# Geophysical Monograph Series
## Including
### Maurice Ewing Volumes
### Mineral Physics Volumes

## Geophysical Monograph Series

1 **Antarctica in the International Geophysical Year,** A. P. Crary, L. M. Gould, E. O. Hulburt, Hugh Odishaw, and Waldo E. Smith (Eds.)

2 **Geophysics and the IGY,** Hugh Odishaw and Stanley Ruttenberg (Eds.)

3 **Atmospheric Chemistry of Chlorine and Sulfur Compounds,** James P. Lodge, Jr. (Ed.)

4 **Contemporary Geodesy,** Charles A. Whitten and Kenneth H. Drummond (Eds.)

5 **Physics of Precipitation,** Helmut Weickmann (Ed.)

6 **The Crust of the Pacific Basin,** Gordon A. Macdonald and Hisashi Kuno (Eds.)

7 **Antarctic Research: The Matthew Fontaine Maury Memorial Symposium,** H. Wexler, M. J. Rubin, and J. E. Caskey, Jr. (Eds.)

8 **Terrestrial Heat Flow,** William H. K. Lee (Ed.)

9 **Gravity Anomalies: Unsurveyed Areas,** Hyman Orlin (Ed.)

10 **The Earth Beneath the Continents: A Volume of Geophysical Studies in Honor of Merle A. Tuve,** John S. Steinhart and T. Jefferson Smith (Eds.)

11 **Isotope Techniques in the Hydrologic Cycle,** Glenn E. Stout (Ed.)

12 **The Crust and Upper Mantle of the Pacific Area,** Leon Knopoff, Charles L. Drake, and Pembroke J. Hart (Eds.)

13 **The Earth's Crust and Upper Mantle,** Pembroke J. Hart (Ed.)

14 **The Structure and Physical Properties of the Earth's Crust,** John G. Heacock (Ed.)

15 **The Use of Artificial Satellites for Geodesy,** Soren W. Henriksen, Armando Mancini, and Bernard H. Chovitz (Eds.)

16 **Flow and Fracture of Rocks,** H. C. Heard, I. Y. Borg, N. L. Carter, and C. B. Raleigh (Eds.)

17 **Man-Made Lakes: Their Problems and Environmental Effects,** William C. Ackermann, Gilbert F. White, and E. B. Worthington (Eds.)

18 **The Upper Atmosphere in Motion: A Selection of Papers With Annotation,** C. O. Hines and Colleagues

19 **The Geophysics of the Pacific Ocean Basin and Its Margin: A Volume in Honor of George P. Woollard,** George H. Sutton, Murli H. Manghnani, and Ralph Moberly (Eds.)

20 **The Earth's Crust: Its Nature and Physical Properties,** John G. Heacock (Ed.)

21 **Quantitative Modeling of Magnetospheric Processes,** W. P. Olson (Ed.)

22 **Derivation, Meaning, and Use of Geomagnetic Indices,** P. N. Mayaud

23 **The Tectonic and Geologic Evolution of Southeast Asian Seas and Islands,** Dennis E. Hayes, (Ed.)

24 **Mechanical Behavior of Crustal Rocks: The Handin Volume,** N. L. Carter, M. Friedman, J. M. Logan, and D. W. Stearns (Eds.)

25 **Physics of Auroral Arc Formation,** S.-I. Akasofu and J. R. Kan (Eds.)

26 **Heterogeneous Atmospheric Chemistry,** David R. Schryer (Ed.)

27 **The Tectonic and Geologic Evolution of Southeast Asian Seas and Islands: Part 2,** Dennis E. Hayes, (Ed.)

28 **Magnetospheric Currents,** Thomas A. Potemra (Ed.)

29 **Climate Processes and Climate Sensitivity** (Maurice Ewing Volume 5), James E. Hansen and Taro Takahashi (Eds.)

30 **Magnetic Reconnection in Space and Laboratory Plasmas,** Edward W. Hones, Jr. (Ed.)

31 **Point Defects in Minerals** (Mineral Physics Volume 1), Robert N. Schock (Ed.)

## Maurice Ewing Volumes

1 **Island Arcs, Deep Sea Trenches, and Back-Arc Basins,** Manik Talwani and Walter C. Pitman III (Eds.)

2 **Deep Drilling Results in the Atlantic Ocean: Ocean Crust,** Manik Talwani, Christopher G. Harrison, and Dennis E. Hayes (Eds.)

3 **Deep Drilling Results in the Atlantic Ocean: Continental Margins and Paleoenvironment,** Manik Talwani, William Hay, and William B. F. Ryan (Eds.)

4 **Earthquake Prediction—An International Review,** David W. Simpson and Paul G. Richards (Eds.)

Geophysical Monograph 32

# The Carbon Cycle and Atmospheric CO₂: Natural Variations Archean to Present

**E. T. Sundquist and W. S. Broecker**
**Editors**

American Geophysical Union
Washington, D.C.
1985

The Carbon Cycle and Atmospheric CO$_2$: Natural
Variations Archean to Present

Library of Congress Cataloging in Publication Data

Main entry under title:

The Carbon cycle and atmospheric CO$_2$.

   (Geophysical monograph, 0065-8448 ; 32)
   Papers presented at the Chapman Conference on
Natural Variations in Carbon Dioxide and the Carbon
Cycle, Tarpon Springs, Fla., Jan. 9-13, 1984,
sponsored by the American Geophysical Union.
   1. Carbon cycle (Biogeochemistry)--Congresses.
2. Atmospheric carbon dioxide--Congresses.
3. Geological time--Congresses.
4. Paleothermometry--Congresses. 5. Geology,
Stratigraphic--Congresses. I. Sundquist, E. T.
(Eric T.) II. Broecker, Wallace S., 1931-
III. Chapman Conference on Natural Variations in
Carbon Dioxide and the Carbon Cycle (1984 : Tarpon
Springs, Fla.). IV. American Geophysical Union.
V. Title: Carbon cycle and atmospheric carbon
dioxide. VI. Series.

QE516.5.C37  1985        574.5'222        85-7353
ISBN  0-87590-060-7
ISSN  0065-8448

# CONTENTS

Preface   ix

Introduction: The Scientific History of Carbon Dioxide   *Roger Revelle*   1

## FROM THE PRESENT TO THE PAST

Geological Perspectives on Carbon Dioxide and the Carbon Cycle   *Eric T. Sundquist*   5

A Reexamination of the Tropospheric Methane Cycle: Geophysical Implications   *Gunnar I. Senum and Jeffrey S. Gaffney*   61

$CO_2$ Driven Equator-to-Pole Paleotemperatures: Predictions of an Energy Balance Climate Model With and Without a Tropical Evaporation Buffer   *Brian P. Flannery, Andrew J. Callegari, Martin I. Hoffert, C. T. Hseih, and Mark D. Wainger*   70

Organic Carbon Preservation in Marine Sediments   *Steven Emerson*   78

Transient Response of the Marine Carbon Cycle   *John R. Southam and William H. Peterson*   89

Ocean Carbon Pumps: Analysis of Relative Strengths and Efficiencies in Ocean-Driven Atmospheric $CO_2$ Changes   *Tyler Volk and Martin I. Hoffert*   99

Detection of El Nino and Decade Time Scale Variations of Sea Surface Temperature From Banded Coral Records: Implications for the Carbon Dioxide Cycle   *Ellen R. M. Druffel*   111

Atmospheric $CO_2$ Variations Based on the Tree-Ring $^{13}C$ Record   *T.-H. Peng*   123

## THE LAST DEGLACIATION

Variations of the $CO_2$ Concentration of Occluded Air and of Anions and Dust in Polar Ice Cores   *H. Oeschger, B. Stauffer, R. Finkel, and C. C. Langway, Jr.*   132

Accelerator Radiocarbon Ages on Foraminifera Separated From Deep-Sea Sediments   *M. Andrée, J. Beer, H. Oeschger, A. Mix, W. Broecker, N. Ragano, P. O'Hara, G. Bonani, H. J. Hofmann, E. Morenzoni, M. Nessi, M. Suter, and W. Wölfli*   143

Changes in Atmospheric $CO_2$: Factors Regulating the Glacial to Interglacial Transition   *Fanny Knox Ennever and Michael B. McElroy*   154

Glacial to Interglacial Changes in Atmospheric Carbon Dioxide: The Critical Role of Ocean Surface Water in High Latitudes   *J. R. Toggweiler and J. L. Sarmiento*   163

The High-Latitude Ocean as a Control of Atmospheric $CO_2$   *T. Wenk and U. Siegenthaler*   185

Last Deglaciation in the Bahamas: A Dissolution Record From Variations of Aragonite Content?   *Andre W. Droxler*   195

Late Holocene Carbonate Dissolution in the Equatorial Pacific: Reef Growth or Neoglaciation?   *R. S. Keir and W. H. Berger*   208

Carbon Cycle Variations During the Past 50,000 Years: Atmospheric $^{14}C/^{12}C$ Ratio as an Isotopic Indicator   *Devendra Lal*   221

Simulation Experiments with Late Quaternary Carbon Storage in Mid-Latitude Forest Communities   *Allen M. Solomon and M. Lynn Tharp*   235

## THE PLEISTOCENE

Carbonate Preservation and Rates of Climatic Change: An 800 kyr Record From the Indian Ocean   *L. C. Peterson and W. L. Prell*   251

Late Quaternary Carbonate Changes in the North Atlantic and Atlantic/Pacific Comparisons   *Thomas J. Crowley*   271

Carbon Deposition Rates and Deep Water Residence Time in the Equatorial Atlantic Ocean Throughout the Last 160,000 Years   *W. B. Curry and G. P. Lohmann*   285

Atmospheric Carbon Dioxide, Orbital Forcing, and Climate   *N. J. Shackleton and N. G. Pisias*   303

Carbon Isotopes in Deep-Sea Benthic Foraminifera: Precession and Changes in Low-Latitude Biomass   *Lloyd D. Keigwin and Edward A. Boyle*   319

Carbon Isotope Variations in Surface Waters of the Gulf of Mexico on Time Scales of 10,000, 30,000, 150,000 and 2 Million Years   *Douglas F. Williams*   329

Carbon Isotope Record of Late Quaternary Coral Reefs: Possible Index of Sea Surface Paleoproductivity   *Paul Aharon*   343


## THE CENOZOIC

Distribution of Major Vegetational Types During the Tertiary   *Jack A. Wolfe*   357

Cenozoic Fluctuations in Biotic Parts of the Global Carbon Cycle   *Jerry S. Olson*   377

An Improved Geochemical Model of Atmospheric $CO_2$ Fluctuations Over the Past 100 Million Years   *Antonio C. Lasaga, Robert A. Berner, and Robert M. Garrels*   397

Oceanic Carbon Isotope Constraints on Oxygen and Carbon Dioxide in the Cenozoic Atmosphere   *N. J. Shackleton*   412

Charcoal Fluxes into Sediments of the North Pacific Ocean: The Cenozoic Record of Burning   *James R. Herring*   419

Changes in Calcium Carbonate Accumulation in the Equatorial Pacific During the Late Cenozoic: Evidence From HPC Site 572   *Nicklas G. Pisias and Warren L. Prell*   443

Carbon Dioxide and Polar Cooling in the Miocene: The Monterey Hypothesis   *Edith Vincent and Wolfgang H. Berger*   455

Oligocene to Miocene Carbon Isotope Cycles and Abyssal Circulation Changes   *Kenneth G. Miller and Richard G. Fairbanks*   469

A "Strangelove" Ocean in the Earliest Tertiary   *Kenneth J. Hsü and Judith A. McKenzie*   487

Mantle Degassing Induced Dead Ocean in the Cretaceous-Tertiary Transition   *Dewey M. McLean*   493


## THE PHANEROZOIC AND PRECAMBRIAN

Variations in the Global Carbon Cycle During the Cretaceous Related to Climate, Volcanism, and Changes in Atmospheric $CO_2$   *M. A. Arthur, W. E. Dean, and S. O. Schlanger*   504

Natural Variations in the Carbon Cycle During the Early Cretaceous   *H. J. Weissert, J. A. McKenzie, and J. E. T. Channell*   531

Warm Cretaceous Climates: High Atmospheric $CO_2$ as a Plausible Mechanism   *Eric J. Barron and Warren M. Washington*   546

Mid-Cretaceous Continental Surface Temperatures: Are High $CO_2$ Concentrations Needed to Simulate Above-Freezing Winter Conditions?   *Stephen H. Schneider, Starley L. Thompson, and Eric J. Barron*   554

Proterozoic to Recent Tectonic Tuning of Biogeochemical Cycles   *T. R. Worsley, J. B. Moody, and R. D. Nance*   561

Potential Errors in Estimates of Carbonate Rock Accumulating Through Geologic Time   *William W. Hay*   573

Nonskeletal Aragonite and $pCO_2$ in the Phanerozoic and Proterozoic   *Philip A. Sandberg*   585

Carbonates and Ancient Oceans: Isotopic and Chemical Record on Time Scales of $10^7$–$10^9$ Years   *Ján Veizer*   595

Carbon Exchange Between the Mantle and the Crust, and Its Effect Upon the Atmosphere: Today Compared to Archean Time   *David J. Des Marais*   602

Photochemical Consequences of Enhanced $CO_2$ Levels in Earth's Early Atmosphere   *James F. Kasting*   612

Are Interpretations of Ancient Marine Temperatures Constrained by the Presence of Ancient Marine Organisms?   *James W. Valentine*   623

Readers of this book will generally fit into two groups. One group is geologists and geochemists, who have studied the global carbon cycle for many decades. These readers will find that the papers in this book present a new view of familiar themes. Whereas much previous work on the carbon cycle, and other geochemical cycles, has emphasized the nature of the steady state maintained by complex networks of feedbacks, recent attention has shifted to the changes implied by the way these feedbacks respond to perturbations.

The other group of readers is the community of scientists who are concerned with anticipating the effects of anthropogenic carbon dioxide. Like the geologists and geochemists, these readers have a long tradition of carbon cycle research within their own disciplines. They, also, have raised subtle but profound questions about the role of the steady state assumption in their studies. As the evidence for preanthropogenic $CO_2$ variations grows, they are no longer satisfied with predictive models that assume ipso facto that the carbon cycle was at a steady state before man's intervention. And as they discover more about the complexities of carbon cycle processes, they seek to know how these processes have behaved in the geologic past.

These groups are not mutually exclusive, and this book is a product of their healthy interaction. The papers in this volume were presented at the Chapman Conference on Natural Variations in Carbon Dioxide and the Carbon Cycle, held in Tarpon Springs, Florida, January 9-13, 1984. This meeting was sponsored by the American Geophysical Union and made possible by support from the U.S. Geological Survey and National Science Foundation. The conference was organized primarily by and for geologists, yet its underlying impetus was a concern for the effects of anthropogenic $CO_2$. It was attended by over 100 scientists, many of whom were not geologists. Its timing was very fortuitous, in that it occurred when many people had many new things to say.

We are delighted that this book begins with a historical introduction by Roger Revelle, who has modestly chosen not to single out his own critical contributions in his account of the scientific history of carbon dioxide. As a leader in the implementation of the atmospheric and oceanic $CO_2$ measurements which began as part of the International Geophysical Year of 1957-1958, he laid the foundation for modern quantitative understanding of what he termed the "vast geophysical experiment" of fossil fuel $CO_2$. His account of earlier scientists' perspectives, as well as his own continuing research contributions, remind us that we are "standing on the shoulders of giants."

The organization of this book is based on subdivisions of the broad range of time scales considered. It is often difficult for non-geologists to jump from human time perspectives to geologic time scales. Likewise, many geologists have found it difficult to relate their work to studies of the modern carbon cycle. The book's first section is an attempt to ease these difficulties. The first paper (by E.T.S.) presents a summary of data for carbon cycle fluxes and reservoirs in the context of a hierarchical series of box models appropriate for a series of time scales ranging from decades to hundreds of millions of years. It is followed by a series of papers on modern carbon cycle and climate processes, concluding with studies of decadal carbon isotope records observed in corals and tree rings. We hope that these papers provide a transition for those who seek geological contexts for knowledge about the modern carbon cycle.

The most impressively documented natural change in atmospheric $CO_2$ occurred during the most recent deglaciation. The book's second section includes papers discussing both the evidence for the deglacial $CO_2$ increase and its possible origins. These themes are extended in the book's third section to the oscillatory Pleistocene record. Together the papers in these sections reflect a very exciting combination of elegant laboratory analysis, mathematical modeling, and reassessment of the sedimentary record.

The book's next section reflects the influence of different processes over longer time scales. $CO_2$ in the Cenozoic atmosphere, according to the papers in this section, has been influenced by processes such as volcanism, weathering, and carbon burial. These processes are generally slower to affect the atmosphere than those hypothesized to account for Pleistocene variations, but they may have caused much larger $CO_2$ variations over time scales of tens of millions of years. The cause-and-effect relationship between $CO_2$ and climate, which appears tantalizingly in the Quaternary record, becomes more conspicuous in scenarios for change over longer time scales. The range of possibilities extends even further in the book's last section, covering the Phanerozoic and Precambrian.

Having opened the Pandora's box of natural $CO_2$ variations, geologists are searching now not only for more evidence of variability, but also for constraints that will narrow the possible down to the probable. In so doing, they are drawn back to a reevaluation of the constraints that tend to return the carbon cycle to a steady state. This reevaluation does not deny the effectiveness of geochemical and geophysical feedback mechanisms, but seeks to better assess them through studies of the system's response to perturbations. The study of geological carbon cycle perturbations will greatly enhance our ability to anticipate future responses to anthropogenic perturbations.

Eric T. Sundquist
Wallace S. Broecker
Editors

# INTRODUCTION
## THE SCIENTIFIC HISTORY OF CARBON DIOXIDE

Roger Revelle

University of California, San Diego
La Jolla, California 92093

Carbon dioxide may be thought of as the most important substance in the biosphere: that part of the earth's atmosphere, hydrosphere, and solid crust in which life exists. It has supported the existence and development of life by serving as the source of carbon, the principal element of which all living beings, including geologists and geochemists, are made. In past times it was a source of the free oxygen in the air and the ocean that makes animal life possible. By absorbing and backscattering the heat radiated from the earth's surface, it maintains, together with atmospheric water vapor, a sufficiently high temperature in the air and the sea to allow liquid water, and therefore life, to exist. Earth's uniquely benign environment for living things depends fundamentally, of course, on its relationship to a small, steady star, the sun; but this relationship is modulated in essential ways by carbon dioxide.

These various roles of carbon dioxide have gradually been elucidated over the past three centuries by a succession of pioneering scientists. In a book devoted to the geologic history of carbon dioxide it seems appropriate to recount the gradual evolution of scientific ideas about this remarkable substance. I shall discuss the course of that evolution during the first three centuries prior to 1900. Many of the ideas on which modern $CO_2$ research is focused arose during this earlier time. By examining this scientific history, however briefly, we should gain perspective on the advances in understanding made in recent years, and the long path ahead of us before full understanding is attained.

A Belgian alchemist and physician, Johannes van Helmont (1579-1644) first used the term "gas" to denote "wild spirits" (spiritus sylvestris) that could be neither constrained by vessels nor reduced to a visible body. He described a number of different gases, notably that produced from the burning of charcoal, the fermentation of wine, and the action of distilled vinegar on limestone, and in emanations from mineral waters. This was, of course, what we now know as carbon dioxide.

Joseph Black (1728-1799) was a distinguished and much loved professor of medicine in the University of Edinburgh. With his two closest friends, Adam Smith and James Hutton, he founded the famous Oyster Club, which met for dinner once a week to discuss scientific and intellectual subjects. Local industrialists often attended these dinners; a favorite topic was the role of science in technological progress. One of Black's principal scientific contributions was his discovery, through careful experiments, of latent heats of melting and vaporization of water and other substances. But his only major publication was a paper based on his doctoral thesis, published by the Philosophical Society of Edinburgh in 1756, entitled "Experiments Upon Magnesia Alba, Quicklime and Some Other Alkaline Substances." In this thesis, Black showed that "Magnesia Alba" (magnesium carbonate) is a compound of an alkaline earth and a gas which he called "fixed air." This was the same gas that was released in the treatment of chalk with acid, in fermenting alcohol, and in burning charcoal. It combined with quicklime to produce a chalky noncaustic material. In studying its properties he found that a burning candle was extinguished when dipped into fixed air, as was the flame of a burning paper. Small animals placed in fixed air did not survive. The air produced in respiration was in part fixed air, which was heavier than common air.

It remained for one of the greatest of all chemists, Antoine Lavoisier (1743-1794) to determine the composition of Black's fixed air. He showed that oxygen, which he at first called "eminently respirable air," combines with carbon to produce a substance he named "chalky aeroform acid." Lavoisier and his younger colleague, Pierre Simon Laplace (1749-1827), studied respiration in animals. They showed that this physiological process consists of combustion of organic compounds with oxygen to produce a measurable quantity of heat plus carbon dioxide and water.

Jan Ingen-Housz (1730-1799) was born in the Netherlands but spent most of his life in England. He was a physician who played an active part in disseminating the practice of innoculation against smallpox throughout western and central Europe. But he was also an active experimental scientist, a friend of Benjamin Franklin, and most famous during his lifetime for his research on electricity. In 1779 he published a book entitled "Experiments Upon Vegetables, Discovering Their Great Powers in Purifying the Common Air in the Sunshine and of Injuring It in the Dark and at Night." In this book he showed that the green parts of plants "restore" the air only when they are in the sunlight and that the action is caused only by the visible part of the sun's radiation. On the other hand, plants exhibit respiration day and night, from all parts of the plant. More important for a beginning of understanding of the carbon cycle in nature, he showed that the carbon in plants is taken up from the carbon dioxide in the air and not from the humus in the soil, as was then generally believed.

Ingen-Housz's work did not become widely known, and the notion that plants derive their sustenance from the humus in the soil persisted for many decades. For example, it was the basis of Thomas Robert Malthus's proposition, in his seminal "Essay on Population" in 1798, that agricultural production can grow only arithmetically in proportion to the increase in the area of cultivated land. (In contrast, Malthus claimed, human populations tend to grow exponentially, and therefore populations will always be limited by the food supply.)

It was not until 1840, when the great organic chemist Justus von Liebig (1803-1873) published "Organic Chemistry and Its Application to Agriculture and Physiology," that the view became widely accepted that the source of carbon assimilated by plants is the carbon dioxide of the atmosphere. Von Liebig pointed out that the relative constancy of the carbon dioxide content of the air, despite the continued exhalation of carbon dioxide by animals, requires that some other process must be removing $CO_2$ from the atmosphere. He showed that even though the percentage of $CO_2$ in the air is relatively small, the atmosphere contains an ample amount to provide for plant growth. Von Liebig made an even more fundamental generalization. Before his time it had been generally believed that both plant and animal life depend on the assimilation of organic, previously living material. He demonstrated that the nutrient substances of plants are almost entirely inorganic. Under the older concept, the potential food production was limited by previously existing organic matter, whereas in the new view, plant production is limited mainly by available water, the quantity and composition of fertilizers applied to the soil, the amount of sunlight received by the plants, or the amount of atmospheric carbon dioxide available to them.

The first measurements of the atmospheric concentration of carbon dioxide were apparently made by the great German polymath Alexander von Humboldt (1769-1859) and the pioneering chemist Joseph Louis Gay-Lussac (1778-1850) during the last decade of the eighteenth century. Gay-Lussac made a balloon ascent up to the then unprecedented height of 8000 meters and took a series of air samples at different altitudes. He later analyzed these samples, and showed that the chemical composition of the air was constant within his experimental accuracy at all levels.

Another polymath, Jean Baptiste-Joseph Fourier (1768-1830), was the first scientist to propose that the gases in the earth's atmosphere act in the same manner as the glass in a "hothouse" to retain heat radiation and thus to raise the earth's temperature considerably above that which would prevail in the absence of an atmosphere. Because of his administrative ability, Fourier was a favorite of Napoleon, who made him a baron and prefect of the French département of Isére. He was also one of the earliest students of Egyptian archeology, supervising the preparation of the great series of volumes on Egypt that resulted from Napoleon's Egyptian expedition. Today he is best remembered as a creative mathematician who developed, among many other contributions, the theory of heat conduction.

We know that Fourier misunderstood the mode of action of a greenhouse. But because of his work on heat conduction in different materials it was natural for him to consider the radiation balance of the atmosphere and the mechanism of heat conduction from the air to outer space. His speculation concerning radiation-absorbing gases in the atmosphere was confirmed by John Tyndall (1820-1893), one of the first physicists to attempt to make a living as a professional scientist. Tyndall measured the absorption of infrared radiation by carbon dioxide and water vapor and showed that the quantities of these substances present in the atmosphere significantly raise the earth's temperature. Although he believed that the major effect was due to water vapor rather than to carbon dioxide, Tyndall was apparently the first to suggest that glacial periods may have been due to a decrease in atmospheric carbon dioxide.

By the end of the nineteenth century it was realized that the carbon dioxide content of the air was probably increasing, because of the large-scale combustion of coal that had begun during previous decades. Svante Arrhenius (1859-1927), who received one of the first Nobel prizes in chemistry for his work on the dissociation of salts in solution, attempted to calculate the effect of a doubling of the atmospheric carbon dioxide concentration on the earth's average surface temperature. As a basis for his calculation he used measurements by Samuel Langley (1834-1906) of infrared radiation received at the earth's surface from the full moon at different altitudes above the horizon, under different levels of atmospheric humidity. Measurements of atmospheric $CO_2$ had been made by a series of workers in the 1880's and 1890's, and Arrhenius used these to estimate the quantity of carbon dioxide and water vapor through which the moon's rays passed. He concluded that a doubling of carbon dioxide would raise the average temperature by 5°-6°C, the higher value occurring at high latitudes. In making this calculation he assumed that the water vapor content of the air would increase with the rise in temperature and thereby contribute to the increased absorption of infrared radiation. Arrhenius thus clearly had the idea of feedback effects in the dynamics of the atmosphere, though the term was not invented until 50 years later.

During Arrhenius's time the annual combustion of coal amounted to nearly one gigaton and was "rapidly increasing." He reasoned that by

> . . . evaporating our coal mines into the air . . . the slight percentage of carbonic acid in the atmosphere may, by the advances of industry, be changed to a noticeable degree in a few centuries . . . [even though] the sea by adsorbing carbonic acid acts as a regulator of huge capacity which takes up about five sixths of the produced carbonic acid. . . . This would imply that there is no real stability in the percentage of carbon dioxide in the air, which is probably subject to considerable fluctuations in the course of time.

Using the same method of calculation, Arrhenius concluded from Langley's data that a 50% reduction in atmospheric carbon dioxide would reduce the average of temperature by about 4°C, which was presumably the temperature during the glacial period. A complete depletion of atmospheric $CO_2$ by itself would cause a fall in temperature by 20°C, and the accompanying diminution of atmospheric water vapor would cause a further, almost equally great drop in temperature.

Arrhenius also pointed out, on the basis of plant growth experiments by the Polish botanist E. Godlewski, that the amount of carbon absorbed by plants in photosynthesis "increases proportionately with the percentage of carbonic acid in the atmosphere up to 1%," and he suggested that the coal deposits of the Carboniferous period were produced during times of "enormous plant growth" resulting from a high concentration of carbon dioxide in the air. He accepted the supposition of other nineteenth century scientists that

> . . . all the oxygen of the air may have been formed at the expense of the carbonic acid in the air. . . . We may take it as established that the masses of free oxygen in the air and of free carbon in the sedimentary strata approximately correspond to each other and that probably all the oxygen in the atmosphere owes its existence to plant life.

Using the geological evidence of vast accumulations of limestone throughout the then known geological past, Arrhenius concluded that very large quantities of carbon dioxide produced by vulcanism had passed through the atmosphere, had been transformed into carbonates in the process of weathering of igneous rocks, and had precipitated as calcium and magnesium carbonates in the ocean. He rejected, on several grounds, the idea that the earth's original atmosphere had contained all or most of the carbon dioxide now in sedimentary carbonates and that this $CO_2$ had been slowly depleted by the process of weathering throughout the history of the earth. He believed instead that volcanic production of carbon dioxide and its depletion by weathering must on the average have been roughly in balance throughout the earth's history, although marked deviations from balance may have occurred because of the varying intensity of volcanic activity in different geologic periods. He thus appealed to an essentially unknown earth process, variations in the volcanic supply of carbon dioxide, which could be balanced by weathering only at variable levels of atmospheric $CO_2$, as a possible explanation of the Pleistocene glaciations.

It remained for the American geologist Thomas C. Chamberlin (1843-1928) to present a "working hypothesis" on the causes of glacial periods which invoked known processes for the depletion of atmospheric carbon dioxide to the levels which Arrhenius had shown could bring on an ice age. He proposed that weathering of silicate rocks would be accelerated whenever the continents stood relatively high above sea level and their area was correspondingly relatively large. During such times the aerated zone above the groundwater table was relatively thick, and weathering could take place throughout this larger volume of material. Also, because of the greater slope of the water table toward the sea, groundwaters could move more rapidly and freely, and their dissolved chemicals could act more effectively in weathering.

Chamberlin supposed that at the same time the accumulation of calcium carbonate skeletons of pelagic organisms in marine deposits would be greatly reduced because of the diminution of shallow water areas in which most such deposits accumulated. Consequently, less carbon dioxide would be released from the sea by precipitation of carbonates. Both increased weathering and reduced calcium carbonate deposition would act in the same direction to lower atmospheric carbon dioxide. This depletion would be accelerated by the drop in temperature of ocean waters that would accompany the lowering of atmospheric temperature; with lower temperature the solubility of carbon dioxide in seawater would be increased, and the ocean would "rob" the atmosphere. Chamberlin thought that production and sedimentary deposition of organic matter in the ocean would also be reduced and that this would act in the opposite direction to the other processes, but he thought this would be a relatively small effect.

Chamberlin recognized that the essential question to explain was the rhythmic alteration of glacial and interglacial periods. He proposed that the increased weathering which had brought about depletion of atmospheric carbon dioxide and consequent cooling would rapidly diminish as the land surfaces became covered with ice. Continuing volcanic supply would then increase the carbon dioxide content and consequently the temperature of the atmosphere until glacial melting would take place, whereupon a high rate of weathering would again occur, atmospheric carbon dioxide would again be depleted, and another glaciation would follow. He thought that his hypothesis of variable rates of weathering depending upon the elevation of the land would also explain the warm and relatively uniform climate of the Cretaceous period when the surfaces of the continents were near or below sea level over large areas. Not only would weathering on land be low in these conditions, but precipitation of calcium carbonate in shallow seas would be relatively high, and large quantities of carbon dioxide released by carbonate deposition would flow from the warm ocean to the atmosphere. He suggested that the Permian glaciation at the end of the Paleozoic epoch could be explained in the same way as Pleistocene glaciation. The vast coal deposits of the preceding Carboniferous period could be explained by reduced weathering during a low stand of the continents, which produced high levels of atmospheric carbon dioxide; these in turn greatly stimulated photosynthetic production of organic matter by plants.

Chamberlin overestimated the rate of depletion of atmospheric carbon dioxide by weathering of silicate rocks, and he underestimated the rate of deposition of organic matter on the seafloor. He was unaware of the cyclic perturbations of the earth's orbit and pole of rotation described by M. Milankovich, which are now believed to have determined the timing of Pleistocene climatic changes. Moreover, he did not realize the relatively poor coupling between rates of weathering and the concentration of atmospheric carbon dioxide, which has been demonstrated in recent years by H. Holland and R. Garrels, among others.

Nevertheless his hypothesis still represents a not unreasonable explanation of climatic variation throughout later geologic time.

Since the time of Chamberlin and that of his student C. F. Tolman, who attempted to estimate the quantities of $CO_2$ that would be taken up or released by the ocean during times of climatic change, there have been important advances in chemical and climatic understanding. These have made possible more quantitative insights into the various roles of $CO_2$ than could be reached by the earlier workers. Among these advances are the discovery and accurate measurement of the carbon isotopes $^{13}C$ and $^{14}C$, and the realization that the distribution of these isotopes in the biota, the ocean, and the atmosphere must be accounted for in any satisfactory model of the carbon cycle. Highly accurate methods for continuous spectrometric measurement of atmospheric carbon dioxide have been developed and used in a network of observing stations to determine the variations of atmospheric carbon dioxide in time and space over the past 25 years. We now know with considerable precision the dissociation constants of carbonic acid in ocean waters, and with much less precision the rates of exchange of $CO_2$ across the sea surface. The development of large, high-speed computers in recent years has made it possible to model the probable climatic effects of the changing atmospheric concentration of carbon dioxide. Studies of tree rings, coral skeletons, lake varves, and deep-sea sediments have yielded proxy records of past climatic and sea level changes. Techniques have been developed for the measurement of carbon dioxide in air trapped in glacial ice; these show that Chamberlin and Arrhenius were right in believing that atmospheric $CO_2$ was much lower during glacial times than at present. These and other advances have given new levels of insight into the carbon dioxide problem. To a very considerable extent, nevertheless, we are standing on the shoulders of the giants who, during the course of three centuries, built the framework of our present understanding.

# GEOLOGICAL PERSPECTIVES ON CARBON DIOXIDE AND THE CARBON CYCLE

Eric T. Sundquist

U.S. Geological Survey, Reston, Virginia 22092

Abstract. A review of global carbon fluxes and reservoirs emphasizes the wide range of time scales that characterize their interactions. These interactions can be organized in a series of box models developed sequentially from short to long time scales. Model structures are developed by adding and lumping boxes, procedures which can be formalized in terms of transformations of the matrix of geochemical exchange coefficients. Time responses of the models are characterized by eigenanalysis, which shows particularly clear patterns when the matrix is arranged to approximate block diagonal form. Over time scales of a few centuries, atmospheric $CO_2$ can be modeled as part of a larger reservoir that includes land plants and part of the oceans. Over time scales of thousands of years, carbon cycle interactions are dominated by "reactive" marine and terrestrial sediments. Over time scales of tens to hundreds of millions of years, the earth surface carbon cycle can be viewed as at "secular equilibrium" controlled by the global cycle of weathering and sedimentation.

## Introduction

Concern for the climatic effects of anthropogenic $CO_2$ has led to substantial worldwide efforts to better understand the carbon cycle. These efforts have focused primarily on a time scale of decades, corresponding to the time scale of the exponential increase in fossil fuel consumption. How relevant are these efforts to geological time scales? Will anthropogenic $CO_2$ have long-term geological effects? This paper is an attempt to provide a transition to geological concerns from the perspective of current $CO_2$ research.

I will first review the carbon cycle reservoirs and processes that can be assessed directly or by inference from historical information. I will then apply this information to the construction of models appropriate to time scales ranging from 10 to $10^8$ years. Particular attention to time scales as frames of reference is necessary because of conspicuous shifts in the relative importance of different processes over different time scales. For example, annual fluctuations of atmospheric $CO_2$ (Figure 1) are caused by seasonal changes that have little bearing on $CO_2$ concentrations 100 years ago or 100 years in the future. Likewise, although the consumption of $CO_2$ during chemical weathering can have but little effect on the atmosphere during the course of a century, weathering is one of the most important processes controlling atmospheric $CO_2$ over millions of years.

Such differences are visualized in this paper in a series of box models, developed sequentially from short to long time scales. This development is given mathematical formalism in terms of matrix algebra and eigenanalysis, which allow relationships among models for different time scales to be shown explicitly. The models are characterized in terms of their time response to perturbations, rather than their behavior at steady state. The effects of basic box modeling approximations can be seen readily, and uncertainties can be evaluated relative to particular time scales.

Many geochemical models, whether global or local in scale, have until recently focused primarily on the description of steady state conditions. These models have contributed greatly to our understanding of geochemical processes. A measure of their success is that they have evolved into essential tools for understanding the nature of geochemical change. However, a new generation of time-dependent geochemical models is emerging, fueled by the pervasive evidence for change observed in the geological record. These models, like their steady state precursors, are characterized by complex systems of feedback processes. An important test for their validity is often to check their ability to return to a steady state after being perturbed. In most cases, they must still rely on steady state approximations as the only available way to estimate some fluxes. Nevertheless, they reflect a new emphasis on the dynamic nature of geochemical cycles. For example, whereas many studies have focused on the constraints that require the oceans and atmosphere to have remained relatively constant in composition throughout much of geological time, recent research has emphasized the significant composi-

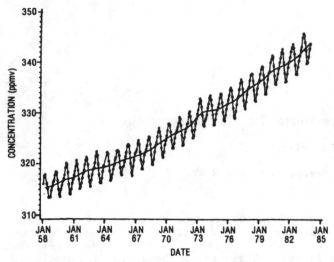

Fig. 1. Atmospheric $CO_2$ concentrations recorded at Mauna Loa Observatory, Hawaii. Data are from files at the Carbon Dioxide Information Center, Oak Ridge National Laboratory, Oak Ridge, Tennessee; data were obtained by C. D. Keeling, Scripps Institution of Oceanography, University of California, La Jolla, California.

tional changes that are possible within those constraints.

In the following descriptions of carbon reservoirs and fluxes, I will point out those estimates that are derived from steady state approximations. I will also attempt to present a representative range of reasonable estimates published to date. I have avoided undocumented and, except where noted, secondary sources. My goal is not only to present a credible framework for relating a broad range of time scales and processes, but also to survey the extent to which significant uncertainties still cloud our understanding of carbon cycle changes.

It is very important to begin to evaluate man's consumption of fossil fuels in the context of the evidence and mechanisms for other carbon cycle changes. For this reason, I choose to begin with an overview of global fossil fuel resources.

### Fossil Fuel Resources

It is a measure of man's influence on global geochemistry that fossil fuel combustion is one of the largest fluxes of carbon dioxide to the atmosphere. Recent estimates of the fossil fuel carbon flux to the atmosphere are close to 5 Gt (1 Gt = $10^9$ metric tons) carbon per year [Marland and Rotty, 1983]. Historically, $CO_2$ production from fossil fuels and cement has grown exponentially. ($CO_2$ from cement manufacturing accounts for 1-2% of the annual $CO_2$ production from fossil fuels [Keeling, 1973a].) From approximately 1860 to 1975, excluding the World War I and II and Depression periods, the annual growth rate was

about 4.3% [Rotty, 1979]. The rate of increase slowed to an estimated 2.25% from 1973 to 1980 [Rotty, 1982], and production decreased slightly in 1980 and 1981 [Marland and Rotty, 1983].

The magnitude of future anthropogenic $CO_2$ perturbations is closely tied to the geology, geography, and technical recoverability of fossil fuel resources. Fossil fuel resource estimates are variable. A significant cause of this variability is the range of definitions of resources and reserves. Key words and phrases in these definitions are subjective. In particular, resource estimation requires incorporating predictions of future technologies and undiscovered occurrences. Although undiscovered resources can often be estimated rationally from geological information and/or exploration statistics, predicting future recovery technologies is very speculative. The only portion of a resource estimate that is known to any reasonable degree of accuracy is known as the "proved reserves," defined as the amount of fossil fuel that is reasonably well measured and can be produced under present economic conditions with existing technologies. Total resource estimates are much higher, because they generally include an estimate of the amount that might be economically exploitable at some time in the future. For purposes of this paper, perhaps the most significant distinction is that between "identified" resources and "total" or "ultimately recoverable" resources. Whereas identified resource estimates are generally based on relatively specific geological information, total resource estimates often involve more vague and regional geological data and technological guesswork. Thus, the differences between identified and total resource estimates provide an indication of the range of geological and technological uncertainties inherent in the total resource data.

Table 1 lists recent estimates of identified and ultimately recoverable fossil fuel resources by fuel type. Equivalent carbon contents are also given.

Coal possesses the greatest potential for carbon dioxide additions to the atmosphere. The amount of carbon contained in coal resources depends on the relative proportions of the various ranks of coal and their corresponding carbon contents. Resource estimates of coal by rank are not commonly reported. An analysis of such data presented by the World Energy Conference (WEC) [WEC, 1980] can be found in the footnotes to Table 1. Other coal resource estimates in Table 1 are converted to equivalent carbon assuming a similar mix of ranks.

Although crude oil reserves are frequently reported and updated (for example, by the Oil and Gas Journal, the International Petroleum Encyclopedia, and the United Nations Institute for Training and Research), estimates of petroleum resources are less common. All of the studies referenced in Table 1 have incorporated an estimate of hypothetical resources. The world's ultimately recoverable oil shale and oil

TABLE 1.  World Remaining Identified and Ultimately Recoverable Fossil Fuel Resources

| Reference | Identified | | Ultimately Recoverable | |
| --- | --- | --- | --- | --- |
| | Quantity | Equivalent Carbon Content, Gt C | Quantity | Equivalent Carbon Content, Gt C |

Coal

| Reference | Quantity | Equiv. C | Quantity | Equiv. C |
| --- | --- | --- | --- | --- |
| WEC [1974] | | | 10,754 | 5381 |
| Peters and Schilling [1978] | | | 11,866[a] | 5994 |
| WEC [1980] | 6447 | 3226 | 13,476 | 6808[b] |
| World Coal Resources and Production [1981] | | | 13,609 | 6810 |

Crude Oil

| Reference | Quantity | Equiv. C | Quantity | Equiv. C |
| --- | --- | --- | --- | --- |
| Despraires [1978] | | | 257 | 216[c] |
| WEC [1980] | | | 301 | 253[c] |
| Moody and Esser [1975] | 101 | 85[c] | 232 | 195[c] |
| Halbouty and Moody [1980] | 115 | 97[c] | 256 | 215[c] |
| Masters et al. [1984] | 103 | 87[c] | 246 | 207[c] |

Natural Gas Liquids

| Reference | Quantity | Equiv. C | Quantity | Equiv. C |
| --- | --- | --- | --- | --- |
| Hubbert [1969] | | | 28-47 | 23-38[d] |
| Linden [1972] | | | 29 | 24[d] |

Natural Gas

| Reference | Quantity | Equiv. C | Quantity | Equiv. C |
| --- | --- | --- | --- | --- |
| Despraires [1978] | | | 246 ($10^{12}$ $m^3$) | 133[e] |
| National Research Council [1979] | | | 210 ($10^{12}$ $m^3$) | 113[e] |
| WEC [1980] | | | 266 ($10^{12}$ $m^3$) | 144[e] |
| Nehring [1981] | 76 ($10^{12}$ $m^3$) | 41[e] | | |
| Grunau [1984] | 82 ($10^{12}$ $m^3$) | 44[e] | | |

Oil Shale (Gigatons Oil Equivalent)

| Reference | Quantity | Equiv. C | Quantity | Equiv. C |
| --- | --- | --- | --- | --- |
| Duncan and Swanson [1965] | 28 | 23[f] | 479 | 407[f] |
| Donnell [1977] | | | 474 | 403[f] |
| WEC [1980] | | | 339 | 288[f] |
| Ovcharenko [1981] | 27 | 23[f] | 450 | 382[f] |

Oil Sands/Heavy Crude (Gigatons Oil Equivalent)

| Reference | Quantity | Equiv. C | Quantity | Equiv. C |
| --- | --- | --- | --- | --- |
| Meyer & Dietzman [1981] | 118[g] | 98[g] | 118[g] | 98[g] |
| WEC [1980] | 116[g] | 97[g] | 116[g] | 97[g] |

Values are in gigatons (Gt) unless otherwise indicated.

[a] Converted to metric tons coal equivalent (T.C.E.); 1 T.C.E. = 29.3 x $10^9$ joules (GJ).  The resource-weighted world average coal energy content for coal is 25 GJ/metric ton [Ion, 1980].

[b] Equivalent carbon content is based on the percentage of carbon in the various ranks of coal, modified from the Synthetic Fuels Data Handbook [Hendrickson, 1975] assuming 10% ash content. Quantities of each rank of coal are from WEC [1980].  Anthracite is assumed to be 6.4% of the anthracite/bituminous value, based on world production figures by rank from 1966-1971 [WEC, 1974]. Anthracite:  81% carbon x 443.9 Gt = 359.6 Gt carbon. Bituminous:  64% carbon x 6492.5 Gt = 4155.2 Gt carbon.  Sub-bituminous:  40% carbon x 4057.2 Gt = 1622.9 Gt carbon.  Lignite: 27% carbon x 2482.6 Gt = 670.3 Gt carbon.  The total is 6808.0 Gt carbon.

[c] Equivalent carbon figures were calculated from the average carbon content of petroleum estimated to be 84% [Rotty, 1983].  Bbl oil x 0.136 = metric tons of oil [Loftness, 1978, p. 734].

[d] Equivalent carbon figures were calculated from the average carbon content of natural gas liquids, estimated to be 82% [Marland and Rotty, 1984].

[e] Average carbon content of natural gas is estimated to be 540 x $10^{-6}$ metric tons C/$m^3$ gas.

[f] Values are for an oil yield of 16-100 gallons per ton of shale.  Equivalent carbon content figures are calculated from the average carbon content of shale oil, estimated to be 85% [Hendrickson, 1975].  Bbl shale oil x 0.146 = metric tons shale oil [Hendrickson, 1975, p. 38].

[g] Bbl heavy oil x 0.156 = metric tons heavy oil, assuming a specific gravity value of 0.98, which is an average value for deposits in Canada and Venezuela [Starr et al., 1981].  Equivalent carbon content is based on average carbon percentage (83.1%) in Canadian heavy oil [Starr et al., 1981]. Values are for known resources only.

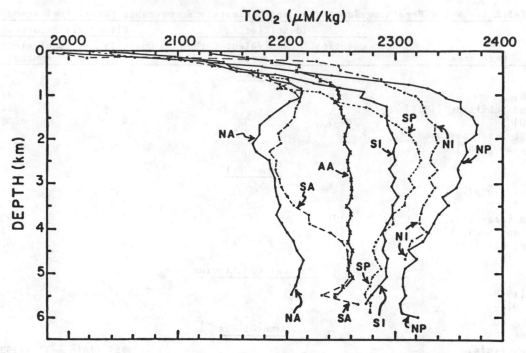

Fig. 2.   Average oceanic concentrations of dissolved inorganic carbon, calculated from the GEOSECS data [from Takahashi et al., 1981].   Data are averages for the following ocean regions:   NA, North Atlantic; SA, South Atlantic; AA, Antarctic region south of latitude 45°S; SI, South Indian; NI, North Indian; SP, South Pacific; and NP, North Pacific.

sand resources are particularly uncertain because recovery technologies are relatively undeveloped. Large quantities of readily available conventional oil supplies have inhibited the study of unconventional oil supplies in many nations.

During the combustion of these resources and the production of $CO_2$, a portion of each of the estimated carbon contents remains unoxidized. Losses of fuel occur in handling and transport, as unoxidized particulates, and in conversion to nonenergy products. The nonoxidized percentages of carbon in coal, oil, and natural gas have been estimated to be 1.8, 8.2, and 2%, respectively [Marland and Rotty, 1983]. These percentages have undoubtedly varied historically as the uses and combustion technologies of fossil fuels have changed.

Energy must be consumed in the refinement of synthetic fuels, with a proportionate amount of carbon released as $CO_2$. As a result of this energy consumption during production, synthetic fuels produce more $CO_2$ per unit of usable energy than do conventional fuels. While conventional oil, gas, and coal produce 14 to 24 kg carbon per $10^9$ joules of energy released as heat, synthetic oil and gas may produce as much as 33 kg carbon per $10^9$ joules [MacDonald, 1978]. However, this effect does not increase the amount of carbon ultimately available for combustion.

A unique aspect of oil shale extraction and mineralogy may significantly add to the amount of $CO_2$ ultimately released to the atmosphere. Oil shales of the Green River Formation, located in Colorado, Utah, and Wyoming, are actually dolomitic marlstones. High-temperature retorting (i.e., retorting at temperatures above 600°C) produces $CO_2$ from the calcination of carbonate minerals. Sundquist and Miller [1980] estimated that high-temperature retorting of the Green River marlstone could produce more than 200 Gt carbon as $CO_2$ from calcination, and about 150 Gt carbon as $CO_2$ from combustion of unrecoverable kerogen. These quantities of $CO_2$ would be added to the $CO_2$ produced by consuming the identified Green River shale oil resources. However, an unknown amount of the $CO_2$ produced by calcination would probably be reabsorbed by relatively rapid weathering of the highly alkaline retorting residues.

From the available data, the total amount of carbon contained in ultimately recoverable fossil fuel resources ranges from 6100 to 7700 Gt carbon. Identified resources account for about 3500 Gt carbon, which might be considered the minimum value for ultimate fossil fuel availability. Thus, the total amount of carbon ultimately available in fossil fuels is probably between 3500 and 7700 Gt carbon. After correcting for typical fractions of unoxidized carbon [Marland and Rotty, 1983] the amount of

TABLE 2. Distribution of Oceanic Dissolved Inorganic Carbon at Time of GEOSECS Measurements (1972-1977)

| Depth, m | Atlantic Ocean | Pacific Ocean | Indian Ocean | Antarctic[a] Ocean | Arctic[b] Ocean | Global Totals |
|---|---|---|---|---|---|---|
| 0-100 | 198 | 382 | 154 | 127 | 20.1 | 880 |
| 100-500 | 776 | 1540 | 630 | 496 | 50.7 | 3490 |
| 500-1000 | 965 | 1970 | 791 | 610 | 46.0 | 4380 |
| 1000-1500 | 944 | 1970 | 787 | 595 | 37.6 | 4330 |
| 1500-2000 | 916 | 1940 | 771 | 582 | 32.2 | 4240 |
| 2000-3000 | 1720 | 3640 | 1450 | 1090 | 46.2 | 7950 |
| 3000-4000 | 1400 | 2940 | 1180 | 888 | 16.6 | 6420 |
| 4000-5000 | 845 | 1730 | 696 | 527 | ... | 3800 |
| 5000-6000 | 247 | 506 | 204 | 155 | ... | 1110 |
| 0-6000 | 8010 | 16,620 | 6660 | 5070 | 250 | 36,600 |

Values are given as gigatons carbon (GtC). Data were calculated from the average concentrations given by Takahashi et al. [1981] and the hypsometric data of Menard and Smith [1966]. The following ocean surface areas were assumed [Baumgartner and Reichel, 1975] (in $10^6$ $km^2$): Atlantic, 80.04; Pacific, 159.023; Indian, 64.173; Antarctic[a], 49.364; Arctic, 8.5. Seawater specific gravity was assumed to be 1.025.

[a] Defined here as ocean south of latitude 50°S.

[b] Calculated from average concentrations given by Anderson and Dyrssen [1981] and the hypsometric data of Gorshkov [1980].

carbon available as $CO_2$ ranges from 3400 to 7500 Gt. (The unoxidized percentage of oil from oil shales and oil sands was assumed to be the same as that for conventional petroleum.) If the marlstones of the Green River Formation are ultimately retorted using high-temperature technologies, the additional release to the atmosphere could be about 350 Gt, increasing the maximum $CO_2$ value to about 8000 Gt carbon.

Attempts to quantify the relationship of fossil fuel consumption to present and past atmospheric $CO_2$ concentrations are hindered by uncertainties in estimates of cumulative production to date. The total industrial $CO_2$ production from 1860 to 1980 can be estimated from the data of Keeling [1973a] for the years 1860-1949 and Marland and Rotty [1983] for the years 1950-1980 as 163 Gt carbon. It should be emphasized that estimates of cumulative $CO_2$ production are subject to an uncertainty of about ±13% [Keeling, 1973a]. It is doubtful that refinements of past annual production estimates will significantly reduce this uncertainty. Thus, there is no reason to expect to "balance" the budget of historical $CO_2$ sources and sinks to better than about ±20 Gt carbon.

It is clear that fossil fuels will continue to play an important role in global carbon geochemistry. However, large uncertainties in the resource estimates and in future energy directions severely limit our ability to predict production, consumption, and $CO_2$ release from fossil fuels. In the short term, the greatest uncertainty with respect to fossil fuels and $CO_2$

production will be the rate of consumption. Over time scales of hundreds of years and longer, the primary uncertainty shifts from the rate to the total amount of consumption of fossil fuels. Over these longer time scales, technological and societal factors will determine how much of the world's ultimately recoverable fossil fuels will actually be exploited.

The Atmosphere

The atmosphere is the best known modern global carbon reservoir. Precise estimates of the mass of carbon as $CO_2$ in the atmosphere can be made from $CO_2$ concentration measurements and the mass and effective molecular weight of the atmosphere. For example, Fraser et al. [1983] give the average atmospheric $CO_2$ concentration as 337.04 parts per million by volume (ppmv) for the year 1980. Taking the effective atmospheric molecular weight as 28.97 (U.S. Standard Atmosphere, 1976) and the mass of the atmosphere as $5.12 \times 10^{21}$ g (Trenberth, 1981), they calculated the size of the 1980 atmospheric $CO_2$ reservoir as

$$(337 \times 10^{-6}) \left(\frac{12.01}{28.97}\right) (5.12 \times 10^{21}) =$$
$$715 \text{ Gt carbon} \qquad (1)$$

(The above values are for dry air. The mass of the total atmosphere, including water vapor, is $5.137 \times 10^{21}$ g [Trenberth, 1981].) Uncertainties in the mass and effective molecular weight of the atmosphere are very small (less than 0.1%

[Trenberth, 1981]).  Thus, the primary uncertainties in estimating the mass of atmospheric $CO_2$ are uncertainties in the average atmospheric $CO_2$ concentration.

These uncertainties are also small, arising primarily from temporal and spatial $CO_2$ variations.  Atmospheric $CO_2$ concentrations vary seasonally, with amplitudes ranging from about 1 ppmv at the south pole [Keeling et al., 1976] to about 15 ppmv at Point Barrow, Alaska [Peterson et al., 1982].  These variations appear to reflect seasonal changes in the balance between terrestrial photosynthesis and respiration, which are more intense in the northern hemisphere because of the greater land area [Machta et al., 1977; Pearman et al., 1983].  Superimposed on the seasonal oscillations, the secular increasing trend ranged from 0.5 to 1.7 ppmv per year between 1960 and 1980 [Keeling, 1983].

Fossil fuel $CO_2$ is also thought to be responsible for spatial atmospheric variations.  Seasonally averaged northern hemisphere concentrations are 2-5 ppmv higher than those at the south pole [Bolin and Keeling, 1963; Mook et al., 1983; Fraser et al., 1983], and this gradient has increased steadily since 1962 [Keeling, 1983].  Likewise, $CO_2$ concentrations measured between 1977 and 1979 in the stratosphere (which constitutes about 15% of the total atmospheric mass [Machta, 1973]) were about 7 ppmv lower than those in the troposphere [Bischof et al., 1980] (see also Ehhalt et al. [1975]).  Taking these variations into account, Fraser et al. [1983] estimated the accuracy of the value calculated in equation (1) to be 0.5-1.0%.  This estimate may be somewhat optimistic, as it is based on a two-dimensional integration of model results which were calibrated against data from predominantly marine locations [Pearman et al., 1983].  $CO_2$ measurements over land are notoriously variable [e.g., Schnell et al., 1981; Verma and Rosenberg, 1981], and it is reasonable to expect zonal gradients from land to sea [Fung et al., 1983].

The seasonal and spatial variations of atmospheric $CO_2$ were used in some of the earliest estimates of global atmospheric mixing rates [Bolin and Keeling, 1963; Bolin and Bischoff, 1970].  Recent refinements of this approach to atmospheric mixing have yielded time constants ranging from a few days to a few years for a two-dimensional atmospheric diffusion simulation model [Pearman and Hyson, 1980].  The atmospheric eddy diffusivities implied by the distribution of trace gases suggest similar time scales [Jacobi and André, 1963; Hidalgo and Crutzen, 1977; Hyson et al., 1980], as do the results of general circulation model studies [Fung et al., 1983].  Together, the rapidity of atmospheric mixing and the subtle magnitude of atmospheric spatial inhomogeneities imply that atmospheric $CO_2$ very closely approximates a homogeneous reservoir over time scales longer than a few years.

The size of the atmospheric $CO_2$ reservoir before human influence is considerably less certain.  Uncertainties in preindustrial atmospheric $CO_2$ estimates have important implications for the calibration of predictive $CO_2$ and climate models.  Until recently, nineteenth century atmospheric $CO_2$ concentrations were thought to have been about 290 ppmv.  This approximation had been inferred from late nineteenth and early twentieth century measurements, as assessed by Callendar [1958] and Bray [1959] and supported by early carbon cycle models [e.g., Bolin and Eriksson, 1959; Keeling, 1973b].  However, current estimates place atmospheric $CO_2$ levels at about 260-280 ppmv during the mid-nineteenth century [Elliott et al., 1984].

The most important historical atmospheric $CO_2$ data are the exquisite records from Mauna Loa and the south pole obtained by Keeling and co-workers (see Figure 1).  Although these data extend only from the late 1950's to the present, their high quality and continuity constrain models for carbon cycle behavior over much longer time scales.  For example, of the 86 Gt of carbon as $CO_2$ released from fossil fuels and cement manufacture between 1958 and 1980, only about 50 Gt were reflected in the measured atmospheric increase [Clark et al., 1982].  If the same proportion of total fossil fuel $CO_2$ production through 1980 (163 Gt; see above) were the only source of the atmospheric $CO_2$ increase since the mid-nineteenth century, the preindustrial atmospheric $CO_2$ concentration would have been about 290-295 ppmv.  Thus, the evidence for lower nineteenth century $CO_2$ levels implies an additional $CO_2$ source and/or a change in the dynamic behavior of the reservoirs that exchange $CO_2$ with the atmosphere.

The strongest evidence for lower preindustrial $CO_2$ concentrations comes from measurements of air trapped in ice samples from Greenland and Antarctica.  During the formation of glacial ice, air originally present in the open porous texture of fresh snow is gradually enclosed in sinuous channels in firn and is finally isolated in bubbles in buried ice.  The time between the deposition of new snow and the occlusion of these bubbles ranges generally from 100 to 3000 years, depending on the accumulation rate.  Because the air in snow and firn appears to mix rapidly with the atmosphere to depths of many meters, the occlusion process occurs over a narrower time range of about 20-600 years [Schwander and Stauffer, 1984].  Thus, the age of the gas is both younger and less precise than the age of the surrounding ice.  Another source of ambiguity is the selective concentration of $CO_2$ during summer melting and refreezing [Neftel et al., 1982; Neftel et al., 1983].  In order to minimize the difficulties caused by melting episodes and age offset, the most desirable locations for ice samples for $CO_2$ measurements must represent a compromise between ice accumulation rates and seasonal temperatures, which are generally inversely related in polar regions.  Other potential problems are the behavior of $CO_2$ during the formation of air clathrates and during solid adsorption or desorption. Experiments and comparisons among cores from different locations suggest that these effects are relatively minor [Neftel et al., 1983; Stauffer et al., 1984].

Measurements in ice from six different locations in both Greenland and Antarctica [World Climate Research Program, 1983], , and a rigorous interlaboratory comparison using ice dated at 800-2500 years B.P. [Barnola et al., 1983], suggest that the atmospheric $CO_2$ concentration was 260 ± 10 ppmv for at least 1000 years prior to the nineteenth century. Although small natural variations ( ~ 10 ppmv) have been inferred from some measurements corresponding to the interval A.D. 1500-1850 [Raynaud and Barnola, 1984], atmospheric $CO_2$ was apparently relatively constant throughout the centuries immediately preceding human influences. However, recent analytical innovations [Moor and Stauffer, 1984] have led Oeschger et al. [this volume] to suggest that a systematic error in previous analyses requires adjustment of the early nineteenth century $CO_2$ concentration to 280 ±5 ppmv. Thus, a prudent range for $CO_2$ levels before the nineteenth century is the "consensus" estimate of 260-280 ppmv [Elliott et al., 1984], corresponding to 550-590 Gt carbon.

This consensus is supported by several other lines of evidence, including carbon isotope measurements in tree rings and $CO_2$ analyses in dated water from the deep ocean. However, these approaches are indirect in the sense that they require implicit or explicit models for the exchange of carbon among the atmosphere, biosphere, and oceans. Preindustrial $CO_2$ estimates based on biosphere and ocean models will be discussed in the appropriate sections below.

It should be noted that $CO_2$ is not the only climatically important carbon compound that has become increasingly abundant in the atmosphere during modern times. Recent measurements have also documented increasing concentrations of methane [Graedel and McRae, 1980; Fraser et al., 1981; Rasmussen and Khalil, 1981a, b; Rasmussen et al., 1981; Blake et al., 1982; Ehhalt et al., 1983; Khalil and Rasmussen, 1983], carbon monoxide [Graedel and McRae, 1980; Khalil and Rasmussen, 1984], and several halocarbon compounds [Lovelock, 1974] (see Logan et al. [1981] for a review of halocarbon data collected in the 1970's). Ice core measurements suggest that the amount of methane in the atmosphere has doubled during the last two centuries [Craig and Chou, 1982; Khalil and Rasmussen, 1982; Rasmussen and Khalil, 1984]. Other studies have examined the potential impact of human activities on the amount of particulate carbon in the atmosphere [e.g., Wolff and Klimisch, 1982; Andreae, 1983; Hansen and Rosen, 1984]. Through their interaction with climate, these atmospheric constituents may influence the global carbon cycle to a degree far exceeding their contribution to the carbon mass balance [Ramanathan, 1975; Wang et al., 1976; Lacis et al., 1981; Chamberlain et al., 1982]. However, because their atmospheric abundances are several orders of magnitude below that of $CO_2$, they will not be considered further in this paper.

## The Oceans

Over time scales of centuries and longer, the effects of fossil fuel $CO_2$ will depend largely on its fate in the oceans. The oceans will not only remove $CO_2$ from the atmosphere by direct dissolution, but also mediate many of the long-term responses of the global carbon cycle.

The distribution of dissolved inorganic carbon in the oceans can be calculated accurately from the results of the GEOSECS expeditions of 1972-1977. Figure 2 shows mean vertical concentration profiles calculated for seven world ocean regions by Takahashi et al. [1981], and Table 2 lists masses calculated from volume-weighted average concentrations. The most conspicuous features of these data are the increase in concentrations between the ocean surface and depths of 1000-2000 m, and the increase in deep water concentrations from the North Atlantic through the southern hemisphere to the North Pacific.

One of the central problems in $CO_2$ research is to determine the abundance and distribution of carbon in the oceans before man's influence. The inhomogeneities apparent in Figure 2 result from the interaction of atmospheric exchange, ocean mixing, and biological cycling. The oceans' response to anthropogenic $CO_2$ has involved interplay among the same basic processes. In attempting to understand these complex processes and their even more complex interactions, $CO_2$ researchers have confronted some of the most fundamental problems in oceanography.

## Air-Sea Exchange

$CO_2$ exchange between the atmosphere and the ocean surface is rapid compared to the time scales of interest in this paper. In the ocean mixed layer, the distribution of dissolved inorganic carbon is close to equilibrium with respect to atmospheric $CO_2$, resulting in a gradient increasing from equator to poles as the solubility of $CO_2$ increases with decreasing temperature [Wattenberg, 1933; Takahashi, 1961; Keeling, 1968; Takahashi et al., 1980; Sundquist and Plummer, 1981]. $CO_2$ gas exchange rates have been quantified using the natural $^{14}C$ levels in the atmosphere and the ocean surface, the changes in those levels caused by anthropogenic effects (specifically, "bomb $^{14}C$" added to the atmosphere by nuclear explosions, and the "Suess effect" caused by $^{14}C$-free $CO_2$ from combustion of fossil fuels), and the application of gas exchange theory to rates determined for the inert radioactive gas $^{222}Rn$. These estimates, summarized in Table 3, can be expressed in terms from the

**TABLE 3.  $CO_2$ Gas Exchange Estimates**

| Reference | $k_{gas}$, $10^{-4}$ moles m$^{-2}$ yr$^{-1}$ atm$^{-1}$ | $(k_{as})^{-1}$, yr | Invasion Flux[a] | | |
|---|---|---|---|---|---|
| | | | moles m$^{-2}$ yr$^{-1}$ | $10^{15}$ moles yr$^{-1}$ | Gt C yr$^{-1}$ |
| *Based on Natural* $^{14}C$ | | | | | |
| Craig [1957] | 4.9–12.2 | 4–10[b] | 13–33 | 4.8–11.9 | 57–143 |
| Revelle and Suess [1957] | 7.0 | 7 | 19 | 6.8 | 82 |
| Arnold and Anderson [1957] | 4.9 | 10 | 13 | 4.8 | 57 |
| Craig [1958] | 4.1–24 | 2–12[c] | 11–66 | 4.0–24 | 48–290 |
| Bolin and Eriksson [1959] | 9.8 | 5 | 26 | 9.6 | 115 |
| Broecker [1963] | 6.1 | 8[d] | 17 | 6.0 | 72 |
| Craig [1963] | 3.3–9.8 | 5–15 | 9–26 | 3.2–9.6 | 38–115 |
| Broecker et al. [1971][e] | | 10–20[e] | | | |
| Keeling [1973b] | 7.0 | 7 | 19 | 6.8 | 82 |
| Broecker [1974] | 5.3 | 9.2 | 14[f] | 5.2 | 62 |
| Oeschger et al. [1975] | 5.4–8.1 | 6–9 | 15–22 | 5.3–8.0 | 64–96 |
| Keeling [1979] | 6.49 | 7.53 | 17.5 | 6.35 | 76.1 |
| Peng et al. [1979] | 5.2–8.9 | 5.5–9.4[g] | 14–24[h] | 5.1–8.7 | 61–104 |
| Siegenthaler et al. [1980] | 6.5 | 7.5 | 18 | 6.4 | 76 |
| Lal and Suess [1983][i] | 2.0–1.6 | 3–25 | 5.3–44 | 1.9–16 | 23–191 |
| Siegenthaler [1983] | 4.6–6.2 | 7.9–10.6 | 12–17 | 4.5–6.0 | 54–73 |
| Kratz et al. [1983] | 7.3 | 6.7 | 20 | 7.1 | 86 |
| *Based on Suess Effect* | | | | | |
| Ferguson [1958] | 24 | 2 | 66 | 24 | 290 |
| Bacastow and Keeling [1973] | 7.0–7.8 | 6.3–7.0 | 19–21 | 6.8–7.6 | 82–91 |
| Stuiver and Quay [1981][e] | | | 21 | | |
| *Based on Bomb* $^{14}C$ | | | | | |
| Bien and Suess [1967] | <4.9 | >10 | <13 | <4.8 | <57 |
| Munnich and Roether [1967] | 9.1 | 5.4 | 24 | 8.9 | 106 |
| Nydal [1968] | 4.9–9.8 | 5–10 | 13–26 | 4.8–9.6 | 57–115 |
| Young and Fairhall [1968] | 8.1–12.2 | 4–6 | 22–33 | 8.0–11.9 | 96–143 |
| Rafter and O'Brien [1970] | 4.1 | 12 | 11 | 4.0 | 48 |
| Broecker et al. [1980a] | 5.5–7.8 | 6.2–8.8 | 15–21[j] | 5.4–7.6 | 65–91 |
| Stuiver [1980] | 7.2 | 6.8 | 19 | 7.0 | 84 |
| Quay and Stuiver [1980] | 6.6 | 7.5 | 18 | 6.4 | 77 |
| Delibrias [1980] | 8.1 | 6.0 | 22 | 7.9 | 95 |
| Druffel and Suess [1983] | 3.9 | 12.5[k] | 11 | 3.8 | 46 |
| Siegenthaler [1983] | 6.48–7.0 | 6.99–7.54 | 17.5–18.9[l] | 6.34–6.84 | 76.0–82.1 |
| *Based on* $^{222}Rn$ | | | | | |
| Broecker and Peng [1974] | 6.1 | 8 | 17 | 6.0 | 72 |
| Peng et al. [1979] | 3.7–6.3 | 7.8–13.2 | 10–17[m] | 3.6–6.2 | 43–74 |
| Peng et al. [1983] | 5.8 | 8.4 | 16[n] | 5.7 | 68 |

[a]Assuming total ocean surface area of $362 \times 10^{12}$ m$^2$, total atmosphere of $1.77 \times 10^{20}$ moles, and atmospheric $pCO_2$ of $270 \times 10^{-6}$ atm.

[b]Given as 7±3.

[c]Given as 7±5.

[d]Recalculated by Broecker and Peng [1974].

[e]$CO_2$ partial pressure not specified.

[f]Given as 17 moles m$^{-2}$ yr$^{-1}$ for a $pCO_2$ of $320 \times 10^{-6}$ atm.

[g]Authors give 7.6 years.

[h]Given as 19 moles m$^{-2}$ yr$^{-1}$ with error of ±5 moles m$^{-2}$ yr$^{-1}$.

[i]Authors used "Suess wiggles" rather than natural $^{14}C$ base levels.

[j]Given as 18–25 moles m$^{-2}$ yr$^{-1}$ for 1955–1973.  Corrected to preindustrial value.

[k]Based on relaxation rate of $0.08°/°° (°/°° \text{-yr})^{-1}$.

[l]Given as 19.3 and 20.6 moles m$^{-2}$ yr$^{-1}$ for $pCO_2$ values of $298.54 \times 10^{-6}$ and $295 \times 10^{-6}$ atm, respectively.

[m]Given as 16 ± 4 moles m$^{-2}$ yr$^{-1}$ for a $pCO_2$ of $320 \times 10^{-6}$ atm.

[n]Given as 17 moles m$^{-2}$ yr$^{-1}$ for a $pCO_2$ of $280 \times 10^{-6}$ atm.

equations for $CO_2$ gas exchange based both on gas exchange theory and global box modeling. From gas exchange theory,

$$\text{invasion flux} = k_{gas}\, pCO_{2a} \qquad (2a)$$

$$\text{evasion flux} = k_{gas}\, pCO_{2s} \qquad (2b)$$

where $pCO_{2a}$ is the partial pressure of $CO_2$ in the atmosphere and $pCO_{2s}$ is the $CO_2$ partial pressure at equilibrium with the ocean surface. For global box models,

$$\text{invasion flux} = k_{as}\, N_a \qquad (3a)$$

$$\text{evasion flux} = k_{sa}\, N_s \qquad (3b)$$

where $N_a$ and $N_s$ are the total inorganic carbon masses in the atmosphere and the ocean surface, respectively. From these equations, it is clear that differences among flux estimates will arise from differences in $CO_2$ masses and partial pressures as well as differences in the estimated constants. In order to facilitate comparison of the estimated constants, the invasion fluxes reported in Table 3 have been corrected to a preindustrial $pCO_{2a}$ of $270 \times 10^{-6}$ atm (corresponding to a $CO_2$ concentration of about 275 ppmv in dry air).

All of the estimation methods point toward a value for $k_{as}^{-1}$ of about 10 years or less. Because the return flux from the ocean mixed layer to the atmosphere must be nearly equal to the global $CO_2$ invasion rate, and because the ocean mixed layer contains an amount of dissolved inorganic carbon close to the amount of carbon as $CO_2$ in the atmosphere (see Table 2), the value of $k_{sa}$ must be comparable to that for $k_{as}$. Thus, $CO_2$ exchange between the atmosphere and the ocean surface is characterized by a time scale of a few years. In fact, it has been shown that, even over time scales of decades, global exchange between the atmosphere and the ocean mixed layer can be accurately modeled as a single box characterized by chemical equilibrium between the atmosphere and the ocean surface [Bolin and Eriksson, 1959; Sundquist and Plummer, 1981]. Although this characterization ignores the significance of disequilibrium in polar surface waters associated with sources of deep water, uncertainties in polar ocean carbon fluxes are derived primarily from uncertainties in ocean mixing and biological processes.

## Ocean Mixing

Unlike the general agreement between gas exchange theory and its representation in $CO_2$ models, there are serious gaps between physical theories of ocean mixing and the parameterizations of ocean mixing in current $CO_2$ models. Transport of $CO_2$ in the oceans is typically modeled using expressions that relate only vaguely to known physical mechanisms. Before a rigorous description of the general circulation of the oceans can be accomplished, fundamental observational and mathematical difficulties must yet be overcome. Many of these difficulties concern the practical scaling of spatial and temporal effects on the transport of heat, salt, and momentum (see, for example, Nihoul, [1981] and Robinson [1983]). Thus, the scaling of effects on $CO_2$ and carbon fluxes is inherently limited by current uncertainties in the physics of ocean mixing.

Because the removal of atmospheric $CO_2$ to the oceans is a fundamentally vertical process, carbon cycle models have traditionally emphasized the vertical component of ocean mixing. The earliest box models to be calibrated using $^{14}C$ estimated the residence time of the deep sea to be in the range of 500-1500 years with respect to exchange with the ocean surface [Craig, 1957; Burling and Garner, 1959; Bien et al., 1960; Broecker et al., 1960; Broecker, 1963]. Recent calculations have put this value at about 1100 years [Keeling, 1979]. More detailed studies have converged on a global value of about 500 years for the residence time of deep ocean water with respect to exchange with water shallower than 1500 m; flushing of the deep Atlantic may occur as rapidly as every 100-250 years [Stuiver, 1976; Broecker, 1979; Stuiver et al., 1983]. If this exchange is accomplished by means of advection, about $50 \times 10^6$ $m^3$ of water must exchange with the global deep sea each second. This flow rate is within the range of global mean upwelling velocities calculated as necessary to balance the downward transport of heat through the thermocline [Stommel, 1958; Robinson and Stommel, 1959; Stommel and Arons, 1960; Wyrtki, 1961; Munk, 1966]. Estimates of deep water formation rates [e.g., Stommel, 1958; Stommel and Arons, 1960; Kuo and Veronis, 1970; Carmack and Foster, 1975; Gordon and Taylor, 1975; Stuiver, 1976; Broecker, 1979; Stuiver et al., 1983] and equatorial upwelling rates [Broecker et al., 1978; Tomczack, 1979; Quay and Stuiver, 1980; Wyrtki, 1981] also lie in the range of tens of millions of cubic meters per second. Thus, although these global and regional upwelling estimates probably oversimplify complex processes, they indicate that vertical ocean mixing is characterized by time scales of hundreds of years, and that its accurate portrayal is critical to modeling carbon cycle changes over these time scales.

Welander [1959] was the first to observe that a diffusive treatment of vertical ocean mixing shows promise of resolving problems encountered in modeling $CO_2$ perturbations over a range of different time scales. This advantage was exploited by Oeschger et al. [1975], whose "box-diffusion" model successfully simulated perturbations associated with both fossil fuel $CO_2$ and bomb $^{14}C$. Welander [1959] demonstrated that a simple box model approximation of diffusion is reasonable when the characteristic time of internal mixing, defined as

TABLE 4.  Apparent Vertical Ocean Diffusivities Based on $^{14}$C

| Reference | $K_v$, cm$^2$ s$^{-1}$ | Location | Depth |
|---|---|---|---|
| | | Based on Natural $^{14}$C | |
| Broecker [1966] | 0.4±0.1 | Pacific-Indian | 0-750 m |
| | 1.5±0.4 | North Atlantic | 0-750 m |
| Craig [1969] | 2.2±0.6 | North Pacific | 1000-4000 m |
| Broecker et al. [1971] | 0.79 | Global, no polar outcrop | 0-1000 m |
| | 0.57 | Global, polar outcrop | 0-1000 m |
| Oeschger et al. [1975] | 0.97-1.77[a] | Global | 75-3800 m |
| Keeling and Bacastow [1977] | 0.9[b] | Global | 1000-4000 m |
| MacIntyre [1980] | 1.3 | Global | 100-2250 m |
| Fiadeiro [1980a] | 0.7 | Pacific | Deepest 3000 m |
| Hoffert et al. [1981] | 0.6 | Global | 100-4100 m |
| Sire et al. [1981][c] | 8.32 | Global | Surface |
| | 1.62 | Global | All depths |
| Crane [1982] | 0.76 | Atlantic | 0-4000 m |
| | 4.0 | Pacific | 3800 m |
| Enting and Pearman [1982][d] | 5.7 | Global | Surface |
| | 19 | Global | 3800 m |
| Enting and Pearman [1983] | 1.6-1.7 | Global | Surface |
| | 1.9-2.0 | Global | 0-1500 m |
| Siegenthaler [1983] | 1.27 | Global, no polar outcrop | 75 m to bottom |
| | 0.70 | Global, polar outcrop | 75 m to bottom |
| Peng et al. [1983] | 1.6[e] | Global | 100-700 m |
| | 0.5 | Global | 700-3800 m |
| Hoffert et al. [1983] | 0.6 | Global | All depths |
| | | Based on Bomb $^{14}$C | |
| Broecker et al. [1978] | 0.5 | Equatorial Atlantic | 100-900 m |
| | 3.0 | North Atlantic | 100-900 m |
| Broecker and Peng [1980][f] | 0.15-0.65 | East Pacific | 0-500 m |
| Stuiver [1980] | 4.0 | Atlantic | 0-1000 m |
| Quay and Stuiver [1980] | 0.6 | Equatorial Atlantic | 0-500 m |
| | 1.5 | Temperate Atlantic | 0-1000 m |
| Broecker et al. [1980b] | 6-7 | North Atlantic | 0-3000+ m |
| Broecker et al. [1980c] | 0.35 | Equatorial Atlantic | 0-1000 m |
| | 2.8 | Temperate Atlantic | 0-1000 m |
| | 1.9-3.3 | Global | Not specified |
| Stuiver and Quay [1981] | 3 | Atlantic and Pacific | Not specified |
| Viecelli et al. [1981] | 5.1 | Northern hemisphere | Thermocline |
| Kratz et al. [1983][g] | 32 | Low latitudes | 75 m |
| | 5.2 | Low latitudes | 1000 m |
| | 2.5 | High latitudes | 75 m |
| | 0.25 | High latitudes | 1000 m |
| Siegenthaler [1983] | 2.2 | Global, no polar outcrop | 75 m to bottom |
| | 1.6 | Global, polar outcrop | 75 m to bottom |

[a]From authors' Table 2.
[b]From authors' Appendix 4.A, in which they argue against a single global value of $K_v$.
[c]$K_v$ assumed to be depth-dependent; 1.62 cm$^2$ s$^{-1}$ is depth-weighted average.
[d]$K_v$ assumed to be depth-dependent; $K_v$ for 3800 m from parameters given by authors.
[e]Chosen to fit tritium penetration data as well as natural $^{14}$C.
[f]Authors note inconsistencies with reasonable gas exchange rates, suggesting importance of horizontal transport.
[g]$K_v$ assumed to be depth-dependent; depth-averaged value given as 2.6 cm$^2$ s$^{-1}$ for high latitudes.

TABLE 5.  Marine Biomass

| Reference | Quantity, Gt C | Notes |
|---|---|---|
| | **Fauna** | |
| Zenkevich et al. [1960] | 0.90[a] | |
| Vinogradov [1965] | 1.80[a] | |
| Bowen [1966] | 0.509 | Includes protozoa, coelenterates, annelids, nematodes, molluscs, echinoderms, arthropods, fishes |
| Bogorov [1967] | 2.9[a] | |
| Bogorov [1967] | 2.4[b] | Zooplankton plus zoobenthos plus nekton from author's Table 4 |
| Bogorov [1969] | 1.5[c] | "Consumers reducers" |
| Moiseev [1969] | 1.55[d] | Zooplankton |
| Bolin [1970] | <5 | |
| Whittaker and Likens [1973] | 0.44 | Excludes estuaries |
| Whittle [1977] | 0.532 | Macrozoobenthos; calculated from data of Menzies [1973] |
| Cauwet [1978] | 0.1 | |
| Nienhuis [1981] | 2.6 | Calculated from data of Conover [1978], who in turn cited Bogorov [1967] and Skopintsev [1971] |
| Nienhuis [1981] | 1.3 | Calculated from data of Rodin et al. [1975] |
| | **Flora** | |
| Bowen [1966] | 2.45 | Plankton and algae |
| Bogorov [1967] | 0.15[a] | Phytoplankton; phytobenthos = 0.02 |
| Bogorov [1967] | 0.091[b] | Phytoplankton plus phytobenthos from author's Table 4 |
| Bogorov [1969] | 0.08[c] | |
| Bolin [1970] | 5 | |
| Whittaker and Likens [1973] | 1.13 | Excludes estuaries |
| Whittaker and Likens [1975] | 1.1[b] | Excludes estuaries |
| Woodwell et al. [1978] | 1.11 | Excludes estuaries |
| Cauwet [1978] | 4 | |
| | **Bacteria** | |
| Bowen [1966] | 0.36 | |
| Sorokin [1978] | 0.23 | |
| Cauwet [1978] | 0.1 | |
| Mopper and Degens [1979] | 0.2 | |
| | **Estuarine Biomass** | |
| Whittaker and Likens [1973] | 0.64 | Flora 0.63; fauna 0.01 |
| Woodwell et al. [1978] | 0.63 | Excludes fauna |
| | **Total or Undifferentiated Biomass** | |
| Bowen [1966] | 3.319 | Includes estuarine biomass and bacteria |
| Bogorov [1967] | 3.05[a] | |
| Bogorov [1967] | 2.536[b] | |
| Bogorov [1969] | 1.79[c] | |
| Bolin [1970] | 5+ | Phytoplankton 5, plus "zooplankton, fish," <5 |
| Reiners [1973b] | 1.5 | |
| Whittaker and Likens [1973] | 2.21 | Includes estuarine biomass |
| Baes et al. [1976] | 1 | |
| Cauwet [1978] | 4.2 | Includes estuarine biomass and bacteria |
| Mopper and Degens [1979] | 3.2 | Plankton and bacteria |

[a]From de Vooys [1979], who cited Moiseev [1971].
[b]Calculated from author's "dry matter" data assuming 45% is carbon [Whittaker and Likens, 1973].
[c]Calculated from data given by Bazilevich et al. [1971] assuming "dry weight" biomass is 45% carbon [Whittaker and Likens, 1973].
[d]From de Vooys [1979].

TABLE 6.  Marine Primary Productivity

| Reference | Flux, Gt C yr$^{-1}$ | Notes |
|---|---|---|
| Bowen [1966] | 29 | |
| Bogorov [1969] | 25[a] | |
| Ryther [1969] | 20 | |
| Whittaker [1970] | 25 | |
| Bazilevich et al. [1971] | 27[b] | |
| SCEP [1970] | 22 | |
| Bolin [1970] | 40 | "Assimilation" |
| Koblentz-Mishke et al. [1970] | 23 | |
| Skopintsev [1971] | 38.4 | |
| Bruyevich and Ivanenkov [1971] | 44 | Revised data of Koblentz-Mishke et al. [1970], assuming radiocarbon data underestimates true productivity |
| Golley [1972] | 25[b] | |
| Sorokin [1974] | 60 to 80 | Assumed radiocarbon data underestimates productivity by a factor of 1.5-2 |
| Lieth [1975] | 25[b] | |
| Whittaker and Likens [1973] | 24.9 | |
| Whittaker and Likens [1975] | 25[b] | |
| Williams [1975] | 36 | |
| Platt and Subba Rao [1975] | 31 | |
| Fogg [1975] | 22-32 | |
| Bolin et al. [1979] | 45 | |
| de Vooys [1979] | 32 | Data of Koblentz-Mishke et al. [1970] corrected for extracellular excretions and vacuum storage |
| de Vooys [1979] | 43.5 | Data of Platt and Subba Rao [1975] corrected for extracellular excretions and vacuum storage |

[a]Calculated from data given by Bazilevich et al. [1971], assuming "dry weight" biomass is 45% carbon [Whittaker and Likens, 1973].

[b]Calculated from authors' "dry matter" data assuming 45% is carbon [Whittaker and Likens, 1973]

$$\tau^* = h^2/2K_{dif} \qquad (4)$$

is much shorter than the response time associated with other processes. Equation (5) provides a rationale for structuring box models with a spatial resolution of

$$h = \sqrt{2K_{dif}\tau^*} \qquad (5)$$

where h is the dimension of a box parallel to the direction of diffusion according to the coefficient $K_{dif}$, and $\tau^*$ is arbitrarily assigned some value much (say, an order of magnitude) shorter than the time scale of interest.

It is therefore tempting to use vertical ocean diffusivities to scale ocean box model structures according to their time response characteristics. Table 4 shows a range of values for the apparent vertical ocean diffusivity ($K_v$) calculated from the distribution of both natural and bomb $^{14}C$. All of these estimates are model-dependent, and thus some of the differences apparent in Table 4 result from the use of different models.  Other differences arise from geographical effects. Nevertheless, estimates derived from globally averaged models cluster well within an order of magnitude, ranging from 0.5 to 2.0 cm$^2$ s$^{-1}$ (based on natural $^{14}C$) and from 1.6 to 3.3 cm$^2$ s$^{-1}$ (based on bomb $^{14}C$). This order of magnitude is consistent with estimates based on the oceans' large-scale thermocline circulation [Wyrtki, 1961; Munk, 1966; Fiadeiro and Craig, 1978] and global tritium penetration [Broecker et al., 1979; Li et al., 1982] and with regional distributions of chlorofluoromethanes [Gammon et al., 1982], tritium [Rooth and Ostlund, 1972; Cline and Ostlund, 1982; Sarmiento, 1983a], and $^{226}Ra$ [Ku et al., 1980; Chung and Craig, 1980]. Assuming that a vertical "mixing length" can be defined by equation (5), with $\tau^*$ assigned a maximum "mixing time" of 50 years, the global $K_v$

estimates based on $^{14}C$ imply that ocean box models for time scales of hundreds of years should be structured with a vertical resolution of at least 400-1000 meters.

Likewise, estimates of apparent horizontal ocean diffusivity ($K_H$) appear to offer a reasonable way to determine the horizontal resolution of ocean box models appropriate to particular time scales. Values for $K_H$ ranging from 0.4 to 6 x $10^7 cm^2 s^{-1}$ have been calculated from the distributions of $^{222}Rn$ [Sarmiento and Rooth, 1980], $^{226}Ra$ [Ku et al., 1980], $^{228}Ra$ [Moore et al., 1980; Sarmiento et al., 1982b], and tritium [Michel and Suess, 1975; Fine and Ostlund, 1980]. This range is consistent with values estimated from thermohaline circulation studies [Kuo and Veronis, 1970; Needler and Heath, 1975; Fiadeiro and Craig, 1978], dye-release experiments [Okubo, 1971], and statistical interpretations of buoy trajectories [Freeland et al., 1975; McWilliams et al., 1983]. In a manner analogous to the derivation of a vertical mixing length from $K_V$, a horizontal resolution of 1000-4000 km can be calculated from the range of $K_H$ estimates and a maximum internal mixing time of 50 years. This resolution might be appropriate for an ocean box model structured for time scales of hundreds of years.

Unfortunately, a number of difficulties limit the value of $K_V$ and $K_H$ in scaling calculations such as those above. Many of the model structures and assumptions used in the estimation of $K_V$ and $K_H$ are vulnerable to serious criticisms. In contrast to mean global models, the mixing processes encompassed by apparent diffusivities may be nonlinear, intermittent, and localized. For example, the $K_V$ values summarized in Table 4 are significantly higher than those derived from measurements and theories of small-scale turbulence [Garrett, 1979]. The $K_V$ values implied by tritium and bomb $^{14}C$ uptake are generally higher than those implied by natural $^{14}C$ levels in the deep sea, leading several modelers to experiment with depth- and/or latitude-dependent $K_V$ functions [Jenkins, 1980; Stuiver, 1980; Broecker et al., 1980c; Sire et al., 1981; Viecelli et al., 1981; Peng et al., 1983; Kratz et al., 1983]. More drastic structural revisions are implied by the observation that the vertical distributions of many ocean properties, particularly tritium, seem to require the dominance of relatively rapid horizontal or isopycnal transport accompanied by mixing at ocean surface or bottom boundaries [e.g., Wunsch, 1970; Welander, 1971; Rooth and Ostlund, 1972; Veronis, 1975; Fine and Ostlund, 1980; Jenkins, 1980, Sarmiento and Rooth, 1980; Fine et al., 1981; Sarmiento et al., 1982a; Sarmiento, 1983b].

## Biological processes

Many current ocean $CO_2$ models do not include explicit terms for the cycling of carbon by the marine biota. While the absence of marine biological terms from $CO_2$ models does not necessarily diminish their predictive value over time scales of decades (see, for example, Keeling [1979] and Peng et al. [1983]), longer-term models must incorporate the fluxes associated with the settling and alteration of particles derived from plants and animals living in ocean surface water.

Tables 5 and 6 show recent estimates of marine biomass and primary productivity. Because of the sparseness and "patchiness" of marine organisms, their biomass is difficult to estimate. The more carefully documented estimates in Table 5 range from about 2 to 4 Gt carbon. Part of the variance among these estimates is associated with the "inverted" ecological pyramid structure represented by the excess of zoomass over phytomass in the estimates of several Soviet scientists [e.g., Bogorov, 1967]. Other scientists maintain that marine ecology follows the classical terrestrial model, in which the producer biomass is larger than the consumer biomass [e.g., Whittaker and Likens, 1973].

The productivity estimates in Table 6 are derived from measurements using the $^{14}C$ uptake technique. According to these data, annual marine primary production exceeds the marine biomass by about an order of magnitude, reflecting the importance of biological recycling in the oceanic carbon budget. Most of this recycling occurs near the ocean surface over very short time scales. Carbon is rapidly recycled via grazing and remineralization [Harris, 1959; Steele, 1972; Menzel, 1974; Harrison, 1980], enabling plankton production to persist even in waters with extremely low bulk nutrient concentrations [McCarthy and Goldman, 1979]. Recent studies have emphasized the importance of localized environments (such as aggregates known as "marine snow") and microheterotrophic organisms in the rapid turnover of carbon and nutrients near the ocean surface [Pomeroy, 1974; Sieburth et al., 1978; Shanks and Trent, 1979; Jackson, 1980; Williams, 1981; Alldredge and Cox, 1982; Knauer et al., 1982; Goldman, 1984; Smith et al., 1984]. Although the relative definitions and roles of primary productivity and recycling are points of contention, particularly in oligotrophic waters [Sorokin, 1974; Shulenberger and Reid, 1981; Jenkins, 1982; Williams et al., 1983; Gieskes and Kraay, 1984; Platt, 1984; Platt et al., 1984], these arguments are not generally pertinent to the time scales considered here. The primary conclusion from the data in Table 6 is the intensity of rapid carbon recycling near the ocean surface.

The relatively small carbon flux leaving the ocean euphotic zone has been equated to "new production," a concept originally derived from marine nitrogen cycling studies [Dugdale and Goering, 1967]. Eppley and Peterson [1979] documented the general agreement between the carbon particulate flux implied by ocean box models and global new production based on extrapolation from

TABLE 7.  Global Particulate Organic Carbon Flux from Surface to Deep Ocean

| Reference | Flux, Gt C yr$^{-1}$ | Notes |
|---|---|---|
| Based on $^{15}$N Uptake | | |
| Eppley and Peterson [1979] | 3.4–4.7 | Excludes polar waters and waters over floor <200 m deep |
| Based on Sediment Trap Data | | |
| Extrapolation from Suess [1980] | 7.3[a] | Open ocean flux at 100 m |
| Extrapolation from Suess [1980] | 3.8[a] | Open ocean flux at 200 m |
| Extrapolation from Betzer et al. [1984a] | 2.4[a] | Open ocean flux at 100 m |
| Extrapolation from Betzer et al. [1984a] | 1.6[a] | Open ocean flux at 200 m |
| Based on Steady State Ocean Models | | |
| Keeling and Bolin [1968] | 1.5–2.4 | Three-box model; Pacific only |
| Pytkowicz [1968] | 2.4 | Based on deep sea oxygen consumption |
| Li et al. [1969] | 2.8[b] | Two-box model |
| Broecker and Li [1970] | 4.4–6.7[c] | Three-box model; carbonate flux included |
| Keeling [1973b] | 2.3–3.4[d] | Two-box model; $\tau_{dm}$ = 1000 to 1500 years; carbonate flux included |
| Broecker [1974] | 1.6[e] | Two-box model |
| Keeling [1979b] | 3.5 | Two-box model; carbonate flux included |
| Bjorkstrom [1979] | 2.4 | Twelve-box model; carbonate flux assumed to be negligible |
| Sundquist [1979] | 2.8 | Two-box model |
| Broecker and Peng [1982] | 2.4[f] | Two-box model |
| Bolin et al. [1983] | 5.0±0.8 | Twelve-box model; authors' reference case |
| Enting and Pearman [1983] | 3.4–3.6 | Upwelling-diffusion model; carbonate flux included |
| Keir [1983a] | 3.7 | Two-box model |
| Keir and Berger [1983] | 4.3–5.0 | Two-box model |
| Broecker and Peng [1984] | 3.3 | Two-box model |
| Knox and McElroy [1984] | 3.0 | Three-box model; carbonate flux included |
| Toggweiler and Sarmiento [1984] | 3.6 | Three-box model; carbonate flux included |
| Siegenthaler and Wenk [1984] | 2.9 | Three-box model; carbonate flux included |
| Viecelli [1984] | 7–12 | One-dimensional diffusion model |
| Keir and Berger [this volume] | 3.3 | Two-box model |
| Toggweiler and Sarmiento [this volume] | 2.8 | Three-box model |
| Volk and Hoffert [this volume] | 2.4 | Upwelling-diffusion model |
| Wenk and Siegenthaler [this volume] | 2.3 | Three-box model |

[a]Calculated from equation (1) of Suess [1980] or equation (1) of Betzer et al. [1984a] and productivities and areas listed in Table 1 of Koblentz-Mishke et al. [1970]. "Neritic" waters omitted to be consistent with calculations of Eppley and Peterson [1979].
[b]Calculated from authors' rate of deep sea $CaCO_3$ dissolution (0.7 Gt C yr$^{-1}$) and organic/carbonate destruction ratio of 4.
[c]Calculated from authors' Table 1, assuming Atlantic Ocean is 23% and Pacific-Indian Ocean is 63% of total ocean area.
[d]Calculated from author's equation (7.9).
[e]Calculated from author's rate of deep sea $CaCO_3$ dissolution (0.8 Gt C yr$^{-1}$) and organic/carbonate destruction ratio of 2.
[f]Calculated from authors' rate of deep sea $CaCO_3$ dissolution (0.6 Gt C yr$^{-1}$) and organic/carbonate destruction ratio of 4.

TABLE 8.   Marine Saprosphere Calculated from Depth-Dependent Data

| Depth, m | Volume[a], $10^{17}$ $m^3$ | Suspended Organic Carbon[b] | | Dissolved Organic Carbon[c] | |
|---|---|---|---|---|---|
| | | $mg/m^3$ | Gt | $g/m^3$ | Gt |
| 0-100 | 0.36 | 30-54 | 1.1-1.9 | 1-2 | 36-72 |
| 100-500 | 1.36 | 11-11.8 | 1.5-1.6 | 0.6-1.3[d] | 82-180 |
| 500-1000 | 1.64 | 4.6-6.3 | 0.75-1.0 | 0.3-0.7 | 49-110 |
| 1000-3000 | 6.10 | 3.0-4.9 | 1.8-3.0 | 0.3-0.7 | 180-430 |
| 3000-6000 | 4.22 | 3.0-3.6 | 1.3-1.5 | 0.3-0.7 | 130-300 |
| 0-6000 | 13.68 | | 6.5-9.0 | | 480-1100 |

[a]From hypsometric data of Menard and Smith [1966].
[b]Concentrations from Wangersky [1976, Tables 3 and 4].  Range shown represents averages for Pacific and South Atlantic samples.
[c]Concentrations from Menzel, 1974, p. 668.
[d]Interpolated from Menzel's surface water and deep (>300 m) water figures.

[15]N uptake studies.  This agreement is further scrutinized in Table 7, which also shows open ocean fluxes calculated from the relationships between organic carbon particle flux and productivity inferred by Suess [1980] and Betzer et al. [1984a] from several sediment trap studies.  These comparisons must be treated cautiously.  Organic carbon in settling particles is rapidly depleted, causing the flux to the deep sea to decrease to about 5-10% of the [14]C-based primary productivity within a few hundred meters of the ocean surface [Bishop et al., 1977, 1978; Knauer et al., 1979; Honjo, 1980; Knauer and Martin, 1981; Wefer et al., 1982; Honjo et al., 1982c; Lorenzen et al., 1983; Angel, 1984].  Particle flux estimates therefore depend strongly on the depth to which they are attributed.  Corrections may also be necessary to account for methodological artifacts in the nitrogen fixation estimates extrapolated by Eppley and Peterson [Glibert et al., 1982], and for these authors' exclusion of Antarctic circumpolar waters, which may be very productive [Platt and Subba Rao, 1975; Jennings et al., 1984].  Whereas the equations of Suess [1980] and Betzer et al. [1984a] are based on mean annual productivities, Eppley and Peterson extrapolated from daily rates.  Flux calculations based on models, of course, depend inherently on modeling structures and assumptions.  Several of the models cited in Table 7 do not distinguish between organic and carbonate particulate carbon.  Nevertheless, most of the data cluster within a range of 2-5 Gt carbon per year.

Most of this flux enters the deep sea in the form of relatively large aggregates such as fecal matter, zooplankton carcasses, and marine snow [Steele, 1972; Menzel, 1974; McCave, 1975; Bishop and Edmond, 1976; Wiebe et al., 1976; Bishop et al., 1977; Bishop et al., 1978; Honjo, 1978, 1980; Dunbar and Berger, 1981; Silver and Alldredge, 1981; Honjo et al., 1982c; Billett et al., 1983].  Settling rates for these aggregates typically range from tens to hundreds of meters per day [Honjo and Roman, 1978; Small et al., 1979; Bruland and Silver, 1981; Komar et al., 1981; Lorenzen et al., 1983a].  Another indication of rapid settling is the pronounced seasonality of the particulate flux, paralleling seasonal surface productivity in many areas [Deuser and Ross, 1980; Deuser et al., 1981; Honjo, 1982; Smith and Baldwin, 1984].

Settling particulate organic matter is oxidized by marine bacteria and other organisms on a time scale that is rapid compared to particle settling and thermocline mixing rates.  This decomposition results in the "oxygen minimum zone" at depths ranging from about 200 to 1500 m, depending on location [Wust, 1935; Wyrtki, 1962].  Deeper waters have relatively high oxygen concentrations because most of the settling organic matter is oxidized before it reaches depths greater than 1500 m [Wiebe et al., 1976; Honjo, 1978; Hinga et al., 1979; Rowe and Gardner, 1979; Deuser and Ross, 1980; Honjo, 1980; Cobler and Dymond, 1980; Honjo et al., 1982c; Wefer et al., 1982; Angel, 1984].  Of the organic matter that reaches the ocean bottom, most is oxidized by benthic organisms before it can be buried [Eadie and Jeffrey, 1973; Wiebe et al., 1976; Hinga et al., 1979; Muller and Suess, 1979; Lorenzen et al., 1983a; Reimers and Suess, 1983; Bender and Heggie, 1984; Reimers et al., 1984; Wangersky, 1984; Emerson, this volume].  Thus, in considering time scales of hundreds of years or longer, the organic carbon particulate flux can be modeled as a direct transfer of carbon from the ocean surface to the depth where it is oxidized.  The flux that remains for burial in the sediments is significant only over time scales longer than hundreds of years.

Biological production and recycling are also associated with the significant reservoir of marine dissolved and suspended organic carbon, termed here the "marine saprosphere."  Estimates

TABLE 9.  Marine Saprosphere

| Reference | Quantity, Gt C | Notes |
|---|---|---|
| Dissolved Organic Carbon | | |
| Williams [1971] | 110 | 0-300 m; assumed 1 $g/m^3$ |
| | 630 | 300-3800 m; assumed 0.5 $g/m^3$ |
| | 740 | Total |
| Menzel [1974] | 665 | Assumed 0.5 $g/m^3$ |
| Williams [1975] | 1000 | Assumed 0.7 $g/m^3$ |
| Cauwet [1978] | 200 | Assumed 0.5 $g/m^3$; apparently a calculation error |
| Mopper and Degens [1979] | 90 | 0-200 m; cited Williams [1975] and Cauwet [1978] |
| | 860 | >200 m; cited Williams [1975] and Cauwet [1978] |
| | 950 | Total |
| See footnote a | 480-1100 | From Table 8 |
| Particulate Organic Carbon | | |
| Williams [1971] | 11 | 0-300 m; assumed 100 $mg/m^3$ |
| | 13 | 300-3800 m; assumed 10 $mg/m^3$ |
| | 24 | Total |
| Menzel [1974] | 14 | Assumed 10 $mg/m^3$ |
| Williams [1975] | 30 | Assumed 20 $mg/m^3$ |
| Cauwet [1978] | 20 | Assumed 100 $g/m^2$ |
| Mopper and Degens [1979] | 30 | Cited Williams [1975] and Cauwet [1978] |
| See footnote a | 6.5-9.0 | From Table 8 |
| Total (Dissolved Plus Particulate) Organic Carbon | | |
| Skopintsev [1971] | 630 | 300-3800 m only; assumed 0.5 $g/m^3$ |
| | 1300-2000 | Assumed 1.0-1.5 $g/m^3$ |
| Williams [1971] | 764 | Sum of dissolved and particulate, as given above |
| Reiners [1973b] | 1000 | |
| Menzel [1974] | 679 | Sum of dissolved and particulate, as given above |
| Williams [1975] | 1030 | Sum of dissolved and particulate, as given above |
| Baes et al. [1976] | 29 | 0-70 m; assumed 1.2 $g/m^3$ |
| | 1620 | >70 m; assumed 1.2 $g/m^3$ |
| | 1649 | Total |
| Cauwet [1978] | 220 | Sum of dissolved and particulate, as given above |
| Mopper and Degens [1979] | 980 | Sum of dissolved and particulate, as given above |
| Skopintsev [1981] | 2000 | Assumed 1.5 $g/m^3$ |
| See footnote a | 486-1109 | From Table 8 |

aCalculated from data of Menzel [1974] and Wangersky [1976].  See Table 8.

of this reservoir, based on the data of Wangersky [1976] and Menzel [1974], are detailed in Table 8. Table 9 compares these calculations to other estimates. There is a wide range of estimates for both dissolved and suspended organic carbon, resulting largely from differences in extraction and analytical techniques [Menzel, 1974; Wangersky, 1984]. The more carefully documented estimates yield a range of about 500-1000 Gt carbon for the total marine saprosphere.

It should be emphasized that the particulate flux described above is not a significant fraction of the "standing crop" of marine suspended organic carbon, which is extracted using conventional sampling and filtration techniques (see, for example, Wangersky [1976]). This distinction between the particulate flux and the suspended standing crop results from the more rapid settling of the large particles that constitute the flux. In the euphotic zone, all particles are subject to rapid biological recycling. Below the euphotic zone, the relationship between the flux and the standing crop is unclear. Some large particles are known to disaggregate easily, leading to the suggestion that the remineralization of carbon in the water column occurs

**TABLE 10.  Marine Carbonate Particulate and Dissolution Fluxes**

| Reference | Particulate Flux From Surface, Gt C yr⁻¹ | Deep Ocean Dissolution Flux, Gt C yr⁻¹ | Notes |
|---|---|---|---|
| **Based on Carbonate Sedimentation Patterns** | | | |
| Broecker [1971] | 1.18 | 0.55-0.63 | Dissolution rate from model of Broecker and Li [1970] (see Table 7 of this paper); author notes discrepancy compared to observed accumulation rates. |
| Berner [1977] | 0.75 | 0.59 | Modified calculations of Broecker [1971] to include aragonite. |
| Broecker [1982] | 0.42 | ... | Calcite accumulation above lysocline assumed to be 1 g cm⁻² kyr⁻¹. |
| Broecker and Peng [1982] | 0.63 | 0.49 | Calcite accumulation above lysocline assumed to be 1.2 g cm⁻² kyr⁻¹. Aragonite dissolution assumed to be 20% of calcite dissolution. |
| **Based on Steady State Ocean Models** | | | |
| Keeling and Bolin [1968] | 0.91-1.88 | 0.91-1.88 | Three-box model; assumed no sedimentation |
| Li et al. [1969] | 0.84±0.24 | 0.72±0.24 | Two-box model |
| Pytkowicz [1973] | 0.64 | 0.38 | Two-box model |
| Broecker [1974] | 1.0 | 0.80 | Two-box model |
| Sundquist [1979] | 0.75 | 0.61 | Two-box model |
| Broecker and Peng [1982] | 0.60 | 0.60 | Two-box model; assumed no sedimentation |
| Bolin et al. [1983] | 0.56±0.26 | 0.56±0.26 | Twelve-box model; authors' reference case; assumed no sedimentation |
| Keir [1983a] | 1.06 | 0.91 | Two-box model |
| Keir and Berger [1983] | 1.30-1.43 | 1.13-1.26 | Two-box model |
| Broecker and Peng [1984] | 0.82 | 0.58 | Two-box model |
| Keir and Berger [1984] | 0.94 | 0.80 | Two-box model; tuned to dissolution model |
| Toggweiler and Sarmiento [this volume] | 0.71 | 0.71 | Three-box model; assumed no sedimentation |
| Wenk and Siegenthaler [this volume] | 0.57 | 0.57 | Three-box model; assumed no sedimentation |

primarily through decomposition of small particles recently shed from larger particles [Honjo and Roman, 1978; Knauer et al., 1979; Honjo et al., 1982c]. On the other hand, studies of the chemical composition of organic material in large particles suggest significant chemical alteration during settling [Wakeham et al., 1980, 1984a, b; Lee and Cronin, 1982; Wefer et al., 1982]. Several studies have suggested a significant contribution from resuspended bottom sediments, which might include a more inert organic component [Spencer et al., 1978; Honjo et al., 1982a, b; Honjo, 1982; Tsunogai et al., 1982].

Similar uncertainties cloud the status of the much larger dissolved component of the marine saprosphere. The reactivity of dissolved organic matter is suggested by its decreasing concentration with depth [Wangersky, 1984]. On the other hand, its carbon isotope composition suggests that it is substantially more inert than suspended organic carbon [Eadie et al., 1978]. This view seems to be supported by $^{14}C$ measurements [Williams et al., 1969] and residence time calculations [Skopintsev, 1971, 1981]. However, the latter approach may understate the extent of recycling and appears to be influenced by the assumption that marine "humus" resembles terrestrial residual organic matter because it can be extracted from seawater using precipitation techniques derived from soil chemistry.

Until the composition and behavior of the marine saprosphere are better known, its role in carbon cycle models will be very uncertain. Because the fluxes of organic carbon settling and decomposition are apparently dominated by rapid transport of large particles, and because the marine saprosphere is dispersed throughout the water column, its treatment as a discrete interactive reservoir, analogous to soil carbon, does not seem to be justified.

Carbonate particles also contribute substantially to the cycling of carbon within the oceans. Calcareous pelagic sediments are abundant throughout the world's oceans, and it is not unusual to find calcareous "ooze" that is more than 70% calcium carbonate by weight (see, for example, Lisitzin and Petelin [1967], Lisitzin [1972], Milliman [1974], Berger et al. [1976], Biscaye et al. [1976], and Kolla et al. [1976]). Modern pelagic carbonates are primarily composed of the calcium carbonate shells of three planktonic organisms: foraminifera, coccolithophorids, and pteropods. Although deep sea sediments may contain very small amounts of the calcareous remains of other shelled organisms, there is no evidence for any nonbiogenic carbonate in the deep oceans.

The shells of foraminifera are the most conspicuous components of most calcareous deep sea sediments. A typical foraminiferal ooze consists mainly of the shells of planktonic foram species (principally of the globigerinid family), with a few benthonic foram tests. Foram shells are generally 50-200 microns in diameter, and most are composed of low-magnesium calcite.

Coccolithophorids are yellow-green algae (chrysophytes) which grow as spheres (coccospheres) composed of numerous calcite plates (coccoliths). The component crystals of coccoliths are usually rhombohedral, and they often form a complex arrangement of plate- and rod-shaped microstructures [Black, 1965]. Diaggregated coccoliths are usually 2-20 microns in diameter and comprise much of the fine-grained component of calcareous oozes.

Pteropods are planktonic gastropods. Those that have shells (belonging to the suborder Euthecosomata) can be up to several millimeters in dimension, although the shells are rarely found intact. They are the only significant aragonitic constituent of modern pelagic sediments.

On the basis of particle size, forams and pteropods would be expected to settle rapidly through the ocean water column, reaching even the deepest seafloor in a matter of days to months [Vinogradov, 1961; Berger and Piper, 1972]. The much smaller size of coccoliths implies a free-settling residence time of about 100 years [Honjo, 1977]. However, Honjo [1975, 1976] has shown that coccoliths may be efficiently transported to the sea floor in fecal pellets, which contribute significantly to the carbonate sedimentation flux as well as to the standing crop of suspended coccoliths found throughout the water column. Another complication in the carbonate particle flux is the fact that many forams and pteropods live at depths substantially below the euphotic zone [Berger, 1969; Bé and Tolderlund, 1971; Bé and Gilmer, 1977]. Forams are known to calcify at depths of up to several hundred meters, as indicated by departures of the carbon and oxygen isotope compositions of their shells from values expected from calcification in surface seawater [Emiliani, 1954; Hecht and Savin, 1972; Savin and Douglas, 1973; Fairbanks et al., 1980; Williams and Healy-Williams, 1980; Curry and Matthews, 1981].

The global carbonate particle flux to the seafloor has been estimated on the basis of sedimentation patterns and steady state ocean models. Current estimates, as shown in Table 10, range from 0.4 to 1.4 Gt carbon per year. The estimates based on sedimentation patterns assume that the average sedimentation rates in areas of minimal dissolution are representative of the global particle flux. Estimates derived from steady state ocean models are based primarily on the dissolution flux implied by marine alkalinity differences. These approaches are independent, and the broad agreement apparent in Table 10 is encouraging. Unfortunately, the carbonate fluxes measured using sediment traps and in situ filtration are not consistent enough to refine these estimates. For example, Honjo et al. [1982c] reported that their PARFLUX sediment traps, deployed at about 400 m depth, collected carbonate fluxes equivalent to 0.17 g C $m^{-2}$ $yr^{-1}$ in the central Pacific and 1.8 g C $m^{-2}$ $yr^{-1}$ in the equatorial Atlantic. For comparison, the above range of global flux estimates is equivalent to about $1.2$-$4.2$ g C m$^{-2}$ yr$^{-1}$ over a deep ocean area of $3.3 \times 10^{14}$ m$^2$. The sediment trap data may be significantly affected by carbonate dissolution during deployment [Thunnell and Honjo, 1981; Betzer et al., 1984b]. Many uncertainties presently obscure both the magnitude of the global carbonate dissolution flux and its potential time response characteristics. These problems are discussed below, with other aspects of carbonate sedimentation.

## Uptake of Anthropogenic CO$_2$

Given the fundamental nature of these uncertainties and the others discussed above, efforts to model the oceanic uptake of anthropogenic CO$_2$ are very crude. Several investigators have attempted to assess oceanic CO$_2$ uptake directly through interpretation of the distribution of dissolved inorganic carbon [Brewer, 1978; Chen and Millero, 1979; Chen, 1982a, b]. However, these calculations rely on back-calculations that require problematic local assumptions equivalent to the problems discussed above in a global context [Shiller, 1981; Chen et al., 1982]. Thus, "direct" estimates of oceanic CO$_2$ absorption appear to be no more reliable than estimates from global models. In order to demonstrate the range of global model CO$_2$ uptake estimates, Table 11 summarizes the CO$_2$ uptake of several models which were run without an interactive biosphere component. The diffusive-type models, which best reproduce carbon isotope effects, absorbed 32-47% of the model fossil fuel CO$_2$ injection. If this range is representative of oceanic absorption of anthropogenic CO$_2$ over the entire period prior to 1980, the total anthropogenic CO$_2$ injection can be estimated by assuming a "preanthropogenic" atmospheric CO$_2$ level of 260-280 ppmv. This estimate is 180-240 Gt carbon, of which about 160 Gt was contributed by fossil fuels. If the preanthropogenic CO$_2$ concentration proves to be close to 280 ppmv, as suggested by Moor and Stauffer [1984], the total anthropogenic injection is estimated at 180-240 Gt carbon. Within this broad range of uncertainties, oceanic CO$_2$ models can accommodate substantial net CO$_2$ contributions from effects on land plants and soils.

### Terrestrial Plants and Soils

The carbon pool in terrestrial plants is generally assessed through analysis of the sizes and characteristics of individual vegetation or ecosystem types. Table 12 shows a convergence among recent estimates for the present size of this pool, reflecting a general agreement among various investigators regarding global classification schemes and the carbon masses associated with classification units. Some classification schemes, particularly those based on climatic variables, yield estimates of "potential" rather than actual biomass [e.g., Bazilevich et al., 1971].

Global terrestrial net primary production is estimated using similar classification schemes, and a similar agreement among estimates is apparent in Table 13. Gross primary production is generally estimated to be about twice the net primary production. It is apparent that land plants cycle carbon rapidly, assimilating their own weight in 10 years or less. Virtually all of the net primary production is thought to be returned to the atmosphere by heterotrophic respiration in soils. However, because the heterotrophic CO$_2$ flux is so difficult to measure, the steady state assumption may obscure significant fluctuations at any given time and location.

Soil carbon (termed here the "terrestrial saprosphere") is widely and unevenly distributed, making its assessment very difficult. Table 14 summarizes recent evaluations. A fraction of this reservoir cycles very slowly, resulting in $^{14}$C ages of 1000 years or more [Jenkinson and Rayner, 1977; Schlesinger, 1977; O'Brien and Stout, 1978; Stout et al., 1981; Newbould, 1982]. However, the rates and mechanisms of soil carbon cycling are very poorly known (e.g., Reiners [1973a]). Schlesinger [1984] points out that "nearly half" of the soil carbon to a depth of 30 cm is lost through oxidation in agricultural studies of a few years' duration, supporting the idea that soil organic matter should be categorized according to its age or refractory nature. Unfortunately, such a categorization does not yet exist on a global scale.

The storage of organic carbon on land has been a major point of controversy in CO$_2$ research because of questions concerning the change in its size over the last 150 years. An expansion of this pool has been postulated by some as a way of resolving discrepancies among model results for CO$_2$ and carbon isotopes [Keeling, 1973b; Oeschger et al., 1975]. Yet trends in land use and large-scale deforestation in the last century suggest that the land carbon pool has probably decreased [Bolin, 1977; Revelle and Munk, 1977; Woodwell et al., 1978]. Some estimates of this decrease are summarized in Table 15. As was shown in the previous section, current ocean models are not necessarily inconsistent with a cumulative non-fossil fuel CO$_2$ flux of up to about 140 Gt carbon (or up to about 80 Gt carbon if the preanthropogenic CO$_2$ level was close to 280 ppmv). Thus, the controversy relates primarily to modern yearly CO$_2$ budgets based on land use data and ocean models [Hampicke, 1979; Houghton et al., 1983; Hobbie et al., 1984].

### Weathering and Sedimentation

Over time scales of thousands of years and longer, atmospheric CO$_2$ is significantly influenced, if not controlled, by the cycle of weathering and sedimentation. This cycle includes the chemical and mechanical erosion of carbonates and organic carbon in continental rocks, the reaction of soil and atmospheric CO$_2$

TABLE 11. Ocean Model Uptake of Fossil Fuel $CO_2$

| Reference | Time Period | Fossil Fuel $CO_2$ Production, Gt C | $CO_2$ Absorbed by Model Ocean, Gt C | Ratio of Absorbed to Produced $CO_2$ | Notes |
|---|---|---|---|---|---|
| | | Preindustrial to Recent Period | | | |
| Broecker et al. [1971] | PI-1960 | 84.5 | 29.6-42.3 | 0.35-0.50 | Outcrop-diffusion model |
| Keeling [1973b] | PI-1954 | 68.6 | 14.5-23.3 | 0.21-0.34 | Two-box model |
| Siegenthaler and Oeschger [1978] | PI-1970 | 114.4 | 43.2 | 0.38 | Box-diffusion model |
| Broecker et al. [1980c] | PI-1973 | 128.5 | 42.1 | 0.33 | Box-diffusion model of Oeschger et al. [1975]. |
| | | 128.5 | 52.4 | 0.41 | |
| MacIntyre [1980] | 1859-1979 | 156.5 | 60.7 | 0.39 | Three-box model |
| Siegenthaler [1983] | 1860-1980 | 162.1 | 63.9 | 0.39 | Diffusion with polar outcrop; calibrated with PI $^{14}C$ |
| Bolin et al. [1983] | 1860-1980 | 162.1 | 76.7 | 0.47 | Calibrated with bomb $^{14}C$ |
| | 1860-1980 | 162.1 | 45.4 | 0.28 | Twelve-box model; authors' version A |
| | 1860-1980 | 162.1 | 41.2 | 0.25 | Twelve-box model; authors' version B |
| | | "Mauna Loa" Period[a] | | | |
| Broecker et al. [1979] | 1958-1975 | 62.5 | 22.8 | 0.36 | Box-diffusion model |
| Crane [1982] | 1959-1978 | 75.0 | 23.7-32.5 | 0.32-0.43 | Box-diffusion model |

Only model runs with no net biosphere flux are included. PI stands for preindustrial.
[a]Period during which models can be calibrated against the continuous atmospheric record from Mauna Loa.

TABLE 12.  Carbon in Terrestrial Plants

| Reference | Carbon in Biomass, Gt C | | Notes |
|---|---|---|---|
| | Present | Past | |
| Bowen [1966] | 510 | | |
| Bolin [1970] | 450 | | |
| Bazilevich et al. [1971] | | 1080[a] | "Reconstructed plant cover"; climatic classification scheme |
| Garrels and Mackenzie [1972] | 1000 | | Includes marine biomass (see Table 5) |
| Keeling [1973b] | 560 | | Adapted from Olson [1970] |
| Whittaker and Likens [1973] | 827 | | Includes continental aquatic plants |
| Reiners [1973b] | 833 | | Includes terrestrial fauna |
| Olson [1974] | | 1070 | "Preagricultural" |
| Garrels et al. [1975] | 480 | | |
| Rodin et al. [1975] | | ~ 1200 | |
| Whittaker and Likens [1975] | 827[a] | 945-990 | "Present" is 1950 |
| Baes et al. [1976] | 680 | | |
| Olson et al. [1978] | 557 | (1)696, (2)1050 | "Present" is 1970; "past" estimates for (1) 1860, (2) "preagricultural" |
| Bolin et al. [1979] | 700±20 | | |
| Bolin et al. [1979] | 592 | | Attributed to Duvigneaud |
| Ajtay et al. [1979] | 560 | | |
| Olson [1981, 1982] | 560±100 | | "Present" is 1980 |
| Olson et al. [1983] | 460-661 | | "Present" is 1980 |

[a]Calculated from dry biomass assuming carbon content is 45%.

with silicate minerals, the transport of bicarbonate and organic carbon by rivers, the deposition of carbonates and organic carbon in oceanic sediments, and the diagenesis of marine sediments. These processes have been summarized in several steady state sediment cycle models [e.g., Garrels and Mackenzie, 1972; Garrels and Perry, 1974; Garrels et al., 1975] and in time-dependent models [Garrels et al., 1976; Garrels and Lerman, 1981; Berner et al., 1983; Lasaga et al., this volume].

Although the analysis in this paper is restricted to simple carbon cycle models, it should be emphasized that the long-term behavior

TABLE 13.  Terrestrial Net Primary Production

| Reference | Flux, Gt C yr$^{-1}$ | Notes |
|---|---|---|
| Whittaker [1970] | 49 | |
| Olson [1970] | 54 | |
| SCEP [1970] | 56 | |
| Bazilevich et al. [1971] | 78 | |
| Lieth [1972, 1975a] | 45[a] | |
| Golley [1972] | 40[a] | |
| Whittaker and Likens [1973] | 45 | Excludes freshwater |
| Reiners [1973b] | 50 | |
| Keeling [1973b] | 56 | Adapted from Olson [1970] |
| Whittaker and Likens [1975] | 53[a] | |
| Lieth [1975b] | 55[a] | |
| Baes et al. [1976] | 56 | |
| Bolin et al. [1979] | 63 | |
| Ajtay et al. [1979] | 60 | |

[a]Calculated from dry matter production assuming carbon content is 45%.

TABLE 14.  Terrestrial Saprosphere

| Reference | Quantity, Gt C | Notes |
|---|---|---|
| Bolin [1970] | 700 | |
| Keeling [1973b] | 1085 | Adapted from Olson [1970] |
| Reiners [1973b] | 700 | Based on Bolin [1970] |
| Keeling [1973b] | 1050 | |
| Bazilevich [1974] | 1392 | Cited by Ajtay et al. [1979]; excluded peat; depth 1 m |
| Kovda [1974] | ~ 1400 | Cited by Buringh [1984]; estimated "humus" as 2400 Gt |
| Baes et al. [1976] | 1080 | Based on Olson [1970] |
| Bohn [1976] | 2946±500 | |
| Schlesinger [1977] | 1456 | |
| Bohn [1978] | 3000 | |
| Bolin et al. [1979] | 2840 | Attributed to Duvigneaud |
| Bolin et al. [1979] | 1672 | |
| Ajtay et al. [1979] | 1636 | Inventoried by ecosystem; depth 1 m |
|  | 2070 | Inventoried by geographic region |
| Bohn [1982] | 2200 | |
| Post et al. [1982a] | 1395±200 | |
| Post et al. [1982b] | 1484 | |
| Zinke et al. [1984] | 1309 | Used Holdridge life zones |
|  | 1728 | Used 1° latitude bands |
| Buringh [1984] | 1477 | Inventoried by soil type and land use |
| Schlesinger [1984] | 1515 | Inventoried by ecosystem |

of the carbon cycle is strongly coupled to the cycling of several other elements, including phosphorus, nitrogen, oxygen, sulfur, calcium, magnesium, and iron. As emphasized by Lasaga [1980, 1981], such coupling implies nonlinear behavior and can significantly affect response times. This aspect of carbon cycling is ignored for the sake of simplicity in this paper, but coupling must be considered in simulating the behavior of the carbon cycle through time.

The Importance of Alkalinity

Several of the coupling relationships that affect carbon cycling occur through effects on ocean alkalinity. The distribution of ocean alkalinity is shown in Table 16, calculated from the GEOSECS averages of Takahashi et al. [1981] in a manner directly analogous to the calculation of Table 2. Alkalinity is a necessary component of carbon cycle models because it completes the seawater charge balance, thereby fulfilling the requirements for calculating the equilibrium speciation of dissolved inorganic carbon.

Although alkalinity need not be modeled explicitly in carbon cycle models, it is always incorporated implicitly if charge balance is maintained. For example, the models of Berner et al. [1983] and Lasaga et al. [this volume] postulate a constant state of oceanic saturation with respect to calcite throughout the last 100 million years, during which their models indicate that the ocean/atmosphere $CO_2$ content has changed significantly. These conditions imply that times of high atmospheric $CO_2$ were also times of high oceanic alkalinity. Alkalinity fluxes occur implicitly in these models through the precipitation and/or dissolution of carbonates invoked to maintain a constant state of saturation and through the interconversion of bicarbonate and $CO_2$ accompanying redox reactions. The coincidence of high alkalinity with high $CO_2$ is consistent with the concept of oceanic carbonate "buffering" of atmospheric $CO_2$ changes; increasing alkalinity, for example by increasing carbonate dissolution (see below), enhances oceanic uptake of atmospheric $CO_2$. Without some technique for estimating alkalinity effects, carbon cycle model calculations cannot reasonably partition $CO_2$ between the oceans and atmosphere.

River Fluxes

The important role of $CO_2$ in chemical weathering is reflected in the flux of bicarbonate carried by rivers to the oceans. As summarized by Holland [1978], nearly two-thirds of the river bicarbonate flux is derived from the atmosphere; the remainder comes mainly from carbonate rock weathering. Table 17 summarizes

TABLE 15. Net Anthropogenic Carbon Transfer From Terrestrial Plants and Soils to Atmosphere

| Reference | Time Period | Transfer, Gt C | | |
| | | From Plants | From Soils | Total |
|---|---|---|---|---|
| | Based on Land-Use Data | | | |
| Bolin [1977] | 1800-mid-1970's | 30-60 | 10-40 | 40-100 |
| Revelle and Munk [1977] | 1860-1970 | 50 | 22 | 72 |
| Bohn [1978] | Prehistory-1870 | ... | 180 | ... |
| | 1870-1970 | ... | 150 | ... |
| | Prehistory-1970 | ... | 330 | ... |
| Pankrath [1979] | 1860-1974 | ... | ... | 120 |
| Chan et al. [1980] | 1860-1970 | ... | ... | 215-217[a] |
| Moore et al. [1981] | 1860-1970 | ... | ... | 148 |
| Shaver et al. [1982] | 1958-1980 | ... | ... | 57 |
| Woodwell et al. [1983] | 1860-1980 | ... | ... | 135-228 |
| Houghton et al. [1983] | 1958-1980 | ... | ... | 38-76 |
| Buringh [1984] | Prehistory-early 1980's | ... | 537 | ... |
| Schlesinger [1984] | mid-1800's-early 1980's | ... | 36 | ... |
| | Based on Tree-Ring $^{13}$C Records | | | |
| Freyer [1978] | 1860-1974 | ... | ... | 70 |
| Stuiver [1978a] | 1850-1950 | ... | ... | 120 |
| Wagener [1978] | 1800-1935 | ... | ... | 170 |
| Siegenthaler et al. [1978] | 1860-1974 | ... | ... | 157-195 |
| Wilson [1978] | 1860-1890 | ... | ... | 110 |
| Emanuel et al. [1982] | 1740-1979 | -54 | 228 | 174 |
| Peng et al. [1983] | 1800-1980 | ... | ... | 274 |
| Emanuel et al. [1984] | 1800-1975 | 172 | 59 | 231 |
| Stuiver et al. [1984] | 1600-1975 | ... | ... | 50-250 |
| Peng [this volume] | 1800-1980 | ... | ... | 144 |

[a]Gross transfer

TABLE 16. Distribution of Oceanic Total Alkalinity at Time of GEOSECS Measurements (1972-1977)

| Depth, m | Atlantic Ocean | Pacific Ocean | Indian Ocean | Antarctic[a] Ocean | Arctic[b] Ocean | Global totals |
|---|---|---|---|---|---|---|
| 0-100 | 192 | 370 | 149 | 114 | 17.9 | 840 |
| 100-500 | 709 | 1390 | 563 | 435 | 45.2 | 3140 |
| 500-1000 | 851 | 1700 | 687 | 531 | 40.9 | 3810 |
| 1000-1500 | 832 | 1680 | 678 | 519 | 33.4 | 3740 |
| 1500-2000 | 813 | 1670 | 666 | 508 | 28.5 | 3690 |
| 2000-3000 | 1530 | 3140 | 1260 | 951 | 41.1 | 6920 |
| 3000-4000 | 1250 | 2550 | 1030 | 775 | 14.7 | 5620 |
| 4000-5000 | 746 | 1510 | 606 | 460 | ... | 3320 |
| 5000-6000 | 219 | 442 | 177 | 135 | ... | 970 |
| 0-6000 | 7140 | 14,450 | 5820 | 4430 | 220 | 32,050 |

Data are given as equivalents x $10^{15}$. Data were calculated from the average concentrations given by Takahashi et al. [1981] and the hypsometric data of Menard and Smith [1966]. The following ocean surface areas were assumed [Baumgartner and Reichel, 1975] (in $10^6$ km$^2$): Atlantic, 80.04; Pacific, 159.023; Indian, 64.173; Antarctic[a], 49.364; Arctic, 8.5. Seawater specific gravity was assumed to be 1.025.

[a]Defined here as ocean south of latitude 50°S.
[b]Calculated from average concentrations given by Anderson and Dryssen [1981] and the hypsometric data of Gorshkov [1980].

TABLE 17. Carbon Fluxes from Rivers to Oceans

| References | DIC[a] | DOC[b] | POC[c] | PIC[d] | Total | Notes |
|---|---|---|---|---|---|---|
| Livingstone [1963] | 0.372 | ... | ... | ... | ... | Inferred from data on many rivers worldwide |
| Garrels and Mackenzie [1971] | 0.37 | 0.32 | ... | 0.16 | ... | From a steady state model of the sedimentary cycle over geologic time |
| Garrels and Mackenzie [1972] | 0.39 | 0.28 | 0.07 | 0.02 | 0.76 | |
| Menzel [1974] | ... | 0.08-0.19 | 0.04-0.068 | ... | ... | Linear extrapolation of data from three rivers |
| Garrels et al. [1975] | 0.450 | 0.128 | 0.070 | ... | ... | Used reasoning similar to that of Menzel [1974] |
| Duce and Duursma [1977] | ... | 0.11 | 0.06 | ... | ... | Assumed larger runoff than did other authors |
| Holland [1978] | 0.528 | ... | ... | ... | ... | |
| Kempe [1979a] | 0.445 | 0.123 | 0.066 | 0.197 | 0.831 | Based on Garrels and Mackenzie's [1972] data |
| Kempe [1979b] | 0.454 | ... | ... | ... | ... | Revision of Livingstone's [1963] data |
| Baumgartner and Reichel [1975] | 0.4 | 0.1 | 0.06 | ... | ... | |
| Richey et al. [1980] | ... | 1.0 (TOC)[e] | ... | ... | ... | Based largely on organic transport patterns observed for Amazon River |
| Meybeck [1981] | 0.400 | 0.215 | 0.180 | 0.150 | 0.945 | |
| U.S. Department of Energy [1981] | ... | 0.792 (TOC) | ... | ... | ... | "Best estimate" of flux |
| Schlesinger and Melack [1981] | ... | 0.39 (TOC) | ... | ... | ... | Average of determination by two different methods |
| Meybeck [1982] | ... | 0.215 (TOC) | 0.180 | 0.170 | ... | Extrapolation based on data from North America |
| Mulholland and Watts [1982] | ... | 0.3 (TOC) | ... | ... | ... | |
| Milliman et al. [1984] | ... | ... | 0.295 | ... | ... | Recalculated according to method of Meybeck [1981], assuming higher POC concentrations for large, turbid rivers; based on data from the Yangtze River |

[a]Dissolved inorganic carbon.
[b]Dissolved organic carbon.
[c]Particulate organic carbon.
[d]Particulate inorganic carbon.
[e]Total (dissolved plus particulate) organic carbon.

TABLE 18. Marine Carbonate Burial Rates

| Reference | Continental Shelves | Continental Slopes and Deep Ocean | Total | Notes |
|---|---|---|---|---|
| **Based on Steady State Ocean Models** | | | | |
| Li et al. [1969] | ... | ... | 0.12 | See Table 10 |
| Pytkowicz [1973] | ... | ... | 0.26 | See Table 10 |
| Broecker [1974] | ... | ... | 0.20 | See Table 10 |
| Sundquist [1979] | ... | ... | 0.14 | See Table 10 |
| Keir [1983a] | ... | ... | 0.15 | See Table 10 |
| Keir and Berger [1983] | ... | ... | 0.17 | See Table 10 |
| Broecker and Peng [1984] | ... | ... | 0.24 | See Table 10 |
| Keir and Berger [this volume] | ... | ... | 0.14 | See Table 10 |
| **Based on Holocene Carbonate Sedimentation Patterns** | | | | |
| Broecker [1971] | ... | ... | 0.55-0.63 | See Table 10 |
| Berner [1977] | ... | ... | 0.16 | See Table 10 |
| Milliman [1974] | 0.089a | 0.19b | 0.28 | Shelf estimates as corrected by Hay and Southam [1977] |
| Hay and Southam [1977] | 0.16-0.48 | 0.19 | 0.35-0.67 | Authors' shelf estimate; deep sea from Milliman [1974] |
| Smith [1978] | 0.072 | ... | ... | Reefs only; based on calcification studies |
| **Based on Pre-Holocene Sedimentation Patterns** | | | | |
| Milliman [1974] | 0.030 | 0.19 | 0.22 | Post-Miocene; shelf estimate as corrected by Hay and Southam [1977] |
| Hay and Southam [1977] | ... | 0.096 | ... | Excludes continental slope |
| Hay [this volume] | 0.020c | 0.526 | 0.546 | Pliocene |
| Ronov [1980, 1982a, b] | ... | ... | 0.145 | Phanerozoic; not corrected for sediment recycling |
| | ... | ... | 0.095 | Phanerozoic; see Hay [this volume] |
| **Based on Steady State Rock Cycle Models** | | | | |
| Garrels and Mackenzie [1972] | ... | ... | 0.160 | Flux to Mesozoic-Cenozoic rocks |
| | ... | ... | 0.056 | Flux to Paleozoic and older rocks |
| Garrels and Perry [1974] | ... | ... | 0.142 | |
| Garrels et al. [1975] | ... | ... | 0.142 | |
| Kempe [1979b] | ... | ... | 0.17 | |
| Kitano [1980] | ... | ... | 0.15 | |
| Garrels and Lerman [1981] | ... | ... | 0.15 | Holocene |
| Okumura et al. [1983] | ... | ... | 0.19 | |

Data are given as gigatons carbon per year.

aIncludes burial in Red Sea, Black Sea, Mediterranean Sea.

bAuthor assumes that this estimate, based on post-Miocene Deep Sea Drilling Project rates, is valid for the Holocene.

cFrom platform and geosyncline data of Ronov [1980, 1982a, b].

flux estimates since the landmark study of Livingstone [1963].

The organic carbon fluxes listed in Table 17 must be interpreted very cautiously. Several studies have shown that the dissolved organic carbon in rivers and estuaries is a product of a rapid local recycling between plankton production and benthic remineralization [Nixon, 1981; Crawford et al., 1974; Sigleo et al., 1983; Sigleo and Helz, 1981]. Thus, the "flow" of organic carbon at a given point near the mouth of a river may merely reflect the dynamics of local nutrient cycling rather than the transport of inert components to the oceans.

## Carbonate Burial

The net sedimentation rate of calcium carbonate depends on the fractionation of carbonate deposition between the continental shelves and the ocean basins [Berger and Winterer, 1974; Hay and Southam, 1977] and on the extent of carbonate dissolution, which occurs mainly in the ocean basins. Shelf-basin fractionation has fluctuated drastically in the past, chiefly in response to sea level changes (see, for example, Milliman [1974] and Wilson [1975]). These fluctuations are reflected in the range of estimates for shelf carbonate burial shown in Table 18, which also lists estimates for the ocean basins. The global sedimentation rate of shelf carbonates during the Holocene (a time of relatively high sea level) has been estimated at 2-20 times the long-term rate based on pre-Holocene sediment sequences.

Dissolution is the most conspicuous factor affecting the distribution of pelagic carbonate sediments. It has long been known that calcareous sediments are absent in the deepest parts of the oceans (see, for example, Murray and Renard [1891]). Dissolution is most complete at the greatest depths. The depth below which calcite cannot be found in modern sediments is known as the calcite compensation depth (CCD). Detailed mapping of the CCD has revealed that it occurs at different depths in different regions [Lisitzin, 1972; Biscaye et al., 1976; Kolla et al., 1976; Berger et al., 1976]. The Pacific CCD is generally between depths of 4000 and 5000 m, while the Atlantic CCD occurs below depths of 5000 m. The Indian Ocean CCD generally lies at depths between those observed in the Atlantic and Pacific. In all the oceans, the CCD shoals significantly in nearshore and polar waters, while equatorial sediments are characterized by exceptionally deep carbonate-rich sediments.

Detailed studies of calcareous sediments have revealed significant dissolution effects at depths much shallower than the CCD. In general, these effects increase with increasing depth, culminating in complete dissolution of calcareous materials below the CCD. Because aragonite is the most soluble pelagic carbonate mineral, it is not surprising that the pteropods dissolve at relatively shallow depths. Aragonitic pteropod

shells are not often found in sediments deeper than 2000-3000 m in the Atlantic Ocean and 500-1500 m in the Pacific Ocean [Berner et al., 1976; Berger, 1978a].

The various species of calcitic foraminifera show differing susceptibilities to dissolution above the CCD. Species that secrete spiny, porous shells generally dissolved at shallower depths than those which produce massive, smooth shells [Berger, 1967; Ruddiman and Heezen, 1967]. Benthonic species usually dissolve at greater depths than planktonic species [Oba, 1969]. These species-dependent differences led Berger [1968] to develop useful quantitative techniques which reflect the extent of dissolution in foraminiferal sediments. He defined the "foraminiferal lysocline" as the depth at which a statistical summary of species assemblages changes from values dominated by solution-susceptible species to values dominated by solution-resistant species. Berger [1981] has shown that the foraminiferal lysocline occurs at depths ranging from about 500 to 1500 m above the CCD throughout the world oceans. Variations in the depth relationship between the lysocline and CCD appear to reflect regional differences in seawater chemistry and surface ocean productivity [Berger, 1977; Takahashi and Broecker, 1977]. Extending the application of foraminiferal dissolution indices to still shallower depths, Berger [1978b] introduced the "$R_0$-level," defined as the maximum depth to which the foram species assemblage remains unaffected by dissolution.

Attempts to quantify the dissolution fluxes associated with these sedimentary criteria have produced equivocal results. Several studies based on dissolution experiments and on observed dissolution-dependent species patterns have suggested that substantial dissolution must take place before significant sediment effects are apparent [Berger, 1971, 1976, 1977, 1978b; Adelseck, 1978; Ku and Oba, 1978]. However, studies based on well-documented net accumulation rates show no evidence for appreciable calcite dissolution above the lysocline [Berger et al., 1982]. The resolution of this conflict may lie in a relatively sudden increase in dissolution at the lysocline [Berger et al., 1982]. This interpretation is supported by evidence relating the lysocline to the thermodynamic stability of calcite in overlying waters.

The relationship between carbonate dissolution and seawater chemistry has been one of marine geochemistry's greatest frustrations. The GEOSECS expeditions have provided abundant data that greatly clarify this problem. Broecker and Takahashi [1978] observed that lysocline depths in the Atlantic and Pacific coincide with "critical carbonate ion" concentrations which are consistent with calcite solubility measurements in seawater [Ingle et al., 1973]. Plummer and Sundquist [1982] showed that the critical carbonate ion curve coincides with a thermodynamically defined calcite equilibrium boundary

in the oceans. A consistent equilibrium interpretation of the lysocline has also been demonstrated in the Indian Ocean [Peterson and Prell,
1984].

In spite of these advances, the quantification
of marine carbonate dissolution remains
uncertain. Experimental dissolution rates show
vast differences between the fluxes from sediment
beds into overlying waters and the fluxes predicted from studies of stirred suspended
sediments [Keir, 1983b]. Oceanic alkalinity
profiles, once thought to reflect only the
influence of carbonate dissolution, are now known
to be affected by the oxidation of organic matter
as well [Brewer et al., 1975; Brewer and Goldman,
1976; Chen, 1978; Tsunogai and Watanabe, 1981].
Steady state models based on alkalinity fluxes
must take this significant effect into account.
Organic decomposition has also been implicated as
the cause, through release of metabolic $CO_2$, of
substantial carbonate dissolution near the sea-
sediment interface at depths above the lysocline
[Berger, 1970; Broecker, 1971; Emerson and
Bender, 1981]. Several studies have suggested
that pteropod aragonite may comprise a significant component of the carbonate particles
settling into the deep sea, and that these
particles may dissolve before they reach the sea
floor [Berner, 1977; Fiadeiro, 1980b; Berner and
Honjo, 1981; Byrne et al., 1984; Betzer et al.,
1984b]. Some carbonate particles may reach
depths below the lysocline through downslope
transport along the seafloor [Thiede, 1977; Land,
1979; Berger et al., 1982]. Thus, there are many
fundamental problems yet to be solved in evaluating the distribution and magnitude of marine
carbonate dissolution.

## Organic Carbon Burial

Table 19 lists recent estimates of the global
rates of organic carbon burial on continental
shelves and in ocean basins. The wide range of
estimates partly reflects uncertainties in the
influence of sea level on shelf sedimentation and
in the influence of nutrient "loading" from
agriculture. In agreement with Gershanovich et
al. [1974] and Volkov and Rozanov [1983], Walsh
[1984] has specified the continental slopes as a
locus for particularly high rates of organic
carbon deposition. Walsh's burial flux is con-
spicuously greater than the other estimates in
Table 19, and his calculations have been criticized by Broecker and Peng [1984] and Emerson
[this volume] on the basis of nutrient budgets.
Walsh's slope flux is 5 times that of the Soviet
authors, who in turn estimate a deep ocean flux
an order of magnitude greater than the estimate
of Berner [1982]. Indeed, Walsh [1984] points
out that his slope carbon burial flux requires a
total sediment flux that is several times the
global river suspended load.

Another uncertainty in the assessment of

organic carbon burial rates is the time scale of
interest. For example, Berner [1982] and Berner
and Raiswell [1983] distinguish three burial
rates: one for surficial sediments, one for
burial to a few meters, and one for burial over
millions of years. In the model of Lasaga et al.
[this volume], the latter rate is properly considered to be a "present" burial rate because the
model is concerned with time scales of millions
of years. Thus, the time scale of interest
determines how the "present" is defined. Similar
considerations have been applied to carbonate
sedimentation [Hay and Southam, 1977].

### Eigenanalysis

The information summarized above concerns
carbon fluxes and reservoirs that interact over
time scales ranging from decades to hundreds of
millions of years. These interactions can be
organized into a series of box models that
describe a series of sequentially longer time
scales. These models are hierarchical in the
sense that the fluxes and reservoirs for shorter
time scales are subsets of those for longer time
scales. However, because my purpose is to
emphasize the importance of particular processes
to particular time scales, the building of this
hierarchy requires not only progressively
expanding the frames of reference, but also selec-
tively subsuming short-term fluxes and reservoirs
as they become relatively unimportant. In box
model parlance, I am concerned with both adding
new boxes and lumping together old boxes.

The criteria used in adding and lumping are
necessarily somewhat arbitrary and subjective. I
have subdivided the continuum of time scales into
frames of reference that are convenient given
current knowledge about the carbon cycle. For
each time scale, I have chosen a structure of
fluxes and reservoirs according to two criteria:
(1) the fluxes represent possibilities for
significant change in the sizes of the reservoirs; and (2) each reservoir can be characterized internally according only to its size.
These considerations are implicit in most box
modeling studies. Rigorous application of these
criteria would require exhaustive sensitivity
analysis of models that accurately represent a
full understanding of the relevant fluxes and
reservoirs. This task will occupy (and probably
elude) geochemical modelers for many years to
come. For this paper, the above criteria
comprise a basis for a quantitative approach to
developing a modeling hierarchy based on time
scales.

This approach relies on the calculation of
eigenvalues for matrices of linear exchange
coefficients, as applied by Southam and Hay
[1976] and by Lasaga [1980, 1981] to the kinetic
treatment of geochemical cycles. Using Lasaga's
notation, the time dependence of the reservoirs
in a linear box model is represented by the
equation

TABLE 19. Marine Organic Carbon Burial Rates

| Reference | Continental Shelves (Including estuaries) | Continental Slopes and Deep Ocean | Total | Notes |
|---|---|---|---|---|
| Bogdanov et al. [1971] | ... | ... | 0.001-0.03 | As cited by Berner [1982] |
| Williams [1971] | 0.092 | 0.0027 | 0.095 | Author's Table 2 corrected to agree with text |
| Garrels and Mackenzie [1972] | ... | ... | 0.03 | Flux to Mesozoic-Cenozoic rocks |
|  | ... | ... | 0.01 | Flux to Paleozoic and older rocks |
| Mopper and Degens [1972] | ... | ... | 0.072-0.216 | As cited by Degens and Mopper [1976] |
| Garrels and Perry [1974] | ... | ... | 0.035 | Phanerozoic average |
| Gershanovich et al. [1974] | 0.022 | 0.200 | 0.222 | Holocene; as cited by Romankevich [1984] |
| Seibold [1974] | ... | ... | 0.04 |  |
| Garrels and Perry [1974] | ... | ... | 0.031 | Flux to "deep rocks" |
| Garrels et al. [1975] | ... | ... | 0.073 | Flux to "recent sediments" |
|  | ... | ... | 0.030 | Flux to "old sediments" |
| Garrels et al. [1976] | ... | ... | 0.03 |  |
| Holland [1978] | ... | ... | 0.120 |  |
| Kempe [1979b] | 0.3-0.4 | ... | ... |  |
| Deuser [1979] | 0.20-0.45 | ... | ... |  |
| Mulholland [1981] | 0.55 | ... | ... | "Potential" |
| Mackenzie [1981] | 0.29 | ... | ... | "Deltaic" accumulation |
| Wollast and Billen [1981] | 0.30 | ... | ... |  |
| Walsh et al. [1981] | 0.75 | ... | ... | Calculated to balance anthropogenic nitrogen; authors assume export to continental slopes |
| Garrels and Lerman [1981] | ... | ... | 0.038 | Holocene |
| Peterson [1981] | ... | ... | 0.018 |  |
| Berner [1982] | 0.137 | 0.020 | 0.157 | Surficial sediments |
|  | 0.110 | 0.016 | 0.126 | Burial to several meters |
| Berner and Raiswell [1983] | ... | ... | 0.046 | Pre-Pleistocene |
| Volkov and Rozanov [1983] | 0.022-0.030 | 0.200-0.270 | 0.222-0.300 |  |
| Romankevich [1984] | 0.0063 | 0.057 | 0.063 |  |
| Walsh [1984] | 0.26 | 1.09 | 1.35 | See note for Walsh et al. [1981] above |
| Lasaga et al. [this volume] | ... | ... | 0.0456 | Pre-agricultural |

Data are given as gigatons carbon per year.

$$\frac{dA}{dt} = KA \qquad (6)$$

where A is the vector of reservoir sizes and K is the matrix of exchange coefficients. If the jth eigenvector and corresponding eigenvalue of K are represented by $\psi_j^o$ and $E_j$, then equation (6) is satisfied by a time-dependent sum of vectors of the form

$$a_j \psi_j^o e^{E_j t}$$

The values of the coefficients $a_j$ are derived from the decomposition of the initial value vector $A^o$ into a sum of multiples of the eigenvectors.

Within this matrix algebra framework, I have broadly interpreted the first criterion above to mean that the reciprocals of the eigenvalues associated with a box model should fall within the time scale range of the model. When this condition is satisfied, the model's responses to perturbations will be significant over the chosen time scale. Of course, the generality of the linear approximation is limited to a restricted range of perturbations, and the model's time dependence may be influenced by external factors (such as climatic forcing or fossil fuel burning) that are not inherent in equation (6). A more exhaustive analysis of time response would include scrutiny of the eigenvectors. This would be particularly important in isolating the behavior of a single reservoir (such as atmospheric $CO_2$) or in analyzing very large departures from steady state. Nevertheless, it is assumed that the eigenvalues are an adequate criterion for assessing response times to within the orders-of-magnitude ranges considered here. In structuring models for progressively longer time scales, new boxes are added when the associated eigenvalues indicate response times of the appropriate order(s) of magnitude.

Likewise, eigenanalysis can be applied to the second criterion for structuring models according to time scale. The internal characterization of a reservoir can be greatly simplified if its internal "mixing time" is significantly shorter than its "transfer time" associated with external fluxes [Welander, 1959]. If the eigenvalues associated with exchange among a particular group of boxes indicate response times that are significantly shorter than the time scale of interest, then these boxes are candidates for lumping. Their relatively rapid response to perturbations can be viewed as internal redistributions within a single box. These internal redistributions occur so quickly that they are essentially independent of the slower changes caused by fluxes to and from the box.

This application requires a departure from the simple box modeling convention of randomly mixed reservoirs. Throughout this development, I assume that carbon may be distributed and redis-tributed unevenly within a box. Thus, the internal characterization of a reservoir encompasses not only the amount defining its size, but also the distribution of that amount and the internal fluctuations in the distribution. For such an internal characterization to depend only on the reservoir's size, any internal fluctuations must be rapid in relation to any changes caused by the external fluxes to and from the reservoir. When this condition is satisfied, the distribution of material within the reservoir should be definable in terms that depend only on the reservoir's total size.

The criteria regarding significant change and rapid internal mixing are not necessarily consistent. In some cases, a model structure implied by the change condition will not conform to the condition for internal mixing. A prominent example is the case of diffusive mixing. This problem has been considered in this paper as it applies to ocean mixing and serves as an illustration of the arbitrary nature of some of the scaling decisions made. In this case, the internal "mixing time" and external "transfer time" are not readily distinguishable. Box model approximations implied by equation (5), with $\tau*$ chosen to be significantly less than the time scale of interest, will be associated with eigenvalue reciprocals significantly less than the time scale of interest. However, such a model will portray diffusive mixing much more accurately than a model with eigenvalues in agreement with the time scale.

In some exchanges, fluxes of carbon may be coupled to fluxes of other elements, causing response times to be shorter than those implied by consideration of carbon alone [Lasaga, 1980]. Another inherent limitation is that short-term oscillations undoubtedly persist for long times. For example, atmospheric $CO_2$ has certainly fluctuated seasonally for millions of years. It is fortunate for students of the carbon cycle that these fluctuations are minor compared to those that appear to have occurred over longer time scales. If, as several papers in this volume seem to indicate, it is true that relatively long-term changes are significantly greater than short-term changes, then the latter can be viewed as insignificant "noise" associated with the long-term "signal." However, it is quite feasible that, in some cases, comparable changes may be characteristic of different time scales. For these cases, the use of time scales as frames of reference may be more appropriate for examining hypotheses than for simulating reality.

Implicit in the use of eigenvalues as criteria for adding and lumping boxes is the assumption that particular eigenvalues or groups of eigenvalues can be associated with particular boxes or groups of boxes. This assumption is often valid for geochemical cycles, because many boxes exchange only with a few other boxes. Consequently, the matrix K in equation (6) typically contains many zeros. Moreover, the adding and lumping of boxes can be treated formally in a matrix algebra context. This treatment, based on block-diagonal matrix approximations, is described as follows:

## Theory

The approach here is motivated by two observations: (1) that geochemical systems never operate completely independently, and thus never operate at absolute steady state, and (2) that the formulation of geochemical models inherently requires decisions about how to aggregate or allocate parts of systems on the basis of their time-response characteristics. The following analysis is an attempt to provide a basis for quantifying these decisions and the extent to which steady state is an approximation.

Using the notation of Lasaga [1980], let the components of the n-by-n-dimensional exchange coefficient matrix K be defined by the following conditions:

$$k_{ij} = F_{ji}/A_j \qquad i \neq j \qquad (7)$$

$$k_{ij} = -\sum_{m \neq j} F_{jm}/A_j \qquad i = j \qquad (8)$$

where $F_{ji}$ is the flux from box j to box i; $A_j$ is the size of the box j; the summation in (8) is over all values of m except j; and the subscripts i, j, and m range from 1 to n.

Matrix representation of lumping boxes. Consider the matrix K partitioned as follows:

$$K = \begin{bmatrix} K_* & \bar{K}_{i(n-1)} & \bar{K}_{in} \\ \underline{K}_{(n-1)j} & k_{(n-1)(n-1)} & k_{(n-1)n} \\ \underline{K}_{nj} & k_{n(n-1)} & k_{nn} \end{bmatrix} \qquad (9)$$

where

$$K_* = \begin{bmatrix} k_{11} & \cdots & k_{1(n-2)} \\ \vdots & & \vdots \\ k_{(n-2)1} & \cdots & k_{(n-2)(n-2)} \end{bmatrix}$$

$$\bar{K}_{i(n-1)} = \begin{bmatrix} k_{1(n-1)} \\ \vdots \\ k_{(n-2)(n-1)} \end{bmatrix}$$

$$\bar{K}_{in} = \begin{bmatrix} k_{1n} \\ \vdots \\ k_{(n-2)n} \end{bmatrix}$$

$$\underline{K}_{(n-1)j} = \begin{bmatrix} k_{(n-1)1} & \cdots & k_{(n-1)(n-2)} \end{bmatrix}$$

$$\underline{K}_{nj} = \begin{bmatrix} k_{n1} & \cdots & k_{n(n-2)} \end{bmatrix}$$

If boxes n and (n-1) are lumped, the resulting box has a size of $(A_{n-1} + A_n)$, and the fluxes to and from it are sums of the fluxes to and from the formerly separated boxes. The new coefficient matrix has dimensions (n-1) and is equal to

$$K_L = \begin{bmatrix} K_* & B_{n-1}\bar{K}_{i(n-1)} + B_n\bar{K}_{in} \\ \underline{K}_{(n-1)j} & B_{n-1}(k_{(n-1)(n-1)} + k_{n(n-1)}) \\ + \underline{K}_{nj} & + B_n(k_{nn} + k_{(n-1)n}) \end{bmatrix}$$

$$(10)$$

where

$$B_{n-1} = A_{n-1}/(A_{n-1} + A_n)$$

$$B_n = A_n/(A_{n-1} + A_n)$$

The relationship between K and $K_L$ can be expressed in terms of the matrix transformation

$$K_L = PKQ \qquad (11)$$

where

$$P = \begin{bmatrix} I_* & \bar{0} & \bar{0} \\ \underline{0} & 1 & 1 \end{bmatrix}$$

$$Q = \begin{bmatrix} I_* & \bar{0} \\ \underline{0} & B_{n-1} \\ \underline{0} & B_n \end{bmatrix}$$

$I_*$ is the (n-2)-dimensional identity matrix, and $\bar{0}$ and $\underline{0}$ are (n-2)-dimensional column and row null vectors. By repeating this transformation, using matrix partitions of appropriate dimensions, the exchange coefficient matrix can be determined after lumping any number of boxes.

Effects of lumping on eigenvalues. It is common for matrices representing geochemical systems to contain many coefficients equal to zero, representing the absence of exchange between particular reservoirs. As a special case, consider a block-diagonal matrix of the form

$$K = \begin{bmatrix} K_* & 0 \\ 0 & K_o \end{bmatrix}$$

where 0 is a null matrix and the dimensions are

$K$:  $n \times n$

$K_*$:  $r \times r$

$0$:  $(n-r) \times r$  or  $r \times (n-r)$

$K_o$:  $(n-r) \times (n-r)$

This matrix represents a system composed of two completely independent subsystems. The eigenvalues for this matrix can be obtained by triangularization, represented by the similarity transformation

$$T = C^{-1}KC \qquad (12)$$

where

$$C = \begin{bmatrix} C_* & 0 \\ 0 & C_o \end{bmatrix}$$

The dimensions for $C_*$ and $C_o$ are the same as for $K_*$ and $K_o$, and the null partitions in $C$ are dimensioned appropriately. Expanding the partitioned form of equation (12),

$$T = \begin{bmatrix} C_*^{-1} K_* C_* & 0 \\ 0 & C_o^{-1} K_o C_o \end{bmatrix} \qquad (13)$$

It is apparent that the similarity transformations $C_*^{-1}K_*C_*$ and $C_o^{-1}K_oC_o$ must represent triangularizations of $K_*$ and $K_o$. The eigenvalues of $K$ will be equal to the diagonal elements of $T$.

Now suppose that the last $r$ boxes in the system represented by $K$ are subjected to a lumping transformation that leaves a smaller total number of boxes $m$. Such a transformation can be represented by the equation

$$K_L = P_L K Q_L \qquad (14)$$

where

$$P_L = \begin{bmatrix} I_* & 0 \\ 0 & P_o \end{bmatrix}$$

$$Q_L = \begin{bmatrix} I_* & 0 \\ 0 & Q_o \end{bmatrix}$$

The dimensions of $I_*$ are the same as those of $K_*$.

The dimensions of $P_o$ and $Q_o$ are $(m-r) \times (n-r)$ and $(n-r) \times (m-r)$, respectively. The resulting partitioned matrix is

$$K_L = \begin{bmatrix} K_* & 0 \\ 0 & P_o K_o Q_o \end{bmatrix} \qquad (15)$$

This matrix can also be triangularized:

$$T_L = D^{-1} K_L D \qquad (16)$$

or, in partioned form,

$$T_L = \begin{bmatrix} D_*^{-1} K_* D_* & 0 \\ 0 & D_o^{-1} P_o K_o Q_o D_o \end{bmatrix} \qquad (17)$$

where the upper left and lower right partitions are triangularizations of $K_*$ and $P_o K_o Q_o$. The eigenvalues of $K_L$ are equal to the diagonal elements of $T_L$.

The eigenvalues of $K$ and $K_L$ can be compared by comparing the diagonal elements of $T$ and $T_L$. Because the upper left partitions of $T$ and $T_L$ are both similar to $K_*$, they must have the same eigenvalues. Thus, in the case of a block-diagonal coefficient matrix, the lumping transformation affects only the eigenvalues associated with the block that is lumped.

In a geochemical modeling context, block-diagonal matrices are trivial in the sense that any geochemically independent systems, whether related in any way or not, may be rigorously grouped together in block-diagonal form. If the fluxes associated with each block are at steady state, there will be one zero eigenvalue associated with each block, contradicting the general stipulation that the matrix for a linear geochemical system have one and only one zero eigenvalue (as shown by Lasaga [1980]. The importance of the block-diagonal matrix to geochemical modeling is as a limiting case in defining models that are appropriate to particular time scales. If an exchange coefficient matrix can be arranged in a way that approximates block-diagonal form, the eigenproblem for that matrix will approach the analysis described above.

In terms consistent with that analysis, the partitioned matrix

$$K = \begin{bmatrix} K_* & F \\ E & K_o \end{bmatrix} \qquad (18)$$

will approach the relationships expressed in equations (13) and (17) if the elements of partitions $E$ and $F$ approach zero relative to the elements of $K_*$ and $K_o$. This condition occurs when the response times characteristic of exchange among boxes $(n-r+1)$ through $n$ and among boxes 1 through $(n-r)$ are significantly shorter than the response times for exchange between these two groups of boxes. Either group of boxes may then be lumped in any way without significantly affecting the eigenvalues associated with the other group.

This approach can be extended to coefficient matrices that approximate block-diagonal form with any number of diagonal blocks. In fact, any geochemical coefficient matrix can be viewed as the product of lumping diagonal blocks, with each block representing a reservoir made up of parts that exchange more rapidly with each other than with other reservoirs. The validity of the box modeling approach depends largely on the extent to which exchange in the "real" system approaches a block-diagonal representation.

Another common situation is for the elements of partitions E and F to approach zero relative to the elements of $K_O$ but not relative to those of $K_*$. This condition occurs when the response times associated with exchange among boxes $(n-r+1)$ through n are significantly shorter than response times throughout the rest of the system. Under these circumstances, rapidly exchanging boxes can generally be lumped without changing the order of magnitude of the slower response times. That is, in a geochemical system characterized by eigenvalues spanning several orders of magnitude, boxes associated with the large eigenvalues can be lumped without significantly affecting the order of magnitude of the smaller eigenvalues.

Another approach to the effects of lumping on eigenvalues is the application of perturbation theory, which has received considerable attention in efforts to analyze the accuracy of computed eigensystems (e.g., Wilkinson [1965]). In particular, Gerschgorin's theorems [Wilkinson, 1965, pp. 71-81] point toward a more rigorous and general treatment than that given above.

Steady state as an approximation. The block-diagonal approximation is also very useful in understanding how the concept of steady state relates to particular time scales. In a truly block-diagonal matrix, each block along the diagonal is associated with one and only one zero eigenvalue. In a geochemical model, this condition would imply that the exchange coefficients within each block determine a steady state for exchange among a particular group of reservoirs without dependence on the behavior of the other reservoirs. In an approximately block-diagonal matrix such as that defined by equation (18), each diagonal block such as $K_O$ or $K_*$ will be associated with one and only one eigenvalue that is near zero relative to the other eigenvalues associated with that block. In geochemical terms, this implies that the exchange coefficients within that block define a near-steady state that is almost independent of the behavior of the other reservoirs. This conclusion can be extended to the case in which the elements of the off-diagonal partitions approach zero relative to the elements of only one diagonal block. The response time represented by

that block's near-zero eigenvalue is so slow that it cannot be associated with any significant changes over time scales comparable to the response times represented by the other eigenvalues within the system represented by the same diagonal block.

Thus, the time scale of interest determines whether such a system can be considered to be effectively independent. When the time scale of interest coincides with the range of all but the near-zero eigenvalue, the eigenvector associated with that value defines an approximate steady state. Over longer time scales, more significant changes may be associated with the near-zero eigenvalue, contradicting the steady state approximation. Although independent steady states can be approximated for each of the diagonal blocks in an approximately block-diagonal matrix of exchange coefficients, the interpretation of those approximations depends entirely on the time scale of interest.

Matrix interpretation of adding boxes. Consider the matrix defined by equation (18). If the elements of partitions E and F approach zero relative to those of $K_O$, and if the time scale of interest coincides with the range of all of the eigenvalues of $K_O$ except the one near zero, the system represented by $K_O$ can be modeled as an independent system over the appropriate time scale of interest. However, if the time scale of interest lengthens to include time scales comparable to the response time associated with the near-zero eigenvalue of $K_O$, the system must be expanded by adding boxes to those represented by $K_O$.

This process can be viewed formally as the inverse of lumping. For shorter time scales, the near-zero eigenvalue of $K_O$ may be viewed as representing very slow exchange with a single external box. As the time scale of interest increases, the "external" exchange becomes more significant, and its model formulation may require more detail. In other words, the single external box can no longer be treated as single or as external.

Thus, when steady state is viewed as an approximation relative to a particular diagonal submatrix and time scale, the process of adding boxes means only that additional rows and columns are considered to be part of the submatrix under study. These rows and columns will be associated with additional eigenvalues that will replace the former "near-zero" eigenvalue and will determine the relevance of the new rows and columns to the new time scale of interest. A new steady state approximation may be associated with a new eigenvalue that is near zero relative to the others.

## Application

Figures 3 through 5 depict a hierarchical series of carbon cycle box models beginning with a short-term $CO_2$ model and concluding with a model having time response characteristics of

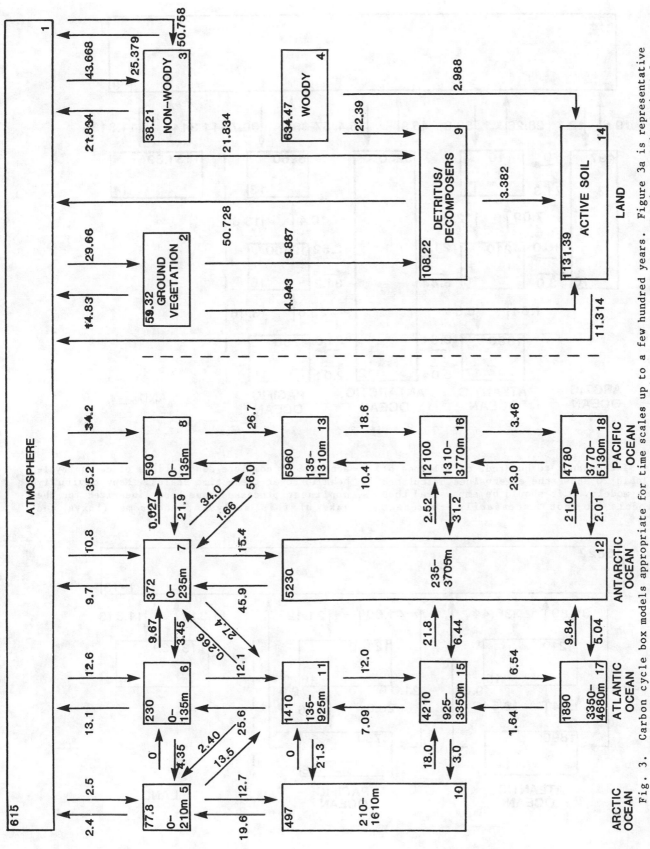

Fig. 3. Carbon cycle box models appropriate for time scales up to a few hundred years. Figure 3a is representative of models used to simulate the short-term effects of anthropogenic CO₂. The ocean part of this model is derived from Bolin et al. [1983] and the land part from Emanuel et al. [1984a, b]. Figures 3b-3d represent box lumping experiments, as described in text. Reservoir sizes are shown in upper left box corners, in gigatons of carbon. Fluxes (arrows) are indicated in gigatons of carbon per year. Box index numbers, in lower right box corners, are consistent for all figures.

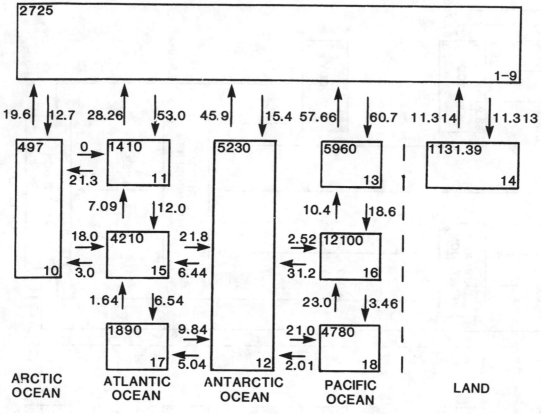

Fig. 3b

hundreds of thousands to hundreds of millions of years.  Table 20 shows the eigenvalues calculated for these models.  It should be emphasized that these models do not necessarily represent accurate or complete descriptions of carbon cycle behavior over these time scales.  Many obviously nonlinear processes have been linearized for the sake of this analysis, and some fluxes and

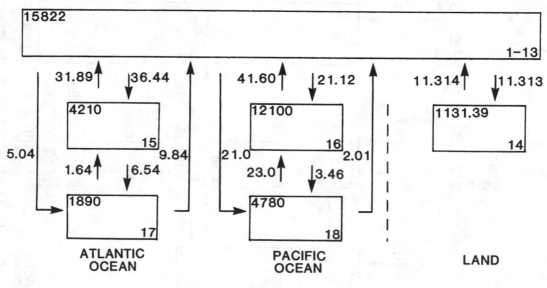

Fig. 3c

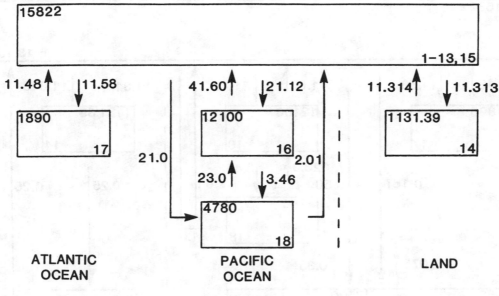

Fig. 3d

reservoirs have been evaluated in spite of large uncertainties. Nevertheless, the models reasonably approximate order-of-magnitude time response characteristics, and the eigenvalues provide a quantitatively consistent tabulation of their interrelationships.

Figures 3a-3d represent an experiment in progressive lumping of boxes. Figure 3a depicts an atmospheric $CO_2$ reservoir coupled to the twelve-box ocean model of Bolin et al. [1983] and the five-box terrestrial carbon model of Emanuel et al. [1984a, b]. The model of Bolin et al. was chosen for this analysis because it depicts both geographic and depth resolution, and because it incorporates explicit particulate carbon fluxes. Bolin et al. rigorously calibrated this model using matrix inversion techniques applied to the distribution of dissolved inorganic carbon, $^{14}C$, oxygen, phosphorus, and alkalinity. Exchange coefficients corresponding to the arrows in Figure 3a were calculated directly from the authors' diffusive and advective flux terms. Additional exchange terms were calculated corresponding to the authors' particulate fluxes, assuming that these fluxes can be represented by direct transfer from surface to deep ocean boxes. The terrestrial model of Emanuel et al. was chosen because it is well documented in linearized form, with reservoirs designated on the basis of time response characteristics. The terrestrial fluxes depicted in Figure 3a were evaluated from the authors' exchange coefficients. The "woody" and "non-woody" reservoirs refer to parts of trees.

Table 20 shows time responses for Figure 3a ranging from less than a year to hundreds of years. Clearly, if the time scale of interest is hundreds of years, many of these boxes can be lumped. In Figure 3b, the atmosphere has been lumped with the surface ocean and with land plants and detritus. In Figure 3c, this aggregate is expanded to include all polar and intermediate ocean waters. Finally, Figure 3d shows that all but the bottom Atlantic, deep and bottom Pacific, and soil carbon boxes have been lumped. The effects of these lumping operations can be observed in the effects on the smaller nonzero eigenvalues, as shown in Table 20. The models represented by Figures 3c and 3d are characterized by response times on the order of 100-200 years, and the eigenvalues defining these response times can be seen to be within 50% of the corresponding eigenvalues for Figure 3a.

In all of the models shown in Figure 3, the carbon cycle is depicted as a closed system encompassing the atmosphere, living and dead plant material, and the oceans. The eigenvalues in Table 20 confirm that, over time scales up to centuries, the most significant natural changes in the carbon cycle must occur as a result of redistributions of carbon within this system.

Until very recently, it was widely assumed that natural changes could not be important over this time scale. However, recent measurements of air trapped in ice cores have shown that significant changes in atmospheric $CO_2$ may have occurred during the most recent glacial epoch over time scales as short as 100 years [see paper by Oeschger et al., this volume]. Oceanographers have also discovered evidence for unexpectedly large changes in the oceans' deep circulation over time scales of decades [Brewer et al., 1983; Roemmich and Wunsch, 1984; Swift, 1984]. Because the oceans contain so much more carbon than the atmosphere, a redistribution of carbon to or from the oceans can have a large effect on the

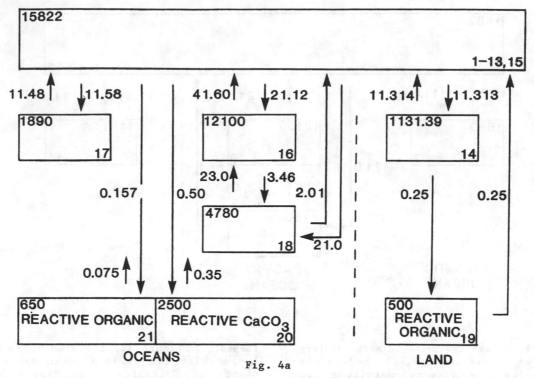

Fig. 4a

Fig. 4. Carbon cycle box models appropriate for time scles of thousands of years. Figure 4a represents addition of "reactive" sediments to the model shown in Figure 3d. Figure 4b represents lumping of all boxes shown in Figure 3. Reservoir sizes, fluxes, and box indices shown are as in Figure 3.

atmosphere. Thus, it is not surprising that several studies [e.g. Wenk and Siegenthaler, this volume; Ennever and McElroy, this volume; Toggweiler and Sarmiento, this volume; Volk and Hoffert, this volume] have hypothesized a link between atmospheric $CO_2$ and the cycling of carbon within the oceans. It is also apparent that heretofore undiscovered mechanisms of change may lie concealed in our lack of knowledge about changes in the exchange of $CO_2$ between the atmosphere and terrestrial organic matter.

In Figure 4, the time scale of interest is

extended by adding boxes to represent exchange with "reactive" sediments. The amount of marine reactive organic carbon is based on the observation that sedimentary organic carbon profiles typically show diagenetic depletion with increasing depth in the uppermost 20-30 cm below the sea-sediment interface [Berner, 1982; Grundmanis and Murray, 1982]. Reactive marine organic carbon was therefore defined as the organic carbon in the uppermost 30 cm of open ocean sediments (assumed to contain 0.5% organic carbon) and the uppermost 50 cm of shelf sedi-

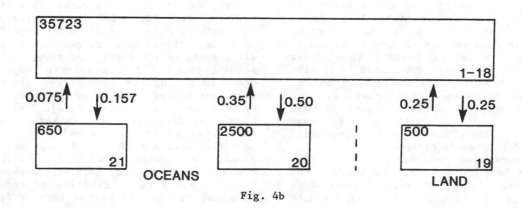

Fig. 4b

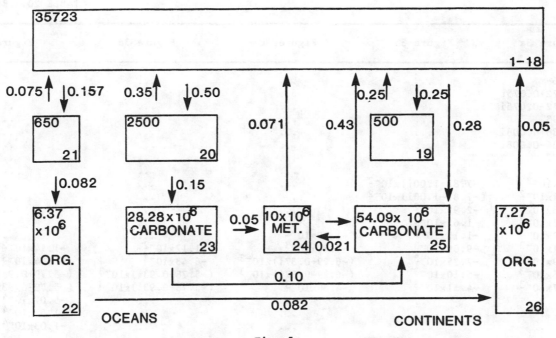

Fig. 5a

Fig. 5.   Carbon cycle box models appropriate for time scales of $10^5$ years and longer. Figure 5a represents addition of all sediments to the model shown in Figure 4b. Abbreviations "org." and "met." stand for the sedimentary organic and active metamorphic carbon reservoirs, respectively.   Figure 5b represents lumping of all boxes shown in Figures 3 and 4.   Reservoir sizes, fluxes, and box indices are as shown in Figure 3.

ments (assumed to contain 1% organic carbon). Reactive marine carbonate carbon was estimated by adapting the carbonate fossil fuel $CO_2$ "neutralizing" capacity calculated by Broecker and Takahashi [1977].   They calculated that a total of 5000 Gt carbon would be available as carbonate to react with fossil fuel $CO_2$. However, this calculation assumed reaction with a static sediment column, and recent models have shown that only about half as much carbonate is available when a realistic sediment flux is assumed [Sundquist, 1983; Keir, 1984].   Terrestrial reactive carbon was estimated from the difference between the "active soil carbon" of Emanuel et al. [1984a, b] and the total soil carbon estimates in Table 14. Fluxes to and from the marine reactive sediments

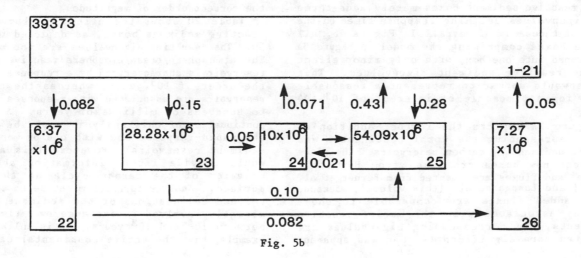

Fig. 5b

TABLE 20. Eigenvalues

| Figure 3a | Figure 3b | Figure 3c | Figure 3d | Figure 4a |
|---|---|---|---|---|
| $-1.16$ | ... | ... | ... | ... |
| $-0.572+0.075i$ | ... | ... | ... | ... |
| $-0.572-0.075i$ | ... | ... | ... | ... |
| $-0.393$ | ... | ... | ... | ... |
| $-0.236+0.008i$ | ... | ... | ... | ... |
| $-0.236-0.008i$ | ... | ... | ... | ... |
| $-0.177$ | ... | ... | ... | ... |
| $-0.129$ | ... | ... | ... | ... |
| $-6.5 \times 10^{-2}$ | $(-7.87+1.00i) \times 10^{-2}$ | ... | ... | ... |
| $-3.51 \times 10^{-2}$ | $(-7.87-1.00i) \times 10^{-2}$ | ... | ... | ... |
| $-2.16 \times 10^{-2}$ | $-2.95 \times 10^{-2}$ | ... | ... | ... |
| $-1.67 \times 10^{-2}$ | $-1.62 \times 10^{-2}$ | ... | ... | ... |
| $-9.77 \times 10^{-3}$ | $-1.11 \times 10^{-2}$ | $-1.30 \times 10^{-2}$ | ... | ... |
| $-7.71 \times 10^{-3}$ | $-9.09 \times 10^{-3}$ | $-9.80 \times 10^{-3}$ | $-1.12 \times 10^{-2}$ | $-1.14 \times 10^{-2}$ |
| $-6.99 \times 10^{-3}$ | $-7.25 \times 10^{-3}$ | $(-6.29+0.37i) \times 10^{-3}$ | $-6.43 \times 10^{-3}$ | $-6.44 \times 10^{-3}$ |
| $-4.45 \times 10^{-3}$ | $-5.10 \times 10^{-3}$ | $(-6.29-0.37i) \times 10^{-3}$ | $(-5.78+0.93i) \times 10^{-3}$ | $(-5.79+0.93i) \times 10^{-3}$ |
| $-4.27 \times 10^{-3}$ | $-4.31 \times 10^{-3}$ | $-4.86 \times 10^{-3}$ | $(-5.78-0.93i) \times 10^{-3}$ | $(-5.79-0.93i) \times 10^{-3}$ |
| ... | ... | ... | ... | $-5.07 \times 10^{-4}$ |
| ... | ... | ... | ... | $-1.55 \times 10^{-4}$ |
| ... | ... | ... | ... | $-1.03 \times 10^{-4}$ |
| ... | ... | ... | ... | ... |
| ... | ... | ... | ... | ... |
| ... | ... | ... | ... | ... |
| ... | ... | ... | ... | ... |
| $6.69 \times 10^{-15}$ | $2.22 \times 10^{-16}$ | $2.22 \times 10^{-16}$ | $1.11 \times 10^{-16}$ | $8.67 \times 10^{-17}$ |

Units are $yr^{-1}$.

were estimated from Tables 10, 18, and 19. The terrestrial reactive reservoir was assigned a residence time of 2000 years.

In Figure 4a, the reactive sediments are simply added to the model depicted in Figure 3d. In Table 20, it is apparent that adding these boxes has little impact on the response times associated with exchange within the atmosphere/ocean/biosphere system. Adding the three reactive sediment boxes merely adds three new eigenvalues defining response times on the order of thousands of years. In Figure 4b, all of the boxes comprising the model in Figure 3a are lumped into one box, with only minor effect on the reactive sediment eigenvalues. This figure would appear to represent a reasonable model for a timescale of interest of $10^3$-$10^4$ years.

Figure 5a depicts the further addition of boxes representing the total continental and oceanic sedimentary carbon reservoirs. The sizes of these new boxes are taken from Hay [this volume] and fluxes are adapted from Berner et al. [1983] and Lasaga et al. [this volume]. Because these added fluxes are comparable to those already associated with the marine reactive sediments, the corresponding eigenvalues are affected more significantly than was apparent

when the reactive sediment boxes were added in Figure 4a. Thus, although Figure 4b adequately approximates carbon cycle behavior over time scales of thousands of years, the time response within that system is affected by the magnitude of the long-term weathering and sedimentation fluxes. The size and interactions of the "metamorphic" reservoir in Figure 5 are very uncertain, but the values shown are probably of the correct order of magnitude.

Table 20 shows the effect of lumping the reactive sediment boxes, as depicted in Figure 5b. The resulting eigenvalues are not uniform. The atmosphere/ocean/biosphere/reactive sediment reservoir is characterized by a response time on the order of $10^5$ years, whereas the sediment reservoirs are associated with responses of tens to hundreds of millions of years. A logical continuation of this analysis would be to lump the rapid-response box with one or more of the sediment reservoirs. However, this approach would provide little information about the behavior of the carbon cycle at the earth surface. Another approach might be to seek more refined delineations of the sediment volumes actually involved in exchange over time scales between $10^5$ and $10^8$ years. It is unlikely, for example, that the entire continental carbonate

for Carbon Cycle Models

| Figure 4b | Figure 5a | Figure 5b | All Reservoirs |
|---|---|---|---|
| ... | ... | ... | -1.16 |
| ... | ... | ... | -0.573+0.075i |
| ... | ... | ... | -0.573-0.075i |
| ... | ... | ... | -0.394 |
| ... | ... | ... | -0.236+0.008i |
| ... | ... | ... | -0.236-0.008i |
| ... | ... | ... | -0.178 |
| ... | ... | ... | -0.129 |
| ... | ... | ... | $-6.51 \times 10^{-2}$ |
| ... | ... | ... | $-3.52 \times 10^{-2}$ |
| ... | ... | ... | $-2.17 \times 10^{-2}$ |
| ... | ... | ... | $-1.67 \times 10^{-2}$ |
| ... | ... | ... | $-9.86 \times 10^{-3}$ |
| ... | ... | ... | $-7.74 \times 10^{-3}$ |
| ... | ... | ... | $-7.06 \times 10^{-3}$ |
| ... | ... | ... | $(-4.40+0.10i) \times 10^{-3}$ |
| ... | ... | ... | $(-4.40-0.10i) \times 10^{-3}$ |
| $-5.07 \times 10^{-4}$ | $-5.08 \times 10^{-4}$ | ... | $-5.03 \times 10^{-4}$ |
| $-1.55 \times 10^{-4}$ | $-2.44 \times 10^{-4}$ | ... | $-2.44 \times 10^{-4}$ |
| $-1.18 \times 10^{-4}$ | $-2.10 \times 10^{-4}$ | ... | $-2.09 \times 10^{-4}$ |
| ... | $-1.33 \times 10^{-5}$ | $-1.30 \times 10^{-5}$ | $-1.24 \times 10^{-5}$ |
| ... | $-1.42 \times 10^{-8}$ | $-1.42 \times 10^{-8}$ | $-1.41 \times 10^{-8}$ |
| ... | $-8.23 \times 10^{-9}$ | $(-7.49+0.09i) \times 10^{-9}$ | $-7.95 \times 10^{-9}$ |
| ... | $-7.33 \times 10^{-9}$ | $(-7.49-0.09i) \times 10^{-9}$ | $-7.07 \times 10^{-9}$ |
| ... | $-7.10 \times 10^{-9}$ | $-6.60 \times 10^{-9}$ | $-5.27 \times 10^{-9}$ |
| $2.38 \times 10^{-18}$ | $-9.30 \times 10^{-18}$ | $-4.24 \times 10^{-22}$ | $2.78 \times 10^{-18}$ |

reservoir contributes to the river bicarbonate flux.

Another interpretation of the Figure 5b eigenvalues is that they define a state of "secular equilibrium" between the slow sedimentary cycle and the relatively rapid earth surface carbon cycle. This is essentially the viewpoint taken by Berner et al. [1983] and Lasaga et al. [this volume]. In these models, the distribution of carbon at the earth's surface is fixed by its slow release and consumption by the sedimentary rock cycle. The secular equilibrium is defined by the maintenance of a constant state of oceanic calcite saturation. Eigenanalysis supports this approach.

For purposes of comparison, the last column in Table 20 shows eigenvalues for a box model defined by all of the discrete fluxes and reservoirs introduced in Figures 3-5. These values confirm the relatively independent nature of the eigenvalues throughout the hierarchical model development process.

## Conclusions

Matrix algebra and eigenanalysis provide a useful basis for examining the interrelationships among carbon cycle processes and reservoirs interacting over a wide range of time scales. By appropriately arranging the matrix of exchange coefficients, and by ranking the eigenvalues in a sequence such as that shown in Table 20, the effects of box modeling approximations can be quantified in terms of changes in particular eigenvalues or groups of eigenvalues. Models appropriate to particular time scales can be constructed, with explicit knowledge of the changes caused by lumping short-term reservoirs.

The carbon cycle can be represented without sediment interactions for time scales up to a few hundred years. Atmospheric $CO_2$ over these time scales can be considered as part of a larger reservoir that includes substantial portions of the oceans and land plants. "Reactive" sediments must be incorporated in models of carbon cycle behavior over thousands of years. Although millennial carbon cycle models need not include interaction with the long-term sedimentary cycle, the time response of millennial models is affected by the magnitudes of the longer-term fluxes. Modeling longer time scales will benefit substantially from a refined understanding of the quantity of sediments actively involved in weathering. Until such an understanding is reached, the earth surface carbon cycle can be viewed as at secular equilibrium with respect to the sedimentary cycle over tens to hundreds of millions of years.

An important observation is the influence of the magnitudes of long-term fluxes on the time response of short-term models, as was observed in adding the long-term sedimentary cycle to a model appropriate to thousands of years. Another example might be the adding of particle fluxes to decadal oceanic $CO_2$ models, many of which ignore particulate transport. In each case, although perturbations of the long-term fluxes are unlikely to influence the short-term cycle, the long-term fluxes affect the response to short-term perturbations. Thus, in certain cases, short-term models must include long-term fluxes.

Within the context of time scales as frames of reference, I have shown how long-term box model structures can be simplified by lumping short-term components. It should be emphasized that more detail in models is always better, as long as the detail is supported by reasonable data or assumptions. In model development, the value of eigenanalysis is that it quantifies compromises in modeling detail.

Acknowledgments. I am grateful to many for their help in preparing this paper. Doug Burns, Deborah Martin, John McGeehin, and Clark Reed were my succession of ambassadors to the U.S. Geological Survey library over a period of several years. Dave Schink, Fred Spilhaus, and the publication staff of the American Geophysical Union provided exceptional management support in spite of my duplicity as diligent editor and laggard author. Tony Lasaga and an anonymous reviewer supplied very useful and timely comments. Typing and formatting were done by Joanne Taylor, and drafting by Doug Burns.

## References

Adelseck, C. G., Jr., Dissolution of deep-sea carbonate: Preliminary calibration of preservational and morphologic aspects, Deep Sea Res., 25, 1167-1185, 1978.

Ajtay, G. L., P. Ketner, and P. Duvigneaud, Terrestrial primary productivity and phytomass, in The Global Carbon Cycle, SCOPE 13, edited by B. Bolin, E. T. Degens, S. Kempe, and P. Ketner, pp. 129-181, John Wiley, New York, 1979.

Alldredge, A. L., and J. L. Cox, Primary productivity and chemical composition of marine snow in surface waters of the Southern California Bight, J. Mar. Res., 40, 517-527, 1982.

Anderson, L., and D. Dyrssen, Chemical constituents of the Arctic Ocean in the Svalbard area, Oceanol. Acta, 4, 305-311, 1981.

Andreae, M. O., Soot carbon and excess fine potassium: Long-range transport of combustion-derived aerosols, Science, 220, 1148-1151, 1983.

Angel, M. V., Detrital organic fluxes through pelagic ecosystems, in Flows of Energy and Materials in Marine Ecosystems, Theory and Practice, edited by M. J. R. Fasham, pp. 475-516, Plenum, New York, 1984.

Arnold, J. R., and E. C. Anderson, The distribution of carbon-14 in nature, Tellus, 9, 28-32, 1957.

Bacastow, R., and C. D. Keeling, Atmospheric carbon dioxide and radiocarbon in the natural carbon cycle, II. Changes from A.D. 1700 to 2070 as deduced from a geochemical model, in Carbon in the Biosphere, AEC Symp. Ser., vol. 30, edited by G. M. Woodwell and E. V. Pecan, pp. 86-135, U.S. Department of Commerce, Springfield, Va., 1973.

Baes, C. F., Jr., H. E. Goeller, J. S. Olson, and R. M. Rotty, The global carbon dioxide problem, Rep. NTIS ORNL-5194, 72 pp., Oak Ridge Natl. Lab., Oak Ridge, Tenn., 1976.

Barnola, J. M., D. Raynaud, A. Neftel, and H. Oeschger, Comparison of $CO_2$ measurements by two laboratories on air from bubbles in polar ice, Nature, 303, 410-413, 1983.

Baumgartner, A., and E. Reichel, The World Water Balance, 179 pp., Elsevier, New York, 1975.

Bazilevich, N. I., Energy flow and biological regularities of the world ecosystems, in Proceedings of the First International Congress of Ecology, edited by A. J. Cave, pp. 47-51, The Hague, Pudoc, Wageningen, 414 pp., 1974.

Bazilevich, N. I., L. Ye. Rodin, and N. N. Rozov, Geographical aspects of biological productivity, Sov. Geogr. Rev. Transl., 12, Engl. Transl., 293-317, 1971.

Bé, A. W. H., and R. W. Gilmer, A taxonomic and zoogeographic review of euthecosomatous pteropods, in Oceanic micropaleontology, edited by A. T. Ramsay, pp. 733-808, Academic Press, Orlando, Fla., 1977.

Bé, A. W. H., and D. S. Tolderlund, Distribution and ecology of living planktonic foraminifera in surface waters of the Atlantic and Indian oceans, in Micropaleontology of Oceans, edited by B. M. Funnel and W. R. Riedel, pp. 105-149, Cambridge University Press, New York, 1971.

Bender, M. L., and D. T. Heggie, Fate of organic carbon reaching the deep sea floor: A status report, Geochim. Cosmochim. Acta, 48, 977-986, 1984.

Berger, W. H., Foraminiferal ooze: Solution at depths, Science, 156, 383-385, 1967.

Berger, W. H., Planktonic foraminifera: selective solution and paleoclimatic interpretation, Deep Sea Res., 15, 31-43, 1968.

Berger, W. H., Ecologic patterns of living planktonic foraminifera, Deep Sea Res., 16, 1-24, 1969.

Berger, W. H., Biogeneous deep sea sediments: Fractionation by deep sea circulation, Geol. Soc. Am. Bull., 81, 1385-1402, 1970.

Berger, W. H., Sedimentation of planktonic foraminifera, Mar. Geol., 11, 325-358, 1971.

Berger, W. H., Biogenous deep sea sediments: Production, preservation and interpretation, in Chemical Oceanography, vol. 5, edited by J. P. Riley and R. Chester, pp. 265-388, Academic Press, Orlando, Fla., 1976.

Berger, W. H., Carbon dioxide excursions and the deep-sea record: Aspects of the problem, in The Fate of Fossil Fuel $CO_2$ in the Oceans, edited by N. R. Anderson and A. Malahoff, pp. 505-542, Plenum, New York, 1977.

Berger, W. H., Deep-sea carbonate: Pteropod distribution and the aragonite compensation depth, Deep Sea Res., 25, 447-452, 1978a.

Berger, W. H., Sedimentation of deep-sea carbonate: Maps and models of variations and fluctuations, J. Foraminiferal Res., 8, 286-302, 1978b.

Berger, W. H., Paleoceanography: The deep-sea record, in The Oceanic Lithosphere, edited by C. Emiliani, pp. 1437-1519, John Wiley, New York, 1981.

Berger, W. H., and R. S. Keir, Glacial-Holocene changes in atmospheric $CO_2$ and the deep-sea record, in Climate Processes and Climate Sensitivity, Geophys. Monogr. Ser. vol. 29, edited by J. E. Hansen and T. Takahashi, pp. 337-351, AGU, Washington, D.C., 1984.

Berger, W. H., and D. J. Piper, Planktonic foraminifera: Differential settling, dissolution and redeposition, Liminol. Oceanogr., 17, 275-287, 1972.

Berger, W. H., and E. L. Winterer, Plate stratigraphy and the fluctuating carbonate line, in Pelagic Sediments on Land and under the Sea, edited by K. J. Hsu and H. C. Jenkyns, pp. 11-48, Blackwell, London, 1974.

Berger, W. H., C. C. Adelseck, Jr., and L. A. Mayer, Distribution of carbonate in surface sediments of the Pacific Ocean, J. Geophys. Res., 81, 2617-2627, 1976.

Berger, W. H., M.-C. Bonneau, and F. L. Parker, Foraminifera on the deep-sea floor: Lysocline and dissolution rate, Oceanol. Acta, 5, 249-258, 1982.

Berner, R. A., Sedimentation and dissolution of Pteropods in the ocean, in The Fate of Fossil Fuel $CO_2$ in the Oceans, edited by N. R. Anderson and A. Malahoff, pp. 243-260, Plenum, New York, 1977.

Berner, R. A., Burial of organic carbon and pyrite sulfur in the modern ocean: Its geochemical and environmental significance, Am. J. Sci., 282, 451-473, 1982.

Berner, R. A., and S. Honjo, Pelagic sedimentation of aragonite: Its geochemical significance, Science, 211, 940-942, 1981.

Berner, R. A., and R. Raiswell, Burial of organic carbon and pyrite sulfur in sediments over Phanerozoic time: A new theory, Geochim. Cosmochim. Acta, 46, 1689-1705, 1983.

Berner, R. A., E. K. Berner, and R. S. Keir, Aragonite dissolution on the Bermuda Pedestal: Its depth and geochemical significance, Earth Planet. Sci. Lett., 30, 169-178, 1976.

Berner, R. A., A. C. Lasaga, and R. M. Garrels, The carbonate-silicate geochemical cycle and its effect on atmospheric carbon dioxide over the past 100 million years, Am. J. Sci., 283, 641-683, 1983.

Betzer, P. R., W. J. Showers, E. A. Laws, C. D. Winn, G. R. DiTullio, and P. M. Kroopnick, Primary productivity and particle fluxes on a transect of the equator at 153°W in the Pacific Ocean, Deep Sea Res., 31, 1-11, 1984a.

Betzer, P. R., R. H. Byrne, J. G. Acker, C. S. Lewis, and R. R. Jolley, The oceanic carbonate system: A reassessment of biogenic controls, Science, 226, 1074-1077, 1984b.

Bien, G., and H. Suess, Transfer and exchange of $^{14}C$ between the atmosphere and the surface water of the Pacific Ocean, in Radioactive Dating and Low-Level Counting, pp. 105-115, International Atomic Energy Agency, Vienna, 1967.

Bien, G. S., N. W. Rakestraw, and H. E. Suess, Radiocarbon concentration in Pacific Ocean water, Tellus, 12, 436-443, 1960.

Billett, D. S. M., R. S. Lampitt, A. L. Rice, and R. F. C. Mantoura, Seasonal sedimentation of phytoplankton to the deep-sea benthos, Nature, 302, 520-522, 1983.

Biscaye, P. E., V. Kolla, and K. K. Turekian, Distribution of calcium carbonate in surface sediments of the Atlantic Ocean, J. Geophys. Res., 81, 2595-2603, 1976.

Bischof, W., P. Fabian, and R. Borchers, Decrease in $CO_2$ mixing ratio observed in the stratosphere, Nature, 288, 347-348, 1980.

Bishop, J. K. B., and J. M. Edmond, A new large volume filtration system for the sampling of oceanic particulate matter, J. Mar. Res., 34, 181, 1976.

Bishop, J. K. B., J. M. Edmond, D. R. Ketten, M. P. Bacon, and W. B. Silker, The chemistry, biology, and vertical flux of particulate matter from the upper 400 m of the equatorial Atlantic Ocean, Deep Sea Res., 24, 511-548, 1977.

Bishop, J. K. B., D. R. Ketten, and J. E. Edmond, The chemistry, biology and vertical flux of particulate matter from the upper 400 m of the Cape Basin in the southeast Atlantic Ocean, Deep Sea Res., 25, 1121-1161, 1978.

Björkstrom, A., A model of $CO_2$ interaction between atmosphere, oceans, and land biota, in The Global Carbon Cycle, SCOPE 13, edited by B. Bolin, E. T. Degens, S. Kempe, and P. Ketner, pp. 403-457, John Wiley, New York, 1979.

Black, M. A., Coccoliths, Endeavor, 24, 131-137, 1965.

Blake, D. R., E. W. Mayer, S. C. Tyler, Y. Makide, D. C. Montague, and F. S. Rowland, Global increase in atmospheric methane concentrations between 1978 and 1980, Geophys. Res. Lett., 9, 477-480, 1982.

Bogdanov, U. A., A. P. Lisitzin, and E. A. Romankevich, Organic matter of suspended and bottom sediments of seas and oceans, in Organic Matter of Recent and Ancient Sediments (in Russian), edited by N. B. Vasseovich, pp. 35-103, Akademiya Nauk SSSR, Moscow, 1971.

Bogorov, V. G., Biological transformation and exchange of energy and matter in the ocean, Oceanology, Engl. Transl., 7, 649-665, 1967.

Bogorov, V. G., Life in the Oceans (in Russian), Biol. Ser., vol. 6, Znaniye, Moscow, 1969.

Bohn, H. L., Estimate of organic carbon in world soils, Soil Sci. Soc. Am. J., 40, 468-470, 1976.

Bohn, H. L., Estimate of organic carbon in world soils: II, Soil Sci. Soc. Am. J., 46, 1118-1119, 1982.

Bolin, B., The carbon cycle, Sci. Am., 223, 125-132, 1970.

Bolin, B., Changes of land biota and their importance for the carbon cycle, Science, 196, 613-615, 1977.

Bolin, B., and W. Bischof, Variations of the carbon dioxide content of the atmosphere in the northern hemisphere, Tellus, 22, 431-442, 1970.

Bolin, B., and E. Eriksson, Changes in the carbon dioxide content of the atmosphere and sea due to fossil fuel combustion, in The Atmosphere and the Sea in Motion, edited by B. Bolin, pp. 130-142, The Rockefeller Institute Press, New York, 1959.

Bolin, B., and C. D. Keeling, Large-scale atmospheric mixing as deduced from the seasonal and meridional variations of carbon dioxide, J. Geophys. Res., 68, 3899-3920, 1963.

Bolin, B., E. T. Degens, P. Duvigneaud, and S. Kempe, The global biogeochemical carbon cycle, in The Global Carbon Cycle, SCOPE 13, edited by B. Bolin, E. T. Degens, S. Kempe, and P. Ketner, pp. 1-53, John Wiley, New York, 1979.

Bolin, B., A. Bjorkstrom, and K. Holmen, The simultaneous use of tracers for ocean circulation studies, Tellus, 35B, 206-236, 1983.

Bowen, H. J. M., Trace Elements in Biochemistry, 241 pp., Academic Press, Orlando, Fla., 1966.

Bray, J. R., An analysis of the possible recent change in atmospheric carbon dioxide concentration, Tellus, 11, 220-230, 1959.

Brewer, P. G., Direct observation of the oceanic $CO_2$ increase, Geophys. Res. Lett., 5, 997-1000, 1978.

Brewer, P. G., and J. C. Goldman, Alkalinity changes generated by phytoplankton growth, Limnol. Oceanogr., 21, 108-117, 1976.

Brewer, P. G., G. T. F. Wong, M. P. Bacon, and D. W. Spencer, An oceanic calcium problem?, Earth Planet. Sci. Lett., 26, 81-97, 1975.

Brewer, P. G., W. S. Broecker, W. J. Jenkins, P. B. Rhines, C. G. Rooth, J. H. Swift, T. Takahashi, and R. T. Williams, A climatic freshening of the deep North Atlantic (north of 50°N) over the past 20 years, Science, 222, 1237-1239, 1983.

Broecker, W. S., Radioisotopes and large scale oceanic mixing, in The Sea, Vol. 2, edited by M. N. Hill, pp. 88-108, John Wiley, New York, 1963.

Broecker, W. S., Radioisotopes and the rate of mixing across the main thermoclines of the ocean, J. Geophys. Res., 71, 5827-5836, 1966.

Broecker, W. S., Calcite accumulation rates and glacial to interglacial changes in oceanic mixing, in The Late Cenozoic Glacial Ages, edited by K. K. Turekian, pp. 239-265, Yale University Press, New Haven, Conn., 1971.

Broecker, W. S., Chemical Oceanography, 214 pp., Harcourt Brace Jovanovich, New York, 1974.

Broecker, W. S., A revised estimate for the radiocarbon age of North Atlantic deep water, J. Geophys. Res., 84, 3218-3226, 1979.

Broecker, W. S., Ocean chemistry during glacial time, Geochim. Cosmochim. Acta, 46, 1689-1705, 1982.

Broecker, W. S., and Y.-H. Li, Interchange of water between the major oceans, J. Geophys. Res., 25, 3545-3552, 1970.

Broecker, W. S., and T.-H. Peng, Gas exchange rates between air and sea, Tellus, 26, 21-35, 1974.

Broecker, W. S., and T.-H. Peng, The distribution of bomb-produced tritium and radiocarbon at Geosecs station 347 in the eastern North Pacific, Earth Planet. Sci. Lett., 49, 453-462, 1980.

Broecker, W. S., and T.-H. Peng, Tracers in the Sea, 690 pp., Lamont-Doherty Geological Observatory, Palisades, N.Y., 1982.

Broecker, W. S., and T.-H. Peng, The climate-chemistry connection, in Climate Processes and Climate Sensitivity, Geophys. Monogr. Ser. vol. 29, edited by J. E. Hansen, and T. Takahashi, pp. 327-336, AGU, Washington, D.C., 1984.

Broecker, W. S., and T. Takahashi, Neutralization of fossil fuel $CO_2$ by marine calcium carbonate, in The Fate of Fossil Fuel $CO_2$ in the Oceans, edited by N. R. Andersen and A. Malahoff, pp. 213-241, Plenum, New York, 1977.

Broecker, W. S., and T. Takahashi, The relationship between lysocline depth and in situ carbonate ion concentration, Deep Sea Res., 25, 65-95, 1978.

Broecker, W. S., and T. Takahashi, Is there a tie between atmospheric $CO_2$ content and ocean circulation?, in Climate Processes and Climate Sensitivity, Geophys. Monogr. Ser. vol. 29, edited by J. E. Hansen and T. Takahashi, pp. 314-326, AGU, Washington, D.C., 1984.

Broecker, W. S., R. Gerard, M. Ewing, and B. C. Heezen, Natural radiocarbon in the Atlantic Ocean, J. Geophys. Res., 65, 2903-2931, 1960.

Broecker, W. S., Y.-H. Li, and T.-H. Peng, Carbon dioxide--Man's unseen artifact, in Impingement of Man on the Oceans, edited by D. W. Hood, pp. 287-324, John Wiley, New York, 1971.

Broecker, W. S., T.-H. Peng, and M. Stuiver,

An estimate of the upwelling rate in the equatorial Atlantic based on the distribution of bomb radiocarbon, J. Geophys. Res., 83, 6179-6186, 1978.

Broecker, W. S., T. Takahashi, H. J. Simpson, and T.-H. Peng, Fate of fossil fuel carbon dioxide and the global carbon budget, Science, 206, 409-418, 1979.

Broecker, W. S., T.-H. Peng, G. Mathieu, R. Hesslein, and T. Torgersen, Gas exchange rate measurements in natural systems, Radiocarbon, 22, 676-683, 1980a.

Broecker, W. S., T.-H. Peng, and T. Takahashi, A strategy for the use of bomb-produced radiocarbon as a tracer for the transport of fossil fuel $CO_2$ into the deep-sea source regions, Earth Planet. Sci. Lett., 49, 463-468, 1980b.

Broecker, W. S., T.-H. Peng, and R. Engh, Modeling the carbon system, Radiocarbon, 22, 565-598, 1980c.

Bruland, K. W., and M. W. Silver, Sinking rates of fecal pellets from gelatinous zooplankton (salps, pteropods, doliolids), Mar. Biol., 63, 295, 1981.

Bruyevich, S. V., and V. N. Ivanenkov, Chemical balance of the world ocean, Oceanology, Engl. Transl., 11, 694-699, 1971.

Buringh, P., Organic carbon in soils of the world, in The Role of Terrestrial Vegetation in the Global Carbon Cycle: Measurement by Remote Sensing, SCOPE 23, edited by G. M. Woodwell, pp. 91-109, John Wiley, New York, 1984.

Burling, R. W., and D. M. Garner, A section of $^{14}C$ activities of sea water between 9°S and 66°S in the southwest Pacific Ocean, N. Z. J. Geol. Geophys., 2, 799-824, 1959.

Byrne, R. H., J. G. Acker, P. R. Betzer, R. A. Feely, and M. H. Cates, Water column dissolution of aragonite in the Pacific Ocean, Nature, 312, 321-326, 1984.

Callendar, G. S., On the amount of carbon dioxide in the atmosphere, Tellus, 10, 243-248, 1958.

Carmack, E. C., and T. D. Foster, On the flow of water out of the Weddell Sea, Deep Sea Res., 22, 711-724, 1975.

Cauwet, G., Organic chemistry of sea water particulates: Concepts and developments, Oceanol. Acta, 1, 99-105, 1978.

Chamberlain, J. W., H. M. Foley, G. J. MacDonald, and M. A. Ruderman, Climate effects of minor atmospheric constituents, in Carbon Dioxide Review, edited by W. C. Clark, pp. 253-277, Oxford University Press, New York, 1982.

Chan, Y.-H., J. S. Olson, and W. R. Emanuel, Land use and energy scenarios affecting the global carbon cycle, Environment International, 4, 189-206, 1980.

Chen, C.-T., Decomposition of calcium carbonate and organic carbon in the deep oceans, Science, 201, 735-736, 1978.

Chen, C.-T., Oceanic penetration of excess $CO_2$ in a cross section between Alaska and Hawaii, Geophys. Res. Lett., 9, 117-119, 1982a.

Chen, C.-T., On the distribution of anthropogenic $CO_2$ in the Atlantic and Southern Oceans, Deep Sea Res., 29, 563-580, 1982b.

Chen, C.-T., and F. J. Millero, Gradual increase of oceanic $CO_2$, Nature, 277, 205-206, 1979.

Chen, C.-T., F. J. Millero, and R. M. Pytkowicz, Comment on calculating the oceanic $CO_2$ increase: A need for caution by A. M. Shiller, J. Geophys. Res., 87, 2083-2085, 1982.

Chung, Y., and H. Craig, $^{226}Ra$ in the Pacific Ocean, Earth Planet. Sci. Lett., 49, 267-292, 1980.

Clark, W. C., K. H. Cook, G. Marland, A. M. Weinberg, R. M. Rotty, P. R. Bell, L. J. Allison, and C. L. Cooper, The carbon dioxide question: A perspective for 1982, in Carbon Dioxide Review 1982, edited by W. C. Clark, pp. 3-44, Oxford University Press, New York, 1982.

Cline, J. D., and H. G. Ostlund, A comparison of the vertical distributions of trichlorofluoromethane (F-11) and tritium in the North Pacific (abstract), EOS Trans. AGU, 63, 972, 1982.

Cobler, R., and J. Dymond, Sediment trap experiment on the Galapagos spreading center, Equatorial Pacific, Science, 209, 801-803, 1980.

Conover, R. J., Transformation of organic matter, in Marine Ecology, Volume IV: Dynamics, edited by O. Kinne, pp. 221-499, John Wiley, New York, 1978.

Craig, H., The natural distribution of radiocarbon and the exchange time of carbon dioxide between atmosphere and sea, Tellus, 9, 1-17, 1957.

Craig, H., A critical evaluation of mixing rates in oceans and atmosphere by use of radiocarbon techniques, Proc. 2nd Int. Conf. Peaceful Uses Atomic Energy, 18, pp. 358-363, United Nations, 1958.

Craig, H., The natural distribution of radiocarbon: Mixing rates in the sea and residence times of carbon and water, in Earth Science and Meteoritics, compiled by J. Geiss and E. D. Goldberg, pp. 103-114, John Wiley, New York, 1963.

Craig, H., Abyssal carbon and radiocarbon in the Pacific, J. Geophys. Res., 74, 5491-5506, 1969.

Craig, H. and C. C. Chou, Methane: The record in polar ice cores, Geophys. Res. Lett., 9, 1221-1224, 1982.

Crane, A. J., The partitioning of excess $CO_2$ in a five-reservoir atmosphere-ocean model, Tellus, 34, 398-405, 1982.

Crawford, C. C., J. E. Hobbie, and K. L. Webb, The utilization of dissolved free amino acids by estuarine microorganisms, Ecology, 55, 551-563, 1974.

Curry, W. B., and R. K. Matthews, Paleooceanographic utility of oxygen isotopic measurements on planktic foraminifera: Indian Ocean core-top evidence, Palaeogeogr. Palaeoclimatol. Palaeoecol., 33, 173-191, 1981.

Degens, E. T., and K. Mopper, Factors controlling the distribution and early diagenesis of organic material in marine sediments, in Chemical Oceanography, vol. 6, edited by J. P. Riley and R. Chester, pp. 59-113, Academic Press, Orlando, Fla., 1976.

Delibrias, G., Carbon-14 in the southern Indian Ocean, Radiocarbon, 22, 684-692, 1980.

Despraires, P., World Energy Resources 1985-2020, vol. 2, Oil and Gas Resources, IPC Science and Technology Press, Guildford, England, 1978.

Deuser, W. G., Marine biota, nearshore sediments, and the global carbon balance, Org. Geochem., 1, 243-247, 1979.

Deuser, W. G., and E. H. Ross, Seasonal change in the flux of organic carbon to the deep Sargasso Sea, Nature, 283, 364-365, 1980.

Deuser, W. G., E. H. Ross, and R. F. Anderson, Seasonality in the supply of sediment to the deep Sargasso Sea and implications for the rapid transfer of matter to the deep ocean, Deep Sea Res., 28A, 495-505, 1981.

de Vooys, C. G. N., Primary production in aquatic environments, in The Global Carbon Cycle, SCOPE 13, edited by B. Bolin, E. T. Degens, S. Kempe, and P. Ketner, pp. 259-292, John Wiley, New York, 1979.

Donnell, J. R., Global oil-shale resources and costs, in The Future Supply of Nature-Made Petroleum and Gas, edited by R. F. Meyer, pp. 843-856, Pergamon, York, New York, 1977.

Duce, R. A., and E. K. Duursma, Inputs of organic matter to the ocean, Mar. Chem., 5, 319-339, 1977.

Dugdale, R. C., and J. J. Goering, Uptake of new and regenerated forms of nitrogen in primary productivity, Limnol. Oceanogr., 12, 196-206, 1967.

Dunbar, R. B., and W. H. Berger, Fecal pellet flux to modern bottom sediment of Santa Barbara Basin (California) based on sediment trapping, Geol. Soc. Am. Bull., Part I, 92, 212-218, 1981.

Duncan, D. C. and V. E. Swanson, Organic-rich shale of the United States and world land areas, U.S. Geol. Surv. Circ., 523, 1965.

Druffel, E. M. and H. E. Suess, On the radiocarbon record in banded corals: Exchange parameters and net transport of $^{14}CO_2$ between atmosphere and surface ocean, J. Geophys. Res., 88, 1271-1280, 1983.

Eadie, B. J., and L. M. Jeffrey, $\delta^{13}C$ analyses of oceanic particulate organic matter, Mar. Chem., 1, 199-209, 1973.

Eadie, B. J., L. M. Jeffrey, and W. M. Sackett, Some observations on the stable carbon isotope composition of dissolved and particulate organic carbon in the marine environment, Geochim. Cosmochim. Acta, 42, 1265-1269, 1978.

Ehhalt, D. H., L. E. Heidt, R. H. Lueb, and E. A. Martell, Concentrations of $CH_4$, CO, $CO_2$, $H_2$, $H_2O$ and $N_2O$ in the upper stratosphere, J. Atmos. Sci., 32, 163-169, 1975.

Ehhalt, D. H., R. J. Zander, and R. A. Lamontagne, On the temporal increase of tropospheric $CH_4$, J. Geophys. Res., 88, 8442-8446, 1983.

Elliott, W. P., R. D. Bojkov, P. Brewer, P. Fraser, R. Gammon, H. Oeschger, G. I. Pearman, J. Peterson, D. Raynaud, U. Sigenthaler, B. Stauffer, G. M. Stokes, and M. Stuiver, The pre-1958 atmospheric concentration of carbon dioxide, EOS, 65, 416-417, 1984.

Emanuel, W. R., G. G. Killough, W. M. Post, and H. H. Shugart, Modeling terrestrial carbon cycling at the global scale, in Global Dynamics of Biospheric Carbon, CONF-8108131, edited by S. Brown, pp. 166-194, U.S. Department of Energy, Washington, D.C., 1982.

Emanuel, W. R., G. G. Killough, W. M. Post, H. H. Shugart, and M. P. Stevenson, Computer implementation of a globally averaged model of the world carbon cycle, Rep. DOE/NBB-0062, TR010, 79 pp., U.S. Department of Energy, Washington, D.C., 1984a.

Emanuel, W. R., G. G. Killough, W. M. Post, and H. H. Shugart, Modeling terrestrial ecosystems in the global carbon cycle with shifts in carbon storage capacity by land-use change, Ecology, 65, 970-983, 1984.

Emerson, S., Organic carbon preservation in marine sediments, this volume.

Emerson, S., and M. Bender, Carbon fluxes at the sediment-water interface of the deep-sea: calcium carbonate preservation, J. Mar. Res., 39, 139-162, 1981.

Emiliani, C., Depth habitats of some species of pelagic foraminifera as indicated by oxygen isotope ratios, Am. J. Sci., 252, 149-158, 1954.

Ennever, F. K., and M. B. McElroy, Changes in atmospheric $CO_2$: Factors regulating the glacial to interglacial transition, this volume.

Enting, I. G., and G. I. Pearman, Description of a one-dimensional global carbon cycle model, Div. Atmos. Phys. Tech. Pap. 42, 95 pp., Commonwealth Sci. and Ind. Res. Organ., Melbourne, Australia, 1982.

Enting, I. G., and G. I. Pearman, Refinements to a one-dimensional carbon cycle model, Div. Atmos. Res. Tech. Paper 3, 35 pp., Commonwealth Sci. and Ind. Res. Organ., Melbourne, Australia, 1983.

Eppley, R. W. and B. J. Peterson, Particulate organic matter flux and planktonic new production in the deep ocean, Nature, 282, 677-680, 1979.

Fairbanks, R. G., P. H. Wiebe, A. W. H. Bé, Vertical distribution and isotopic composition of living planktonic foraminifera in the western North Atlantic, Science, 207, 61-63, 1980.

Fergusson, G. J., Reduction of atmospheric radiocarbon concentration by fossil fuel carbon dioxide and the mean life of carbon dioxide in the atmosphere, Proc. R. Soc. London, Ser. A, 243, 561-574, 1958.

Fiadeiro, M., Carbon cycling in the ocean, in Primary Productivity in the Sea, edited by P. G. Falkowski, pp. 487-496, Plenum, New York, 1980a.

Fiadeiro, M., The alkalinity of the deep Pacific, Earth Planet. Sci. Lett., 49, 499-505, 1980b.

Fiadeiro, M., and H. Craig, Three-dimensional modeling of tracers in the deep Pacific Ocean, I., Salinity and oxygen, J. Mar. Res., 36, 323-355, 1978.

Fine, R. A. and H. G. Ostlund, Exchange times in the Pacific equatorial current system, Earth Planet. Sci. Lett., 49, 447-452, 1980.

Fine, R. A., J. L. Reid, and H. G. Ostlund, Circulation of tritium in the Pacific Ocean, J. Phys. Oceanogr., 11, 3-14, 1981.

Fogg, G. E., Primary productivity, in Chemical Oceanography, vol. 2, edited by J. P. Riley and G. Skirrow, pp. 386-453, Academic Press, Orlando, Fla., 1975.

Fraser, P. J., M. A. K. Khalil, R. A. Rasmussen, and A. J. Crawford, Trends of atmospheric methane in the southern hemisphere, Geophys. Res. Lett., 8, 1063-1066, 1981.

Fraser, P. J., G. I. Pearman, and P. Hyson, The global distribution of atmospheric carbon dioxide, 2., A review of provisional background observations, 1978-1980, J. Geophys. Res., 88, 3591-3598, 1983.

Freeland, H. J., P. B. Rhines, and T. Rossby, Statistical observations of the trajectories of neutrally buoyant floats in the North Atlantic, J. Mar. Res., 33, 383-404, 1975.

Fung, I., K. Prentice, E. Matthews, J. Lerner, and G. Russell, Three-dimensional tracer model study of atmospheric $CO_2$; Response to seasonal exchanges with the terrestrial biosphere, J. Geophys. Res., 88, 1281-1294, 1983.

Gammon, R. H., J. Cline, and D. Wisegarver, Chlorofluoromethanes in the northeast Pacific Ocean: Measured vertical distributions and application as transient tracers of upper ocean mixing, J. Geophys. Res., 87, 9441-9454, 1982.

Garrels, R. M., and A. Lerman, Phanerozoic cycles of sedimentary carbon and sulfur, Proc. Natl. Acad. Sci., 78, 4652-4656, 1981.

Garrels, R. M., and F. T. Mackenzie, Evolution of Sedimentary Rocks, 397 pp., W. W. Norton, New York, 1971.

Garrels, R. M., and F. T. Mackenzie, A quantitative model for the sedimentary rock cycle, Mar. Chem., 1, 27-41, 1972.

Garrels, R. M., and E. A. Perry, Jr., Cycling of carbon, sulfur, and oxygen through geologic time, in The Sea, vol. 5: Marine Chemistry, edited by E. D. Goldberg, pp. 303-336, John Wiley, New York, 1974.

Garrels, R. M., F. T. Mackenzie, and C. Hunt,

Chemical Cycles and the Global Environment, Assessing Human Influences, 206 pp., W. Kaufmann, Los Altos, Calif., 1975.

Garrels, R. M., A. Lerman, and F. T. Mackenzie, Controls of atmospheric $O_2$ and $CO_2$: Past, present, and future, Am. Sci., 64, 306-315, 1976.

Garrett, C., Mixing in the ocean interior, Dyn. Atmos. Oceans, 3, 239-265, 1979.

Gershanovich, D. E., T. I. Gorshkova, and A. I. Koniukhov, Organic matter in recent sediments of continental margins, in Organic Matter in Recent and Fossil Sediments and Methods of its Investigation, (in Russian), Nauka, Moscow, 63 pp., 1974.

Gieskes, W. W. C., and W. Kraay, State-of-the-art in the measurement of primary production, in Flows of Energy and Materials in Marine Ecosystems, Theory and Practice, edited by M. J. R. Fasham, pp. 171-190, Plenum, New York, 1984.

Glibert, P. M., F. Lipschultz, J. J. McCarthy, and M. A. Altabet, Isotope dilution models of uptake and remineralization of ammonium by marine plankton, Limnol. Oceanogr., 27, 639-650, 1982.

Goldman, J. C., Oceanic nutrient cycles, in Flows of Energy and Materials in Marine Ecosystems, Theory and Practice, edited by M. J. R. Fasham, pp. 137-170, Plenum, New York, 1984.

Golley, F. B., Energy flux in ecosystems, in Ecosystem Structure and Function, edited by J. A. Weins, pp. 69-90, Oregon State University, Corvallis, 1972.

Gordon, A. L., and H. W. Taylor, Heat and salt balance within the cold waters of the world ocean, Proceedings of Symposium, Numerical Models of Ocean Circulation, pp. 54-56, National Academy of Sciences, Washington, D.C., 1975.

Gorshkov, S. G., Ocean atlas reference tables, (in Russian), Department of Navigational Oceanography, Ministry of Defense, USSR, 156 pp., 1980.

Graedel, T. E. and J. E. McRae, On the possible increase of the atmospheric methane and carbon monoxide concentrations during the last decade, Geophys. Res. Lett., 7, 977, 1980.

Grunau, H. R., Natural gas in major basins worldwide attributed to source rock type, thermal history and bacterial origin, in Proc. World Pet. Congr. 11th, vol. 2, 293-302, John Wiley, New York, 1984.

Grundmanis, V., and J. W. Murray, Aerobic respiration in pelagic marine sediments, Geochim. Cosmochim. Acta, 46, 1101-1120, 1982.

Halbouty, M. T. and J. D. Moody, World ultimate reserves of crude oil, Proc. World Pet. Congr. 10th, vol. 2, Heyden and Son, Philadelphia, Penn., 291-301, 1980.

Hampicke, V., Net transfer of carbon between

the land biota and the atmosphere, induced by man, in The Global Carbon Cycle, SCOPE 13, edited by B. Bolin, E. T. Degens, S. Kempe, and P. Ketner, pp. 219-236, John Wiley, New York, 1979.

Hansen, A. D. A., and H. Rosen, Vertical distributions of particulate carbon, sulfur, and bromine in the Arctic haze and comparison with ground-level measurements at Barrow, Alaska, Geophys. Res. Lett., 11, 381-384, 1984.

Harris, E., Bull. Bingham Oceanogr. Collect., 17, 31, 1959.

Harrison, W. G., Nutrient regeneration and primary production in the sea, in Primary Productivity in the Sea, edited by P. G. Falkowski, pp. 433-460, Plenum, New York, 1980.

Hay, W. W., Potential errors in estimates of carbonate rock accumulating through geologic time, this volume.

Hay, W. W., and J. R. Southam, Modulation of marine sedimentation by the continental shelves, in The Fate of Fossil Fuel $CO_2$ in the Oceans, edited by N. R. Anderson and A. Malahoff, pp. 569-604, Plenum Press, New York, 1977.

Hecht, A. D., and S. M. Savin, Phenotypic variation and oxygen isotope ratios in recent planktonic foraminifera, J. Foraminiferal Res., 2, 55-67, 1972.

Hendrickson, T. A., Synthetic Fuels Data Handbook, 308 pp., Cameron Engineers, Denver, Colo., 1975.

Hidalgo, H. and P. J. Crutzen, The troposphere and stratospheric composition perturbed by $NO_x$ emissions of high-altitude aircraft, J. Geophys. Res., 82, 5833-5866, 1977.

Hinga, K. R., J. M. Sieburth, and G. R. Heath, The supply and use of organic material at the deep-sea floor, J. Mar. Res., 37, 557-579, 1979.

Hobbie, J., J. Cole, J. Dungan, R. A. Houghton, and B. Peterson, Role of biota in global $CO_2$ balance: The controversy, BioScience, 34, 492-498, 1984.

Hoffert, M. I., A. J. Callegari, and C.-T. Hsieh, A box-diffusion carbon cycle model with upwelling, polar bottom water formation and a marine biosphere, in Carbon Cycle Modelling, SCOPE 16, edited by B. Bolin, pp. 287-305, John Wiley, New York, 1981.

Hoffert, M. I., Tyler Volk, and C.-T. Hsieh, A two-dimensional ocean model for climate and tracer studies, Rept. U.S. Department of Energy Contract No. DE-ACO2-81EV10610, 23 pp., New York University, New York, 1983.

Holland, H. D., The Chemistry of the Atmosphere and Oceans, 351 pp., John Wiley, New York, 1978.

Honjo, S., Dissolution of suspended coccoliths in the deep-sea water column and sedimentation of coccolith ooze, in Dissolution of Deep-Sea Carbonates, Cushman Found. Foraminiferal Res., edited by W. V. Sliter, A. W. H. Bé, and W. H.

Berger, pp. 114-128, Spec. Pub. 13, U.S. National Museum, Washington, D.C., 1975.

Honjo, S., Coccoliths: Production, transportation and sedimentation, Mar. Micropaleontol., 1, 65-79, 1976.

Honjo, S., Biogenic carbonate particulates in the ocean; Do they dissolve in the water column?, in The Fate of Fossil Fuel $CO_2$ in the Oceans, edited by N. R. Anderson and A. Malahoff, pp. 269-294, Plenum, New York, 1977.

Honjo, S., Sedimentation of materials in the Sargasso Sea at a 5,367 m deep station, J. Mar. Res., 36, 469-492, 1978.

Honjo, S., Material fluxes and modes of sedimentation in the mesopelagic and bathypelagic zones, J. Mar. Res., 38, 53-97, 1980.

Honjo, S., Seasonality and interaction of biogenic and lithogenic particulate flux at the Panama Basin, Science, 218, 883-884, 1982.

Honjo, S., and M. R. Roman, Marine copepod fecal pellets: Production, preservation, and sedimentation, J. Mar. Res., 36, 45-57, 1978.

Honjo, S., S. J. Manganini, and L. J. Poppe, Sedimentation of lithogenic particles in the deep ocean, Mar. Geol., 50, 199-220, 1982a.

Honjo, S., D. W. Spencer, and J. W. Farrington, Deep advective transport of lithogenic particles in Panama Basin, Science, 216, 516-518, 1982b.

Honjo, S., S. J. Manganini, and J. J. Cole, Sedimentation of biogenic matter in the deep ocean, Deep Sea Res., 29, 609-625, 1982c.

Houghton, R. A., J. E. Hobbie, J. M. Melillo, B. Moore, B. J. Peterson, G. R. Shaver, and G. M. Woodwell, Changes in the carbon content of terrestrial biota and soils between 1860 and 1980: A net release of $CO_2$ to the atmosphere, Ecol. Monogr., 53, 235-262, 1983.

Hubbert, M. K., Energy resources, in Resources and Man, National Academy of Sciences, W. H. Freeman, San Francisco, pp. 157-200, 1969.

Hyson, P., P. J. Fraser, and G. I. Pearman, A study of the F-11 concentration, using a 2-D transport model, J. Geophys. Res., 85, 1980.

Ingle, S. E., C. H. Culberson, J. Hawley, and R. M. Pytkowicz, The solubility of calcite in sea water at atmospheric pressure and 35°/°° salinity, Mar. Chem., 1, 295-307, 1973.

Ion, D. C., Availability of World Energy Resources, Graham and Trotman, London, England, 1980.

Jackson, G. A., Phytoplankton growth and zooplankton grazing in oligotrophic oceans, Nature, 284, 439-440, 1980.

Jacobi, W., and K. Andre, The vertical distribution of radon 222, radon 220, and their decay products in the atmosphere, J. Geophys. Res., 68, 3799-3814, 1963.

Jenkins, W. J., Tritium and $^3$He in the Sargasso Sea, J. Mar. Res., 38, 533-569, 1980.

Jenkins, W. J., Oxygen utilization rates in North Atlantic subtropical gyre and primary production in oligotrophic systems, Nature, 300, 246-248, 1982.

Jenkinson, D. S., and J. H. Rayner, The turnover of soil organic matter in some of the Rothamsted classical experiments, Soil Sci., 123, 298-305, 1977.

Jennings, J. C., Jr., L. I. Gordon, and D. M. Nelson, Nutrient depletion indicates high primary productivity in the Weddell Sea, Nature, 309, 51-54, 1984.

Keeling, C. D., Carbon dioxide in surface ocean waters, 4., Global distribution, J. Geophys. Res., 73, 4543-4553, 1968.

Keeling, C. D., Industrial production of carbon dioxide from fossil fuels and limestone, Tellus, 25, 174-198, 1973a.

Keeling, C. D., The carbon dioxide cycle: Reservoir models to depict the exchange of atmospheric carbon dioxide with the oceans and land plants, in Chemistry of the Lower Atmosphere, edited by S. Rasool, pp. 251-329, 1973b.

Keeling, C. D., The Suess Effect: $^{13}$Carbon-$^{14}$Carbon Interrelations, Environ. Int., 2, 229-300, 1979.

Keeling, C. D., The global carbon cycle: What we know and could know from atmospheric, biospheric, and oceanic observations, Proceedings of Carbon Dioxide Research Conference: Carbon Dioxide, Science and Consensus, Berkeley Springs, West Virginia, pp. II.4-II.62, U.S. Department of Energy, Washington, D.C., 1983.

Keeling, C. D., and R. B. Bacastow, Impact of industrial gases on climate, in Energy and Climate, pp. 72-95, National Academy of Sciences, Washington, D.C., 1977.

Keeling, C. D., and B. Bolin, The simultaneous use of chemical tracers in oceanic studies, II., A three-reservoir model of the North and South Pacific Oceans, Tellus, 20, 17-54, 1968.

Keeling, C. D., J. A. Adams, Jr., C. A. Ekdahl, Jr., and P. R. Guenther, Atmospheric carbon dioxide variations at the South Pole, Tellus, 28, 552-564, 1976.

Keir, R. S., Reduction of thermohaline circulation during deglaciation: The effect on atmospheric radiocarbon and $CO_2$, Earth Planet. Sci. Lett., 64, 445-456, 1983a.

Keir, R. S., Variation in the carbonate reactivity of deep-sea sediments: Determination from flux experiments, Deep Sea Res., 30, 279-296, 1983b.

Keir, R. S., Recent increase in Pacific $CaCO_3$ dissolution: A mechanism for generating old $^{14}$C ages, Mar. Geol., 59, 227-250, 1984.

Keir, R. S., and W. H. Berger, Atmospheric $CO_2$ content in the last 120,000 years: The phosphate-extraction model, J. Geophys. Res., 88, 6027-6038, 1983.

Keir, R. S., and W. H. Berger, Late Holocene carbonate dissolution in the equatorial Pacific: Reef growth or neoglaciation?, this volume.

Kempe, S., Carbon in the freshwater cycle, in The Global Carbon Cycle, SCOPE 13, edited by B. Bolin, E. T. Degens, S. Kempe, and P. Ketner, pp. 317-342, John Wiley, New York, 1979a.

Kempe, S., Carbon in the rock cycle, in The Global Carbon Cycle, SCOPE 13, edited by B. Bolin, E. T. Degens, S. Kempe, and P. Ketner, pp. 343-377, John Wiley, New York, 1979b.

Khalil, M. A. K., and R. A. Rasmussen, Secular trends of atmospheric methane, Chemosphere, 11, 877-883, 1982.

Khalil, M. A. K., and R. A. Rasmussen, Sources, sinks, and seasonal cycles of atmospheric methane, J. Geophys. Res., 88, 5131-5144, 1983.

Khalil, M. A. K., and R. A. Rasmussen, Carbon monoxide in the earth's atmosphere: Increasing trend, Science, 224, 54-56, 1984.

Kitano, Y., Water-rock interaction and geochemical balance of major chemical elements, with emphasis on geochemical balance of calcium and carbon dioxide, Acta Oceanogr. Taiwan., 11, 40-48, 1980.

Knauer, G. A., and J. H. Martin, Primary production and carbon-nitrogen fluxes in the upper 1,500 m of the northeast Pacific, Limnol. Oceanogr., 26, 181-186, 1981.

Knauer, G. A., J. H. Martin, and K. W. Bruland, Fluxes of particulate carbon, nitrogen, and phosphorus in the upper water column of the northeast Pacific, Deep Sea Res., 26A, 97-108, 1979.

Knauer, G. A., D. Hebel, and F. Cipriano, Marine snow: Major site of primary production in coastal waters, Nature, 300, 630-631, 1982.

Knox, F., and M. B. McElroy, Changes in atmospheric $CO_2$: Influence of the marine biota at high latitude, J. Geophys. Res., 89, 4629-4637, 1984.

Koblentz-Mishke, O. J., V. V. Volkovinsky, and J. G. Kabanova, Plankton primary production of the world ocean, in Symposium on Scientific Exploration of the South Pacific, edited by W. S. Wooster, pp. 183-193, National Academy of Sciences, Washington, D.C., 1970.

Kolla, V., A. W. H. Bé, and P. E. Biscaye, Calcium carbonate distribution in the surface sediments of the Indian Ocean, J. Geophys. Res., 81, 2605-2616, 1976.

Komar, P. D., A. P. Morse, L. F. Small, and S. W. Fowler, An analysis of sinking rates of natural copepod and euphausiid fecal pellets, Limnol. Oceanogr., 26, 172-180, 1981.

Kovda, V. A., Soil loss: An overview, Agro-Ecosystems, 3, 205-224, 1977.

Kratz, G., G. H. Kohlmaier, E. O. Sire, U. Fischbach, and H. Brohl, Carbon exchange between atmosphere and oceans in a latitude-dependent advection-diffusion model, Radiocarbon, 25, 459-471, 1983.

Ku, T.-L., and T. Oba, A method for quantitative evaluation of carbonate dissolution in deep-sea sediments and its application to paleoceanographic reconstruction, Quat. Res., 10, 112-129, 1978.

Ku, T.-L., C. A. Huh, and P. S. Chen, Meridional distribution of $^{226}$Ra in the eastern Pacific along Geosecs cruise tracks, Earth Planet. Sci. Lett., 49, 293-308, 1980.

Kuo, H.-H., and G. Veronis, Distribution of tracers in the deep oceans of the world, Deep Sea Res., 17, 29-46, 1970.

Lacis, A., J. Hansen, P. Lee, T. Mitchell, and S. Lebedeff, Greenhouse effect of trace gases, 1970-1980, Geophys. Res. Lett., 8, 1035-1038, 1981.

Lal, D. and H. E. Suess, Some comments on the exchange of $CO_2$ across the air-sea interface, J. Geophys. Res., 88, 3643-3646, 1983.

Land, L. S., The fate of reef-derived sediment on the North Jamaican Island slope, Mar. Geol., 29, 55-71, 1979.

Lasaga, A. C., The kinetic treatment of geochemical cycles, Geochim. Cosmochim. Acta, 44, 815-828, 1980.

Lasaga, A. C., Dynamic treatment of geochemical cycles: Global kinetics, in Kinetics of Geochemical Processes, edited by A. C. Lasaga and R. J. Kirkpatrick, pp. 69-110, Reviews in Mineralogy, vol. 8, Mineralogical Society of America, 1981.

Lasaga, A. C., R. A. Berner, and R. M. Garrels, An improved geochemical model of atmospheric $CO_2$ fluctuations over the past 100 million years, this volume.

Lee, C., and C. Cronin, The vertical flux of particulate organic nitrogen in the sea: Decomposition of amino acids in the Peru upwelling area and the equatorial Atlantic, J. Mar. Res., 40, 227-251, 1982.

Li, Y.-H., T. Takahashi, and W. S. Broecker, Degree of saturation of $CaCO_3$ in the oceans, J. Geophys. Res., 74, 5507-5525, 1969.

Li, Y.-H., T.-H. Peng, W. S. Broecker, and H. G. Ostlund, The average vertical eddy diffusion coefficient of the ocean (abstract), EOS Trans. AGU, 63, 973, 1982.

Lieth, H., Uber die Primärproduktion der Erde, Z. Angew Bot., 46, 1-37, 1972.

Lieth, H., Primary production of the major vegetation units of the world, in Primary Productivity of the Biosphere, edited by H. Lieth and R. H. Whittaker, pp. 203-215, Springer-Verlag, New York, 1975a.

Lieth, H. Primary productivity in ecosystems: Comparative analysis of global patterns, in Unifying Concepts in Ecology, edited by W. H. Van Dobben, and R. H. Lowe-McConnell, pp. 67-98, Junk, The Hague/Pudoc, Wageningen, 1975b.

Linden, H. R., The future development of energy supply systems, paper presented at Fuel Conference in Commemoration of the Golden Jubilee of the Fuel Society of Japan, Tokyo, Oct. 31 to Nov. 2, 1972.

Lisitzin, A. P., Sedimentation in the World Ocean, Spec. Publ. 17, 218 pp., Society Economic Paleontologists and Mineralogists, Tulsa, Okla., 1972.

Lisitzin, A. P., and V. P. Petelin, Features of distribution and modification of $CaCO_3$ in bottom sediments of the Pacific Ocean, Lithol. Miner. Resour., 5, Engl. Transl., 50-65, 1967.

Livingstone, D. A., Chemical composition of rivers and lakes, in Data of Geochemistry, U.S. Geol. Surv. Prof. Pap. 440-G, G1-G61, 1963.

Loftness, R. L., Energy Handbook, 741 pp., Van Nostrand-Reinhold, New York, 1978.

Logan, J. A., M. J. Prather, S. C. Wofsy, and M. B. McElroy, Tropospheric chemistry: A global perspective, J. Geophys. Res., 86, 7210-7254, 1981.

Lorenzen, C. J., N. A. Welschmeyer, A. E. Copping, and M. Vernet, Sinking rates of organic particles, Limnol. Oceanogr., 28, 766-769, 1983a.

Lorenzen, C. J., N. A. Welschmeyer, and A. E. Copping, Particulate organic carbon flux in the subarctic Pacific, Deep Sea Res., 30, 639-643, 1983b.

Lovelock, J. E., Atmospheric halocarbons and stratospheric ozone, Nature, 252, 292-294, 1974.

MacDonald, G. J. F., An overview of the impact of carbon dioxide on climate, Rept. M78-79, 17 pp., The MITRE Corporation, McLean, Va., 1978.

Machta, L., Prediction of $CO_2$ in the atmosphere, in Carbon and the Biosphere, AEC Symp. Ser. vol. 30, edited by G. M. Woodwell and E. V. Pecan, pp. 21-31, U.S. Department of Commerce, Springfield, Va., 1973.

Machta, L., K. Hanson, and C. E. Keeling, Atmospheric carbon dioxide and some interpretations, in The Fate of Fossil Fuel $CO_2$ in the Oceans, edited by N. R. Andersen and A. Malahoff, pp. 131-144, Plenum, New York, 1977.

MacIntyre, F., The absorption of fossil-fuel $CO_2$ by the ocean, Oceanol. Acta, 3, 505-516, 1980.

Mackenzie, F. T., The global carbon cycle: Minor sinks for carbon dioxide, in Carbon Dioxide Effects Research and Assessment Program, CONF-8009140, pp. 360-397, U.S. Department of Energy, Washington, D.C., 1981.

Marland, G., and R. M. Rotty, Carbon dioxide emissions from fossil fuels: A procedure for estimation and results for 1950-1981, Rep. DOE/NBb-0036, 75 pp., U.S. Department of Energy, Washington, D.C., 1983.

Marland, G., and R. M. Rotty, Carbon dioxide emissions from fossil fuels: A procedure for estimation and results for 1950-1982, Tellus, 36B, 232-261, 1984.

Masters, C. D., D. H. Root, and W. D. Dietzman, Distribution and quantitative assessment of world crude oil reserves and resources, in Proc. World Pet. Congr. 11th, vol. 2, John Wiley, New York, pp. 229-237, 1984.

McCarthy, J. J., and J. C. Goldman, Nitrogenous nutrition of marine phytoplankton in nutrient-depleted waters, Science, 203, 670-672, 1979.

McCave, I. N., Vertical flux of particles in the ocean, Deep Sea Res., 22, 491-502, 1975.

McWilliams, J. C., E. D. Brown, H. L. Bryden, C. C. Ebbesmeyer, B. A. Elliott, R. H. Heinmiller, B. Lien-Hua, K. D. Leaman, E. J. Lindstrom, J. R. Luyten, S. E. McDowell, W.

Breckner-Owens, W. Perkins, J. F. Price, L. Reiger, S. C. Riser, H. T. Rossby, T. B. Sanford, C. Y. Shen, B. A. Taft, and J. C. Van Leer, The local dynamics of eddies in the Western North Atlantic, in Eddies in Marine Science, edited by A. R. Robinson, pp. 92-112, Springer-Verlag, New York, 1983.

Menard, H. W. and S. M. Smith, Hypsometry of ocean provinces, J. Geophys. Res., 71, 4305-4325, 1966.

Menzel, D. W., Primary productivity, dissolved and particulate organic matter, and the sites of oxidation of organic matter, in The Sea, vol. 5, Marine Chemistry, pp. 659-678, John Wiley, New York, 1974.

Menzies, R. J., R. Y. George, and G. T. Rowe, Abyssal Environment and Ecology of the World Oceans, Wiley-Interscience, New York, 1973.

Meybeck, M., River transport of organic carbon to the ocean, in Carbon Dioxide Effects Research and Assessment Program, CONF-8009140, pp. 219-269, U.S. Deptartment of Energy, Washington, D.C., 1981.

Meybeck, M., Carbon, nitrogen, and phosphorus transport by world rivers, Am. J. Sci., 282, 401-450, 1982.

Meyer, R. F. and W. D. Dietzman, World geography of heavy crude oils, in The Future of Heavy Crude and Tar Sands, edited by R. F. Meyer and C. T. Steele, pp. 16-26, McGraw-Hill, New York, 1981.

Michel, R. L., and H. E. Suess, Bomb tritium in the Pacific Ocean, J. Geophys. Res., 80, 4139-4152, 1975.

Milliman, J. D., Marine Carbonates, 375 pp., Springer-Verlag, New York, 1974.

Milliman, J. D., X. Qinchun, and Y. Zuosheng, Transfer of particulate organic carbon and nitrogen from the Yangtze River to the ocean, Am. J. Sci., 284, 824-834, 1984.

Moiseev, P. A., The Living Resources of the World Ocean, Moscow, Translated from Russian by Israel Program for Scientific Translations, 339 pp., Jerusalem, 1971.

Moody, J. D. and R. W. Esser, An estimate of the world's recoverable crude oil resources, Proc. World Pet. Congr. 9th, vol. 3, Applied Sciences, London, 1975.

Mook, W. G., M. Koopmans, A. F. Carter, and C. D. Keeling, Seasonal, latitudinal, and secular variations in the abundance and isotopic ratios of atmospheric carbon dioxide, J. Geophys. Res., 88, 10,915-10,933, 1983.

Moor, E., and B. Stauffer, A new dry extraction system for gases in ice, J. Glaciology, in press, 1984.

Moore, W. S., H. W. Feely, and Y.-H. Li, Radium isotopes in sub-arctic waters, Earth Planet. Sci. Lett., 49, 329-340, 1980.

Moore, B., R. D. Boone, J. E. Hobbie, R. A. Houghton, J. M. Melillo, B. J. Peterson, G. R. Shaver, C. J. Vorosmarty and G. M. Woodwell, A simple model for analysis of the role of terrestrial ecosystems in the global carbon

budget, in Carbon Cycle Modelling, SCOPE 16, pp. 365-385, John Wiley, New York, 1981.

Mopper, K., and E. T. Degens, Tech. Rep. Woods Hole Oceanogr. Inst., 72-68, pp. 1-117, Woods Hole, Mass., 1972.

Mopper, K., and E. T. Degens, Organic carbon in the ocean: Nature and cycling, in The Global Carbon Cycle, SCOPE 13, edited by B. Bolin, E. T. Degens, S. Kempe, and P. Ketner, pp. 293-316, John Wiley, New York, 1979.

Mulholland, P. J., Deposition of riverborne organic carbon in floodplains, wetlands and deltas, in Carbon Dioxide Effects Research and Assessment Program, CONF-8009140, pp. 142-172, U.S. Department of Energy, Washington, D.C., 1981.

Mulholland, P. J., and J. A. Watts, Transport of organic carbon to the oceans by rivers of North America: A synthesis of existing data, Tellus, 34, 176-186, 1982.

Muller, P. J., and E. Suess, Productivity, sedimentation rate, and sedimentary organic matter in the oceans, I., Organic carbon preservation, Deep Sea Res., 26A, 1347-1362, 1979.

Munk, W. H., Abyssal recipes, Deep Sea Res., 13, 707-730, 1966.

Munnich, K. O., and W. Roether, Transfer of bomb $^{14}$C and tritium from the atmosphere to the ocean. Internal mixing of the ocean on the basis of tritium and $^{14}$C profiles, in Radioactive Dating and Low-Level Counting, pp. 93-104, International Atomic Energy Agency, Vienna, 1967.

Murray, J., and A. F. Renard, Report on deep-sea deposits based on the specimens collected during the voyage of H. M. S. Challenger in the years 1872-1876, in Challenger Reports, 525 pp., Dulau, London, 1891.

National Research Council, U.S. Energy Supply Prospects to 2010, National Academy of Sciences, Washington, D.C., 1979.

Needler, G. T. and R. A. Heath, Diffusion coefficients calculated from the Mediterranean salinity anomaly in the North Atlantic Ocean, J. Phys. Oceanogr., 5, 173-182, 1975.

Neftel, A., H. Oeschger, J. Schwander, B. Stauffer, and R. Zumbrunn, Ice core sample measurements give atmospheric $CO_2$ content during the past 40,000 years, Nature, 295, 220-223, 1982.

Neftel, A., H. Oeschger, J. Schwander, and B. Stauffer, Carbon dioxide concentration in bubbles of natural cold ice, J. Phys. Chem., 87, 4116-4120, 1983.

Nehring, R., The outlook for conventional petroleum resources, in Long Term Energy Resources, UNITAR, pp. 315-327, Pitman, Marshfield, Mass., 1981.

Newbould, P., Losses and accumulation of organic matter in soils, in Soil Degradation, edited by D. Boels, D. B. Davies, and A. E. Johnston, pp. 107-131, A. A. Balkema, Rotterdam, 1982.

Nienhuis, P. H., Distribution of organic matter in living marine organisms, in Marine Organic Chemistry, Evolution, Composition, Interactions and Chemistry of Organic Matter in Seawater, Elsevier Oceanogr. Ser. vol. 31, edited by E. K. Duursma and R. Dawson, pp. 31-69, Elsevier, New York, 1981.

Nihoul, J. C. J., The turbulent ocean, in Marine Turbulence, Elsevier Oceanogr. Ser. vol. 28, edited by J. C. J. Nihoul, pp. 1-19, Elsevier, New York, 1981.

Nixon, S. W., Remineralization and nutrient cycling in coastal marine ecosystems, in Estuaries and Nutrients, edited by B. J. Neilson and L. E. Cronin, pp. 111-138, Humana, Clifton, N.J., 1981.

Nydal, R., Further investigation on the transfer of radiocarbon in nature, J. Geophys. Res., 73, 3617-3635, 1968.

Oba, T., Biostratigraphy and isotopic paleo-temperature of some deep-sea cores from the Indian Ocean, Tohuku Univ. Sci. Repts., 41, 129-195, 1969.

O'Brien, B. J., and J. D. Stout, Movement and turnover of soil organic matter as indicated by carbon isotope measurements, Soil Biol. Biochem., 10, 309-317, 1978.

Oeschger, H., U. Siegenthaler, U. Schotterer, and A. Gugelmann, A box diffusion model to study the carbon dioxide exchange in nature, Tellus, 27, 168-192, 1975.

Oeschger, H., B. Stauffer, R. Finkel, and C. Langway, Jr., Variations of the $CO_2$ concentration of occluded air and of anions and dust in polar ice cores, this volume.

Okubo, A., Oceanic diffusion diagrams, Deep Sea Res., 18, 789-802, 1971.

Okumura, M., Y. Kitano, and M. Idogaki, Note: Removal of anions by carbonate sedimentation from seawater, Geochem. J., 17, 105-110, 1983.

Olson, J. S., Carbon cycles and temperate woodlands, in Ecological Studies, vol. 1, edited by D. E. Reichle, pp. 73-85, Springer-Verlag, New York, 1970.

Olson, J. S., Terrestrial Ecosystems, in Encyclopedia Britannica, 15th ed., pp. 144-149, 1974.

Olson, J. S., Carbon balance in relation to fire regimes, Fire Regimes and Ecosystem Properties, pp. 327-378, Gen. Tech. Rep. WO-26, U.S. Department of Agriculture, Forest Service, Washington, D.C., 1981.

Olson, J. S., Earth's vegetation and atmospheric carbon dioxide, in Carbon Dioxide Review: 1982, edited by W. C. Clark, pp. 388-398, Oxford University Press, New York, 1982.

Olson, J. S., H. A. Pfunderer, and Y.-H. Chan, Changes in the global carbon cycle and the biosphere, Environ. Sci. Div. Publ. 1050, ORNL/EIS-109, 169 pp., Oak Ridge National Laboratory, Oak Ridge, Tenn., 1978.

Olson, J. S., J. A. Watts, and L. J. Allison, Carbon in live vegetation of major world ecosystems, Rep. ORNL-5862, 152 pp. and map, Oak Ridge National Laboratory, Oak Ridge, Tenn., 983.

Ovcharenko, V. A., Reassessment of oil shale prospects, in Long Term Energy Resources, UNITAR, pp. 451-484, Pitman, Marshfield, Mass., 1981.

Pankrath, J., The global carbon cycle and possible disturbances due to man's intervention, Environ. Int., 2, 357-377, 1979.

Pearman, G. I., and P. Hyson, Activities of the global biosphere as reflected in atmospheric $CO_2$ records, J. Geophys. Res., 85, 4457-4467, 1980.

Pearman, G. I., P. Hyson, and P. J. Fraser, The global distribution of atmospheric carbon dioxide, 1., Aspects of observations and modeling, J. Geophys. Res., 88, 3581-3590, 1983.

Peng, T.-H., Atmospheric $CO_2$ variations based on the tree-ring $^{13}C$ record, this volume.

Peng, T.-H., W. S. Broecker, G. G. Mathieu, and Y.-H. Li, Radon evasion rates in the Atlantic and Pacific Oceans as determined during the Geosecs Program, J. Geophys. Res., 84, 2471-2486, 1979.

Peng, T.-H., W. S. Broecker, H. D. Freyer, and S. Trumbore, A deconvolution of the tree ring based $\delta^{13}C$ record, J. Geophys. Res., 88, 3609-3620, 1983.

Peters, W., and H.-D. Schilling, World Energy Resources, 1985-2020: Coal Resources, 139 pp., IPC Science and Technology, New York, 1978.

Peterson, B. J., Perspectives on the importance of the oceanic particulate flux in the global carbon cycle, Ocean Sci. Eng., 6, 71-108, 1981.

Peterson, L. C., and W. L. Prell, Carbonate dissolution in recent sediments of the eastern equatorial Indian Ocean: Preservation patterns and carbonate loss above the lysocline, Mar. Geol., in press, 1984.

Peterson, L. C., and W. L. Prell, Carbonate preservation and rates of climatic change: An 800 kyr record from the Indian Ocean, this volume.

Peterson, J. T., W. D. Komhyr, and T. B. Harris, Atmospheric carbon dioxide measurements at Barrow, Alaska, 1973-1979, Tellus, 34, 166-175, 1982.

Platt, T., Primary productivity in the central North Pacific: Comparison of oxygen and carbon fluxes, Deep Sea Res., 31, 1311-1319, 1984.

Platt, T., and D. V. Subba Rao, Primary production of marine microphytes, in Photosynthesis and Productivity in Different Environments, edited by J. P. Cooper, pp. 249-280, Cambridge Univ. Press., New York, 1975.

Platt, T., M. Lewis, and R. Geider, Thermodynamics of the pelagic ecosystem: Elementary closure conditions for biological production in the open ocean, in Flows of Energy and Materials in Marine Ecosystems, Theory and Practice, edited by M. J. R. Fasham, pp. 49-84, Plenum, New York, 1984.

Plummer, L. N., and E. T. Sundquist, Total individual ion activity coefficients of calcium and carbonate in seawater at 25°C and 35°/°°

salinity, and implications to the agreement between apparent and thermodynamic constants of calcite and aragonite, Geochim. Cosmochim. Acta, 46, 247-258, 1982.

Pomeroy, L. R., The ocean's food web, a changing paradigm, BioScience, 24, 499, 1974.

Post, W. M., W. R. Emanuel, P. J. Zinke, and A. G. Stangenberger, Soil carbon pools and world life zones, Nature, 298, 156-159, 1982a.

Post, W. M., P. J. Zinke, A. G. Stangenberger, W. R. Emanuel, H. Jenny, and J. S. Olson, Summaries of soil carbon storage in world life zones, in Global Dynamics of Biospheric Carbon, CONF-8108131, edited by S. Brown, pp. 131-139, U.S. Department of Energy, Washington, D.C., 1982b.

Pytkowicz, R. M., Carbon dioxide-carbonate system at high pressures in the oceans, in Mar. Biol. Ann. Rev. 6, edited by H. Barnes, pp. 83-135, George Allen and Unwin, London, 1968.

Pytkowicz, R. M., The carbon dioxide system in the oceans, Swiss J. Hydrol., 35, 8-28, 1973.

Quay, P. D. and M. Stuiver, Vertical advection-diffusion rates in the oceanic thermocline determined from $^{14}C$ distributions, Radiocarbon, 22, 607-625, 1980.

Rafter, T. A., and B. J. O'Brien, Exchange rates between the atmosphere and the ocean as shown by recent $^{14}C$ measurements in the South Pacific, in Proceedings of the 12th Nobel Symposium, Radiocarbon Variations and Absolute Chronology, edited by I. V. Olsson, pp. 355-377, Wiley-Interscience, New York, 1970.

Ramanathan, V., Greenhouse effect due to chlorofluorocarbons: Climate implications, Science, 190, 50, 1975.

Rasmussen, R. A., and M. A. K. Khalil, Increase in the concentration of atmospheric methane, Atmos. Environ., 15, 883-886, 1981a.

Rasmussen, R. A., and M. A. K. Khalil, Atmospheric methane ($CH_4$): Trends and seasonal cycles, J. Geophys. Res., 86, 9826-9832, 1981b.

Rasmussen, R. A., and M. A. K. Khalil, Atmospheric methane in the recent and ancient atmospheres: Concentrations, trends, and interhemispheric gradient, J. Geophys. Res., 89, 11,599-11,605, 1984.

Rasmussen, R. A., M. A.K. Khalil, and R. W. Dalluge, Atmospheric trace gases in Antarctica, Science, 211, 285-287, 1981.

Raynaud, D. and J. M. Barnola, The $CO_2$ record in ice cores: A reconstruction of the atmospheric evolution between 18 ka and 1850 AD (abstract), Ann. Glaciol., 5, 224, 1984.

Reimers, C. E., and E. Suess, The partitioning of organic carbon fluxes and sedimentary organic matter decomposition rates in the ocean, Mar. Chem., 13, 141-168, 1983.

Reimers, C. E., S. Kalhorn, S. R. Emerson, and K. H. Nealson, Oxygen consumption rates in pelagic sediments from the central Pacific: First estimates from microelectrode profiles, Geochim. Cosmochim. Acta, 48, 903-910, 1984.

Reiners, W. A., Terrestrial detritus and the carbon cycle, in Carbon and the Biosphere, AEC Symp. Ser., vol. 30, edited by G. M. Woodwell and E. V. Pecan, pp. 303-327, U.S. Department of Commerce, Springfield, Va., 1973a.

Reiners, W. A., Appendix: Summary of world carbon cycle and recommendations for critical research, in Carbon and the Biosphere, AEC Symp. Ser., vol. 30, edited by G. M. Woodwell and E. V. Pecan, pp. 368-382, U.S. Department of Commerce, Springfield, Va., 1973b.

Revelle, R., and W. Munk, The carbon dioxide cycle and the biosphere, in Energy and Climate, Studies in Geophysics, pp. 140-158, National Research Council, National Academy of Sciences, Washington, D.C., 1977.

Revelle, R., and H. E. Suess, Carbon dioxide exchange between atmosphere and ocean and the question of an increase of atmospheric $CO_2$ during the past decades, Tellus, 9, 18-27, 1957.

Richey, J. E., J. T. Brock, R. J. Naiman, R. C. Wissmar, and R. F. Stallard, Organic carbon: Oxidation and transport in the Amazon River, Science, 207, 1348-1351, 1980.

Robinson, A. R., Overview and summary of eddy science, in Eddies in Marine Science, edited by A. R. Robinson, pp. 1-14, Springer-Verlag, New York, 1983.

Robinson, A. R., and H. Stommel, The oceanic thermocline and the associated thermohaline circulation, Tellus, 3, 295-308, 1959.

Rodin, L. E., N. I. Bazilevich, and N. N. Rozov, Productivity of the world's main ecosystems, in Productivity of World Ecosystems, edited by D. E. Reichle, J. F. Franklin, and D. W. Goodall, pp. 13-26, National Academy of Sciences, Washington, D.C., 1975.

Roemmich, D. and C. Wunsch, Apparent changes in the climatic state of the deep North Atlantic Ocean, Nature, 307, 447-450, 1984.

Romankevich, E. A., Geochemistry of Organic Matter in the Ocean, 334 pp., Springer-Verlag, New York, 1984.

Ronov, A. B., The earth's sedimentary shell: Quantitative patterns of its structures, compositions and evolution (in Russian), Osadochnaya obolochka Zemli (Kolichestvenniye zakonomersti stroyeniga, softavi i evolyutsiya Izd.-vo Nauka), edited by A. A. Yaroshivskiy, Nauka, Moscow, 1980.

Ronov, A. B., The earth's sedimentary shell. The 20th V. I. Vernadskiy Lecture, 1., Int. Geol. Rev., 24, Engl. Transl., 1313-1363, 1982a.

Ronov, A. B., The earth's sedimentary shell. The 20th V. I. Vernadskiy Lecture, 2., Int. Geol. Rev., 24, 1365-1388, 1982b.

Rooth, C. G., and H. G. Ostlund, Penetration of tritium into the Atlantic thermocline, Deep Sea Res., 19, 481-492, 1972.

Rotty, R. M., Global carbon dioxide production from fossil fuels and cement, A.D. 1950 - A.D. 2000, in The Fate of Fossil Fuel $CO_2$ in the Oceans, edited by N. R. Andersen and A. Malahoff, pp. 167-181, Plenum, New York, 1977.

Rotty, R. M., Distribution of and changes in

industrial carbon dioxide production, Rep. 81-23, 23 pp., Institute for Energy Analysis, Oak Ridge Associated Universities, Oak Ridge, Tenn., 1982.

Rowe, G. T., and W. D. Gardner, Sedimentation rates in the slope water of the northwest Atlantic Ocean measured directly with sediment traps, J. Mar. Res., 37, 581-600, 1979.

Ruddiman, W. F., and B. C. Heezen, Differential solution of planktonic foraminifera, Deep Sea Res., 14, 801-808, 1967.

Ryther, J. H., Photosynthesis and fish production in the sea, Science, 166, 72-76, 1969.

Sarmiento, J. L., A simulation of bomb tritium entry into the Atlantic Ocean, J. Phys. Oceanogr., 13, 1924-1939, 1983a.

Sarmiento, J. L., A tritium box model of the North Atlantic thermocline, J. Phys. Oceanogr., 13, 1269-1294, 1983b.

Sarmiento, J. L., and C. G. H. Rooth, A comparison of vertical and isopycnal mixing models in the deep sea based on radon 222 measurements, J. Geophys. Res., 85, 1515-1518, 1980.

Sarmiento, J. L., and J. R. Toggweiler, A new model for the role of the oceans in determining atmospheric $P_{CO_2}$, Nature, 308, 621-624, 1984.

Sarmiento, J. L., C. G. H. Rooth, and W. Roether, The North Atlantic tritium distribution in 1972, J. Geophys. Res., 87, 8047-8056, 1982a.

Sarmiento, J. L., C. G. H. Rooth, and W. S. Broecker, Radium 228 as a tracer of basin wide processes in the abyssal ocean, J. Geophys. Res., 87, 9694-9698, 1982b.

Savin, S. M., and R. G. Douglas, Stable isotope and magnesium geochemistry of Recent planktonic foraminifera from the South Pacific, Geol. Soc. Am. Bull., 84, 2327-2342, 1973.

SCEP (Report of the Study of Critical Environmental Problems), Man's Impact on the Global Environment, 319 pp., MIT Press, Cambridge, Mass., 1970.

Schlesinger, W. H., Carbon balance in terrestrial detritus, Annu. Rev. Ecol. Syst., 8, 51-81, 1977.

Schlesinger, W. H., Soil organic matter: A source of atmospheric $CO_2$, in The Role of Terrestrial Vegetation in the Global Carbon Cycle: Measurement by Remote Sensing, SCOPE 23, edited by G. M. Woodwell, pp. 111-127, John Wiley, New York, 1984.

Schlesinger, W. H., and J. M. Melack, Transport of organic carbon in the world's rivers, Tellus, 33, 172-187, 1981.

Schnell, R. C., S.-A.Odh, and L. N. Njau, Carbon dioxide measurements in tropical East African biomes, J. Geophys. Res., 86, 5364-5372, 1981.

Schwander, J., and B. Stauffer, Age difference between polar ice and the air trapped in its bubbles, Nature, 311, 45-47, 1984.

Seibold, E., Der Meeresboden, 183 pp., Springer-Verlag, New York, 1974.

Shanks, A. L., and J. D. Trent, Marine snow: Microscale nutrient patches, Limnol. Oceanogr., 24, 850, 1979.

Shaver, G. R., J. E. Hobbie, R. A. Houghton, J. M. Melillo, B. Moore, B. J. Peterson, and G. M. Woodwell, The role of terrestrial biota and soils in the global carbon budget, in Global Dynamics of Biospheric Carbon, CONF-8108131, pp. 160-165, U.S. Department of Energy, Washington, D.C., 1982.

Shiller, A. M., Calculating the oceanic $CO_2$ increase: A need for caution, J. Geophys. Res., 86, 11,083-11,088, 1981.

Shulenberger, E., and J. L. Reid, The Pacific shallow oxygen maximum, deep chlorophyll maximum, and primary productivity, reconsidered, Deep Sea Res., 28A, 901-919, 1981.

Sieburth, J. M., V. Smetacek, and J. Lenz, Comment: Pelagic ecosystem structure: Heterotrophic compartments of the plankton and their relationship to plankton size fractions, Limnol. Oceanogr., 23, 1256-1263, 1978.

Siegenthaler, U., Uptake of excess $CO_2$ by an outcrop-diffusion model of the ocean, J. Geophys. Res., 88, 3599-3608, 1983.

Siegenthaler, U., and H. Oeschger, Predicting future atmospheric carbon dioxide levels, Science, 199, 388-395, 1978.

Siegenthaler, U., and T. Wenk, Rapid atmospheric $CO_2$ variations and ocean circulation, Nature, 308, 624-625, 1984.

Siegenthaler, U., M. Heimann, and H. Oeschger, Model responses of the atmospheric $CO_2$ level and $^{13}C/^{14}C$ ratio to biogenic $CO_2$ input, in Carbon Dioxide, Climate and Society, edited by J. Williams, pp. 79-87, Pergamon, New York, 1978.

Siegenthaler, U., M. Heimann, and H. Oeschger, $^{14}C$ variations caused by changes in the global carbon cycle, Radiocarbon, 22, 177-191, 1980.

Sigleo, A. C., and G. R. Helz, Composition of estuarine colloidal material: Major and trace elements, Geochim. Cosmochim. Acta, 45, 2501-2509, 1981.

Sigleo, A. C., P. E. Hare, and G. R. Helz, The amino acid composition of estuarine colloidal material, Estuarine, Coastal Shelf Sci., 17, 87-96, 1983.

Silver, M. W., and A. L. Alldredge, Bathypelagic marine snow: Deep-sea algal and detrital community, J. Mar. Res., 39, 501-530, 1981.

Siré, E. O., G. H. Kohlmaier, G. Kratz, U. Fischbach, and H. Brohl, Comparative dynamics of atmosphere-ocean-models within the description of the perturbed global carbon cycle, Z. Naturforsch., 36a, 233-250, 1981.

Skopintsev, B. A., Recent advances in the study of organic matter in oceans, Oceanology, Engl. Transl., 11, 775-789, 1971.

Skopintsev, B. A., Decomposition of organic matter of plankton, humification and hydrolysis, in Marine Organic Chemistry, Evolution, Composition, Interactions and Chemistry of Organic Matter in Seawater, Elsevier Oceanogr. Ser., vol. 31, edited by

E. K. Duursma and R. Dawson, pp. 125-177, Elsevier, New York, 1981.

Small, L. F., S. W. Fowler, and Y. Unlu, Sinking rates of natural copepod fecal pellets, Mar. Biol., 51, 233, 1979.

Smith, K. L., Jr., and R. J. Baldwin, Seasonal fluctuations in deep-sea sediment community oxygen consumption: Central and eastern North Pacific, Nature, 307, 624-625, 1984.

Smith, R. E. H., R. J. Geider, and T. Platt, Microplankton productivity in the oligotrophic ocean, Nature, 311, 252-254, 1984.

Smith, S. V., Coral-reef area and the contributions of reefs to processes and resources of the world's oceans, Nature, 273, 225-226, 1978.

Sorokin, Y. I., Bacterial populations as components of oceanic ecosystems, Mar. Biol., 11, 101-105, 1971.

Sorokin, Y. I., The primary production of the seas and oceans, in General Ecology, Biocenology, Hydrobiology, vol. 1 (Biology series), Translated from Russian, pp. 3-35, G. K. Hall, Boston, Mass., 1974.

Sorokin, Y. I., Decomposition of organic matter and nutrient regeneration, in Marine Ecology, Volume IV, Dynamics, edited by O. Kinne, pp. 501-616, John Wiley, New York, 1978.

Southam, J. R., and W. W. Hay, Dynamical formulation of Broecker's model for marine cycles of biologically incorporated elements, Math. Geol., 8, 511-527, 1976.

Spencer, D. W., P. G. Brewer, A. Fleer, S. Honjo, S. Krishnaswami, and Y. Nozaki, Chemical fluxes from a sediment trap experiment in the deep Sargasso Sea, J. Mar. Res., 36, 493-523, 1978.

Starr, J., J. M. Prats, and S. A. Messulam, Chemical properties and reservoir characteristics of bitumen and heavy oil from Canada and Venezuela, in The Future of Heavy Crude and Tar Sands, edited by R. F. Meyer and C. T. Steele, pp. 168-173, McGraw-Hill, New York, 1981.

Stauffer, B., H. Hofer, H. Oeschger, J. Schwander, and U. Siegenthaler, Atmospheric $CO_2$ concentration during the last glaciation, Ann. Glaciol., 5, 160-164, 1984.

Steele, J. H., Factors controlling marine ecosystems, in Changing Chemistry of the Oceans, Proceedings of the 20th Nobel Symposium, edited by D. Dryssen and D. Jagner, pp. 209-221, Almquist and Wiksell, Stockholm, 1972.

Stommel, H., The abyssal circulation, Deep Sea Res., 5, 80-82, 1958.

Stommel, H., and A. B. Arons, On the abyssal circulation of the world ocean, II., An idealized model of the circulation pattern and amplitude in oceanic basins, Deep Sea Res., 6, 217-233, 1960.

Stout, J. D., K. M. Goh, and T. A. Rafter, Chemistry and turnover of naturally occurring resistant organic compounds in soil, in Soil Biochemistry, edited by E. A. Paul and J. N. Ladd, pp. 1-73, Marcel Dekker, New York, 1981.

Stuiver, M., The $^{14}C$ distribution in west Atlantic abyssal waters, Earth Planet. Sci. Lett., 32, 322-330, 1976.

Stuiver, M., Atmospheric carbon dioxide and carbon reservoir changes, Science, 199, 253-258, 1978a.

Stuiver, M., Atmospheric carbon dioxide in the 19th century - A reply, Science, 202, 1109, 1978b.

Stuiver, M., $^{14}C$ distribution in the Atlantic Ocean, J. Geophys. Res., 85, 2711-2718, 1980.

Stuiver, M., and P. D. Quay, Atmospheric $^{14}C$ changes resulting from fossil fuel $CO_2$ release and cosmic ray flux variability, Earth Planet. Sci. Lett., 53, 349-362, 1981.

Stuiver, M., P. D. Quay, and H. G. Ostlund, Abyssal water carbon-14 distribution and the age of the world oceans, Science, 219, 849-851, 1983.

Stuiver, M., R. L. Burk, and P. D. Quay, $^{13}C/^{12}C$ ratios in tree rings and the transfer of biospheric carbon to the atmosphere, J. Geophys. Res., 89, 11,731-11,748, 1984.

Suess, E., Particulate organic carbon flux in the oceans, Surface productivity and oxygen utilization, Nature, 288, 260-263, 1980.

Sundquist, E. T., Carbon dioxide in the oceans: Some effects on sea water and carbonate sediments, Ph.D. thesis, 215 pp., Harvard University, Cambridge, Mass., 1979.

Sundquist, E. T., Geologic analogs: Their value and limitations in carbon dioxide research, paper presented at the Sixth ORNL Annual Life Sciences Symposium, U.S. Department of Energy, Energy, Knoxville, Tenn., November 1983.

Sundquist, E. T., and G. A. Miller, Oil shales and carbon dioxide, Science, 208, 740-741, 1980.

Sundquist, E. T., and L. N. Plummer, Carbon dioxide in the ocean surface layer: Some modelling considerations, in Carbon Cycle Modeling, SCOPE 16, edited by B. Bolin, pp. 259-269, John Wiley, New York, 1981.

Swift, J. H., A recent Θ-S shift in the deep water of the northern North Atlantic, in Climate Processes and Climate Sensitivity, Geophys. Monogr. Ser., vol. 29, edited by J. E. Hansen and T. Takahashi, pp. 39-47, AGU, Washington, D.C., 1984.

Takahashi, T., Carbon dioxide in the atmosphere and in Atlantic Ocean water, J. Geophys. Res., 66, 477-494, 1961.

Takahashi, T., and W. S. Broecker, Mechanisms for calcite dissolution on the sea floor, in The Fate of Fossil Fuel $CO_2$ in the Oceans, edited by N. R. Anderson and A. Malahoff, pp. 455-477, Plenum, New York, 1977.

Takahashi, T., W. S. Broecker, and S. R. Werner, Carbonate chemistry of the surface waters of the world oceans, in Isotope Marine Chemistry, edited by E. G. Goldberg, Y. Horibe, and K. Suruhashi, pp. 291-326, Uchida Rokakuho, Tokyo, 1980.

Takahashi, T., W. S. Broecker, and A. E.

Bainbridge, The alkalinity and total carbon dioxide concentration in the world oceans, in Carbon Cycle Modeling, SCOPE 16, edited by B. Bolin, pp. 271-286, John Wiley, New York, 1981.

Thiede, J., Textural variations of calcareous coarse fractions in the Panama Basin (eastern equatorial Pacific Ocean), in The Fate of Fossil Fuel $CO_2$ in the Oceans, edited by N. R. Anderson and A. Malahoff, pp. 673-692, Plenum, New York, 1977.

Thunell, R. C., and S. Honjo, Planktonic foraminiferal flux to the deep ocean: Sediment trap results from the tropical Atlantic and the central Pacific, Mar. Geol., 40, 237-253, 1981.

Toggweiler, J. R., and J. L. Sarmiento, Glacial to interglacial changes in atmospheric carbon dioxide: The critical role of ocean surface water in high latitudes, this volume.

Tomczak, M., Jr., Equatorial upwelling rates, Nature, 278, 307-308, 1979.

Trenberth, K. E., Seasonal variations in global sea level pressure and the total mass of the atmosphere, J. Geophys. Res., 86, 5238-5246, 1981.

Tsunogai, S., and Y. Watanabe, Calcium in the North Pacific water and the effect of organic matter on the calcium-alkalinity relation, Geochem. J., 15, 95-107, 1981.

Tsunogai, S., M. Uematsu, S. Noriki, N. Tanaka, and M. Yamada, Sediment trap experiment in the northern North Pacific: Undulation of settling particles, Geochem. J., 16, 129-147, 1982.

U.S. Department of Energy, Carbon Dioxide Effects Research and Assessment Program, CONF-8009140, 397 pp., Washington, D.C., 1981.

U.S. Standard Atmosphere (1976), published by Natl. Oceanic & Atmos. Admin., Natl. Aeronautics & Space Admin., and the U.S. Air Force, available from Superintendent of Documents, U.S. Government Printing Office, Washington, D.C. 20402.

Verma, S. B. and N. J. Rosenberg, Further measurements of carbon dioxide concentration and flux in a large agricultural region of the Great Plains of North America, J. Geophys. Res., 86, 3258-3261, 1981.

Veronis, G., The role of models in tracer studies, in Proceedings of Symposium, Numerical Models of Ocean Circulation, pp. 133-146, National Academy of Sciences, Washington, D.C., 1975.

Viecelli, J. A., Analysis of a relationship between the vertical distribution of inorganic carbon and biological productivity in the oceans, J. Geophys. Res., 89, 8194-8196, 1984.

Viecelli, J. A., H. W. Ellsaesser, and J. E. Burt, A carbon cycle model with latitude dependence, Clim. Change, 3, 281-301, 1981.

Vinogradov, M. Y., Food sources of the deep-water fauna: Speed of decomposition of dead Pteropoda, Soviet Oceanogr., Engl. Transl., 136/141, 39-42, Trans. from Russian, Dokl. Akad. Nauk SSSR Okeanol., 1961.

Volk, T., and M. I. Hoffert, Ocean carbon pumps: Analysis of relative strengths and efficiencies in ocean-driven atmospheric $CO_2$ changes, this volume.

Volkov, I. I., and A. G. Rozanov, The Sulfur cycle in oceans: Reservoirs and fluxes, in The Global Biogeochemical Sulphur Cycle, SCOPE 19, edited by M. V. Ivanov and J. R. Freney, pp. 357-423, John Wiley, New York, 1983.

Wagener, K., Total anthropogenic $CO_2$ production during the period 1800-1935 from carbon-13 measurements in tree rings, Radiat. Environ. Biophys., 15, 101-111, 1978.

Wakeham, S. G., J. W. Farrington, R. B. Gagosian, C. Lee, H. DeBaar, G. E. Nigrelli, B. W. Tripp, S. O. Smith, and N. M. Frew, Organic matter fluxes from sediment traps in the equatorial Atlantic Ocean, Nature, 286, 798-800, 1980.

Wakeham, S. G., R. B. Gagosian, J. W. Farrington, and E. A. Canuel, Sterenes in suspended particulate matter in the eastern tropical North Pacific, Nature, 308, 840-843, 1984a.

Wakeham, S. G., C. Lee, J. W. Farrington, and R. B. Gagosian, Biogeochemistry of particulate organic matter in the oceans: Results from sediment trap experiments, Deep Sea Res., 31, 509-528, 1984b.

Walsh, J. J., The role of ocean biota in accelerated ecological cycles: A temporal view, BioScience, 34, 499-507, 1984.

Walsh, J. J., G. T. Rowe, R. L. Iverson, and C. P. McRoy, Biological export of shelf carbon is a sink of the global $CO_2$ cycle, Nature, 291, 196-201, 1981.

Wang, W., Y. Yung, A. Lacis, T. Mo, and J. Hansen, Greenhouse effects due to man-made perturbations of trace gases, Science, 194, 685, 1976.

Wangersky, P. J., Particulate organic carbon in the Atlantic and Pacific oceans, Deep Sea Res., 23, 457-465, 1976.

Wangersky, P. J., Organic material in sea water, in The Natural Environment and the Biogeochemical Cycles, The Handbook of Environmental Chemistry, vol. 1, part C, edited by O. Hutzinger, pp. 25-62, Springer-Verlag, New York, 1984.

Wattenberg, H., Kalziumkarbonat und Kohlensauregehalt des Meerwassers, Wiss. Ergeb. Dtsch. Atlantischen Exped., 1925-1927, 8, 233-331, 1933.

Wefer, G., E. Suess, W. Balzer, G. Liebezeit, P. J. Muller, C. A. Ungerer, and W. Zenk, Fluxes of biogenic components from sediment trap deployment in circumpolar waters of the Drake Passage, Nature, 299, 145-147, 1982.

Welander, P., On the frequency response of some different models describing the transient exchange of matter between the atmosphere and the sea, Tellus, 11, 348-354, 1959.

Welander, P., The thermocline problem, Philos. Trans. R. Soc. London, Ser. A., 270, 415-421, 1971.

Wenk, T., and U. Siegenthaler, The high-latitude ocean as a control of atmospheric $CO_2$, this volume.

Whittaker, R. H., Communities and Ecosystems, 162 pp., Macmillan, New York, 1970.

Whittaker, R. H., and G. E. Likens, Carbon in the biota, in Carbon and the Biosphere, AEC Symp. Ser., vol. 30, edited by G. M. Woodwell and E. V. Pecan, pp. 281-302, U.S. Department of Commerce, Springfield, Va., 1973.

Whittaker, R. H., and G. E. Likens, The biosphere and man, in Primary Productivity of the Biosphere, edited by H. Lieth and R. H. Whittaker, pp. 305-328, Plenum, New York, 1975.

Whittle, K. J., Marine organisms and their contribution to organic matter in the ocean, Mar. Chem., 5, 381-411, 1977.

Wiebe, P. H., S. H. Boyd, and C. Winget, Particulate matter sinking to the deep-sea floor at 2000 m in the Tongue of the Ocean, Bahamas, with a description of a new sedimentation trap, J. Mar. Res., 34, 341, 1976.

Wilkinson, J. H., The Algebraic Eigenvalue Problem, 662 pp., Oxford University Press, New York, 1965.

Williams, D. F., and N. Healy-Williams, Oxygen isotopic-hydrographic relationships among recent planktonic foraminifera from the Indian Ocean, Nature, 283, 848-852, 1980.

Williams, D. F., M. A. Sommer, II, and M. L. Bender, Carbon isotopic compositions of recent planktonic foraminifera of the Indian Ocean, Earth Planet. Sci. Lett., 36, 391-403, 1977.

Williams, P. J. le B., Biological and chemical aspects of dissolved organic material in sea water, in Chemical Oceanography, vol. 2, edited by J. P. Riley and G. Skirrow, pp. 301-363, Academic Press, Orlando, Fla., 1975.

Williams, P. J. le B., Incorporation of microheterotrophic processes into the classical paradigm of the planktonic food web, Kiel. Meeresforsch., Sonderh., 5, 1-28, 1981.

Williams, P. J. le B., K. R. Heinemann, J. Marra, and D. A. Purdie, Comparison of [14]C and $O_2$ measurements of phytoplankton production in oligotrophic waters, Nature, 305, 49-50, 1983.

Williams, P. M., The distribution and cycling of organic matter in the ocean, in Organic Compounds in Aquatic Environments, edited by S. J. Faust and J. V. Hunter, pp. 145-163, Marcel Dekker, New York, 1971.

Williams, P. M., H. Oeschger, and P. Kinney, Natural radiocarbon activity of the dissolved organic carbon in the northeast Pacific Ocean, Nature, 224, 256-258, 1969.

Wilson, A. T., Pioneer agriculture explosion and $CO_2$ levels in the atmosphere, Nature, 273, 40-41, 1978.

Wilson, J. L., Carbonate Facies in Geologic History, 471 pp., Springer-Verlag, New York, 1975.

Wolff, G. T. and R. L. Klimisch (Eds.), Particulate Carbon Atmospheric Life Cycle, 411 pp., Plenum, New York, 1982.

Wollast, R., and G. Billen, The fate of terrestrial organic carbon in the coastal area, in Carbon Dioxide Effects Research and Assessment Program, CONF-8009140, pp. 331-359, U.S. Department of Energy, Washington, D.C., 1981.

Woodwell, G. M., R. H. Whittaker, W. A. Reiners, G. E. Likens, C. C. Delwiche, and D. B. Botkin, The biota and the world carbon budget, Science, 199, 141-146, 1978.

Woodwell, G. M., J. E. Hobbie, R. A. Houghton, J. M. Melillo, B. Moore, B. J. Peterson, G. R. Shaver, Global deforestation: Contribution to atmospheric carbon dioxide, Science, 222, 1081-1086, 1983.

World Climate Research Program, Report of the WMO (CAS) meeting of experts on the $CO_2$ concentrations from pre-industrial time to IGY (WCP-53), Rep. 10, World Meteorol. Organ., Geneva, 1983.

World Coal Resources and Production, World Coal, 7, 50, 1981.

WEC (World Energy Conference), Survey of Energy Resources, The United States National Committee of the WEC, New York, 1974.

WEC (World Energy Conference), Survey of Energy Resources, Federal Institute for Geosciences and Natural Resources, Hanover, Federal Republic of Germany, 1980.

Wunsch, C., On oceanic boundary mixing, Deep Sea Res., 17, 293-301, 1970.

Wüst, G., Schichtung und Zirkulation des Atlantischen Ozeans. Die Stratosphere, Wiss. Ergeb, Dtsch. Atlantischen. Exped. Meteor, 6, 1, 1935.

Wyrtki, K., The thermocline circulation in relation to the general circulation in the oceans, Deep Sea Res., 8, 39-64, 1961.

Wyrtki, K., The oxygen minima in relation to ocean circulation, Deep Sea Res., 9, 11-23, 1962.

Wyrtki, K., An estimate of equatorial upwelling in the Pacific, J. Phys. Oceanogr., 11, 1205-1214, 1981.

Young, J. A., and A. W. Fairhall, Radiocarbon from nuclear weapons tests, J. Geophys. Res., 73, 1185-1200, 1968.

Zenkevich, L. A., N. G. Barsanova, and G. M. Belyaev, Kolichestvennoe raspredelenie donnoi fauny v abissali Mirovogo okeana (Quantitative Distribution of Zoobenthos in the Oceanic Abyss), Dokl. Akad. Nauk. SSSR, 130, 1960.

Zinke, P. J., A. G. Stangenberger, W. M. Post, W. R. Emanuel, and J. S. Olson, Worldwide Organic Soil Carbon and Nitrogen Data, 141 pp., Rep. ORNL/TM-8857, Oak Ridge Natl. Lab., Oak Ridge, Tenn., 1984.

A REEXAMINATION OF THE TROPOSPHERIC METHANE CYCLE:
GEOPHYSICAL IMPLICATIONS

Gunnar I. Senum and Jeffrey S. Gaffney

Environmental Chemistry Division, Department of Applied Science
Brookhaven National Laboratory, Upton, New York 11973

Abstract. Existing data concerning sources, sinks, and natural isotopic ($^{13}C$, D, $^{14}C$) abundances of atmospheric methane, a greenhouse gas, are reexamined in light of recent indications of an increasing global atmospheric methane burden. This increase has important geophysical implications with regard to greenhouse effects. Previous evaluations of Ehhalt have been revised by updating bacteria source terms and the hydroxyl sink terms. Microorganism-derived source rates have been lowered from $700 \pm 130$ to $300 \pm 140 \times 10^{12}$ g yr$^{-1}$. Sink rates for atmospheric methane are dropped from $635 \pm 30$ to $280 \pm 60 \times 10^{12}$ g yr$^{-1}$. The revised budget is examined with relation to the known $\delta^{13}C$, $\delta D$ and $^{14}C$ methane abundances and found to be in reasonable agreement, with recent biogenic source terms accounting for 80-90% of the observed increase. Although there are limited data, existing $\delta^{13}C$ and $\delta D$ yearly trends support this observation. Future needs for measurements are discussed in light of this budget.

Introduction

Recent observations by Rasmussen and co-workers have indicated that the global tropospheric methane burden is increasing at an appreciable rate in both northern and southern hemispheres (N.H. and S.H.) [Fraser et al., 1981; Rasmussen and Khalil, 1981a, b; Rasmussen et al., 1983], the present concentrations of $CH_4$ being 1.67 and 1.51 parts per million by volume (ppmV) and increasing at rates of $1.9 \pm 0.4$ and $1.4 \pm 0.4\%$ yr$^{-1}$ in the N.H. and S.H., respectively. This observed interhemispheric difference is due to the slow transport or mixing rate of air between the hemispheres (1.1 to 1.2 years) and to the preponderance of land-based sources terms in the northern hemisphere [Rasmussen et al., 1983]. These observations have been confirmed by the work of Rowland and colleagues indicating an ~1.5% yr$^{-1}$ increase in methane [Blake et al., 1982]. Using a tropospheric mass of $4.1 \times 10^{21}$ g based on existing mixing volume of known releases of conservative perfluorocarbon tracers (R.N. Dietz, personal communication, 1983), these concentrations lead to a tropospheric methane burden of

$3600 \times 10^{12}$ g and a rate of increase of $60 \pm 10 \times 10^{12}$ g yr$^{-1}$. Recent data from polar ice cores [Craig and Chou, 1982] support the observed increase in $CH_4$ as far back as A.D. ~1580 with a strong increase indicated around the beginning of the industrial revolution. Their data extrapolated to the 1982 level of 1.67 ppmV ($CH_4$) in the atmosphere yield a 2.0% increase over that time period, in good agreement with the more recent air measurements (see Figure 1). These data strongly suggest that natural as well as anthropogenic variations may be important in greenhouse effects.

Methane is produced from bacterial (recent), fossil, and possibly abiogenic sources and is removed in the troposphere by reaction with hydroxyl radical [OH] and by transport across the tropopause into the stratosphere (where it is subsequently oxidized). The stratospheric (N.H) methane concentration is ~1.67 ppm at the tropopause and decreases at a nearly uniform rate of 0.062 ppm/km from the tropopause [Ehhalt, 1972, 1974, 1976, 1979, Ehhalt and Schmidt, 1978].

The observed increases in the atmospheric methane burden have important geophysical implications in relation to the greenhouse effect [Wang et al., 1976; and Lacis et al., 1981]. This is due to the fact that methane and other trace atmospheric gases have reasonably strong infrared absorption bands in windows that allow them to behave in a linear (Beer's law) fashion. That is, their tropospheric absorption is not approaching saturation similar to $CO_2$ (see Figure 2), and they absorb in spectral regions where they do not compete with the dominant absorption of $CO_2$ or $H_2O$. Table 1 gives the rate of increase in the greenhouse effect due to the rate of increase in the concentration of atmospheric trace gases. From these calculated greenhouse warming rates, it is apparent that approximately 30% of the current increase in the greenhouse warming rate can be attributed to the yearly rate of increase in atmospheric methane. Also, $CH_4$ reaching the stratosphere may be an important mechanism for $ClO_x$ species sinking as HCl, thus affecting the freon-ozone stratosphere depletion [Blake et al., 1982]. Consequently, it becomes especially important to understand the sources and sinks of atmospheric methane.

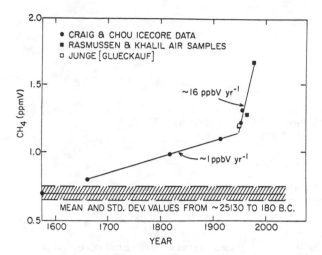

Fig. 1. Existing data showing evidence from ice core data [Craig and Chou, 1982] and direct atmospheric measurements for atmospheric methane increases over the past 200 years [Rasmussen and Khalil, 1981a, b, 1983; Junge, 1984].

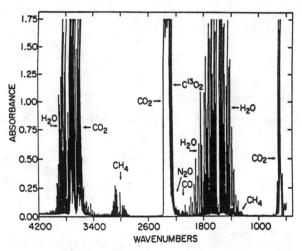

Fig. 2. Fourier transform infrared spectrum of tropospheric air, showing absorption bands of $CO_2$, $^{13}CO_2$, $CH_4$, $H_2O$, $N_2O$, and CO (100-m path, 0.24-$cm^{-1}$ resolution, air partially dried, sample taken at Brookhaven National Laboratory using Nicolet 7199 spectrometer and long-path gas cell).

In this work we have reexamined the existing methane budgets of Ehhalt [1972, 1974, 1976, 1979] and Ehhalt and Schmidt [1978] and have revised these data in light of updated estimates of bacterially derived methane source rates and revised methane sink rates. Existing natural isotopic data ($^{13}C$, D, $^{14}C$) for bacterial, abiogenic, and fossil sources are examined for the limited atmospheric isotopic methane determinations and are discussed in light of the presented budget.

### Methane Budget Revisions

The sources of atmospheric methane can be divided into bacterial (recent) and fossil and abiogenic (ancient) source terms using their $^{14}C$ abundance; i.e., $^{14}C$ is either at present-day levels or absent.

### Bacterial Sources (Recent)

The bacterial sources of methane as identified and reviewed previously [Ehhalt, 1972, 1974, 1976, 1979, Ehhalt and Schmidt, 1978] are given in the first column of Table 2. Since Ehhalt's most recent review [Ehhalt, 1979], additional evidence has been produced which necessitates a revision of the biogenic (bacterial) estimates; these are as follows:

1. Recent work by Zimmerman et al. [1982] indicates that termites are a significant source of atmospheric methane, contributing ~150 x $10^{12}$ g $CH_4$ $yr^{-1}$ to the atmosphere. These authors indicate an uncertainty in their estimates of a factor of 2, leading to a source of 190 ± 120 x $10^{12}$ g $CH_4$ $yr^{-1}$. This source has been reexamined by Rasmussen and Khalil [1983], who have also shown that ter-

TABLE 1. Predicted Global Heating Effects of Methane
Compared to Other Greenhouse Gases

| Trace Gas | Present Ambient Concentration | Yearly Rate of Increase | Increase in Greenhouse Effect mK/yr | |
|---|---|---|---|---|
| | | | Wang et al. [1976] | Lacis et al. [1981] |
| $CH_4$ | 1.65 ppm | 1.7% | 1.7 | 6.2 |
| $CO_2$ | 330 ppm | 0.3% | 1.3 | 2.4 |
| $N_2O$ | 335 ppb | 0.4% | 0.9 | 3.5 |
| $CF_2Cl_2$ | 300 ppt | 9% | 1.8 | 4.9 |
| $CF_2Cl_3$ | 190 ppt | 10% | | |

TABLE 2. Bacterial Tropospheric Methane Sources - Recent Biogenic

| Source | Source Rate[a] $10^{12}$ g $CH_4$ $yr^{-1}$ | Revised Source Rate[b] $10^{12}$ g $CH_4$ $yr^{-1}$ |
|---|---|---|
| Enteric fermentation | | |
| Animals | $150 \pm 50$ | $150 \pm 50$ |
| Termites | --- | $50 \pm 40$ |
| Humans | | $4 \pm 2$ |
| Rice paddy fields | 280 | $55 \pm 25$ |
| Swamps and marshes | $245 \pm 55$ | $30 \pm 10$ |
| Freshwater sediments | $13 \pm 12$ | $6 \pm 5$ |
| Tundras | $1.7 \pm 1.4$ | $1.7 \pm 1.4$ |
| Oceans | $9 \pm 7.7$ | $9 \pm 7.7$ |
| Total recent biogenic | $700 \pm 130$ | $300 \pm 140$ |

[a]Ehhalt [1979].
[b]This work.

mites are indeed a significant source of methane. Their estimate, of $50 \pm 40 \times 10^{12}$ g $yr^{-1}$, however, is lower than the previous work. The uncertainties in estimation of the world termite population lead to some of the discrepancy. In the revised budget we have adopted $50 \pm 40 \times 10^{12}$ g $yr^{-1}$ until further data are evaluated.

2. Cicerone and Shetter [1981] have remeasured the methane fluxes from both fertilized and unfertilized rice paddies. They concluded that Ehhalt's estimate, which is based on earlier laboratory rice paddy measurements, is too large and should be reduced to $60 \times 10^{12}$ g $CH_4$ $yr^{-1}$. They noted that this value is for generously fertilized rice paddies and perhaps should be further reduced to account for lesser usage of fertilizers in less developed nations. An unfertilized rice paddy would have ~20% of the methane flux of a fertilized paddy; consequently, we can estimate the rice paddy methane flux to be $55 \pm 25 \times 10^{12}$ g/$yr^{-1}$.

3. There have been additional measurements on methane fluxes from swamps, marshes, and freshwater lakes [Cicerone and Shetter, 1981] that indicate a lower methane flux than used by Ehhalt in his estimations. Ehhalt used a derived $CH_4$ flux value of 200 g $CH_4$ $m^{-2}$ $yr^{-1}$ or 400% of the recent measurements for fresh water lakes. Cicerone and Shetter caution about using their flux value to extrapolate the total $CH_4$ flux due to the large variability in their data; nonetheless, the earlier data used by Ehhalt would clearly be no better, and thus we derive a freshwater $CH_4$ flux of $6 \pm 5 \times 10^{12}$ g $CH_4$ $yr^{-1}$. The recent measurement by Hariss and Sebacher [1981] of $CH_4$ flux from swamps in the southern United States similarly indicated smaller $CH_4$ fluxes from freshwater swamps and marshes, i.e., by approximately an order of magnitude. Ehhalt [1979] used the Baker-Blocker et al. [1977] $CH_4$ flux for Michigan swamps and ponds,

i.e., $CH_4$ fluxes ranging from 33 to 401 g $m^{-2}$ $yr^{-1}$. This lower range of $CH_4$ flux is in agreement with Harris' values found in Virginia swamps. Thus we have revised Ehhalt's estimate by an order of magnitude to $30 \pm 10 \times 10^{12}$ g $CH_4$ $yr^{-1}$ for swamps and marshes.

4. An additional methane source not considered by Ehhalt is the formation and release of $CH_4$ from humans by fermentation in the large intestine. Wolin [1981] has calculated that approximately 4 g $CH_4$ $d^{-1}$ per person is formed and released to the ambient. This then extrapolates to a yearly $CH_4$ flux of $4 \pm 2 \times 10^{12}$ g $CH_4$ $yr^{-1}$. These revisions of the recent biogenic source terms are summarized in Table 2. Sheppard et al. [1982] have estimated that $9.10 \times 10^{14}$ g $yr^{-1}$ of methane is emitted from natural ecosystems, leading to an overall biogenic source rate of $12.10 \times 10^{14}$ g $yr^{-1}$. This source rate is ~4 times larger than the flux recommended here and will be discussed later in detail.

Fossil and Abiogenic Sources (Ancient)

In addition to the recent biogenic source of $CH_4$, there are fossil and abiogenic sources (ancient), which are given in Table 3 according to Ehhalt's [1979] estimate. There are no further data regarding fossil and abiogenic methane sources to our knowledge, and consequently, Ehhalt's estimates have not been revised here.

It is interesting to note that recently Gold [1979] has proposed that large amounts of abiogenic primordial methane have been trapped in the earth's crust. This methane, it is argued, is released from deep within the crust by outgassing, principally along fault lines and during earthquakes. Although this hypothesis has been criticized in several quarters [Rawls, 1981; American Chemical Society, 1980], if it is correct, then the impact

TABLE 3.    Fossil and Abiogenic (Ancient)
Troposheric Methane Sources

| | Source Rate, $10^{12}$ g $CH_4$ $yr^{-1}$ |
|---|---|
| Fossil sources[a] | |
| Coal fields | 14 ± 8 |
| Lignite coal | 3.7 ± 2 |
| Natural gas losses | 14 ± 7 |
| Automotive | 0.5 |
| Abiogenic sources | |
| Volcanos | 0.2 |
| Gold methane[b] | 0.1 |
| | |
| Total ancient sources | 33 ± 17 |
| Total recent biogenic sources[c] | 300 ± 140 |
| | |
| Total Methane Sources | 330 ± 150 |

[a]Ehhalt [1979].
[b]Welhan and Craig [1979].
[c]From Table 2 [Welhan and Craig, 1979].

of this methane should be apparent in the ultimate
reservoir for this outgassed material, i.e., the
atmosphere. Existing literature data concerning
the measured $^{14}C$, $^{13}C$, and deuterium natural
isotopic abundances of atmospheric methane appear
to indicate the possibility that unknown fossil or
abiogenic methane deposits exist on the present-day
earth [Ehhalt, 1979; Rust, 1981; Rust and Stevens,
1980]. However, as is pointed out in our discussion on the existing natural isotopic abundances
with respect to this budget, simpler more orthodox
explanations can be invoked regarding the observed
methane isotopic ratios which require no added
abiogenic source terms. Welhan and Craig [1979]

have noted methane evolution along with enriched
$^{3}He$ from certain geothermal vents in the east
Pacific. From their data an upper limit of 0.1 x
$10^{12}$ g $yr^{-1}$ can be made for this possible deep-
earth-sourced methane, which is a trivial correction to the fossil source.

Tropospheric Methane Sinks

The known tropospheric sinks of atmospheric
methane as tabulated by Ehhalt [1979] are summarized in Table 4. Again, there have been additional
data, which necessitates a revision of Ehhalt's
methane sinks, that is, specifically Ehhalt's
assumption that [OH] = 1 x $10^6$ $cm^{-3}$ as a mean
tropospheric OH concentration. This may be now
regarded as slightly high, with a more realistic
value being 3-4 x $10^5$ $cm^{-3}$ based on fluorocarbon
sink rates [Singh et al., 1979]. Indeed, if the
recent work of Rasmussen et al. [1983] is used to
estimate the [OH] concentration using the Junge
relationship [Junge, 1974], one estimates again
that the mean tropospheric lifetime of methane is
8-15 years. This compares well with the lifetime
estimated by comparison with $CH_3CCl_3$ of 10.5 ± 1.8
years [Mayer et al., 1982]. These data lead to a
mean tropospheric OH radical concentration of ~3-4
x $10^5$ molecules $cm^{-3}$. This then reduces the
methane sink rate by [OH] to 205 ± 30 x $10^{12}$ g
$yr^{-1}$.

Recent computer simulations [Crutzen and Gidel,
1983] have estimated the methane sink rate by OH to
be 320 x $10^{12}$ g $yr^{-1}$, corresponding to an average
annual average [OH] of ~5-6 x $10^5$ molecules $cm^{-3}$.
Recent work by Lowe and Schmidt [1983] examining
nonurban atmospheric formaldehyde levels indicates
that mean global [OH] is probably lower than this
value. Now recognized $NO_x$ sinks such as $NO_3$ reaction with trace organics and sulfur compounds were
not considered in the model [Crutzen and Gidel,
1983]. Due to this uncertainty and others (i.e.,
light intensity (cloudiness), nitrogen oxide, and
reactive organic emission inventory discrepancies)

TABLE 4.    Sinks of Tropospheric Methane

| Sinks | Sink Rate,[a] $10^{12}$ g $CH_4$ $y^{-1}$ | Revised Sink Rate,[b] $10^{12}$ g $CH_4$ $yr^{-1}$ |
|---|---|---|
| Microorganisms | <1 | <1 |
| Stratospheric transport | 55 ± 30 | 55 ± 30 |
| OH depletion | 580 | 205 ± 30 |
| $O(^1D)$ depletion | ---- | 1.5 ± 1 |
| $NO_3$ depletion[c] | ---- | 20 ± 3 |
| | | |
| Total sink rate | 635 ± 30 | 280 ± 60 |

[a]Ehhalt [1979].
[b]This work.
[c]Assuming rate of 1 x $10^{-18}$ $cm^3$ $molecule^{-1}$ $s^{-1}$ and a $[NO_3]_{ss}$ of 25 ppt.

we have chosen to use the lower estimate based on the integrated chemical sink measurements.

Comparison of our overall budget with that of Crutzen and Gidel [1983] shows excellent agreement between the overall source term and sink term. However, this is somewhat fortuitous because of their neglect of stratospheric loss of methane due to transport and their somewhat higher estimate of the loss of methane due to reaction by OH.

An additional methane sink is the reaction of methane with $O(^1D)$. An ambient $O(^1D)$ concentration of 0.01-0.05 molecules $cm^{-3}$ can be estimated [Logan et al., 1981]. A reaction rate constant of 2.4 x $10^{-10}$ $cm^3$ molecule$^{-1}$ $s^{-1}$ over the temperature range 200 to 300 K is given by Baulch et al. [1980], which leads to a trivial methane sink rate of (0.5-2.5) x $10^{12}$ g $CH_4$ yr$^{-1}$. Another possible sink for methane is the reaction with nitrate radical. If nitrate reacts at a rate of 1 x $10^{-18}$ $cm^3$ molecule$^{-1}$ $s^{-1}$, then a 250 parts per trillion (ppt) steady state of $NO_3$ would give an equivalent loss rate. $NO_3$ concentrations will most likely be less. An upper estimate of this sink assuming this rate and a steady state $NO_3$ concentration of 25 ppt leads to a sink of 20 ± 3 x $10^{12}$ g yr$^{-1}$. If this reaction rate is slower (or if [$NO_3$] is less) it will lead to a correspondingly smaller sink term. This reaction rate should be studied to evaluate this proposed sink.

## Discussion

Since the atmosphere is increasing in concentration with respect to $CH_4$, the observed net 60 ± 10 x $10^{12}$ g $CH_4$ increase in the atmospheric $CH_4$ burden should be the difference of the source and sink rates, and it is in reasonable agreement with this budget.

An apparent discrepancy in the above interpretation, however, is indicated by the measured $^{14}C$ content of atmospheric $CH_4$. Ehhalt [1979] has reviewed these measurements and has concluded that ~83% of atmospheric methane must be derived from recent biogenic sources. This conclusion is based on the fact that bacterially produced methane will have a $^{14}C$ content consistent with recently photosynthesized plant matter, whereas fossil and abiogenic methane will be devoid of any $^{14}C$, since it has a relatively short lifetime ($t_{1/2}$ = 5570 years) compared to the age of fossil and abiogenic methane. Thus the level of $^{14}C$ observed in atmospheric methane is apparently not consistent with our present understanding of methane sources; i.e., the $^{14}C$ observation implies an 83% recent, 17% ancient source ratio, whereas the revised compilation gives a 90% recent, 10% ancient source ratio, if one believes this difference to be significant. Nonetheless, this apparent discrepancy can be resolved by noting that the quoted $^{14}C$ content measurements of $CH_4$ are all from the period 1949 through 1960, whereas the sources are tabulated for (1982) values. The yearly increase in the $CH_4$ burden is ascribed to increased agricultural activity [Rasmussen and Khalil, 1981a, b; Fraser et al.,

1981], which implies that most of the added methane is of recent biogenic origin. Consequently, we would expect to find at present a larger recent/ancient ratio than that determined for the 1949 through 1960 period. This viewpoint will be further supported in the discussion of $\delta(^{13}C)$ yearly trends.

Sheppard et al. [1982] have argued that humic substances are the source of anaerobic bacterial production of methane, and that these carbon sources are older and therefore depleted in carbon 14. They use their data along with a derived source term of 1260 x $10^{12}$ g $CH_4$ yr$^{-1}$ to determine a 3.3-year half-life for methane and a 2.6 x $10^6$ molecule $cm^{-3}$ mean concentration of OH. Following their example, if one assumes that all the biogenic methane is derived from humic substances, then our budget leads to 10% fossil carbon. If one assumes 23.5% dead carbon in methane [Sheppard et al., 1982] then one is lead to a mean residence time (MRT) of humic material of 1200 years as compared to their estimate of 1365 years. Assuming 17% dead carbon leads to a MRT for the earth's humus of 605 years. It is clear that more data are required for the $^{14}C$ content of atmospheric methane to better determine these source terms, and also to determine if post-nuclear-test $^{14}C$ (bomb carbon) has been incorporated into the methane. Both budgets are in reasonable agreement with regard to the $^{14}C$ data. Sheppard et al.'s [1982] estimate for the MRT of methane in the atmosphere is 3.3 years. Our MRT value for $CH_4$ in the troposphere, based on a value of 3.6 x $10^{15}$ g of $CH_4$ in the troposphere and a source rate of 3.3 x $10^{14}$ g yr$^{-1}$ is 11 years, which is in good agreement with the independently measured residence times of Rasmussen and Khalil [1981a, b] and Rowland and co-workers [Blake et al., 1982; Mayer et al., 1982]. Recent measurements of formaldehyde over the Pacific [Lowe and Schmidt, 1983] also indicate that modeling the mean OH concentration at 0.5 to 2 x $10^6$ molecules $cm^{-3}$ is too large a value. Thus, if one takes the existing data on OH and uses Sheppard et al.'s [1982] source rate along with the present tropospheric budget one would expect to see a 10-20% increase in atmospheric methane in 1 year's time. This has not been the case; although there is evidence for an increase in the rate, it is much smaller than this. As has been pointed out by Cicerone et al. [1983], term seasonal fluctuations and source term variability can be substantial. More source emission inventory measurements are required over a number of annual cycles to determine seasonal, temperature, and relative humidity effects before these apparent discrepancies can be clarified.

Another discrepancy, also previously noted [Ehhalt, 1979; Rust and Stevens, 1980; Rust, 1981; Stevens and Rust, 1982], is that of $\delta(^{13}C)$ and $\delta D$ between the sources of methane and present atmospheric methane. A compilation of the $\delta(^{13}C)$ and $\delta D$ (in parts per thousand) for methane sources as derived from the work of Schoell [1980] and Rust [1981] is presented in Table 5. From this listing

TABLE 5.  Methane $\delta(^{13}C)$ and $\delta(D)$ Values According to Sources

| | $\delta(^{13}C)_{PDB}$[b] | $\delta D_{SMOW}$[c] |
|---|---|---|
| **Recent Biogenic** | | |
| Terrestrial | | |
| Glacial drift deposit gases | −78 ± 7 | −243 ± 18 |
| Marsh gases | −74 ± 4 | −222 ± 19 |
| Marine sediments | −68.7 ± 2.5 | −202 ± 8 |
| Enteric Fermentation Gases | | |
| $C_3$ plants (alfalfa, soybean grasses) | −63.7 ± 7.0 | --- |
| $C_4$ plants (corn) | −50.3 ± 2.7 | --- |
| **Fossil** | | |
| Natural gases associated with crude oils | | |
| N.W. Germany | −45.4 ± 1.6 | −202 ± 1 |
| S. Germany | −40.0 ± 4.7 | −197 ± 24 |
| North Sea | −49.2 ± 2.6 | −231 ± 18 |
| N. Italy | −42.5 ± 5.8 | −166 ± 15 |
| Egypt | −45.8 ± 0.7 | −239 ± 24 |
| Thermogenically produced $CH_4$ | | |
| Marine sapropelic source rock, Delaware−Val Verde Basin, Texas | −38.5 ± 2.6 | −147 ± 10 |
| Terrigenic (humic source rock) coal gas, N.W. Germany | −27.6 ± 2.0 | −145 ± 8 |

[a]Data are from Schoell [1980] and Rust [1981].
[b]Referenced to the Pee Dee Belemnite (PDB) limestone standard.
[c]Referenced to Standard Mean Ocean Water (SMOW).

together with the estimated methane source rates given in Tables 2 and 3, we should be able to calculate the expected $\delta(^{13}C)$ for atmospheric methane, which is given as follows:

| | $\delta(^{13}C)$ |
|---|---|
| Recent biogenic sources | −64 ± 2 |
| Fossil sources | −38 ± 1 |
| Total sources | −61 ± 3 |

In order to compare this total value with the measured atmospheric methane $\delta(^{13}C)$ values (see Table 6) we have to correct the measured value for the kinetic isotope effect (KIE) for the major methane sink:  abstraction reaction by hydroxyl radical. Thus, the natural isotopic fractionation reactions of concern are

$$^{12}CH_4 + OH \xrightarrow{k_{12}} {}^{12}CH_3 + H_2O \qquad (1)$$

$$^{13}CH_4 + OH \xrightarrow{k_{13}} {}^{13}CH_3 + H_2O \qquad (2)$$

i.e., differing isotopes of methane will be depleted at differing rates and consequently the steady state isotopic concentration of methane will differ from the isotopic ratio of the source terms.  The KIE for equations (1) and (2) has been measured by Rust and Stevens [1980] and found to be 3 ppt (i.e., $k_{12}/k_{13} = 1.003$).

Examining the atmospheric methane isotopic data as given in Table 6, it appears that the $\delta(^{13}C)$ value has been decreasing, at a value of ~0.44 ± 0.05 per mill per year (See Figure 3).  Consequently, we can estimate the current (1983) northern hemisphere methane $\delta(^{13}C)$ value to be −48.4 ± 1.0 per mill.  This is in good agreement with recent work of C.M. Stevens (personal communication, 1984).  When the KIE correction is applied, this adjusts the $\delta(^{13}C)$ value for atmospheric methane to −50.4 ± 1 per mill.  This can be compared to the calculated value of −61 ± 3 per mill as derived from estimates of methane sources, and apparently there is a discrepancy of approximately 10 per mill.

An explanation of part of this discrepancy is apparent in the work of Barker and Fritz [1981] and Coleman et al. [1981] who have observed a kinetic

TABLE 6.   Atmospheric Methane $\delta^{13}C$ and $\delta D$ Measurements

| Year of Measurement | $\delta(^{13}C)_{PDB}$ per mill | $\delta(D)_{SMOW}$ per mill |
|---|---|---|
| 1960[a,b] | $-39 \pm 0.1$ | $-76$ |
| 1970[c] | $-41.3$ | $-86$ |
| 1970[c] | $-43.1$ | $-94$ |
| 1980[d] | $-47.5 \pm 0.3$ | $---$ |
| 1982[e] | $-47.6$ | |

[a]Begemann and Friedman [1968].
[b]Bainbridge et al. [1961].
[c]Ehhalt [1972].
[d]Stevens and Rust [1982]. Mean of eight samples taken from April through July 1980.
[e]C.M. Stevens (personal communication, 1984).

isotope effect for methane when it is subjected to microbial oxidation. That is, when methane is released into aerobic environments (i.e., underlying methane deposits seeping into overlying groundwater or methane formed in deposits which are released into overlaying lake, swamp, or marsh waters), it is partially oxidized by bacteria to $CO_2$, with the $^{12}C$ isotope being oxidized to a greater extent than the $^{13}C$ isotope. Thus as Barker and Fritz observed, when a large proportion of the dissolved methane was oxidized, then the remaining methane was more depleted in $^{12}C$ and had the highest $\delta(^{13}C)$ values. They observed a $\delta(^{13}C)$ shift of 5 to 30 per mill in an experimental situation and a $\delta(^{13}C)$ shift of 8 to 16 per mill in a study of Ontario groundwater which had an underlying source of methane. Ehhalt [1972] in his review reported earlier work which determined that methane gas bubbles which had formed at the bottom of a 10-m-deep lake lost 76 to 80% of the original methane after ascent to the surface due to bacterial oxidation in the water. This mechanism has also been proposed and discussed by Stevens and Rust [1982]. Thus taking this bacterial oxidation KIE into account we can reestimate the biogenic $\delta(^{13}C)$ contribution as $-59$ to $-64$ and the overall estimated $\delta(^{13}C)$ as $-57$ to $-61$, reducing the disagreement. Indeed, there are additional uncertainties which may further resolve this discrepancy. For example, there is a need to determine the $\delta(^{13}C)$ values for termite-derived methane and other source terms. Further experimental $\delta(^{13}C)$ determinations are needed which will hopefully clarify these discrepancies.

Altogether these explanations given in an attempt to resolve the apparent $\delta(^{13}C)$ discrepancy are more plausible than those advanced by Gold [1979], which are partially based on the apparent $\delta(^{13}C)$ and $\delta(^{14}C)$ discrepancies between atmospheric methane and its sources and sinks. However, recent work indicates that in the southern hemisphere the opposite $\delta(^{13}C)$ trend is occurring (C.M. Stevens, personal communication, 1984). Other sources of unknown heavy isotopic methane need to be examined here (such as methane clathrates) to explain this reversal in the southern hemisphere.

Examination of the atmospheric methane $\delta D$ data reveals a possible trend in the values, i.e., $\delta D$ has been decreasing at a rate of $14 \pm 4$ per mill per decade. This reflects Begemann and Friedman's [1968] comments that their methane $\delta D$ analyses are possibly contaminated by local methane sources and Ehhalt's [1981] comment that his samples [Ehhalt, 1972] taken in Boulder, Colorado, were not contaminated by local sources. This apparent decrease in tropospheric methane $\delta D$ is consistent with an increasing portion of the methane burden being derived from biogenic sources which have more negative $\delta D$ values than abiogenic sources. Consequently the present atmospheric methane $\delta D$ can be extrapolated to $-104 \pm 4$ per mill. The measured KIE for the reactions [Gorden and Mulac, 1975]

$$CH_3D + OH \xrightarrow{k_D} products \qquad (3)$$

$$CH_4 + OH \xrightarrow{k_H} products \qquad (4)$$

is $k_H/k_D = 1.50$. This corrects the measured $\delta D$ to $-322$ per mill, a value more depleted in deuterium than the known abiogenic sources of methane. An estimate of the biogenic methane $\delta D$ can be obtained as follows. It has been experimentally observed by Schoell [1980] that municipal sewage sludge can produce methane with $\delta D$ of $-323$ per mill when the $H_2O$ present in the sludge has a $\delta D$ of 0, which is reasonable for $H_2O$. Consequently,

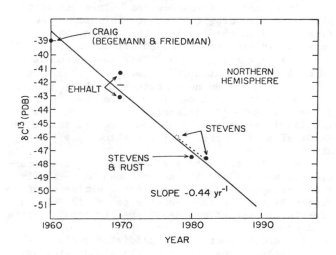

Fig. 3.  Carbon 13 isotope data for atmospheric methane in the northern hemisphere. Data are from Craig [1982], Begemann and Friedman [1968], Ehhalt [1974], Stevens, private communication [1984], and Stevens and Rust [1982].

we can assume this value to be indicative of methane produced by biogenic processes. Coleman et al. [1981] have also noted that the hydrogen isotopes are also enriched by methane-oxidizing bacteria, with the change in the $\delta D$ value of partially oxidized methane being 8-14 times the change in the $\delta^{13}C$ value. Therefore, there may be a discrepancy in the apparent magnitude of the $\delta D$ tropospheric methane trend. Nonetheless, there exist far fewer data for $\delta D$ compared to $\delta(^{13}C)$ for atmospheric methane and its sources, and as a result it is not certain if there is any discrepancy in the methane $\delta D$ value until further experimental determinations are made.

## Conclusions

As has been previously noted, the atmospheric methane burden is increasing, due to increased biological activity, that is, the proportion of recent biogenic methane sources compared to fossil or abiogenic sources is increasing. This proportional increase appears to be supported by trends in the $\delta(^{13}C)$ and $\delta D$ of atmospheric methane i.e., atmospheric methane is being depleted in both $^{13}C$ and D due to the proportionally increasing $^{13}C$- and D-depleted microorganism-derived sources of methane. However, this trend is based on few isotopic methane measurements, and further isotopic measurements should be made in order to confirm these trends. Likewise, there are several bacterial and fossil sources of methane for which further isotopic, $\delta(^{13}C)$ and $\delta D$, measurements should be made. This would help in the comparison between the isotopic ratio of atmospheric methane and its sources.

The proportional increase in bacterial methane sources based on new experimental data has helped in resolving the $^{14}C$ discrepancy between atmospheric methane and its sources. This was accomplished by reducing the recent biogenic source rate, based on the latest data, leading to an increase in the fractional contribution of bacterial methane sources. This resolution is far more plausible than Gold's [1979] hypothesis, which argues for large unknown sources of abiogenic terrestial methane in order to resolve the $^{14}C$ discrepancy. More $^{14}C$ determinations are needed. Low-level gas proportional counting and Tandem Van de Graaf mass spectrometry may allow this [Gaffney et al., 1984].

Similarly, the $\delta(^{13}C)$ discrepancy between atmospheric methane and its sources can be partially resolved by noting (1) the reduced estimates of the recent biogenic methane source rates, (2) the kinetic isotope effect manifested by aerobic oxidation of methane as it is transported from its source through bacterially oxidizing media, i.e., groundwater, lakes, etc., to the atmosphere, and (3) the uncertainties in the isotopic composition of methane released from termites and other uncharacterized sources. Again, the need for further isotopic measurements is apparent and would help in resolving any discrepancy.

The $\delta D$ isotopic ratios for atmospheric methane and its sources do not presently appear to be in discrepancy; however, this is due more to a lack of $\delta D$ measurements for various sources. Further measurements would hopefully further our understanding of atmospheric methane and its sources. This is becoming more necessary as it is becoming more apparent that the increasing methane burden will provide 30% of the future increase in greenhouse warming effect, thus insuring that a knowledge of atmospheric methane and its sources and sinks will become more important in the future.

We would also point out here that methane is only one of a number of trace atmospheric gases that could contribute to climatic impacts. As Kerr [1983] points out, the chemistry (including trace gases and aerosols) and temperature of the overall system (including natural and anthropogenic perturbations) must be considered if the overall effects are to be understood and predicted.

This is not only the case with regard to our present-day and future tropospheric greenhouse effects, but must also be considered when reconstructing paleoclimates. Methane released from clathrates in permafrost or other traps could cause short-term effects which may more reasonably explain glacier recession and other apparently shorter time period phenomena. Also, not only trace gases but aerosols, especially carbonaceous soot and sulfuric acid aerosols from forest fires and volcanic activity could play key roles and should not be overlooked when examining the carbon cycle and climatic effects.

Acknowledgments. We wish to thank D. Ehhalt, R. Dietz, C.M. Stevens, and B. Manowitz for valuable discussions. The assistance of J. Bennett in typing and editing the manuscript is gratefully acknowledged. This work was supported under the auspices of the United States Department of Energy under contract DE-AC02-76CH00016.

## References

American Chemical Society, Deep reservoir natural gas has big potential, Chem. Eng. News, 58 (35), 7, 1980.

Bainbridge, A.E., H.E. Suess, and I. Friedman, Isotopic composition of atmospheric hydrogen and methane, Nature, 192, 648-649, 1961.

Baker-Blocker, A., T.M. Donahue, and K.H. Mancy, Methane flux from wetlands areas, Tellus, 29, 245-259, 1977.

Barker, J.F. and P. Fritz, Carbon isotope fractionation during microbial methane oxidation, Nature, 293, 289-291, 1981.

Baulch, D.L., R.A. Cos, R.F. Hampson, Jr., J.A. Kerr, J. Troe, and R.T. Watson, Evaluated kinetic and photochemical data for atmospheric chemistry, J. Phys. Chem. Ref. Data, 9, 295, 1980.

Begemann, F. and I. Friedman, Tritium and deuterium in atmospheric methane, J. Geophys. Res., 73, 1149-1153, 1968.

Blake, D.R., E.W. Mayer, S.C. Tyler, Y. Makiye, D.C. Montague, and F.S. Roweland, Global increase in atmospheric methane concentrations between 1978 and 1980, Geophys. Res. Lett., 9, 477-480 (1982).

Cicerone, R.J., and J.D. Shetter, Sources of atmospheric methane: Measurements in rice paddies and a discussion, J. Geophys Res., 86, 7203, 1981.

Cicerone, R.J., J.D. Shetter, and C.C. Delwicke, Seasonal variation of methane flux from a California rice paddy. J. Geophys. Res., 88 (C15), 11,022-11,024, 1983.

Coleman, D.D., J.B. Risatti, and M. Schoell, Fractionation of carbon and hydrogen isotopes by methane oxidizing bacteria, Geochim. Cosmochim. Acta, 45, 1033-1037, 1981.

Craig, H., and C.C. Chou, Methane: The record in polar cores, Geophys. Res. Lett., 9, 1221-1224, 1982.

Crutzen, P.J., and L.T. Gidel, A two dimensional photochemical model of the atmosphere, J. Geophys. Res. 88, (C11), 6641-6661, 1983.

Ehhalt, D.H., Methane in the atmosphere, in Proceedings of Symposium on Carbon and the Biosphere, AEC Symp. Ser. 30, vol. 30, pp. 144-158, U.S. Atomic Energy Commission, Oak Ridge, Tenn., 1972.

Ehhalt, D.H., The atmospheric cycle of methane, Tellus, 26, 58-70, 1974.

Ehhalt, D.H., The Atmospheric cycle of methane, Proc. Symp. Microb. Prod. Util. Gases, Göttingen, Germany (1975), 13-22, 1976.

Ehhalt, D.H., Der atmosphärische Krieslauf von Methan, Naturwissenschaften, 66, 307-311, 1979.

Ehhalt, D.H., and U. Schmidt, Sources and sinks of atmospheric methane, Pure Appl. Geophys., 116, 452-464, 1978.

Fraser, P.J., M.A.K. Khalil, R.A. Rasmussen, and A.J. Crawford, Trends of atmospheric methane in the southern hemisphere, Geophys. Res. Lett., 8, 1063-1066, 1981.

Gaffney, J.S., R.L. Tanner, and M. Phillips, Separating carbonaceous aerosol source terms using thermal evolution, carbon isotopic measurements and C/N/S determinations. The Science of the Total Environment, 36, 53-60, 1984.

Gold, T., Terrestial sources of carbon and earthquake outgassing, J. Pet. Geol., 1, 3-19, 1979.

Gorden, S., and W.A. Mulac, Reactions of the OH (X $2_\pi$) radical produced by the pulse radiolysis of water vapor, Int. J. Chem. Kinet., 1, 289-299, 1975.

Hariss, R.C., and D.I. Sebacher, Methane flux in forested freshwater swamps of the southeastern United States, Geophys. Res. Lett., 8, 1002-1004, 1981.

Junge, C.E., Residence time and variability of tropospheric trace gases, Tellus, 26, 477-488, 1974.

Kerr, R.A., Trace gases could double climate warming. Science, 220, 1364-1365, 1983.

Lacis, A., J. Hansen, P. Lee, T. Mitchell, and S.

Lebedeff, Greenhouse effect of trace gases, 1970-1980, Geophys. Res. Lett., 8, 1035-1038, 1981.

Logan, J.A., M.J. Prather, S.C. Wofsy, and M.B. McElroy, Tropospheric chemistry: a global perspective, J. Geophys. Res, 86 (C8), 7210-7254, 1981.

Lowe, D.C., and U. Schmidt, Formaldehyde (HCHO) measurements in the nonurban atmosphere. J. Geophys. Res., 88, 10,844-10,858, 1983.

Mayer, E.W., D.R. Blake, S.C. Tyler, T. Makide, D.C. Montague, and F.S. Rowland, Methane interhemispheric concentration gradient and atmospheric residence time, Proc. Natl. Acad. Sci. USA, 79, 1366-1370, 1982.

Rasmussen, R.A., and M.A.K. Khalil, Increase in the concentration of atmospheric methane, Atmos. Environ., 15, 883-886, 1981a.

Rasmussen, R.A., and M.A.K. Khalil, Atmospheric methane ($CH_4$): Trends and seasonal cycles, J. Geophys. Res., 86 (C10), 9826-9832, 1981b.

Rasmussen, R.A., and M.A.K. Khalil, Global production of methane by termites, Nature, 301, 704-705, 1983.

Rasmussen, R.A., M.A.K. Khalil, and R.W. Dalluge, Atmospheric trace gases in Antarctica, Science, 211, 285-287, 1983.

Rawls, R., Controversial natural has theory being tested, Chem. Eng. News, 59, (28), 17-18, 1981.

Rust, F., Ruminant methane ($\delta C^{13}/C^{12}$) values: relation to atmospheric methane, Science, 211, 1044-1046, 1981.

Rust, F. and C.M. Stevens, Carbon kinetic isotope effect in the oxidation of methane by hydroxyl, Int. J. Chem. Kinet., 12, 371-377, 1980.

Schoell, M., The hydrogen and carbon isotopic composition of methane from natural gases of various origins, Geochim. Cosmochim. Acta, 44, 649-661, 1980.

Sheppard, J.C., H. Westberg, J.F. Hopper, K. Ganesan, and P. Zimmerman, Inventory of global methane sources and their production rates, J. Geophys. Res., 87, 1305-1312, 1982.

Singh, H.B., L.J. Salas, H. Shieishi, and E. Scribner, Atmospheric halocarbons, hydrocarbons, and sulfur hexafluoride: Global distribution, sources and sinks, Science, 203, 899-903, 1979.

Stevens, C.M., and R.E. Rust, The carbon isotopic composition of atmospheric methane, J. Geophys. Res., 87, 4879-4882, 1982.

Wang, W.C., Y.L. Yung, A.A. Lacis, T. Mo, and J.E. Hansen, Greenhouse effect due to man-made perturbation of trace gases, Science, 194, 685-690, 1976.

Welhan, J.A., and H. Craig, Methane and hydrogen in east Pacific Rise hydrothermal fluids, Geophys. Res. Lett., 6, 829-831, 1979.

Wolin, W.J., Fermentation in the rumen and human large intestine, Science, 213, 1463-1468, 1981.

Zimmerman, P.R., J.P. Greenberg, S.O. Wandiga, and P.J. Crutzen, Termites: A potentially large source of atmospheric methane, carbon dioxide and molecular hydrogen, Science, 218, 563-565, 1982.

# $CO_2$ DRIVEN EQUATOR-TO-POLE PALEOTEMPERATURES: PREDICTIONS OF AN ENERGY BALANCE CLIMATE MODEL WITH AND WITHOUT A TROPICAL EVAPORATION BUFFER

Brian P. Flannery and Andrew J. Callegari

Corporate Research-Science Laboratories, Exxon Research and Engineering Co.
Annandale, New Jersey 08801

Martin I. Hoffert, C. T. Hseih, and Mark D. Wainger

Department of Applied Science, New York University
New York, New York 10003

Abstract. Direct measurements of $CO_2$ trapped in polar ice cap cores indicate levels some two-thirds current values during the last ice age. In addition, other estimates, based on geochemical models for weathering over the Phanerozoic, indicate $CO_2$ variations as large as 10 or more times current values. Here we investigate the influence of such carbon dioxide variation through the greenhouse effect on the horizontally averaged equator-to-pole temperature distribution of the earth's surface. The influence for long-wave radiation to space and from the atmosphere to the surface is represented by functions derived from the LOWTRAN radiation code. Surface temperatures are computed using a multireservoir energy balance climate model with separate land, water, and atmosphere components. A range of "climates" are computed as a function of carbon dioxide amounts for 0.5 to 20 times current value using both constant and variable eddy thermal diffusivities. Model results indicate that $CO_2$ variations alone account for only about half the "observed" climate variations during the Cretaceous. Other effects, such as a different distribution of land and sea, might also be of importance but were not considered here.

## 1. Introduction

Direct and inferred evidence for natural variations in the atmospheric concentration of $CO_2$ suggests that the $CO_2$ greenhouse effect might be responsible for some or all of the differences between ancient and modern climates [Crowley, 1983]. For example, analysis of air bubbles trapped in glacial ice cores at both poles indicates that atmospheric $CO_2$ was lower (200-220 ppm) than modern values (300-340 ppm)

when the Wisconsin Ice Age was at its peak (18 kyr B.P.) and that the concentration rose to near its modern, or preindustrial, value at the time of the most rapid retreat of continental glaciation [Neftel et al., 1982; Oeschger et al., this volume]. Over time scales of plate tectonic motion, seafloor spreading, etc. (10-100 Myr), the sedimentary record indicates that atmospheric carbon dioxide levels are controlled by the shifting balance between surface degassing and the carbonate-silicate geochemical weathering cycle [Berner et al., 1983]. Model simulations of this effect suggest that atmospheric carbon dioxide may have been much higher, by a factor of 10 or more, during the Cretaceous, when the global climate was warmer [Lasaga et al., this volume]. Over still longer periods, even higher, greater than 100-1000, concentrations of $CO_2$ have been proposed, consistent with current views of planetary formation and evolution, to resolve the so-called "faint early sun paradox"; namely, the apparent maintenance of above-freezing planetary temperatures when insolation was only about 70% of today's value. Owen et al. [1979], and others, have computed from a radiative-convective climate model that a compensatory greenhouse warming from more abundant carbon dioxide might resolve this apparent discrepancy [Thompson, 1984].

Apart from the global warming implicit in higher carbon dioxide amounts, it is of interest to climate theory to understand the equator-to-pole temperature distributions associated with different climatic regimes. These constitute a first-order refinement of zero-dimensional (or radiative-convective) climate models insofar as they incorporate information on meridional heat transfer by atmospheric and ocean circulations. Poleward heat flux is an essential feature of the planetary surface temperature distribution; its effects are represented in simplified (parameterized) form by Fourier heat conduction laws with eddy thermal diffusivities in latitude-resolved energy balance climate models (EBMS) of the type pioneered by Sellers [1969] and Budyko [1969] and reviewed recently by North et al. [1981]. However, when an EBM was used by Barron et al. [1981] to study the reconstructed paleotemperature distribution of the Cretaceous, these authors found "total heat transport must be maintained at close to present day values despite the fact that meridional temperature gradient was considerably reduced." This is a problem for EBM's, since the poleward heat flux to first order is proportional to the meridional temperature gradient, although there is a partially compensatory effect from latent heat flux toward the poles which increases as the planet warms [Manabe and Wetherald, 1980; Gal-Chen and Schneider, 1976; Coakley et al., 1983; Cess and Wronka, 1979; Flannery, 1984]. Another possible explanation of increased heat flux with diminished temperature gradient is horizontal "advective adjustments" by the atmosphere-ocean system, whereby sufficient horizontal transport processes are generated by the fluid dynamic system to satisfy some thermodynamic or radiative constraint [Stone, 1978; Lindzen and Farrell, 1980; Hoffert et al., 1983]. However, a more detailed analysis of this effect during earlier, warmer climates requires consideration of the infrared radiation flux at each latitude leaving the atmosphere and that radiated downward from the atmosphere to the surface. In addition, the possible redistribution of energy between land, ocean, and atmosphere during climatic change suggests that a multireservoir climate model with distinct temperatures for these components should be employed.

In this paper we use a multireservoir EBM to examine the latitudinal dependence of temperature in the steady state, annual mean, for climate forcing by variations in $CO_2$ relative to its modern value. The governing equations are presented in section 2. The individual reservoir energy balance equations contain terms for the infrared cooling to space at the top of the atmosphere as well as the back radiation to the surface at the bottom of the atmosphere. In section 3, a parameterization of these terms is developed based on the Air Force Geophysics Laboratory LOWTRAN5 computer code [Kneizys et al., 1980]. In section 4, we describe our model results for different atmospheric $CO_2$ amounts. Finally, in section 5, we present the conclusions of this study.

## 2. An Energy Balance Model With Reservoirs for Air, Sea and Land

The energy balance model described below evaluates annual, mean, steady state surface temperatures for three, latitudinally resolved reservoirs, air, sea, and land, heated by sunlight averaged to give the correct annual mean insolation. The three reservoirs exchange energy in the form of radiation, latent heat, and sensible heat, and the air and sea reservoirs transport energy poleward diffusively. These processes can be expressed by the following equations:

Sea

$$-\frac{d}{dx}(1-x^2)\, D_S \frac{dT_S}{dx} = S_S(x,T_S) - F_{SA} + BOA \quad (1a)$$

Air

$$-\frac{d}{dx}(1-x^2)\, D_A \frac{dT_A}{dx}(1+\psi_L) = S_A(x,T_S,T_L) \quad (1b)$$

$$+ fF_{SA} + (1-f)F_{LA} - BOA - TOA$$

Land

$$F_{LA} = S_L(x,T_L) + BOA \quad (1c)$$

where $x = \sin\theta$, $\theta$ is latitude. Here insolation $S(x,T)$ is given as the product of two factors: $s(x)$, the local insolation at the top of the atmosphere, and $(1-\omega)$, an absorbed fraction where $\omega(x,T)$ is the local albedo. We represent the albedo as did Hoffert et al. [1983] with a dependence on latitude associated with mean solar zenith angle, and a sudden change to higher albedo when surface ice forms at 0°C on land and -2°C in the surface ocean. Locally absorbed insolation is partitioned so that 70% directly heats the surface and 30% heats the air. This fractional allocation remains constant during perturbations. The fraction of surface covered by sea, $f$, is assumed to be constant with latitude. $F_{SA}$ and $F_{LA}$ represent the transfer from sea (S) or land (L) to air (A) of infrared radiation, latent heat and sensible heat:

$$F_{SA} = \varepsilon\sigma T_S^4 + C_{LH}[q_{sat}(T_S) - r_a q_{sat}(T_A)]$$

$$+ C_{SH}(T_S - T_A) \quad (2a)$$

$$F_{LA} = \varepsilon\sigma T_L^4 + C_{LH}[r_L q_{sat}(T_L) - r_a q_{sat}(T_A)]$$

$$+ C_{SH}(T_L - T_A) \quad (2b)$$

where $r_a$ ($r_L$) is the relative humidity of air (land), and $q_{sat}$ is the saturation water vapor pressure at the temperature of the reservoir. Air transfers heat to surface and space by infrared radiation, which we discuss in more

detail in the next section 3. BOA and TOA are
fluxes through the bottom and top of atmosphere,
respectively. Our formulation includes the
dependence of BOA and TOA on surface temperatures, latitude, and the concentration of CO$_2$ in
the atmosphere.

Finally, air and sea transport energy poleward
diffusively. The left-hand sides of equations
(1a) and (1b) are standard diffusion equations on
a sphere. We use separate diffusion coefficients
in air, $D_A$, and sea $D_S$. The values of $D_A$ and $D_S$
were "tuned" along with other adjustable
parameters of the model to match temperature
distributions on modern earth. For the sea the
diffusive flux is simply proportional to a temperature gradient, as in a heat conduction law of
the Fourier type. However, in air we also
include a term that causes transport to occur
when there exists a gradient of latent heat
energy. Such transport is known to be important
in nature, and also has been found to be
important in general circulation model (GCM)
simulations, both in an absolute sense and in
forcing experiments [Manabe and Wetherald, 1980].
The factor $\psi_L = r_a\, q_{sat}\, L/c_p T$ is the ratio of
latent heat energy to thermal energy. As
described by Flannery [1984] this formulation
accounts reasonably well, within the framework of
an EBM, for the important effect of transport of
the energy of latent heat in water vapor. Since
the model contains no hydrological cycle, we
simply assume that climate varies so as to
preserve constant relative humidity, as has been
found to be roughly true in simulations using
GCM's. Thus we fix both $r_a$ and $r_L$ at 0.6.

With annual mean insolation and a constant
ratio of land to sea surface coverage, the model
is symmetric about the equator. At the equator,
temperature gradients with respect to latitude
vanish, by symmetry. At the pole we express
boundary conditions as in the work by Flannery
[1984] so that heat flux into a small annular
region about the pole matches surface heat losses
from the region. However, when the pole is ice
covered, we simply require the sea surface temperature to be -2°C and let the land fraction be
unity.

We solve the nonlinear EBM equation using a
relaxation technique on a finite difference grid.
The grid contains 99 points equally spaced by
surface area between the equator and a point
nearly at the pole. Locating the ice line correctly produces some difficulty, but we determine
its location consistently using an iterative
approach. In the figures presented later in the
paper the "discretization" introduced by the grid
will be seen in some figures. As coded, a zone
is either ice free or ice covered; this produces
some discontinuity in the results as the ice
shifts across the mesh.

Figure 1 shows the temperature distributions
of the model as tuned to match the climate of
modern earth. The ice line, $T_S$ = -2°C, is at 72°
latitude. The global mean sea surface tempera-

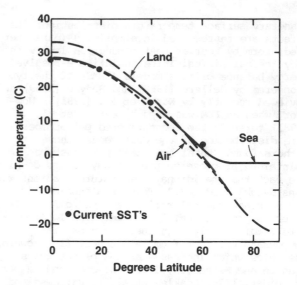

Fig. 1.   Three-component model temperature
distributions tuned to match current climate.
Data from Oort [1983] for annual mean sea surface
temperature (SST) are shown.

ture is 17.5°C. The figure contains several
values for annual mean sea surface temperature
which indicate that the model matches modern
earth quite accurately [Oort, 1983].

3.    Radiation Parameterizations

To develop parameterizations for the dependence of atmospheric radiative fluxes TOA and
BOA, we made use of the computer code LOWTRAN5,
obtained from the Air Force Geophysics Laboratory
[Kneizys et al., 1980]. The code treats atmospheric radiative transfer using band model
approximations for the opacity of important atmospheric constituents with resolution of 20 cm$^{-1}$,
over a frequency domain from 350 cm$^{-1}$ into the
visible. The code also includes a set of five
model atmospheres appropriate to summer and
winter in mid-latitudes and subarctic regions,
and to the tropics. We refer to these as the
five "standard atmospheres."

The LOWTRAN5 absorptance formulation lumps the
effects of infrared absorber gases (CO$_2$, N$_2$O,
CH$_4$, O$_3$, etc.) in a given wave number interval
together. We adapted the code to the present
carbon dioxide greenhouse problem by isolating,
with the knowledgeable assistance of W. Wiscombe,
the bands where CO$_2$ is the dominant contributor,
and increased the absorptance with carbon dioxide
amount accordingly in those bands only. The
errors associated with this approximation should
be small. In addition, opacity coefficients were
added to extend the models to frequencies below
350 cm$^{-1}$, where LOWTRAN5 terminates, but which
still contain an important fraction of the total
IR flux (15%). We ran the code in a manner that

integrated the radiance calculation over a set of
Gaussian quadrature points in angle to obtain
total fluxes. Change in the concentration of
atmospheric $CO_2$ was accounted for by adjusting
the uniformly mixed trace gas absorber amount in
those frequency bands where $CO_2$ contributes to
opacity. Finally, we perturbed the model atmospheres, as discussed below, to develop
parameterizations for regimes differing from the
standard models.

We used LOWTRAN5 with perturbations to calculate results for specific situations, and then
fit the particular solutions to convenient
algebraic relations that allow us to interpolate
smoothly to intermediate values of the
parameters. The total set of parameters for
which we evaluated dependence are (1) surface
temperature, (2) latitude, (3) temperature difference between the planetary surface and the
bottom of the atmosphere (TOA only), (4) time of
year, (5) concentration of atmospheric $CO_2$, (6)
fractional cloudiness, (7) temperature difference
from bottom of atmosphere to bottom of clouds
(BOA only), and (8) temperature difference from
bottom of atmosphere to top of clouds (TOA only).
In the annual mean models we always assumed time
of year to be 0.25, i.e., at an equinox, and we
held the cloud parameters fixed with fractional
cloudiness at 0.5, cloud bottoms $10°C$ cooler than
the surface, and cloud tops $15°C$ cooler than the
surface. We developed latitudinal interpolation
formulas by assigning latitude $15°$, $45°$, $60°$ to
the tropical, mid-latitude, and subarctic models.

Several examples of our results for the magnitude of the fluxes BOA and TOA and their seasonal
dependence and sensitivity to clouds and $CO_2$
changes are given in Figures 2-5. In Figure 2,
we show the results for BOA and TOA as a function
of latitude, time of year, and especially surface
temperature. In the figure, symbols represent
specific models calculated with LOWTRAN5, while
the curves were derived from our fitting functions. We calculated results with surface temperature different from the five supplied model
atmospheres by altering the model atmosphere as
follows. For each model we retained the same
temperature difference $T(z)-T(0)$ with height, z,
but varied $T(0)$ in steps of $5°C$ over an interval
from $-20°C$ to $+20°C$ from the fiducial case.
Also, at each height we adjusted the perturbed
model so that its relative humidity distribution
remained identical to that of the original model.
Thus, the perturbation maintains the same temperature gradient and humidity as the base model.
In this way, the mid-latitude models are extended
into a regime where they overlap with subarctic
and tropical models. Note that TOA basically
shows linear temperature dependence, with an
intercept and slope that are similar to values
often used in simple EBM models where TOA = A+BT
with A = 200 W m$^{-2}$ and B = 2 W m$^{-2}$ $(°C)^{-1}$ [North
et al., 1981]. However, BOA depends nonlinearly
on temperature.

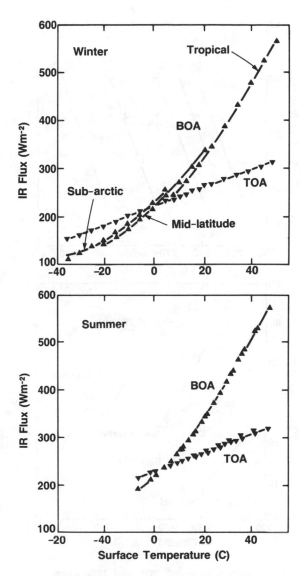

Fig. 2.    Dependence of IR flux on planetary
surface temperature for tropical, mid-latitude
and subarctic model atmospheres during winter
(top) and summer (bottom) seasons. In summer the
model atmosphere results virtually coincide.

Our radiative parameterizations also consider
effects of clouds. The models in LOWTRAN5 were
evaluated by placing "blackbody" clouds at height
levels in the five standard models corresponding
to being located at temperatures 5, 10, 15 and
$20°C$ cooler than the surface and evaluating the
change in BOA and TOA for a totally cloudy sky.
Some results of those models are shown in
Figure 3.

Finally, we evaluated sensitivity to $CO_2$ concentration by calculating models with $CO_2$ varied
from its present concentration in the range from
0.5 to 20 times, but with the standard model

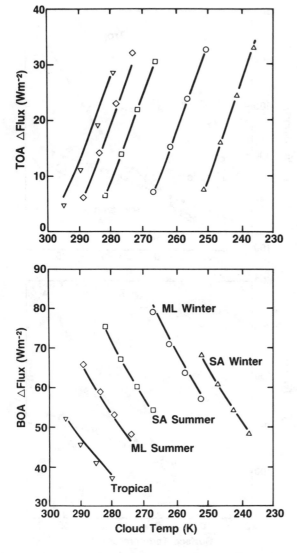

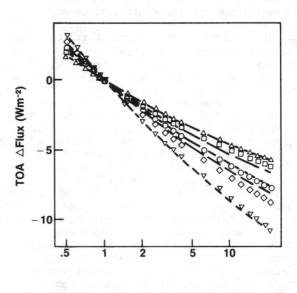

the atmospheric temperature will have changed, so that at a given latitude the atmosphere will be warmer (cooler) for increased (decreased) CO$_2$. Third, and simultaneously with the previous effect, the absolute humidity changes with atmospheric temperature. Here we kept relative humidity constant, which agrees with many previous findings. It is in fact the change in absolute humidity which most influences the total effect of CO$_2$ changes.

Thus, our EBM plus radiative parameterization accounts for most of the effects of altering CO$_2$: the direct effect, the temperature change, and the absolute humidity change. However, there is one more effect that our "surface" model cannot evaluate, namely, the change in vertical

Fig. 3.    Dependence of TOA and BOA fluxes on "blackbody" cloud temperature for a totally cloudy sky.  Results from radiation model calculations (symbols) and interpolated curves are presented for each of the five model atmospheres.

atmospheres.  Those results are shown in Figure 4.  Note that this is not the total CO$_2$ greenhouse effect, which can only be evaluated in a model where all feedback effects come into play [Ramanathan, 1981].  Our EBM does include most, but not all of the additional effects.  First, when CO$_2$ changes, the net atmospheric flux TOA + BOA changes slightly.  For example, when CO$_2$ is doubled in tropical regions, TOA decreases by 3 W m$^{-2}$, while BOA increases by 1.3 W m$^{-2}$, causing a net decrease in flux to space of 1.7 W m$^{-2}$.  Second, as a result of heating or cooling, when the entire model reequilibrates,

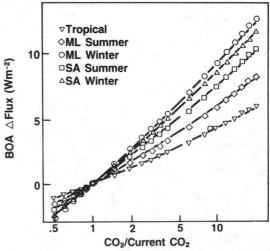

Fig. 4.    Modified LOWTRAN results for changes in atmospheric radiative fluxes due to changes in atmospheric CO$_2$ concentration.

temperature distribution that accompanies variation of $CO_2$. As discussed by Ramanathan [1981], for enhanced $CO_2$ the troposphere warms and the stratosphere cools; thus, in some general sense, the atmospheric lapse rate increases. We believe that this additional effect explains why our model shows somewhat lower sensitivity to changing $CO_2$ than has been found in other studies. For doubled $CO_2$ we find a global mean temperature rise of about 1.4°C, compared with results from vertically resolved radiative-convective models in the range 1.5-2°C. Thus, although our model's sensitivity to radiative perturbations is reasonable, the computed radiative perturbation for the 2 x $CO_2$ case is somewhat less than currently cited values ($\Delta T \approx 2.5$°C for a $CO_2$ doubling).

Even without evaluating a detailed model, a general measure of the "sensitivity" of our radiative $CO_2$ forcing can be found as follows. At some intermediate latitude, typically about 30°, net heating by meridional transport vanishes, so that latitude behaves as though it were in radiative equilibrium, with local radiative heat loss equaling local insolation. In Figure 5 we show the surface temperature at 30° latitude at which TOA equals net insolation. Note that the lower axis scales logarithmically, so that the effect of $CO_2$ is essentially linear for logarithmic increments of $CO_2$. The magnitude of the change is about +1.2°C when the concentration of atmospheric $CO_2$ doubles. (In the complete model the change in global mean temperature is about 1.4°C for doubling.)

### 4. Results for Steady State Models as a Function of $CO_2$ Concentration

Several of our results for models as a function of the concentration of atmospheric $CO_2$, relative to the present $CO_2$ value, are described here. In Figure 6 we show the variation in temperature change with latitude for 0.5, 2, and 20 times present $CO_2$. Note that enhanced $CO_2$ causes warming to amplify in polar regions. This occurs in part from ice albedo feedback, but also in part from the increased effect of meridional transport of latent heat with warmer climate. This is discussed in more detail by Flannery [1984]. Note that $CO_2$-induced changes are highly non-linear. Increasing $CO_2$ concentrations by a factor of 20 only enhances warming by a factor of 4 (not 10) over warming induced by doubling $CO_2$. However, our neglect of possible changes in lapse rate could be seriously in error for such large changes in $CO_2$.

- In Figure 7 we show the variation of ice line (TS = -2°C), the equator-to-pole temperature difference, and the mean sea surface temperature as a function of $CO_2$ concentration. Note that sea ice disappears when $CO_2$ increases about 20 times, and that the temperature difference between equator and pole decreases monotonically

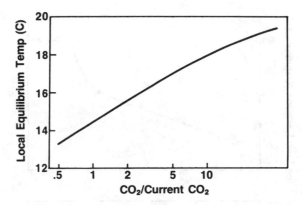

Fig. 5. Sensitivity of radiative model to changes in atmospheric $CO_2$ estimated by considering surface temperature at 30° latitude calculated by equating flux out of top of atmosphere to net insolation (local radiative equilibrium).

with enhanced $CO_2$ and warmer mean temperature. In Figure 7c we also indicate, by tick marks, estimates for the mean sea surface temperature during the Wisconsin Ice Age 18 kyr B.P., the Cretaceous 100 Myr B.P., and a measured modern value [Crowley, 1983; Barron et al., 1981].

As a final result in Figure 8 we show the distribution of sea surface temperature with latitude for a model with 20 x $CO_2$, compared with data for the Cretaceous [Barron et al., 1981].

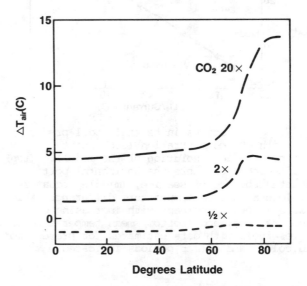

Fig. 6. Variation of atmospheric temperature with latitude for atmospheric $CO_2$ concentration of 1/2, 2, and 20 times present value (330 ppm) as calculated by energy balance model. Polar amplification is due to a combination of ice albedo feedback and poleward transport of latent heat.

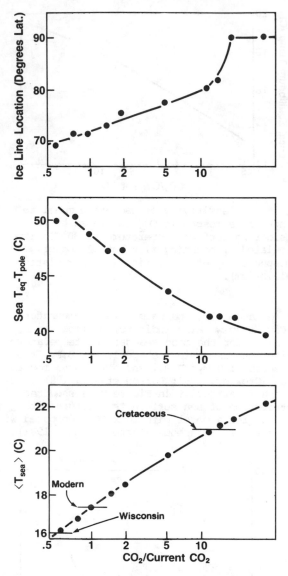

Fig. 7.    Variations in several model predictions with $CO_2$ concentration (each point corresponding to a solution of the model): (top) ice line location where the horizontal portion indicates absence of sea ice, (middle) equator-to-pole sea surface temperature difference, which shows a monotone decrease with increasing $CO_2$ concentrations, and (bottom) mean temperature of sea reservoir with data estimates for present and both colder and warmer periods indicated.

Note that the Cretaceous data appear to indicate that the equator-to-pole temperature change was even smaller than predicted by the model. Thus, Cretaceous temperatures are not only warmer but also more isothermal than modern temperatures.

The dashed line labeled Tec corresponds to a model where we impose a prescribed variation in equatorial temperature with changed $CO_2$. Tec stands for Tropical Evaporation Constraint. Depending on the required temperature at the equator, poleward transport must change in a way that enhances or limits heat removal. In the model this means that the diffusion coefficients $D_S$, $D_A$ must vary with forcing. As discussed by Hoffert et al. [1983] for a simpler class of EBM's, solutions are possible for a range of values of mean temperature. For the particular model shown here we achieved a temperature distribution closer to the observational data by requiring the diffusion coefficients to increase, from 0.22 to 0.40 W m$^{-2}$ ($^{\circ}$C)$^{-1}$ in air and from 0.55 to 0.57 W m$^{-2}$ ($^{\circ}$C)$^{-1}$ in sea. This suggests that changes in circulation patterns might also be required to explain the observed tendency toward isothermality in the Cretaceous.

## 5.    Conclusions

Our EBM results suggest that $CO_2$ variations of the magnitude suggested in recent discussions of paleoclimate would contribute significantly to the differences between ancient and modern climates. However, in the case of the ice ages, $CO_2$ forcing, alone, seems to account for only about half the measured temperature change. Measurements by Neftel et al. [1982] show that atmospheric $CO_2$ during the ice age was about 70% of the modern, preindustrial value (190 ppm com-

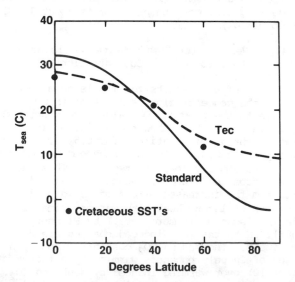

Fig. 8.    Latitudinal dependence of sea surface temperature for two versions of the model compared with proxy Cretaceous data [Barron et al., 1981]. The TEC model, which constrains equatorial temperatures and involves $CO_2$-concentration dependent diffusion coefficients, produces a more isothermal climate as indicated by the data. The standard model, where diffusion coefficients are constant, yields a larger equator-to-pole gradient.

pared to 270 ppm). Figure 7c shows that $CO_2$ would need to decrease to about 40% of the pre-industrial value to account for the "known" temperature decrease. However, we know that other important climatological differences might also influence ice age temperatures. In particular, the role of Milankovitch cycles should be investigated using a seasonal model. If the altered seasonal insolation affected the extent of ice cover, the additional influence of ice albedo feedback might account for the remaining cooling.

Similarly, although values for the $CO_2$ concentration in the Cretaceous are tentative, if $CO_2$ rose by a factor of 20 or more, then temperature does rise much of the way toward inferred levels in the Cretaceous. However, again there are significant additional changes that cannot be ascribed to $CO_2$ alone. The pole-to-equator temperature difference seems to be too small to be explained without additional mechanisms. In the Cretaceous we know that the distribution of land and sea differed from modern times. The absence of land or ice at the poles does tend to cause a shallower global temperature gradient [Barron and Washington, this volume].

Acknowledgments. The authors want to thank W. Wiscombe of NYU for his assistance in modifying LOWTRAN5 to more accurately treat the $CO_2$ effect.

References

Barron, E. J., S. L. Thompson, and S. H. Schneider, An ice-free Cretaceous?, results from climate model simulations, Science, 212, 501-508, 1981.

Barron, E. J., and W. M. Washington, Warm Cretaceous Climates: High atmospheric $CO_2$ as a plausible mechanism, this volume.

Berner, R. A., A. C. Lasagna, and R. M. Garrels, The carbonate-silicate geochemical cycle and its effect on atmospheric carbon dioxide over the past 100 million years, Am. J. Sci., 283, 641-683, 1983.

Budyko, M. I., The effect of solar radiation variations on the climate of the earth, Tellus, 21, 616-619, 1969.

Cess, R. D., and J. C. Wronka, Ice ages and the Milankovich theory: A study of interactive climate feedback mechanisms, Tellus, 31, 185-192, 1979.

Coakley, J. A., Jr., R. D. Cess, and F. M. Yurevich, The effect of tropospheric aerosols on the earth's radiation budget: A parameterization for climate models, J. Atmos. Sci. 40, 116-138, 1983.

Crowley, T. J., The geologic record of climatic change, Rev. of Geophys. Space Phys., 21, 828-877, 1983.

Flannery, B. P., Energy balance models incorporating transport of thermal and latent energy, J. Atmos. Sci., 41, 414-421, 1984.

Gal-Chen, T., and S. H. Schneider, Energy balance climate modeling: Comparison of radiative and dynamic feedback mechanisms, Tellus, 28, 108-121, 1976.

Hoffert, M. I., B. P. Flannery, A. J. Callegari, C. T. Hsieh and W. Wiscombe, Evaporation-limited tropical temperatures as a constraint on climate sensitivity, J. Atmos. Sci., 40, 1659-1668, 1983.

Kneizys, F. X., E. P. Shettle, W. O. Gallery, J. H. Chetwynd, Jr., L. W. Abrue, J. E. A. Selby, R. W. Fenn, and R. A. McClatchy, Atmospheric transmittance/radiance: Computer code LOWTRAN5, Rep. AFGL-TR-80-0067, Envir. Res. Pap., 697, Air Force Geophys. Lab., Hanscom AFB, Bedford, Mass., 1980.

Lasaga, A. C., R. A. Berner, and R. M. Garrels, An improved geochemical model of atmospheric $CO_2$ fluctuations over the past 100 million years, this volume.

Lindzen, R. S., and B. F. Farrell, The role of polar regions in global climate and a new parameterization of global heat transport, Mon. Weather Rev., 108, 2064-2079, 1980.

Manabe, S., and R. T. Wetherald, On the distribution of climate change resulting from an increase in the $CO_2$ content of the atmosphere, J. Atmos. Sci. 37, 99-118, 1980.

Neftel, A., H. Oeschger, J. Schwandor, B. Stauffer, and R. Zumbrun, Ice core sample measurements give atmospheric $CO_2$ content during the past 40,000 years, Nature, 295, 220-223, 1982.

North, G. R., R. F. Cahalan, and J. A. Coakley, Energy balance climate models, Rev. Geophys. Space Phys., 19, 91-121, 1981.

Oeschger, H., B. Stauffer, R. C. Finkel, and C. C. Langway, Jr., Variations of the $CO_2$ Concentration of Occluded Air and of Anions and Dust in Polar Ice Cores, this volume.

Oort, A. H., Global atmospheric circulation statistics, 1958-1973, NOAA Prof. Pap. 14, Nat. Oceanic and Atmos. Admin., Washington, D.C., 1983.

Owen, T., R. D. Cess and V. Ramanathan, Early earth: An enhanced carbon dioxide greenhouse to compensate for reduced solar luminosity, Nature, 277, 640-642, 1979.

Ramanathan, V., The role of ocean-atmosphere interactions in the $CO_2$ climate problem, J. Atmos. Sci., 38, 918-930, 1981.

Sellers, W. D., A global climate model based on the energy balance of the earth-atmosphere system, J. Appl. Meteorol., 8, 329-400, 1969.

Stone, P. H., Constraints on dynamical transports of energy on a spherical planet, Dyn. Atmos. Oceans, 2, 123-139, 1978.

Thompson, S. L., The faint early sun-climate paradox: Role of atmospheric carbon dioxide, paper presented at Chapman Conference on Natural Variations in Carbon Dioxide and the Carbon Cycle, AGU, Tarpon Springs, Fla, Jan. 9-13, 1984.

# ORGANIC CARBON PRESERVATION IN MARINE SEDIMENTS

Steven Emerson

School of Oceanography, University of Washington, Seattle, Washington 98195

Abstract. Mechanisms that control the preservation of organic carbon in the ocean are the key to interpreting the paleoceanographic carbon signal. This process is investigated here utilizing published data from sediment trap experiments in a model that couples organic carbon and oxygen reactions in marine sediments. The carbon content of sediments is most strongly influenced by the particulate organic carbon flux to the bottom, the bottom water oxygen content, and the organic matter degradation rate constant. The familiar correlation between sediment organic carbon content and sedimentation rate is likely a result of the direct relationship between the total particulate flux and organic carbon flux rather than a preservation effect of rapid sedimentation. The data indicate that the degradation rate "constant" for organic material oxidation in marine sediments increases with increasing particulate organic carbon flux to the bottom. The positive correlation of these parameters enhances the relative importance of the bottom water oxygen concentration to organic matter preservation. Cycles of the carbon concentration over the past 100,000 years in marine sediments may be influenced by changes in bottom oxygen concentration. Recent arguments that organic carbon burial on the continental slope is an effective sink for anthropogenic fossil fuel $CO_2$ require 50% of the particulate algal carbon raining from the photic zone of the continental shelves to escape degradation and undergo burial. Calculations presented here suggest that the organic matter degradation rate is much too rapid for this amount to be buried and that the potential sink is overestimated by roughly a factor of 5.

## Introduction

The vast majority of marine sediments have a carbon content that is between 0.2 and 2%. This is observed over a large spectrum of environmental conditions: in water depths ranging from a few hundred meters to abyssal depths, in regions where the surface water primary productivity varies from that characteristic of upwelling to values typical of central gyres, in locations where bottom water oxygen concentrations range from near saturation to almost one-tenth this value, and in sedimentation regimes with accumulation rates varying from 0.1 to 30 cm per 1000 years. In this paper I examine the mechanisms of organic carbon preservation with the goal of demonstrating the controlling processes and the reason for the relatively small variability in concentration. The procedure is to adapt a model that describes the relationship between sedimentary carbon and pore water oxygen concentrations [see Jahnke et al., 1982; Emerson et al., 1984] to a wide range of marine carbon conditions. The model is fueled with literature data that include sediment trap organic carbon fluxes, bottom water oxygen concentration, sediment carbon content, and the associated sediment accumulation and bioturbation rates. The sensitivity of organic carbon content to model variables is evaluated based on the data, and then some educated guesses are forwarded concerning the implications of these results with regard to the process of organic carbon preservation in the oceans, present and past. I shall attempt to determine the reason for the correlations between sedimentation rate and organic carbon content (and organic matter degradation rate) noted in the literature, the value of sedimentary organic carbon content as a paleoproductivity indicator and other potential paleontological interpretations, and the likelihood for organic matter in surface sediments on the continental shelf and slope to be a sink for anthropogenic fossil fuel carbon.

## The Model

The model that I utilize is a somewhat simplified analytical version of the numerical algorithm presented by Jahnke et al. [1982]. It is identical to that used by Emerson et al. [1984] except here I include the sedimentation rate term. In the latter paper the process of sedimentation was deleted because it plays a minor role in deep-sea carbon preservation; however, this assumption is inadequate for application of the model to carbon fluxes on the continental margins. The previous tests of the model, referenced above, support its accuracy in the description of pore water-sediment carbon chemistry in marine sediments.

Assumptions incorporated into formulation of the terms depicting biological mixing and organic matter degradation, and arguments for and against carbon degradation within the sediment, are explained in the paper by Emerson et al. [1984]. One of the most serious model simplifications for the purpose here is that I assume oxygen is the most important electron acceptor for carbon degradation and that the other oxidizing agents ($NO_3^-$, $MnO_2$, $FeO(OH)$, and $SO_4^{2-}$) can be neglected. In deep-sea sediments where sulfate reduction is not important it has been shown that $NO_3^-$, $MnO_2$, and $FeO(OH)$ reduction account for only a few percent of the carbon degradation [Bender and Heggie, 1984; Jahnke et al., 1982]. However, this has not been argued in continental slope and borderland regions where sulfate reduction can occur very near the sediment-water interface. In the following I suggest that the model should be applicable in regions where the bottom waters have at least 30 μmol/kg oxygen year-round and where the sedimentation rate does not exceed 30 cm per 1000 years (1000 years = 1 kyr). Berner [1980a] has shown that the interface sulfate pore water gradients are empirically correlated with sedimentation rate in such a way that a value of 30 cm/kyr would imply a gradient, $dSO_4^{2-}/dz$), equal to $3.0 \times 10^{-8}$ moles $cm^{-4}$. Using a $SO_4^{2-}$ diffusion rate corrected for porosity and tortuosity of $2 \times 10^{-6}$ at 2°C, the corresponding sulfate flux into the sediments would be $6.0 \times 10^{-14}$ moles $SO_4 cm^{-2} s^{-1}$, which is equivalent to a predicted carbon utilization of $\sim 12 \times 10^{14}$ moles C $cm^{-2} s^{-1}$ assuming a stoichiometric reaction with "Redfield" organic matter. Oxygen gradients at the interface of the sediments of San Clemente Basin of the California borderlands (a location that approaches the 30 μmolar, 30 cm/kyr limits (see Table 1) measured with microelectrodes indicate that the $O_2$ gradient reaches near zero a few millimeters below the interface (C. Reimers, personal communication, 1984). Thus, a minimum $O_2$ flux (assuming a bottom water value of 30 μmoles/ℓ) would be $6 \times 10^{-6} \times (30 \times 10^{-9}/0.3) = 60 \times 10^{-14}$ $\times 106/130 = 49 \times 10^{-14}$, or about 4 times the carbon utilization by the sulfate reduction. Furthermore, in the model analysis we use carbon measurements from the surface sediments which should have been under the influence of only a very small fraction of the depth-integrated sulfate reduction.

While it is clear that, to a first approximation, the dynamics of oxygen reduction are more important than those of the secondary oxidants ($NO_3$, Mn(IV), and $SO_4^{2-}$) in determining the rate of organic carbon preservation in the environments considered here, we should not lose sight of the fact that these electron acceptors could play an important role in this process.

Since the preservation rate of carbon in the open ocean is usually less than 10% of the particulate flux to the bottom [Suess and Müller, 1980], oxidants that degrade a relatively small fraction of the carbon may be a factor in determining the concentration buried in the sediments [Bender and Heggie, 1984]. There are two observations which lead me to believe that oxidation by oxygen is the dominant reaction controlling carbon preservation in locations considered here. In an analysis of the carbon content of equatorial Pacific sediments where the particulate organic carbon rain rate to the sediments is roughly equal (sites C, H and M of the manganese nodule project, MANOP, Emerson et al., 1984), the location with highest bottom water oxygen content (site C) had a much lower carbon content. Furthermore, there are dramatic carbon cycles on the time scale of glacial periods preserved in deep sea sediments with abundant sulfate in the pore waters (for example, MANOP site H, M. Lyle and S. Emerson, unpublished results, 1984). It is unlikely that the reactivity of this carbon, which in all cases has been degraded to the point where it represents only a few percent of that which left the euphotic zone, could be very different between the glacial stages. If, in spite of these arguments, sulfate reduction becomes more important in determining the nearsurface carbon content as bottom water oxygen decreases, then the effect of oxygen on carbon preservation will be overestimated in this model.

Another potential problem of applying the deep-sea model to the nearshore environment is the influence of relatively nondegradable terrestrial organic matter near land [Hedges and Parker, 1976]. In the model we assume that all of the carbon exceeding a concentration of a few tenths of one percent has a planktonic origin and is homogeneously degradable. Hedges and Mann [1979] showed that on the Washington continental shelf, terrestrial organic carbon becomes a small fraction (<10%) of the total beyond 50 km from shore, which is a justification for our assumption with regard to the continental slope locations but not in the nearshore borderland basins. If a significant fraction of the carbon preserved in the borderland environments is of a terrestrial origin, it implies that the "degradable" carbon content of the sediments in these locations is even smaller than that indicated here. Thus, the sedimentary carbon concentrations in these locations should be considered maxima for the purpose of this discussion.

The model begins with the steady state, constant porosity equation for the "degradable" carbon, C (moles $cm^{-3}_{bulk}$), distribution with depth z (cm, positive downward) in the sediments:

$$0 = K\frac{d^2C}{dz^2} - s\frac{dc}{dz} - kC \qquad (1)$$

$$-K\frac{dc}{dz}\Big)_0 + sC = F_c \qquad \text{at } z = 0$$

$$\frac{dc}{dz}\Big)_{\bar{z}} = 0 \qquad \text{at } z = \bar{z} \tag{2}$$

where $K$ ($cm^2_{bulk}$ $s^{-1}$) is the bioturbation coefficient, $k$ ($s^{-1}$) is the first-order organic matter degradation rate constant, $s$ ($cm\ s^{-1}$) is the sedimentation rate, $F_c$ (moles $cm^{-2}$ $s^{-1}$) is the particulate organic carbon flux to the bottom, and $\bar{z}$ is the depth in the sediments at which oxygen reaches zero. The solution to this equation relates the carbon content in the sediments to the values for $k$, $K$, $F_c$, and $s$:

$$C = \frac{F_c}{k}\left[\frac{r_1 e^{-(r_2 z - r_1 \bar{z})} + r_2 e^{(r_1 z - r_2 \bar{z})}}{(\frac{r_1}{r_2})e^{r_1 \bar{z}} - (\frac{r_2}{r_1})e^{-r_2 \bar{z}}}\right] \tag{3}$$

at $z = \bar{z}$

$$C_{\bar{z}} = F_c\left[\frac{\frac{r_1 + r_2}{k}e^{(r_1 - r_2)\bar{z}}}{(\frac{r_1}{r_2})e^{r_1 \bar{z}} - (\frac{r_2}{r_1})e^{-r_2 \bar{z}}}\right] \tag{4}$$

with

$$r_1 = s/2K + (\frac{s^2}{4K^2} + \frac{k}{K})^{1/2} \tag{5}$$

$$r_2 = -s/2K + (\frac{s^2}{4K_2} + \frac{k}{K})^{1/2}$$

The equivalent of $C_{\bar{z}}$ (m $cm^{-3}_{bulk}$) in units of $g/cm^3_s$ is derived from the porosity $\phi$ and the molecular weight of carbon, $M$; $C^+ = C_{\bar{z}}(M)/(1 - \phi)$. The relationship between the measured carbon concentration $\bar{C}$ ($g_C/g_s$) and $C^+$ is

$$\bar{C} = C^+/[C^+\alpha + (1-(C^+\alpha/\rho_{o.m.})\rho_m]$$

where $\alpha$ is the concentration of carbon in organic matter (assuming a "Redfield" composition for the organic material, $\alpha = 2.7$ g organic matter/g C), and $\rho_{o.m.}$ and $\rho_m$ (1.1 and 2.7, respectively) are the densities of organic matter and the mineral fraction of the sediments.

The equation for oxygen, $O_2$ (moles/$cm^3_{porewater}$) is

$$0 = \phi D\frac{d^2 O}{dz^2} - \gamma k C \tag{6}$$

$$(O_2)_0 = O_2^o \qquad \text{at } z = 0$$

$$(O_2)_{\bar{z}} = 0 \qquad \text{at } z = \bar{z} \tag{7}$$

where $\gamma$ is the assumed "Redfield" ratio between moles oxygen reduced to moles carbon oxidized ($\gamma = 1.3$ mol $O_2$/mol C). Finally, the carbon oxygen mass balance in the region $z = 0$ to $\bar{z}$ is

$$k\int_{z=0}^{z=\bar{z}} Cdz = (1/\gamma)F_{O_2} \tag{8}$$

By combining equations (3), (6), (7), and (8) we arrive at an expression for the relationship between oxygen concentration in the bottom waters, $O_2^o$ (moles $cm^{-3}$), particulate organic carbon flux $F_c$, and the $O_2$ zero depth, $\bar{z}$:

$$\frac{\phi O_2^o D_{O_2}}{F_c} =$$

$$\frac{e^{-r_1 \bar{z}}(\frac{r_2}{r_1})\left[\bar{z}\,e^{r_1 \bar{z}} - (1/r_1)e^{r_1 \bar{z}} + 1/r_1\right]}{(\frac{r_1}{r_2})e^{r_1 \bar{z}} - (\frac{r_2}{r_1})e^{-r_2 \bar{z}}}$$

$$-\frac{e^{r_1 \bar{z}}(\frac{r_1}{r_2})\left[\bar{z}\,e^{-r_2 \bar{z}} + \frac{e^{-r_2 \bar{z}}}{r_2} - \frac{1}{r_2}\right]}{(\frac{r_1}{r_2})e^{r_1 \bar{z}} - (\frac{r_2}{r_2})e^{-r_2 \bar{z}}} \tag{9}$$

where $D_{O_2}$ is the diffusion coefficient for oxygen at 1°C corrected for tortuosity ($6 \times 10^{-6}$ $cm^2$ $s^{-1}$). Since $O_2^o$, $F_c$, $s$, $K$, and $C$ are known from the data, the variables $\bar{z}$ and $k$ can be derived from equations (4) and (9).

## Results

### The Data

The data used in the model and their source are presented in Table 1. The carbon fluxes, $F_c$, are tabulated for deep sediment trap moorings only. Yearlong deployments are favored in the table because of the observed seasonal cyclicity of the flux [Deuser and Ross, 1980]; however, in the cases of the northwest Atlantic [Rowe and Gardner, 1979; Hinga et al, 1979] this criterion was not met. The associated information in the table ($O_2$, $C$, and $s$ data) was derived from the references with the exception of some of the sedimentation rates which were taken from the sources cited in the footnotes. The borderlands basin information is from the tabulation of Smith and Hinga [1983], who cite A. Soutar, K. Bruland

TABLE 1.    Particulate Organic Carbon Fluxes and Associated Data From
Different Locations in the Oceans

| Location | Depth, | $F_c$, | $O_2$, | s, | $\bar{C}$, | $\dfrac{s\bar{C}}{F_c} \times 100$ |
|---|---|---|---|---|---|---|
| | m | $\mu$molC cm$^{-2}$ yr$^{-1}$ | $\mu$mol $\ell^{-1}$ | cm kyr$^{-1}$ | % | % Burial |
| E equatorial Pacific [Emerson et al., 1984] | | | | | | |
| MANOP site H | 3500 | 12 | 110 | 0.5 | 0.6 | 0.8 |
| MANOP site M | 3100 | 14 | 110 | 1.0 | 1.2 | 3 |
| E equatorial Pacific [Cobler and Dymond, 1980] | 2690 | 19 | 110 | 5 | 1.0 | 8 |
| NW Atlantic [Deuser and Ross, 1980] | 5000-5200 | 6 | 272 | (2)[b] | 0.5 | 6 |
| NW Atlantic, Rise [Rowe and Gardner, 1979] | 3650 | 35 | 291 | 7 | 1.3 | 8 |
| NW Atlantic, slope [Hinga et al., 1979] | 1345 | 45 | 246 | 7 | 1.6 | 8 |
| Patton Escarpment | 3815 | 82[a] | 172 | 10-30[d] | 1-2 | 11 |
| San Clemente Basin | 1800 | 82 | 57 | 25[c] | 5 | 57 |
| Santa Catalina Basin [Smith and Hinga, 1983] | 1300 | 82 | 18 | 20[c] | 6 | 45 |
| San Diego Trough [Smith and Hinga, 1983] | 1230 | 82 | 31 | 10-20[c] | 3 | 34 |

[a] The carbon flux is assumed to be the same as that in the other borderland basins (see text).
[b] From Broecker and Peng [1982].
[c] From Emery [1960].
[d] Assumed to be the same as that for the borderland basins.

and W. T. Reed (unpublished data) for the sediment trap results. Smith and Hinga [1983] show that the benthic community respiration at the Patton Escarpment and in the San Diego Trough and Santa Catalina Basin are different by only about 10%, supporting the assumption of consistency of the rain rate of particulate organic matter to the bottom in this region. The organic carbon flux to the sediments of San Clemente Basin is assumed to be the same as that for the other borderland basins. Table 1 is not intended to be a complete review of sediment trap experiments which have the appropriate ancillary data, but rather a tabulation of results that are representatiave of a wide range of oceanic conditions. The carbon content of the sediments in the last three locations, those that are within 100 km of shore, may include a substantial portion of recalcitrant terrestrial organic matter.

## Factors Controlling Carbon Preservation

The sedimentation rate s and bioturation rate K. Emerson et al. [1984] suggested that sedimentation rates in the range of a few tenths to 1 cm/kyr played very little role in the preservation of carbon in the deep-sea. In

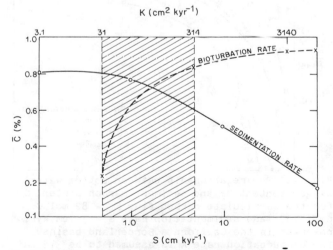

Fig. 1. The carbon content of sediments, $\bar{C}$, predicted by the model as a function of sedimentation rate s and bioturbation rate K. The hatched area is the range of s and K observed in deep-sea sediments. Other model parameters are those observed for the eastern equatorial Pacific: $F_c$ = 10 $\mu$mol C cm$^{-2}$ yr$^{-1}$; $O_2^0$ = 110 $\mu$m; k (the organic matter degradation rate constant) = 4 x 10$^{-1}$ s$^{-1}$.

Figure 1, percent carbon is plotted against sedimentation rate for a constant particulate organic carbon flux. The model-predicted role of the sedimentation rate is indeed minor for carbon fluxes and sedimentation rates typical of the deep-sea. Only at sedimentation rates approaching 10 cm/kyr does the value of s significantly affect carbon preservation. Notice that model-derived carbon concentrations decrease with increasing sedimentation rate because of the dilution by the noncarbon fraction. The ramification of this result with regard to observed trends is discussed later.

The influence of bioturbation is far more dramatic than that of sedimentation, particularly in the region of lower bioturbation rates (Figure 1). A typical deep-sea value for K is 150 cm²/kyr [see Cochran, 1984], but nearshore values can be more than 100 times this rate [Carpenter et al., 1984]. The model-derived carbon content increases with bioturbation rate and is influenced most in the lower range of bioturbation, with the effect diminishing as K increases. In fact, the greatest effect is at bioturbation rates lower than those most commonly observed. The reason for the increase in carbon content with increasing K is that bioturbation introduces carbon to deeper portions of the sediment column where it more effectively

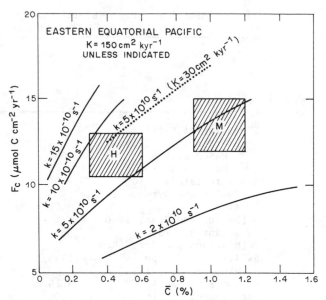

Fig. 3. The variation in the percent carbon preserved, $\bar{C}$, as a function of the particulate organic carbon rain rate $F_C$ for different organic carbon degradation rate constants k. The conditions for the solutions were $O_2^0 = 110$ µm kg⁻¹, s = 1 cm kyr⁻¹, and K = 150 cm² kyr⁻¹ except where indicated. The hatched fields are those occupied by MANOP sites M and H (Table 1). (Modified from Emerson et al. [1984].

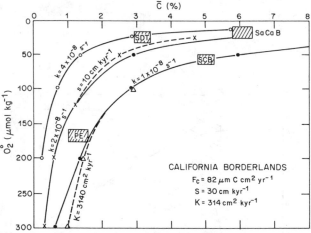

Fig. 2. The relationship between bottom water oxygen content $O_2^0$ and sediment carbon content $\bar{C}$ for the particulate carbon flux $F_C$ = 82 mol C cm⁻² yr⁻¹ and sedimentation rate $\bar{s}$ = 30 cm kyr, observed in the California Borderland basins. The bioturbation rate K is assumed to be 314 cm² kyr⁻¹. A range for the organic matter degradation rate k of 1-4 x 10⁻⁸ s⁻¹ brackets the measurements (Table 1) which are denoted by the symbols. SaCaB, Santa Catalina Basin; SCB, San Clemente Basin; SDT, San Diego Trough; PE, Patten Escarpment. The effects of deceasing the sedimentation rate s to 10 cm kyr⁻¹ and increasing the bioturbation rate K to 3140 cm² kyr⁻¹ are shown by the dashed lines.

depletes the pore waters of oxygen and thus enhances carbon preservation. As the oxygen zero level, z, becomes shallower, the effect of increasing the bioturbation rate diminishes. (One should bear in mind that the trends here are those predicted for a constant particulate carbon rain rate. In the environment both the sedimentation rate and bioturbation rate may be a function of this value.)

The model predicts that the most dramatic effects of the sedimentation and bioturbation rates on the carbon concentration occur at opposite extremes of the range of observed values in the ocean. We will see later that this has the effect of rendering the carbon content relatively insensitive to these parameters except for very rapid sedimentation or low bioturbation rates.

The oxygen boundary value $O_2^0$. The carbon content of sediments is dependent on the bottom water oxygen concentration because the flux of oxygen into the sediment is directly proportional to this value. Decreasing the bottom water oxygen concentration decreases the depth to which oxygen penetrates the sediments, thus increasing the probability that carbon will escape oxidation by $O_2$. Arguments for the inefficiency of alternative electron acceptors in the environments considered here were presented in the previous section. One of the most difficult problems in interpreting the factors controlling the carbon content of sediments is separating the influence of the flux of carbon to the bottom and the concentration of oxygen in the overlying waters. I attempt to make this distinction by examining the carbon content of sediments of the California Borderlands (Table 1), where there is a large range in bottom water oxygen but a fairly constant rain rate of particulate organic matter. In Figure 2, model-derived curves for the variation of organic carbon as a function of bottom water oxygen content are compared with the borderland data (from Table 1). The measurements are bracketed by a narrow range for the organic matter degradation rate constant, $k = 1-4 \times 10^{-8}$ $s^{-1}$, and sedimentation rate and bioturbation rate uncertainties are unimportant (I assume a minimum value for K of 314 $cm^2 kyr^{-1}$ in these nearshore waters).

The results in Figure 2 appear to be convincing evidence for the effect of oxygen on carbon preservation and for consistency between the data and the model; however, the uncertainty of the fraction of terrestrially derived organic carbon in the sediments of these basins is a caveat for adopting the trends totally at face value. Since there are no major rivers entering the ocean in this region, it may not be a serious problem. However, if the degradable marine-derived organic carbon content is less than half of the total sedimentary organic carbon, then the trends would be less dramatic, and the effect of oxygen less convincing. It is clear from the figure, though, that the organic matter degradation rate k required to fit the data cannot be much greater than $4 \times 10^{-8}$ $s^{-1}$.

The particulate carbon flux F and the organic matter degradation rate k. Emerson et al. [1984] determined the organic matter degradation rate constant necessary to explain the preservation of carbon at MANOP sites in the equatorial Pacific. Figure 3 is modified from their paper. The value of C is plotted as a function of the particulate organic carbon flux $F_c$ for the appropriate oxygen bottom water value and bioturbation rate constant. A fairly narrow range for the organic matter degradation rate constant, $k = 2-8 \times 10^{-10}$ $s^{-1}$, is required to explain the data at sites M and H. This value is about 100 times less than that derived for the California Borderlands in Figure 2. Interpretations similar to those presented in Figures 2 and 3 were carried through for the rest of the data in Table 1. The resulting organic matter degradation constants are presented in Table 2 and Figure 4 along with the value determined by Grundmanis and Murray [1982] for oxygenated west equatorial Pacific sediments. There is a clear trend of increasing degradation rate constant with increasing rain rate of organic matter. This is consistent with the concept of multiple rate constants for degrading organic matter suggested by Jorgensen [1979] and Berner [1980b]

and tested by Westrich and Berner [1984]. The degradation rate constant for fresh Long Island Sound plankton determined in the last of these references is included in Table 2. The observed trend is precisely the reason that the organic carbon content of sediments does not vary as strongly as the variation in organic carbon flux to the ocean bottom.

Discussion

Correlations Between Organic Carbon Content and Sedimentation Rate

The relationship between organic carbon content of marine sediments and sedimentation rate has been frequently observed [e.g., Heath et al., 1977; Müller and Suess, 1979], and it is often suggested that rapid sedimentation provides a mechanism for enhancing the preservation of organic carbon in the marine environment. Given the organic matter degradation rates observed in Table 2, it hardly seems feasible that this is the real mechanism. The predicted depth in the sediment to the oxygen zero level for the cases listed in Table 2 is 0.2 cm to 3 cm. With these values and the measured sedimentation rates (Table 1), one can readily see that the residence time for organic carbon with respect to degradation in the sediments (Table 2) is several orders of magnitude shorter than the equivalent value for transport through the oxygenated zone by burial. There is no reason then to expect that the sedimentation rate would play any role other than that of diluting the organic carbon concentration with nonorganic material as the rate increases. The decrease in C with more rapid sedimentation rate is indeed the model prediction for a constant carbon rain rate (Figure 1). A more likely explanation for the

TABLE 2. Organic Matter Degradation Rate Constant k and Residence Time τ at Different Locations in the Oceans

| Location | k, $S^{-1} \times 10^{10}$ | τ years |
|---|---|---|
| W equatorial Pacific[a] | 0.5-4 | 80-630 |
| E equatorial Pacific | 2-8 | 40-160 |
| W atlantic continental slope | 10-30 | 11-32 |
| California Borderlands | 100-400 | 0.8-3.2 |
| Fresh plankton[b] (laboratory study) | 5700 | 0.06 (21d) |

[a]Degradation rates determined by Grundmanis and Murray [1982] using $O_2$ profiles. The value for k has been recalculated for a bioturbation rate of 150 $cm^2$ $kyr^{-1}$ to be consistent.

[b]Westrich and Berner [1984].

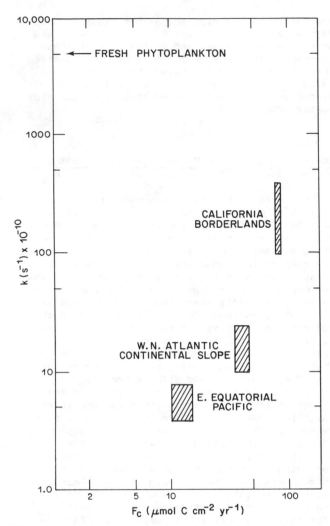

Fig. 4. The organic carbon degradation rate constant k determined by the model as a function of the particulate organic carbon flux $F_c$, measured by sediment traps for three locations in the ocean (see Tables 1 and 2).

The proportionality between organic carbon content and sedimentation rate observed by Müller and Suess [1979] for northwest Africa, the Oregon coast, and equatorial Pacific is also indicated in the figure. The reason for the offset between the model-derived and measured values has to do with the magnitude of the particulate carbon flux used to generate the curves and the content of refractory carbon in the sediments, both of which are unknown. The importance of the comparison is in the trends rather than the absolute values.

The explanation proposed here for the observed correlation between sedimentary carbon content and sedimentation rate is consistent with correlations between the organic matter degradation rate constant, both oxic and anoxic, and sedimenation rate [Toth and Lerman, 1977; Reimers and Suess, 1983]. If the rate of sediment accumulation is a measure of organic carbon flux to the sediments, then according to the relationship between carbon flux and rate constant k in Figure 4, there must be a correlation between sedimentation rate and k.

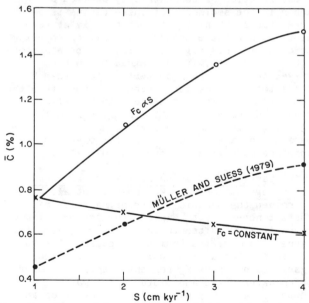

Fig. 5. The model-predicted relationship between sediment carbon content $\bar{C}$ and sedimentation rate s (solid lines). The line labeled "$F_c$ = constant" is the trend predicted if the carbon rain rate is constant. The line labeled "$F_c \propto s$" illustrates the result if $F_c$ increases at the same rate as s; i.e., a doubling of s causes a doubling of $F_c$. The dashed line is the trend observed by Müller and Suess [1979] for the equatorial Pacific, Argentine Basin, and Oregon coast. The model-derived curves are for the conditions $O_2^o$ = 110 μm/kg, K = 150 cm² s⁻¹, $F_c$ = 10 μmC cm⁻² yr, and s = 1 cm kyr⁻¹. The significance of the comparison between the model curves and the data from Müller and Suess is in the change of $\bar{C}$ as a function of s rather than the absolute values (see text).

observed increase in sedimentary carbon content with increasing sedimentation rate is the concomitant increase in the rain rate of organic matter to the sediments. Since the observed fraction of organic carbon reaching bottom sediment traps in the deep ocean appears to be rather consistent at about 5-10% regardless of the total flux [e.g., Honjo, 1980; Fisher, 1983; Deuser and Ross, 1980], it is reasonable to assume a direct relationship between particulate organic carbon flux and total sediment accumulation rate. The exact value of this relationship is, of course, unknown and probably not constant. The relationship between organic carbon content of sediments and sedimentation rate for carbon rain that increases at the same rate as sedimentation is presented in Figure 5.

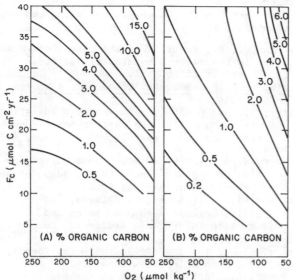

Fig. 6.    Lines of constant carbon content as a function of particulate organic carbon flux F and bottom water oxygen concentration $O_2^=$. Model solutions were derived using a sedimentation rate of 1 cm kyr$^{-1}$. (a) The solutions for a constant organic matter degradation rate of 6 x 10$^{-10}$ s$^{-1}$ (the value for the eastern equatorial Pacific (Table 2). (b) The solution for an organic carbon degradation rate constant that increases with $F_c$ in the manner indicated in Figure 4 and Table 2.

## Interpretation of Carbon Cycles in Deep-Sea Sediments

From the discussion in the results section it is clear that the most important factors controlling the preservation of organic carbon in deep-sea sediments are the carbon flux to the bottom, the overlying oxygen content, and the degradation rate constant. Variation of the predicted carbon concentration in sediments as a function of the particulate organic carbon flux and bottom water $O_2$ concentration is presented in Figure 6. The difference between Figure 6a and Figure 6b illustrates the effect of constant and variable degradation rate constants on the carbon content. The dependency of the preserved carbon content on the input flux is roughly cut in half by specifying that the degradation rate constant increase with $F_c$ in the manner suggested in Figure 4. The conditions for Figure 6b are those observed for the equatorial Pacific and the western Atlantic continental slope; to derive the trend for carbon fluxes greater than 40μm cm$^{-2}$ yr$^{-1}$ would require the incorporation of rate constants observed for the California Borderlands, and the trend in the constant carbon lines would become even more vertical. Using conditions similar to those observed today in the eastern equatorial Pacific ($O_2^=$ = 112 μM, $F_c$ ~ 15

μmolC cm$^{-2}$ yr and $\bar{C}$ ~ 1%) as a point of reference, both a doubling of the carbon flux and a decrease in the bottom water oxygen content from 100 to 50 μM/kg would increase the carbon content by about a factor of two.

Müller and Suess [1979] derived an empirical relationship between present-day estimates of productivity, sedimentation rate, and carbon content and used it to suggest that the carbon cycles observed in core 12392-1 off northwest Africa indicate an approximately three-fold increase in primary productivity for oxygen isotope stages 3 and 6. Some of the data used to construct the empirical relationship came from the Baltic Sea, Peru margin, and northwest African continental slope, all areas with dramatically less bottom water oxygen than at the other locations (the Argentine Basin, central North Pacific and Oregon margin). I believe that Müller and Suess omitted one of the most important variables in considering their relationship: the bottom water oxygen content. The high values of carbon observed in glacial stages of core 12392-1 off northwest Africa and in the eastern equatorial Pacific [Pedersen, 1983] could be a result of low concentration of bottom water oxygen rather than (or in addition to) a productivity increase. There is abundant evidence for a more stagnant bottom water circulation during glacial periods than today [e.g., Boyle and Keigwin, 1982; Curry and Lohmann, 1983], and one would expect this condition to expand the extent of oxygen minimum layers in the ocean. The trends in Figure 6b indicate that regions with present-day low oxygen content would be most susceptible to preservation of a high carbon signal during these times.

The present analysis suggests that carbon cycles in deep-sea sediments can be caused by a number of processes, the most likely of which are a change in bottom water oxygen content and/or carbon flux to the sediments. By itself the carbon signal is insufficient to separate these two potential causes. Paleoceanographers will have to search for independent evidence to distinguish between a decrease in bottom water oxygen content and an increase in particulate carbon flux as explanations for the observed carbon signal.

## Preservation of Fossil Fuel Carbon as Organic Matter in Marine Sediments

Walsh and collaborators [Walsh et al., 1981a, b] have suggested that up to 15% of the $CO_2$ that has entered the atmosphere by fossil fuel burning and deforestation could reside as organic carbon in the surface sediments of the continental slope. To provide a sink of this magnitude, roughly one half of the carbon presently produced by primary production on the continental shelves must be exported to the slope area, where 50% of that one half must be stored in the sediments. Thus, half of the organic matter that escapes recycling in the photic zone of the continental shelves has to be buried on the slope.

The fraction of the particulate organic carbon flux that escapes oxidation based on sediment trap studies on the continental slope and deep sea (Table 1) is about 10% (the value is higher in the nearshore waters of the California Borderlands). The residence time for degradation of carbon at the sediment-water interface of the continental slope is determined in this paper (Table 2) to be less than 30 years. Based on the model presented here and the available sediment trap studies, I would suggest that 90% of the particulate organic carbon delivered to the continental slopes is degraded on a time scale of less than 30 years. Thus, the estimates by Walsh et al. [1981a, b] of the potential for carbon burial on the continental shelves are probably in excess by at least a factor of 5. For the estimates here to be drastically in error there would have to be either vast areas of slope sediments with higher carbon content (much greater than 1%) than generally measured or a dramatic recent increase in sedimentation rate on the continental slope. Neither of these phenomena has been observed, and until they are, one must view the fossil fuel carbon storage hypothesis as one that is counter to the present-day estimates of carbon cycling in the ocean.

Acknowledgments. I would like to thank John Hedges and James Murray for insightful comments on the original manuscript. Financial support was provided by National Science Foundation grant OCE 83-15820. University of Washington, School of Oceanography contribution 1409.

## References

Bender, M. L., and D. T. Heggie, Fate of organic carbon reaching the sea floor: A status report, Geochim. Cosmochim. Acta, 48, 977-986, 1984.

Berner, R. A., Early Diagenesis, 241 pp., Princeton University Press, Princeton, N.J., 1980a.

Berner, R. A., A rate model for organic matter decomposition during bacterial sulfate reduction in marine sediments, in Biogeochimie de la Matière Organique à l'Interface Eau Sediment Marin, Colloq. Int. CNRS, No. 293, pp. 34-44, 1980b.

Boyle, E., and L. Keigwin, Deep circulation of the North Atlantic over the last 200,000 years: Geochemical evidence, Science, 218, 784-787, 1982.

Broecker, W. S., and T.-H. Peng, Tracers in the Sea, 690 pp., LDGO Press, Columbia University, Palisades, N.Y., 1982.

Carpenter, R., M. L. Peterson, J. T. Bennett, and B. L. K. Somayajulu, Mixing and cycling of uranium, thorium and $^{210}$Pb in Puget Sound sediments, Geochim. Cosmochim. Acta, 48, 1949-1963.

Cobler, R., and J. Dymond, Sediment trap experiment on the Galapagos spreading center, equatorial Pacific, Science, 209, 801-804, 1980.

Cochran, J. K., Particle mixing rates from MANOP sites in the eastern equatorial Pacific: Evidence from $^{210}$Pb, $^{239,240}$Pu and $^{137}$Cs distributions, Geochim. Cosmochim. Acta, in press, 1984.

Curry, W., and G. P. Lohmann, Reduced advection into Atlantic Ocean deep eastern basins during the last glaciation maximum, Nature, 306, 577-579, 1983.

Deuser, W. G., and E. H. Ross, Seasonal change in the flux of organic carbon to the deep Sargasso Sea, Nature, 283, 364-365, 1980.

Emerson, S., K. Fischer, C. Reimers, and D. Heggie, Organic carbon dynamics and preservation in deep sea sediments, Deep Sea Res., in press, 1984.

Emery, K. O., The Sea off Southern California, 360 pp., John Wiley, New York, 1960.

Fischer, K., Particle fluxes to the eastern tropical Pacific Ocean-Sources and proceses, Ph.D. dissertation, 225 pp., Oregon State University, Corvallis, 1983.

Grundmanis, J., and J. W. Murray, Stoichiometry of decomposing organic matter in aerobic marine sediments, Geochim. Cosmochim. Acta, 46, 1101-1121, 1982.

Heath, G. R., T. C. Moore, and J. P. Dauphin, Organic carbon in deep sea sediments, in The Fate of Fossil Fuel $CO_2$ in the Oceans, edited by N. R. Andersen and A. Malahoff, pp. 605-628, Plenum, New York, 1977.

Hedges, J., and D. Parker, Land-derived organic matter in surface sediments from the Gulf of Mexico, Geochim. Cosmochim. Acta, 40, 1019-1029, 1976.

Hedges, J., and D. C. Mann, The lignin geochemistry of marine sediments from the southern Washington coast, Geochim. Cosmochim. Acta, 43, 1809-1818, 1979.

Hinga, K. K., J. M. Sieburth, and G. R. Heath, The supply and use of organic material at the deep-sea floor, J. Mar. Res., 37, 557-579, 1979.

Honjo, S., Material fluxes and modes of sedimentation in the mesopelagic and bathypelagic zones J. Mar. Res. 38, 53-97, 1980.

Jahnke, R., D. Heggie, S. Emerson, and V. Grundmanis, Pore waters of the central Pacific Ocean: Nutrient results, Earth Planet. Sci. Lett., 61, 233-256, 1982.

Jorgensen, B. B., A comparison of methods for quantification of bacterial sulfate reduction in coastal marine sediments, II, Calculation from mathematical models, Geomicrobiol. J., 1, 29-97, 1979.

Müller, P. J., and E. Suess, Productivity sedimentation rate and sedimentary organic matter in the oceans, I, Organic carbon preservation, Deep Sea Res., 26, 1347-1367, 1979.

Pedersen, T. F., Increased productivity in the eastern Equatorial Pacific during the last glacial maximum (19,000-14,000 yr B.P.), Geology, 11, 16-19, 1983.

Reimers, C. E., and E. Suess, The partitioning of organic carbon fluxes and sedimentary organic matter decomposition rates in the ocean, Mar. Chem., 13, 141-168, 1983.

Rowe, G. T., and W. D. Gardner, Sedimentation rates in the slope water of the northwest Atlantic Ocean measured directly with sediment traps, J. Mar. Res., 37, 581-600, 1979.

Smith, K. L., and K. H. Hinga, Sediment community respiration in the deep sea, in The Sea, vol. 8, edited by G. T. Rowe, pp. 331-370, John Wiley, New York, 1983.

Suess, E., and P. J. Müller, Productivity, sedimentation rate and sedimentary organic matter in the oceans, II, Elemental fractionation, in Biogeochimie de la Matière Organique à l'Interface Eau-Sediment Marin,

Colloq. Int. CNRS, No. 293, pp. 17-26, Centre National de la Recherche Scientifique, Paris, 1980.

Toth, D. J., and A. Lerman, Organic matter reactivity and sedimentation rates in the ocean, Am. J. Sci. 277, 465-485, 1977.

Walsh, J. J., G. T. Rowe, R. L. Iverson, and C. P. McRoy, Biological export of shelf carbon is a sink of the global $CO_2$ cycle, Nature, 291, 196-201, 1981a.

Walsh, J. J., E. T. Premuzic, and T. E. Whitledge, Fate of nutrient enrichment on continental shelves as indicated by the C/N content of bottom sediments, in Ecohydrodynamics, edited by J. C. J. Nihoul, pp. 13-49, Elsevier, New York, 1981b.

Westrich, J. T., and R. A. Berner, The role of sedimentary organic matter in bacterial sulfate reduction:  The G model tested, Limnol. Oceanogr., 29, 236-249, 1984.

# TRANSIENT RESPONSE OF THE MARINE
## CARBON CYCLE

John R. Southam and William H. Peterson

Rosenstiel School of Marine and Atmospheric Science (MGG)
University of Miami, Miami, Florida 33149

Abstract. Recent measurements of $pCO_2$ in ice cores indicate fluctuation of the natural atmospheric $CO_2$ on several hundred year time scales, which are much shorter than previously envisioned. Recent isotope measurements suggest $CO_2$ fluctuations may have led rather than lagged ice volume changes. The results argue for climatic fluctuations being forced by changes in the marine carbon cycle. A simple one-dimensional model of the ocean's coupled carbon, oxygen, and phosphorus (COP) system is developed. An eigenfunction expansion of the Green's function is presented as the solution to an impulsive change in the surface nutrient concentration. It is shown that the system reaches its new equilibrium state on a time scale of several hundred years. Changes in the COP system with this time scale are commensurate with the $pCO_2$ variability observed in the ice record.

## Introduction

Recent research (much of which was presented at the Chapman Conference on Natural Variations in Carbon Dioxide and the Carbon Cycle [AGU, 1984] and appears in this volume [e.g. Oeschger et al., this volume] indicates that the natural atmospheric $CO_2$ concentration has varied much more rapidly than previously measured. Data [Shackleton and Pisias, this volume] suggest that $CO_2$ variations may have led rather than lagged ice volume changes, contrary to prevailing ideas. These results indicate that $CO_2$ may have played a significant role in the forcing of the glacial-Holocene climate transition. This argues for climatic fluctuations being forced by changes in the marine carbon cycle and hence by changes in nutrient cycles and/or thermohaline circulation.

Assessing the role of the oceans in forcing atmospheric $CO_2$ fluctuations requires a modeling approach. The experience of modelers trying to understand the role of the oceans in controlling atmospheric $CO_2$ levels in the face of the anthropogenic $CO_2$ transient suggests this is not an easy task. The nature of the problem dictates

that the model be time dependent. It is advantageous therefore to keep the spatial dependence as simple as possible, so that considerable iteration and experimentation with the model is feasible. Large numerical models are too costly to allow the necessary experimentation. However, we suggest that the other extreme of complexity, the simple box model, is not adequate. For example, a box model does not have the correct frequency response to a forcing whose time scale is of the order of or less than the characteristic mixing time scale of the boxes. One could envision a box model with a large number of small boxes that would have the correct frequency response. However, such models require rather cumbersome numerical solutions from which it is difficult to draw general inferences on the dynamics of the system. We prefer to construct a simple continuous model that yields analytical solutions from which dynamical implications are more transparent. Evidence presented at this meeting [Stauffer et al., 1984; Oeschger et al., this volume] indicates natural atmospheric $CO_2$ excursions have occurred on the order of a few hundred years, which is less than the oceanic mixing time scale which is usually taken to be of the order of a thousand years. Difficulties with box models may also arise from spatial considerations when large gradients in property fields exist. Again a large number of boxes is unsatisfactory because profiles and their gradients should be a model result not an input. The appropriate choice of boxes prior to a perturbation is not guaranteed to be the most appropriate choice after the perturbation.

In this paper we show that perturbations in the surface phosphate level of a simple one-dimensional model of the ocean's coupled carbon oxygen, and phosphate system (COP model) can force significant excursions in the marine carbon system with a time scale on the order of several hundred years. A perturbation in surface phosphate can be considered to be a consequence of either a sudden change in the delivery of nutrients to the surface ocean by river input or a sudden change in the flux from the rate of vertical circulation. Specifically, by deriving an eigenfunction expansion of the Green's function, we show that the oxygen profile will reach its new equilibrium state in response to a step function perturbation in the surface phosphate within a thousand years.

Since both dissolved oxygen and phsophate are coupled to the degradation of organic carbon particles settling through water column, this calculation implies the dynamics of the phosphate system respond on this same scale. Furthermore, this implies the same time scale for response of the $\Sigma CO_2$ system to the same perturbation in the surface nutrient levels. The imposition of a step function perturbation in surface nutrient level was made for mathematical convenience. However, our model, as formulated for the oxic case, is a linear system. Linear system analysis demonstrates that the time scale for the natural response of the system (as opposed to the forced response) is completely determined by the time dependence of the Green's function. This is a preliminary result, chosen because it was the simplest analytical result available from the model. Nonetheless, we feel it is an important step to show that the marine carbon, oxygen and nutrient systems can respond on several hundred year time scales. The mathematical techniques used are applicable to perturbations in the surface nutrient levels with a more physically realistic time dependence than the step function.

## The Marine Carbon Cycle

### Processes

Because most of the dissolved carbon in the sea is in the form of bicarbonate and carbonate, the ocean is a much larger reservoir of carbon than is the atmosphere, unlike the case for virtually all other gases. Hence, the ocean ultimately determines the atmospheric $CO_2$ content rather than vice versa. At the air-sea interface the $CO_2$ concentration in the overlying atmosphere is controlled primarily by the $pCO_2$ of the surface layer and the layer's temperature. Changes in the total dissolved carbon content ($\Sigma CO_2$) and/or alkalinity ($A_T$) of these surface waters result in changes of the surface layer $pCO_2$ and hence of atmospheric $CO_2$.

Two major types of biologically produced particles containing carbon fall from the surface ocean into the deep: organic carbon and calcium carbonate particles. On the average, for every four carbon atoms which fall from the surface ocean in the form of organic tissue, one carbon atom falls in the form of $CaCO_3$, according to Broecker and Peng [1982]. The rates at which these particles are produced in the surface ocean are functions of the availability of limiting nutrient, among other variables.

Differences in the destruction mechanisms of these two types of particles have important implications for their chemical signature or distribution in the water column. Even though the flux of organic carbon out of the surface layer exceeds the flux of $CaCO_3$, a very small fraction of this flux reaches the bottom, and even less is preserved in the sediment. Organic carbon is scavenged high in the water column. Sediment trap data are consistent with an exponential decrease with a length scale of approximately 250-350 m [Honjo et al., 1982]. This is also consistent with the location of the oxygen minimum and phosphate maximum at a depth of 300-1000 m. The $CaCO_3$ particles sink further in the water column before experiencing the corrosive effect of undersaturated waters. A much larger fraction of the flux of $CaCO_3$ (than of organic carbon) originating in the overlying surface water survives settling through the water column and diagenetic processes at the seafloor to become incorporated into the sediments.

The destruction of both these types of particles has important implications for water chemistry. However, as the effect of the destruction of organic C is felt higher up in the water column, it has a more immediate effect on the chemistry of the surface ocean and atmospheric $CO_2$ content than carbonate destruction. In addition, recall that the organic C flux from the surface layer is larger than the $CaCO_3$ flux. However, the carbonate component is important as a recorder of changes in the chemistry of the overlying water column and in controlling the marine carbon cycle on long time scales.

A primary focus of this paper is the role organic carbon particle creation and destruction play in the marine carbon cycle. The flux of organic carbon to the seafloor is intimately connected to the marine nutrient cycle. We take phosphate to be the ultimate limiting nutrient. In the steady state, the $PO_4$ incorporated into the sediments must exactly balance the $PO_4$ flux delivered to the oceans by rivers. In the surface waters carbon is incorporated into particles in the ratio of approximately 100 carbon atoms to every phosphorus atom. However, on time scales shorter than the residence time of $PO_4$ (about $10^5$ years) perturbations or imbalances in the $PO_4$ system lead to perturbations in the carbon system. Consideration of phosphorus is interesting for another reason.

Consider a hypothetical ocean free of any pelagic carbonate-secreting organisms and which is maintained at constant alkalinity by some unspecified mechanisms operating on the shelves. The distribution of $\Sigma CO_2$ in the deep sea would be analogous to that of $PO_4$. The distributions of both in the deep sea would be controlled by the same processes. The only differences would be a consequence of the differences in the flux conditions describing their rates of supply to the surface reservoir and the appearance of the Redfield ratio in the particle destruction term. In the deep reservoir

## Steady-State COP Model

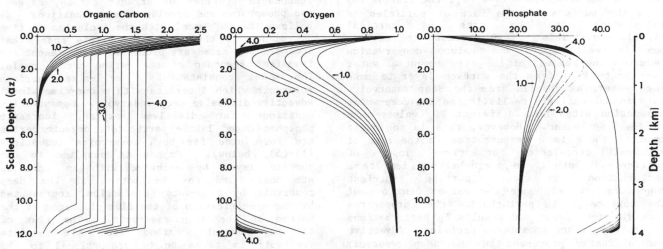

Fig. 1. Profiles versus scaled depth of organic carbon, dissolved oxygen and inorganic phosphate concentrations (all normalized by their surface concentrations) for the COP model. Each profile corresponds to a fixed Wyrtki number in the range from 1.0 to 4.0 in increments of 0.2

of this hypothetical ocean, $PO_4$ is an analog for $\Sigma CO_2$.

An additional constraint on the description of processes in the deep reservoir is provided by the $O_2$ system. One of the controls on oxygen in the deep ocean is the number of carbon particles available for oxidation by bacteria and other organisms. Hence a perturbation in the $PO_4$ system will also cause a perturbation in oxygen in addition to the carbon system. The processes operating in the deep reservoir which determine the time scale for the response of an oxygen profile to a perturbation in surface nturients are the same processes which contribute to the determination of the time for the response of $PO_3$ and $\Sigma CO_2$ to the same perturbation. Thus the magnitude of the time scale for the response of the oxygen system has implication for $\Sigma CO_2$, and hence $pCO_2$ transients.

## Steady State Models of the Marine Carbon, Oxygen, and Phosphorus System

Before considering perturbations in the carbon, oxygen, phosphorus system, several aspects of the steady state system pertaining to the problem must be considered. Figure 1 shows COP model profiles. These profiles are from a steady state model of the coupled carbon, oxygen, and phosphorus system [J.R. Southam et al., unpublished manuscript, 1984] that consists of a well-mixed surface reservoir that interacts with a one-dimensional advective-diffusive deep reservoir. A more complete description of the model can be found in the next section. The profiles are laterally averaged and are to be interpreted as mean property profiles. In the hypothetical ocean in which alkalinity remains constant the $PO_4$ profile is the analog of the $\Sigma CO_2$ profile.

The carbon profile corresponds to a consumption rate at each depth that is proportional to the concentration of organic C at that depth. The initial flux to the deep reservoir is taken to be proportional to the nutrient level in the surface ocean. The consumption of organic carbon removes oxygen and releases nutrients at depth. The distribution in the interior is a consequence of the balance of advective and diffusive processes and the destruction of organic particles. The structure of these profiles reflects the two length scales involved in the vertical balance, namely, the particle consumption rate scale ($1/\alpha \sim 250-350$ m) and the pycnocline scale ($K/w \sim 500-1000$ m.) Details of the structure of all three coupled profiles, or the state of the system, can be specified by the introduction of two dimensionless parameters: the ventilation parameter and the Wyrtki number (see the notation list at the end of this paper for formal definitions of these parameters). The ventilation parameter ($\rho$) is the ratio of the length scales mentioned above. This parameter is proportional to the upward advective velocity and hence the intensity of vertical circulation. The Wyrtki number ($\nu_w$) is interpreted as the ratio of the carbon flux produced in the surface to the downward diffusive flux of oxygen in the absence of advection. In the steady state this number is proportional to the river input of $PO_4$. Both $\rho$ and $\nu_w$ are of order one in today's ocean (for justification of scaling, see below). If in addition the model conserves $PO_4$, the flux of $PO_4$ from the surface in the form of particles is independent of the rate of vertical circulation. This is true because the surface box is well mixed and volume conservation requires the removal of the same amount of water (and hence $PO_4$) from the surface layer to make deep water, as upwells from the deep reservoir. This is not entirely realistic, as the deep-water production sites have different $PO_4$ values than in the open ocean. However, this is no longer true if there is a perturbation in the rate of vertical circulation or river input of nutrients. Both these perturbations lead to a perturbation in the surface nutrient concentration and hence the carbon flux out of the surface. The perturbation then propagates into the deep layer and results in perturbations of the oxygen and phosphorus profiles. Advective and diffusive processes in the deep reservoir provide a feedback mechanism between deep and surface nutrient levels, and after some time the profiles reach new equilibrium states, and the flux of $PO_4$ to the sediments again equals the river influx.

Modeling of the $CO_2$ system requires the introduction of two additional variables to the COP model. Total dissolved inorganic carbon ($\Sigma CO_2$) and alkalinity ($A_T$) are the most convenient choice. In the hypothetical ocean where the alkalinity remains constant, $\Sigma CO_2$ profiles are characterized by a parameter analogous to the Wyrtki number.

The interesting question from the perspective of variability in the natural carbon cycle is: "What processes are responsible for the observed rapid increase in the rates of change in atmospheric $CO_2$, observed in ice core records?" In an attempt to understand the causal mechanisms for variability in $CO_2$ and the marine carbon cycle we have extended the COP model to include time dependence. The model is used to study the effect of perturbing the system from an initial steady state. In particular, we will perturb oxygen and phosphorus profiles by perturbing the flux of carbon produced in the surface reservoir. Because of the intimate coupling between $CO_2$ and the carbon, oxygen, and phosphorus system, understanding the nature of transient responses of this system has important implications for the natural carbon cycle and fossil fuel problems.

### Time Dependent COP Model Description and Solutions

#### The Processes and Model Equations

The time dependent model discussed here is an extention of the steady state model of the coupled carbon, oxygen and phophorus model described by J.R. Southam et al., (unpublished manuscript, 1984). The model was developed to be

the simplest possible model that would produce reasonable profiles of organic carbon, oxygen, and phosphorus and develop anoxic conditions. We believe the COP model with the inclusion of time dependence is the simplest possible model that includes the necessary processes to represent the transient response of the marine carbon cycle. The model consists of a well-mixed surface reservoir which interacts with a one-dimensional advective-diffusive deep reservoir. Conservation equations for dissolved oxygen, inorganic phosphate, and labile particulate organic carbon are formulated for both reservoirs (equations (1)-(5), below). Oxygen is supplied to the surface layer by exchange with an infinite atmosphere and to the bottom of the deep reservoir by deep-water production, represented by the specification of the flux of oxygen at the bottom of the deep reservoir. The flux of settling organic carbon from the surface reservoir is taken to be proportional to the surface phosphate concentration, thus coupling the phosphate, and carbon equations. The organic carbon particles fall into the deep reservoir with a constant settling velocity. This rain of organic particles is oxidized in the deep reservoir at a rate that is first order in organic carbon and independent of oxygen until the oxygen is exhausted, at which point oxidation ceases. Phosphate is supplied to the surface reservoir by a small river input and to the deep reservoir by deep-water production, represented by the specification of the flux of phosphate at the bottom of the deep reservoir. For a detailed description and justification of the processes included or excluded from the model the reader is referred to J.R. Southam et al. (unpublished manuscript, 1984). This paper will be concerned only with transitions from an initial oxic state to a final oxic state. Hence the processes of nitrate and sulfate reduction are not included.

The well-mixed surface reservoir is described by conservation equations for dissolved $PO_4$ and organic carbon particles (see the notation list for definition of variables and parameters). For dissolved $PO_4$,

$$\frac{dP(t)}{dt} + \beta P(t) = \frac{K}{h_s} \frac{\partial P}{\partial z}\bigg|_{z=0} + \frac{I(t)}{h_s} \quad (1)$$

where the diffusive flux from the deep reservoir (there is no advective term due to conservation of mass in the surface layer), plus the river input (I) is balanced by the net uptake of $PO_4$ in organic particles and storage in the surface reservoir. For organic carbon particles,

$$\frac{dC(t)}{dt} + \frac{v}{h_s} C(t) = r \beta P(t) \quad (2)$$

where storage of organic particles plus the flux of organic particles settling out of the surface reservoir is balanced by the net particle production rate in this reservoir. The oxygen concentration is assumed to be maintained at near saturation by exchange with the atmosphere. This is justified because the atmospheric oxygen reservoir is so large compared with the ocean's that it can be treated as infinite.

In the deep reservoir the distributions of particulate organic carbon, dissolved oxygen, and inorganic phosphate are governed by the conservation equations,

$$\frac{\partial C}{\partial t} = -v\,\frac{\partial C}{\partial z} - \Omega\,C\,\Theta\,(0) \tag{3}$$

$$\frac{\partial O}{\partial t} = K\,\frac{\partial^2 O}{\partial z^2} + w\,\frac{\partial O}{\partial z} - \eta\,\Omega\,C\,\Theta\,(0) \tag{4}$$

$$\frac{\partial P}{\partial t} = K\,\frac{\partial^2 P}{\partial z^2} + w\,\frac{\partial P}{\partial z} + \frac{\Omega}{r}\,C\,\Theta\,(0) \tag{5}$$

We are interested in transitions away from an initial steady state solution so that the following notation is introduced for the surface reservoir:

$$P(t) = P_o + p(t) \tag{6}$$
$$C(t) = C_o + c(t)$$

where $C_o$ and $P_o$ denote steady state surface reservoir concentrations, and the initial conditions for the surface reservoir become

$$P(t=o) = P_o \Rightarrow p(t=o) = 0 \tag{7}$$
$$C(t=o) = C_o \Rightarrow c(t=o) = 0$$

Clearly, the perturbations from the steady state satisfy

$$c(t) = \int_o^t dt'e^{\frac{-v}{h_s}(t-t')}\,r\beta p\,(t') \tag{8}$$

and

$$p(t) = \int_o^\infty dt'e^{-\beta(t-t')}\Big(\frac{K}{h_s}\frac{\partial p}{\partial z}\,(z=o^+,\,t') \tag{9}$$
$$+ \frac{I(t')}{h_s}\Big)$$

where $p(z,t)$ is the perturbation from the steady state $P_o(z)$ in the deep reservoir. If the surface reservoir phosphate concentration increases or decreases abruptly, i.e. a step function perturbation away from the steady state, the resulting perturbation in the particulate organic carbon concentration is

$$C(t) = \frac{r\,\beta h_s}{v}\,\Delta p\,(1 - e^{-\frac{v}{h_s}t}) \tag{10}$$

where $\Delta p$ is the magnitude of the perturbation in the phosphate concentration. This perturbation in surface phosphate can be considered to be a consequence of either a sudden change in river input or a sudden change in the flux from the deep reservoir due to a sudden change in the rate of vertical circulation. Note that the time scale $(h_s/v)$ is associated with particles sinking out of the surface reservoir ($h_s/v$ is order of magnitude $10^{-2}$–$10^{-1}$ years).

Next we consider perturbations from an initial steady state solution. The response of the ocean's chemistry system to this abrupt perturbation in $PO_4$ can be important because it establishes the time scale dictated by internal processes in the deep reservoir. The following notation is introduced for convenience

$$C(z,t) = C_o(z) + c(z,t)$$
$$P(z,t) = P_o(z) + p(z,t) \tag{11}$$
$$O(z,t) = O_o(z) + o(z,t)$$

where again the subscript o denotes the initial steady state solution prior to perturbation. The initial conditions become

$$C(z,t=o) = C_o(z) \Rightarrow c(z,t=o) = 0$$
$$P(z,t=o) = P_o(z) \Rightarrow p(t,t=o) = 0 \tag{12}$$
$$O(z,t=o) = O_o(z) \Rightarrow o(z,t=o) = 0$$

### Organic Carbon Solution

Equation (3), together with initial conditions from (12), yields a solution for the perturbation in the particulate organic carbon profile in the deep reservoir, for the oxic case:

$$c(z,t) = c(t - \frac{z}{v})e^{-\frac{\Omega}{v}z}\,\Theta(t - \frac{z}{v}) \tag{13}$$

where the step function insures causality. The perturbation in the surface carbon concentration propagates into the deep reservoir at the particle settling velocity, where it is attenuated with depth by bacterial degradation with the length scale $\alpha = \Omega/v$ determined by the organic carbon consumption rate.

### Oxygen and Phosphate Solutions

The complete formulation of the time dependent model of oxygen and phosphate profiles requires in addition to the conservation equations (4) and (5), and initial conditions, the specification of an appropriate set of boundary conditions. The conditions at the top of the deep reservoir are that O$_2$ and P$_{O_4}$ are continuous so that

$$o(z=o^+, t) = 0 \tag{14}$$

$$p(z=o^+,t) = p(t)$$

As discussed by J.R. Southam et al. (unpublished manuscript, 1984) the imposition of flux conditions rather than fixed conditions at the bottom of the deep reservoir leads to physically more realistic behavior within the context of a steady state model. The O$_2$ and P$_{O_4}$ fluxes to the deep reservoir by deep-water production were taken to be proportional to the vertical advective velocity, which is determined by the rate at which deep water is produced. Thus it seems reasonable to impose the following conditions at the bottom of the deep reservoir to describe the above perturbations in our system:

$$K\frac{\partial}{\partial z} p(z=h,t) + w\, p(z=h,t)$$
$$= w\, p(t-\frac{h}{w}\, \varepsilon)\, \theta(t-\frac{h}{w}\, \varepsilon) \tag{15}$$

and

$$K\frac{\partial}{\partial z} o(z=h,t) + w\, o(z=h,t) = 0 \tag{16}$$

Perturbations in the surface phosphate concentrations perturb the flux to the bottom, but with a time delay $(h/w)\varepsilon$, the time required for a parcel of water to sink from the surface to the bottom of the deep reservoir ($\varepsilon = A_s/A \gg 1$ is the ratio of the deep water-source area to the total surface area of the ocean). For a more complete description and justification of the mixed or flux type boundary condition the reader is referred to J.R. Southam et al. (unpublished manuscript, 1984).

### Solutions: Time Scale of the Response of Oxygen Profiles

In this section we determine the time scale required for oxygen and phosphate profiles to respond to perturbations within the system. To accomplish this, we consider a problem in which the oxygen profile is perturbed by a step function change in surface phosphate and, hence, a perturbation in organic carbon produced in the surface reservoir. Within the context of the results of the steady state model the perturbation in surface phosphate corresponds to perturbing the Wyrtki number.

For the step function perturbation in the surface phosphate, the perturbed oxygen equation becomes

$$\frac{\partial o}{\partial t} = K\frac{\partial^2}{\partial z^2} o + w\frac{\partial o}{\partial z} - \eta r \frac{\Omega}{v}\beta h_s \Delta p \cdot$$
$$(1 - e^{-\frac{v}{h_s}(t - \frac{z}{v})}) e^{-\frac{\Omega}{v}z} \theta(t - \frac{z}{v}) \tag{17}$$

$$\frac{\delta p}{\delta t} = K\frac{\delta^2 p}{\delta z^2} + w\frac{\delta p}{\delta z} + \Omega\,\beta\,h_s\,\Delta p \cdot$$
$$(1-e^{-\frac{v}{h_s}(t-\frac{z}{v})}) e^{-\frac{\Omega}{v}z} \theta(t-\frac{z}{v}) \tag{18}$$

These equations contain a large number of parameters which characterize the processes which collectively determine the oxygen and phosphate profiles. The values of most of these parameters are not well known. To reduce the number of parameters in the solution the equations are scaled, and nondimensional variables are introduced. This scaling results in certain nondimensional groups occurring in the solutions. This allows for the most compact presentation of the dependence of the solutions on the imposed parameters and explicitly shows the relative importance of these parameters. By not introducing specific values into the solutions the reader can take what he thinks are appropriate values for the imposed parameters and substitute them into the definitions of the nondimensional numbers and then examine the solution for the appropriate nondimensional numbers. More specifically, in the COP model (as was discussed in the section on steady state models) we were able to reduce the number of parameters in the solution to two parameters: the ventilation parameter and the Wyrtki number. The scaling for the time dependent model is similar to the COP model scaling except that we define a perturbation to the Wyrtki number and introduce a time scale:

Ventilation parameter,

$$\rho = \frac{w}{K\frac{\Omega}{v}} \tag{19}$$

Perturbation in the Wyrtki number,

$$\Delta\nu_w = \frac{\eta r\,\beta h_s\,\Delta p}{K\frac{\Omega}{v}\,o} \tag{20}$$

Perturbation in the nutrient number,

$$\Delta\nu_n = \frac{\beta\,h_s\,\Delta p}{K\frac{\Omega}{v}\,P_o} \tag{21}$$

$$z' = \frac{\Omega}{v} z = \alpha z \tag{22}$$

$$t' = K \frac{\Omega^2}{v^2} t = K \alpha^2 t \tag{23}$$

Under these scalings and normalizing the oxygen and phosphate profiles by their respective surface reservoir concentrations the model equations become

$$\frac{\partial o}{\partial t'} = \frac{\partial^2}{\partial z'^2} o + \rho \frac{\partial}{\partial z'} o - \Delta \nu_w e^{-z}$$

$$\left(1 - e^{\frac{-v}{h_s \Omega} \left( \frac{t'v^2}{K\Omega} - z' \right)} \right) \theta\left( \frac{t'v^2}{K\Omega} - z' \right) \tag{24}$$

$$\frac{\partial p}{\partial t'} = \frac{\partial^2}{\partial z^2} p + \rho \frac{\partial}{\partial z'} p + \Delta \nu_n e^{-z'}$$

$$\left(1 - e^{-\frac{v}{h_s \Omega} \left( \frac{t'v^2}{K\Omega} - z' \right)} \right) \theta\left( \frac{t'v^2}{K\Omega} - z' \right) \tag{25}$$

The auxillary equations appropriate to this scaling and normalization are as follows:

Boundary conditions

$$O(z' = o^+, t') = 0$$

$$\left( \frac{\partial o}{\partial z'} + \rho o \right)_{z'=\alpha h} = 0$$

$$p(z' = o^+, t') = p(t')$$

$$\left( \frac{\partial p}{\partial z'} + \rho p \right)_{z'=\alpha h} = \rho \frac{p}{p_o} \left( t' - \kappa \alpha^2 \frac{h}{w} \varepsilon \right)$$

$$\theta\left( t' - \kappa \alpha^2 \frac{h}{w} \varepsilon \right) \tag{26}$$

Initial conditions

$$o\left( \frac{z'}{\alpha}, \quad t' = o \right) = 0$$

$$p\left( \frac{z'}{\alpha}, \quad t' = o \right) = 0 \tag{27}$$

Mathematically the simplest problem to consider is the oxygen system because of the appearance of the inhomogeneous term appearing in the bottom $PO_4$ flux condition. This term introduces feedback between the surface and deep reservoirs through the process of deep-water production. To explore the consequences of the dynamics expressed by the above formulation, we first consider the consequences of this term being small, i.e., $p \ll P_o$, the perturbation being small compared to the steady state value. Calculations of time scales depending on this assumption represent lower limits.

To determine the time scale required to perturb an oxygen profile, we first consider the problem in which the carbon production is perturbed by a step function increase in surface phosphate concentration. This leads to a change, $\Delta \nu_w$, in the Wyrtki number.

For $t \gg (h + h_s)/v$ (i.e. $t \gg 10^o$–$10^1$ years, the time required for a carbon particle to settle to the bottom) the perturbation for the normalized oxygen profiles has the space and time dependence represented by the following Green's function expansion. The reader uncomfortable with the validity of this result may verify it by direct substitution into equation (25). Clearly, boundary conditions (26) and initial conditions (27) are satisfied.

$$o(z', t') = -\Delta \nu_w \, e^{-\frac{\rho}{2} z'} \sum_{n=1}^{\infty} \left\{ \left( 1 - e^{-(1 - \frac{\rho}{2})\alpha h} \right) \cdot \right.$$

$$\left( \frac{1 - \left(1 - \frac{\rho}{2}\right) \sin \sqrt{\lambda_n}\, \alpha h}{\frac{\alpha h}{2} \left( 1 - \frac{\sin 2\sqrt{\lambda_n}\, \alpha h}{2 \sqrt{\lambda_n}\, \alpha h} \right)} \right)$$

$$+ \left( \frac{\sqrt{\lambda_n}\, \alpha h \, \cos \sqrt{\lambda_n}\, \alpha h}{\frac{\alpha h}{2} \left( 1 - \frac{\sin 2\sqrt{\lambda_n}\, \alpha h}{2 \sqrt{\lambda_n}\, \alpha h} \right)} \right)$$

$$\left. \frac{\left(1 - e^{-(\lambda_n + \frac{\rho^2}{4}) t'} \right)}{\left( \frac{\rho^2}{4} + \lambda_n \right)\left( \left(1 - \frac{\rho}{2}\right)^2 + \lambda_n \right)} \sin \sqrt{\lambda_n}\, z' \right\} \tag{28}$$

where the eigenvalues $\lambda_n$ satisfy

$$\tan \sqrt{\lambda_n}\, \alpha h = - \frac{2}{\rho} \sqrt{\lambda_n} \tag{29}$$

This eigenvalue equation is a consequence of applying the mixed or flux type boundary condition at the bottom of the deep reservoir. If the fixed value type boundary condition, discussed by J.R. Southam et al. (unpublished manuscript, 1984), had been imposed, the corresponding eigenvalues would have been

$$\lambda_n = \left( \frac{n\pi}{\alpha h} \right)^2 \tag{30}$$

TABLE 1. Eigenvalues From Equation (28)

| ρ/n | 1 | 2 | 3 | 4 | 5 |
|------|---------|---------|---------|---------|---------|
| 0.25 | 0.03284 | 0.17386 | 0.44873 | 0.86019 | 1.40859 |
| 0.5  | 0.04188 | 0.19016 | 0.46746 | 0.81985 | 1.42869 |
| 0.75 | 0.04749 | 0.20299 | 0.48399 | 0.89813 | 1.44787 |
| 1.00 | 0.05124 | 0.21297 | 0.49824 | 0.91478 | 1.48259 |
| 1.25 | 0.05390 | 0.22078 | 0.51039 | 0.92973 | 1.48259 |

The eigenvalue equation (28) is a consequence of imposing a mixed or flux type boundary condition to the bottom of the deep reservoir. The first five eigenvalues from this equation for a deep reservoir of depth αh = 12 (i.e., h = 3-4 km) are listed here.

Thus a different time dependence or characteristic time response results, depending on the type of boundary condition imposed. It is easy to show that from consideration of equation (28),

$$\lambda_n > \left(\frac{n\pi}{\alpha h}\right)^2 \qquad \text{for all n} \qquad (31)$$

Hence, the system adjusts more rapidly to perturbation from the initial steady state when the flux conditions are imposed at the bottom of the deep reservoir.

The features of this Green's function expansion for the response of the oxygen profile are as follows:

1. This expansion converges rapidly for large values of t (see discussion below).

2. For large t the series approaches a new steady state oxygen profile in response to a perturbation in the surface nutrient levels.

3. A positive perturbation in the Wyrtki number results in a negative perturbation in the oxygen profile (i.e., higher surface productivity leads to increased oxygen consumption.)

4. The effect of the perturbation is more pronounced high in the water column because of the factor $\exp(-\rho\alpha z/2)$ in the expansion.

5. Because the system is linear, the space-time behavior of the perturbation is independent of the initial state (however, this may no longer be true if the perturbation drives the system to anoxia).

The term corresponding to the lowest eigenvalues (i.e., n = 1) dominates the expansion for large values of t. The solution (equation (28)) and the time scaling (equation (23)) indicate that the time required to make the transition from the initial steady state to the final steady state is determined by the time scale

$$\tau = \frac{1}{K\alpha^2\left(\lambda_1 + \frac{\rho^2}{4}\right)} = \frac{1}{K\alpha^2\lambda_1 + \frac{w^2}{4K}} \qquad (32)$$

Two contributions to this time scale appear in the denominator: (1) a purely advective-diffusive time scale $4K/w^2$ and (2) a time scale containing the lowest order eigenvalue $1/K\alpha^2\lambda_1$. The advective-diffusive time scale $4K/w^2$ is the time scale appropriate for the response of any conservative tracer to a perturbation of its steady state distribution. The process of oxygen consumption reduces the time scale appropriate for the response of an oxygen profile. The consumption of oxygen in the deep reservoir enters the expression through the scaling parameter $\alpha$, which is the reciprocal of length scale of carbon oxidation, and the eigenvalue $\lambda_1$. The value of $\tau$ is an essential result of this paper in that the oxygen profile will have changed by $\sim(1-e^{-1})$ from its initial to its final shape in time $\tau$. Estimates of the magnitude of $\tau$ are made by choosing convenient but typical parameter values appropriate to present-day oceans. Specifically, the following values are used:

Depth,

$$h = 4 \text{ km}$$

Thermohaline scale depth,

$$\frac{K}{w} = \frac{2}{3} \text{ km}$$

Consumption scale,

$$\frac{1}{\alpha} = \frac{1}{3} \text{ km} \qquad O_2$$

Vertical eddy diffusivity

$$K = 1 \frac{\text{cm}^2}{\text{sec}}$$

The model yields results which should be interpreted as horizontally averaged profiles. It would be unjustified to use parameter values determined from a single profile from a single station to infer global implications. The above choice of parameters is appropriate for the following reasons. The residence time (h/w) is ~850 years, which is in agreement with the much used $10^3$ years mixing time scale for the ocean (see, for example, Broecker and Peng, [1982] for the derivation of such an estimate using a box model of the ocean's radiocarbon distribution).

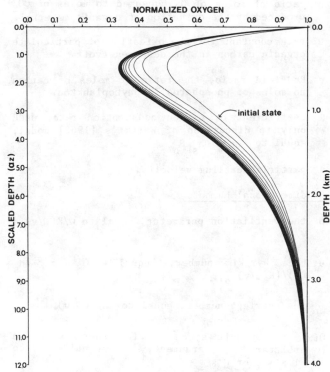

Fig. 2. Response of oxygen profiles (normalized to surface concentration) from the time dependent COP model to a perturbation of the initial Wyrtki number of 0.2  Each profile corresponds to a time step of $\tau/2$ (see equation (32)).    Assuming convenient but typical modern-day values of the $O_2$ consumption length scale $(1/\alpha \sim 1/3$ km) and the pycnocline length scale $(K/w \sim 2/3$ km) gives a real depth scaling as shown on the right side of the depth axis.    These length scales and a diffusivity $(\kappa)$ of 1 $cm^2/s$ give a time step of ~150 years between each profile.

The value of K = 1 $cm^2/s$ is a rather conventional value for global scale models.    The parameter choices also place the thermocline, oxygen minimum, and phosphate maximum at reasonable depths (in comparison to mean GEOSECS profiles).  The choice of $1/\alpha$ is also consistent with the expontential scale of the organic carbon profile from sediment trap data [Honjo et al., 1982].    Finally, the above choices of parameters yield

$$\tau \sim 350 \text{ years} \qquad (33)$$

whereas the time scale for a conservative tracer, $4K/w^2$, is about 550 years.

The model solutions in Figure 2 show the evolution of an oxygen profile resulting from a perturbation in the surface phosphate concentration which is equivalent to a perturbation in the Wyrtki number.  The profiles represent a transition from an original steady state characterized by an initial Wrytki number

to a new, final state with a larger Wyrtki number.  Each profile corresponds to a scaled time increment of $\tau/2$ (~150 years).  By the time t = $3\tau$ (~1000 years), the oxygen profile has essentially reached its final state.    The time scale on which an $O_2$ profile can respond to a perturbation is determined by the time constant $\tau$.  This is a dynamically determined quantity as opposed to the residence time of oxygen, which is a steady state concept.

The solutions are produced by taking the first 10 terms of the expansion in equation (29) with $\Delta \nu_w$ = 0.2.    There is virtually no difference in the profiles for 5 or 10 terms.  The first five eigenvalues for various values of $\rho$ are presented in Table 1.    Such a perturbation can be viewed as forced by a change in delivery of nutrients to the surface ocean by rivers or nutrients previously sequestered on the shelves during high sea level stands [Broecker, 1982] or the delivery of nutrients to the surface layer by a sudden increase in the rate of vertical circulation. The profiles in Figure 2 show that the $O_2$ content of the deep reservoir can change dramatically on a time scale of several hundred years in response to a change in surface $PO_4$. .This is especially true high in the water column (200-2000 m) because the carbon flux from the surface is largest at these depths.  This is contrasted with the residence time of $O_2$, which is of the order of $10^5$ years.    The reason for this dramatic difference between the response time of the profiles and residence time is that the calculation of the residence time involves the volume of the entire ocean, where changes in oxygen profiles are restricted to the depth range containing the $O_2$ minimum.

## Conclusions

Because of the intimate connection between atmospheric $pCO_2$ and organic carbon particle production in the surface ocean it is necessary to consider the coupling between the carbon and nutrient systems.  This is especially important when considering rapid transients such as the glacial-Holocene transition.  During a transient induced by changes in thermohaline circulation, for example, it is possible to increase or decrease the supply of nutrients to the surface and hence change the flux of carbon out of the mixed layer.    The model calculation presented above identified two time scales: (1) a purely advective-diffusive time scale applicable to a conservative tracer $4K/w^2$ (O(500 years)) and (2) a time scale for oxygen and phosphate which depends on the consumption of organic carbon in the deep reservoir.  Because of the importance of organic carbon consumption on the $CO_2$ system, $pCO_2$ can also be expected to respond on this same time scale to perturbations in the surface nutrient level.  The simple calculations showing the evolution of a mean oxygen profile in response to changes in mixed layer nutrient level demonstrate dramatic changes on time scales of several hundred years. Changes in the COP system with this time scale are reasonable and consistent with variability in $pCO_2$ observed in the ice core record.

Two interesting questions which can be addressed with this formulation of COP system are (1) "How does the feedback between the surface and deep reservoir by deep-water production affect the time scales involved?" and (2) "What is the magnitude and time dependence of the surface nturient level in response to changes in the rate of vertical circulation?"

## Notation

### Times

t   time

### Lengths

(Depths are frequently scaled by $\alpha$, e.g. $\alpha z$ see Figure 1 for sketch of the model domain.)

z   depth coordinate (positive down, 0 at boundary between surface reservoir and deep reservoir).

$h_s$   depth of well-mixed surface reservoir.

h   depth of deep reservoir.

L   width of domain.

### Dependent Variables

C   particulate organic carbon concentration in deep reservoir.

$O_o$   dissolved oxygen concentration in surface reservoir (as $O_o$ is determined by the overwhelming abundance of $O_2$ in the atmosphere it functions more as a boundary condition).

O   dissolved oxygen concentration in deep reservoir.

$P_o$   inorganic phosphate concentration in the surface reservoir.

P   inorganic phosphate concentration in deep reservoir.

### Physical Transport Parameters

K   vertical eddy diffusion coefficient.

w   vertical velocity.

### Biological Parameters

$\alpha$   reciprocal length scale of oxygen consumption in the deep reservoir, equal to $\Omega/v$.

$\beta$   rate constant for the production of organic particles in the surface reservoir.

$\eta$   ratio of moles of $O_2$ consumed to moles of $CO_2$ produced by the oxidation of organic material.

$\Omega$   rate constant for the oxidation of particulate organic carbon in the deep reservoir.

r   Redfield ratio, the ratio of moles of carbon to moles of phosphorus in phytoplankton.

$R_o$   coefficient of oxygen consumption rate, used only in discussion of Wyrtki's [1962] model, equal to $\eta r \frac{w}{v} \beta h_s P_o$

v   particle settling velocity.

### Nondimensional Numbers

$\rho$   the ventilation parameter, equal to $w/K(\Omega/v) = w/\kappa\alpha$.

$\nu_w$   the Wyrtki number, equal to $(\eta r \beta h_s P_o)/[K(\Omega/v)O_o]$.

$\nu_n$   the nutrient number, equal to $\beta h/K(\Omega/v)$.

$0(x)$   unit step function where x is an arbitrary argument, equal to $1 \ x < 0$ or $0 \ x \geqslant 0$.

## References

AGU, Chapman Conference on Natural Variations in Carbon Dioxide and the Carbon Cycle, Tarpon Springs, Fla., Jan. 9-13, 1984.

Broecker, W.S., Glacial to interglacial changes in ocean chemistry, Prog. Oceanogr. 11, 151-197, 1982.

Broecker, W.S., and T.H. Peng, Tracers in the Sea, 690 pp., Lamont-Doherty Geological Observatory, Palisades, N.Y., 1982.

Honjo, S., S. Manganini, and J. Cole, Sedimentation of biogenic matter in the deep sea, Deep Sea Research, 29, 609-625, 1982.

Oeschger, H., B. Stauffer, R. Finkel, and C. Langway, Jr., Variations of the $CO_2$ concentration of occluded air and of anions and dust in polar ice cores, paper presented at Chapman Conference on Natural Variations in Carbon Dioxide and the Carbon Cycle, Tarpon Springs, Fla., Jan. 9-13, 1984.

Stauffer, B., U. Siegenthaler, and H. Oeschger, Atmospheric $CO_2$ concentration during the last 50,000y from ice core analysis, paper presented at Chapman Conference on Natural Variations in Carbon Dioxide and the Carbon Cycle, Tarpon Springs, Fla., Jan. 9-13, 1984.

Shackleton, N.J., and N.G. Pisias, Atmospheric carbon dioxide, orbital forcing, and climate, paper presented at Chapman Conference on Natural Variations in Carbon Dioxide and the Carbon Cycle, Tarpon Springs, Fla., Jan. 9-13, 1984.

Wyrtki, K, The oxygen minima in relation to ocean circulation, Deep Sea Research, 9, 11-23, 1962.

# OCEAN CARBON PUMPS: ANALYSIS OF RELATIVE STRENGTHS AND EFFICIENCIES IN OCEAN-DRIVEN ATMOSPHERIC $CO_2$ CHANGES

Tyler Volk and Martin I. Hoffert

Department of Applied Science, New York University
New York, New York 100003

Abstract. An ocean carbon pump is defined as a process that depletes the ocean surface of $\Sigma CO_2$ relative to the deep-water $\Sigma CO_2$. Three pumps are recognized: a carbonate pump, a soft-tissue pump, and a solubility pump. The first two result from the the biological flux of organic and $CaCO_3$ detritus from the ocean's surface. The third results from the increased $CO_2$ solubility in downwelling cold water and is demonstrated by a one-dimensional upwelling-diffusion model of an abiotic ocean. In the soft-tissue and solubility pumps, working strengths are defined in terms of the $\Delta\Sigma CO_2$ each creates between surface and deep-water. Efficiencies of each pump are quantified as a ratio of working strength to potential maximum strength. Using alkalinity, nitrate, and $\Sigma CO_2$ to remove the carbonate pump signal from ocean or model data, the individual working strengths of the soft-tissue and solubility pumps can be calculated by scaling the soft-tissue's $\Delta\Sigma CO_2$ to the surface-to-deep $\Delta PO_4$. This technique is applied to a three-box ocean model known to demonstrate high-latitude control of atmospheric $CO_2$ through a variety of circulation and biological changes. Considering each pump separately reveals that the various changes which lower $pCO_{2atm}$ in the model are caused primarily by an increased solubility pump. Analysis of global ocean data indicates a positive solubility pump signal, subject to uncertainties in the C:P Redfield ratio and in the preindustrial $pCO_{2atm}$. If C:P = 105 and $pCO_{2atm}$ = 270 µatm, the efficiency of the solubility pump is about 0.5. We suggest that this type of analysis of relative carbon pump strengths will be an effective method for inter-model and intra-model comparison and diagnosis of underlying oceanic mechanisms for $pCO_{2atm}$ changes.

## Introduction

Models that simulate the steady-state distribution of total carbon dioxide ($\Sigma CO_2$) and various other tracers, such as carbon 14, temperature (T), phosphate ($PO_4$), nitrate ($NO_3$), oxygen ($O_2$), salinity (S), in the ocean can potentially be utilized for studying the variation of atmospheric $CO_2$ ($pCO_{2atm}$) with ocean changes. A hierarchy of such models exists, consisting of simple box models [Keeling and Bolin, 1968], multi-box models [Bolin et al., 1983], one-dimensional models [Hoffert et al. 1981], two-dimensional models [Baes and Killough, 1985], and three-dimensional models [Maeir-Reimer, 1984].

Employing primarily the simpler models in quantifying the relationship between ocean changes and $pCO_{2atm}$ is a relatively recent interest fueled by the discovery of lowered $pCO_{2atm}$ during the last glacial period [Neftel et al., 1982; Oeschger and Stauffer, 1985]. In general, $pCO_{2atm}$ is a function of the state of the ocean's surface properties, specifically, $\Sigma CO_2$, alkalinity (TA), T, S, $PO_4$, and silicate (SI) (see the appendix). Changes in natural $pCO_{2atm}$ in time scales in which it is in equilibrium with the ocean must correlate with changes in what is a rather complex distribution of these properties.

There are two general types of changes that the ocean's surface properties can undergo. One is a change in the total ocean-atmosphere inventory of a property. An example is the higher salt/water ratio during glacial times in which the assumption might be that the change in the average equals the change in the surface. The second type of change is where the total inventory remains constant, but the distribution of a property changes. An example here would be a change in the C:P Redfield ratio of organic detritus. Hybrids are also possible, such as a change in the $PO_4$ inventory that alters the vertical distribution of $\Sigma CO_2$. Broecker [1981] explored these possibilities in developing the phosphate-extraction hypothesis for the glacial-interglacial $pCO_{2atm}$ levels.

Recent work in the direction of the second type is of interest because it examines changes in the processes within the ocean. For example, Broecker and Takahashi [1984] discussed changes in ocean circulation upon $pCO_{2atm}$, explored with a two-box ocean model put into two end-states, that of Redfield-ocean and a thermodynamic ocean. There is also the important group of three-box models which look at the role of various circulation magnitudes and high-latitude biology, temperature, and other properties on $pCO_{2atm}$ [Siegenthaler and Wenk, 1984; Knox and McElroy, 1984; Sarmiento and Toggweiler, 1984; Wenk and Siegenthaler this volume; Toggweiler and Sarmiento, this volume; Ennever and McElroy, this volume].

A problem with these models that change the distribution of $CO_2$ within a constant ocean-atmosphere inventory is that although they represent the global ocean in an exceedingly simple manner, they are fairly complex systems. Combining a simple marine biosphere with a simple circulation scheme and carbonate chemistry results in a model requiring a computer for experimentation. Understanding reasons for different states of a model is difficult enough, and comparing effects across various models is more so; we can reasonably expect this difficulty to magnify as models further up the hierarchy are employed for such studies.

The goal of this paper is to develop a quantitative vocabulary enabling us to compare different models. A key questions is: can the numerous and potentially infinite model property variations that cause $pCO_{2atm}$ changes be reduced to a smaller and finite set of more fundamental causes? One answer lies in the definition of what will be termed carbon pumps. Our approach will be to describe the pumps and formalize expressions for their relative working strengths and efficiencies. Applying these strengths and efficiencies will be shown to be valuable in analyzing both outputs of models and ocean data.

Carbon Pump Types, Efficiencies
and Strengths

Although the term "pump" is most commonly associated with mechanical devices that create a pressure gradient, it also applies to processes that create or maintain a concentration gradient, such as the sodium ion pump in the membrane of a cell. The second sense is appropriate for the ocean carbon system. One of the distinctive features of $\Sigma CO_2$ in the world ocean is that the deep-water $\Sigma CO_2$ ($\Sigma CO_{2d}$ as global mean) is higher than the surface mixed layer $\Sigma CO_2$ ($\Sigma CO_{2m}$ as global mean). This is true of all ocean regions, with the global means of $\Sigma CO_{2m}$ = 2012 µmol/kg compared to $\Sigma CO_{2d}$ (below 1200 m) = 2284 µmol/kg [Takahashi et al., 1981a]. What maintains this vertical carbon gradient? Since diffusive ocean mixing is always actively destroying differences in concentration, there must be a continuously counteracting process that is maintaining the vertical carbon gradient. The active process might be termed a global ocean carbon pump; its action is important because of the sensitivity of $pCO_{2atm}$ to $\Sigma CO_{2m}$.

The global ocean carbon pump is actually a system of components, each of which is a pump itself. Three such components may be distinguished.

Two of the pumps are biological, and will be termed the soft-tissue pump and the carbonate pump. They are due to the action of organisms at the ocean's surface removing $\Sigma CO_2$ and incorporating it into organic and calcium carbonate particles, respectively. These particles are transported into the depths through gravitational settling and active biotransport [Redfield et al., 1963] where the carbon and other components are returned to their dissolved, inorganic forms.

The important effect of the soft-tissue pump on the vertical distribution of $\Sigma CO_2$, $PO_4$, $NO_3$, and $O_2$ is well recognized [Broecker, 1983]. $PO_4$ and $NO_3$ affect TA; the surface depletion of $PO_4$ and $NO_3$ by the soft-tissue pump raises TA relative to what it would be otherwise. The reduction of $\Sigma CO_2$ and elevation of TA by the soft-tissue pump both serve to lower $pCO_{2atm}$ in equilibrium with the surface at a constant temperature and salinity.

The carbonate pump has competing effects on $pCO_2$: the lowering of TA due to the removal of calcium ions raises $pCO_2$, while the lowering of $\Sigma CO_2$ through the carbonate ion removal lowers $pCO_2$. The TA effect is dominant in the vicinity of present surface conditions, and so a stronger carbonate pump raises $pCO_2$. A full treatment of the carbonate pump in the terminology of strength and efficiency, as will be done for the soft-tissue pump, must be deferred to a later work. The primary concern here will be to remove the carbonate pump signal from the $\Sigma CO_2$ data.

The third pump refers to a process which we term the solubility pump and is driven by the $CO_2$ solubility differences of warm and cold water, i.e., the surface and deep-water. The solubility pump can be examined through a modeling experiment.

We use the one-dimensional upwelling diffusion ocean model of Hoffert et al. [1981] to produce a vertical $\Sigma CO_2$ distribution in the absence of a marine biosphere. See Figure 1a and the appendix for the model description. In the absence of the soft-tissue and carbonate pumps, the solubility pump alone determines the $\Sigma CO_2$ profile, unlike the real ocean, which is an as yet undetermined mixture of the three. Without biology, the model's solution reduces to a particularly simple form, with the profile determined by $\Sigma CO_{2w}$ (the warm mixed layer concentration), $\Sigma CO_{2u}$ (the concentration of the upwelling water at the base of the one-dimensional column), and a depth parameter, $z^*$ = K/w, where K is the average vertical eddy diffusivity and w the upwelling velocity. Setting $z^*$ = 500 m, calibrated from the globally-averaged vertical temperature profile [Hoffert et al., 1980], and requiring the inventory of $CO_2$ in the ocean-atmosphere system to be the same as measured by the Geochemical Ocean Sections Study (GEOSECS) of $3.16 \times 10^{18}$ moles of $CO_2$ for the ocean [Takahashi et al., 1981a] reduces the problem to specifying a relationship between $\Sigma CO_{2u}$ and $\Sigma CO_{2w}$. Their relation can develop a measure of the efficiency of the solubility pump.

As the overturning water moves from the warm surface box to the cold surface box, its $pCO_2$ will drop due to cooling, putting it below $pCO_{2atm}$ and,

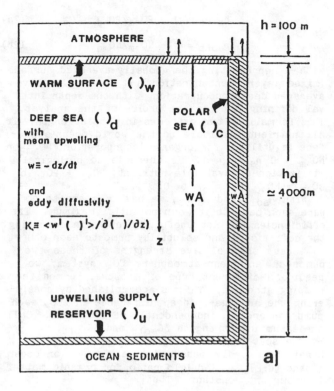

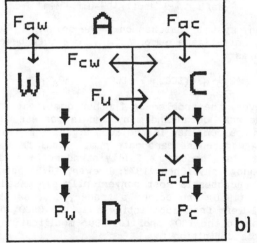

Fig. 1. (a) Schematic of the upwelling-diffusion model of Hoffert et al. [1980; 1981]. (b) Schematic of the Siegenthaler and Wenk [1984] model. Also see Wenk and Siegenthaler [this volume].

depending upon its residence time, will gain $CO^2$ from the atmosphere through the cold air-sea interface, determining its concentration, $\Sigma CO_{2C} = \Sigma CO_{2u}$. Two end-cases are possible.

Case 1. Cold water sinks from the surface in equilibrium with the atmosphere, i.e., $pCO_{2C} = pCO_2$ atm.

Case 2. Cold water sinks from the surface at the $\Sigma CO_2$ concentration of the warm surface, i.e., $\Sigma CO_{2C}$

$= \Sigma CO_{2W}$. In this case of isochemical cooling, no $CO_2$ is picked up from the atmosphere.

Since case 1 is the maximum possible difference between $\Sigma CO_{2C}$ and $\Sigma CO_{2W}$, it will create the largest possible gradient in an abiotic ocean's vertical $\Sigma CO_2$ profile and can be considered the case in which the solubility pump is the strongest. In case 2 the solubility pump is the weakest. We therefore define a solubility pump efficiency, $\eta_{solu}$, as

$$\eta_{solu} = (\Delta C_{solu})/(\Delta C_{solu,max}) \qquad (1)$$

where $\Delta C_{solu}$ and $\Delta C_{solu,max}$ are the actual and maximum surface-to-deep $\Sigma CO_2$ differences, respectively, due to the solubility pump. $\Delta C_{solu,max}$ is calculated as the amount of $\Sigma CO_2$ added to $\Sigma CO_{2W}$ if the $\Sigma CO_{2W}$ water is cooled to the cold region's temperature and equilibrated with $pCO_{2atm}$. In case 1 above, $\eta_{solu} = 1.0$ and in case 2, $\eta_{solu} = 0$. In one-dimensional abiotic model, if $\Sigma CO_{2C},eq$ represents the $\Sigma CO_{2C}$ for case 1, $\eta_{solu} = (\Sigma CO_{2C} - \Sigma CO_{2W})/(\Sigma CO_{2C}, eq - \Sigma CO_{2W})$. For any given $\eta_{solu}$, there is a unique steady-state $\Sigma CO_2$ distribution that satisfies the constraint of constant $CO_2$ in the atmosphere-ocean system and with gas exchange simplified so that $pCO_{2W} \approx pCO_{2atm}$ (see appendix).

Figure 2a shows the abiotic $\Sigma CO_2$ profile when $\eta_{solu} = 1$ and $\eta_{solu} = 0$ against the GEOSECS globally averaged data. Two points are apparent. First, with the solubility pump at its maximum efficiency, the surface-to-deep $\Sigma CO_2$ difference is less than that of the actual ocean. This demonstrates that the solubility pump alone could not produce the present ocean's $\Sigma CO_2$ profile. Second, the solubility pump can maintain a $\Sigma CO_2$ profile that is similar to the ocean's in respect to its causing a definite depletion of the surface $\Sigma CO_2$ relative to its deep value, i.e., $\Sigma CO_{2W} < \Sigma CO_{2d}$, thus demonstrating the pumping effect.

Figure 2b shows $pCO_{2atm}$ as a function of $\eta_{solu}$ for $0 < \eta_{solu} < 1$. The resulting $pCO_{2atm}$ varies from 720 µatm to 460 µatm, respectively, again demonstrating the pumping power of the solubility pump from its absence, $\eta_{solu} = 0$, to its maximum strength, $\eta_{solu} = 1$. With the soft-tissue and carbonate pumps on, the model produces a $pCO_{2atm}$ of 260 µatm [Volk, 1984], therefore, assuming that $\eta_{solu} = 1$ in the abiotic state, this model calculates the presence of the marine biosphere maintains $pCO_{2atm}$ 200 µatm lower than it would be otherwise. Although one could define the $\Delta pCO_{2atm}$ from pump minimum to pump maximum as a measure of strength, it will the more fruitful to define the pump's strength in terms of the surface-to-deep $\Sigma CO_2$ difference it creates, as discussed below, and which is different than the efficiency.

Note that the concept is unchanged if the ocean model has deep water upwelling into the polar sea, instead of being transported from the warm surface box. Typically, models such as Siegenthaler and Wenk's (1984) consider the polar surface as a mixture of transport between warm surface and cold surface and between deep and cold surface regions.

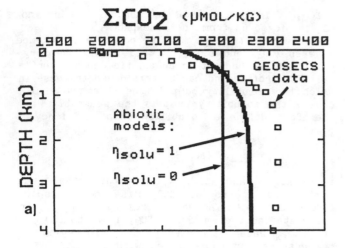

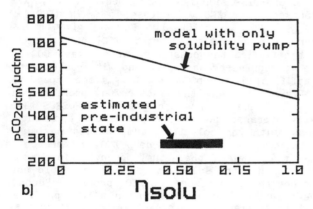

Fig. 2. (a) $\Sigma CO_2$ model profile of an abiotic ocean using the upwelling-diffusion model for a full and absent solubility pump, $\eta_{solu} = 1$ and $\eta_{solu} = 0$, respectively. The $CO_2$ in the model ocean-atmosphere system is a constant; the salinity-normalized GEOSECS data is from Takahashi et al. [1981b]. (b) Plot of $pCO_{2atm}$ as a function of the solubility pump's efficiency, $\eta_{solu}$, in an abiotic model. The $\eta_{solu}$ for the preindustrial state is taken from Table 2.

The upwelling into the cold region from deep in an abiotic ocean could still be considered as between two extremes, reaching equilibrium with the atmosphere ($\eta_{solu} = 1$), or alternatively, upwelling under sea ice and having no gas exchange with the atmosphere ($\eta_{solu} = 0$).

Understanding the solubility pump by itself is relatively straightforward, but the real ocean and often models are a mixture of the three pumps. In order to begin a comparison, we first develop an efficiency for the soft tissue pump, similarly as was done for the solubility pump.

A soft-tissue efficiency, $\eta_{soft}$, is obtained by assuming that it is tied to the ocean's nutrient ($PO_4$ or $NO_3$) content. Then

$$\eta_{soft} = 1 - PO_{4m}/PO_{4d} \tag{2a}$$

or

$$\eta_{soft} = 1 - NO_{3m}/NO_{3d} \tag{2b}$$

with m representing the globally-averaged surface (mixed layer) concentration and d the globally-averaged deep concentration, with the depth interval defining d being the choice of the analyst. This formalizes two extremes in pump strength. If all nutrient available to the world ocean's surface is utilized for organic incorporation, $PO_{4m}$ or $NO_{3m} = 0$ $\eta_{soft} = 1$. If there is no biological utilization of available nutrient, $PO_{4m} = PO_{4d}$ or $NO_{3m} = NO_{3d}$ and $\eta_{soft} = 0$.

With $\eta_{solu}$ and $\eta_{soft}$, we have a means to compare each pump to its own potential maximum. The efficiencies are not useful, however, in comparing the soft-tissue and solubility pump to each other. To assess the relative strengths of these two pumps in an ocean-atmosphere $CO_2$ system, we require a different type of measure, termed the pumping strength. This is accomplished by considering the surface-to-deep difference in $\Sigma CO_2$ each pump can create, independently of the other. Call these pumping strengths $\Delta C_{solu}$ and $\Delta C_{soft}$.

The soft-tissue strength, $\Delta C_{soft}$ can be straightforwardly measured in any model or ocean if the $[C:P]_{soft}$ Redfield ratio for organic matter is known and constant. Then,

$$\Delta C_{soft} = [C:P]_{soft} (PO_{4d} - PO_{4m}) \tag{3}$$

In any system that has only the solubility pump and the soft-tissue pump, $\Delta C_{solu}$ can then be calculated as

$$\Delta C_{solu} = (\Sigma CO_{2d} - \Sigma CO_{2m}) - \Delta C_{soft} \tag{4}$$

However, the problem is further complicated in that the real ocean and many models possess all three pumps combined in the $\Sigma CO_2$ signal. A means to disaggregate the carbonate pump signal from the $\Sigma CO_2$ data is a procedure fairly standard in chemical oceanography (Chen, 1978; Brewer, 1978) but for present purposes is most conveniently expressed by Fiadeiro (1980) who coined a tracer, $C_{ox}$, as the residual left from subtracting away the $CaCO_3$ contribution to the $\Sigma CO_2$ profile. Our modification of Fiadeiro's definition is

$$C_{ox} = \Sigma CO_2 - (1/2)(TA - TA_m + NO_3 - NO_{3m} + PO_4 - PO_{4m}) \tag{5}$$

We have modified Fiadero's original formulation by adding the constants $TA_m$, $NO_{3m}$, and $PO_{4m}$ so that $C_{ox,m} = \Sigma CO_{2m}$, for interpretive convenience. Equation (5) assumes that 1 equivalent of TA is lost for every mole of $NO_3$ or $PO_4$ created. Although the ocean's $PO_4$ variation is within the range of uncertainty of the $NO_3$ measurements and essentially can be ignored in analyzing ocean data (Fiadeiro, 1980), in a model the TA change due to $PO_4$ as well as $NO_3$ can be exactly specified.

Equation (5) eliminates the carbonate pump

signal from $\Sigma CO_2$ data and therefore leaves only the solubility and soft-tissue pumps in the generated tracer of $C_{OX}$. Then we define a $\Delta C_{OX}$, such that

$$\Delta C_{OX} = C_{OX,d} - C_{OX,m} \qquad (6)$$

$\Delta C_{OX}$ must contain both remaining pump signals, so $\Delta C_{OX} = \Delta C_{soft} + \Delta C_{solu}$

$$\text{or } \Delta C_{solu} = \Delta C_{OX} - \Delta C_{soft} \qquad (7)$$

Thus in any system consisting of all three pumps, the working strengths of the solubility and soft-tissue pump can be found knowing $\Sigma CO_2$, TA, $NO_3$, $PO_4$, and $[C:P]_{soft}$ (or $[C:N]_{soft}$).

With the resulting $\eta_{solu}$, $\eta_{soft}$, $\Delta C_{soft}$, and $\Delta C_{solu}$ we have information about the solubility and soft-tissue pumps' efficiencies ($\eta$'s) and strengths ($\Delta C$'s). The efficiencies are useful for comparing each pump to its own maximum and the strengths for comparing the pumps to each other. The next step is to use these formulations in analysis.

### Analysis of Models

With the technique of pump strength analysis we make the following suggestions. Since the combined pump strengths were defined in terms of the surface-to-deep $\Sigma CO_2$ difference, any changes in a closed atmosphere-ocean carbon system that involves change in this difference should be a change in one or more of the pumps. The magnitude of change for each pump should be calculable. Second, since this kind of analysis has not been performed on models before, its use should give insight into the operation of models, both between different experiments with a particular model and between different models. The possibility for intermodel comparison is potentially most important, since up to this point the community has only had the rise or fall of $pCO_{2atm}$ as a common effect to use in model comparison. The pumps should allow distinctions of similarities and differences regarding the causal connection between $pCO_{2atm}$ and the processes within the models.

The first model to examine is that of Broecker and Takahashi (1984). There is a warm surface box and a cold deep box which outcrops in a polar sea. Thus both warm and deep boxes have contact with the atmosphere. A soft tissue pump is present only in the warm box; the alkalinity is constant, there is neither a carbonate pump nor TA changes associated with the soft-tissue pump.

Two ends states are calculated. In the first, the so-called Redfield-ocean, the warm-to-deep $\Sigma CO_2$ difference is fully determined by the particulate flux associated with the phosphate difference. Since $\Delta C_{solu} = 0$, by equation (1) $\eta_{solu} = 0$. This is an ocean with a soft-tissue pump only.

The other end state is the thermodynamic-ocean. Here an infinitely fast gas exchange brings both deep and warm boxes into equilibrium: $pCO_{2w} =$

$pCO_{2d} = pCO_{2atm}$. Now $\Delta C_{OX}$ is less than $\Delta C_{soft}$. The $\Delta C_{soft}$ has remained identical to that of the Redfield-ocean because the $PO_4$ distribution is unchanged. From equation (7), this means $\Delta C_{solu}$ is less than 0. The solubility pump is only positive when the cold region is a sink for the atmospheric $CO_2$ flow. In this thermodynamic ocean, the cold region is a source. The rapid gas exchange reduces the $\Delta C_{OX}$ from the Redfield-ocean to that of $\Delta C_{solu,max}$. The soft-tissue pump remained constant in the two cases, while the solubility pump changed from 0 to a negative value between the Redfield- and thermodynamic-oceans. Accordingly, $pCO_{2atm}$ increased in this change. A decrease in the sum of the strengths of the soft-tissue and solubility pumps increases $pCO_{2atm}$.

This shows we can understand the operation of the Broecker and Takahashi model through the carbon pumps. Furthermore, the absence of a positive solubility pump in their model suggests that the Redfield-ocean is not a true end state in the creation of the largest $\Delta C_{OX}$ possible. The maximum $\Delta C_{OX}$ in this model occurs when the $\Sigma CO_{2d} - \Sigma CO_{2w}$ is obtained solely from the soft-tissue pump. Can a positive solubility pump exist in addition to a soft tissue pump? A simple thought-experiment shows this to indeed be possible.

Consider an ocean with a warm surface box that shuttles water to a cold polar box. There it picks up $CO_2$ from the atmosphere; it then descends into a deep box and there receives the oxidation products from the soft-tissue pump. Therefore, its $\Delta C_{OX}$ is greater than $\Delta C_{soft}$. This addition of a cold polar box with some exchange from a warm surface box is exactly the structure of the so-called high-latitude models of Siegenthaler and Wenk (1984), Knox and McElroy (1984), and Sarmiento and Toggweiler (1984).

These high-latitude models consist of three ocean boxes, a warm surface, a cold surface, and a deep box. Soft-tissue and carbonate pumps exist in both surface boxes, and various vertical and horizontal advective and exchange fluxes exist between the boxes. The Wenk and Siegenthaler (1985, this volume) model finds that decreased vertical exchange between the cold and deep boxes, increased horizontal exchange between the warm and cold boxes, and an increased biological flux from the cold box all individually reduce $pCO_{2atm}$. The causes for these $pCO_{2atm}$ changes are not immediately comprehensible, and if the carbon pumps analysis is useful, then it should be able to elucidate new information from the high-latitude models.

We have constructed the Siegenthaler and Wenk model and run their standard-case and two circulation-change cases. See Figure 1b and the appendix for the model description. The soft-tissue pump's contribution to TA changes was treated more explicitly than in the Siegenthaler and Wenk documentation of the model so that equations (1), (2), (3), and (7) could be used to construct the efficiencies and strengths for the solubility and soft-tissue pumps for each model run.

TABLE 1. Three Box Model Analysis

| Property | Box[a] | Standard Case (st) | $1/2 \times F_{cd,st}$ | $2 \times F_{cw,st}$ | Notes |
|---|---|---|---|---|---|
| $PCO_{2atm}$[b] | a | 283 | 230 | 256 | |
| | w | 0.30 | 0.30 | 0.30 | |
| $PO_4$[c] | c | 1.43 | 1.06 | 1.23 | |
| | m | 0.42 | 0.38 | 0.40 | |
| | d | 2.24 | 2.24 | 2.24 | |
| | w | 2324 | 2324 | 2324 | |
| $TA$[c] | c | 2364 | 2351 | 2357 | |
| | m | 2329 | 2327 | 2328 | |
| | d | 2392 | 2392 | 2392 | |
| | w | 1958 | 1913 | 1937 | |
| $CO_2$[c] | c | 2158 | 2101 | 2127 | |
| | m | 1980 | 1934 | 1958 | |
| | d | 2284 | 2286 | 2285 | |
| | m | 1980 | 1934 | 1958 | Equation (5) |
| $C_{ox}$[c] | d | 2236 | 2238 | 2237 | |
| $\Delta C_{ox}$[c] | | 256 | 304 | 279 | $C_{ox,d} - C_{ox,m}$ |
| $\Delta C_{solu,max}$[c] | | 170 | 178 | 173 | Note d |
| $\Delta C_{soft}$[c] | | 236 | 242 | 239 | Equation (3) |
| $\eta_{soft}$ | | 0.81 | 0.83 | 0.82 | Equation (2) |
| $\Delta C_{solu}$[c] | | 20 | 62 | 40 | Equation (7) |
| $\eta_{solu}$ | | 0.12 | 0.35 | 0.23 | Equation (1) |

Output of the Siegenthaler and Wenk [1984] model of Figure 1b, analyzed in terms of carbon pumps for each case.

a   a,w,c and d refer to the atmosphere, warm surface, cold surface, and deep reservoirs of Figure 1b. m is the area-weighted average surface concentration; for this model $( )_m = 0.89 ( )_w + 0.11 ( )_c$

b   µatm

c   µmol/kg

d   $C_{solu,max}$ is the amount of $CO_2$ that would be gained by $C_{ox,m}$ if its $pCO_2$ equaled $pCO_{2atm}$ at the polar temperature (3°C in this model), keeping $TA_m$ and $PO_{4m}$ constant.

Results are displayed in Table 1. It reveals that both the case of halving the cold-deep vertical exchange and that of doubling the warm-cold horizontal exchange reduce $pCO_{2atm}$ primarily by increasing the solubility pump over the standard-case. Take the first case. The $\Delta C_{soft}$ does increase slightly, from 236 µmol/kg to 242 µmol/kg, representing a change in $\eta_{soft}$ from 0.81 to 0.83. However, the $\Delta C_{solu}$ increase is more substantial, from 20 µmol/kg to 62 µmol/kg, with the corresponding increase in $\eta_{solu}$ from 0.12 to 0.35. The change in $\Delta C_{solu}$ is 7 times the change in $\Delta C_{soft}$, showing that the pulling of additional $CO_2$ into the ocean in the lowering of $pCO_{2atm}$ case is due to an increase in strength of solubility pump. The case of doubling the warm-cold flux reveals the same underlying cause for the $pCO_{2atm}$ change, an increase in solubility pump.

Knox and McElroy focused on changes in the marine biosphere in the cold box. By allowing the simulated glacial state's cold box $PO_4$ and $NO_3$ to go to zero from an interglacial state where $PO_4$ = 1.1 µmol/kg and $NO_3$ = 17 µmol/kg in the cold box, the model reduced $pCO_{2atm}$ from 288 µatm to 161 µatm. It might superficially appear that this change was due to an increased soft-tissue pump. However, our carbon pump analysis of the model calculated that the $\Delta C_{soft}$ only slightly increased from an interglacial 225 μmol/kg to a glacial 231 μmol/kg while $\Delta C_{solu}$ went from 22 μmol/kg to 149 μmol/kg. It is actually the solubility pump, brought into full action in their glacial case, that caused the dramatic $pCO_{2atm}$ reduction. Although the soft-tissue pump in the cold box became stronger in the glacial state, it could do no more than contribute to reducing the entire model's surface nutrients to zero. For example, with a $NO_{3d}$ = 33 μmol/kg and a $[C:N]_{soft}$ = 7, their soft-tissue pump acting at full strength, i.e., $\eta_{soft}$ = 1, could contribute only 231 μmol/kg to a $\Delta C_{ox}$. The actual $\Delta C_{ox}$ in this case was 380 μmol/kg, a surface-to-deep $C_{ox}$ (or $\Sigma CO_2$) difference impossible to achieve with the soft-tissue pump changes alone.

The result is that two circulation changes and one biological change of the high-latitude models are acting in a similar manner in that they each cause an increase in the solubility pump. This means that the carbon pumps analysis allows a level of understanding to be brought to bear on a model's specific causes and the $pCO_{2atm}$ results. Without a comparison of the strengths of the pumps, it is difficult to relate $pCO_{2atm}$ changes to their underlying causes. With such a comparison, numerous specific experiments may be related directly to changes they create in the carbon pumps.

Before moving on to an analysis of the GEOSECS ocean data, we note that in both interglacial cases of Siegenthaler and Wenk and Knox and McElroy, the $\Delta C_{ox}$ is greater than $\Delta C_{soft}$, demonstrating the presence of a positive solubility pump that added to the $\Sigma CO_2$ difference made by the soft-tissue pump (and carbonate pump). Thus the Redfield-ocean of Broecker and Takahashi is an end state only in that particular model, not an end state for other models or for the ocean itself. One last point is that our doubling the $CaCO_3:C_{soft}$ ratio in the particulate flux from 0.2 (the standard case) to 0.4, a 100% increase in the carbonate pump, raised $pCO_{2atm}$ from 283 μatm to 309 μatm. This is relatively small compared to a typical reduction of 70 μatm in $pCO_{2atm}$ with an approximately 50% increase in the soft-tissue pump (Broecker, 1981).

Analysis of the Present Ocean State

What is the state of the carbon pumps in the ocean today; and can we deduce the former steady-state of the pre-industrial atmosphere-ocean system?

The GEOSECS data from Takahashi et al (1981b) for $PO_4$, $NO_3$, TA, and $\Sigma CO_2$ for ten ocean regions were area-weighted with data from Levitus (1982) to construct global surface (< 75 m) and deep (> 2000 m) water values. These are listed in Table 2 for four cases, the global values, the global values excluding the Antarctic Circumpolar area south of 50°S, the global values normalized to S = 35°/oo, and the global values excluding the Antarctic Circumpolar and normalized to S = 35°/oo. These four will show the differences that

arise in the analysis between including or excluding the Antarctic Circumpolar with its unusual surface properties and between the actual and salinity-normalized cases.

From the $PO_4$, $NO_3$, TA, and $\Sigma CO_2$ data, $C_{ox}$ was constructed using equation (5). The problem that now arises is that the $C_{ox,m}$ ($\Sigma CO_{2m}$) measured during the GEOSECS was already somewhat increased over the preindustrial value by absorption of $CO_2$ from the $pCO_{2atm}$ transient driven by fossil fuel burning. The $pCO_{2atm}$ during GEOSECS was approximately 320 μatm, about 8 μ atm above the globally-averaged $pCO_{atm}$ of 312 μatm (Takahashi, personal communication, 1984). In a preindustrial steady-state, $pCO_{2m}$ = $pCO_{2atm}$. The preindustrial $pCO_{2atm}$ is not known to perhaps less than 20 μatm. Moor and Stauffer (1984) indicate at preindustrial range of 270-280 μatm, while previous estimates gave a value closer to 260 μatm (Oeschger and Stauffer, 1984). We used the values of $PO_{4m}$, $TA_m$, $\Sigma CO_{2m}$ along with S = 35°/oo and SI = 5 μmol/kg and adjusted the $T_m$ for each of the four cases until a $pCO_{2m}$ = 312 μatm was returned from the equilibrium carbonate chemistry (see appendix). The $pCO_{2m}$ was then changed to 270 ± 10 μatm for the four cases. The $\Sigma CO_{2m}$ differences between the $pCO_{2m}$ = 312 μatm and $pCO_{2atm}$ = 270 μatm varied by less than 3% in the four cases. The correction to $C_{ox,m}$ for a preindustrial $pCO_{2atm}$ = 270 ± 10 μatm was determined to be -31 ± 9 μmol/kg. This correction is applied to obtain a preindustrial $C_{ox,m}$ in Table 2.

Another assumption in this analysis is that of $[C:P]_{soft}$ = 105, as in Redfield et al (1963). Recent work by Takahashi et al (1985) using isopycnal surfaces indicates $[C:P]_{soft}$ = 103 if $C_{ox}$ and $PO_4$ are examined directly, but $[C:P]_{soft}$ = 140 if $O_2$ and $NO_3$ are used to deconvolve the $C_{ox}$ variation. Table 2 shows the $\Delta C_{soft}$ using equation (3) and $[C:P]_{soft}$ = 105.

Using the preindustrial $C_{ox,m}$ and the GEOSECS $C_{ox,d}$, $\Delta C_{ox}$ is obtained and is significantly larger than $\Delta C_{soft}$. This indicates the presence of a positive solubility pump signal in the globally-averaged ocean. The $\Delta C_{solu}$ is obtained using equation (7).

Equation (2) calculates $\eta_{soft}$, which is higher in the cases that excluded the Antarctic circumpolar area ($\eta_{soft}$ = 0.85) than in the global value ($\eta_{soft}$ = 0.77), because the high nutrient values in the circumpolar represent an inefficiency in the soft-tissue pump.

To obtain an ocean value of $\Delta C_{solu,max}$, water with properties of $PO_{4m}$, $TA_m$, SA = 35 °/oo, SI = 5 μmol/kg, and T = 1.5° C was equilibrated with $pCO_{2atm}$ = 312 μatm using the carbonate chemistry routines to calculate a maximum $C_{ox}$ that could be created by the solubility pump alone. The difference between this value and $C_{ox,m}$ (GEOSECS) gives a $\Delta C_{solu,max}$ and finally $\eta_{solu}$ from equation (1). The range of $\eta_{solu}$ for the four cases for the $pCO_{2atm}$ pre-industrial value of 270 ± 10 μatm is 0.43 - 0.68. Efficiency values for the S-normalized cases were about 0.06 higher than the non-normalized cases and the cases omitting the

TABLE 2.   GEOSECS Data Analysis[a]

| Property[b] | Depth (m) | Global | Global without Antarctic Circumpolar Area | Global S=35 /oo | Global without Antarctic Circumpolar Area, S = 35°/oo | Notes |
|---|---|---|---|---|---|---|
| $PO_4$ | <75 | 0.50 | 0.33 | 0.51 | 0.33 | |
| | >2000 | 2.20 | 2.20 | 2.21 | 2.22 | |
| $NO_3$ | <75 | 5.48 | 2.60 | 5.62 | 2.59 | |
| | >2000 | 32.60 | 32.66 | 32.87 | 32.93 | |
| TA | <75 | 2324 | 2327 | 2322 | 2314 | |
| | >2000 | 2395 | 2400 | 2414 | 2419 | |
| $TCO_2$ | <75 | 2016 | 1997 | 2016 | 1985 | |
| | >2000 | 2288 | 2292 | 2306 | 2310 | |
| $C_{ox}$ | <75 | 2016 | 1997 | 2016 | 1985 | Equation (5) |
| | >2000 | 2238 | 2240 | 2246 | 2242 | |
| $C_{ox}$ | <75 | 1985 ±9 | 1966 ±9 | 1985 ± 9 | 1954 ± 9 | Preindustrial |
| | >2000 | 2238 | 2240 | 2246 | 2242 | (see text) |
| $\Delta C_{ox}$ | | 253 ±9 | 274 ±9 | 260 ± 9 | 288 ± 9 | Preindustrial ($C_{ox,d}-C_{ox,m}$) |
| $\Delta C_{solu,max}$ | | 136 | 159 | 135 | 159 | See text |
| $\Delta C_{soft}$ | | 178 | 196 | 178 | 198 | Equation (3) |
| $\eta_{soft}$ | | 0.77 | 0.85 | 0.77 | 0.85 | Equation (2) |
| $\Delta C_{solu}$ | | 74 ±9 | 78 ±9 | 82 ±9 | 90 ±9 | Equation (7) Preindustrial |
| $\eta_{solu}$ | | 0.54 ±0.07 | 0.49 ±.06 | 0.61 ± .07 | 0.56 ± .06 | Equation (1) |

a    Global data from GEOSECS [Takahashi et al. 1981b] used to calculate the global properties of the soft-tissue and solubility pumps, using area-weighting from Levitus [1982].
b    μmol/kg, except for the non-dimensional $\eta_{solu}$ and $\eta_{soft}$.

Antarctic circumpolar about were 0.05 lower than the global cases.

According to this analysis, there is a solubility pump operating in the ocean's $\Sigma CO_2$ signal at about 50% efficiency. It creates a global $\Sigma CO_2$ surface-to-deep difference ($\Delta C_{solu}$) of 74 μmol/kg for the global, non-S-normalized case, compared to the difference created by the soft-tissue pump ($\Delta C_{soft}$) of 178 μmol/kg. The solubility pump's strength appears therefore to be about 40% that of the soft tissue pump.

This analysis is subject to uncertainty in the preindustrial $pCO_{2atm}$ and in the $[C:P]_{soft}$ ratio. For example, the solubility pump signal disappears from the analysis if $[C:P]_{soft} = 140$ and if $pCO_{2atm}$ was somewhat above 280 μatm. More certain analysis of the ocean data with the technique of pump strengths would have to wait for more accuracy in these values.

Finally, the 1-D upwelling-diffusion model is used with an internal source term to simulate a global $PO_4$ profile in Figure 3a. See the appendix for details. Using the $PO_4$ model as a calibration, fluxes associated with the internal source term and the change of warm surface water into bottom upwelling water are scaled by N:P = 16:1, and a $NO_3$ model is produced in Figure 3b. This scaling of fluxes then is put at 105:1 to create a $C_{ox}$

## PHOSPHATE (μMOL/KG)

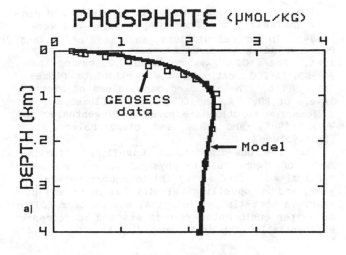

a)

## NITRATE (μMOL/KG)

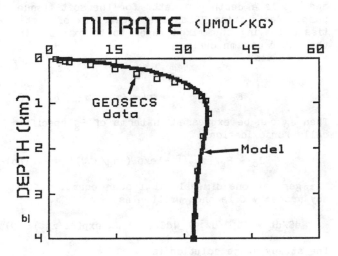

b)

## $C_{OX}$ (μMOL/KG)

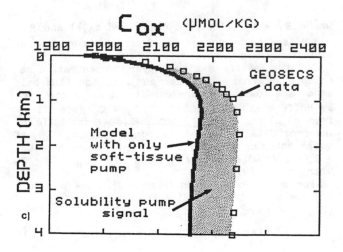

c)

profile with $C_{OX,m}$ (preindustrial) = 1966 μmol/kg. This $C_{OX}$ model profile does not match the actual $C_{OX}$ profile, generated from the GEOSECS data. The difference between the data and the model is the visual representation of the presence of the solubility pump as quantified in Table 2.

### Discussion

A technique for analyzing a three-component carbon pump system has been developed. Treatment of the carbonate pump was primarily to subtract it from the data to show a combined signal of the solubility and soft-tissue pumps. These can be isolated through nutrient-carbon coupling.

The technique has proven useful in a variety of instances. The Broecker and Takahashi [1984] ocean circulation change study apparently centers around the variation of the solubility pump between zero and a negative value. That the solubility pump can be negative has to do with what will now be shown to be a relationship between the soft-tissue and solubility pumps.

With the present ocean $PO_4$ average content of about 2 μmol/kg, if $\eta_{soft}$ = 1 a maximum $\Delta C_{soft}$ can be maintained that is 105 x 2 = 210 μmol/kg. For today's values of $\Sigma CO_{2m}$, $TA_m$, and $S$, the difference in equilibrium $\Sigma CO_2$ between 20°C and 2°C at a constant and approximately current $pCO_2atm$ is a $\Delta C_{solu,max}$ of about 140-170 μmol/kg. That the present $\Delta C_{soft,max}$ is larger than $\Delta C_{solu,max}$ is important because it means that if deep water such that $(\Sigma CO_{2d} - \Sigma CO_{2m}) > \Delta C_{solu,max}$ upwells in the polar regions, its $pCO_2$ will be greater than $pCO_2atm$, and the region will outgas $CO_2$ to the atmosphere and be a source. In this case the ocean would not be able to take advantage of the solubility pump. On the other hand, if the average $PO_4$ content was only 1 μmol/kg, then the polar regions could only be sinks.

Our analysis indicates that the three-box high-latitude models exhibit variations primarily in their solubility pump. How does this work? One example is the case of decreased exchange between

Fig. 3. (a) Model-generated $PO_4$ profile using the upwelling-diffusion model with a calibrated internal source term (J(z) and a parameterization ($F_{wu}$) of preformed $PO_4$ and entrained $PO_4$. (b) Model-generated $NO_3$ scaling the J(z) and $F_{wu}$ from the $PO_4$ model by N:P = 16:1. (c) Model-generated $C_{OX}$ profile scaling J(z) and $F_{wu}$ to $PO_4$ by $[C:P]_{soft}$ = 105. This profile, like that of $PO_4$ and $NO_3$ contains only a soft-tissue pump, while the $C_{OX}$ data will have both solubility and soft-tissue pumps (the carbonate pump was eliminated in forming $C_{OX}$, see equation (5)). The possible presence of a positive solubility pump in the data is shown by the gap between model and data. $C_{OX,w}$ was taken as 1966 μmol/kg, for a preindustrial $pCO_2atm$ of 270 μatm (see Table 2). The GEOSECS data is for in situ salinity from Takahashi et al. [1981b].

the deep and cold surface boxes. The deep box can be thought of as contaminating the cold box's possibility to use the solubility pump, since exchange with the deep box brings water into the cold box such that $(C_{ox,d} - C_{ox,m}) > \Delta C_{solu,max}$ and its presence will therefore reduce the effectiveness of the solubility pump. It appears that a slight inefficiency in the global biological pump, if it occurs in the polar regions, can clog the solubility pump. This is a linkage between the two. This clogging of the solubility pump is due to its dependence on a small region of the world's ocean surface for operation, while the soft-tissue pump acts vertically everywhere. The solubility pump's downward vertical action of water potentially enriched in $\Delta C_{solu}$ occurs only in the polar regions in these models. In the ocean, intermediate waters also may act as parts of the solubility pump. This should be examined.

These demonstrations of the pump analysis have been with the simplest models. Further tests of the technique should apply the analysis to more detailed models, such as the twelve-box global model of Bolin et al. [1983] and the two-dimensional model of Baes and Killough [1985]. Do solubility pump signals exist in the steady-states of these models? Can the analysis, which depends here upon a global aggregation of concentrations, be useful in more complex models?

Hopefully, yes, for the technique appears to be of use in looking at the aggregated GEOSECS data. If any model's $\Delta C_{ox}$ is greater than $\Delta C_{soft}$ then the solubility pump is present. It should not matter how many horizontal boxes have to be averaged to obtain the aggregated data for the pump analysis. In some ways, the models are more attractive domains for the application of the technique than the actual ocean, because the Redfield ratios in models are specified exactly, unlike the situation in the ocean.

The main accomplishment has been to add a level in the causal hierarchy. Specific ocean changes can cause changes in the pumps which affect $pCO_{2atm}$. This should facilitate a classification of model types. Perhaps all ocean-driven $CO_2$ changes can be put into the pump scheme. For example, the phosphate extraction model [Broecker, 1981] is clearly a variation of the soft tissue pump. Furthermore, new questions can be asked. For instance, does an increased soft-tissue pump due to elevated $PO_4$ levels aid in clogging the polar regions and therefore decrease the solubility pump? Distinguishing the strengths and efficiencies of the pumps may help formulate questions about models and possible ocean changes in the future.

### Appendix: Model Calculations

Routines to calculate $pCO_2$ from $\Sigma CO_2$ or $\Sigma CO_2$ from $pCO_2$, knowing TA, T, S, $PO_4$, and SI were provided to us by T. Takahashi. The calculations essentially follow those presented by Takahashi et al. [1982].

The three-box ocean model in Figure 1b and results in Table 1 are from Siegenthaler and Wenk [1984]. In our calculations, each mole of $NO_3$ and $PO_4$ created is taken to decrease TA by 1 equivalent. The $CaCO_3:C_{soft}$ ratio is 0.2, making the $PO_4:NO_3:TA:\Sigma CO_2$ ratio in the particulate fluxes 1:16:35:156. The standard case values of average levels of $PO_4$, TA, and $\Sigma CO_2$ and the fluxes and conditions for solution are given in Siegenthaler and Wenk [1984] and Wenk and Siegenthaler [this volume].

For the one-dimensional upwelling-diffusion model of Figure 1a, the physical parameters are a vertical eddy diffusivity (K) in square meters per year, and an upwelling velocity (w) in meters per year. A soft-tissue biological source term (J) in moles per cubic meter year is assumed to decrease exponentially with depth according to

$$J(z) = J_0 \exp(-z/z_{ox})$$

where $J_0$ is the soft-tissue pump source at $z = 0$ and $z_{ox}$ is a depth parameter for the soft-tissue pump. Designating $F_g$ as the total flux of soft-tissue leaving the mixed layer to be oxidized in the water column during a year's time,

$$F_g = J_0 \int_0^{h_d} \exp(-z/z_{ox})dz \qquad (8a)$$

Then $J_0$ can be expressed in terms of $F_g$ specifically for oxidation

$$J_0 = F_g/[z_{ox}(1 - \exp(-h_d/z_{ox})] \qquad (8b)$$

The general one-dimensional tracer equation for any property C is then written as

$$dC/dt = Kd^2C/dz^2 + wdC/dz + J_0 \exp(-z/z_{ox}) \qquad (9)$$

The steady state solution is

$$C(z) = B_1 + B_2 \exp(-zw/K) + B_3 \exp(-\frac{z}{z_{ox}}) \qquad (10)$$

where $B_3 = F_g/[(1-\exp(-h_d/z_{ox}))(w-K/z_{ox})]$ and $B_1$ and $B_2$ are to be solved for from two boundary conditions.

We can specify a mixed layer concentration as an upper boundary condition and a bottom flux balance as a lower boundary condition, as in the work by Hoffert et al. [1980]: (1) $C_0 = C_w$ (specified from data) and (2) $wC_u = wC_{hd} + K(dC_{hd}/dz)$ estimating from data the upwelling concentration, $(C_u)$, of water entering the diffusive field. Then, $B_1 = C_u - B_3(1-z^*/z_{ox}) \exp(-h_d/z_{ox})$ and $B_2 = C_w - B_1 - B_3$ where $z^* = K/w$.

For the abiotic ocean (Figure 2), $F_g = B_3 = 0$, and the solution simplifies to

$$C(z) = C_u + (C_w - C_u) \exp(-z/z^*) \qquad (11)$$

For the abiotic solution, the following properties were assumed to be well mixed, equal to their GEOSECS global averages normalized to S = 35°/oo [Takahashi et al. 1981a, 1981b] and therefore the values for the warm and cold surface boxes: TA=2387 µeq/kg, S = 35°/oo, PO$_4$ = 2.1 µmol/kg, SI = 100 µmol/kg. Also, T$_W$ = 20°C and T$_C$ = 1.5° C. The area of the world ocean was 3.34 x 10$^{14}$m$^2$ and total ocean-atmosphere inventory of 3.22 x 10$^{18}$ moles of CO$_2$. The calculation at a specified η$_{solu}$ begins by guessing ΣCO$_{2W}$, obtaining pCO$_{2W}$ by Takahashi et al. [1982] routines, setting pCO$_{2atm}$ = pCO$_{2W}$ and obtaining ΣCO$_{2C}$ = ΣCO$_{2u}$ from equation (1):

$$ΣCO_{2u} = ΣCO_{2W} + η_{solu}(ΣCO_{2c,eq} - ΣCO_{2W})$$

Using K = 2000 m$^2$/yr, w = 4 m/yr and z$^*$ = 500 m [Hoffert et al. 1980], equation (11) produces the ΣCO$_2$(z) profile. The amount of CO$_2$ in the atmosphere-ocean model is compared to the desired inventory. An iterative procedure follows by guessing a new ΣCO$_{2W}$ until convergence to the inventory.

This calculation does not require specification of a cold surface area or gas exchange coefficient, although this could be done. The interest is on the solubility pump efficiency, so area and gas exchange rate could be adjusted to obtain any particular ΣCO$_{2C}$. Since A$_W$ < A$_C$, generally in models and in the ocean, pCO$_{2W}$ ≈ pCO$_{2atm}$. With the cold surface as a sink, pCO$_{2W}$ would be several microatmospheres above pCO$_{atm}$ for a steady state in the atmosphere. This error in obtaining Figure 2b is much less than the uncertainty in the other parameters.

For the one-dimensional upwelling-diffusion model with soft-tissue pump (Figure 3), we use equation (10). A fit to PO$_4$ GEOSECS data (Figure 3a) was obtained with F$_g$ = 4500 µmol/$^2$yr and z$_{ox}$ = 770 m. PO$_{4u}$ and PO$_{4W}$ were set at 2.18 µmol/kg and 0.33 µmol/kg, respectively. This mathematically implies a PO$_4$ flux (F$_{wu}$) between warm surface layer and bottom upwelling water of F$_{wu}$ = ρw(PO$_{4u}$ - PO$_{4W}$) = 7600 µmol/m$^2$yr, where ρ is the density of seawater. The F$_{wu}$ flux may physically represent preformed properties that upwell through the one-dimensional column directly into the cold surface and/or entrainment of properties by the downwelling plume of bottom water formation [see Southam and Peterson, this volume]. The point is that the PO$_4$ model is a calibration. The NO$_3$ and C$_{ox}$ models are produced from the PO$_4$ model calibration by scaling the F$_g$ and F$_{wu}$ fluxes with Redfield ratios: F$_g$, PO$_4$: F$_g$, NO$_3$: F$_g$, C$_{ox}$ = F$_{wu}$, PO$_4$: F$_{wu}$, NO$_3$: F$_{wu}$, C$_{ox}$ = 1:16:105.

Note that this C$_{ox}$ model contained the soft-tissue pump only. The NO$_3$ model scales quite well from PO$_4$, but the C$_{ox}$ with soft-tissue pump only does not make the surface-to-deep difference shown by the GEOSECS data of Figure 3c, implying the existence of a solubility pump if the Redfield ratios are correct.

Acknowledgements. This research was supported by the Department of Energy, Carbon Dioxide Research Division, Office of Basic Energy Sciences, under contract DE-AC02-81EV10610 to New York University. Many thanks to Taro Takahashi for numerous invaluable discussions and for supplying the chemical formulations determining the pCO$_2$/ΣCO$_2$ system.

## References

Baes, C.F., Jr., and G.G. Killough, Chemical and biological processes in CO$_2$-ocean models, Oak Ridge Natl. Lab. Life Sci. Symp. Global Carbon Cycle, 6th, in press, 1985

Bolin, B.A. Bjorkstrom, K. Holmen, and B. Moore, The simultaneous use of tracers for ocean circulation studies, Tellus, 35B, 206-236, 1983.

Brewer, P.G., Direct observation of the oceanic CO$_2$ increase, Geophys. Res.Let., 5, 997-1000, 1978.

Broecker, W.S., Glacial to interglacial changes in ocean and atmosphere chemistry, in Climatic Variations and Variability: Facts and Theories, edited by A. Berger, pp. 111-121 D. Reidel, Hingham, Mass., 1981.

Broecker, W.S., The ocean, Sci. Am., 249, 146-161, 1983.

Broecker, W.S., and T. Takahashi, Is there a tie between atmospheric CO$_2$ content and ocean circulation?, in Climate Processes and Climate Sensitivity, Geophys. Monogr. Serv., Vol. 29, edited by J.E. Hansen and T. Takahashi, pp. 314-326, AGU, Washington, D.C., 1984.

Chen, C-T. A., Decomposition of calcium carbonate and organic carbon in the deep oceans, Science, 210, 735-736, 1978.

Ennever, F.K., and M.B. McElroy, Changes in atmospheric CO$_2$: Factors regulating the glacial to interglacial transition, this volume.

Fiadeiro, M.E., Carbon cycling in the ocean, in Primary productivity in the Sea, edited by P, Falkowski, pp. 487-496, Plenum, New York, 1980.

Hoffert, M.I., A.J. Callegari, and C.T. Hsieh, The role of deep sea heat storage in the secular response to climatic forcing, J. Geophys. Res., 85, 6667-6679, 1980.

Hoffert, M.I., A.J. Callegari, and C.T. Hsieh, A box-diffusion carbon cycle model with upwelling, polar bottom water formation and a marine biosphere, in Carbon Cycle Modeling; Scope 16, edited by B. Bolin pp. 287-305, John Wiley, New York, 1981.

Keeling, C.D., and B. Bolin, The simultaneous use of chemical tracers in oceanic studies: A three-reservoir model of the North and South Pacific oceans, Tellus, 20, 17-54, 1968.

Knox, F., and M.B. McElroy, Changes in atmospheric CO$_2$: Influence of the marine biota at high latitude, J. Geophys. Res., 89, 4629-4637, 1984.

Levitus, S., Climatological Atlas of the World Ocean,NOAA Prof. Pap. 13, U.S. Dept. of Commer., Washington, D.C., 1982.

Maier-Reimer, E., A three-dimensional ocean carbon cycle model, paper presented at Chapman

Conference on Natural Variations in Carbon Dioxide and the Carbon Cycle, AGU, Tarpon Springs, Fla., Jan. 9-13, 1984.

Moor, E., and B. Stauffer, A new dry extraction system for gases in ice, J. Glaciol., in press, 1984.

Neftel, A., H. Oeschger, J. Schwander, B. Stauffer, and R. Zumbrunn, Ice core measurements give atmospheric $CO_2$ content during the last 40,000 years, Nature, 295, 220-223, 1982.

Oeschger, H., and B. Stauffer, Review of the history of the atmospheric $CO_2$ recorded in ice cores, Oak Ridge Natl. Lab. Life Sci. Symp. Global Carbon Cycle, 6th. in press, 1985.

Redfield, A.C., B.H. Ketchum, and F.A. Richards, The influence of organisms on the composition of sea-water, in The Sea, Vol.2, edited by M.N. Hill, pp. 26-77, Interscience, New York, 1963.

Sarmiento, J.L., and J.R. Toggweiler, A new model for the role of the oceans in determining atmospheric $pCO_2$, Nature, 308, 620-624, 1984.

Siegenthaler, U., and T. Wenk, Rapid atmospheric $CO_2$ variations and ocean circulation, Nature, 308, 624-626, 1984.

Southam, J. R., and W. H. Peterson, Transient response of the marine carbon cycle, this volume.

Takahashi, T., W. S. Broecker, and A.E. Bainbridge, The alkalinity and total carbon dioxide concentration in the world oceans, in Carbon Cycle Modeling; Scope 16, edited by B. Bolin, pp. 271-286, John Wiley, New York, 1981a.

Takahashi, T., W.S. Broecker, and A. E. Bainbridge, Supplement to the alkalinity and total carbon dioxide concentration in the world oceans, in Carbon Cycle Modeling; Scope 16, edited by B. Bolin, pp. 159-200, John Wiley, New York, 1981b.

Takahashi, T., R.T. Williams, and D.L. Bos, Carbonate chemistry, in GEOSECS Pacific Expedition, Vol. 3, Hydrographic Data, 1973-1974, Chap. 3, U.S.Government Printing Office, Washington, D.C., 1982.

Takahashi, T., W.S. Broecker, and S. Langer, Redfield ratio based on chemical data from isopycnal surfaces, J. Geophys.Res., in press, 1985.

Toggweiler, J.R., and J.L. Sarmiento, Glacial to interglacial changes in atmospheric carbon dioxide: The critical role of ocean surface waters in high latitudes, this volume.

Volk, T., Multi-property modeling of the marine biosphere in relation to global climate and carbon cycles, Ph.D. thesis, 348 pp., New York Univ., New York, May 1984.

Wenk, T., and U. Siegenthaler, The high-latitude ocean as a control of atmospheric $CO_2$, this volume.

# DETECTION OF EL NINO AND DECADE TIME SCALE VARIATIONS OF SEA SURFACE TEMPERATURE FROM BANDED CORAL RECORDS: IMPLICATIONS FOR THE CARBON DIOXIDE CYCLE

Ellen R. M. Druffel

Woods Hole Oceanographic Institution, Woods Hole, Massachusetts  02543

Abstract. Stable oxygen isotope ratios from annually banded corals are correlated with historical records of sea surface temperature in the central and eastern tropical Pacific Ocean. El Nino events between 1929 and 1976 are detected using this method, but there are discrepancies between the records of El Ninos from corals and those determined using historical hydrographic and meteorologic data. The average annual depletion of $\delta^{18}O$ during El Nino events is greater at the Galapagos Island sites (0.45°/oo) than at the Fanning and Canton Island sites in the mid-Pacific (0.20-0.30°/oo and <0.2°/oo, respectively). Of prime importance is evidence of decade time scale variability of sea surface temperature (SST) in the tropical Pacific. In particular, annually averaged SST appears to have been 0.5°-1°C higher in the eastern tropical Pacific during the 1930's than during subsequent years. A significant net flux of $CO_2$ from the surface ocean to the atmosphere is envisioned during these periods of higher SST.

## Introduction

Prior to the intervention of man, the pressure of $CO_2$ in the atmosphere (p$CO_2$) varied throughout geologic time. Neftel et al. [1982] provided evidence that during the height of the last glaciation, p$CO_2$ was about 210 ppm or about one-third less than the 19th century value (270-280 ppm). Stauffer et al. [1984] showed strong evidence of 50 ppm variations during the glacial period over very short time periods, of the order of 100 years or more. It seems unlikely that the shelf hypothesis proposed by Broecker [1982] could explain excursions on so short a time scale, as it is envisioned that significant storage of organic matter on the continental shelves could only occur over a period of thousands of years. A more plausible mechanism to explain rapid changes in p$CO_2$ is change in upper ocean circulation, which would cause changes in the physical and chemical properties of the surface ocean and most likely induce changes in nutrient utilization [Wenk and Siegenthaler, this volume; Toggweiler and Sarmiento, this volume].

Recent evidence linking shifts in p$CO_2$ to short-term changes in ocean circulation (in particular, SST) have been presented by several authors [Bacastow et al., 1980, 1981a, b; Schnell et al., 1981; Gammon et al., 1984]. These observations imply that net fluxes of $CO_2$ from ocean to atmosphere (1-3 metric gigatons (Gt) C/yr) are the result of widespread changes in SST (+1°-2°C) throughout the Pacific.

Influence of SST excursions on the flux of $CO_2$ may be particularly important in the equatorial regions of the world oceans. Preanthropogenic $\Delta^{14}C$ levels in the tropical Pacific were 15-30°/oo lower than at higher latitudes (Figure 1) due to upwelling of $^{14}C$-depleted waters from subsurface depths. Keeling et al. [1965] showed that p$CO_2$ levels in these equatorial surface waters are 40-80 ppm above atmospheric equilibrium values. This is the result of an imbalance between sources (upwelling of $CO_2$-enriched water) and sinks (photosynthetic production of biomass) of $CO_2$ to the surface ocean and long air/sea equilibration times. High p$CO_2$ indicates that the equatorial surface ocean acts as a net source of $CO_2$ to the atmosphere, whereas temperate and most polar surface oceans are either in equilibrium or act as a net sink for $CO_2$. If the nature of upwelling in the equatorial ocean were to change, this delicate balance would be altered, thus changing the net flux of $CO_2$ to the atmosphere and causing perturbations in atmospheric p$CO_2$ on a time scale of months to years. If the equatorial ocean warms up markedly as a result of reduced upwelling, such as that which occurs during El Nino, opposing effects on the $CO_2$ balance are expected: (1) the solubility of $CO_2$ will be lower in warmer water, (2) rates of primary production will be lower as a result of reduced nutrient levels, and (3) excess $CO_2$ supplied from subsurface waters will

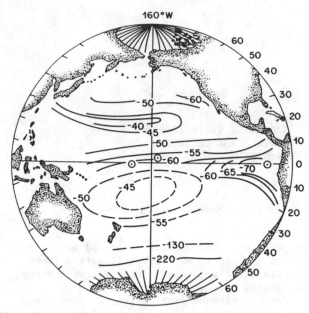

Fig. 1.  Isolines of preanthropogenic radiocarbon levels in surface waters of the Pacific
Ocean displayed in $\Delta^{14}C$ units (per mil). Distributions are based on early seawater measurements
[Rafter and Fergusson, 1957; Rafter, 1968; Bien
et al., 1960] and analyses of coral bands
[Druffel, 1981a, b; Konishi et al., 1983;
Toggweiler, 1983].  Dashed isolines indicate
estimates based on post-bomb distributions
[Linick, 1975].  Circles show sampling locations for this study.

be reduced due to decreased upwelling.  The first
two effects will cause an increased release of
$CO_2$ from surface ocean to atmosphere, whereas
the latter effect would cause a reduction of the
flux of $CO_2$ into the atmosphere.  Equatorial
warming, such as that experienced during El Nino,
causes the post-bomb $^{14}C/^{12}C$ ratios in surface
seawater to rise [Druffel, 1981a, and unpublished manuscript, 1984]. Reduced upwelling increases the residence time of waters at the surface, allowing more of the $^{14}CO_2$ admitted from
the atmosphere to be retained within the equatorial zone.  On these time scales, however,
oceanic $^{14}CO_2/^{12}CO_2$ ratios would still be depleted with respect to that in the atmosphere.
This is due to a long isotopic equilibration
time for $^{14}CO_2$ of 10–14 years [Druffel and
Linick, 1978], in comparison to the chemical
equilibration time for $CO_2$ of only 1 year
[Broecker et al., 1980].

El Nino is the response of the ocean and
climate to a forcing function that is not yet
understood.  Wyrtki [1965] determined that a
high Southern Oscillation Index (SOI), which is
a measure of the strength of the southeast
trade winds over the southern Pacific, preceded

El Nino by 1 year.  McCreary [1976] proposed
that subsequent reduction of the SOI generates
an eastward propagating equatorial Kelvin wave
front, which causes the water accumulated in
the western Pacific to "slosh back" toward the
eastern Pacific.  These El Nino/Southern Oscillation (ENSO) events are associated with a
dramatic sea level lowering at Truk Island
(7.5°N, 151.9°E) [Meyers, 1982] and considerable thickening of the pycnocline, as well as
high sea level and heavy rainfall in the
eastern tropical Pacific.  The Peru Current is
virtually absent during these periods, and upwelling is suppressed to depths well below the
mixed layer.  As a result, warm, nutrient-poor,
low-salinity waters invade the eastern and
central tropical Pacific, causing massive
deaths of various fishes and guano birds that
normally feed on the fish.  El Nino is an economic and ecological disaster to the western
coast of South America.

Statistics on the periodicities of the SST
anomalies associated with ENSO are relatively
unknown prior to 1960, as there is no uniform
index available in which these anomalies have
been recorded. Until recently, researchers have
relied on records of water temperature, rainfall, and disturbances to the anchoveta indus-
try, mostly from the coasts of Peru and Ecuador.
However, El Nino is not only a coastal phenomenon; in fact, some weak and moderate events occur only at sea and do not extend as far east
as the coast [Wyrtki, 1975].  Long-term records
from offshore locations, such as the Galapagos
or the Line Islands, would provide the index
necessary to assess the periodicity of temporal
variations of SST in the eastern tropical Pacific.  Initial efforts to obtain such a uniform
index from stable oxygen isotope ratios in annual coral bands are presented here.  These
data will be compared to the record of ENSO occurrences reconstructed by Quinn et al. [1978]
from a combination of environmental, hydrographic, and meteorological data.

The purpose of this study is threefold:  (1)
to establish the usefulness of coral skeletons
as integrators of SST in the eastern and central
tropical Pacific region, (2) to identify a
coral-inhabited island that is particularly sensitive to weak as well as strong ENSO events,
and (3) to reconstruct an accurate record of
past ENSO events, as well as long-term (several
year) changes in SST and salinity for the past
50 years.

Corals as Integrators of SST

It is generally believed that hermatypic
corals accrete aragonite from an internal pool
of carbon and oxygen within the calcioblastic
layer according to the following equation:

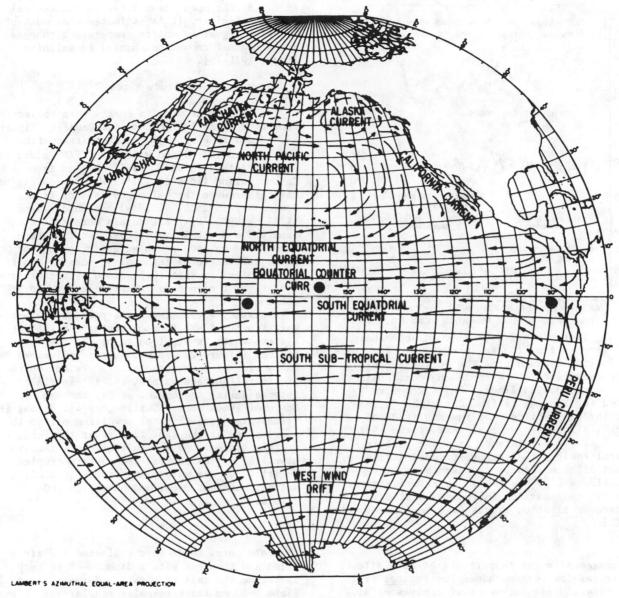

Fig. 2.  Surface currents of the Pacific Ocean [from Michel, 1974].  Circles show sampling locations:  Canton Island (2°S, 171°W), Fanning Island (4°N, 159°W), and the Galapagos Islands (1°S, 90°W).

$$Ca^{2+} + 2HCO_3^- \rightleftharpoons CaCO_3 + H_2O + CO_2 \qquad (1)$$

Sources and sinks of C and O in this internal pool are controlled by three major processes: (1) input of dissolved inorganic carbon (DIC) and $H_2O$ from surrounding seawater, (2) uptake and release of $CO_2$ by metabolic processes such as coral respiration and algal photosynthesis, and (3) accretion of aragonite.  The effect of metabolic processes on the $^{13}C/^{12}C$ ratio of accreted aragonite appears to be more important than that on the $^{18}O/^{16}O$ skeletal ratios

(see Swart [1983] for a review).  Metabolic processes contribute toward a constant overall depletion in coral skeletal $\delta^{18}O$ values, with respect to that expected for equilibrium precipitation, however, variations of the $^{18}O/^{16}O$ ratio (in areas of constant salinity and water composition) are controlled by fluctuations of SST [Weber and Woodhead, 1972].  Using high-resolution sampling of Montastrea annularis from three sites in the North Atlantic, Fairbanks and Dodge [1979] obtained a direct correlation between $\delta^{18}O$ measurements and month-

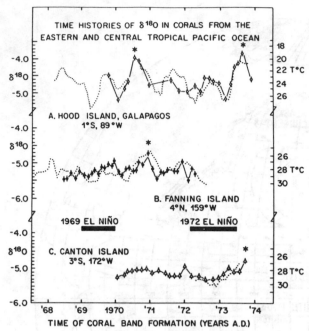

TIME HISTORIES OF $\delta^{18}O$ IN CORALS FROM THE
EASTERN AND CENTRAL TROPICAL PACIFIC OCEAN

A. HOOD ISLAND, GALAPAGOS
1°S, 89°W

B. FANNING ISLAND
4°N, 159°W

1969 EL NIÑO    1972 EL NIÑO

C. CANTON ISLAND
3°S, 172°W

TIME OF CORAL BAND FORMATION (YEARS A.D.)

Fig. 3.  Time histories of $\delta^{18}O$ in subannual
coral samples from the eastern and central tropical Pacific Ocean during the period 1968-1974.
Sea surface temperature for each location is
shown by the dotted lines.  The SST record
plotted for Fanning is from Christmas Island
(2°N, 157°W).  Notice that the amplitude of the
seasonal signal, as well as that of the 1972 El
Nino, is lower as distance from the eastern
tropical Pacific (Galapagos Islands) increases.
One out of nine results represents an average
of duplicate $\delta^{18}O$ analyses.  An asterisk marks
unusually cool periods following El Nino/
Southern Oscillation events (see text for
detail).

ly averages of water temperature at each site.
Similar results were obtained for Porites from
the northern Great Barrier Reef [Chivas et al.,
1983], and for Pocillopora damicornis from Oahu
[Weil et al., 1981) and from the Gulf of Panama
[Dunbar and Wellington, 1981].

These studies revealed that $\delta^{18}O$ is shifted
by +0.21-0.23°/oo for every 1.0°C drop in
temperature, the same as that observed by
Epstein et al. [1953] for the inorganic precipitation of calcite.  In this way, changes of
<0.5°C are easily detected.  Some authors have
obtained conflicting results, due to additional
variables that affect the $\delta^{18}O$ signature, such
as salinity changes due to high rates of evaporation and precipitation in surrounding seawater
[Swart and Coleman, 1980].  A shift of +0.06°/oo
in $\delta^{18}O$ is encountered for every 0.10°/oo rise
in salinity.  Thus, the $\delta^{18}O$ variation in coralline aragonite is a combined signal reflective
of changes in SST, salinity, and water composi-

tion.  At the open ocean sites in the central
tropical Pacific, it is estimated that most of
the $\delta^{18}O$ signal is due to temperature changes
and a minimal amount to changes in salinity
[Love, 1971-75].

Study Sites

Two species of massive corals were collected
from five sites in the tropical Pacific (Figure
2).  Heads of Porites lobata were taken from
the leeward (CTFN) and windward (CFAN) sides of
Fanning Island (3°52'N, 159°15'W) and from
Canton Island (2°48'S, 171°43'W).  Pavona clavus
was gathered in the Galapagos Islands from
Urvina Bay (0°25'S, 91°17'W) and Hood Island
(1°23'S, 89°37'W).  All corals were collected
from <10 m depth in outer reef locations and
were laved by open ocean waters.  Lagoons or
areas influenced by lagoonal or coastal runoff
were avoided so that temperature records representative of open ocean water masses could be
obtained.

The study sites lie in the path of the westward flowing South Equatorial Current (SEC)
system, which is supplied mainly by the Peru
Current (Figure 2).  Upwelling is influenced by
the action of the southeast trade winds on the
equatorial surface waters, and is especially
intense during the months of May through
November [Wooster and Guillen, 1974].  Strong SE
trades also cause coastal upwelling off South
America and at the same time power the Peru
Current system.  As a result of the extensive
upwelling in the central and eastern tropical
Pacific, surface waters are typically cool,
saline, nutrient-rich, and have high $pCO_2$
levels.

Methods

Whole coral heads were collected and air
dried and then cut with a diamond-edged rock
saw along the axis of growth.  X-radiographs of
slabs 5-10 mm thick revealed regular variations
in skeletal density.  Alizarin staining techniques, long-term field observations [Glynn and
Wellington, 1983], bomb-radiocarbon distributions [Druffel, 1981a, and unpublished manuscript, 1984] and stable oxygen isotope studies
(this work) revealed that Pavona from the Galapagos accreted one band pair per year of high-
and low-density aragonite.  In contrast, most
Porites coral from the Fanning and Canton sites
accreted two density band pairs per year, probably the result of reduced seasonal variation
in some parameters controlling coral growth
(SST, ambient light, etc.).  An exception was
the Porites head collected by J. R. Toggweiler
from the NW coast of Fanning Island (CTFN)
which had only one band pair per year.

The Galapagos specimens were sectioned into
annual bands; radiocarbon measurements were re-

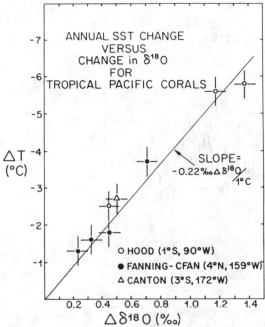

Fig. 4. Correlation between annual range in historical sea surface temperature (SST) records (ΔT°C) and annual range in $\delta^{18}O$ values ($\Delta\delta^{18}O$) in coral bands from Hood Island (Galapagos Islands), Fanning Island, and Canton Island. The line shown represents a slope of 0.22°/oo per 1°C decrease in SST predicted by the temperature-dependent aragonite-water fractionation factor. This slope is the same as that defined for the equilibrium precipitation of calcite [Epstein et al., 1953].

ported previously [Druffel, 1981a]. Small aliquots of material used for $^{14}C$ measurements were saved and analyzed for stable oxygen and carbon isotopes. Subannual samples were drilled (1 mm diameter) from Fanning, Canton, and Hood Island coral slabs. As the upward growth rates of these corals were not the same, sampling resolution varied among specimens. A growth rate of ~1.5-2.0 cm/yr for Fanning (CFAN), Canton, and Hood Island heads allowed sampling of 9-12 samples/yr whereas the second Fanning specimen (CTFN) grew less than 1.0 cm/yr, which reduced the sampling resolution to 6 samples/yr.

In preparation for stable isotope analyses, samples were crushed in methanol and roasted at 375°C under vacuum for 1 hour. They were acidified with 100% orthophosphoric acid at 50°C, and the evolved $CO_2$ was measured on a V.G. Micromass 602E mass spectrometer. Results are reported in the standard δ (per mil) notation relative to the PDB-1 Chicago Standard, assuming $\delta^{18}O$ (NBS-20) = -4.18°/oo and $\delta^{13}C$ (NBS-20) = -1.06°/oo. The precision of $\delta^{18}O$ measurements determined from numerous duplicate analyses of the same aragonite is ±0.07 °/oo standard deviation (SD).

The $\delta^{13}C$ signal in corals is sufficiently complicated, and the controls on it are so numerous, that adequate treatment of the $\delta^{13}C$ data collected as part of this study is not possible in this brief a forum. These data, as well as the radiocarbon measurements of the mid-Pacific corals, are discussed in a separate paper [E. M. Druffel, unpublished manuscript, 1984].

### Intrannual $\delta^{18}O$ Results: Central and Eastern Tropical Pacific

Stable oxygen isotope analyses of subannual coral samples that grew from 1968-1974 in the tropical Pacific are shown in Figure 3. The chronology of each isotope curve was established by fitting the $\delta^{18}O$ measurements to records of average monthly SST from each location (dotted lines). Temperature records from Christmas Island (2°N, 157°W) are plotted with the Fanning data (4°N, 159°W), as no records from Fanning were available for this time period. The temperature axis (right side of Figure 3) is scaled relative to $\delta^{18}O$ (left side of Figure 3), assuming that temperature alone affects $\delta^{18}O$ (-0.22°/oo per 1°C rise). Despite the use of different species in this study, $\delta^{18}O$ values from Hood Island are proportionately greater (by 0.8°/oo, 4°C cooler) than those from the warmer locations in the central tropical Pacific.

At all three locations, temperature appears to be the predominant factor controlling the variations of the $\delta^{18}O$ signal. In Figure 4, the annual range in SST (ΔT) during each year is plotted versus the observed annual variation in $\delta^{18}O$ ($\Delta\delta^{18}O$) at each site. There is close agreement between these values and the line representing a direct correlation between SST and $\Delta\delta^{18}O$ (slope is 0.22°/oo per 1°C). From this agreement, it appears that changes in salinity that occurred at these tropical locations were either small (<0.3°/oo) or short-lived.

Even during the ENSO periods of 1969 and 1972, reductions in salinity are not apparent in the $\delta^{18}O$ records. We observe no large negative excursions of $\delta^{18}O$ beyond those predicted by the observed temperature changes. It appears, however, that despite the good correlation between annual ΔT and $\Delta\delta^{18}O$ in the Fanning record, there is a discrepancy between absolute $\delta^{18}O$ values and SST. Values of $\delta^{18}O$ seem too high during most of 1969 and low during most of 1970. The SST record used for comparison, however, was taken from Christmas Island, which is located 2° closer to the equator (2°N, 157°W) than Fanning Island (4°N, 159°W) and is influenced by different water masses during part of the year.

It might be expected that differences in salinity accompany ENSO events in the vicinity of the Galapagos Islands, due to the influx of fresher waters from the Panama Basin [Wyrtki et

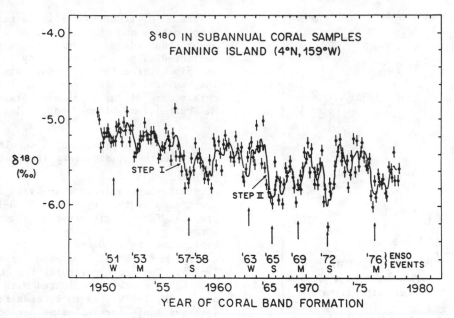

Fig. 5. Time history of $\delta^{18}O$ in subannual coral samples from a specimen collected from the leeward side of Fanning Island (CTFN). The line is a 6-month (3 point) running mean of the values. Note the two-step trend toward lighter values for 1957 (step I) and 1965 (step II). Correlation of low $\delta^{18}O$ values with ENSO events is consistent throughout the record, with the strong event of 1972 being the most impressive. The severity of each ENSO event is indicated by (W) for weak, (M) for moderate, and (S) for strong. Results of duplicate analyses are plotted as a single averaged point.

al., 1976]. However, there is no evidence of this effect during the 1972 El Nino at Hood Island, as $\delta^{18}O$ changed in proportion to the observed average temperature rise for this period (Figure 3). Hood is the southernmost island in the Galapagos archipelago, apparently far enough south to escape the influence of low salinity waters accompanying El Nino.

A prominent feature in each of the three stable isotope records is the appearance of pulses of cool water (high $\delta^{18}O$ values) with the return of upwelling after ENSO events. These pulses are noticed during mid-1970 and mid-1973 at Hood, late 1970 at Fanning, and late 1973 at Canton (marked by asterisks in Figure 3). Occurrence of cool water pulses is in phase with high $pCO_2$ anomalies found by Bacastow et al. [1980] at several of the $CO_2$ monitoring sites in the Pacific. This suggests that the renewal of upwelling following an ENSO event is related to the net flux of $CO_2$ into the atmosphere.

### Thirty-Year Subannual $\delta^{18}O$ Record at Fanning Island

Stable oxygen isotope results are reported for the second Fanning coral, CTFN, sampled subannually (6 samples/yr) over the period

1950-1979 (Figure 5). A 6-month (3 point) running mean is plotted as the solid curve. It is important to point out that $\delta^{18}O$ results from the two Fanning corals are similar, with the exception of a 1½-year period when there was a 0.2°/oo offset (1969-1970) (Figures 3 and 5). The reason for this may be the fact that CTFN grew off the leeward (NW) side of the island, which would be influenced by slightly warmer waters, especially during ENSO events (i.e., 1969), than those at the windward (SE) edge, where CFAN was collected.

An annual temperature signal is discernable during most years in the CTFN results. However, its amplitude appears attenuated by about a factor of 2 during the 1950's. This is due to the fact that the coral slab deviated from the vertical axis of growth during this period, causing contamination in the drilled samples due to incorporation of earlier and subsequently accreted material.

However, this does not account for the overall greater values apparent during the earlier part of the record. A least squares fit of the data for non-ENSO periods reveals an overall decline of 0.45°/oo from 1950 to 1979. In particular, two periods of rapid decline in $\delta^{18}O$ (steps I and II in Figure 5) appear during the onset of two strong ENSO events (1957 and

1965), after which there is a partial recovery toward higher values. If this $\delta^{18}O$ record is representative of the area surrounding Fanning Island, sudden changes in the combined SST/salinity signal are indicated and appear to be induced by major ENSO events. The trends apparent in the 30-year Fanning Island record do not appear to correlate with the surface air temperatures reported by Angell and Korshover [1983] for the tropics, most likely because the sites from which air temperature data were chosen were remote with respect to the eastern and central tropical Pacific Ocean. It is difficult to ascertain the significance of high $\delta^{18}O$ values (indicative of lower SST) during 1963-1965, due to the scatter. There is evidence in the literature of slower growth rate of $pCO_2$ in the atmosphere [Bacastow, 1979] and lower world air temperatures [Angell and Korshover, 1983] during the period following the Agung eruption in 1963.

Superimposed on the long-term changes are the presence of strong (1957, 1965, 1972), moderate (1953, 1969, 1976), and weak (1951, 1963) ENSO events marked by unusually low $\delta^{18}O$ values. There are also lower $\delta^{18}O$ values during 1959 and 1974, presently considered non-ENSO periods. The average offset during strong events is about 0.3°/oo, which reflects a maximum of 1.5°C warmer SST (assuming no salinity change) averaged over 12-month periods. The severe event of 1972 stands out among the other strong ENSO episodes because of the very low $\delta^{18}O$ value of -6.25°/oo revealed at the height of the event. Moderate and weak events are marked by a decrease of 0.1-0.2°/oo and are somewhat proportional to the length of each event, which ranges from 4 to 10 months.

## Interannual $\delta^{18}O$ Results: Eastern Tropical Pacific

Results of Pavona clavus samples from Urvina Bay (UB), Isabella Island and from Hood Island (HI) are presented in Figure 6. An overall enrichment of $^{18}O$ by 0.3°/oo is apparent in the UB results relative to those for HI, due to the enhanced influence of Ekman divergence and/or the Cromwell Current at this location nearer the equator (0°25'S). The UB results average -4.08°/oo during the 1930's and -3.96°/oo during the 1940's, with the exception of a few low values that represent coral growth during ENSO events. The major event of 1941 is displayed by a reduction in $\delta^{18}O$ of ~0.33°/oo. The moderate (1939-1940 and 1953) and weak (1932, 1943-1944, and 1951) events described by Quinn et al. [1978] show a smaller reduction (0.10-0.20°/oo). Overall, there is agreement between the historical record compiled by Quinn et al. [1978] and the $\delta^{18}O$ record from banded corals. This corre-

lation supports the suspicion that the low $\delta^{18}O$ values obtained for coral growth during 1937 and 1952 also represent warm water/low salinity conditions in the eastern tropical Pacific that have not previously been recognized as such in the literature. These values are significantly depleted (by 0.15-0.20°/oo) with respect to the baseline values established for the 1930's and 1940's. Evidence that the weak ENSO event of 1951 lasted through 1952 is provided by the sea level atmospheric pressure anomaly between Santiago, Chile, and Darwin, Australia [Quinn et al., 1978], which lasted throughout the first half of 1952. There was no similar SOI anomaly apparent during 1937. Nonetheless, the lower $\delta^{18}O$ value indicates the presence of higher SST or lower-salinity waters, or a combination of both, during 1937. This suggests that a weak ENSO event, not associated with an SOI anomaly, may have occurred several hundred kilometers west of the South American coast, and thus went unnoticed in the historical records. The high value for 1950 growth (-3.87°/oo) indicates an unusually cold period, which agrees with records of SST along the Peruvian coast.

Most important is the long-term trend toward lower $\delta^{18}O$ values apparent in the 1930's (Figure 6), which is evidence of higher SST and/or less saline waters for this period. A least squares fit of the data from non-ENSO years reveals a significant decrease from -4.01 in 1930 to -4.14°/oo by 1938 (Figure 6). Least squares analyses of the 1940's data indicate an insignificant slope, with values that range from -4.01°/oo in 1942 to -3.92°/oo by 1950. The average offset of $\delta^{18}O$ values between the 1930's and 1940's, therefore, is 0.10-0.15°/oo, which is equivalent to the signal from a weak ENSO event. Whether the 1930's was a period of consecutive weak ENSO events or local or widespread warming cannot be absolutely determined at this point. This area was tectonically active, which led to the death of the reef in 1954 when a portion of the shoreline was uplifted [Richards, 1957]. However, there is no evidence of warm water inputs from lava pools or other tectonically associated water bodies [P. Glynn, personal communication, 1980]. In view of this and the exposed location of Urvina Bay to equatorial currents (i.e., Cromwell Current), it is likely that this coral was accurately recording a widespread warming event (or a reduction in the countercurrent flow) in the eastern tropical Pacific during the 1930's.

El Nino events recorded by Hood Island coral during 1965, 1969 and 1972 displayed roughly the same depletion in $^{18}O$ (0.2-0.5°/oo) as that observed in the UB sample. It is important to note that results of annual sampling of the 1972 event (-4.85°/oo) agree with the results obtained from subannual sampling (Figure 3) where the values ranged from -4.6 to -5.0°/oo

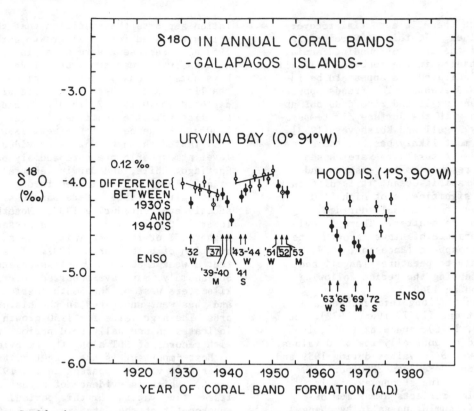

Fig. 6. Stable isotope measurements in annual samples from corals collected in the Galapagos Islands. Each point represents an average of two to four analyses. During El Nino years (solid circles), $\delta^{18}O$ is lower, indicating the presence of warmer and/or less saline waters. Lower $\delta^{18}O$ values are also obtained for 1937 and 1952 (open circles with dot), periods which are not considered ENSO events in the literature. Values from non-ENSO years (open circles) are lower during the 1930's than during the 1940's, suggestive of overall warmer sea surface temperature during the earlier period (see text for details).

with an average of -4.8°/oo. The baselines obtained using both sampling procedures for normal years are also similar (-4.3 to -4.5°/oo).

Unlike the UB record, values from the years preceding ENSO events (1964, 1968, and 1971) in the Hood Island record are also depleted in $^{18}O$. This is probably a function of the sectioning procedure which splits a given El Nino year (December-November) into two separate annual bands which are from March-February. Thus, 30% of the El Nino signal (December-February) should appear in the preceding year. This suggests that the $\delta^{18}O$ signal associated with ENSO events is stronger at Hood Island than at Urvina Bay. Sectioning of the HI samples was more accurate than for the UB samples, due to the very high growth rate (1-2 cm/yr) and the well-defined annual density bands. In view of this difference, no viable conclusions can be made regarding the suspected trend toward longer El Nino events from the 1930's-1940's to the 1960's-1970's.

The level of noise in the $\delta^{18}O$ values surrounding the HI baseline (-4.35±0.13°/oo SD) is significantly higher than that surrounding the UB baseline (-3.97±0.06°/oo SD, after removing the 0.12°/oo depletion during the 1930's) (see Figure 6). It is difficult to use the CTFN data to test for this change in noise level, as the results from the 1950's are attenuated due to a sampling artifact. A larger noise level in the HI $\delta^{18}O$ results may indicate that occurrences of weak ENSO events had become more frequent during the 1960s and 1970s, or that site HI is more sensitive to weak El Nino events than site UB. However, without $\delta^{18}O$ records from the same time periods for both sites, we cannot distinguish between these two possibilities as explanations of the apparently higher levels of noise.

Discussion

This study has revealed a direct correlation between $\delta^{18}O$ in banded corals and historical records of SST in the extensive region affected by ENSO. There appears to be no noticeable dependence of $\delta^{18}O$ on salinity changes that may have occurred. This provides a valuable integrator of the ocean's response to atmospheric forcing, records that have been incomplete up to this point.

The largest differential between annually averaged $\delta^{18}O$ for normal years and that for ENSO years was found at the two Galapagos Island sites (0.2-0.5°/oo). The amplitude of this signal at the three mid-Pacific sites was about half the size (0.1-0.3°/oo). It is desirable to obtain records from both general locations, however, in order to determine the spatial extent of each ENSO event as well as the temporal relationship between changes in the eastern and central equatorial Pacific. Christmas Island may be a better mid-Pacific location than Fanning, in that there is less influence from currents other than the SEC (i.e., North Equatorial Countercurrent, NECC), which is the predominant current flowing through the region affected by ENSO. Canton Island seems to be located at the westernmost edge of "El Nino country"; however, it will be very important to study the effect that the catastrophic event of 1982-1983 had at this location. If Canton is indeed sensitive only to very intense events, records from this site will be important for corroborating records of these events from more sensitive locations to the east.

It is desirable to use monthly sampling intervals from coral bands to determine accurately the severity and length of individual ENSO events. However, this involves an enormous number of analyses. Also, the data may not represent monthly averaged SST values if the upward growth rate of the coral is less than 1 cm/yr. Calcification takes place on the calcioblastic layer surrounding the individual cup-shaped polyps, which extend down into the skeleton. Thus calcification occurs simultaneously over a distance equal to the depth of polyp (1-5 mm). Should corals with high growth rate and small polyp size not be available from a given location, $\delta^{18}O$ from seasonal samples would at least allow us to distinguish between weak and strong ENSO events.

There is evidence from the Urvina Bay $\delta^{18}O$ record that ENSO conditions prevailed during 1937 and 1952, despite the absence of these events in historical records from the South American coast. This incongruity suggests that records from the South American coast are not necessarily indicative of warming events in the eastern tropical Pacific. Isotopic records from a location off the coast may be a better integrator of the incidence of ENSO. A revised time history of ENSO events based on these UB data appears in Figure 6. The reliability of the record can be improved by obtaining higher precision (±0.03°/oo) and by implementing more careful sampling procedures on a high growth rate coral that exhibits minimal intra-annual banding.

A major conclusion of this study is that there appears to have been decade time scale variability of SST in the eastern and central tropical Pacific over the past half century. Warmer (and/or less saline) waters, by about 0.5°-1°C, apparently predominated in the Galapagos Islands during the 1930's than during the subsequent decade. In addition, sea surface temperature appears to have increased by about 1°-2°C from 1956 to 1965 at the Fanning Island site. If these changes are representative of large-scale warmings over a substantial area of the tropical Pacific, significant effects on the short-term $CO_2$ cycle are likely.

Bacastow [1976] observed decreased $pCO_2$ at the south pole and at Mauna Loa during strong SOI and increased levels when the winds relaxed, marking the onset of an ENSO. More recently, he has extended this correlation to include ocean station PAPA (50°N, 145°W) and Fanning Island [Bacastow et al., 1980]. Schnell et al. [1981] calculated an exchange factor of 0.64 ± 0.07 ppm atmospheric $CO_2$ per 1°C change in SST for two sites in the North Pacific. From these studies, it was shown that atmospheric $CO_2$ was stored in cooler than normal surface waters and that when SST was higher than normal, there was a net flux of $CO_2$ to the atmosphere. However, maximum $pCO_2$ levels are not reached until the year following an ENSO event [Bacastow et al., 1980], which is coincident with the renewal of upwelling and the appearance of unusually cool waters in the tropical Pacific (Figure 3). Newell et al. [1978] demonstrated that the SST changes in the Pacific equatorial upwelling region are correlated with these $CO_2$ changes, with the Indian and Atlantic oceans contributing only minor fractions to the variance. Gammon et al. [1984] described a slower than normal rise in atmospheric $CO_2$ during the first part of the severe ENSO of 1982-1983 and a subsequent net release of about 3 Gt C from the surface ocean to the atmosphere during the warmer second year.

Bacastow et al. [1980] used a simple equilibrium model to calculate that a change in SST representative of a major ENSO event would correspond to an atmospheric $CO_2$ rise of only 0.4 ppm, approximately 40% of the observed change. However, it is difficult to understand how a rise in SST could cause a higher rate of $CO_2$ flux into the atmosphere, as cool upwelling waters of the SEC/Peru Current regime have very high $pCO_2$ values [Keeling et al., 1965] and are believed to be a net source of $CO_2$ to the atmosphere during non-ENSO conditions. Decreased upwelling might be expected to reduce $pCO_2$ in ocean water and as a result reduce the net flux of $CO_2$ to the atmosphere. On the other hand, increased primary production during upwelling (cold) periods may act to decrease $pCO_2$ in surface waters and thus reduce their capacity as a $CO_2$ source to the atmosphere, in contrast to nonupwelling periods when nutrients are low.

Nonetheless, the net flux of $CO_2$ from ocean to atmosphere occurs during the latter part of an ENSO event and during the following 6–12 month period. This suggests that the appearance of warm waters associated with ENSO, as well as the return of cool, $CO_2$-rich upwelling waters to the surface, both contribute to a higher net flux of $CO_2$ from ocean to atmosphere.

When we consider the decade time scale variability of SST revealed by corals, the impact on the carbon cycle will most likely be even greater. The decade-long perturbation in SST noticed during the 1930s at the Urvina Bay site was approximately 0.5°–1°C. This SST anomaly is about half of the perturbation noticed during a major ENSO event (Figure 6). If the impact on the carbon cycle during a 12–18 month ENSO event is taken as a net addition of 2–3 Gt C from surface ocean to atmosphere during the second part of the event, and the 0.5°–1°C rise at the Galapagos is symptomatic of a widespread warming throughout the equatorial Pacific, then it is estimated that there was a net flux of 1–2 Gt C from surface ocean to atmosphere during the early 1930's, in addition to the net release of 4–6 Gt C during the ENSO events of 1932, 1939–1940, and 1941. It is likely that most of the net $CO_2$ input from sea to air as a result of the long-term warming was restricted to the early 1930's after which quasi-steady state conditions were attained despite the retention of high SST. However, if the $CO_2$ release was dominated by the failure of the marine biota to resume a high level of biomass production in the equatorial zone, then net $CO_2$ release to the atmosphere over the entire interval of high SST would be possible. Had the oceans continued to supply net $CO_2$ over the entire period, a measurable effect would be expected in the $\delta^{13}C$ record from tree rings.

Similarly, the two-step warming indicated by the Fanning Island (4°N, 159°W) record from 1956 to 1970 may have caused a net input of carbon to the atmosphere. It is difficult to quantitatively assess the spatial extent of this SST anomaly, however, because of its location; it may reflect temperature changes in the NECC rather than in the SEC. An indication of this is borne out by the temperature records from Christmas Island, which do not indicate a warming over this time period. In the event that these data represent a widespread warming, an upper limit for the net input of $CO_2$ from surface ocean to atmosphere during this 1°–2°C warming is estimated to be 2 Gt C during the

late 1950's (step I) and 1 Gt C during the mid-1960's (step II). This correlates with the overall increase in $pCO_2$ (corrected for fossil fuel addition and seasonal variation) of ~1 ppm seen in the record from the south pole [Keeling, 1973] during these periods.

## Conclusions

The major purpose of this study was to corroborate the usefulness of coral bands as integrators of SST in the tropical Pacific and subsequently to reveal the presence or absence of ENSO episodes prior to the time for which good historical SST records are available.

Major conclusions of this tropical Pacific study are as follows:
1. The $^{18}O/^{16}O$ ratios in coralline aragonite from monthly and annually averaged growth increments are directly correlated with historical SST records from the same time periods. Salinity variations appear to have been minimal during the period 1968–1973, even during ENSO events.
2. The $\delta^{18}O$ signal during strong ENSO events is greatest at the Galapagos Island sites (0.45°/oo), less at Fanning Island (0.20–0.30°/oo) and very weak at Canton Island (<0.20°/oo). Therefore, the best location for tracking past ENSO events is the Galapagos Islands. It is anticipated that Christmas Island would be the most desirable mid-Pacific location, due to its position with respect to the SEC.
3. Decade time scale variations of SST were noticed in the tropical Pacific: (1) the eastern tropical Pacific was warmer by ~0.5°–1°C during the 1930's than during the subsequent decade, and (2) the mid-tropical Pacific underwent a warming trend from 1956 to 1970, which occurred rapidly in two stages coincident with the two major ENSO episodes of 1957 and 1965. The onset of each of these warming periods most likely accompanied increases in the net input of $CO_2$ from surface ocean to atmosphere of 1–3 Gt C.

Acknowledgements. I am indebted to P. Glynn, G. Wellington, R. Schneider and S. V. Smith for providing the coral samples. Many thanks to S. M. Griffin and D. Kaminski for their technical expertise, and to M. Harvey for preparing the manuscript. Comments from R. Bacastow, L. Keigwin, E. Boyle, P. Brewer, S. M. Griffin, R. Dunbar, P. Glynn, L. Benavides and an anonymous reviewer are appreciated. Support from the National Science Foundation under grants nos. OCE81-11954 and OCE83-15260 is gratefully acknowledged. This is Woods Hole Oceanographic Institution contribution no. 5735.

## References

Angell, J. K., and J. Korshover, Global temperature variations in the troposphere and stratosphere, 1958-1982, Month. Weather Rev., 111, 901-921, 1983.

Bacastow, R. B., Modulation of atmospheric carbon dioxide by the Southern Oscillation, Nature, 261, 116-118, 1976.

Bacastow, R. B., Dip in the atmospheric CO₂ level during the mid-1960s, J. Geophys. Res., 84, 3108-3114, 1979.

Bacastow, R. B., J. A. Adams, C. D. Keeling, D. J. Moss, T. P. Whorf, and C. S. Wong, Atmospheric carbon dioxide, the Southern Oscillation and the weak 1975 El Nino, Science, 210, 66-68, 1980.

Bacastow, R. B., C. D. Keeling, and T. P. Whorf, Seasonal amplitude in atmospheric CO₂ concentration at Mauna Loa, Hawaii, 1959-1980, in Proceedings of the WMO/ICSU/ UNEP Scientific Conference on Analysis and Interpretation of Atmospheric CO₂ Data, Bern, 14-18 Sept. 1981, pp. 169-176, World Meterological Organization, Geneva, 1981b.

Bacastow, R. B., C. D. Keeling, and T. P. Whorf, Seasonal amplitude in atmospheric CO₂ concentration at Canadian Weather Station P, 1970-1980, in Proceedings of the WMO/ICSU/UNEP Scientific Conference on Analysis and Interpretation of Atmospheric CO₂ Data, Bern, 14-18 Sept. 1981, pp. 163-168, World Meterological Organization, Geneva, 1981a.

Bien, G. S., N. W. Rakestraw, and H. E. Suess, Radiocarbon concentration in Pacific Ocean water, Tellus, 12(4), 436-443, 1960.

Broecker, W. S., Glacial to interglacial changes in ocean and atmosphere chemistry, Prog. Oceanogr., 11, 151-197, 1982.

Broecker, W. S., T.-H. Peng, and T. Takahashi, A strategy for the use of bomb-produced radiocarbon as a tracer for the transport of fossil fuel CO₂ into the deep-sea source regions, Earth Planet. Sci. Lett., 49, 463-468, 1980.

Chivas, A. R., P. Aharaon, J. Chappell, C. Veastuin, and E. Kiss, Trace elements and stable-isotope ratios of annual growth bands as environmental indicators, in Proceedings of the Inaugural Great Barrier Reef Conference, Townsville, Australia, edited by J. T. Baker et al., pp. 77-81, JCU Press, Townsville, Qld., Australia, 1983.

Druffel, E. M., and T. W. Linick, Radiocarbon in annual coral rings of Florida, Geophys. Res. Lett., 5(11), 913-916, 1978.

Druffel, E. M., Radiocarbon in annual coral rings from the eastern tropical Pacific, Geophys. Res. Lett., 8, 59-62, 1981a.

Druffel, E. M., Bomb-produced radiocarbon in surface waters of the world's oceans as seen in annual coral rings (abstract), EOS Trans. AGU, 62(8), 82, 1981b.

Dunbar, R. B., and G. M. Wellington, Stable isotopes in a branching coral monitor sea-sonal temperature variation, Nature, 293, 453-455, 1981.

Emiliani, C., J. H. Hudson, E. A. Shinn, and R. Y. George, Oxygen and carbon isotopic growth record in a reef coral from the Florida Keys and a deep-sea coral from Blake Plateau, Science, 202, 627-630, 1978.

Epstein, S., R. Buchsbaum, H. A. Lowenstam, and H. C. Urey, Revised carbonate-water isotopic temperature scale, Geol. Soc. Am. Bull. 64, 1315-1326, 1953.

Fairbanks, R. G., and R. E. Dodge, Annual periodicity of the ¹⁸O/¹⁶O and ¹³C/¹²C ratios in the coral Montastrea annularis, Geochim. Cosmochim. Acta, 43, 1009-1020, 1979.

Gammon, R. H., W. D. Komhyr, L. S. Waterman, T. Conway, and K. Thoning, Estimating the natural variation in atmospheric CO₂ since 1860 from interannual changes in tropospheric temperature and the history of major El Nino events, paper presented at Chapman Conference on Natural Variations in Carbon Dioxide and the Carbon Cycle, AGU, Tarpon Springs, Fla., Jan. 9-13, 1984.

Glynn, P. W., and G. M. Wellington, Corals and Coral Reefs of the Galapagos Islands, 331 pp., University of California Press, Berkeley, 1983.

Keeling, C. D., Industrial production of carbon dioxide from fossil fuels and limestone, Tellus, 25, 174-198, 1973.

Keeling, C. D., N. W. Rakestraw, and L. S. Waterman, CO₂ in the surface waters of the Pacific Ocean, 2, Measurements of the distribution, J. Geophys. Res., 70, 6087-6097, 1965.

Konishi, K., T. Tanaka, and M. Sakanoue, Secular variation of radiocarbon concentration in seawater, Proc. Int. Coral Reef Symp., 4th, 1983.

Linick, T. W., Uptake of bomb-produced radiocarbon in the surface water of the Pacific Ocean, Ph.D. dissertation, 255 pp., Univ. of Calif., San Diego, 1975.

Love, C. M. (Ed.), Eastropac Atlas, Circ. 330, vols 1, 2, 5, 6, 9, and 10, NOAA, U. S. Dep. of Commer., Washington, D.C., 1972-1975.

McCreary, J., Eastern tropical ocean response to changing wind systems: With application to El Nino, J. Phys. Oceanogr., 6, 632-645, 1976.

Meyers, G., Interannual variation in sea level near Truk Island - A bimodal seasonal cycle, J. Phys. Oceanogr., 12, 1161-1168, 1982.

Michel, R. L., Uptake of bomb-produced tritium by the Pacific Ocean, Ph.D. thesis, Univ. of Calif. San Diego, 1974.

Neftel, A., H. Oeschger, J. Schwander, G. Stauffer, and R. Zumbrunn, Ice core sample measurements give atmospheric CO₂ content during the past 40,000 yr, Nature, 295, 220-223, 1982.

Newell, R. E., A. R. Navato, and J. Hsiung, Long-term global sea surface temperature fluctuations and their possible influence on atmospheric $CO_2$ concentrations, Pure Appl. Geophys., 116, 351-371, 1978.

Quinn, W. H. D. O. Zopf, K. S. Short, and R. T. W. Kuo Yang, Historical trends and statistics of the Southern Oscillation, El Nino and Indonesian droughts, Fish. Bull., 76, 663-678, 1978.

Rafter, T. A., Carbon-14 variations in nature, 3, [14]C measurements in the South Pacific and Antarctic oceans, N. Z. J. Sci., 11, 551-588, 1968.

Rafter, T. A., and G. J. Fergusson, The atom bomb effect: Recent increase in the [14]C content of the atmosphere, biosphere and surface waters of the oceans, N. Z. J. Sci. Technol., Sect. B, 38, 871-883, 1957.

Richards, A. F., Volcanism in eastern Pacific Ocean basin, 1954-1955, Proc. Int. Geol. Congr., 10th, 1, 19-31, 1957.

Schnell, R. C., M. N. Harris, and J. A. Schroeder, A relationship between Pacific Ocean temperatures and atmospheric carbon dioxide concentrations at Barrow and Mauna Loa, in Proceedings of the WMO/ICSU/UNEP Scientific Conference on Analysis and Interpretation of Atmospheric $CO_2$ data, Bern, 14-18 Sept. 1981, pp. 155-162, World Meteorological Organization, Geneva, 1981.

Stauffer et al., Ann. Glaciol., in press, 1984.

Swart, P. K., Carbon and oxygen isotope fractionation in scleractinian corals: A review, Earth Sci. Rev., 19, 51-80, 1983.

Swart, P. K., and M. L. Coleman, Isotopic data for scleractinian corals explain their paleotemperature uncertainties, Nature, 283, 557-559, 1980.

Toggweiler, J. R., A six zone regionalized model for bomb radiotracers and $CO_2$ in the upper kilometer of the Pacific Ocean, Ph.D. dissertation, 421 p., Columbia University, New York, 1983.

Toggweiler, J. R., and J. W. Sarmiento, Glacial to interglacial changes in atmospheric carbon dioxide: The critical role of ocean surface water in high latitudes, this volume.

Weber, J. N., and P. M. J. Woodhead, Temperature dependence of oxygen-18 concentration in reef coral carbonates, J. Geophys. Res., 77, 463-473, 1972.

Weil, S. M., R. W. Buddemeier, S. V. Smith, and P. M. Kroopnick, The stable isotopic composition of coral skeletons: Control by environmental variables, Geochim. Cosmochim. Acta, 45, 1147-1153, 1981.

Wenk, T., and U. Siegenthaler, The high-latitude ocean as a control of atmospheric $CO_2$, this volume.

Wooster, W. S., and O. Guillen, Characteristics of El Nino in 1972, J. Mar. Res., 32, 387-403, 1974.

Wyrtki, K., Surface currents of the eastern tropical Pacific Ocean, Inter Am. Trop. Tuna Comm. Bull., 9, 271-294, 1965.

Wyrtki, K., El Nino - The dynamic response of the equatorial Pacific Ocean to atmospheric forcing, J. Phys. Oceanogr., 5, 572-584, 1975.

Wyrtki, K., E. Stroup, W. Patzert, R. Williams, and W. Quinn, Predicting and observing El Nino, Science, 191, 343-346, 1976.

# ATMOSPHERIC $CO_2$ VARIATIONS BASED ON THE TREE-RING [13]C RECORD

T.-H. Peng

Environmental Sciences Division, Oak Ridge National Laboratory
Oak Ridge, Tennessee 37831

Abstract. The reconstruction of atmospheric $CO_2$ concentrations over the last 150 years, based on a deconvolution of the tree-ring-based [13]C record, is reviewed. Assuming that Freyer's latest composite [13]C record for the northern hemisphere represents global changes in the [13]C/[12]C ratio of atmospheric $CO_2$ induced by changes in atmospheric $CO_2$ concentration due to deforestation, soil manipulation, and combustion of fossil fuels, the deconvolution, using the modified box-diffusion model, gives the following results: (1) the magnitude of the integrated $CO_2$ release from the terrestrial biosphere since 1800 is about 90% of that from fossil fuel; (2) over the two-decade period covered by the Mauna Loa atmospheric $CO_2$ record, the input from the forest-soil source is about 15% of that from fossil fuels; (3) the [13]C/[12]C trend and the atmospheric $CO_2$ anomaly over the last two decades have been dominated by the input of fossil fuel $CO_2$; and (4) the pre-1850 atmospheric $CO_2$ content is estimated to be about 266 ppm. The integrated amount of $CO_2$ released from the terrestrial biosphere between A.D. 1800 and 1980 is estimated to be $12 \times 10^{15}$ mol which is to be compared with the previous estimate of $22 \times 10^{15}$ mol on the basis of Freyer and Belacy's (1983) [13]C data. This indicates that tree-ring [13]C-based estimates are volatile and further changes are likely.

## Introduction

Release of biogenic $CO_2$ (both from fossil fuel combustion and from deforestation and soil manipulation) into the atmosphere has changed atmospheric $CO_2$ levels over the past few hundred years. But the time history of atmospheric $CO_2$ variations was not known until 1958, when the first precise $CO_2$-monitoring system was established at Mauna Loa [Keeling et al., 1982]. Therefore, the lack of direct observed atmospheric $CO_2$ data before 1958 is a major problem when attempting to predict future increases in the $CO_2$ level due to anthropogenic carbon input into the atmosphere; that is, the geochemical carbon cycle models used for these predictions cannot be properly validated with the complete time series of atmospheric $CO_2$ variations.

To reconstruct past atmospheric $CO_2$ levels, we must know the release history of biogenic $CO_2$ inputs. Although the amount of $CO_2$ released due to combustion of fossil fuel can be estimated from fossil fuel production records [Keeling, 1973; Rotty, 1981], insufficient data make estimating biospheric $CO_2$ releases very difficult. Based on limited biomass change data, some estimates of annual input between 1960 and 1975 and estimates of cumulative input up to recent years were made [Bolin, 1977; Adams et al., 1977; Woodwell et al., 1978; Wong, 1978; Hampicke, 1979; Seiler and Crutzen, 1980; Moore et al., 1981]. Houghton et al. [1983] estimated a time series of this type for $CO_2$ releases, using land use and demographic data.

Stuiver [1978] was the first to derive the magnitude of historical forest-soil $CO_2$ releases from the [13]C/[12]C record in tree rings. Wagener [1978], Freyer [1978], and Siegenthaler et al. [1978] also utilize the tree-ring [13]C data for estimating the total $CO_2$ emission from terrestrial ecosystems. Summarizing all available [13]C records for tree-ring series, Peng et al. [1983] compared composite [13]C trends based on sample material (whole wood or cellulose), site of tree growth (freestanding or forested), and climate environments. Using the decade-averaged composite [13]C trend of Freyer and Belacy [1983] for the northern hemisphere and the modified box-diffusion model of Oeschger et al. [1975], Peng et al. [1983] reconstructed past atmospheric $CO_2$ levels. Stuiver et al. [1984] employed different deconvolution schemes to derive the history of biospheric $CO_2$ releases and atmospheric $CO_2$ variations since A.D. 235 from long tree-ring [13]C records for the Pacific coast of North and South America.

After receiving more tree-ring [13]C results [Leavitt and Long, 1983], Freyer [1984] recalculated the [13]C/[12]C trend of the northern hemisphere (based mainly on [13]C data presented by Freyer and Belacy [1983]). He concluded that no long-term trend exists for a period of a few centuries before A.D. 1800. By contrast, a clear trend of decreasing $^{13}C$ is observed after A.D. 1800. The overall decrease of $\delta^{13}C$ from A.D. 1800 to 1980 is estimated to be about $-1.5°/°°$, which is $0.5°/°°$ less than that estimated from the composite $^{13}C$ trend used for the previous deconvolution schemes [Peng et al., 1983]. The terrestrial biospheric contribution to the lowering of $\delta^{13}C$ in atmospheric $CO_2$ was reevaluated by Peng and Freyer [1984]. The objective of this paper is to present past atmospheric $CO_2$ variations, based on the deconvolution of composite tree-ring $\delta^{13}C$ records.

## Method

The basic principle used in relating the $^{13}C/^{12}C$ ratio in tree rings to the amount of biogenic $CO_2$ released to the atmosphere is as follows. Biogenic $CO_2$ has $\delta^{13}C$ values averaging about $-26°/°°$ [Schwarz, 1970] ($\delta^{13}C$ per mil = $[(R_{sample} - R_{PDB})/R_{PDB}] \times 1000$, where R is the $^{13}C/^{12}C$ isotopic ratio and PDB (Pee Dee belemnite) is the standard used for measuring the $^{13}C/^{12}C$ isotopic ratio). The preanthropogenic atmosphere had a value of about $-6°/°°$ [Bolin et al., 1981]. Therefore, the addition of biogenic $CO_2$ to the atmosphere will reduce the $\delta^{13}C$ value of atmospheric $CO_2$. The time history of the variation of atmospheric $\delta^{13}C$ is recorded in tree rings when the trees assimilate atmospheric $CO_2$ by photosynthesis to form plant materials. The situation is complicated by several factors. The carbon atoms from fossil fuel and from forest-soil sources will exchange with carbon atoms in the ocean and in the terrestrial biosphere reservoir. The $\delta^{13}C$ dilution effects caused by this exchange must be properly accounted for if the tree-ring $^{13}C$ record is to be converted to a $CO_2$ production record. We will need a global carbon model to resolve this problem.

In addition to atmospheric $\delta^{13}C$ changes, the observed $\delta^{13}C$ variations in tree rings could also be caused by a number of other factors. The local fluctuation in the $^{13}C/^{12}C$ ratio, generated by the input of $^{13}C$-depleted $CO_2$ from soil respiration, may influence the $^{13}C$ content of trees. The $^{13}C/^{12}C$ record carried by such trees would not be a reliable representation of $\delta^{13}C$ variations on a global scale. The $^{13}C/^{12}C$ records of oak trees in the Spessart Mountains in southern Germany [Freyer and Belacy, 1983] show sudden jumps and nonsystematic variations not synchronous with respect to time. The canopy effects of forest trees and the partial uptake of $CO_2$ from soil respiration are considered to be the main reasons for such $\delta^{13}C$ variations. Therefore, the selection of freestanding trees far away from urban areas for $\delta^{13}C$ analysis should minimize these effects.

Ring-to-ring variations in the chemical makeup of plant material will lead to changes in the $\delta^{13}C$ of whole wood. By analyzing only cellulose, this effect of variations in the ratio of cellulose to lignin and extractives can be eliminated.

The $\delta^{13}C$ variations in tree rings can also be affected by environmental variability, due to the effects of changing irradiance, relative humidity, and growing season temperature on stomatal geometry, all of which cause varying isotopic fractionation during photosynthesis.

A recent model of carbon isotope fractionation by C3 plants [Farquhar et al., 1982] envisions atmospheric $CO_2$ diffusing into intercellular leaf spaces and causing a kinetic fractionation. This is followed by further isotopic separation during carboxylation of RuDP (Ribulose Di-Phosphate). $^{13}C$ diffuses more slowly in air than does $^{12}C$ ($\alpha_{diffusion} = 0.9955$) ($\alpha = R_{product}/R_{source}$, where R is the $^{13}C/^{12}C$ ratio). The magnitude of the fractionation by RuDP ($\alpha_{RuDP}$) is not well determined; G. D. Farquhar (personal communication, 1983) places it at 0.97.

The rate of diffusion of $CO_2$ through leaf stomata is controlled by the $CO_2$ partial pressure ($pCO_2$) gradient between the leaf interior and the atmosphere. The sizes of the stomatal openings adjust according to variations in relative humidity, irradiance, and temperature, thereby controlling the size of the $CO_2$ reservoir within the leaf. RuDP carboxylation determines how much of that internal reservoir will be converted to glucose. The final isotopic composition of the plant tissue will be determined by diffusion or carboxylation, which limits the amount of $CO_2$ assimilated by the plant. If diffusion alone limits the assimilation rate, all of the $CO_2$ that diffuses into the leaf is eventually fixed as glucose, and the diffusion fractionation will dominate:

$$\left. \frac{^{13}C}{^{12}C} \right)_{plant} = \left. \frac{^{13}C}{^{12}C} \right)_{atm} \alpha_{diff}$$

However, if diffusion is rapid when compared with carboxylation, no $CO_2$ concentration gradient will exist across the stomata, and the $\delta^{13}C$ of the plant will depend only on enzyme fractionation:

$$\left. \frac{^{13}C}{^{12}C} \right)_{plant} = \left. \frac{^{13}C}{^{12}C} \right)_{atm} \alpha_{RuDP}$$

Trees, as C3 plants, have $\delta^{13}C$ values that fall between the two extremes. Thus, environmental factors that change the ratio of diffusion resistance to carboxylation resistance change the $^{13}C/^{12}C$ ratio of the plant (see Figure 1). Farquhar et al. [1982] and Francey and Farquhar [1982] attribute much of the $\delta^{13}C$ variability observed in C3 plants to the effects on stomatal geometry of changing irradiance, relative humidity, and temperature.

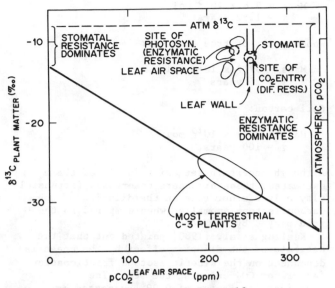

Fig. 1. Relationship between the $\delta^{13}C$ for cellulose in C3 plants as a function of leaf space $CO_2$ partial pressure (as documented by Farquhar et al. [1982]). If this pressure is near the atmospheric value, then the enzymatic fractionation dominates. If this pressure is very low, the fractionation associated with diffusion through the stomatal openings into the leaf air space must dominate. Although the exact position of the line has yet to be pinned down, the approximate trend has been established.

Ideally, freestanding trees that have an adequate supply of moisture and that experience relatively consistent temperatures throughout their growing season should be used in order to reduce environmental effects. The $^{13}C$ record from cellulose of such trees would best represent the global atmospheric $\delta^{13}C$ record. However, in selecting trees in Tasmania largely according to the above criteria, Francey [1981] found a $^{13}C$ change smaller than that expected from fossil fuels alone.

Leavitt and Long [1983] found a correlation between $\delta^{13}C$ of the cellulose fraction of juniper trees in Arizona and the mean December temperature or precipitation at the tree sites. Freyer and Belacy [1983] also demonstrated a correlation between the $\delta^{13}C$ record of the Scots pine trees and autumn temperature and spring precipitation. But the relationship does not hold for other trees at different sites and in different climatic zones. In his recalculation of the global $\delta^{13}C$ trend of tree rings, Freyer [1984] does not correct for environmental effects. Trees used for reconstruction of past atmospheric $\delta^{13}C$ variations are mostly freestanding. Cellulose is the only material used in most of the analyses of $\delta^{13}C$. Efforts were made to reduce the local fluctuation and material effects, but it should be kept in mind

that the reconstructed global $^{13}C$ trend of tree rings contains environmental noises. In addition, if the $^{13}C$ fractionation in trees correlates with atmospheric $pCO_2$ (partial pressure of $CO_2$ gas), no amount of averaging will give the right temporal record for atmospheric $CO_2$ content.

The global carbon cycle model used for reconstruction of past atmospheric $CO_2$ concentration (see Figure 2) is the modified box-diffusion ocean model of Peng et al. [1983], coupled with living biosphere and soil carbon reservoirs [Emanuel et al., 1981]. Major features of this modified box-diffusion model are as follows:

1. The oceanic photosynthesis-respiration cycle is included.

2. The formation of deep water is considered. A loop is inserted that brings intermediate water (∿1000-m depth) to the surface in the polar region, increases its density, and sends it back to the bottom of the model ocean. From there it upwells to the intermediate depth, completing the cycle.

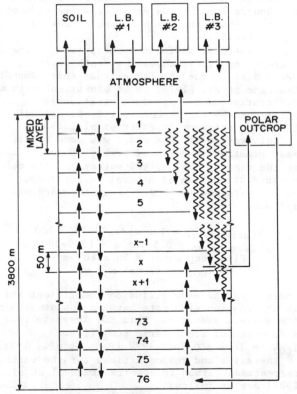

Fig. 2. Multibox model used to calculate $CO_2$ uptake and carbon isotope dilution. A detailed description is given by Peng et al. [1983]. Some special features are briefly described in the text. (Living biomass (L.B.)1, ground vegetation; L.B.2, nonwoody parts of tree plus detritus and decomposers; L.B.3, woody parts of tree.)

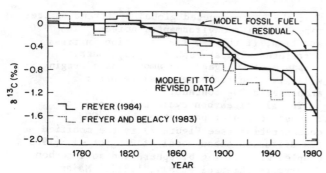

Fig. 3. Terrestrial biosphere CO$_2$ contribution (curve marked "residual") to the decline in atmospheric $^{13}$C/$^{12}$C ratio, obtained by subtracting the model-derived contribution of fossil fuel CO$_2$ (curve marked "model fossil fuel") from the model curve that fits Freyer's [1984] revised $^{13}$C record (solid step curve). The dotted step curve is the composite $^{13}$C record of Freyer and Belacy [1983].

Because this cycle mixes the deep ocean, the eddy diffusion coefficient below the level from which the source for deep water is drawn is reduced to about one-half the value chosen by Oeschger et al. [1975].

3. The eddy diffusion coefficient in the main thermocline (i.e., above the depth of the deep-water source) is larger than that adopted by Oeschger et al. [1975] to be consistent with the penetration of bomb-produced tritium as determined by the Geochemical Section Studies (GEOSECS) program [Broecker et al., 1980].

Calibration of this model was made to fit the mean penetration depth of bomb-produced tritium at the time of the GEOSECS survey and the mean depth distribution of natural radiocarbon. The values obtained for these parameters are as follows:

$$K_{thermocline} = 1.6 \ cm^2 \ s^{-1} \ (100-700 \ m)$$
$$K_{deep \ ocean} = 0.5 \ cm^2 \ s^{-1} \ (700-3800 \ m)$$
$$Flux \ deep \ water = 50 \times 10^6 \ m^3 \ s^{-1}$$
$$I_{CO2} = 17 \ mol \ m^{-2} \ yr^{-1}$$

where K is the eddy diffusion coefficient and $I_{CO2}$ is the CO$_2$ invasion rate between the atmosphere and the surface seawater. This CO$_2$ invasion rate is a function of atmospheric pCO$_2$ (i.e., $I_{CO2} = 16 \times pCO_2/pCO_2^{1800}$ [Peng et al., 1983]).

The sizes and response times for the various terrestrial carbon reservoirs [Emanuel et al., 1981] are as follows:

Ground vegetation (living biomass (L.B.)1]:

Mass = 5.8 x 10$^{15}$ mol
$\tau$ = 4 years

Nonwoody parts of tree plus detritus and decomposers (L.B. 2):

Mass = 9.8 x 10$^{15}$ mol
$\tau$ = 1.92 years

Woody parts of tree (L.B. 3):

Mass = 37.7 x 10$^{15}$ mol
$\tau$ = 25 years

Soil carbon:

Mass = 93.4 x 10$^{15}$ mol
$\tau$ = 100 years.

Although uncertainties exist in all of these estimates, these biosphere reservoirs fortunately play only a minor role in the dilution of the carbon isotope anomalies when compared to the role played by the ocean.

Keeling et al. [1980] pointed out that the estimate of the dilution effect by the ocean is dependent on the kinetic isotope fractionation factor for CO$_2$ entering the sea. The fractionation factor of 0.998 suggested by Siegenthaler and Munnich [1981] is adopted for computations.

Tans [1981] showed that the $\delta^{13}$C in fossil fuel CO$_2$ changes with time because of variations in the proportions of fossil fuels used (e.g., coal, oil, natural gas). This variation in $\delta^{13}$C is also included in the model computations.

Once the dilution model has been set, the $^{13}$C/$^{12}$C time history for atmospheric CO$_2$ is calculated, assuming that the only perturbation is attributed to the addition of fossil fuel CO$_2$ to the system. The $\delta^{13}$C change obtained this way is subtracted from the composite tree-ring $\delta^{13}$C record. The residual $^{13}$C/$^{12}$C anomaly is then assumed to represent the forest-soil CO$_2$ contribution. An estimate of the time history of

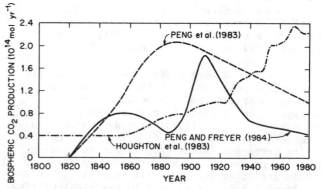

Fig. 4. Comparison of the terrestrial biosphere CO$_2$ input function derived by Peng and Freyer [1984] using Freyer's [1984] revised $^{13}$C record with that derived by using the Freyer and Belacy [1983] composite $^{13}$C record. The time history of CO$_2$ release from the terrestrial biosphere, as derived from land use data [Houghton et al., 1983] is also shown for comparison.

TABLE 1.  Results of $\delta^{13}C$ Deconvolution Using the Peng et al. [1983]
Modified Box-Diffusion Ocean Model of Oeschger et al. [1975]

|  | Freyer's [1984] Revised $^{13}C$ Data | Freyer and Belacy's [1983] $^{13}C$ Data |
|---|---|---|
| Integrated fossil fuel $CO_2$ input as of 1980 | $14 \times 10^{15}$ mol | $14 \times 10^{15}$ mol |
| Integrated terrestrial biosphere $CO_2$ input as of 1980 | $12 \times 10^{15}$ mol | $22 \times 10^{15}$ mol |
| $\Delta pCO_2$) fossil fuel (1958-1980) | $24 \times 10^{-6}$ atm* | $24 \times 10^{-6}$ atm |
| $\Delta pCO_2$) biosphere (1958-1980) | $1 \times 10^{-6}$ atm | $4 \times 10^{-6}$ atm |
| Pre-1850 atmospheric $pCO_2$ | $266 \times 10^{-6}$ atm | $243 \times 10^{-6}$ atm |
| Ice core $pCO_2$ [Barnola et al.,1983] | $258-266 \times 10^{-6}$ atm | |

*1 atm = $1.013 \times 10^5$ Pa.

forest-soil $CO_2$ input is then obtained by iteration.  The $^{13}C/^{12}C$ anomaly generated by this history is calculated and compared with the residual $^{13}C/^{12}C$ anomaly.  The history is then adjusted in such a way as to improve the match between the calculated and the observed residual.  The atmospheric $CO_2$ variations with time can thus be estimated with the input of $CO_2$ derived from forest-soil and known fossil fuel $CO_2$ inputs.

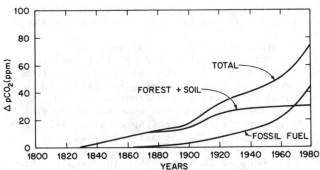

Fig. 5.  Increase in atmospheric $pCO_2$ versus time, obtained from the combined forest plus soil and fossil fuel $CO_2$ inputs (curve marked "total").  Also shown are the individual contributions of the two components.  As can be seen over the 20-year period for which we have an atmospheric $CO_2$ record, the trend is dominated by the fossil fuel $CO_2$ contribution.  These curves are based on Peng and Freyer's [1984] results.

## Results

A comparison of the tree-ring-based $\delta^{13}C$ record determined by Freyer [1984] with that by Freyer and Belacy [1983] is shown in Figure 3.  Three model-computed curves are also presented.  The model residual curve represents the $^{13}C$ changes needed to make the $^{13}C$ variations, in addition to those variations caused by fossil fuel $CO_2$ inputs, fit the observed $^{13}C/^{12}C$ record.  We attribute this residual to the input of $CO_2$ generated by deforestation and soil manipulation.

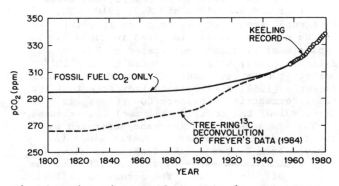

Fig. 6.  Time history of atmospheric $pCO_2$ derived from the deconvolution of Freyer's [1984] tree-ring-based $^{13}C$ record.  The model-based curve for fossil fuel input only is also shown. The Keeling record is from Keeling et al. [1982]; Freyer's data are given by Freyer [1984].

TABLE 2.   Results of Deconvolution on $^{13}C$ Records of Freyer and Belacy [1983]
and of Stuiver et al. [1984]

| | Freyer and Belacy [1983] Data | | Stuiver et al. [1984] Data | |
| --- | --- | --- | --- | --- |
| | Peng et al. [1983] Model | Stuiver et al. [1984] Model | Peng et al. [1983] Model | Stuiver et al. [1984] Model |
| Integrated biospheric $CO_2$ releases (x $10^{15}$ mol) | 22.8 (as of 1975) | 28.8 (as of 1975) | 2.0 (1860-1975) | 3.3 (1860-1975) |
| Fossil fuel $CO_2$ release 1860-1975 (x $10^{15}$ mol) | 11.3 | 11.3 | 11.3 | 11.3 |
| Preindustrial atmospheric $CO_2$ partial pressure (x$10^{-6}$ atm)* | 243 | 230 | 283 | 276 |
| $CO_2$ partial pressure ($10^{-6}$ atm) of air bubble in ice core [Barnola et al., 1983] | 258-266 | 258-266 | 258-266 | 258-266 |

*1 atm = 1.013 x $10^5$ Pa.

The estimate of the time history of forest-soil $CO_2$ input needed to produce the residual curve shown in Figure 3 is presented in Figure 4.  Also shown in Figure 4 is the time history of the $CO_2$ input from the same sources deconvolved from the previous Freyer and Belacy [1983] composite $^{13}C$ record and the terrestrial biospheric $CO_2$ release time history based on land use data [Houghton et al., 1983].  The major difference between the two methods is that the $CO_2$ release based on the land use method increases with time since early 1800, while that based on tree-ring $^{13}C$ decreases with time since early 1900.  Two maximum periods of $CO_2$ releases are determined by the deconvolution of Freyer's [1984] revised $^{13}C$ record, and these appear around A.D. 1860 and A.D. 1910.  The integrated amount of $CO_2$ released from this source as of A.D. 1980 is given in Table 1.  The estimate of 22 x $10^{15}$ mol (based on Freyer and Belacy's [1983] data) is reduced to 12 x $10^{15}$ mol after data revision.  The estimate based on Freyer's [1984] revised data indicates that the total terrestrial biosphere $CO_2$ release is about 90% of the total fossil fuel $CO_2$ release.

The atmospheric $CO_2$ increase generated by the terrestrial biosphere scenario alone (as derived from Freyer's [1984] data) is shown in Figure 5.  As can be seen, its contribution has leveled off since 1940.  The actual contribution over the last 20 years (i.e., the period of time for which a reliable atmospheric $CO_2$ record exists) is only about 0.1 Pa (1 x $10^{-6}$ atm) (see Table 1).  This is smaller than the estimate of about 0.4 Pa (about 4 x $10^{-6}$ atm) based on the Freyer and Belacy [1983] data.  The fossil fuel contribution over this interval of time computed by the same model is about 2.4 Pa (24 x $10^{-6}$ atm).  Obviously, the atmospheric $CO_2$ anomaly over the last 20 years is predominantly generated by the release of fossil fuel $CO_2$.  This leads to the conclusion that oceanic uptake of $CO_2$ released from the terrestrial biosphere in earlier years has almost exactly balanced the new production from this source during the last 20 years.  Hence, as long as the relaxation of the anomalies generated by the emissions of terrestrial biosphere $CO_2$ in the past balances the inputs over the last two decades, releases of forest-soil $CO_2$ do not aggravate the modeling strategy for fossil fuel $CO_2$ used by the ocean modelers.

The preanthropogenic atmospheric $CO_2$ concentration was calculated by using the time history of forest-soil $CO_2$ release derived from the deconvolution of tree-ring $^{13}C$ data.  In the model calculation, the atmospheric $pCO_2$ curve is forced through the observed $pCO_2$ value of 315.5 ppm for 1958 [Keeling et al., 1982].  As shown in Figure 6, the scenario derived from Freyer's [1984] revised $^{13}C$ data yields a pre-1800 $pCO_2$ of 26.9 Pa (266 x $10^{-6}$ atm).  This is to be compared with the previous estimate of 24.6 Pa (243 x $10^{-6}$ atm), using Freyer and Belacy's [1983] $^{13}C$ data.

The $\delta^{13}C$ records of Stuiver et al. [1984] from the Pacific coast of North and South America are not included in Freyer's [1984] revised composite $^{13}C$ trend.  Stuiver et al. chose six trees (one tree from 40° south and five trees from 37° north to 57° north) to make up a composite $\delta^{13}C$ record for the purpose of deconvolution for land biospheric $CO_2$ release. The model they use is the box-diffusion model of Oeschger et al. [1975], coupled with two biospheric reservoirs. The size of these reservoirs changes with time as carbon comes in and out of these boxes as required by deconvolution of their $^{13}C$ records. The vertical mixing rate of the ocean is taken to be 2.2 $cm^2 s^{-1}$ instead of the value (1.6 $cm^2 s^{-1}$) used by Peng et al. [1983]. The $^{13}C$ changes from one decade to the next, as measured in tree rings, are taken to be the sums of $^{13}C$ changes due to (1) $CO_2$ exchange with the ocean reservoir, (2) fossil fuel $CO_2$ releases, and (3) biogenic $CO_2$ flux. They deconvolved the $^{13}C$ data individually for each decade-averaged sector. Hence, the derived time histories of atmospheric $CO_2$ variations and of biospheric $CO_2$ fluxes are decade-averaged step functions with shapes similar to those of $^{13}C$ data. This is to be contrasted with smooth curve fitting to the step function of $^{13}C$ data employed by Peng et al. [1983].

Comparisons of the results of the deconvolution of Stuiver et al. [1984] with those of Peng et al. [1983] are shown in Table 2. Using the Freyer and Belacy [1983] $^{13}C$ record, Stuiver et al. [1984] derived an integrated biospheric $CO_2$ release of 28.8 x $10^{15}$ mol for the period between 1760 and 1975. For the same set of data, the Peng et al. [1983] model gave an estimate of 22.8 x $10^{15}$ mol for the same period. Based on these estimates of biospheric $CO_2$ flux, the preindustrial atmospheric $CO_2$ partial pressure was calculated to be 230 ppm by the Stuiver et al. [1984] deconvolution and 243 ppm by the Peng et al. [1983] smooth-curve-fitting method.

To make a further comparison, Peng and Freyer [1984] deconvolved the $^{13}C$ record of Stuiver et al. [1984] using the Peng et al. [1983] model. This comparison is also shown in Table 2. Due to the smaller magnitude of the $\delta^{13}C$ decrease, Stuiver et al.'s data led to a smaller land biospheric $CO_2$ flux than did those based on Freyer's [1984] $^{13}C$ record. In addition, it is clear that the Stuiver et al. [1984] deconvolution method gives consistently higher estimates (i.e., 3.3 x $10^{15}$ mol) of biospheric $CO_2$ flux for the period between 1860 and 1975 than does the smooth-curve-fitting method. The corresponding estimates of preindustrial atmospheric $pCO_2$ are 283 ppm for the Peng et al. [1983] model and 276 ppm for the Stuiver et al. [1984] model. This difference is caused mainly by the higher ocean-mixing rates used in the Stuiver et al. [1984] model (i.e., 2.2 $cm^2 s^{-1}$ versus 1.6 $cm^2 s^{-1}$). The ocean dilutes more $^{13}C$ released from the land biosphere in the Stuiver et al. [1984] model. Other major contributing factors include the differences between smooth curve fitting and direct-step curve fitting and between keeping the size of biospheric reservoirs constant [Peng et al.,

1983] and varying the size with biospheric changes.

As mentioned earlier, the preindustrial $CO_2$ concentration in the atmosphere is estimated to be about 266 ppm using the Peng et al. [1983] model and the Freyer [1984] revised $^{13}C$ record. Although this value is consistent with recent results of ice core studies (i.e., $pCO_2^{1800} \simeq 258$ to 266 ppm [Barnola et al., 1983]), the new $^{13}C$ data from the Pacific coast of North and South America [Stuiver et al., 1984] indicate an estimate of 283 ppm when the same model is used for deconvolution. These estimates are also different from those based on Freyer and Belacy's [1983] data (i.e., 243 ppm). The implication of these differences is that the environment- and climate-induced noises in the tree-ring $^{13}C$ record are not clearly defined and filtered out. To make a significant linkage between the tree-ring $^{13}C$ record and the past atmospheric $CO_2$ content, the true relationship between atmospheric $^{13}C$ variations and tree-ring $^{13}C$ changes will have to be established. The recent work of Farquhar et al. [1982] and Francey and Farquhar [1982] toward developing a simple theory to explain carbon isotope fractionation by plants is an important step toward understanding the meaning of the $^{13}C/^{12}C$ record in tree rings.

## Conclusions

If the tree-ring-based $^{13}C/^{12}C$ record constructed by Freyer [1984] is representative of the fossil fuel and terrestrial biospheric perturbation-induced $^{13}C/^{12}C$ change for atmospheric $CO_2$, the deconvolution of this $^{13}C$ record using the Peng et al. [1983] modified box-diffusion ocean model of Oeschger et al. [1975] leads to the following conclusions: (1) The total release of $CO_2$ from the terrestrial biosphere (as of 1980) is estimated to be about 12 x $10^{15}$ mol, which is only 55% of the previous estimate using the Freyer and Belacy [1983] data. Compared to total fossil fuel $CO_2$ released during the same period, the biospheric $CO_2$ input is about 90% of the fossil fuel $CO_2$ input. (2) The uptake by the ocean of $CO_2$ released from the terrestrial biosphere in earlier years has almost exactly balanced the new production by this source during the last 20 years. (3) The atmospheric $CO_2$ anomaly over the last 20 years is predominantly caused by the release of fossil fuel $CO_2$. (4) The pre-1850 atmospheric $pCO_2$ is estimated to be about 266 ppm, which is consistent with the recent results of ice core studies.

Considering the changes in the averages of the global composite $^{13}C$ record from tree rings when more $^{13}C$ data became available, as demonstrated here, it is clear that $^{13}C$-based estimates of atmospheric $CO_2$ variations are highly volatile and further changes are likely.

Acknowledgments. Research at Oak Ridge National Laboratory (ORNL) was supported jointly by the National Science Foundation's Ecosystem Studies Program under Interagency Agreement BSR-8115316, A03 and by the Carbon Dioxide Research Division, Office of Energy Research, U.S. Department of Energy, under contract DE-AC05-84OR21400 with Martin Marietta Energy Systems, Inc. Publication 2390, Environmental Sciences Division, ORNL.

## References

Adams, J. A. S., M. S. M. Mantovani, and L. L. Lundell, Wood versus fossil fuel as a source of excess carbon dioxide in the atmosphere: A preliminary report, Science, 196, 54-56, 1977.

Barnola, J. M., D. Raynaud, A. Neftel, and H. Oeschger, Comparison of $CO_2$ measurements by two laboratories on air from bubbles in polar ice, Nature, 303, 410-413, 1983.

Bolin, B., Changes of land biota and their importance for the carbon cycle, Science, 196, 613-615, 1977.

Bolin, B., C. D. Keeling, R. B. Bacastow, A. Björkstrom, and U. Siegenthaler, Carbon cycle modelling, in Carbon Cycle Modelling, SCOPE 16, edited by B. Bolin, pp. 1-28, John Wiley, New York, 1981.

Broecker, W. S., T.-H. Peng, and R. Engh, Modelling the carbon system, Radiocarbon, 22, 565-598, 1980.

Emanuel, W. R., G. G. Killough, and J. S. Olson, Modeling the circulation of carbon in the world's terrestrial ecosystems, in Carbon Cycle Modelling, SCOPE 16, edited by B. Bolin, pp. 335-353, John Wiley, New York, 1981.

Farquhar, G. D., M. H. O'Leary, and J. A. Berry, On the relationship between carbon isotope discrimination and the intercellular carbon dioxide concentration in leaves, Aust. J. Plant Physiol., 9, 121-137, 1982.

Francey, R. J., Tasmanian tree rings belie suggested anthropogenic [13]C/[12]C trends, Nature, 290, 232-235, 1981.

Francey, R. J., and G. D. Farquhar, An explanation of the [13]C/[12]C variations in tree rings, Nature, 297, 28-31, 1982.

Freyer, H. D., Preliminary evaluation of past $CO_2$ increase as derived from [13]C measurements in tree rings, in Carbon Dioxide, Climate and Society, edited by J. Williams, pp. 69-78, Pergamon, New York, 1978.

Freyer, H. D., Interpretation of the northern hemispheric record of [13]C/[12]C trends of atmospheric $CO_2$ in tree rings, in Proceedings of the Sixth ORNL Life Sciences Symposium - The Global Carbon Cycle: Analysis of the Natural Cycle and Implications of Anthropogenic Alterations for the Next Century, edited by J. R. Trabalka and D. E. Reichle, Oak Ridge National Laboratory, Oak Ridge, Tennessee, in press, 1984.

Freyer, H. D., and N. Belacy, [13]C/[12]C records in northern hemisphere trees during the past 500 years: Anthropogenic impact and climatic superpositions, J. Geophys. Res., 88, 6844-6852, 1983.

Hampicke, U., Net transfer of carbon between the land biota and the atmosphere, induced by man, in The Global Carbon Cycle, SCOPE 13, edited by B. Bolin, E. T. Degens, S. Kempe, and P. Ketner, pp. 219-236, John Wiley, New York, 1979.

Houghton, R. A., J. E. Hobbie, J. M. Melillo, B. Moore, B. J. Peterson, G. R. Shaver, and G. M. Woodwell, Changes in the carbon content of terrestrial biota and soils between 1860 and 1980: A net release of $CO_2$ to the atmosphere, Ecol. Monogr., 53(3), 235-262, 1983.

Keeling, C. D., Industrial production of carbon dioxide from fossil fuel and limestone, Tellus, 25, 1174-1198, 1973.

Keeling, C. D., R. B. Bacastow, and P. Tans, Predicted shift in the [13]C/[12]C ratio of atmospheric carbon dioxide, Geophys. Res. Lett., 7, 505-508, 1980.

Keeling, C. D., R. B. Bacastow, and T. P. Whorf, Measurements of the concentration of carbon dioxide at Mauna Loa Observatory, Hawaii, in Carbon Dioxide Review, edited by W. C. Clark, pp. 377-385, Oxford University Press, New York, 1982.

Leavitt, S. W., and A. Long, An atmospheric [13]C/[12]C reconstruction generated through removal of climatic effects from tree-ring [13]C/[12]C measurements, Tellus, 35B, 92-102, 1983.

Moore, B., R. D. Boone, J. E. Hobbie, R. A. Houghton, J. M. Melillo, B. J. Peterson, G. R. Shaver, C. J. Vorosmarty, and G. M. Woodwell, A simple model for analysis of the role of terrestrial ecosystems in the global carbon budget, in Carbon Cycle Modelling, SCOPE 16, edited by B. Bolin, pp. 365-385, John Wiley, New York, 1981.

Oeschger, H., U. Siegenthaler, U. Schotterer, and A. Gugelmann, A box diffusion model to study the carbon dioxide exchange in nature, Tellus, 27, 168-192, 1975.

Peng, T.-H., and H. D. Freyer, Revised estimates of atmospheric $CO_2$ variations based on the tree-ring [13]record, in Proceedings of the Sixth ORNL Life Sciences Symposium - The Global Carbon Cycle: Analysis of the Natural Cycle and Implications of Anthropogenic Alterations for the Next Century, edited by J. R. Trabalka and D. E. Reichle, Oak Ridge National Laboratory, Oak Ridge, Tennessee, in press, 1984.

Peng, T.-H., W. S. Broecker, H. D. Freyer, and S. Trumbore, A deconvolution of the tree ring based $\delta$[13]C record, J. Geophys. Res., 88, 3609-3620, 1983.

Rotty, R. M., Data for global $CO_2$ production from fossil fuels and cement, in Carbon Cycle Modelling, SCOPE 16, edited by B. Bolin, pp. 121-123, John Wiley, New York, 1981.

Schwarz, H. P., The stable isotopes of carbon, in *Handbook of Geochemistry*, edited by K. H. Wedepohl, pp. 1-16, Springer-Verlag, New York, 1970.

Seiler, W., and P. J. Crutzen, Estimates of gross and net fluxes of carbon between the biosphere and the atmosphere from biomass burning, *Clim. Change*, 2, 207-247, 1980.

Siegenthaler, U., and K. O. Munnich, $^{13}C/^{12}C$ fractionation during $CO_2$ transfer from air to sea, in *Carbon Cycle Modelling*, *SCOPE 16*, edited by B. Bolin, pp. 249-258, John Wiley, New York, 1981.

Siegenthaler, U., M. Heimann, and H. Oeschger, Model responses of the atmospheric $CO_2$ level and $^{13}C/^{12}C$ ratio to biogenic $CO_2$ input in carbon dioxide, in *Climate and Society*, edited by J. Williams, pp. 79-84, Pergamon, New York, 1978.

Stuiver, M., Atmospheric $CO_2$ increases related to carbon reservoir changes, *Science*, 199, 253-258, 1978.

Stuiver, M., R. L. Burk, and P. D. Quay, $^{13}C/^{12}C$ ratios in tree rings and the transfer of biospheric carbon to the atmosphere, *J. Geophys. Res.*, in press, 1984.

Tans, P., $^{13}C/^{12}C$ of industrial $CO_2$, in *Carbon Cycle Modelling*, *SCOPE 16*, edited by B. Bolin, pp. 127-129, John Wiley, New York, 1981.

Wagener, K., Total anthropogenic $CO_2$ production during the period 1800-1935 from carbon-14 measurements in tree rings, *Radiat. Environ. Biophys.*, 15, 101-111, 1978.

Wong, C. S., Atmospheric input of carbon dioxide from burning wood, *Science*, 200, 197-200, 1978.

Woodwell, G. M., R. H. Whittaker, W. A. Reiners, G. E. Likens, C. C. Delwiche, and D. B. Botkin, The biota and the world carbon budget, *Science*, 199, 141-146, 1978.

# VARIATIONS OF THE $CO_2$ CONCENTRATION OF OCCLUDED AIR AND OF ANIONS AND DUST IN POLAR ICE CORES

H. Oeschger and B. Stauffer

Physics Institute, University of Bern, Switzerland

R. Finkel and C. C. Langway, Jr.

Ice Core Laboratory, State University of New York at Buffalo
Amherst, New York 14226

Abstract. Analysis of impurities entrapped in natural ice is the most promising method for reconstructing the history of atmospheric composition before the period of direct measurement and offers the possibility of extending the record to at least 100,000 years B.P. We report here the present state of work in this field, with special emphasis on atmospheric $CO_2$ concentration. After discussing the mechanism by which atmospheric gases are entrapped in ice, we report $CO_2$ concentrations in ice core samples, up to 100,000 years old, from deep drilling projects in Greenland and the Antarctic. Results from ice deposited during the last 2000 years allow us to estimate the preindustrial atmospheric $CO_2$ level, an important boundary condition for modelling the anthropogenic $CO_2$ increase. Using older samples from a deep ice core drilled at Dye 3, Greenland, we show that the $CO_2$ concentration was 180 to 200 ppmv at the end of the Wisconsin and increased during the transition to the Holocene to values in the 260 to 300 ppmv range. Detailed $CO_2$ measurements on sections of the Wisconsin part of the Dye 3 core which, based on $\delta^{18}O$, were deposited during times of significant climatic variation, show that the $\delta^{18}O$ variations were accompanied by simultaneous correlated rapid $CO_2$ variations. Other parameters, including micro-particle concentration and $Cl^-$, $NO_3^-$ and $SO_4^{2-}$ concentrations also showed significant variations which correlate with the measured $\delta^{18}O$ shifts.

## Introduction

The atmospheric $CO_2$ concentration is a valuable parameter for understanding the carbon cycle. Changes in atmospheric $CO_2$ since the beginning of industrialization give information about the extent to which the carbon cycle has been perturbed by anthropogenic influences, while changes in the more distant past reflect the natural variability of the carbon cycle and provide information for estimating the sensitivity of climate to atmospheric $CO_2$ variations. In spite of the importance of atmospheric $CO_2$, continuous and precise measurements date back only to 1958, when sampling at Mauna Loa was begun.

Gas analysis of air entrapped in natural ice is the most promising method for reconstructing the history of the atmospheric $CO_2$ concentration before the period of direct measurement and offers the possibility of extending the record to at least 100,000 years B.P. [World Meteorological Organization, 1983]. In this paper we describe the present state of work in this field. Especially crucial to the use of ice cores for the reconstruction of past $CO_2$ levels is understanding the mechanisms by which air is occluded in ice and possible processes which could lead to deviations of the composition of the trapped air from that of the atmosphere. Such deviations could be produced both during the occlusion process and during the long interval while the air bubbles are in contact with the surrounding ice matrix in situ in the glacier or during core recovery.

After discussing this point, we give an overview of $CO_2$ measurements from the recent preindustrial period and from the period including the last glaciation and its termination. These measurements show fairly stable $CO_2$ concentrations during the Holocene and evidence for shifts in atmospheric $CO_2$ concentration during and at the end of the Wisconsin. The existence of an atmospheric $CO_2$ shift at the end of the glaciation has recently been supported by $^{13}C$ measurements on foraminifera in ocean sediments. Rapid $CO_2$ shifts during the Wisconsin are correlated with other parameters measured in the ice, such as $\delta^{18}O$, $Cl^-$, $NO_3^-$, $SO_4^{2-}$ and dust concentrations [Hammer et al., 1984; Finkel and Langway, 1985].

At the end of the paper, we propose future research programs which will help to increase confidence in the results obtained hitherto and which promise to give more precise and detailed information on carbon cycle fluctuations and the mechanisms involved.

This paper partly corresponds to the paper by Oeschger and Stauffer [1983].

## The Experimental Method

The $CO_2$ concentration measurements reported here have been performed on ice core samples, up to 100,000 years old, from deep drilling projects in Greenland and Antarctica.

Two methods have been used for collecting the occluded air from ice samples: melting the sample or crushing the ice. Both methods have advantages and disadvantages. In the melting procedure, any $CO_2$ which might have been preferentially dissolved in the ice is also extracted so that fractionation of $CO_2$ from other gases is less likely. However, contamination with $CO_2$ from carbonates, either in the ice or at the surface of the sample, is more likely to occur. The dry extraction procedure, which is based on crushing the still frozen samples, avoids such contamination problems, but may lead to fractionation of $CO_2$ from the other air gases. Experiments carried out by the groups in Bern [Zumbrunn et al., 1982] and in Grenoble [Delmas et al., 1980] have now clearly indicated that the dry extraction technique leads to more reliable results. In this paper we will therefore limit ourselves to the discussion of data obtained by the dry extraction method in these two laboratories.

In the most recent studies, gas analysis has been performed by either gas chromatography or laser infrared spectroscopy. The experimental errors of the gas analysis itself are only of the order of ±1%, ∿3 ppm, while the overall error is significantly larger (±3%, ∿10 ppm). It is still unclear whether the large overall error is due to natural small-scale variability of the $CO_2$ content of the air in "adjacent" natural ice samples, e.g., seasonal effects, or is experimentally introduced during the extraction procedure. To compare different dry extraction and measuring procedures, the laboratories in Bern and Grenoble studied

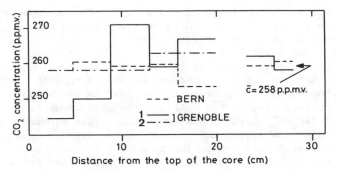

Fig. 1. $CO_2$ concentrations measured in Bern and Grenoble on a core section from Dome C (East Antarctica, depth 132.9 m below the surface). Experimental precisions are 3% for the Bern measurements, 10% for set 1, measured in Grenoble, and 3 % for set 2, also measured in Grenoble. The results obtained in both laboratories agree within the limits of experimental error. The air was trapped by the ice during a time interval lying inside the period 800-2500 years B.P. The average $CO_2$ value is 258 ppmv [Barnola et al., 1983].

samples from the same East Antarctic ice core (Dome C) [Barnola et al., 1983]. The results agree nicely within the error limits (Figure 1), and suggest that the mean $CO_2$ level recorded by Antarctic ice for the period 800-200 years B.P. is about 260 parts per million by volume (ppmv).

Recently, in Bern a new grinder for processing larger ice samples (0.5 to a few liters) has been constructed [Moor and Stauffer, 1984]. After grinding the samples, the released gases are transferred to a trap cooled by a cryocooler to 15 K. All the air gases, except helium, are frozen out in this cold trap. The $CO_2$ concentration is measured with this system, but, interestingly enough, on preindustrial samples from the south pole, $CO_2$ concentrations in the range of 280 to 285 ppmv have been observed, 20 to 25 ppmv higher than previous results. These new measurements raise doubts about the accuracy of the absolute values obtained in earlier measurements in which the air released from the ice is passed through cold traps at -80°C to trap water vapor before laser or gas chromatographic analysis. Tests in the Bern laboratory indicate that during gas transport and diffusion, $CO_2$ shows an affinity for water vapor which can lead to a demixing of $CO_2$ relative to the main air components $N_2$, $O_2$ and Ar.

Based on these experiences, we believe that the absolute $CO_2$ concentration values obtained in the Bern-Grenoble interlaboratory comparison tests are too low by 15 to 20 ppm. We conclude that in the early 19th century the concentration was in the range of 280±5 ppmv.

The problem of demixing of CO$_2$ from other air components is not the only problem encountered during the measuring procedure. In the measurements of small samples, in which the surface area of the walls of the instrument is relatively large compared to the sample volume, we have observed the release of CO$_2$ from instrument surfaces in the presence of water vapor. This effect leads to initial values for the CO$_2$ concentration which are too high [Zumbrunn et al., 1982]. For the measurements presented here, this effect can be kept relatively small by keeping the water vapor pressure low and the extraction time short and can partly be corrected for. It should have little influence on the final data.

The most direct way to investigate the source of the variability in the absolute values of the CO$_2$ concentrations measured in gases extracted from ice cores would be calibration by the measurement of ice samples which contain air, the composition of which is known from direct measurements. Unfortunately, at present such samples are not available.

### Trapping of Air in Natural Ice

Figure 2 is a schematic of the process by which gases are trapped in ice. In the uppermost centimeters of an ice cap or glacier, newly deposited snowflakes transform by diagenesis to firn grains of essentially ellipsoidal shape. With increasing pressure from subsequent snow deposition, the firn grains sinter together, reducing the porous air-filled space in the firn. Within a given depth zone (see Figure 2), the sinuous channels between the firn grains begin to close off, and the channels become separated into isolated elongate bubbles with no further interaction with the atmosphere [Stauffer, 1981].

The analysis of the N$_2$/O$_2$/Ar ratios in ice originating from very cold areas with no summer melting shows that, within experimental uncertainty, the measured ratios agree with those in air [Raynaud and Delmas, 1977], indicating that the entrapment of air in ice is essentially a mechanical process of collection of air samples, which occurs with no differentiation of the gas components. However, CO$_2$ has different physicochemical properties from other atmospheric gases, including a much higher solubility and different adsorption properties. It is, for example, known that CO$_2$ is significantly enriched in melt layers [Stauffer et al., 1984a]. It has also been observed that, in cold and dry accumulation areas, freshly fallen snow contains air

in microbubbles with an increased CO$_2$ content [Stauffer et al., 1981]. However, measurements on snow a few days after its fall [Oeschger et al., 1982] show a decrease of the CO$_2$ excess toward the atmospheric value. CO$_2$ enrichment might also be caused by CO$_2$ adsorbed on firn grains. Therefore, one cannot automatically conclude that the CO$_2$ content of the occluded air in the ice also corresponds to its atmospheric content. Estimates, however, indicate that for ice from very cold accumulation areas the processes mentioned here could lead only to relatively insignificant CO$_2$ enrichments of not more than 10 ppm [Oeschger et al., 1984].

As Figure 2 shows, air occlusion occurs at depth in a glacial deposit. The value of the occlusion depth depends on temperature and snow accumulation rate. Therefore the gas and the ice in a sample of glacial ice are of different age, the gas being somewhat younger than the precipitation which encompasses it. At Dye 3, Greenland, where between 1979 and 1981 a 2037-m-deep ice core was drilled [Langway et al., 1984], the air is occluded in firn layers which are about 100 years older. Gas occlusion is a continuous process, taking place during a time interval corresponding to one fifth to one third of the age difference between the occluded

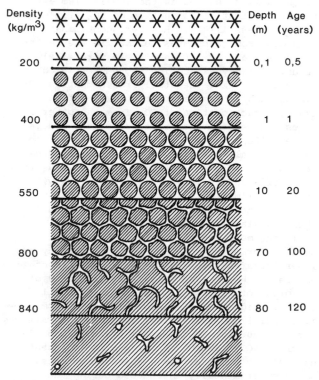

Fig. 2. Two-dimensional schematic of the metamorphosis of snow to firn and ice; depth-age relationships, typical for Greenland ice cores.

TABLE 1.    Characteristics of Drill Sites in Greenland and Antarctica
(Temperature, Accumulation and Information on Gas Enclosure)

| Site | Location | Mean Annual Air Temperature, °C | Annual Accumulation, m | Width of Age Distribution, years | Difference Between Age of Ice and Mean Age of Air, years |
|------|----------|----------------|-------------------|------------------|-----------------|
| Measured Data | | | | | |
| Siple Station | 75°55'S,  83°55'W | -24 | 0.5 | 22 | 95 |
| Calculated Data | | | | | |
| Vostok | 78°28'S, 106°48'E | -57 | 0.022 | 590 | 2800 |
| Dome C | 74°39'S, 124°10'E | -53 | 0.036 | 370 | 1700 |
| South Pole | 90°00'S | -51 | 0.084 | 220 | 950 |
| Byrd Station | 79°59'S, 120°01'W | -28 | 0.16 | 54 | 240 |
| North Central | 74°37'N,  39°36'W | -31.7 | 0.11 | 76 | 350 |
| Crête | 71°07'N,  37°19'W | -30 | 0.265 | 46 | 200 |
| Camp Century | 77°11'N,  61°09'W | -24 | 0.34 | 31 | 130 |
| Dye 3 | 65°11'N,  43°50'W | -19.6 | 0.5 | 22 | 90 |

air and the ice.  An estimate of the age difference between the air and precipitation can be made by assuming that the air is well mixed down to the depth where the air occlusion mechanism starts.  To estimate the time interval for the air occlusion process, a model based on measured firn/ice density profiles has been used.  The data obtained [Schwander and Stauffer, 1984] for firn samples from Dye 3, Greenland, and Siple Station, Antarctica (Table 1), show that gas enclosure in summer firn layers occurs earlier and more rapidly than in the winter layers.  This observation shows that ice cores from very cold regions with high accumulation might enable us to improve the time resolution of the $CO_2$ information in young samples.  This is especially important for reconstruction of the atmospheric $CO_2$ increase since A.D. 1850, the time period to which at present the youngest air samples extracted from very cold regions date back.

### Interaction Between the Air in the Bubbles and the Surrounding Ice

As mentioned above, good evidence exists that on occlusion the $CO_2$ content of the trapped air is close to that of the atmosphere.  A further question concerns the possible change of the gas composition of the air in the bubbles during the long storage times involved.  The $CO_2$ concentration might be affected by interaction with the ice itself or with occluded impurities (carbonates).

First, we discuss the direct interaction of the air in the bubbles with the ice.  Density measurements of newly recovered ice indicate that the bubble size shrinks faster than would be expected from the increasing hydrostatic pressure [Gow and Williamson, 1975].  Indeed, at a certain depth the bubbles start to disappear, and probably air hydrates are formed [Miller, 1969; Shoji and Langway, 1982].  Once deep ice cores are raised to the surface, a decompression process occurs, and air bubbles begin to regenerate.  One might question whether the gas composition of the air in these reformed bubbles might differ from the original composition.  To investigate this possibility, the $CO_2$ content of the air in ice cores as a function of time after core recovery was studied [Neftel et al., 1983].  Measurements were made on adjacent samples 1 week, 2 months and 1 year after recovery.  Within the experimental errors, the same $CO_2$ concentrations were observed, although in the 1-week-old samples the air pressure rose significantly more slowly after crushing than in the older samples, indicating that the ice relaxation and bubble reformation processes had not yet come to an end.

A second possible cause for a shift in the $CO_2$ concentration of the occluded air is interaction with the chemical impurities contained in the ice.  Indeed, the increase of the atmospheric $CO_2$ concentration at the end of the last glaciation coincides with a significant change of the chemical constituents of the ice.  Although this

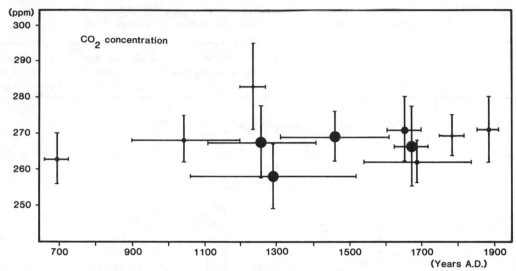

Fig. 3.  CO$_2$ concentration of air occluded in ice cores from Greenland and Antarctica. The errors are the standard deviations of the single measurements.  The measurements have been performed at Bern using the crushing system for 1-cm$^3$ ice samples.

is still an open question, experimental observations tend to suggest that this interaction does not play an important role, since the CO$_2$ shifts for the glacial and postglacial transition have been observed in both Greenland and Antarctic ice cores which show stronlgy different impurity contents.

Data

Data on the Last 2000 Years of the Preindustrial Atmospheric CO$_2$ Concentration

The CO$_2$ concentration data obtained on air extracted from ice cores and measured in the laboratories in Bern and Grenoble have been discussed and compared at the WMO (CAS) "Meeting of Experts on the CO$_2$ Concentrations From Preindustrial Times to 1958" at Boulder, Colorado, June 22-25, 1983 [World Meteorological Organization, 1983].  The two laboratories reported data from six different locations in Greenland and Antarctica for the last 2000 years.  The data from the interlaboratory comparison are shown in Figure 3.  Based on the data given in Figures 1 and 3, one would conclude that the preindustrial concentration lay in the range of 260 to 270 ppmv, but, as mentioned above, more recent data, obtained using a new gas extraction system, suggest somewhat larger preindustrial values (∿280 ppm) [Moor and Stauffer, 1984]. Additional laboratory tests in Bern as well as intercomparisons with the Grenoble laboratory will be required to show whether the new results are experimentally more correct.

In any case, the ice core data suggest that the preindustrial CO$_2$ concentration was significantly (10 to 30 ppmv) lower than the 295 ppmv obtained by back extrapolation of the Mauna Loa record assuming only a fossil CO$_2$ input and a constant airborne fraction [World Meteorological Organization, 1983].  This discussion underscores the experimental difficulties of obtaining an absolute calibration of the method and of reconstructing the CO$_2$ increase from the beginning of the last century.

As mentioned before, we have greater confidence in the relative accuracy of the CO$_2$ concentrations than in the absolute ones, since systematic sources of error, such as fractionation of gases during occlusion in the ice or during measurement, should affect all the data equally.  The time series shown in Figure 3 therefore deserves special attention.  Measurements were made with the 1 cm$^3$ crushing system. The CO$_2$ concentrations covering the period A.D. 900 to 1900 in samples from Greenland and Antarctic ice cores lie almost without exception in the 260- to 270-ppmv concentration band.  The one sample set from Byrd Station, covering the period A.D. 1180 to 1260, is probably contaminated due to bad core quality, i.e., cracks, which allowed the introduction of more recent air with higher CO$_2$ concentration.  This data set clearly shows that during the last 1000 years the atmospheric CO$_2$ concentration, averaged over the 30- to 500-year (depending on the accumulation rates) interval required for occlusion of air in the ice, did not show CO$_2$ excursions as large as that which we observe at pre-

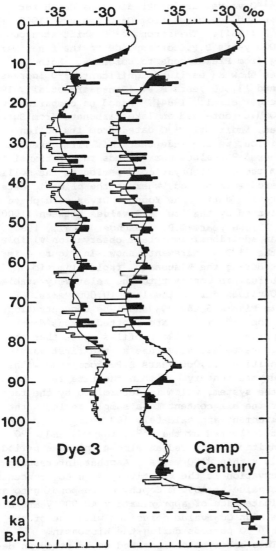

Fig. 4.  Profiles of $\delta^{18}O$ measured in Copenhagen along the Dye 3 (0 to 1982 m depth) and the Camp Century (0 to 1370 m depth) ice cores plotted on a common linear time scale based on considerations discussed by Dansgaard et al. [1982].  In the time interval 40,000 to 30,000 years B.P. on the left side of the Dye 3 $\delta^{18}O$ profile, the core increments analyzed in more detail regarding $CO_2$ concentrations in this paper are indicated.

sent.  We consider this as strong indication that the presently observed increase (>340 ppmv) is an anthropogenic phenomenon.

A data set obtained by the Grenoble group (D. Raynaud, personal communication, 1984), not available for presentation here, seems to indicate variations of the $CO_2$ concentration of about 10 ppm over several centuries.  Based on

our data set as well as on their information, we cannot exclude the possibility that during the last 1000 preindustrial years the atmospheric $CO_2$ concentration might have shown fluctuations of the order of ±10 ppmv.

## Data on Atmospheric $CO_2$ Concentrations, Period 50,000 to 1 B.C.

Measurements of $CO_2$ concentrations in ice from the glacial-postglacial transition, and also from parts of the Wisconsin glaciation, have given very surprising results.  In 1979, measurements carried out in the laboratories in Bern and Grenoble on several ice cores from both hemispheres showed that at the end of the Wisconsin the $CO_2$ concentration was 180 to 200 ppmv and increased at the transition to the Holocene to values in the 260- to 300-ppmv range [Berner et al., 1980; Delmas et al., 1980; Neftel et al., 1982].

The new deep ice core from Dye 3, Greenland [Langway et al., 1984], has enabled a much more detailed study of the time dependence of the $CO_2$ variations to be carried out.  In particular, this core has allowed detailed $CO_2$ measurements on sections of the Wisconsin part of the ice core which, based on other information, were deposited during times of significant rapid climatic variation [Dansgaard et al., 1984; Herron and Langway, 1984].  Although the Holocene part of this core shows significant $CO_2$ enrichments with respect to the atmospheric concentrations [Stauffer et al., 1984] and is therefore not included in Figure 3, we believe that during the Wisconsin, due to the significantly lower temperature, relatively little summer melting occurred

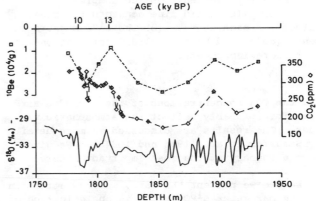

Fig. 5.  The $^{10}Be$ concentrations ($10^4$ atoms per gram of ice), $CO_2$ concentrations (ppm) and $\delta^{18}O$ data obtained for the Dye 3 ice core.  The tentative time marks are suggested by the comparison with $^{14}C$-dated European lake sediments [Oeschger et al., 1983].

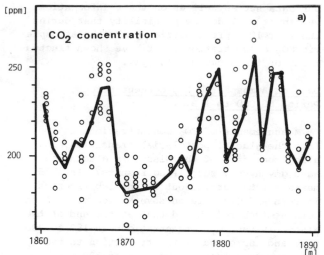

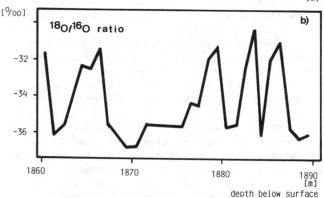

depth below surface

Fig. 6. CO$_2$ and $\delta^{18}$O values measured on ice samples from Dye 3 (the 30-m increment corresponds to about 10,000 years of accumulation; it is indicated in Figure 4). (a) The results of single measurements of the CO$_2$ concentration of air extracted from ice samples (circles). The solid line connects the mean values for each depth. (b) The $\delta^{18}$O measurements (connected by the solid line) done on 0.1-m core increments.

and the CO$_2$ concentrations of the occluded air therefore probably reflect the atmospheric values. The results are discussed in more detail by Stauffer et al. [1984 and 1984a]. We give here a summary of the most important observations.

Before we present the data, we draw attention to the parameter $\delta^{18}$O, which is an indication of the $\delta^{18}$O content of the water molecules which make up the snow. A dense data set over the entire Dye 3 core length has been measured by Dansgaard et al. [1982]; $\delta^{18}$O is the per mil deviation of the $^{18}$O/$^{16}$O ratio in the ice compared to that of standard mean ocean water. In Figure

4, $\delta^{18}$O values are plotted versus depth for Greenland ice cores from Camp Century (1966) and Dye 3 (1981). The strong $\delta^{18}$O shift at around 10,000 years B.P. corresponds to the final warming from Pleistocene to Holocene. Both ice cores show an earlier significant $\delta^{18}$O increase around 13,000 years B.P. [Dansgaard et al., 1982; Oeschger et al., 1984]. Based on comparison with $\delta^{18}$O data obtained on lake carbonate from Swiss lakes, indirectly $^{14}$C dated from its pollen composition, we conclude that this first trend to higher $\delta^{18}$O values corresponds to the transition from the Oldest Dryas cold period to the Bølling-Allerød warm period, whereas the climatic deterioration during the Younger Dryas cold phase is reflected by the low $\delta^{18}$O values between 11,000 and 10,000 years B.P. [Oeschger et al., 1984].

An additional important observation visible in the $\delta^{18}$O measurements shown in Figure 4 is that during the Wisconsin, rapid climatic shifts occurred, in contrast to the relatively stable $\delta^{18}$O values during the last 10,000 years.

In Figure 5 the CO$_2$ and $^{10}$Be concentrations and the $\delta^{18}$O values are plotted versus depth. Comparison of the two profiles shows that the increase to higher values at the first warming transition, 13,000 years B.P., occurred within about one century. Within the time resolution of the system, which is determined by the fact that the air content in a sample reflects the environment approximately 100 years later than that reflected by the $\delta^{18}$O, the CO$_2$ and $\delta^{18}$O transitions occurred at almost the same period, i.e., within ±100 years. Another interesting observation is the relatively high CO$_2$ concentration value at 1890 m depth, corresponding to an estimated age of approximately 40,000 years B.P. This high CO$_2$ value coincides with one of the high $\delta^{18}$O periods during the Wisconsin.

This observation inspired us to study in detail the relationship between $\delta^{18}$O and CO$_2$ in the approximately 30,000- to 40,000-year-old section of the Dye 3 ice core. The results are shown in Figure 6. The surprising result is that all the rapid $\delta^{18}$O oscillations are accompanied by simultaneous, in general well correlated CO$_2$ oscillations [Stauffer et al., 1984a]. Again, even very detailed measurements showed no evidence of a phase shift between CO$_2$ and $\delta^{18}$O. At present we tend to conclude that the CO$_2$ and $\delta^{18}$O changes occurred almost simultaneously.

Other parameters have been observed to vary in a similar fashion over this same time interval. Dansgaard et al. [1984] have shown that microparticle concentrations as measured by light scattering show strong $\delta^{18}$O-correlated variations (Figure 7). Concentrations of particles are more than a factor of 6 higher during cold periods than during warm.

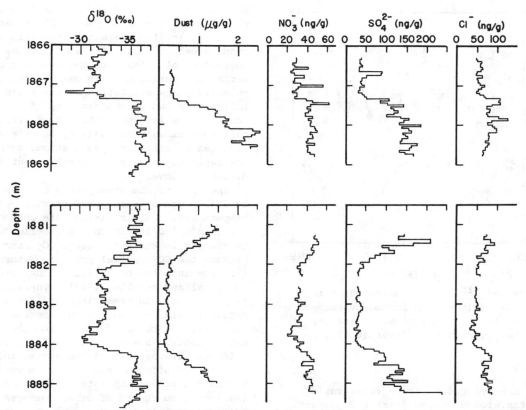

Fig. 7.   Dust, anion and $\delta^{18}O$ concentrations in a section of the Dye 3 ice core which show evidence of rapid climate change from $\delta^{18}O$ variations.   The dust concentrations and $\delta^{18}O$ values are from Dansgaard et al. [1982].   Chloride, nitrate and sulfate concentrations are from Finkel and Langway [1985].

R. C. Finkel and C. C. Langway [1985] have observed that $Cl^-$, $NO_3^-$ and $SO_4^{2-}$ concentrations all showed significant variations which correlate with the measured $\delta^{18}O$ shifts (Figure 7). The cold/warm ratios are about 1.5 for $Cl^-$ and $NO_3^-$ and about 4 for $SO_4^{2-}$.   Beer et al. [1984a] have observed similar $\delta^{18}O$-correlated variations in $^{10}Be$ at the same depths (Figure 8). Because of the larger sample requirements for the $^{10}Be$ analyses, measurements could not be made with the same resolution as for the other species.   $^{10}Be$ is also more concentrated, by about a factor of 2.5, during periods of high $\delta^{18}O$.   In general, nonvolatile constituents have lower concentrations during periods of high $\delta^{18}O$ (warm periods) than during periods of low $\delta^{18}O$ (cold periods).   However, because the relative concentration shifts are not the same for all species, the variations cannot be the result of a simple dilution modulation of a constant impurity flux by a variable water vapor flux system.

What is remarkable in the set of data from the Dye 3 core is the apparent bistable nature of the observed variations.   Cold and warm period values of all parameters tend to occur in narrow bands relative to the difference between the bands.   The correlated variation of these parameters suggests the existence of a bistable climate system during the Wisconsin defined by two sets of the above parameters, one describing the warm and the other the cold climate state. What might be the nature of such rapid climate shifts?   Although we are not yet able to answer this question, a clue may lie in the results of Ruddiman and McIntyre [1977], who have used microfaunal studies of sediment cores to trace the position of the North Atlantic polar front between 18,000 and 6000 years B.P. and who have shown that "during major climatic changes the North Atlantic polar front moved as a line hinged in the western Atlantic south-southeast of Newfoundland, sweeping out an arc larger than 45°".   The position of this polar front as determined in the sediment record is closely correlated with the climate changes recorded by the $^{18}O$ ice core record in the late Wisconsin, after 18,000 years B.P.   Whether changes in the

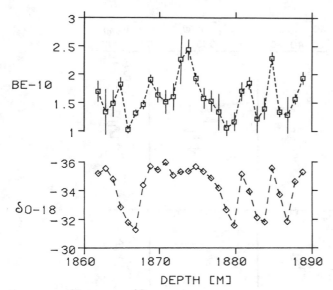

Fig. 8.   $^{10}Be$ and $\delta^{18}O$ values measured on ice samples from Dye 3 (the 30-m increment corresponds to about 10,000 years of accumulation; it is indicated in Figure 4) [Beer et al., 1984b].

polar front are related to the earlier rapid climate shifts around 40,000 years B.P. is not known.

This behavior of the North Atlantic polar front during past climate shifts between glacial and interglacial conditions suggests a strong relation between North Atlantic surface seawater temperatures and climate.  Changes in North Atlantic circulation will affect sea ice extent and, additionally, by changing surface seawater temperature, might lead to changes in seawater biology and chemistry which could lead to some of the changes in constituents observed in the ice core record.

### Independent (Isotopic) Information on Atmospheric $CO_2$ Variations

Besides the direct information on the history of the atmospheric $CO_2$/air ratio from ice cores, independent isotopic information on $CO_2$ system changes can be derived from the carbon isotopic information ($^{13}C/^{12}C$ and $^{14}C/^{12}C$) in natural records like tree rings, corals, sediments and $CO_2$ occluded in natural ice.

Broecker [e.g., Broecker, 1982] has shown that the difference between the $^{13}C/^{12}C$ ratios of benthic and planktonic foraminifera in ocean sediments gives information about chemical differences between the surface and the deep ocean arising from biological activity which might reflect atmospheric $CO_2$ changes.  Indeed,

Shackleton et al. [1983] have used such data to reconstruct an atmospheric $CO_2$ record which, for the glacial-postglacial transition, corresponds nicely to that obtained from the ice cores.  These data also indicate the presence of $CO_2$ variations during the Wisconsin.  Due to the relatively low resolution of the ocean sediment information and the uncertainty of the time scale of the ice core information, it is difficult to decide if the data by Shackleton et al. also can be considered as a strong support for the rapid $CO_2$ changes during the last glaciation discussed above.

The size of the atmospheric $CO_2$ reservoir is influenced by such parameters as sea surface temperature, oceanic mixing rate and ocean volume.  $^{14}C/^{12}C$ levels in atmospheric $CO_2$ respond to changes in the atmospheric $CO_2$ concentration because the $CO_2$ partial pressure determines the $^{14}C$ flux into the ocean [Siegenthaler et al., 1980; Lal and Revelle, 1983].  Therefore the $^{14}C/^{12}C$ record in tree rings and ice cores should provide an independent check on the atmospheric $CO_2$ record derived from the ice core measurements.  However, it is difficult to differentiate between carbon system induced effects and effects resulting from variations in the $^{14}C$ production rate.  Measurements of the long-term record of other cosmogenic radionuclides such as $^{10}Be$ and $^{36}Cl$ will permit correction for production rate changes [Beer et al., 1984b] and allow comparison of the $^{14}C$ record in tree rings and ice cores with the long-term $CO_2$ record.

### Conclusion and Recommendations

At present, based on the agreement between conclusions derived from $CO_2$ measurements in Greenland and Antarctic ice cores and from carbon isotopic results from sea sediment cores, we believe that the existence of the $CO_2$ transition at the end of the last glaciation can be considered as confirmed.  Because of the possible influence of other effects on $CO_2$ content in ice cores and the present lack of confirmatory data, we cannot be so confident about the interpretation of the rapid $CO_2$ changes which we have observed at around 40,000 years B.P.

The importance of paleoinformation for understanding the complex environmental system is underlined by the ice core, ocean and lake sediment studies discussed above.  The continuation of such experimental efforts should be given high priority.  In addition, model calculations relating isotopic information ($^2H/H$ and $^{18}O/^{16}O$ in $H_2O$) to general circulation models (GCM's) should be developed, as should further efforts to relate GCM's to fluxes of dust, sea salts, reduced nitrogen and sulfur and other geochemical species which are valuable climatic indicators. There is increased evidence that the North Atlantic is a crucial area for triggering broad changes in total oceanic circulation. Study of the propagation of changes in North Atlantic circulation to the entire ocean system should not be neglected.

## References

Barnola, J. M., D. Raynaud, A. Neftel, and H. Oeschger, Comparison of $CO_2$ measurements by two laboratories on air from bubbles in polar ice, Nature, 302, 401-413, 1983.

Beer, J., H. Oeschger, M. Andrée, G. Bonani, M. Suter, W. Wölfli, and C. C. Langway, Temporal variations in the [10]Be concentration levels found in the Dye 3 ice core, Greenland, Ann. Glaciol., 5, 16-17, 1984a.

Beer, J., M. Andrée, H. Oeschger, U. Siegenthaler, G. Bonani, H. Hofmann, E. Morenzoni, M. Nessi, M. Suter, W. Wölfli, R. Finkel, and C. C. Langway, Jr., The Camp Century [10]Be record: Implications for long-term variations of the geomagnetic dipole moment. Nucl. Instrum. Methods, B5, 380-384, 1984b.

Berner, W., H. Oeschger, and B. Stauffer, Information on the $CO_2$ cycle from ice core studies, Radiocarbon, 22, 227-235, 1980.

Broecker, W. S., Ocean chemistry during glacial time, Geochim. Cosmochim. Acta, 46, 1689-1705, 1982.

Dansgaard, W., H. B. Clausen, N. Gundestrup, C. U. Hammer, S. J. Johnsen, P. M. Kristinsdottir, and N. Reeh, A new Greenland deep ice core, Science, 218, 1273-1277, 1982.

Dansgaard, W., H. B. Clausen, D. Dahl-Jensen, N. Gundestrup, C. U. Hammer, S. J. Johnsen, N. Reeh, and H. Oeschger, North Atlantic climatic oscillations revealed in deep Greenland ice cores, in Climate Processes and Climate Sensitivity, Geophys. Monogr. Ser., vol. 29, edited by J. E. Hansen and T. Takahashi, pp. 288-298, AGU, Washington, D.C., 1984.

Delmas, R. J., J. M. Ascencio, and M. Legrand, Polar ice evidence that atmospheric $CO_2$ 20,000 y B.P. was 50% of present, Nature, 284, 155-157, 1980.

Finkel, R. C., and C. C. Langway, Jr., Global and local influences on the chemical composition of snowfall at Dye 3, Greenland: The record between 10 kaBP and 40 kaBP, Earth Planet. Sci. Lett., in press, 1985.

Gow, A. J., and T. Williamson, Gas inclusions in the Antarctic ice sheet and their glaciological significance, J. Geophys. Res., 80, 5101-5108, 1975.

Hammer, C. U., H. B. Clausen, W. Dansgaard, A. Neftel, P. Kristinsdottir, and E. Johnson, Continuous impurity analysis along the Dye 3 deep core, in The Greenland Ice Sheet Program, Geophys. Monogr. Ser., edited by C. C. Langway et al., AGU, Washington, D.C., in press, 1984.

Herron, M. M., and C. C. Langway, Jr., Chloride, nitrate and sulfate in the Dye 3 and Camp Century, Greenland ice cores, in The Greenland Ice Sheet Program, Geophys. Monogr. Ser., edited by C. C. Langway et al., AGU, Washington, D.C., in press, 1984.

Lal, D., and R. Revelle, Atmospheric $pCO_2$ changes recorded in lake sediments, Nature, 308, 344-346, 1983.

Langway, C. C., Jr., H. Oeschger, and W. Dansgaard, The Greenland ice sheet in perspective, in The Greenland Ice Sheet Program, Geophys. Monogr. Ser., edited by C. C. Langway et al., AGU, Washington, D.C., in press, 1984.

Miller, S. L., Clathrate hydrates of air in Antarctic ice, Science, 165, 489-490, 1969.

Moor, E., and B. Stauffer, A new dry extraction system for gases in ice, J. Glaciol., in press, 1984.

Neftel, A., H. Oeschger, J. Schwander, B. Stauffer, and R. Zumbrunn, Ice core sample measurements give atmospheric $CO_2$ content during 40,000 y, Nature, 295, 220-223, 1982.

Neftel, A., H. Oeschger, J. Schwander, and B. Stauffer, $CO_2$ concentration in bubbles of natural cold ice, J. Phys. Chem., 87, 4116-4120, 1983.

Oeschger, H., and B. Stauffer, Review of the history of the atmospheric $CO_2$ recorded in ice core, paper presented at the Sixth ORNL Life Sciences Symposium on the Global Carbon Cycle, Oak Ridge Natl. Lab., Knoxville, Tenn., Oct. 31 to Nov. 2, 1983.

Oeschger, H., B. Stauffer, A. Neftel, J. Schwander, and R. Zumbrunn, Atmospheric $CO_2$ content in the past deduced from ice-core analyses, Ann. Glaciol., 3, 227-232, 1982.

Oeschger, H., J. Beer, U. Siegenthaler, B. Stauffer, W. Dansgaard, and C. C. Langway, Jr., Late-glacial climate history from ice cores, in Climate Processes and Climate Sensitivity, Geophys. Monogr. Ser., vol. 29, edited by J. E. Hansen and T. Takahashi, pp. 299-306, AGU, Washington, D.C., 1984.

Raynaud, D., and R. Delmas, Composition des gaz contenu dans la glace polaire, IAHS AISH Publ., 188, 377-381, 1977.

Ruddiman, W., and A. McIntyre, Late Quaternary surface ocean kinematics and climatic change in the high-latitude North Atlantic, J. Geophys. Res., 82, 3877-3887, 1977.

Schwander, J., and B. Stauffer, Age difference between polar ice and the air trapped in its bubbles, Nature, 311, 45-47, 1984.

Shackleton, N. J., M. A. Hall, and Cang Shuxi, Carbon isotope data in core V19-30 confirms reduced carbon dioxide content in ice age atmosphere, Nature, 306, 319-322, 1983.

Shoji, H., and C. C. Langway, Jr., Air hydrate inclusions in fresh ice core, Nature, 298, 548-550, 1982.

Siegenthaler, U., M. Heimann, and H. Oeschger, $^{14}C$ variations caused by changes in the global carbon cycle, Radiocarbon, 22, 177-191, 1980.

Stauffer, B., Mechanismen des Lufteinschlusses in natürlichem Eis, Z. Gletscherk. Glazial-geol., 17, 17-56, 1981.

Stauffer, B., W. Berner, H. Oeschger, and J. Schwander, Atmospheric $CO_2$ history from ice core studies, Z. Gletscherk. Glazialgeol., 17, 1-15, 1981.

Stauffer, B., H. Hofer, H. Oeschger, J. Schwander, and U. Siegenthaler, Atmospheric $CO_2$ concentration during the last glaciation, Ann. Glaciol., 5, 160-164, 1984a.

Stauffer, B., A. Neftel, H. Oeschger, and J. Schwander, $CO_2$ concentration in air extracted from Greenland ice samples, in The Greenland Ice Sheet Program, Geophys. Monogr. Ser., edited by C. C. Langway et al., AGU, Washington, D.C., in press, 1984.

World Meteorological Organization, Report of the WMO (CAS) meeting of experts on the $CO_2$ concentration from pre-industrial times to 1958, I.G.Y. WMO Project on Research and Monitoring of Atmospheric $CO_2$, Rep. 10, WCP-53, Geneva, 1983.

Zumbrunn, R., A. Neftel, and H. Oeschger, $CO_2$ measurements on 1-$cm^3$ ice samples with an IR laser spectrometer (IRLS) combined with a new dry extraction device, Earth Planet. Sci. Lett., 60, 318-324, 1982.

ACCELERATOR RADIOCARBON AGES ON FORAMINIFERA
SEPARATED FROM DEEP-SEA SEDIMENTS

M. Andrée, J. Beer, and H. Oeschger

Physics Institute, University of Bern, Switzerland

A. Mix, W. Broecker, N. Ragano, and P. O'Hara

Lamont-Doherty Geological Observatory, Columbia University, Palisades, New York

G. Bonani, H.J. Hofmann, E. Morenzoni, M. Nessi, M. Suter, and W. Wölfli

Institut für Mittelenergiephysik, ETH Hönggerberg, Zürich, Switzerland

Abstract. A first set of accelerator radio-carbon dates for foraminifera shells separated from a deep-sea core from the western equatorial Pacific is reported. While the ultimate object-ive of this work is to obtain evidence for changes in the rate of deep-sea ventilation over the last 20,000 years, this preliminary study concentrates on illuminating some of the pos-sible biases which will surely complicate such studies. The results reveal that while whole shells and shell fragments of a single species give ages which agree within experimental error, there are significant differences among the ages for coexisting whole shells of different plank-tonic species. It is not possible as yet to pin down the source of these differences. Because of this, the finding that the benthic-plankto-nic age difference was greater 6000 to 12,000 years ago than over the last 5000 years does not necessarily mean that the ventilation rate for the deep sea was significantly slower during late glacial and early Holocene times than it is today. Other equally plausible explanations are possible. Much has yet to be learned about the origin and seafloor history of the material in deep-sea cores before any firm answers regarding paleocirculation rates can be obtained by this approach. Such studies should initially be con-centrated on cores from areas of the seafloor characterized by a higher ratio of sedimentation rate to bioturbation depth than is found for typical open sea sediments.

## Introduction

The discovery by Oeschger and his co-workers [Neftel et al., 1982] of sharp changes in the $CO_2$ content of the air extracted from polar ice cores has led to speculation regarding processes which might induce sudden changes in the $CO_2$ con-tent of the atmosphere. So far the promising candidates all involve changes in the distribu-tion of $\Sigma CO_2$ between the surface and deep ocean. As pointed out by several authors [Broecker and Takahashi, 1984; Toggweiler and Sarmiento, this volume; Siegenthaler and Wenk, 1984; Knox and McElroy, this volume], such redistributions could be accomplished by changing the pattern and rate of deep-sea ventilation or by changing the nu-trient content of surface waters in Antarctic regions. Thus, the ice core results have spurred interest in reconstructing paleocirculation rates in the ocean.

The advent of accelerator mass spectrometry (AMS) offers a new way to address this question. Radiocarbon dates can now be made on about 1 mg of carbon (as opposed to about 1000 mg using the conventional decay counting technique) [Nelson et al., 1977; Muller, 1977]. This makes it pos-sible to obtain ages on foraminifera shells hand-picked from deep-sea cores. Ages can be obtained separately on benthic (i.e., bottom dwelling) and planktonic (i.e., surface dwelling) foraminifera picked from the same core slice. The difference between these ages should carry information about the rate of deep-sea ventila-tion at the time the sediment layer was deposi-ted. The potential of such measurements is made clear by the map shown in Figure 1 of the pre-sent-day difference between the apparent radio-carbon ages of dissolved inorganic carbon from surface water and from 3 km depth water.

## Methods and Results

To date we have made 42 accelerator [14]C mea-surements on samples separated from deep-sea se-

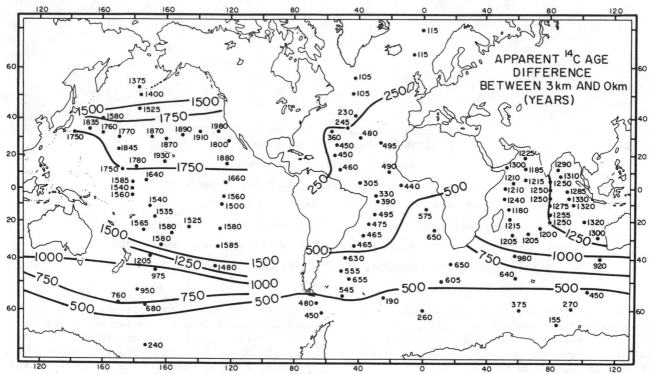

Fig. 1. Map showing the $^{14}$C age difference calculated from the $\Delta^{14}$C values for prenuclear surface water (see Broecker and Peng [1982] for summary) and the $\Delta^{14}$C values
for water at 3 km depth as measured during the GEOSECS program [Ostlund and Stuiver,
1980; Stuiver and Ostlund, 1980, 1983].

diments. The samples were hand-picked and converted to $CO_2$ gas by reaction with 100% $H_3PO_4$ at
Lamont-Doherty Geological Observatory. The $CO_2$
gas was converted to carbon targets at the University of Bern, using the zinc reduction method
[Andrée et al., 1984]. The $^{14}$C/$^{12}$C ratio in the
target carbon was measured at the ETH Hönggerberg
accelerator facility [Suter et al., 1984]. The
results are listed in Table 1. The errors given
are the 1σ errors including the errors of the
sample measurement, the standard measurements
and the background measurements. For details,
see Andrée et al. [1984].

All our results to date were obtained from
core V28-238 (3.12 km depth, 1.0°N latitude,
160.5°E longitude). An $^{18}$O/$^{16}$O stratigraphy
for this core has been published by Shackleton
and Opdyke [1973]. Their results are reproduced in Figure 2. The site of the core lies
above the present calcite lysocline (3.5 km)
but below that of aragonite [Berger et al.,
1977; Broecker and Takahashi, 1978]. This is
important because dissolution effects can
seriously bias the results [Peng and Broecker,
1984]. The number of individual shells needed
for one sample is about 150 for Pulleniatina
obliquiloculata, 300 for Globigerinoides saccu-

lifer, 300 for Neogloboquadrina dutertrei, 700
for Globigerinoides ruber, and 300-600 for the
mixed benthics. In addition to measuring the
radiocarbon ages on the hand-picked samples, the
abundances of the foraminifera in the sediment
were measured (see Table 1). This is important
because bioturbation coupled with abundance
change can produce serious age biases [Peng and
Broecker, 1984].

In a separate paper [Broecker et al., 1984]
the potential sources of bias which must be considered when interpreting age differences between benthic and planktonic foraminifera are
discussed. Hence, they will only briefly be re-
iterated here.

Habitat-related biases. Although there is a
growing consensus among marine geologists that
the planktonic shells found in marine sediment
calcify in the upper 100 m of the water column
[Fairbanks et al., 1980, 1983], claims to the
contrary cannot be totally disregarded [Be,
1977; Shackleton and Vincent, 1978]. As shown
in Figure 3, the preanthropogenic gradient of
the $^{14}$C/C ratio in the upper water column of the
western equatorial Pacific is sufficiently small
that if the foraminifera of interest formed their
shells in the upper 100 m of the water column,

the assumption that their carbon is representative of surface water would be met. It is also clear from this figure that contamination with $^{14}C$ from nuclear testing has greatly enhanced the water column $^{14}C/C$, offering a rather foolproof means of checking this assumption. Foraminifera caught in deep sediment traps could be analyzed for their $^{14}C/C$ ratio by AMS. In this way, debates regarding the mean depth habitat of various species of planktonic foraminifera could be largely resolved.

While it is known that benthic foraminifera live on the seafloor, no one knows to what extent and to what depth they burrow into the sediments. If benthic foraminifera live a few centimeters below the sediment/water interface, they form their shells from carbonate dissolved in sediment pore waters. Pore water $\Sigma CO_2$ is a mixture of bottom water $\Sigma CO_2$, $CO_3^{2-}$ derived from the dissolution of $CaCO_3$, and $CO_2$ derived from respiration. That derived from sediment carbonate will be lower in $^{14}C$ than bottom water to an extent dependent on the bioturbation depth and accumulation rate. Because organic debris is rapidly consumed, that derived from respiration will generally be higher in $^{14}C$ concentration than that for bottom water $\Sigma CO_2$. The available pore water data [McCorkle et al., 1983; Emerson et al., 1984] suggest that these effects are not large. If 90% of the pore water $\Sigma CO_2$ were of bottom water origin (i.e., $\Delta^{14}C = -225$ per mil), 5% of $CaCO_3$ origin (i.e., $\Delta^{14}C = -400$ per mil), and 5% of organic origin (i.e., $\Delta^{14}C = -80$ per mil), then the composite would have a $\Delta^{14}C$ value of $-227$ per mil, giving the benthics an age only about 20 years older than had they grown in bottom water. However, before firm conclusions can be drawn, more information must be obtained about the $\Sigma CO_2$, alkalinity, and $^{13}C/^{12}C$ profiles in sediment pore water and the habitats of the benthic foraminifera of interest.

Bioturbation-related biases. Radiocarbon studies by Peng et al. [1979] and by Berger and Killingley [1982] suggest that the sediments from the Ontong Java Plateau in the western Pacific are stirred to a depth of about 9 cm on a time scale of less than 1000 years. As shown by Sundquist [1979] and by Broecker and Peng [1982], the coupling between bioturbation and partial dissolution of $CaCO_3$ (in steady state condition of accumulation and dissolution) decreases the radiocarbon age of the material in the bioturbated zone (below that which would be found had no dissolution occurred). If the various taxa selected for dating were subjected to different extents of dissolution-induced shell breakage, then the age differences between them would be altered. It is for this reason that cores from above the lysocline should be selected, as they should show minimal dissolution effects.

Bioturbation coupled with temporal changes in the abundance of the foram fractions selected for dating can also produce biases in the age difference between these fractions [see Peng and Broecker, 1984]. This is why careful track is kept of the abundances of the species selected for dating.

Contamination-related biases. Contamination of the foraminifera used for dating could occur after burial on the seafloor if the pore water became supersaturated with respect to calcite, it could occur during storage of the core after it was retrieved from the seafloor, and it could occur during the preparation of the samples. We have checked the latter two of these possibilities by running hand-picked foraminifera samples from a depth of 12 m (age about 700,000 years) in core V28-238. The results obtained are indistinguishable from those for the Bern system blanks. Hence, it is not likely that contamination with younger carbon occurs during storage or preparation. While this test also suggests that contamination during burial on the seafloor is also not important, it is not conclusive in this regard.

Biases introduced by along-bottom transport. Like snow in a blizzard, deep-sea sediment can be picked up by bottom currents and dropped downstream. Sorting according to grain size and density would be expected during both erosion and deposition. If the sediment subject to this transport were Holocene in age, this process would be hard to detect. As benthic forams are generally larger and more dense than planktonic forams, it is possible that they would be less subject to along-bottom transport. If so, then the age difference between benthics and planktonics in a sediment might be reduced by this process. The planktonics at a given site could in part have been purged and then be suspended and moved to the site of interest, while the benthics would all have grown on site.

Age Differences Among Planktonic
Fractions

The results show that the pattern of age differences for core V28-238 is very complex. No simple explanation accounts for all the age differences. Because of this, no firm conclusions with regard to temporal changes in the $^{14}C/C$ ratio between the surface and deep waters of the equatorial Pacific can be drawn from these preliminary results. Biases clearly exist which obscure the message. It will take much work on a number of cores to read through these biases.

The complications can be demonstrated when the differences between the ages of various planktonic pairs are examined. Four planktonic

TABLE 1. Summary of Accelerator $^{14}$C Ages Obtained to Date on Core V28-238
Together with the Abundances of the Foraminifera in the Sediment

| Depth, cm | Material | | Abundance, mg/g | $^{14}$C Age, Years B.P. |
| | Type | Size | | |
|---|---|---|---|---|
| 2.5-4.5 Piston Core | G. sacculifer | Whole | ... | 5,500±230 |
| | P. obliquiloculata | Whole | ... | 4,330±100 |
| | | | | |
| 3-8 Trigger Weight Core | G. sacculifer | Whole | 8.1 | 4,640±160 |
| | G. sacculifer | Fragments | ... | 4,650±100 |
| | G. ruber | Whole | 7.7 | 5,680± 90 |
| | P. obliquiloculata | Whole | 58.2 | 4,760±160 |
| | P. obliquiloculata | Fragments | ... | 4,930±230 |
| | N. dutertrei | Whole | 3.6 | 5,410± 80 |
| | Benthics | Whole | 0.29 | 6,150±180 |
| | | | | |
| 10-12 Piston Core | G. sacculifer | Whole | 6.8 | 5,880±100 |
| | G. ruber | Whole | 23.1 | 7,670±100 |
| | P. obliquiloculata | Whole | 82.2 | 6,390±160 |
| | N. dutertrei | Whole | 3.7 | 9,070±120 |
| | Benthics | Whole | 0.38 | 9,530± 80 |
| | | | | |
| | Bulk | <25 μm | ... | 6,320± 60 |
| | Bulk | 25-63 m | ... | 6,620± 70 |
| | Bulk | >63 μm | ... | 6,870±140 |
| | | | | |
| 13-14 Piston Core | G. sacculifer | Whole | 9.0 | 8,350±100 |
| | P. obliquiloculata | Whole | 100.0* | 8,620±100 |
| | Benthics | Whole | 1.01 | 12,080±120 |
| | | | | |
| 18-20 Piston Core | G. sacculifer | Whole | 6.2* | 9,730±220 |
| | G. sacculifer | Whole | 6.2* | 8,490±150 |
| | G. ruber | Whole | 27.6 | 9,580±110 |
| | P. obliquiloculata | Whole | 111.0 | 9,300±220 |
| | P. obliquiloculata | Whole | 111.0 | 9,680±170 |
| | N. dutertrei | Whole | 6.7 | 11,230±130 |
| | Benthics | Whole | 0.73 | 11,660±260 |
| | | | | |
| 30-31 Piston Core | G. sacculifer | Whole | 4.1 | 11,650±260 |
| | P. obliquiloculata | Whole | 78.6 | 12,680±460 |
| | Benthics | Whole | 0.92 | 16,140±390 |
| | | | | |
| 34-35 Piston Core | G. sacculifer | Whole | 2.7* | 13,560±220 |
| | | | | |
| 41-43 Piston Core | G. sacculifer | Whole | 1.0* | 17,780±390 |
| | P. obliquiloculata | Whole | 47.4 | 19,620±190 |
| | Benthics | Whole | 1.29 | 20,650±220 |
| | | | | |
| | Bulk | <25 μm | ... | 17,800±160 |
| | Bulk | 25-63 μm | ... | 19,440±260 |

TABLE 1.   (continued)

| Depth, cm | Material Type | Size | Abundance, mg/g | $^{14}C$ Age, Years B.P. |
|---|---|---|---|---|
| 45-47 Piston Core | G. sacculifer | Whole | 1.3* | 19,620±240 |
| | G. ruber | Whole | 5.6 | 19,380±260 |
| | N. dutertrei | Whole | 5.8 | 21,000±250 |
| | Benthics | Whole | 1.36 | 22,110±350 |
| 1200 Piston Core | G. sacculifer | Whole | ... | >39,000 |
| | P. obliquiloculata | Whole | ... | >39,000 |

*For these samples the abundance was calculated from values for adjacent samples.

species were selected for this preliminary study. Two, G. sacculifer and G. ruber, are solution prone, and two, P. obliquiloculata and N. dutertrei, are solution resistant. All are thought to calcify in the upper 100 m of the water column [Fairbanks et al., 1980, 1983]. The age differences between P. obliquiloculata and the other three planktonic species are summarized in Table 2. Large differences are found. For the levels of 3-8 and 10-12 cm the ages for G. ruber and N. dutertrei are much greater than those for P. obliquiloculata and G. sacculifer. For the 18-20 cm level, G. ruber has the same age as P. obliquiloculata, and N. dutertrei a much older age than P. obliquiloculata. At this level, two separate runs were made on G. sacculifer. They yield ages differing by 1200 years. One of these ages is close to that of P. obliquiloculata, and the other is much younger. We have no explanation for this difference. It alerts us that other "fliers" might be present in the data set.

There are three explanations for the above differences. The first is that there are errors of unknown origin associated with the measurements which are much larger than those taken into account in computing the age errors given in Table 1 (included in these errors are the errors of blank and standard runs, counting statistics, and instabilities in the ratio measurements). The general reproducibility of the $^{14}C$ measurements made using the Zurich accelerator on samples prepared in Bern argues against this possibility [Suter et al., 1984]. However, until more duplicate samples have been prepared and run, this possibility cannot be entirely discounted.

The second explanation has to do with dissolution effects. As discussed by Broecker et al. [1984], breakage of shells resulting from partial dissolution may decrease the mean age of shells in the mixed layer. The bioturbated layer includes a mixture of shells with an age spectrum spanning several thousand years. The older shells in this mixture resided in the mixed layer longer and therefore have an increased probabi-

lity of breakage. As we generally sample only whole shells, the probability that younger shells are overrepresented in the mixture is increased. There are two reasons to believe that this is not the dominant effect. First, the ages for whole shells and for fragmented shells agree within measurement error for both P. obliquiloculata and G. sacculifer (Table 1). If the fragments were generated by partial dissolution during residence in the mixed layer, the age of the fragmented shells would be expected to be significantly older than that of the whole shells. As we see no difference in the ages, the suggestion is that the shells were fragmented before being stirred into the sediment or during sample washing.

The other argument against dissolution is that the core is situated above the lysocline, where little dissolution would be anticipated. This is consistent with the following check: If there is no dissolution, the mixed layer age $t_{ml}$ (in thousands of years) is given by the following relationship [see Broecker and Peng, 1982]:

$$t_{ml} = 8.27 \ln[1 + (h/8.27A)]$$

where h (centimeters) is the thickness of the bioturbated layer, A (centimeters per kiloyear) the accumulation rate of the sediment, and 8.27 (kiloyears) the mean life of radiocarbon. Radiocarbon studies on nearby box cores from the western equatorial Pacific [Berger and Killingley, 1982] suggest bioturbation zones of about 9 cm in thickness. Our results yield a mean sedimentation rate of about 2.3 cm/ky. Thus the expected age for mixed layer calcite would be 3200 years. Partial dissolution would reduce this age [Broecker and Peng, 1982]. For the sample from 2.5-4.5 cm depth in the V28-238 piston core we get ages for the planktonic foraminifera of 5000 years (for G. sacculifer) and 4330 years (for P. obliquiloculata). Thus, none of the ages are younger than the expected value. Rather than shedding light on the disso-

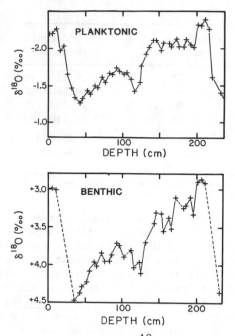

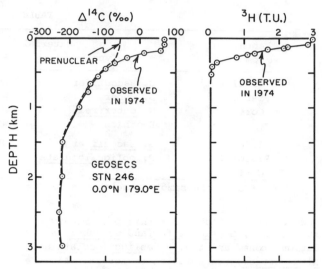

Fig. 3. Plot of radiocarbon [Ostlund and Stuiver, 1980] and tritium [Ostlund and Brescher, 1982] as a function of water depth at a GEOSECS station located on the equator in the western Pacific. A knowledge of the prenuclear $\Delta^{14}C$ value for surface water ($\simeq$ -60 per mil) and of the vertical distribution of tritium (dominantly from bomb tests) allows reconstruction of the prenuclear vertical distribution of $^{14}C/C$ ratios.

Fig. 2. Measurements of $\delta^{18}O$ for planktonic and benthic forams from core V28-238 (3.12 km, 1.0°N latitude and 160.5°E longitude) as published by Shackleton and Opdyke [1973]. The planktonic analyses are all on G. sacculifer and hence are subject to the influence of bioturbation. Because of major changes in ecology, the benthic analyses are made on one species in glacial sections and on another species in interglacial sections. Thus bioturbation does not suppress the amplitude of the benthic $\delta^{18}O$ change across the glacial-to-interglacial boundary.

lution question, this raises a new question: why is the core top so old?

One way to explain the high core top age is to assume that about 9 cm of sediment was lost during the coring procedure. Such a large loss

from a piston core is not expected but cannot be excluded. Another possibility is that part of the sediment is being derived from along-bottom transport.

The calculation of the core top age involves the assumption that the sedimentation is in steady state; i.e., the sum of dissolution and accumulation balances the carbonate input. Berger and Killingley [1982] suggest that the carbonate chemistry of the Pacific Ocean has not been in steady state in the recent past. If so, the situation is far more complex.

A third possible cause for the discrepancy

TABLE 2. Age Differences Between Planktonic Species

| Depth of Sample, cm | Age of G. sacculifer Minus Age of P. obliquiloculata, Years | Age of G. ruber Minus Age of P. obliquiloculata, Years | Age of N. dutertrei Minus Age of P. obliquiloculata, Years |
|---|---|---|---|
| 2.5-4.5 | +1170±250 | | |
| 3-8 | -120±230 | +920±180 | +650±180 |
| 10-12 | -510±190 | +1280±190 | +2680±200 |
| 13-14 | -270±140 | | |
| 18-20 | -380±390 | +90±300 | +1740±310 |
| 30-31 | -1030±530 | | |

TABLE 3. Summary of Foraminifera Abundance Data for Core V28-238

| Depth, cm | Abundance, mg/g | | | | |
| --- | --- | --- | --- | --- | --- |
| | P. obliquiloculata | G. sacculifer | G. ruber | N. dutertrei | Benthics |
| 3-8 | 58 | 8.1 (0.14) | 7.7 (0.13) | 3.6 (0.06) | 0.3 |
| 10-12 | 82 | 6.8 (0.08) | 23.1 (0.28) | 3.7 (0.05) | 0.4 |
| 13-14 | 100 | 9.0 (0.09) | | | 1.0 |
| 18-20 | 111 | 6.2 (0.06) | 27.6 (0.25) | 6.7 (0.06) | 0.7 |
| 30-31 | 79 | 4.1 (0.5) | | | 0.9 |
| 41-43 | 47 | 1.0 (0.2) | | | 1.3 |
| 45-47 | | 1.3 | 5.6 | 5.8 | 1.4 |

The numbers in parentheses are the ratios of the abundance to the abundance of P. obliquiloculata.

among the ages for the planktonic fractions is related to the variation with depth of the abundances of the species of interest (see Table 3 and Figure 4). If the abundance decreases up-core, bioturbation would mix more older shells from the lower sediment layers up into younger sediments. This would result in higher core top ages than in the case of constant abundance. Conversely, the age of the core top would be decreased for a species which has an increasing

abundance up-core. In Table 3 it can be seen that G. sacculifer has a nearly constant abundance over the upper 20 cm. By contrast, P. obliquiloculata, G. ruber and N. dutertrei all show decreases over this depth interval. It should be kept in mind that the abundances observed are those after smoothing by bioturbation has occurred. Had not bioturbation occurred, these changes would be even more pronounced. Tables 2 and 3 show that there is no apparent

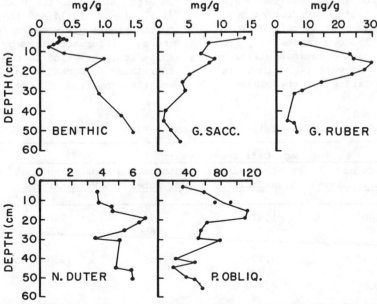

Fig. 4. Abundance (in milligrams of foraminifera per gram of sediment) versus depth (in centimeters) plots for the four planktonic species and for mixed benthics in core V28-238 obtained as part of this work.

TABLE 4.   Stable Isotope Results
for the Holocene From Core V28-238

|                          | $\delta^{18}O$ | $\delta^{13}C$ |
|--------------------------|------|------|
| P. obliquiloculata       | -1.50 | 0.98 |
| G. sacculifer            | -2.2* |      |
| G. ruber                 | -2.65 | 0.51 |
| N. dutertrei             | -1.20 | 1.19 |
| Benthics (Uvigerina)     | +3.0* |      |

The values given for P. obliquiloculata,
G. sacculifer, and N. dutertrei are averages
for samples from 3-8 (trigger weight core),
10-12, 13-14 and 15-16 cm depth analyzed at
Lamont-Doherty Geological Observatory.
   *From Shackleton and Opdyke [1973].

correlation between the pattern of abundances
and the age anomalies.  So while abundance-indu-
ced age difference biases must be present, they
are unlikely to be the whole story.

If experimental errors, dissolution-induced
shell breakage, abundance variation, and along-
bottom transport are not the major culprits,
then what is?  Possibly, the whole shells of
G. ruber and N. dutertrei that reach the sediment come from a population living well below
the surface.  This possibility must be explored
by doing $^{14}C/C$ determinations on shells caught
in deep sediment traps.  We should mention that
such a direct $^{14}C/C$ ratio approach is far supe-
rior to the $^{18}O/^{16}O$ or $^{13}C/^{12}C$ approaches which
have already been explored [e.g., Fairbanks et
al., 1983].  The stable isotope approaches can
be highly misleading because of so-called vital
effects which shift the isotope ratios away from
the value representing thermodynamic equilibrium
between the calcite and the water.  As shown in
Table 4, stable isotope ratios for the plank-

tonic species from the Holocene section of core
V28-238 do not point to the habitat depths
needed to explain the $^{14}C$ anomalies observed.
For example, G. ruber gives the appearance from
its $\delta^{18}O$ values of living much shallower than
P. obliquiloculata.  The radiocarbon age difference suggests just the opposite.

Another possible explanation involves differential bioturbation.  Are particles differing in
size, texture and shape moved during bioturbation with equal probability?  While also not a
likely prospect [Ruddiman et al., 1980], this
possibility cannot be ignored.  Other possible
explanations will surely be unearthed.  To sort
out all the possibilities, our knowledge as to
the origin and history of particles making up
deep-sea sediments has to be greatly improved.

### Age Difference Between Benthic
### and Planktonic Fractions

With the caution that the inability to interpret the planktonic data demands, the mixed
benthic-planktonic age differences are now considered.  These age differences are summarized
in Table 5.  The differences found between the
ages for the mixed benthic samples and those
found for P. obliquiloculata and G. sacculifer
from the 3-8 cm sample are about as expected
from the present-day deep water/surface water
age difference (i.e., about 1600 years).  Those
for G. ruber and N. dutertrei are far too small.
Surprising is the large increase in the benthic-
planktonic age difference in going from the 3-8
cm depth sample to the next three deeper layers
(i.e., 10-12, 13-14 and 18-20 cm).  For three
of the planktonic species (not for N. dutertrei)
the age difference rises by about 1500 years.
While it is tempting to conclude that this jump
is related to a change in the rate of deep-sea
ventilation, such an inference is not warranted

TABLE 5.   Summary of the Mixed Benthic-Planktonic Age Differences

| Depth of Sample, cm | Age Difference, years | | | | Age of P. obliquiloculata, years |
|---|---|---|---|---|---|
| | Benthics and P. obliquiloculata | Benthics and G. sacculifer | Benthics and G. ruber | Benthics and N. dutertrei | |
| 3-8 | 1,390 | 1,510 | 470 | 740 | 4,800 |
| 10-12 | 3,140 | 3,650 | 1,860 | 460 | 6,400 |
| 13-14 | 3,460 | 3,730 | | | 8,600 |
| 18-20 | 2,360* | 3,170* | 2,080 | 430 | 9,500 |
| | 1,980* | 1,930* | | | |
| 30-31 | 3,460 | 4,490 | | | 12,700 |

The present-day observed 3 km minus surface water apparent age difference is 1600 years.
*Two samples of the same species have been measured at this depth.

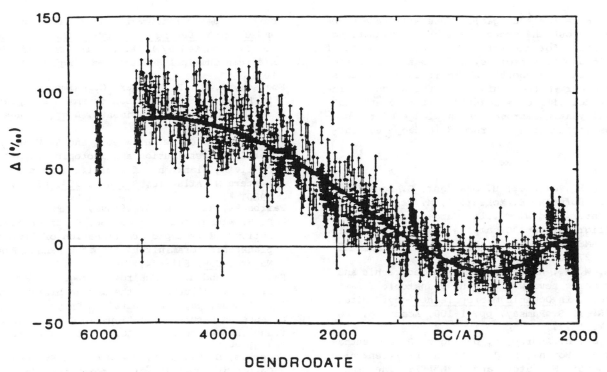

Fig. 5.   $\Delta^{14}C$ for atmospheric $CO_2$ over the last 7000 years as summarized by Damon [1982].

because of the many questions raised by the planktonic age difference anomalies.

As it is not productive to go further with a discussion of this preliminary data set, we will turn to another question.  If further measurements were to show that the rate of deep-sea ventilation was slower over the time interval from about 12,000 to about 6000 years ago, would this violate anything already known about the earth's $^{14}C$ cycle?  Specifically, would it be consistent with what is known about the time history of $^{14}C/C$ ratios for atmospheric $CO_2$ from measurements on tree rings [Suess, 1980; Damon, 1982]?  This record is reproduced in Figure 5.  These data indicate that the $^{14}C$ content of atmospheric carbon was about 9% higher 7000 years ago than it is today.  Until very recently these changes have been attributed by most workers in the field to changes in cosmic ray production of $^{14}C$ resulting from changes in the earth's dipole field.  Archeomagnetic results suggest that changes in the earth's dipole field over the last 7000 years have about the magnitude and timing needed to explain the observed changes.  However, recent studies of the $^{10}Be$ record in ice cores by the Bern group [Beer et al., 1984] suggest that the required cosmic ray production changes may not have occurred.  If so, then the logical causal candidate becomes changes in the rate of deep-sea ventilation.

If the $^{14}C$ record is to be explained in this way, then a slowing down in the rate of exchange between the deep sea and upper ocean-atmosphere reservoir between 7000 and 3000 years ago would be required.  Our results are in the right direction to accommodate the observed atmospheric $^{14}C/C$ changes.  However, were the results to prove more or less valid as they stand, then even larger atmospheric changes would be expected unless the age increase were restricted to about half the volume of the ocean.

Conclusions

Our measurements show that it is technically possible to obtain $^{14}C$ ages of sufficient accuracy on foraminifera separated from deep-sea sediments to detect changes in the rate of deep-sea ventilation over the last 18,000 years.  They also show that serious biases are present in this record which will have to be understood if the age differences are to be used to reliably estimate paleocirculation rates. Modeling studies show that biases can be generated through the interaction between bioturbation and abundance change and the interaction between bioturbation and dissolution-induced shell breakage.  Thus, attempts should be made to find cores with the least dissolution effects and with the highest possible ratio of sedimentation rate to bioturbation depth (i.e., with the lowest core top $^{14}$C ages). More will have to be learned about the growth habitats of planktonic and benthic foraminifera, about the mechanics of the bioturbation process, and about redeposition via lateral transport. While it is not at this point clear whether this quest for paleocirculation rates will be successful, it is clear that it will vastly improve our knowledge of the processes influencing the record in deep-sea sediments.

## References

Andrée, M., J. Beer, H. Oeschger, G. Bonani, H. J. Hofmann, E. Morenzoni, M. Nessi, M. Suter, and W. Wölfli, Target preparation for milligramsized $^{14}$C samples and data evaluation for AMS measurements, Nucl. Instrum. Methods, 233, 274-279, 1984.

Be, A. W. H., An ecological, zoogeographic and taxonomic review of recent planktonic foraminifera, in Oceanic Micropaleontology, edited by A. T. S. Ramsay, pp. 1-100, Academic, New York, 1977.

Beer, J., M. Andrée, H. Oeschger, U. Siegenthaler, G. Bonani, H. J. Hofmann, E. Morenzoni, M. Nessi, M. Suter, and W. Wölfli, The Camp Century $^{10}$Be record: Implications for long-term variations of the geomagnetic dipole moment, Nucl. Instrum. Methods, 233, 380-384, 1984.

Berger, W. H., and J. S. Killingley, Box cores from the equatorial Pacific: $^{14}$C sedimentation rates and benthic mixing, Mar. Geol., 45, 93-125, 1982.

Berger, W. H., T. C. Johnson, and E. L. Hamilton, Sedimentation on Ontong Java Plateau: Observations on a classic "carbonate monitor", in The Fate of Fossil Fuel $CO_2$ in the Oceans, edited by N. R. Andersen and A. Malahoff, pp. 543-567, Plenum, New York, 1977.

Broecker, W. S., and T. H. Peng, Tracers in the Sea, Eldigio Press, Palisades, New York, 1982.

Broecker, W. S., and T. Takahashi, The relationship between lysocline depth and in situ carbonate ion concentration, Deep Sea Res., 25, 65-95, 1978.

Broecker, W. S., and T. Takahashi, Is there a tie between atmospheric $CO_2$ content and ocean circulation?, in Climate Processes and Climate Sensitivity, Geophys. Monogr. Ser. vol. 29, edited by J. E. Hansen and T. Takahashi, pp. 314-326, AGU, Washington, D.C., 1984.

Broecker, W. S., A. Mix, M. Andrée, and H. Oeschger, Radiocarbon measurements on co-existing benthic and planktonic foraminifera shells: Potential for reconstructing ocean ventilation times over the past 20,000 years, Nucl. Instrum. Methods, 233, 331-339, 1984.

Damon, P. E., Fluctuation of atmospheric radio-carbon and the radiocarbon timescale, in Nuclear and Chemical Dating Techniques, ACS Symp. Ser., no. 176, edited by L. A. Currie, pp. 233-244, American Chemical Society, Washington, D.C., 1982.

Emerson, S. R., K. Fischer, C. Reimers, and D. Heggie, Organic carbon dynamics and preservation in deep-sea sediments, Deep Sea Res., in press, 1984.

Fairbanks, R. G., P. H. Wiebe, and A. W. H. Be, Vertical distribution and isotopic composition of living planktonic foraminifera in the western N. Atlantic, Science, 207, 61-63, 1980.

Fairbanks, R. G., M. Sverflove, R. Free, P. H. Wiebe, and A. W. H. Be, Vertical distribution and isotopic fractionation of living planktonic foraminifera from the Panama Basin, Nature, 298, 841-844, 1983.

Knox, F., and M. B. McElroy, Changes in atmospheric $CO_2$: Influence of the marine biota at high latitudes, paper presented at Chapman Conference on Natural Variations in Carbon Dioxide and the Carbon Cycle, AGU Tarpon Springs, Fla., Jan. 9-13, 1984.

McCorkle, D. C., S. R. Emerson, and P. Quay, Carbon isotopes in marine porewaters (abstract), Eos Trans AGU, 64, 721, 1983.

Muller, R. A., Radioisotope dating with a cyclotron, Science, 196, 489-494, 1977.

Neftel, A., H. Oeschger, J. Schwander, B. Stauffer, and R. Zumbrunn, Ice core sample measurements give atmospheric $CO_2$ content during the past 40,000 years, Nature, 295, 220-223, 1982.

Nelson, D. E., R. G. Korteling, and W. R. Stott, Carbon-14: Direct-detection at natural concentrations, Science, 198, 507-508, 1977.

Ostlund, H. G., and R. Brescher, GEOSECS tritium, Tritium Lab. Data Rep. 12, Univ. of Miami, Florida, 1982.

Ostlund, H. G., and M. Stuiver, GEOSECS Pacific radiocarbon, Radiocarbon, 22, 25-53, 1980.

Peng, T. H., and W. S. Broecker, The impacts of bioturbation on the age difference between benthic and planktonic foraminifera in deep-sea sediments, Nucl. Instrum. Methods, 233, 346-352, 1984.

Peng, T. H., W. S. Broecker, and W. H. Berger, Rates of benthic mixing in deep-sea sediment as determined by radioactive tracers, Quat. Res., 11, 141-149, 1979.

Ruddiman, W. F., G. A. Jones, T. H. Peng, L. K. Glover, B. P. Glass, and P. J. Liebertz, Tests for size and shape dependency in deep-sea mixing, Sediment. Geol., 25, 257-276, 1980.

Shackleton, N. J., and N. D. Opdyke, Oxygen isotope and paleomagnetic stratigraphy of equatorial Pacific core V28-238: Oxygen isotope temperatures and ice volumes on a $10^5$ and $10^6$ year scale, Quat. Res., 3, 39-55, 1973.

Shackleton, N. J., and E. Vincent, Oxygen and carbon isotope studies in recent planktonic foraminifera from S. W. Indian Ocean, Mar. Micropaleontol., 3, 1-13, 1978.

Siegenthaler, U., and T. Wenk, Rapid atmospheric $CO_2$ variations and ocean circulation, Nature, 308, 624-625, 1984.

Stuiver, M., and H. G. Ostlund, GEOSECS Atlantic radiocarbon, Radiocarbon, 22, 1-24, 1980.

Stuiver, M., and H. G. Ostlund, GEOSECS Indian Ocean and Mediterranean radiocarbon, Radiocarbon, 25, 1-29, 1983.

Suess, H. E., The radiocarbon record in tree rings of the last 8000 years, Radiocarbon, 22, 200-209, 1980.

Sundquist, E. T., Carbon dioxide in the ocean: Some effects on sea water and carbonate sediments, Ph.D. thesis, Harvard Univ. Cambridge, Mass., 1979.

Suter, M., R. Balzer, G. Bonani, H. J. Hofmann, E. Morenzoni, M. Nessi, W. Wölfli, J. Beer, M. Andrée, and H. Oeschger, Precision measurement of rare isotopes - Results and prospects, Nucl. Instrum. Methods, 233, 117-122, 1984.

Toggweiler, J. R., and J. L. Sarmiento, Glacial to interglacial changes in atmospheric carbon dioxide: The critical role of ocean surface water in high latitudes, this volume.

# CHANGES IN ATMOSPHERIC $CO_2$:
## FACTORS REGULATING THE GLACIAL TO INTERGLACIAL TRANSITION

Fanny Knox Ennever and Michael B. McElroy

Center for Earth and Planetary Physics, Harvard University
Cambridge, Massachusetts 02138

Abstract. An acceptable model for the glacial ocean must be consistent with available data for $\delta^{13}C$ in both deep and shallow waters, and must account also for recent observations indicating that the level of atmospheric $CO_2$ was lower during glacial times than in the preindustrial interglacial environment by about 100 ppm. Glacial deep-sea carbon was isotopically light compared to present, suggesting that the glacial to interglacial transition was accompanied by transfer of about $5 \times 10^{16}$ moles of carbon from the ocean to an organic reservoir, most probably a combination of the terrestrial biosphere and coastal sediments. The associated change in $pCO_2$ was investigated using a five-box model for the ocean-atmosphere system and was found to be insufficient on its own to account for the ice core data. It is argued that the low values for glacial $pCO_2$ require higher deep-sea nutrient concentrations and/or lower levels of preformed nutrients. Reduction in preformed nutrient concentrations could reflect a change in the supply of water to zones of downwelling, specifically a larger relative contribution in glacial times due to advection of surface waters from lower latitude, or an increased efficiency for biological utilization of nutrients at high latitude, or both.

## 1. Introduction

Studies of air trapped in polar ice indicate that the level of atmospheric $CO_2$ was significantly lower during the last ice age than it is today, by more than 100 ppm [Berner et al., 1979; Delmas et al., 1980; Neftel et al., 1982]. The transition from glacial to interglacial conditions was accompanied by a rise in $CO_2$, from about 200 to 300 ppm.

The extent to which change in $CO_2$ served as a causative factor in climate change is unclear. It is apparent though that the observed variation in $CO_2$ must reflect a significant difference in either the physical or the chemical state of the ocean in glacial times as compared to the present environment. An excursion in $pCO_2$ as large as 100 ppm in a time interval as brief as a few thousand years can only be attributed to enhanced release or uptake of $CO_2$ by the ocean [Broecker, 1981].

There are two problems, in some sense separable. One concerns the nature of the physical and chemical processes responsible for the abrupt changes in $CO_2$ and climate which took place during the last termination. The other relates to the difference in the physical and chemical states of the ocean-atmosphere system during glacial conditions as compared to the subsequent interglacial environment. Analysis of the glacial-interglacial transition requires a time dependent model and is of necessity more demanding. We shall focus here on differences between the steady state environments. An adequate definition of the glacial environment is obviously prerequisite to a satisfactory discussion of the processes responsible for the transition, and there are important constraints which must be satisfied by an acceptable model for the glacial ocean.

Average values for sea surface temperature were lower in glacial times, by about 2°C [CLIMAP Project Members, 1976]. Sea level was also lower, by 75 to 150 m [Denton and Hughes, 1981], and salinity was correspondingly higher, by about 1°/oo. The implied variations in temperature and salinity have opposite effects on $pCO_2$. A 2°C drop in temperature would result in a drop in $pCO_2$ by about 20 ppm, while an increase in salinity by 1°/oo would lead to an increase in $pCO_2$ by about 12 ppm, resulting in little net change in the concentration of gas in the atmosphere [Broecker, 1981].

Inferences concerning changes in the isotopic composition of oceanic carbon carry more important, if less direct, implications for $pCO_2$. The isotopic composition of carbon in the deep sea is recorded in the carbonate shells of benthic foraminifera preserved in marine sediments. Analyses by Shackleton [1977], who pioneered this technique, indicate that marine carbon was isotopically light during glacial time. The early data suggested a shift in $\delta^{13}C$ of magnitude 0.7°/oo [Shackleton, 1977; Broecker, 1981, 1982a, b]. More recent analyses favor a smaller value, between 0.2 and 0.4°/oo [Shackleton et al., 1983a]. The change in the relative abundance of carbon isotopes, from light to heavy, between glacial and interglacial time suggests that carbon was removed from the atmosphere-ocean system over the intervening period, sequestered in some organic reservoir. If we assume that the terrestrial biosphere provides the dominant sink, and adopt a value of -26°/oo for $\delta^{13}C$ characteristic of this environment, we require that the ocean lost between $2.5 \times 10^{16}$ and $5 \times 10^{16}$ moles C during the glacial-interglacial transition to account for the observed shift in deep-sea $\delta^{13}C$, between 0.2 and 0.4°/oo. The quantity of carbon transferred out of the ocean would be somewhat larger, between $3 \times 10^{16}$ and $6 \times 10^{16}$ moles C if, as suggested by Broecker [1981], we attribute the removal to formation of coastal organic sediments with a $\delta^{13}C$ of -21°/oo.

The $\delta^{13}C$ of surface water should reflect the $\delta^{13}C$ of deep water modified to account for selective removal of light C by the marine biosphere. The mean difference between $\delta^{13}C$ in surface waters today as compared with the deep waters is about 2°/oo. We shall describe this difference by the quantity

$$\Delta = \delta^{13}C_{surface} - \delta^{13}C_{deep} \qquad (1)$$

The sediment core analyzed by Shackleton et al. [1983a] showed that during glacial times, $\Delta$ was higher, by as much as 0.75°/oo, a result which might be attributed to a higher concentration of nutrients in the glacial ocean. It requires in any event a larger rate for removal of carbon from the surface water as organic material during glacial time as compared with the present if the data presented by Shackleton et al. [1983a] are considered representative. Acceptable models for the glacial ocean must account not only for the apparent change in $\delta^{13}C_{deep}$ but also for the variation in $\Delta$. Differences between the glacial and interglacial environments are summarized in Table 1. As we shall see, it is no easy task to develop a model which accounts simultaneously for all these constraints.

Removal of carbon from the ocean in organic form should result in a decrease in the acidity of the sea and a corresponding drop in the position of the lysocline. The so-called calcium carbonate preservation spike marking the termination records this event [Berger, 1977]. The phenomenon is temporary, however, in that enhanced removal of $CaCO_3$ depletes alkalinity, restoring acidity, thus raising the position of the lysocline. Evidence from deep-sea cores suggests that the mean position of the lysocline in the glacial ocean was approximately the same as today. This result follows as a natural consequence of the requirement that removal of carbonate to sediments balance the input from rivers.

We begin in section 2 with a discussion of models for oceanic carbon, emphasizing the important role of preformed nutrient in setting the level of atmospheric $CO_2$. We continue in section 3 with an evaluation of changes in $CO_2$, $\delta^{13}C$, and $\Delta$ expected with various assumptions regarding the nature of the glacial environment, seeking to limit the range of possible scenarios. Summary remarks and brief comments on the nature of the glacial-interglacial transition are given in section 4.

2. Models for Oceanic Carbon

The simplest model for oceanic carbon would involve two well-mixed reservoirs, one representing the mean surface regime in contact with the atmosphere and the second describing the bulk of ocean water below the thermocline. The surface reservoir contains relatively warm water and is the site for biological productivity. Carbon enters the surface reservoir from below and is removed either by the return flow or as particulate material formed by biological processes near the surface. A modification of the two-box model would allow for outcropping of cold deep water at high latitudes, providing a direct connection between the atmosphere and the deep.

The loss of carbon from the surface to deep in particulate form must be balanced in the modified two-box approach by a source provided by transfer through the atmosphere, from the cold outcrop to the remainder of the ocean's surface [Broecker and Takahashi, 1984]. This illustrates a difficulty with two-box models: carbon in the present system is in fact transferred through the atmosphere in the opposite sense [Broecker and Peng, 1982]. A more elaborate model is required to simulate the real situation. To this end we developed [Knox and McElroy, 1984] three- and five-box models which allow for advection of water and chemical compounds as part of the overall ocean circulation, treating the chemistry of the polar surface ocean explicitly. The importance of the polar ocean has also been recognized by Sarmiento and Toggweiler [1984] and Siegenthaler and Wenk [1984].

Two-box models encounter further difficulties if one attempts to predict the level of oxygen dissolved in deep water. It is customary to assume that nutrient concentrations are reduced to zero in the warm surface box. The flux of organic carbon from the warm surface to depth in the original two-box formulation then exceeds the capacity of the return flow to supply the quantity of oxygen necessary for respiration. In this case the deep water would be anoxic, and the flux of organic carbon to sediments would be unrealistically high. The modified two-box model avoids this problem, since deep-water oxygen in this case is equilibrated with the atmosphere through the polar outcrop. But oxygen is then too high, and the assumptions of the model, specifically the requirement that oxygen be con-

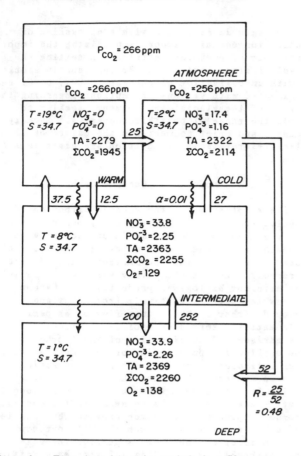

Fig. 1. Five-box kinetic model for $CO_2$. Quantities in italics represent inputs to the model. Concentrations are given in micromoles per kilogram, salinities in per mill, and advective fluxes of water in sverdrups (1 Sv = $10^6$ m$^3$ s$^{-1}$). The parameter $\alpha$ defines the fraction of incoming nutrients consumed by photosynthetic organisms in cold surface water. Average concentrations for the whole ocean: TA = 2365; $\Sigma CO_2$ = 2250; $NO_3^-$ = 33; $PO_4^{-3}$ = 2.2. Film thicknesses (for $CO_2$ and $O_2$ invasions) in warm and cold surface waters are 25 and 20 µm, respectively. For details see Knox and McElroy [1984]. Standard case: $\alpha$ = 0.01, R = 0.48.

stant through the deep reservoir, preclude any treatment of oxygen utilization at depth. These difficulties are removed in the higher-resolution models.

The quantity of oxygen at depth in more realistic simulations depends critically on the level of preformed nutrients, i.e. the quantities of N and P returned from the surface to depth unused by the biota. Preformed nutrients also play an important role in controlling the concentration of atmospheric $CO_2$. They provide a simple measure of the efficiency of the biological pump which sets the gradient in carbon between the atmosphere-surface ocean system on the one hand

and the deep sea on the other. We define the efficiency of the biological pump as the fractional contribution made by organic detritus to the transfer of carbon between the surface and depth. A high concentration of preformed nutrient reflects an inefficient pump, resulting in high levels of atmospheric $CO_2$ and elevated levels of dissolved oxygen at depth. Low concentrations of preformed nutrient lead to low values of $pCO_2$ and correspondingly low values for dissolved oxygen. We shall illustrate these concepts in what follows using the five-box formulation given by Knox and McElroy [1984] (see Figure 1).

The concentration of preformed nutrients, equivalently the concentration of nutrients in the high-latitude surface box, is determined by a combination of physical and biological effects. Waters entering the high-latitude surface regime from below are initially high in nutrients, in contrast to the source from surface waters at lower latitude. In the absence of biological activity at high latitudes, the concentration of preformed nutrients would simply reflect the relative importance of these two sources, specified in our model by the parameter R defining the fraction of waters in the high-latitude surface

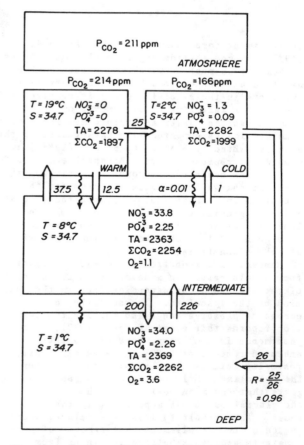

Fig. 2. Same as Figure 1, but with R = 0.96.

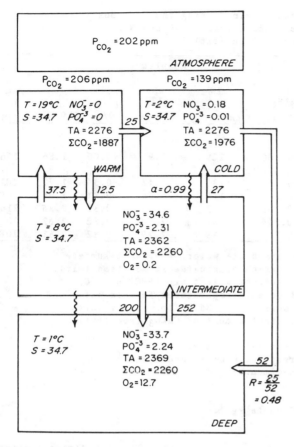

Fig. 3.  Same as Figure 1, but with α = 0.99.

box derived from surface currents connecting this regime to the nutrient-depleted water at low latitude.  Nutrients may also be removed from the high-latitude surface reservoir by in situ biological activity, described by a parameter α defining the fraction of incoming nutrients removed in particulate form by high-latitude biota.  Sensitivity of results to R and α is illustrated in Figures 1-3.

Figure 1 is designed to simulate the preindustrial environment prior to recent large additions of $CO_2$ associated with burning of fossil fuel and clearance of land for agriculture.  The physical parameters of the model are the same as those used by Knox and McElroy [1984] except that the temperature of the warm surface layer is set at 19°C rather than 22°C.  With the lower temperature, the value for atmospheric $pCO_2$ is 266 ppm [Barnola et al., 1983], as compared with 282 ppm given in the earlier study.  We assumed that transfer of water from the warm surface reservoir to the high-latitude box accounts for a flux of 25 Sv (1 Sv equals $10^6$ $m^3$ $s^{-1}$) with a further flux of 27 Sv supplied from below, corresponding to a value of R equal to 0.48.  Utilization of nutrients is inefficient at high latitudes in this model: α is set equal to 0.01.  Preformed

nutrients account for approximately half the concentration of nutrients in the deep ocean, in accord with observation [Redfield et al., 1963; Takahashi et al., 1981].  Carbon is transferred through the atmosphere from the low-latitude ocean to the polar environment.  Levels of $O_2$ in intermediate and deep waters are 129 μmoles $kg^{-1}$ and 138 μmoles $kg^{-1}$, respectively.

A change in R from 0.48 to 0.96 leads to a reduction in atmospheric $CO_2$ from 266 ppm to 211 ppm, as illustrated in Figure 2.  As in Figure 1, the flux of carbon through the atmosphere proceeds from low to high latitude.  The concentration of preformed nutrient is reduced by more than an order of magnitude with the increase in R:  the concentration of $NO_3^-$ in the high-latitude surface reservoir drops from 17.4 to 1.3 μmoles $kg^{-1}$.  Deep-sea oxygen is also reduced, to 1.1 μmoles $kg^{-1}$ and 3.6 μmoles $kg^{-1}$ in the intermediate and deep reservoirs, respectively.

Similar results are obtained by changing α, as shown in Figure 3.  The value of α in this case was set equal to 0.99, with the magnitude of R identical to that used in Figure 1.  Concentrations of preformed nutrients in Figure 3 are comparable to those in Figure 2, and as expected, results for atmospheric $CO_2$ and deep-sea oxygen are similarly depressed.  The low values for $CO_2$ and $O_2$ in this case may be attributed to reductions in preformed nutrients imposed by efficient biological uptake at high latitudes.  Suppression of nutrients at high latitude in Figure 2 is a consequence of a reduction in the transport of high-nutrient water from below.  Results in Figures 1-3 confirm the importance of the concentration of preformed nutrients as the master variable controlling both atmospheric $pCO_2$ and deep-sea oxygen.

There are suggestions that formation of bottom water in the North Atlantic was much less efficient in the glacial environment than today [Boyle and Keigwin, 1982; Curry and Lohmann, 1983; Shackleton et al., 1983b].  Reduction in the net rate of bottom water formation would have little effect on atmospheric $CO_2$ if R and α were

TABLE 1.    Changes Between Glacial and Interglacial
Environments

| Parameter | Change |
|---|---|
| **Ocean** | |
| Temperature | 2°C warming |
| Average depth | 100 m rise |
| Salinity | 1°/oo lowering |
| $\delta^{13}C$ of deep water | 0.4°/oo increase |
| $\delta^{13}C$ of surface water | 0.4°/oo decrease |
| Surface – Deep $\delta^{13}C$ | 0.8°/oo decrease |
| Mean lysocline depth | change of less than 500 m |
| **Atmosphere** | |
| $pCO_2$ | 100 ppm rise |

TABLE 2.  Results for the Glacial Ocean Obtained Using the Five-Box
Model Assumptions Concerning the Concentration of C and N
and the Parameters R and α (See Text)

| | | Inputs (Average Ocean) | | | | | Results | | | | |
|---|---|---|---|---|---|---|---|---|---|---|---|
| | R | [TA], μmole/kg | [ΣCO$_2$], μmole/kg | [NO$_3^-$], μmole/kg | $\delta^{13}$C, °/oo | IW [O$_2$], μmole/kg | WSW $\delta^{13}$C, °/oo | DW $\delta^{13}$C, °/oo | Δ, °/oo | Atm. CO$_2$, ppm |
| **Present** | | | | | | | | | | |
| Case A | 0.48 | 2365 | 2250 | 33.0 | +0.100 | 129 | 1.92 | +0.06 | 1.86 | 266 |
| **Glacial** | | | | | | | | | | |
| Case B | 0.48 | 2482 | 2376 | 33.9 | −0.314 | 137 | 1.42 | −0.35 | 1.77 | 283 |
| Case C | 0.48 | 2476 | 2374 | 37.1 | −0.266 | 119 | 1.65 | −0.31 | 1.96 | 268 |
| Case D | 0.86 | 2476 | 2374 | 37.1 | −0.266 | 2 | 2.03 | −0.32 | 2.35 | 216 |
| Case E | 0.48 | 2460 | 2374 | 52.0 | −0.266 | 37 | 2.50 | −0.32 | 2.82 | 202 |
| Case F | 0.62 | 2466 | 2373 | 45.0 | −0.266 | 19 | 2.29 | −0.32 | 2.61 | 208 |

Abbreviations are:  TA, titration alkalinity; IW, intermediate water; WSW, warm surface
water; DW, deep water; R, proportion of sinking water which originates in WSW (see text).
For the present ocean, temperatures of warm and cold surface waters are 19°C and 2°C,
respectively; salinity is 34.7°/oo.  For the glacial ocean, temperatures fo warm water and
cold surface waters are 17°C and 0°C, respectively; salinity is 35.7°/oo.  The results
assume that ocean chemistry adjusts to preserve a constant amount of CaCO$_3$ (see text).
Case A:  standard interglacial conditions.
Case B:  removal of 5 x 10$^{16}$ moles organic C to terrestrial biomass.
Case C:  removal of 2 x 10$^{16}$ moles organic C to terrestrial biomass and 3 x 10$^{16}$ moles
   organic C along with 4.5 x 10$^{15}$ moles N to shelf sediments.
Case D:  as in case C, with R = 0.86.
Case E:  as in case C, with [NO$_3^-$] = 52 μmole kg$^{-1}$.
Case F:  as in case C, with R = 0.62 and [NO$_3^-$] = 45 μmole kg$^{-1}$.

fixed, according to the present model.  On the
other hand, a change in the mechanisms or location of bottom water formation would undoubtedly
be accompanied by significant variations in
either or both R and α with associated implications for pCO$_2$.

### 3.  Chemistry of the Glacial Ocean

As a first step in evaluating the change in
oceanic and atmospheric chemistry at the glacial
to interglacial transition, consider Shackleton's
original suggestion, that an increase in terrestrial biomass was responsible for the observed
increase in $\delta^{13}$C of the deep ocean [Shackleton,
1977].  We adopt the physical parameters listed
in Table 1, keeping R and α fixed at the values
used in our standard interglacial reference, Figure 1.  Results for the glacial ocean are summarized in Table 2.
The concentration of deep-sea nutrients is
somewhat higher in the glacial model, reflecting
the smaller volume of ocean water.  Removal of
5x10$^{16}$ moles C in organic form from the ocean at
the end of glacial time would tend to lower the
level of atmospheric CO$_2$ by about 50 ppm.  There
is an associated drop in the position of the
lysocline, by about 1 km, reflecting the
decreased acidity of the new environment.  Subse-

quent accumulation of CaCO$_3$ in sediments restores
the lysocline to near its original position,
releasing CO$_2$ to the atmosphere.  We calculated
the effect of lysocline adjustments using a CaCO$_3$
preservation model similar to that of Keir and
Berger [1983].  We allowed, however, for the
influence of ocean topography (taken from Menard
and Smith [1966]), in contrast to the Keir and
Berger study, which assumed a flat-bottomed
ocean.  The amount of CaCO$_3$ preserved in our
model for the present ocean, 2.75x10$^{13}$ moles
yr$^{-1}$, was used as the standard, and values of
ΣCO$_2$ and alkalinity in the deep ocean were
adjusted in a 1:2 ratio following changes in C$_{org}$
and/or N content so as to reach this level of
calcite preservation in all glacial ocean
scenarios.  Adjustment in the lysocline depth
acts to raise atmospheric pCO$_2$, by about 20 ppm,
offsetting partially the 50-ppm drop due to removal of organic C.  Adjustments in ocean volume
and temperature cause a further small rise in
pCO$_2$, by about 10 ppm, resulting in the preindustrial standard of 266 ppm.  As shown in Table 2,
the ultimate result of the removal of 5x10$^{16}$
moles of organic C is a net drop in pCO$_2$ of about
17 ppm.  If growth of the terrestrial biosphere
were the only factor of importance in determining
the glacial to interglacial change in ocean chemistry, we would expect the level of atmospheric
CO$_2$ to have been somewhat higher during glacial

time, in contrast to the lower values implied by the ice core data [Berner et al., 1979; Delmas et al., 1980; Neftel et al., 1982; Barnola et al., 1983].

Our choice of magnitude for the organic carbon sink insures an acceptable value for the change in deep-sea $\delta^{13}C$. The $\delta^{13}C$ of surface waters was calculated using mass balance considerations. Physical fractionation factors were taken from Siegenthaler and Munnich [1981], Mook et al. [1974], and Thode et al. [1965]. Biological fractionation factors were taken as 1.0 for formation of $CaCO_3$ and 0.979 for synthesis of organic material. Note that $\Delta$ for the glacial ocean, case B, is about the same, or slightly lower, than $\Delta$ for the preindustrial standard, in contrast to observation, which suggests that $\Delta$ decreased from glacial to interglacial time by about 0.75°/oo [Shackleton et al., 1983a]. The relatively small variation in $\Delta$ between cases A and B is as expected, since there is little change in the productivity of the two oceans.

Case C provides an alternate way to account for the observed change in deep-sea $\delta^{13}C$. We assume that removal of organic carbon reflects, in part, growth of the terrestrial biosphere and, in part, deposition of carbon in coastal sediments. We adopt a value of $2 \times 10^{16}$ moles C for the terrestrial component, based on an estimate for the quantity of carbon contained now in the biomass and soils of North America which were earlier covered by ice [Denton and Hughes, 1981]. The remainder, $3 \times 10^{16}$ moles C, is ascribed to coastal sediments and is assumed to have been deposited with nutrients, N and P, represented by the Redfield ratios [Redfield et al., 1963]. Case C differs from case B in the important respect that the content of oceanic nutrients is allowed to vary during the glacial to interglacial transition. Correcting for the change in ocean volume, the concentration of $NO_3^-$ decreases from 37 $\mu$moles kg$^{-1}$ to 33 $\mu$moles kg$^{-1}$. This fluctuation is significantly less than that investigated by Broecker [1981], who considered a variation in the value of $\delta^{13}C$ in the deep ocean of 0.7°/oo, which he attributed exclusively to exchange of carbon with coastal sediments. The change in the ratio C/N for the deep ocean associated with the glacial-interglacial transition described by cases C and A is thus much less than that assumed by Broecker [1981]. The impact for atmospheric $CO_2$ is correspondingly reduced. As shown in Table 2, $pCO_2$ is predicted to drop by 2 ppm between glacial and interglacial times: $\Delta$ decreases by about 0.10°/oo, less than the observed change by about 0.6°/oo.

Case D differs from case C in that it allows in addition for a variation in R, from 0.48 to 0.86. It is consistent as before with the observed change in deep-sea $\delta^{13}C$. The higher value of R leads to significant reduction in the concentration of preformed nutrients, and the concentration of atmospheric $CO_2$ is lowered accordingly, to 216 ppm. This is accomplished

however at a cost: the concentration of oxygen in intermediate waters is reduced to 2 $\mu$moles kg$^{-1}$. Given the inevitable heterogeneity of the ocean, this suggests that oxygen would be essentially absent in extensive regions of the ocean, and there is no evidence for such sustained glacial anoxia. The value for $\Delta$ in case D is higher than in case A by about 0.5°/oo. Thus, case D is in better accord with the constraints which must be satisfied by an acceptable model. It yields a value for $pCO_2$, however, somewhat higher than one might wish and a value for $\Delta$ which is too low. The implications for $O_2$ are even more serious: denitrification [McElroy, 1983] and perhaps phosphorite formation [Froelich et al., 1982] would be widespread in the environment described by case D and might be expected to result in rapid loss of N and P, calling into question the validity of the assumption made regarding the existence of a steady state. It raises the possibility that the concentrations of N and P may have been significantly higher in the glacial ocean and that the glacial to interglacial transition was accompanied by large-scale loss of oceanic nutrients.

This possibility is considered further in cases E and F. In both instances the requirements for deep-sea $\delta^{13}C$ are satisfied by the combination of terrestrial and coastal exchanges of organic carbon explored in case C. Case E allows for an increase in the concentration of deep-sea nutrients in the glacial ocean of almost 60% relative to the interglacial environment. It yields a value for $pCO_2$ consistent with observation, about 200 ppm, and a value for $\Delta$ relative to our interglacial standard which is slightly higher than the result reported by Shackleton et al. [1983a]. The concentration of oxygen in intermediate waters is low, but perhaps acceptable. Case F involves an enhancement in nutrient concentration approximately half that involved in case E, adjusting R to derive satisfactory results for both $pCO_2$ and $\Delta$. The concentration of oxygen is lower in case F than in case E but probably reasonable.

It is clear that a range of models can be constructed to agree with observation using suitable choices of values for $\alpha$ and R and for the concentration of nutrients. It is difficult, however, if we are to account for the low level of $pCO_2$, to avoid the conclusion that levels of oxygen in the glacial ocean must have been significantly less than values prevalent today.

## 4. Summary and Concluding Remarks

Available observations on $pCO_2$, $\delta^{13}C$ of the deep sea, and $\delta^{13}C$ of shallow waters provide important constraints on the chemistry of the glacial ocean.

The estimate for $\delta^{13}C$ in the deep glacial ocean is less than the contemporary value by between 0.2 and 0.4°/oo. Part of the difference

must be attributed to a smaller reservoir of terrestrial organic carbon during glacial time.  The
balance may reflect reduced storage of organic
carbon in coastal sediments.

Variations in the concentration of dissolved
inorganic carbon result in, at most, small
changes in the concentration of atmospheric $CO_2$.
The markedly reduced values for $pCO_2$ in glacial
times implied by the ice core data require that
the efficiency of the marine biological pump
regulating the difference in carbon between the
surface and deep ocean must have been significantly higher in the past.  If we assume that
nutrients are taken up according to the Redfield
ratio and that this ratio remains fixed in time,
the value for $pCO_2$ should reflect the concentration of deep-sea N, the ratio C/N, and the concentration of preformed N.  The ratio C/N controls the effective flux of carbon from the deep
sea to the atmosphere:  large concentrations of N
allow high productivity in surface waters where
nutrients are used to depletion, ensuring that a
large fraction of upwelling C is returned to the
deep in organic form.  A high concentration of
preformed nutrients would imply that much of the
upwelling nutrient is returned to the deep
unused.  For fixed C/N, $pCO_2$ should respond
directly to changes in the concentration of preformed nutrients expressed as a ratio with
respect to the abundance of deep-sea N.

The concentration of preformed nutrients in
the model developed here reflects the concentration of N at depth, the efficiency with which N
is supplied to the cold surface water by exchange
with the intermediate layer, the extent to which
nutrients in the cold surface regime are diluted
by low-nutrient waters from lower latitude, and
the efficiency with which nutrients are utilized
by the local cold water biota.  The complexity of
the actual situation is expressed in the model by
three simple quantities: the concentration of
deep-sea N and parameters R and $\alpha$ summarizing the
effects of physics and biology, respectively.  We
showed that acceptable results for the glacial
environment can be obtained with certain combinations of values for N, R and $\alpha$:  the value for
total oceanic C was fixed by the requirement that
we reproduce the observed value for deep-sea
$\delta^{13}C$.

It is important to develop procedures to
discriminate between these possible models, to
determine which, if any, of the various cases
studied may be considered to yield a satisfactory
first-order description of the glacial ocean.
Further observations of $\delta^{13}C$ in surface waters,
specifically data defining the value of $\delta^{13}C$ in
downwelling regions, would be valuable.  Data
bearing on the concentration of oxygen in the
deep sea would also be useful and could aid in
distinguishing the relative importance of
physical and biological processes.  A better
understanding of the processes regulating R and $\alpha$
is also imperative.  One would expect surface
currents to have been somewhat stronger in the

glacial environment, given the higher value for
the meridional temperature gradient which applied
at that time [CLIMAP Project Members, 1976].
This could contribute to a larger value of R.
The value of $\alpha$ could be similarly elevated, since
zones of bottom water formation would be expected
to shift to lower latitudes characterized by
higher levels of ambient light.  Further work,
specifically, directed studies of deep-water formation in the present environment, is required to
place these qualitative considerations on a more
quantitative footing.

We paid scant attention here to the factors
influencing the transition from glacial to
interglacial conditions.  To the extent that
climatic oscillations on time scales of 22,000
and 41,000 years appear to be correlated with
changes in the orbital parameters of the earth,
it would seem necessary that mechanisms responsible for the transition should be related ultimately to a trigger driven by small changes in
the seasonal distribution of incident radiation.
The last termination, for which relatively good
data exist, merits special attention in this
regard.  It was associated apparently with a
rapid and significant change in the concentration
of atmospheric $CO_2$.  Warming occurred simultaneously in both hemispheres [Dreimanus, 1977;
Shackleton, 1978; Porter, 1981; Mercer, 1983;
Broecker, 1984], even though the simple albedo
theory [Suarez and Held, 1979] predicts that
warming in the northern hemisphere should have
been accompanied by compensatory cooling in the
south.  Warming associated with a rise in $CO_2$
could account for the global character of the
observed change [Broecker, 1984].  Did the change
in $CO_2$ arise as a consequence of changes in climate stimulated by a small change in the distribution of insolation?  Alternatively, was the
change in $CO_2$ itself the result of changing
illumination, and is the changing level of $CO_2$ in
fact the trigger which drives subsequent changes
in both $CO_2$ and climate?

It is apparent that a variety of positive
feedback mechanisms exist to reinforce climate
change once it begins, and there may be similar
influences at work for $CO_2$.  Climatic warming
could cause a retreat in the extent of sea ice,
and could be accompanied by reductions in R and $\alpha$
as discussed above, with further increases in
$pCO_2$, causing additional warming.  Knox and McElroy [1984] raised the possibility that small
changes in $\alpha$ stimulated by variations in illumination could provide the necessary connection
between changes in the orbital parameters of the
earth and the observed changes in climate.  Relatively small changes in illumination could be
amplified to provide large modifications to the
radiation budget of the earth, modulated for the
most part by varying levels of atmospheric $CO_2$
and $H_2O$.  Additional work is required to carry
this hypothesis further.  The need exists in particular for a time dependent model of ocean chemistry and for better data on the chronology of

chemical and isotopic changes during the glacial-interglacial transition. Information on the relative timing of the variations in $CO_2$ and climate would be particularly valuable in this regard.

Acknowledgments. We are indebted to W. S. Broecker and S. C. Wofsy for helpful discussions. This work was supported by National Science Foundation grant ATM-81-17009 and National Aeronautics and Space Administration grant NAGW-359.

## References

Barnola, J. M., D. Raynud, A. Neftel, and H. Oeschger, Comparison of $CO_2$ measurements by two laboratories on air from bubbles in polar ice, Nature, 303, 410-412, 1983.

Berger, W. H., Deep-sea carbonate and the deglaciation preservation spike in pteropods and foraminifera, Nature, 269, 301-304, 1977.

Berner, W., B. Stauffer, and H. Oeschger, Past atmospheric composition and climate, gas parmeters measured on ice cores, Nature, 275, 53-55, 1979.

Boyle, E. A., and L. D. Keigwin, Deep circulation of the North Atlantic over the last 200,000 years: Geochemical evidence, Science, 218, 784-787, 1982.

Broecker, W. S., Glacial to interglacial changes in ocean and atmospheric chemistry, in Climate Variations and Variability: Facts and Theories, edited by A. Berger, pp. 111-121, D. Reidel, Hingham, Mass., 1981.

Broecker, W. S., Glacial to interglacial changes in ocean chemistry, Prog. Oceanogr., 11, 151-197, 1982a.

Broecker, W. S., Ocean chemistry during glacial time, Geochim. Cosmochim. Acta, 46, 1689-1705, 1982b.

Broecker, W. S., Terminations in Milankovich and Climate, Part 2, edited by A. L. Berger et al., pp. 687-698, D. Reidel, Hingham, Mass., 1984.

Broecker, W. S., and T.-H. Peng, Tracers in the Sea, 690 pp., Eldigio Press, Palisades, N.Y., 1982.

Broecker, W. S., and T. Takahashi, Is there a tie between atmospheric $CO_2$ content and ocean circulation?, in Climate Processes and Climate Sensitivity, Geophys. Monogr. Ser., vol. 29, edited by J. E. Hansen and T. Takahashi, pp. 314-326, AGU, Washington, D. C., 1984.

CLIMAP Project Members, The surface of the ice-age earth, Science, 191, 1131-1138, 1976.

Curry, W. B., and G. P. Lohmann, Reduced advection into Atlantic Ocean deep eastern basins during last glacial maximum, Nature, 306, 577-580, 1983.

Delmas, R. J., J.-M. Ascencio, and M. Legrand, Polar ice evidence that atmospheric $CO_2$ 29,000 BP was 50% of the present, Nature, 282, 155-157, 1980.

Denton, G. H., and T. J. Hughes (Eds.), The Last Great Ice Sheets, 484 pp., John Wiley, New York, 1981.

Dreimanus, A., Late Wisconsin glacial retreat in the Great Lakes region, North. Amer. Ann. N.Y. Acad. Sci. 228, 70-89, 1977.

Froelich, P. N., M. L. Bender, N. A. Luedtke, G. R. Heath, and T. DeVries, The marine phosphorus cycle, Am. J. Sci., 282, 474-511, 1982.

Keir, R. S., and W. H. Berger, Atmospheric $CO_2$ content in the last 120,000 years: The phosphate extraction model, J. Geophys. Res., 88, 6027-6038, 1983.

Knox, F., and M. B. McElroy, Changes in atmospheric $CO_2$: Influence of the marine biota at high latitude, J. Geophys. Res., 89, 4629-4637, 1984.

McElroy, M. B., Marine biology: Controls on atmospheric $CO_2$ and climate, Nature, 302, 328-329, 1983.

Menard H. W., and S. M. Smith, Hypsometry of ocean basin provinces, J. Geophys. Res., 71, 4305-4325, 1966.

Mercer, J. H., Cenozoic glaciation in the southern hemisphere, Annu. Rev. Earth Planet. Sci., 11, 99-132, 1983.

Mook, W. G., J. C. Bommerson, and W. H. Staverman, Carbon isotope fractionation between dissolved bicarbonate and gaseous carbon dioxide, Earth and Planet. Sci. Lett., 22, 169-176, 1974.

Neftel, A., H. Oeschger, J. Schwander, B. Stauffer, and R. Zumbrunn, Ice core sample measurements give atmospheric $CO_2$ contents during the past 40,000 years, Nature, 295, 220-223, 1982.

Porter, S. C., Pleistocene glaciation in the southern lake district of Chile, Quat. Res., 16, 263-292, 1981.

Redfield, A. C., B. H. Ketchum, and F. A. Richards, The influence of organisms on the composition of sea-water, in The Sea, vol. 2, edited by M. N. Hill, pp. 26-77, John Wiley, New York, 1963.

Sarmiento, J. L., and J. R. Toggweiler, A new model for the role of the oceans in determining atmospheric $pCO_2$, Nature, 308, 621-624, 1984.

Shackleton, N. J., Tropical rainforest history and the equatorial Pacific carbonate dissolution cycles, in The Fate of Fossil Fuel $CO_2$ in the Ocean, edited by N. R. Anderson and A. Malahoff, pp. 355-374, Plenum, New York, 1977.

Shackleton, N. J., Some results of the CLIMAP project, in Climatic Change and Variability: A Southern Perspective, edited by A. B. Pittock et al., pp. 69-76, Cambridge University Press, New York, 1978.

Shackleton, N. J., M. A. Hall, J. Line, and C. Chuxi, Carbon isotope data in core V19-30 confirm reduced carbon dioxide concentration in the ice age atmosphere, Nature, 306, 319-322, 1983a.

Shackleton, N. J., J. Imbrie, and M. A. Hall,

Oxygen and carbon isotope record of east Pacific core V19-30: Implications for the formation of deep water in the late Pleistocene North Atlantic, <u>Earth Planet. Sci. Lett</u>., <u>65</u>, 233-244, 1983b.

Siegenthaler, U., and K. O. Munnich, $^{13}C/^{12}C$ fractionation during $CO_2$ transfer from air to sea, in <u>Carbon Cycle Modelling</u>, <u>SCOPE 16</u>, edited by B. Bolin, pp. 249-257, John Wiley, New York, 1981.

Siegenthaler, U., and T. Wenk, Rapid atmospheric $CO_2$ variations and ocean circulation, <u>Nature</u>, <u>308</u>, 624-626, 1984.

Suarez, M. J., and I. M. Held, The sensitivity of an energy balance climate model to variations in orbital parameters, <u>J. Geophys. Res</u>., <u>84</u>, 4825-4836, 1979.

Takahashi, T., W. S. Broecker, and A. E. Bainbridge, Supplement to the alkalinity and total carbon dioxide concentration in the world oceans, in <u>Carbon Cycle Modelling</u>, <u>SCOPE 16</u>, edited by B. Bolin, pp. 159-199, John Wiley, New York, 1981.

Thode, H. G., M. Shima, C. E. Rees, and K. V. Krishnamurty, Carbon-13 isotope effects in systems containing carbon dioxide, bicarbonate, carbonate, and metal ions. <u>Can. J. Chem</u>., <u>43</u>, 582-595, 1965.

# GLACIAL TO INTERGLACIAL CHANGES IN ATMOSPHERIC CARBON DIOXIDE: THE CRITICAL ROLE OF OCEAN SURFACE WATER IN HIGH LATITUDES

J.R. Toggweiler and J.L. Sarmiento

Geophysical Fluid Dynamics Program, Princeton University
Princeton, New Jersey 08542

Abstract. Recent measurements of the $CO_2$ content of air bubbles trapped in glacial ice have shown that the partial pressure of atmospheric $CO_2$ during the last ice age was about 70 ppm lower than during the interglacial. Isotopic measurements on surface- and bottom-dwelling forams living during the ice age have shown that the $^{13}C/^{12}C$ gradient between the ocean's surface and bottom layers was 25% larger during the last ice age than at present. Broecker (1982) proposed that an increase in the phosphate content of the deep sea could explain these observations. We follow up here on a proposal by Sarmiento and Toggweiler (1984) that glacial to interglacial changes in $P_{CO2}$ are related to changes in the nutrient content of high-latitude surface water. We develop a four-box model of the ocean and atmosphere which includes low- and high-latitude surface boxes, an atmosphere, and a deep ocean. In simplest form the model equations show that the $CO_2$ content of high-latitude surface water is directly connected to the huge reservoir of $CO_2$ in deep water through the nutrient content of high-latitude surface water. The relationship between the $CO_2$ content of low latitude surface water and the deep sea is more indirect and depends to a large extent on transport of $CO_2$ through the atmosphere from high latitudes. We illustrate how the $^{14}C$ content of the atmosphere and that of high-latitude surface water constrain model solutions for the present ocean and how ice age $^{13}C$ observations constrain ice age parameters. We propose that the low ice age $P_{CO2}$ can be produced by a reduction in local exchange between high-latitude surface water and deep water. The model requires that the current exchange rate of about 50 Sv be reduced to about 10 Sv. We review evidence in the geologic record for widespread changes in deep convection around Antarctica about 14,000 years ago which are synchronous with the change in atmospheric $P_{CO2}$.

## Introduction

Three important facts about the chemistry of the ocean and atmosphere during the geologically recent past have come to light in recent years.

$CO_2$ measurements on gas bubbles trapped in glacial ice have shown that the partial pressure of $CO_2$ in the atmosphere during the last ice age (~18,000 years ago) was about 70 ppm lower than it has been during the interglacial period (beginning ~10,000 years ago) [Neftel et al., 1982]. This is significant because it suggests that global climate changes have been induced, or at least amplified, by changes in the $CO_2$ content of the atmosphere. Climate projections for the future predict a 2°C rise in temperature with a doubling of atmospheric $CO_2$ levels [Manabe and Stouffer, 1980]. Lower $CO_2$ levels in the past are therefore consistent with cooler climates. A $CO_2$-induced cooling during the ice age explains how glaciations might occur simultaneously in both northern and southern hemispheres in spite of the fact that the Milankovitch orbital forcing favors the growth of ice sheets in one hemisphere but not the other.

The glacial ice also tells us that the $P_{CO2}$ of the atmosphere prior to the period of rapid industrialization was about 270 ± 10 ppm compared to about 345 ppm today [Stauffer et al., 1984]. Therefore, the change in atmospheric $CO_2$ content since the middle of the last century to the present is about the same as the change between glacial and interglacial periods. Although no obvious climatic changes have been linked to the recent rise in $CO_2$, most students of climatic change hold that the effects of $CO_2$ on climate operate through feedbacks in the climatic system involving, for example, changes in water vapor transport or changes in the ice albedo feedback in high latitudes. Because of the flywheel effect of the oceans these processes may take some time to become established [Bryan et al., 1982].

Shackleton [1977], Vincent et al. [1981], and Shackleton et al. [1983a, b] have shown using carbon isotope measurements in deep-dwelling benthic forams that the $^{13}C/^{12}C$ ratio in the deep sea was lower during the ice age than at present. At the same time the $^{13}C$ content of surface-dwelling planktonic forams seems not to have changed [Shackleton et al., 1983b]. While all the data on isotopic changes in benthic forams seem to support some glacial to interglacial increase in the $\delta^{13}C$

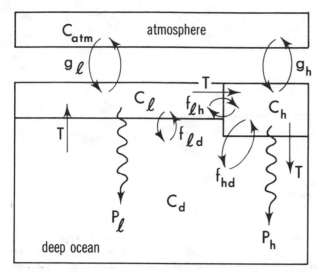

Fig. 1.  Schematic diagram of the four-box model.

content of the deep sea, there is uncertainty with regard to the absolute magnitude.  Estimates of the magnitude range from barely significant to as much as 0.7 per mil.  We assume that the highly resolved 0.5 per mil increase recorded by Shackleton et al. [1983a, b] for core V19-30 in the eastern equatorial Pacific is a representative value.  This means that the gradient in $^{13}C/^{12}C$ between the surface and the deep sea increased by 25% over the preindustrial gradient of about 2.0 per mil.  This observation suggests that the ice age ocean was more effective at sequestering biologically fixed $CO_2$ in the deep sea.  One can also infer that the oxygen content of the deep sea was lower during the ice age because of the close association between $\delta^{13}C$ and oxygen in the ocean's photosynthesis/respiration cycle.

Broecker [1982] proposed that an increase in the phosphorus to carbon ratio in the glacial deep sea could explain the lower atmospheric $P_{CO2}$ and the carbon isotope changes.  He proposed that erosion of organic sediments from exposed continental shelves during the ice age might add extra phosphorus to the ocean.  In simplest terms, this hypothesis says that a unit of nutrient-rich deep water upwelling to the surface during the ice age sustains more biological productivity.  This enhances the flux of detrital organic carbon from the surface to the deep sea and pulls $CO_2$ out of the atmosphere.  The excess organic carbon in the deep sea, when oxidized by deep-sea organisms, reduces deep-sea oxygen levels and increases the $^{13}C/^{12}C$ gradient between the surface and deep ocean.  One consequence of Broecker's shelf erosion hypothesis is that lower atmospheric $CO_2$ levels must follow the growth of the ice sheets; the proposed ocean chemistry changes can only occur after the sea level has been lowered.

What we propose here is a different model for

the $CO_2$ lowering which shows how changes in the ocean's large-scale circulation can reduce the $CO_2$ content of the atmosphere independently of continental glaciation and sea level changes.  Like Broecker's model, this one calls on changes in ocean nutrient content but not in the deep sea or over the ocean as a whole.  This model requires only that changes in the nutrient content in high-latitude surface water be altered.  Nutrient levels in present-day high-latitude surface water are quite high, particularly in the Antarctic.  We will show that these high nutrient levels are consistent with a well-oxygenated deep ocean.  Lower $CO_2$ levels in the ice age atmosphere are consistent with lower nutrient levels in high-latitude surface water and a poorly oxygenated deep sea.  We propose that glacial to interglacial changes in atmospheric $CO_2$ are probably driven by changes in the dynamics of convection in the Antarctic.  Time scales for this transition are short enough that changes in atmospheric $P_{CO2}$ might have played an active role in forcing the end of the last ice age.

The model to be discussed here has been presented in preliminary form by Sarmiento and Toggweiler [1984].  Two other groups, working independently, have published papers using very similar models [Siegenthaler and Wenk, 1984; Knox and McElroy, 1984].  These authors also have papers appearing in this volume [Wenk and Siegenthaler, this volume; Ennever and McElroy, this volume].  We will detail here how the model works and how carbon isotopes constrain model solutions.  In the first section below we present

## SURFACE TEMPERATURE (°C)

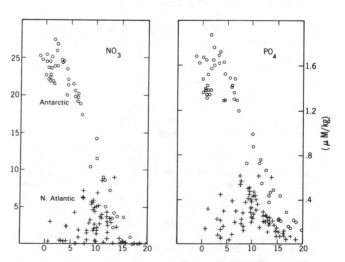

Fig. 2.  Nitrate and phosphate content of high-latitude surface water plotted against surface temperature.  Antarctic points are plotted as open circles.  North Atlantic points are plotted as crosses.  GEOSECS and TTO data.

TABLE 1.   Stoichiometric Relationships Between
Model Variables

|  | Standard Redfield Ratios[a] | Ratios Used in Model |
|---|---|---|
| $\Delta O_2/\Delta P$ | -138 | -169 |
| $\Delta C_{org}/\Delta P$ | 106 | 130 |
| $\Delta \Sigma CO_2/\Delta P$[b] | 132.5 | 162.5 |
| $\Delta Alk/\Delta P$ | 38 | 50 |

[a]Redfield et al. [1962].

[b]Assuming for both cases that 20% of $\Sigma CO_2$ increase in deep sea is due to $CaCO_3$ dissolution [Li et al., 1969; Kroopnick, 1974].

a set of idealized model equations which provide the simplest and most illustrative description of the model. In the second and third sections we present model results for the present ocean and ice age ocean, respectively. In the discussion to follow we attempt to relate the model's predictions to what one observes in the real world and examine the evidence for predicted ocean chemistry changes. An appendix follows, in which technical aspects of the model's construction are presented.

The Idealized Model Equations

The model presented here is diagrammed in Figure 1. It differs from a standard geochemical two box ocean model by including a box which represents high-latitude surface water, combining North Atlantic and Antarctic deep water formation regions. $C_{atm}$, $C_\ell$, $C_h$, and $C_d$ represent the tracer contents of the atmosphere, low-latitude surface water, high-latitude surface water, and the deep ocean, respectively. This model configuration is similar to one developed by Keeling and Bolin [1968] for the Pacific Ocean. Using a data set of much poorer quality than those presently available, Keeling and Bolin attempted with great difficulty to derive fluxes between the boxes from averaged data. In our approach here we will make some simplifying assumptions which Keeling and Bolin did not make. Also, in producing a global model we use the overall atmospheric balances of $CO_2$ and $^{14}C$ as model constraints. At the time when Keeling and Bolin developed their model no one had reason to believe that the $CO_2$ content of the atmosphere had varied so widely in the recent past. Keeling and Bolin, therefore, did not explore the possibilities which we consider here.

We include in the high-latitude box the area

of the North Atlantic poleward of 60°N and the areas of the South Atlantic, South Pacific, and Indian Oceans poleward of 50°S. Approximately 15% of the ocean's surface is included in this definition; the area of the ocean south of 50°S is seven times larger than the area of the North Atlantic north of 60°N. The surface of the Arctic Ocean is ignored because ice coverage cuts off exchange between the ocean and atmosphere. The surface boxes $\ell$ and $h$ are assumed to have average mean annual temperatures of 21.5° and 2.5°C, respectively, as determined from the Levitus [1982] data set. We assume that the low-latitude box is 100 m deep and the high-latitude box is 250 m deep.

The bidirectional arrows labeled $g_\ell$ and $g_h$ represent gas exchange between the surface boxes and the atmosphere. Peng et al. [1979] found that an average gas exchange piston velocity of about 3 m/day applies over most of the ocean, and we adopt this value for both surface boxes. This does not mean that the gas exchange rates are the same; the effect of temperature on $CO_2$ solubility makes the high-latitude $CO_2$ gas exchange rate approximately twice that of the low-latitude box. The wiggly lines labeled $P_\ell$ and $P_h$ represent particulate phosphorus fluxes from the surface layers to the deep sea.

The bidirectional arrows labeled $f_{hd}$, $f_{\ell d}$, and $f_{\ell h}$ represent simple first-order water exchange between the boxes indicated by the subscripts. The arrows labeled with a T represent a one-way cycle in which water flows from the low-latitude surface box to the high-latitude box, to the deep ocean box, and back to the low-latitude surface box again. The variable T represents the large-scale thermohaline overturning of the ocean in which warm water flows poleward, cools, sinks into the deep sea, and then upwells back to the warm surface through the ocean's thermocline. Of the bidirectional fluxes, the $f_{hd}$ term connecting high-latitude surface water with the deep sea is by far the most important. It can be thought of as deep convection in high latitudes or as isopycnal mixing along deep density surfaces outcropping at high latitudes. The preliminary report mentioned above [Sarmiento and Toggweiler, 1984] discusses model solutions for ordinary $^{12}CO_2$ in which $f_{\ell d}$ and $f_{\ell h}$ are set equal to zero. We will maintain this simplification throughout this section.

Photosynthesizing ocean plants living in warm ocean surface water are capable of taking up virtually all available nitrate and phosphate and of incorporating these vital nutrients into their body tissues. For this reason the biological productivity of warm surface water can be said to be nutrient limited. When the plants are eaten by grazing zooplankton, the nutrients ultimately find their way into excreted particles which sink out of the photic upper layers. When a unit of deep water upwells into the warm surface ocean, all of the upwelled nitrate and phosphate thus returns to the deep sea in organic form as a component of

TABLE 2.　State Variables for the Ocean/Atmosphere System

| | Preindustrial Value | Source | Ice Age Value[a] |
|---|---|---|---|
| Volume of the ocean | $1.292 \times 10^{18}$ m$^3$ | Levitus [1982] | $1.249 \times 10^{18}$ m$^3$ |
| Area of the ocean's surface | $3.49 \times 10^{14}$ m$^2$ | Sverdrup et al. [1942] | same |
| Mole volume of the atmosphere | $1.773 \times 10^{20}$ moles | Weast and Astle [1979] | same |
| Average AOU content of the deep sea below upper 100 m (AOU$_d$) | 154 µmol/kg | GEOSECS/TTO | ---- |
| Total phosphate content of the ocean | $2.77 \times 10^{15}$ moles (avg = 2.09 µmol/kg) | GEOSECS/TTO | same |
| Total alkalinity content of the ocean | $3.140 \times 10^{18}$ eq. (avg = 2371 µeq/kg) | GEOSECS/TTO | $3.267 \times 10^{18}$ equivalents |
| Total ΣCO$_2$ content of the ocean | $2.990 \times 10^{18}$ moles (avg = 2258 µmol/kg) | GEOSECS/TTO | ---- |
| Average δ$^{13}$C of the ocean | 0.55 per mil | Broecker and Peng [1982][b] | -- |
| Average [$^{14}$C] of the ocean | $2.27 \times 10^{-15}$ moles/kg (85.1% of atmospheric equilibrium) | Stuiver et al. [1981][c] | -- |
| Total ΣCO$_2$ content of the ocean and atmosphere | $3.0245 \times 10^{18}$ moles | ---d | $3.1517 \times 10^{18}$ moles |
| Total Σ$^{13}$CO$_2$ content of ocean and atmosphere | $3.0263 \times 10^{18}$ moles $\cdot^{13}C/^{12}C_{stnd}$ | ---e | $3.1521 \times 10^{18}$ moles $\cdot^{13}C/^{12}C_{stnd}$ |
| Total Σ$^{14}$CO$_2$ content of ocean and atmosphere | $2.5578 \times 10^{18}$ moles $\cdot^{14}C/^{12}C_{stnd}$ | ---f | $2.6182 \times 10^{18}$ moles $\cdot^{14}C/^{12}C_{stnd}$ |
| Depth of the low-latitude surface box | 100 meters | | same |
| Depth of the high-latitude surface box | 250 meters | | same |
| Temperature and salinity of the low-latitude surface box | 21.5°C, 34.7 mg/g | | 20.0°C, 35.9 mg/g |
| Temperature and salinity of the high-latitude surface box | 2.5°C, 34.7 mg/g | | 2.0°C, 35.9 mg/g |

[a]Ice age state variables have been determined as follows: for the ocean volume we assume that enough fresh water leaves the ocean to increase salinity from 34.7 to 35.9 per mil following Broecker [1982]; we add $1.272 \times 10^{17}$ moles of isotopically light (-23 per mil) carbon to the system (this is sufficient to decrease the deep ocean δ$^{13}$C by 0.5 per mil); we then add an equal amount of alkalinity and isotopically neutral (0 per mil) ΣCO$_2$ to the system to simulate CaCO$_3$ dissolution from the seafloor; we assume that half of the added ΣCO$_2$ is terrestrial

sinking particles. In terms of the diagram in Figure 1 we can write

$$P_\ell = PO4_d \cdot T \qquad (1)$$

where only the major water flux terms, $f_{hd}$ and T, are considered. To avoid confusion regarding subscripts, we will not subscript molecular formulas in expressions in which model subscripts appear. For example, we will use $PO4_d$ in place of $PO4_{d}$.)

When the same unit of deep water upwells to the surface, the $CO_2$ content of the water is reduced in proportion to the phosphorus removal as organic tissues are formed. The molar ratio of C to P taken up is known as the Redfield ratio and is traditionally given the value 106:1 [Redfield et al., 1962]. Following Broecker [1982], we will use phosphate as a representative, idealized limiting nutrient. We could easily choose to use nitrate in the same manner. Nitrate is thought to be more limiting in the ocean than phosphate; however, it has some small non-Redfieldian sources and sinks within the ocean which are not well quantified.

The Redfield ratio which links phosphorus and nitrogen to carbon in fixed proportions is based on measurements of C:P and C:N ratios in live ocean plants sieved from surface water. It is normally assumed that these ratios hold for organic particulate matter throughout the water column. Evidence from particle traps, however, suggests that organic phosphorus and nitrogen are remineralized more rapidly than organic carbon [Knauer et al., 1979; Honjo, 1980]. Therefore, the ratios of C:P or C:N are larger in particles which are destroyed in the deep sea, as opposed to those destroyed in the upper ocean. Indeed, Takahashi et al. [1984] show that relative changes in oxygen and phosphorus along intermediate water density surfaces follow an average slope of $-172 \pm 6:1$. Given the Redfield ratio of 1.3 moles $O_2$ consumed per mole organic carbon oxidized, this slope translates into a C:P ratio of about 130:1.

For the high latitude surface box we can write the following equation:

$$P_h = PO4_d \cdot f_{hd} - PO4_h \cdot (T+f_{hd}) \qquad (2)$$

where again the particle flux $P_h$ refers to the downward flux of phosphorus and the flux terms $f_{\ell d}$ and $f_{\ell h}$ are ignored.

If we assume that the oxygen content of the two surface boxes is always in equilibrium with the atmosphere, we can write an equation describing the deep-ocean oxygen content as a function of $P_\ell$ and $P_h$. Here, we express the oxygen content in terms of the apparent oxygen utiliztion (AOU), which is the difference between the dissolved oxygen concentration at saturation (a function of water temperature) and the measured oxygen concentration. We assume here that the AOU of the two surface boxes is zero.

$$AOU_d = r_{O2:P} \cdot (P_\ell + P_h)/(T+f_{hd}) \qquad (3)$$

---

Table 2 (continued)

biomass with a $\Delta^{14}C$ of -50 per mil; the other half is assumed to be dead with respect to $^{14}C$.

[b]The average $\delta^{13}C$ for the world ocean is estimated using the correlation between $\delta^{13}C$ and $PO_4$ given by Broecker and Peng [1982] Figure 6-12, p. 309 and the average $PO_4$.

[c]A subtle, but important, error exists in the Stuiver et al. [1981] compilation. Although the reported average $^{14}C$ concentration units for the various ocean basins are given as moles per cubic meter, the figures and tables actually have units of moles per 1000 kg. One must multiply Stuiver et al.'s [1981] $^{14}C$ concentrations by 1.025 before comparing $^{14}C$ with $^{12}C$ in units of moles per cubic meter. Otherwise an error of 10-15% will result with respect to the amount of $^{14}C$ decay or "aging" one would predict for the ocean relative to the atmosphere.

[d]The total $\Sigma CO_2$ content of the preindustrial ocean and atmosphere is the sum of the total $\Sigma CO_2$ for the ocean given above plus the atmospheric $CO_2$ content in 1973 (320 x $10^{-6}$ atm times 1.773 x $10^{20}$ moles of air) less twice the fossil fuel $CO_2$ input through 1973 (11 x $10^{15}$ moles [Broecker and Peng, 1982]). We have assumed that a terrestrial biomass release equal to the fossil fuel release has entered the atmospheric during the last century.

[e]The total $\Sigma^{13}CO_2$ content of the preindustrial ocean and atmosphere is calculated as in note d above, where the ocean $\delta^{13}C$ is +0.55 per mil, the atmosphere $\delta^{13}C$ is -7.2 per mil, and the subtracted fossil fuel/biomass $\delta^{13}C$ is -26 per mil. The number of significant digits reported in the carbon isotope totals reflects relatively small isotopic shifts between the tracers.

[f]The total $\Sigma^{14}CO_2$ content of the preindustrial/prebomb ocean and atmosphere is computed in similar manner to $^{13}C$ above where the 1973 atmosphere is 450 per mil enriched in $^{14}C$ and the released biomass $^{14}C$ is assumed to be -60 per mil (the fossil fuel component contains no $^{14}C$). We subtract 8.42 x $10^4$ moles of bomb $^{14}C$ from the total. This total is derived from the 314 x $10^{26}$ bomb $^{14}C$ atoms reported by Broecker et al. [1980] to be residing in the ocean in 1973 and from the 480 per mil of bomb $^{14}C$ in the atmosphere (450 per mil plus 30 per mil for compensation of the Seuss effect).

The term $r_{O2:P}$ is the ratio of oxygen consumed to phosphate regenerated when organic matter is oxidized. We also assume in equation (3) that all particulate organic matter falling into the deep box is oxidized. In other words, none of the particulate flux is lost to the sediments.

We can now combine equations (1), (2), and (3) and write an expression for the phosphate content of high-latitude surface water:

$$PO4_h = PO4_d - AOU_d/r_{O2:P} \qquad (4)$$

We see from equation (4) that the nutrient content of high-latitude surface water is directly linked to the oxygen content of the deep sea; i.e., a well-oxygenated deep sea coexists with high nutrient levels in high-latitude surface water. Since the average phosphate concentration of the deep sea is about 2.15 µmol/kg and the average AOU is about 154 µmol/kg, we calculate a $PO4_h$ of 1.24 µmol/kg. Nutrients in newly formed deep water are often referred to as "preformed nutrients"; here, $PO4_h$ is the same as preformed phosphate.

Figure 2 shows the nitrate and phosphate contents of high-latitude (summer) surface water plotted against temperature. The circles represent Antarctic locations, while the pluses represent the North Atlantic. The nutrient contents of Antarctic and North Atlantic surface waters climb above near-zero mid-latitude values at temperatures below 15°C. Antarctic nutrient values increase to much higher values than those found in the North Atlantic. Our preformed phosphate concentration of 1.24 µmol/kg plots just below the cluster of Antarctic points.

In high latitudes, far more nutrients are delivered to the surface than can be utilized by plants. High-latitude surface water is not underlain by the intense thermocline found at lower latitudes; in winter the water column is often able to convect to great depth, bringing nutrient-rich deep water to the surface. Over much of the year there is insufficient light to support plant growth, so nutrients go unutilized. Because we define our high-latitude box to be that region of the surface ocean in which nutrient levels are nonzero, we can use the information in Figure 2 as a rough guide in demarcating the box's latitudinal extent. The midpoint of the Antarctic nutrient rise is found at about 10°C. If we take the 10° summer isotherm as our boundary, the high-latitude box is defined as the ocean area poleward of 50°S in the Antarctic and 60°N in the North Atlantic.

It is instructive to consider what would happen if there were no high-latitude region in the ocean. If, somehow, warm surface water (with preformed $PO4 = 0$) were to be taken directly to the deep ocean and allowed to upwell back to the surface, phosphate could enter into the deep sea only in the form of organic particles. Looking at the oxygen balance, one finds that the oxygen demand of such a particle flux depletes the deep ocean of oxygen, even if one allows the downwelling water to become saturated with oxygen at deep-ocean temperatures. Expressed mathematically,

$$O2_d = O2_{sat}(T) - r_{O2:P} \cdot PO4_d \qquad (5)$$

The saturation oxygen content at the average temperature of the deep sea is approximately 325 µmol/kg. Given $PO4_d = 2.15$ µmol/kg and $r_{O2:P} = 169$, one calculates that the oxygen demand from the phosphate flux is greater than the oxygen content of newly formed deep water. This situation is not observed in the ocean today; the average oxygen content of the deep sea is about 170 µmol/kg. Therefore, the deep sea is not depleted in oxygen to even half its saturation content. The real ocean shunts much of the upward flux of phosphate to high-latitude regions, allowing phosphate to reach the surface outside of the domain of the warm surface ocean. In high latitudes the deep sea's oxygen content is replenished without upwelled phosphate generating a tremendous organic particle demand on the deep-ocean oxygen content.

For the other tracers we want to study with this model ($\Sigma CO_2$, alkalinity, $^{13}C$, and $^{14}C$) the equations describing their steady state distributions among the four boxes in Figure 1 will be more complicated. There are no simplifying assumptions to be made regarding surface water as there are with phosphate and oxygen. One must also contend with exchanges between the ocean and atmosphere. There is, however, one simple relationship we can construct. Again ignoring $f_{\ell d}$ and $f_{\ell h}$, we can write the following equation for total $CO_2$ in the deep box:

$$\Sigma CO2_d \cdot (T+f_{hd}) = \Sigma CO2_h \cdot (T+f_{hd}) + r_{\Sigma C:P} \cdot (P_\ell + P_h) \qquad (6)$$

Here, $r_{\Sigma C:P}$ represents the ratio of total carbon (including $CaCO_3$) to phosphate in sinking particulate matter. We assume a constant fraction of four parts organic carbon to one part carbonate in particulate matter as originally proposed by Li et al. [1969] and later supported by Kroopnick [1974]. We then express $r_{\Sigma C:P}$ in terms of the C:P ratio in organic matter as follows:

$$r_{\Sigma C:P} = \frac{r_{C_{org}:P}}{(1-f_{Ca})} = \frac{130}{0.8} = 162.5 \qquad (7)$$

where $f_{Ca}$ is the carbonate fraction of the total carbon flux. Rearranging and substituting equations (1) and (2) into (6), we arrive at the following result:

$$\Sigma CO2_h = \Sigma CO2_d - r_{\Sigma C:P} \cdot (PO4_d - PO4_h) \qquad (8)$$

Rearranging again, we find

$$\frac{\Sigma CO2_d - \Sigma CO2_h}{PO4_d - PO4_h} = r_{\Sigma C:P} \qquad (9)$$

Equation (9) states that the relative differences in $\Sigma CO_2$ and phosphate between high-latitude surface water and deep water are given simply by a constant ratio. One can derive a similar relationship for alkalinity, where the constant relating alkalinity to phosphorus is given by

$$r_{Alk:P} = 2 \cdot f_{Ca} \cdot r_{\Sigma C:P} - r_{N:P} \qquad (10)$$

$r_{Alk:P}$ has a value of 50 when we substitute the constants cited above. We assume $r_{N:P}$ has a value of 15. Table 1 summarizes the stoichiometric relationships between oxygen, alkalinity, carbon, and phosphorus used in this study.

Equation (9) is a very significant result. The high-latitude $\Sigma CO_2$ and alkalinity together determine the $P_{CO2}$ of high-latitude surface water. Because both of these quantities are anchored to deep-sea values, this result says, in effect, that the $CO_2$ content of the atmosphere is anchored to the deep sea through the nutrient content of high-latitude surface water. The relationship between the $P_{CO2}$ of low-latitude surface water and the deep sea is more indirect, depending on transport of $CO_2$ through the atmosphere from high-latitudes.

One can calculate a value of $r_{\Sigma C:P}$ using equation (9) in order to test the applicability of our C:P ratio of organic matter. We have computed average $\Sigma CO_2$ and phosphate values for the deep sea of 2257 µmol/kg and 2.15 µmol/kg, respectively (Table 2). Averaging all the surface Geochemical Ocean Sections Study (GEOSECS) data for the Antarctic south of 50° while normalizing to the deep-sea salinity, we find $\Sigma CO2_h$ = 2178 ± 10 µmol/kg and $PO4_h$ = 1.51 µmol/kg. We assume that Antarctic surface water contains ~25 µmol/kg of fossil fuel $CO_2$ and subtract this from $\Sigma CO2_h$. (A 50-ppm excess of fossil fuel $CO_2$ in the atmosphere becomes a $\Sigma CO_2$ excess of 25 µmol/kg in surface water using a buffer factor of 14.) Substituting into equation (9), we find that $r_{\Sigma C:P}$ = 150-175. This result roughly confirms the number predicted, 162.5, from Takahashi et al.'s [1984] observations of $O_2$:P. The standard Redfield ratio of 106 units organic carbon to one unit phosphorus represents a total carbon to phosphorus ratio of only 132.5 using equation (7).

Before solving the full set of equations for $\Sigma CO_2$, alkalinity, and the carbon isotopes for all four boxes, we write a set of conservation equations which express the deep-box tracer concentrations as functions of the total amount of tracer in the ocean and atmosphere. As an example we write out the conservation equation for $\Sigma CO_2$ below:

$$\Sigma CO2_d = (\Sigma CO2_T - \Sigma CO2_h \cdot V_h - \Sigma CO2_\ell \cdot V_\ell - P_{CO2} \cdot V_{atm})/V_d$$
$$(11)$$

$\Sigma CO2_T$ is the total amount of carbon in the ocean and atmosphere in units of moles C. The terms $V_h$, $V_\ell$, $V_{atm}$, and $V_d$ represent the volumes of the subscripted boxes. The volume of the atmosphere is expressed as moles of gas molecules such that the product of the $P_{CO2}$ (a mixing ratio) and the

mole volume of the atmosphere has units of moles $CO_2$. Conserving total tracer content ensures that the present-day partitioning of tracers will not influence model solutions for other possible scenarios.

We have averaged all the GEOSECS and Transient Tracers in the Ocean (TTO) data within 5° latitude belts at standard National Oceanographic Data Center (NODC) levels and added the belts together to produce total tracer contents for the ocean. The results appear in Table 2. The validity of using conservation expressions like equation (11) in our model is compromised by the fact that the ocean and atmosphere are not presently at steady state with respect to $\Sigma CO_2$, $^{13}C$, and $^{14}C$; modern additions of fossil fuel $CO_2$ and bomb $^{14}C$ have perturbed the presumed steady state which existed prior to industrialization. We account for this by simply subtracting the added quantities from the totals. The addition of unknown amounts of terrestrial biomass carbon to the atmosphere makes this correction somewhat imprecise; the maximum errors, however, are not large.

Before proceeding we would like to introduce an additional observation into the model regarding the degree of oxygen saturation in newly formed deep-water masses. Weiss et al. [1979] report that the oxygen content of remnant Weddell Winter Water and Weddell Shelf Water during the summer is about 40 µmol/kg undersaturated. Gordon et al. [1984] report that the oxygen content of Antarctic surface water under sea ice at the end of the austral winter is 45 µmol/kg undersaturated (86% of saturation). Gordon et al. argue that the 14% undersaturation has not been appreciably altered by biological processes over the winter period. These observations are characteristic of oceanographic observations in deep-water formation regions: nowhere are subsurface water masses in these regions less than 10% undersaturated. Because the process of oxygen equilibration takes a finite amount of time (one month for a 100-m mixed layer) it is not hard to imagine that a convectively unstable water mass might not aquire its full measure of oxygen.

The above observations are suggestive that newly formed deep water is not fully saturated with oxygen. If newly formed deep water carries an initial oxygen deficit into the deep sea, the oxygen deficit of average deep water due to oxidation of organic matter is reduced accordingly. We will make an assumption that the model's $AOU_h$ is 10% undersaturated and will consider this to be a much better choice than $AOU_h$ = 0. A 10% undersaturation in newly formed deep water reduces the deep-sea oxygen deficit caused by the oxidation of organic matter by 30 µmol/kg, nearly a 20% reduction. A 10% undersaturation in $AOU_h$ also increases the predicted $PO4_h$ or preformed phosphate to 1.41 µmol/kg (see equation (12) below). Referring back to Figure 2, we find that the predicted $PO4_h$ is inside the field of Antarctic data. The model equations in the appendix include an $AOU_h$ term.

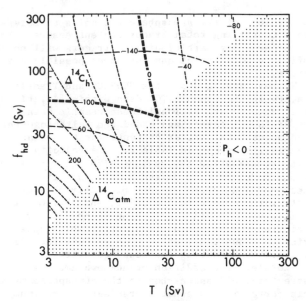

T (Sv)

Fig. 3. Contour plot of model results for the $^{14}C$ contents of the atmosphere, $\Delta^{14}C_{atm}$, and high-latitude surface water, $\Delta^{14}C_h$, in the parameter space of high-latitude convection ($f_{hd}$) versus thermohaline overturning (T). Contours for atmospheric $\Delta^{14}C$ are the diagonally trending dashed lines labeled as -80, -40, etc. starting from the upper right. Contours for high-latitude surface water are the horizontally trending dashed lines labeled -140, -100, and -60. The heavy contour lines represent contours for $\Delta^{14}C_{atm} = 0$ and $\Delta^{14}C_h = -100$ per mil. Horizontal and vertical axes are log scales. The contours are constructed from 440 model runs evenly spaced over the parameter range. The model run in this figure represents the simple case in which $f_{\ell d}$ and $f_{\ell h}$ are set to zero. $AOU_d = 154$ µmol/kg, $AOU_h = 30$ µmol/kg, and $r_{\Sigma C:P} = 162.5$. The stipled region represents the region of the parameter space in which $P_h$ is less than zero (see text).

A Model Solution For The Present Ocean

Our point of departure for describing the present ocean is to use the observed $AOU_d$ and the assumed $AOU_h$ to determine the particle fluxes into the deep sea. We can then use the partitioning of $^{14}C$ between the various boxes to limit the flux terms, $f_{hd}$ and T. Later in this section we will investigate the role of $f_{\ell d}$. As we see in equation (12),

$$PO4_h = \frac{(T + f_{hd} + f_{\ell d})}{(T + f_{hd})} \cdot (PO4_d - \frac{AOU_d}{r_{O2:P}}) + \frac{AOU_h}{r_{O2:P}} \quad (12)$$

knowing $AOU_d$ and $AOU_h$ allows one to predict the phosphate content of high-latitude surface water

as a function of T, $f_{hd}$, and $f_{\ell d}$. Knowing $PO4_h$ in turn yields $P_\ell$ and $P_h$, the particulate phosphate fluxes (equations (1) and (2)). Equation (12) is simplified from equation (A4) in the appendix. We have left out the conservation and $PO4_\ell$ terms, which have little impact in the present illustration.

Two facts are known about the preindustrial and prebomb ocean and atmosphere with regard to $^{14}C$. We know that the $\Delta^{14}C$ content of the atmosphere was zero per mil. We also know that the average $^{14}C$ content of Antarctic surface water was $-100 \pm 20$ per mil from oceanic $^{14}C$ measurements made prior to the bomb tests [Broecker and Peng, 1982, p. 415]. (We also know the $^{14}C$ content of low-latitude surface water from coral measurements; however, low-latitude $\Delta^{14}C$ mainly tracks the atmosphere and proves redundant in the present applications.) Throughout the remainder of the paper we will plot model results as contours in a parameter space defined by the major flux terms, $f_{hd}$ and T. High-latitude convection ($f_{hd}$) and thermohaline overturning (T) affect the partitioning of $^{14}C$ between the model boxes in different ways such that contours of atmospheric $\Delta^{14}C$ and high-latitude surface water $\Delta^{14}C$ intersect and define a solution for today's ocean.

Model results for the preindustrial ocean and atmosphere appear in Figures 3, 4a-4d, and 5a-5c. Figures 3 and 4 present the simple case in which $f_{\ell d}$ and $f_{\ell h}$ are set equal to zero. Figure 5 examines the effect of varying $f_{\ell d}$ with $f_{\ell h}$ held constant. Each figure represents over 400 model runs in which $f_{hd}$ and T are varied over two orders of magnitude. The scales for $f_{hd}$ and T are identical; each is a log scale in transport extending from 3 to 300 Sv (1 Sv = $10^6$ m/s).

In Figure 3, isolines for the preindustrial $^{14}C$ contents of the atmosphere and high-latitude surface water are drawn in the $f_{hd}$ versus T parameter space. $\Delta^{14}C_{atm}$ values in the upper right-hand corner of the diagram are close to -80 per mil, whereas those in the lower left are greater than +400 per mil. We see that values of $\Delta^{14}C_{atm}$ decrease away from the lower left-hand corner as both $f_{hd}$ and T increase. The flux terms $f_{hd}$ and T act together to ventilate the deep sea. When both are small (lower left-hand part of Figure 3), $^{14}C$ produced in the atmosphere tends to remain there. When both are large (upper right), the great reservoir of low $^{14}C$ in the deep sea is able to dilute the atmosphere's $^{14}C$ production and bring the atmospheric $\Delta^{14}C$ down.

We show three contours of $\Delta^{14}C$ in high-latitude surface water in Figure 3: -60, -100, and -140 per mil. As one might expect, the $\Delta^{14}C_h$ is especially sensitive to exchange between high-latitude surface water and deep water. When $f_{hd}$ is large, the deep ocean is able to pull $\Delta^{14}C_h$ further from atmospheric values.

The point where the $\Delta^{14}C_h = -100$ isoline meets the $\Delta^{14}C_{atm} = 0$ isoline defines today's ocean in the $f_{hd}$ versus T parameter space. In Figure 3 the intersection of the $^{14}C$ lines occurs at the boun-

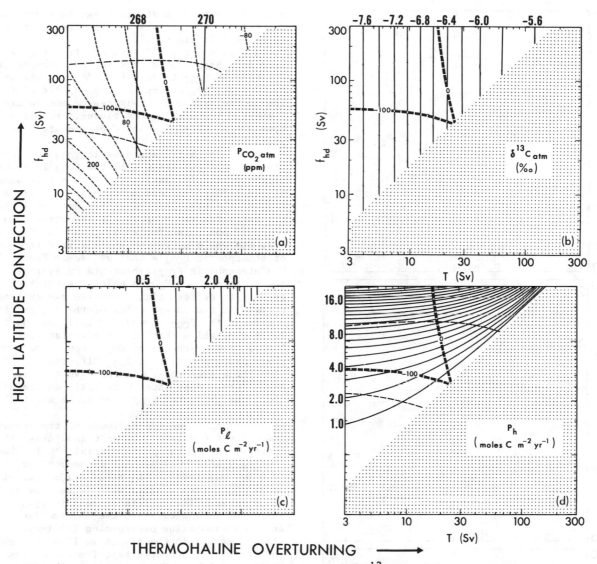

Fig. 4. Contour plots for (a) atmospheric $P_{CO2}$, (b) $\delta^{13}C_{atm}$, (c) $P_\ell$, and (d) $P_h$ overlying the same parameter space and $^{14}C$ contours shown in Figure 3. Phosphate particle fluxes, $P_\ell$ and $P_h$, have been multiplied by $r_{\Sigma C:P}$ (162.5) in Figures 4c and 4d in order to present results in units of moles C per square meter per year (organic plus $CaCO_3$).

dary of the shaded region where $f_{hd}$ is 43 Sv and T is 24 Sv. Given the uncertainty in our knowledge of the preindustrial $\Delta^{14}C$ value for average high-latitude surface water, $f_{hd}$ for today's ocean could be as low as 30 Sv or as high as 75 Sv.

A little more than half of the diagram is shaded, indicating that the high-latitude phosphate particle flux, $P_h$, in that part of the parameter space is less than zero. It is not difficult to understand why $P_h$ becomes negative over much of the parameter space given the $AOU_d$ restriction. $P_\ell$ and $P_h$, when converted to organic carbon fluxes, must sum to produce the deep-sea oxygen deficit (equation (3)). Because the

low-latitude area is much larger than the high-latitude area, $P_\ell$ represents a much bigger drain on the deep-sea oxygen levels than $P_h$. $P_\ell$ increases as T increases (equation (1)), and even though T brings oxygen into the deep sea, the particle flux from low-latitude surface water negates it. High-latitude exchange with deep water adds much more oxygen to the deep sea than the high-latitude particle flux can remove. Therefore, when T is large relative to $f_{hd}$, the particle flux from low-latitude surface water can, by itself, account for more than the observed oxygen deficiency in the deep sea, and $P_h$ is forced to be negative. Note that when $f_{\ell d} = 0$ in equation (12)

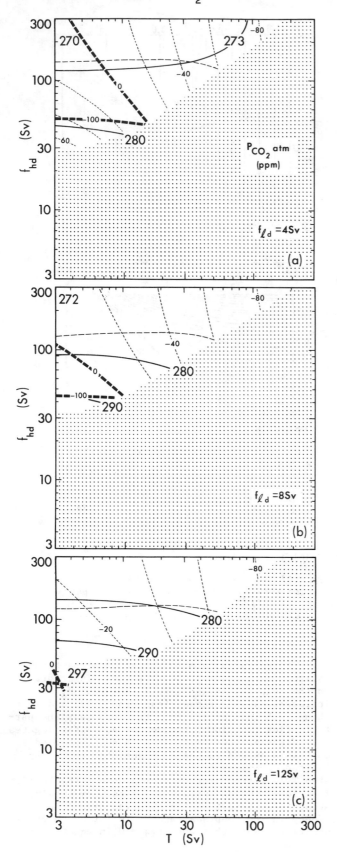

and AOU$_d$ is held constant, PO4$_h$ is independent of f$_{hd}$ and T. PO4$_h$ equals 1.41 μmol/kg at every point on this diagram.

In Figure 4 the $^{14}$C isolines from Figure 3 are used as a base plot over which contours of the atmospheric P$_{CO2}$ (Figure 4a), Δ$^{14}$C$_{atm}$ (Figure 4b), P$_\ell$ (Figure 4c), and P$_h$ (Figure 4d) are plotted. The particle fluxes in Figures 4c and 4d are here converted to total carbon fluxes (organic + carbonate) in units of moles C per square meter per year.

In Figure 4a the atmospheric P$_{CO2}$ isolines plot as vertical lines with the P$_{CO2}$ increasing slightly with increasing T. At the intersection of the Δ$^{14}$C$_{atm}$ = 0 and Δ$^{14}$C$_h$ = -100 isolines, the P$_{CO2}$ of the atmosphere is 268.8 ppm, well within the estimated range of the preindustrial P$_{CO2}$, 270 ± 10 ppm [Stauffer et al., 1984]. Fixing AOU$_d$ over the whole parameter space limits the range of atmospheric P$_{CO2}$'s that we see in Figure 4a. The atmospheric P$_{CO2}$ represents an average of the two surface ocean P$_{CO2}$'s which are weighted with respect to differential areas and gas exchange rates. At the intersection of the $^{14}$C isolines in Figure 4a the P$_{CO2}$ of the low-latitude box is 269.6 ppm, while that of the high-latitude box is 266.5. Free exchange of CO$_2$ through the atmosphere dampens potential P$_{CO2}$ differences larger than those seen here. If the oxygen content of newly formed deep water were fully saturated (AOU$_h$ = 0), the predicted P$_{CO2}$'s would be about 25 ppm lower.

In Figure 4b we see contours of the atmospheric $^{13}$C content overlying the Δ$^{14}$C isolines. At the intersection of the preindustrial $^{14}$C isolines, δ$^{13}$C$_{atm}$ equals -6.3 per mil as opposed to about -7.6 per mil today. The atmospheric $^{13}$C/$^{12}$C composition becomes heavier with increasing T, but shows no sensitivity with respect to f$_{hd}$. This situation is analogous to the one in Figure 4a. As the thermohaline overturning (T) increases, the flux of carbon in the advective flow through both surface boxes increases relative to the gas exchange fluxes. The effect is more limiting in the high-latitude box because its volume is so much smaller. Cooler temperatures in the high-latitude box favor an equilibrium in which the gas phase is isotopically lighter; warmer temperatures in the low-latitude box favor an isotopically heavier gas phase. Therefore, the atmosphere's

Fig. 5. Contours of atmospheric P$_{CO2}$ (solid lines), Δ$^{14}$C$_{atm}$ (short-dashed lines), and Δ$^{14}$C$_h$ (longer-dashed lines) in the the same parameter space shown in Figures 3 and 4, but where f$_{\ell d}$ varies from 4 to 12 Sv. Here f$_{\ell h}$ is held fixed at 10 Sv. The Δ$^{14}$C contours for 0 per mil in the atmosphere and -100 per mil in high-latitude surface water are accentuated as before. In Figure 5c (f$_{\ell d}$ = 12 Sv) the intersection of the $^{14}$C isolines occurs in the stipled region along the left-hand side of the figure. AOU$_d$ = 154 μmol/kg, AOU$_h$ = 30 μmol/kg, and r$_{\Sigma C:P}$ = 162.5.

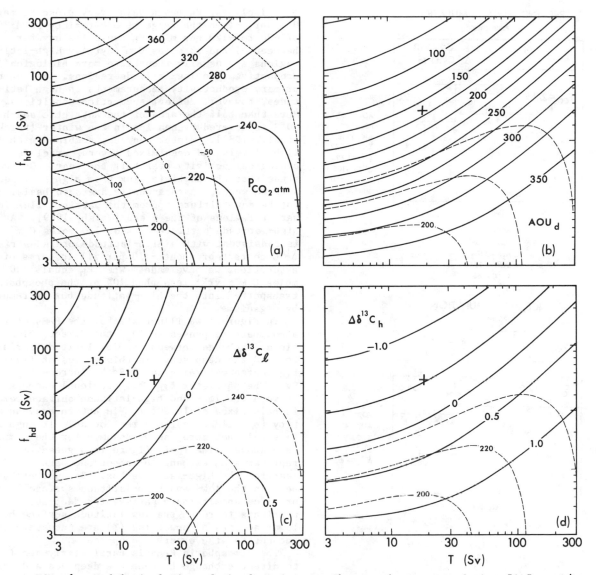

Fig. 6. Model simulation of the last ice age where we have assumed that 51.5 μmol/kg of isotopically light carbon (−26 per mil) has invaded the deep ocean. This input of carbon is sufficient to lower the $\delta^{13}C$ of the deep sea by about 0.5 per mil. We assume that an equal number of moles of $CaCO_3$ dissolves from the seafloor to balance the pH. In this figure the AOU of the deep sea is allowed to vary while the particle flux from high-latitude surface water is fixed at 0.5 moles C m$^{-2}$ yr$^{-1}$. The plus in each figure part marks the position in $f_{hd}$ versus T parameter space which yields the optimal model solution for the present ocean discussed in the text. The contours shown in Figure 6a are atmospheric $P_{CO2}$ (solid lines) and $\Delta^{14}C_{atm}$ (dashed lines). Contours of $AOU_d$ are plotted in Figure 6b. In Figures 6c and 6d are plotted the difference in $\delta^{13}C$ between the ice age and present (preindustrial) for the low-latitude and high-latitude surface boxes, respectively. In Figures 6b through 6d contours of atmospheric $P_{CO2}$ at 200, 220, and 240 ppm (dashed lines) are carried over from Figure 6a.

$\delta^{13}C$ is driven toward heavier values when T is large. Because the time required to equilibrate the $^{13}C$ isotopic composition between surface ocean and atmosphere is 10 times longer than the time required to equilibrate $CO_2$ chemically [Broecker and Peng, 1974], the effect of surface layer

flushing with respect to the parameter T is much larger for $^{13}C$ in Figure 4b than it is for $P_{CO2}$ in Figure 4a.

Figures 4c and 4d show the particulate carbon fluxes out of the surface boxes as functions of $f_{hd}$ and T. As we expect from equation (1), $P_\ell$,

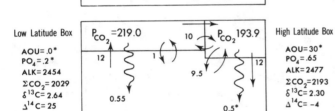

Fig. 7. Steady state concentrations of model variables in the ocean and atmospheric boxes for the preindustrial and ice age simulations. Units are micromoles per kilogram for AOU, $PO_4$, and $\Sigma CO_2$; microequivalents per kilogram for alkalinity, parts per million for $P_{CO2}$; and parts per thousand (per mil) for $\delta^{13}C$ and $\Delta^{14}C$. Model transports between boxes are given in units of sverdrups ($10^6$ $m^3/s$). Particle fluxes are given next to the wiggly arrows in units of moles C per square meter per year. Asterisks indicate values that are initial assumptions.

expressed here as a carbon flux, increases proportionately with T. At the intersection of the preindustrial $\Delta^{14}C$ isolines, $P_l$ has a value of 0.82 moles C m$^{-2}$ yr$^{-1}$. $P_h$ in Figure 4d is sensitive primarily to $f_{hd}$. At the intersection of the bold

$\Delta^{14}C$ isolines, $P_h$ has a value very close to zero. However, given the range of uncertainty in $\Delta^{14}C_h$ of $-80$ to $-120$ per mil, $P_h$ could lie anywhere between $-1$ and $+3$ moles C m$^{-2}$ yr$^{-1}$. High-latitude regions are usually thought to more biologically productive than low-latitude regions. All of the primary productivity measurements in high latitudes, however, reflect summertime conditions. More than half the area of the Antarctic south of 50°S is covered by sea ice in the winter [Burckle et al., 1982]. Winter sea ice, combined with low light levels and deep mixed layers, precludes biological activity over half the year. One could argue that the particle flux from high-latitude surface water, on an annually averaged basis, may not be much different from the particle flux in warmer regions of the ocean [Walsh, 1969]. A high-latitude particle flux of 1.0 moles C m$^{-2}$ yr$^{-1}$ is consistent with a high-latitude preindustrial $\Delta^{14}C$ content of about $-110$ per mil in terms of the assumptions we have made. When $P_h$ equals 1.0 moles C m$^{-2}$ yr$^{-1}$ less than 10% of the phosphate transported into the high-latitude box is removed by organisms.

In Figure 5 we illustrate how the $P_{CO2}$ of the atmosphere responses to $f_{ld}$, the direct, bidirectional exchange between the low-latitude surface box and the deep sea. In this series of figures, $f_{ld}$ increases from 4 to 12 Sv by increments of 4 Sv. The parameter $f_{lh}$, bidirectional exchange between the low- and high-latitude surface boxes, is held fixed at 10 Sv; $f_{lh}$ is not an easy quantity to estimate for the real ocean, although it is surely not zero, as we assumed for the simple case above. The 10-Sv figure is a guess; it represents an exchange of about one part in 40 per year for the high-latitude box. Because exchange between the low- and high-latitude surface boxes provides an alternate pathway for deep-sea nutrients to reach the low-latitude surface box, lower amounts of upwelling (T) are necessary to account for the observed deep-sea AOU.

The atmospheric $P_{CO2}$ is particularly sensitive to direct exchange between the deep sea and low-latitude surface water. As $f_{ld}$ increases from 0 in Figure 4a through 12 Sv in Figures 5a-5c, the atmospheric $P_{CO2}$ increases from 269 to 297 ppm at the intersection of the $\Delta^{14}C$ isolines. The $P_{CO2}$ isolines no longer respond only to T in the $f_{hd}$ versus T parameter space, but instead swing over to respond primarily to $f_{hd}$, particularly in the lower range of T.

The $\Delta^{14}C$ isolines also respond to $f_{ld}$. Because direct mixing between low-latitude surface water and the deep sea increases contact between the deep ocean and atmosphere, an increase in $f_{ld}$ must be offset by a drop in $f_{hd}$ or T. Hence, the $\Delta^{14}C_{atm}$ isolines tend to shift downward and to the left in the $f_{hd}$ versus T parameter space. The isolines of $\Delta^{14}C_h$, on the other hand, are not affected by $f_{ld}$. Therefore, only T decreases in the model solution as direct exchange between the deep sea and low-latitude surface water increases. When $f_{ld}$ is equal to 12 Sv, only 3 Sv of T are required to meet the $^{14}C$ constraints. The particle flux from low-latitude surface water increases with $f_{\ell d}$ as more phosphate is delivered to low-latitude surface water. Less high-latitude particle flux is needed to bring $AOU_d$ up to 154 $\mu$mol/kg, and we see that the shaded area of the diagrams ($P_h$ less than zero) expands, especially at low values of T.

The rise in atmospheric $P_{CO2}$ in response to direct exchange between the deep sea and low-latitude surface water reveals an important point about how the model works. We see from equation (12) that when T is small, $PO4_h$ increases as $(f_{hd}+f_{\ell d})/f_{hd}$ increases. The increase in $PO4_h$ is mirrored by an increase in high-latitude $\Sigma CO_2$ (recall the simplified equation (9)). Less nutrient-depleted water is flushed through the high-latitude box as $f_{\ell d}$ substitutes for T in the $^{14}C$ balance. Therefore, we might say that the $CO_2$ content of the atmosphere rises in response to $f_{\ell d}$ because high-latitude surface water becomes more like deep water. We will see in the model simulation for the ice age ocean below that the atmospheric $P_{CO2}$ was lower during the ice age because high-latitude surface water was less like deep water then is observed at present.

We conclude from Figure 5 that only a minimal amount ($<4$ Sv) of direct exchange between the deep sea and low-latitude surface water is consistent with the preindustrial atmospheric $P_{CO2}$ of 270 $\pm$ 10 ppm. This conclusion is, of course, dependent on the assumption which we made about the oxygen content of newly formed deep water. The size of the $AOU_h$ effect on atmospheric $P_{CO2}$ is about the same as the effect of 12 Sv of direct exchange between the deep sea and low-latitude surface water. We will adopt as an optimal solution for the present ocean a set of parameters in which $f_{\ell d}$ equals a token 1 Sv and $f_{\ell h}$ = 10 Sv. (A minimal amount of direct exchange between low-latitude surface water and the deep sea is important when $f_{hd}$ and T are both small.) For our optimal case, $f_{hd}$ is between 45 and 60 Sv and T is about 19 Sv. $P_\ell$ equals 0.90 moles C m$^{-2}$ yr$^{-1}$. The preindustrial $\delta^{13}C$ of the atmosphere is -6.18 per mil, while the $\delta^{13}C$ contents of low- and high-latitude surface water are -2.64 and -1.72 per mil, respectively. We will consider $P_h$ to be limited to a range 0 to 1.0 moles C m$^{-2}$ yr$^{-1}$, which is consistent with the 45-60 Sv range in $f_{hd}$ noted above and with values of $\Delta^{14}C_h$ between -100 and -110 per mil.

## The Ice Age Ocean And Atmosphere

We have adapted the model slightly to perform a simulation of ice age conditions. Given Shackleton et al.'s [1983a, b] observation that the $\delta^{13}C$ of the deep sea became lighter by about 0.5 per mil during the ice age, we calculate that approximately 50 $\mu$mol/kg of isotopically light (-26 per mil) carbon must have been added to the deep sea. Because the depth of the lysocline seems to have remained the same between glacial

and interglacial periods [Broecker, 1982], we assume that an equal amount of $CaCO_3$ dissolved from the seafloor to hold the pH of the deep sea in balance. These changes increase the alkalinity and $\Sigma CO_2$ of the deep sea by about 100 $\mu$eq/kg and $\mu$mol/kg, respectively. Removal of water from the ocean to form continental ice shrinks the ocean volume by 3% and increases the ocean's salinity, phosphate, alkalinity, and $\Sigma CO_2$ contents proportionally. We assume that the surface temperature of the low-latitude box decreases to 20.0°C and that of the high-latitude box to 2.0°. Added together, these changes increase the $P_{CO2}$ of the model by 21 ppm at the position in the $f_{hd}$ versus T parameter space which defines today's ocean.

Because we do not know the AOU of the deep sea during the ice age, we allow $AOU_d$ to become a variable quantity. In order to maintain the same degree of determinacy we fix the particulate carbon flux from high-latitude surface water at 0.5 moles C m$^{-2}$ yr$^{-1}$. All of the other quantities, including $AOU_h$, the C to P ratio in organic matter, $f_{\ell d}$, and $f_{\ell h}$, remain at the levels for the present ocean as given above. In Figure 6 we present the model results for the ice age simulation. We retain the same $f_{hd}$ versus T parameter space shown in previous results. The pluses mark the position where the $\Delta^{14}C$ = 0 and the $\Delta^{14}C_h$ = -100 isolines meet in the present-day solution.

In figure 6a we plot contours of atmospheric $P_{CO2}$ and $\Delta^{14}C_{atm}$. Atmospheric $P_{CO2}$ values increase from less than 190 ppm at the bottom of the figure to 400 ppm in the upper left-hand corner. The appearance of Figure 6a is radically different from previous $P_{CO2}$ plots because there are no restrictions on $AOU_d$. Figure 6b shows contours of the deep-box oxygen content overlying a few of the atmospheric $P_{CO2}$ contours from Figure 6a. Model $AOU_d$ values decrease from nearly 375 $\mu$mol/kg in the lower right-hand corner of the figure to less than 50 $\mu$mol/kg in the upper left. $AOU_d$ increases with increasing T because particle fluxes from low-latitude surface water increase with T. $AOU_d$ decreases with increasing $f_{hd}$ because the ventilation of the deep sea through high latitudes introduces new oxygen without increasing the flux of organic material to the deep sea. Over most of the parameter domain the atmospheric $P_{CO2}$ contours parallel those of $AOU_d$. In the lower right-hand section of the figures, where the $P_{CO2}$ and $AOU_d$ deviate from one another, the more limited gas exchange capability of $CO_2$ causes the $P_{CO2}$ to rise with increasing T.

Because we know that the $P_{CO2}$ of the atmosphere during the ice age lies between 200 and 220 ppm [Neftel et al., 1982], we expect to find a model solution for the ice age between the 200- and 220-ppm $P_{CO2}$ isolines in Figure 6a. To reach this region of parameter space, a substantial reduction in $f_{hd}$ is required. The deep-sea oxygen deficit in this region of parameter space lies between 280 and 325 $\mu$mol/kg. Because the oxygen content of the deep sea at full saturation would be about 325 $\mu$mol/kg at today's average deep-sea temperature, we see that the model's ice age prediction lies very close to a condition of full anoxia in the deep sea. Although it is not shown here, the phosphate content of high-latitude surface water parallels the $AOU_d$ isolines. To reach the 200-220 ppm $P_{CO_2}$ region in Figure 6a requires a reduction in $PO_{4_h}$ from 1.41 μmol/kg to about 0.6 to 0.7 μmol/kg.

From observations by Shackleton et al. [1983b] and others cited by Broecker [1982], we know that the $\delta^{13}$ of planktonic forams did not change appreciably between the ice age and recent times. This is equivalent to saying that the $\delta^{13}$ content of the low-latitude surface box did not change. In Figure 6c we have plotted the differences between the $\delta^{13}C$ content of low-latitude surface water as obtained from the ice age simulation, and the preindustrial $\delta^{13}C_\ell$ determined above (+2.64 per mil). We see that the 0 isoline, which shows no difference between the ice age and preindustrial, passes through the 200-220 ppm $P_{CO_2}$ field directly below the modern solution and at somewhat smaller values of T. If we choose as a representative ice age solution the intersection of the $P_{CO_2}$ = 210 ppm and $\Delta\delta^{13}C_\ell$ = 0 isolines in Figure 6c, the model predicts that $f_{hd}$ was only 10 Sv during the ice age, one fifth of its present value, and T was about 12 Sv, a little more than half its present value. The particle flux from low-latitude surface water was lower along with T at 0.55 moles C m$^{-2}$ yr$^{-1}$ down from 0.9 moles C m$^{-2}$ yr$^{-1}$. In Figure 6a we predict that the $\Delta^{14}C$ content of the ice age atmosphere was about +100 per mil. More significant, perhaps, is the prediction that the $\Delta^{14}C$ gradient between low-latitude surface water and the deep sea was about 200 per mil compared to about 110 per mil for the modern ocean.

In Figure 6d we have plotted the difference in $\delta^{13}C$ between ice age and preindustrial high-latitude surface water ($\delta^{13}C_h$ preindustrial = +1.72 per mil). The model predicts that the $\delta^{13}C$ of high-latitude surface water was more than 0.5 per mil heavier during the ice age.

Figure 7 summarizes our results for the preindustrial and ice age scenarios. Steady state concentrations of oxygen, phosphate, alkalinity, and $\Sigma CO_2$ are given for the three ocean boxes. Isotopic ratios for $^{13}C$ and $^{14}C$ are given for the ocean boxes and atmosphere. Water fluxes in Sverdrups are indicated alongside the transport arrows, and particle fluxes in moles C per square meter per year are given next to the wiggly arrows. Asterisks indicate that concentrations or fluxes are fixed values.

Sarmiento and Toggweiler [1984] illustrate an ice age scenario in which $f_{hd}$ is held constant and the high-latitude particle flux is allowed to vary. This scenario predicts a 70-ppm decrease in atmospheric $P_{CO_2}$ with an increase in the high-latitude particle flux to 5 moles C m$^2$/yr$^{-1}$. The effect of a change in high-latitude particle flux on the model is virtually identical to a change in $f_{hd}$ except that the ventilation time for the deep sea would remain at modern values in the former

case. Sarmiento and Toggweiler examine the possibility that atmospheric $CO_2$ changes might be driven by changes in high-latitude insolation acting to increase or decrease the high-latitude particle flux. According to this scenario, high-latitude insolation and particle fluxes may have been enhanced during periods of greater tilt in the earth's axis, e.g., 10,000-15,000 years ago. This possibility has been analyzed in more detail by Knox and McElroy [1984].

We find it hard to imagine, however, that a 10 or 15% change in insolation could produce a 500% change in the organic particle flux from high-latitude surface water. It seems much more physically reasonable that lower ice age $P_{CO_2}$'s were caused by a change in Antarctic circulation or convection which produced a large reduction in $f_{hd}$, as will be discussed below. This conclusion is shared by Siegenthaler and Wenk [1984]. The important feature common to both of these lower $P_{CO_2}$ scenarios is a large reduction in high-latitude nutrient concentrations. We have also adapted the model to Broecker's [1982] shelf erosion hypothesis by increasing the model's deep-ocean phosphate content and have found an interesting result: the $P_{CO_2}$ of the atmosphere does not go down with an increase in deep ocean phosphate unless the particle flux from high-latitude surface water is also increased to pull down the surface nutrient content.

An important consequence of the model's ice age scenario is its illustration of how the deep ocean's oxygen content varies with respect to $f_{hd}$ and T (Figure 6b). It is commonly thought that a sluggish ocean circulation leads to deep-sea anoxia. If "sluggish" refers to smaller amounts of both $f_{hd}$ and T, this conception is completely wrong; when $f_{hd}$ and T vary together, the oxygen content of the deep sea remains the same. The oxygen content of the deep sea declines when either (1) the thermohaline overturning becomes more vigorous and $f_{hd}$ remains the same or (2) exchange between the deep sea and surface in high latitudes decreases and T remains the same. Changes in relative quantities of $f_{hd}$ and T alter the deep ocean's oxygen content.

### Discussion: What Is $f_{hd}$?

The combined quantities of $f_{hd}$ and T which are required to account for today's partitioning of $^{14}C$ amount to some 70 Sv. Literature estimates [Warren, 1981] of the rates of bottom water formation are substantially less. Most agree that the rate of formation of North Atlantic Deep Water (NADW) is about 10 Sv. Amounts of Antarctic Bottom Water (AABW) sinking from Antarctic continental shelves add up to less than 10 Sv.

One might characterize these types of deep and bottom water as extreme end members on potential temperature versus salinity (T/S) diagrams. They are relatively easy to locate in ocean surveys and command the most attention as a result. However, if these extreme types of deep and bottom water represent the principal means by which the deep ocean is ventilated, the $^{14}C/^{12}C$ ratio in the modern atmosphere would be substantially higher than we observe. Broecker [1979] has estimated that the time scale over which the deep western Atlantic is ventilated is very short (not much more than 100 years) because there is simply no evidence that much $^{14}C$ decay has occurred. Ten sverdrups of NADW formation will not do the trick.

As a simplification we might think of the formation of extreme T/S water masses as the thermohaline overturning or 'T' process in this model. As described by Warren [1981]

The sinking that is known to take place seems not to be merely a concomitant of the overall meridional density gradient, because most of it occurs from sheltered, semi-enclosed regions (Antarctic continental shelf, Norwegian Sea, low-latitude marginal seas) where near surface water is drawn in, contained long enough to become exceptionally dense, and then is forced back to the open ocean, sinking to depth because of its high density.

High-density bottom water displaces existing deep water upward, leading to the upwelling component of the T circulation. Because the formation of dense bottom water is generally a high-latitude process, the surface water which is "drawn in" to semienclosed basins must come from lower latitudes. This is especially well illustrated in the North Atlantic, where warm, salty surface water is found at quite high latitudes and gives up great quantities of heat when it cools. We see this process occurring in the GEOSECS surface water $P_{CO_2}$'s [Broecker et al., 1979] which drop dramatically below atmospheric values in the North Atlantic north of 40°. North Atlantic surface water flows poleward and cools at such a rate that the exchange of $CO_2$ with the atmosphere cannot re-equilibrate the upper layers fast enough. We also see evidence for the thermohaline nature of NADW formation in the fact that the North Atlantic is relatively impoverished in nutrients; it maintains a low nutrient inventory because it imports nutrient-poor surface water and exports deep water.

The process which we call $f_{hd}$ is another matter. We have been alluding to $f_{hd}$ as an Antarctic process because the high-latitude Southern Ocean covers such an immense area, and because the high nutrient content of Antarctic surface water seems to reflect frequent contact with deep water. From a global ocean heat blance Gordon [1975] has estimated that 38 Sv of Antarctic Bottom Water must be forming to account for the ocean's heat loss to the atmosphere. In an earlier paper, Gordon [1971] estimated that an upwelling of 60 Sv of upper Circumpolar Deep Water with a salinity of 34.6 per mil is necessary to balance the input of fresh water to the Antarctic surface. This volume of upwelling is consistent with that theoretically

generated by Ekman divergence south of the polar front. Gordon [1981] contrasts conditions in the Arctic and Southern oceans. In the Arctic, large amounts of fresh water input produce a strong, sharp pycnocline. Very low nutrient concentrations are found in the low-salinity upper layer. The Antarctic pycnocline is characterized by low stability. A relatively small fresh water input nearly balances an upward thermal bouyancy flux from warm Circumpolar Deep Water.

Although the mass balances require large-scale overturning in the Antarctic of the right magnitude to comply with our $f_{hd}$ requirement, the actual overturning is not easy to observe. Gordon [1982] reports on observations made during 1977 and 1978 in the Weddell Sea in which an extensive patch of anomalously cold deep water was found between 200 and 2700 m. The patch of cold water was about a half degree colder than water in the same area during the early 1970's. Gordon attributes the cold patch to convective mixing between deep water and wintertime surface water during the Weddell Polynya which persisted from 1974-1976 in the same area. The rate of overturning required to produce this feature is estimated to be as high as 15 Sv during the Polynya years. Gordon [1978] has also observed a narrow (14 km radius) cold core eddy east of the Weddell Sea which extended from 200 to 4000 m. This eddy is also thought to have been a remnant feature of wintertime convection.

If 50 Sv of deep water does in fact upwell to the surface, where does it go? Some small fraction of it sinks as bottom water from the Weddell Shelf. Some part flows equatorward and sinks at the Antarctic Convergence to form intermediate water. The remainder must cool and sink back into the mass of Circumpolar Deep Water. Apart from the few verifiable episodes of open ocean convection and perhaps 10-15 Sv of intermediate water formation, it is not easy to find 50 Sv of water sinking around the Antarctic today. We therefore leave the search for $f_{hd}$ partially unfulfilled.

Evidence for Changes in High-Latitude Convection in Antarctic Sediments

From the distribution of opal in Antarctic sediments Cooke and Hays [1982] have claimed that sea ice extended out from the continent on a year-round basis during the ice age. Winter sea ice coverage 18,000 years ago was probably double the present winter sea ice cover. Cooke and Hays claim that summer sea ice covered slightly more area than present winter ice. Although this latter conclusion has been refuted by more recent work [Burckle et al., 1982], it is clear that the zone of diatomaceous ooze which characterizes modern Antarctic sediments in the zone of seasonally varying sea ice was displaced hundreds of kilometers to the north 18,000 years ago. In two of the cores studied by Cooke and Hays, sedimentation rates averaged more than 30 cm/1000 years; Cooke and Hays are able to show that the pattern of sedimentation changed within 200 years at these sites about 14,000 years ago, just before or during the beginning stages of important changes in northern hemisphere ice volume [Ruddiman and McIntyre, 1981]. Oeschger et al. [1984] have dated the rise of atmospheric $CO_2$ levels at the end of the last ice age at about 14,000 years B.P.

Hays et al. [1976a] and Cooke and Hays [1982] have also shown that the abundance of the radiolarian Cycladophora davisiana in Antarctic sediments shows an abrupt decline in conjunction with the southward displacement of silicious ooze at the close of the last ice age. C. davisiana abundance as a percentage of the total radiolarian fauna drops from 30-40% to less than 5% over a wide expanse of the Southern Ocean. Morely and Hays [1983] have shown that high abundances of C. davisiana in modern sediments are found only in the Sea of Okhotsk. Morely and Hays propose that the Sea of Okhotsk represents a modern analogue of oceanographic conditions that existed in high-latitude regions during the late Pleistocene. The present Sea of Okhotsk is capped by a very low salinity upper layer and strong pycnocline, in stark contrast to Antarctic conditions today [Gordon, 1981].

Following Morely and Hays [1983], we propose that surface water around the Antarctic 18,000 years ago was capped by a stable pycnocline which inhibited deep convection. This surface layer was relatively depleted in nutrients and, as a result, the $CO_2$ content of the atmosphere was lower. At present, upwelling of warm, Circumpolar Deep Water prevents salinities in Antarctic surface water from falling much below 33.8 per mil and provides a substantial amount of the heat needed to melt back Antarctic sea ice in the spring [Gordon, 1981]. This upwelling is driven by Ekman divergence in the cyclonic gyres between the Antarctic Circumpolar Current and the Antarctic coast. The sedimentary evidence suggests that the circulation in these gyres was radically disrupted during the ice age.

What, then, would stir up the cyclonic gyres and restart deep Antarctic convection 14,000 years ago? Changes in the earth's orbital parameters may have increased the shear in the zonal wind field near Antarctica, increased the local Ekman suction, and forced a general meltback of Antarctic sea ice. Or, a decrease in local precipitation or meltwater runoff from the continent might have reduced the stability of the Antarctic water column. Detailed examination of these possibilities is beyond the scope of this paper. We can, however, suggest an oceanographic factor which may have played an important role. The buoyancy flux which causes the present Antarctic water column to be so close to neutral stability comes from deep water which is about 3°C warmer than winter surface water. The warmth of present-day deep water is due to its component of relatively warm, salty North Atlantic Deep Water. Streeter and Shackleton [1979] and Boyle and Keigwin [1982], among others, have shown that the production of

NADW was reduced during periods of maximal ice sheet extent and ice sheet growth. The cessation of NADW formation would tend to stabilize the Antarctic water column and lower atmospheric $P_{CO2}$. Conversely, the return of NADW formation would tend to reduce stability in the Antarctic water column and open up the possibility of deep convection and $CO_2$ release from the deep ocean.

## Other Evidence

The best evidence in favor of our hypothesis, that a change in high-latitude convection is responsible for the glacial to interglacial changes in the $CO_2$ content of the atmosphere, is the synchroneity of Antarctic sedimentary changes and the atmospheric $P_{CO2}$ change. What other evidence do we have which might be brought to bear on the predictions of this model? The most obvious change predicted for the ice age ocean is the near anoxia of the deep sea. Broecker's [1982] shelf extraction hypothesis predicts an oxygen depletion for the deep sea of the same magnitude. If the average oxygen content of the deep sea was 50 $\mu$mol/kg or less, we should expect to see abundant evidence of anoxic sediments in areas where oxygen minima are present today. Vast areas of the deep Pacific should have been in contact with a greatly intensified oxygen minimum zone.

We can find no evidence of an anoxia of this scale. We can, however, offer several reasons for its apparent absence. First, the most oxygenated water in any ocean will be found near the bottom where most of the ocean's sediments are located. Second, if the duration of the intense anoxia were about 10,000 years, we might expect to see only 10 cm of sediment accumulate during this period of time. When oxygen returns to the deep sea and benthic organisms recolonize formerly anoxic areas, their bioturbation may erase the record by oxidizing the 10 cm of anoxic sediment. Areas most likely to preserve a large-scale anoxic event would therefore be (1) in the Pacific at 1000-3000 m depth and (2) in regions of high sediment accumulation. Areas of high accumulation in the Pacific in this depth range would mainly be on continental slopes. Such areas are the least desirable areas to study from a sedimentological point of view because turbidites and slumping make dating very difficult, if not impossible. Therefore, we conclude that although it is unlikely that a major anoxic event could have gone undetected, there are reasons why this might be so. A systematic search in the deep Pacific for evidence of lower oxygen levels is clearly needed.

The model also predicts that the $\delta^{13}C$ content of high-latitude surface water was higher during glacial time by at least 0.5 per mil and that the phosphate content was lower by about 0.7 $\mu$mol/kg. Either of these predictions might be verified by measuring the isotopic content of Antarctic planktonic forams or their Cd to Ca ratios [Boyle and Keigwin, 1982]. A fundamental difficulty is that planktonic foraminifera do not thrive in cold

Antarctic waters and are not abundant in Antarctic sediments south of the polar front. Diatoms are much more successful in the Antarctic, where silica is available in abundance.

A better test might be measurement of Cd to Ca ratios in ice age Antarctic Bottom Water. If ice age surface water was dramatically lower in nutrients, the cadmium content should have been lower also [Boyle and Keigwin, 1982]. Therefore, benthic forams living in areas bathed by Antarctic Bottom Water might record the surface water signal. One problem with this idea is that present-day Weddell Shelf Water entrains surrounding deep water as it sinks to the bottom. This will attenuate whatever surface signal is present.

The model predicts that the $\Delta^{14}C$ content of the ice age atmosphere was about +100 per mil, assuming there has been no change in the $^{14}C$ production rate. The oldest varved lake sediments, taken from Lake of the Clouds, Minnesota, date back to 10,000 years B.P. and have an age-corrected $^{14}C$ activity of +90-100 per mil [Stuiver, 1971]. Variations in atmospheric $^{14}C$ have been commonly attributed to changes in the earth's magnetic dipole field. Radiocarbon measurements, like those of the Lake of the Clouds record, are often held out as evidence of a long-term (8000-9000 year) sinusoidal variation in the magnetic field which has caused the $^{14}C$ production rate to vary. Keir [1983] has investigated the possibility that higher radiocarbon levels in the atmosphere 10,000 years ago might be evidence for a slowdown in deep sea ventilation in combination with a magnetic field oscillation. Longer, independently dated $^{14}C$ records are needed before atmospheric $^{14}C$ changes at the glacial-interglacial boundary can be assessed.

Finally, the model predicts that the gradient in $\Delta^{14}C$ between surface water and deep water doubled from present values. The difference between surface and deep water $^{14}C$ activity should be independent of uncertainties regarding atmospheric $^{14}C$ production. A promising new technique was unveiled recently to measure this difference in handpicked planktonic and benthic forams which are subjected to accelerator $^{14}C$ dating [Andree et al., this volume]. Even with the passage of three $^{14}C$ half-lives a glacial to interglacial signal of 100 per mil should be detectable.

## Conclusions

Overturning of the low-stability Antarctic water column makes surface water over a broad expanse of the Southern Ocean much like average deep water with relatively high nutrient and $\Sigma CO_2$ contents. As a result, the modern deep sea is well oxygenated, and the atmospheric $P_{CO2}$ is relatively high. We have proposed that during the ice age this overturning was disrupted, leading to a less oxygenated deep sea and a reduced atmospheric $P_{CO2}$. This proposal is consistent with sedimentological evidence which suggests that the ice age Southern Ocean was capped with an intense pycno-

cline and a low-salinity upper layer. An abrupt change in Antarctic sedimentation about 14,000 years ago occurs at about the same time that atmospheric $CO_2$ levels began to rise. We have demonstrated that a fivefold reduction in the rate at which high-latitude surface water exchanges with the deep sea can produce a 70-ppm lowering of atmospheric $P_{CO2}$ and can reproduce the late Pleistocene-Holocene $\delta^{13}C$ changes recorded in ocean sediments.

The four-box model developed here offers a number of predictions about the ice age ocean and atmosphere which may be verifiable in future work. There are as follows:
1. The average oxygen content of the deep sea was within 50 $\mu mol/kg$ of zero oxygen.
2. The average phosphate content of newly formed deep water was 0.6-0.7 $\mu mol/kg$, or about half that of today's Antarctic surface water.
3. The $\delta^{13}C$ content of high-latitude surface water was more than 0.5 per mil heavier than preindustrial values.
4. The $^{14}C$ content of the atmosphere was about +100 per mil, assuming there has been no change in the $^{14}C$ production rate in the upper atmosphere.
5. The $\Delta^{14}C$ difference between warm surface water and average deep water increased by 100 per mil over the present difference of about 110 per mil.

Because the time scales over which these glacial to interglacial changes might occur are the same as those which goven the overturning of the ocean (500-1000 years), the climatic forcing of atmospheric $CO_2$ probably played a pivotal role in amplifying the orbital forcing which paced the climatic changes [Hays et al., 1976b].

## Appendix

One can think of the model as a set of simple linear equations which can be solved by algebraic substitution. The actual solution requires a nested iteration scheme, however, because of the nonlinear equation through which the $P_{CO2}$ values of the two surface boxes are computed [Takahashi et al., 1980].

The model equations are given below in their complete form. In contrast to the equations in the text we include the terms $f_{\ell d}$, $f_{\ell h}$, $PO4_\ell$, and $AOU_h$. The latter two are assumed to be constants which are given to the model as input parameters. We start below with the equations for $P_\ell$, $P_h$ describing the phosphorus particle fluxes and proceed to list a full set of equations for $PO4_h$, $PO4_d$, $Alk_h$, $Alk_d$, and $Alk_\ell$.

$$P_\ell = PO4_d \cdot (T+f_{\ell d}) + PO4_h \cdot f_{\ell h} - PO4_\ell \cdot (T+f_{\ell d}+f_{\ell h}) \tag{A1}$$

$$P_h = PO4_d \cdot f_{hd} - PO4_h \cdot (T+f_{hd}+f_{\ell h}) + PO4_\ell \cdot (T+f_{\ell h}) \tag{A2}$$

$$r_{O2}:P \cdot (P_\ell + P_h) = AOU_d \cdot (T+f_{hd}+f_{\ell d}) - AOU_h \cdot (T+f_{hd}) \tag{A3}$$

Combining equations (A1), (A2), and (A3) and substituting the conservation expression for $PO4_d$ (see equation (11) in text) we find

$$PO4_h = \left[ \frac{T+f_{hd}+f_{\ell d}}{T+f_{hd}} \cdot \left( \frac{PO4_T}{V_d} - \frac{AOU_d}{r_{02:P}} \right) + \frac{AOU_h}{r_{02:P}} \right.$$

$$\left. - PO4_\ell \cdot \left( \frac{f_{\ell d}}{T+f_{hd}} + \frac{T+f_{hd}+f_{\ell d}}{T+f_{hd}} \cdot \frac{V_\ell}{V_d} \right) \right]$$

$$\cdot \left[ 1 + \frac{T+f_{hd}+f_{\ell d}}{T+f_{hd}} \cdot \frac{V_h}{V_d} \right]^{-1} \tag{A4}$$

In producing Figures 3, 4, and 5, for the pre-industrial ocean we start with equation (A4) because $AOU_d$ is a known quantity. We then derive $P_\ell$ and $P_h$ as in (A1) and (A2).

For an ice age simulation we do not know the AOU of the deep sea. We assume instead that $P_h$ is the same as that given by the preindustrial solution. The phosphate balance for the high-latitude box is then rewritten as in equation (A5):

$$PO4_h = \left[ PO4_T \cdot \frac{f_{hd}}{V_d} - P_h + PO4_\ell (T + f_{\ell h} - f_{hd} \cdot \frac{V_\ell}{V_d}) \right]$$

$$\cdot \left[ T + f_{\ell h} + f_{hd}(1 + \frac{V_h}{V_d}) \right]^{-1} \tag{A5}$$

$P_\ell$ can then be calculated using equation (A1) and $AOU_d$ using (A3).

For alkalinity we have

$$Alk_h = \frac{\dfrac{Alk_T}{V_d} \cdot A - r_{Alk:P} \cdot \left( \dfrac{B \cdot P_\ell + P_h}{M} \right)}{1 + A \cdot \dfrac{V_h}{V_d} - f_{\ell h} \cdot \dfrac{B}{M}} \tag{A6}$$

where

$$M = T + f_{hd} + f_{\ell h}$$

$$N = T + f_{\ell d} + f_{\ell h}$$

$$A = \frac{1}{M} (f_{hd} + B (T+f_{\ell d}))$$

$$B = \frac{T + f_{\ell h} - f_{hd} \cdot \dfrac{V_\ell}{V_d}}{N + (T+f_{\ell d}) \cdot \dfrac{V_\ell}{V_d}}$$

$$Alk_d = \left[ Alk_h \cdot (M - (T + f_{\ell h}) \cdot \frac{\ell h}{N} \right.$$

$$+ r_{Alk:P} \left( \frac{T+f_{\ell h}}{N} \cdot P_\ell + P_h \right) \Bigg]$$

$$\cdot \left[ f_{hd} + \frac{(T+f_{\ell h}) \cdot (T+f_{\ell d})}{N} \right]^{-1} \tag{A7}$$

$$Alk_\ell = \left[ Alk_d \cdot (T+f_{\ell d}) + Alk_h \cdot f_{\ell h} - r_{Alk:P} \cdot P_\ell \right]/N \tag{A8}$$

The term $Alk_T$ which appears in equation (A6) represents the total number of equivalents of alkalinity in the ocean see (see Table 2).

Solving the $\Sigma CO_2$ equations requires iterating to a stable solution. The procedure consists of a $\Sigma CO2_\ell$ loop nested within a $\Sigma CO2_h$ loop. Each loop involves an iterative convergence in which the $P_{CO2}$ and $\Sigma CO_2$ are successively recalculated. After each $\Sigma CO2_h$ (outer) loop is completed, a new $\Sigma CO2_d$ and $P_{CO2atm}$ is determined. When the difference between successive values of $\Sigma CO2_h$ reaches a specified minimum, the procedure stops. The $\Sigma CO_2$ equations are given below followed by a detailed outline of the entire procedure.

$$\Sigma CO2_h = \left[ (\Sigma CO2_T - P_{CO2atm} \cdot V_a) \frac{A}{V_d} - r_{\Sigma C:P} \cdot \left( \frac{B \cdot P_\ell + P_h}{M} \right) \right.$$

$$+ \frac{g_h \cdot \beta_h}{M} \cdot (P_{CO2atm} - P_{CO2_h})$$

$$+ \left. g_\ell \cdot \beta_\ell \cdot \frac{B}{M} \cdot (P_{CO2atm} - P_{CO2_\ell}) \right]$$

$$\cdot \left[ 1 + A \frac{V_h}{V_d} - B \frac{f_{\ell h}}{M} \right]^{-1} \tag{A9}$$

$$\Sigma CO2_d = \left[ \Sigma CO2_h \cdot (M - (T+f_{\ell h}) \cdot \frac{f_{\ell h}}{N}) \right.$$

$$+ r_{\Sigma C:P} \cdot \left( \frac{T+f_{\ell h}}{N} \cdot P_\ell + P_h \right)$$

$$- g_h \cdot \beta_h \cdot (P_{CO2atm} - P_{CO2_h})$$

$$- \left. ( \frac{T+f_{\ell h}}{N} ) \cdot g_\ell \cdot \beta_\ell \cdot (P_{CO2atm} - P_{CO2_\ell}) \right]$$

$$\cdot \left[ f_{hd} + \frac{(T+f_{\ell h}) \cdot (T+f_{\ell d})}{N} \right] \tag{A10}$$

$$\Sigma CO2_\ell = \left[\Sigma CO2_d \cdot (T+f_{\ell d}) + \Sigma CO2_h \cdot f_{\ell h} - r_{\Sigma C:P} \cdot P_\ell \right.$$

$$\left. + g_\ell \cdot \beta_\ell \cdot (P_{CO2atm} - P_{CO2_\ell}) \right] / N \qquad (A11)$$

where A, B, M, and N are the same as for alkalinity above and $\beta_\ell$ and $\beta_h$ are the $CO_2$ solubilities according to Takahashi et al. [1980]. The term $\Sigma CO2_T$ which appears in equation (9) represents the total number of moles of $CO_2$ in the ocean-atmosphere system (see Table 2). The $P_{CO2}$ of the atmosphere is a simple weighted average of $P_{CO2_\ell}$ and $P_{CO2_h}$:

$$P_{CO2atm} = \frac{g_\ell \cdot \beta_\ell \cdot A_\ell \cdot P_{CO2_\ell} + g_h \cdot \beta_h \cdot A_h \cdot P_{CO2_h}}{g_\ell \cdot \beta_\ell \cdot A_\ell + g_h \cdot \beta_h \cdot A_h}$$

$$(A12)$$

where $A_\ell$ and $A_h$ are the respective areas of the low- and high-latitude ocean boxes.
A complete outline of the solution procedure follows in schematic form.

For a given set of flux parameters (T, $f_{hd}$, $f_{\ell d}$, and $f_{\ell h}$)

-determine $PO4_h$ using equation (A4) or (A5).
-determine $P_\ell$ and $P_h$ using $PO4_h$ and (A1) and (A2).
-determine $Alk_h$, $Alk_d$, and $Alk_\ell$ using equations (6A) - (8A).

Guess initial values for $\Sigma CO2_\ell$ and $\Sigma CO2_h$.

-determine $P_{CO2_\ell}$ and $P_{CO2_h}$ using $Alk_\ell$, $Alk_h$, the initial values of $\Sigma CO2_\ell$ and $\Sigma CO2_h$, and the $P_{CO2}$ subroutine of Takahashi et al. [1980].
-determine $P_{CO2atm}$ using equation (A12).
-determine an initial value of $\Sigma CO2_d$ using (A10).

Begin iteration

-calculate $P_{CO2_\ell}$ using $Alk_\ell$ and $\Sigma CO2_\ell$.
-calculate a new value of $\Sigma CO2_\ell$ ("$\Sigma CO2_\ell^n$") using $\Sigma CO2_h$ and $\Sigma CO2_d$ and equation (A11).
-test new $\Sigma CO2_\ell$ against old $\Sigma CO2_\ell$ ("$\Sigma CO2_\ell^o$").
-if 0.002 μmol/kg > ABS ($\Sigma CO2_\ell^o - \Sigma CO2_\ell^n$), jump out.
-$\Sigma CO2_\ell = \Sigma CO2_\ell^o + (\Sigma CO2_\ell^n - \Sigma CO2_\ell^o)/100$
-calculate $P_{CO2_h}$ using $Alk_h$ and $\Sigma CO2_h$.
-calculate a new value of $P_{CO2atm}$ using (A12).
-calculate a new value of $\Sigma CO2_h$ ("$\Sigma CO2_h^n$") using equation (A9).
-test new $\Sigma CO2_h$ against old $\Sigma CO2_h$ ("$\Sigma CO2_h^o$").
-if 0.0005 μmol/kg > ABS ($\Sigma CO2_h^o - \Sigma CO2_h^n$), jump out.
-$\Sigma CO2_h = \Sigma CO2h^o + (\Sigma CO2_h^n - \Sigma CO2_h^o)/5$
-calculate a new $\Sigma CO2_d$ using $\Sigma CO2_h$ and equation (A10).

For the carbon isotopes the same iterative procedure is followed except that values of $P_{CO2atm}$, $P_{CO2_\ell}$, and $P_{CO2h}$ are carried over from the ordinary $^{12}C$ solutions. Because the $P_{CO2}$'s for $^{12}C$ do not have to be recalculated for each iteration, the convergence is much faster.

The $^{13}C$ and $^{14}C$ contents of the ocean and atmosphere are expressed in the model in pseudo-$\Sigma CO2$ (concentration) form following the $^{14}C*$ notation of Craig [1969]. This means that the fractional isotopic deviation from the isotopic reference (i.e., $1 - \delta^{13}C/1000$ or $1 - \Delta^{14}C/1000$) is simply multiplied by the $\Sigma CO2$ or $P_{CO2}$. Because the isotopic results are converted back into delta notation at the end of a model run, the isotopic ratio of the reference never enters the calculation.

The biological fractionation factors for $^{13}C$ and $^{14}C$, $^{13}\varepsilon$ and $^{14}\varepsilon$, respectively, only apply to the organic fractions of the particle fluxes. Therefore each appears in an expression, $(1-f_{ca})^i\varepsilon + f_{ca}$, when multiplied by $P_\ell$ and $P_h$ in the isotopic $\Sigma CO2$ equations. Here $f_{ca}$ is the fraction of the carbon flux which is $CaCO_3$; the isotopic fractionation associated with $CaCO_3$ formation or dissolution is assumed to be negligible. $^{13}\varepsilon$ is equal to 0.977; i.e., the per mil fractionation for $^{13}C/^{12}C$ in organic matter is -23 per mil. $^{14}\varepsilon$ is equal to 0.954.

The atmosphere-ocean partial pressure difference for carbon isotope i is written according to Siegenthaler and Munnich [1981] as follows:

$$^iKF \cdot (^if_{as}(T) \cdot {}^iR_{atm} \cdot P_{CO2atm} - {}^if_{sa}(T) \cdot R_s^i \cdot P_{CO2_s})$$

where the subscript s denotes ocean surface, KF is the kinetic fractionation factor, and $^if_{as}$ and $^if_{sa}$ are the temperature dependent thermodynamic fractionation factors for air to sea and sea to air transfer, respectively. The $^iR$ terms represent the isotopic ratio of isotope i to isotope 12 in the reservoir in question; in our model formulation the $^iR$ terms appear as $\Sigma^iCO2/\Sigma^{12}CO2$, where the $\Sigma^{12}CO2$ concentration is carried over from the $^{12}C$ solutions. Table A1 lists the kinetic and air-sea fractionation factors for the modern and ice age temperatures used in the model.

Decay for radiocarbon is expressed by substituting M' and N' below for M and N in the $\Sigma CO2$ equations:

$$M' = T + f_{hd} + f_{\ell h} + \lambda \cdot V_h \qquad (A13)$$

$$N' = T + f_{\ell d} + f_{\ell h} + \lambda \cdot V_\ell \qquad (A14)$$

where $\lambda$ is the decay constant for $^{14}C$, $1.2097 \times 10^{-4}$ $yr^{-1}$, and $V_h$ and $V_\ell$ are box volumes. The decay of radiocarbon in all of the model reservoirs must be balanced by the input of new $^{14}C$ to the atmosphere. To account for the addition of new $^{14}C$ in the equations, we add the expression $\lambda/M' \cdot \Sigma CO2_T$ to the

TABLE A1.  Fractionation Factors for Gas Exchange

| | Ocean Temperature | $^{13}f_{as}$ | $^{13}f_{sa}$ | $^{14}f_{as}$ | $^{14}f_{sa}$ | $^{13}KF$ | $^{14}KF$ |
|---|---|---|---|---|---|---|---|
| | | | *Modern* | | | | |
| Low-latitude box | 21.5 | 0.99893 | 0.99091 | 0.99786 | 0.98182 | 0.9995 | 0.9990 |
| High-latitude box | 2.5 | 0.99884 | 0.98860 | 0.99768 | 0.97720 | 0.9995 | 0.9990 |
| | | | *Ice Age* | | | | |
| Low-latitude box | 20.0 | 0.09893 | 0.99076 | 0.99786 | 0.98152 | 0.9995 | 0.9990 |
| High-latitude box | 2.0 | 0.99884 | 0.98854 | 0.99768 | 0.97708 | 0.9995 | 0.9990 |

numerator of equation (A9) for $\Sigma CO2_h$ and add $\lambda \cdot \Sigma^{14}CO2_T$ to the numerator of equation (A12) for $P_{CO2atm}$. We subtract $\lambda \cdot \Sigma^{14}CO2_T$ from the numerator of equation (A10) for $\Sigma CO2_d$.

*Acknowledgments.*  This work has been supported through the Visiting Scientist Program at the Geophysical Fluid Dynamics Laboratory (NOAA grant 04-7-022-44017) and by ARL/NOAA grant NA83RAC00052 and NSF grant OCE-8110155.  We would like to thank Arnold Gordon, Robert Stallard, Mitsuhiro Kawase, and Robert Gardiner-Garden for critically reading the manuscript and Susan Leigh for editorial suggestions.  We would like to acknowledge Phil Tunison for drafting the figures and Johann Callan for typing the manuscript and preparing the camera-ready copy.  Special thanks are due Sol Hellerman and Marty Jackson for their help in data analysis.

## References

Andree, M., et al., Accelerator radiocarbon ages on foramimifera separated from deep-sea sediments, this volume.

Boyle, E.A., and L.D. Keigwin, Deep circulation of the N. Atlantic over the last 200,000 yrs: Geochemical evidence, Science, 218, 748-787, 1982.

Broecker, W.S., A revised estimate of the radiocarbon age of North Atlantic Deep Water, J. Geophys. Res., 84, 3218-3226, 1979.

Broecker, W.S., Ocean chemistry during glacial time, Geochim. Cosmochim. Acta, 46, 1689-1705, 1982.

Broecker, W.S., and T.-H. Peng, Gas exchange rates between air and sea, Tellus, 26, 21-35, 1974.

Broecker, W.S., and T.-H. Peng, Tracers in the Sea, Eldigio Press, Lamont-Doherty Geological Observatory, Palisades, N.Y. 1982.

Broecker, W.S., and T.-H. Peng, and R.E. Engh, Modelling the carbon system, Radiocarbon, 22, 565-598, 1980.

Broecker, W.S., T. Takahashi, H.J. Simpson, and T.-H. Peng, Fate of fossil fuel carbon dioxide and the global carbon budget, Science, 206, 409-418, 1979.

Bryan, K., F.G. Komro, S. Manabe, and M.J. Spelman, Transient climate response to increasing atmospheric carbon dioxide, Science, 215, 56-58, 1982.

Burckle, L.H., D. Robinson, and D. Cooke, Reappraisal of sea ice distribution in Atlantic and Pacific sectors of the Southern Ocean at 18,000 yr BP, Nature, 299, 435-437, 1982.

Cooke, D.W., and J.D. Hays, Estimates of Antarctic Ocean seasonal sea-ice cover during glacial intervals, in Antarctic Geoscience, Int. Union of Geol. Sci., Ser. B, No. 4, edited by C. Craddock, pp. 1017-1025, University of Wisconsin Press, Madison, 1982.

Craig, H., Abyssal carbon and radiocarbon in the Pacific, J. Geophys. Res., 74, 5491-5506, 1969.

Ennever, F.K., and M.B. McElroy, Changes in atmospheric $CO_2$: Factors regulating the glacial to interglacial transition, this volume.

Gordon, A.L., Oceanography of Antarctic waters, in Antarctic Oceanology I, Antarct. Res. Ser., vol. 15, edited by J.L. Reid, pp. 169-203, AGU, Washington, D.C., 1971.

Gordon, A.L., General ocean circulation, in Numerical Models of Ocean Circulation, pp. 39-53, National Academy of Sciences, Washington, D.C., 1975.

Gordon, A.L., Deep Antarctic convection west of Maud Rise, J. Phys. Oceanogr., 8, 600-612, 1978.

Gordon, A.L., Seasonality of Southern Ocean sea ice, J. Geophys. Res., 86, 4193-4197, 1981.

Gordon, A.L., Weddell Deep Water variability, J. Mar. Res., 40, suppl., 199-217, 1982.

Gordon, A.L., C.T.A. Chen, and W.G. Metcalf, Winter mixed layer entrainment of Weddell Deep Water, J. Geophys. Res., 89, 637-640, 1984.

Hays, J.D., J.A. Lozano, N. Shackleton, and G. Irving, Reconstruction of the Atlantic and western Indian Ocean sectors of the 18,000 B.P. Antarctic Ocean, in Investigation of Late Quaternary Paleoceanography and Paleoclimatology, Mem. 145, edited by R.M. Cline and J.D. Hays, pp. 337-372, Geological Society of America, Boulder, Colo., 1976a.

Hays, J.D., J. Imbrie, and N.J. Shackleton, Variations in the earth's orbit: Pacemaker of the ice ages, Science, 194, 1121-1132, 1976b.

Honjo, S., Material fluxes and modes of sedimentation in the mesopelagic and bathypelagic zones, J. Mar. Res., 38, 53-97, 1980.

Keeling, C.D., and B. Bolin, The simultaneous use of chemical tracers in oceanic studies, Tellus, 20, 17-54, 1968.

Keir, R.S., Reduction of thermohaline circulation during deglaciation: The effect on atmospheric radiocarbon and $CO_2$, Earth Planet. Sci. Lett., 64, 445-456, 1983.

Knauer, G.A., J.H. Martin, and K.W. Bruland, Fluxes of particulate carbon, nitrogen, and phosphorus in the upper water column of the northeast Pacific, Deep Sea Res.,26A,97-108,1979.

Knox, F., and M. McElroy, Changes in atmospheric $CO_2$: Influence of marine biota at high latitudes J. Geophys. Res., 89, 4629-4637, 1984.

Kroopnick, P., Correlations between $\delta^{13}C$ and $\Sigma CO_2$ in surface waters and atmospheric $CO_2$, Earth Planet. Sci. Lett., 22, 397-403, 1974.

Levitus, S., Climatological Atlas of the World Ocean, NOAA Prof. Pap. 13, U.S. Government Printing Office, Washington, D.C., 1982.

Li, Y.-H., T. Takahashi, and W.S. Broecker, Degree of saturation of $CaCO_3$ in the oceans, J. Geophys. Res., 74, 5507-5525, 1969.

Manabe, S., and R.J. Stouffer, Sensitivity of a global climate model to an increase of $CO_2$ concentration in the atmosphere, J. Geophys. Res., 85, 5529-5554, 1980.

Morely, J.J., and J.D. Hays, Oceanographic conditions associated with high abundances of the radiolarian Cycladophora davisiana, Earth Planet. Sci. Lett., 66, 63-72, 1983.

Neftel, A., H. Oeschger, J. Schwander, B. Stauffer, and R. Zumbrunn, Ice core measurements give atmospheric $P_{CO_2}$ content during the past 40,000 yrs., Nature, 295, 220-223, 1982.

Oeschger, H., J. Beer, U. Siegenthaler, B. Stauffer, W. Dansgaard, and C.C. Langway, Late glacial climate history from ice cores, in Climate Processes and Climate Sensitivity, Geophys. Monogr. Ser., vol. 29, edited by J.E. Hansen and T. Takahashi, pp. 299-306, AGU, Washington, D.C., 1984.

Peng, T.-H., W.S. Broecker, G.G. Mathieu, and Y.-H. Li, Radon evasion rates in the Atlantic and Pacific oceans as determined during the GEOSECS program, J. Geophys. Res., 84, 7839-7845, 1979.

Redfield, A.C., B.H. Ketchum, and F.A. Richards, The influence of organisms on the composition of sea water, in The Sea, vol. 2, edited by M.N. Hill, pp. 26-77, Wiley-Interscience, New York, 1962.

Ruddimen, W.F., and A. McIntyre, The mode and mechanism of the last deglaciation: Oceanic evidence, Quat. Res., 16, 125-134, 1981.

Sarmiento, J.L., and J.R. Toggweiler, A new model for the role of the oceans in determining atmospheric $P_{CO_2}$, Nature, 308, 621-624, 1984.

Shackleton, N.J., Carbon-13 in Uvigerina:

Tropical rainforest history and the equatorial Pacific carbonate dissolution cycles, in The Fate of Fossil Fuel $CO_2$ in the Oceans, edited by N. Anderson and A. Malahoff, pp. 401-428, Plenum, New York, 1977.

Shackleton, N.J., J. Imbrie, and M.A. Hall, Oxygen and carbon isotope record of east Pacific core V19-130: Implications for the formation of deep water in the late Pleistocene North Atlantic, Earth Planet. Sci. Lett., 65, 233-244, 1983a.

Shackleton, N.J., M.A. Hall, J. Line, and Cang Shuxi, Carbon isotope data in core V19-30 confirm reduced carbon dioxide concentration in the ice age atmosphere, Nature, 306, 319-322, 1983b.

Siegenthaler, U., and T. Wenk, Rapid atmospheric $CO_2$ variations and ocean circulation, Nature, 308, 624-626, 1984.

Siegenthaler, U., and K.O. Munnich, $^{13}C/^{12}C$ fractionation during $CO_2$ transfer from air to sea, in Carbon Cycle Modelling, SCOPE 16, edited by B. Bolin, pp. 249-257, John Wiley, New York, 1981.

Stauffer, B., H. Hofer, H. Oeschger, J. Schwander, and U. Siegenthaler, Atmospheric $CO_2$ concentration during the last glaciation, Proc. Symposium on Ice and Climate Modelling. Annals of Glaciology, in press, 1984.

Streeter, S.S., and N.J. Shackleton, Paleocirculation of the deep North Atlantic: 150,000 year record of benthic foraminifera and oxygen-18, Science, 203, 168-171, 1979.

Stuiver, M., Evidence for the variation of atmospheric $^{14}C$ content in the late Quaternary, in The Late Cenozoic Glacial Ages, edited by K. Turekian, pp. 57-70, Yale University Press, New Haven, Conn., 1971.

Stuiver, M., H.G. Ostlund, and T.A. McConnaughey, GEOSECS Atlantic and Pacific $^{14}C$ distribution, in Carbon Cycle Modelling, SCOPE 16, edited by B. Bolin, pp. 201-221, John Wiley, New York, 1981.

Sverdrup, H.U., M.W. Johnson, and R.H. Fleming, The Oceans, Prentice-Hall, Englewood Cliffs, N.J., 1942.

Takahashi, T., W.S. Broecker, A.E. Bainbridge, and R.F. Weiss, Carbonate chemistry of the Atlantic, Pacific, and Indian oceans: The results of the GEOSECS expeditions, 1972-1978, Tech. Rep. 1, CU-1-80, Lamont-Doherty Geol. Obs., Palisades, N.Y. 1980.

Takahashi, T., W.S. Broecker, and S. Langer, Redfield ratio based on chemical data from isopycnal surfaces, J. Geophys. Res., in press, 1984.

Vincent, E., J.S. Killingley, and W.H. Berger, Stable isotopes in benthic foraminifera from Ontong-Java Plateau, box cores ERDC 112 and 123, Palaeogeogr. Palaeoclimatol. Palaeoecol., 33, 221-230, 1981.

Walsh, J.J., Vertical distribution of Antarctic phytoplankton, II, A comparison of phytoplankton standing crops in the Southern Ocean with

that of the Florida Strait, Limnol. Oceanogr.,
14, 86-94, 1969.

Warren, B.A., Deep circulation of the world ocean,
in Evolution of Physical Oceanography,
Scientific Surveys in Honor of Henry Stommel,
edited by B. Warren and C. Wunsch, pp. 6-41,
MIT Press, Cambridge, Mass., 1981.

Weast, R.C., and M.J. Astle (Eds.), CRC Handbook

of Chemistry and Physics, 60th ed., p. F-198,
CRC Press, Boca Raton, Fla., 1979.

Weiss, R.F., H.G. Ostlund, and H. Craig, Geochemical studies of the Weddell Sea, Deep Sea Res.,
26, 1093-1120, 1979.

Wenk, T., and U. Siegenthaler, The high-latitude
ocean as a control of atmospheric CO$_2$, this
volume.

# THE HIGH-LATITUDE OCEAN AS A CONTROL OF ATMOSPHERIC $CO_2$

T. Wenk and U. Siegenthaler

Physics Institute, University of Bern, Switzerland

**Abstract.** It is suggested that the rapid natural atmospheric $CO_2$ variations during and at the end of the last glaciation which are indicated by ice core studies may have been caused by changes in the high-latitude oceans, particularly in the Antarctic. Concentrations of nutrients (N, P) in surface water are near zero in large ocean areas, but relatively high in high-latitude oceans. A circulation change could lead to more complete nutrient utilization and thus to a lower $pCO_2$ of surface waters in these regions. Possible changes are discussed, and their effects on atmospheric $CO_2$ concentrations, carbon isotope ratios and dissolved oxygen in the deep sea are estimated by means of a simple box model. Time-dependent calculations show that after a sudden change of circulation rate, the atmospheric $CO_2$ concentration would approach its new steady state value with a relaxation time of about 200 years.

## Introduction

Natural ice contains air bubbles that are assumed to be samples of the atmosphere at the time of ice formation. Measurements on the occluded air show an increase of the atmospheric carbon dioxide concentration from about 200 ppm to about 270 ppm at the end of the last glaciation, 10,000 years ago [Neftel et al., 1982] and rapid changes of 50 to 70 ppm during the last glaciation, 30,000 to 40,000 years ago [Stauffer et al., 1984]. A good correlation with variations of $\delta^{18}O$ and of concentrations of trace constituents in the ice [Oeschger et al., this volume] supports the possibility of a relation to climatic changes. It is important to understand the causes of these natural $CO_2$ variations, also in connection with the question of whether similar changes could occur also under present interglacial conditions. Of special interest are the size and the rate of such changes.

The ocean contains about 60 times as much carbon as the atmosphere and effectively controls the atmospheric $CO_2$ level on a time scale of up to $10^4$ years. It is therefore reasonable to look for oceanic causes of the observed atmospheric $CO_2$ variations. The equilibrium partial pressure of carbon dioxide in the surface ocean, $pCO_2$, is determined by the physical and chemical properties of the surface water. Variations of each of the properties determining $pCO_2$ could lead in principle to atmospheric $CO_2$ variations.

During the maximum of the last glaciation, 18,000 years ago, the mean surface temperature of the ice-free ocean was about 1.5°C lower than it is today [CLIMAP, 1981]. All other parameters being constant, this temperature change would have lowered $pCO_2$ of the surface water by some 20 ppm. Because of the large amounts of water stored in ice caps, the mean ocean salinity was at that time about 1 per mil higher than today. This increase in salinity corresponds to a rise in $pCO_2$ by some 10 ppm, which partly compensates the effect of lower temperatures. For this reason, mean temperature and salinity changes cannot explain why atmospheric $CO_2$ concentrations were lower by 70 ppm during the last glaciation. Rather, some change of the chemical parameters determining $pCO_2$ (i.e., of the total concentration of inorganic carbon, $\Sigma CO_2$, and alkalinity) must have been responsible for these long-term variations. $\Sigma CO_2$ and alkalinity are significantly lower in warm surface waters than in the deep sea. The difference is due to the biological activity in the surface ocean: buildup of organic tissue and carbonate shells lowers $\Sigma CO_2$ and alkalinity of surface water. Organic and shell particles sink to deeper ocean layers where they are oxidized and dissolved. Availability of the nutrients phosphate and nitrate limits the biological activity in most regions where surface concentrations of these nutrients are low. Broecker [1982] proposed that a higher phosphate content of the glacial ocean might have induced a decrease of $\Sigma CO_2$ and $pCO_2$ in surface water. During interglacial times, the additional phosphate would be stored in sediments on the continental shelf, exposed to erosion during glacial periods of low sea level. This explanation of atmospheric $CO_2$ variations involves sea level changes as well as interactions with deep-sea sediments. These processes require rather long time spans and can therefore not explain the rapid atmospheric $CO_2$ variations suggested by ice core results.

Nutrients are not depleted everywhere in the surface ocean. They are in fact rather abundant in Antarctic waters and the equatorial Pacific Ocean, regions where the vertical circulation is strong. There, other factors are limiting biological activity. The combination of inhomogeneous nutrient depletion, water temperature, and surface circulation produces a complex pattern of $pCO_2$ of surface waters in the world ocean and induces a flow of $CO_2$ from regions of high $pCO_2$ to regions of low $pCO_2$ through the atmosphere. The atmospheric $CO_2$ level is a result of this balance.

We try to study the sensitivity of the atmospheric $CO_2$ concentration to changes related to high-latitude oceans, especially in the Antarctic, by means of a simple model of the atmosphere-ocean system. The model is not intended to be a detailed representation of the global carbon cycle, but a means to judge the importance of processes in the high-latitude oceans for atmospheric $CO_2$ and to get a rough estimate of the size and rates of possible changes. Basic considerations have already been published elsewhere [Siegenthaler and Wenk, 1984]. They are extended here by also considering dissolved oxygen, by including the mass of atmospheric $CO_2$ in the carbon balance, and by time-dependent model calculations.

### The Model

Our four-box model, shown in Figure 1, is an extension of a model proposed by Broecker and Takahashi [1984]. A similar model was first used by Keeling and Bolin [1968] to study the circulation of the Pacific Ocean. One box represents the atmosphere (A). The ocean is subdivided into a deep-sea box (D) and a warm (W) and a cold (C) surface box. The surface boxes have relative areas of 89% (W) and 11% (C), corresponding to the ocean areas north and south of 50°S. The depth of the surface boxes is fixed as 200 m.

Properties considered with this model are phosphate ($PO_4$), alkalinity, carbon ($\Sigma CO_2$ and $pCO_2$), oxygen ($O_2$), and carbon isotope ratios ($^{13}C/^{12}C$ and $^{14}C/^{12}C$). Different fluxes transport the properties between the boxes (cf.

Figure 1). $F_{cw}$ and $F_{cd}$ are exchange fluxes of water, $F_{cw}$ between cold and warm surface and $F_{cd}$ between cold surface and deep sea. $F_{cw}$ represents the effect of surface currents and $F_{cd}$ of deep overturning in the high-latitude ocean. $F_u$ stands for the upwelling (deep water formation) circulation: Dense water formed at the cold surface sinks to the deep sea and flows at depth to lower latitudes where it rises back to the surface. Since the cold surface is the only source of deep water, preformed values of nutrients in deep-sea water are identical with the concentrations in the cold surface.

The fluxes of biogenic particles, $P_w$ and $P_c$, transport properties from the surface to the deep sea in the following ratios.

$$\Delta PO_4 : \Delta\Sigma CO_2 : \Delta \text{ alkalinity} : \Delta O_2$$
$$= 1 : 155 : 40 : -175$$

(In the preceding paper [Siegenthaler and Wenk, 1984], a wrong value (35) was given for the ratio of alkalinity to phosphate changes, while 40 had been used for the calculations.)

These numbers are based on the following assumptions. The particles are composed of organic matter and inorganic shell parts of $CaCO_3$. About 80% of the carbon is in the reduced organic form [Broecker and Peng, 1982, p. 12]. The oxidation of organic matter utilizes oxygen dissolved in deep-sea water, hence the negative sign for the oxygen change. The ratios of properties in the organic matter

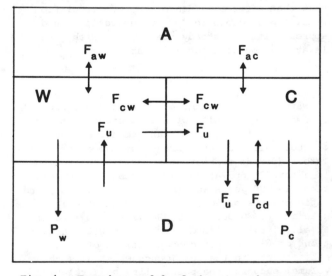

Fig. 1. Four-box model of the atmosphere-ocean system, consisting of warm surface ocean (W), cold surface ocean (C), deep sea (D), and atmosphere (A). Properties are transported between the boxes by water fluxes $F_{cd}$, $F_{cw}$, and $F_u$, by gas exchange fluxes $F_{aw}$, $F_{ac}$, and by particle fluxes $P_w$ and $P_c$.

(Redfield ratios) are assumed to be

$$\Delta PO_4 : \Delta NO_3 : \Delta \Sigma CO_2 : \Delta O_2 \\ = 1 : 17 : 130 : -175$$

according to W.S. Broecker et al. (Redfield Ratio Estimates Based on Chemical Data from Isopycnal Surfaces, preprint, 1983). Alkalinity is changed mainly by the dissolution of $CaCO_3$, but also by the oxidation of nitrogen bound in organic matter to nitrate, which accounts for 30% of the alkalinity change.

The flux of $CO_2$ from the atmosphere to the sea surface, $F_{as}$, is proportional to the difference of partial pressures at the interface and can be written as

$$F_{as} = k \, (pCO_2 \, (A) - pCO_2 \, (S))$$

The partial pressure of $CO_2$ in the surface water, $pCO_2$, can be calculated from the physical and chemical properties, using chemical equilibrium equations [Bacastow, 1981]. The transfer coefficient k equals the product (transfer velocity) x ($CO_2$ solubility). Peng et al. [1979] determined transfer velocities from $^{222}Rn/^{226}Ra$ disequilibria in surface water of the Atlantic and Pacific oceans. From their data (transfer velocities at actual water temperatures, not normalized to 20°C) and $CO_2$ solubilities [Weiss, 1974] we determined k values and found that the mean k for the region >50°S is about 1.7 times as large as for the rest of the ocean. This is essentially an effect of the temperature dependence of the solubility; mean transfer velocities for the warm and cold surface, as estimated in this way, are equal to a few percent.

For each property and each box we can formulate a balance equation. This yields a system of nonlinear equations; the nonlinearity arises because of the complex dependence of $pCO_2$ on $\Sigma CO_2$ and alkalinity.

We are interested in two kinds of results from this model. First, we want to know how the steady state properties are affected by some change of the model parameters, e.g., fluxes, and second, how fast such changes occur. The steady state can be calculated by solving the model equations by linearization and iteration. The study of the time dependence requires the numerical integration of the full system of balance equations.

### The Standard Case

For working with the model, we have to fix the values of several model parameters while others are treated as unknowns. These parameters (see Table 1) approximately define today's

state of the ocean, assumed to be a steady state. As far as possible they have been chosen in agreement with mean GEOSECS data as given by Takahashi et al. [1981]. All concentrations are normalized to 35 per mil salinity. Temperatures are 23°C and 3°C for the warm and cold surface, respectively. The total amounts of $\Sigma CO_2$ and alkalinity in the system are chosen according to Takahashi et al. [1981].

In the warm surface the $PO_4$ concentration is assumed to have the fixed value of 0.3 µmol $kg^{-1}$, independent of circulation changes. If, for example, the upwelling flux $F_u$ becomes stronger, more phosphate is transported to the surface, which leads to an increased biological productivity. Such a change does not directly influence $\Sigma CO_2$ or $pCO_2$, because the increased upwelling of $CO_2$-rich water is com-

TABLE 1. Model Parameters for the Standard Case

| Property | Warm Surface | Cold Surface | Deep Sea |
|---|---|---|---|
| Relative area | 0.89 | 0.11 | 1 |
| Depth, m | 200 | 200 | 3500 |
| Temperature, °C | 23 | 3 | – |
| Salinity, per mil | 35 | 35 | 35 |
| $\Sigma CO_2$, µmol $kg^{-1}$ | 1952 | 2158 | 2285 |
| Alkalinity, µeq $kg^{-1}$ | 2315 | 2360 | 2393 |
| Phosphate, µmol $kg^{-1}$ | 0.30 | 1.43 | 2.25 |
| Oxygen, µmol $kg^{-1}$ | 219 | 332 | 188 |

| Flux | Value |
|---|---|
| **Water** | |
| $F_u$ | 15 Sv |
| $F_{cd}$ | 44 Sv |
| $F_{cw}$ | 10 Sv |
| **Particles** | |
| $P_w$ | 0.64 mol C $m^{-2}$ $yr^{-1}$ |
| $P_c$ | 0.93 mol C $m^{-2}$ $yr^{-1}$ |
| **Gas exchange (for 300 ppm)** | |
| $F_{aw}$ | 19.5 mol C $m^{-2}$ $yr^{-1}$ |
| $F_{ac}$ | 33.0 mol C $m^{-2}$ $yr^{-1}$ |
| **Net gas exchange for standard case** | |
| W → A | 0.078 mol C $m^{-2}$ $yr^{-1}$ |
| A → C | 0.638 mol C $m^{-2}$ $yr^{-1}$ |

1 sverdrup (Sv) = $10^6$ $m^3$ $s^{-1}$.

pensated by the higher particle flux; only an indirect influence is possible, via the coupling to the cold surface. In the cold surface, the productivity is assumed constant, so that changes in the water fluxes lead to changes of $PO_4$, $\Sigma CO_2$, and $pCO_2$.

Water fluxes and high-latitude particle flux are determined to fulfill best the following requirements for the standard case:

1. The phosphate concentration in the cold surface has to be about 1.6 $\mu$mol kg$^{-1}$, compared to the deep-sea value of 2.3 $\mu$mol kg$^{-1}$. According to Takahashi et al. [1981], the measured mean surface concentration of phosphate in Antarctic waters is 1.5 $\mu$mol kg$^{-1}$ (normalized to a salinity of 35 per mil) in summer. This value is estimated to be about 0.2 $\mu$mol kg$^{-1}$ higher in winter, based on observation of the subsurface core of cold winter water (W.S. Broecker, personal communication, 1984). New measurements on winter water [Jennings et al., 1984] confirm this observation. The annual mean

value, relevant for the exchange of $CO_2$ between atmosphere and surface water, is assumed to be 16 $\mu$mol kg$^{-1}$.

2. Measured $^{14}C$ concentrations impose strong constraints on the fluxes: In the atmosphere, $\Delta^{14}C$ should be 0 per mil, in the warm and cold surface about -50 and -100 per mil, respectively, and -160 per mil in the deep sea [Broecker and Peng, 1982, pp. 245-252].

3. The atmospheric $CO_2$ level should lie in the pre-industrial range, 270-280 ppm [Moor and Stauffer, 1984].

4. The particle fluxes should agree with values gained from direct observations, 0.3-1.1 mol C m$^{-2}$ yr$^{-1}$ [Mopper and Degens, 1979; Eppley and Peterson, 1979].

The set of parameters for the standard case is given in Table 1. As can be seen from the results in Table 2, the requirements 1-4 are generally fulfilled in a satisfactory way. $\Delta^{14}C$ in the warm surface water is -35 per mil,

TABLE 2.   Results of Sensitivity Test

| | Box | Standard Case | $1/2 \times F_{cd}$ | $2 \times F_u$ | $2 \times F_{cw}$ | $2 \times P_c$ | $1/2 \times S_c$ | $T_c$ Lower | $T_w$ Lower |
|---|---|---|---|---|---|---|---|---|---|
| $pCO_2$, ppm | A | 280 | 235 | 258 | 264 | 266 | 250 | 254 | 274 |
| | W | 281 | 238 | 261 | 266 | 267 | 252 | 257 | 274 |
| | C | 274 | 221 | 243 | 252 | 258 | 226 | 244 | 272 |
| $\Sigma CO_2$, $\mu$mol kg$^{-1}$ | W | 1952 | 1915 | 1936 | 1940 | 1941 | 1928 | 1931 | 1974 |
| | C | 2158 | 2106 | 2129 | 2138 | 2143 | 2112 | 2162 | 2157 |
| $PO_4$, $\mu$mol kg$^{-1}$ | C | 1.43 | 1.05 | 1.23 | 1.29 | 1.32 | 1.10 | 1.43 | 1.43 |
| P, mol C m$^{-2}$ yr$^{-1}$ | W | 0.64 | 0.58 | 1.08 | 0.78 | 0.63 | 0.59 | 0.64 | 0.64 |
| $O_2$, $\mu$mol kg$^{-1}$ | D | 188 | 123 | 154 | 163 | 169 | 130 | 212 | 188 |
| $\delta^{13}C$, per mil | A | -6.83 | -6.45 | -6.39 | -6.51 | -6.72 | -6.02 | | |
| | W | 1.65 | 2.03 | 2.13 | 2.00 | 1.76 | 2.31 | | |
| | C | 1.39 | 1.77 | 1.56 | 1.51 | 1.50 | 1.68 | | |
| | D | 0.46 | 0.43 | 0.43 | 0.44 | 0.45 | 0.41 | | |
| $\Delta^{14}C$, per mil | A | 0 | 35 | -31 | -1 | 3 | 49 | | |
| | W | -35 | -9 | -73 | -41 | -33 | 2 | | |
| | C | -96 | -58 | -107 | -93 | -96 | -61 | | |
| | D | -162 | -163 | -159 | -161 | -161 | -163 | | |

Sensitivity of the steady state to changes of model parameters. $1/2 \times Y$ indicates a reduction of parameter Y by 50%, $2 \times Y$ a doubling. Temperatures are reduced by 3°C, i.e., to 0°C and 20°C for the cold ($T_c$) and the warm ($T_w$) surface, respectively. Only one parameter is changed in each case, except for the reduction of the cold ocean area ($1/2 \times S_c$, see text). The change of oxygen solubility is taken into account in the case of changing temperatures.

somewhat high compared to the observed values. This is a consequence of the assumed mean gas exchange flux, which was chosen according to the evaluation of bomb-$^{14}$C data by Broecker et al. [1980]. The influence of this minor inconsistency should not be important. The observed mean values of $\Sigma CO_2$ are 1978 μmol kg$^{-1}$ and 2200 μmol kg$^{-1}$ for the warm and the cold surface, respectively. These values have to be corrected for the influence of anthropogenic $CO_2$, which yields estimated pre-industrial values of 1950 μmol kg$^{-1}$ and 2175 μmol kg$^{-1}$. These numbers are in good agreement with the values resulting for our standard case (1952 μmol kg$^{-1}$), taking the uncertainties arising from seasonal influences, different salinities, and Redfield ratios into account.

## Different Steady States

For testing the influence of changes of model parameters on the dependent variables, like atmospheric $CO_2$ level or isotopic composition of carbon, we have considered several cases where a single flux was changed by a factor of 2. As a boundary condition the total amounts of carbon, phosphorus, and alkalinity in the system were kept constant. The results of these tests are shown in Table 2. Some numbers are slightly different from those published previously [Siegenthaler and Wenk, 1984]. The difference arises because we are no longer neglecting the atmospheric $CO_2$ capacity, as described in the introduction.

A reduction by 50% of the water exchange flux between the cold surface and the deep sea, $F_{cd}$, lowers the atmospheric $CO_2$ concentration by 44 ppm. The cold water remains at the surface for a longer time than in the standard case, so that more phosphate and carbon is utilized by the biological activity and $\Sigma CO_2$ and hence also $pCO_2$ are lower. The $pCO_2$ in the warm surface ocean is also decreased; this is due to transfer of $CO_2$ to the cold surface via the atmosphere. Such a reduction diminishes the ventilation of the deep sea, and less $^{14}$C, produced in the atmosphere, is transported to the deep sea. Its concentration in the atmosphere and surface ocean rises by about 35 per mil.

An increase of the large-scale circulation $F_u$ lowers the atmospheric $CO_2$. A doubling of $F_u$ results in a decrease of 21 ppm. The effect is due to the dilution of phosphate-rich cold water by water depleted in phosphate. $\Sigma CO_2$ and $pCO_2$ in the cold surface decrease. In this case, deep sea ventilation is increased, and the steady state concentration of $^{14}$C in the atmosphere and surface ocean is lowered. The cold surface water can also be diluted with water depleted in phosphate by increasing the water exchange $F_{cw}$ between warm and cold surfaces. For a doubling of $F_{cw}$ the atmospheric $CO_2$ level is decreased by 22 ppm. Carbon 14 is not significantly affected, because no variation of the exchange with the deep sea occurs. The phosphate concentration of the cold surface ocean cannot only be influenced by circulation changes. A higher biological productivity in the cold surface ocean implies a higher phosphate and carbon consumption, lowering $\Sigma CO_2$ and $pCO_2$. Carbon 14 is only slightly affected; the small effect in the atmosphere is due to the lower $CO_2$ concentration, i.e., less dilution of the newly produced $^{14}$C by stable carbon.

In all the scenarios discussed, phosphate and $\Sigma CO_2$ concentrations in the surface ocean are lower than in the standard case, which is equivalent to a greater fraction of $\Sigma CO_2$ being transported to the deep sea by biogenic particles and less by newly formed deep-sea water. The preformed concentrations of nutrients in the deep water are changed. Because all the deep water is formed in the cold surface ocean, the preformed concentrations of nutrients in deep-sea water are equal to the nutrient concentrations in the cold surface ocean. Concentrations of preformed nutrients are not affected in the case of $CO_2$ variations induced by changes of surface temperature.

A cooling of the high-latitude surface ocean by 3°C lowers the atmospheric $CO_2$ level by as much as 25 ppm. The same decrease of temperature in the warm surface ocean induces a decrease of only 6 ppm. The difference indicates the importance of the high-latitude ocean for the atmospheric $CO_2$ concentration, which is mainly determined by the properties of the cold surface water. This can be explained by the combined effect of gas exchange and circulation. The coupling between the cold surface and the deep sea and between the cold and the warm surface is much stronger than the coupling of the warm surface to the deep sea. For this reason, the properties of the cold surface ocean control the total amount of carbon in the system surface ocean-atmosphere.

The importance of atmospheric $CO_2$ transfer is illustrated by changing the gas exchange rates (cf. Figure 2). A hypothetical reduction of the gas exchange flux to 1% of its standard value causes a rise of the atmospheric $CO_2$ level by 20 ppm to about 300 ppm. In this "Redfield Ocean" [Broecker and Takahashi, 1984], $\Sigma CO_2$ and $pCO_2$ in surface water are determined by the interplay of biological activity and ocean circulation, but hardly affected by gas exchange. A multiplication of the gas exchange rate by a factor of 100 ("thermodynamic ocean"), on the other hand, induces a decrease of the atmosphe-

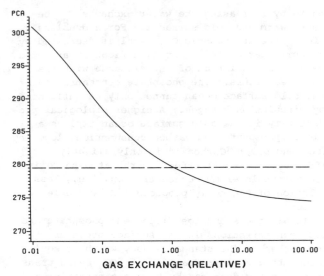

Fig. 2. Dependence of atmospheric $CO_2$ concentration (PCA, in parts per million) on the strength of gas exchange. Horizontal axis: gas exchange rates, in relative units (1 corresponds to standard gas exchange rates). The dashed line indicates the value for standard gas exchange rates.

ric $CO_2$ concentration by 5 ppm. The exchange of $CO_2$ between different oceanic surface regions via the atmosphere thus proves to be an important factor determining the atmospheric $CO_2$ level.

Because of the importance of the cold surface ocean, the effect changes in the area of this region has been considered, too. In case $1/2 \times S_c$ of Table 2, the cold surface area, $S_c$, is reduced by 50% (and the warm surface increased by 6%), while the flux densities (flux per unit area) are kept constant. This means that the total exchange between cold surface and deep sea, $F_{cd}$, is diminished by 50%. The total amount of water upwelling in low latitudes, $F_u$, becomes larger by 6%, because it is kept proportional to the warm surface area. The reduction of the cold surface involves a dilution of the high-latitude surface water by water of low latitudes, depleted in $\Sigma CO_2$. This leads to a considerable decrease of the atmospheric $CO_2$ concentration. In this case, the assumption of constant flux densities is important. Other assumptions which may be as plausible, e.g., constant absolute fluxes, would have different consequences. Therefore it is not possible to answer the question regarding the influence of cold surface area changes without specifying the changed conditions in more detail. T. Volk (unpublished manuscript, 1984) independently pointed to the fact that larger sea ice cover 18,000 years ago involved an areal reduction of the cold Antarctic surface water and that this may have led to an atmospheric $CO_2$

change. He estimated from CLIMAP data that the area of water with temperatures <4°C was reduced to about 44% of its present size on an annual mean basis. Our assumed reduction of the cold surface area by 50% approximately corresponds to this change.

One problem in changing the rate of deep water formation is the supply of oxygen to the deep sea: If deep water formation is reduced, more carbon from oxidation of organic particles is added per volume of deep-sea water, and more dissolved oxygen is utilized. The figures in Table 2 show, however, that our model assumptions do not involve anoxic conditions in the whole deep sea. Because we keep the total amount of carbon in the system constand and because of the large volume of the deep sea, $\Sigma CO_2$ and alkalinity in the deep sea are changed only little. For this reason and for the rather short times we consider, we can neglect the interaction with carbonates in the sediments.

The cases discussed above are only a choice of many different possibilities. The model does not at all behave in a linear way. The possible effect of changing horizontal mixing, for example, will lead to a decrease of the atmospheric $CO_2$ level only until the surface boxes are well mixed and have nearly the same chemical properties. Different changes do not combine in a linear way, either. A reduction of the cold-deep exchange $F_{cd}$ by 50% decreases the atmospheric $CO_2$ level by 44 ppm, and a doubling of the large-scale circulation $F_u$ by 21 ppm; however, both changes occurring simultaneously will not decrease the atmospheric $CO_2$ level by 65 ppm but only by 55 ppm. An approximate lower limit for the atmospheric $CO_2$ concentration is obtained by assuming that the exchange between cold surface and deep sea, $F_{cd}$, is reduced to about 10% of its standard value, so that the phosphate concentration in the cold surface is 0.3 μmol $kg^{-1}$, as in the warm surface. In this case, atmospheric $CO_2$ drops to 178 ppm.

Time-Dependent Calculations

It is of interest to know how fast the atmospheric $CO_2$ level is adjusted to new oceanic conditions. Therefore we have carried out time-dependent calculations, assuming a sudden change of some flux. The results of a typical case are shown in Figures 3-5. At time 0, the exchange flux between cold surface and deep sea, $F_{cd}$, is reduced by 50%. Figure 3 shows the decrease of the atmospheric $CO_2$ during the first 1000 years. A logarithmic plot of the transient atmospheric $CO_2$ excess, i.e., the difference to the new steady state value (Figure 4), yields a straight line, which shows that the change proceeds

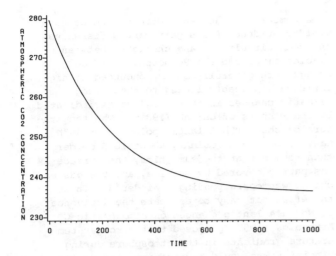

Fig. 3. Decrease of atmospheric $CO_2$ concentration (in parts per million) after a reduction of the water exchange between cold surface and deep sea, $F_{cd}$, from 44 Sv to 22 Sv. Time in years after the change.

strictly exponentially with a relaxation time of 225 years. The value of this relaxation time can be estimated and understood as follows: As discussed, for example, by Broecker and Peng [1974], the partial pressures of $CO_2$ in atmosphere and surface ocean equilibrate within a few years. Therefore we can consider these three reservoirs A, W, and C as one combined "upper" reservoir.

Atmosphere and surface ocean then contain

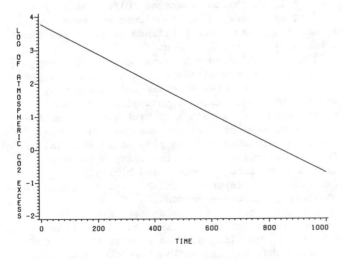

Fig. 4. Plot of the natural logarithm of the difference of the $CO_2$ concentration after a reduction of $F_{cd}$ to the new steady state value. The difference decreases exponentially with a relaxation time of $\tau = 225$ years. Time in years.

excess amounts of carbon, $n_a$ and $n_s$ (expressed in moles of C per square meter of ocean surface), which must be transported to the deep sea by the flux of newly formed deep water, $F_{cd} + F_u$. The relaxation time for reaching the new steady state is therefore given by

$$\tau = \frac{(n_a + n_s) \cdot S_{oc}}{(F_{cd} + F_u) \cdot \Delta\Sigma CO_2}$$

where $S_{oc}$ is the ocean surface and $\Delta\Sigma CO_2$ the $\Sigma CO_2$ excess in newly formed deep water. For a given $\Delta pCO_2$, the excess amount of atmospheric $CO_2$ per square meter of ocean surface is given by

$$n_a = 0.488 \text{ mol m}^{-2} \text{ ppm}^{-1} \cdot \Delta\Sigma CO_2$$

The $\Sigma CO_2$ excess in the warm and cold surface ocean is on the average

$$\Delta\Sigma CO_2 = 0.74 \text{ }\mu mol \text{ kg}^{-1} \text{ ppm}^{-1} \cdot \Delta pCO_2$$

and $n_s$ is

$$n_s = 0.74 \text{ }\mu mol \text{ C kg}^{-1} \text{ ppm}^{-1} \cdot \rho \cdot d \cdot \Delta$$
$$= 0.152 \text{ mol C m}^{-2} \text{ ppm}^{-1} \cdot \Delta pCO_2$$

where d is depth of surface ocean (= 200 m) and $\rho$ is density of ocean water (= 1024 kg m$^{-3}$).

For a given $\Delta pCO_2$ the excess amount of $CO_2$ in the atmosphere, $n_a$, is about 3 times as large as that in the layer of surface water, $n_s$, although the total amount of carbon is 3 times less, due to the chemical buffer effect accompanying $CO_2$ changes. The capacity of the atmosphere for excess $CO_2$ corresponds to that of an ocean layer of about 640 m depth. For a total deep-water production of 22 Sv + 15 Sv = 37 Sv in the case studied, we obtain $\tau = 264$ years. If the atmospheric $CO_2$ capacity were neglected, the relaxation would correspond to the flushing time of the warm plus surface, i.e., $\tau \cong 62$ years.

The value $\tau = 264$ years is in reasonable agreement with the value obtained from numerical calculations. If the effect of the finite volume of the deep sea is taken into account, the agreement becomes even better. The above considerations indicate that the time constant depends only little on the depth of the surface ocean layers which we chose, somewhat arbitrarily, as 200 m.

The relaxation time for achieving new steady state is not strongly dependent on the way the $CO_2$ change is produced. It varies inversely with the total flux of water to the deep sea after the circulation change. Therefore in the reverse transition, from $1/2 \times F_{cd}$ to $F_{cd}$, atmospheric $CO_2$ increases with a time constant of about 160 years.

The induced change of $\delta^{13}$C is, as the CO$_2$ change, exponential in time and has, apart from minor corrections, the same relaxation time as atmospheric CO$_2$. This is not true for the change of the concentration of dissolved oxygen in the deep sea, as can be seen from Figure 5. It is controlled by the residence time of water in the deep sea, which is about 700 years. This means that significant atmospheric CO$_2$ variations could occur within a few centuries after a change of the oceanic conditions, while the oxygen concentration of deep-sea water would react much slower.

## Discussion and Conclusions

Originally, Oeschger et al. [1984] suggested that changes in the high-latitude ocean surface, where nutrients are relatively abundant and therefore do not limit biological productivity, may have led to the CO$_2$ variations of 50 to 70 ppm, indicated by ice core studies. The simple model we have used here shows that, in principle, oceanic changes can indeed produce atmospheric CO$_2$ variations of this magnitude, and that these variations may proceed rather fast, within a few centuries, provided of course that the oceanic changes are so rapid.

An interesting result is also that temperature changes only in high-latitude regions may have a significant influence, even if the temperature in the rest of the ocean surface remains constant. Although the high-latitude regions cover only about 10% of the world ocean, they seem to be of great importance for the atmospheric CO$_2$ concentration.

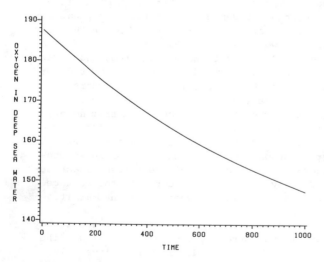

Fig. 5. Time dependence of the concentration of dissolved oxygen (in micromoles per kilogram) in deep-sea water after a reduction of F$_{cd}$ to 50%. Time in years.

A number of deep-sea sediment studies have yielded evidence for significant differences in ocean circulation and chemistry between glacial times and the Holocene. Surface changes are generally most pronounced at high latitudes, probably linked to the strong climatic changes in these regions as indicated by the results of CLIMAP [1981]. The temperature of the North Atlantic poleward of 40°N was, according to CLIMAP, about 10°C colder than now, but at the same time, the surface was partly covered by sea ice, and the gas exchange was correspondingly hindered. It is therefore not easy to estimate the influence of North Atlantic changes on atmospheric CO$_2$. Arrhenius [1952] proposed that stronger temperature gradients in the atmosphere during glacial times could have caused higher wind stress at the sea surface and thus have affected oceanic circulation. Sediment studies clearly point to changed oceanic circulation. The formation of North Atlantic deep water appears to have been significantly reduced during glacial periods [Streeter and Shackleton, 1979; Boyle and Keigwin, 1982; Shackleton et al., 1983]. Sea ice cover changed in the North Atlantic [Ruddiman and McIntyre, 1981] as well as in the Antarctic [Cooke and Hays, 1982], accompanied by changes in biological productivity. The magnitude of the changes in Antarctic waters is not yet known, and it is not clear whether the formation of Antarctic bottom water was different in glacial times (L.H. Burckle, personal communication, 1984).

For a better understanding of the history of the global carbon cycle, more data are needed, especially on carbon isotopes. Different series of $\delta^{13}$C measurements on planktonic and benthic foraminifera have been published [e.g., Curry and Lohmann, 1983; Shackleton et al., 1983].

The $\delta^{13}$C difference between surface and deep water is a measure of the carbon transport to the deep sea by particles relative to transport by water, and a variation of this difference may be the result of a change of biological productivity or of some circulation change. The horizontal gradient of $\delta^{13}$C in planktonic foraminifera is determined by the interplay between gas exchange, surface currents, and biological franctionation [Broecker and Peng, 1982, pp. 306-311]. For these reasons, $\delta^{13}$C measurements on sediment cores can provide key information in the search for the explanation of the atmospheric CO$_2$ variations. The results [Shackleton et al., 1983; Keigwin and Boyle, this volume] indicate $\delta^{13}$C values in the ocean lower by about 0.2-0.5 per mil during the glacial than during the Holocene. This difference cannot be explained by changed circulation or producitivity but seems to be a consequence of an interaction with a carbon reservoir outside the atmosphere-ocean system, affecting $\delta^{13}C$ and probably the total amount of carbon in the ocean [Shackleton, 1977; Keigwin and Boyle, this volume]. In this laboratory, a program for measuring $\delta^{13}C$ of atmospheric $CO_2$ trapped in old polar ice has been started (H. Friedli et al., unpublished manuscript, 1984). The $\Delta^{14}C$ difference between surface water and deep sea depends mainly on the rate of deep water formation. Attempts are now being made to reconstruct the value of this difference for glacial time by separately measuring $\Delta^{14}C$ of planktonic and benthic foraminifera stored in sediments [Andrée et al., this volume].

In summary, this model, and similar work by Knox and McElroy [1984] and Sarmiento and Toggweiler [1984], shows that changes in ocean circulation might have produced atmospheric $CO_2$ variations of the order of 50 ppm and that these changes might have occurred within a few centuries. According to these ideas, a lower atmospheric $CO_2$ level would have been accompanied by a lower concentration of dissolved oxygen in the deep sea, but the decrease of its mean value would not be so large as to produce widespread anoxia (which is not observed). In order to conform or discard the suggested mechanism, more observational data are needed, especially on carbon isotope variations in the past. The simple box models as discussed here can obviously only yield qualitative indications of what might have to be studied by more realistic models. Finally, changes of the ocean circulation may not only have happened in the past but could possibly also occur in connection with man-made environmental changes and should therefore also be studied from this point of view.

Acknowledgments.  We wish to thank H. Oeschger for valuable discussions.  This work was financially supported by the Swiss National Science Foundation.

## References

Andrée, M., J. Beer, H. Oeschger, A. Mix, W. S. Broecker, N. Ragano, P. O'Hara, G. Bonani, H. J. Hofmann, E. Morenzoni, M. Nessi, M. Suter, and W. Woelfli, Accelerator radiocarbon ages on foraminifera separated from deep-sea sediments, this volume.

Arrhenius, G. O., Sediment cores from the east Pacific, Rep. Swed. Deep Sea Exped., 5 (1), 227 pp. 1952.

Bacastow, R., Numerical evaluation of the evasion factor, Carbon Cycle Modelling, SCOPE 16, edited by B. Bolin, pp. 95-101, John Wiley, New York, 1981.

Boyle, E. A., and L. D. Keigwin, Deep circulation of the North Atlantic over the last 200,000 years: Geochemical evidence, Science, 218, 784-787, 1982.

Broecker, W. S., Ocean chemistry during glacial time, Geochim. Cosmochim. Acta, 46, 1689-1705, 1982.

Broecker, W. S., and T.-H. Peng, Gas exchange rates between air and sea, Tellus, 26, 21-35, 1974.

Broecker, W. S., and T.-H. Peng, Tracers in the Sea, Eldigio, Palisades, N.Y., 1982.

Broecker, W. S., and T. Takahashi, Is there a tie between atmospheric $CO_2$ content anc ocean circulation?, in Climate Processes and Climate Sensitivity, Geophys. Monogr. Ser., vol. 29, edited by J.E. Hansen and T. Takahashi, pp. 324-326, AGU, Washington, D.C., 1984.

Broecker, W. S., T.-H. Peng, and R. Engh, Modeling the carbon system, Radiocarbon, 22, 565-598, 1980.

CLIMAP Project Members, Seasonal reconstruction of the earth's surface at the last glacial maximum, Geol. Soc. Am. Map Chart Ser., MC-36, 1981.

Cooke, D. W., and J. D. Hays, Estimates of Antarctic Ocean seasonal sea-ice cover during glacial intervals, Antarctic Geoscience, edited by C. Craddock, pp. 1017-1025, University of Wisconsin Press, Madison, 1982.

Curry, W. B., and G. P. Lohmann, Reduced advection into Atlantic Ocean deep eastern basins during last glaciation maximum, Nature, 306, 577-580, 1983.

Eppley, R. W., and B. J. Peterson, Particulate organic matter flux and planktonic new production in the deep ocean, Nature, 282, 677-680, 1979.

Jennings, J. C. Jr, L. I. Gordon, and D. M. Nelson, Nutrient depletion indicates high primary productivity in the Weddell Sea, Nature, 309, 51-54, 1984.

Keeling, C. D., and B. Bolin, The simultaneous use of chemical tracers in oceanic studies, 2, Tellus, 20, 17-54, 1968.

Keigwin, L. D., and E. A. Boyle, Carbon isotopes in deep-sea benthic foraminifera: Precession and changes in low-latitude biomass, this volume.

Knox, F., and M. B. McElroy, Changes in atmospheric $CO_2$: Influence of the marine biota at high latitude, J. Geophys. Res., 89, 4629-4637, 1984.

Moor, E., and B. Stauffer, A new dry extraction system for gases in ice, J. Glaciol., in press, 1984.

Mopper, K., and E. T. Degens, Organic carbon in the ocean: Nature and cycling, The Global Carbon Cycle, SCOPE 13, edited by B. Bolin,

E. T. Degens, S. Kempe, and P. Ketner, pp. 293-316, John Wiley, New York, 1979.

Neftel, A., H. Oeschger, J. Schwander, B. Stauffer, and R. Zumbrunn, Ice core sample measurements give atmospheric CO$_2$ content during the past 40,000 years, Nature, 295, 220-223, 1982.

Oeschger, H., J. Beer, U. Siegenthaler, B. Stauffer, W. Dansgaard, and C.C. Langway, Late glacial climate history from ice cores, in Climate Processes and Climate Sensitivity, Geophys. Monogr. Ser., vol. 29, edited by J. E. Hansen and T. Takahashi, pp. 299-306, AGU, Washington, D.C., 1984.

Oeschger, H., B. Stauffer, R. Finkel, and C. C. Langway, Variations of the CO$_2$ concentration of occluded air and of anions and dust in polar ice cores, this volume.

Peng, T.-H., W. S. Broecker, G. G. Mathiew, and Y.-H. Li, Radon evasion rates in the Atlantic and Pacific oceans as determined during the GEOSECS program, J. Geophys. Res., 84, 2471-2486, 1979.

Ruddiman, W. F., and A. McIntyre, The North Atlantic Ocean during the last deglaciation, Palaeogeogr. Palaeoclimatol. Palaeoecol., 35, 145-214, 1981.

Sarmiento, J. L., and J. R. Toggweiler, A new model for the role of the oceans in determining atmospheric pCO$_2$, Nature, 308, 621-624, 1984.

Shackleton, N. J., Carbon-13 in Uvigerina: Tropical rainforest history and the equatorial

Pacific carbonate dissolution cycles, in The Fate of Fossil Fuel CO$_2$ in the Oceans, edited by N. R. Anderson and A. Malahoff, pp. 401-428, Plenum, New York, 1977.

Shackleton, N. J., J. Imbrie, and M. A. Hall, Oxygen and carbon isotope record of East Pacific core V19-30: Implications for the formation of deep water in the late Pleistocene North Atlantic, Earth Planet. Sci. Lett., 65, 233-244, 1983.

Siegenthaler, U., and T. Wenk, Rapid atmospheric CO$_2$ variations and ocean circulation, Nature, 308, 624-625, 1984.

Stauffer, B., H. Hofer, H. Oeschger, J. Schwander, and U. Siegenthaler, Atmospheric CO$_2$ concentration during the last glaciation, Ann. Glaciol., 5, 160-164, 1984.

Streeter, S. S., and N. J. Shackleton, Paleocirculation of the deep North Atlantic: A 150,000 year record of benthic foraminifera and $^{18}$O, Science, 203, 168-170, 1979.

Takahashi, T., W. S. Broecker, and A. E. Bainbridge, Supplement to the alkalinity and total carbon dioxide concentration in the world oceans, Carbon Cycle Modelling, SCOPE 16, edited by B. Bolin, pp. 159-200, John Wiley, New York, 1981.

Weiss, R. F., Carbon dioxide in water and seawater: The solubility of a non-ideal gas, Mar. Chem., 2, 203-215, 1974.

# LAST DEGLACIATION IN THE BAHAMAS:
## A DISSOLUTION RECORD FROM VARIATIONS OF ARAGONITE CONTENT?

Andre W. Droxler

Rosenstiel School of Marine and Atmospheric Science, MGG
University of Miami, Miami, Florida   33149

Abstract. Detailed studies of three core tops and corresponding box cores display the last 20,000-year history of the Bahamian basins and the adjacent Atlantic Ocean. The cores are from three different areas, the northern part of Tongue of the Ocean at 1,900 m, NE Providence Channel at 2,390 m, and the crest of Eleuthera Ridge at 4,000 m. Variations of aragonite content are occurring in concert with the variations of carbonate and quartz content, and also parallel the oxygen isotopic record on Globigerinoides rubra known to be related to the glacial rhythm of the earth's climate. The most drastic mineralogical and lithological changes are recorded during the transition interval between the last glacial maximum at 18,000 years B.P. and the hypsithermal at 6,000 years B.P. of the current interglacial. During the deglaciation interval, the mineralogical and lithological variations and oxygen isotopic records seem to have recorded together regression events of 1,000 years duration that can be correlated with established events in ice cores from Greenland and Antarctica, in North Atlantic cores, and in the southwest of France. In addition, the lag time of several thousand years between bank flooding and aragonite content in the basins has been confirmed to exist in several cores from different Bahamian basins and from the adjacent Atlantic Ocean. Based on these observations, the aragonite variations seem to be related to a more climatic, thus oceanic signal than the regional alternate flooding and exposure of the Bahamas Banks. We propose that aragonite dissolution cycles analogous to the calcite dissolution glacial cycles observed in deeper waters are the main cause of the aragonite variations.

## Introduction

The sediments deposited between graded beds of lime sand and mud from sediment gravity flows are either a periplatform carbonate ooze (60-95%)

carbonate) in the Bahamian basins or a marl to hemipelagic mud (2-72% carbonate) in the adjacent Atlantic Ocean. The carbonate is a mixture of aragonite, calcite, magnesian calcite (calcite with more than 4% $MgCO_3$), and the noncarbonate a combination of clays, quartz, and feldspars. This sediment deposited by continuous settling has three main sources. The calcite (tests of planktonic foraminifera and plates of coccoliths) and some of the aragonite (tests of pteropods) are pelagic in origin. The fine aragonite (needles), organically (calcareous algae) or inorganically precipitated, and some of the magnesian calcite are derived from either the margin or the top of the shallow carbonate banks, and thus are neritic in origin. Some of the magnesian calcite is also precipitated on the subsurface sediment as in situ cement. Finally, the noncarbonate fraction (clays, quartz, and feldspar) is transported into the region by oceanic currents, today mainly through the Western Boundary Undercurrent.

Along the cores from the Bahamian basins, the content of aragonite and calcite varies in antithetical cycles coupled with more irregular variation of magnesian calcite [Supko, 1963; Pilkey and Rucker, 1966; Rucker, 1968; Kier and Pilkey, 1971; Lynts et al., 1973; Boardman, 1978; Droxler et al., 1983; Boardman et al., 1983; Kiefer, 1983; Droxler, 1984; Mullins et al., 1984]. Also, carbonate cycles are observed down the cores, and they parallel the aragonite cycles; low carbonate content intervals systematically occur during aragonite lows and correspond to quartz maxima [Droxler et al., 1983; Droxler, 1984]. These cycles are closely correlated with the oxygen isotope record of planktonic foraminifera, and together with additional nannostratigraphy and radiocarbon evidence, clearly relate the aragonite and carbonate cycles to the glacial/interglacial rhythm of the earth's climate during the late Quaternary [Droxler et al., 1983].

The origin of the aragonite cycles remains a

controversy today.  In view of the detailed correspondence between the oxygen isotope record and the aragonite curve in a core from the northern part of Tongue of the Ocean and because of the lag time between the bank flooding and the aragonite increase, Droxler et al. [1983] have rejected the widely accepted idea that the aragonite cycles reflect the intermittent exposure and flooding of the banks [Supko, 1963; Kier and Pilkey, 1971; Boardman, 1978; Boardman et al., 1983].  Instead, we have suggested that the aragonite cycles are the product of shallow dissolution cycles, analogous to the Pleistocene calcite dissolution cycles established in deeper waters of the Atlantic Ocean.

This paper presents further evidence that the aragonite cycles are related to more global climatic, and thus oceanic, variation than to regional carbonate bank flooding.  It shows a detailed study of the record of the last 20,000 years in several cores from the Bahamian basins and the adjacent Atlantic Ocean.  The study of this time interval is certainly critical for trying to solve the controversial origin of the aragonite cycles.  This time interval includes the transition between the last glacial maximum at 18,000 years B.P. and the present interglacial hypsithermal at 6,000 years B.P.; drastic changes have occurred during this transition in the sediments.

## Core Material and Analyses

We have selected for analyses the top of three piston cores, GS 7705-34 from a slope terrace in the northern Tongue of the Ocean, CI 8108-54 from Abaco Knoll in the northeast Providence Channel, and GS 7705-17 from the crest of Eleuthera Ridge in the Atlantic Ocean, just adjacent to the Bahamian escarpment (Figure 1).  (These cores will be referred to from now on by their last two digits).  Because of the elevated location of these areas, away from the major axes of the basins where gravity flows are channeled through, the sediments have been mainly deposited by continuous settling.  The three cores, 34, 54, and 17, were collected in water depths of 1,900, 2,390, and 4,000 m, respectively, and represent an intermediate environment between the open deep Atlantic Ocean and the restricted, relatively shallow Bahamian basins.  We have also used three box cores I, IV, and VIII, collected in the same three areas as the piston cores.  Core 34 and correspondent box core I as well as core 17 and box core VIII match sufficiently well to be considered together as two single cores (Figure 2).  Box core IV, in spite of being from the same area as core 54, displays a lower average sedimentation rate and therefore was considered as an individual core.

Samples for analyses were taken every 2.5 cm and in critical intervals every 1 cm.  All or several of the following analyses have been carried out on closely spaced samples:  (1)

size fraction separation, smaller and larger than 62 µm by wet sieving, (2) carbonate mineralogy (aragonite, calcite and, magnesian calcite content) by X-ray diffraction, (3) carbonate content by Karbonat-Bombe, (4) quartz content by X-ray diffraction, (5) oxygen isotope analyses on G. rubra, (6) radiocarbon dates on bulk sediment, (7) appearance of the Globoratalia menardii complex.

## Glacial/Interglacial Variation in Mineralogy and Lithology:  Correlation and Timing

Figure 3 compares the variations of aragonite, carbonate, and quartz content and also $\delta^{18}O$ measured on Globigerinoides rubra over the last 20,000 years in core 34 reported relative to the PDB standard .  The correlation between the different analyses is striking, even if the negative correlation (mirror image) between quartz and carbonate content was expected.  An age model, based on 15 radiocarbon dates, demonstrates that the drastic changes in the different curves occur between 16,000 and 8,000 years.  This is the transition between the last glacial and the current interglacial maximum, in good agreement with the $\delta^{18}O$ curve described in the Atlantic sediments by Ruddiman and McIntyre [1981] and Duplessy et al. [1981] and in the Antarctic and Greenland ice cores by Johnsen et al. [1972].  The glacial interval (end of isotopic stage 2) occurs between 130 and 90 cm in the core; it is characterized by low aragonite (30%) and carbonate (~85%) content and by a relatively high percent of quartz.  During the glacial to interglacial transition (90-55 cm), the aragonite content first reaches a minimum (note the quartz maximum at the same time) and then sharply increases.  The present interglacial (55-0 cm) is characterized by a high aragonite and carbonate content, both reaching a maximum at 6,000 years B.P., and by the absence of quartz.  There is a slight decrease of aragonite and carbonate at the very top of the core.  The record of the last 16,000 years B.P. was also studied in three other cores, box core IV and core tops 54 and 17 (Figure 4).  These cores display, as does core 34, good correlations between aragonite, carbonate, and quartz content.  The amplitude of the variations for both aragonite and carbonate content increases toward deeper water and the more open Atlantic.  The age model for each core is based on at least four radiocarbon dates.  In box IV, in spite of the very low sedimentation rate, the aragonite increase is dated by $^{14}C$ between 10,000 and 5,000 years B.P.; in core 54, from the same setting at Abaco Knoll as box core IV but with a sedimentation rate twice as high, the aragonite increase is dated between 12,000 and 6,000 years B.P..  The aragonite increase in core 17, the deepest core, during the last deglaciation is more complex.  After being totally absent during the last glacial, aragonite first appears around

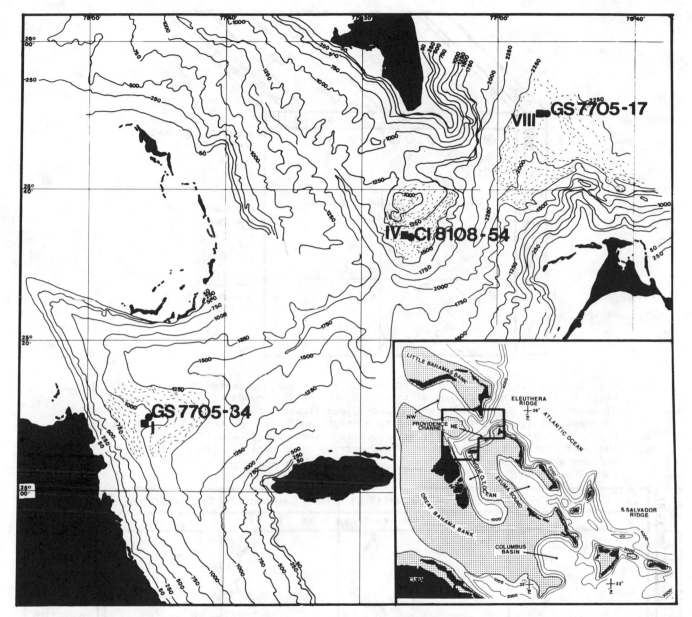

Fig. 1.  Location map for cores and and corresponding box cores, contours in fathoms
(1 fm = 1.829 m) (modified from Andrews et al. [1970]).  Each core was collected from
an elevated area, several hundred meters above the main axes of canyons, in order to
recover the most continuous sedimentary records (these appropriate locations act as a
shelter during the occurrence of catastrophic events as gravity flows transiting along
the canyons.)  Inset shows the location of the map in the Bahamian Archipelago; the
dotted pattern delineates carbonate banks shallower than 200 m; all contours in
meters.

15,000 years B.P.; between 15,000 and 12,000
years, the aragonite content never reaches more
than 30% of the total carbonate.  At around
14,000 years B.P., during a quartz maximum,
aragonite drops to zero, followed by a sharp
increase of aragonite occurring between 12,000
and 7,000 years B.P..  During the present
interglacial, since 5,000 years B.P., cores IV,
54, and 17 display, as it was observed in core
34, a slight decrease of aragonite and carbonate
content, corresponding to a slight increase of
quartz content.

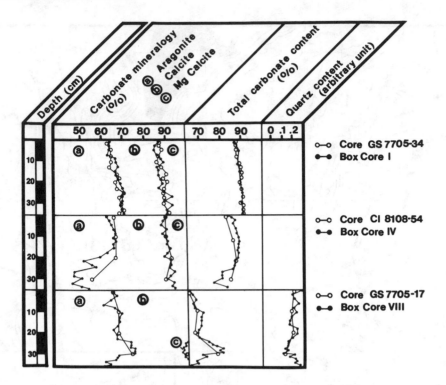

Fig. 2.  Matching tops of three piston and core boxes, collected from three different
areas in the northern Tongue of the Ocean, in the NE Providence Channel, and on
Eleuthera Ridge in the adjacent Atlantic (see Figure 1 for their locations), based on
carbonate mineralogy, total carbonate content, and quartz content. Note the excellent
match for box cores I and VIII with corresponding cores 34 and 17, respectively, and
the poor match between box core IV and corresponding core 54, due to different
sedimentation rates.

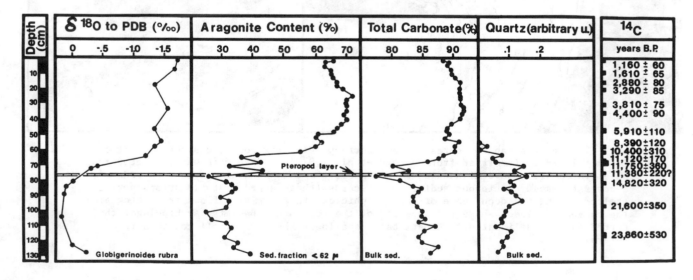

Fig. 3.  Detailed analyses in core top 34 of oxygen isotopes recorded on
*Globigerinoides rubra*, aragonite, total carbonate, and quartz content, for the last
glacial maximum and the current interglacial, placed in a time frame based on 15
radiocarbon dates. Note the good correlation between the different analyses. The
drastic changes in each case occurred during the glacial to interglacial transition.

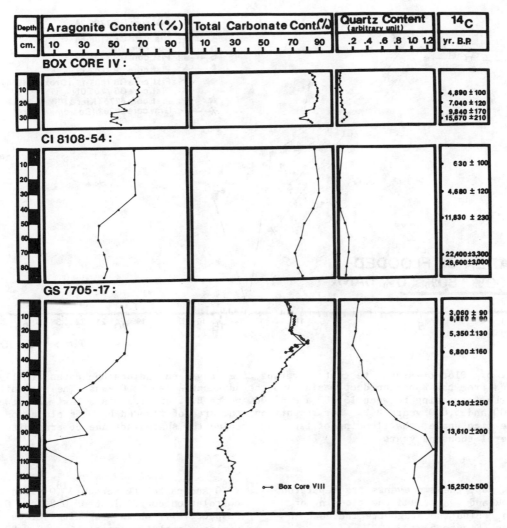

Fig. 4. Variations of aragonite, carbonate, and quartz content in three core tops from the NE Providence Channel (box core IV and gravity core CI 8108-54) and from Eleuthera Ridge (GS 7705-17). Note the good correlations between the three curves during the last deglaciation. In spite of different sedimentation rates, the glacial/interglacial transition is well defined. The aragonite sharp increase is dated between 12,000 and 6,000 years B.P. In core 17, the aragonite disappeared for a short while at around 13,500 years B.P. and then increased until 7,000 years B.P.

In order to illustrate the general timing of the aragonite increase during the last deglaciation, we have plotted aragonite content of the bulk sediment versus time based on 43 radiocarbon dates in six cores from this study, in three cores from Droxler [1984], and in two cores from Kier and Pilkey [1971] and Boardman [1978] (Figure 5 and Table 1). The plot clearly shows that the sharp increase in aragonite started both in the Bahamian basins and in the adjacent deep Atlantic Ocean between 15,000 and 11,000 years B.P. The aragonite maximum is reached between 7,000 and 5,000 years B.P. in the different cores.

The combination of a sea level curve for the last deglaciation established in Bermuda by Neumann [1969], with the hypsographic curves of two representative regions of the shallow banks, Joulters Cays [Harris, 1979] and southern Tongue of the Ocean [Palmer, 1979], gives a good estimate of the timing of bank flooding (Figure 6). Seventy percent of the Bahama Banks must have been flooded between 7,000 and 4,000 years B.P. (Figure 6). Since 4,000 years B.P., sea level has risen very slowly or remained stationary. When the timing of the aragonite increase during the last glacial to interglacial transition is compared to the flooding of the

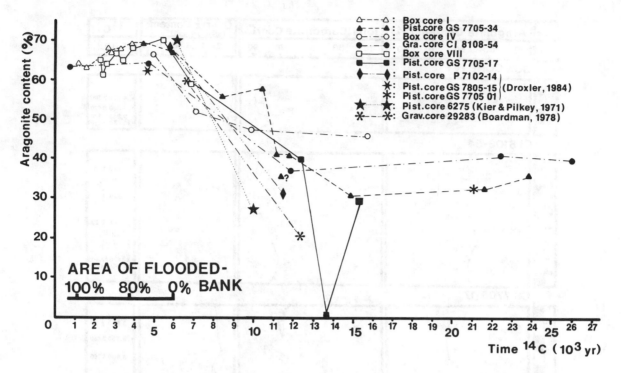

Fig. 5.  Plot of aragonite content versus time (based on radiocarbon dates) in 11 cores from the Bahamian deep basins and the adjacent Atlantic Ocean.  The aragonite starts increasing between 15,000 and 11,000 years B.P. and reaches a maximum between 7,000 and 5,000 years B.P.  For comparison, the area of flooded bank is placed on the same time scale; the flooding of bank lags behind the sharp increase of aragonite by several thousand years.

carbonate bank top, the two events are separated by several thousand years, and the flooding of the banks lags behind the aragonite increase in the deep basins (Figure 5).

### The Deglaciation in Cores 34 and 54

More detailed curves, based on 1-cm-spaced samples, were established for the deglaciation period between 15,000 and 8,000 years B.P. in cores 34 and 54.  Because the aragonite variations are present in both bulk sediment and silt-clay fraction, and because this latter fraction makes, on the average, up to 80% of the bulk sediment, variations of aragonite content must be related to the silt-clay fraction.  We therefore will rely on the fine fraction to study in detail the transition between the last glacial maximum and the current Holocene interglacial.

The aragonite, carbonate, and quartz content for the silt-clay fraction was analyzed every 1 cm in cores 34 and 54.  The oxygen isotopic composition of G. rubra was measured in core 34 with almost the same 1-cm-spaced samples.  Again the covariance of the curves is obvious, and the curves display good similarities between the two cores (Figure 7).  In core 34, aragonite versus

$\delta^{18}O$ and carbonate versus $\delta^{18}O$ yield a negative correlation coefficient of 0.81 and 0.93, respectively, while aragonite versus carbonate yields a positive correlation of 0.80.

The sharp aragonite increase occurs in core 34 just after the first appearance of the G. menardii complex at 63 cm (Figure 7).  The interpolated age of that level, 10,500 to 11,000 years B.P., matches well with the dated appearance of the G. menardii complex at 11,000 years in the Atlantic Ocean by Ericson et al. [1961].  The G. menardii stratigraphy confirms well the general timing of core 34 based on $^{14}C$ and leads us to reject the reversed radiocarbon date of 11,380 years B.P. at 73-74 cm as too young.  The appearance of the G. menardii complex in core 54 is more difficult to place; both G. menardii and G. tumida are present between 48 and 52 cm, in agreement with the $^{14}C$ age of 11,800 years B.P., and then disappear for the next 3 cm; they reappear at 45 cm, a level corresponding to an interpolated age of 10,000 to 10,500 years B.P.

The curves in core 34 display three well-defined regression periods during the general deglaciation: (III) at 10,750, (II) at 12,750, and (I) at 14,250 years B.P. (Figure 7).  The

TABLE 1   Radiocarbon Dates

| Core Type | Core Number | Core Depth cm | Uncorrected Years B.P. | Corrected Years B.P. |
|---|---|---|---|---|
| Piston core | P 7102-14 | 76-77 | 5730 ± 120 | |
| | | 99-100 | 11.430 ± 180 | |
| Piston core | GS 7705-01 | 69-70 | 21.000 ± 1400 | |
| Box core | I | 0-2.5 | 1160 ± 60 | 1630 ± 60 |
| | | 2.5-5.0 | 1210 ± 65 | 1670 ± 70 |
| | | 5.0-7.5 | 1540 ± 65 | 2000 ± 65 |
| | | 7.5-10.0 | 1610 ± 65 | 2060 ± 65 |
| | | 12-13 | 2630 ± 90 | |
| | | 14-16 | 2880 ± 90 | 3330 ± 80 |
| | | 19-21 | 3290 ± 85 | 3750 ± 85 |
| | | 29-31 | 3810 ± 75 | 4270 ± 75 |
| Piston core | GS 7705-34 | 34-35 | 4400 ± 90 | |
| | | 47-49 | 5910 ± 110 | |
| | | 57-59 | 8390 ± 120 | |
| | | 60-62 | 10.400 ± 110 | |
| | | 66-67 | 11.200 ± 170 | 10.720 ± 170 |
| | | 66.5-69 | 11.750 ± 360 | |
| | | 73-74 | 11.380 ± 220 | |
| | | 80-82 | 14.820 ± 320 | |
| | | 95.5-96.5 | 21.610 ± 350 | 21.220 ± 350 |
| | | 114-116 | 23.860 ± 530 | — |
| Gravity core | CI 8108-54 | 9-10 | 630 ± 100 | 1140 ± 100 |
| | | 29-30 | 4680 ± 120 | 5130 ± 130 |
| | | 49-50 | 11.830 ± 230 | 12.260 ± 240 |
| | | 72-73 | 22.400 ± 3300 | 22.780 ± 3300 |
| | | 75-76 | 26.600 ± 3000 | 26.970 ± 3000 |
| Box core | IV | 14-15 | 4890 ± 100 | 4530 ± 100 |
| | | 22-23 | 7040 ± 120 | — |
| | | 27-28 | 9840 ± 170 | 9440 ± 170 |
| | | 32-33 | 15.670 ± 210 | 15.290 ± 210 |
| Piston core | GS 7805-15 | 18-19 | 4550 ± 150 | |
| Box core | VIII | 0-1 | 2240 ± 100 | |
| | | 2-3 | 2800 ± 90 | |
| | | 3-4 | 3400 ± 90 | |
| | | 4-5 | 2570 ± 105 | |
| | | 6-7 | 2360 ± 90 | |
| | | 9-10 | 3060 ± 90 | |
| | | 13-14 | 3850 ± 90 | |
| | | 23-24 | 5350 ± 130 | |
| | | 35-36 | 6800 ± 160 | |
| Piston core | GS 7705-17 | 70-72 | 12.330 ± 250 | |
| | | 89-91 | 13.610 ± 200 | |
| | | 127-131 | 15.250 ± 500 | |

same three periods seem to be recorded in core 54; however, the record is not as clear, probably owing to the low sedimentation rates in core 54 and the scarcity of [14]C dates. The level in core 17 without any aragonite and a maximum value of quartz (Figure 4) probably corresponds to the regression period (I). When the variations in core 34 are plotted on a time scale instead of on core depth, the three regression periods observed in core 34 correlate crudely with the timing of

the regression periods on both Antarctic and Greenland ice cores of Byrd and Camp Century stations, respectively [Johnsen et al., 1972] (Figure 8). These three regressions may be related to the three Dryas cold events observed in Aquitaine (France) by Laville et al. [1983]. The last regression period, at 10,750 years B.P. (Younger Dryas?) is the best marked in the aragonite, carbonate, and $\delta^{18}O$ curves in core 34. Duplessy et al. [1981] and Rudimann and

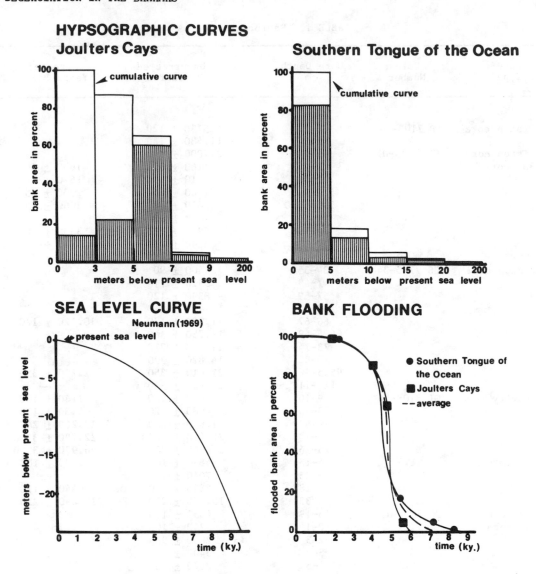

Fig. 6.  Combinations between the hypsographic curves of two areas on the Great Bahama Bank and a sea level curve from Bermuda in order to produce a flooding curve for the shallow banks.  According to this curve, the flooding of the banks started at around 7,000 years B.P., and the major part of the shallow banks was flooded by 3,000 years B.P.

McIntyre [1981] have both documented this cold period in the North Atlantic (Figure 8).  If we accept this correlation, then the major aragonite increase would correspond to the warming period just after the Younger Dryas, the Termination $I_B$ of Duplessy et al. [1981].  The major increase of carbonate content (quartz decrease) and the sharp depletion in $\delta^{18}O$ would correspond to the warming just after the Dryas event (II), indicated by a rapid change of $\delta^{18}O$ in Greenland ice (Figure 8).  This event probably marks the transition from a glacial to an interglacial mode of climatic fluctuation [Dansgaard et al., 1971].

If our interpretation is correct, events as

short as 1,000 years must have been recorded in the sediment of the Bahamian basins.  To our knowledge, such a resolution has not been observed before in cores with similar sedimentation rates.

Discussion

Our detailed analyses of the last 20,000 years in the Bahamian basins and adjacent Atlantic Ocean lead us to two major observations.

1.  The lag time between bank flooding and aragonite increase in the basins, already observed in core 34 [Droxler et al., 1983], has

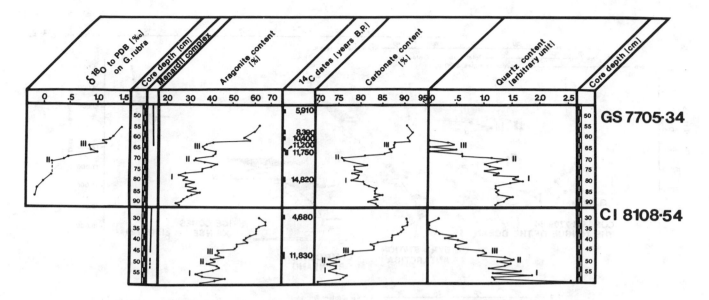

Fig. 7.  Detailed analyses in cores 34 and 54 of the glacial/interglacial transition between 16,000 and 8,000 years B.P., known as the deglaciation period.  Note the overall good correlations between the different curves; in core 34, correlation coefficients of 0.81 between aragonite and $\delta^{18}O$ and 0.80 between aragonite and total carbonate.  Note also the close similarities between the two cores; both display a step-like transition with three reversal events I, II, and III, at 14,250, 12,750, and 10,750 years B.P., respectively.

been confirmed here to be a general fact in several cores from different Bahamian basins and the adjacent Atlantic Ocean.  Thus, the hypothesis that flooding and exposure of the banks have caused the aragonite cycles is not easily reconciled with our data.

2.  The aragonite variations seem to have occurred in concert with the other well-established drastic changes during the last deglaciation.  They seem to have also recorded global events of 1,000 years duration during the deglaciation interval.  The aragonite cycles therefore should also be related to a more global oceanic cause than the regional cause of flooding of the shallow banks.  How conclusive are these deductions?

The bank flooding curve has some limitations because it is based on one sea level curve and two hypsographic curves of specific areas from the Great Bahama Bank.  An updated version (A.C. Neumann, personal communication, 1982) of a sea level curve from Bermuda [Neumann, 1969] was chosen because of its proximity to our study area and its general agreement with several other curves published by Morner [1971].  We have to remain cautious, however, because no general consensus has yet been reached on sea level changes in the Holocene, especially for the last 6,000 years [Kidson, 1982].  The areas of Joulters Cay and southern Tongue of the Ocean were selected as representative of the Great

Bahama Bank for the reason that they are the only ones where the Pleistocene morphology of the upper slope (previous to the Holocene transgression) is known with some detail.  The Joulters area is also the immediate hinterland of core 34.  The resulting bank flooding curve appears to be the best one that can be compiled with the data available at the present time.  As an indirect backup, the beginning of bank flooding at around 6,000 years B.P. fits well with the $^{14}C$ dates (6,600 and 5,500 years B.P.) of the oldest Holocene sediment deposited in the Bight of Abaco [Neumann and Land, 1975].  The sharp aragonite increase at around 10,000 years B.P. (see deglaciation section), if explained by bank flooding, would correspond to a sea level position at around 30 m below present-day sea level, on a hypothetical slope terrace.  Such a terrace, to our knowledge, has never been identified in the Bahamas.

On the other hand, the lag time between bank flooding and the increase of aragonite cannot really be explained by "old" contaminated radiocarbon dates, caused by the addition of old sediment eroded from the upper slope and the bank top during the Holocene transgression.  The radiocarbon dates fit well with the worldwide established $\delta^{18}O$ stratigraphy during the last glacial to interglacial transition, and also with the stratigraphy based on the G. menardii complex, at least in core 34.  In conclusion, the

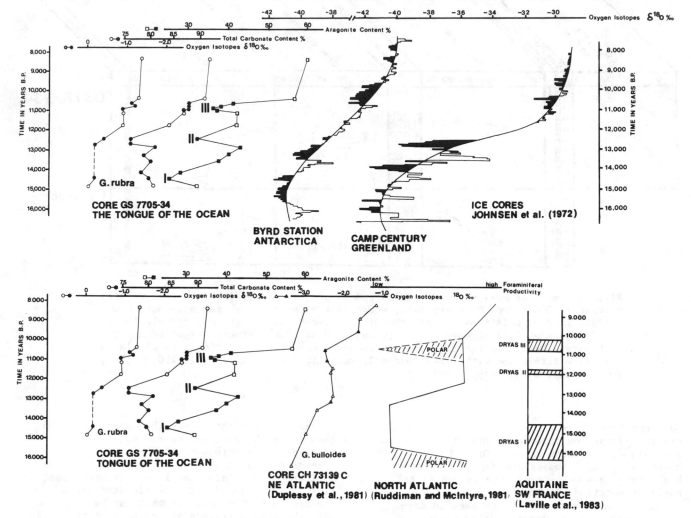

Fig. 8. In core top GS 7705–34, variations of $\delta^{18}O$, total carbonate, and aragonite content during the last deglaciation plotted versus a time scale instead of core depth. Comparisons with North Atlantic data and ice core data from Antarctica and Greenland are shown. Note the general agreement between the different studies; see text for details. (The open symbols correspond to levels of existing radiocarbon dates, the solid symbols to interpolated dates.)

observations of the lag time of several thousand years between bank flooding and aragonite increase in the basins seem to be general and real according to the present available data.

In addition, the aragonite variations are occurring in concert with the other variations of carbonate and quartz content, and to some extent of $\delta^{18}O$. These latter variations have global oceanic origin. The noncarbonate fraction had to be transported during glacial intervals through oceanic currents into the Bahamian basins, because the latter are surrounded by a pure carbonate environment. Several distinct events are recorded in each core of this study: the

noncarbonate input reached a peak sometime between 14,000 and 15,000 years B.P., a minimum at around 6,000 years B.P., and since then displays a slight increase according to the curves of carbonate and quartz content. A similar trend was observed by Balsam [1981] in cores from the western margin of the North Atlantic ranging in latitude between 32° and 42° and water depth between 3,000 and 4,000 m. Balsam [1981] and, more recently, M.T. Ledbetter and W.L. Balsam (unpublished manuscript, 1985) also suggested that carbonate dissolution was intense during late glacial time, lessened in the early interglacial, reaching a minimum around

6,000 years, and increased again in the late Holocene. This dissolution pattern recalls, maybe coincidentally, the curve of the aragonite content in the cores of this study. There, the aragonite maximum is always observed between 5,000 and 6,000 years, the hypsithermal of the present interglacial. In addition, we saw that the aragonite content sharply increased during the second part of the deglaciation interval, when the carbonate preservation is at its maximum in the western margin of the North Atlantic based on percent foraminiferal fragments (M.T. Ledbetter and W.L. Balsam, unpublished manuscript, 1985). Also, the aragonite content slightly decreased since 6,000 years B.P., when Balsam [1981] and M.T. Ledbetter and W.L. Balsam (unpublished manuscript, 1985) reported a decrease in preservation.

Due to the close correlations between the variations of aragonite content and both the known oceanic global variations of $\delta^{18}O$ of planktonic foraminifera and the carbonate and quartz content, the causes of the aragonite cycles should also be related to the open ocean and should be more general than just the flooding of the surrounding carbonate bank tops. Based on these observations and deductions, and mainly on the good correlation between the variation of the aragonite content and the preservation pattern of M.T. Ledbetter and W.L. Balsam (unpublished manuscript, 1985), we are proposing again [Droxler et al., 1983] that the pattern of aragonite content in the cores of this study reflects mainly the fluctuations of the much shallower aragonite dissolution levels. Due to the greater solubility of aragonite, these cycles would be the shallow replicates of the Pleistocene glacial calcite dissolution cycles, established in deeper waters of the Atlantic Ocean by several authors [Gardner, 1975; Bé et al., 1976; Thunell, 1976; Damuth, 1977; Crowley, 1983; Balsam, 1981, 1983; M.T. Ledbetter and W.L. Balsam, unpublished manuscript, 1985]. The preservation of aragonite, minimum during the last glacial maximum, would have increased to reach a maximum at 6,000 years B.P. and since then slightly decreased again. To the primary dissolution factor modulating the aragonite cycles, variations in parallel of aragonite input from the shallow banks can explain the observed pattern of high aragonite accumulation rates during interglacial time (Figure 9). The aragonite accumulation rates in core 34 are low during the last glacial maximum and slightly increase during the deglaciation interval; however, the striking pattern of the aragonite accumulation rate is its sudden increase since 6,000 years B.P. by a factor of 4 to 6. According to our interpretation, these high aragonite accumulation rates during the last 6,000 years would reflect both maximum preservation of aragonite and maximum aragonite input from the shallow banks. A low compaction effect for these recent sediments may also

partially explain the high aragonite accumulation rates since 6,000 years B.P.

## Conclusions

1. Detailed analyses of the last 20,000-year record in several cores from the Bahamian basins and the adjacent Atlantic Ocean show that variations of aragonite content are occurring in concert with the variations of carbonate and quartz content, and also with the oxygen isotopic record of Globigerinoides rubra; these latter variations are known to be related to the glacial rhythm of the earth's climate.

2. The most drastic changes occurred during the deglaciation interval, between 16,000 and 8,000 years B.P. During that time, the aragonite content in the cores sharply increased and displayed good correlations with the carbonate content and oxygen isotopic record. Together, the different mineralogical and lithological variations seem to have recorded regression events of 1,000 years duration that crudely correlated with established events in ice cores from Greenland and Antarctica, in North Atlantic cores, and in SW France.

3. The lag time of several thousand years between bank flooding and aragonite content, already observed in one core by Droxler et al. [1983], has been confirmed to be a general fact in several cores from different Bahamian basins and from the adjacent Atlantic Ocean.

4. Based on these observations, the aragonite variations seem to be related to more global climatic, thus oceanic, fluctuations than the regional alternate flooding and exposure of the Bahama Banks. We are proposing instead aragonite dissolution cycles analogous to the calcite dissolution cycles observed in deeper waters and tied to the glacial rhythm of the earth's climate.

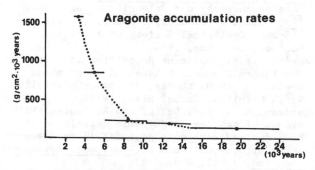

**Core GS 7705 34 & Box Core I**

Fig. 9. Variations of aragonite accumulation rates in core 34 and box core I. The aragonite accumulation rates, low between 24,000 and 15,000 years B.P. and slightly increasing between 15,000 and 6,000 years B.P., suddenly jumped since 6,000 years B.P. by a factor of 4 to 6.

5. The interpretation of the aragonite cycles to be controlled by dissolution offers opportunities to monitor back in time carbonate dissolution in the water column well above the dissolution levels recorded by calcitic plankton. These shallow dissolution cycles may be sensitive indications of basic variations in the cycling of carbon and carbon dioxide between the atmosphere and the oceans. Also, the high resolution of the aragonite record (a 1,000-year event) opens up the prospect of understanding better the last deglaciation interval, when drastic changes occurred within the earth's climate and thus within the oceans.

Acknowledgments. Supported by La Compagnie Française des Péroles - TOTAL (France), by National Science Foundation grants OCE 80-19266 and 82-15101 (principal investigator, W. Schlager) and by a contribution from the Rosenstiel School of Marine and Atmospheric Science. I thank W. Schlager and C. Emiliani for helpful discussions and advice, D. Price for help during the oxygen isotope analyses and D. Stipp and his graduate students D. Weston, N. Samter, and A. Castellanos for radiocarbon dates.

## References

Andrews, J.E., F.P. Shepard, and R.J. Hurley, Great Bahama Canyon, Geol. Soc. Am. Bull., 81, 1061-1078, 1970.

Balsam, W.L., Late Quaternary sedimentation in the western North Atlantic: Stratigraphy and paleoceanography, Palaeogeogr. Palaeoclimatol. Palaeoecol., 35, 215-240, 1981.

Balsam, W.L., Carbonate dissolution on the Muir Seamount (western North Atlantic), interglacial and glacial changes, J. Sediment. Petrol., 53(3), 719-731, 1983.

Bé, A.W.H., J.E. Damuth, L. Lott, and R. Free, Late Quaternary climatic record in western equatorial Atlantic sediments, in Investigation of Late Quaternary Paleoceanography and Paleoclimatology, Mem. 145, edited by R.M. Cline and J.D. Hay, pp. 165-200, Geological Society of America, Boulder, Colorado, 1976.

Boardman, M.R., Holocene deposition in northwest Providence Channel, Bahamas: A geochemical approach, Ph.D. thesis, 155 pp., University of North Carolina, Chapel Hill, 1978.

Boardman, M.R., L.A. Dulin, and R. Kentes, High stands of sea level: Rhythmic deposition of bank-derived carbonate sediment in the deep periplatform environment, Geol. Soc. Am. Abstr. Programs, 16(6), 528, 1983.

Crowley, T.J., Calcite dissolution patterns in the central North Atlantic during the last 150,000 years, Mar. Geol., 51, 1-14, 1983.

Damuth, J.E., Late Quaternary sedimentation in the western equatorial Atlantic, Geol. Soc. Am. Bull., 88, 695-710, 1977.

Dansgaard, W., S.J. Johnsen, H.B. Clausen, and C.C. Langway, Jr., Climatic record revealed by the Camp Century ice core, in The Late Cenozoic Glacial Ages, edited by K.K. Turekian, pp.37-56, Yale University Press, New Haven, Conn., 1971.

Droxler, A.W., Late Quaternary glacial cycles in the Bahamian deep basins and in the adjacent Atlantic Ocean, Ph.D. thesis, 165 pp., University of Miami, Miami, Fla., 1984.

Droxler, A.W., W. Shlager, and C.C. Whallon, Quaternary aragonite cycles and oxygen-isotope record in Bahamian carbonate ooze, Geology, 11, 235-239, 1983.

Duplessy, J.C., G. Delibrias, J.L. Turon, C. Pujol, and J. Duprat, Deglacial warming of the northeastern Atlantic Ocean: Correlation with the paleoclimatic evolution of the European continent, Palaeogeogr. Palaeoclimatol. Palaeoecol., 35, 121-144, 1981.

Ericson, D.B., M. Ewing, G. Wollin, and B.C. Heezen, Atlantic deep-sea cores, Geol. Soc. Am. Bull., 72, 193-286, 1961.

Gardner, J.V., Late Pleistocene carbonate dissolution cycles in the eastern equatorial Atlantic, in Dissolution of Deep-Sea Carbonates Spec. Publ. Cushman Found. Foraminiferal Res., vol. 13, edited by W.V. Sliter, A.W.H. Bé, and W.H. Berger, pp. 129-141, U.S. National Museum, Washington, D.C., 1975.

Harris, P.M., Facies anatomy and diagenesis of a Bahamian ooid shoal, Sedimenta VII, 163 pp., University of Miami, Coral Gables, Fla., 1979.

Johnsen, S.J., W. Dansgaard, H.B. Clausen, and C.C. Langway, Jr., Oxygen isotope profiles through the Antarctic and Greenland ice sheets, Nature, 235, 429-434, 1972.

Kidson, C., Sea-level changes in the Holocene, Quat. Sci. Rev., 1(2), 121-151, 1982.

Kiefer, K.B., Quaternary climatic cycles recorded in the isotopic record of periplatform pelagic deposition: Northwest Providence Channel, Bahamas, M.S. thesis, 107 pp., Duke University, Durham, N.C., 1983.

Kier, J.S., and O.H. Pilkey, The influence of sea level changes on sediment carbonate mineralogy, Tongue of the Ocean, Bahamas, Mar. Geol., 11, 189-200, 1971.

Laville, H., J.L. Turon, J.P. Texier, J.P. Raynal, F. Delpech, M.M. Paquereau, F. Prat, and A. Debenath, Upper Pleistocene paleoclimatic history of Aquitaine and Bay of Biscay since the last interglacial, Bull. Inst. Geol. Bassin Aquitaine, 34, 219-241, 1983.

Ledbetter, M.T. and W.L. Balsam, Paleoceanography of the deep Western Boundary Undercurrent on the North American Continental margin for the last 25,000 years, Geology, in press, 1985.

Lynts, G.W., J.B. Judd, and C.F. Stehman, Late Pleistocene history of Tongue of the Ocean, Bahamas, Geol. Soc. Am. Bull., 84, 2605-2684, 1973.

Morner, N.A., The Holocene eustatic sea-level

problem, Geol Mijnbouw, 50, 699-702, 1971.

Mullins, H.T., K.C. Heath, H.M. Van Buren, and C.R. Newton, Anatomy of a modern open-ocean carbonate slope: Northern Little Bahama Bank, Sedimentology, 31, 141-168, 1984.

Neumann, A.C., Quaternary sea-level data from Bermuda, Congr. Int. Union Quat. Res. Abstr. Paris 8th, 228-229, 1969.

Neumann, A.C., and L.S. Land, Lime mud deposition and calcareous algae in the Bight of Abaco, Bahamas: A budget, J. Sediment. Petrol., 45, 763-786, 1975.

Palmer, M.S., Holocene facies geometry of the leeward bank margin, Tongue of the Ocean, Bahamas, M.S. thesis, 200 pp., University of Miami, Coral Gables, Fla., 1979.

Pilkey, O.H., and J. Rucker, Mineralogy of Tongue of the Ocean sediments, J. Mar. Res., 24, 276-285, 1966.

Rucker, J.B., Carbonate mineralogy of sediments of Exuma Sound, Bahamas, J. Sediment. Petrol., 38, 68-72, 1968.

Ruddiman, W.F., and A. McIntyre, The North Atlantic Ocean during the last deglaciation, Palaeogeogr. Palaeoclimatol. Palaeoecol., 35, 145-214, 1981.

Supko, P.R., A quantitative X-ray diffraction method for the mineralogical analysis of carbonate sediments from Tongue of the Ocean, M.S. thesis, 159 pp., University of Miami, Coral Gables, Fla., 1963.

Thunell, R.C., Calcium carbonate dissolution history in late Quaternary deep-sea sediments western Gulf of Mexico, Quat. Res., 6, 281-297, 1976.

# LATE HOLOCENE CARBONATE DISSOLUTION IN THE EQUATORIAL PACIFIC: REEF GROWTH OR NEOGLACIATION?

R.S. Keir and W.H. Berger

Scripps Institution of Oceanography, University of California, San Diego
La Jolla, California 92093

Abstract. Equatorial Pacific sediments in water depths greater than 4 km exhibit increased calcium carbonate dissolution and anomalously old $^{14}$C ages in the upper ~5-10 cm of sediment. This suggests that that these sediments have been "chemically" eroded by unusually high dissolution rates within the recent past. An atmosphere-ocean-sediment model is employed to evaluate the effect of Holocene carbonate reef growth on deep sea sediment accumulation/erosion rate as well as the resulting carbonate fraction and radiocarbon stratigraphies. This model also predicts the temporal atmospheric $CO_2$ content. We assume that about 6 g $CaCO_3$ $cm^{-2}$ of ocean bottom is extracted to produce lowered carbonate values and anomalous $^{14}$C ages. The model assumes a continuous supply of settling particles throughout the Holocene, and that the anomalously old sediment is produced by a sudden increase in reef growth within the last 3 kyr. This time interval corresponds to the turnover time of the particulate carbonate influx in the upper mixed layer of sediment. In cases where a burst of reef growth ends before 3 kyr B.P. the aging effect on the mixed layer is subsequently eradicated by the settling young carbonate. The result is a "normal" mixed layer age of 3 to 5 kyr overlying a discontinuous age jump or hiatus. Slow continuous reef growth throughout the Holocene produces a similar result. This type of $^{14}$C profile is not observed. However, a burst of reef growth in the last 3 kyr would not only produce the peculiar radiocarbon profile which is observed, but also a pulse-like increase of greater than 70 ppm in the atmospheric $CO_2$ concentration. This is not seen in published data on interstitial air trapped in ice cores. Accordingly, it appears that basin-to-shelf transfer by reef growth is not the cause of the recent erosional event in the Pacific. Rather, a mechanism must be found which produces undersaturation of deep waters in some other fashion, for example, by an increase in productivity caused by global cooling ("neoglaciation").

## Introduction

Holocene calcareous sediments deposited in the equatorial Pacific Ocean below water depths of 4 km show a rather peculiar profile regarding carbonate values and $^{14}$C ages. Calcium carbonate percentages are reduced in the upper decimeter relative to adjacent values deeper in the sediment (Figure 1). This observation is generally valid for the entire Indo-Pacific region [Berger, 1970]. The depletion reflects the Holocene increase in dissolution of carbonate, which represents the most recent phase in the well-known late Pleistocene dissolution cycles in this region [Olausson, 1965; Oba, 1969; Broecker, 1971; Berger, 1973; Thompson and Saito, 1974; Luz and Shackleton, 1975; Thompson, 1976]. In addition, the lowest carbonate values tend to occur at the very top of the cores. This is especially striking in the box cores taken on the EURYDICE (ERDC) Expedition, Leg 9 [Berger and Killingley, 1982].

The radiocarbon ages of carbonate sediments in the top several centimeters of sublysoclinal cores are abnormally old (Figure 2). Both in the western and eastern equatorial Pacific, radiocarbon ages in surface sediments of the shallower cores are between 3000 and 4000 years (3 to 4 kyr), while ages exceed 6 kyr in the cores taken below 4 km depth. In the deepest cores, there is a steep age-versus-depth gradient in the upper part of the sediment, despite the expectation that this portion of the sediment is mixed by benthic organisms. The association of depleted carbonate content with old radiocarbon exhibited in these core tops suggests that recent erosion has occurred rather than partial dissolution only [Sundquist et al., 1977; Keir, 1984]. In the steady state condition, increased dissolution causes the mixed layer carbonate to become younger in age [Broecker and Peng, 1982; Keir, 1984]. Given that erosion is necessary to explain simultaneously the sublysoclinal carbonate and radiocarbon profiles, it is of

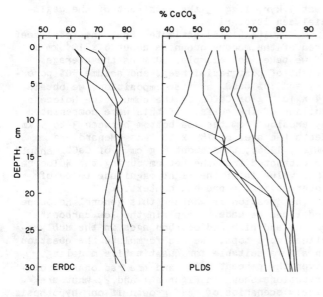

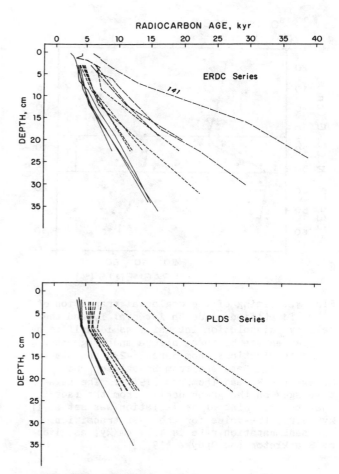

Fig. 1. Calcium carbonate fraction versus depth for sediments below 4 km water column in western (ERDC) and eastern (PLDS) equatorial Pacific. For location of cores, see Berger and Killingley [1982].

interest to explore possible mechanisms by which a late Holocene dissolution pulse could be produced. One mechanism which comes to mind is the buildup of shelf carbonate at the expense of deep-sea carbonate. Here we formulate this idea within the framework of a quantitative model. The question which we attempt to answer is this: could late Holocene growth of coral reefs have been the cause of the erosional event, or must other mechanisms be called upon, such as recent global cooling ("neoglaciation").

### Post-Termination Dissolution and Reef Hypothesis

We have earlier suggested that increased reef growth during the Holocene could be responsible for the observation that strong carbonate dissolution follows deglaciation [Berger, 1982; Berger and Keir, 1984]. The observation apparently is valid for all late Pleistocene deglaciation events. The preservation data of Thompson [1976], from core V28-238, when combined with the oxygen-isotope data of Shackleton and Opdyke [1973], indicate that this is so (Figure 3). Seven terminations have been combined in the graph, by positioning the midpoints on top of each other. The scale is that appropriate for the stage 1/stage 2 boundary. The boundaries were picked as follows (the values in parentheses being those given by Shackleton and Opdyke [1973]: 1/2, 23 cm (22 cm); 5/6, 216 (220); 7/8, 426 (430); 9/10, 593 (595); 11/12, 756 (755); 15/16, 1021 (1015); 19/20, 1215 (1210). Only

Fig. 2. Radiocarbon age versus depth for equatorial Pacific cores below 2 km depth [Berger and Killingley, 1982]. For ERDC, dashed lines are cores below 3400 m water depth. For PLDS, dashed lines are cores below 4 km water depth.

distinct terminations with a large amplitude were chosen for plotting.

Results show considerable scatter but suggest some general features, as follows: (1) The terminations connect periods of $\delta^{18}O$ maxima and minima, emphasizing the suddenness of the transition. The pre-deglaciation maximum may derive from long-term changes in ice composition [Olausson, 1981] as well as from maximum cooling. The post-deglaciation minimum may derive from a number of effects, including climatic overshoot phenomena (incomplete mixing of the ocean during meltwater input, unusual warming due to high $pCO_2$ from decrease of ocean productivity). (2) The transition periods are characterized by unusually good preservation of carbonate. The deglaciation preservation spike is well known for Termination I [Berger, 1977; Shackleton, 1977], and has equivalents for earlier termination events, as is clear from the inspection of the data of Thompson [1976]. The reasons for the deglacial preserva-

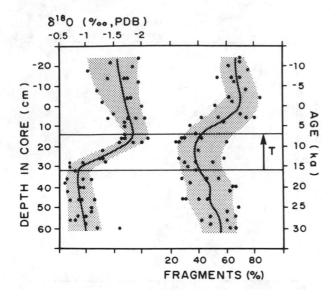

Fig. 3. Timing of the deglacial transition of δ¹⁸O, identified as depth interval T, and the relative dissolution intensity as identified by the percentage of foraminifera and fragments. Seven terminations from core 28-238 are stacked (see text). Data are from Shackleton and Opdyke [1973] and Thompson [1976]. The negative ages on the graph derive from the fact that the midpoint of deglaciation was set at 11 kyr B.P., the value for the last transition. The sedimentation rate is 1.71 cm/ky, as given by Shackleton and Opdyke [1973].

tion event are unknown. Shackleton [1977] suggested buildup of the terrestrial biosphere, which extracted $CO_2$ from the ocean-atmosphere system, thus increasing carbonate saturation. The locking-up of organic carbon on shelves (phosphate extraction model of Broecker [1982]) also would produce such a peak in preservation [Keir and Berger, 1983]. (3) The fragmentation profile shows a rapid increase in carbonate dissolution in post-transition time, with maximum dissolution values a few thousand years after the transition. It seems reasonable to assume that this dissolution pulse is caused, at least in part, by buildup of shallow-water carbonate.

The reef hypothesis can be supported by a number of arguments, as follows:

1. Modern coral reefs commonly grow on an unconformity produced by the erosion of older Pleistocene reefs during the previous low stand of sea level [Daly, 1915; Thurber et al, 1965; Lalou et al., 1966; Purdy, 1974; Thom et al, 1978; Hopley et al, 1978; Marshall and Davies, 1984]. Radiocarbon analysis of borehole samples of modern reefs yield virtually no dates in the range of 9 to 30 kyr B.P. (see, for example, Hopley [1982], Table 8.3] for references. Thus it seems plausible that the bulk of modern carbonate reef accumulation took place within the

last 9 kyr, lagging the main part of the deglacial sea level rise.

2. Smith [1978] estimated that the total reef area of the modern ocean is about $6 \times 10^5$ km². If we take, for example, 20 m as the average height of postglacial reef, and assume 50% porosity and 80% $CaCO_3$ as its composition, we obtain $14 \times 10^{18}$ g of $CaCO_3$ as the cumulative Holocene buildup of the reefs. If this were compensated by erosion from the sea bottom between 3 to 5 km depth (an area of $185 \times 10^6$ km² [Menard and Smith, 1966], then about 8 g cm⁻² of $CaCO_3$ should be extracted from the bottom during the Holocene. This value is of the right magnitude to be of interest in the present context.

The question is whether this general Holocene model can be made to explain the low carbonate values and high radiocarbon ages in the sublysoclinal core tops. We can formulate the question in a form suitable for quantitative modeling. Given the percent $CaCO_3$ and the radiocarbon distributions shown in Figures 1 and 2, what are likely scenarios of reef growth if our hypothesis is to explain the core top data? We shall show that the necessary scenarios are, in fact, unlikely. Thus, it appears that the reef growth hypothesis, which seems quite satisfactory for post-transition dissolution in general, cannot explain an additional late Holocene dissolution effect as indicated in our ERDC core top data.

## Model Assumptions and Input

The model we use is a combination of two previously published models. The first [Keir, 1984] relates percent $CaCO_3$ and radiocarbon stratigraphies to changing carbonate saturation in the deep ocean. According to this "diagenetic" model, an increase from 20% to 40% in undersaturation (with respect to calcite) over a 1500-year period results in the erosion of about 6 g cm⁻² $CaCO_3$. The resulting radiocarbon stratigraphies are similar to those seen in cores ERDC 129 and 141.

The second model [Broecker, 1982] contains a two-box ocean configuration for total $CO_2$ and alkalinity. With a suitable dissolution feedback from the seafloor, this model has been used to relate changes in atmospheric $pCO_2$ to records of stable isotopes ($\delta^{18}O$, $\delta^{13}C$) and of carbonate preservation in deep-sea cores [Keir and Berger, 1983; Broecker and Peng, 1984]. In the present study, we use the "diagenetic" model as the dissolution feedback to the box model. Here we are interested in the effect of coral reef growth scenarios, with a wide range of time scales on predicted percent $CaCO_3$ and radiocarbon stratigraphies (and also on predicted $pCO_2$ changes). Note that the previously used sediment feedback algorithum is not suitable for rapidly occurring deep-sea chemical changes because it was assumed that carbonate accumulation plus dissolution balances sedimentation input. This will not be the case, for example, if a rapid increase in acidity occurs in the water overlying calcareous sediment.

We shall ignore the effect of extracting organic carbon from the ocean. Shackleton [1977] and Broecker [1982] have proposed that such extraction occurs maximally during the deglaciation, approximately 13 to 9 kyr B.P. In this paper, we shall be largely concerned with reef growth episodes which might have occurred much more recently than this date.

The model only considers the response of a flat-bottomed, single-depth ocean with an area of about $3 \times 10^8$ km$^2$. This limitation prevents consideration of the different regional and depth-related responses. Thus, the model presented here represents a gross overview of the behavior of the ocean-atmosphere-sediment system. For example, the time scale of the dissolution response would be longer if only a portion of the seafloor is sensitive to the increased undersaturation. The uncertainty is compounded by the fact that the chemical kinetics of the sediment response are virtually unknown. Thus, introducing geographic boundary conditions without better definition of kinetics would not necessarily improve results.

We assume that the influx of dissolved carbonate from rivers and the transfer of particulate organic carbon and carbonate from the surface ocean to the deep sea occur at a constant rate over the period of interest, ranging from 0.5 to 9 kyr B.P. Also, it is assumed that the production of $^{14}$C by cosmic rays in the atmosphere remains constant. Forcing of the system is assumed to be solely the consequence of extraction of CaCO$_3$ from the surface ocean by reef growth. This results in a change in the chemical equilibrium of the surface water characterized principally by the reactions

$$Ca^{++} + 2HCO_3^- \rightarrow CaCO_3 \text{ (s)} + CO_2 \text{ (aq.)} + H_2O$$

$$Ca^{++} + HCO_3^- + H_2BO_3^- \rightarrow CaCO_3 \text{ (s)} + H_3BO_3 + H_2O$$

which lower the seawater pH and increase the pCO$_2$. As a consequence, the dissolved carbonate ion concentration will also be lowered, and when this signal is transferred to the deep ocean by the thermohaline circulation, dissolution of the calcareous sediment will increase, particularly in sensitive areas below the lysocline, where partial undersaturation has previously existed.

Generally, we will consider cases where reef growth occurs at a constant rate after a pre-existing no-growth steady state condition. Also, we postulate a constant flux of 2.6 mg cm$^{-2}$ yr$^{-1}$ of particulate CaCO$_3$ to the deep ocean together with a particulate organic carbon flux 3.5 times greater than the carbonate flux on a molar basis (91 $\mu$mole C cm$^{-2}$ yr$^{-1}$). This is similar to the P/2/1 case of Keir and Berger [1983]. However,

in this previous paper we assumed that 10% of the particulate carbonate flux dissolved prior to burial, while we here take about 44% (1.14 mg cm$^{-2}$ yr$^{-1}$) to dissolve before burial. Primarily, this is because we wish to match our results against ERDC 141, discussed later, which is estimated to have a carbonate input of 1.46 mg cm$^{-2}$ yr$^{-1}$. The overall 2.6 mg cm$^{-2}$ yr$^{-1}$ flux is necessary to preserve the observed alkalinity difference between the surface and deep ocean against a thermohaline circulation carrying a 1000-year deep water residence time. Alternatively, one could use a particulate carbonate flux of 1.62 mg cm$^{-2}$ yr$^{-1}$ against a 1500-year deep ocean residence time, retaining the assumption of 10% dissolution before burial. Our reading of the literature indicates that the shorter time is probably more nearly correct [e.g., Broecker, 1979]. At present, there is no consensus on how much of the particulate CaCO$_3$ flux dissolves before incorporation into the sediment, and a discussion of this matter is beyond the scope of this paper. Figure 4 summarizes the steady and temporal carbonate fluxes in the model.

## Modifications to Previous Models

Details of the calculations are given in two previous papers [Keir and Berger, 1983; Keir, 1984]. The original quasi-stationary response of the sediment in the first paper [Keir and Berger, 1983, Appendix B] is replaced with the model presented in the second one. Given the assumptions described in the previous section, equations (6) and 7) of Keir and Berger [1983] for total inorganic carbon and alkalinity change in the upper reservoir ($\Sigma_T$ and $A_m$) become

$$\frac{d\Sigma_T}{dt} = \frac{1}{\tau_m}(\Sigma_d - \Sigma_m) - \frac{1}{V_m}[(1 + \nu_1) I_{carb} + E_{carb}(t) - H_{riv}] \quad (1)$$

$$\frac{dA_m}{dt} = \frac{1}{\tau_m}(A_d - A_m) - \frac{1}{V_m}[(2 - \nu_0\nu_1) I_{carb} + 2(E_{carb} + I_{carb} - H_{riv})] \quad (2)$$

Notation is given in Table 1. The only change to the equations governing properties of the deep box is that the total dissolution flux (in milligrams per square centimeter per year, of calcium carbonate to the deep sea is given by

$$\zeta = 1.14 + J(t)$$

where $J(t)$ is the sediment dissolution response. As before, this is assumed to be a function of

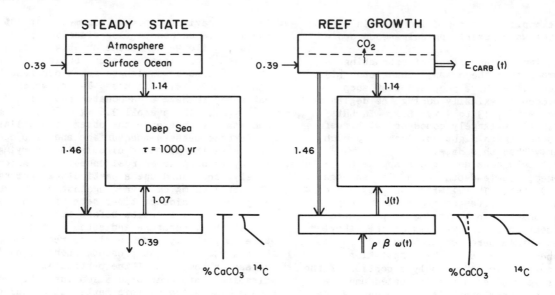

MODEL PARTICULATE CaCO$_3$ FLUXES

Fig. 4. Model calcium carbonate particulate and dissolution fluxes, in milligrams per square centimeters per year compared for initial steady-state and transient conditions.

the carbonate fraction, $B_1$, of sediment near the overlying water interface, and of the dissolved carbonate ion concentration of the deep ocean, $c(t)$. We assume this function to be

$$J = [\kappa B_1 (1 - \frac{c}{c_{eq}})^{(n+1)}]^{1/2} \qquad (3)$$

where $\kappa$ and $n$ are constants, and $c_{eq}$ is the dissolved carbonate ion concentration at equilibrium with calcite. Table 1 lists the initial steady state conditions. The initial value for $J$ is 1.07 mg cm$^{-2}$ yr$^{-1}$, leaving 0.39 mg cm$^{-2}$ yr$^{-1}$ to accumulate, which balances the river supply. The initial surface ocean $pCO_2$ is 270 ppm.

The diagenetic model for the sediment as described previously [Keir, 1984] was driven by a prescribed dissolved carbonate ion concentration function, $c(t)$. Values for the bioturbation diffusion coefficient and mixed layer thickness are assumed to be 20 cm$^{-2}$ kyr$^{-1}$ and 8 cm. Discussion of the effects due to variation of mixing rates is given elsewhere [Keir, 1984]. In this work, c at any instant in time is a function of the total $CO_2$ and alkalinity of the deep ocean. Certain modifications were necessary because different numerical techniques were used for the ocean-atmosphere model and for the diagenetic model. The sediment dissolution flux given in Equation 3 is a slowly varying function of the carbonate content, $B_1$. Thus, from the known value of $c(t)$ at time t, an estimate of $B_1^{(p)}$ $(t+\Delta t)$ can be obtained with the Crank-Nicolson

calculation. The result is combined with an average estimate of $B_1^{(p)}(t+1/2\Delta t)$ for the Runge-Kutta calculation of the ocean-atmosphere total $CO_2$, alkalinity, etc. at $t + \Delta t$. A corrected value of $B_1(t+\Delta t)$ is then obtained from a second Crank-Nicolson calculation employing $c(t)$, $c(t+\Delta t)$, $B_1(t)$, and $B_1^{(p)}(t+\Delta t)$.

We take this opportunity to correct an error in our earlier paper (S. Richardson, Iowa State University, personal communication, 1984). Two of the equations in Appendix A [Keir and Berger, 1983] have not been printed correctly. The equations (A5) and (A12) of that paper are

$$g(y, c_m) = py^3 + qy^2 + ry + s$$

$$\frac{\partial g}{\partial y} = 3py^2 + 2qy + r$$

where the coefficients q, r, and s are functions of the dissolved $CO_2$ concentration, $c_m$. Using the notation of Keir and Berger [1983],

$$q = q_0 (\Sigma_T - \gamma c_m) - pA_m + q_1$$

$$r = -\beta_2 [r_0 A_m - (\Sigma_T - \gamma c_m)] (\Sigma_T - \gamma c_m)$$
$$- r_1 A_m^2 + r_3 [A_m + 2(\Sigma_T - \gamma c_m)]$$

$$s = -s_0 A_m^2 (\Sigma_T - \gamma c_m)$$

TABLE 1. Notation and Numerical Values

| Symbol | Description | Value |
|---|---|---|
| | Constants | |
| $\tau_m$ | surface ocean residence time | 57 years |
| $\tau_d$ | deep ocean residence time | 1000 years |
| $\upsilon_1$ | ratio of organic to carbonate particulate flux | 3.5 |
| $\upsilon_0$ | ratio of alkalinity titrated to $CO_2$ produced by organic matter oxidation | 0.17 |
| $H_{riv}$ | river supply of dissolved $CO_3^-$ | 0.39 mg $cm^{-2}$ $yr^{-1}$ |
| $I_{carb}$ | rate of particulate carbonate transfer from surface to deep ocean | 2.6 mg $cm^{-2}$ $yr^{-1}$ |
| $\kappa$ | flux rate constant | $10^4$ $mg^2$ $cm^{-4}$ $yr^{-2}$ |
| $c_{eq}$ | equilibrium $CO_3$ = with calcite | 0.095 mmole $kg^{-1}$ |
| | Time Dependent Coefficient | |
| $E_{carb}$ | Coral reef growth rate per unit area of ocean bottom | 0* |
| | Dependent Variables | |
| $\Sigma_T$ | total dissolved inorganic carbon in the atmosphere and surface ocean per volume of surface ocean | 2.708* mmole $kg^{-1}$ |
| $\Sigma_m$ | concentration of dissolved carbon in surface ocean | 2.005* mmole $kg^{-1}$ |
| $A_m$ | total alkalinity of surface ocean | 2.341* meq $kg^{-1}$ |

*Steady state value.

## Box Core ERDC 141

Of the sublysoclinal cores, ERDC 129 and ERDC 141 have the most detailed radiocarbon profiles (Figure 2). Two age reversals appear in ERDC 129, suggesting that the record may be slightly disturbed. We chose ERDC 141, therefore, as target for our analysis.

This core lies 2.5° north of the equator in the western equatorial Pacific. It is the second-deepest core (4325 m) in the ERDC box core series. The detailed radiocarbon stratigraphies of ERDC 129 and ERDC 141 were published by Krishnamurthy et al. [1979] and are also listed by Berger and Killingley, [1982]. The accumulation rate of calcium carbonate based on the age-versus-depth profile below 8 cm is about 0.4 mg $cm^{-2}$ $yr^{-1}$.

The depth of ERDC 141 is close to the average for the Indo-Pacific (4100 m), and in this sense it may be considered "typical" for the carbonate dissolution rate and its stratigraphic pattern. The average depth of carbonate dissolution on the Indo-Pacific seafloor is greater. We can estimate it by successively multiplying the relative abundances of depth intervals [Menard and Smith, 1966] with the corresponding proportion of the settling carbonate dissolving within the interval

[Berger et al., 1982]. The result of this procedure is shown in Figure 5. The average depth at which carbonate dissolves appears to be near 4900 m, roughly the elevation of the carbonate compensation depth (CCD).

Primarily, we shall consider the effect of extracting 6.3 g of coral $CaCO_3$ per square centimeter of ocean bottom at various rates during the last 9 kyr. It has been shown that this amount of excess dissolution can result in a radiocarbon distribution similar to that of ERDC 141 [Keir, 1984]. Because the model used in this paper simply considers a flat-bottomed ocean, the corresponding reef accumulation would be near 20 x $10^{18}$ g of $CaCO_3$.

## Results: Radiocarbon and Percent $CaCO_3$ Stratigraphy

In Figure 6, predicted radiocarbon and %$CaCO_3$ stratigraphies that result from three different reef growth functions are compared. These contrast slow continuous growth with a recent, relatively intense, short pulse of reef growth. We have also compared the different effects that result from a reef growth pulse centered at about 5 kyr B.P. as opposed to 1 kyr B.P. After the reefs begin to grow, the $^{14}C$ age versus depth profiles in the bioturbated layer become progressively older until at some point a "maximum $^{14}C$ age profile" is reached. After this time the $^{14}C$ ages in the mixed layer generally decrease. It will be shown below that the aging of the mixed sediment greatly depends on the reef growth rate.

The case in Figure 6a is taken fom previous work [Keir, 1984]. Here the erosion is due to a prescribed change in the chemistry of the seawater overlying the sediment. Results from forcing by coral reef growth are shown in Figures 6b-6d. For these cases, the cumulative growth of the corals is set equal to the integrated excess dissolution,

$$\int_0^2 [J(t) - .39]\, dt$$

found originally for the bottom-water forcing (Figure 6a).

The cases in Figures 6a and 6b are similar in that the perturbation, whether a 20% increase in the undersaturation of the bottom water or excess growth of coral reef, is presumed to occur over 1500 years followed by a reversion to normal condition for the last 500 years. Despite the difference in the shape of the resulting sediment accumulation rate versus time relation, the final radiocarbon age versus depth profile is quite similar in these two cases. The model-predicted profile is systematically about 1.5 kyr younger than the measured $^{14}C$ ages in the upper 4 cm. A slow continuous extraction of 6.3 g $cm^{-2}$ over 9000 years (Figure 6c) produces about the same

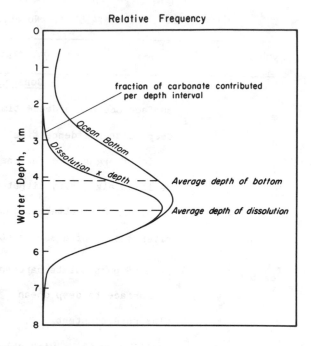

Fig. 5. Indo-Pacific hypsography (excluding marginal seas) [Menard and Smith, 1966] and relative percentage of total dissolution versus water depth (see text).

$^{14}C$ age at the base of the erosional discontinuity as observed in the first two cases (Figures 6a and 6b). Below the base of the mixed layer, the $^{14}C$ age jumps to about 15 to 17 kyr B.P. with the historical 0.63 cm $kyr^{-1}$ accumulation recorded beneath. However, as shown in Figure 6c, slow reef growth of 0.7 mg $cm^{-2}$ $yr^{-1}$ produces a negative accumulation of only -0.2 cm $kyr^{-1}$, which does not pump sufficient old carbonate against the young sedimentation flux to cause appreciable aging of the carbonate in the mixed layer. The result is a "normal", relatively young 3 to 5 kyr old mixed layer, with a sudden transition to sediment 10 kyr older at the base of the mixed layer. This type of profile is not observed in ERDC 141 and, to our knowledge, is not observed in any other dated deep-sea core top.

Figure 6d illustrates the age versus depth profile evolution as a result of mid-Holocene maximal reef growth centered about 5 kyr B.P., which would be the case, for example, if the fringing-reef crest growth for Hayman Island [Hopley et al., 1978; Hopley, 1982] were representative of the global postglacial reef accumulation. While a maximum old-age anomaly of the mixed layer would have occurred about 4 kyr B.P., by the present time this profile would have relaxed to near "normal." This occurs because the continuous sedimentation influx of young carbonate after the erosional episode resupplies radiocarbon to the mixed layer, and the sediment

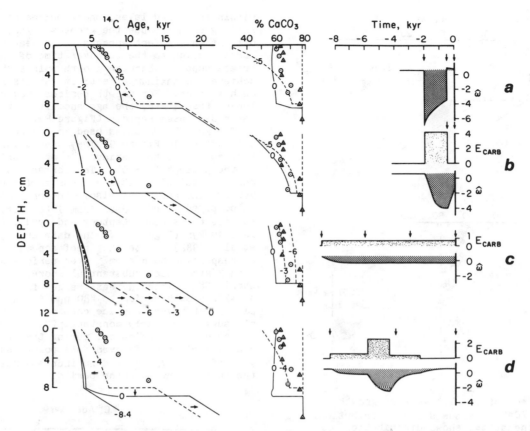

Fig. 6.  Predicted $^{14}$C age and %CaCO$_3$ versus depth distributions according to accumulation rate, $\omega$, and coral reef growth function (b - d) E$_{CARB}$ is shown on the right. Negative numbers on depth distribution curves indicate time of occurrence in kiloyears relative to present (t = 0).  Position of the depth distributions in time is indicated by arrows on right side of Figure.  Case a is from Keir [1984].

mixing is sufficiently strong to distribute the particles downward before their dissolution near the interface.

A rough estimate of the time necessary to rejuvenate the mixed layer is given by its carbonate capacity (about 3.8 g cm$^{-2}$) divided by the supply (1.46 mg cm$^{-2}$ yr$^{-1}$), which comes to about 2600 years.  Thus, erosional episodes which have ended more than 3000 years ago should produce erosional hiatuses at the base of the mixed layer, but show normal mixed-layer ages.  The present $^{14}$C profile (Figure 6d) would look similar to the case of slow continuous reef growth (Figure 6c), although the past historical development of the $^{14}$C profiles would have been different.

The "diagenetic" sediment model will not correctly predict both the radiocarbon and %CaCO$_3$ profiles quantitatively.  In general, the extraction of 6.3 g cm$^{-2}$ of CaCO$_3$ results in a CaCO$_3$ depletion of about 18 to 20% on an average, although the gradient of the CaCO$_3$ profile depends on whether erosion has subsided by the present time (compare Figure 6b to Figures 6a,

6c, and 6d).  The observed depletion of the carbonate fraction in the upper 8 cm of ERDC 141 is less than that predicted above, being about 11% on the average.  If the cumulative reef growth is taken as 3.75 g cm$^{-2}$ (approximately 60% of 6.33), the resulting depth profile of CaCO$_3$ is similar to that observed (Figure 7).  However, it can be seen that whereas 6.3 g cm$^{-2}$ extracted over 1500 years causes an average increase in the $^{14}$C age of about 3.5 kyr in the mixed layer, an extraction of 3.75 g cm$^{-2}$ gives disproportionately less of an age increase, only about 1.3 kyr (Figure 7).  Also, the $^{14}$C age just below the mixed layer at 8$^+$ cm is about 10 kyr B.P. in the 3.75 g cm$^{-2}$ case.  As a result, the calculated $^{14}$C profile below 8 cm appears systematically 5 kyr younger than observed (Figure 7).

The model study shows that the lower cumulative erosion of 3.75 g cm$^{-2}$ produces a good fit for the carbonate stratigraphy, but that the fit for the radiocarbon stratigraphy is poor.  The situation is reversed for the higher cumulative erosion.  This discrepancy also appears when the model is applied to the carbonate and radiocarbon

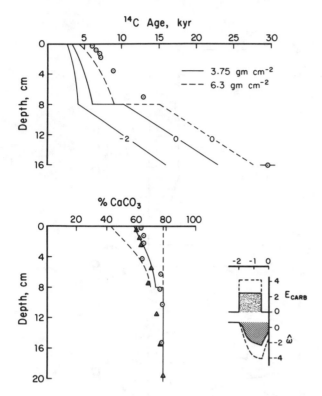

Fig. 7. Effect of cumulative reef growth on [14]C age and %CaCO3 versus depth distribution. The solid line at t=0 shows distribution resulting from 3.75 gm cm$^{-2}$ reef growth over 1.5 kyr (lower right). The dashed line indicates distribution resulting from 6.3 g cm$^{-2}$ cumulative growth as shown in Figure 6b.

stratigraphies of ERDC 129 [Keir, 1984]. There are several conceivable explanations for the dilemma, including missing core tops, an excessively resistant (and therefore old) carbonate component, and removal of fine carbonate by winnowing. The available data cannot resolve the problem. Grain-size distributions, and grain-size-specific radiocarbon values would be necessary.

### Reef Growth and pCO2 of the Atmosphere

The model indicates that cumulative reef growth of 6.3 g cm$^{-2}$, during a relatively short period of 1.5 kyr, produces a significant aging of carbonate in the mixed layer, but reef growth of the same amount spread over a time interval of 9 kyr does not (Figures 6b and 6c). Effects of the rate of reef growth on radiocarbon stratigraphy are illustrated in Figure 8. It shows the maximum [14]C age profiles in the mixed layer that develop for a cumulative growth of 6.3 g cm$^{-2}$ over several time periods ranging from 0.5 to 9 kyr. Below the mixed layer the resulting age profiles are very similar in all cases. However,

within the mixed layer, short pulses of rapid reef growth produce the greatest ages. The cut-off for this effect appears to be near 3 kyr, corresponding to the turnover time of the particulate supply. Shorter growth pulses than 3 kyr achieve the maximum effect; it matters little how much shorter. For growth periods progressively longer than 3 kyr, the age anomaly in the mixed layer decreases rapidly (Figure 8).

The growth functions used in Figure 8 are illustrated in Figure 9. They are positioned in time such that the maximum [14]C age profile occurs at the present (t = 0). The corresponding sediment accumulation rates and curves for pCO2 versus time are given also. The predicted pCO2 is compared to pCO2 data from interstitial air in two ice cores (Camp Century and Byrd), for the last 10 kyr (Figure 9). The data are from Neftel et al. [1982]. Note that reef accumulation corresponding to 6 g cm$^{-2}$ over periods of less than 3 kyr causes substantial increase in atmospheric CO2. This increase ranges from 70 ppm for a 3-kyr growth period to 480 ppm for a 0.5-kyr period. However, the ice core data of Neftel et al. suggest a nearly constant pCO2 for the entire Holocene. In particular, during the last 1500 years, a cluster of seven data points suggests a pCO2 of near 270 ppm and no increase relative to the rest of the Holocene period.

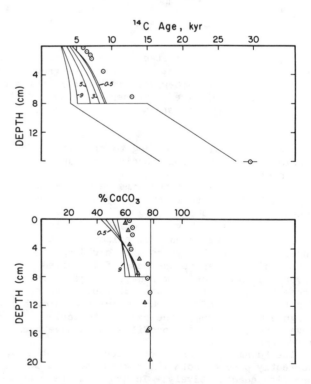

Fig. 8. Maximum age versus depth distribution resulting from 6.3 g cm$^{-2}$ cumulative growth over periods ranging from 0.5 to 9 kyr (number indicated on curve). Reef growth functions are shown in Figure 9.

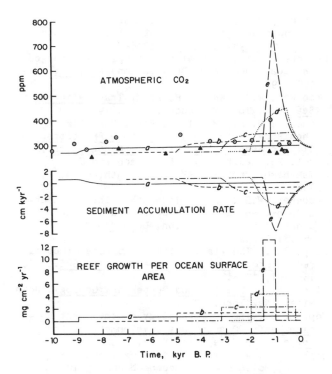

Fig. 9. Sediment accumulation rate and atmospheric $CO_2$ content versus time corresponding to reef growth functions shown in lower figure. In all cases, cumulative growth is 6.3 g $cm^{-2}$. Data points indicate ice core analysis by Neftel et al. [1982]. Circles indicate Camp Century (Greenland); triangles, Byrd Station (Antarctica). Data points which may have been affected by drill fluid contamination have been removed.

## Discussion

Assuming that the ice core data are representative of the Holocene atmospheric $CO_2$, the result in Figure 9 implies that substantial growth (i.e. corresponding to about 6 g per square centimeter of ocean bottom) of coral reefs did not occur within the last 3 kyr, or for that matter within any 3 kyr of the Holocene. Otherwise, some sort of pulse-like increase in atmospheric $CO_2$ should be evident in the ice cores. The ice core data do not prohibit slow and continuous reef growth throughout the Holocene, however. For example, a growth rate corresponding to 0.7 mg $cm^{-2}$ $yr^{-1}$ increases the $pCO_2$ by only 20 ppm. However, if such growth is the only function driving dissolution, then the expected $^{14}C$ distribution has a young mixed layer over a large hiatus (Figure 6c). This type of profile has not been observed in deep sea sediments.

Assuming the ERDC 141 tells a typical story (which is supported by the available information), we must then conclude that reef buildup is

not the reason for the late Holocene dissolution pulse. Our results do not negate the idea that Holocene reef growth contributes to the overall increase in Holocene dissolution observed in the Indo-Pacific, but it does appear that reef growth can be excluded as the probable cause of the late Holocene dissolution pulse. There remain two other obvious candidates for possible causes: an increase in the age difference between the deep waters of the Atlantic and Indo-Pacific and a global increase in ocean productivity.

At present, the age of deep water increases going from the Atlantic to the Indian and Pacific oceans, reflecting the advective pattern of water movement. Due to the continuous oxidation of particulate organic carbon, supplied from surface waters, the dissolved $CO_2$ concentration increases in the older water downstream in the Indo-Pacific. The result is that carbonate ion concentration (and hence saturation) decreases [Broecker and Takahashi, 1978]. If this downstream effect were increased, and the "diffusive" interchange of water between the Atlantic and Pacific water decreased, we would then expect increased carbonate dissolution on the Indo-Pacific but decreased dissolution in the Atlantic. In fact, inspection of a number of box cores from the North and South Atlantic, for aragonite preservation, indicates decreased dissolution in the Atlantic for the last few thousand years. Thus, a change in deep-sea circulation, or more specifically, a strengthening of North Atlantic bottom water production, is a viable hypothesis to be considered in explaining the late Holocene Pacific dissolution pulse.

The position of dissolution levels in the world ocean is closely tied to the overall carbonate productivity of the ocean. Keir [1983] has shown that a decrease in vertical circulation, postulated for the period of deglaciation, produces a pronounced preservation maximum. Conversely, an increase in mixing (and productivity) will produce a corresponding dissolution maximum. This relationship was first pointed out by Li et al. [1969], and was subsequently used by van Andel et al. [1975] to reconstruct the productivity history of the ocean in the Cenozoic. It is likely that circulation and mixing intensity were stimulated by global cooling of the climate in the last few thousand years. This cooling, known as "neoglaciation" [Flint, 1971, p. 524] spans roughly the last 2500 years, and culminated in the Little Ice Age, between about 500 and 100 years B.P. In addition to growth of glaciers, it has also found expression in increasing desertification [see Lamb, 1982]. An intensification of ocean mixing due to cooling at high latitudes would have produced a decrease in atmospheric $pCO_2$, through "biological pumping" of carbon into deep water. Such an effect is more in line with the ice core data than the increase predicted by the coral reef hypothesis. For the last 8000 years, the lowest $pCO_2$ values are found within the last 1000 years.

Because of the absence of a recent $pCO_2$ increase in the ice core data, we suggest that an increase in North Atlantic bottom water production, and in overall productivity, should be investigated as a likely driving factor rather than reef growth.  In late Pleistocene sediments, periods of dissolution in the Indo-Pacific straddle periods of increasing $\delta^{18}O$ values [Berger and Vincent, 1981; Keir and Berger, 1983].  In this light, the late Holocene dissolution pulse appears as the first indication that the new glacial age is fast approaching.  If these suggestions are valid, it would mean that changes in the carbon cycle (carbonate dissolution) precede changes in the ice reservoirs (sea level) rather than vice versa.  A similar conclusion, that the carbon cycle leads rather than lags sea level change, was put forward by Shackleton et al. [1983] on the basis of $\delta^{13}C$ stratigraphy.

Acknowledgements.  Financial support for this work was provided by the Electric Power Research Institute and NSF grant OCE82-19553.  We also are grateful to the reviewers for helpful comments.

## References

Berger, W.H., Planktonic foraminifera: Selective solution and the lysocline, Mar. Geol., 8, 111-138, 1970.

Berger, W.H., Deep-sea carbonates: Pleistocene dissolution cycles, J. Foraminiferal Res., 3, 187-195, 1973.

Berger, W.H., Deep-sea carbonate and the deglaciation preservation spike in pteropods and foraminifera, Nature, 269, 301-304, 1977.

Berger, W.H., Deglacial $CO_2$ buildup: Constraints on the coral-reef model, Palaeogeogr. Palaeoclimatol. Palaeoecol., 40, 235-253, 1982.

Berger, W.H., and R.S. Keir, Glacial-Holocene changes in atmospheric $CO_2$ and the deep-sea record, in Climate Processes and Climate Sensitivity, Geophys. Monogr. Ser., vol. 29, edited by J. Hansen and T. Takahashi, pp. 337-351, AGU, Washington, D. C., 1984.

Berger, W.H., and J.S. Killingley, Box cores from the equatorial Pacific:  $^{14}C$ sedimentation rates and benthic mixing, Mar. Geology, 45, 93-125, 1982.

Berger, W.H., and E. Vincent, Chemostratigraphy and biostratigraphic correlation:  Exercises in systemic stratigraphy, Oceanol. Acta, Proc. Int. Geol. Cong. 26th, 115-127, 1981.

Berger, W.H., M.C. Bonneau, and F.L. Parker, Foraminifera on the deep-sea floor: Lysocline and dissolution rate, Oceanol. Acta, 5, 249-258, 1982.

Broecker, W.S., Calcite accumulation rates and glacial to interglacial changes in ocean mixing, in The Late Cenozoic Glacial Ages, edited by K. K. Turekian, pp. 239-265, Yale University Press, New Haven, Conn., 1971.

Broecker, W.S., A revised estimate for the radiocarbon age of North Atlantic deep water, J. of Geophys. Res., 84, 3218-3226, 1979.

Broecker, W.S., Glacial to interglacial changes in ocean chemistry, Prog. Oceanogra., 11, 151-197, 1982.

Broecker, W.S., and T.-H. Peng, Tracers in the Sea, Eldigio,, Palisades, N.Y., 1982.

Broecker, W. S., and T.-H. Peng, The climate-chemistry connection, in Climate Processes and Climate Sensitivity, Geophys. Monogr. Ser., vol. 29, edited by J.E. Hansen and T. Takahashi, pp. 327-336, AGU, Washington, D. C., 1984.

Broecker, W.S., and T. Takahashi, The relationship between lysocline depth and in-situ carbonate ion concentration, Deep Sea Res., 25, 65-95, 1978.

Daly, R.A., The glacial-control theory of coral reefs, Proc. Am. Acad. the Arts and Sci., 51, 155-251, 1915.

Flint, R.F., Glacial and Quaternary Geology, John Wiley, New York, 1971.

Hopley, D., The Geomorphology of the Great Barrier Reef:  Quarternary Development of Coral Reefs, John Wiley, New York, 1982.

Hopley, D., R.F. Mclean, J. Marsh, and A.S. Smith, Holocene-Pleistocene boundary in a fringing reef:  Hayman Island, North Queensland, Search, 9, 323-325, 1978.

Keir, R.S., Reduction of thermohaline circulation during deglaciation: The effect on atmospheric radiocarbon and $CO_2$, Earth Planet. Sci. Lett., 64, 445-456, 1983.

Keir, R.S., Recent increase in Pacific $CaCO_3$ dissolution:  A mechanism for generating old $^{14}C$ ages, Mar. Geol., 59, 227-250, 1984.

Keir, R.S., and W.H. Berger, Atmospheric $CO_2$ content in the last 120,000 years:  The phosphate-extraction model, J. Geophys. Res., 88, 6027-6038, 1983.

Krishnamurthy, R.V., D. Lal, L. K. Somayajulu, and W.H. Berger, Radiometric studies of box cores from the Ontong-Java Plateau, Proc. Indian Acad. Sci., Sect. A 88, 273-283, 1979.

Lalou, C., J. Labeyrie, and G. Delebrias, Datations des calcaires coralliens de l'atoll de Muroroa (archipel des Tuamotu) de l'époque actuelle jusqu'à - 500,000 ans., C.R. Hebd. Seances Acad. Sci., Ser. D 263, 1946-1949, 1966.

Lamb, H.H., Climate History and the Modern World, Methuen, London, 387 pp., 1982.

Li, Y.H., T. Takahashi, and W.S. Broecker, Degree of saturation of $CaCO_3$ in the oceans, J.Geophys. Res., 74, 5507-5525, 1969.

Luz, B., and N.J. Shackleton, $CaCO_3$ solution in the tropical east Pacific during the past 130,000 years, in Dissolution of Deep-Sea Carbonates, Cushman Found. Foraminiferal Res., Spec. Publ., 13, edited by W.V. Sliter, A.W. Be, and W.H. Berger, pp. 142-150, U.S. National Museum, Washington, D.C., 1975.

Marshall, J.F., and P.J. Davies, Last interglacial reef growth beneath modern reefs in the southern Great Barrier Reef, Nature, 307, 44–46, 1984.

Menard, H.W., and S.M. Smith, Hypsometry of ocean basin provinces, J. Geophys. Res., 71, 4305–4325, 1966.

Neftel, A., H. Oeschger, J. Schwander, B. Stauffer, and R. Zumbrunn, Ice core sample measurements give atmosphere $CO_2$ content during the past 40,000 yr., Nature, 295, 220–223, 1982.

Oba, T., Biostratigraphy and isotopic paleotemperatures of some deep-sea cores from the Indian Ocean, Sci. Rep. Tohoku Univ. Ser. 2, 41, 221–252, 1969.

Olausson, E., Evidence of climatic changes in North Atlantic deep-sea cores, with remarks on isotopic paleotemperature analysis, Prog. Oceanogr., 3, 221–252, 1965.

Olausson, E., On the isotopic composition of late Cenozoic sea water, Geogr. Ann., 63A, 311–312, 1981.

Purdy, E.G., Reef configurations, cause and effect, in Reefs in Time and Space, Society of Economic Paleontologists and Mineralogists, Spec. Publ., edited by L.F. Laporte, pp. 9–76, Tulsa, Okla., 1974.

Shackleton, N.J., Carbon-13 in Uvigerina: Tropical rain forest history and the equatorial Pacific carbonate dissolution cycles, in The Fate of Fossil Fuel in the Oceans, edited by N.R. Andersen and A. Malahoff, pp. 401–427, Plenum, New York, 1977.

Shackleton, N.J., and N.D. Opdyke, Oxygen isotope and Paleomagnetic stratigraphy of equatorial Pacific core V28-238: Oxygen isotope temperatures and ice volumes on a $10^5$ year and $10^6$ year scale, Quart. Res., 3, 39–55, 1973.

Shackleton, N.J., J. Imbrie, and M.A. Hall, Oxygen and carbon isotope record of east Pacific core V19-30: Implications for the formation of deep water in the late Pleistocene North Atlantic, Earth Planet. Sci. Lett., 65, 233–244, 1983.

Smith, S.V., Coral-reef area and the contributions of reefs to processes and resources of the world's oceans, Nature, 273, 225–226, 1978.

Sundquist, E., D.K. Richardson, W.S. Broecker, and T.-H. Peng, Sediment mixing and carbonate dissolution in the southeast Pacific Ocean, in The Fate of Fossil Fuel in the Oceans, edited by N.R. Andersen and A. Malahoff, pp. 429–454, Plenum, New York, 1977.

Thom, B.G., G.R. Orme and H.R. Polach, Drilling investigation of Bewick Islands. Philos. Trans. Ry. Soc. London, 291, 37–54, 1978.

Thompson, P.R., Planktonic foraminiferal solution and progress toward a Pleistocene equatorial transfer function, J. of Foraminiferal Res., 6, 203–227, 1976.

Thompson, P.R., and T. Saito, Pacific Pleistocene sediments; Plankonic Foraminifera: Dissolution cycles and geochronology, Geology, 2, 333–335, 1974.

Thurber, D.L., W.S. Broecker, R.L. Blanchard, and H.A. Potratz, Uranium-series ages of Pacific atoll coral, Science, 149, 55–58, 1965.

van Andel, T.H., G.R. Heath, and T.C. Moore, Cenozoic history and paleoceanography of the central equatorial Pacific Ocean, Geol. Soc. Am. Mem., 143, 1975.

# CARBON CYCLE VARIATIONS DURING THE PAST 50,000 YEARS: ATMOSPHERIC $^{14}C/^{12}C$ RATIO AS AN ISOTOPIC INDICATOR

Devendra Lal[1]

Institute of Geophysics and Planetary Physics, University of California
Los Angeles, California 90024

**Abstract.** We explore the conditions under which the paleorecord of the atmospheric $^{14}C/^{12}C$ ratio can serve as an isotopic tracer for changes in the carbon cycle during the past 50,000 years. Climatic changes on time scales of $10^3$-$10^4$ years are expected to cause appreciable changes in the carbon cycle, both in the sizes of the dynamic carbon reservoirs and in the rates of mixing between them. The atmospheric $^{14}C$ paleorecord shows appreciable variations with amplitudes in the range of 2-10% at frequencies of $5 \times 10^{-3}$-$10^{-4}$ cycles per year. In order to study the information contained in atmospheric $^{14}C$ specific activity, we have carried out sensitivity studies theoretically by varying the carbon cycle parameters over a range of frequencies. These studies show that whereas atmospheric $^{14}C$ activity responds sensitively to certain changes, with the magnitude of change being in the range observed, the task of deducing changes in carbon cycle parameters from the $^{14}C$ paleorecord is a complex one. It involves delineating changes in the global production rate of $^{14}C$ due to terrestrial and extraterrestrial causes. An interesting conclusion of this work is that the observed slow decrease in the atmospheric $^{14}C/^{12}C$ ratio during the past 8000 years probably originated (in large part) from changes in the carbon cycle parameters. An interesting implication of this deduction to geomagnetic dipole field change during this period is discussed.

## Introduction

Geophysical and biogeochemical paleorecords eloquently speak of the great variability in the dynamic processes occurring in the atmosphere-ocean system during the glacial and interglacial periods [Imbrie and Imbrie, 1979; Bolin, 1980]. The biogeochemical cycle of carbon, which operates via the fluidic atmosphere-hydrosphere system, is therefore expected to be a very sensitive indicator of the climatic forcing on the system. Thus the studies of paleofluctuations in the carbon cycle constitute an entry into the problem of climatic variations in the past.

Changes in the marine carbon cycle were first recognized by Arrhenius in the equatorial Pacific [Arrhenius, 1952] from studies of marine sediments. It is now well known from such studies that the ocean chemistry changes on geologically short time scales, and that there have been several major transitions, described as commotion in the sea [cf. Berger, 1982a], during the (oscillatory) chemical evolutionary history of the oceans. The atmospheric endowment of carbon dioxide has also changed appreciably in the past $(30-40) \times 10^3$ years [Berner et al., 1980; Delmas et al., 1980], by as much as about 50% on time scales of the order of half a millenium [Stauffer et al., 1984]. These inferences are based on analyses of air trapped in ice cores. Independent support of a large variation in the atmospheric $pCO_2$ during $(15-20) \times 10^3$ years B.P. has been provided [Lal and Revelle, 1984] from analysis of $^{14}C/^{12}C$ ratio variation in a lake sediment. (Thus $CO_2$ can no longer be regarded as a constant constituent of the atmosphere, even on time scales of the order of a century.)

Based on the ice core data, it appears that appreciable changes have probably occurred in the carbon content of the oceans on short time scales, $10^3$-$10^4$ years, at least during the glacial and late glacial periods. (Comparable data series are not yet available for the Holocene.) Such changes are an integral part of some of the models proposed recently to explain the 90 parts per million by volume increase in atmospheric $pCO_2$ during transition from glacial to Holocene time (see Broecker and Peng [1984] for implications of the various models). The first attempt to explain the $pCO_2$ change was due to Broecker [1982], who suggested a scenario for deposition (extraction) of C and P on the shelf during the Holocene transgression. Berger [1982b], on the other hand, suggested a coral growth scenario during the same epoch. In such mechanisms, the

---

[1] On leave from Physical Research Laboratory, Ahmedabad 380 009, India.

total change in the dissolved carbon content of the ocean amounts to -3 to +5%, with accompanying changes in $^{13}C$ content of the ocean in some of the models.

The radionuclide $^{14}C$, along with the stable isotopes $^{12}C$ and $^{13}C$, can be used as a tracer to delineate changes in the carbon cycle during the past 50,000 years. This is an interesting period of time, as it includes the last glacial and several interstadial stages [Imbrie and Imbrie, 1979; Broecker and van Donk, 1970; Ruddiman and McIntyre, 1981]. It seems reasonable to assume that climatic changes must alter the distribution of $^{14}C$ and $^{12}C$ in the different carbon cycle reservoirs. However, most studies have emphasized that $^{14}C$ changes in the atmosphere (and in other reservoirs) are related primarily to changes in the global production rate of $^{14}C$, that is, $Q(^{14}C)$. Variations in $Q(^{14}C)$ have been believed to be due to either the solar modulation [Stuiver, 1961; Castagnoli and Lal, 1980], or the geomagnetic screening of the incident cosmic ray flux [Elsasser et al., 1956; Bucha, 1970], or changes in the carbon cycle [Stuiver, 1970; Lal and Venkatavaradan, 1970]. In this paper, we treat in some detail how the $^{14}C/^{12}C$ ratio changes in the carbon reservoirs for hypothetical changes in the carbon cycle parameters. These sensitivity studies have been carried out to determine whether a paleorecord of $^{14}C$ activity in any of the carbon reservoirs can be deconvoluted to delineate changes in the carbon cycle parameters.

Any hypothetical carbon cycle variation will probably be associated with changes in the marine sedimentary record. The character of this record would be related to causative changes, e.g., in the productivity, degree of carbonate saturation in the oceans as a function of depth, rise in sea level and the associated biogeochemical processes in the coastal region. Therefore, whereas the $^{14}C/^{12}C$ ratios in the carbon reservoirs represent an integration of changes produced by a number of processes and mechanisms, including the global changes in $Q(^{14}C)$, there is the hope that by combining several kinds of sedimentary data, it may be possible to delineate principal causes of $^{14}C/^{12}C$ changes in the reservoirs. This paper represents an attempt in this direction, building on several previous studies [Lal and Venkatavaradan, 1970; Siegenthaler et al., 1980; Keir, 1983; Lal and Revelle, 1984]. Several studies have been carried out earlier on the response of the different carbon reservoirs to changes in $Q(^{14}C)$, the different reservoirs being treated as internally well mixed. Reference is made here to papers by Houtermans [1966], Kigoshi and Hasegawa [1966], and Ekdahl and Keeling [1973].

### Model Studies of Sensitivity of $^{14}C/^{12}C$ Ratios in Carbon Reservoirs to Changes in Carbon Cycle Parameters

#### Carbon Cycle Models

For sensitivity studies of the carbon cycle, we adopt for simplicity the box model treatment which has been found to be quite sufficient to judge the response of the system to transients [Houtermans, 1966; Kigoshi and Hasegawa, 1966; Ekdahl and Keeling, 1973]. We, however, consider "open" box models. By "open," we imply that the carbon in the atmosphere-ocean system leaves the system or receives carbon from other reservoirs, e.g., the shelf regions or the sediments. Most of the $^{14}C$ model calculations have adopted a "closed" carbon cycle, since the residence time of carbon in the oceans, ~ $5 \times 10^4$ years (see section on changes in the carbon content of the reservoirs), is long compared to the mean life of $^{14}C$. However, as we have seen, differences of the order of 5% may be expected in the carbon inventory of the oceans between glacials. The same order of differences may arise over periods of $10^3$–$10^4$ years due to climate forcing, changes in sea level, etc., as discussed earlier. We, therefore, have to use an open model to be realistic.

We have considered two types of box models in our work. These are shown in Figures 1a and 1b. Figure 1a shows a five-box model including the biosphere. The ocean is divided into two layers, the mixed and the deep. The deep layer is in exchange with the sediment as well as with the atmosphere. The model is equivalent to the southern outcrop model proposed by Craig [1963] to take into account the fact that the atmosphere exchanges its carbon dioxide directly with the southern ocean, which has a $^{14}C/^{12}C$ ratio much lower than that of the rest of the ocean surface.

With a proper choice of the parameters $K_{ad}$ and $K_{da}$, the southern ocean is simulated. Similarly, the episodes of deposition of carbon in the shelf are simulated by proper choice of $K_{sd}$ and $K_{ds}$.

The model adopted in Figure 1b was primarily designed to test the sensitivity of the real ocean to changes in vertical mixing rates by adding an intermediate layer between the mixed layer and the deep ocean. To simplify computations, the biosphere reservoir was deleted.

The computations were carried out by solving the transient coupled first-order equations using the fourth-order Runge-Kutta method. Transients were imposed, starting from a steady state, by changing either the exchange parameters or the contents of the boxes. The particular merit of this approach lies in the fact that nonlinear situations, as for example, arising from a time dependent change in mixing rate, can be treated. The disadvantage, however, is the absence of an explicit solution. But in any case, the solutions become too cumbersome when one considers cyclic models as in Figure 1a.

The sizes of the boxes adopted are as follows: atmosphere (58 m), mixed layer (75 m), whole ocean (3875 m); the numbers within parentheses are the ocean equivalent depths of the boxes. In Figure 1b, the depths of the mixed and the intermediate layers were taken to be 100 m and 500 m, respectively. The amount of carbon in the biosphere was taken to be 2.4 times the amount in the atmosphere at present.

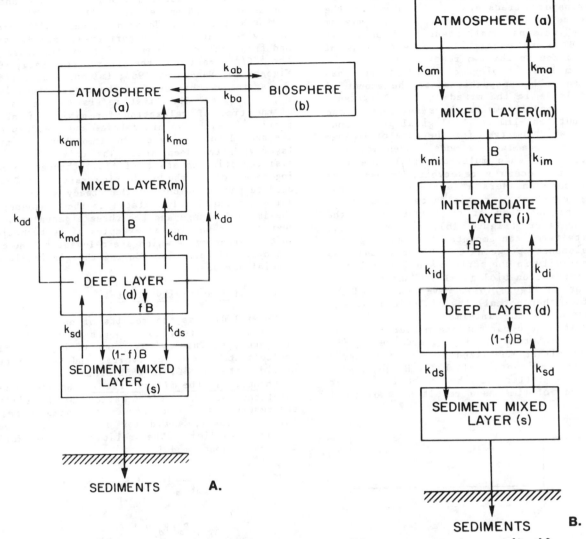

Fig. 1. "Open" type box models used for considering perturbations in $^{14}C/^{12}C$ ratios in the carbon cycle reservoirs. The ocean is divided into two layers in Figure 1a. This is a cyclic model, as there exists a direct exchange between the atmosphere and the deep sea. The chain model in Figure 1b has no biosphere, but the ocean is subdivided into three layers.

Exchange coefficients were fixed at $K_{am} = 0.1$ yr$^{-1}$, $k_{dm} = 10^{-3}$ yr$^{-1}$, and $K_{ba} = 1.67 \times 10^{-2}$ yr$^{-1}$. In Figure 1b, $K_{di}$ was taken as $10^{-3}$ yr$^{-1}$ and $K_{mi} = 5 \times 10^{-2}$ yr$^{-1}$. These parameters, with the global average amounts of carbon in the oceans as a function of depth [Takahashi et al., 1981], agree well with the measured $^{14}C/^{12}C$ ratios in the oceans [Broecker and Peng, 1982]. There is some flexibility in the choice of the exchange parameters, but since we are only concerned with the transients in $^{14}C/^{12}C$ ratios in the different reservoirs for any hypothetical change in the carbon cycle parameters, we will not discuss this issue further.

Besides mixing of dissolved carbon within the oceans by transfer across the boxes by first-order

rate constants (K terms in Figures 1a and b), carbon is transferred unidirectionally by transfer of particulate biological carbon from the mixed layer. This produces a vertical gradient in the concentration of dissolved carbon, with the lowest concentration in the mixed layer. The effect of biological transfer of carbon was first considered by Lal (1962). The available data on dissolved concentrations at that time led to values of mean biological removal time of 260 and 170 years for the Atlantic and Pacific Oceans, respectively. It should, however, be noted that the relative $^{14}C/^{12}C$ ratios in the mixed layer and the deep sea are only a function of the rate of exchange of water, and not a function of the biological rate constant B (see also Lal and

Venkatavaradan [1970]). The reason is that biological transport transfers $^{14}C$ and $^{12}C$ in the same ratio as in the mixed layer, and the mixing "resistance" remains unaffected. The principal effect of B is to change the concentrations of dissolved carbon in the two reservoirs.

In the models, biological productivity has been assumed to be proportional to the amount of carbon available in the mixed layer. The result for a two-box model remains the same, even if one considers nutrient-limited biological production. The picture which emerges for the case of biological transfer in a two-box model loses its simplifying features when multibox models are considered for the oceanic reservoir, irrespective of whether one considers a first-order or zero-order rate constant for biological production. Even with the addition of one layer, e.g., the intermediate layer (Figure 1b), the biological productivity term tantamounts to changing the mixing resistance between the layers, and the specific activity in the surface and deeper layers becomes dependent on biological production.

Note that in Figure 1b we have considered the observed fact that not all the biological production is added to the intermediate or the deep layers. Most of the particulate matter is remobilized in the intermediate layer, and the remaining part is either mobilized in the deep layer or lost to the sediments [Lal, 1977; Honjo, 1980]. A part of the latter may be remobilized over longer periods of time as a result of biological

activity at the water-sediment interface and in the sediments [Hinga et al., 1979].

For a three-box "closed" model, with oceans subdivided into two compartments, steady state and transient equations have been discussed for the $^{14}C/^{12}C$ ratios in the reservoirs [Craig, 1957; Kigoshi and Hasegawa, 1966; Lal and Suess, 1983]. The system is characterized by three eigenvalues, $\mu_1$, $\mu_2$ and $\lambda$. The first refers to the inverse of the time of relaxation for mixing of atmospheric $^{14}C$ with the mixed layer and the deep sea, the second for mixing of the atmosphere and mixed layer with the deep sea. The third, $\lambda$, is the disintegration constant of $^{14}C$. Changes in mixing rates or in the production rate of $^{14}C$ will produce perturbations in the steady state ratios which are explicitly related to the frequency of the imposed change and the three eigenvalues. We next discuss particular examples of perturbations in the carbon cycle which are relevant to the carbon cycle perturbations during the glacial-interglacial periods.

## Results of Sensitivity Studies

We will discuss the results of our model calculations separately for changes in (1) the mixing rates, (2) biological productivity, (3) carbon content of the reservoirs, and (4) the production rate of $^{14}C$, in that order.

**Changes in the mixing rates.** Steady state equations for the $^{14}C/^{12}C$ ratios $R_i$ in different reservoirs (subscripts a for atmosphere, m for mixed layer, and d for deep sea) have been discussed earlier. The relative ratios for a three-box closed model

$$\frac{R_a}{R_d} = 1 + \frac{\lambda}{K_{ma}K_{dm}}[K_{ma} + K_{md} + K_{dm} + B + \lambda] \quad (1)$$

$$\frac{R_m}{R_d} = 1 + \frac{\lambda}{K_{dm}} \quad (2)$$

are functions only of the first-order rate constants [Lal and Revelle, 1984]. It is convenient to express the ratios relative to the deep reservoir, which contains more than 90% of $^{12}C$ and $^{14}C$. Any changes in the rate constants will produce changes in the specific activities $R_a$ and $R_m$ but only inappreciably in $R_d$ (see Figure 2). The fractional changes in the specific activities in the atmosphere and the mixed layer can be obtained by differentiating equations (1) and (2) with respect to the rate constant of interest. For example, differentiating (1) with respect to $K_{dm}$, we obtain

$$\frac{d(R_a/R_d)}{R_a/R_d} = -\frac{dK_{dm}}{K_{dm}} \cdot \frac{\lambda}{\lambda + K_{dm}}$$

$$\simeq -\frac{dK_{dm}}{K_{dm}} \cdot \frac{\lambda}{K_{dm}} \quad (3)$$

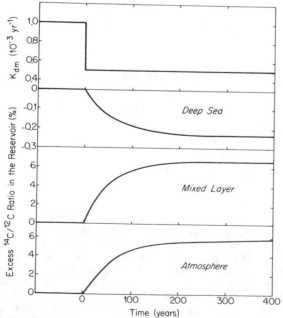

Fig. 2. The steady state $^{14}C/^{12}C$ ratios in the carbon cycle (two-layer ocean) are perturbed at t=0 due to a step function change in the vertical mixing rate ($K_{dm}$). Resulting changes in the specific activity of $^{14}C$ in the deep sea, the mixed layer, and the atmosphere are shown.

which implies that the fractional change in $R_a/R_d$ would be about one-eighth of the fractional change in $K_{dm}$; an increase in $K_{dm}$ decreases the ratio. Similarly, one can derive the functional dependence for other rate constants.

$K_{dm}$, $K_{di}$ and $K_{im}$ variation: A 50 % decrease in $K_{dm}$ produces about 6% increase in specific activity of the atmosphere [Lal and Venkatavaradan, 1970]. This is also seen easily from equation (3). The time to reequilibrium for a step function change in $K_{dm}$ (decrease by 50%) is $\lesssim$ 50 years (Figure 2). This is valid over a wide range of parameters for the rate constants. However, longer equilibration times would be expected for a box-diffusion model or a thicker mixed layer [Siegenthaler et al., 1980].

The results of a simulation for sinusoidal variations in the rate of mixing between the mixed layer, intermediate layer, and deep sea are shown in Figure 3, with an amplitude of ±50%. Note that the phase difference is small, as expected, since the relaxation period for mixing between the atmosphere and the mixed layer with the rest of the ocean is small ($\lesssim$ 50 years). The amplitude in $R_a$ is not sensitive to the frequency of variation in mixing rates for slower changes. The apparent asymmetric behavior is an artifact of the simulation wherein the mixing rates vary by factors of 0.5 and 1.5 in the extreme cases.

$K_{ad}$ variation: As was pointed out earlier, $K_{ad}$ and $K_{da}$ represent direct exchange between the atmosphere and the deep sea (southern ocean outcrop), and correspond to the cyclic model first produced by Craig [1963]. The effect of changes in the rate of mixing between the atmosphere and the deep sea outcrop was discussed by Lal and Venkatavaradan [1970] who showed that appreciable changes, at the level of several percent, can occur due to changes in the effective rate of exchange $K_{ad}$, $K_{da}$. Assuming a 12% outcrop area, we find using the model in Figure 1a that the atmospheric $^{14}C/^{12}C$ ratio decreases by 2.5% for a doubling of $K_{ad}$, and increases by 3% if $K_{ad}$ is set equal to zero. These results are in agreement with those of Lal and Venkatavaradan [1970] and indicate that changes in the exchange rate, $K_{ad}$ (due to changes in wind velocities), and changes in the exposed surface area of the ocean (due to changes in sea ice cover) over time periods of the order of 1000 years or larger can easily cause decreases or increases in the atmospheric $^{14}C/^{12}C$ ratios by up to ±5%.

$K_{am}$ variation: The effect of variations in the air-sea exchange rate over most of the ocean, $K_{am}$, is important, particularly if the exchange rate were to decrease from its present value. An increase in the exchange rate can decrease the atmospheric $^{14}C/^{12}C$ ratio by at most 6%, which is the (deduced) difference between the atmosphere and the mixed layer, corresponding to $K_{am}$ = 0.1 yr$^{-1}$. For small differences in $K_{am}$, the atmospheric $^{14}C/^{12}C$ ratio changes by 0.8% for a 10% change in $K_{am}$. From observations of the rate of

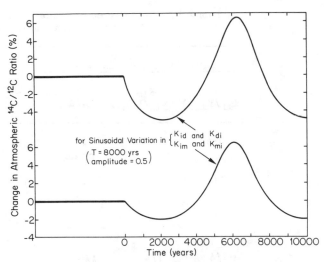

Fig. 3. The temporal variation in atmospheric $^{14}C/^{12}C$ ratios for sinusoidal variation in the mixing parameters, $K_{id}$ and $K_{im}$ with a period of 8000 years for a three-layer ocean (Figure 1b).

rise of $^{14}C/^{12}C$ ratios during the 200-year period oscillations, Lal and Suess [1983] deduced that the residence time of carbon dioxide in the atmosphere has remained within the range of 3 to 25 years during the past 8000 years.

Changes in the biological productivity. We have modeled the effect of changes in biological productivity for a three-layer ocean model (Figure 1b). The results for hypothetical changes in B and exchange constants are summarized in Table 1, for periods 50 and 200 years after start of perturbation. We present only the changes in the mixed layer and the intermediate layer. The atmosphere keeps in step with the mixed layer except for any change that occurs in its $CO_2$ content.

Shifts in biological productivity change atmospheric $^{14}C$ specific activity in a sense opposite to that due to changes in vertical mixing. Increased vertical mixing, by itself, would increase atmospheric $pCO_2$ and decrease the atmospheric $^{14}C/^{12}C$ ratio. However, vertical mixing would bring more nutrients to surface waters and increase biological productivity. This constitutes a negative feedback on the effect of vertical mixing on atmospheric $pCO_2$ and $^{14}C$ specific activity. The complex interplay of biological productivity and vertical mixing (see Table 1) necessitates considering more realistic ocean models. We will later consider the manner in which changes in B and in mixing rates gang up or oppose during glacial/interglacial periods.

Biological productivity has been identified as one of the principal mechanisms of change in atmospheric $pCO_2$ [Broecker, 1982; Broecker and Peng, 1984; McElroy, 1982; Lal and Revelle, 1984]. The availability of nutrients is an important question which is related to changes in vertical mix-

TABLE 1.    Calculated Changes in Atmospheric and Mixed Layer Reservoir
Due to Changes in $K_{mi}$ and B from Steady State

| Step Function Change | | | Percentage Change in $^{14}C/^{12}C$ Ratio | |
|---|---|---|---|---|
| $K_{mi}$ ($K_{mi}/K_{mio}$) | $K_{im}$ ($K_{im}/K_{imo}$) | B ($B/B_0$) | Mixed Layer | Intermediate Layer |
| 1/4 | 1/4 | 1 (no change) | +8.6 (+18.4) | −2.0 (−3.0) |
| 1/2 | 1/2 | 1 | +4.4 (+6.7) | −1.1 (−1.2) |
| 4 | 4 | 1 | −5.3 (−5.2) | +1.4 (+1.1) |
| 1/4 | 1/4 | 1/4 | +8.3 (+17.4) | −2.1 (−2.7) |
| 4 | 4 | 4 | −5.0 (−5.25) | +1.2 (+0.7) |

Subscript zero in the exchange coefficient implies the steady state value.
Percentage figures give the calculated change after 50 years of the step
function change in exchange coefficients; the numbers within parentheses
give the values after 200 years.

ing and also to changes in sea level [Broecker, 1982]. Therefore, any discussions of B must necessarily involve modeling changes in nutrients and carbonate chemistry. We are concerned at present only with the changes in $^{14}C/^{12}C$ ratios in the main carbon reservoirs.

Changes in the carbon content of the reservoirs. Changes in the atmospheric and oceanic reservoirs will be considered separately.

Changes in atmospheric carbon reservoir: This aspect has been discussed earlier [Lal and Venkatavaradan, 1970; Siegenthaler et al., 1980]. More recently, Lal and Revelle [1984] have pointed out that the paleorecord of atmospheric $^{14}C/^{12}C$ ratios could be used to study variations in atmospheric $pCO_2$, combining it with supplementary data to take into account other causative mechanisms of the change. A simple theoretical treatment of the problem is given by Lal and Revelle [1984]. We present here the results of simulations for the carbon cycle model as in Figure 1a, for hypothetical changes in atmospheric $pCO_2$ beginning from a steady state at t=0. For a 50% decrease in atmospheric $pCO_2$, the atmospheric $^{14}C/^{12}C$ ratios increase by ~6% (Figure 4). The time to equilibration is less than 20 years for a range of carbon cycle parameters.

The causal mechanism is the change in the mixing resistance across the air-sea interface. The difference in specific activities between the atmosphere and the mixed layer for the present amount of $CO_2$ in the atmosphere in the model is

~6%. Halving the amount of $CO_2$ in the atmosphere doubles the mixing resistance across the air-sea interface and causes an increase in atmospheric $^{14}C/^{12}C$ ratio by 6%. Doubling the atmospheric $pCO_2$ would decrease the mixing resistance by a factor of 2, and this would cause a reduction in atmospheric $^{14}C/^{12}C$ ratio by 3%. A decrease in atmospheric $CO_2$ content, therefore, leads to a larger effect than an increase by the same factor.

Changes in oceanic carbon inventory: The residence time of carbon in the oceans is an order of magnitude longer than the mean life of $^{14}C$ [Bolin et al., 1979]. However, as was mentioned earlier, appreciable fluctuations (possibly at the 5% level) in the carbon inventory of the oceans probably occurred at least during the glacial and late glacial periods. The problem of estimating a meaningful value of the residence time of carbon in the oceans is in fact complicated by temporal imbalances in input and output. The best present estimates for the rates of supply and removal of carbon to/from the ocean are $(5.6-9.6) \times 10^{14}$ g/yr [Garrels et al., 1975; Bolin et al., 1979; F. T. Mackenzie, personal communication, 1984]. This range of values corresponds to values of $4 \times 10^4$ to $7 \times 10^4$ years for the residence time of carbon in the oceans. The corresponding range of the rate constant K is $(1.4-2.5) \times 10^{-5}$ yr$^{-1}$.

For our sensitivity studies, we have considered removal of carbon from the oceans under two

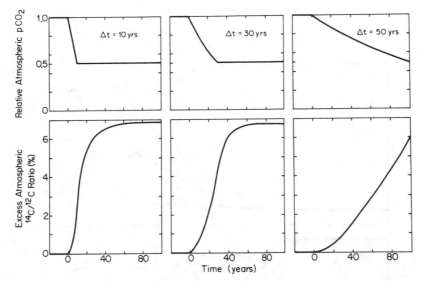

Fig. 4. (Top) The atmospheric $CO_2$ inventory reduced exponentially by half, starting at t=0, when a steady-state situation prevailed in time periods $\Delta t$=10, 30, and 50 years. (Bottom) The corresponding changes in the atmospheric $^{14}C/^{12}C$ ratios shown as a function of time for the three cases.

situations: the carbon inventory in the ocean remains constant or decreases. In the former case, the loss of carbon is balanced by an input of carbon from streams.

The results of model calculations for different values of K are shown in Figure 5. As expected, the $^{14}C/^{12}C$ ratio in the atmosphere in Figure 5a keeps on increasing with time, because in this case the $^{12}C$ inventory decreases with time as $e^{-Kt}$. However, in the case of constant carbon reservoirs, the oceanic specific activity of $^{14}C$ decreases because of addition of dead carbon from streams, contributed from weathering of rocks (Figure 5b). (In practice, the stream carbon has a high $^{14}C/^{12}C$, nearly the same as in the atmosphere, derived by exchange from the atmosphere. But the net result is an addition of dead carbon to the ocean-atmosphere system.) The time to equilibration in the case of constant carbon reservoir (Figure 5b) is the mean life of $^{14}C$.

We have also considered a possibly more realistic situation (Figure 6) where the oceanic carbon inventory is assumed to oscillate sinusoidally as a result of removal to sediments and input by dissolution of sediments. The mean age of sediment carbon, relative to surface ocean carbon, was taken to be 5000 years [Broecker, 1982; Keir, 1984]. The age of the surface sediments, which dissolve with the changing state of the carbonate equilibrium in the deep sea, is experimentally as well as theoretically determined to lie in this bracket; the effect is produced as a result of complex mixing by bioturbation and dissolution. The results in Figure 6 correspond to the onset of a removal/ addition cycle beginning with steady state at t=0. The time

evolution of the atmospheric $^{14}C/^{12}C$ ratio is followed in Figure 6. The phase difference between the oceanic carbon inventory and $^{14}C/^{12}C$ ratios in the atmosphere is small, as expected. Also, the time to equilibration is large (mean life of $^{14}C$). The evolution curve depends sensitively on the phase of the carbon inventory variation.

Changes in global production rate of $^{14}C$, $Q(^{14}C)$. This aspect has been treated extensively in several publications [Stuiver, 1961; Suess, 1980; Castagnoli and Lal, 1980]. It is well known that the changes, if any, in the cosmic ray flux as incident on the heliosphere are small on time scales of the order of $10^5$ years, but that the incident flux of cosmic rays on the earth can vary appreciably [cf. Reedy et al., 1983]. Principal changes are due to (1) solar modulation of cosmic ray flux streaming in the heliosphere and (2) variations in the geomagnetic screening of cosmic ray flux due to changes in the geomagnetic field.

The changes in atmospheric $^{14}C/^{12}C$ ratios with $Q(^{14}C)$ depend on the frequency ($\omega$) with which $Q(^{14}C)$ changes. The amplitude of the $^{14}C/^{12}C$ ratio in the deep ocean, relative to that in $Q(^{14}C)$ is attenuated by a factor of $(1 + \omega^2/\lambda^2)^{1/2}$ (see, for example, Houtermans [1966] and Kigoshi and Hasegawa [1966]). The amplitude in the atmosphere is larger, and depends partly on the particular carbon cycle model adopted. Atmospheric attenuation factors are about 200, 25 and 10 for periodicities in $Q(^{14}C)$ of 10, 100 and 1000 years, respectively.

Global $Q(^{14}C)$ changes with the geomagnetic dipole field, M, and changes in solar modulation

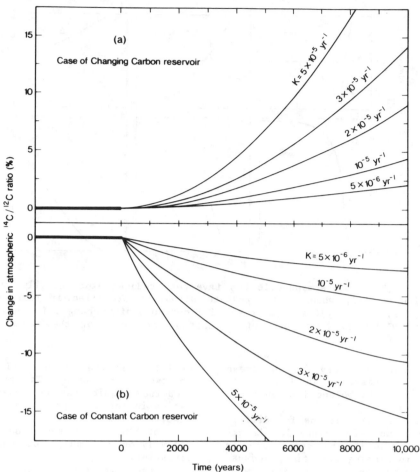

Fig. 5. A steady state closed (K=0) atmosphere-ocean model (Figure 1a) assumed to begin losing its oceanic carbon inventory at t=0, with different assumed rate constants, K. (Note that the residence time of carbon in the oceans exceeds $10^{-5}$ $yr^{-1}$; see text.) In Figure 5a, the carbon inventory decreases due to removal of carbon. However, in Figure 5b, the carbon losses are compensated by introduction of carbon from streams so that the ocean inventory of carbon remains constant.

have been recalculated using the recent cosmic ray data [Castagnoli and Lal, 1980]. The tree ring $^{14}$C/$^{12}$C record for the past 8000 years is believed to contain contributions from both these causes [Suess, 1980]. The record has two prominent frequencies: (1) a high-frequency wiggle with an amplitude of ±(1-2)%, period of ~200 years, and (2) a slow change with an amplitude of about 10%, period of about 12,000 years [Suess, 1980]. These variations seem to have natural explanations.

A recent analysis by Castagnoli and Lal [1980] showed that solar modulation in flux would cause a change in Q($^{14}$C) by about ±20% during periods of unusual solar activity, namely, the Maunder Minimum and Solar Maximum epochs. The high-frequency oscillations, known as Suess wiggles, are explainable since the attenuation factor for a period of 200 years is about 10. The observed 1-2% magnitude in wiggles can thus be understood.

Apparently, high-frequency changes ($>10^{-2}$ cycles per year) have not occurred during the present interglacial period. Otherwise, they would have been detected, particularly if they occurred in the past 8000 years, since the tree ring data are quite suited for the study of these changes. Suess [1980] has documented a prominent cycle in $^{14}$C/$^{12}$C ratios of about 200 years period, this, as discussed above, is due to solar modulation of Q($^{14}$C). Other cycles noted by Suess [1980] correspond to frequencies of $<10^{-3}$ cycles per year.

The slow variation, on the other hand, is canonically believed to be due to a variation in the geomagnetic dipole field [Bucha, 1970].

Discussion and Conclusions

We have considered the possible influence of climatic changes in the $^{14}$C/$^{12}$C ratios in the car-

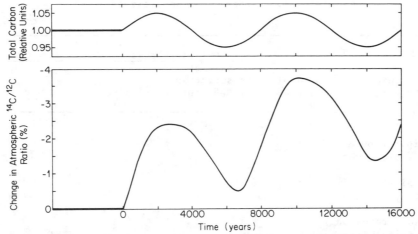

Fig. 6. Model simulation for change in the atmospheric $^{14}C/^{12}C$ ratio for a hypothetical change in the oceanic carbon inventory. Beginning with a "closed" box model (Figure 1a with no transfer or exchange with sediments) at t=0, the oceanic inventory of carbon is modulated sinusoidally with a period of 8000 years. During gains from the surface sediments, the mean age of the carbon is assumed to be 5000 years, with regard to the mixed layer.

bon cycle to see if paleodata on $^{14}C/^{12}C$ in the carbon reservoirs can be used to delineate carbon cycle changes during the past 50,000 years. We will now make an attempt to put together our results of sensitivity calculations in a proper perspective of the expected changes due to particular climatic changes which occurred during the recent glacial and interglacial periods. We will then examine the evidence and see what, if any, conclusions can be drawn from the paleodata on $^{14}C$ (and $^{13}C$).

All changes in the carbon cycle are interrelated, and therefore, in the final analysis one would have to consider the various carbon cycle changes in the particular combination they occur. But we will first consider the changes separately and their effects on $^{14}C/^{12}C$ ratios, particularly in the atmosphere. The reason for the latter is that paleodata are usually available for the atmosphere alone. It is absolutely necessary in such studies to determine $^{14}C/^{12}C$ ratios in all the carbon cycle reservoirs to deconvolve the effects of the different changes considered, and the necessity of obtaining such data can never be overemphasized. Only recently, a bold attempt has been made [Andree et al., 1984] to determine $^{14}C/^{12}C$ ratios in surface and deep waters by analyzing pelagic and benthic organisms.

The present ocean exhibits a characteristic vertical mixing time of $(1-1.5) \times 10^3$ years ($K_{dm} \simeq 10^{-3} y^{-1}$). This produces a gradient of ~15% within the oceans in the $^{14}C/^{12}C$ ratio. A difference of about 6% exists between the atmosphere and the ocean due to the mixing resistance across the air-sea interface. Thus the atmospheric $^{14}C/^{12}C$ ratio can be lowered by a maximum of 20% due to changes (speeding up) in the mixing rates. There is no theoretical limit to the magnitude by which the ratio can be increased!

The data in Tables 2 and 3 show the manner in which carbon cycle changes gang up or oppose changes in the atmospheric $^{14}C/^{12}C$ ratios; the magnitudes of changes are discussed in earlier sections. The main question before us is how the carbon cycle parameters vary with climate.

During warm periods, meridional temperature gradients would be smaller; the atmosphere would be more sluggish (smaller wind stress curl), and the thermohaline circulation would be reduced. On the other hand, during glacial periods, the oceanic mixing, in particular upwelling of intermediate waters, would be expected to be enhanced over warmer periods. A special situation arises from the spread of meltwater during the Holocene transgression. The meltwater lid proposed by Worthington [1968] for the entire ocean would cause a stable stratification and possibly inhibit ocean turnover for thousands of years [Berger and Killingley, 1982]. In the meltwater situation, our sensitivity study calculations may not hold, because the global ocean circulation model itself changes radically. However, it is obvious that this would raise the $^{14}C/^{12}C$ ratios in the atmosphere and in the surface and intermediate waters. We also note that the Holocene transgression was a complex one, with marked oscillatory changes in ice volume during 13,000-10,000 years B.P. [Ruddiman and McIntyre, 1981].

During the glacial period, higher mixing of surface and intermediate waters would lead to lower atmospheric $^{14}C/^{12}C$ ratios. Large $pCO_2$ excursions occurred during the most recent glacial epoch [Stauffer et al., 1984]. It is interesting to note that the increases in atmospheric $^{14}C/^{12}C$ ratios due to decreases in atmospheric $pCO_2$ might be counteracted by changes in vertical mixing. However, the situation may be quite the reverse. For instance, the decrease in $pCO_2$ may

TABLE 2.  Carbon Cycle Sensitivity Studies

| Oceanic Process | Sign of Change | Oceanic Change | | Atmospheric Change | |
|---|---|---|---|---|---|
| | | State of Preservation of Sediments | $^{13}C/^{12}C$ | $pCO_2$ | $^{14}C/^{12}C$ |
| Precipitation of organic carbon | ↑ | ↑ | ↑ | ↓ | ↑ |
| Precipitation of $CaCO_3$ | ↑ | ↓ | − | ↑ | ↓ |
| Exchange rates only Vertical($K_{dm}$, $K_{di}$) | ↑ | − | − | ↑ | ↓ |
| Air-sea exchange with the southern ocean ($K_{ad}$) | ↑ | − | − | ↑ | ↓ |
| Biology and exchange rates* $K_{dm}/B$ | ↑ | − | − | ↑ | ↓ |
| $K_{dm}/B$ | ↓ | ↓ | ↑ | ↓ | ↑ |

Changes in the carbon cycle parameters are given due to increase (upward arrows) or decrease in the process considered.  A dash means that the effect is small or zero.

*Here we consider hypothetically a change in the ratio of vertical mixing and biological productivity.  The two are, however, coupled in the real ocean.

itself be a result of decreased upwelling in the equatorial Pacific, i.e., lower vertical mixing.

To conclude, our sensitivity analyses show that appreciable changes are expected to occur in the $^{14}C/^{12}C$ ratios in the carbon cycle reservoirs, at levels of several percent, due to plausible changes in the carbon cycle.  In the absence of changes in $Q(^{14}C)$, the ratio in the mixed layer of the ocean changes primarily due to variations in the vertical mixing rates or in the carbon content of the deep ocean.  The atmospheric $^{14}C/^{12}C$ ratio keeps in step with this change with a lag of time period of ~(50-100) years.  In addition, the atmospheric ratio also changes due to variations in the atmospheric $pCO_2$ and in the rate of exchange of $CO_2$ across the air-sea inter-

TABLE 3.  Expected Variations in Atmospheric $^{14}C/^{12}C$ ratios During Glacial/Interglacial Periods

| | State of Preservation of Deep-Sea Carbonates | Vertical Mixing (Intermediate Waters) | Atmospheric $pCO_2$ | Net Expected Change in Atmospheric $^{14}C/^{12}C$ |
|---|---|---|---|---|
| Interglacial to Glacial | ↓ | ↑ | ↓ | ↓ |
| Glacial to Interglacial | ↑ | ↓ | ↑ | ↑ |
| Intraglacial | ? | ↓ | ↓ | ↑ |

face, including that in the southern ocean. A number of complex causal mechanisms have been identified for the changes in atmospheric $pCO_2$ [c.f. Broecker and Peng, 1984]. Each of these would directly or indirectly entail changes in the atmospheric $^{14}C/^{12}C$ ratio.

From a Fourier analysis of the variation of the atmospheric $^{14}C/^{12}C$ ratios, it may be possible to delineate the principal frequencies with which the carbon cycle has been perturbed. However, over the relatively short time period of interest, 50,000 years, the system is not expected to exhibit several cycles. Herein lies the complexity of the task. Clearly, one would have to make use of all the supplementary data, besides utilizing the paleodata on $^{14}C/^{12}C$ ratios in the mixed layer and the deep sea [Andree et al., 1984]. Examples of useful data are $^{13}C/^{12}C$ ratios in pelagic and benthic organisms, [Shackleton et al., 1983] and age distribution of carbonate sediments as a function of depth [Keir, 1984].

It seems especially important to at least ask what type of changes one expects during the past 8000 years for which we have accurate tree ring data [Suess, 1980]. A number of prominent effects, leading to changes in the carbon inventory of the oceans, or in the air-sea exchange or vertical mixing parameters, could have caused the atmospheric $^{14}C/^{12}C$ ratio to continuously decrease from 10,000 years B.P.:

1. The state of preservation of sediments was highest during ~10,000 years B.P. [Berger, 1977; Broecker and Peng, 1982; Kent, 1982; Droxler et al., 1983; Crowley, 1983], when the aragonite compensation depth was considerably depressed. Subsequently, this state declined, but the manner in which this happened cannot be ascertained unambiguously. The coral reef growth event was not synchronous with the last deglacial sea level rise, and the growth may have occurred $(5-9) \times 10^3$ years ago (see Keir [1984] for a discussion).

2. The vertical mixing rates are expected to be higher today than during the deglacial sea level rise epoch.

3. The direct exchange between the atmosphere and the deep sea (southern ocean outcrop) was certainly lower during the glacial period. Worthington's [1968] meltwater hypothesis, discussed earlier, would not only inhibit direct exchange between the atmosphere and the deep sea, but also air-sea exchange in general, during the Holocene transgression.

It is unfortunate that exact calculations cannot be carried out for the expected variation in $^{14}C/^{12}C$ ratios during the past 10,000 years, since the supplementary data available to date do not allow quantifying the changes for the global ocean. The carbonate chemistry effects are recognized to be complex with appreciable differences between the Pacific and the Atlantic. Besides Andree et al. [1984], no one else has attempted to determine vertical mixing rates in the past. The interpretation of $^{13}C/^{12}C$ ratios in pelagic and benthic organisms in the sea water

[Shackleton et al., 1983] for this purpose is complicated due to the complex histories of water masses [c.f. Broecker and Peng, 1982]. However, if we make the plausible assumptions that during the last 10,000 years, the carbon content of the oceans increased by 3% due to dissolution of surface sediments of age 5000 years, that the vertical exchange rate changes (increased) by a factor of 2, and that the southern ocean outcrop exchange is now effective, one would expect a compounded decrease of about 10% in $^{14}C/^{12}C$ ratios in the past 10,000 years. This would be consistent with the observed decrease for the past 8000 years.

The above argument implies that an appreciable part of the slow variation in atmospheric $^{14}C/^{12}C$ ratios could in fact be due to carbon cycle changes rather than changes in the geomagnetic dipole field. A recent reexamination of the archeomagnetic data by D. Lal (unpublished manuscript, 1984) argues that there is in fact no strong case for accepting most of the geomagnetic field change as postulated [Bucha, 1970]. The principal problems in accepting the reality of the geomagnetic field change as being responsible for most or all of the slow $^{14}C/^{12}C$ change are (D. Lal, unpublished manuscript, 1984) (1) the large scatter in the archeomagnetic data [c.f. Barton et al., 1979]; (2) the inadequate data base used to construct global dipole field changes, (3) the appreciably shorter period in the archeomagnetic data (9000 years), compared to the 12,000 years $^{14}C$ periodicity, and (4) the lack of phase difference between the geomagnetic field changes and $^{14}C/^{12}C$ ratios (a difference of about 1000 years would be expected for a 10,000-year variation in the field).

We therefore believe in the emerging view that possibly the slow $^{14}C/^{12}C$ changes are largely due to changes in the carbon cycle during the past 8000 years, rather than due to variation in the geomagnetic dipole field. If the atmospheric $^{14}C/^{12}C$ record can be extended to the past, using lake sediments, it will be possible to ascertain the existence of the proposed sinusoidal variation in geomagnetic dipole field. Some aspects of this have been considered (D. Lal, unpublished manuscript, 1984). We stress here that it is the dipole field we are interested in, because that is what the cosmic rays "see"; local changes are not important for the purpose of calculations of $Q(^{14}C)$.

Subsequent to the presentation of these results, we learned of the work of Oeschger and his collaborators, who have examined the record of fallout of cosmic ray-produced $^{10}Be$ in ice samples from Camp Century [Beer et al., 1984]. They do not find any evidence for a geomagnetic field change large enough to explain the observed slow change in the atmospheric $^{14}C/^{12}C$ ratio during the last 8000 years. Thus, $^{10}Be$ data seem to support our conclusion regarding the dominating effect of carbon cycle changes in controlling the atmospheric $^{14}C$ specific activity, and the need to reexamine the archeomagnetic data on which the proposed geomagnetic change was based.

Acknowledgements. I am very grateful to B. Bolin, W. S. Broecker, W. H. Berger, R. S. Keir, H. Oeschger, U. Seigenthaler, and E. Sundquist for helpful discussions on a variety of problems considered here. This work was carried out while I was on sabbatical leave at the Institute of Geophysics and Planetary Physics, University of California, Los Angeles, and I would like to gratefully acknowledge the support and facilities provided.

## References

Andree, M., J. Beer, H. Oeschger, A Mix, W. Broecker, N. Ragano, P. O'Hara, G. Bonani, H. J. Hofmann, E. Morenzoni, M. Nessi, M. Suter, and W. Woelfli, Accelerator radiocarbon measurements on foraminifera separated from deep-sea sediments, this volume.

Arrhenius, G., Sediment cores from the East Pacific, Rept. Swed. Deep Sea Exped. 1947-1948, 5(1-4), 1-227, 1952.

Barton, C. E., R. T. Merrill, and M. Barbetti, Intensity of the earth's magnetic field over the last 10,000 years, Phys. Earth Planet. Inter., 20, 96-110, 1979.

Beer, J., M. Andree, H. Oeschger, U. Siegenthaler, G. Bonani, H. Hofmann, E. Morenzoni, M. Nessi, M. Suter, W. Wolfli, R. Finkel, and C. Langway, Jr., The Camp Century $^{10}Be$ record: Implications for long-term variations of the geomagnetic dipole moment, Nucl. Instrum. Methods, 233, 380-384, 1984.

Berger, W. H., Deep-sea carbonate and the deglaciation preservation spike in pteropods and foraminifera, Nature, 269, 301-304, 1977.

Berger, W. H., Climate steps in ocean history - lessons from the Pleistocene, in Climate in Earth History, Stud. Geophys., pp. 43-54, National Academy Press, Washington, D.C., 1982a.

Berger, W. H., Deglacial $CO_2$ buildup: Constraints on the coral reef model, Palaeogeogr. Palaeoclimatol. Palaeoecol., 40, 235-253, 1982b.

Berger, W. H., and J. S. Killingley, The Worthington effect and the origin of the Younger Dryas, J. Mar. Res., 40, Suppl., 27-38, 1982.

Berner, W., H. Oeschger, and B. Stauffer, Information on the $CO_2$ cycle from ice core studies, Radiocarbon, 22, 227-235, 1980.

Bolin B., Climatic changes and their effects on the biosphere, Rep. WMO 542, 49 pp., World Meterorological Organization, Geneva, 1980.

Bolin, B., E. T. Begens, P. Duvigneaud, and S. Kempe, The global biogeochemical carbon cycle, in The Global Carbon Cycle, edited by B. Bolin, E. E. Egens, S. Kempe, and P. Ketner, pp. 1-11, John Wiley, New York, 1979.

Broecker, W. S., Glacial to interglacial changes in ocean chemistry, Prog. Oceanogr., 11, 151-197, 1982.

Broecker, W. S., and T.-H. Peng, Tracers in the Sea, 690 pp., Lamont-Doherty Geological Observatory, Columbia University, Palisades, New York, 1982.

Broecker, W. S., and T.-H. Peng, The climate chemistry connection, in Climate Processes and Climate Sensitivity, edited by J. E. Hansen and T. Takahashi, pp. 327-336, Geophysical Monograph 29, American Geophysical Union, Washington, D.C., 1984.

Broecker, W. S., and J. van Donk, Insolation changes, ice volumes, and the $O^{18}$ record in deep sea cores, Rev. Geophys. Space Physics, 8, 169-198, 1970.

Bucha, V., Influence of the earth's magnetic field on radiocarbon dating, in Radiocarbon Variations and Absolute Chronology, edited by I. U. Olsson, pp. 501-511, Wiley Interscience, New York, 1970.

Castagnoli, G., and D. Lal, Solar modulation effects in terrestrial production of carbon-14, Radiocarbon, 22, 133-158, 1980.

Craig, H., The natural distribution of radiocarbon and the exchange of carbon dioxide between atmosphere and sea, Tellus, 9, 1-17, 1957.

Craig, H., The natural distribution of radiocarbon: Mixing rates in the sea and residence times of carbon and water, in Earth Science and Meteoritics, edited by J. Geiss and E. D. Goldberg, pp. 109-120, North Holland, Amsterdam, 1963.

Crowley, T. J., Calcium-carbonate preservation patterns in the Central North Atlantic during the last 150,000 years, Mar. Geol., 151, 1-14, 1983.

Delmas, R. J., J. M. Ascencio, and M. Legrand, Polar ice evidence that atmospheric $CO_2$ 20,000 year B.P. was 50% of present, Nature, 284, 155-157, 1980.

Droxler, A. W., W. Schlager, and C. C. Whallon, Quaternary aragonite cycle and oxygen-isotope records in Bahamian carbonate ooze, Geology, 11, 235-239, 1983.

Ekdahl, C. A., and C. D. Keeling, Atmospheric carbon dioxide and radiocarbon in the natural carbon cycle, I, Quantitative deductions from records at Mauna Loa Observatory and at the south pole, in Carbon and the Biosphere, edited by G. M. Woodwell and E. V. Pecan, pp. 51-85, U.S. Atomic Energy Commission, Springfield, Virginia, 1973.

Elsasser, W., E. P. Ney, and J. R. Wickler, Cosmic-ray intensity and geomagnetism, Nature, 178, 1226-1227, 1956.

Garrels, R. M., F. T. Mackenzie, and C. Hunt (Eds.), Chemical cycles and the global environment, 206 pp., W. Kaufmann, Los Altos, California, 1975.

Hinga, K. R., J. M. Seiburth, G. Ross Heath, The supply and use of organic material at the deep sea floor, J. Mar. Res., 37, 557-579, 1979.

Honjo, S., Material fluxes and modes of sedimentation in the mesopelagic and bathypelagic zones, J. Mar. Res., 38, 53-97, 1980.

Houtermans, J., On the quantitative relationships between geophysical parameters and the natural $C^{14}$ inventory, Z. Phys., 193, 1-12, 1966.

Imbrie, J., and K. P. Imbrie, Ice Ages: Solving the Mystery, 224 pp., Enslow, Hillside, N. J., 1979.

Keir, R. S., Reduction of thermohaline circulation during deglaciation: The effect on atmospheric radiocarbon and $CO_2$, Earth Planet. Sci. Lett., 64, 445-456, 1983.

Keir, R. S., Recent increase in Pacific $CaCO_3$ dissolution: A mechanism for generating old $^{14}C$ ages, Mar. Geol., 59, 227-250, 1984.

Kent, D. V., Apparent correlation of palaeomagnetic intensity and climatic records in deep-sea sediments, Nature, 299, 538-539, 1982.

Kigoshi, K., and H. Hasegawa, Secular variation of atmospheric radiocarbon concentration and its dependence on geomagnetism, J. Geophys. Res., 71, 1065-1071, 1966.

Lal, D., Cosmic-ray-produced radionuclides in the sea, J. Oceanogr. Soc. Jpn. 20th Anniv. Vol., 600-614, 1962.

Lal, D., The oceanic microcosm of particles, Science, 198, 997-1009, 1977.

Lal, D., and R. Revelle, Atmosperic $pCO_2$ changes recorded in lake sediments, Nature, 308, 344-346, 1984.

Lal, D., and H. E. Suess, Some comments on the exchange of $CO_2$ across the airsea interface, J. Geophys. Res., 88, 3643-3646, 1983.

Lal, D., and V. S. Venkatavaradan, Analysis of the causes of $C^{14}$ variations in the atmosphere, in Radiocarbon Variations and Absolute Chronology, Nobel Symp. 12, edited by I. U. Olsson, pp. 549-569, John Wiley, New York, 1970.

McElroy, M. B., Marine biology: Controls on $CO_2$ and climate, Nature, 302, 328-329, 1982.

Reedy, R. C., J. R. Arnold, and D. Lal, Cosmic-ray record in solar system matter, Annu. Rev. Nucl. Sci., 33, 505-537, 1983.

Ruddiman, W. F., and A. McIntyre, The North Atlantic Ocean during the last deglaciation, Palaeogeogr. Palaeoclimatol. Palaeoecol., 35, 145-214, 1981.

Shackleton, N. J., M. A. Hall, J. Line, and C. Shuxi, Carbon isotope data in core V 19-30 confirm reduced carbon dioxide concentration in the ice age atmosphere, Nature, 306, 319-322, 1983.

Siegenthaler, U., M. Heimann, and H. Oeschger, $C^{14}$ variations caused by changes in the global carbon cycle, Radiocarbon, 22, 177-191, 1980.

Stauffer, B., H. Hofer, H. Oeschger, J. Schwander, and U. Siegenthaler, Atmospheric $CO_2$ concentration during the last glaciation, Ann. Glaciol., 5, 160-164, 1984.

Stuiver, M., Variations in radiocarbon concentration and sunspot activity, J. Geophys. Res., 66, 273-276, 1961.

Stuiver, M., Tree ring, varve and carbon-14 chronologies, Nature, 228, 454-455, 1970.

Suess, H. E., The radiocarbon record in tree rings of the last 8000 years, Radiocarbon, 22, 200-209, 1980.

Takahashi, T., W. S. Broecker, and A. E. Bainbridge, Supplement to the alkalinity and total carbon dioxide concentration in the world oceans, in Carbon Cycle Modelling, SCOPE 16, edited by B. Bolin, pp. 159-199, John Wiley, New York, 1981.

Worthington, L. V., Genesis and evolution of water masses, Meteorol. Monogr., 8, 63-67, 1968.

# SIMULATION EXPERIMENTS WITH LATE QUATERNARY CARBON STORAGE IN MID-LATITUDE FOREST COMMUNITIES

Allen M. Solomon

Environmental Sciences Division, Oak Ridge National Laboratory
Oak Ridge, Tennessee 37831

M. Lynn Tharp

Computer Services Division, Oak Ridge National Laboratory
Oak Ridge, Tennessee 37831

Abstract. The assumption was tested that forest biomass in communities on the modern landscape is equivalent to that in similar communities on the late Quaternary landscape. Forest carbon storage dynamics during the past 16,000 yers were derived from a mathematical model of forest processes and individual tree species behavior. Modern pollen and climate data sets provided pollen-climate transfer functions to generate model driving variables from fossil pollen records. Climate variables were estimated from fossil pollen stratigraphies in Tennessee, Ohio, and Michigan. Only simulated early postglacial warming produced the large carbon gains one would expect in mixed deciduous-coniferous forests from unglaciated regions. The simulated mid-Holocene warming generated little carbon storage response by temperate deciduous forests and large carbon gains in northern hardwood-conifer forests, unlike the linear relationship expected when equivalence is assumed between modern and prehistoric forests. Late-glacial, mid-latitude forests may have contained more biomass than would be expected from equivalent forests on the modern landscape. Simulations of alternate hypotheses to explain the enhanced late glacial biomass cannot distinguish effects of reduced seasonal temperature extremes from effects of changing species' temperature tolerances.

## Introduction

In attempting to determine how future $CO_2$-induced climate changes will affect carbon stored in forest biomass, we examined the behavior of biomass of late Pleistocene and Holocene forests. Prehistoric climate changes as rapid as those expected in the future cannot be documented (although they may have occurred), but other relevant prehistoric climate characteristics and vegetation responses existed.

For example, late Pleistocene seasonal temperature extremes apparently were reduced in nonglaciated portions of eastern North America [Bryson and Wendland, 1967; Moran, 1972; Williams and Barry, 1974] even more than they are expected to be in the future [Manabe and Wetherald, 1980]. The magnitude of mid-latitude temperature differences between the full glacial and the Holocene (5°-15°C; [Lamb, 1982]) is similar to that projected for a $CO_2$ quadrupling [Manabe and Stouffer, 1980; Manabe, 1983]. Mid-latitude temperatures 7500 to 4000 years ago were 1-3°C higher than today's, [Lamb et al., 1966; Bartlein and Webb, 1984], as they may be under a $CO_2$ doubling [Manabe and Wetherald, 1975]. Reduced precipitation during this period is documented in the midwestern United States [Bartlein et al., 1984], and enhanced precipitation may have occurred along the east coast [Davis et al., 1975] similar to geographic precipitation shifts possible under increased atmospheric $CO_2$ [Manabe et al., 1981; Wigley et al. 1980].

Grove [1984] concluded that the carbon stored in global terrestrial biomass was relatively low during full glacial time, increasing considerably to a maximum between 9500 to 4500 years ago and declining to an intermediate amount by the present day. His conclusions were drawn by tabulating biomass for various times and areas in the world. Each entry was based on some combination of geological evidence of absence or presence of vegetation, pollen-analytical evidence of the kind of vegetation if present, and modern biomass values from the equivalent vegetation types. Grove's conclusions are reasonable, assuming that specific prehistoric vegetation types were indeed equivalent to modern ones. The assumption may not be valid due to differences between late Quaternary environmental variables (e.g., climate, natural vegetation disturbance, lack of anthropomorphic vegetation disturbance) and those of today. The work described below is an attempt to test the assumption, i.e., whether differences between late Quaternary and modern climatic or other variables could induce carbon storage in prehistoric forests significantly different from that in modern equivalent forests.

## Approach

The vegetation that responded to the prehistoric climatic variations is reflected by fossil pollen grains preserved in lake and bog sediments. This relationship might be used to calculate a modern pollen and biomass transfer function, which then could be applied to fossil pollen to estimate "fossil" biomass. In fact, pollen profiles possess most characteristics needed to so reconstruct forest biomass (i.e., carbon storage) on ancient landscapes. Wind-transported pollen is ubiquitous, being produced and deposited every year, thereby providing evidence of taxon absence as well as taxon presence. Sedimentary pollen composition is determined by species composition of surrounding forests [Davis, 1963; Webb et al., 1978, 1981]. Pollen deposited in lakes and bogs is preserved almost indefinitely in the contemporaneous sediments [Sangster and Dale, 1964]. Most pollen can be identified at the genus level [Faegri and Iverson, 1975], and some at the species level [e.g., Solomon, 1983a,b].

One required attribute that fossil pollen (absolute or relative measure) lacks is a documented numerical relationship with the quantities of biomass in nearby forests. Theoretically, the amount of pollen deposited per annual unit of sediment can be calibrated with the number or density of trees on the landscape. However, Davis et al. [1973] found that lakes surrounded by landscapes forested as little as 40% and as much as 100% received about the same annual pollen influx. Pollen proportions are even less well suited to reconstructing biomass amounts. For example, tree pollen composition of closed forests is little different from that in adjacent treeless areas of tundra [Ritchie and Lichti-Federovich, 1967] or prairie [McAndrews, 1966].

In addition, modern pollen and biomass data are derived from tree associations controlled not only by climate but also by disturbance from recent land use. Frequency and intensity of anthropogenic disturbances are unlikely to match those of prehistoric natural disturbances. Differences between biomass in disturbed and in undisturbed forests [Olson et al., 1983] can be great, while pollen composition may reflect few or none of these differences [Janssen, 1967; Davis, 1967].

Pollen and biomass regressions (transfer functions [e.g., Webb et al., 1981; Bradshaw and Webb, 1984]) must be applied to fossil pollen samples that contain values in the calibration data sets. This presents no difficulty for samples of the past 10 millenia but is important to use of late glacial pollen samples, which are frequently unlike those of the present day [Overpeck et al., 1984].

A different approach to reconstructing ancient forest biomass uses known principles of ecological and biological limits to forest tree growth in a simulation model designed to directly mimic environmental cause and ecological effect. Forests are "grown," based on each species' growth potential and on the climate-modulated competitive capabilities, relationships, and interactions among the tree species [Botkin et al., 1972; Shugart and West, 1977]. The individual tree species responses to climate variables can be determined from trees in scattered forest stands rather than from forested landscapes, avoiding the difficulties induced by recent land use. Like the dynamics of actual forests, the model output depends on the different capabilities and behavior of individual tree species to determine collective forest biomass, reducing the importance of unique environmental conditions in applications of the model to prehistoric landscapes. Tests of this model on modern forests of eastern North America [Solomon et al., 1984] indicate that it reliably reproduces forest biomass.

The difficulty in applying such a model of cause and effect to estimating biomass on late Quaternary landscapes is that quantitative estimates of cause (e.g., climate) are required before the model can simulate effect (biomass). The present version of the model requires estimates of annual growing season length, monthly temperature and precipitation and their standard deviations, severely reducing the kinds of paleoclimatological data that can be used.

The driving variables can be provided by multivariate regressions of modern pollen and climate data sets applied to fossil pollen records. The pollen-climate relationships appear to be quite robust [Webb and Bryson, 1972; Webb et al., 1983], particularly when violations of regression assumptions are minimized. In addition, effects of land use on the pollen-climate relationships are minimal, based on tests with pollen from pre-agricultural sediments [Bartlein et al., 1984]. This may be logical, considering that tree pollen percentages appear to reflect land use little and to reflect forest biomass even less (see discussion above). Thus the two approaches (first principles simulation and pollen-environment regression) can provide complementary elements of an approach to estimating past biomass, which neither element is capable of generating in isolation.

We applied the approach to simulating late Quaternary biomass in three distinct steps (Figure 1). First, late Quaternary climatic variables (Figure 1, BCD) were reconstructed from fossil pollen records (D), based upon

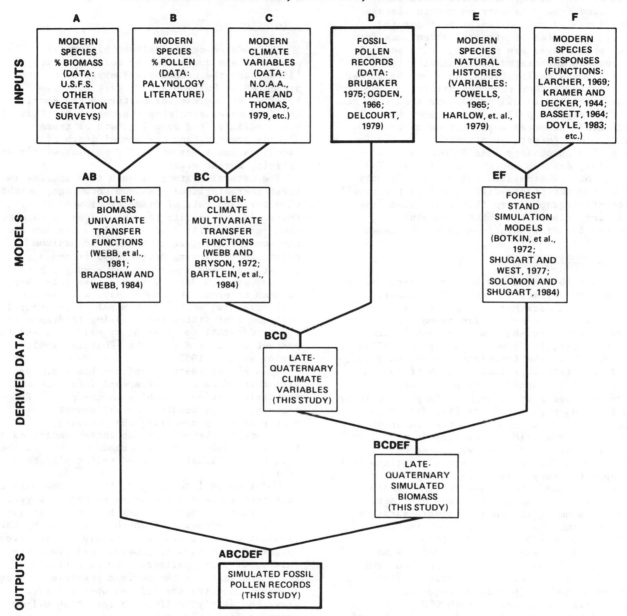

Fig. 1. Diagram of approach that integrates forest-stand simulation modeling with fossil pollen data via modern forest biomass, modern climate, and modern pollen data sets. Box letter designations are provided to ease visual comprehension of relationships within the diagram.

multivariate relationships (BC) between pollen (B) and climate (C) on the modern landscape, as detailed below. Second, the climate reconstructions were used to drive the simulation model (EF), which integrates tree growth capabilities (E) and responses of growth

to environmental variables (F), also detailed below.

The objective of these operations was to simulate forest biomass (Figure 1, BCDEF). Before the simulations can be interpreted, a third step requires a consistency test. Here, the model output was transformed into "simulated fossil pollen percentages" (ABCDEF) with percent pollen-percent biomass transfer functions (AB) constructed from data sets on modern biomass proportions (A) and pollen proportions (B). Although estimates of biomass percentages from pollen percentages are not subject to some of the problems associated with quantitative pollen-biomass calibrations, we view the former as indicative rather than definitive.

The simulated fossil pollen was then compared with the original pollen stratigraphy (D). The model rarely produces a perfect match for reconstructed forests at all stratigraphic levels in a pollen chronology [see Solomon et al., 1980, 1981; Solomon and Shugart, 1984]. The portions of the simulated pollen stratigraphy that were unacceptably different from the fossil pollen stratigraphy were then eliminated (see discussion below), leaving the remaining long-term forest simulations to be examined.

## Climate Reconstruction

The most objective and best documented method for deriving climate variables from fossil pollen spectra uses multivariate statistics to relate a set of modern pollen taxon frequencies to the climate variables at adjacent weather stations [Cole, 1969; Webb and Bryson, 1972]. Deficiencies of the technique for our purposes, include the fact that combinations of fossil pollen taxon frequencies which are not in the modern data set produce unreliable proxy climate results; that e.g., soils or fire frequency variations in both time and space can produce pollen taxon variations indistinguishable from the effects of the climate variables that are assumed to be the only variance source; and that the separate proxy climate variables (e.g., July and January temperature) are not independent, but rather depend on the same set of limited differences among the modern climate variables and on the same characteristic combinations of modern pollen variables. As a result, proxy climate data represent estimates of the most likely climate, rather than the unknown true climate values. They are the best available proxy data, and are probably adequate, particularly during the past 10,000 years.

Transfer functions and derived estimates of July and January mean temperature and of annual precipitation and growing season length for all three sites were provided by P. J. Bartlein and T. Webb III, using the methods of Bartlein and Webb [1984] and Bartlein et al. [1984]. Temperature and precipitation and their standard deviations in all 12 months are required to parameterize the simulation model. These were estimated from today's variance and Bartlein's data as described by Solomon and Shugart [1984]. The four climate variables for each of 12 months, and growing season length for each year, were calculated from every pollen sample analyzed by

the original authors [Brubaker, 1975; Ogden, 1966; Delcourt, 1979] at the three fossil pollen sites.

## Simulation Model

The modeled ecological and biological processes are documented by Solomon et al. [1984], and the mathematical expressions are presented in Solomon and Shugart [1984]. Briefly, the forests of eastern North America (FORENA) model simulates the annual addition of new seedlings, the annual growth of trees already present, and the annual death of seedlings and trees on a 1/12-ha plot within an infinite forest canopy.

The natural history of each tree species is first characterized: maximum known age, height, diameter, and rate of diameter growth; shade tolerance; capability to sprout from roots when above-ground parts die, as well as minimum and maximum sprouting age; minimum and maximum tolerance to growing degree-days (temperature summations above a minimum temperature, here 5.5°C); minimum January temperature tolerance; maximum tolerance to annual days with inadequate soil moisture; seedling germination on mineral soil and leaf litter and seedling tolerance to summer drought, to browsing by wildlife, and to percent of maximum sunlight [Fowells, 1965; Harlow et al., 1979].

This set of features defines how each species will respond to a set of annual intrinsic and extrinsic forcing variables on the plot. For example, annual seedling establishment is accomplished by comparing the seedling requirements listed above to those conditions on the plot. Subsequently, a random number of the eligible seedlings are added to the plot each year.

Under optimal conditions, the modeled size of each tree increases a maximum amount each year according to a logistic growth function [Doyle, 1983], the end-point of which is defined by the greatest diameter and age possible for the tree (Figure 2a). However, growing conditions are rarely, if ever, perfect. The probability of death increases as the maximum possible tree age is approached (Figure 2b) and when the tree grows too slowly, defined as less than 0.1-cm annual radial increment.

Shading differentially reduces the optimum growth of shade-tolerant and shade-intolerant species (Figure 2c). Shading is calculated from extinction of light through the simulated leaf area above each tree on the plot [Monsi et al., 1973] generated by taller trees on the plot and averaged across the area of the plot. Optimum growth is also reduced by greater stand density to simulate nutrient competition (Figure 2d).

These five features (seedling generation, optimum growth, death from age or slow growth, growth lost to shading, and growth lost to increased stand density) define intrinsic

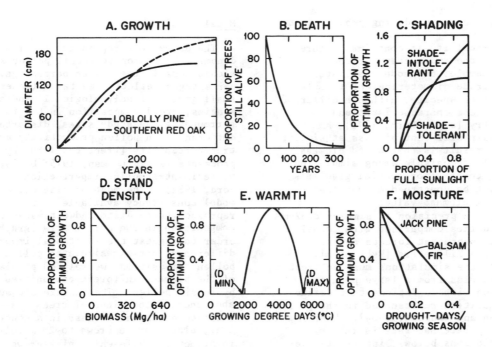

Fig. 2.  Tree responses to major intrinsic and extrinsic forcing variables in the
FORENA model.  (a) Optimum growth of two species as a function of maximum age (years)
and diameter (diameter at breast height) of each.  Note that about two-thirds the
maximum diameter is achieved in about one-half the maximum age.  (b) Mortality as a
function of a tree's age and the maximum age (AGEMX) a tree can achieve, using
loblolly pine as an example.  (c) Portion of optimum tree growth lost due to
nonoptimal light intensities.  Different growth equations are used for shade-tolerant
and shade-intolerant species.  (d) Proportion of optimum tree growth lost due to
nonoptimal crowding.  (e) Proportion of optimum tree growth lost due to nonoptimal
annual growing degree-days.  D MIN and D MAX are the lowest and highest degree-day
values, respectively, within which the species can grow.  Degree-day requirements of
sycamore are used as an example.  (f) Proportion of optimum tree growth of two tree
species lost due to nonoptimal growing season days in which soil moisture is below
the wilting point.

competitive interactions on the plot.  Extrinsic
forces (warmth, coldness, and moisture) modify
the interactions by shifting the competitive
advantages among species.  Growing degree-day
(heat sums, i.e., cumulative temperature above a
specified threshold temperature) maxima and
minima found within the geographic range of each
species define the limits to a parabolic respone
curve [Larcher, 1969; Figure 2e] used to decrease
the annual optimum growth of each tree on the
plot.  This occurs when each species' response
curve is compared with a stochastic annual
growing degree-day value that the model
calculates from the monthly temperature means
and standard deviations.  No seedlings are
selected and no annual growth occurs when the
stochastic January temperature is below the
lowest January temperature found within the
geographic range of each tree to simulate
effects of winter low temperatures [Sakai and
Weiser, 1973].
    Moisture response is simulated by reducing

annual growth in proportion to the days in each
yearly growing season during which soil moisture
is below the wilting point [Bassett, 1964] of
-15 bars (drought-days, Figure 2f).  Drought-days
are stochastic variables, calculated annually
from a soil column of predefined soil moisture
capacity, into which the stochastic monthly
precipitation enters and from which it leaves
either as runoff (when soil moisture capacity is
exceeded) or as evapotranspiration, determined
by the stochastic monthly temperature
[Thornthwaite and Mather, 1957].
    Like the proxy climate data, the simulation
model represents the best available approach for
the task, but it also contains certain
limitations for our purposes.  First, the model
can only simulate forest tree growth as a
function of the proxy climate driving
variables.  Errors in the latter are translated
into the former.  The consistency checks (below)
permit us to avoid the worst effects of this
difficulty, but the problem will only be solved

with future advances in producing proxy climate chronologies.

Second, the model itself contains features which are either too rudimentary or too precise. For example, the model estimates biomass from a crude diameter:height formula applied to all tree species, although different tree species possess characteristic and different geometry [e.g., Horn, 1971], and therefore, contain different biomass at similar diameters. The result is that stand biomass may be quantitatively comparable among simulation sequences along temporal or spatial gradients, but it cannot yet be construed as a reliable measure of actual biomass values.

In contrast, the precision of simulated 32-m diameter plots defies comparison with fossil pollen data that represent forests covering hundreds of square kilometers [Webb et al., 1978]. Instead, the simulations must be treated as vegetation samples from a large population, and should be run with conditions that define the different habitats represented in the pollen records [Solomon and Shugart, 1984]. When simulation of separate habitats is lacking (e.g., the simulations below), inconsistencies are generated between correctly modeled simulations and pollen records.

## Long-Term Simulations

The model was run with climate variables from the pollen record at Camp 11 Lake, Marquette County, Michigan (46°40'N, 88°01'W [Brubaker, 1975]), Silver Lake, Logan County, Ohio (40°26'N, 83°40'W [Ogden, 1966]), and Anderson Pond, White County, Tennessee (36°02'N, 85°30'W [Delcourt, 1979]). The model results are averages of ten runs at each site. Soil moisture capacity for a 100-cm-deep soil at each site was from the most common nearby upland soil (Michigamme sandy loam, Camp 11 Lake [Berndt, 1984]; Miamian silt loam, Silver Lake [Waters and Siegenthaler, 1979]; Waynesboro loam, Anderson Pond [Moore et al., 1982]).

The $^{14}C$ dating control appears adequate at Camp 11 Lake and at Anderson Pond but is inadequate for lowermost sediments at Silver Lake. Assuming that the sedimentation rate between the oldest two dates (6150 $^{14}C$ years at 5.25 m, 9800 $^{14}C$ years at 6.1 m) applies to the undated lowermost 2.4 m of sediment, basal sediments at 8.5 m would be 18,100 years old. Yet, the absence of a distinctive pollen-stratigraphic event (replacement of tundra by spruce forest) which occurs in nearby areas of Ohio between 15,000 and 14,000 $^{14}C$ years ago [Shane, 1980] indicates the basal Silver Lake sediments are younger than 15,000 $^{14}C$ years. The first pollen-stratigraphic event identifiable at Silver Lake (spruce forest decline, 7.0 m) is dated at 13,000-12,500 $^{14}C$ years elsewhere [Shane, 1980]).

## Model Consistency

The datum that represents the lowest denominator common to fossil pollen and model simulations is the pollen percentage. Pollen percentages reflect tree taxon (genera, species, small groups of palynologically indistinguishable species, etc.) composition in the forests surrounding the study sites, distorted by taxon differences in pollen productivity and transportability [Prentice, 1984] by limnological processes [e.g., Lehman, 1975; Davis, 1973], and by sediment-sampling imperfections [Davis and Ford, 1982]. The most precise criterion of model consistency available is its ability to reproduce these distorted estimates of forest composition in the pollen stratigraphy. In order to suggest the functional importance of differences between simulated pollen and fossil pollen composition, we rely on a relative index of the biomass in forest communities reflected by fossil pollen. The stature and apparent abundance of the kinds of trees present can indicate potential biomass in a general sense, i.e., black spruce grows to 28 m tall and 1/2 m in diameter, while white pine grows to 61 m tall and 3 1/2 m in diameter [Fowells, 1965]. Thus, a white pine-dominated forest would be expected to contain more biomass than a spruce-dominated forest by virtue of the greater forest depth associated with the white pine.

Pollen profiles from Camp 11 Lake (Figure 3a) can ben subdivided into three zones. From 10,350 years before present (B.P.) to 8000 years B.P. pollen stratigraphy is dominated by jack/red pine with spruce, fir, birch, and oak. The mid-Holocene interval of increased warmth at Camp 11 Lake is characterized by the appearance of white pine at about 8000 years B.P. and its dominance from 7000 to 3000 years B.P., decline of jack/red pine, absence of spruce and fir, and steadily increasing oak and sugar maple pollen percentages. Pollen stratigraphy of the most recent 3000 years is marked by dominance of birch pollen, decreased but plentiful white pine pollen, and replacement of the thermophilous hardwoods (oak and sugar maple) by spruce and fir.

To compare simulation output with the fossil pollen, we converted simulated biomass densities to biomass percentages and then calculated "simulated pollen percentages" from published regressions of modern pollen and biomass percentages [Webb et al., 1981; Bradshaw and Webb, 1984]. The resulting simulated pollen diagram (Figure 3b) is similar to the measured pollen diagram (Figure 3a). The initial period of jack/red pine dominance with spruce, fir, birch, and oak is accurately simulated, although the simulations appear to extend the period by about 1000 years. As in the pollen record, the simulated mid-Holocene interval is characterized by dominance of white pine, decline of jack/red pine, absence of spruce and fir, and enhanced

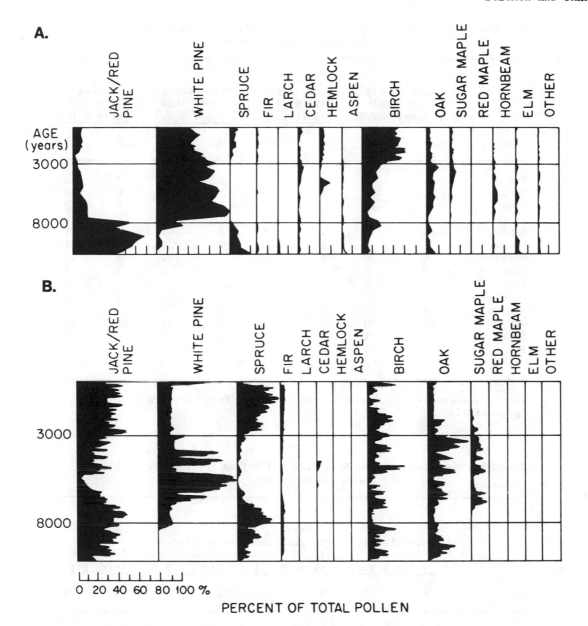

Fig. 3.  Pollen stratigraphy at Camp 11 Lake.  (a) Measured pollen stratigraphy
[Brubaker, 1975].  (b) Simulated pollen stratigraphy.

percentages of oak and sugar maple.  Unlike the
pollen record, jack/red pine expands about 5000
simulated years ago at the expense of white pine.

During the most recent 3000 years, the
simulation clearly misses the mark.  It correctly
simulates the return of small-stature spruce and
fir (though in too great abundance) and the
decline of the thermophilous hardwoods, oak, and
sugar maple.  However, it continues to simulate
large percentages of small-stature jack/red
pine, more like the pollen percentages in a
nearby lake surrounded by coarse outwash sands
(Yellow Dog Pond [Brubaker, 1975]) than like

Camp 11 Lake pollen.  Not apparent in simulations
(Figure 3b) is the generous amount of white pine
and the distinctive increase and dominance of
birch (both large-stature trees) during the
final 3000 years of pollen record at Camp 11
Lake (Figure 3a).  As a result, this portion of
the simulation likely contains lower carbon
storage than the forests which pollen actually
reflects.

The measured and simulated pollen frequencies
from Silver Lake and Anderson Pond are presented
as average values (Tables 1 and 2) during periods
defined for this paper:  200-4000 years B.P.

TABLE 1.  Measured (M) and Simulated (S) Pollen Percentages at Anderson
Pond, White County, Tennessee

| Tree | Late Glacial (>12,000 years B.P.) | | Early Postglacial (12,000-8000 years B.P.) | | Mid-Holocene (8000-4000 years B.P.) | | Recent (4000-200 years B.P.) | |
|---|---|---|---|---|---|---|---|---|
| | M | S | M | S | M | S | M | S |
| Fir | 2 | 2 | 0 | 0 | 0 | 0 | 0 | 0 |
| Spruce | 21 | 13 | 8 | 0 | 1 | 0 | 0 | 0 |
| Pine | 41 | 45 | 7 | 0 | 4 | 0 | 8 | 0 |
| Birch | 1 | 6 | 1 | 0 | 2 | 0 | 2 | 0 |
| Maple | 0 | 0 | 2 | 1 | 2 | 1 | 2 | 1 |
| Hickory | 1 | 0 | 4 | 6 | 8 | 12 | 7 | 14 |
| Chestnut | 0 | 0 | 0 | 3 | 2 | 7 | 4 | 12 |
| Beech | 0 | 0 | 1 | 2 | 1 | 2 | 1 | 5 |
| Ash | 6 | 2 | 6 | 1 | 11 | 1 | 5 | 1 |
| Sweetgum | 0 | 0 | 0 | 0 | 1 | 1 | 1 | 1 |
| Ironwood | 4 | 0 | 18 | 5 | 2 | 7 | 1 | 5 |
| Oak | 21 | 30 | 49 | 76 | 65 | 64 | 66 | 61 |
| Basswood | 0 | 1 | 0 | 0 | 0 | 0 | 0 | 0 |
| Elm | 1 | 0 | 2 | 2 | 1 | 4 | 1 | 1 |
| Walnut | 0 | 0 | 1 | 4 | 1 | 0 | 1 | 0 |

Data are averages of individual sample values from Delcourt [1978,
Appendix II].


TABLE 2.  Measured (M) and Simulated (S) POllen Percentages at Silver
Lake, Logan County, Ohio

| Tree | Late Glacial (>12,000 years B.P.) | | Early Postglacial (12,000-8000 years B.P.) | | Mid-Holocene (8000-4000 years B.P.) | | Recent (4000-200 years B.P.) | |
|---|---|---|---|---|---|---|---|---|
| | M | S | M | S | M | S | M | S |
| Fir | 7 | 1 | 4 | 0 | 0 | 0 | 0 | 0 |
| Spruce | 70 | 35 | 14 | 0 | 0 | 0 | 0 | 0 |
| Pine | 12 | 53 | 24 | 43 | 4 | 7 | 3 | 4 |
| Birch | 3 | 9 | 3 | 0 | 1 | 0 | 1 | 0 |
| Maple | 1 | 1 | 1 | 9 | 1 | 2 | 2 | 1 |
| Hickory | 0 | 0 | 4 | 0 | 12 | 1 | 14 | 1 |
| Chestnut | 0 | 0 | 0 | 0 | 1 | 0 | 1 | 0 |
| Beech | 0 | 0 | 1 | 1 | 4 | 1 | 4 | 2 |
| Ash | 0 | 1 | 4 | 0 | 2 | 0 | 4 | 1 |
| Sweetgum | 0 | 0 | 0 | 0 | 0 | 0 | 1 | 0 |
| Ironwood | 3 | 0 | 11 | 3 | 4 | 4 | 4 | 6 |
| Oak | 4 | 0 | 14 | 36 | 57 | 68 | 59 | 72 |
| Elm | 0 | 0 | 17 | 3 | 10 | 10 | 5 | 9 |
| Walnut | 0 | 0 | 3 | 5 | 4 | 7 | 2 | 4 |

Data are estimated from Ogden [1966, Plate I].

(recent), 4000-8000 years B.P. (mid-Holocene), 8000-12,000 years B.P. (early postglacial), and older than 12,000 years B.P. (late glacial). Measured pollen frequencies were averaged from individual pollen counts at Anderson Pond [Delcourt, 1978, Appnedix II] and were visually estimated from the Silver Lake pollen diagram [Ogden, 1966, Plate I]. The simulated pollen values were derived from simulated biomass density as described for Camp 11 Lake above.

The Anderson Pond late glacial, mid-Holocene, and recent pollen stratigraphy appears well matched by the simulations (Table 1). Both measured and simulated late-glacial pollen spectra are dominated by similar proportions of pine, spruce, and oak, with lesser amounts of fir, birch, and ash. Measured and simulated pollen of the past 8000 years is primarily from oak, hickory, and chestnut, with minor amounts of maple, beech, ash, sweetgum, ironwood, and elm.

The early postglacial pollen from Anderson Pond is not as well matched by simulated pollen (Table 1). The pollen stratigraphy contains abundant ash, spruce, pine, and ironwood. The model did not reproduce the spruce and pine abundance or the open forests that promote ironwood abundance [Heide, 1981] in late glacial and early postglacial time. The poor match between pollen and the model results in the early postglacial constrains the use of the model output, because the modeled abundance of large oak trees coupled with the absence of smaller simulated spruce and pine enhances total simulated stand biomass compared with forests probably reflected by measured pollen.

Simulated pollen adequately matches measured pollen at Silver Lake during the mid-Holocene and recent time (Table 2). Oak dominates, with minor amounts of elm, walnut, ironwood, beech, maple, and pine. Hickory is poorly represented in simulations, and oak is too abundant, but the overall effect on stand biomass is probably not great.

The simulated late glacial is dominated by boreal trees, and the early postglacial by boreal and temperate deciduous trees at Silver Lake. However, the proportions of taxa involved match fossil pollen poorly. Late glacial spruce pollen outnumbers pine by 6 to 1 in the sediments, but pine dominates spruce 3 to 2 in the simulations (Table 2). Because white spruce is a larger tree than jack pine, actual closed canopy forests may have stored more carbon than the simulated ones during the late glacial. In contrast, early postglacial simulated trees appear to represent greater total biomass than do the trees reflected by measured pollen frequencies. Pine, elm, oak, spruce, and ironwood, in that order, compose most of the measured early postglacial pollen, compared with simulated abundance of pine, oak, maple, and walnut.

The foregoing examination of consistency

between modeled and measured forests suggests that the following portions of the simulated pollen records correspond well enough to fossil pollen records to allow interpretation of the simulated biomass: 16,000 to 12,000 years B.P., and younger than 8000 years B.P. at Anderson Pond; younger than about 10,000 years B.P. at Silver Lake; and 10,000 to 3000 years B.P. at Camp 11 Lake.

Simulation Results

Recent (Prelogging Time ~200-4000 Years B.P.)

The high-frequency variance in the biomass profiles precludes serious consideration of any differences between Silver Lake and Anderson Pond biomass (Figure 4). The Camp 11 Lake biomass clearly declines during recent time, from 130-140 Mt/ha at 4000 years B.P. to 80-100 Mt/ha at 1500 years B.P. The magnitude of the biomass decline is uncertain because of the poor match between fossil and simulated pollen (Figure 3) that probably results in simulations which underestimate the actual biomass of forests surrounding Camp 11 Lake.

Mid-Holocene (4000-8000 Years B.P.)

This period is reliably simulated at all three sites. It is also of particular interest in considering possible future effects of $CO_2$-induced warming, because mid-latitude temperatures were 1°-2°C greater than today's [Lamb, 1982]. Grove [1984] suggests that carbon storage in global forests reached maximum Holocene values during this period, based on analogy with temperature and carbon storage in modern forests. Reconstructions of forest geography indicated reduced mid-latitude forest cover in some areas [Webb et al., 1983] but northward invasion of tundra by boreal elements [Lamb, 1982, Figure 14], with the net change in area remaining uncertain. The simulations examined the forest density change in response to mid-Holocene climatic shifts.

Although hidden by large-amplitude, high-frequency variance, average simulated biomass values at Anderson Pond (190-200 Mt/ha) and Silver Lake (150-170 Mt/ha) are not particularly different from biomass during periods which either preceded or followed the mid-Holocene warm period (Figure 4). Biomass change is evident only at Camp 11 Lake, where large-stature deciduous tree species replaced small boreal conifers. There, biomass increased by about 50%, from 75-80 Mt/ha at 8000-7000 years B.P. to 115-120 Mt/ha at 5000-4000 years B.P. Had a fossil pollen record been simulated in boreal forests to the north of Camp 11 Lake, the biomass increase probably would have disappeared, as it does in simulated warming of modern boreal forests [Solomon et al., 1984].

The simulations suggest nonlinear forest

Scott W. Starratt
Dept. of Paleontology
U. C. Berkeley
Berkeley, Ca. 94720

**ANDERSON POND, TN.**

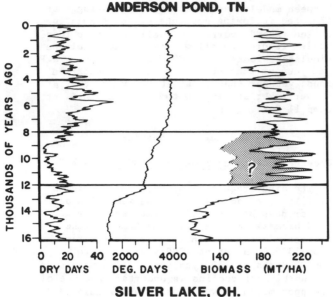

**CAMP 11 LAKE, MI.**

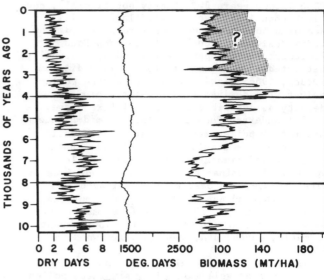

Fig. 4.    (continued)

**SILVER LAKE, OH.**

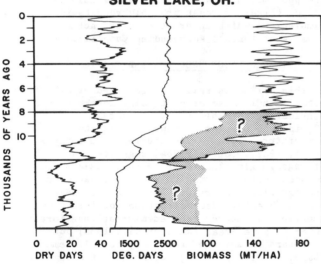

Fig. 4.  Dry days, growing degree-days, and total above-ground stand biomass simulated at Anderson Pond, Tennessee; Silver Lake, Ohio; and Camp 11 Lake, Michigan.  Less reliably simulated portions of the chronologies and likely values of expected errors are indicated by shading. Silver Lake age estimates older than 10,000 years B.P. were deleted for reasons discussed in text.

responses to the same amount of mid-Holocene warming at all three sites.  Biomass shifts are essentially absent at Silver Lake and Anderson Pond and are of greater magnitude at Camp 11 Lake than might be expected from the warming of 1°-2°C.  In contrast, simulated modern vegetation response to warming [Solomon et al., 1984] and the sparse data on biomass correlation with warmth on the modern landscape both appear

to indicate more linear relationships [e.g., Merz, 1978; Olson et al., 1983 and Cohen, 1973, p. 93-96; Delcourt et al., 1981, Figure 3; Nelson and Zillgitt, 1969, p. 16-25].

Early Postglacial (8000-12,000 Years B.P.)

Both the Anderson Pond and Silver Lake simulations may have overestimated the actual stand biomass during this period, indicating that biomass dynamics portrayed in Figure 4 are not reliable.  These dynamics include simulated biomass which reached the greatest values in the 16,000-year chronology at Anderson Pond.  Biomass at Silver Lake rapidly increased from 80 Mt/ha to 150 Mt/ha sometime before 10,000 simulated years ago.  Dating uncertainties in Silver Lake sediments and the poor match of simulated and fossil pollen preclude further consideration of the Silver Lake simulated biomass during this period.

Camp 11 Lake was occupied by a buried ice block until about 10,350 years B.P. [Brubaker, 1975].  When organic sediments began accumulating, a spruce and pine forest already surrounded the lake (Figure 3).  Simulated biomass during deposition of early postglacial stratigraphy varied about a mean value of 100 Mt/ha (Figure 4).

Despite the simulation uncertainties at Silver Lake and Anderson Pond, it still seems probable that carbon storage during this interval was greatest in the diverse forests simulated at Anderson Pond and least in the species-depauperate forests simulated at Camp 11 Lake.  Forest carbon storage values near Silver Lake may have been somewhere between the two extremes.

Late Glacial (Older than 12,000 Years B.P.)

The presence or absence of late-glacial pollen at Silver Lake is unknown because of the [14]C-dating problems discussed above. Inconsistencies between simulated and measured Silver Lake pollen composition also render biomass estimates unusable there. The Camp 11 Lake site was covered by late glacial ice sheets. Thus, Anderson Pond sediments provide the only reliable biomass estimates during this time period.

Stand biomass in the Anderson Pond simulations remained at 110-120 Mt/ha from 16,000 to 14,000 years B.P. Then it began to increase, reaching a maximum of about 190 Mt/ha from 12,500 years B.P., coincident with extirpation of the boreal tree species [Delcourt, 1979]. Earlier simulations of forests at this site 16,000-14,000 years B.P. [Solomon et al., 1980, 1981] were based on a simpler model which generated only 25-35 Mt/ha, similar to the 20-80 Mt/ha stand biomass in modern northern Taiga woodland [Olson et al., 1983, Table 5]. Subsequent model improvements (routines to generate soil moisture and tree responses; routines to generate winter low temperatures and tree responses) resulted in the higher biomass estimates, which are more characteristic of modern deciduous-coniferous forests [Olson et al., 1983].

A similar mixture apparently grew over wide areas of the late glacial landscape (e.g., Minnesota [Wright, 1968]; Illinois [King, 1981]; Indiana [Williams, 1974]; Ohio [Ogden, 1966]; Tennessee [Delcourt, 1979]; Virginia [Craig, 1969]), unlike the limited areal extent of equivalent modern forests. The important difference in conditions that separated late glacial mixed forests from Holocene forests in which spruce rarely grew with the other taxa (e.g., mid-Holocene at Camp 11 Lake) is hypothesized to be winter temperatures like today's coupled with much colder late glacial summer [Bryson and Wendland, 1967; Wright, 1968, 1983; Moran, 1972].

Are decreased seasonal temperature extremes both sufficient and necessary [e.g., Botkin, 1981] to account for the unique late-glacial forest composition? Reduced seasonality may explain (sufficiency) the enhanced late glacial biomass [Solomon, 1982], but it is not an exclusive requirement (necessity) for the phenomenon. Other processes that might generate the same result have been suggested e.g., effects on boreal species of stress from reduced atmospheric $CO_2$ during full glacial (C. F. Cooper, personal communication 1983); effects of mid-latitude daylength on high-latitude boreal species, (M. B. Davis, personal communication 1982).

Changing ecological amplitudes of plant populations presents both a testable and a credible cause of late glacial temperate and boreal species admixtures and associated increased carbon storage. The implications of changing ecological amplitudes may be unwelcome to any paleoecologists who require equivalence of modern and Pleistocene species to reconstruct prehistoric forests or climates [e.g., Solomon et al., 1980; Davis, 1983; Webb et al., 1983; Birks, 1981], but it also may be a natural outcome of oscillating climates.

A species' population consists of intergrading subpopulations (ecotypes) each possessing differing environmental requirements [Antovonics, 1971]. Of particular relevance here, ecotypes frequently have been identified which possess tolerance to features either of cold winters or of warm summers, but not to both (e.g., jack pine [Rudolph and Yeatman, 1982]; white pine [Bourdeau, 1963; Mergen, 1963]; hemlock [Olson et al., 1959; Kessell, 1979]; beech [Camp, 1951]; red oak [Overlease, 1975; McDougal, 1982]; red maple [Perry and Wu, 1960]). Response of any of these species to, e.g., a warming, would involve (1) decreased success and perhaps the extinction of local northern subpopulations possessing low tolerance to warmth (greatest cold tolerance), (2) increased success of southern subpopulations which tolerate warmth (but not cold), and (3) no immediate change in the warmest temperatures that members of the population can tolerate. The population as a whole, therefore, would possess a narrower temperature range and a warmer mean temperature after the warming than it did before the warming.

This hypothetical situation was simulated for red oak (the northernmost oak species in eastern North America). The simulated oak species possessed the modern tolerance to maximum warmth (4590 growing degree-days) but a lower tolerance to minimum warmth (880 growing degree-days) than it does on the modern landscape (1100 growing degree-days). The other 71 simulated species were provided with their modern environmental requirements.

Under late glacial temperatures inferred from nonpalynological evidence at Anderson Pond [Solomon and Shugart, 1984] but with modern seasonal temperature extremes, the most successful species were white spruce and the cold-tolerant red oak (Figure 5), much like the composition of the 15,000-year pollen spectrum [Delcourt, 1979] and unlike red oak behavior on the modern landscape. The spruce-oak forest also contained about twice the biomass as forests which were simulated under the same climatic conditions, but with red oak defined by its modern temperature requirements, i.e. with six percent less degree-day range. Thus, in addition to decreased seasonality, another sufficient cause of the late glacial forest characteristics could be warming-induced shifts in population temperature tolerances.

Discussion

The foregoing simulation experiments reconstructed shifts in carbon storage of forest

## SIMULATED FULL-GLACIAL VEGETATION: INCORRECT RED OAK SPECIES PARAMETERS

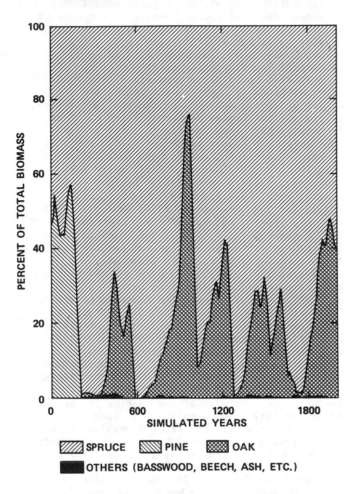

Fig. 5. Simulated late-glacial vegetation at Anderson Pond, with lowest growing degree-day tolerance of northern red oak decreased (growing degree-day range increased) by 6%.

stands during the late Quaternary. The purpose of the reconstructions was to examine the validity of the assumption that prehistoric forest carbon storage variations can be projected from modern forest biomass. These variations include responses to mid-Holocene temperatures, 1°-2°C warmer than today's, to large temperature increases that characterized the shift to postglacial (Holocene) climate, and to reduced seasonal temperature extremes that accompanied late glacial climates of unglaciated eastern North America.

Simulations of mid-Holocene warming at all three sites appear adequately consistent with pollen evidence to allow inferring that simulated forest behavior could mimic actual forest dynamics near the pollen sampling sites. If this is so, the above-ground carbon storage did not significantly change in response to mid-Holocene warming in the deciduous forests simulated (Anderson Pond and Silver Lake). The increased biomass in mixed coniferous-deciduous forests (Camp 11 Lake) appears to be localized as a function of shifting, climate-mediated competitive advantages between the small-stature spruce and pine, growing at the south edges of their geographic ranges, and large-stature oak and maple at the north edge of their ranges [e.g., Little, 1971].

In contrast, the effects of early postglacial warming were quite apparent in the elevated carbon densities at the two southern sites. The simulated biomass values are of unknown reliability because of the poor match between simulated and fossil pollen. However, the direction and magnitude of the simulated change are essentially as suggested from analysis of modern climate and biomass relationships [Grove, 1984].

Simulations of late glacial climate and forest responses may be of greater interest because conditions depart from modern climate-forest associations. Due to an admixture of boreal conifers with temperate deciduous species, forests surrounding Anderson Pond may have contained 40 to 50% more biomass than would be simulated (e.g., early postglacial at Camp 11 Lake) or suggested from forest types [Olson et al., 1983, Figure 4] under the most similar modern climates. However, the enhanced carbon storage probably could have occurred in all mid-latitude forests without change in the overall conclusion that global terrestrial carbon storage was much lower during the full and late glacial than during most of the Holocene, up to the centuries of greatest deforestation [Grove, 1984].

The reason for enhanced carbon storage is relevant not only for extrapolating modern carbon storage to that of the late glacial, but also for extrapolating late glacial forest dynamics onto a future landscape on which atmospheric $CO_2$ concentrations increase. If reduced seasonality produced enhanced carbon storage in mid-latitude forests, we might expect modern forests also to store additional atmospheric carbon when rising atmospheric $CO_2$ concentrations begin to reduce seasonality. If, on the other hand, the late glacial biomass enhancement reflected no real vegetation response to seasonality, but instead was due to wider ecological amplitudes than species now possess, the opposite effect could occur. Future warming induced by $CO_2$ would decrease ecological amplitudes, probably decreasing the potential stand biomass as well.

Presently, we know of no documented way to determine (1) whether reduced seasonality or shifting environmental requirements among tree populations were the central or only cause of temperate-boreal species mixtures and enhanced carbon storage in the late glacial or (2) which, if either, process will dominate a future warmer landscape. Portions of the answers may involve simulation modeling akin to that used above. This is the only available approach with which hypotheses can be tested, as illustrated, for example, by simulating the potential changes in ecological amplitude that may be required to explain prehistoric vegetation assemblages, or simulating the unique combinations of climate variables that could generate those assemblages. Yet, at best, simulation can only verify or disqualify ideas on what could have occurred. They cannot demonstrate, conclusively, what did occur in the past or what should occur in the future. Instead, entirely new kinds of physiochemical data may be required to reduce the number of independent variables that knowledge, intuition, and modeling exercises indicate are relevant to each problem.

Acknowledgments. We thank P. J. Bartlein and T. Webb III for providing the pollen-climate transfer functions and the resultant climate estimates used in this research. T. Webb III, J. J. Pastor, and J. S. Olson contributed thorough and constructive reviews of the manuscript. This research was sponsored by the National Science Foundation's Ecosystem Studies Program under Interagency Agreement No. BSR8115316, A03 with Martin Marietta Energy Systems, Inc., under contract No. DE-ACO5-84OR21400 with the U.S. Department of Energy. Publication No. 2397, Environmental Sciences Division, Oak Ridge National Laboratory.

### References

Antovonics, J., The effects of a heterogeneous environment on the genetics of natural populations, Am. Sci. 59, 593-599, 1971.

Bartlein, P. J., T. Webb, III, and E. Fleri, Holocene climatic change in the northern Midwest: Pollen-derived estimates, Quat. Res., in press, 1984.

Bartlein, P. J., and T. Webb, III, Mean July temperature at 6000 yr B.P. in eastern North America: Regression equations for estimates from fossil pollen data, in Syllogeus National Museum of Canada, Ottawa, in press, 1984.

Bassett, J. R., Tree growth as affected by soil moisture availability, Soil Sci. Soc. Am. Proc., 28, 436-438. 1964.

Berndt, L. W., Soil Survey of Baraga County, Michigan, U.S. Department of Agriculture, Washington, D.C., in press, 1984.

Birks, H. J. B., The use of pollen analysis in the reconstruction of past climates: A review, in Climate and History, edited by T. M. L. Wigley, M. J. Ingram and G. Farmer, pp. 111-138, Cambridge University Press, New York, 1981.

Botkin, D. B., Causality and succession, in Forest Succession: Concepts and Application, edited by D. C. West, H. H. Shugart, and D. B. Botkin, pp. 36-55, Springer-Verlag, New York, 1981.

Botkin, D. B., J. F. Janak, and J. R. Wallis, Some ecological consequences of a computer model of forest growth, J. Ecol., 60, 849-872, 1972.

Bourdeau, P. F., Photosynthesis and respiration of Pinus strobus L. seedlings in relation to provenance and treatment, Ecology, 44, 710-716, 1963.

Bradshaw, R. H., and T. Webb, III, Relationships between contemporary pollen and vegetation data from Wisconsin and Michigan, U.S.A., Ecology, in press, 1984.

Brubaker, L. B., Postglacial forest patterns associated with till and outwash in northcentral upper Michigan, Quat. Res., 5, 499-527, 1975.

Bryson, R. A., and W. M. Wendland, Tentative climatic patterns for some late glacial and post-glacial episodes in central North America, in Life, Land and Water, edited by W. J. Mayer-Oakes, pp. 271-298, University of Manitoba Press, Winnipeg, 1967.

Camp, W. H., A biogeographic and paragenetic analyses of the American beech (Fagus), Am. Philos. Soc. Yearb. 1950, 166-169, 1951.

Cohen, S. B., Oxford World Atlas, 190 pp., Oxford University Press, New York, 1973.

Cole, H. S., Objective reconstruction of the paleoclimatic record through the application of eigenvectors of present-day pollen spectra and climate to the late-Quaternary pollen stratigraphy, Ph.D. Dissertation, 110 pp., University of Wisconsin, Madison, August 1969.

Craig, A. J., Vegetational history of the Shenandoah Valley, Virginia, Geol. Soc. Am. Spec. Pap., 123, 283-296, 1969.

Davis, M. B., On the theory of pollen analysis, Am. J. Sci., 261, 897-912, 1963.

Davis, M. B., Redeposition of pollen grains in lake sediments, Limnol. Oceanogr., 18, 44-52, 1973.

Davis, M. B., Holocene vegetational history of the eastern United States, in Late Quaternary Environments of the United States, vol. 2, The Holocene, edited by H. E. Wright, Jr., pp. 166-181, University of Minnesota Press, Minneapolis, 1983.

Davis, M. B., and M. S. Ford, Sediment focusing in Mirror Lake, New Hampshire, Limnol. Oceanogr., 27, 137-150, 1982.

Davis, M. B., L. B. Brubaker, and T. Webb, III, Calibration of absolute pollen influx, in Quaternary Plant Ecology, edited by H. J. B. Birks and R. G. West, pp. 9-27, Halsted, New York, 1973.

Davis, R. B., Pollen studies of near-surface sediments in Maine lakes, in Quaternary Paleoecology, edited by E. J. Cushing and H. E. Wright, Jr., pp. 143-173, Yale University Press, New Haven, Connecticut, 1967.

Davis, R. B., T. E. Bradstreet, R. Stuckenrath, and H. W. Borns, Vegetation and associated environments during the past 14,000 years near Moulton Pond, Maine, Quat. Res., 5, 435-465, 1975.

Delcourt, H. R., Late Quaternary vegetation history of the Eastern Highland Rim and adjacent Cumberland Plateau of Tennessee, Ph.D. Dissertation, 210 pp., University of Minnesota, Minneapolis, August 1978.

Delcourt, H. R., Late Quaternary vegetation history of the Eastern Highland Rim and adjacent Cumberland Plateau of Tennessee, Ecol. Monogr., 49, 255-280, 1979.

Delcourt, H. R., D. C. West, and P. A. Delcourt, Forests of the southeastern United States: Quantitative maps for above-ground woody biomass, carbon, and dominance of major tree taxa, Ecology, 62, 879-887, 1981.

Doyle, T. W., Competition and growth relationships in a mixed-age, mixed-species forest community, Ph.D. Dissertation, 85 pp., University of Tennessee, Knoxville, June 1983.

Faegri, K., and J. Iverson, Textbook of Pollen Analysis, 295 pp., Hafner, New York, 1975.

Fowells, H. A., Silvics of forest trees of the United States, USDA For. Serv. Agric. Handb., 271, 762 pp., 1965.

Grove, A. T., Changing climate, changing biomass, and changing atmospheric $CO_2$. Prog. Biometeorol., 3, 5-10, 1984.

Hare, F. K., and M. K. Thomas, Climate Canada, 2nd ed., 230 pp., John Wiley and Sons Canada Limited, Toronto, 1979.

Harlow, W. M., E. S. Harrar, and F. M. White, Textbook of Dendrology, 6th ed., 510 pp., McGraw-Hill, New York, 1979.

Heide, K., Late Quaternary vegetational history of northcentral Wisconsin, U.S.A.: Estimating forest composition from pollen data, Ph.D. Dissertation, 189 pp., Brown University, Providence, R.I., May 1981.

Horn, H. S., The adaptive geometry of trees, 144 pp., Princeton University Press, Princeton, New Jersey, 1971.

Janssen, C. R., A comparison between the recent regional pollen rain and the subrecent vegetation in four major vegetation types in Minnesota (U.S.A.). Rev. Palaeobot. Palynol., 2, 331-342, 1967.

Kessell, S. R., Adaptation and dimorphism in eastern hemlock, Tsuga canadensis (L.). Carr., Amer. Nat., 113, 333-350, 1979.

King, J. E., Late Quaternary vegetation history of Illinois, Ecol. Monogr., 51, 43-62, 1981.

Kramer, P. J., and J. P. Decker, Relation between light intensity and rate of photosynthesis of loblolly pine and certain hardwoods, Plant Physiol., 19, 350-358, 1944.

Lamb, H. H., Reconstruction of the course of postglacial climate over the world, in Climatic Change in Later Prehistory, edited by A. Harding, pp. 11-32, Edinburgh University Press, Edinburgh, Scotland, 1982.

Lamb, H. H., R. P. W. Lewis, and A. Woodroffe, Atmospheric circulation and the main climatic variables between 8000 and 0 B.C.: Meteorological evidence, in World Climate From 8000 to 0 B.C., Proceedings of an International Symposium, pp. 174-217, R. Meterological Society, London, 1966.

Larcher, W., The effect of environmental and physiological variables on the carbon dioxide gas exchange of trees, Photosynthetica, 3, 167-198, 1969.

Lehman, J. T., Reconstructing the rate of accumulation of lake sediment: The effect of sediment focusing, Quat. Res., 5, 541-550, 1975.

Little, E. L., Atlas of United States trees, Vol. I, Conifers and important hardwoods, USDA For. Serv. Misc. Publ., 1146, 9 pp., 1971.

Manabe, S., Carbon dioxide and climate change, Adv. Geophys., 25, 39-82, 1983.

Manabe, S., and R. J. Stouffer, Sensitivity of a global climate model to an increase of $CO_2$ concentration in the atmosphere, J. Geophys. Res., 85, 5529-5554, 1980.

Manabe, S., and R. T. Wetherald, The effects of doubling the $CO_2$-concentration on the climate of a general circulation model, J. Atmos. Sci., 32, 3-15, 1975.

Manabe, S., and R. T. Wetherald, On the distribution of climate change resulting from an increase in $CO_2$ content of the atmosphere, J. Atmos. Sci., 37, 99-118, 1980.

Manabe, S., R. T. Wetherald, and R. J. Stouffer, Summer dryness due to an increase of atmospheric $CO_2$ concentration, Climatic Change, 3, 347-386, 1981.

McAndrews, J. H., Postglacial history of prairie, savanna, and forest in northwestern Minnesota, Mem. Torrey Bot. Club, 22, 1-72, 1966.

McDougal, K., Geographic variability in flavonoid profiles of red oak Quercus rubra L. (abstract), Am. J. Bot., 70 (5, part 2), 123, 1983.

Mergen, F., Ecotypic variation in Pinus strobus L., Ecology, 44, 716-727, 1963.

Merz, R. W., Forest Atlas of the Midwest, 48 pp., U.S. Department of Agriculture Forest Service, North Central Forest Experiment Station, St. Paul, Minnesota, 1978.

Monsi, M., Z. Uchijima, and T. Oikawa, Structure of foliage canopies and photosynthesis, Annu. Rev. Ecol. Syst., 4, 301-327, 1973.

Moore, R. K., J. F. Campbell, and W. C. Moffitt, Soil survey of White and Van Buren Counties, Tennessee, 86 pp., U.S. Department of Agriculture Soil Conservation Service, Washington, D.C., 1981.

Moran, J. M., An analysis of periglacial climatic indicators of late glacial time in North America, Ph.D. Dissertation, 160 pp., Univ. of Wisconsin, Madison, January 1972.

Nelson, T. C., and W. M. Zillgitt, A Forest

Atlas of the South, 27 p., U.S. Department of Agriculture Southeast Forest Experiment Station, Asheville, North Carolina, 1969.

Ogden, J. G., III, Forest history of Ohio: I, Radiocarbon dates and pollen stratigraphy of Silver Lake, Logan County, Ohio, Ohio J. Sci., 66, 387-400, 1966.

Olson, J. S., F. W. Stearns, and H. Nienstaedt, Eastern hemlock seeds and seedlings: Response to photoperiod and temperature, Connecticut Agricultural Experiment Station, Bull., 620, 70 pp., New Haven, 1959.

Olson, J. S., J. A. Watts, and L. J. Allison, Carbon in live vegetation of major world ecosystems, ORNL/TM-5862, 164 pp., Oak Ridge Nat. Lab., Oak Ridge, Tennessee, 1983.

Overlease, W. R., Population studies of red oak (Quercus rubra L.) and northern red oak (Quercus rubra var. borealis (Michx. F.) Farw.)., Proc. Pa. Acad. Sci., 49, 138-140, 1975.

Overpeck, J. T., T. Webb, III, and I. C. Prentice, Quantitative interpretation of fossil pollen spectra: Dissimilarity coefficients and the method of modern analogs, Quat. Res., in press, 1984.

Perry, T. O., and W. C. Wu, Genetic variation in the winter chilling requirement for date of dormancy break for Acer rubrum, Ecology, 41, 790-794, 1960.

Prentice, I. C., Pollen representation, source area and basin size: Towards a reunified theory of pollen analysis, Quat. Res., in press, 1984.

Ritchie, J. C., and S. Lichti-Federovich, Pollen dispersal phenomena in arctic-subarctic Canada, Rev. Palaeobot. Palynol., 3, 255-266, 1967.

Rudolph, T. D., and C. W. Yeatman, Genetics of jack pine, USDA For. Serv. Res. Pap., WO-38, 64 pp., 1982.

Sakai, A., and C. J. Weiser, Freezing resistance of trees in North America with reference to tree regions, Ecology, 54, 118-126, 1973.

Sangster, A. G., and H. M. Dale, Pollen grain preservation of underrepresented species in fossil spectra, Can. J. Bot., 42, 437-449, 1964.

Shane, L. K., Detection of a late-glacial climatic shift in central midwestern pollen diagrams (abstract), Abstracts, Am. Quat. Assoc., 6th Biennial Meet., pp. 171-172, 1980.

Shugart, H. H., and D. C. West, Development of an Appalachian deciduous forest succession model and its application to assessment of the impact of the chestnut blight, J. Environ. Manage., 5, 161-179, 1977.

Solomon, A. M., Plant community response to decreased seasonality during full-glacial time (abstract), Abstracts, Am. Quat. Assoc. 7th Biennial Meet., p. 18, 1982.

Solomon, A. M., Pollen morphology and plant taxonomy of red oaks in eastern North America, Am. J. Bot., 70, 495-507, 1983a.

Solomon, A. M., Pollen morphology and plant taxonomy of white oaks in eastern North America, Am. J. Bot., 70, 481-494, 1983b.

Solomon, A. M., and H. H. Shugart, Integrating forest-stand simulations with paleoecological records to examine long-term forest dynamics, in Proceedings of the European Science Foundation Workshop on Forest Dynamics, Uppsala, Sweden, edited by F. Andersson, Elsevier, Holland, in press, 1984.

Solomon, A. M., H. R. Delcourt, D. C. West, and T. J. Blasing, Testing a simulation model for reconstruction of prehistoric forest-stand dynamics, Quat. Res., 14, 275-293, 1980.

Solomon, A. M., D. C. West, and J. A. Solomon, Simulating the role of climate change and species immigration in forest succession, in Forest Succession: Concepts and Application, edited by D. C. West, H. H. Shugart, and D. B. Botkin, pp. 154-177, Springer-Verlag, New York, 1981.

Solomon, A. M., M. L. Tharp, D. C. West, G. E. Taylor, J. W. Webb, and J. L. Trimble, Response of unmanaged forests to $CO_2$-induced climate change: Available information, initial tests, and data requirements, Rept. TR009, 93 pp., U.S. Department of Energy, Washington, D.C., 1984.

Thornthwaite, C. W., and J. R. Mather, Instructions and tables for computing potential evapotranspiration and the water balance, Publ. Climatol., 10, 183-311, 1957.

Waters, D. D., and V. L. Siegenthaler, Soil Survey of Logan County, Ohio, 181 pp., U.S. Department of Agriculture Soil Conservation Service, Washington, D.C., 1979.

Webb, T., III, and R. A. Bryson, The late- and post-glacial sequence of climatic events in Wisconsin and east-central Minnesota: Quantitative estimates derived from fossil pollen spectra by multivariate statistical analysis, Quat. Res., 2, 70-115, 1972.

Webb, T., III, R. A. Laeski, and J. C. Bernabo, Sensing vegetational patterns with pollen data: Choosing the data, Ecology, 59, 1151-1163, 1978.

Webb, T., III, S. Howe, R. H. W. Bradshaw, and K. M. Heide, Estimating plant abundances from pollen percentages: The use of regression analysis, Rev. Palaeobot. Palynol., 34, 269-300, 1981.

Webb, T., III, E. J. Cushing, and H. E. Wright, Jr., Holocene changes in the vegetation of the midwest, in Late-Quaternary Environments of the United States, vol. 2, The Holocene, edited by H. E. Wright, Jr., pp. 142-165, University of Minnesota Press, Minneapolis, 1983.

Wigley, T. M. L., P. D. Jones, and P. M. Kelly, Scenario for a warm, high-$CO_2$ world, Nature, 283, 17-21, 1980.

Williams, A. S., Late-glacial-postglacial history of the Pretty Lake Region, northeastern Indiana, U.S. Geol. Surv., Prof.

Pap., 686-13, 23 pp., 1974.

Williams, J., and R. G. Barry, Ice age experiments with the NCAR general circulation model: Conditions in the vicinity of the northern continental ice sheets, in Climate of the Arctic, edited by G. Weller and S. A. Bowling, pp. 143-149, Geophysical Institute, University of Alaska, Fairbanks, 1974.

Wright, H. E., Jr., The roles of pine and spruce in the forest history of Minnesota and adjacent areas, Ecology, 49, 937-955, 1968.

Wright, H. E., Jr., Introduction, in Late-Quaternary Environments of the United States, Vol. 2, The Holocene, edited by H. E. Wright, Jr., pp. xi-xvii, University of Minnesota Press, Minneapolis, 1983.

# CARBONATE PRESERVATION AND RATES OF CLIMATIC CHANGE: AN 800 KYR RECORD FROM THE INDIAN OCEAN

L. C. Peterson[1] and W. L. Prell

Department of Geological Sciences, Brown University, Providence, Rhode Island 02912

Abstract. In the eastern equatorial Indian Ocean, a foraminiferal-based Composite Dissolution Index (CDI) identifies the present level of the lysocline (3800 m) and reflects the modern carbonate saturation gradient in the water column. Piston cores from water depths of 2900 to 4400 m on the Ninetyeast Ridge at 6°S provide an 800 kyr record of bathymetric and temporal variations in carbonate preservation as measured by the CDI. Dissolution of carbonate is out of phase with glacial-interglacial $\delta^{18}O$ cycles. Cross-spectral analysis reveals high coherency (>0.85) between the CDI, the $\delta^{18}O$ record, and its derivative over the 100 kyr and 41 kyr Milankovitch frequencies, indicating a simple linear response of overall dissolution intensity to climate change. Variations in preservation are proportional to and in phase with maximum rates of change in the $\delta^{18}O$ record, suggesting a rapid response of the local carbonate system to climatic forcing. Both the Broecker shelf deposition model and circulation models calling for changes in the production or chemistry of deep waters provide plausible mechanisms for explaining the observed dissolution patterns. However, the lack of a well-defined precessional (23 and 19 kyr) response in the CDI and loss of coherency with $\delta^{18}O$ over the same frequencies tends to favor a circulation explanation, as strict interpretation of a shelf model would require coherency at all of the dominant Milankovitch periods. Alternatively, shelf-basin carbon inventory models may be too simple and may produce a nonlinear response in the deep sea carbonate reservoir for small changes in sea level.

## Introduction

The carbon chemistry of the ocean is intimately related to the balance between the precipitation of calcium carbonate in supersaturated surface waters and its dissolution in undersaturated deep waters. Temporal variations in this balance are recorded in deep-sea sediments and provide important historical information about the global geochemical cycle of $CO_2$.

Available evidence suggests that carbonate accumulation and preservation patterns have varied between and within ocean basins through time. In the equatorial Pacific, Arrhenius [1952] was the first to observe that deep-sea cores contained lower percentages of carbonate in sediments deposited during periods of warm climate than in those deposited during inferred periods of cold climate. Subsequent workers have generally attributed the cause of these carbonate fluctuations to increased dissolution intensity during interglacial intervals [e.g., Berger, 1973; Luz, 1973; Thompson and Saito, 1974; Thompson, 1976; Adelseck, 1977]. In the North Atlantic, Berger [1973] presented early foraminiferal evidence suggesting that late Pleistocene dissolution patterns were similar to those observed in the Pacific. Later Atlantic studies, however, generally indicate that dissolution intensified during glacial periods [e.g., Gardner, 1975; Damuth, 1975; Bé et al., 1976; Balsam, 1981; Crowley, 1983], in apparent opposition to Pacific dissolution patterns. In addition, a number of studies have noted the possibility of a distinct phase shift between dissolution intensity and glacial-interglacial variations in the oxygen isotopic composition of foraminifera [e.g., Luz and Shackleton, 1975; Shackleton and Opdyke, 1976; Moore et al., 1977].

Far fewer dissolution records have up to now been available from the Indian Ocean, although both Atlantic-type and Pacific-type patterns have been reported in single cores [Oba, 1969; Williams and Keany, 1978; Corliss and Thunell, 1983]. This study attempts to clarify and further document Indian Ocean patterns by giving detailed consideration to a suite of piston cores recovered from a small region of the eastern equatorial Indian Ocean. Results presented here provide additional constraints which must be taken into consideration in the evaluation and testing of models of the global carbon cycle.

[1]Now at Rosenstiel School of Marine and Atmospheric Science, University of Miami, Miami, Florida 33149.

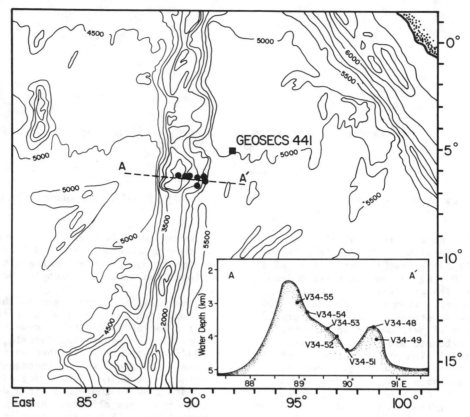

Fig. 1.  Regional bathymetry of Ninetyeast Ridge showing location of piston cores used in this study.

## Strategy and Modern Setting

Seven piston cores ranging in water depth from 2992 m to 4382 m were selected for study from the eastern flank of the Ninetyeast Ridge at 6°S (Figure 1; Table 1).  The bathymetric interval straddled by these seven cores encompasses the present-day level of the foraminiferal lysocline (3800 m); hence, they are in position to act as a particularly sensitive monitor of temporal changes in dissolution intensity.

The modern dissolution setting of this area has been recently evaluated from a study of recent sediment samples from the crest and flanks of the Ninetyeast Ridge [Peterson and Prell, 1984].  In this previous study, we utilized a combination of six commonly measured dissolution indicators (Table 2) to describe the preservation of carbonate in 45 Holocene core-top samples. Because each of these indicators can also be influenced by ecologic and sedimentologic factors, principal components analysis was applied to these data in order to partition out the common variance among the six indicators attributable to the dissolution process.  The first principal component from our core-top

TABLE 1.  Location and Water Depth of Indian Ocean Piston Cores

| Core | Latitude | Longitude | Water Depth, m |
|------|----------|-----------|----------------|
| V34-55 | 06°02.4'S | 88°57.4'E | 2992 |
| V34-54 | 06°05.0'S | 89°10.0'E | 3254 |
| V34-48 | 06°15.0'S | 90°33.0'E | 3656 |
| V34-53 | 06°07.0'S | 89°35.0'E | 3812 |
| V34-52 | 06°10.0'S | 89°48.0'E | 3984 |
| V34-49 | 06°22.0'S | 90°36.0'E | 4088 |
| V34-51 | 06°11.5'S | 89°58.0'E | 4382 |

TABLE 2.  Dissolution Indices Used to Calculate the Composite Dissolution Index (CDI)

| Index | Method of Calculation |
|---|---|
| Weight percent calcium carbonate | - - - |
| Percent coarse fraction (>63 μm) | - - - |
| Percent whole planktic foraminifera (>150 μm) | $\dfrac{\text{no. whole planktic foram. tests}}{(\text{no. whole planktic tests + no. planktic fragments})}$ |
| Percent benthic foraminifera (>150 μm) | $\dfrac{\text{no. benthic foram. tests}}{(\text{no. benthic foram. tests + no. whole planktic foram. tests})}$ |
| Percent whole G. menardii (>150 μm) | $\dfrac{\text{no. whole G. menardii}}{(\text{no. whole G. menardii + no. G. menardii fragments})}$ |
| Percent radiolarians (>150 μm) | $\dfrac{\text{no. radiolarian tests}}{(\text{no. radiolarian tests + no. whole planktic foram. tests})}$ |

analysis explained 83% of the variability in the raw dissolution data and was adopted as a unified measure of relative dissolution intensity, or Composite Dissolution Index (CDI).  The bathymetric distribution of the values of the CDI on the core-top samples reflects the preservation gradient within the modern sediments (Figure 2a).  The foraminiferal lysocline presently lies at a depth of 3800 m, at a level closely approximating the calcite saturation horizon in the water column predicted by the critical carbonate ion model of Broecker and Takahashi [1978] (Figure 2b).

In order to simplify the presentation of our downcore results, we have chosen to further utilize the CDI as a descriptive measure of relative preservation state.  Raw data for the six dissolution indices which go into its calculation and further details of their measurement in all downcore samples are presented elsewhere [Peterson, 1984].

### Stratigraphy

Given its sensitive location near the depth of the present-day lysocline, core V34-53 (3812 m) was sampled at 5-cm intervals.  All other cores were sampled at 10-cm intervals.  Oxygen isotope data provide the principal stratigraphy for this study.  The oxygen isotope data for all cores but V34-49 are plotted against estimated age in Figure 3.  The chronology used for our age models was developed by Imbrie et al. [1984].  All $\delta^{18}O$ data from cores V34-55, -54, -53, -51, and -48 were gathered on the benthic foraminifer Planulina wuellerstorfi.  Oxygen isotope data for the shortest core, V34-52, were obtained on the deep-dwelling planktic foraminifer, Pulleniatina

obliquiloculata.  The $\delta^{18}O$ data for V34-53 are presented in Table 3; all other isotopic data are reported by Peterson [1984].  An isotopic stratigraphy for V34-49 is not yet available.  Stratigraphy for this core is currently based on variations in the different dissolution indices.  Additional stratigraphic information for the set of cores comes from the presence of a distinctive volcanic ash layer and the identification of several key biohorizons in each core (Figure 3).

The amount of detail visible in the isotopic records is a function of sampling density, bioturbation, and sediment accumulation rates.  Together, bioturbation and generally low accumulation rates (1.0-1.7 cm/kyr) tend to smooth the records somewhat.  The results of this smoothing process can be seen in the amplitude of the glacial-interglacial changes in $\delta^{18}O$.  The magnitude of the $\delta^{18}O$ difference between climatic extremes generally ranges between about 1.1 and 1.6°/$_{oo}$ over most of these records.  This range is somewhat less than the full 1.6-1.9°/$_{oo}$ signal recognized in cores with high accumulation rates [Peng et al., 1977; Shackleton, 1977a], but is close enough to suggest that the isotopic and dissolution records have not been significantly distorted by the vagaries of the depositional process.

### Late Pleistocene Preservation Patterns

Downcore preservation states were described relative to the profile of the modern CDI model (Figure 2a) by simple postmultiplication of the standardized downcore dissolution indicator data matrix by the vector of principal component factor scores for the core-top CDI [Peterson and

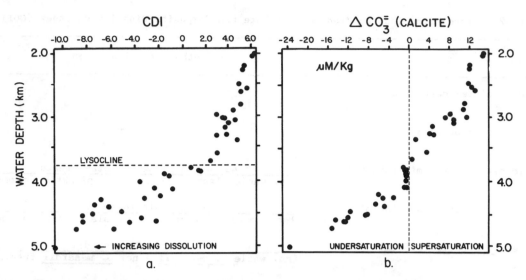

Fig. 2. Modern carbonate preservation gradient in local sediments compared to modern carbonate saturation gradient in water column. (a) Baseline values of the Composite Dissolution Index (CDI) from the core-top study of Peterson and Prell [1984], illustrating the modern preservation gradient in the Ninetyeast Ridge area. The position of the modern lysocline approximately corresponds to the cross-over from positive- to negative-valued loadings. (b) Bathymetric profile of water column carbonate saturation, expressed as $\Delta CO_3^=$ (calcite), extrapolated to core-top positions from profiles calculated for local GEOSECS stations [Takahashi et al., 1980]. The local saturation horizon for calcite ($\Delta CO_3^= = 0$) falls at about 3700-m water depth in the study area.

Prell, 1984, Table 5]. The results are shown in Figure 4, with CDI time series for the individual cores arranged in order of increasing water depth. Well-preserved sediments are indicated by more positive CDI values, whereas poorly preserved sediments are indicated by more negative values.

Given this array of records (Figure 4), the systematically greater variability at greater water depths strongly implies changes in the depth-dependent dissolution gradient through time. However, the regular variations in preservation can be clearly traced from core to core, with a first-order periodicity of approximately 100 kyr. Intervals of poor preservation tend to occur either during periods of interglacial climate (odd-numbered isotope stages) or near the transitions from interglacial to glacial climate.

Figure 5 is an alternate representation of all of the time and bathymetric variation observed in the same set of data. Rather than plotting the CDI values as a sequence of time series, the data were first sampled at 10 kyr intervals, plotted as a function of estimated age and depth of the core site, and then contoured. The result is a more easily visualized portrayal of bathymetric variations in carbonate preservation through time. The most striking feature of this plot is once again the strong evidence for a periodicity approximating 100 kyr in the preservation

record. In addition, there is evidence of a much longer period supercycle in the data which manifests itself as the shoaling of poor preservation levels between about 300-600 kyr ago. In the Pacific, Adelseck [1977] has previously drawn attention to a long-period "Brunhes Dissolution Cycle", which appears in the carbonate records of Arrhenius [1952], Hays et al. [1969], and within his own data from the eastern equatorial region. Crowley [this volume] has also observed this mid-Brunhes dissolution increase in North Atlantic records, suggesting that the Brunhes supercycle is a global phenomenon.

Though convenient, display of the downcore preservation data as a depth-age plot (Figure 5) makes it difficult to recognize more than the dominant mode of variability. In order to help determine which parts of the ocean-atmosphere-climate system ultimately force changes in deep-sea carbonate preservation, a more detailed look at the patterns and timing of events in the preservation record is necessary. To some extent this can be done visually by examining the most detailed record. Figure 6 shows the records of $\delta^{18}O$ and the CDI from core V34-53 (3812 m) plotted together for comparison. The CDI values for this core are listed with the $\delta^{18}O$ data in Table 3. Inspection of Figure 6 suggests that dissolution variations are not synchronous with glacial-interglacial intervals, but appear to be

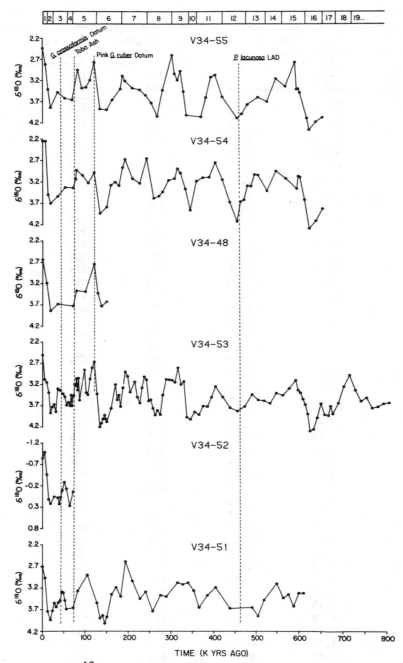

Fig. 3.  Variations in $\delta^{18}O$ as a function of estimated age for six of the transect cores.  Key biohorizons (coiling direction change in <u>G. crassaformis</u>, local last appearance of pink <u>G. ruber</u>, <u>P. lacunosa</u> LAD), and the stratigraphic level of the Toba Ash are identified.  For details on the derivation of the age models and the source of the time scale, see text and Peterson [1984].

somewhat offset.  Periods of maximum preservation show a tendency to occur on the isotopic Terminations, or periods of rapid deglaciation. A good example of this is the preservation peak across the most recent Termination I.  Similar preservation spikes found at this time are well documented globally [e.g., Chen, 1968; Diester-Haass et al., 1973; Berger, 1977; Berger and Killingley, 1977], and have also been observed across older Terminations [Thompson and Saito, 1974].  Intervals of maximum dissolution (i.e., poorest preservation) generally coincide with

TABLE 3.  Estimated Age, Measured $\delta^{18}$ and Calculated Values of the Composite Dissolution Index (CDI) for Samples From Core V34-53

| Core Depth, cm | Estimated Age, kyr | $\delta^{18}O$,[a] °/°° | CDI |
|---|---|---|---|
| 0 | 0 | +2.51 | 1.67 |
| 5 | 5 | +3.08 | 2.69 |
| 10 | 10 | +3.15 | 3.25 |
| 15 | 15 | +3.39 | 2.26 |
| 20 | 19 | +3.87 | 0.76 |
| 25 | 23 | +3.73 | −0.42 |
| 30 | 27 | +3.67 | −0.43 |
| 35 | 32 | +3.84 | 1.22 |
| 40 | 36 | +3.30 | 0.40 |
| 45 | 40 | +3.33 | 1.18 |
| 50 | 44 | ---[b] | 0.86 |
| 55 | 48 | +3.42 | −0.03 |
| 60 | 52 | +3.49 | 0.74 |
| 65 | 57 | +3.69 | −0.34 |
| 70 | 61 | +3.62 | −1.50 |
| 75 | 65 | +3.70 | −3.87 |
| 80 | 67 | +3.45 | −4.39 |
| 85 | 69 | +3.70 | −4.20 |
| 90 | 71 | +3.47 | −4.11 |
| 95 | 74 | +3.46 | −5.90 |
| 100 | 76 | +3.20 | −6.68 |
| 105 | 78 | +3.23 | −6.12 |
| 110 | 80 | +3.06 | −1.94 |
| 115 | 82 | +3.33 | 0.29 |
| 120 | 85 | +3.05 | 0.13 |
| 125 | 87 | +3.57 | 0.23 |
| 130 | 99 | +2.86 | −0.25 |
| 135 | 103 | +3.39 | −0.64 |
| 140 | 107 | +3.44 | −0.59 |
| 145 | 112 | +3.07 | −0.77 |
| 150 | 117 | +2.81 | 0.09 |
| 155 | 122 | +2.67 | 1.75 |
| 160 | 128 | +3.42 | 2.33 |
| 165 | 135 | +4.20 | 1.72 |
| 170 | 139 | +4.11 | 1.49 |
| 175 | 142 | +4.01 | 1.28 |
| 180 | 146 | +3.92 | 1.16 |
| 185 | 148 | +4.01 | 0.92 |
| 190 | 151 | +4.08 | 0.95 |
| 195 | 161 | +3.76 | 0.45 |
| 200 | 171 | +3.20 | 0.43 |
| 205 | 175 | +3.56 | 0.85 |
| 210 | 179 | +3.45 | −0.53 |
| 215 | 183 | +3.72 | 0.39 |
| 220 | 188 | +3.28 | −4.42 |
| 225 | 194 | +2.90 | −1.68 |
| 230 | 200 | +3.01 | −2.00 |
| 235 | 205 | +3.38 | −0.48 |
| 240 | 216 | +3.14 | −0.18 |
| 245 | 222 | +3.50 | −0.46 |
| 250 | 228 | +3.64 | −1.56 |
| 255 | 233 | +3.28 | −0.71 |
| 260 | 238 | +3.02 | −0.18 |
| 265 | 243 | +3.09 | 2.13 |
| 270 | 248 | +3.59 | 1.43 |
| 275 | 254 | +3.56 | 2.04 |
| 280 | 259 | +3.76 | 0.55 |

TABLE 3.  (continued)

| Core Depth, cm | Estimated Age, kyr | $\delta^{18}O$,[a] °/°° | CDI |
|---|---|---|---|
| 285 | 264 | +3.92 | 0.71 |
| 290 | 269 | +3.81 | −0.72 |
| 295 | 276 | +3.91 | −1.09 |
| 299 | 282 | +3.45 | −3.81 |
| 305 | 290 | +3.08 | −2.46 |
| 310 | 297 | +3.09 | −3.98 |
| 315 | 304 | +3.10 | −2.53 |
| 320 | 310 | +3.15 | −4.62 |
| 325 | 318 | +2.81 | −1.79 |
| 330 | 324 | +3.20 | 0.00 |
| 335 | 331 | +3.13 | 1.89 |
| 340 | 338 | +3.97 | −1.12 |
| 345 | 347 | +4.02 | −0.28 |
| 350 | 356 | +3.85 | −3.68 |
| 355 | 366 | +3.91 | −3.62 |
| 360 | 376 | +3.71 | −4.98 |
| 365 | 386 | +3.72 | −3.86 |
| 370 | 395 | +3.50 | −6.35 |
| 375 | 405 | +3.25 | −2.53 |
| 380 | 422 | +3.50 | −1.57 |
| 385 | 439 | +3.75 | −1.36 |
| 390 | 457 | +3.83 | −2.99 |
| 395 | 474 | +3.72 | −5.71 |
| 400 | 491 | +3.44 | −3.55 |
| 405 | 506 | +3.57 | −0.28 |
| 410 | 521 | +3.59 | −1.15 |
| 415 | 536 | +3.65 | −0.85 |
| 420 | 551 | +3.41 | −4.35 |
| 425 | 566 | +3.46 | −3.27 |
| 430 | 581 | +3.30 | −4.57 |
| 435 | 596 | +3.11 | −2.43 |
| 440 | 601 | +3.34 | −2.52 |
| 445 | 607 | +3.40 | −1.24 |
| 450 | 612 | +3.55 | −2.23 |
| 455 | 617 | +3.68 | 2.15 |
| 460 | 623 | +3.90 | 2.25 |
| 465 | 628 | +4.30 | 2.29 |
| 470 | 636 | +4.26 | 1.48 |
| 475 | 645 | +3.99 | 1.67 |
| 480 | 653 | +3.66 | 1.31 |
| 485 | 662 | +3.92 | 2.30 |
| 490 | 670 | +3.94 | 0.02 |
| 495 | 675 | +3.72 | 0.71 |
| 500 | 680 | +3.90 | −4.35 |
| 505 | 693 | +3.65 | −2.17 |
| 510 | 706 | +3.26 | −1.65 |
| 515 | 719 | +2.99 | 2.58 |
| 520 | 731 | +3.34 | 2.44 |
| 525 | 744 | +3.60 | 2.66 |
| 530 | 757 | +3.52 | 1.43 |
| 535 | 770 | +3.76 | −0.47 |
| 540 | 783 | +3.73 | 0.63 |
| 545 | 796 | +3.66 | 1.37 |
| 550 | 809 | +3.64 | 2.55 |

[a]All isotopic analyses performed on _Planulina wuellerstorfi_ and referred to the PDB Standard.

[b]Not enough sample for isotopic analysis.

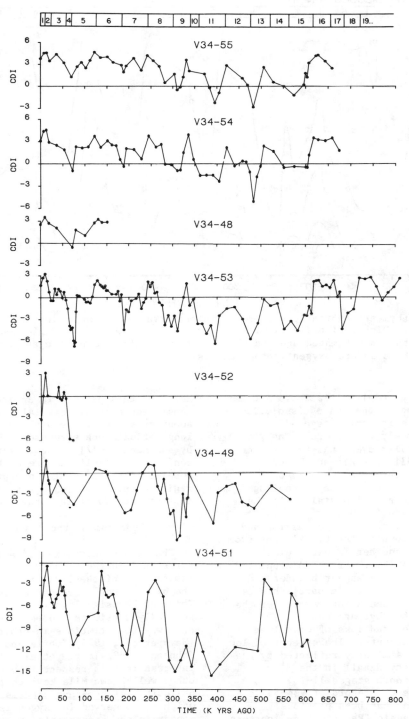

Fig. 4.  Variations in the calculated values of the Composite Dissolution Index (CDI) for the seven transect cores through time.  Positive values indicate better relative preservation; negative values indicate poorer relative preservation.  Cores are arranged in order of increasing water depth.  Standard oxygen isotope stages are indicated across the top.

COMPOSITE DISSOLUTION INDEX

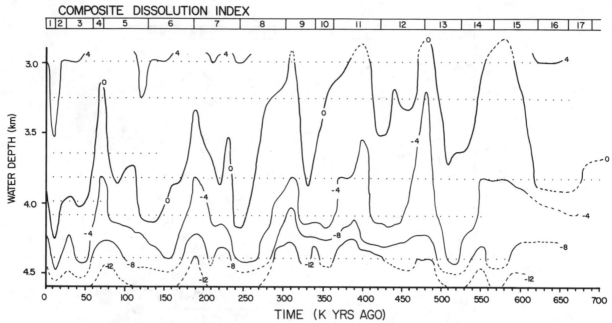

Fig. 5. Bathymetric variations in relative preservation through time, as expressed by the Composite Dissolution Index. The CDI values from Figure 4 were plotted according to core depth and estimated age and then contoured. The numbered legend across the top shows the standard oxygen isotope stages.

those times of major glacial initiation. A clear example of this timing is the sharp dissolution pulse found across the isotopic stage 4/5 boundary at approximately 75 kyr ago. The global nature of this particular event has been recently noted by Crowley [1983]. Berger and Vincent [1981] point out that the co-occurrence of dissolution maxima and times of major cooling over longer time spans are also visible in the Pacific data of Thompson [1976].

This pattern whereby preservation maxima and minima tend to coincide with the transitions from one climatic state to another is particularly clear over the last 250 kyr. Prior to about 250 kyr ago, dissolution events appear broader and less spikelike in the record. In part, this is an artifact of lower accumulation rates over the interval from 300 to 600 kyr ago due to intensified dissolution and loss of dissolved carbonate from the sediments. The effects of low accumulation rate are similarly reflected by the loss of higher-frequency detail in the $\delta^{18}O$ record throughout isotopic stages 11-15.

Also evident in the CDI record of V34-53 is a weak suggestion of what has been previously referred to as the classic "Pacific-type" pattern of intensified dissolution during interglacials and enhanced preservation during glacials. Preservation in glacial stage 6, for example, is generally better than preservation in adjacent interglacial stages 5 and 7. Similarly, preservation in glacial stage 16 is generally better than that in interglacials 15 and 17.

However, this pattern is apparently not consistent over all glacial intervals. Finally, superimposed over the whole record is the very long period Brunhes supercycle previously noted by Adelseck [1977]. The longer record of V34-53 confirms the cyclic nature of this phenomenon, as opposed to the possible interpretation of a regime change at 250-300 kyr if based only on the shorter records of the other cores.

Spectrum of the Dissolution Record

The strong suggestion of a low-frequency 100 kyr signal in the preservation records is very reminiscent of the 100 kyr cycle which dominates the oxygen isotope record of the last 900 kyr. This 100 kyr signal in the isotope record is generally considered to be causally related in some way to variations in orbital eccentricity [Hays et al., 1976; Imbrie and Imbrie, 1980; Johnson, 1982; Imbrie et al., 1984]. Of great interest in this respect is whether or not the other well-known Milankovitch periods of obliquity (41 kyr) and precession (23 and 19 kyr) which are observed in oxygen isotope records also appear in the preservation record of V34-53.

Compared to visual observations, the techniques of cross-spectral analysis provide a more rigorous means of analyzing the relationships between the $\delta^{18}O$ and preservation (CDI) signals. Cross-spectral analysis of the paired records in V34-53 gives an indication of the frequencies over which they are most highly

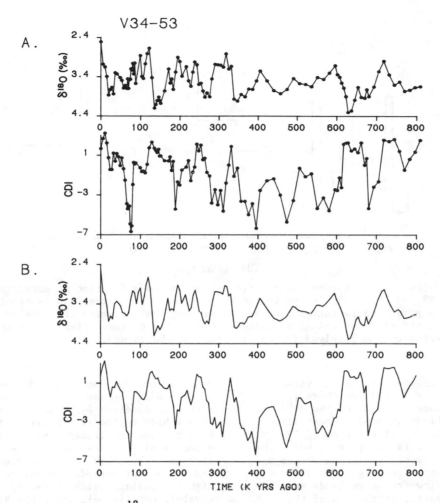

V34-53

A.

B.

Fig. 6. (a) Variations in $\delta^{18}O$ and the calculated Composite Dissolution Index as a function of estimated age for the more detailed record of V34-53. (b) Time series of $\delta^{18}O$ and the CDI for core V34-53 after interpolation at intervals of 5 kyr. Note that the interpolated time series preserves the original character of the two records.

correlated (coherent). More important, the method makes it possible to examine the relative phasing between the two signals across the entire range of statistically significant frequencies. Detailed descriptions of the cross-spectral technique and examples of applications can be found in work by Jenkins and Watts [1968], Pisias et al. [1973], Moore et al. [1977], Dunn [1982], and Imbrie et al. [1984].

In all calculations the sampling interval for the time series was fixed at 5 kyr (Figure 6b); hence the spectral estimates ideally cover a frequency range up to the Nyquist frequency of 0.1 cycles per thousand years. Because of generally low accumulation rates, this level of resolution is unlikely to be approached. Pisias [1983] has found that oxygen isotope records from cores with accumulation rates less than 2 cm/kyr show a significant reduction in variance at higher frequencies as predicted by simple

bioturbation models. However, given known accumulation rates, an estimate of the resolution obtainable (Nyquist period) can be made using the formula:

$$NP = 2(SI/SR)$$

where NP is the Nyquist period, SI is the sampling interval, and SR is the sedimentation rate [Balsam and Posmentier, 1984]. The results of this calculation of the Nyquist period for V34-53 are plotted as a function of age in Figure 7. The figure shows that not all periodicities of potential interest are capable of being resolved over all intervals of the core. Specifically, accumulation rates over the interval from about 400 to 600 kyr are too low to preserve evidence of the precessional periodicities of 23 kyr and 19 kyr predicted by the astronomical theory.

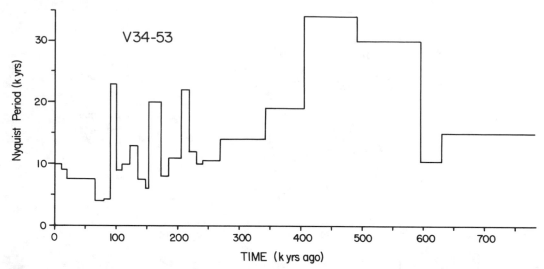

Fig. 7.  Calculated variations in the Nyquist period as a function of estimated age down core V34-53.  The Nyquist period is the shortest period that can be resolved in a core with a given sedimentation rate and sample interval assuming the core is otherwise unaffected by geological disturbances.  Only periodicities longer than the Nyquist period can be resolved for any interval of the record.

The results of a cross-spectral analysis of $\delta^{18}O$ versus the CDI for V34-53 are shown in Figure 8.  The variance spectrum of the $\delta^{18}O$ record shows the well-documented concentrations of power at the orbital periodicities of 100 kyr, 41 kyr, and near 23 and 19 kyr (Figure 8a).  The spectrum of the CDI reveals that much of the variance in the preservation signal is also concentrated in the low-frequency bands near 100 kyr and 41 kyr.  However, resolution of the precessional bands in the CDI is much less clear, although power is found near 19 kyr and in a broad frequency band from about 27 to 30 kyr.

In spite of low accumulation rates downcore, the presence of all four orbital periods in the $\delta^{18}O$ spectrum indicates that the overall record of V34-53 is clearly adequate to resolve the higher precessional frequencies.  The absence of discrete spectral peaks at both 23 and 19 kyr in the spectrum of the CDI suggests that preservational variations do not record a strong precessional response.  Further experiments with the most detailed portion of the record (0-350 kyr; see Figure 7) also failed to clearly resolve the precessional bands in the CDI.

Of great interest with regards to understanding the possible origin of the observed dissolution cycles is the significant coherency between the two records over the lower orbital frequencies.  The coherency spectrum across the bottom of Figure 8a shows that $\delta^{18}O$ and the CDI record are significantly coherent over the eccentricity (100 kyr) and obliquity (41 kyr) frequency bands. Coherency is a measure of the correlation between two signals over a given frequency band after the effects of any phase

differences have been removed [Jenkins and Watts, 1968].  High coherencies imply not only the mutual presence of significant periodicities, but also a high degree of similarity between their time-varying amplitudes, and provide strong evidence of a causal link between the two signals.  The measured coherencies at the 100 kyr and the 41 kyr periods are not only surprisingly high in absolute value (0.84 and 0.90, respectively), but lie well above the 95% significance level.  Since the coherency coefficient has all of the properties of the Pearsonian coefficient of linear correlation [Jenkins and Watts, 1968], the square of any coherency represents that fraction of the variance observed in one signal that is linearly related to the variance in the other signal over a given frequency band.  Squaring the measured coherencies suggests that as much as 71% of the variance in the CDI over the 100 kyr frequency band and 81% of the variance in the CDI over the 41 kyr frequency band can be explained as a simple linear response to climatic forcing as represented by the $\delta^{18}O$ curve.  The lack of significant coherency at 19 kyr and the absence of a discrete peak in the CDI spectrum at 23 kyr suggests that the CDI and $\delta^{18}O$ are, at most, weakly linked over the precessional frequency bands.

An additional product of the analysis is the phase spectrum between $\delta^{18}O$ and the CDI shown in Figure 8b.  The phase spectrum is simply a listing of frequency-specific estimates of average phase shifts over the 800 kyr study interval.  The results quantify general impressions gained from the previous visual examination of the two records.  At the 100 kyr frequency band   changes

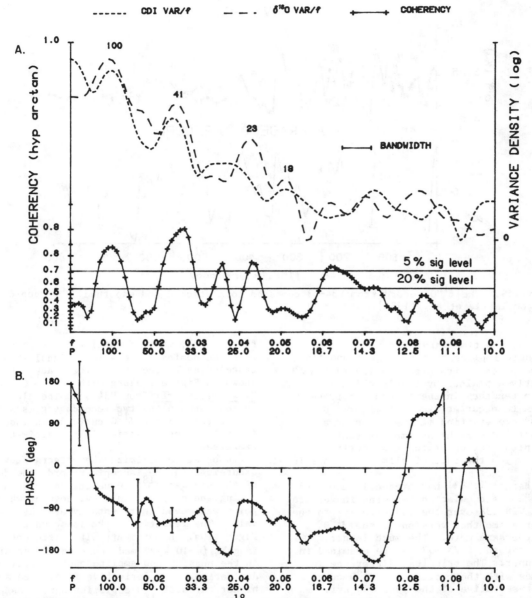

Fig. 8. (a) Variance spectra of the $\delta^{18}O$ (long-dashed line) and the CDI (short-dashed line) records from V34-53, and the coherency spectrum between the two records (solid line with pluses) for the interval from 0 to 805 kyr. Variance spectra are plotted on arbitrary log scales, while the coherency spectrum is plotted on a hyperbolic arctangent. The sample interval is 5 kyr. Significance levels for the coherency spectrum are shown (0.80 and 0.95 levels). (b) Phase spectrum of $\delta^{18}O$ versus the CDI for V34-53. Negative phase indicates that isotopic maxima and minima lead preservation maxima and minima, respectively.

in preservation lag behind changes in $\delta^{18}O$ by approximately 70°, or about 19 kyr. At the 41 kyr band, this lag is on the order of 110°, or about 12 kyr. In fact, over the whole range of frequencies encompassed by the orbital periods of 100 kyr and 19 kyr, the sense of the phase difference is such that preservation maxima lag behind isotopic maxima (glacials) and preser-

vation minima lag behind isotopic minima (interglacials).

Carbonate Preservation and Rates of Change in the Climate System

The statistical evidence of a close relationship between the time-varying amplitudes of the

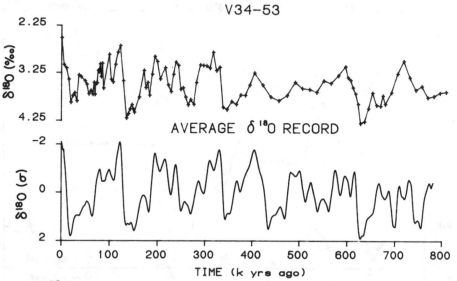

Fig. 9. The $\delta^{18}O$ record from V34-53 compared to a stacked and smoothed $\delta^{18}O$ record from Imbrie et al. [1984].

$\delta^{18}O$ record and the time-varying amplitudes of the CDI strongly implies that a large proportion of the record of carbonate preservation in V34-53 can be directly explained by simple climatic forcing. The tendency in the record for preservation maxima to occur at times of major $^{18}O$ depletion and preservation minima to occur at times of major $^{18}O$ enrichment implies that extremes in preservation state occur during intervals in which the rate of climatic change is greatest.

Keir and Berger [1983] have recently compared preservation and accumulation patterns in western Pacific core V28-238 over the last 120 kyr to the derivative of a smoothed version of the $\delta^{18}O$ record from the same core. The much longer preservation record of V34-53 can be examined in a similar manner. The principal difference in strategy comes with the choice of a $\delta^{18}O$ curve taken as representative of the last 800 kyr. Pleistocene variations in the $\delta^{18}O$ record are generally interpreted to reflect changes in oceanic isotopic composition due to the waxing and waning of the great continental ice sheets [Shackleton, 1967; Shackleton and Opdyke, 1973]. We have chosen to use a stacked and smoothed $\delta^{18}O$ record constructed by Imbrie et al. [1984] as the best available approximation of global ice volume, rather than the benthic $\delta^{18}O$ record from our own core. The stacked $\delta^{18}O$ curve is considered to be a more realistic model because it consists of the $\delta^{18}O$ records from shallow-dwelling planktic foraminifera in five cores from low- and mid-latitude sites which have been normalized and averaged together. Matthews and Poore [1980] have emphasized that the $\delta^{18}O$ of tropical, shallow-dwelling planktics provides the best record of global ice volume. By stacking

the records, most of the local influences other than the dominant ice volume signal should have canceled each other out. The stacked record is shown in Figure 9 along with the $\delta^{18}O$ record of P. wuellerstorfi from V34-53. The close correlation between the two records suggests that the use of the stacked $\delta^{18}O$ curve as an independent ice volume approximation is a legitimate exercise.

In order to minimize the effects of very small excursions in the isotopic record, the derivative of the stacked $\delta^{18}O$ curve was calculated by taking the slope of an 8 kyr segment whose midpoint was moved incrementally at 1 kyr intervals. The resulting curve is shown by itself in Figure 10a. Experiments with different segment lengths (4-10 kyr) made no significant difference in the phasing and amplitude of the calculated derivative. The derivative is plotted such that higher rates of $^{18}O$ depletion appear as peaks and higher rates of $^{18}O$ enrichment appear as valleys. The highest rates of $^{18}O$ change are associated with the major isotopic depletions which identify the glacial Terminations; these can be easily identified in Figure 10a from their chronology. Because of the characteristic sawtooth nature of the $\delta^{18}O$ record [Broecker and Van Donk, 1970], the periods of isotopic enrichment which signal the initiation of glacial conditions are less pronounced in the derivative curve than the Terminations.

The derivative of the stacked $\delta^{18}O$ record is shown in Figure 10b, this time superimposed over the CDI from V34-53. One's initial impression from this comparison is that a remarkable similarity exists between variations in preservation state and the times of maximum rate of change in the $\delta^{18}O$ record. The degree of

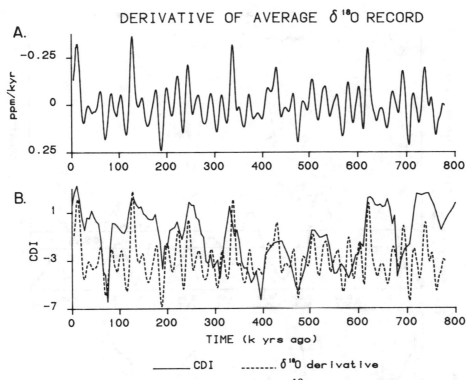

Fig. 10. (a) Calculated derivative of the stacked $\delta^{18}O$ record in units of parts per million per kiloyear. The derivative is plotted such that rapid $^{18}O$ depletions appear as peaks, and periods of rapid $^{18}O$ enrichment appear as valleys. (b) Plot of the $\delta^{18}O$ derivative (dashed line) superimposed over the CDI record from V34-53.

similarity is particularly striking over the last 250-300 kyr, where higher accumulation rates give much better resolution of events in the CDI.

A cross spectrum calculated between the CDI and the derivative of the $\delta^{18}O$ record (the presumed forcing function) appears in Figure 11a. Given that the $\delta^{18}O$ derivative contains the same periodic components as the actual measured $\delta^{18}O$ record, the high coherencies observed between the CDI and $\delta^{18}O$ derivative curve are expected. Coherencies at the 100 kyr and 41 kyr periods are nearly identical to those calculated earlier between the CDI and the $\delta^{18}O$ record in V34-53 (0.85 and 0.90, respectively).

The phase spectrum calculated between the CDI and the $\delta^{18}O$ derivative is shown in Figure 11b with 95% confidence intervals. The record of carbonate preservation (CDI) and the inferred $\delta^{18}O$ forcing function are essentially in phase over the 100 kyr and 41 kyr frequency bands. Of particular significance here is the statistical confirmation in a much longer record of the general observation made by Keir and Berger [1983] that preservation maxima and minima are nearly coincident with the times of maximum rate of change in the $\delta^{18}O$ record. Conceptually, this view differs from previous work which has tended to view offsets between preservation extremes and measured $\delta^{18}O$ extremes as a lead-lag phenomenon

reflecting differing response times within the system. Luz and Shackleton [1975], for example, suggested that the preservation lag they observed in the eastern equatorial Pacific reflected a delayed response of the ocean carbonate ion system to changes in $CaCO_3$ deposition elsewhere, such as at higher latitudes or in the North Atlantic. The observations of Keir and Berger [1983] and from this study, in part, relieve us of the necessity of postulating significant delays in the carbonate system. On the contrary, the high correlation and close phase relationship between extremes in preservation state and the derivative of the $\delta^{18}O$ record suggest a rapid response of the carbonate system to simple climatic forcing. Somehow this forcing must translate into changes in the carbonate saturation state of the local deep ocean.

## Discussion

In general, most models put forth to account for the temporal and spatial variations in carbonate dissolution have tended to fall into two broad categories. Inventory models generally invoke temporary changes in the partitioning of carbonate and organic carbon between various oceanic and terrestrial reservoirs to explain dissolution patterns. Circulation models are

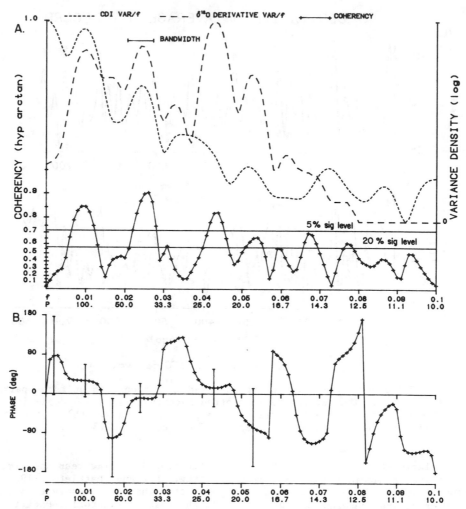

Fig. 11. (a) Variance and coherency spectra calculated between the CDI record of V34-53 and the derivative of the stacked $\delta^{18}O$ record for the interval from 0 to 805 kyr. Variance spectra of the CDI (short-dashed line) and the $\delta^{18}O$ derivative (long-dashed line) are plotted as arbitrary log scales. The coherency spectrum (solid line with pluses) shows significant coherency (0.95 level) between the two records over the 100 kyr and 41 kyr frequency bands. (b) Phase spectrum of the CDI from V34-53 versus the derivative of the $\delta^{18}O$ forcing function. The 95% confidence interval is shown. Negative phase indicates that the $\delta^{18}O$ derivative leads the CDI index.

those which call for changes in the circulation patterns of the deep ocean, with the resulting redistribution or fractionation of $CO_2$, $CaCO_3$, and nutrients, to explain records of carbonate accumulation and preservation.

In the Indian Ocean, the high coherencies found between the CDI from V34-53 and the $\delta^{18}O$ record strongly imply that the bulk of the preservation record can be explained as a simple linear response to changes in climate. Whatever process is ultimately driving local dissolution intensity is in phase with rates of change in the $\delta^{18}O$ signal. Finally, the CDI and $\delta^{18}O$ records are strongly linked over the 100 kyr and 41 kyr frequency bands, but not over those at 23 and 19

kyr. Given these observations as boundary conditions, we now consider the potential explanatory power of both inventory and circulation models.

## Inventory Models

Shackleton [1977b] suggested that organic carbon exchange between ocean and continent during the climatically induced growth and decay of the terrestrial biosphere was the principal determinant of deep-sea preservation patterns. Other inventory models invoke some form of shelf-basin fractionation and are presumed to be directly driven by glacio-eustatic sea level excursions. Possibly the most widely discussed of these shelf inventory models is that of Broecker [1981, 1982]. Broecker proposed that during a postglacial sea level rise, deposition of organic-rich sediments on newly flooded shelves and in estuaries would result in a reduction of dissolved carbon and phosphorus in the ocean. In particular, the removal of $CO_2$ from the open-ocean reservoir would temporarily reduce the degree of $CaCO_3$ undersaturation in the deep sea, resulting in a pulse of excess preservation of carbonate. Similarly, the destruction of these same shelf sediments during the early phases of glacial-induced sea level fall would inject $CO_2$ back into the ocean, temporarily increasing the degree of under-saturation until neutralized by the dissolution of carbonate. Alternatively, other sea level driven inventory models call for simple mass balance considerations with respect to the build-up and erosion of shelf carbonates to explain dissolution patterns [Olausson, 1965; Milliman, 1974; Berger and Winterer, 1974; Hay and Southam, 1977; Berger, 1982].

As an indicator of global ice volume, the $\delta^{18}O$ record is also an indirect measure of changing sea level. Hence, periods of rapid depletion and enrichment of $^{18}O$ can be inferred to reflect periods of rapid sea level rise and fall. The strong tendency in the record of V34-53 for preservation maxima to occur at times of major $^{18}O$ depletion (sea level rise) and preservation minima to occur at times of major $^{18}O$ enrichment (sea level fall) would seem to support a shelf-basin type of inventory mechanism to explain the local preservation record. If one assumes, as did Keir and Berger [1983], that the rate of transfer of organic carbon to and from shelf reservoirs is linearly proportional to the rate of sea level rise and fall, then the results described here are consistent with preservation patterns predicted by the Broecker model. On the other hand, those sea level driven inventory models calling solely for $CaCO_3$ mass balance changes due to the build-up and erosion of shelf carbonates cannot explain the observed preservation record. Rather than producing a preservation peak, for example, rapid carbonate precipitation or deposition on the shelves during a postglacial rise of sea level would instead compete with the deep sea for available carbonate.

An obvious drawback of the Broecker shelf hypothesis is that it requires large amounts of shallow-water deposition of organic-rich sediments over relatively short periods of time. Recent measurements of glacial atmospheric $CO_2$ concentrations from ice cores [Oeschger et al., this volume] suggests that some past $pCO_2$ variations were far too rapid (a few hundreds of years) to be explained by a shelf deposition model. However, one mechanism is unlikely to explain all variability on all time scales. Over the longer time scales considered in this study,

sea level changes, and the resulting organic carbon flux onto and off of the shelf, may well have an important effect on deep-sea preservation patterns.

The Broecker model, by its very nature, calls for changes in the $CO_2$ budget of the whole-ocean reservoir. Given that all ocean basins have shelves, the model provides a mechanism for producing a global response in the carbonate system. As noted earlier, similar observations of preservation maxima across glacial Terminations (trangressions) and preservation minima across major cooling events (regressions) have now been made from widespread localities. The apparently global and in-phase nature of many of these short-duration preservation phenomena is also consistent with a shelf-basin type of model.

Although the above relationships are quite suggestive, there are nevertheless a number of problems and uncertainties associated with the Broecker shelf scenario. For example, obser-vations of the more general out-of-phase tendency between dissolution fluctuations in the Atlantic and Pacific through at least the late Pleistocene imply an overall fractionation between basins that is difficult to explain by simple whole-ocean inventory changes. In terms of frequency-domain considerations, a strict interpretation of the Broecker model, coupled with the assumption of Keir and Berger [1983] that organic carbon transfer to and from the shelf is proportional to rates of sea level change, would require coherency in our data at all of the dominant frequencies present in the $\delta^{18}O$ (i.e., sea level) record. As noted before, the absence of significant coherency between the CDI and $\delta^{18}O$ over the 23 kyr and 19 kyr frequency bands suggests that these two records are, at most, weakly linked over the precessional frequencies. However, a possible caveat to the latter argument may simply be that the Broecker shelf model, as presently formulated, is not linear for the smaller changes in sea level associated with higher-frequency (23 and 19 kyr) oscillations in global ice volume.

Circulation Models

An alternative interpretation of our data is that the bulk of the local preservation record is the product of past changes in deep ocean circulation patterns. Abyssal waters in the Indian and Pacific oceans are largely a mixture of waters of both Atlantic and Antarctic origin. Because of their high-latitude source areas, production of these deep and bottom waters is apt to be particularly sensitive to climatic forcing. Our data suggest that deep-water production rates, or at least changes in deep-water chemistry, would have to vary directly in proportion to rates of change in the $\delta^{18}O$ record in order to produce the preservation record we observe in the Indian Ocean.

Considerable evidence now points to reduced

production of North Atlantic Deep Water (NADW) during the most recent glacial maximum [Duplessy et al., 1975; Streeter and Shackleton, 1979; Boyle and Keigwin, 1982; Curry and Lohmann, 1982, 1983]. A diminished glacial production of NADW relative to the deep-water input from other sources has often been proposed to explain observations of the general see-saw tendency between dissolution fluctuations in the Atlantic and Pacific [e.g., Berger, 1973; Volat et al., 1980]. The weak suggestion of a "Pacific-type" dissolution pattern in parts of the record of V34-53 may in fact reflect such longer-term circulation changes. Whether or not simple circulation models can explain both this overall see-saw tendency between basins and the apparently global and in-phase preservation extremes associated with the most rapid climatic transitions remains to be seen.

Tentative support, however, for a circulation explanation comes from the frequency-domain response of high-latitude climatic change. In the North Atlantic, Ruddiman and McIntyre [1981, 1984] have shown that downcore foraminiferal estimates of past sea surface temperatures contain mostly 23 kyr power south of about 50°N. North of this latitude, in the principal source regions of NADW, equivalent foraminiferal-based sea surface temperature records contain mostly 41 kyr and 100 kyr power. If deep-water changes are indeed driven by climatic processes active in their areas of formation, we would tend to expect records of NADW behavior to preserve responses over the latter frequency bands, at the expense of the higher precessional frequencies. Boyle [1984], for example, has attributed a dominant 41 kyr periodicity in a South Atlantic record of Cd/Ca ratios in benthic foraminifera to past fluctuations in NADW flow. In the Indian Ocean, the strong concentrations of power in the 100 kyr and 41 kyr frequency bands in the CDI from V34-53, and the lack of clearly defined power and coherency with $\delta^{18}O$ over the precessional frequency bands, are also consistent with such a hypothesis. At this time, given our data and this simple interpretation of frequency-domain relationships, we view the weak linkage of the CDI and $\delta^{18}O$ over the precessional frequencies and the high coherencies at 100 kyr and 41 kyr as evidence that deep-water circulation changes, driven by high-latitude climatic forcing in the North Atlantic, are the principal source of variability in the preservation record of the equatorial Indian Ocean.

## Conclusions

The preservation of calcareous microfossils fluctuates markedly in a regular and periodic manner in a suite of bathymetrically distributed piston cores from the eastern flank of the Ninetyeast Ridge. The dominant cycles visible in a measured Composite Dissolution Index are of about 100 kyr and 41 kyr duration. In addition,

there is evidence of a much longer period super-cycle in the data which manifests itself as a shoaling of poor preservation levels between about 300 and 600 kyr ago.

Dissolution variations are not synchronous with glacial-interglacial intervals as determined from $\delta^{18}O$ stratigraphy. However, cross-spectral analysis reveals high coherency (>0.85) between the CDI from core V34-53, the $\delta^{18}O$ record, and the derivative of the $\delta^{18}O$ record over the 100 kyr and 41 kyr Milankovitch frequencies, indicating a simple linear response of overall dissolution intensity to changing climate. Variations in carbonate preservation are proportional to and in phase with maximum rates of change in the $\delta^{18}O$ record, suggesting a rapid response of the local carbonate system to climatic forcing.

Both the shelf-basin inventory model of Broecker [1981, 1982] and circulation models provide plausible mechanisms for explaining the origins of the observed Indian Ocean dissolution cycles. However, to the degree that one is willing to accept a simple linear version of the Broecker shelf model and to assume that changes in deep-water production or chemistry are directly driven by high-latitude climatic variations, we believe our results tend to implicate a circulation explanation. We base our choice largely upon the lack of a well-defined precessional response in the CDI from V34-53 and the absence of significant coherency between preservation variations and the 23 and 19 kyr frequency bands which show up clearly in foraminiferal $\delta^{18}O$. We note that a strict interpretation of the shelf inventory model would require coherency with all of the dominant periodicities in the $\delta^{18}O$ record. Our one qualification to this argument is that the Broecker mechanism, as presently formulated, may not be linear for the smaller changes in sea level generally associated with the low-amplitude precessional cycles in the $\delta^{18}O$ record.

Acknowledgments. The authors are grateful to John Imbrie and Wally Broecker for helpful discussion and to Angeline Duffy for her programming assistance. We would also like to thank Taro Takahashi for originally taking the Vema-34 series of cores and for providing preliminary carbonate chemistry data at an early stage of this study. This research was supported by National Science Foundation grants ATM78-25629 and ATM80-18897 (SPECMAP) through the Climate Dynamics Section, Division of Atmospheric Sciences, and the Division of Polar Programs. We thank the National Science Foundation for its continued support of the Lamont-Doherty Geological Observatory core collection.

## References

Adelseck, C. G., Jr., Recent and late Pleistocene sediments from the eastern equatorial Pacific

Ocean: Sedimentation and dissolution, Ph.D. thesis, 192 pp., University of California, San Diego, 1977.

Arrhenius, G., Sediment cores from the east Pacific, Rep. Swed. Deep Sea Exped. 1947-1948, 5, 228 pp., 1952.

Balsam, W. L., Late Quaternary sedimentation in the western North Atlantic: Stratigraphy and paleoceanography, Palaeogeogr. Palaeoclimatol. Palaeoecol., 35, 215-240, 1981.

Balsam, W. L., and E. S. Posmentier, Comments on spectral analysis of core data, Palaeogeogr. Palaeoclimatol. Palaeoecol., 45, 369-372, 1984.

Bé, A. W. H., J. E. Damuth, L. Lott, and R. Free, Late Quaternary climatic record in western equatorial Atlantic sediments, in Investigations of Late Quaternary Paleoceanography and Paleoclimatology, Geol. Soc. Am. Mem. 145, edited by R. M. Cline and J. D. Hays, pp. 165-200, Geological Society of America, Boulder, Colorado, 1976.

Berger, W. H., Deep-sea carbonates: Pleistocene dissolution cycles, J. Foraminiferal Res., 3, 187-195, 1973.

Berger, W. H., Deep-sea carbonate and the deglacial preservation spike in pteropods and foraminifera, Nature, 269, 301-304, 1977.

Berger, W. H., Increase of carbon dioxide in the atmosphere during deglaciation: The coral reef hypothesis, Naturwissenschaften, 69, 87-88, 1982.

Berger, W. H., and J. S. Killingley, Glacial-Holocene transition in deep-sea carbonates: Selective dissolution and the stable isotope signal, Science, 197, 563-566, 1977.

Berger, W. H., and E. Vincent, Chemostratigraphy and biostratigraphic correlation: Exercises in systemic stratigraphy, Oceanol. Acta, Proc. Int. Geol. Congr. 26, C(4), Geol. Oceans Symp., 115-127, 1981.

Berger, W. H., and E. L. Winterer, Plate stratigraphy and the fluctuating carbonate line, in Pelagic Sediments on Land and Under the Sea, Spec. Publ. Inter. Assoc. Sedimentol., vol. 1, edited by K. J. Hsu and H. Jenkyns, pp. 11-48, Blackwell, Oxford, England, 1974.

Boyle, E. A., Cadmium in benthic foraminifera and abyssal hydrography: Evidence for a 41 kyr obliquity cycle, in Climate Processes and Climate Sensitivity, Geophys. Monogr. Ser., vol. 29, edited by J. E. Hansen and T. Takahashi, pp. 360-368, AGU, Washington, D. C., 1984.

Boyle, E. A., and L. D. Keigwin, Deep circulation of the North Atlantic over the last 200,000 years: Geochemical evidence, Science, 218, 784-787, 1982.

Broecker, W. S., Glacial to interglacial changes in ocean and atmosphere chemistry, in Climatic Variations and Variability: Facts and Theories, edited by A. Berger, pp. 109-120, D. Reidel, Hingham, Mass., 1981.

Broecker, W. S., Glacial to interglacial changes in ocean chemistry, Prog. Oceanogr., 11, 151-197, 1982.

Broecker, W. S., and T. Takahashi, The relationship between lysocline depth and in situ carbonate ion concentration, Deep Sea Res., 25, 65-95, 1978.

Broecker, W. S., and J. Van Donk, Insolation changes, ice volumes, and the $^{18}O$ record in deep-sea cores, Rev. Geophys. Space Phys., 8, 169-198, 1970.

Chen, C., Pleistocene pteropods in pelagic sediments, Nature, 219, 1145-1147, 1968.

Corliss, B. H., and R. C. Thunell, Carbonate sedimentation beneath the Antarctic Circumpolar Current during the late Quaternary, Mar. Geol., 51, 293-326, 1983.

Crowley, T. J., Calcium carbonate preservation patterns in the central North Atlantic during the last 150,000 years, Mar. Geol., 51, 1-14, 1983.

Crowley, T. J., Late Quaternary carbonate dissolution changes in the North Atlantic and Atlantic/Pacific comparisons, this volume.

Curry, W. B., and G. P. Lohmann, Carbon isotopic changes in benthic foraminifera from the western South Atlantic: Reconstruction of glacial abyssal circulation patterns, Quat. Res., 18, 218-235, 1982.

Curry, W. B., and G. P. Lohmann, Reduced advection into Atlantic Ocean deep eastern basins during last glaciation maximum, Nature, 306, 577-580, 1983.

Damuth, J. E., Quaternary climate changes as revealed by calcium carbonate fluctuations in western equatorial Atlantic sediments, Deep Sea Res., 22, 725-743, 1975.

Diester-Haass, L., H. J. Schrader, and J. Thiede, Sedimentological and paleoclimatological investigations of two pelagic ooze cores off Cape Barbas, North-West Africa, Meteor Forschungsergeb. Reihe C, 16, 19-66, 1973.

Dunn, D. A., Change from "Atlantic-type" to "Pacific-type" carbonate stratigraphy in the middle Pliocene equatorial Pacific Ocean, Mar. Geol., 50, 41-60, 1982.

Duplessy, J. C., L. Chenouard, and F. Vila, Weyl's theory of glaciation supported by isotopic study of Norwegian core K11, Science, 188, 1208-1209, 1975.

Gardner, J. V., Late Pleistocene carbonate dissolution cycles in the eastern equatorial Atlantic, in Dissolution of Deep-Sea Carbonates, Spec. Publ. Cushman Found. Foraminiferal Res., vol. 13, edited by W. V. Sliter, A. W. H. Bé, and W. H. Berger, pp. 129-141, U. S. National Museum, Washington, D. C., 1975.

Hay, W. W., and J. R. Southam, Modulation of marine sedimentation by the continental shelves, in The Fate of Fossil Fuel $CO_2$ in the Oceans, edited by N. R. Andersen and A. Malahoff, pp. 569-604, Plenum, New York, 1977.

Hays, J. D., T. Saito, N. D. Opdyke, and L. H. Burckle, Plio-Pleistocene sediments of the equatorial Pacific: Their paleomagnetic, biostratigraphic, and climatic record, Geol.

Soc. Am. Bull., 80, 1481-1514, 1969.

Hays, J. D., J. Imbrie, and N. J. Shackleton, Variations in the earth's orbit: Pacemaker of the ice ages, Science, 194, 1121-1132, 1976.

Imbrie, J., and J. Z. Imbrie, Modeling the climatic response to orbital variations, Science, 207, 943-953, 1980.

Imbrie, J., J. D. Hays, D. G. Martinson, A. McIntyre, A. C. Mix, J. J. Morley, N. G. Pisias, W. L. Prell, and N. J. Shackleton, The orbital theory of Pleistocene climate: Support from a revised chronology of the marine $\delta^{18}O$ record, in Milankovitch and Climate: Understanding the Response to Astronomical Forcing, Part I, edited by A. Berger, J. Imbrie, J. Hays, G. Kukla, B. Saltzman, pp. 296-306, D. Reidel, Hingham, Mass., 1984.

Jenkins, G. M., and D. G. Watts, Spectral Analysis and Its Applications, 525 pp., Holden-Day, San Francisco, California, 1968.

Johnson, R. G., Brunhes-Matuyama magnetic reversal dated at 790,000 yr B. P. by marine-astronomical correlations, Quat. Res., 17, 135-147, 1982.

Keir, R. S., and W. H. Berger, Atmospheric $CO_2$ content in the last 120,000 years: The phosphate-extraction model, J. Geophys. Res., 88, 6027-6038, 1983.

Luz, B., Stratigraphic and paleoclimatic analysis of late Pleistocene tropical southeast Pacific cores, Quat. Res., 3, 56-72, 1973.

Luz, B., and N. J. Shackleton, $CaCO_3$ solution in the tropical east Pacific during the past 130,000 years, in Dissolution of Deep-Sea Carbonates, Spec. Publ. Cushman Found. Foraminiferal Res., vol. 13, edited by W. V. Sliter, A. W. H. Bé, and W. H. Berger, pp. 142-150, U. S. National Museum, Washington, D. C., 1975.

Matthews, R. K., and R. Z. Poore, Tertiary $\delta^{18}O$ record and glacio-eustatic sea-level fluctuations, Geology, 8, 501-504, 1980.

Milliman, J. D., Marine Carbonates, 375 pp., Springer-Verlag, New York, 1974.

Moore, T. C., Jr., N. G. Pisias, and G. R. Heath, Climate changes and lags in Pacific carbonate preservation, sea surface temperature and global ice volume, in The Fate of Fossil Fuel $CO_2$ in the Oceans, edited by N. R. Andersen and A. Malahoff, pp. 145-165, Plenum, New York, 1977.

Oba, T., Biostratigraphy and isotopic paleotemperatures of some deep-sea cores from the Indian Ocean, Sci. Rep. Tohoku Univ., Ser. 2, 41, 129-195, 1969.

Oeschger, H., B. Stauffer, R. Finkel, and C. Langway, Jr., Variations of the $CO_2$ concentration of occluded air and anions and dust in polar ice cores, this volume.

Olausson, E., Evidence of climatic changes in North Atlantic deep-sea cores, with remarks on isotopic paleotemperature analysis, Prog. Oceanogr., 3, 221-252, 1965.

Peng, T. H., W. S. Broecker, G. Kipphut, and N.

J. Shackleton, Benthic mixing in deep sea cores as determined by $^{14}C$ dating and its implications regarding climate stratigraphy and the fate of fossil fuel $CO_2$, in The Fate of Fossil Fuel $CO_2$ in the Ocean, edited by N. R. Andersen and A. Malahoff, pp. 355-374, Plenum, New York, 1977.

Peterson, L. C., Late Quaternary deep-water paleoceanography of the eastern equatorial Indian Ocean: Evidence from benthic foraminifera, carbonate dissolution, and stable isotopes, Ph.D. thesis, 439 pp., Brown Univ., Providence, R. I., May 1984.

Peterson, L. C., and W. L. Prell, Carbonate dissolution in Recent sediments of the eastern equatorial Indian Ocean: Preservation patterns and carbonate loss above the lysocline, Mar. Geol., in press, 1984.

Pisias, N. G., Geologic time series from deep-sea sediments: Time scales and distortion by bioturbation, Mar. Geol., 51, 99-113, 1983.

Pisias, N. G., J. P. Dauphin, and C. Sancetta, Spectral analysis of late Pleistocene-Holocene sediments, Quat. Res., 3, 3-9, 1973.

Ruddiman, W. F., and A. McIntyre, Oceanic mechanisms for amplification of the 23,000-year ice-volume cycle, Science, 212, 617-627, 1981.

Ruddiman, W. F., and A. McIntyre, Ice-age thermal response and climatic role of the surface Atlantic Ocean, 40°N to 63°N, Geol. Soc. Am. Bull., 95, 381-396, 1984.

Shackleton, N. J., Oxygen isotope analyses and Pleistocene temperatures reassessed, Nature, 215, 15-17, 1967.

Shackleton, N. J., Tropical rainforest history and the equatorial Pacific carbonate dissolution cycles, in The Fate of Fossil Fuel $CO_2$ in the Oceans, edited by N. R. Andersen and A. Malahoff, pp. 401-427, Plenum, New York, 1977a.

Shackleton, N. J., The oxygen isotope stratigraphic record of the late Pleistocene, Philos. Trans. R. Soc. London, Ser. B, 280, 169-182, 1977b.

Shackleton, N. J., and N. D. Opdyke, Oxygen isotope and paleomagnetic stratigraphy of equatorial Pacific core V28-238: Oxygen isotope temperatures and ice volumes on a $10^5$ and $10^6$ year scale, Quat. Res., 3, 39-55, 1973.

Shackleton, N. J., and N. D. Opdyke, Oxygen-isotope and paleomagnetic stratigraphy of Pacific core V28-239, late Pliocene to latest Pleistocene, in Investigations of Late Quaternary Paleoceanography and Paleoclimatology, Geol. Soc. Am. Mem. 145, edited by R. M. Cline and J. D. Hays, pp. 449-464, Geological Society of America, Boulder, Colo., 1976.

Streeter, S. S., and N. J. Shackleton, Paleocirculation of the deep North Atlantic: 150,000 year record of benthic foraminifera and oxygen-18, Science, 203, 168-171, 1979.

Takahashi, T., W. S. Broecker, A. E. Bainbridge, and R. F. Weiss, Carbonate Chemistry of the

Atlantic, Pacific, and Indian Oceans: The Results of the GEOSECS Expeditions, 1972-1978, Lamont-Doherty Geological Observatory Technical Report No. 1., CU-1-80, 1980.

Thompson, P. R., Planktonic foraminiferal dissolution and the progress towards a Pleistocene equatorial Pacific transfer function, J. Foraminiferal Res., 6, 208-227, 1976.

Thompson, P. R., and T. Saito, Pacific Pleistocene sediments: Planktonic foraminifera,

dissolution cycles and geochronology, Geology, 2, 333-335, 1974.

Volat, J. L., L. Pastouret, and C. Vergnaud-Grazzini, Dissolution and carbonate fluctuations in Pleistocene deep-sea cores: A review, Mar. Geol., 34, 1-28, 1980.

Williams, D. F., and J. Keany, Comparison of radiolarian/planktonic foraminiferal paleoceanography of the subantarctic Indian Ocean, Quat. Res., 9, 71-86, 1978.

# LATE QUATERNARY CARBONATE CHANGES IN THE NORTH ATLANTIC AND ATLANTIC/PACIFIC COMPARISONS

Thomas J. Crowley[1]

Climate Dynamics Program, National Science
Foundation, Washington, D.C. 20550

Abstract. This paper discusses North Atlantic carbonate dissolution fluctuations during the last 400,000 years (oxygen 18 stages 1-11). Dissolution is inferred based on fluctuations of percent benthonic foraminifera and planktonic foraminiferal fragments. Stratigraphic control is based on percent carbonate fluctuations correlated to oxygen 18 records. A summary of results for the last 400,000 years indicates that (1) North Atlantic dissolution is relatively low during interglacial maximum events and increases with increasing ice volume; however, the relationship between dissolution and ice volume is not linear; (2) the lysocline changed by about 200-300 m in response to the dissolution fluctuations; (3) abyssal regions of the eastern Atlantic have a different dissolution pattern than bathyal regions; (4) comparison of Atlantic and Pacific carbonate dissolution records indicates that sometimes dissolution changes in the two basins are in phase, sometimes out of phase; (5) there is a tendency for a mid-Brunhes (200-450,000 years BP) increase in dissolution intensity in both the Atlantic and the Pacific. Interpretation of the results suggests that out-of-phase dissolution events between the Atlantic and Pacific can be explained in terms of changing deep water production rates. Abyssal/bathyal variations may be due to decreased deep water production rates coupled with increased residence times of deep waters below local sill depth in the eastern basins. In-phase Atlantic/Pacific dissolution events occur during intervals of ice growth. This result implies that shelf/basin transfer of carbon during sea level lowerings is in part responsible for whole-ocean changes in the carbon budget and global carbonate dissolution events. Long-term (100,000 years) changes in dissolution correlate in part with orbital insolation variations. At present, four processes primarily operating on 1,000-year time scales have been identified that may contribute to lower-frequency (100,000 year)
dissolution variations: changes in the rate of sea level lowering, global weathering rates, high-latitude carbonate productivity, and the rate of oceanic mixing. The relative importance of these mechanisms has not yet been quantitatively determined.

## Introduction

Deep-sea carbonate deposits are an important reservoir of carbon, and fluctuations in the preservation of biogenous carbonate sediments in part reflect changes in the carbon budget. This paper discusses features of North Atlantic carbonate dissolution variations during the last 400,000 years (through oxygen 18 stage 11). Results will be compared with Pacific dissolution records. It will be demonstrated that joint use of Atlantic/Pacific records enables a more complete interpretation of the dissolution fluctuations. The whole-ocean carbonate dissolution record can be summarized by two main patterns: (1) an out-of-phase Atlantic/Pacific response that is probably due to changes in deep-water production rates and (2) an in-phase response indicative of whole-ocean changes in the carbon budget. The amplitude of the in-phase response varies with time.

## Study Area

Dissolution patterns were examined in 15 cores in the North Atlantic between 15 and 45 N (Figure 1 and Table 1). Core depths vary from 3370 to 5029 m. Shallower cores were excluded because an earlier paleoecological study [Crowley, 1976] detected only minor and irregular variations in carbonate dissolution in cores from water depths less than 3000 m. Cores are from both the western and the eastern Atlantic basins. Berger [1968] has shown that the lysocline varies between the two basins, probably because the Mid-Atlantic Ridge influences the relative amounts of Antarctic Bottom Water (AABW) contribution to the deep

[1] On leave from Department of Physics, University of Missouri-Saint Louis.

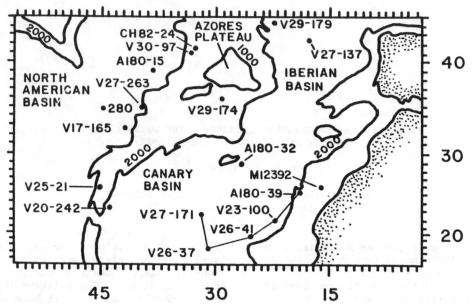

Fig. 1. North Atlantic cores examined for dissolution fluctuations, plotted against bathymetry in fathoms (1000 fm = 1830 m). Core locations are listed in Table 1. Transect in the southeast Canary Basin illustrates cores used to construct Figure 5. Dissolution information is not available for cores M12392, V29-179, and V30-97. Their locations are plotted for reference only. See Crowley [1983a] for additional details.

waters in each basin. In the eastern basin the AABW contribution is not that of a discrete water mass. Rather, some AABW mixes with North Atlantic Deep Water (NADW) at the Romanche Fracture Zone on the equator [Broecker et al.,

TABLE 1. Locations of North Atlantic Cores Illustrated in Figure 1

| Core | Latitude, N | Longitude, W | Depth, m |
|------|-------------|--------------|----------|
| 280 | 34 57' | 44 16' | 4256 |
| A180-15 | 39 36' | 36 42' | 4710 |
| A180-32 | 29 07' | 26 15' | 5029 |
| A180-39 | 25 50' | 19 18' | 3470 |
| CH82-24 | 41 40' | 32 50' | 3370 |
| M12392 | 25 10' | 16 50' | 2573 |
| V17-165 | 32 45' | 41 54' | 3924 |
| V20-232 | 23 22' | 43 39' | 4565(?) |
| V23-100 | 21 18' | 22 41' | 4579 |
| V25-21 | 26 24' | 45 27' | 3693 |
| V26-37 | 16 38' | 31 06' | 4898 |
| V26-41 | 19 19' | 26 08' | 4341 |
| V27-137 | 42 42' | 17 04' | 4883 |
| V27-171 | 21 44' | 32 34' | 4693 |
| V27-263 | 35 00' | 40 55' | 3704 |
| V29-174 | 36 18' | 29 22' | 3420 |
| V29-179 | 44 00' | 24 32' | 3331 |
| V30-97 | 41 00' | 32 55' | 3371 |

1980]. The predominance of NADW in the lower water column of the eastern basins results in a marked depression of dissolution levels [Broecker and Takahashi, 1980]. In the western basin the lysocline is at about 4100 m, whereas in the Canary Basin it is at almost 5000 m [Berger, 1968; Biscaye et al., 1976].

Variations in surface sediment preservation are in part due to different ecological provinces having different amounts of solution-susceptible species [Berger, 1968]. Sediments in the study area underlie a diverse range of ecological provinces (Figure 2). In addition to the four provinces illustrated, a fifth lies along a narrow 200-km coastal strip off the upwelling region of West Africa. A sixth province (polar organisms) is found in glacial-level samples, as the northern part of the study area (about 42 N) was the southernmost boundary of the polar front during the last glacial maximum [McIntyre et al., 1976]. No attempt was made to incorporate coastal upwelling dissolution records [e.g., Diester-Haass, 1976, 1983] into the North Atlantic open-ocean dissolution record. Observations of Parker and Berger [1971] indicate that carbonate preservation is strongly modified beneath upwelling areas, and late Pleistocene changes in West African upwelling might impose a large local signal on the dissolution records.

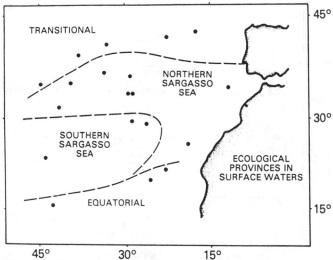

Fig. 2. Ecological provinces in the study area
[e.g., Kipp, 1976]. Figure after Crowley [1981].

## Methods

### Stratigraphy

Variations of percent calcium carbonate were
used as the principal stratigraphic tool in the
study. For the time interval spanning the last
170,000 years (oxygen 18 stages 1-6), carbonate
records from 15 cores were tied into two oxygen
18 records for the purposes of correlation [cf.
Shackleton and Opdyke, 1973]: core 280
[Emiliani, 1958] and V23-100 [Parkin and
Shackleton, 1973]. Although carbonate and
isotope variations are not completely
synchronous [Crowley, 1976], the timing of the
offsets are sufficiently close (a few thousand
years) that a reasonably accurate stratigraphy
can be achieved. However, the lack of
extremely precise time control, coupled with
generally coarse sampling resolution (usually
10 cm intervals in records with sedimentation
rates less than 2 cm/1000 years), places
some limits on detailed stratigraphic control.
This problem can sometimes be mitigated by
examination of cores with higher sedimentation
rates.

Generalizations drawn from the study of the
last 170,000 years were applied to a long core,
V26-41, (Figure 1), in the Canary Basin.
Stratigraphic control was based on fluctuations
of carbonate that were correlated to a nearby
core with carbonate and oxygen 18 control.

### Dissolution Indices

Dissolution variations were interpreted
based on fluctuations of percent planktonic
foraminiferal fragments and of percent
benthonic foraminifera. Results utilizing

these two techniques show a fair degree of
internal consistency. Theoretical
considerations precluded the use of other
possible indices. For example, Berger [1973
cf. Thompson and Saito, 1974] has demonstrated
that percent resistant foraminifera appears to
be a good index of Pacific dissolution
fluctuations. However, resistant foraminifera
vary by ecological province. Application of
such a technique to the Atlantic encounters the
problem of large temporal changes in surface
ecology during the late Pleistocene in the
North Atlantic [Crowley, 1981]. Variations in
percent calcium carbonate have also been
utilized with some success to interpret
dissolution variations in deep Pacific cores
[Luz and Shackleton, 1975]. In the Atlantic,
however, there are large terrestrial sources of
noncarbonate [Broecker et al., 1958; Hays and
Perruzza, 1972; Kolla et al., 1979];
fluctuations of percent carbonate may therefore
not correlate well with changes in deep-sea
dissolution.

Percent fragments were computed based on
the number of fragments in a sample of 350-400
whole planktonic foraminifera (>149μ).
Percent benthonic foraminifera were calculated
for the same sample. Parker and Berger [1971]
suggested that benthonic percentages greater
than 1% indicate increased seafloor
dissolution. Inspection of numerous North
Atlantic records indicates that this value also
holds for the North Atlantic. Variations below
that number appear to be random, whereas
variations greater than 1% can often be traced
from core to core and/or coincide with
intervals of increased fragments. The fragment
records indicate reliable intercore
correlations over a water depth range from
about 3000 to 4500 m. Preliminary tests on the
reproducibility of the fragment index indicates
a precision of 2% (1 σ) for the index (15
samples).

It should be noted that variations of both
indices provide only an estimate of relative
changes in preservation states. Quantitative
estimates of dissolution loss require more
work. However, Thunell and Honjo [1981] have
shown that in situ comparisons of percent
fragments with weight percent loss of calcium
carbonate from a moored array in the Panama
Basin reveal close parallels between the two
indices.

## Results

### Last 170,000 Years

Figures 3-6 illustrate the major results of
earlier dissolution studies on fifteen cores in
the central North Atlantic [Crowley, 1983a, b]
during the last 170,000 years. This time
interval comprises the last glacial cycle and
the penultimate glacial maximum. Studies in

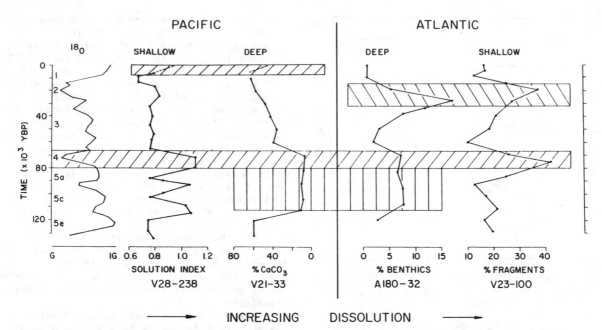

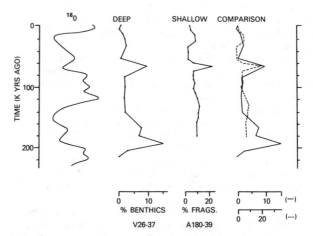

Fig. 3. Comparison of carbonate preservation patterns in Atlantic/Pacific records for the last 130,000 years, plotted against generalized oxygen 18 record with stage numbers. Pacific records are from Luz and Shackleton [1975] and Thompson [1976]; Atlantic records are from Crowley [1983a]. Dashed lines in Holocene Pacific sections indicate evidence for increased dissolution from other cores [Berger, 1977a]. The 4600-m-deep V23-100 record is plotted as a "shallow" core in order to demonstrate the depth range over which this pattern holds [Figure from Crowley, 1983a].

other parts of the North Atlantic indicate a similar pattern in those regions [Gardner, 1975; Be et al., 1976; Thunell, 1976; Droxler et al., 1983]. Two typical Atlantic records from "shallow" and "deep" sites (3000–4500 m and >4500 m) are compared with two Pacific records. The segregation by water depth was required because previous studies indicated different patterns of dissolution between the two depth intervals [cf. Luz and Shackleton, 1975]. The principal results are as follows:

1. Dissolution increased during glacial oxygen 18 stages 2 and 4 in both deep and shallow cores (Figure 3). Shallow cores consistently indicate that stage 4 dissolution is stronger than stage 2.

2. North Atlantic dissolution is low during interglacial oxygen 18 stages 1 and 5e and during the lower part of interstadial stage 3. There is also an indication of an intermediate level of dissolution in stages 5a–5d; however, this pattern is less consistent than the other patterns.

3. Abyssal cores (>4500 m) in the Canary Basin (cf. Figure 1) show a different dissolution pattern than shallower cores [cf. Crowley, 1983b]. Figure 4 illustrates the pattern for stage 6, when severe dissolution in abyssal cores occurs at the same time as minor

Fig. 4. A comparison of deep and shallow dissolution records for the last 170,000 years. Scales are adjusted so that fluctuations in both cores for the upper 100,000 years have unit length. The divergence of records in the lower part of the record is interpreted to be significant: benthonic percentages of 7–8% require removal of 90% of the carbonate, whereas fragment values of 12% require transfer of only about 3% of the forams from a coarse size fraction to a finer size fraction, with actual carbonate dissolution being perhaps minimal.

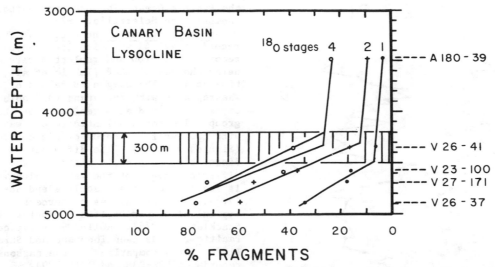

Fig. 5. The lysocline at three points in time. Results are from the southeast Canary Basin transect illustrated in Figure 1. See Figure 3 for stratigraphic levels. The alternate slope to the lysocline during stage 4 is drawn to illustrate uncertainty in interpretation. Note that this uncertainty amounts to only about 100 m change in the interpretation of the magnitude of the lysocline fluctuation [from Crowley, 1983a].

dissolution in shallow cores. Other deep records in the eastern basin suggest that intense dissolution persisted through the entire interval of stage 6 (A180-32 [Crowley, 1983b]; V27-178 [Gardner, 1975]).

4. The lysocline changed by about 200-300 m in response to the increased dissolution (Figure 5). This amount is about the same as that estimated by Boyle [1983] from theoretical considerations. There is also an indication that during the intense dissolution event of stage 4, dissolution increased above the lysocline. Similar trends have been found in records from stage 4 in the South Atlantic [Tappa and Thunell, 1984].

5. North Atlantic dissolution patterns primarily reflect changes in deep-sea chemistry. This conclusion is supported by two lines of evidence: (1) The pattern of dissolution response is similar for almost all open-ocean sites between 0 and 40 N (the seafloor is sufficiently shallow north of 45 N that little dissolution occurs even during glacial times [McIntyre et al., 1972]. This area encompasses five ecological provinces (the four from Figure 2 plus the glacial polar assemblage), and it is unlikely that surface ecology changes could conspire everywhere to produce the same seafloor pattern. (2) A core from a region with only very minor faunal changes shows the same general pattern as other North Atlantic cores (Figure 6), thereby implying that surface ecology effects do not bias the open-ocean record.

6. An Atlantic/Pacific dissolution comparison (Figure 3) indicates that sometimes

dissolution fluctuations in the two basins are out of phase (e.g., last 30,000 years) and sometimes in phase (e.g., about 70,000 years ago).

Last 400,000 Years

Generalizations drawn from the last 170,000 years were used as a guide to interpreting a longer North Atlantic record. The dissolution record of a core (V26-41) used in the study of the last 130,000 years (cf. Figure 1) was extended back to 400,000 years (oxygen 18 stage 11). Stratigraphic control was based on fluctuations of carbonate that were correlated (Figure 7) to a nearby core with carbonate and oxygen 18 control. The carbonate records for the two cores are very similar; dashed lines in Figure 7 indicate levels of particularly confident correlation. There are three major new results:

1. The model of low North Atlantic dissolution during interglacials needs to be modified. Only during the early part of interglacials, when most oxygen 18 records indicate minimum ice volume, is dissolution consistently at a minimum. There is still a tendency for dissolution to increase with increasing ice volume; however, the relationship between the two indices is not linear. Relatively small changes in inferred ice volume are sometimes associated with large increases in dissolution.

2. Dissolution is low during some glacial intervals. In the lower part of stage 6, a carbonate high corresponds to a dissolution

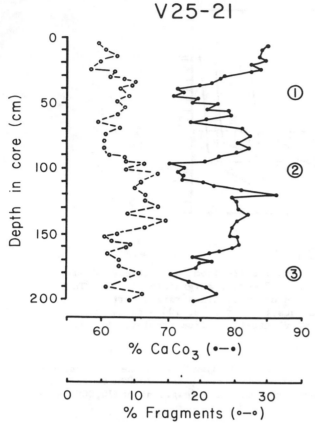

Fig. 6. Comparison of percent fragments and
carbonate in a core that has only minor downcore
variations in foraminiferal populations (see
Crowley and Matthews, [1983, Figure 2] for a plot
of abundance variations in the dominant
population). Numbered intervals for the carbonate
record refer to carbonate minima as defined by
McIntyre et al. [1972]. These intervals correlate
with oxygen 18 stages 2, 4, and 6.

low, even though oxygen 18 values indicate a
large amount of ice volume. In the upper part
of stage 8, there is another interval of high
carbonate/low dissolution. This interval
immediately follows a sharp carbonate minimum
that coincides with the lower of two stage 8
oxygen 18 maxima (estimated age of 270,000
years). The carbonate minimum can be found in
a number of North Atlantic cores and coincides
with the final phase of a long interval of
increased dissolution.

3. There is a down core increase in the
intensity of dissolution. Dissolution in upper
stages 9 and 11 is particulary intense, and
certain intervals are almost entirely composed
of foraminiferal fragments. There is also
evidence for increased "mid-Brunhes"
dissolution in other regions of the world
ocean: the South Atlantic [Johnson et al.,
1977; Briskin and Schreiber, 1978], the Indian
Ocean [Peterson and Prell, this volume], and

the Pacific Ocean [Thompson and Saito, 1974;
Thompson and Sciarrillo, 1978].

Comparisons with the Pacific. The fragment
record of V26-41 was tied into the oxygen 18
record of V23-100 and replotted versus time
using the new SPECMAP [Imbrie et al., 1984]
time scale. The oxygen 18 record of V23-100
was replaced with the standard "whole-ocean"
oxygen 18 record also generated by the SPECMAP
group. The North Atlantic record was compared
with two Pacific records of dissolution
(Figures 8 and 9). The different Pacific
dissolution records in Figures 8 and 9 are from
different levels of the water column. V28-238
is located near the lysocline and does not show
large, coherent changes of percent carbonate
typical of deeper Pacific cores [cf. Luz and
Shackleton, 1975]. RC11-209 is typical of deep
Pacific records (see Thompson and Sciarrillo
[1978] for a comparison of the carbonate
records of V28-238 and RC11-209).

The Atlantic/Pacific comparison in Figure 8
indicates trends similar to the last 170,000
years. Sometimes dissolution in the two basins
is out of phase, sometimes in phase. The
in-phase events (indicated by a dashed line)
are particulary noteworthy because they tend to
occur near intervals of inferred ice volume
increase. In addition to the stage 4/5
boundary, there are in-phase events associated
with the stage 6/7 boundary and in the middle
of stage 7. This latter event corresponds to a
time of rapid glacial growth documented by
Ruddiman and McIntyre [1982]. Other in-phase
dissolution events occurred during two inferred
glacial growth events at the end of stage 9 and
one near the end of stage 11. The stage 11
event (estimated age of 390,000 years) is
particularly intense and can also be found in
the Caribbean [Prell, 1982], the equatorial
Indian Ocean [Peterson and Prell, this volume],
and the southern ocean [Corliss and Thunell,
1983]. The stage 4 event is also very
widespread; taken together, these two events
represent the best evidence for a whole-ocean
change in the carbonate reservoir. Figure 9
compares the V26-41 North Atlantic record with
an eastern equatorial Pacific record from below
the lysocline [Hays et al., 1969]. The
carbonate record of dissolution in this core is
similar to that of other cores from the same
region [Hays et al., 1969; Thompson and
Sciarrillo, 1978]. An extremely accurate
chronology for this core is not available,
because the sampling rate is low and oxygen
isotope control is lacking. An absolute
chronology was based on two assumptions: (1)
Pacific dissolution coincides approximately
with interglacials; and (2) dissolution lags
the isotope record by 5,000-10,000 years [Luz
and Shackleton, 1975; Moore et al., 1977].

Comparison of North Atlantic and deep
Pacific records indicates that dissolution
intensity increased during a mid-Brunhes
interval from approximately 200,000-450,000

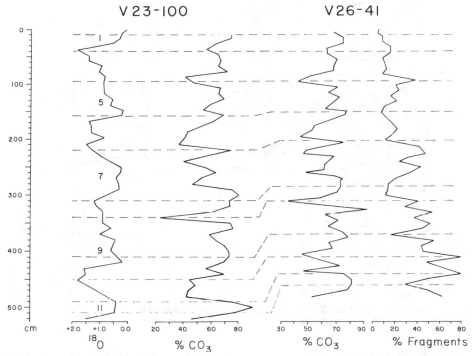

Fig. 7. Dissolution, carbonate, and oxygen 18 records, plotted versus core depth, of two cores from the Canary Basin (cf. Figure 1). Similar carbonate records for the two cores enable correlation of the dissolution record of one core with the 0-18 record of the other core (interglacial stages identified in the 0-18 record). Dashed lines represent intervals of particularly confident correlation. V23-100 data are from Parkin and Shackleton [1973].

years BP (evidence discussed below suggests that Atlantic dissolution returned to low values in stage 13). There is some indication that this pattern also occurs in the South Atlantic and the equatorial Indian Ocean [Johnson et al., 1977, Peterson and Prell, this volume]. The mid-Brunhes change appears to be global in scope and indicates a long-term excursion of the deep-sea carbonate reservoir.

It might be argued that diagenetic effects cause downcore increases in dissolution. However, the return to relatively low dissolution values in the lower Brunhes in the Pacific (Figure 9) and the South Atlantic [Johnson et al., 1977] would not be expected if diagenesis were the origin of the pattern. High carbonate values in stages 13 and 14 of low-latitude North Atlantic cores [Ruddiman, 1971; Parkin and Shackleton, 1973] suggest that dissolution also returns to relatively low values in the lower Brunhes of the North Atlantic.

Discussion

Interpretation of Results for the Last 170,000 Years

Since Broecker and Takahashi [1978] have shown that dissolution changes

reflect changes in the carbonate ion concentration of the deep waters, the discussion below will focus on mechanisms resonsible for changing the level of the deep-water carbonate ion concentration. The analysis of results will address three topics: differences in response between the Atlantic and Pacific basins (basin/basin fractionation), differences in response by water depth (depth variations), and the origin of in-phase dissolution responses for the Atlantic and Pacific.

Basin/basin fractionation. The out-of-phase dissolution patterns between the Atlantic and Pacific during the last 30,000 years strongly suggest control by changes in deep-water production rates [cf. Berger, 1970]. Production of large volumes of NADW results in a net export of deep water from the Atlantic to the Pacific. Because deep waters acquire lower carbonate ion concentration as they age, Pacific deep waters have a lower carbonate ion concentration than the Atlantic. Pacific preservation is correspondingly worse than the Atlantic.

A reversal in the basin/basin preservation tendencies occurred during the last glacial maximum, when Atlantic dissolution increased and Pacific dissolution decreased. Since the

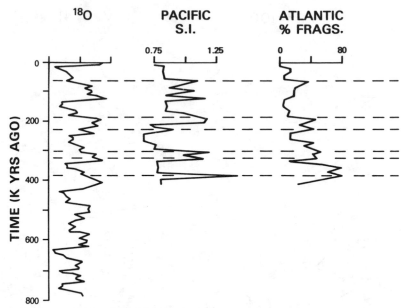

Fig. 8. Correlation of the V26-41 North Atlantic dissolution record in Figure 7 with Pacific core V28-238 [Thompson, 1976].  See text for details.

Pacific lysocline remained shallower than the Atlantic lysocline [Broecker, 1982a], the general pattern of deep-water flow was probably the same, but NADW production rates may have decreased.  The older deep waters in the North Atlantic would result in increased dissolution. Pacific deep waters would not necessarily be younger, but preservation would still improve because the excess carbonate dissolved in the Atlantic would be credited to the Pacific deep-sea budget in order to maintain a whole-ocean balance.

The model of decreased NADW production rates during the last glacial maximum is supported by geochemical evidence from the western North Atlantic, where both Cd/Ca and carbon 13/carbon 12 ratios in benthonic foraminifera can be interpreted as indicating decreased NADW production [Boyle and Keigwin, 1982; Keigwin and Boyle, this volume].  The carbon 13 values for the lower part of stage 3 in these studies suggest interglacial-level production rates of NADW.  The interpretation is consistent with high levels of carbonate

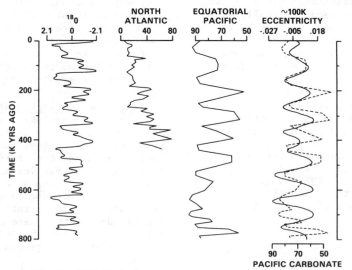

Fig. 9. Correlation of the V26-41 North Atlantic dissolution record in Figure 7 with Pacific core RC11-209 [Hays et al., 1969].  The Pacific record is compared with the 100,000-year eccentricity record.  See text for details.

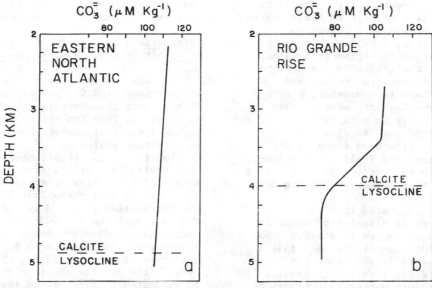

Fig. 10. Carbonate ion/water depth profiles for (a) eastern North Atlantic and (b) Rio Grande Rise of the Vema Channel in the South Atlantic. GEOSECS carbonate ion data as illustrated by Broecker and Takahashi [1980]. Figure from Crowley [1983b].

preservation during the lower part of stage 3. In the Canary Basin, lower stage 3 preservation is good at depths as great as 5000 m (Figure 3).

Depth variations. The results illustrated in Figures 3 and 4 and by Crowley [1983b] also indicate that the dissolution response in the deep sea varies by water depth. Stated in another way, increased dissolution at abyssal depths does not necessarily imply increased dissolution at shallower depths. Since the illustrated cores are only from the Canary Basin, the following interpretation will be restricted to that region. However, depth variations have also been observed for the western North Atlantic [Balsam, 1983].

Canary Basin cores indicate that the level separating deep and shallow dissolution patterns is approximately 4500 m [Crowley, 1983b]. Examination of the carbonate ion profiles for different parts of the Atlantic (Figure 10) suggests one possible interpretation for depth variations. The present carbonate ion concentration in the eastern North Atlantic shows a linear change with depth. The pattern reflects a single water mass: the NADW [Broecker and Takahashi, 1980]. In the western South Atlantic (Figure 10b), the carbonate ion profile is quite different. The discontinuity at about 4000 m marks the approximate boundary between the NADW and the underlying AABW.

The pronounced depth variation of dissolution patterns that occurs during some glacial intervals may possibly be interpreted in terms of transformation of the Canary Basin carbonate ion profile into a pattern similar to that occurring at present in the western South Atlantic. This is not to imply that hydrographic conditions were analogous, although Crowley [1983b] suggested the possible presence of a greater AABW influence during some of the glacial intervals [cf. Berger, 1968; Gardner, 1975; Ledbetter and Johnson, 1976; Ledbetter, 1979].

An alternative interpretation for depth variations is suggested by studies of Curry and Lohmann [1983, this volume]. Benthic carbon 13 variations in the Sierra Leone Basin (eastern North Atlantic, south of the Canary Basin) also indicate a depth gradient, with cores below local sill depth (connecting the western basin) recording significantly lighter benthic carbon 13 values during glacial intervals. Curry and Lohmann interpreted the pattern in terms of reduced advection of oxygen-rich deep waters into the eastern basins. The increased residence time of waters below sill depth contributes to the observed carbon 13 patterns.

An interpretation analogous to that for the Sierra Leone Basin can be applied to the Canary Basin to the north, except that the local sill depth (postulated from carbonate preservation) would be 4500 m between the two eastern Atlantic basins and the index monitored is carbonate dissolution rather than benthic carbon 13. This interpretation is supported by bathymetric evidence indicating that the sill depth between the two basins is about 4500 m.

In-phase Atlantic/Pacific dissolution changes. Previous work [Berger, 1977b] has demonstrated the existence of an apparently global carbonate preservation spike during the termination of the last glaciation (event dated about 11,000 years B.P. by W.H. Berger (personal communication, 1984)). This event is not recorded in most of the Atlantic studies, probably because of the generally deeper core locations, lower sedimentation rates, and coarser sampling intervals. However, Figure 3 illustrates that sometimes dissolution also increased simultaneously in both basins - oxygen 18 stages 5a-5d and especially stage 4. The latter event has also been found in the Indian Ocean [Peterson and Prell, this volume]. The in-phase dissolution event implies a whole-ocean change in the deep-sea carbonate reservoir.

Crowley [1983a] suggested that the coincidence of the major dissolution event in stage 4 with the major change in climate from interglacial to glacial may indicate a causal connection. One possibility is that the cooling/regression event resulted in erosion and transport of organic matter from transient shallow reservoirs (shelf sediments, marshes, etc.) into the deep sea [cf. Berger and Vincent, 1981]. The corresponding change in total inorganic carbon ( $\Sigma$ carbon dioxide) would result in increased seafloor dissolution. The above model is analogous to that suggested by Broecker [1982a, b]. However, Broecker focused on the transgression event associated with the last deglaciation in order to explain the observed ice core variations in atmospheric carbon dioxide [e.g., Neftel et al., 1982].

The coincidence of global-scale dissolution events with major sea level falls during the last 400,000 years can be considered as additional evidence in support of the shelf/basin fractionation model. The model is also supported by unusually light benthic carbon 13 values in Atlantic and Pacific sediments during oxygen 18 stage 4 [Boyle and Keigwin, 1982; Shackleton et al., 1983; Keigwin and Boyle, this volume].

The above model is not meant to imply exlusive control of global dissolution events by shelf/basin fractionation. Other evidence clearly points to the influence of additional factors on the global carbon budget [e.g., Keigwin and Boyle, this volume; Shackleton and Pisias, this volume; Oeschger et al., this volume]. But the consistent coincidence of global-scale dissolution events with sea level falls suggests that marine regressions have an important influence on the global carbon budget.

An alternative model for global dissolution changes involves an increased rate of oceanic mixing [Broecker, 1971]. A rapid recycling rate of deep water would return nutrients to surface waters more rapidly. In order to balance the resultant change in carbonate flux, dissolution would have to increase on the seafloor. Although plausible, there is at present no evidence in support of this model,

nor is there a priori justification for linking it to major times of ice growth.

## Fluctuations of the Global Carbonate Reservoir on 100,000 Year Time Scales

The discussion below addresses two aspects of long-term (100,000 years) changes in dissolution: (1) the role of orbital variations as a forcing function and (2) possible physical/chemical changes that may be occurring on this time scale.

Imbrie et al., [1984] have demonstrated that orbital variations account for a substantial fraction of total ice volume fluctuations during the last 400,000 years. The correlation of the dissolution records in Figure 9 with the 100,000-year eccentricity record also indicates some significant first-order similarities in both period and amplitude of response during the same time period (the precise phase relationship cannot be evaluated because of the low temporal resolution in the Pacific record.) A spectral analysis of the entire Pleistocene record of this Pacific core by Moore et al., [1982] verifies the significant coherence between the 100,000 year carbonate and eccentricity records. There is also a suggestion of significant 400,000-year variance in the carbonate record. A. L. Berger [1977] has noted that the most important term in the series expansion for eccentricity occurs at this period. The 100,000- and 400,000-year eccentricity cycles account for 47% of the total variance in the carbonate record. However, there are still some significant discrepancies between the carbonate and eccentricity records, e.g., stage 11 (high dissolution/low eccentricity) and stages 15 and 17 (low dissolution/high eccentricity). Other factors must play an important role in the evolution of the carbon cycle on this time scale.

The physical processes responsible for low frequency (100,000 years) carbonate fluctuations have not received much attention. Some 100,000-year signals may reflect amplified versions of millennial scale processes. Listed below are four mechanisms that might contribute to changes in the whole-ocean carbonate budget.

The rate of change of sea level. Peterson and Prell [this volume] present some evidence indicating a correspondence between the rate of sea level lowering and the intensity of Indian Ocean dissolution (the rate at which organic matter is added to the deep-sea reservoir affects the saturation level of the deep sea).

Changes in the rate of oceanic mixing. Long-term changes in this 1000-year process [Broecker, 1971] could have produced the observed pattern.

Changes in the area of high-latitude productivity. Luz and Shackleton [1975] proposed that increased high-latitude carbonate productivity during interglacials should cause a compensating increase of seafloor dissolution. Boyle [1983] estimated that North Atlantic high latitude productivity changes are responsible for about 10% of the changes in carbonate accumulation rate in a Pacific core. The area of high-latitude surface waters has not been the same for each interglacial. For example, very high carbonate values occur during stage 11 in the North Atlantic [Ruddiman and McIntyre, 1976] and southern ocean [Hays et al., 1976]. Warm North Pacific waters during stage 11 [Sachs, 1973] also suggest higher productivity in that region. The very intense dissolution in Stage 11 (Figure 8) may in part reflect these high-latitude events.

Changes in global weathering rates. Kukla [1977] has shown that there is a significant difference in interglacial weathering horizons between the lower and middle Brunhes. The approximate transition is around 450,000 years B.P.  Prior to 450,000 years B.P., interglacial horizons in Europe and western Russia were drier (more Mediterranean) than during later times. G.J. Kukla (personal communication, 1983) suggests that this pattern may extend across Eurasia to China. Thus, a very large fraction of the total land area may have undergone a shift in weathering that would presumably change the river input of carbon, phosphorus, and alkalinity to the oceanic reservoir. Since the duration of the low-dissolution regime exceeds the residence time of phosphorus (100,000 years), global weathering changes may have affected the carbonate reservoir. The approximate coincidence of weathering and dissolution horizons indicates that the connection warrants more scrutiny.

## Summary and Conclusions

Results based on a study of the North Atlantic dissolution record for the last 400,000 years indicate the following conclusions:

1.  Dissolution is low during interglacial maximum conditions and increases with increasing ice volume; however, the relationship between ice volume and dissolution is not linear.

2.  Dissolution changes are primarily due to changes in deep-water chemistry.

3.  The lysocline shallowed by about 300 m during glacial dissolution events.

4.  Atlantic dissolution events are sometimes out of phase with Pacific dissolution events. This pattern probably reflects changes in North Atlantic Deep Water production rates. There is a high degree of internal consistency among three indices (dissolution, benthic carbon 13, and Cd/Ca ratios) that can be used in part to infer changes in NADW production rates.

5.  Abyssal cores sometimes have different dissolution patterns than bathyal cores. This pattern can be explained by increased residence times of deep-water masses below local sill depth in the eastern Atlantic basins.

6.  Correlation of the North Atlantic record with Pacific cores indicates a tendency for dissolution to increase in both basins during times of ice volume increase. The coincidence with times of major sea level lowering implies that shelf/basin fractionation processes contribute to global fluctuations of dissolution.

7.  The intensity of dissolution increases in both basins in the middle Brunhes (approximately 200,000-450,000 years ago). There is also a tentative indication that dissolution returns to low levels in both basins in the lower Brunhes.

8.  Changes of dissolution on 100,000 year time scales correlate only in part with orbital insolation variations. The 100,000 year signals may be amplified versions of processes operating on millennial time scales. There is some evidence for three mechanisms that might be affecting dissolution on this time scale: the rate of change of sea level lowering, variations in global weathering rates, and the area of high-latitude carbonate productivity. A fourth mechanism (changes in the rate of oceanic mixing) is possible, but evidence for the process is lacking at the present time.

Acknowledgments. The Lamont core collection is supported by NSF grant OCE 78-25448. Further support for this work was provided by funds from the University of Missouri-Saint Louis. The analysis of results has benefited from input by William B. Curry, Wolfgang H. Berger, and Eric T. Sundquist. I thank Celeste Savattere for processing the samples and Janice C. Kriegel for typing the manuscript.

## References

Balsam, W.B., Carbonate dissolution on the Muir Seamount (western North Atlantic): Interglacial/glacial changes, J.Sediment.Petrol., 53, 719-731, 1983.

Be, A.W.H., J.E. Damuth, L. Lott, and R. Free, Late Quaternary climatic record in western equatorial Atlantic sediments, in Investigation of Late Quaternary Paleoceanography and Paleoclimatology, Geol. Soc. Am. Mem.145, edited by R.M. Cline and J.D. Hays, pp. 165-200, Geological Society of America, Boulder, Colo., 1976.

Berger, A.L., Support for the astronomical theory of climatic change, Nature, 269, 44-45, 1977.

Berger, W.H., Planktonic foraminifera: Selective solution and paleoclimatic interpretation, Deep Sea Res., 15, 31-43, 1968.

Berger, W.H., Biogenous deep-sea sediments: Fractionation by deep-sea circulation, Deep Sea Res., 81, 1385-1402, 1970.

Berger, W.H., Deep-sea carbonates: Pleistocene dissolution cycles, J. Foraminiferal Res., 3, 187-195, 1973.

Berger, W.H., Carbon dioxide excursions and the deep sea record: Aspects of the problem, in The Fate of Fossil Fuel $CO_2$ in the Oceans, edited by N.R. Andersen and A. Malahoff, pp., 505-542, Plenum, New York, 1977a.

Berger, W.H., Deep-sea carbonates and the deglaciation preservation spike in pteropods and foraminifera, Nature, 269, 301-304, 1977b.

Berger, W.H., and E. Vincent, Chemostratigraphy and biostratigraphic correlation: Exercises in systematic stratigraphy, Geol. Oceans Symp., Proc. Int. Geol. Cong., 26th ,Oceanol.Acta., xx, 115-127, 1981.

Biscaye, P.E., V. Kolla, and K.K. Turekian, Distribution of calcium carbonate in surface sediments of the Atlantic Ocean, J.Geophys. Res., 81, 2595-2603, 1976.

Boyle, E.A., Chemical accumulation variations under the Peru Current during the past 130,000 years, J.Geophys. Res., 88, 7667-7680, 1983.

Boyle, E.A., and L.D. Keigwin, Deep circulation of the North Atlantic over the last 200,000 years: Geochemical evidence, Science, 218, 784-787, 1982.

Briskin, M., and B.C. Schreiber, Authigenic gypsum in marine sediments, Mar. Geol., 28, 37-49, 1978.

Broecker, W.S., Calcite accumulation rates and glacial to interglacial changes in ocean mixing, in The Late Cenozoic Glacial Ages, edited by K.K. Turekian, pp. 239-265, Yale University Press, New Haven, Conn., 1971.

Broecker, W.S., Glacial to interglacial changes in ocean chemistry, Prog. Oceanogr., 11, 151-197, 1982a.

Broecker, W.S., Ocean chemistry during glacial time, Geochim. Cosmochim. Acta, 46, 1689-1705, 1982b.

Broecker, W.S., and T. Takahashi, The relationship between lysocline depth and in-situ carbonate ion concentration, Deep Sea Res., 25, 65-95, 1978.

Broecker, W.S., and T. Takahashi, Hydrography of the central Atlantic, III, The North Atlantic Deep-Water complex, Deep Sea Res., 27, 591-613, 1980.

Broecker, W.S., T. Takahashi, and M. Stuiver, Hydrography of the central Atlantic, II, Waters beneath the two-degree discontinuity, Deep Sea Res., 27, 397-419, 1980.

Broecker, W.S., K.K. Turekian, and B.C. Heezen, The relation of deep-sea sedimentation rates to variations in climate, Am. J. Sci., 256, 503-517, 1958.

Corliss, B.H., and R.C. Thunell, Carbonate sedimentation beneath the Antarctic Circumpolar Current during the Late Quaternary, Mar. Geol., 51, 293-326, 1983.

Crowley, T.J., Fluctuations of the eastern North Atlantic gyre during the last 150,000 years, Ph.D. dissertation, 282 pp., Brown Univ., Providence, R.I., 1976.

Crowley, T.J., Temperature and circulation changes in the eastern North Atlantic during the last 150,000 years: Evidence from the planktonic foraminiferal record, Mar.Micropaleontol., 6, 97-129, 1981.

Crowley, T.J., Calcium-carbonate preservation patterns in the central North Atlantic during the last 150,000 years, Mar. Geol., 51, 1-14, 1983a.

Crowley, T.J., Depth-dependent carbonate dissolution changes in the eastern North Atlantic during the last 170,000 years, Mar. Geol., 54, M25-M31, 1983b.

Crowley, T.J., and R.K. Matthews, Isotope-plankton comparisons in a late Quaternary core with a stable temperature history, Geology, 11, 275-278, 1983.

Curry, W.B., and G.P. Lohmann, Reduced advection into Atlantic Ocean deep eastern basins during last glacial maximum, Nature, 306, 577-580, 1983.

Curry, W.B., and G.P. Lohmann, Carbon deposition rates and deep water residence time in the equatorial Atlantic Ocean throughout the last 160,000 years, this volume.

Diester-Haass, L., Late Quaternary climatic variations in northwest Africa deduced from east Atlantic sediment cores, Quat.Res., 6, 299-314, 1976.

Diester-Haass, L., Late Quaternary sedimentation processes on the west African continental margin and climatic history of west Africa (12-18 N), "Meteor"Forschungsergeb.,Reihe C, 16, 19-66, 1983.

Droxler, A.W., W. Schlager, and C.C. Whallon, Quaternary aragonite cycles and oxygen-isotope record in Bahamian carbonate ooze, Geology, 11, 235-239, 1983.

Emiliani, C., Paleotemperature analysis of core 280 and Pleistocene correlations, J.Geol., 66, 264-275, 1958.

Gardner, J.V., Late Pleistocene carbonate dissolution cycles in the eastern equatorial Atlantic, in Dissolution of Deep-Sea Carbonates, Cushman Found.Foraminiferal Res. Spec. Publ., 13, edited by W.V. Sliter, A.W.H. Be, and W.H. Berger, 1975., pp. 129-141, U.S. National Museum, Washington D.C., 1975.

Hays, J.D., and A. Perruzza, The significance of calcium carbonate oscillations in eastern equatorial Atlantic deep-sea sediments for the end of the Holocene warm interval, Quat. Res., 2, 355-362, 972.

Hays, J.D., J. Imbrie, and N.J. Shackleton, Variations in the earth's orbit: Pacemaker of the ice ages, Science, 194, 1121-1132, 1976.

Hays, J.D., T. Saito, N.D. Opdyke, and L.H. Burckle, Pliocene-Pleistocene sediments of the equatorial Pacific: Their paleomagnetic, biostratigraphic, and climatic record, Geol.Soc. Am.Bull., 8, 1481-1514, 1969.

Imbrie, J., et al, The orbital theory of Pleistocene climate: Support from a revised

chronology of the marine 0-18 record, in
Milankovitch and Climate, Part I, edited by A.
Berger et al., pp. 269-305, D. Reidel,
Hingham, Mass., 1984.

Johnson, D.A., M. Ledbetter, and L.H. Burckle,
Vema Channel paleo-oceanography: Pleistocene
dissolution cycles and episodic bottom water
flow, Mar. Geol., 23, 1-33, 1977.

Keigwin, L.D., and E.A. Boyle, Carbon isotopes in
deep-sea benthic foraminifera: Precession and
changes in low-latitude biomass, this volume.

Kipp, N.G., New transfer function for estimating
sea-surface conditions from sea-bed
distribution of planktonic foraminiferal
assemblages in the North Atlantic, in
Investigation of Late Quaternary
Paleoceanography and Paleoclimatology, Geol.
Soc. Am. Mem., edited by R.M. Cline and J.D.
Hays, 145, pp. 3-42, Geological Society of
America, Boulder, Colo., 1976.

Kolla, V., P.E. Biscaye, and A.F. Hanley,
Distribution of quartz in late Quaternary
atlantic sediments in relation to climate,
Quat.Res., 11, 261-277, 1979.

Kukla, G.J., Pleistocene land-sea correlations, I,
Europe, Earth Sci.Rev., 13, 307-374, 1977.

Ledbetter, M.T., Fluctuations of Antarctic Bottom
Water velocity in the Vema Channel during the
last 160,000 years, Mar.Geol., 33, 71-89,
1979.

Ledbetter, M.T., and D.A. Johnson, Increased
transport of Antarctic Bottom Water in the
Vema Channel during the last ice age, Science,
194, 837-839, 1976.

Luz, B., and N.J. Shackleton, $CaCO_3$ solution in
the tropical east Pacific during the past
130,000 years, in Dissolution of Deep-Sea
Carbonates, Cushman Found. Foraminiferal
Res.Spec. Publ. 13, edited by W.V. Sliter,
A.W.H. Be, and W.B. Berger, pp. 142-150, U.S.
National Museum, Washington D.C., 1975.

McIntyre, A., W.F. Ruddiman, and R. Jantzen,
Southward penetrations of the North Atlantic
polar front: Faunal and floral evidence of
large scale surface water mass movements over
the past 225,000 years, Deep Sea Res., 19,
61-77, 1972.

McIntyre, A., et al., The glacial North Atlantic
18,000 years ago: A CLIMAP reconstruction, in
Investigation of Late Quaternary
Paleoceanography and Paleoclimatology,
Geol.Soc. Am. Mem.145, edited by R.M. Cline
and J.D. Hays, pp. 43-76, Geological Society
of American, Boulder, Colo., 1976.

Moore, T.C., N.G. Pisias, and G.R. Heath, Climate
changes and lags in Pacific carbonate
preservation, sea surface temperature, and
global ice volume, in The Fate of Fossil Fuel
$CO_2$ in the Oceans, edited by N.R. Andersen and
A. Malahoff, pp. 145-166, Plenum, New York,
1977.

Moore, T.C., N.G. Pisias, and D.A. Dunn, Carbonate
time series of the Quaternary and late Miocene
sediments in the Pacific Ocean: A spectral

comparison, Mar. Geol., 46, 217-234, 1982.

Neftel, A., H. Oeschger, J. Schwander, B.
Stauffer, and R. Zumbrunn, Ice core sample
measurements give atmospheric $CO_2$ content
during the past 40,000 years, Nature, 295,
220-223, 1982.

Oeschger, H., B. Stauffer, R. Finkel, and C.
Langway, Jr., Variation of the $CO_2$
concentration of occluded air and of anions
and dust in polar ice cores, this volume.

Parker, F.L., and W.H. Berger, Faunal and solution
patterns of planktonic foraminifera in surface
sediments of the South Pacific, Deep Sea Res.,
18, 73-107, 1971.

Parkin, D.W., and N.J. Shackleton, Trade wind and
temperature correlations down a deep-sea core
off the Saharan Coast, Nature, 245, 455-457,
1973.

Peterson, L.C., and W.L. Prell, Carbonate
preservation and rates of climatic change: An
800 kyr record from the Indian Ocean, this
volume.

Prell, W.L., Oxygen and carbon isotope
stratigraphy for the Quaternary of hole 502B:
Evidence for two modes of isotopic
variability, in Initial Re.Deep Sea
Drill.Proj., 68, 455-464, 1982.

Ruddiman, W.F., Pleistocene sedimentation in the
equatorial Atlantic: Stratigraphy and faunal
paleoclimatology, Geol. Soc. Am. Bull., 82,
283-302, 1971.

Ruddiman, W.F., and A. McIntyre, Northeast
Atlantic paleoclimatic changes over the past
600,000 years, in Investigation of Late
Quaternary Paleoceanography and
Paleoclimatology, Geol. Soc. Am. Mem. 145,
edited by R.M. Cline and J.D. Hays, pp.
111-146, Geological Society of America,
Boulder, Colo., 1976.

Ruddiman, W.F., and A. McIntyre, Severity and
speed of northern hemisphere glaciation
pulses: The limiting case?, Geol. Soc.
Am. Bull., 93, 1273-1279, 1982.

Sachs, H.M., Late Pleistocene history of the North
Pacific: Evidence from a quantitative study of
radiolaria in core V21-173, Quat. Res., 3,
89-98, 1973.

Shackleton, N.J., and N.D. Opdyke, Oxygen isotope
and paleomagnetic stratigraphy of equatorial
Pacific core V28-238: Oxygen isotope
temperatures and ice volume on a 100,000 and
1,000,000 year scale, Quat.Res., 3, 39-55,
1973.

Shackleton, N.J.,and Pisias, N.G., Atmospheric
carbon dioxide, orbital forcing and climate
change in the Pleistocene, this volume.

Shackleton, N.J., J. Imbrie, and M.A. Hall, Oxygen
and carbon isotope record of east Pacific core
V19-30: Implications for the formation of deep
water in the late Pleistocene North Atlantic,
Earth Planet.Sci. Lett., 65, 233-244, 1983.

Tappa, E., and R.C. Thunell, Late Pleistocene
glacial/interglacial changes in planktonic
foraminiferal biofacies and carbonate

dissolution patterns in the Vema Channel, Mar. Geol., 58, 101-122, 1984.

Thompson, P.R., Planktonic foraminiferal dissolution and the progress towards a Pleistocene equatorial Pacific transfer function, J. Foraminiferal Res., 6, 208-227, 1976.

Thompson, P.R., and T. Saito, Pacific Pleistocene sediments: Planktonic foraminifera dissolution cycles and geochronology, Geology, 2, 333-335, 1974.

Thompson, P.R., and J.R. Sciarrillo, Planktonic foraminiferal biostratigraphy in the equatorial Pacific, Nature, 275, 29-33, 1978.

Thunell, R.C., Calcium carbonate dissolution history in late Quaternary deep-sea sediments, western Gulf of Mexico, Quat.Res., 6, 281-297, 1976.

Thunell, R.C., and S. Honjo, Calcite dissolution and the modification of planktonic foraminiferal assemblages, Mar.Micropaleontol., 6, 169-182, 1981.

# CARBON DEPOSITION RATES AND DEEP WATER RESIDENCE TIME IN THE EQUATORIAL ATLANTIC OCEAN THROUGHOUT THE LAST 160,000 YEARS

W. B. Curry and G. P. Lohmann

Department of Geology and Geophysics, Woods Hole Oceanographic Institution
Woods Hole, Massachusetts 02543

Abstract. In the deep, silled basins of the equatorial Atlantic Ocean, large changes have occurred in the rates of carbonate production and dissolution, deep water residence time, and rates of organic carbon accumulation over the last 160,000 years. Accumulation of carbonate in shallow, relatively undissolved sediments decreased by one half during glacial maxima, reflecting reduced surface water production of carbonate. The most severe dissolution occurred during stage 4 (64,000 to 75,000 years B.P.) when carbonate accumulation decreased abruptly below 3750 m. Benthic foraminiferal $\delta^{13}C$ decreased below 3750 m during all glacial maxima, reflecting greater remineralization of organic carbon and reduced dissolved oxygen concentrations below sill depth (~3750 m) within the eastern Atlantic. While the reduced $[O_2]$ inferred during stage 2 (13,000 to 32,000 years B.P.) can be attributed predominantly to increased residence time of the deep water, the lower $[O_2]$ during stage 4 resulted largely from increased delivery of organic carbon to the deep eastern basins. No bathymetric gradient in benthic foraminiferal $\delta^{13}C$ is detectable during stage 5e (115,000 to 125,000 years B.P.), implying that the eastern Atlantic was as well ventilated with deep water during the last interglacial period as it is today. During stages 1 and 5, organic carbon accumulation was constant below 3000 m. During stages 2, 3 and 4, the accumulation of organic carbon increased below 3750 m, reflecting in part better preservation because of reduced oxygen conditions.

## Introduction

Bathymetric gradients of deep water physical and chemical properties reflect the large-scale mixing that results from abyssal circulation. Deep-sea sediments deposited down the slopes of submarine rises record, in their lithology and chemistry, these gradients. The preservation of carbonate sediments, the chemistry of benthic foraminifera, and the preservation of organic carbon are determined to a large extent by the chemistry of the overlying deep water. We have reconstructed bathymetric profiles of these parameters in sediments deposited during the last 160,000 years on the Sierra Leone rise (5°N, 20°W) in the eastern equatorial Atlantic. Our data are measurements of benthic foraminiferal $\delta^{13}C$ and the carbonate and organic carbon accumulation in the sediments. Our results focus on the changes in deep water $\Sigma CO_2$, the bathymetric gradient of dissolved oxygen concentration, and the intensity of carbonate dissolution that accompanied glacial-interglacial climate change during the late Quaternary.

## Strategy

The eastern basins of the Atlantic Ocean are bounded on the west by the Mid-Atlantic ridge and on the south by the Walvis ridge. Constrained by this topography, water found below 3750 m in the east originates in the western Atlantic, flows across low-latitude fracture zones into the eastern basins, and fills them to the sill depth of the ridge [Metcalf et al., 1964; Warren, 1981]. Western and eastern Atlantic profiles of temperature and salinity diverge at approximately 3750 m, marking the sill depth of the Mid-Atlantic ridge near the equator [Metcalf et al., 1964]. Deep water in the western Atlantic exhibits a temperature decrease below 4000 m because of the presence of Antarctic Bottom Water (AABW). As the cold, southern origin AABW flows across the equator into the northern hemisphere, it acts as an eastern boundary current along the western flank of the Mid-Atlantic ridge [Warren, 1981]. Because AABW is blocked from the eastern basins by the Mid-Atlantic ridge, no benthic thermocline is formed in the eastern Atlantic basins. For AABW to enter the eastern Atlantic, it must shoal above the sill of the Mid-Atlantic ridge while maintaining this unusual position as an eastern boundary current. Today, the deep water

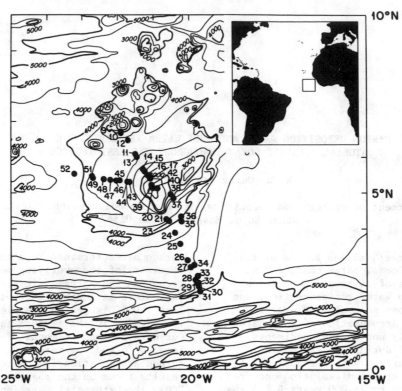

Fig. 1.    Location map of cores recovered on cruise 066 of R/V ENDEAVOR in 1981.
Cores were recovered in three transects of the rise, covering the complete depth
range of 2800 to 5300 m.

below the sill depth in the eastern Atlantic is
a mixture of North Atlantic Deep Water (NADW)
and only 10–20% AABW [Edmund, 1974].

In 1981, we collected gravity cores from the
slopes of the Sierra Leone rise within the
eastern Atlantic, sampling the entire depth
range of the rise from 2800 to 5300 m (see
Figure 1).  Fifteen cores with complete stratigraphic records (Table 1) were chosen on the
basis of extensive biostratigraphic, lithologic,
and geochemical analysis.  Our characterization
of past changes of the deep ocean carbon system
is derived from lithologic (carbonate, organic
carbon), geochemical (stable isotopic composition of benthic foraminifera), and physical
property analysis (wet and dry bulk density) of
the sediments recovered in these cores.

Measurements of carbonate concentration,
organic carbon concentration, and sediment bulk
density, when combined with an appropriate
chronology, provide a record of carbonate accumulation and dissolution and organic carbon accumulation and remineralization on the Sierra Leone
rise.  Because of the close geographical proximity of our cores, we assume that the input from
the surface water is equal in all cores.  Net
carbonate accumulation then equals the gross
production of carbonate in surface water minus

the loss to dissolution.  Assuming no advective
input nor any down-slope reworking of carbonate,
changes in the shape of the carbonate accumulation profile result from changes in the corrosiveness of deep water.  Temporal changes in the
accumulation of carbonate in shallow cores
reflect the production of carbonate in surface
water because little dissolution occurs
shallower than the lysocline [Berger et al.,
1982].  In contrast, the flux of organic carbon
arriving at the sediment is only a fraction of
the total carbon produced in the euphotic zone
because most of the carbon is remineralized
above 1000 m.  The flux of organic carbon
measured at sediment trap moorings is relatively
constant [Honjo et al., 1982a, b] below 2000 m,
suggesting that the input rate of carbon to the
sediment in the deep ocean can be assumed
constant below this depth.  Bathymetric gradients
of organic carbon accumulation on the Sierra
Leone rise then reflect the extent to which this
carbon is preserved in sediments at different
water depths.  The bathymetric profile of organic
carbon accumulation in the deep ocean reflects
in part the availability of dissolved oxygen at
different depths in the water column, and the
shape of the profile provides a qualitative
record of $O_2$ gradients in the water column.

TABLE 1.  Location, Water Depth, and Length of Recovery for
15 Cores Collected in 1981 at the Sierra Leone Rise, Eastern
Equatorial Atlantic Ocean

| CORE | Latitude, deg/min | Longitude, deg/min | Depth, m | Length, cm |
|------|-------------------|--------------------|----------|------------|
| EN066-10GGC | 6 38.72 N | 21 53.84 W | 3527 | 309 |
| EN066-16GGC | 5 27.28 N | 21 08.64 W | 3152 | 196 |
| EN066-17GGC | 5 22.05 N | 21 05.33 W | 3050 | 278 |
| EN066-21GGC | 4 14.00 N | 20 37.58 W | 3995 | 274 |
| EN066-23GGC | 4 10.85 N | 20 32.15 W | 4105 | 270 |
| EN066-24GGC | 3 50.85 N | 20 22.05 W | 4504 | 298 |
| EN066-26GGC | 3 05.14 N | 20 00.97 W | 4745 | 300 |
| EN066-29GGC | 2 27.63 N | 19 45.70 W | 5104 | 226 |
| EN066-32GGC | 2 28.38 N | 19 44.09 W | 5003 | 155 |
| EN066-36GGC | 4 18.38 N | 20 12.71 W | 4270 | 294 |
| EN066-38GGC | 4 55.19 N | 20 29.94 W | 2931 | 247 |
| EN066-39GGC | 5 04.34 N | 20 52.05 W | 2818 | 280 |
| EN066-43GGC | 5 15.87 N | 21 38.36 W | 3197 | 283 |
| EN066-44GGC | 5 15.81 N | 21 42.74 W | 3428 | 311 |
| EN066-47GGC | 5 18.88 N | 22 11.80 W | 3951 | 296 |

GGC: Giant Gravity Core.  These cores contain the most complete
stratigraphic sections over the widest possible depth range.

Certain benthic foraminifera have been shown
to faithfully record deep water $\delta^{13}C$
[Belanger et al., 1981; Graham et al., 1981],
and past changes in $\delta^{13}C$ of oceanic $\Sigma CO_2$
reflect at least two changing aspects of the
carbon system.  First, transfer of organic carbon
between shallow water marine or terrestrial
reservoirs and the oceanic reservoir causes
changes in mean oceanic $\delta^{13}C$ because the
remineralized organic carbon is approximately 20
to 25 parts per mil ($^o/_{oo}$) depleted with
respect to the isotopic composition of oceanic
$\Sigma CO_2$ [Shackleton, 1977; Broecker, 1982].
Changes in the concentration of $\Sigma CO_2$ in the
ocean caused by transfers to and from the
oceanic reservoir produce records of benthic
foraminiferal $\delta^{13}C$ which are parallel and of
equal magnitude at all locations in all oceans.

Superimposed on these global effects are changes
in benthic foraminiferal $\delta^{13}C$ caused by
changes in the patterns of abyssal circulation.
Today, production of nutrient-depleted deep
water in the North Atlantic is an important
source of deep water for the world ocean.  As
NADW flows from the Atlantic to the Pacific,
input and remineralization of organic carbon
reduces deep water $[O_2]$ and $\delta^{13}C$ while
increasing deep water $\Sigma CO_2$ and dissolved
nutrient ($PO_4$, $NO_3$) concentration.  The net
flow of deep water from the Atlantic to the
Pacific produces a gradient in $\delta^{13}C$ of
approximately $1^o/_{oo}$ between the ocean basins
[Kroopnick et al., 1972; Kroopnick, 1974, 1980].
Changes in the distribution of $\delta^{13}C$ in the
deep ocean can result from changes in abyssal
circulation patterns and deep water production

TABLE 2. Stable Isotope Composition of the Benthic Foraminifer Planulina wuellerstorfi During the Last 160,000 years

| Core | Depth | Holocene | | | | | 18,000–30,000 years B.P. | | | | | 45,000–60,000 years B.P. | | | | |
|---|---|---|---|---|---|---|---|---|---|---|---|---|---|---|---|---|
| | m | 18 O Mean | 18 O s.d. | 13 C Mean | 13 C s.d. | n | 18 O Mean | 18 O s.d. | 13 C Mean | 13 C s.d. | n | 18 O Mean | 18 O s.d. | 13 C Mean | 13 C s.d. | n |
| 38 | 2931 | 2.64 | --- | 1.05 | --- | 1 | 4.04 | 0.13 | 0.55 | 0.12 | 4 | 3.67 | 0.04 | 0.67 | 0.17 | 2 |
| 17 | 3050 | 2.65 | --- | 1.12 | --- | 1 | 4.13 | 0.08 | 0.53 | 0.08 | 2 | 3.77 | 0.02 | 0.70 | 0.21 | 2 |
| 16 | 3152 | 2.54 | --- | 0.94 | --- | 1 | 4.23 | 0.11 | 0.44 | 0.08 | 7 | 3.68 | 0.06 | 0.62 | 0.06 | 3 |
| 44 | 3428 | 2.59 | --- | 1.05 | --- | 1 | 4.29 | 0.15 | 0.49 | 0.04 | 4 | 3.60 | 0.06 | 0.48 | 0.06 | 2 |
| 10 | 3527 | 2.25 | --- | 0.82 | --- | 1 | 3.99 | 0.10 | 0.37 | 0.14 | 4 | 3.57 | 0.01 | 0.36 | 0.20 | 2 |
| 21 | 3995 | 2.50 | --- | 0.78 | --- | 1 | 3.96 | 0.21 | 0.10 | 0.11 | 3 | 3.69 | 0.09 | 0.48 | 0.08 | 3 |
| 36 | 4270 | 2.75 | --- | 0.75 | --- | 1 | 4.23 | 0.06 | 0.12 | 0.08 | 5 | 3.69 | 0.09 | 0.29 | 0.19 | 3 |
| 24 | 4504 | 2.50 | --- | 0.59 | --- | 1 | 4.05 | 0.25 | 0.19 | 0.03 | 3 | 3.56 | 0.12 | 0.25 | 0.06 | 3 |
| 26 | 4745 | 2.64 | --- | 0.84 | --- | 1 | 4.13 | 0.07 | -0.11 | 0.14 | 5 | 3.59 | 0.05 | 0.11 | 0.12 | 3 |
| 32 | 5003 | 2.76 | --- | 0.86 | --- | 1 | 4.08 | 0.04 | -0.15 | 0.19 | 3 | --- | --- | --- | --- | 1 |
| 29 | 5104 | 2.57 | --- | 0.85 | --- | 1 | 3.75 | 0.35 | -0.20 | 0.05 | 2 | --- | --- | --- | --- | - |

| Core | Depth | 64,000–75,000 years B.P. | | | | | 115,000–125,000 years B.P. | | | | | 146,000–160,000 years B.P. | | | | |
|---|---|---|---|---|---|---|---|---|---|---|---|---|---|---|---|---|
| | m | 18 O Mean | 18 O s.d. | 13 C Mean | 13 C s.d. | n | 18 O Mean | 18 O s.d. | 13 C Mean | 13 C s.d. | n | 18 O Mean | 18 O s.d. | 13 C Mean | 13 C s.d. | n |
| 38 | 2931 | 3.67 | 0.18 | 0.37 | 0.12 | 3 | 2.85 | --- | 0.87 | --- | 1 | 4.00 | --- | -0.07 | --- | 1 |
| 17 | 3050 | 3.72 | 0.20 | 0.71 | 0.18 | 3 | 2.36 | --- | 0.65 | --- | 1 | 3.98 | 0.04 | 0.13 | 0.05 | 2 |
| 16 | 3152 | 3.58 | 0.07 | 0.40 | 0.06 | 3 | --- | --- | --- | --- | - | --- | --- | --- | --- | - |
| 44 | 3428 | 3.46 | 0.18 | 0.35 | 0.36 | 2 | 2.66 | --- | 0.76 | --- | 1 | 3.97 | 0.01 | 0.30 | 0.16 | 1 |
| 10 | 3527 | 3.41 | 0.37 | 0.51 | 0.59 | 2 | 3.00 | 0.32 | 0.52 | 0.18 | 2 | 3.75 | 0.16 | -0.30 | 0.09 | 2 |
| 21 | 3995 | 3.67 | 0.03 | -0.02 | 0.20 | 2 | 2.74 | 0.12 | 0.71 | 0.15 | 3 | 3.94 | --- | -0.32 | --- | 3 |
| 36 | 4270 | 3.77 | 0.16 | 0.21 | 0.18 | 2 | 2.71 | 0.04 | 0.55 | 0.19 | 4 | 4.08 | 0.11 | -0.38 | 0.10 | 1 |
| 24 | 4504 | 3.83 | --- | -0.10 | --- | 1 | 2.59 | 0.12 | 0.51 | 0.09 | 2 | 3.59 | 0.11 | -0.60 | 0.13 | 2 |
| 26 | 4745 | --- | --- | --- | --- | - | 2.63 | 0.31 | 0.42 | 0.07 | 4 | --- | --- | --- | --- | - |
| 32 | 5003 | --- | --- | --- | --- | - | --- | --- | --- | --- | 2 | --- | --- | --- | --- | - |
| 29 | 5104 | --- | --- | --- | --- | - | --- | --- | --- | --- | - | --- | --- | --- | --- | - |

The mean and standard deviation (s.d.) are presented for all analyses which fall within the specified stratigraphic interval; n is the number of observations.

TABLE 3.  Average Calcium Carbonate Accumulation for 15 Cores From the Sierra
Leone Rise, Eastern Equatorial Atlantic

| Core | Depth, m | Interval, $10^3$ Years B.P. | | | | | |
|------|----------|------|------|------|------|------|------|
|      |          | 0–7 | 18–30 | 45–60 | 64–75 | 115–125 | 146–160 |
| 39 | 2818 | 0.81 (3) | 0.32 (3) | 0.61 (6) | 0.26 (4) | 0.78 (5) | 0.22 (4) |
| 38 | 2931 | 0.87 (5) | 0.31 (7) | 0.62 (8) | 0.58 (9) | 0.71 (5) | ---- |
| 17 | 3050 | 0.77 (3) | 0.38 (4) | 0.63 (7) | 0.44 (7) | 0.71 (5) | ---- |
| 16 | 3152 | 0.82 (3) | 0.50 (8) | 0.66 (10) | 0.56 (10) | ---- | ---- |
| 43 | 3197 | 0.90 (3) | 0.14 (1) | 0.58 (5) | 0.59 (6) | 0.49 (3) | 0.15 (2) |
| 44 | 3428 | 0.55 (2)[a] | 0.31 (4) | 0.57 (6) | 0.39 (5) | 0.67 (5) | 0.24 (5) |
| 10 | 3527 | 0.97 (4) | 0.36 (4) | 0.53 (7) | 0.49 (8) | 0.72 (5) | ---- |
| 47 | 3951 | 0.86 (4) | 0.32 (6) | 0.51 (7) | 0.29 (9) | 0.65 (5) | ---- |
| 21 | 3995 | 0.97 (4) | 0.25 (4) | 0.57 (10) | 0.25 (9) | 0.71 (4) | ---- |
| 23 | 4105 | 0.71 (3) | 0.19 (3) | 0.57 (8) | 0.32 (8) | 0.70 (8) | ---- |
| 36 | 4270 | 0.67 (3) | 0.32 (5) | 0.56 (11) | 0.29 (14) | 0.68 (6) | ---- |
| 24 | 4504 | ---- [b] | 0.32 (8) | 0.50 (9) | 0.14 (9) | 0.65 (4) | 0.29 (9) |
| 26 | 4745 | 0.80 (4) | 0.29 (6) | 0.51 (9) | 0.24 (15) | 0.68 (5) | ---- |
| 32 | 5003 | 0.91 (4) | 0.25 (9) | 0.78 (13) | ---- | ---- | ---- |
| 29 | 5104 | 0.91 (5) | 0.13 (9) | 0.73 (17) | 0.21 (21) | ---- | ---- |

The accumulation rate (in grams per square centimeter per 1000 years) is
calculated as the product of the $CaCO_3$ concentration, the sedimentation
rate, and a calculated value of bulk density based on an empirical
relationship observed in these cores.  The number in parentheses is the number
of carbonate determinations within each interval which determined the average
rate of accumulation.

[a] Core top is probably missing because of overpenetration of the gravity
core.

[b] Profile of $^{14}C$ in the core documents that ~6 cm of the mixed layer is
missing (L. G. DuBois, personal communication, 1983).

rates and are recorded as changes in regional
gradients of benthic foraminiferal $\delta^{13}C$.
Synoptic patterns of benthic foraminiferal
$\delta^{13}C$ reflect the gradients in $[O_2]$,
$\Sigma CO_2$ and nutrient concentration between
locations.

Unfortunately, measurements of benthic foraminiferal $\delta^{13}C$ cannot be directly converted
to measurements of the dissolved oxygen concentration in the overlying deep water.  Although
the assumption that a deep water mass is saturated with dissolved oxygen at its site of formation is reasonable and is applicable throughout
the period of time examined in this paper, the
preformed $\delta^{13}C$ of the deep water cannot be
assumed to be constant.  Different rates of
exchange between the atmosphere and the high-
latitude surface ocean for $CO_2$ and $O_2$, and
formation of deep water from high-latitude
surface water that is not nutrient depleted, can

TABLE 4.    Average Organic Carbon Accumulation for Seven Cores From the Sierra Leone Rise, Eastern Equatorial Atlantic

| Core | Depth, m | Interval, $10^3$ Years B.P. | | | | | |
|------|------|--------|--------|--------|--------|--------|--------|
| | | 0–7 | 18–30 | 45–60 | 64–74 | 115–125 | 146–160 |
| 38 | 2931 | 3.7 (3) | 1.6 (4) | 2.1 (8) | 3.0 (9) | 0.5 (1) | --- |
| 10 | 3527 | 5.2 (3) | 1.9 (4) | 3.6 (6) | 5.4 (7) | 3.0 (5) | --- |
| 47 | 3951 | 5.2 (3) | 2.8 (6) | 4.3 (7) | 6.2 (9) | 3.1 (5) | 6.2 (1) |
| 21 | 3995 | 6.6 (1) | 3.3 (1) | 6.4 (5) | 6.1 (5) | 4.4 (4) | 12.8 (1) |
| 36 | 4270 | 3.8 (1) | 3.5 (5) | 7.9 (8) | 12.0 (11) | 4.0 (5) | --- |
| 24 | 4504 | --- [a] | 3.7 (4) | 7.5 (9) | 7.0 (8) | 3.7 (2) | 7.4 (8) |
| 29 | 5104 | 5.8 (3) | 5.9 (6) | 9.0 (12) | 12.0 (16) | --- | --- |

The accumulation rate (in milligrams per square centimeter per 1000 years) is calculated as the product of the organic carbon concentration, the sedimentation rate and a calculated value of bulk density based on an empirical relationship observed in these cores.  The number in parentheses is the number of organic carbon determinations within the specified stratigraphic interval which determined the average rate of accumulation.

[a] Profile of $^{14}C$ in the core documents that ~6 cm of the mixed layer is missing (L. G. DuBois, personal communication, 1983).

produce different initial ratios of $[O_2]/\delta^{13}C$ in a deep water mass.  Lacking specific knowledge of the preformed $\delta^{13}C$ in the deep water, isolated measurements of benthic foraminiferal $\delta^{13}C$ cannot be translated directly into deep water dissolved oxygen concentration.

### Analytical Methods

Carbonate concentration in sediment is measured using a differential pressure technique based on the design of Jones and Kaiteris [1983].  Samples of known dry weight are reacted with HCl in vessels of known volume, and the pressure increase is measured.  The ratio of $CO_2$ produced by the sample to $CO_2$ produced by an equivalent weight of reagent-grade calcium carbonate is the proportion of calcium carbonate in the sample, assuming that all carbonate in the sample is $CaCO_3$.  Based on replicate analyses of both standards and unknowns, reproducibility of this measurement is better than ±2%.  For measurement of organic carbon, raw samples of known dry weight are reacted with $H_3PO_4$ to remove all inorganic carbon.  The residues are analyzed with a LECO WR12 carbon analyzer.  Analytical precision based on replicate analyses of unknowns is ±0.1%.  Benthic foraminiferal samples for isotopic analysis were picked from the >250 μm fraction of the washed sediment sample.  Five to 10 individuals of Planulina wuellerstorfi were reacted in $H_3PO_4$ at 50°C under vacuum.  The evolved $CO_2$ and $H_2O$ were separated by a series of three distillations utilizing frozen isopropyl alcohol slush and liquid nitrogen.  The $CO_2$ was analyzed on-line in a VG Micromass 602E mass spectrometer at the Woods Hole Oceanographic Institution using standard techniques [Curry and Lohmann, 1982, 1983].  The data are referred to the PDB (Peedee Belemnite) isotopic standard.

Organic carbon ($C_{org}$) and carbonate ($CaCO_3$) accumulation rates are calculated as the product of their concentration in the sediment, the bulk density of the sample and the sedimentation rate.  The accumulation rate data are presented in terms of milligrams $C_{org}$ per square centimeter per 1000 years and grams $CaCO_3$ per square centimeter per 1000 years.  The complete methodology is presented in the appendix, including examples of stratigraphic control and the effects of different chronologies.

### Data

The isotopic data and carbonate and organic carbon accumulation rate data are presented in Tables 2-4 as averages for six discrete glacial maxima and minima within the last glacial-interglacial cycle. In the following discussion, we consider each interval separately.

### Late Holocene (Present to 7,000 years B.P.)

Consistent with the local physical properties and chemistry of today's deep water, bathymetric gradients of benthic foraminiferal $\delta^{18}O$ and $\delta^{13}C$ during the Holocene are small (Figures 2a and 2b). In the eastern Atlantic basins deeper than 3000 m, temperature decreases by less than 0.5°C, and salinity decreases by 0.05°/₀₀ [Bainbridge, 1981]. The calculated equilibrium $\delta^{18}O$ gradient for calcite [Epstein et al., 1953] that results from these bathymetric gradients falls within the expected range of analytical, biological, and sedimentological errors. No variation with water depth in benthic foraminiferal $\delta^{13}C$ is expected given the present distribution of $[PO_4]$ and $[NO_3]$ [Bainbridge, 1981], and measurements of $\delta^{13}C$ of deep water $\Sigma CO_2$ exhibit little change with depth in the water column today [Kroopnick, 1980].

Carbonate accumulation in the eastern equatorial Atlantic during the late Holocene does not exhibit a significant decrease in the water column (Figure 2c). Calcite dissolution, as marked by increasing fragmentation of planktonic foraminifera, increases below 4800 m, which is considered the level of the lysocline in this region [Thunell, 1982]. No bathymetric decrease in carbonate accumulation occurs in the eastern Atlantic for at least two reasons: First, the deep water in the deepest parts of the eastern basins is only slightly undersaturated with respect to calcite. Destruction of the larger (>250 μm) planktonic foraminifera at depth, seen as a decrease in accumulation rate and an increase in fragmentation, indicates that dissolution must be occurring in deeper water. However, this weakly corrosive water is apparently limited to fragmenting the planktonic foraminifera without removing much calcite from the sediment. Second, if down-slope transport of carbonate is an important process, the additional carbonate input into the deeper sites would balance the removal of carbonate by dissolution. We have no evidence that down-slope transport of carbonate is a significant process in this region at this time, although observations of the accumulation rate of insoluble sediments suggests that active down-slope reworking occurred in the past.

In the eastern basins, organic carbon accumulation does not exhibit any significant varia-tion with water depth during the Holocene (Figure 2d). Because Holocene sedimentation rates are relatively constant with depth (1.2-1.7 cm/1000 years), differential burial rates do not affect the preservation of organic carbon [Muller and Suess, 1979]. Since there are no bathymetric changes in deep water $[O_2]$, oxygen availability does not affect the depth distribution of organic carbon oxidation.

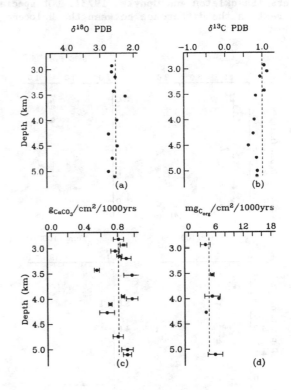

*O to 7ky BP*

Fig. 2. Depth profiles of (a) $\delta^{18}O$ of <u>Planulina wuellerstorfi</u>, (b) $\delta^{13}C$ of <u>Planulina wuellerstorfi</u>, (c) carbonate accumulation in grams $CaCO_3$ per square centimeter per 1000 years and d) organic carbon accumulation in milligrams $C_{org}$ per square centimeter per 1000 years. The isotopic data are Holocene samples from cores with validated Holocene sections, and the accumulation rate data represent averages of the accumulation rate within the interval 0 to 7000 years B.P. The mean value of each measurement in cores from depths shallower than 3750 m is plotted on each graph as a dotted line. The data points are plotted as mean value (±1σ) within this stratigraphic interval and reflect variability within the interval, not precision of the measurement or significance of the mean. No large gradients in the water column are present at this time in any of the measured parameters. (Note: The accumulation of carbonate at 3428m is low because part of the mixed layer was lost during coring.)

## Stage 2 Glaciation (18,000 to 30,000 years B.P.)

Benthic foraminiferal $\delta^{18}O$ during the last glacial maximum exhibited no significant change with depth in the water column, although the average $\delta^{18}O$ value was ~1.6°/°° greater (Figure 3a). This enrichment of oxygen isotopic composition reflects the preferential, temporary storage of $H_2^{16}O$ in northern hemisphere ice sheets [Shackleton and Opdyke, 1973]. Of special interest is the difference between the Holocene

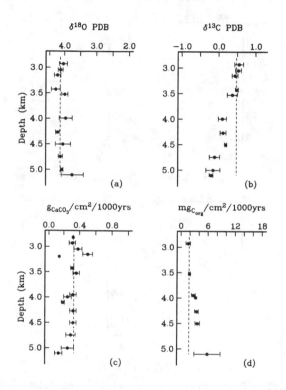

*18 to 30ky BP*

Fig. 3. Depth profiles of (a) $\delta^{18}O$ of <u>Planulina wuellerstorfi</u>, (b) $\delta^{13}C$ of <u>Planulina wueller-storfi</u>, (c) carbonate accumulation in grams $CaCO_3$ per square centimeter per 1000 years and (d) organic carbon accumulation in milligrams $C_{org}$ per square centimeter per 1000 years. The data are averages of all data falling within the interval 18,000 to 30,000 years B.P.. At this time a large decrease in $\delta^{13}C$ occurred below 3750 m which corresponds to a decrease in $[O_2]$ of 90 to 100 µmole/kg and an increase in $\Sigma CO_2$ of 70 to 75 µmoles/kg. Overall the rate of carbonate accumulation was about one half of the interglacial rate. No decrease with depth can be detected in carbonate accumulation, although fragmentation of planktonic foraminifera below 3750 m was greater than during the Holocene. A small increase with depth in organic carbon accumulation occurred below 3750 m.

and stage 2 profiles of $\delta^{13}C$. In contrast to the Holocene the profile of $\delta^{13}C$ during the stage 2 glaciation exhibited a significant decrease below 3750 m: Shallower than 3750 m, average $\delta^{13}C$ values were ~0.5°/°° (PDB); between 3750 m and 4500 m, $\delta^{13}C$ values decreased to ~0.1°/°° (PDB); and below 4500 m, $\delta^{13}C$ values were −0.1 to −0.2°/°° (PDB). Assuming Redfield stoichiometry in the remineralized organic carbon [Redfield et al., 1963], this decrease in $\delta^{13}C$ with depth corresponds to a bathymetric decrease in $[O_2]$ of ~90 to 100 µmoles/kg and bathymetric increases in $\Sigma CO_2$ of ~70 to 75 µmoles/kg and in $[PO_4]$ of ~0.7 µmoles/kg.

Carbonate accumulation was reduced at all depths in the water column during stage 2 (Figure 3c). At depths shallower than 3750 m, the preservation of sediment carbonate (indicated by the extent of fragmentation) was excellent and equaled the preservation in Holocene sediments. The overall decrease in carbonate accumulation at all depths resulted from decreased production of calcium carbonate in the equatorial Atlantic surface ocean during stage 2. No decrease in carbonate accumulation occurred below 3750 m despite the sharp increase in $\Sigma CO_2$ that is inferred from the $\delta^{13}C$ bathymetric profile. Greater fragmentation of planktonic foraminifera occurred below 3750 m, indicating increased dissolution, but as in the Holocene, dissolution did not increase to a degree that removed significant amounts of carbonate from the sediment.

A small bathymetric increase in organic carbon accumulation occurred during stage 2 (Figure 3d). At sites deeper than 3750 m, black mud deposition occurred stratigraphically at the base of stage 2. The reduced oxygen environment below 3750 m, indicated by the bathymetric profile of benthic foraminiferal $\delta^{13}C$, enhanced preservation of organic carbon in the sediments. The fossilized remains of many aerobic, benthic organisms can be found preserved in the black sediments, implying that the anoxia was postdepositional.

## Early Stage 3 (45,000 to 60,000 years B.P.)

Benthic foraminiferal $\delta^{18}O$ (Figure 4a) exhibited no variation with depth in the water column at this time. The profile of benthic foraminiferal $\delta^{13}C$ (Figure 4b) decreased with depth in the water column, but not to the extent seen during stage 2. Cores located at depths shallower than 3750 m were slightly enriched in $^{13}C$ relative to their respective stage 2 values. Of the cores located below 3750 m, four contributed data to both stage 2 and stage 3 profiles. During stage 2, benthic foraminifera in these cores were on average 0.4°/°° (range: 0.3 to 0.6°/°°) depleted in

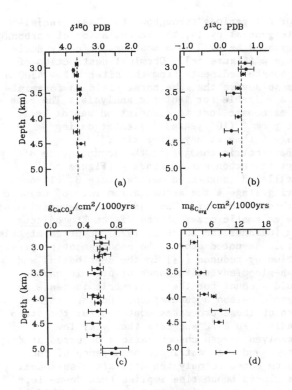

*45 to 60ky BP*

Fig. 4. Depth profiles of (a) δ$^{18}$O of <u>Planulina wuellerstorfi</u>, (b) δ$^{13}$C of <u>Planulina wuellerstorfi</u>, (c) carbonate accumulation in grams CaCO$_3$ per square centimeter per 1000 years and (d) organic carbon accumulation in milligrams C$_{org}$ per square centimeter per 1000 years. The data are averages of all data falling within the interval 45,000 to 60,000 years B.P. No variation with depth in the water column occurred in the δ$^{18}$O profile, and only a small decrease in δ$^{13}$C occurred. Carbonate accumulation increased below 5000 m suggesting that advective sources contributed to the accumulation in the deep sites. Hiatuses present in other cores support this hypothesis. A large increase with depth in the accumulation rate of organic carbon may have resulted from advective input.

δ$^{13}$C relative to the average δ$^{13}$C value shallower than 3750 m. During stage 3 the average difference between the shallow sites and the four deep cores was only 0.3°/oo (range: 0.1 to 0.5°/oo), so that over this depth interval the average gradient was slightly less steep.

Carbonate accumulation decreased gradually down to 4800 m during stage 3 (Figure 4c). Below 4800 m carbonate accumulation increased sharply. This increase in accumulation below 4800 m is the only clear evidence that our assumption of equal initial input of carbonate from the surface into each site was ever violated. Without an additional source of car-

bonate for deeper sites not available to shallower sites, only physical transport mechanisms exist to increase carbonate accumulation in deeper water. Down-slope transport of carbonate is one possible process. Transport and redeposition of carbonate by bottom currents is another and, in this case, seems likely. Many cores recovered in this region between 4500 and 5000 m contain hiatuses as well as coarse sand layers within stage 3 sediments. Anomalously high sedimentation rates occurred at this time in other cores. Such sedimentary features are not seen at other water depths. The presence of hiatuses and the abrupt increase in carbonate accumulation rate in the deepest sites suggests that bottom current activity increased during stage 3, creating areas of erosion and preferential accumulation of carbonate.

Organic carbon accumulation increased markedly with water depth during stage 3 (Figure 4d). At shallow sites, organic carbon accumulation was 2 to 4 mg cm$^{-2}$ (10$^3$ years)$^{-1}$, comparable to or slightly greater than the rates observed at shallow depths during stages 1 and 2. Below 3750 m, accumulation of organic carbon increased to 6 to 8 mg cm$^{-2}$ (10$^3$ years)$^{-1}$ during stage 3, compared to only 4 to 6 mg cm$^{-2}$ (10$^3$ years)$^{-1}$ during stage 2. The steeper bathymetric gradient observed during stage 3 may have resulted from two factors: (1) lower dissolved oxygen concentrations in the basin and (2) increased advective input of organic carbon in the deep sites. We cannot determine deep water [O$_2$] during stage 3, although we have noted that the bathymetric gradient was less steep than during stage 2. The bathymetric increase in organic carbon accumulation is consistent with the hypothesis of the increased advective input of sediments into the deep sites at this time. Sediment trap studies of particle fluxes attribute large increases in the lithogenic particulate flux [Honjo et al., 1982b] and smaller increases in the flux of organic carbon [Honjo et al., 1982a] to resuspension and advection processes. The patterns of carbonate and organic carbon accumulation in this region suggest that advective transport of particulates was an active process at this time.

Stage 4 (64,000 to 75,000 years B.P.)

While there was no bathymetric gradient of benthic foramininferal δ$^{18}$O (Figure 5a), benthic foraminiferal δ$^{13}$C decreased sharply below 3750 m during stage 4 (Figure 5b). In the five cores shallower than 3750 m, average δ$^{13}$C values during stages 2 and 4 were equal. Three cores below 3750 m contributed isotopic data to both stages 2 and 4; the average difference in δ$^{13}$C between these deep cores and the shallow cores averaged 0.3°/oo during stage 2 and 0.4°/oo during stage 4. Although this implies that the bathymetric gradients in $[O_2]$ and $\Sigma CO_2$ during stage 2 were only slightly less than during stage 4, we suspect that the gradients during stage 4 were much greater. The average value of $\delta^{13}C$ below 3750 m measured during stage 4 is biased both by dissolution and sampling. Because as much as 30 cm of sediment in the middle of stage 4 often contain very low carbonate and no benthic foraminifera, the calculated average $\delta^{13}C$ may not be representative of the most severe oxygen depletion.

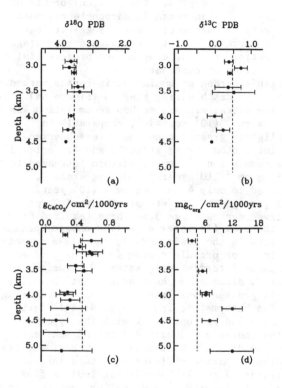

δ¹⁸O PDB    δ¹³C PDB

*64 to 75ky BP*

Fig. 5. Depth profiles of (a) $\delta^{18}O$ of _Planulina wuellerstorfi_, (b) $\delta^{13}C$ of _Planulina wuellerstorfi_, (c) carbonate accumulation in grams $CaCO_3$ per square centimeter per 1000 years and (d) organic carbon accumulation in milligrams $C_{org}$ per square centimeter per 1000 years. The data are averages of all data falling within the interval 64,000 to 75,000 years B.P. A large decrease in $\delta^{13}C$ occurred below 3750 m at this time. The gradient is similar in magnitude to that observed during stage 2. No isotopic data are available from below 4500 m because of severe dissolution. Increased dissolution is reflected in the carbonate accumulation rate profile by a large decrease below 3750 m. High organic carbon accumulation is associated with the widespread depostion of _Ethmodiscus rex_ oozes.

Of all periods throughout the last glacial-interglacial cycle, the accumulation of carbonate decreased below 3750 m most dramatically during stage 4 (Figure 5c). Chemical destruction of carbonate sediments from the sites below 4500 m was so severe that the cores yield no foraminifera suitable for isotopic analysis. The rate of carbonate lost to dissolution was at least $0.1 \text{ g cm}^{-2} (10^3 \text{ years})^{-1}$ greater during stage 4 than at any other time.

Bathymetric changes in the accumulation of organic carbon during stage 4 (Figure 5d) paralleled those seen during stage 3, although during stage 4 the accumulation rate of organic carbon was higher. The widespread deposition of the enigmatic oozes of the diatom _Ethmodiscus rex_ [Gardner and Burckle, 1975] occurred at this time. As noted above, the enhancement of preservation by reduced $[O_2]$ in the deep basins and down-slope/advective input of organic carbon could account for the bathymetric increase in organic carbon accumulation. During stage 4 both of these processes contributed: the steep gradient in $\delta^{13}C$ suggests that very low dissolved oxygen concentrations occurred at deep sites, and the restricted occurrence of the diatom oozes to only the deep sites surrounding the Sierra Leone rise implies that down-slope transport was an important process at this time.

### Early Stage 5 (115,000 to 125,000 years B.P.)

Bathymetric profiles of benthic foraminiferal $\delta^{18}O$ and $\delta^{13}C$ during this interval (Figures 6a and 6b) exhibit only small gradients and parallel the profiles observed during the Holocene. The average $\delta^{18}O$ value for stage 5e (2.7 to 2.8°/oo PDB) was similar to that observed during the Holocene, and its low value reflects the low volume of northern hemisphere ice at this time. There was no decrease in benthic foraminiferal $\delta^{13}C$ below 3750 m, indicating that there was no gradient in $[O_2]$ or $\Sigma CO_2$. The average $\delta^{13}C$ value shallower than 3750 m was 0.3°/oo less than during the Holocene. This difference in isotopic composition between the Holocene and stage 5e is observed at other locations in the Atlantic Ocean [Boyle and Keigwin, 1982; Curry and Lohmann, 1982], but not at the best studied site in the Pacific Ocean (V19-30, eastern equatorial Pacific [Shackleton et al., 1983]).

A small decrease in carbonate accumulation in deeper water occurred during stage 5e (Figure 6c). Overall, the rate of carbonate accumulation was lower during this interval than during the Holocene, but note that the sedimentation rate used for the accumulation rate calculation is an average for all of stage 5 and not just stage 5e. As a result, the average accumulation rate of carbonate at this time may be underestimated. No bathymetric changes

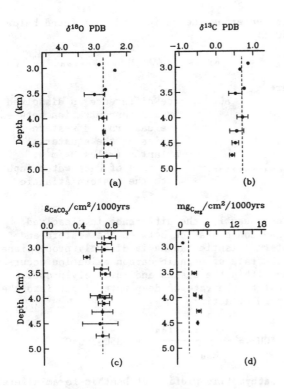

*115 to 125ky BP*

**Fig. 6.** Depth profiles of (a) $\delta^{18}O$ of <u>Planulina wuellerstorfi</u>, (b) $\delta^{13}C$ of <u>Planulina wueller-storfi</u>, (c) carbonate accumulation in grams CaCO₃ per square centimeter per 1000 years and (d) organic carbon accumulation in milligrams C_org per square centimeter per 1000 years. The data are averages of all data falling within the interval 115,000 to 125,000 years B.P. The profiles of each measured parameter parallel the profiles observed during stage 1 (see Fig. 2). No gradient in isotopic composition is observed. Carbonate accumulation was high at all depths in the water column while organic carbon accumulation was low and did not increase below 3750 m.

occurred in organic carbon accumulation (Figure 6d). Bathymetric increases in organic carbon accumulation were glacial phenomena.

## Late Stage 6 (146,000 to 160,000 years B.P.)

From our limited data for stage 6, the bathymetric variations in the measured properties paralleled the profiles observed for the other glacial stages (Figures 7a-7d). Carbon isotopic compositions, in particular, exhibited a sharp decrease in the water column, indicating a sharp decrease in [O₂] in the deep sites. The limited data on carbonate accumulation suggest that its average rate was low (0.2 g cm⁻² (10³ years)⁻¹) and that no large decrease

occurred below 3750 m. Organic carbon accumulation rates are available for only three cores during stage 6, insufficient for study of their depth distribution.

### Discussion

At no time during the last glacial-interglacial cycle was there a bathymetric gradient in $\delta^{18}O$ of the benthic foraminifera in the eastern Atlantic. Absence of such a gradient implies either that there were no bathymetric variations in temperature or isotopic composition of the deep water or that, if there were any such variations, the changes in temperature and $\delta^{18}O$ of the deep water affected the $\delta^{18}O$ of the calcite in opposite ways and cancelled each other out. Because of the silled geometry of the eastern basins, we prefer the

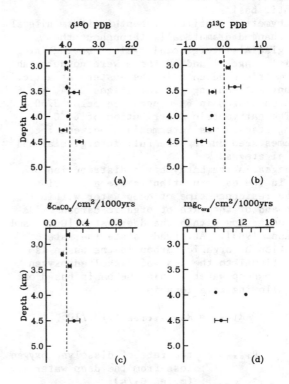

*146 to 160ky BP*

**Fig. 7.** Depth profiles of (a) $\delta^{18}O$ of <u>Planulina wuellerstorfi</u>, (b) $\delta^{13}C$ of <u>Planulina wueller-storfi</u>, (c) carbonate accumulation in grams CaCO₃ per square centimeter per 1000 years and (d) organic carbon accumulation in milligrams C_org per square centimeter per 1000 years. The data are averages of all data falling within the interval 146,000 to 160,000 years B.P. The profiles, although composed of fewer data, are similar to those observed during stage 2 (see Fig. 3).

first explanation. If input of deep water from the western into the eastern basins has always been across silled, low-latitude fracture zones (as it is today), then only small bathymetric gradients in temperature and isotopic composition are expected below the sill. Assuming no gradient in the isotopic composition of the deep water, the precision of our isotopic measurements limits resolution of a temperature difference to 0.4°C. A benthic thermocline as steep as that seen today in the western Atlantic (greater than 2°C in 1500 m) would have been detected, and its existence can be ruled out for the last glacial-interglacial cycle. In this discussion, we proceed under the assumption that the input of deep water into the eastern Atlantic throughout the last glacial-interglacial cycle followed today's route and that no sources of deep water other than NADW, AABW or their glacial counterparts entered the eastern Atlantic basins.

Bathymetric profiles of benthic foraminiferal $\delta^{13}C$ changed dramatically throughout the last glacial-interglacial cycle. During interglacial stages 1 and 5, there were no detectable bathymetric gradients in the eastern Atlantic. In contrast, during glacial stages 2, 4, and 6, large decreases in $\delta^{13}C$ occurred below 3750 m. The bathymetric distribution of $\delta^{13}C$ during stage 3 was intermediate between the extremes seen during the full interglacial and glacial stages.

Changes in the bathymetric distribution of $\delta^{13}C$ in the eastern Atlantic are a function of the residence time of deep water within the basin and of the rate of organic carbon oxidation occurring within the deep water [Curry and Lohmann, 1983]. At steady state the rate of oxidation of organic carbon in the basin is proportional to the loss of dissolved oxygen from the deep water within the basin based on the following relationship:

$$R_{loss} = R_{oxidation} \cdot 135/105$$

where

| | |
|---|---|
| $R_{loss}$ | the rate of dissolved oxygen loss from the deep water (moles $O_2$/s); |
| $R_{oxidation}$ | the rate of organic carbon oxidation in the deep water (moles C/s); |
| 135/105 | the Redfield ratio of 135 moles $O_2$ consumed by the oxidation of 105 moles of carbon. |

For today's abyssal circulation pattern, the rate of dissolved oxygen loss in the eastern basin equals the product of the deep water input rate across the fracture zones and the difference in dissolved oxygen concentration between the deep

water entering the basins and that found below 3750 m:

$$R_{loss} = AOU \cdot R_{advection}$$

where

| | |
|---|---|
| AOU | the difference in dissolved oxygen concentration between the eastern and western basins of the equatorial Atlantic (moles $O_2$/m$^3$); |
| $R_{advection}$ | the rate of deep water input into the eastern Atlantic (m$^3$/s). |

Consequently, the difference in dissolved oxygen concentration between the western and eastern Atlantic basins is directly proportional to the rate of organic carbon oxidation occurring within the basins and inversely proportional to the rate of deep water input into the eastern Atlantic:

$$AOU = \frac{R_{oxidation}}{R_{advection}} \cdot 135O_2/105C$$

The bathymetric profile of benthic foraminiferal $\delta^{13}C$, which measures the bathymetric gradient of dissolved oxygen concentration, is quantitatively related to the difference in dissolved oxygen concentration (AOU) between the eastern and western Atlantic basins. Assuming that no water enters the deep eastern Atlantic from sources other than the western basins, deep water at sill depth in the eastern basins provides an estimate of the initial $\delta^{13}C$ of the deep water flowing across the fracture zones. The contrast in $\delta^{13}C$ (and the inferred contrast in [$O_2$]) above and below the sill depth provides an estimate of the loss, to organic carbon remineralization, of dissolved oxygen from the deep water.

Today, there is no bathymetric gradient in $\delta^{13}C$ or [$O_2$] in the deep eastern Atlantic, suggesting that the rate of advection into the eastern basins is large compared to the local rate of organic carbon oxidation. Similarly, during stage 5e, only small bathymetric variations in $\delta^{13}C$ occurred, so the ratio of oxidation to advection at that time was similar to today's. A unique solution for the advective rate is not possible from measurements of $\delta^{13}C$ gradients without specific knowledge of the rate of carbon oxidation.

The bathymetric gradient of $\delta^{13}C$ was generally much greater throughout the last glacial-interglacial cycle than during the Holocene and stage 5e. During stage 2, $\delta^{13}C$ decreased below 3750 m by 0.7°/oo, corresponding to a decrease in [$O_2$] of 90 to 100 µmoles/kg. During stage 4, the gradient was similar, even though data from the deepest sites are not available because of severe dissolution. The gradient (based on even less data) was also large during stage 6. These comparatively steep bathymetric gradients indicate that the ratio of organic carbon oxidation to advection was much greater during the last three glaciations than it is today. We have argued that the increase in this ratio during stage 2 must have been due in part to a decrease in the advective rate into the eastern Atlantic [Curry and Lohmann, 1983]. We calculated that the rate of organic carbon oxidation would have to increase by approximately a factor of 6 for the $\delta^{13}C$ gradient to result from increased carbon remineralization alone. Based on other estimates of the glacial increase in rain rate of organic carbon from increased surface productivity [Muller and Suess, 1979; Pedersen, 1983], we concluded that the advective rate into the eastern Atlantic must have decreased by a factor of 2.

The large differences between the profiles of organic carbon accumulation for stages 2 and 4 suggest that the model of reduced advection proposed for stage 2 may not apply to stage 4. The high rates of carbon accumulation at this time imply that the overall flux and oxidation rates may have increased. The bathymetric profile of organic carbon accumulation during stage 4 shows a large increase below the sill which may have resulted from increased preservation in low oxygen conditions. Down-slope reworking of organic carbon associated with the Ethmodiscus rex oozes also occurred. If more organic carbon was delivered to the deep sites than the shallow sites and the gradient in $\delta^{13}C$ reflected unequal rates of organic carbon remineralization above and below sill depth, then the rate of advection into the eastern Atlantic during stage 4 may have exceeded the rate observed during stage 2.

The bathymetric profile of benthic foraminiferal $\delta^{13}C$ during stage 3 exhibited a gradient that was intermediate between the interglacial profiles of stages 1 and 5e and the glacial profiles of stages 2 and 4. From the large increase in the accumulation of organic carbon below 3750 m, we suspect that much of the gradient resulted from increased carbon oxidation. Independent measurements of the nutrient concentration of deep water in the North Atlantic indicate that the production rate of NADW during stage 3 was the same as observed during the Holocene and stage 5 [Boyle and Keigwin, 1982]. Additionally, the increase in carbonate accumulation in the two deepest sites during stage 3 (see Figure 4c) and the associated erosional hiatuses and winnowed sediments imply that deep water flow through the eastern Atlantic exceeded today's rates. We consider it unlikely that the decrease in $\delta^{13}C$ below 3750 m during stage 3 was caused by decreased advection.

The bathymetric profiles of carbonate accumulation demonstrate that the production of carbonate in surface water parallels glacial-interglacial climate change. Assuming that in situ carbonate dissolution within pore water and down-slope reworking are negligible, accumulation of carbonate above the lysocline reflects changes in production at the sea surface. Locally, the interglacial accumulation rate of carbonate averaged 0.7 to 0.8 g $cm^{-2}$ ($10^3$ years)$^{-1}$, twice the rate observed during glacial stages 2 and 4. In shallow sites the accumulation rate of carbonate was greater during stage 4 than during stage 2. Only during stage 4 did carbonate accumulation decrease significantly with water depth in a manner parallel to the observed steep bathymetric gradients in $\delta^{13}C$ and organic carbon accumulation. Delivery of large amounts of organic carbon, its increased oxidation, and the consequent increase in the corrosiveness of the deep water produced the greatest losses of carbonate to dissolution during stage 4. During stages 1 and 5e, no detectable bathymetric decreases in carbonate accumulation occurred. And while no decreases are detected in stages 2 and 3, greater fragmentation of planktonic foraminifera suggests that the deep water was slightly more corrosive to carbonate during those times. The limited data for stage 6 suggest that the profile of carbonate accumulation closely resembles the profile for stage 2. Clearly, the most severe dissolution of carbonate sediments in the eastern Atlantic during the last glacial-interglacial cycle occurred during stage 4.

Crowley [1983] and Peterson and Prell [1983] each have noted that dissolution was most severe during times of falling sea level, presumably as a result of the transfer of large amounts of organic carbon from terrestrial to oceanic reservoirs. Our observation that the most severe dissolution in the eastern Atlantic occurred during the major sea level fall following stage 5 is consistent with their observations. Crowley [1983] also noted that the dissolution event during stage 4 is the only time throughout the last glacial-interglacial cycle when dissolution occurred in phase between the Atlantic and Pacific oceans. He recognized two modes of dissolution: One mode resulted from sea level change and was synchronous and in phase between ocean basins. The other mode resulted from changes in deep water production in the North Atlantic and was exactly out of phase between oceans. The differences in the shape of the carbonate accumulation profiles between stages 2 and 4 reflect the differences in dissolution intensity between these two mechanisms. While there is evidence of only a slight increase in dissolution when production of deep water decreased during stage 2 [Boyle and Keigwin, 1982; Curry and Lohmann, 1983, this paper;

Shackleton et al., 1983], stage 4 recorded severe dissolution as delivery and oxidation of large amounts of organic carbon occurred in the eastern Atlantic.

## Conclusions

1. Throughout the last glacial-interglacial cycle, there were no gradients in the bathymetric profile of $\delta^{18}O$ in the eastern Atlantic. It is, therefore, unlikely that any gradients occurred in deep water temperature or oxygen isotopic composition related to water mass boundaries. The bathymetric gradients observed in other geological and geochemical parameters preserved in the sediments indicate that ridge topography has controlled abyssal circulation in the equatorial Atlantic throughout the last glacial-interglacial cycle much as it does today: The sill depth of the Mid-Atlantic ridge (3750 m) often coincided with the water depth at which changes in sedimentation and sediment properties occurred. Increases and decreases in the intensity of today's pattern of deep water advection are the only changes that occurred in the deep water circulation of the eastern Atlantic during the last 160,000 years.

2. Large changes in the bathymetric profile of $\delta^{13}C$ occurred throughout the last 160,000 years. While no bathymetric gradients occurred during the Holocene and stage 5e, large decreases in $\delta^{13}C$ occurred below 3750 m during stages 2 and 4. These gradients in $\delta^{13}C$ reflect both the rate of oxidation of organic carbon in the deep water and the residence time of the deep water below 3750 m in the basin. Although both factors are undoubtedly important, high amounts of organic carbon in the sediments suggest that the steep gradient observed during stage 4 resulted from increased oxidation of organic carbon below the sill. Comparatively low amounts of organic carbon in the sediments during stage 2 suggest that its steep $\delta^{13}C$ gradient was caused by an increase in residence time of the deep water below 3750 m.

3. Carbonate sedimentation patterns record changes in deep water chemistry by changing the bathymetric gradient of carbonate accumulation. During the Holocene and stage 5e, well-oxygenated deep water, low in $\Sigma CO_2$, was found below 3750 m in the eastern Atlantic. As a result, dissolution effects in the sediments were minor, and no loss of carbonate from the deep sites can be detected. During stage 2, the decrease in $\delta^{13}C$ below the sill reflected an increase in $\Sigma CO_2$ that increased the corrosiveness of the deep water. A slight increase in carbonate dissolution was recorded by increased fragmentation of planktonic foraminifera, but no large loss of carbonate was observed from the deep sites. The steepest gradient in $\delta^{13}C$ probably occurred during stage 4, and only at

this time did carbonate accumulation decrease significantly below 3750 m. Severe fragmentation of planktonic foraminifera occurred at this time at all water depths. The accumulation rate of carbonate in the shallowest cores provides our best measure of changes in carbonate production in the surface water of the equatorial Atlantic. Carbonate accumulation was ~0.8 g cm$^{-2}$ (10$^3$ years)$^{-1}$ during the Holocene and ~0.7 g cm$^{-2}$ (10$^3$ years)$^{-1}$ during stage 5e. During glacial maxima, carbonate accumulation decreased to ~0.4 g cm$^{-2}$ (10$^3$ years)$^{-1}$.

## Appendix

Calculation of accumulation rate for a sediment component relies on three separate measurements: (1) the concentration of the component in the sediment (grams per gram of raw sediment), (2) the bulk density of the sediment (grams of raw sediment per cubic centimeter of raw sediment), and (3) the sedimentation rate of the core (centimeters per 1000 years). The measurement of concentration has been discussed above, and we will not elaborate further. We determine bulk density from an empirical relationship of bulk density and carbonate concentration that we observe in eastern equatorial Atlantic sediments. Seventy-three measurements of bulk density and carbonate concentration, from all depths in the water column below 2800 m and distributed throughout the upper 3 m of sediment column, produced the following linear least squares relationship:

$$\rho_{dry} = 0.0066(\%CaCO_3) + 0.29$$

The correlation coefficient (r) is +0.76, and the standard error of estimate for prediction of dry bulk density ($\rho_{dry}$ = grams dry weight per cubic centimeter of wet volume) is ±0.09 g/cm$^3$.

Sedimentation rate is sensitive to two factors: the identification of synchronous stratigraphic levels within each core and the proper assignment of an age to each level. Our identification of stratigraphic levels is based on carbonate concentration changes, presence or absence of Globorotalia menardii, and oxygen isotopic changes in the benthic foraminifer Planulina wuellerstorfi. In Figure A1, we present typical records for cores shallower and deeper than 3750 m. On the basis of high-resolution sampling of both carbonate concentration and stable isotopic composition at termination 1, no detectable lead or lag occurred in this region. We assign stage boundaries on the basis of changes in both carbonate concentration and isotopic composition, although within some stratigraphic levels the frequency of sampling is greater for carbonate concentration. At these levels, identification of the isotopic stage boundaries relies

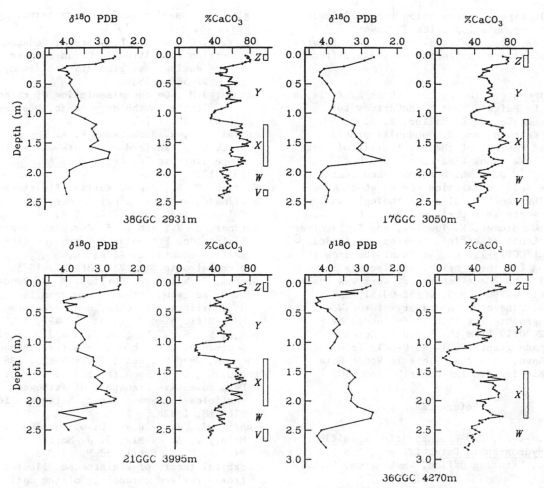

Fig. A1.  Four examples of the stratigraphic records used for correlation and assignment of chronology.  Stable isotopic, lithologic, and biostratigraphic records are presented for two cores recovered from shallower than 3750 m and two from deeper.  The zonation of Ericson and Wollin [1968] is identified for each core.  On the basis of these records, correlation to the Shackleton and Opdyke [1973] chronology was achieved.

heavily on the changes in carbonate concentration after the approximate location has been identified using stable isotope stratigraphy. On the basis of sampling frequency, the largest error expected in the identification of a stratigraphic level is less than 3 cm.  The shortest stratigraphic intervals which we identify are about 20 cm thick, so the error in the accumulation rate calculation based on misidentification of a stratigraphic level is about ±10%. For longer stratigraphic intervals (e.g., stages 3 and 5) the error is closer to ±5% or less. The effect of this random error is to introduce noise into the profiles of accumulation rate rather than change their shape.  Our results concerning the overall glacial-interglacial changes in accumulation would be unaffected.

Different chronologies introduce systematic

errors that affect both the shape of bathymetric profiles of accumulation and the overall differences between glacial and interglacial intervals. We present our results based on sedimentation rates determined from the Shackleton and Opdyke [1973] time scale.  If we applied one of the Spectral Mapping Project (SPECMAP) time scales (D. G. Martinson, personal communication, 1984), our results would differ systematically by the following amounts: (1) stage 1 accumulation rates would be 18% higher; (2) stage 2 accumulation rates 3% lower; (3) stage 3 accumulation rates 8% higher; (4) stage 4 accumulation rates 18% lower; and (5) stage 5 accumulation rates 4% lower.  Another SPECMAP time scale [Imbrie et al., 1984] changes our results by the following amounts: (1) stage 1 accumulation rates would be 8% higher; (2) stage 2 accumulation rates 37%

higher; (3) stage 3 accumulation rates 9% lower; (4) stage 4 accumulation rates 9% lower; and (5) stage 5 accumulation rates 8% lower. Our results and conclusions are unaffected by these differences.

Acknowledgments. The authors thank E. Boyle, R. Fairbanks, L. Keigwin, and M. McCartney for helpful discussions. K. Miller, E. Boyle, W. Berger, L. Keigwin, and E. Sundquist provided constructive reviews of the first draft of this manuscript. We thank Lisa Dubois for her preliminary $^{14}$C data which established that part of the Holocene section for EN066-24GGC was missing. The geochemical and lithologic analyses of the sediments were performed by D. R. Ostermann, C. Blumer, N. Tousley, and T. Fleming. The Sierra Leone rise field program was funded by NSF grant OCE78-19768. We thank the crew of R/V ENDEAVOR for their help in collecting the cores. All subsequent laboratory work was funded by NSF grants OCE80-10935, OCE82-00151, and OCE83-09124. The core laboratory of the Woods Hole Oceanographic Institution is funded by grant OCE82-00119 from the National Science Foundation and grant N00014-82-C-0019 from the Office of Naval Research. This is Woods Hole Oceanographic Institution contribution 5807.

# References

Bainbridge, A. E., GEOSECS Atlantic Expedition, Vol. I, Hydrographic Data, 121 pp., U.S. Government Printing Office, Washington, D. C., 1981.

Belanger, P. E., W. B. Curry, and R. K. Matthews, Core-top evaluation of benthic foraminiferal isotopic ratios for paleo-oceanographic interpretation, Palaeogeogr. Palaeoclimatol. Palaeoecol., 33, 205-220, 1981.

Berger, W. H., Planktonic foraminifera: Selective solution and paleoclimatic interpretation, Deep Sea Res., 15, 31-43, 1968.

Berger, W. H., M.-C. Bonneau, and F. L. Parker, Foraminifera on the deep-sea floor: Lysocline and dissolution rate, Oceanol. Acta, 5, 249-258, 1982.

Boyle, E. A., and L. D. Keigwin, Deep circulation of the North Atlantic over the last 200,000 years: Geochemical evidence, Science, 218, 784-787, 1982.

Broecker, W. S., Glacial to interglacial changes in ocean chemistry, Progr. Oceanogr., 11, 151-197, 1982.

Crowley, T. J., Secular changes in North Atlantic carbonate dissolution during the last 400,000 years (abstract), Eos Trans. AGU, 64, 738, 1983.

Curry, W. B., and G. P. Lohmann, Carbon isotopic changes in benthic foraminifera from the western South Atlantic: Reconstruction of glacial abyssal circulation patterns, Quat. Res., 18, 218-235, 1982.

Curry, W. B., and G. P. Lohmann, Reduced advection into Atlantic Ocean deep eastern basins during last glaciation maximum, Nature, 306, 577-580, 1983.

Edmund, J.M., On the dissolution of carbonate and silicate in the deep ocean, Deep Sea Res., 21, 455-480, 1974.

Epstein, S., R. Buchsbaum, H. A. Lowenstam, and H. C. Urey, Revised carbonate-water isotopic temperature scale, Geol. Soc. Am. Bull., 64, 1315-1326, 1953.

Ericson, D. B., and G. Wollin, Pleistocene climates and chronology in deep-sea sediments, Science, 162, 1227-1234, 1968.

Gardner, J. V., and L. H. Burckle, Upper Pleistocene Ethmodiscus rex oozes from the eastern equatorial Atlantic, Micropaleontology, 21, 236-242, 1975.

Graham, D. W., B. H. Corliss, M. L. Bender, and L. D. Keigwin, Carbon and oxygen isotopic disequilibria of Recent deep-sea foraminifera, Mar. Micropaleontol., 6, 483-497, 1981.

Honjo, S., S. J. Manganini, and J. J. Cole, Sedimentation of biogenic matter in the deep ocean, Deep Sea Res., 29, 609-625, 1982a.

Honjo, S., D. W. Spencer, and J. W. Farrington, Deep advective transport of lithogenic particles in Panama Basin, Science, 216, 516-518, 1982b.

Imbrie, J., J. D. Hays, D. G. Martinson, A. McIntyre, A. C. Mix, J. J. Morley, N. G. Pisias, W. L. Prell and N. J. Shackleton, The orbital theory of Pleistocene climate: Support from a revised chronology of the marine $\delta^{18}$O record, in Milankovitch and Climate, edited by A. Berger et al., pp. 269-305, D. Reidel Co., Dordrecht, Holland, 1984.

Jones, G. A., and P. Kaiteris, A vacuum-gasometric technique for rapid and precise analysis of calcium carbonate in sediments and soils, J. Sediment. Petrol., 53, 655-660, 1983.

Kroopnick, P., Modeling the $O_2$-$CO_2$-$^{13}$C system in the eastern equatorial Pacific, Deep Sea Res., 21, 211-217, 1974.

Kroopnick, P., The distribution of $^{13}$C in the Atlantic Ocean, Earth Planet. Sci. Lett., 49, 469-484, 1980.

Kroopnick, P., R. F. Weiss, and H. Craig, Total $CO_2$, $^{13}$C and dissolved oxygen-$^{18}$O at GEOSECS II in the North Atlantic, Earth Planet. Sci. Lett., 16, 103-110, 1972.

Metcalf, W. G., B. C. Heezen, and M. C. Stalcup, The sill depth of the Mid-Atlantic ridge in the equatorial region, Deep Sea Res., 11, 1-10, 1964.

Muller, P. J. and E. Suess, Productivity, sedimentation rate, and sedimentary organic matter in the oceans - I. Organic carbon preservation, Deep Sea Res., 26A, 1347-1362, 1979.

Pedersen, T. F., Increased productivity in the eastern equatorial Pacific during the last glacial maximum (19,000 to 14,000 yr B.P.), Geology, 11, 16–19, 1983.

Peterson, L. C. and W. L. Prell, Changes in carbonate saturation and circulation of Indian Deep Waters during the Pleistocene (abstract), Geol. Soc. Am., Abstr. Programs, 15, 660, 1983.

Redfield, A. C., B. H. Ketchum and F. A. Richards, The influence of organisms on the composition of sea-water, in The Sea, Volume 2, edited by M. N. Hill, pp. 26–77, Wiley-Interscience, New York, 1963.

Shackleton, N. J., Tropical rainforest history and the equatorial Pacific carbonate dissolution cycles, in Fate of Fossil Fuel $CO_2$ in the Oceans, edited by N. R. Andersen and A. Malahoff, pp. 401–427, Plenum, New York, 1977.

Shackleton, N. J., and N. D. Opdyke, Oxygen isotope and paleomagnetic stratigraphy of equatorial Pacific core V28-238: Oxygen isotope paleotemperatures and ice volumes on a $10^5$ year and $10^6$ year scale, Quat. Res., 3, 39–55, 1973.

Shackleton, N. J., J. Imbrie, and M. A. Hall, Oxygen and carbon isotope record of east Pacific core V19-30: Implications for formation of deep water in the Pleistocene North Atlantic, Earth Planet. Sci. Lett., 65, 233–244, 1983.

Thunell, R. C., Carbonate dissolution and abyssal hydrography in the Atlantic Ocean, Mar. Geol., 47, 165–180, 1982.

Warren, B. A., Deep circulation of the world ocean, in Evolution of Physical Oceanography, edited by B. A. Warren and C. Wunsch, pp. 6–41, MIT Press, Cambridge, Mass., 1981.

# ATMOSPHERIC CARBON DIOXIDE, ORBITAL FORCING, AND CLIMATE

N. J. Shackleton

Godwin Laboratory for Quaternary Research, University of Cambridge
Cambridge CB2 3RS, England

Lamont-Doherty Geological Observatory, Palisades, New York 10964

N. G. Pisias

College of Oceanography, Oregon State University, Corvallis, Oregon 97331

Abstract. We have analyzed a 340,000-year record of benthic and planktonic oxygen and carbon isotope measurements from an equatorial Pacific deep-sea core. The data provide estimates of both global ice volume and atmospheric carbon dioxide concentration over this period. The frequencies characteristic of changes in the earth-sun orbital geometry dominate all the records. Examination of phase relationships shows that atmospheric carbon dioxide concentration leads ice volume over the orbital bandwidth, and is forced by orbital changes through a mechanism, at present not fully understood, with a short response time. Changes in atmospheric $CO_2$ are not primarily caused by glacial-interglacial sea level changes, which had been hypothesized to affect atmospheric $CO_2$ through the effect on ocean chemistry of changing sedimentation on the continental shelves. Instead, variations in atmospheric $CO_2$ should be regarded as part of the forcing of ice volume changes.

## Introduction

There are now two independent sets of data available which indicate that there have been climatically significant changes in the atmospheric carbon dioxide concentration during the Pleistocene. One the one hand, Berner et al., (1978), Delmas et al., (1980), and Neftel et al., (1982) have made measurements of the carbon dioxide in air bubbles trapped in the Antarctic and Greenland ice sheets. These measurements demonstrate that during the last glacial, the atmospheric carbon dioxide concentration was significantly lower than its pre-anthropogenic recent level. On the other hand, Shackleton et al., (1983a), following up reasoning put forward by Broecker (1982) have obtained a record of the [13]C difference between ocean surface and deep

waters from which they derived an atmospheric $CO_2$ record. While the ice core measurements provide an estimate that is both more direct and more accurate, the detailed analysis of deep-sea sediment cores is capable of providing longer records. Our objective here is to examine an extension of the record given by Shackleton et al., (1983a) with a view to obtaining a better understanding of the underlying mechanisms associating $CO_2$ and climate changes.

## Carbon Isotopes and Atmospheric Carbon Dioxide

Broecker (1982) pointed out that if the ice age atmospheric $CO_2$ concentration was indeed about 200 ppm as compared with a pre-anthropogenic level of about 280 ppm, the cause can only be sought in the ocean. Changes in the surface temperature of the ocean could be invoked, but published data (CLIMAP, 1976) suggest that these are far too small to explain the observations. On the other hand, there are several mechanisms by which the $CO_2$ pressure over the ocean surface might become reduced: in fact, any mechanism which extracts more carbon from the ocean surface would have this effect.

The removal of carbon from surface waters results from photosynthesis and from the production of calcareous skeletons. The amount of organic carbon removed is limited by the nutrient content of the upwelling deep water; the nature of the organisms present determines the ratio of carbonate to organic matter that sinks from the surface into deep water. Since photosynthesis results in a substantial carbon isotope fractionation, [13]C measurements may be used to investigate possible variations in the removal of organic matter.

Broecker (1982) was primarily concerned with evaluating all the possible oceanic contributions to the $CO_2$ record observed in ice cores. Since then, Shackleton et al., (1983a) have shown that

303

the [13]C record indicates that variations in the removal of organic (i.e., isotopically depleted) carbon from surface waters have been of a sufficient magnitude to give rise to atmospheric $CO_2$ variations similar in scale to those observed in ice cores. Thus we may for the present ignore those mechanisms which remove carbonate carbon (Berger, 1982) since this would not significantly alter the [13]C value of the surface water and thus cannot explain the [13]C data shown by Shackleton et al., (1983a). Each of the several mechanisms discussed by Broecker (1982) has been simply modeled by Broecker and Peng (1984) with a view to demonstrating testable differences between them. In fact, of the models tested by Broecker and Peng (1984), only that one which combines the extraction of carbon from the ocean to be stored in the continental biomass (Shackleton, 1977) with changes in the carbon/phosphorus ratio in organic residues falling to the deep sea (the mechanism suggested to Broecker by P. K. Weyl) is able to explain the data published by Shackleton et al., (1983a). In this paper we will be concerned with evaluating that model by making a more detailed investigation of an extended [13]C record. We do not deny the possible operation of other mechanisms, but since the independent record of atmospheric $CO_2$ variations available from ice cores is similar to that predicted by the [13]C data (Shackleton et al., 1983a) is seems reasonable to assume that these other mechanisms have been of secondary importance.

## Orbital Forcing

The climatically important parameters of the earth's orbital system are the eccentricity of the orbit and the orientation of the earth's rotational axis, which both varies in its alignment with respect to the major axis of the orbit (the precession of the equinoxes, henceforth "precession") and changes its angle relative to the normal to the orbit (henceforth "tilt"). Berger (1981) has outlined the history of astronomical investigations into the time series of these parameters; we have used the calculations of Berger (1978) in this work. Eccentricity has varied with a period of about 100 kyr over the time interval investigated in this paper, although spectral analysis of a longer eccentricity record reveals a complex spectrum with an important term having a 413-kyr period. Precession is also a complex function; over the time interval of our study it is dominated by terms with periods of about 23 kyr and about 19 kyr. Tilt has a much more regular variation with a mean period of about 41 kyr. Hays et al., (1976) demonstrated that all these periods can be extracted by time series analysis of appropriate climate records from deep sea sediments.

It is generally agreed that the variation in summer insolation at about 65°N might be a plausible forcing for Pleistocene continental ice

sheet variation (Imbrie and Imbrie, 1980); spectral analysis of the 65°N July insolation record shows about 20% variance at the tilt frequency and the remainder in precession; there is essentially no power at the eccentricity period. At higher latitudes, tilt becomes relatively more important. Time series analysis of temperature records from North Atlantic deep-sea sediment cores has shown a spatial pattern reflecting this transition through latitude in the relative importance of tilt and precession (Ruddiman and McIntyre, 1984). The dominant 100 kyr period in many Pleistocene climate records is thought to derive from the response of ice sheets with long, non-linear response times to forcing by precession (Hays et al., 1976; Imbrie and Imbrie, 1980).

We consider that it is now well established that orbital forcing is strongly impressed on long climatic records. Thus in order to evaluate the role of atmospheric $CO_2$ variations in climate change the $CO_2$ record must first be determined over a long enough time interval for its relationship to the orbital variations to be determined. To this end we have extended the record obtained by Shackleton et al., (1983a), placed it on an appropriate time scale, and examined its time and frequency domain characteristics in comparison with those of the forcing function and of the well-known ice volume response as monitored by [18]O variations in the same core.

## Analytical Methods

Core V19-30 was taken at 03°21'S, 83°21'W in 3091-m water depth; it was discussed briefly by Ninkovich and Shackleton (1975), who showed a record of carbonate content in the core and compared this with the carbonate record of nearby core V19-29, in which they obtained an oxygen isotope record. This comparison suggested that V19-30 would yield a detailed isotope record to late Stage 10 at about 340 kyr BP (thousand years before present). We have sampled the core for isotope studies at 3-cm intervals; since the average accumulation rate is over 5 cm/kyr, this represents an average sampling interval of a little over 500 years. Oxygen and carbon isotope analysis was performed using a VG Isogas triple collector mass spectrometer. The carbon dioxide was released using 100% orthophosphoric acid at 50°C; the water that is also released in the reaction was removed in a single pass through a 10-turn stainless steel spiral trap cooled to about -94°C using a commercially available cooler; we have found that the efficient removal of water is particularly important if high-precision [13]C data are to be obtained. Analytical reproducibility of [13]C measurements for replicate analysis of homogeneous carbonates was about ±0.04 per mil when these measurements were made. Virtually every sample contained sufficient specimens of the benthic <u>Uvigerina senticosa</u> for

TABLE 1. $^{18}$O and $^{13}$C Data for Uvigerina senticosa, $^{13}$C Data for Neogloboquadrina dutertrei, and $^{13}$C Difference Between N. dutertrei and U. senticosa in Core V19-30 Interpolated at 1.05 kyr Intervals, Assuming Uniform Accumulation Between the Control Points Listed in Table 3

| Age kyr | $\delta^{18}$O Uvig. | $\delta^{13}$C Uvig | $\delta^{13}$C N.dut. | $\Delta\delta^{13}$C |
|---|---|---|---|---|
| 0.0 | 3.48 | 0.04 | 1.50 | 1.46 |
| 1.05 | 3.46 | 0.08 | 1.46 | 1.38 |
| 2.10 | 3.31 | 0.06 | 1.42 | 1.36 |
| 3.15 | 3.29 | -0.05 | 1.56 | 1.61 |
| 4.20 | 3.40 | -0.07 | 1.57 | 1.64 |
| 5.25 | 3.30 | -0.07 | 1.57 | 1.64 |
| 6.30 | 3.56 | 0.01 | 1.54 | 1.53 |
| 7.35 | 3.53 | 0.03 | 1.42 | 1.39 |
| 8.40 | 3.47 | -0.12 | 1.21 | 1.33 |
| 9.45 | 3.83 | -0.01 | 1.28 | 1.29 |
| 10.50 | 3.89 | -0.13 | 1.11 | 1.24 |
| 11.55 | 4.18 | -0.23 | 1.18 | 1.41 |
| 12.60 | 4.42 | -0.26 | 1.11 | 1.37 |
| 13.65 | 4.86 | -0.26 | 1.24 | 1.50 |
| 14.70 | 4.81 | -0.36 | 1.23 | 1.59 |
| 15.75 | 5.03 | -0.36 | 1.28 | 1.64 |
| 16.80 | 4.98 | -0.44 | 1.36 | 1.80 |
| 17.85 | 4.97 | -0.41 | 1.46 | 1.87 |
| 18.90 | 4.89 | -0.29 | 1.25 | 1.54 |
| 19.95 | 4.89 | -0.39 | 1.62 | 2.01 |
| 21.00 | 4.96 | -0.45 | 1.53 | 1.98 |
| 22.05 | 4.97 | -0.34 | 1.50 | 1.84 |
| 23.10 | 4.96 | -0.34 | 1.44 | 1.78 |
| 24.15 | 4.92 | -0.33 | 1.68 | 2.01 |
| 25.20 | 4.82 | -0.19 | 1.54 | 1.73 |
| 26.25 | 4.68 | -0.18 | 1.53 | 1.71 |
| 27.30 | 4.65 | -0.11 | 1.66 | 1.77 |
| 28.35 | 4.47 | -0.19 | 1.44 | 1.63 |
| 29.40 | 4.54 | -0.28 | 1.43 | 1.71 |
| 30.45 | 4.56 | -0.20 | 1.42 | 1.62 |
| 31.50 | 4.64 | -0.18 | 1.61 | 1.79 |
| 32.55 | 4.61 | -0.31 | 1.59 | 1.90 |
| 33.60 | 4.58 | -0.13 | 1.75 | 1.88 |
| 34.65 | 4.68 | -0.10 | 1.64 | 1.74 |
| 35.70 | 4.60 | -0.15 | 1.60 | 1.75 |
| 36.75 | 4.59 | -0.13 | 1.66 | 1.79 |
| 37.80 | 4.54 | -0.06 | 1.70 | 1.76 |
| 38.85 | 4.50 | -0.05 | 1.52 | 1.57 |
| 39.90 | 4.41 | -0.26 | 1.56 | 1.82 |
| 40.95 | 4.53 | -0.18 | 1.72 | 1.90 |
| 42.00 | 4.63 | -0.18 | 1.74 | 1.92 |
| 43.05 | 4.63 | 0.14 | 1.65 | 1.51 |
| 44.10 | 4.52 | -0.20 | 1.54 | 1.74 |
| 45.15 | 4.57 | -0.15 | 1.66 | 1.81 |
| 46.20 | 4.61 | -0.25 | 1.62 | 1.87 |
| 47.25 | 4.58 | -0.25 | 1.64 | 1.89 |
| 48.30 | 4.56 | -0.19 | 1.53 | 1.72 |

TABLE 1. (continued)

| Age kyr | $\delta^{18}$O Uvig. | $\delta^{13}$C Uvig. | $\delta^{13}$C N.dut. | $\Delta\delta^{13}$C |
|---|---|---|---|---|
| 49.35 | 4.54 | -0.13 | 1.61 | 1.74 |
| 50.40 | 4.52 | -0.18 | 1.39 | 1.57 |
| 51.45 | 4.56 | -0.09 | 1.38 | 1.47 |
| 52.50 | 4.32 | 0.04 | 1.47 | 1.43 |
| 53.55 | 4.28 | -0.44 | 1.33 | 1.77 |
| 54.60 | 4.51 | -0.21 | 1.56 | 1.77 |
| 55.65 | 4.54 | -0.15 | 1.26 | 1.41 |
| 56.70 | 4.53 | -0.12 | 1.49 | 1.61 |
| 57.75 | 4.15 | -0.45 | 1.26 | 1.71 |
| 58.80 | 4.40 | -0.56 | 1.35 | 1.91 |
| 59.85 | 4.45 | -0.74 | 1.41 | 2.15 |
| 60.90 | 4.65 | -0.64 | 1.47 | 2.11 |
| 61.95 | 4.67 | -0.63 | 1.46 | 2.09 |
| 63.00 | 4.65 | -0.64 | 1.29 | 1.93 |
| 64.05 | 4.67 | -0.68 | 1.20 | 1.88 |
| 65.10 | 4.58 | -0.69 | 1.34 | 2.03 |
| 66.15 | 4.56 | -0.73 | 1.49 | 2.22 |
| 67.20 | 4.60 | -0.59 | 1.53 | 2.12 |
| 68.25 | 4.60 | -0.49 | 1.47 | 1.96 |
| 69.30 | 4.46 | -0.25 | 1.46 | 1.71 |
| 70.35 | 4.46 | -0.46 | 1.45 | 1.91 |
| 71.40 | 4.31 | -0.25 | 1.56 | 1.81 |
| 72.45 | 4.21 | -0.40 | 1.64 | 2.04 |
| 73.50 | 4.28 | -0.17 | 1.56 | 1.73 |
| 74.55 | 4.35 | -0.10 | 1.82 | 1.92 |
| 75.60 | 4.19 | -0.15 | 1.65 | 1.80 |
| 76.65 | 4.20 | -0.15 | 1.61 | 1.76 |
| 77.70 | 4.13 | -0.09 | 1.54 | 1.63 |
| 78.75 | 4.05 | -0.38 | 1.58 | 1.96 |
| 79.80 | 4.04 | -0.08 | 1.46 | 1.54 |
| 80.85 | 3.98 | -0.25 | 1.45 | 1.70 |
| 81.90 | 3.97 | -0.05 | 1.37 | 1.42 |
| 82.95 | 4.03 | -0.19 | 1.35 | 1.54 |
| 84.00 | 4.17 | -0.17 | 1.38 | 1.55 |
| 85.05 | 4.20 | -0.30 | 1.47 | 1.77 |
| 86.10 | 4.31 | -0.22 | 1.38 | 1.60 |
| 87.15 | 4.34 | -0.18 | 1.60 | 1.78 |
| 88.20 | 4.37 | -0.19 | 1.78 | 1.97 |
| 89.25 | 4.40 | 0.01 | 1.67 | 1.66 |
| 90.30 | 4.26 | -0.06 | 1.53 | 1.59 |
| 91.35 | 4.34 | -0.16 | 1.52 | 1.68 |
| 92.40 | 4.33 | 0.00 | 1.62 | 1.62 |
| 93.45 | 4.20 | 0.05 | 1.60 | 1.55 |
| 94.50 | 4.10 | 0.07 | 1.45 | 1.38 |
| 95.55 | 4.07 | 0.03 | 1.50 | 1.47 |
| 96.60 | 4.12 | 0.03 | 1.62 | 1.59 |
| 97.65 | 4.18 | 0.04 | 1.59 | 1.55 |
| 98.70 | 4.12 | -0.03 | 1.55 | 1.58 |
| 99.75 | 4.02 | 0.04 | 1.60 | 1.56 |
| 100.80 | 4.01 | -0.01 | 1.59 | 1.60 |
| 101.85 | 4.04 | -0.00 | 1.55 | 1.55 |
| 102.90 | 4.07 | -0.00 | 1.57 | 1.57 |

TABLE 1. (continued)

| Age kyr | $\delta^{18}$O Uvig. | $\delta^{13}$C Uvig. | $\delta^{13}$C N.dut. | $\Delta\delta^{13}$C |
|---------|-------|-------|-------|-------|
| 103.95 | 4.01 | -0.15 | 1.50 | 1.65 |
| 105.00 | 4.23 | -0.25 | 1.33 | 1.58 |
| 106.05 | 4.27 | -0.23 | 1.33 | 1.56 |
| 107.10 | 4.18 | -0.39 | 1.30 | 1.69 |
| 108.15 | 4.35 | -0.23 | 1.38 | 1.61 |
| 109.20 | 4.40 | -0.33 | 1.47 | 1.80 |
| 110.25 | 4.26 | -0.36 | 1.50 | 1.86 |
| 111.30 | 4.03 | -0.13 | 1.50 | 1.63 |
| 112.35 | 3.99 | -0.13 | 1.27 | 1.40 |
| 113.40 | 3.67 | -0.04 | 1.11 | 1.15 |
| 114.45 | 4.05 | 0.01 | 1.19 | 1.18 |
| 115.50 | 3.53 | -0.02 | 1.28 | 1.30 |
| 116.55 | 3.42 | 0.05 | 1.34 | 1.29 |
| 117.60 | 3.57 | 0.09 | 1.33 | 1.24 |
| 118.65 | 3.43 | 0.09 | 1.34 | 1.25 |
| 119.70 | 3.26 | -0.02 | 1.21 | 1.23 |
| 120.75 | 3.23 | 0.01 | 1.27 | 1.26 |
| 121.80 | 3.37 | 0.06 | 1.15 | 1.09 |
| 122.85 | 3.36 | -0.12 | 1.06 | 1.18 |
| 123.90 | 3.45 | 0.0 | 0.89 | 0.89 |
| 124.95 | 3.57 | -0.16 | 1.05 | 1.21 |
| 126.00 | 3.61 | -0.02 | 0.81 | 0.83 |
| 127.05 | 4.19 | -0.10 | 0.73 | 0.83 |
| 128.10 | 4.28 | -0.40 | 0.82 | 1.22 |
| 129.15 | 4.45 | -0.23 | 1.05 | 1.28 |
| 130.20 | 4.49 | -0.16 | 1.08 | 1.24 |
| 131.25 | 4.01 | -0.33 | 1.15 | 1.48 |
| 132.30 | 4.54 | -0.29 | 1.19 | 1.48 |
| 133.35 | 4.77 | -0.30 | 1.14 | 1.44 |
| 134.40 | 4.85 | -0.37 | 1.11 | 1.48 |
| 135.45 | 5.17 | -0.36 | 1.32 | 1.68 |
| 136.50 | 4.97 | -0.47 | 1.27 | 1.74 |
| 137.55 | 4.91 | -0.51 | 1.23 | 1.74 |
| 138.60 | 4.92 | -0.55 | 1.18 | 1.73 |
| 139.65 | 5.03 | -0.44 | 1.20 | 1.64 |
| 140.70 | 4.92 | -0.47 | 1.15 | 1.62 |
| 141.75 | 4.85 | -0.48 | 1.06 | 1.54 |
| 142.80 | 4.88 | -0.40 | 1.21 | 1.61 |
| 143.85 | 4.87 | -0.52 | 1.13 | 1.65 |
| 144.90 | 4.85 | -0.43 | 1.16 | 1.59 |
| 145.95 | 4.83 | -0.37 | 1.18 | 1.55 |
| 147.00 | 4.81 | -0.51 | 1.07 | 1.58 |
| 148.05 | 4.77 | -0.50 | 1.11 | 1.61 |
| 149.10 | 4.87 | -0.48 | 1.14 | 1.62 |
| 150.15 | 4.90 | -0.47 | 1.24 | 1.71 |
| 151.20 | 5.10 | -0.57 | 1.06 | 1.63 |
| 152.25 | 4.75 | -0.71 | 1.12 | 1.83 |
| 153.30 | 4.78 | -0.59 | 1.11 | 1.70 |
| 154.35 | 4.86 | -0.60 | 1.12 | 1.72 |
| 155.40 | 4.73 | -0.66 | 1.11 | 1.77 |
| 156.45 | 4.63 | -0.67 | 1.04 | 1.71 |
| 157.50 | 4.54 | -0.69 | 1.10 | 1.79 |

TABLE 1. (continued)

| Age kyr | $\delta^{18}$O Uvig. | $\delta^{13}$C Uvig | $\delta^{13}$C N.dut. | $\Delta\delta^{13}$C |
|---------|-------|-------|-------|-------|
| 158.55 | 4.51 | -0.61 | 1.22 | 1.83 |
| 159.60 | 4.85 | -0.51 | 1.15 | 1.66 |
| 160.65 | 4.81 | -0.49 | 1.11 | 1.60 |
| 161.70 | 4.55 | -0.68 | 1.09 | 1.77 |
| 162.75 | 4.60 | -0.67 | 1.19 | 1.86 |
| 163.80 | 4.66 | -0.66 | 1.08 | 1.74 |
| 164.85 | 4.63 | -0.68 | 1.14 | 1.82 |
| 165.90 | 4.61 | -0.70 | 1.18 | 1.88 |
| 166.95 | 4.52 | -0.64 | 1.17 | 1.81 |
| 168.00 | 4.42 | -0.58 | 1.13 | 1.71 |
| 169.05 | 4.37 | -0.53 | 1.14 | 1.67 |
| 170.10 | 4.37 | -0.56 | 1.25 | 1.81 |
| 171.15 | 4.37 | -0.59 | 1.19 | 1.78 |
| 172.20 | 4.32 | -0.51 | 1.17 | 1.68 |
| 173.25 | 4.51 | -0.48 | 1.22 | 1.70 |
| 174.30 | 4.55 | -0.51 | 1.04 | 1.55 |
| 175.35 | 4.58 | -0.61 | 1.17 | 1.78 |
| 176.40 | 4.56 | -0.60 | 1.19 | 1.79 |
| 177.45 | 4.70 | -0.58 | 1.16 | 1.74 |
| 178.50 | 4.55 | -0.62 | 1.06 | 1.68 |
| 179.55 | 4.48 | -0.56 | 1.08 | 1.64 |
| 180.60 | 4.51 | -0.54 | 1.05 | 1.59 |
| 181.65 | 4.60 | -0.58 | 1.09 | 1.67 |
| 182.70 | 4.67 | -0.54 | 1.29 | 1.83 |
| 183.75 | 4.71 | -0.49 | 1.25 | 1.74 |
| 184.80 | 4.64 | -0.50 | 1.35 | 1.85 |
| 185.85 | 4.61 | -0.58 | 1.39 | 1.97 |
| 186.90 | 4.67 | -0.31 | 1.41 | 1.72 |
| 187.95 | 4.49 | -0.33 | 1.34 | 1.67 |
| 189.00 | 4.43 | -0.23 | 1.36 | 1.59 |
| 190.05 | 4.38 | -0.21 | 1.04 | 1.25 |
| 191.10 | 4.13 | -0.19 | 1.28 | 1.47 |
| 192.15 | 4.05 | -0.22 | 1.40 | 1.62 |
| 193.20 | 4.02 | -0.14 | 1.50 | 1.64 |
| 194.25 | 4.02 | -0.02 | 1.40 | 1.42 |
| 195.30 | 3.95 | -0.06 | 1.56 | 1.62 |
| 196.35 | 3.88 | 0.18 | 1.42 | 1.24 |
| 197.40 | 3.68 | -0.03 | 1.60 | 1.63 |
| 198.45 | 3.85 | 0.05 | 1.19 | 1.14 |
| 199.50 | 4.25 | 0.18 | 1.32 | 1.14 |
| 200.55 | 4.18 | 0.19 | 1.32 | 1.13 |
| 201.60 | 3.99 | 0.09 | 1.50 | 1.41 |
| 202.65 | 4.00 | 0.03 | 1.31 | 1.28 |
| 203.70 | 3.94 | -0.02 | 1.42 | 1.44 |
| 204.75 | 3.90 | 0.02 | 1.49 | 1.47 |
| 205.80 | 3.86 | 0.05 | 1.44 | 1.39 |
| 206.85 | 3.77 | -0.09 | 1.51 | 1.60 |
| 207.90 | 3.83 | 0.00 | 1.46 | 1.46 |
| 208.95 | 3.75 | -0.02 | 1.71 | 1.73 |
| 210.00 | 3.71 | -0.01 | 1.36 | 1.37 |
| 211.05 | 3.48 | -0.14 | 1.44 | 1.58 |
| 212.10 | 3.76 | -0.12 | 1.60 | 1.72 |

TABLE 1.    (continued)

| Age kyr | $\delta^{18}O$ Uvig. | $\delta^{13}C$ Uvig. | $\delta^{13}C$ N.dut. | $\Delta\delta^{13}C$ |
|---------|------|------|------|------|
| 213.15 | 3.74 | −0.14 | 1.55 | 1.69 |
| 214.20 | 3.80 | −0.21 | 1.37 | 1.58 |
| 215.25 | 3.89 | −0.17 | 1.38 | 1.55 |
| 216.30 | 3.97 | −0.24 | 1.17 | 1.41 |
| 217.35 | 4.02 | 0.02 | 1.10 | 1.08 |
| 218.40 | 3.97 | −0.15 | 1.14 | 1.29 |
| 219.45 | 4.10 | −0.30 | 1.24 | 1.54 |
| 220.50 | 4.41 | −0.42 | 1.25 | 1.67 |
| 221.55 | 4.58 | −0.48 | 1.17 | 1.65 |
| 222.60 | 4.60 | −0.46 | 1.15 | 1.61 |
| 223.65 | 4.70 | −0.45 | 1.16 | 1.61 |
| 224.70 | 4.67 | −0.39 | 1.20 | 1.59 |
| 225.75 | 4.62 | −0.40 | 1.38 | 1.78 |
| 226.80 | 4.53 | −0.47 | 1.38 | 1.85 |
| 227.85 | 4.51 | −0.36 | 1.36 | 1.72 |
| 228.90 | 4.50 | −0.47 | 1.21 | 1.68 |
| 229.95 | 4.49 | −0.52 | 1.15 | 1.67 |
| 231.00 | 4.46 | −0.34 | 1.33 | 1.67 |
| 232.05 | 4.42 | −0.09 | 1.38 | 1.47 |
| 233.10 | 4.29 | −0.22 | 1.37 | 1.59 |
| 234.15 | 4.03 | −0.18 | 1.34 | 1.52 |
| 235.20 | 3.91 | −0.13 | 1.43 | 1.56 |
| 236.25 | 3.74 | −0.05 | 1.41 | 1.46 |
| 237.30 | 3.62 | −0.18 | 1.34 | 1.52 |
| 238.35 | 3.67 | −0.31 | 1.17 | 1.48 |
| 239.40 | 3.98 | −0.17 | 1.13 | 1.30 |
| 240.45 | 3.80 | −0.41 | 1.07 | 1.48 |
| 241.50 | 4.00 | −0.33 | 1.10 | 1.43 |
| 242.55 | 4.35 | −0.33 | 1.01 | 1.34 |
| 243.60 | 4.51 | −0.30 | 1.14 | 1.44 |
| 244.65 | 4.59 | −0.38 | 1.18 | 1.56 |
| 245.70 | 4.80 | −0.48 | 1.50 | 1.98 |
| 246.75 | 4.81 | −0.61 | 1.23 | 1.84 |
| 247.80 | 4.79 | −0.62 | 1.36 | 1.98 |
| 248.85 | 4.85 | −0.53 | 1.38 | 1.91 |
| 249.90 | 4.75 | −0.70 | 1.31 | 2.01 |
| 250.95 | 4.83 | −0.63 | 1.32 | 1.95 |
| 252.00 | 4.66 | −0.64 | 0.96 | 1.60 |
| 253.05 | 4.65 | −0.81 | 1.06 | 1.87 |
| 254.10 | 4.73 | −0.70 | 1.04 | 1.74 |
| 255.15 | 4.72 | −0.70 | 1.06 | 1.76 |
| 256.20 | 4.70 | −0.71 | 1.18 | 1.89 |
| 257.25 | 4.69 | −0.80 | 0.96 | 1.76 |
| 258.30 | 4.65 | −0.86 | 0.97 | 1.83 |
| 259.35 | 4.64 | −0.85 | 1.04 | 1.89 |
| 260.40 | 4.69 | −0.79 | 1.13 | 1.92 |
| 261.45 | 4.74 | −0.92 | 0.96 | 1.88 |
| 262.50 | 4.71 | −0.94 | 1.10 | 2.04 |
| 263.55 | 4.72 | −0.93 | 1.02 | 1.95 |
| 264.60 | 4.72 | −1.00 | 1.21 | 2.21 |
| 265.65 | 4.67 | −0.86 | 1.23 | 2.09 |

TABLE 1.    (continued)

| Age kyr | $\delta^{18}O$ Uvig. | $\delta^{13}C$ Uvig. | $\delta^{13}C$ N.dut. | $\Delta\delta^{13}C$ |
|---------|------|------|------|------|
| 266.70 | 4.76 | −0.82 | 1.18 | 2.00 |
| 267.75 | 4.67 | −0.80 | 1.36 | 2.16 |
| 268.80 | 4.79 | −0.76 | 1.45 | 2.21 |
| 269.85 | 4.70 | −0.61 | 1.12 | 1.73 |
| 270.90 | 4.60 | −0.60 | 1.28 | 1.88 |
| 271.95 | 4.50 | −0.59 | 1.37 | 1.96 |
| 273.00 | 4.51 | −0.76 | 1.29 | 2.05 |
| 274.05 | 4.51 | −0.78 | 1.40 | 2.18 |
| 275.10 | 4.61 | −0.78 | 1.44 | 2.22 |
| 276.15 | 4.61 | −0.76 | 1.45 | 2.21 |
| 277.20 | 4.45 | −0.81 | 1.25 | 2.06 |
| 278.25 | 4.46 | −0.67 | 1.26 | 1.93 |
| 279.30 | 4.25 | −0.60 | 1.21 | 1.81 |
| 280.35 | 4.29 | −0.59 | 1.39 | 1.98 |
| 281.40 | 4.10 | −0.43 | 1.43 | 1.86 |
| 282.45 | 4.02 | −0.36 | 1.46 | 1.82 |
| 283.50 | 3.94 | −0.38 | 1.52 | 1.90 |
| 284.55 | 3.93 | −0.47 | 1.19 | 1.66 |
| 285.60 | 4.01 | −0.48 | 1.28 | 1.76 |
| 286.65 | 4.02 | −0.37 | 1.20 | 1.57 |
| 287.70 | 3.89 | −0.36 | 1.22 | 1.58 |
| 288.75 | 3.91 | −0.38 | 1.27 | 1.65 |
| 289.80 | 4.03 | −0.39 | 1.19 | 1.58 |
| 290.85 | 4.21 | −0.47 | 1.44 | 1.91 |
| 291.90 | 4.33 | −0.57 | 1.37 | 1.94 |
| 292.95 | 4.29 | −0.56 | 1.46 | 2.02 |
| 294.00 | 4.35 | −0.51 | 1.46 | 1.97 |
| 295.05 | 4.40 | −0.64 | 1.35 | 1.99 |
| 296.10 | 4.41 | −0.52 | 1.26 | 1.78 |
| 297.15 | 4.47 | −0.46 | 1.25 | 1.71 |
| 298.20 | 4.46 | −0.41 | 1.31 | 1.72 |
| 299.25 | 4.48 | −0.39 | 1.29 | 1.68 |
| 300.30 | 4.46 | −0.40 | 1.32 | 1.72 |
| 301.35 | 4.28 | −0.53 | 1.58 | 2.11 |
| 302.40 | 4.24 | −0.52 | 1.50 | 2.02 |
| 303.45 | 4.12 | −0.47 | 1.58 | 2.05 |
| 304.50 | 4.11 | −0.51 | 1.48 | 1.99 |
| 305.55 | 4.17 | −0.50 | 1.62 | 2.12 |
| 306.60 | 4.17 | −0.36 | 1.88 | 2.24 |
| 307.65 | 4.09 | −0.25 | 1.70 | 1.95 |
| 308.70 | 4.14 | −0.37 | 1.70 | 2.07 |
| 309.75 | 3.91 | −0.42 | 1.53 | 1.95 |
| 310.80 | 3.95 | −0.24 | 1.45 | 1.69 |
| 311.85 | 3.87 | −0.35 | 1.28 | 1.63 |
| 312.90 | 4.07 | −0.28 | 1.43 | 1.71 |
| 313.95 | 4.04 | −0.24 | 1.56 | 1.80 |
| 315.00 | 3.94 | −0.25 | 1.42 | 1.67 |
| 316.05 | 4.03 | −0.34 | 1.34 | 1.68 |
| 317.10 | 4.11 | −0.29 | 1.34 | 1.63 |
| 318.15 | 4.01 | −0.23 | 1.42 | 1.65 |
| 319.20 | 3.91 | −0.25 | 1.21 | 1.46 |

TABLE 1.  (continued)

| Age kyr | $\delta^{18}O$ Uvig. | $\delta^{13}C$ Uvig. | $\delta^{13}C$ N.dut. | $\Delta\delta^{13}C$ |
|---------|------|-------|-------|------|
| 320.25 | 3.76 | -0.33 | 1.31 | 1.64 |
| 321.30 | 3.64 | -0.07 | 1.31 | 1.38 |
| 322.35 | 3.58 | -0.18 | 1.31 | 1.49 |
| 323.40 | 3.50 | -0.17 | 1.35 | 1.52 |
| 324.45 | 3.45 | -0.08 | 1.24 | 1.32 |
| 325.50 | 3.46 | 0.02 | 1.28 | 1.26 |
| 326.55 | 3.29 | -0.11 | 1.30 | 1.41 |
| 327.60 | 3.32 | -0.11 | 1.34 | 1.45 |
| 328.65 | 3.34 | -0.06 | 1.10 | 1.16 |
| 329.70 | 3.43 | -0.28 | 1.03 | 1.31 |
| 330.75 | 3.35 | -0.32 | 0.93 | 1.25 |
| 331.80 | 3.37 | -0.16 | 0.95 | 1.11 |
| 332.85 | 3.54 | -0.19 | 0.98 | 1.17 |
| 333.90 | 3.80 | -0.34 | 0.99 | 1.33 |
| 334.95 | 3.87 | -0.34 | 0.85 | 1.19 |
| 336.00 | 4.04 | -0.41 | 0.84 | 1.25 |
| 337.05 | 4.23 | -0.37 | 0.91 | 1.28 |
| 338.10 | 4.28 | -0.25 | 0.82 | 1.07 |
| 339.15 | 4.35 | -0.30 | 0.71 | 1.01 |
| 340.20 | 4.46 | -0.39 | 0.89 | 1.28 |
| 341.25 | 4.53 | -0.39 | 0.90 | 1.29 |
| 342.30 | 4.68 | -0.42 | 0.95 | 1.37 |

*For the 355-425 micron fraction where available. When this size was not available, measurements of other sizes were normalized using Table 2.

isotopic analysis; the majority also contained sufficient of the planktonic Neogloboquadrina dutertrei for an analysis to be made of specimens in a controlled size range (failure to control the size of planktonic species analyzed leads to unacceptably large random isotopic variations, especially for $^{13}C$). Since it was not possible to select the preferred size range, 355-425 microns, in every sample, a size-dependent adjustment was made for analyses made in other size fractions in order to provide a time series. Time series of isotopic data are listed in Table 1; the raw data will be published elsewhere and are available from the first author.  Table 2 gives the adjustment factors applied for different size fractions. The deep-water $^{13}C$ time series are shown with the addition of 0.90 per mil to the Uvigerina measurements; this provides values which approximate the $^{13}C$ content of dissolved CO$_2$ (Graham et al., 1981; Duplessy et al., 1984).

No adjustment has been made to the planktonic $^{13}C$ data, which are probably also isotopically light with respect to $^{13}C$ values in surface waters.

## Time Series Development

In order to display the various data sets as time series, we worked with the benthic $^{18}O$ record. We have made use of the time scale developed by D. G. Martinson et al. (unpublished manuscript, 1984) for the interval 280 kyr BP to present, as well as direct correlation to radiometriclally dated marine terraces (Bloom et al., 1974; Chappell and Veeh, 1978). We used the time scale of Imbrie et al., (1984) for the earlier part of our record to 340 kyr BP. The first of these time scales was developed using a stacked and averaged suite of oxygen isotope records obtained from benthic foraminifera which was derived by Pisias et al., (1984); transfer of this time scale proved very straightforward, which is perhaps not surprising, as the suite of cores used by Pisias et al. includes core V19-29, close to core V19-30, as an important ingredient. The longer time scale of Imbrie et al., (1984) is also based on a stacked and averaged set of oxygen isotope records, in this case from planktonic foraminifera and necessarily based on cores with lower accumulation rates and hence poorer stratigraphic resolutions. The age control points that we have used are listed in Table 3, from which it may be seen that these control points are distributed at approximately 10-kyr intervals. In the Panama Basin area it is well established that accumulation rates are strongly modulated by climatic change (Arrhenius, 1952; Pedersen, 1983; Boyle, 1983), and we have sought to control the time scale at points of rapid change of climate (as indicated by the $^{18}O$ record) in the hope that these also represent points at which accumulation did in fact change. However, we attempted at the same time to keep to a minimum the number of control points used so as

TABLE 2. Adjustment Factors Used to Normalise $^{13}C$ Analyses of Neogloboquadrina dutertrei to Values for the 355-425 micron Size Range.

| Size Range, microns | Adjustment per mil |
|---------------------|--------------------|
| 500-425 | -0.27 |
| 415-355 | 0.0 |
| 355-300 | 0.27 |
| 300-255 | 0.51 |

TABLE 3.   Age Model Used in Figure 2
and in Spectral Analysis

| Depth, m | Age, kyr |
|----------|----------|
| 0.0 | 0.0 |
| 0.3225 | 6.09 |
| 0.6675 | 9.935 |
| 1.16 | 15.0 |
| 2.795 | 28.0 |
| 3.365 | 40.0 |
| 4.1 | 53.0 |
| 4.8 | 60.4 |
| 5.74 | 71.6 |
| 6.27 | 85.0 |
| 6.49 | 93.3 |
| 6.72 | 102.0 |
| 7.35 | 115.0 |
| 7.74 | 126.2 |
| 8.25 | 135.2 |
| 9.45 | 151.0 |
| 11.34 | 190.0 |
| 11.75 | 198.5 |
| 12.34 | 220.0 |
| (ash layer L, 12.69-12.82 m) | |
| 12.82 | 233.0 |
| 13.16 | 243.0 |
| 14.83 | 279.0 |
| 15.49 | 291.0 |
| 17.64 | 335.0 |
| 18.04 | 339.0 |

to avoid affecting the high-frequency characteristics of the record. In fact, the density of control points used is small compared with the number used in creating the averaged records on which the time scale is based. In one section below we show a result that was obtained using a slightly more complex time scale. The underlying assumption of this refined time scale is that it may be preferable to assume a uniform accumulation of noncarbonate between age control points rather than simply assume a uniform total sedimentation rate between the same points. This would be a considerable improvement if it were the case that carbonate dissolution were the dominant cause of changes in carbonate accumulation rate; in reality, changes in productivity, which also affected the substantial flux of biogenic silica, intervened as well (Boyle, 1983), so that the model represents only a minor improvement.

The basis of the time scales developed by D. G. Martinson et al. (unpublished manuscript, 1984) and Imbrie et al., (1984) is that Pleistocene changes in global ice volume have been controlled by changes in the earth's orbital geometry (Hays et al., 1976). The result of applying these time scales to the present core, a data set not available to Martinson et al. (unpublishd manuscript, 1984) or Imbrie et al.

(1984) is a striking vindication of this approach; all the records shown in Figure 1, as well as $^{18}O$ in N. dutertrei (not shown here) and combinations of the data sets, show striking concentrations of variance in the bandwidth of the orbital forcing functions as well as coherencies with the orbital elements that are in some cases significantly higher than given by any core record in the data set used by Imbrie et al. (1984). This improvement stems, we believe, from the closer sampling interval and significantly improved analytical accuracy in the present data set.

Finally, in order to perform various statistical tests it is necessary to develop time series with equally spaced sample intervals in time. We have linearly interpolated the raw data sets at 0.35-kyr intervals for the present study; this is less than the actual sampling interval in a few short sections of the core (in the lowest accumulation-rate section from 6.49 to 6.72 m the sampling interval of 3 cm would represent about 1.2 kyr), and greater than the actual sampling interval in other parts (between 1.16 and 2.80 m, our 3 cm sampling interval represents about 0.25 kyr). For numerical convenience, statistical tests have been conducted on a subset consisting of every third interpolated data point (i.e., 326

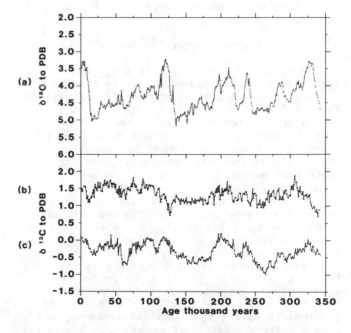

Fig. 1.   (a) $^{18}O$ record in Uvigerina senticosa from V19-30, per mil to PDB.   (b) $^{13}C$ record in Neogloboquadrina dutertrei, per mil to PDB, normalized to 355-425 micron fraction.   (c) $^{13}C$ record in Uvigerina senticosa from V19-30, per mil to PDB plus 0.9 to scale for ocean total $CO_2$. Figures 1a-c are shown for 326 data points at 1.05-kyr intervals, data from Table 1 and time scale from Table 3..

values spaced every 1.05 kyr). Critical tests have been replicated using the two other sets of 326 values to provide three estimates that are in part independent. Figure 1 shows the subset of 326 data points on which the majority of the statistical tests were performed, for each variable used. Table 1 shows this subset of the interpolated data.

## Statistical Techniques

Standard time series analysis procedures as outlined by Jenkins and Watts (1968) have been used to estimate variance distribution as a function of frequency as well as coherency and phase relationships between records. All the calculations in this paper were made using time series with 326 data points; spectral and cross spectral estimates were made using 120 lags of the autocovariance and cross-covariance functions. Confidence limits shown in spectral and coherency plots are for the 80% level, and in the phase plots at the 95% level. In those figures which show phase estimates, confidence limits are only shown for phase estimates in frequency bands over which the signals are coherent. The nonzero 80% confidence level for coherency estimates is based on the statistical analysis of Julian (1975); all other confidence limits are as defined by Jenkins and Watts (1968).

## The Deep-Water $^{18}O$ Record

The record of $^{18}O$ changes in _Uvigerina senticosa_ shown in Figure 1a closely resembles the best previously available records. However, the sampling resolution is unusually good for a record of this length. We will show elsewhere that this unusual resolution may be exploited to learn more of the operation of the mechanism by which climate changes; here it is sufficient to state that we propose to use the $^{18}O$ record as a monitor of global ice volume, and hence also as an approximate monitor of sea level change, as has become standard practice. Mix and Ruddiman (1984) have discussed the extent to which the slow turnover of ice within major ice sheets may lead to a lag between the true ice-volume record and its $^{18}O$ proxy. They conclude that a lag in the range 1-3 kyr is possible. We prefer the lower end of this range for the reason that factors that increase this lag also exacerbate problems in making glaciological sense of the oxygen isotope record, for example, by increasing the rate at which ice would be postulated to accumulate after the last interglacial.

In Figure 2 the variance distribution of the $^{18}O_{ben}$ record is shown on a logarithmic scale; this spectrum resembles those previously published for $^{18}O$ records covering this time interval (Hays et al., 1976). The variance distribution is given in Table 4, which may be regarded as a good estimate of the $^{18}O$ variance

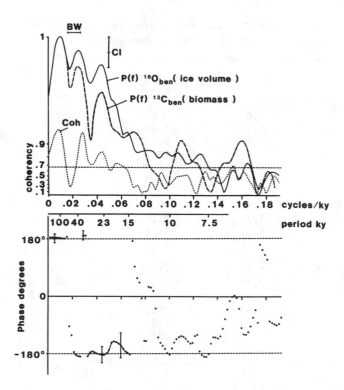

Fig. 2. Log spectra and coherence and phase relationships between $^{13}C_{ben}$ and $^{18}O_{ben}$ in core V19-30. The two records are coherent in all orbital bandwidths and in antiphase. Thus there is no significant lead or lag between positive $^{18}O$ (high ice volume) and negative $^{13}C$ (low continental biomass). In the top part of the figure, coherency estimates which lie above the dashed line are significant at the 20% level. The 80% confidence interval (CI) is shown for the variance estimates of each time series. In the bottom part, the 95% confidence interval is shown only for phase estimates in those portions of the spectra in which there is significant coherency between the two records.

associated with the bandwidth of each of the known orbital forcing elements: eccentricity, tilt, and precession.

## The Deep-Water $^{13}C$ Record

The deep-water $^{13}C$ record shown in Fugure 1c is obtained after adding 0.9 per mil to the analytical values for _Uvigerina senticosa_ to take account of isotopic deviation from the $^{13}$ content of the dissolved $CO_2$ (Graham et al., 1981; Duplessy et al., 1984). Only a small number of detailed deep-water $^{13}C$ records have been published to date. Shackleton et al., (1983b) have already discussed the upper part of the V19-30 record, comparing it with the North Atlantic record published by Shackleton (1977). They concluded by examining the two $^{13}C$ records that

TABLE 4. Summarized Variance Distributions
Expressed as Equivalent Amplitude

| | Bandwidth | | |
| Variable | 100 kyr | 40 kyr | 23 kyr |
| --- | --- | --- | --- |
| $^{13}C_{ben}$ | 0.35 | 0.24 | 0.13 |
| $^{13}C_{plk}$ | 0.25 | 0.19 | 0.14 |
| $^{13}C_{plk-ben}$ | 0.35 | 0.25 | 0.15 |
| $^{18}O_{ben}$ | 0.54 | 0.46 | 0.37 |

Equivalent amplitude (twice square root of
variance) is given in per mil.

one can distinguish two separate processes: a
globally operative change in the mean $^{13}C$ of
total dissolved $CO_2$, and changes in the between-
ocean $^{13}C$ gradient. Curry and Lohmann (1982) have
also demonstrated changes in local $^{13}C$ gradients.
Shackleton et al., (1983b) argued that the
location of V19-30 would be a reasonable single
site at which to monitor the global deep-water
$^{13}C$ signal. To date we have been unable to
identify a better locality, although we have been
able to obtain confirmatory data for parts of the
record from other areas. In the future it would
be desirable both to obtain records from other
parts of the Pacific and to obtain a weighted
mean global record taking account of those North
Atlantic (and perhaps North Pacific) water masses
that underwent more dramatic $^{13}C$ fluctuations
than are witnessed in V19-30. For the present, we
make the assumption that such a summation would
not greatly affect our conclusions.

Figure 2 compares the variance spectrum of
this $^{13}C$ record with that of $^{18}O$ in the same
benthic specimens, and shows their coherency and
phase relationships. The $^{13}C$ record shows
variance at all the orbital frequencies and high
coherency with $^{18}O$ in these frequency bands;
the phase plot shows that there is no significant
lead or lag between the $^{18}O_{ben}$ record (ice
volume) and the $^{13}C_{ben}$ record (carbon in
continental biomass and/or shelf sediments).
Since the residence time of carbon in the
biosphere is quite short (less than 1 kyr) a very
small phase lag between climate and continental
biomass is to be expected. On the other hand, if
the chief reservoir for the isotopically light
carbon removed from and returned to the ocean
were continental shelf sediments, one might
expect a significant lag; one might expect the
maximum rate of extraction of carbon to occur at
times when the greatest area of the continental

shelves was covered by water rather than at times
of most rapid sea level rise. Thus we consider
that the chief contribution to the variations in
$^{13}C$ shown in Figure 1c and summarized in Table 4
may be changes in the continental biosphere as
proposed by Shackleton (1977) rather than changes
in the mass of organic-rich sediments stored on
the shelves as suggested by Broecker (1982),
following a suggestion by N. Niitsuma and T.-L
Ku.

The Surface Water $^{13}C$ Record

It is crucial to the arguments developed here
that the $^{13}C$ record that we have derived from the
analysis of Neogloboquadrina dutertrei and shown
in Figure 1b is also a reasonable depiction of
global changes in the $^{13}C$ content of ocean
surface waters. Broecker (1982) showed low-
resolution $^{13}C$ records from planktonic
foraminifera that were rather erratic but argued
that there was no systematic difference between
glacial and interglacial values. The only data
set available to Broecker (1982) that was derived
from size-controlled planktonic specimens was
from western equatorial Pacific core RC17-177
(data provided to Broecker by Shackleton). Given
the very poor stratigraphic resolution of this
core, the data are consistent with the high-
resolution record shown in Figure 1b. The $^{13}C$
minimum at about 12 kyr BP, missed in the 10-cm
sampling of core RC17-177, is present in 5-cm
sampling of nearby core V28-235 (unpublished
measurements, 1984).

Figure 3 shows data from cores RC11-120 and
E49-18 in the subantarctic Indian Ocean (Hays et
al., 1976, also unpublished data, 1981); although
this is not an ideal area from which to derive a
surface water $^{13}C$ record, it does in fact provide

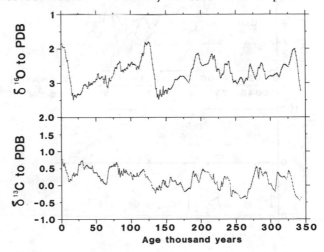

Fig. 3a. $^{18}O$ and $^{13}C$ records in Globigerina
bulloides from cores RC11-120 and E49-18 (using
the patch approach of Hays et al., 1976, but
using the entire RC11-120 record and only a short
section of E49-18).

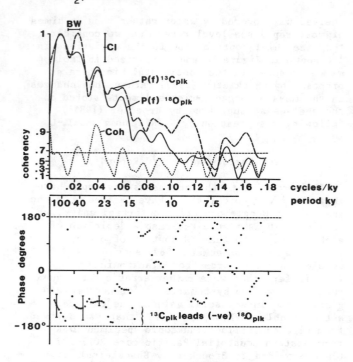

Fig. 3b. Log spectra and coherency and phase plot for $^{13}$C and $^{18}$O in subantarctic G. bulloides from Figure 3a. Symbols as in Figure 2.

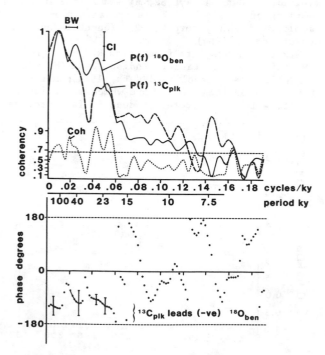

Fig. 3c. Log spectra and coherency and phase plot for $^{13}$C$_{plk}$ and $^{18}$O$_{ben}$ in core V19-30 (East Pacific) from Figure 1. Symbols as in Figure 2. The phase plots in figures 3b and 3c indicate that in both areas the surface water $^{13}$C record has the same phase relationship with ice volume.

a remarkably similar record to V19-30; over the orbital frequency band the phase relationship between $^{13}$C and $^{18}$O in this record is statistically indistinguishable from the analogous phase relationship for core V19-30 (Figure 3).

On the basis of these data, we suggest that the record of N. dutertrei in V19-30 is in fact a fair approximation to a global surface water $^{13}$C record, and thus by subtracting from these the values obtained from U. senticosa we can estimate the varying surface-to-deep water $^{13}$C difference ($^{13}$C$_{plk-ben}$) in the ocean. Figure 4 shows the record of changing $^{13}$C difference between N. dutertrei and U. senticosa in V19-30. In the future, it will perhaps be appropriate to develop an averaged surface water $^{13}$C record using cores from several regions so as to minimize local effects, but at present this would only lead to a degradation of the V19-30 record, which has very good stratigraphic resolution. Also a very high degree of accuracy in stratigraphic correlation would be needed to obtain a difference record using a pair of cores, whereas this problem does not arise if the two data sets are from the same core. Table 4 contains the variance data for $^{13}$C in N. dutertrei. It may be seen that the variance is broadly similar to the variance in the benthic $^{13}$C data; however, the time series are significantly different (Figure 1) as a result of the appreciable phase lag at all frequencies.

It should be noticed here that we propose to use two measurable variables ($^{13}$C$_{ben}$, $^{13}$C$_{plk}$) to estimate two oceanographically meaningful variables ($^{13}$C of mean ocean CO$_2$, mean $^{13}$C enrichment of surface water with reference to deep water). Since variations in $^{13}$C$_{plk}$ in the ocean arise from a combination of these two oceanographic parameters, we are not discouraged by the observation that the $^{13}$C$_{plk}$ data set does not show such a clear variance spectrum as $^{13}$C$_{ben}$ does (Figure 5), nor such high coherencies with orbital elements.

The Atmospheric Carbon Dioxide Record

Figure 4 shows, using scale 1, variations in the vertical $^{13}$C difference between the surface ocean and the deep ocean. Three additional scales have been attached to the figure. First, scale 2 shows the changes in atmospheric CO$_2$ concentration predicted by scale 1. This estimate is based on Broecker's (1982) Table 3; it assumes that changes in the $^{13}$C difference measure changes in the amount of carbon extracted as organic matter, with an unchanging ratio of organic carbon to carbonate carbon (in Table 3 of Broecker (1982), this scenario is found as change in PO$_4$, holding other variables constant; this is the only one of Broecker's scenarios that does significantly change the carbon isotope gradient). As Shackleton et al., (1983a) already noted, this shows a similar range to that observed in ice cores (Neftel et al., 1982),

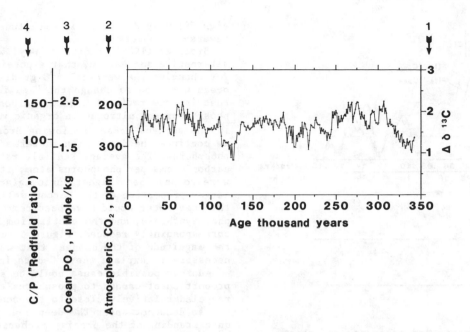

Fig. 4.    Surface-to-deep water $^{13}C$ difference obtained by subtracting data in Figures 1c and 1b, shown at a 1.05 kyr interpolation interval. Scales are shown (1) in per mil (by subtraction); (2) in parts per million by volume atmospheric $CO_2$ concentration implied by scale 1; (3) in mean phosphorus content of ocean deep water that would be required to produce scale 1; (4) in mean Redfield ratio that would be required to explain scale 1(a) (note that scales 3 and 4 are presented as alternatives).

supporting the notion that this is the primary mechanism giving rise to atmospheric $CO_2$ variations. Scale 3 shows the changes in the phosphorus content (micromoles per kilogram) of upwelling deep water that would be required to explain scale 1, again from Table 3 in Broecker (1982). Scale 4 shows as an alternative explanation, changes in the Redfield Ratio (the ratio of C to P in organic matter leaving surface water and oxidizing in deep water) that would be required to explain (a). Broecker (1982) included this alternative only as a footnote, after a suggestion made to him by P. K. Weyl; the mechanism makes each phosphorus atom more effective, rather than increasing the number of phosphorus atoms, in order to reduce atmospheric $CO_2$.

As Shackleton et al., (1983a) have already pointed out with reference to a part of the data shown in Figure 4, the scale of change in atmospheric $CO_2$ concentration implied by the $^{13}C$ data is entirely consistent with the independent data provided by Neftel et al., (1982). The extended record now available supports the existence of a systematic relationship between atmospheric $CO_2$ and climate similar to that observed for the last glacial cycle. Our objective in the remainder of this paper is to evaluate, so far as the present data permit, both the origin of these variations and the part that

they may have played in causing climatic change.

Broecker (1982) argued that a plausible mechanism for changing the phosphorus content of upwelling oceanic deep water would be storage of phosphorus in sediments on the continental shelves as these were flooded. This mechanism was investigated in more detail by Keir and Berger (1983), who modeled the system on the assumption that the rate of phosphorus extraction was proportional to the rate of sea level change, which they represented by the time derivative of the oxygen isotope record. In order to test this model, we have examined the phase relationship between $^{18}O_{ben}$ (proxy for sea level) and $^{13}C_{plk-ben}$ (proxy for ocean phosphorus). Figure 6 shows coherency and phase plots comparing $^{18}O_{ben}$ and $^{13}C_{plk-ben}$ (ice volume and atmospheric $CO_2$). The two spectra are remarkably similar and the coherency estimates massively significant over the whole orbital bandwidth. It may be seen from the phase plot that the data are not consistent with a model in which sea level change causes phosphorus extraction. Instead, the phase relationship shows that change in $^{13}C_{plk-ben}$ leads change in ice volume. This suggests that atmospheric carbon dioxide concentration may be in some manner implicated in the forcing of ice volume changes. We also compared the time derivatives of these two quantities, so as to test the model of Keir and Berger (1983) in

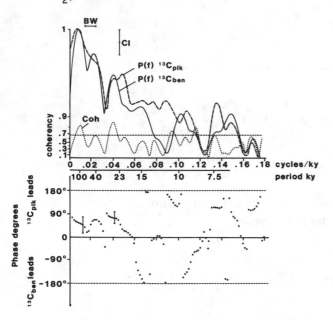

Fig. 5.   Log spectra and coherency and phase plots showing the relationship between $^{13}$C in Uvigerina and in N. dutertrei (Figure 1). Symbols as in Figure 2.

precisely their form; as might be expected, the result is essentially identical to that shown in Figure 6. The true lag between CO$_2$ and sea level may have been slightly overestimated, if we take account of the factors evaluated by Mix and Ruddiman (1984) as mentioned previously.

In order to confirm this important result we estimated the phase lag between $^{13}$C$_{plk-ben}$ and $^{18}$O$_{ben}$ over each of the three orbital bandwidths using each of the three partly independent data sets. The mean overall lag is about 2.5 kyr; estimated separately in each orbital band, one obtains (10±6 kyr at 100 kyr; 2.7±2.6 kyr at 40 kyr; 1.8±1.7 kyr at 23 kyr and 4.5±1.6 kyr at 19 kyr (all 95% confidence limits). We also estimated the phase lag between orbital forcing (represented by June insolation at 65°N) and atmospheric CO$_2$. These estimates show that atmospheric CO$_2$ variations lag the solar insolation forcing, and lead ice volume response, by approximately equal intervals, over all the orbital forcing frequencies. We conclude from this discussion that neither the model proposed by Broecker (1982) nor the differently formulated version investigated by Keir and Berger (1983) has operated over the past 300 kyr in the manner that they envisaged. In fact, the phase relationship between the surface water and deep water $^{13}$C records is itself a demonstration of this; Broecker (1982) envisaged phosphorus as being extracted from the ocean to the shelves together with carbon. If this were the case the two $^{13}$C records would be in phase but

significantly different in amplitude, whereas the reverse is observed.

Broecker (1982) credits P. Weyl with an alternative suggestion, that a possible mechanism for changing the vertical $^{13}$C gradient in the ocean would be to change the "Redfield Ratio", that is, the ratio of organic carbon to phosphorus and nitrogen in organic matter formed in the surface ocean. Following Broecker (1982) we continue the discussion in terms of phosphorus. The average Redfield ratio is 106 carbon atoms per phosphorus atom; if this number were to have been higher during glacial times, a larger proportion of the carbon welling up from the deep water would have been removed by photosynthesis, and the equilibrium CO$_2$ pressure correspondingly reduced. Figure 4 scale 4 shows the magnitude of the changes that would be necessary to explain the $^{13}$C data in this way.

Another possible cause could be sought in the proportion of carbon to phosphorus that is recycled relatively close to the ocean surface and so does not enter the deep sea. A better understanding of the precise mechanism will probably require a more detailed evaluation of the cycling of phosphorus in today's ocean, perhaps by means of sediment trap data. However, our data do provide support for a controlling mechanism that is forced by orbital variations in

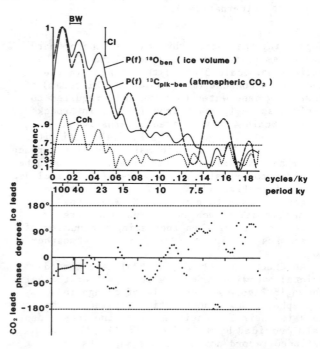

Fig. 6. Log spectra and coherency and phase plots showing the relationship between $^{18}$O$_{ben}$ and $^{13}$C$_{plk-ben}$. Symbols as in Figure 2. The phase plot shows that $^{13}$C$_{plk-ben}$ (atmospheric CO$_2$) leads $^{18}$O$_{ben}$ (ice volume) over the whole of the orbital frequency bandwidth.

a manner that is independent of ice sheet growth.

We examined the time differentials of $^{18}O_{ben}$ and of $^{13}C_{plk-ben}$ in order to test the model of Keir and Berger (1983). Examining the time differentials has the effect of clarifying the display of the higher-frequency components in the record. Here the two records are significantly different; in particular, the $^{13}C$ record shows a spectral peak at a period of about 14 kyr that is not evident in the $^{18}O$ record. In order to investigate the significance of this peak we examine the record using a slightly more sophisticated time scale in which it is assumed that total noncarbonate, instead of total sediment, accumulates uniformly between the control points in Table 4. To obtain time derivatives, the data were first smoothed with a low-pass filter (half-amplitude cutoff frequency 0.16 cycles/kyr); the derivative at each point was estimated from the slope between the two adjacent values. The refined time scale, analyzed in Figure 6, slightly enhances the spectral peak near 13 ky and emphasizes its separation from that in the $^{18}O$ record at 14.7 kyr, which is probably a combination ("difference tone") of tilt and precession $(1/23-1/40)^{-1}$ kyr. In fact, the frequency of this spectral peak in the $^{13}C_{plk-ben}$ record corresponds to a third harmonic of the tilt frequency, suggesting that a nonlinear response to an insolation signature dominated by tilt (i.e., a high-latitude insolation pattern) is responsible for forcing the $CO_2$ signal. In support of this origin for the spectral peak near 13 kyr, we show (Figure 7) that this peak has very high coherency with a third harmonic of tilt, which we generated by producing $(tilt)^3$, and that there is a phase shift that is indistinguishable from 180° between the fundamental tilt record in the atmospheric $CO_2$ record and its third harmonic. Possibly, a significant part of the variance in the precessional bandwidth (23 kyr to 19 kyr periods) may derive from the second harmonic of the tilt period, as is suggested by the significant coherency estimate; it would be difficult to demonstrate this rigorously in such a short record, since this second harmonic lies between the two dominant peaks in the precession spectrum.

The signal that we are interpreting as a record of atmospheric carbon dioxide therefore has an origin at a latitude where tilt is relatively more important than precession, as compared with the ice volume record which has been most successfully modeled with a forcing function with a strong precession component. The mechanism by which tilt variations force $CO_2$ changes is obscure but clearly involves a nonlinear response (to explain the strong harmonic content shown in Figure 7) but with a shorter time constant than that of ice sheet growth, to explain the small lag between tilt and $CO_2$ that is also shown in Figure 7. Moreover $CO_2$ leads ice volume over the whole of the orbital

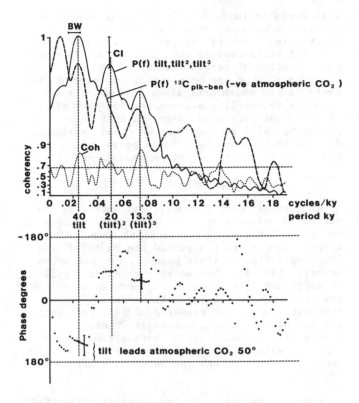

Fig. 7. Log spectra and coherency and phase plot showing the relationship between the time derivative of $^{13}C_{ben} - ^{13}C_{plk}$ and a record consisting of tilt together with its second and third harmonics. Symbols as in Figure 2.

bandwidth. For both of these reasons we conclude that atmospheric $CO_2$ is not forced by climate change or by associated sea level change, but is forced independently by high latitude insolation. Thus changes in atmospheric $CO_2$ may represent part of the mechanism by which orbital variations induce changes in climate.

We suggest, therefore, that either the effective Redfield ratio or the mode of cycling of phosphorus in the ocean is controlled by high-latitude solar insolation, probably through its effect on deep-water circulation. An interesting corollary of this finding is that it will almost certainly also affect the P/C ratio of the component that becomes incorporated in the sediment. This in turn will tend to change the P/C ratio in the entire ocean, at a rate determined by the oceanic residence times of C and P (of the order $10^5$ years); thus a low-frequency component may enter the atmospheric $CO_2$ system through this mechanism.

Phase plots show that over all orbital forcing periods, changes in atmospheric $CO_2$ lead changes in ice volume; that is, these changes must be regarded as contributing to the forcing of the cryosphere component of the climate record, rather than acting as a passive amplifier. This also seems to imply that even in the absence of ice volume changes, which may well enhance $CO_2$ fluctuations, orbital variations are capable of inducing variations of the order of 30% in the concentration of carbon dioxide in the atmosphere. This may be an important part of the mechanism by which orbital variations caused climatic variability without the feedback effect of ice sheet growth and decay. A test of this prediction will require that we obtain analogous data to those shown here for a period prior to the onset of large ice volume variations.

## Conclusions

Atmospheric carbon dioxide concentration has varied in response to orbital forcing. Changes in the tilt of the earth's rotational axis with respect to the orbit have had the major effect, implying a high-latitude control. The mechanism probably involves changes in the internal cycling of nutrients within the ocean, perhaps as a result of changes in ocean deep-water circulation, and not transfer of nutrients into and out of the oceanic reservoir. These atmospheric carbon dioxide changes contributed, together with direct insolation variations, to controlling the growth and decay of continental ice sheets.

Acknowledgments.   We are especially indebted to M. A. Hall for his care of the mass spectrometer and its operation, and to Mike Tabecki and Cang Shuxi for  help in processing the samples. Mark Broadbent saved us a great deal of time by his understanding of the idiosyncracies of the Cambridge University IBM computer. We are grateful to Wolfgang Berger, John Imbrie, Ted Moore, and Tom Pedersen for careful reviews of the manuscript. Laboratory work in Cambridge was supported by NERC grant GR3/3606. This study is a part of the SPECMAP project, supported by NSF grant ATM 80-19253. LDGO contribution 3679.

## References

Arrhenius, G., Sediment cores from the East Pacific. Rep. Swed. Deep Sea Exped. 5, 6-227, 1952.

Berger, A. L., The astronomical theory of paleoclimates, in Climatic Variations and Variability: Facts and Theory, edited by A. L. Berger, pp. 501-525, D. Reidel, Hingham, Mass., 1981.

Berger, A. L., Long-term variations of daily insolation and Quaternary climatic changes, J. Atmos. Sci., 35, 2362-2367, 1978.

Berger, W. H., Increase of carbon dioxide in the atmosphere during deglaciation: The coral reef hypothesis, Naturwissenschaften, 69, 87-88. 1982.

Berner, W., B. Stauffer, and H. Oeschger, Past atmospheric composition and climate, gas parameters measured in ice cores. Nature, 275, 53-55, 1978.

Bloom, A. L., W. S. Broecker, J. M. A. Chappell, R. K. Matthews, and K. J. Mesolella, Quaternary sea level fluctuations on a tectonic coast: New $^{230}Th/^{234}U$ dates from the Huon Peninsula, New Guinea. Quaternary Research, 4, 185-205, 1974.

Boyle, E. A., Chemical accumulation variations under the Peru Current during the past 130,000 years, J. Geophys. Res., 88, 7667-7680, 1983.

Broecker, W. S., Glacial to interglacial changes in ocean chemistry. Prog. Oceanogr, 11, 151-197, 1982.

Broecker, W. S. and T.-H. Peng, The climate-chemistry connection, in Climate Processes and Climate Sensitivity, Geophys. Monogr. Ser., vol. 29, edited by J. E. Hansen and T. Takahashi, pp. 327-336, AGU, Washington, D. C., 1984.

Chappell, J. and H. H. Veeh, $^{230}Th/^{234}U$ age support of an interstadial sea level of -40m at 30,000 yr. BP, Nature, 276, 602-603, 1978.

CLIMAP Project Members, The surface of the ice-age earth, Science, 191, 1131-1137, 1976.

Curry, W. B. and G. P. Lohmann, Carbon isotopic changes in benthic foraminifera from the western equatorial South Atlantic: Reconstruction of glacial abyssal circulation patterns. Quat. Res. 18, 218-235, 1982.

Delmas, R. J., J.-M. Ascenio, and M. Legrand, Polar ice evidence that atmospheric $CO_2$ 20,000 years ago was 50% of present. Nature, 284, 155-157, 1980.

Duplessy, J.-C., N. J. Shackleton, R. K. Matthews, W. Prell, W. F. Ruddiman, M. Caralp, and C. H. Hendy,  Carbon-13 record of benthic foraminifera in the last interglacial ocean: Implications for the carbon cycle and the global deep water circulation. Quat. Res., 21, 225-243, 1984.

Graham, D. W., B. H. Corliss, M. L. Bender  and L. D. Keigwin,  Carbon and oxygen isotopic disequilibria of recent deep-sea benthic foraminifera. Mar. Micropaleontol., 6, 483-497, 1981.

Hays, J. D., J. Imbrie  and N. J. Shackleton, Variations in the earth's orbit: pacemaker of the ice ages. Science, 194, 1121-1132, 1976.

Imbrie, J. and J. Z. Imbrie,  Modelling the climatic response to orbital variations. Science, 207, 943-954, 1980.

Imbrie, J., J. D. Hays, D. G. Martinson, A. McIntyre, A. C. Mix, J. J. Morley, N. G. Pisias,  W. Prell and  N. J. Shackleton, The orbital theory of Pleistocene climate: support from a revised chronology of the marine $^{18}O$ record, in Milankovich and Climate, edited by A. Berger, J. Imbrie, J. Hays, G. Kukla and B. Saltzman. pp. 269-305, D. Reidel, Hingham, Mass, 1984.

Jenkins, G. M. and D. G. Watts, Spectral Analysis and Its Applications, Holden-Day, San Francisco, Calif., 1968.

Julian, P. R., Comments on the determination of significance levels of the coherence statistic. J. Atmos. Sci., 32, 836-838, 1975.

Keir, R. and W. H. Berger, Atmospheric $CO_2$ content in the last 120,000 years: the phosphate-extraction model. J. Geophys. Res., 88, 6027-6038, 1983.

Mix, A. and W. F. Ruddiman, Oxygen-isotope analyses and Pleistocene ice volumes, Quat. Res., 21, 1-20, 1984.

Neftel, A., H. Oeschger, J. Schwander, B. Stauffer, and R. Zumbrunn, Ice core measurements give atmospheric $CO_2$ content during the past 40,000 yr. Nature, 295, 220-223, 1982.

Ninkovich, D. and N. J. Shackleton, Distribution, stratigraphic position and age of ash layer "L", in the Panama Basin region. Earth Planet. Sci. Lett., 27, 20-34, 1975.

Pedersen, T. F., Increased productivity in the eastern equatorial Pacific during the last glacial maximum (19,000 to 14,000 yr B.P.). Geology, 11, 16-19, 1983.

Pisias, N. G., D. G. Martinson, T. C. Moore, N. J. Shackleton, W. L. Prell, J. D. Hays and G.

Boden, High resolution stratigraphic correlation of benthic oxygen isotope records spanning the last 300,000 years, Mar. Geol., 56, 119-136 (1984).

Ruddiman, W. F. and A. McIntyre, Ice-age thermal response and climatic role of the surface Atlantic Ocean, 40°N to 63°N. Bull. Geol. Soc. Am., 95, 381-396, 1984.

Shackleton, N. J., Carbon-13 in Uvigerina: tropical rainforest history and the Equatorial Pacific carbonate dissolution cycles, In The Fate of Fossil Fuel $CO_2$ in the Oceans, edited by N. R. Andersen and A. Malahoff, pp. 401-427. Plenum, New York, 1977.

Shackleton, N. J., M. A. Hall, J. Line, and Cang Shuxi, Carbon isotope data in core V19-30 confirm reduced carbon dioxide concentration of the ice age atmosphere. Nature, 306, 319-322, 1983a.

Shackleton, N. J., J. Imbrie and M. A. Hall, Oxygen and carbon isotope record of East Pacific core V19-30: Implications for the formation of deep water in the late Pleistocene North Atlantic, Earth Planet. Sci. Lett., 65, 233-244, 1983b.

CARBON ISOTOPES IN DEEP-SEA BENTHIC FORAMINIFERA:
PRECESSION AND CHANGES IN LOW-LATITUDE BIOMASS

Lloyd D. Keigwin

Woods Hole Oceanographic Institution, Woods Hole, Massachusetts 02543

Edward A. Boyle

Massachusetts Institute of Technology, Cambridge Massachusetts 02139

Abstract. Detailed carbon isotope records have been obtained from benthic foraminifera in the North Atlantic (CHN 82 Sta 24 Core 4PC, 42°N, 32°W, 3427 m, 244 kyr) and central equatorial Pacific (KNR 73 Sta 4 Core 3PC, 0°S, 106°W, 3606 m, 410 kyr). These data demonstrate that the North Atlantic site has always been nutrient depleted relative to the Pacific site. Although down-core carbon isotope data in the Atlantic and Pacific reflect changes in deep-ocean circulation patterns, a larger fraction of $\delta^{13}C$ variability in both oceans is due to changes in the inventory of continental reduced carbon. The magnitude of this carbon isotope biomass signal is nearly twice as large during oxygen isotope stages 4 and 6 as it is during stage 2. Carbon isotope variability in both oceans contains significant power at the 23-kyr frequency band which is coherent with insolation changes caused by precession of the earth's rotational axis. We propose that this correlation is due to orbitally driven variations in low-latitude biomass, and that the lower amplitude of the carbon isotope signal during the last 60 kyr is due to the reduction in precession parameter amplitude during this period.

Introduction

Shackleton [1977] proposed that the carbon isotope composition of deep-sea benthic foraminifera changes in response to variations in the storage of carbon enriched in $^{13}C$ in the trees and soils on the continents. He argued that destruction of the northern hardwood forests (by the glacial ice mass) and reduction in the size of tropical rain forests (due to increased aridity on the continents during the most recent glacial maximum) transfers a sufficient quantity of carbon enriched in $^{12}C$ and reduces the average $\delta^{13}C$ of the ocean. The magnitude of oceanic average $\delta^{13}C$ change has been shown to be less than originally estimated, however, because a reduced input of water enriched in $^{13}C$ into the deep North Atlantic during glacial periods alters the carbon isotopic signal of Atlantic cores relative to the global mean [Streeter and Shackleton, 1979; Duplessy et al., 1975; Boyle and Keigwin, 1982; Curry and Lohmann, 1983; Shackleton et al., 1983; Berger and Keir, 1984]. The change in the oceanic average carbon isotopic composition between the present and the 18-kyr (1 kyr=$10^3$ years) glacial maximum is now thought to be about 0.4°/oo [Shackleton et al., 1983; E.A. Boyle and L.D. Keigwin, unpublished manuscript, 1984], corresponding to a mass of about 50 x $10^{15}$ moles of reduced carbon, about one-quarter of the present-day mass of carbon stored in plants and soils.

The carbon isotopic records of Shackleton [1977] and Shackleton et al., [1983] only covered the last 140,000 years, so it was not possible from these data to determine if this same pattern of change has occurred over several glacial/interglacial transitions. In addition, the North Atlantic record is based on analysis of three species and core M12392 is located in the eastern basin of the tropical Atlantic Ocean, which has been shown to have an anomalously large component due to ocean circulation changes [Curry and Lohmann, 1983]. In order to examine the variability of deep-sea carbon isotopic composition over time, we have obtained two benthic carbon isotope records from the western North Atlantic Ocean (over the last 244,000 years) and from the equatorial Pacific (over the last 410,000 years). Together with the recent Pacific carbon isotope data of Shackleton et al., [1983], these data allow us to examine the temporal variability of the continental biomass.

These data also provide information on changes in deep-ocean circulation and chemical composition. These aspects of the data, as well as a more detailed discussion of the stratigraphy of the cores, will be presented elsewhere; in this report we emphasize the aspects of the data related to changes in the continental biomass.

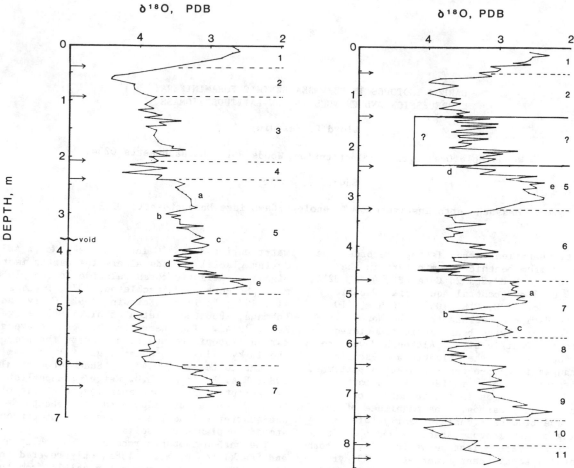

Fig. 1.  Oxygen isotope stratigraphy for our two cores.  All data are from
C. wuellerstorfi.  There was a 32-cm gap in core CHN 82 Sta 24 Core 4PC; this is
presumably a coring artifact, since there is no evidence for any real stratigraphic
gap.  The arrows on the depth scale indicate the chronological marker horizons chosen
as indicated in the text.  The numbers to the right of each plot indicate the oxygen
isotope stage boundaries.

### Analytical Methods

Benthic foraminifera (Cibicidoides
wuellerstorfi) were handpicked from the wet-sieved
>150-μm fraction of 15-g dry samples of the cores.
The portion of the samples adjacent to the core
liner was discarded prior to disaggregation.
Isotopic analyses of the foraminifera were done
using minor modifications of methods reported
elsewhere [Boyle and Keigwin, 1982].

### Chronology

Stratigraphy for these cores is based on the
oxygen isotopic composition of benthic foraminifera
shown in Figure 1.  These data reflect dominantly
the globally synchronous changes in the average

oxygen isotopic composition of the ocean due to
storage of water depleted in $^{18}O$ in glacial ice
sheets [Shackleton and Opdyke, 1973].  The record
from the North Atlantic core (with an average
sedimentation rate of 3.1 cm/kyr) preserves at
least 90% of the full amplitude of the Atlantic
oxygen isotope signal reported by others
[Shackleton, 1977; Streeter and Shackleton, 1979;
Duplessy et al., 1980] and displays all of the
reported stages and substages.  The Pacific record
has a lower sedimentation rate (2.0 cm/kyr) and a
correspondingly less detailed isotopic record,
although it still records 85% of the Pacific $\delta^{18}O$
amplitude reported by Ninkovich and Shackleton
[1975].  Nonetheless, the major stages and
substages are clearly recognizable, with the
exception of the interval between early stage 2 and

TABLE 1.  Assumed Chronology for Sediment Cores

| Depth in Core, cm | Assumed Age, kyr |
|---|---|
| CHN 82 Sta 24 Core 4PC (Atlantic) | |
| 0 | 0 |
| 40 | 11 |
| 91 | 28 |
| 212 | 62 |
| 240 | 73 |
| 475-32* | 127 |
| 606-32* | 187 |
| 643.5-32* | 205 |
| 663-32* | 214 |
| KNR 73 Sta 4 Core 3PC (Pacific) | |
| 0 | 0 |
| 52 | 11 |
| 132 | 28 |
| 241 | 107 |
| 325 | 127 |
| 470 | 187 |
| 587 | 245 |
| 642 | 281 |
| 750 | 339 |
| 805 | 362 |
| 832 | 405 |
| 842 | 410 |

*All depths below 350 cm have 32 cm subtracted to to correct void interval.

substage 5d. This interval is not recognizable in terms of the standard oxygen isotope stratigraphy and must reflect a sedimentary disturbance in this section. For this reason we exclude this portion of the core from further consideration.

The absolute chronology is based on radiometric dates for the upper isotopic stages: the stage 1/2 boundary ("termination I") is dated at approximately 11 kyr by $^{14}C$ [Broecker et al., 1958]; the 4-5 boundary is dated at 73 kyr using K-Ar dates from the Toba tuff [Ninkovich et al., 1978]; the 5-6 boundary is set at 127 kyr through correlation with the 124-kyr Barbados I coral high stand dated by $^{230}Th$ ingrowth [Broecker et al., 1968]. Other dates derive from linear interpolation from the 730-kyr K-Ar-dated Brunhes/Matuyama magnetic reversal [Mankinen and Dalrymple, 1979] in core V28-238 [Shackleton and Opdyke, 1973] and the tune-up chronology of Imbrie et al. [1984]: 28 kyr for the 2-3 boundary; 107 kyr for the stage 5d isotope maximum; 187 kyr for the 6-7 boundary; the "7.2" local isotope minimum at 205 kyr; 245 kyr for the 7-8 boundary; 281 kyr for the midpoint of the "8.4-8.5" isotope fluctuations; 339 kyr for the 9-10 boundary; 362

kyr for the 10-11 boundary; 405 kyr for the "11.3" isotope minimum. The placement of these markers on the isotopic curves is shown in Figure 1 as specified in Table 1. We will also make use of the carbon isotope data from core V19-30 [Shackleton et al., 1983]. Our chronology follows theirs with two minor modifications: we specify a 4-5 boundary at 569 cm and place the 5-6 boundary at 792 cm (which is closer to the midpoint of the isotopic range). These modifications are minor and not significant in terms of absolute time, but result in a better match of the isotopic data from the two Pacific records. All other dates are derived from linear interpolation between these marker points.

Results

The carbon isotope data show that deep water at the Atlantic site has always had higher $\delta^{13}C$ than at the Pacific site (Figure 2). This evidence

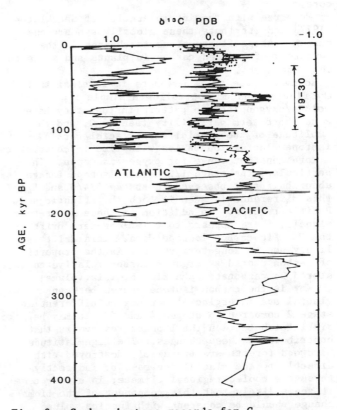

Fig. 2. Carbon isotope records for C. wuellerstorfi plotted versus age. The V19-30 data of Shackleton et al. [1983] have been shifted by adding 1.1°/∘∘ to the published data; this compensates for species fractionation and interlaboratory offsets. In the intervals where our core has reliable data, we have drawn the line connecting our points, and plotted the V19-30 data as individual points. In the interval where our core is not reliable, we have drawn the line between the V19-30 points; this section is indicated by the line near the right margin.

verifies the conclusion that basin-basin nutrient fractionation has existed between the Atlantic and Pacific for the last 225,000 years [Boyle and Keigwin, 1982]. Our Pacific carbon isotope data show fluctuations similar to those in core V19-30 in the intervals where the data overlap, and we use V19-30 data in the interval where we have evidence of sedimentation disturbance. Our Atlantic data are quite different from the data of core M12392, a difference we ascribe to anomalous eastern basin hydrography [Curry and Lohmann, 1983]. Although the carbon isotope difference between the oceans changes in accordance with the North Atlantic deep-water fluctuations seen in cadmium data [Boyle and Keigwin, 1982], it appears that the largest fraction of the carbon isotope variability is due to globally synchronous changes in the average $\delta^{13}C$ of the ocean. The amplitude of $\delta^{13}C$ fluctuations is enhanced in the Atlantic due to deep-water circulation changes and, in this case, also due to the higher sedimentation rate of the Atlantic core.

We agree with the interpretation of Shackleton [1977] and attribute these globally synchronous carbon isotope fluctuations to changes in the inventory of reduced carbon in plants and soils on the continents.

However, with these new data it is clear that the fluctuations do not follow a simple ice volume/oxygen isotope pattern in that they have more short-term variability than $\delta^{18}O$, and the amplitude of the glacial/interglacial carbon isotope fluctuation varies with time in contrast to a more constant range for oxygen isotopes. In particular, the stage 1/2 carbon isotope change is about half that observed in stages 3/4/5 and 5/6/7; this difference is seen in both the Atlantic and Pacific records. In addition to these short-term signals, there appears to be a long-term drift in the Pacific carbon isotope data. This drift is likely due to long-term changes in the proportion of carbon stored as organic carbon relative to that stored as carbonate [Shackleton, this volume].

Why is the carbon isotope change between glacial and interglacial periods so different in stage 2 compared to stages 4 and 6? It may help to distinguish between the biomass reservoirs that contribute to these changes. The high-latitude hardwood forests are presumably destroyed either directly by the glacial ice mass or indirectly through a cooled regional climate; in either case it seems likely that this component of the biomass change should be coherent with the ice volume signal. On the other hand, it is not clear that the size of the low-latitude tropical rain forests need be linked so closely to glacial mass, since their sizes are presumably controlled by the tropical water balance [Sarnthein, 1978; Street and Grove, 1976, 1979; Street-Perrot and Harrison, 1984]. Although there plausibly may be a global "teleconnection" between ice cover and tropical aridity, other processes may dominate low-latitude paleoclimatic variability. Kutzbach and Otto-Bliesner [1982] utilized a low-resolution

atmospheric general circulation model (GCM) to infer an intensified African-Asian monsoon caused by precession-induced radiation distribution differences 9 kyr ago. Prell [1984] reports an increase in upwelling-correlated foraminifera in early Holocene sediments off the Arabian peninsula during this period. Rossignol-Strict [1983] provided evidence that Mediterranean sapropel deposition is related to an index based on low-latitude seasonal insolation changes, and argued that the mechanism causing the correlation was increased African rainfall caused by variations in the monsoon intensity. Tropical rainfall variations should lead to variations in tropical forest biomass. In order to examine this possibility, we now turn to time and frequency domain analyses of the carbon isotope signals.

## Spectral and Cross-Spectral Analytical Methods

Hays et al., [1976], Pisias [1976], Ruddiman and McIntyre [1981], and Imbrie et al. [1984] have applied spectral and cross-spectral analyses to deep-sea sediment records. These investigators have demonstrated the utility of these methods, and we follow their lead in applying these methods to our data. Our methods differ slightly from the methods used by these investigators, however, being based on the fast Fourier transform (FFT) technique. This technique is well known to physical oceanographers and is computationally efficient. It is also easier for nonspecialists to interpret the physical significance of the FFT output and its transformations. Although the theory behind these methods has been discussed in detail elsewhere [Brigham, 1974; Koopmans, 1974], we present an account of our use of these methods in the appendix so that others may attempt to reproduce our results. Although differences between computational methods should lead to no significant differences in the final outcome, they may be relevant for strict quantitative comparisons.

Some marine geologists regard spectral analysis with reservation. Partly, this skepticism stems because the method seems indirect and less than intuitive. However, the signals picked up by this technique are usually obvious in the raw data, and spectral analysis may be best regarded as an objective technique for testing whether fluctuations in the raw data are indeed due to statistically significant periodicities.

## Spectral and Cross-Spectral Analysis of Carbon Isotope Records

In order to fill in the stage 3-4-5c gap in our Pacific record, we have used the V19-30 Uvigerina record of Shackleton et al. [1983]. The two records were converted to equally spaced (in time) records. We then added $1.1°/oo$ to the V19-30 data. Of this correction, about $0.9°/oo$ is due to a systematic offset between Uvigerina and Cibicidoides wuellerstorfi [Duplessy et al., 1984].

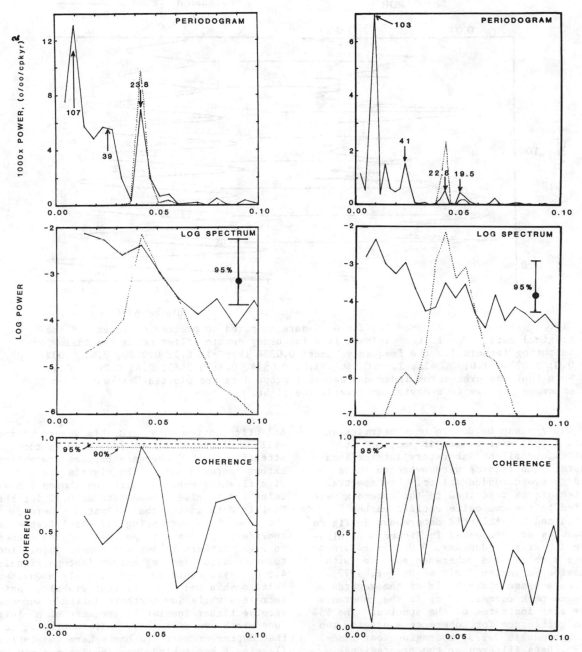

Fig. 3.  Periodograms, log power spectra, and coherence spectra for carbon isotope records (solid lines) and precession (dotted lines) (e sin ω) calculated using the formula of Berger [1977].  This index is proportional to the June earth-sun distance and anticorrelated to low-latitude summer insolation in the northern hemisphere. Power and coherence spectra are based on averaging two periodogram bandwidths to give 4 equivalent degrees of freedom (confidence limits are based on the probability that two uncorrelated white noise records will have spurious peaks or coherences greater than a given value).  The units for precession spectra are arbitrary and not marked, although the scalings for $\delta^{13}C$ and e sin ω cover the same decadal ranges.

## NORTH ATLANTIC

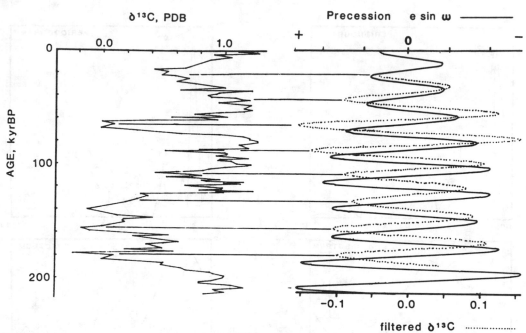

Fig. 4. Frequency-filtered Pacific $\delta^{13}C$ data compared to precession index and the original data. The filter employed is a frequency domain filter using the following weighting factors for the frequency bands: 0.0234 (kyr$^{-1}$), 0.1; 0.0280, 0.4; 0.0327, 0.8; 0.0374, 1.0; 0.0421, 1.0; 0.0467, 1.0; 0.0514, 0.8; 0.0561, 0.4; 0.0607, 0.1. Note that the precession index and carbon isotope data are plotted "backwards" so that enlarged biomass is a deviation towards the right.

The other 0.2°/₀₀ may be due to uncertainties in the offset, slight differences in bottom $\delta^{13}C$ between sites, or slight laboratory intercalibration offsets. The records were averaged where they overlapped to avoid sudden shifts. The spectral analysis techniques described in the appendix were then applied to the composite Pacific carbon isotope data and the Atlantic data shown in Figure 2. The results are presented in Figure 3 as high-resolution linear periodograms, 4 degree of freedom logarithmic spectra, and coherence spectra (with the precession parameter e sin ω [Berger, 1977]. The computed average phase angle of the records at the coherence peak corresponding to the coherence maximum is also indicated on the spectra. The 95% level of significance for coherence averaging two periodogram bands is 0.975; averaging four bands gives 0.795. Data filtered at the precessional frequency are displayed in Figures 4 and 5 along with the original data and the precession index.

The spectra (Figure 3) are dominated by the "100 kyr" periodicity which is observed in most paleoclimatic spectra [Imbrie et al., 1984]. Peaks corresponding to obliquity and precession periods occur in the transformed data.

We believe that the 41-kyr peak is likely to be due to deep-ocean circulation and high-latitude biomass variability; a similar peak appears in the

Atlantic cadmium data, and this aspect of the data will be discussed elsewhere. Here we turn our attention to the coherence between precession and carbon isotopic data. The signals are significantly coherent with the chosen bandwidths; using n=4 to give a bandwidth of 0.10 for the Pacific data leaves the estimated coherence about the same while increasing the significance level considerably. The average phase angles correspond to high northern hemisphere summer insolation (due to precession) leading carbon isotope signals by 4.5-6.7 kyr. This is approximately a quarter-phase relationship and suggests that global oceanic $\delta^{13}C$ increases while low northern latitude summers receive higher insolation (because of a minimum June earth-sun distance). Alternatively, it may be that September northern hemisphere insolation (lagging 6 kyr behind June insolation) may be the relevant parameter. If a global carbon isotope increase is ascribed to increasing tropical continental biomass, then this would imply higher tropical humidity during periods of higher northern hemisphere summer insolation. This certainly seems to have been the case during the last 20 kyr where a tropical "pluvial" period in intertropical Africa is documented at 8.6 kyr, lagging 3.6 kyr behind a precession-induced low northern latitude summer insolation maximum at 12.2 kyr [Street and Grove,

## PACIFIC

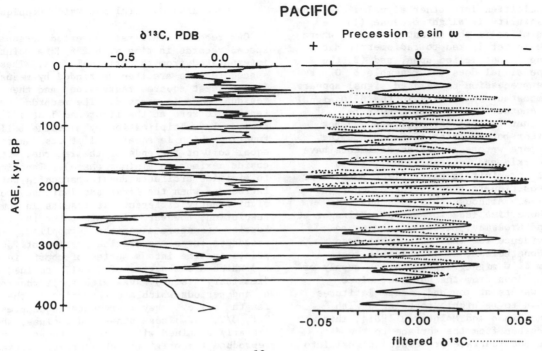

Fig. 5.  Frequency-filtered Atlantic $\delta^{13}$C data compared to precession index and the original data.  The filter employed is a frequency domain filter using the following weighting factors for the frequency bands:  0.0341, 0.1; 0.0366, 0.4; 0.0390, 0.8; 0.0415, 1.0; 0.0439, 1.0; 0.0463, 1.0; 0.0488, 1.0; 0.0512, 0.8; 0.0537, 0.4; 0.0561, 0.1.  Note that the precession index and carbon isotope data are plotted "backwards" so that enlarged biomass is a deviation towards the right.

1979].  There also is some indication of an earlier pluvial at about 26 kyr which would lag the insolation peak by about 8 kyr.  We propose that precession control of tropical humidity and biomass is the mechanism by which coherence between global oceanic $\delta^{13}$C and precession is produced.  The amplitude of the $\delta^{13}$C signal in this frequency band is about 0.1°/∘∘ in the Pacific core and 0.3°/∘∘ in the Atlantic core.  The Pacific amplitude is likely to be lower because of bioturbational smoothing; the Atlantic signal may be enhanced by ocean circulation changes.  A value of 0.1°/∘∘ corresponds to a transfer of 12 x $10^{15}$ moles of carbon.  Whitaker and Likens [1973] estimated the tropical forest biomass (stemwood, branches, leaves, and roots) as 38 x $10^{15}$ moles.  Brown and Lugo [1984] recently estimated this mass at 17 x $10^{15}$ moles.  If soils are included, it appears that the amplitude of our precession-linked $\delta^{13}$C variations are comparable to the mass of tropical carbon.  This magnitude seems reasonable in view of the large variations in tropical vegetation during the most recent glaciation [Shackleton, 1977, and references therein].  If the amplitude of the humidity cycle is roughly proportional to the precession index, then this mechanism may explain why the most recent glacial maximum has a low carbon isotope change relative to stages 4 and 6: the precession parameter has been at about half the

value over the last 60 kyr compared to the previous 300 kyr.  If this explanation is correct, then a similar reduction should be apparent in the filtered carbon isotope data during the last period when eccentricity was similarly low around 400 kyr.  Although our Pacific data cover this range, the frequency filtering techniques are not useful for cycles near the ends of the data series.  Nonetheless, this reduction in amplitude may be left as a prediction which can be tested by studies of older sediment cores.

T.J. Crowley (personal communication, 1984) suggests that the high amplitudes of the $\delta^{13}$C change in early stage 6 and 4 may reflect excess carbon destruction upon glaciation occurring after a period of long climatic stability.  While this mechanism may play a role, the large $\delta^{13}$C fluctuations during stage 6 (a long period of glaciation) suggest that our proposed precession mechanism is dominant.

There are two possible artifacts which we believe are not significant but cannot prove to be so.  We mention them here as possible qualifications on our interpretation.  One objection that could be raised is our use of the "SPECMAP" dates for the older intervals.  Since these dates are "tuned" so as to produce maximum coherence between orbital parameters and $\delta^{18}$O, it may be objected that this procedure could induce orbital periodicities into other signals. We argue that this possiblity is slight because (1) the results are similar in the Atlantic record, where the dates are better linked to radiometric dates and the tuning is not relied upon, and (2) the carbon isotope signal does not resemble $\delta^{18}O$ very much, and the precession-correlated signal appears to be more important than it is in the $\delta^{18}O$ signal (Figure 3). Another artifact that must be mentioned is that benthic foraminifera may not strictly monitor bottom water $\delta^{13}C$. Since the near-surface pore waters of marine sediments have low $\delta^{13}C$ [McCorkle and Emerson, 1983], it is possible that foraminifera may reflect pore water isotopic composition to some extent depending on where they live. Although this idea has been debated for some time (see for example Belanger et al. [1981] and Grossman, [1984]), to date no convincing evidence for this effect has been presented. Nonetheless, it is possible that benthic foram $\delta^{13}C$ might respond to the supply of carbon raining down from the surface; since near-surface waters at low and middle latitudes have been shown to be highly correlated with precession [Ruddiman and McIntyre, 1981], this transfer of carbon from the surface to the deep might carry a near-surface precession signal into the deep sea. Although we expect that neither of these possible artifacts is significant, they must be kept in mind.

## Conclusions

1. The benthic carbon isotope signal in both western North Atlantic and Pacific records is dominated by fluctuations in global average $\delta^{13}C$. We follow Shackleton [1977] in attributing this signal to changes in the continental biomass (reduced carbon in trees and soils).

2. The magnitude of the glacial/interglacial carbon isotope change is less during the most recent deglacial hemicycle than that observed in the isotope stage 3/4/5 and 5/6/7 intervals. This difference is observed in both oceans.

3. The carbon isotope fluctuations are significantly coherent with the earth's precession parameter. Low-latitude summer insolation in the northern hemisphere leads maximum oceanic carbon isotope values (maximum continental biomass) by 4-6 kyr. We propose that this correlation is due to tropical humidity variations driven by insolation distribution changes as seen in the GCM study of Kutzbach and Otto-Bliesner [1982]. Variations in the amplitude of the precession-induced insolation changes caused by changes in orbital eccentricity may vary the magnitude of the global carbon isotope signal. We propose this explanation for the reduced amplitude of the carbon isotope signal during the most recent deglacial hemicycle and predict that a similar reduction should be observed during the last interval of low precession around 400 kyr.

## Appendix: Spectral Analysis Techniques

Our records are first converted to equally spaced records in time with 256 data points interpolated from the original data. These equally spaced records are then detrended by means of a linear least squares regression, and then any residual mean is removed. The records are then tapered to zero at the first and last 10% of the records by multiplication by a cosine bell (a function which is equal to 1 at its inner end and equal to 0 at the ends of the records, with a cosine maximum to minimum shape). Tapering is done to minimize artifacts due to harmonics which can be introduced when the first and last data points are significantly different; it results in some loss of resolution, however. At this point the fast Fourier transform algorithm is applied. This procedure turns the original series of 256 real points in time into a series of coefficients ($a_j$, $b_j$) which are a series of "real" (cosine) and "imaginary" (sine) waves with amplitudes of $a_j$ and $b_j$ and periods which are harmonics of the record length T; i.e., they represent frequencies $1/T$, $2/T$, $3/T$,.... These sines and cosines, when linearly combined with appropriate amplitudes, reproduce the original data series. These harmonics give a complete and unique representation of the data (i.e., all 'reasonable' data in time can be represented by these components in only one way), and the component cosines and sines are orthogonal (i.e., cosines of any one of the harmonic frequencies cannot be reproduced as a linear combination of the other component frequencies).

This Fourier transform is converted into a periodogram (a representation of the power at each frequency) by multiplying:

$$(a + ib) * (a - ib) = a^2 + b^2 \qquad (1)$$

(where i is $\sqrt{-1}$). It is possible that frequencies that are not resolved by the periodogram are present in the data, and nonlinear spectral estimations can provide a better estimate in such cases. These techniques cannot prove that the resolved frequencies are necessarily in the data, however. The periodogram itself has poor statistical properties with only 2 equivalent degrees of freedom such that white noise (variability random with time) can produce peaks of magnitude comparable to that of real periodicities. In order to improve stability of the periodogram we average periodogram bands using the Daniel estimator [Koopmans, 1974].

$$\overline{P}_j = [ \sum_{k=-n}^{+n} P_{j+k} ]/(2n+1) \qquad (2)$$

This procedure improves the statistical properties at the expense of frequency resolution. In order to keep the averaged frequencies the same as those of the original periodogram, when averaging an even number of bands n we take n-1 bands centered on the periodogram frequency and add on half of the power of the bands on either side of these n-1 bands. The equivalent degrees of freedom are given as 2n, and the confidence limits are then computed from a chi-square table [Koopmans, 1974]. This averaging procedure converts the periodogram into a spectrum, which is then plotted on a logarithmic scale with an error bar which signifies the probable level of spectral noise expected from white noise. In this work we have taken n=2 to maximize frequency resolution. In the Atlantic core this gives a bandwidth of 0.008 $kyr^{-1}$, and in the Pacific core the bandwidth is 0.005 $kyr^{-1}$.

Given two time series that we would like to compare for coherent variation as a function of frequency, we compute the coherence magnitude of two series with Fourier coefficients $a_j, b_j$ and $c_j, d_j$ as the product

$$(a + ib) * (c - id) \qquad (3)$$

This procedure returns complex numbers $C_j + iQ_j$ which are then band-averaged as before; the coherence (the correlation coefficient in frequency space) is computed by normalization to the power of each series:

$$\text{coherence} = \left[ \frac{(\overline{C}^2 + \overline{Q}^2)}{\overline{P}_1 * \overline{P}_2} \right]^{1/2} \qquad (4)$$

where P is the power of each series at the frequency of interest. The average phase angle between the two series at that frequency is computed as

$$\text{phase} = \tan^{-1}(\overline{Q}/\overline{C}) \qquad (5)$$

The level at which coherence is significantly different from zero is computed as outlined by Koopmans [1974]. The statistical test is based on the probability that the coherence of two white noise records will exceed a given level.

An additional procedure that can be applied is frequency filtering. Hays et al., [1976] utilized a time domain moving average filter; if a reasonably wide bandwidth is chosen, similar results are obtained using a frequency domain filter and the inverse Fourier transform. The Fourier coefficients are multiplied by zero at

frequency bands out of the range of the frequency filter, and they are multiplied by symmetrical weighting factors $w_i$ within the filter, where $w_i$ is chosen to vary smoothly from 1 at the center of the frequency band to zero at its limits. In this work we specify the weighting factors for the frequency bands used in filtering operations in the figure captions.

Acknowledgments. We thank Charlie Eriksen for helpful advice on spectral analysis and Dave Aubrey for his comments on our technical description. Core collection and curation is supported by NSF grant OCE 820019 to the Woods Hole Oceanographic Institution (WHOI). Thanks to Jim Broda for his help in obtaining these core samples and to C. Eben Franks for operating the mass spectrometer. This research was funded by NSF grant OCE 8217644 to L.D.K. and grant OCE 8209362 to E.A.B. This is WHOI contribution 5837.

References

Belanger, P.E., W.B. Curry and R.K. Matthews, Core-top evaluation of benthic foraminiferal isotopic ratios for paleo-oceanographic interpretations, Palaeo., Palaeo., Palaeo., 33, 205-220, 1981.
Berger, A., Support for the astronomical theory of climate change, Nature, 269, 44-45, 1977.
Berger, W.H. and R.S. Keir, Glacial-Holocene changes in Atmospheric $CO_2$ and the deep sea record, in Climate Processes and Climate Sensitivity, Geophys. Monogr. Ser., Vol. 29, edited by J.E. Hansen and T. Takahashi, 337-351, AGU, Washington, D.C. 1984.
Boyle, E.A. and L.D. Keigwin, Deep circulation of the North Atlantic over the last 200,000 years: Geochemical evidence, Science, 218, 784-787, 1982.
Brigham, E.O. (1974), The Fast Fourier Transform, Prentice-Hall, Englewood Cliffs, N.J., 1974.
Broecker, W.S., K.K. Turekian, and B. Heezen, The relation of deep-sea sedimentation rate changes to variations in climate, Am. J. Sci., 256, 503, 1958.
Broecker, W.S., et al., Milankovitch hypothesis supported by precise dating of coral reefs and deep sea sediments, Science, 159, 297-300, 1968.
Brown, S., and A.E. Lugo, Biomass of tropical forces: A new estimate based on forest volumes, Science, 233, 1288-1293, 1984.
Curry, W., and G.P. Lohmann, Reduced advection into Atlantic Ocean deep eastern basins during last glacial maximum, Nature, 306, 577-580, 1983.
Duplessy, J.C., L. Chenoiard and F. Vila, Weyl's theory of glaciation supported by study of Norwegian Sea core K11, Science, 188, 1208, 1975.
Duplessy, J.C., J. Moyes, and C. Pujal, Deep water formation in the North Atlantic during the last ice age, Nature, 286, 479-482, 1980.

Duplessy, J.C., N.J. Shackleton, R.K. Matthews, W. Prell, W.F. Ruddiman, M. Caralp, and A. Hendy, $^{13}$C record of benthic foraminifera in the last interglacial ocean: Implications for the carbon cycle and the global deep water circulation, Quat. Res., 21, 225-243, 1984.

Grossman, E.L., Stable isotope fractionation in live benthic foraminifera from the southern California borderland, Palaeo., Palaeo., Palaeo., 47, 301-327, 1984.

Hays, J.D., J. Imbrie, and N.J. Shackleton, Variations in the earth's orbit: Pacemaker of the ice ages, Science, 194, 1121-1132, 1976.

Imbrie, J., J.D. Hays, D.G. Martinson, A. McIntyre, A. Mix, J.J. Morley, N. Pisias, W. Prell, and N.J. Shackleton, The orbital theory of Pleistocene climate: Support from a revised chronology of the late marine $\delta^{18}$O record, in Milankovitch and Climate, edited by A. Berger et al., D. Reidel, Hingham, Mass., 1984.

Koopmans, L.H., The Spectral Analysis of Time Series, Academic, New York, 1974.

Kutzbach, J.E., and B.L. Otto-Bliesner, The sensitivity of the African-Asian monsoonal climate to orbital parameter changes for 9000 years B.P. in a low-resolution general circulation model, J. Atmos. Sci., 39, 1177-1188, 1982.

Mankinen, E.A. and G.B. Dalrymple, Revised geomagnetic polarity time scale for the interval 0-5 m.y. B.P., J. Geophys. Res., 84, 615-626, 1979.

McCorkle, D. and S. Emerson, Carbon isotopes in marine porewaters (abstract), EOS Trans. AGU, 64, 721, 1983.

Ninkovich, D., and N.J. Shackleton, Distribution, stratigraphic position, and age of ash layer "L" in the Panama Basin region, Earth Planet. Sci. Lett., 27, 20-34, 1975.

Ninkovich, D., N.J. Shackleton, A.A. Abdel-Monem, J.D. Obradovich, and G. Izett, K-Ar age of the late Pleistocene eruption of Toba, north Sumatra, Nature, 276, 574-577, 1978.

Pisias, N.G., Late Quaternary sediment of the Panama Basin: Sedimentation rates, periodigitres, and controls of carbonate and opal accumulation, in Investigation of Late Quaternary Paleoceanography and Paleoclimatology, edited by R. Cine and J.D. Hays), Mem. 145, pp. 375-391, Geological Society of America, Boulder, Colo.,1976.

Prell, W.L., Variation of monsoonal upwelling: A response to changing solar radiation, in Climate Processes and Climate Sensitivity, Geophys. Monogr. Ser., vol. 29, edited by J.E. Hansen and T. Takahashi, pp. 48-57, AGU, Washington, D.C., 1984.

Rossignol-Strict, M., African monsoons, an immediate climate response to orbital insolation, Nature, 304, 46-49, 1983.

Ruddiman, W., and A. McIntyre, Oceanic mechanisms for amplification of the 23,000 year ice volume cycle, Science, 212, 617-627, 1981.

Sarnthein, M., Sand deserts during glacial maximum and climate optimum, Nature, 272, 43-46, 1978.

Shackleton, N.J., Carbon-13 in Uvigerina: Tropical rainforest history and the equatorial Pacific carbonate dissolution cycles, in Fate of Fossil Fuel $CO_2$ in the Oceans, edited by N. Andersen and A. Malahof, pp. 401-427, Plenum, New York, 1977.

Shackleton, N.J., Oceanic carbon isotope constraints on oxygen and carbon dioxide in the Ceneozoic atmosphere, this volume.

Shackleton, N.J. and N.D. Opdyke, Oxygen isotope and paleomagnetic stratigraphy of equatorial Pacific core V28-238: Oxygen isotope temperatures and ice volumes on a $10^5$ and $10^6$ year time scale, Quat. Res., 3, 39-55, 1973.

Shackleton, N.J., J. Imbrie, and M. Hall, Oxygen and carbon isotope record of east Pacific core V19-30: Implications for the formation of deep water in the late Pleistocene North Atlantic, Earth Planet. Sci. Lett., 65, 233-244, 1983.

Street, F.A., and A.T. Grove, Environmental and climatic implications of late Quaternary lake-level fluctuations in Africa, Nature, 261, 385-390, 1976.

Street, F.A., and A.T. Grove, Global maps of lake-level fluctuations since 30,000 yr. B.P., Quat. Res., 12, 83-118, 1979.

Street-Perrot, F.A., and S.P. Harrison, Temporal variations in lake levels since 30,000 yr. B.P. - An index of the global hydrological cycle, in Climate Processes and Climate Sensitivity, Geophys. Monogr. Ser., vol. 29, edited by J. Hansen and T. Takahashi, pp. 118-129, AGU, Washington, D.C., 1984.

Streeter, S.S., and N.J. Shackleton, Paleocirculation of the deep North Atlantic: 150,000 year record of benthic foraminifera and oxygen 18, Science, 203, 168-204, 1979.

Whitaker, R.H. and G.E. Likens, Tropical Forest Biomass, in Carbon and the Biosphere, edited by G.M. Woodwell and E.V. Pecan, p. 281, 1973. (Available as CONF-720510 from the National Technical Information Service, Springfield, VA.)

# CARBON ISOTOPE VARIATIONS IN SURFACE WATERS OF THE GULF OF MEXICO ON TIME SCALES OF 10,000, 30,000, 150,000 AND 2 MILLION YEARS

Douglas F. Williams

Department of Geology, University of South Carolina, Columbia, South Carolina 29208

Abstract. Detailed carbon isotope records from surface-dwelling planktonic foraminifera appear to represent $\delta^{13}C$ variations in the surface ocean mixed layer of the Gulf of Mexico on different time scales. Correlation of late Pleistocene planktonic $\delta^{13}C$ records from the Gulf of Mexico, Panama Basin and southeast Indian Ocean suggests that the surface water $\delta^{13}C$ record monitors global changes in the carbon cycle, perhaps changes in atmospheric $CO_2$ levels, and not local effects due to variations in productivity or water temperature. Variations in $\delta^{13}C$ of $0.5^o/oo$ occur with a frequency of <500 years in the Holocene and may record background variations in atmospheric $pCO_2$. The meltwater effect on the $\delta^{13}C$ of surface waters is largest during the most rapid portion of the deglaciation from isotope stage 2 into stage 1 (from approximately 17 to 12.5 kybp). On time scales representing the last 150,000 years, the most prominent negative $\delta^{13}C$ excursion begins in late stage 6 and culminates in the earliest part of isotope stage 5e, a time representing the warmest global temperatures and highest sea levels of the last 150 ky B.P. Major changes in the benthic $\delta^{13}C$ record may lead those in the planktonic $\delta^{13}C$ record by as much as 10,000 years, but more work is needed to fully explore this observation. A small but significant offset of $-0.3^o/oo$ occurs at 900 ky B.P. in the $\delta^{13}C$ records of the Gulf of Mexico (E67-135) and Caribbean (DSDP Site 502B (Prell, 1982)), in good agreement with the positive $^{18}O$ stepwise shift observed in globally distributed DSDP records.

## Introduction

The objectives of this study are (1) to examine the surface water $\delta^{13}C$ signal of the Gulf of Mexico as a possible record of atmosphere-surface water interactions during different time scales of the Pleistocene and (2) to determine the timing and frequency of the Pleistocene $\delta^{13}C$ excursions in the Gulf of Mexico as they differ or correspond with similar changes in other ocean basins. It is hoped that detailed $\delta^{13}C$ records from an oceanographically stable region like the Gulf of Mexico will provide important information on the geographic variability of surface water $\delta^{13}C$ changes in terms of potential lead-lag relationships between atmospheric $pCO_2$, glacial-interglacial climatic change and meltwater input during deglaciations. The last deglaciation is examined to determine the effects of surface salinity changes from meltwater on the $\delta^{13}C$ record of the surface waters. Since the chemical and physical oceanography of the Gulf of Mexico is responsive to north and equatorial Atlantic surface and deep water circulation, its surface water $\Sigma CO_2$ content should be sensitive to changes in northern hemisphere $CO_2$ concentrations. At the same time, the Gulf of Mexico should be relatively insensitive to variations in sea surface temperature and in upwelling rate, since the gulf is known to remain fairly stable during glacial-interglacial cycles [Brunner and Cooley, 1976].

Currently, there are several models which attempt to explain the changes in the $\delta^{13}C$ composition of the ocean in response to glacial-interglacial variations. Shackleton [1977] proposed that changes in the $\delta^{13}C$ benthic foraminifera during the late Pleistocene reflected large-scale changes in the terrestrial biomass and soil carbon (humus), which in turn introduced $^{13}C$-depleted $CO_2$ into the atmosphere. Broecker [1982] alternatively accounted for these same $\delta^{13}C$ changes through the burial of carbon 12 and phosphorus organic matter in shelf sediments during rapid rises in sea level. Recently, Shackleton et al. [1983] proposed that the vertical $\delta^{13}C$ contrast between planktonic and benthic foraminifera independently corroborate the evidence of decreased atmospheric $CO_2$ concentrations during the last glacial ice age from studies of the carbon dioxide content in air bubbles in polar ice cores [Delmas et al., 1980; Neftel et al., 1982].

The isotopic data in this report are primarily based on monospecific analyses of the surface-dwelling planktonic foraminiferal species Globigerinoides ruber, from a restricted size fraction, according to well-established procedures [Williams et al., 1977]. The isotopic records representing the Holocene and last

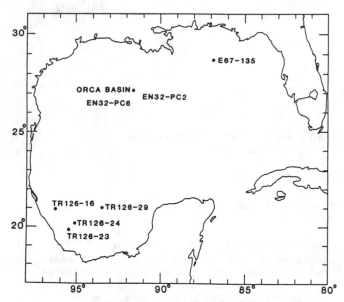

Fig. 1. Bathymetric map of the Gulf of Mexico showing the locations of the Pleistocene borehole Eureka 67-135 near the DeSoto Canyon; the anoxic hypersaline Orca Basin, from which piston cores EN32-PC2 and PC6 are located; and the late Pleistocene piston cores TR126-16, -23, -24 and -29 from the southwest Gulf of Mexico.

glacial-interglacial transition come from piston cores EN32-PC2 and EN32-PC6 from the intraslope Orca Basin (Figure 1) [Trabant and Presley, 1978]. The absolute time frame for the Holocene isotopic records is based on radiocarbon dates as discussed by Leventer [1980] and Leventer et al. [1983]. The chronology for the detailed late Pleistocene records from the southwestern Gulf of Mexico (Figure 1) is established using the ages for the Pleistocene oxygen isotopic stage boundaries [Shackleton and Opdyke, 1973]. An isotopic record representing approximately the last 1.7 million years is from the DeSoto Canyon borehole Eureka 67-135. The chronology for E67-135 is based in part on the oxygen isotope stratigraphy and the foraminiferal and calcareous-nannofossil biostratigraphy as reported by Brunner and Keigwin [1981], Gartner et al. [1983] and Neff [1983].

Before proceeding with a description of the surface water $\delta^{13}C$ data from the Gulf of Mexico, a brief discussion of the assumptions and complications in the use of biogenic carbonate foraminiferal $\delta^{13}C$ compositions to infer oceanic and atmospheric $\delta^{13}C/pCO_2$ changes is perhaps warranted. While $\delta^{13}C$ compositions of calcitic shells of planktonic foraminifera and calcareous nannofossils (primarily coccolithophorids) are complicated by interspecific $\delta^{13}C$ differences [Savin and Douglas, 1973; Williams et al., 1977;

Berger et al., 1978; Dudley et al., 1982] and some intraspecific differences due to size [Vincent et al., 1980; Curry and Matthews, 1979], these two groups (1) live primarily in the euphotic zone of the upper surface waters, (2) are the most important fixers of $CaCO_3$ from the upper waters, (3) closely reflect the physiochemical properties of those waters to varying degrees (temperature, $\delta^{13}C$ of $\Sigma CO_2$, etc.), and (4) represent our best hope of reconstructing the buffering action and exchange between the atmosphere-surface ocean reservoir system in the past. Previous studies have relied primarily on analyses of benthic foraminifera or on the $\delta^{13}C$ difference (contrast) between benthic and planktonic species as a measure of the mean oceanic $\delta^{13}C$ composition or surface to deep water $\delta^{13}C$ gradient. These data have been used as proxy information for inferring past changes in (1) atmospheric $pCO_2$ [Shackleton et al., 1983], (2) productivity [Broecker, 1982; Berger et al., 1978], (3) longterm shifts in the $\delta^{13}C$ gradients of the oceans [Keigwin, 1982; Vincent et al., 1980; Keigwin and Bender, 1979], and (4) the rate of deep water production and circulation [Boyle and Keigwin, 1982; Curry and Lohmann, 1982]. Recently, Shackleton [1984, this volume] has agreed that $\delta^{13}C$ data from total (bulk) carbonate are most appropriate for evaluating the long-term history of the global organic carbon reservoir.

While each of these approaches is valid, none is without its own set of mitigating factors, i.e., large interspecific $\delta^{13}C$ differences among benthic foraminifera, continuing questions regarding the source of the $^{13}C$ utilized by benthic foraminifera, changing carbonate sources (the proportion of nannofossil-foraminiferal-detrital $CaCO_3$) through time, and diagenetic and recrystallization effects, among other potential factors. In this study, the approach is taken to closely partition and monitor the surface water $\delta^{13}C$ signal in an oceanographically stable region using a single planktonic foraminiferal species (G. ruber) from a restricted size fraction to minimize complicating factors as mush as possible. G. ruber is well known to inhabit the upper water column and to reflect quite accurately the seasonal (intraannual) [Bé and Tolderlund, 1971; Williams et al., 1977, 1981] and glacial-interglacial signal [e.g., Broecker and van Donk, 1970; Leventer et al., 1983]. I cannot provide at this time unequivocal evidence that the $\delta^{13}C$ of this species, or any other planktonic microfossil, reflects atmospheric changes in $pCO_2$ on a one-to-one basis, but this assumption is not uniquely restricted to this study.

The purpose of this paper, therefore, is not to argue for or against the use of a particular species, microfossil group, or $CaCO_3$ fraction, but instead to present an internally consistent data set which I believe contains information

## HOLOCENE RECORDS (EN32-PC2)

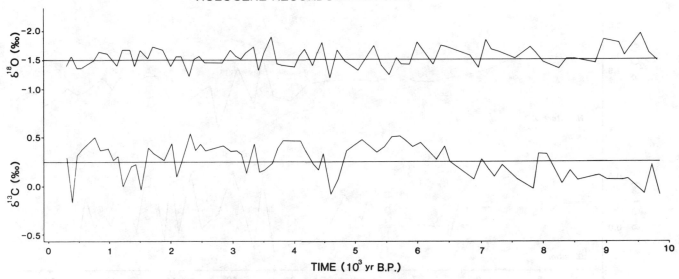

Fig. 2. Holocene oxygen and carbon isotopic records for Orca Basin core EN32-PC2 based on analyses of <u>Globigerinoides ruber</u>. The time scale is based on two radiocarbon determinations on the total carbon and should be regarded as tentative at this time.

regarding the background signals of $\delta^{13}C_{\Sigma CO2}$ they may reflect surface water buffering of atmospheric $pCO_2$ changes during various portions of the Pleistocene epoch.

### Results and Discussion

#### The Holocene Signal

The oxygen and carbon isotopic records for EN32-PC2 represent approximately the last 10,000 years (Figure 2). The $\delta^{18}O$ record is characterized by rapid fluctuations with an amplitude of 0.5 to 0.3⁰/oo (equal to approximately 1.5° to 2.5°C) and with an approximate frequency of 100 to 200 years, which is nearly equivalent to the sampling interval in the core. The chronology is based only on three radiocarbon dates of the total carbon and therefore must be considered tentative at this time. Extremely good resolution is permitted in this core because the sediments are black, unbioturbated anoxic muds nearly throughout the core's length. Although no clear evidence exists in the $\delta^{18}O$ record for a well-expressed thermal maximum, the $\delta^{18}O$ values from approximately 6000 to 9500 years before present (6 to 9.5 ky B.P.) are on the average more negative (warmer) than during the other parts of the record. Like the $\delta^{18}O$ record, the $\delta^{13}C$ record contains numerous short-lived (<500-year duration) positive and negative $\delta^{13}C$ excursions with an amplitude of 0.3 to 0.5⁰/oo (Figure 2). The most consistent change in the $\delta^{13}C$ record is the 0.3 to 0.4⁰/oo

depleted interval from approximately 6.5 to 9.8 ky B.P. No obvious correlation exists between the $\delta^{18}O$ and $\delta^{13}C$ records in EN32-PC2, suggesting that surface water temperature and salinity changes are not the primary controlling factors of the carbon 13 excursions.

Data are available for the time interval from approximately 4 to 13 ky N.P. in the Holocene portion of EN32-PC6 from two species of planktonic foraminifera (<u>Globigerinoides ruber</u> and <u>Neogloboquadrina dutertrei</u>) which inhabit slightly different average water depths (Figure 3). The absolute values for the two species reflect the well-documented $\delta^{13}C$ differences that exist between these two species [Williams et al., 1977; Berger et al., 1978; Williams et al., 1981]. Despite this difference and that due to sample availability of N. dutertrei, the $\delta^{13}C$ records from the two species indicate that a basic similarity exists between the two records. As in EN32-PC2, the carbon isotope variations in EN32-PC6 are on the order of 0.3 to 0.7⁰/oo (Figure 3). The surface water $^{13}C$ variations are particularly large and rapid around the Pleistocene/Holocene boundary (from 11.5 to 9 ky B.P.).

A comparison of the last 10,000 years in EN32-PC2 and PC6 reveal significant similarities in the positive and negative $\delta^{13}C$ excursions with time (Figure 4). Broad, approximately 1,000-year negative $\delta^{13}C$ excursions occur during the interval from 6.5 to 9.8 ky B.P. in both records and, despite uncertainties in the chronology at this time, it is possible that correlation of the

## HOLOCENE RECORDS (EN32-PC6)

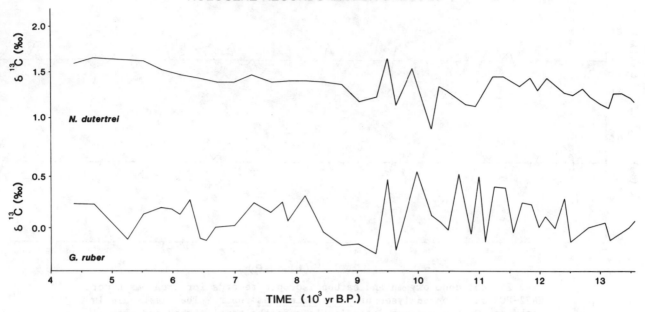

Fig. 3. Carbon isotopic records based on <u>Globigerinoides</u> <u>ruber</u> and <u>Neogloboquadrina</u> <u>dutertrei</u> from Orca Basin core EN32-PC6 from approximately 4 to 13 ky B.P. [from Leventer et al., 1983].

## HOLOCENE RECORDS

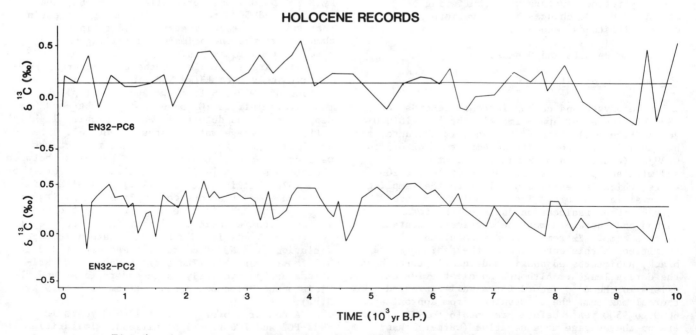

Fig. 4. An approximate comparison between the timing and magnitude of carbon isotopic events recorded in the Holocene records of <u>Globigerinoides</u> <u>ruber</u> from the Orca Basin piston cores EN32-PC2 and PC6. Both records represent relatively undisturbed and unbioturbated sequences, since much of the sediment since 8,500 years B.P. is deposited under the anoxic Orca Basin Brine.

## STAGE 2/1 RECORD (EN32-PC6)

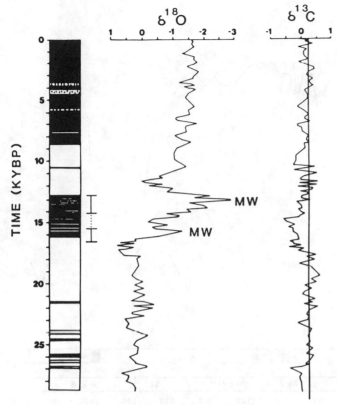

Fig. 5. A comparison of the carbon and oxygen isotopic records for Orca Basin core EN32-PC6 during the deglaciation from isotope stage 2 through stage 1. The left hand column and the chronology are based on the work by Leventer et al. [1982; 1983]. The largest intervals of meltwater input to the $\delta^{18}O$ record are indicated by the capitalized MW symbols. The line drawn through the carbon isotope record is meant to provide visual assistance in seeing negative and positive $\delta^{13}C$ intervals.

short-term (less than 200 year) negative $^{13}C$ excursions may exist during the middle to late Holocene portions of the records.

### The Glacial-Interglacial Signal of the Last 30,000 Years

The oxygen isotope record of EN32-PC6 has provided to date the best documented evidence for the magnitude and timing of the input of $^{18}O$-depleted glacial meltwater during the last deglaciation [Leventer et al., 1982, 1983] (Figure 5). The sedimentary record of EN32-PC6 contains alternating grey hemipelagic muds and black laminated organic-rich intervals during both the maximum periods of meltwater input between 16.5

and 12.5 ky B.P. and the establishment of the Orca Basin brine, approximately 8.5 m.y. B.P., respectively (Figure 5). The carbon isotopic record for EN32-PC6 reveals a pronounced negative $\delta^{13}C$ excursion beginning approximately 17.5 ky B.P., preceding the initial deglaciation of the southern periphery of the Laurentide Ice Sheet by approximately 1,000 years. The most negative $\delta^{13}C$ values observed in this record occur approximately 14.8 ky B.P. immediately before the second major meltwater pulse and continue through to where the most negative $\delta^{18}O$ values occur at 13 ky B.P. The $\delta^{13}C$ variations during the glacial maximum conditions between 26 to 18 ky B.P. are similar to those $\delta^{13}C$ values found in the late Holocene interval.

The impact of the meltwater on the surface water $\delta^{13}C$ values therefore appears to be significant but transitory. The negative $\delta^{13}C$ excursion actually precedes the onset of the meltwater, last approximately 4,000 years and is then followed by $0.5^{O}/oo$ $\delta^{13}C$ fluctuations on the order of less than 250 years in the interval between 12,000 and 10,000 ky B.P.

### The Late Pleistocene Signal of the Last 150,000 Years

The oxygen isotopic record of TR126-23 shows a characteristic isotope stage 6 through stage 1 $\delta^{18}O$ signal. After Termination II at the stage 5/6 boundary, the substages of stage 5 can be discerned. Isotope stage 3 appears as a broad interval of intermediate $\delta^{18}O$ values, with a gradual enrichment in $^{18}O$ occurring across the stage 3/2 boundary. The transition from stage 2 into isotope stage 1 (Holocene) contains evidence of meltwater spikes as a result of the last deglaciation.

The most striking feature about the carbon isotope record of TR126-23 for the last 150,000 years (Figure 6) is the large ($>1^{O}/oo$) negative $\delta^{13}C$ event spanning from 150 to 120 ky B.P. The most negative $\delta^{13}C$ values correspond with the earliest part of isotope stage 5e. After the $\delta^{13}C$ values recover during substages 5d, 5c and 5b, rapid excursions with an amplitude of 0.3 to $-0.5^{O}/oo$ characterize the record during stage 5a, immediately preceding the glacial advance during stage 4. A broad negative $\delta^{13}C$ excursion occurs across the isotope stage 3/4 boundary. The $\delta^{13}C$ values then become broadly positive during isotope stage 3, and rapid, negative $\delta^{13}C$ events characterize the transition from isotope stage 2 to 1 as shown in the more detailed record of EN32-PC6.

The $\delta^{13}C$ records from three additional cores from the same region of the southwestern Gulf of Mexico as TR126-23 (TR126-16, -24, -29) indicate that the large negative $\delta^{13}C$ event of late stage 6 and isotope stage 5e may indeed be two separate events, with the earlier negative event around 140-135 ky B.P. and the other centered around

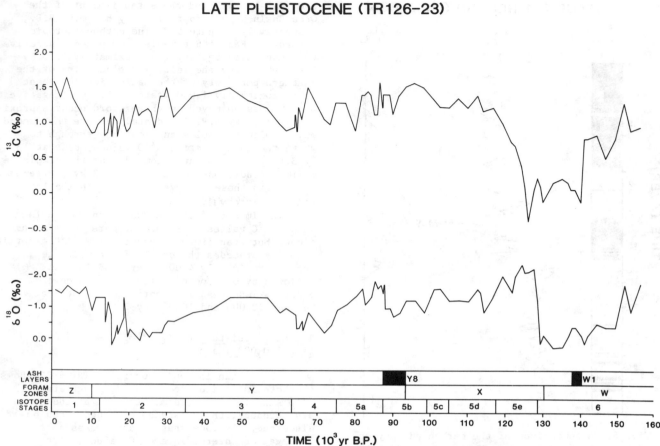

**Fig. 6.** The oxygen and carbon isotopic record for the late Pleistocene piston core TR126-23 based on monospecific analyses of <u>Globigerinoides</u> <u>ruber</u>. The oxygen isotopic record represents a continuous sequence for the last 150,000 years and shows the well-recognized oxygen isotope stage boundaries, the foraminiferal biostratigraphic Ericson zones, and the well-dated and geochemically fingerprinted Y8 and W1 ash horizons. The line drawn through the carbon 13 record is an approximate mean value for the last 120,000 years, and the vertical arrows indicate time periods of particularly negative $\delta^{13}$C excursions.

stage 5e (Figure 7). It is not known at this time, however, how widespread this stage 6/5e $\delta^{13}$C event is in the Gulf of Mexico, or how this event is translated into the deep water. A limited amount of benthic foraminiferal $\delta^{13}$C data from TR126-16 [Falls, 1980] suggests the existence of a coeval deep water change of lesser magnitude than the surface water change, and work is in progress to document the geographic variability of this particularly large $\delta^{13}$C event.

### The Pleistocene Record of the Last 1.7 Million Years Before Present

The longest available oxygen and carbon isotopic records for the Gulf of Mexico are from E67-135, a borehole located near the DeSoto Canyon to the southeast of the Mississippi River

drainage system (Figure 8). Although the $\delta^{18}$O record and biostratigraphy of E67-135 indicate the presence of several hiatuses [Neff, 1983], at least the majority of the Pleistocene glacial-interglacial intervals are present, and the record should therefore provide an approximation of the long-term carbon isotope excursions in the surface waters of the Gulf of Mexico. The $\delta^{18}$O record shows (1) a stepwise 1o/oo shift to more positive values during the middle Pleistocene at approximately 900 ky B.P. and (2) a change to higher-amplitude (>2o/oo) and longer-period (100,000 year) isotopic fluctuations in the interval from 800 ky B.P. to the present (Figure 8). These shifts in mean value, amplitude and frequency in E67-135 are similar to those observed in Caribbean Site 502B [Prell, 1982], in V28-239 in the Pacific [Shackleton and Opdyke, 1976] and in several Mediterranean

# STAGE 5/6 (110-150 KY B.P.)

## $\delta^{13}C_{PDB}$(‰)  <u>G. ruber</u>

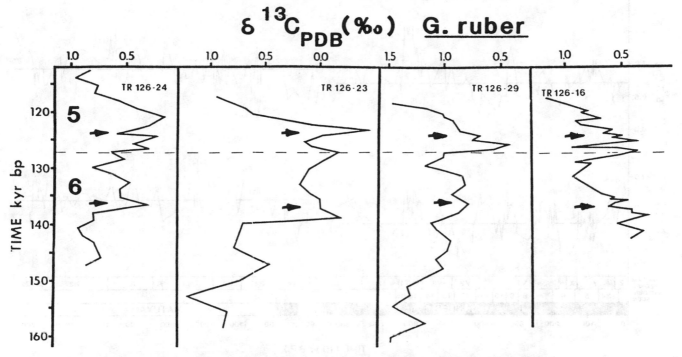

Fig. 7. Four $\delta^{13}C$ records for the time period from approximately 155 to 115 ky B.P. in four piston cores from the southwestern Indian Ocean (Figure 1). All data are based on monospecific analyses of <u>Globigeri-noides ruber</u>, and the horizontal dashed line indicates the position of the isotope stage 6/5 boundary [after Falls, 1980]. The horizontal arrows in each record represent the possible indication of two negative $\delta^{13}C$ events near the isotope stage 6/5 boundary.

Deep Sea Drilling Project (DSDP) sites [Thunell and Williams, 1983]. Interestingly, the carbon isotope record of E67-135 also shows a small (~0.3°/oo) difference between these two climatic periods (Figure 8). The carbon isotope variations prior to 900 ky B.P. are characterized by high-frequency fluctuations with an amplitude on the order of 0.5 to 1°/oo. After 800 ky B.P. the $\delta^{13}C$ record changes to longer period negative and positive $\delta^{13}C$ intervals, with an excursion of particularly positive values (>1.5°/oo) corresponding approximately to isotope stages 15-14 and extending into stage 13. A negative $\delta^{13}C$ excursion is then recorded during stages 11 and 10, followed by a positive excursion during isotope stage 9. Another distinct interval is a broad negative $\delta^{13}C$ event within the Ericson biostratigraphic T zone, the approximate equivalent to the Jaramillo and Brunhes Chron. This negative $\delta^{13}C$ interval is also seen in Caribbean Site 502B (Figure 9), as are the positive and negative excursions between the intervals of isotope stages 15 through 9 in E67-135 and Site 502B. Within the approximate stratigraphy from both the for-

aminiferal zonation and isotope stages in E67-135 and Site 502B, therefore, an approximate correspondence seems to exist between the Gulf of Mexico and the Caribbean, not only between individual positive and negative carbon isotope events, but also between long-term shifts in the $\delta^{13}C$ record of the Pleistocene. However, before point-for-point correlations can be attempted between the Gulf of Mexico and Caribbean, further refinements in the isotopic records and respective chronologies are necessary.

## Climatic and Oceanographic Implications of the Surface Water $\delta^{13}C$ Records

A complete review of the factors affecting the $\delta^{13}C$ of surface water total $CO_2$ ($\Sigma CO_2$) and atmosphere-surface ocean $CO_2$ exchange is totally beyond the scope of this study, but a brief summary of some of the more important factors that relate to the Gulf of Mexico (as a tropical-subtropical low-productivity basin) is warranted. The $\delta^{13}C$ record of the last 150,000 years will then be discussed and used as a focus for inter-

## PLEISTOCENE RECORD (E67-135)

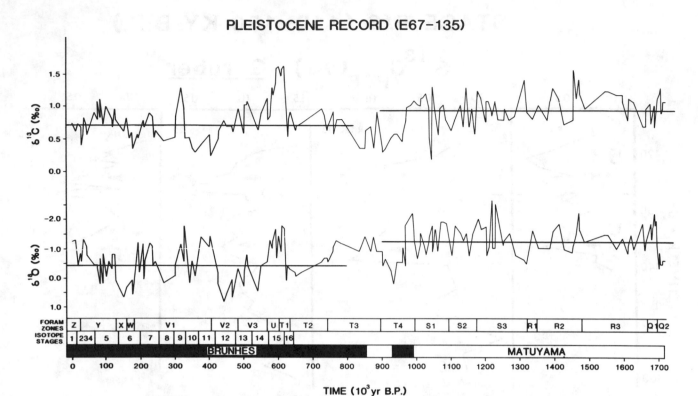

Fig. 8.  The oxygen and carbon isotopic records for the Pleistocene bore-hole Eureka 67-135 based on monospecific isotopic analyses of <u>Globigeri-noides ruber</u>.  The time scale and inferred paleomagnetic polarity boundaries are based on interpolated age dates for the oxygen isotopic stages and for the Ericson foraminiferal faunal zones [after Neff, 1983].  The horizontal lines drawn through the oxygen and carbon isotopic records represent approximate mean values and illustrate the stepwise depletion in $^{13}$C and enrichment in $^{18}$O occurring at approximately 900 ky B.P.

preting the surface $\delta^{13}$C changes during the Holocene and Pleistocene time scales.

In terms of inorganic carbon, the atmosphere and surface ocean reservoirs above the thermocline are nearly equal (approximately 0.0068 and 0.0075 x $10^{20}$ g C, respectively) [Keeling, 1973; Skirrow, 1975].  While the atmosphere and surface ocean reservoirs are in a quasi-steady state equilibrium, pronounced disequilibrium exists between the partial pressures of $CO_2$ in the atmosphere and surface ocean due to a combination of many factors: (1)  productivity variations which produce local/regional differences in the biological fixation of $CO_2$; (2)  upwelling rates which introduce $CO_2$-rich ($^{13}$C depleted) water from below the thermocline; (3)  surface circulation and temperature patterns [Takahashi, 1961; Keeling, 1968; Gordon et al., 1971; Kroopnick, 1974, 1984; Skirrow, 1975; Kroopnick et al., 1977].  The ocean basically handles a given input of atmospheric $CO_2$ via fixation as organic matter and calcium carbonate in the surface waters, oxidation of the organic matter in the water column, and

dissolution or burial of the $CaCO_3$ at the sediment-water interface [Kroopnick, 1974; Berger, 1977]. The amount of atmospheric $CO_2$ which the surface ocean mixed layer can absorb depends on the buffering capacity of seawater and chemical equilibria among the species of dissolved inorganic carbon ($CO_2$, $HCO_3$, $CO_3^{--}$) [Sundquist et al., 1979]. Detailed studies of the anthropogenic increase in atmospheric $pCO_2$ and carbon 13-carbon 14 inter-relations have shown that the atmosphere and the surface ocean are closely linked [Keeling, 1973, 1979; Siegenthaler et al., 1978].  Assuming that the atmosphere-surface ocean remain linked on geologically meaningful time scales, then the $\delta^{13}$C of biogenic carbonate precipitated from the surface ocean mixed layer (i.e., foraminifera, coccolithophorids, molluscs, corals, etc.) should serve as proxy data to model $\delta^{13}$C and $pCO_2$ changes in the atmosphere-surface ocean system.

Recently, Shackleton et al. [1983] showed that surface water $\delta^{13}$C changes over the last 150,000 years are similar in magnitude to those recorded by benthic foraminifera and that the $\delta^{13}$C records

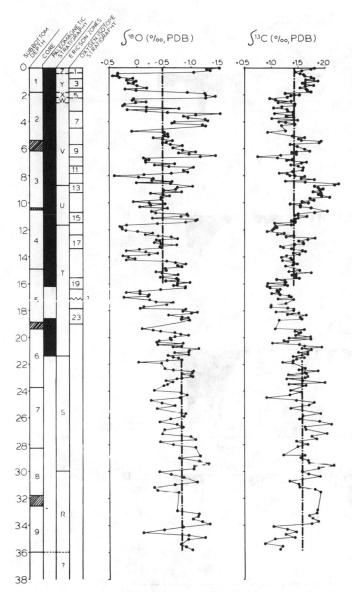

Fig. 9. The oxygen and carbon isotopic records for hydraulically piston cored DSDP Site 502B from the Caribbean based on monospecific analyses of Globigerinoides sacculifer [from Prell, 1982]. Also shown are the Ericson foraminiferal faunal zones, oxygen isotope stages, and paleomagnetic stratigraphy for the site. The vertical dash-dotted lines indicate the mean isotopic value in each of the records and illustrate the negative $^{13}C$ and positive $^{18}O$ shifts which occur at the approximate level between the Jaramillo subchron and Brunhes normal chron.

for the planktonic species Neogloboquadrina dutertrei from the Panama Basin and Globigerina bulloides from the southern Indian Ocean were similar in character. In Figure 10, a comparison of the planktonic $\delta^{13}C$ record from the Gulf of

Mexico (G. ruber, TR126-23) clearly argues that the surface $\delta^{13}C$ records are reflecting temporal variations of the average $\delta^{13}C$ of the surface ocean and not changes in local factors. The Gulf of Mexico planktonic record provides the low-productivity, low-dissolution mid-gyre-like environment needed to minimize the potential effects of productivity, temperature and upwelling changes in the $\delta^{13}C$ of the surface waters [Shackleton et al., 1983]. The three most easily recognizable surface water $\delta^{13}C$ events occur from approximately 150 to 120 ky B.P. (stage 6 to 5e), 60 to 55 ky B.P. (stage 4 to 3) and 27 to 8 ky B.P. (stage 2 to 1), with perhaps other minor events in stage 4 and late stage 5. These minor events and the differences in the timing of the three major events may be either more locally produced or due to sampling errors and nonlinear sedimentation rates. In fact, the largest differences among the planktonic $\delta^{13}C$ records occur in the Panama Basin record which is based on a species (N. dutertrei) which inhabits depths closer to the base of the seasonal thermocline than the mixed surface layer and which was analyzed as a dissolution-resistant species to minimize the effects of severe dissolution [Shackleton et al., 1983].

Regardless of these minor differences, however, the basic similarity between the planktonic $\delta^{13}C$ records suggests that the surface water signal is a very sensitive monitor of changes in atmospheric $pCO_2$. In fact, the rapid rise in atmospheric $CO_2$ content recorded in ice cores at the last deglaciation [Neftel et al., 1982] correlates equally as well with the decrease in the surface-to-deep water $^{13}C$ gradient [Shackleton et al., 1983]. Earlier in the record, Shackleton et al. [1983] used the deep sea gradient to infer that the period of 130-120 ky B.P. witnessed higher atmospheric $CO_2$ levels than the pre-anthropogenic levels during the Holocene. Again, Figure 10 illustrates that this major atmospheric $CO_2$ change is recorded in a most pronounced fashion in the planktonic records, with the benthic $^{13}C$ depletion actually leading the peak planktonic depletion in stage 6-5e by approximately 10,000 years. In fact, rather than subtracting the surface and deep $\delta^{13}C$ records, examination of the timing of particular $^{13}C$-depleted events in the two records reveals an intriguing suggestion of a lead-lag relationship between the surface and deep water $^{13}C$ signals at other times (Figure 10). Clearly, we are dealing with a limited amount of data, but a preliminary comparison of the three major $^{13}C$ depletion events in the planktonic records with the benthic record (Figure 10) suggests that the negative benthic events in stages 6, 4 and 2 lead the surface water events at stages 6/5e, 4/3 and 2/1 by possibly as much as 6,000 to 10,000 years. An isotopic study of very high latitude cores (>60°S) by L. D. Labeyrie and J.-C. Duplessy (unpublished manuscript, 1984) is revealing that changes in the surface and deep water $\delta^{13}C$ records are nearly synchronous, so what could possibly produce

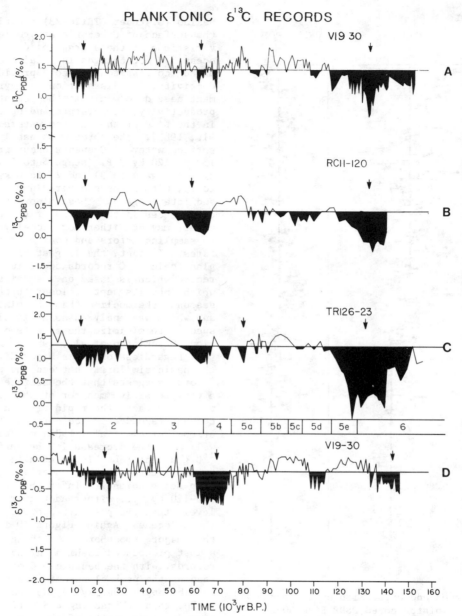

Fig. 10.  Comparison of planktonic foraminiferal $\delta^{13}C$ records from
(a)  the Panama Basin (V19-30, N. dutertrei) [Shackleton et al., 1983],
(b)  the southeast Indian Ocean (RC11-120, G. ruber) [Shackleton et al.,
1983] and (c)  the Gulf of Mexico (TR126-23, G. ruber) (this study),
with (d)  the benthic $\delta^{13}C$ record of Uvigerina from RC11-120 [Shackleton
et al., 1983].  Vertical arrows indicate the most prominent negative
$\delta^{13}C$ events in each of the records.  The chronology is based on the
established oxygen isotope stages in each of the cores.

such a lead-lag effect in the mid- to low-latitude
$\delta^{13}C$ records shown here?  Shackleton [1977]
admitted the possibility that negative $\delta^{13}C$
events in benthic forams could be related to
enhanced oxidation of $^{13}C$-depleted organic matter
on the seafloor but discounted this possibility

as a widespread occurrence.  Recently, Curry and
Lohmann [1983] reported evidence for increased
delivery of organic matter into the eastern
basins of the Atlantic coupled with times of
reduced advection of oxygen-rich deep water.
Perhaps, therefore, the apparent lag between the

surface and deep water $\delta^{13}C$ events reflects some increase in the response time of the carbon cycle or the rate of organic carbon delivery from the surface to deep water in mid- to low-latitude regions sufficiently large to decouple the surface and deep water carbon reservoirs. Obviously, much more work needs to be done before any of these preliminary ideas can be tested, and I conclude that a rigorous study is needed of surface water $\delta^{13}C$ records as an important means of partitioning the oceanic reservoir into separate vertical and horizontal components, as shown to be important in box models for the transfer of $CO_2$ (and $^{13}C$) between natural reservoirs [e.g., Broecker et al., 1960, 1971; Broecker and Li, 1970; Keir and Berger, 1983; Broecker and Peng, 1982].

In summary, therefore, the close agreement between the character of the $\delta^{13}C$ record of TR126-23 from the Gulf of Mexico and timing of the negative surface $^{13}C$ excursions in the Panama Basin and southern Indian Ocean argues that global changes in the level of atmospheric $CO_2$ and not local factors, are driving the surface water $\delta^{13}C$ record. The Gulf of Mexico record further corroborates the interesting observation made by Shackleton et al. [1983] that the highest levels of atmospheric $CO_2$ may have coincided with the warm global temperatures and high sea levels during stage 5e. The results presented here also suggest an association between negative surface water $\delta^{13}C$ (high atmospheric $CO_2$ levels) immediately preceding the rapid changes in global ice volume which occurred at the isotopic stage 6/5e, 4/3 and 2/1 boundaries. Until evidence is presented to the contrary, therefore, I further suggest that the surface $\delta^{13}C$ record for the Holocene and the lack of correlation between it and the $\delta^{18}O$ record most probably represent the response of the surface waters to high-frequency background fluctuations in atmospheric $pCO_2$ on time scales of less than 1,000 years. The meaning of the long-term $\delta^{13}C$ record for the Pleistocene (i.e., E67-135 in the Gulf of Mexico and Site 502B in the Caribbean) is even more problematic. However, the fact that the magnitude and frequency of the surface water $\delta^{13}C$ fluctuations in the Gulf of Mexico and Caribbean are similar within the available stratigraphic framework suggests that the basic controls on the surface water records have not appreciably changed during the Pleistocene. More Pleistocene records with better chronological control will have to become available, however, before it is possible to determine if the long-term $\delta^{13}C$ shift at approximately 900 ky B.P. is related to differences in climatic states associated with the stepwise shift in $\delta^{18}O$.

## Conclusions

1. The $\delta^{13}C$ signal in the surface ocean mixed layer, as recorded in the biogenic carbonate of planktonic foraminifera, from such disparate regions as the Gulf of Mexico, Panama Basin and southeast Indian Ocean, may serve as a reliable monitor of past fluctuations in atmospheric $CO_2$ content and the degree of linkage between the atmosphere and ocean and between the surface and deep water. Isotopic studies of planktonic foraminifera should therefore play a large role in future investigations to document the timing and magnitude of surface water variations as a record of $pCO_2$ and global changes in the carbon budget.

2. The surface water $\delta^{13}C$ signal does not appear for all practical purposes to be correlated with sea surface temperature changes as recorded by the glacial-interglacial oxygen isotopic signal on either 10,000-year or 100,000-year cycles during the Pleistocene. A possible lead-lag relationship may exist between negative $\delta^{13}C$ changes in the surface and deep waters during the late Pleistocene.

3. The input of freshwater into the Gulf of Mexico from the southern periphery of the Laurentide Ice Sheet had a significant but transitory effect on the surface water $\delta^{13}C$. The largest negative isotopic shift occurred during the initial input of the meltwater and lasted for a total duration of approximately 4,000 years.

4. A small (<0.4°/oo) but significant shift occurs in the $\delta^{13}C$ record for the Pleistocene at approximately 900 ky B.P. The timing and magnitude of the early to late Pleistocene stepwise shift in $\delta^{13}C$ (1) occurs in conjunction with a shift in mean $\delta^{18}O$ values and (2) occurs in the Gulf of Mexico at approximately the same time as in the Caribbean Site 502B.

5. The Holocene surface water carbon record reveals many high-frequency, negative $\delta^{13}C$ excursions on the order of 0.3 to 0.5°/oo on a time scale varying between 100 and 500 years, depending on the accumulation rate of the sedimentary section and the sample density. Further work may or may not corroborate the suggestion that these $\delta^{13}C$ changes reflect background changes in atmospheric $CO_2$ levels.

Acknowledgements. The author would like to acknowledge the criticisms of Wolf Berger and an anonymous reviewer and discussions with Dick Fillon, Jean-Claude Duplessy and Laurent Labeyrie for improving the manuscript. I thank the American Geophysical Union for financial assistance to present our Gulf of Mexico work at the Chapman Conference. Thanks to Charlotte Brunner and Stephan Gartner for providing samples from Eureka 67-135. I also thank Amy Leventer, Mary Evans, David Mucciarone and Donna Black of the Stable Isotope Laboratory at the University of South Carolina for technical and drafting assistance. This work was supported through NSF grants ATM-8200424 and OCE-8208911.

## References

Be, A.W.H., and D.S. Tolderlund, Distribution and ecology of living planktonic foraminifera in surface waters of the Atlantic and Indian

Ocean, in The Micropaleontology of Oceans edited by B.M. Funnell and W.R. Riedel, pp. 105-149, Cambridge University Press, 1971.

Berger, W.H., Carbon dioxide excursions and the deep sea record: Aspects of the problem, in The Fate of Fossil Fuel $CO_2$ in the Ocean, edited by N.R. Anderson and A. Malahoff, pp. 505-542, Plenum, New York, 1977.

Berger, W.H., J.S. Killingley and E. Vincent, Stable isotopes in deep-sea carbonates: Box core ERDC-92, west equatorial Pacific, Oceanol. Acta, 1, 203-216, 1978.

Boyle, E.A., and L.D. Keigwin, Deep circulation of the North Atlantic over the last 200,000 years: Geochemical evidence, Science, 218, 784-787, 1982.

Broecker, W.S., Glacial to interglacial changes in ocean chemistry, Prog. Oceanogr., 11, 151-197, 1982.

Broecker, W.S., and Y-H. Li, Interchange of water between the major oceans, J. Geophys. Res., 75, 3545-3552, 1970.

Broecker, W.S., and T-H. Peng, Tracers in the Sea, 690 pp., Lamont-Doherty Geological Observatory, Palisades, N.Y., 1982.

Broecker, W.S., and J. van Donk, Insolation changes, ice volumes, and the $^{18}O$ record in deep-sea cores, Rev. Geophys. Space Phys., 8, 169-198, 1970.

Broecker, W.S., R. Gerard, M. Ewing, and B. C. Heezen, Natural radiocarbon in the Atlantic Ocean, J. Geophys. Res., 65, 2903-2931, 1960.

Broecker, W.S., Y-H. Li, and T-H. Peng, Carbon dioxide--Man's unseen artifact, in Impingement of Man on the Oceans, edited by D.W. Hood, pp. 297-324, John Wiley, New York, 1982.

Brunner, C.A., and J.F. Cooley, Circulation in the Gulf of Mexico during the late glacial maximum, 18,000 years ago, Geol. Soc. Amer. Bull., 87, 681-686, 1976.

Brunner, C.A., and L.D. Keigwin, Late Neogene biostratigraphy and stable isotope stratigraphy of a drilled core from the Gulf of Mexico, Mar. Micropaleontol., 6, 397-418, 1981.

Curry, W.B., and G.P. Lohmann, Carbon isotopic changes in benthic foraminifera from the western South Atlantic: Reconstruction of glacial abyssal circulation patterns, Quat. Res., 18, 218-235, 1982.

Curry, W.B., and G.P. Lohmann, Reduced advection into Atlantic Ocean deep eastern basins during last glaciation maximum, Nature, 306, 577-580, 1983.

Curry, W.B., and R.K. Matthews, Isotopic fractionation in recent fossil planktonic foraminifera from the Indian Ocean: Analyses of equilibrium and disequilibrium patterns, Geol. Soc. Amer. Abstr. Programs, 11, 408, 1979.

Delmas, R.J., J.-M. Ascencio, and M. Legrand, Polar ice evidence that atmospheric $CO_2$ 20,000 years bp was 50% of present, Nature, 284, 155-157, 1980.

Dudley, W., et al., A reevaluation of calcareous nannofossil oxygen isotope data, Geol. Soc. Amer. Abstr. Programs, 14, 478, 1982.

Falls, W.F., Glacial meltwater in-flow into the Gulf of Mexico during the last 150,000 years: Implications for isotope stratigraphy and sea level studies, M.S. thesis, University of South Carolina, Columbia, 1980.

Gartner, S., M.P. Chen, and R.J. Stanton, Late Neogene nannofossil biostratigraphy and paleoceanography of the northeastern Gulf of Mexico and adjacent areas, Mar. Micropaleontol., 8, 17-50, 1983.

Gordon, A.L., Antarctic polar front zone in Antarctic Oceanology I, edited by J.L. Reid, Ant. Res. Ser., 15, Am. Geophys. Union, pp. 205-211.

Keeling, C.D., Carbon dioxide in surface waters, 4, Global distribution. J. Geophys. Res., 73, 4543-4553, 1968.

Keeling, C.D., The carbon dioxide cycle: Reservoir models to detect exchange of atmosphere carbon dioxide with the oceans and land plants, in Chemistry of Lower Atmosphere, edited by S. I. Rasool, pp. 251-329, Plenum, New York, 1973.

Keeling, C.D., The Suess effect: 13 carbon-14 carbon interrelations, Environ. Int., 2, 299-300, 1979.

Keigwin, L.D., Stable isotope stratigraphy and paleoceanography of sites 502 and 503, Initial Rept. Deep Sea Drill. Proj., 68, 1982.

Keigwin, L.D., and M.L. Bender, Speculation about the upper Miocene change in abyssal Pacific dissolved bicarbonate $\delta^{13}C$, Earth Planet. Sci. Lett., 45, 383-393, 1979.

Keir, R.S., and W.H. Berger, Atmospheric $CO_2$ content in the last 120,000 years: The Phosphate-extraction model, J. Geophys. Res., 88, 6027-6038, 1983.

Kroopnick, P.M., Modeling of $O_2$-$CO_2$-$^{13}C$ system in the eastern equatorial Pacific, Deep Sea Res., 21, 211-217, 1974.

Kroopnick, P.M., The distribution of $\Sigma CO_2$ and $\delta^{13}C$ in the world ocean, Deep Sea Res., in press, 1984.

Kroopnick, P.M., S.V. Margolis, and C.S. Wong, $\delta^{13}C$ variations in marine carbonate sediment as indicators of the $CO_2$ balance between the atmosphere and oceans, in The Fate of Fossil Fuel $CO_2$ in the Oceans, edited by N.R. Andersen and A. Malahoff, pp. 295-321, Plenum, New York, 1977.

Labeyrie, L.D., and J.-C. Duplessy, Changes in the oceanic C13/C12 ratio during the last 150,000 years, Palaeogeogr., Palaeoclimatol., Palaeoecol., in press, 1984.

Leventer, A., The oxygen isotopic record for the last 20,000 years from anoxic sediments of the Orca Basin, Gulf of Mexico, M.S. thesis, University of South Carolina, Columbia, 1980.

Leventer, A., D.F. Williams, and J.P. Kennett, Dynamics of the Laurentide ice sheet during the last deglaciation: Evidence from the Gulf of Mexico, Earth Planet. Sci. Lett., 59, 11-17, 1982.

Leventer, A., D. F. Williams, and J.P. Kennett, Relationships between anoxia, glacial meltwater and microfossil preservation/productivity in the Orca Basin, Gulf of Mexico, Mar. Geol., 53, 23-40, 1983.

Neff, E., Pleistocene planktonic foraminiferal biostratigraphy and paleoclimatology of the Gulf of Mexico, M.S. thesis, University of South Carolina, Columbia, 1983.

Neftel, A., H. Oeschger, J. Schwander, B. Stauffer, and R. Zumbrunn, Ice core sample measurements give atmospheric $CO_2$ contents during the past 40,000 years, Nature, 295, 220-223, 1982.

Prell, W.L., Oxygen and carbon isotope stratigraphy for the Quaternary of Hole 502B: Evidence for two modes of isotopic variability, Initial Rep. of the Deep Sea Drill. Proj., 68, p. 592, 1982.

Savin, S.M., and R.G. Douglas, Stable isotope and magnesium geochemistry of Recent planktonic foraminifera from the South Pacific, Geol. Soc. Amer. Bull, 84, 2327-2342, 1973.

Shackleton, N.J., Carbon-13 in Uvigerina: Tropical rain forest history in the equatorial Pacific carbonate dissolution cycles, in The Fate of Fossil Fuel $CO_2$ in the Ocean, edited by N.R. Andersen and A. Malahoff, pp. 401-427, Plenum, New York, 1977.

Shackleton, N.J., The carbon isotope record of the Cenozoic, in Petroleum Source Rocks, edited by J. Brooks and A.J. Fleet, Blackwell, Oxford, in press, 1984.

Shackleton, N.J., Oceanic carbon isotope constraints on oxygen and carbon dioxide in the Cenozoic atmosphere, this volume.

Shackleton, N.J., and N.D. Opdyke, Oxygen isotope and paleomagnetic stratigraphy of equatorial Pacific core V28-238: Oxygen isotope temperatures and ice volumes on a $10^5$ year and $10^6$ year scale, Quat. Res., 3, 39-55, 1973.

Shackleton, N.J., and N.D. Opdyke, Oxygen isotope and paleomagnetic stratigraphy of Pacific Core V28-239: Late Pliocene and latest Pliocene, in Investigations of Late Quaternary Paleoceanography and Paleoclimatology, edited by R.M. Cline and J.D. Hayes, pp. 449-464, Geol. Soc. Amer. Bull. Mem., 145, 1976.

Shackleton, N.J., M.A. Hall, J. Line, and C. Chuxi, Carbon isotope data in core V19-30 confirm reduced carbon dioxide concentration in the ice age atmosphere, Nature, 306, 319-322, 1983.

Siegenthaler, U., M. Heimann, and H. Oeschger, Model responses of the atmospheric $CO_2$ level and $^{13}C/^{12}C$ ratio to biogenic $CO_2$ input, in Carbon Dioxide Climate and Society, edited by J. Williams, pp. 79-87, Pergamon, New York, 1978.

Skirrow, G., The dissolved gases-carbon dioxide, in Chemical Oceanography, edited by J.P. Riley and G. Skirrow, pp. 1-192, Academic, New York, 1975.

Sundquist, E.T., L.N. Plummer, and T.M.L. Wigley, Carbon dioxide in the ocean surface: the homogeneous buffer factor, Science, 204, 1203-1205, 1979.

Takahashi, T., Carbon dioxide in the atmosphere and in the Atlantic Ocean water. J. Geophys. Res., 66, 477-494, 1961.

Thunell, R.C., and D.F. Williams, The stepwise development of Pliocene-Pleistocene paleoclimatic and paleoceanographic conditions in the Mediterranean: Oxygen isotopic studies of DSDP sites 125 and 132, Utrecht Micropaleont. Bull., 30, 111-127, 1983.

Trabant, P.K., and B.J. Presley, Orca Basin: An anoxic depression on the continental slope, northwest Gulf of Mexico, in Framework, Facies and Oil-Trapping Characteristics of the Upper Continental Margin, Stud. Geol., 7, edited by A.H. Bouma, G.T. Moore, and J. Coleman, pp. 289-303, Amer. Assoc. Petrol. Geol., Tulsa, Okla., 1978.

Vincent, E., J.S. Killingley, and W.H. Berger, The magnetic epoch-6 carbon shift: a change in the ocean's $^{13}C/^{12}C$ ratio 6.2 million years ago. Mar. Micropaleontol., 5, 185-203, 1980.

Williams, D.F., M.A. Sommer, and M.L. Bender, Carbon isotopic compositions of Recent planktonic foraminifera of the Indian Ocean, Earth Planet. Sci. Lett., 36, 391-403, 1977.

Williams, D.F., A.W.H. Bé, and R.C. Fairbanks, Seasonal stable isotopic variations in living planktonic foraminifera from Bermuda plankton tows, Palaeogeogr., Palaeoclimatol., Palaeoecol., 33, 71-102, 1981.

# CARBON ISOTOPE RECORD OF LATE QUATERNARY CORAL REEFS: POSSIBLE INDEX OF SEA SURFACE PALEOPRODUCTIVITY

Paul Aharon

Department of Geology, Louisiana State University
Baton Rouge, Louisiana 70803

Abstract. Giant clams (Tridacna gigas) associated with the raised coral reefs in New Guinea yield a $\delta^{13}C$ record spanning the last $10^5$ years which indicates a conspicuous $^{13}C$ depletion during the interglacials and a pronounced $^{13}C$ enrichment during the late ice age interstadials. The distinct positive correlation between the sea level and the $\delta^{13}C$ records from the same stratigraphic sequence implies a coupling between climatic changes and the carbon cycle. The relationship between the metabolic rhythm of coral reef biota and $\delta^{13}C$ variations, established from modern coral reefs, serves as an important guide for the quantitative assessment of the glacial-interglacial $\delta^{13}C$ changes. Model estimates suggest that coral reefs formed during isotope stage 3 were about 2 to 3 times more productive than the interglacial coral reefs from the same sequence. The shifts in the metabolic performance of coral reefs are comparable with the productivity changes of the open ocean biota. The synchronous fertility changes in distinct compartments of the marine biota probably reflect a widespread feature of the ice age ocean and are attributed to contemporary variations in the nutrient chemistry of the ocean.

## Introduction

Recent documentation of $^{13}C/^{12}C$ variations in benthic foraminifera from late Pleistocene deep-sea sediments led Shackleton [1977] and Broecker [1982a, b] to suggest that temporal and spatial redistribution of carbon occurred between different compartments of the global carbon system. Cycling changes in the terrestrial biomass and/or in organic-rich shelf sediments, triggered by the periodic glaciations, were proposed by Shackleton and Broecker, respectively. Broecker's model [1982a, b] has attracted particular attention because it predicts (1) an increase in the nutrient level (e.g., $PO_4$) of the glacial ocean and (2) a proportional enhancement of $CO_2$ consumption by the oceanic biomass resulting in the depletion of $CO_2$ in the surface ocean-atmosphere system.

The significant role that the oceanic biomass might have played in the past by regulating the $CO_2$ level of the surface ocean-atmosphere warrants an investigation of changes in oceanic paleoproductivity through the glacial cycle. The validity of Broecker's postulated changes in oceanic fertility during glaciations can be tested either in the pelagic fraction of deep-sea sediments, as first suggested by Arrhenius [1950], or in coral reefs.

In the present paper I restrict the study to the discussion of organic carbon production changes over the past $10^5$ years in the coral reef environment. The coral reef system conceals several readily observable advantages for pursuing the question of climate-related fertility changes of the oceanic biomass. The coral reef community with its association of plants and animals is one of the most productive natural ecosystems [Aharon, 1982]. From the data summarized in Table 1 it also becomes apparent that coral reefs constitute a quantitatively significant ecosystem within the marine biosphere compartment. In addition, the coral reefs are so narrowly adapted to their tropical, nutrient-deficient, oceanic environment as to be very sensitive to change. It seems only logical to expect, therefore, that this sensitivity to ambient change persisted through the glacial cycle and that these changes are imprinted in the coral reef sedimentary record. The present study takes advantage of a sequence of uplifted coral reef terraces in New Guinea which, although discontinuous, provides closely spaced stratigraphic windows into the past $10^5$ years.

## Geological Setting of the Materials Studied

Narrow, sublinear fringing and barrier reefs enclosing lagoons form a detailed sequence of emergent terraces along 80 km of the present northeast coast of Huon Peninsula, New Guinea

TABLE 1. Distribution of Organic Carbon Between Ecosystem Compartments Making up the Marine Biosphere in the Modern Ocean.

| | Standing Crop,[a] $10^6$ g C km$^{-2}$ | Percent of Total Marine Biomass | Net Primary Production, g C m$^{-2}$ d$^{-1}$ | Mean Turnover Time,[b] Years |
|---|---|---|---|---|
| Open ocean | 1.4 | 25.9 | 0.15 | 0.024 |
| Upwelling zones | 10 | 0.2 | 0.68 | 0.04 |
| Continental shelves | 4.5 | 6.9 | 0.44 | 0.028 |
| Algal beds and reefs[c] | 900 | 31 | 3.2 | 0.77 |
| Estuarines | 450 | 36 | 2.0 | 0.63 |
| Total marine | 4.8 | 100 | 0.19 | 0.07 |

Estimates are calculated from Woodwell [1978].
[a]    Biomass per unit area.
[b]    Ratio of biomass to net primary production.
[c]    According to DeVooys [1979], tropical coral reefs make up over 50% of the algal beds and reef biomass.

(Figure 1, top). Previous studies by Chappell [1974] and Bloom et al. [1974] have shown that the terraces are constructional stratigraphic features associated with eustatic sea level fluctuations superimposed on an upward tectonic movement. Chappell [1974] recognized twenty uplifted coral reef terraces in the sequence, of which only the lower seven are discussed here (Figure 1, bottom). Radiometric dating [Veeh and Chappell, 1970; Bloom et al., 1974] demonstrated that these coral reef terraces from sea level landwards are spaced between the Holocene reef I and the last interglacial interval represented by reefs VIIa and VIIb (Table 2). The terraces in between (II, III, IV, V, VI) represent a quasi-regular recurrence of interstadial sea level maxima at intervals of approximately 20,000 years throughout the entire time span of the Wisconsinan (Würm) glaciation.

The distribution of reef facies (i.e., forereef sediments-reef crest-backreef) as indicated by field observations served as a guide for the sampling of well-preserved giant clams, species Tridacna gigas. In order to provide an unbiased sampling of the seven coral reef complexes, three separate traverses were taken across the coastal strip and the first steep flight of narrow terraces up to reef VII [Aharon, 1980].

Sialum lagoon and barrier reef constitutes a modern analog of the raised coral reef terraces because it is geographically close to the main sampling areas and hence represents a local continuity of the sequence. Figure 2 illustrates the bathymetric map and the coral reef standing crop at Sialum based on observations from October 1977. The lagoon, typical for a tectonically active coast, is narrow (mean 280 m) and elongated (5.2 km) with a mean water depth of 3 m in the areas of active

coral growth and a maximum depth of 17 m. Water circulation on the reef is affected by the seasonal predominance of the southeast trade and the northwest monsoon reflecting the twice-yearly movement of convergence zones across the equator [Aharon, 1980]. Current velocities in the Vitiaz Strait are high (up to 4 m s$^{-1}$) due to the funnelling effect of the narrow strait on water originating in the equatorial water currents. Water temperature varies seasonally between 27°C and 29°C and surface water salinities between 34.5°/oo and 36°/oo [Aharon, 1980]. The tidal range across the Sialum reef is relatively small, not exceeding 1 m but considering the shallowness of the barrier, the tides are an important factor in the water balance of the reef environment. During high tide, rapid exchange of water takes place between the lagoon and the open sea, whereas at low tide the water refluxes from the lagoon through the barrier gaps to the open sea.

Field and Laboratory Experimental Methods

A wide variety of molluscs and corals are present on the modern reef, yet only the giant clams and the low-porosity species of corals (e.g., Porites) are still preserved on the raised fossil reefs. For the present stable isotope study the giant clams were preferred over corals for the following reasons:
1. They are sedentary reef inhabitants of enormous weight (up to 150 kg) and therefore often preserved "in situ" in the growth orientation.
2. The living habitat of the giant clams is restricted to the shallow water of the reef crest and front facies, not exceeding 10 m depth. The consistency of their habitat makes less difficult the comparison of stable isotope

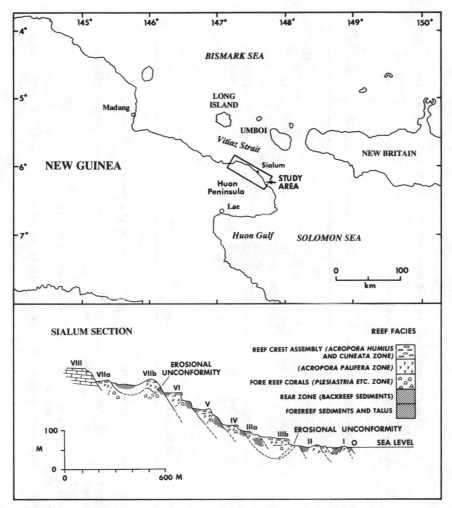

Fig. 1. (Top) Location map of the late Quaternary coral reef terraces along Huon Peninsula. (Bottom) cross section of the uplifted sequence at Sialum (lower).

compositions from reef terraces separated in time.

3. The annual zonation of the giant clam exoskeleton provides a convenient key for locating summer and winter seasonal deposition [Aharon and Chappell, 1983].

4. The giant clams seem to deposit the exoskeleton in apparent isotopic equilibrium with the ambient water [Aharon, 1983].

5. The massive aragonitic material, readily checked for diagenesis by microscope and X ray diffraction methods, is shown by $^{14}C$ and $^{13}C$ studies [Chappell and Polach, 1972; Aharon, 1980] generally to remain immune from carbon exchange and diagenesis.

6. By way of contrast, coral recrystallization and diagenetic exchange of isotopes on the raised reefs are extensive [Aharon and Veeh, 1984], and, more important, the hermatypic corals fractionate stable carbon isotopes in a manner related to their metabolic activity (vital effects [Erez 1978]). These two factors

render corals less suitable for probing the coral reef environment.

Yearly average stable isotope analyses are based on powders crushed from slices cut across the full thickness of the massive internal hinge region of each giant clam shell. Spot drilling along the individual growth increments, according to the method described by Emiliani [1956], yielded subsamples which resolve seasonally based isotopic compositions.

The preparation and purification of $CO_2$ from carbonates is essentially that described by McCrea [1950] using a vacuum extraction line. Prior to the $CO_2$ extraction, the powders were roasted at 400°C for half an hour in vacuum in order to eliminate any isobaric interferences at mass 45 arising from traces of organic material. The dissolved total carbon ($\Sigma CO_2$) in water samples from Sialum was fixed by $BaCO_3$ precipitation immediately after the water sample recovery [see Gleason et al., 1969]. This

TABLE 2.  Carbon Isotope Data for the Late Quaternary Coral Reefs in New Guinea and the Corresponding Organic Productivity Estimates

| Coral Reef | Age,[a] $10^3$ Years | Number of Specimens[b] | $\delta^{13}C$,[c] °/oo PDB | $d\delta^{13}C_r$,[d] °/oo PDB | $d\delta^{13}C_{sw}$,[e] °/oo PDB | $dt$,[f] °C | $P/R$[g] | $P$,[h] $gCm^{-2}d^{-1}$ |
|---|---|---|---|---|---|---|---|---|
| Modern | 0 | 10(30) | 2.28±0.08 | 0 | 0 | 0 | 1.0±0.1 | 5.9±1.2 |
| Reef I | 7±0.1 | 6(13) | 2.75±0.11 | 0.47 | 0.34 | -2.7 | 1.16 | 7.8 |
| Reef II | 28.5±0.6 | 1(4) | 3.12±0.26 | 0.84 | 0.10 | -2.9 | 1.59 | 12.9 |
| Reef IIIb | 40 | 6(18) | 2.93±0.11 | 0.65 | 0.12 | -2.0 | 1.42 | 10.9 |
| Reef IIIa | 45 | 3(9) | 2.98±0.15 | 0.70 | 0.15 | -2.5 | 1.45 | 11.1 |
| Reef IV | 60±4 | 6(18) | 3.06±0.11 | 0.78 | -0.08 | -3.0 | 1.67 | 13.8 |
| Reef V | 85±4 | 10(17) | 2.82±0.09 | 0.54 | 0.19 | 0 | 1.24 | 8.7 |
| Reef VI | 107±6 | 5(14) | 2.78±0.12 | 0.50 | 0.10 | 0 | 1.28 | 9.2 |
| Reef VIIb | | | | | | | | |
| Lagoon | 117 | 7(16) | 2.55±0.10 | 0.27 | 0.0 | 0 | 1.19 | 8.1 |
| Barrier | 120±3 | 7(20) | 2.36±0.10 | 0.08 | 0.0 | -3.0 | 1.13 | 7.4 |
| Reef VIIa | 133±4 | 10(24) | 2.48±0.08 | 0.20 | 0.02 | 0 | 1.13 | 7.4 |

a  Age estimates according to Aharon [1983].
b  Giant clam individuals analyzed per coral reef unit.  Figures in parentheses indicate the number of isotope determinations, including chemistry.
c  Yearly average isotope compositions.  Uncertainties are best estimates of the pooled standard error ($1\sigma_{\bar{x}}$), allowing for the instrumental precision and for the number of giant clams analyzed per reef.
d  $d\delta^{13}C_r = \delta^{13}C_{ancient} - \delta^{13}C_{modern}$ (coral reefs).
e  $d\delta^{13}C_{sw} = \delta^{13}C_{ancient} - \delta^{13}C_{modern}$ (deep-sea core data of Crowley and Matthews [1983]).
f  Isotope temperature difference from present values from Aharon [1983].
g,h  Metabolic performance and net productivity estimates of coral reefs according to equations (8) and (9) (see text).

# SIALUM LAGOON, NEW GUINEA
## (6°5′S, 147°36′E)

### a. BATHYMETRIC MAP

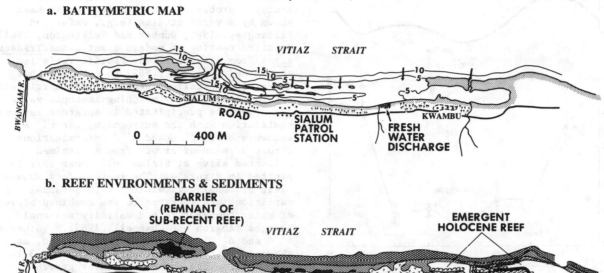

### b. REEF ENVIRONMENTS & SEDIMENTS

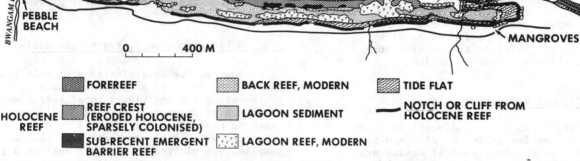

Fig. 2. The modern lagoon and barrier reef at Sialum. (a) Bathymetric map with the depth contours in meters. Arrows indicate the water circulation on the reef under the influence of the southeast trade winds. (b) Distribution of the active coral reef and sediments.

analytical method is generally inferior to the acid stripping method described by Kroopnick [1974], yet it was the only practical method under the field conditions experienced in New Guinea. The extraction of $CO_2$ from the $BaCO_3$ was performed in the same manner as that for $CaCO_3$ samples. The $^{18}O/^{16}O$ ratios of water samples were determined by the $CO_2$–$H_2O$ equilibration method of Epstein and Mayeda[1953].

The unknown $^{13}C/^{12}C$ and/or $^{18}O/^{16}O$ isotope ratio of the sample $CO_2$ was compared on a gas source mass spectrometer with the known ratio of a reference $CO_2$ standard. The isotope data from the carbonates are reported in the conventional delta notation (δ) as deviations from the (PDB) reference standard according to Craig [1957]. Based on multiple successive analyses performed on an internal standard, the overall reproducibility of the stable isotope method is estimated to be ±0.1°/oo.

## Carbon Isotope Shifts in Coral Reefs on Variable Time Scales

$^{13}C/^{12}C$ isotope ratios in a marginal oceanic environment supporting coral reefs are regulated at two different levels:

1. The ratios of carbon isotopes can vary through time for reasons related to global changes in seawater carbon.

2. The local carbon reservoir might change for reasons related exclusively to local conditions.

The two levels of the problem will be discussed in turn, first documenting $^{13}C/^{12}C$ variations in the coral reef environment from New Guinea on time scales ranging from diurnal to secular. A quantitative assessment of the documented isotope variations follows thereafter.

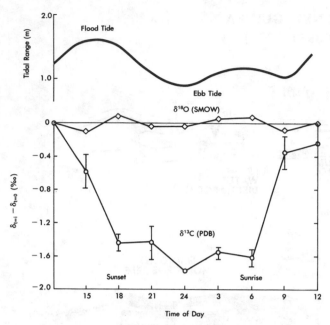

Fig. 3. Oxygen and carbon isotope diurnal variations during the period of sampling (October 18-19, 1977) in the water column overlying the coral reef at Sialum.

## $^{13}C/^{12}C$ Diurnal Variations

The $^{18}O/^{16}O$ and $^{13}C/^{12}C$ diurnal variations in seawater samples collected at 3 m depth on top of the active coral reef at Sialum are illustrated in Figure 3. Whereas no detectable variations beyond the analytical limits of precision are evident from the $\delta^{18}O$ record, significant shifts are shown by the $\delta^{13}C$ record, in particular at the transition between daylight and darkness and vice versa. The $\delta^{13}C$ compositions of the dissolved $\Sigma CO_2$ in the water column exhibit marked $^{13}C$ depletions of up to $1.8^{\circ}/oo$ during darkness relative to the starting values during daylight. Similar diurnal profiles were obtained by Weber and Woodhead [1971] from six different coral reef environments in the Pacific and Atlantic provinces and by Smith and Kroopnick [1981] from a Hawaiian coral reef and tank experiments with corals. These two studies report a $\delta^{13}C$ shift varying between $0.6^{\circ}/oo$ and $1.5^{\circ}/oo$ over a range of circulation intensities. It is also noted that the ubiquitous diurnal rhythm shown by the carbon isotopes is always accompanied by shifts in the levels of $\Sigma CO_2$ and dissolved $O_2$ [Smith and Kinsey, 1978; Kinsey, 1978].

## $^{13}C/^{12}C$ Seasonal Periodicity

Unlike the diurnal isotope variations, the seasonal isotope variability is more difficult to ascertain because of the complex logistics

involved in the long-term monitoring. However, long-lived organisms with skeletal growth bands could be used in favorable circumstances as isotopic probes of their ambient environment as shown by several studies [e.g., Wefer and Killingley, 1980; Dunbar and Wellington, 1981]. Detailed testing of modern giant clams Tridacna gigas from different Indo-Pacific reefs led Aharon and Chappell [1983] to conclude that these offer excellent probes of the coral reef environment, closely matching isotopic values for aragonite precipitated in apparent isotopic equilibrium with the surrounding water. A sequence of 34 $\delta^{13}C$ and $\delta^{18}O$ determinations across 5 years of growth from a specimen collected alive at Sialum in October 1977 is plotted in Figure 4. The isotope data show a clear seasonal periodicity; $\delta^{18}O$ seasonal variations are ascribed to the combined effect of water temperature and salinity seasonal changes [Aharon and Chappell, 1983]. Values of $\delta^{13}C$ and $\delta^{18}O$ are positively correlated, and there is a trend of $^{13}C$ enrichment of 0.3 to $0.6^{\circ}/oo$ from the summer to winter growth increments.

## $^{13}C/^{12}C$ Variations on $10^5$ Years Time Scale

Carbon isotope determinations of multiple giant clams (numbers as shown in Table 2) representing the individual coral reef complexes yield a discontinuous record of $^{13}C/^{12}C$ changes over the past $10^5$ years. The $\delta^{13}C$ data, supported by $\delta^{18}O$ measurements from the same samples, are plotted in Figure 5 against a time scale established by the radiometric determinations (Table 2). The climatic significance of the oxygen isotope record and the quantitative estimates of temperature and ice volume changes derived from $\delta^{18}O$ data have been reported elsewhere [Aharon, 1983]. Features of the carbon isotope record in Figure 5 relevant to the present discussion can be summarized as follows.

The $\delta^{13}C$ compositions of coral reefs representing the last interglacial episode (i.e., fringing reef VIIa and barrier reef and lagoon VIIb) are statistically indistinguishable from the modern reef values ($2.39^{\circ}/oo \pm 0.14$ and $2.23^{\circ}/oo \pm 0.11$) derived from a composite of fringing and lagoon reefs, respectively. By contrast, the coral reefs that grew during the early and late interstadial culminations are up to $0.5^{\circ}/oo$ and $0.8^{\circ}/oo$, respectively, enriched in $^{13}C$ relative to modern values. The distinct positive correlation shown in Figure 6 between the sea levels and $\delta^{13}C$ records from the same stratigraphic sequence implies a coupling between climatic changes and the carbon cycle. Evaluation of this apparent coupling requires a quantitative assessment of the factors affecting the temporal variations of $^{13}C/^{12}C$ ratios in the water column overlying coral reefs.

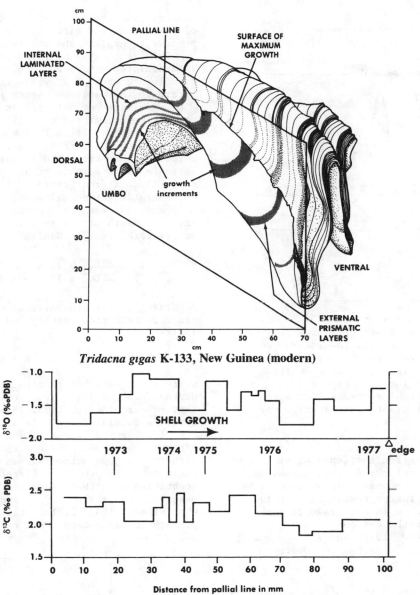

*Tridacna gigas* K-133, New Guinea (modern)

Fig. 4.  Seasonal periodicities of $\delta^{18}O$ and $\delta^{13}C$ from analyses of growth increments of a modern giant clam (T. gigas).

### A Model of Organic Carbon Production and Consumption by Coral Reefs

Coral reefs, marginal to the tropical ocean, tend to develop unique structures through intimate interaction of biological and physical processes. Two of the most important of these processes are the organic carbon production and consumption and the carbonate deposition described by the equations below:

$$CO_2 + H_2O = CH_2O + O_2 \qquad (1)$$

$$Ca^{2+} + 2HCO_3 = CaCO_3 + H_2O + CO_2 \qquad (2)$$

Production/consumption of organic matter and deposition/dissolution of carbonate could cause changes in $\delta^{13}C$ of the $\Sigma CO_2$. However, metabolic processes in coral reefs (equation (1)) would have a much greater effect on the $^{13}C$ of the overlying seawater than the calcification process (equation (2)) does. This is because organic compounds processed by the coral reef biota contain a large fraction of the cycling carbon and are deficient in the $^{13}C$ content [Land et al., 1975], while reef biogenic carbonates yield $\delta^{13}C$ compositions near those of the dissolved $\Sigma CO_2$ [Weber, 1974; Wefer and Berger, 1981]. Hence temporal changes in the

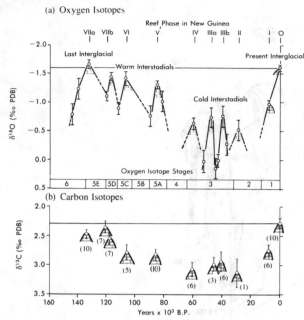

Fig. 5. Discontinuous isotope records for the past $10^5$ years yielded by the lower coral reef sequence in New Guinea. Bars represent the $1\sigma$ standard error estimate. (a) The $\delta^{18}O$ record [Aharon, 1983]. (b) The $\delta^{13}C$ record (see Table 2). Figures in parentheses indicate the number of individual giant clams averaged per coral reef unit.

organic carbon production and consumption of the reef biota would affect the $\delta^{13}C$ of dissolved $\Sigma CO_2$ in a predictable manner. If we denote by $\varepsilon$ the enrichment factor resulting from the process by which carbon is withdrawn from the water, the change in the carbon isotope composition of the $\Sigma CO_2$ left can be described by the Rayleigh equation [Broecker and Oversby, 1971]:

$$R_i/Ro = (N_i/N_o)^\varepsilon \qquad (3)$$

were $R_i$ and $Ro$ stand for the actual and initial isotope ratios and $Ni$ and $No$ for the actual and initial amount of dissolved carbon. This relation can be sufficiently approximated by

$$\Delta\delta^{13}C = \delta^{13}C_i - \delta^{13}C_o = 10^3\varepsilon \ln(\Sigma CO_2{}^i/\Sigma CO_2{}^o) \quad (4)$$

The isotopic enrichment factor, $10^3\varepsilon$, during formation of organic matter in coral reefs, has been estimated to be $-19^o/oo$ [Smith and Kroopnick, 1981].

## Estimates of Coral Reef Productivity at Sialum

Investigation of the factors affecting the diurnal $\delta^{13}C$ variations in seawater overlying coral reefs [Smith and Kroopnick, 1981;

Aharon, 1982] has singled out the metabolism of the coral reef biota as the most likely process to explain the observations in Figure 3. This is because daylight photosynthesis is associated with the $^{13}C$ enrichment in the dissolved carbon pool, while respiration will return to the water $CO_2$ which is depleted in $^{13}C$. Estimates of net organic carbon production integrated over 12 hours of daylight and net organic carbon consumption integrated over 12 hours of darkness for the Sialum experiment are tabulated in Table 3. The estimates were calculated using (1) equation (4); (2) data from Figure 3; (3) an initial $\Sigma CO_2$ concentration of normal seawater (i.e., 2 mmole $L^{-1}$) and, (4) the two expressions below, h being the mean depth of the seawater column above the active coral reef at Sialum:

$$\begin{aligned} r_n &= \Delta\Sigma CO_2 \times 12 \times h \times 12.01 \\ p &= \Delta\Sigma CO_2 \times 12 \times h \times 12.01 \end{aligned} \qquad (5)$$

According to the conventional definition of gross primary productivity and respiration [Smith, 1973, 1974],

$$P = p + r_d \qquad R = r_n + r_d \qquad (6)$$

where $r_n$, $r_d$ are net respiration rates during darkness and daylight, respectively. Assuming that $r_n = r_d$ [Smith, 1973, 1974] and using r and p values tabulated in Table 3, we obtain $P = 11$ g C m$^{-2}$ d$^{-1}$, $R = 11.8$ g C m$^{-2}$ d$^{-1}$, and a P/R ratio of 0.93. The P/R estimate obtained with the isotope method agrees well with estimates from other coral reef environments quantified by either dissolved $O_2$ or by $\Sigma CO_2$ observations [Aharon, 1982] and is very near to what Kinsey [1983] calls "standard metabolic performance of modern coral reefs" at

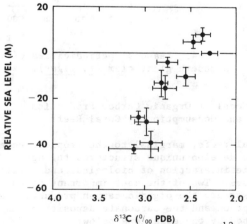

Fig. 6. Inverse relationship shown by $\delta^{13}C$ and Paleosea level data obtained from the coral reef terraces in New Guinea. Sea level data from Table 1 [Aharon, 1983].

TABLE 3. Estimates of Net Organic Carbon Production and Consumption for
the Modern Coral Reef at Sialum Using the Isotope Method

| Diurnal Cycle | $d\delta^{13}C$,[a] $^o$/oo PDB | $d\delta^{13}C/dt$,[b] $^o$/oo hr$^{-1}$ | $d\Sigma CO_2/dt$,[c] mmole hr$^{-1}$ | r,[d] g C m$^{-2}$ d$^{-1}$ | p,[e] g C m$^{-2}$ d$^{-1}$ |
|---|---|---|---|---|---|
| Darkness | -1.57 | -0.13 | 0.0137 | 5.9 | |
| Daylight | 1.33 | 0.11 | -0.0117 | 5.9 | 5.1 |

[a] Data from Figure 3.
[b] Observed change in $\delta^{13}C$ of dissolved carbon per unit of time.
[c] Inferred change in the $\Sigma CO_2$ per unit of time from equation (4) (see text) assuming an initial concentration of normal seawater (i.e., 2 mmole L$^{-1}$).
[d,e] Net productivity estimates from equation 5 (see text).

$P/R = 1.0 \pm 0.1$. The best estimate of the biomass effect on the $\delta^{13}C$ of reef water is derived from the expression below:

$$f = (\Delta\Sigma CO_2/\Sigma CO_2^o) = (\Delta\delta^{13}C/10^3\epsilon) \qquad (7)$$

At Sialum, the biomass effect is therefore about $8^o$/o.

## Constraints on the Secular $\delta^{13}C$ Variations in Coral Reefs

Three variables are likely to explain the secular $\delta^{13}C$ shifts shown in Figure 5: (1) the thermodynamic effect of temperature change on the isotopic enrichment factor $\epsilon$, (2) changes in the $\delta^{13}C$ of the surface ocean, and (3) changes in the metabolic performance of the coral reefs. These three factors and their combined effect on the $\delta^{13}C$ values are examined in turn.

### The Temperature Factor

The ambient water temperature is a prime physical factor which is climate related and could affect the $^{13}C$ content of carbonate. The maximum cooling around New Guinea during the late interstadial culminations was about 3°C below present (Table 2); the temperature dependency of $\epsilon$, as established experimentally by Emrich et al. [1970], is $0.035^o$/oo °C$^{-1}$. The temperature change would tend to lower the $\delta^{13}C$ composition by $0.11^o$/oo relative to present while the documented $\delta^{13}C$ changes suggest enrichments of up to $0.8^o$/oo. On thermodynamic grounds, therefore, the temperature variations are inadequate to explain the recorded $\delta^{13}C$ shifts, and other factors must have been involved.

### Changes in the $^{13}C$ of the Surface Ocean

Relatively few studies have investigated the $\delta^{13}C$ changes of the late Pleistocene surface ocean. Broecker [1982a, b] summarized the existing $\delta^{13}C$ data from planktonic foraminifera of deep-sea sediments and concluded that (1) the planktonic record is invariably noisy and (2) a negative shift of $0.1^o$/oo from the last glacial maximum to the present could be discerned. More recent reports by Shackleton et al. [1983] and Crowley and Matthews [1983] have extended the planktonic $\delta^{13}C$ record to the last interglacial event.

Crowley and Matthews' study of a deep-sea core located beneath the North Atlantic mid-gyre deserves particular attention because their $\delta^{13}C$ record is more likely to reflect the effect of chemical changes on the $\delta^{13}C$ of surface ocean due to the extreme deficiency of nutrients there. A comparison of the coral reef $\delta^{13}C$ record with the planktonic $\delta^{13}C$ record of Crowley and Matthews [1983] reveals that there is a distinct dissimilarity between the two, the coral reefs being significantly enriched in $^{13}C$, particularly during isotope stage 3. As the isotope trends observed in the coral reef $\delta^{13}C$ record cannot be fully explained by chemical changes in the surface ocean, it must be concluded that shifts in the metabolic performance of ice age coral reefs hold the key.

### Organic Productivity Effects on $\delta^{13}C$

Any long-term imbalance in the metabolic performance (P/R ratio) leading to shifts in the organic productivity of coral reefs would affect the $\delta^{13}C$ composition of the water column over the reef, and the isotopic change would in turn be registered by the carbonate extracted from the reef water. On the premise that there is a link between the documented $\delta^{13}C$ variations on different time scales, the carbon isotopes in the parental carbon reservoir and consequently in the reef material could be scaled to the metabolic performance of the coral reefs using the relationship established from the diurnal

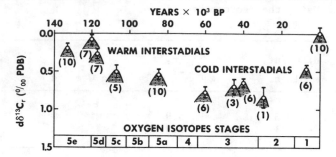

**(a) CARBON ISOTOPE SHIFTS FROM PRESENT VALUES**

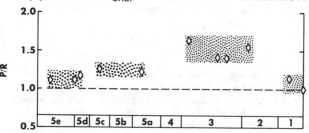

**(b) RATIO OF C_ORG. PRODUCTION TO OXIDATION**

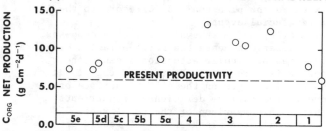

**(c) PALEO-PRODUCTIVITY OF ICE-AGE CORAL REEFS**

Fig. 7.  Model estimates of metabolic performance (P/R) and net organic carbon production (p) of coral reefs through the last glacial cycle.  Note that higher productivities correspond with the observed $^{13}$C enrichments of the late interstadial reefs.

variations study at Sialum.  Viewed this way, a change of 0.93 P/R units corresponds to a 1.33 °/oo change of $\delta^{13}$C (see Table 3 and text).

Organic Productivity Changes
of Ice-Age Coral Reefs

P/R estimates for the coral reef units present in the New Guinea sequence, listed in Table 2, are derived from the productivity parametrization of $\delta^{13}$C changes above, corrected for the temperature effect and changes in the $\delta^{13}$C of the surface ocean:

$$d\ (P/R) = [d\ \delta^{13}C_r - d\ ^{13}C_{sw} - d\epsilon/dt \times dt]/1.43 \tag{8}$$

The corresponding net organic productivity changes are computed using the derivation below and keeping the organic oxidation rate constant ($r_n$), an assumption often used in studies concerned with biological productivity (see, for example, Smith [1973]).

$$p = r_n(2P/R - 1) \tag{9}$$

The results plotted in Figure 7 together with the $\delta^{13}$C record suggest that (1) unlike the present and the last interglacial units, the coral reefs which grew late during the last ice age stages performed at metabolic rates at least twice as high and (2) while the modern reef acts as an organic sink zone (P/R < 1), the ice age reefs were net exporters of organic carbon (P/R > 1).

Discussion

A productivity parametrization of $\delta^{13}$C shifts contains the explicit assumption that the effective exchange of water between the reef environment and the open ocean remained essentially unchanged over different time scales.  Although the rate of mixing between the open ocean and the reef water is unknown, two lines of evidence seem to indicate that paleocirculation through the ancient reefs did not differ significantly from present.

First, the amplitude and periodicity of the seasonal $\delta^{18}$O and $\delta^{13}$C variations of the premodern reefs, typified by the record shown in Figure 8, are compatible with the modern record in Figure 4.  The seasonal productivity changes inferred from the $\delta^{13}$C variations are corroborated by the results of Kinsey [1983], who established through direct monitoring of dissolved $O_2$ that seasonal variability is a major factor in community metabolism of coral

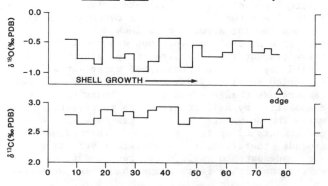

Fig. 8.  Seasonal periodicities of $\delta^{18}$O and $\delta^{13}$C from analyses of growth increments of a 60,000-year-old giant clam associated with the coral reef crest IV.

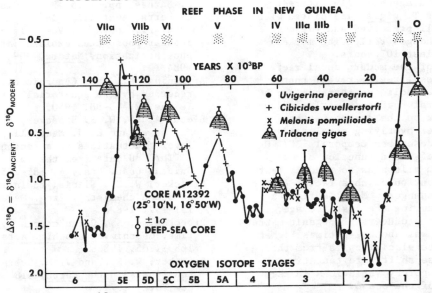

Fig. 9. Comparison of $\delta^{18}O$ of coral reefs (this study) and deep-sea benthics from Meteor core 12392 [Shackleton, 1977]. The isotopic sequences are plotted to the same scale of change from present values and are adjusted to yield best agreement for the modern samples. Deep-sea $\delta^{18}O$ data are adjusted to match the chronology of the coral reef terraces.

reefs. Alternatively, if the ancient reefs were more isolated from the open ocean, then the $\delta^{18}O$ record derived from the same coral reefs would be far more sensitive to such a disturbance than $\delta^{13}C$ is. However, the agreement between the coral reefs and the deep ocean $\delta^{18}O$ records, shown in Figure 9, seems to rule out this possibility. Having established that linking the productivity to $\delta^{13}C$ has some validity, next I discuss the factors which could affect the productivity of the surface ocean biota on $10^5$ years time scale.

Regarding coral reefs, the most important factors which limit the productivity of the biomass are stirring, light intensity, temperature, inorganic carbon supply, oxygen concentration, and nutrient supply [Larkum, 1983]. Although there has been little work on the effect of these limiting factors on coral reef fertility, it seems unlikely that any of the factors above, with the notable exception of the nutrients, were significantly different from present levels during the growth of the ice age coral reefs. To the contrary, the oxygen isotope work indicates water temperatures of only 3°C cooler during isotope stage 3 (Table 2), yet the inferred productivity levels were about 2 to 3 times higher than present.

Nutrient supply, especially nitrogen and phosphorus, is probably the major factor which limits productivity in coral reefs [Kinsey and Domm, 1974; Crossland, 1983; Larkum, 1983]. The levels of these two essential nutrients in reef

water are low (0.03 to 0.1 $\mu m$ $L^{-1}$ for N and 0.1 to 0.3 $\mu m$ $L^{-1}$ for P) and practically do not differ from the measured levels in the adjacent tropical ocean [Crossland, 1983]. Were the nutrient delivery to the reef water not increased during episodes of higher organic production levels, the coral reef biomass and the associated organics would have been consumed, and the system would have collapsed. As we hold evidence to the contrary, then additional nutrients must have been imported into the reefs either from the adjacent land or from the adjacent tropical ocean. The correspondence between the paleoproductivity shifts derived from deep-sea sediments [Muller and Suess, 1979; Pedersen, 1983; Morris et al., 1984] and coral reefs seem to favor changes in the nutrient chemistry of the ocean. However, in view of the ongoing discussion on the likely climate-related mechanisms of oceanic fertility, I leave the discussion open whether the higher productivity of the glacial ocean reflects a shelf-sediment dynamics [Broecker, 1982a, b] or changes in the rate of ocean mixing [Newell et al., 1978; Sarmiento and Toggweiler, 1984].

## Summary

A late Quaternary sequence of raised coral reef terraces along the Huon Peninsula, New Guinea, offers an excellent opportunity to study the carbon cycle through the last ice age in a marginal oceanic environment. While the reef limestones are diagenetically altered, giant clams, species $\underline{T.\ gigas}$, associated with the terraces are well preserved and yield a record of $^{13}C/^{12}C$ changes reflecting marked contrasts over the last $10^5$ years.

Isotope studies of the modern coral reef, analogous to the fossil reefs, reveal that variations in the metabolic performance are a major feature of coral reef communities and have a significant effect on the carbon isotope chemistry of seawater overlying coral reefs. A simple model which describes temporal $^{13}C/^{12}C$ changes in coral reefs as a function of community metabolism allows us to readily test the relative contribution of this factor to the documented isotope changes. Model estimates suggest that, late during the last ice age, coral reefs were net producers of organic matter and their fertility was about 2-3 times higher than that of the interglacial units from the same sequence. These shifts are tentatively attributed to contemporary variations in the nutrient chemistry of the ocean.

Reconstruction of productivity shifts from carbon isotope studies on a more detailed level than given here is necessary for a better understanding of the links between climate, ocean chemistry, and marine biota at low latitudes. For such research, the context of a flight of similar raised coral reef terraces would provide an ideal testing ground.

Acknowledgments. I thank W. Berger, E. Sundquist, and an anonymous reviewer for offering valuable comments and Faye Dedon, Jana Kloss and Sissie Gum, who patiently typed several versions of the manuscript. AGU is acknowledged for providing financial support to attend the Chapman Conference on Natural Variations in Carbon Dioxide and the Carbon Cycle.

## References

Aharon, P., Stable isotope geochemistry of a late Quaternary coral-reef sequence, New Guinea: Application of high resolution data to Palaeoclimatology, Ph.D. thesis, 259 pp., Australian National University, Canberra, October 1980.

Aharon, P., $^{13}C/^{12}C$ isotope ratio variations over the last $10^5$ years in a New Guinea coral-reef environment: Implications for the fertility shifts of the tropical ocean, in The Cycling of C, N, S and P in Terrestrial and Aquatic Ecosystems, edited by I. E. Galbally and J. R. Freney, pp. 119-132, Australian Academy of Science, Canberra, 1982.

Aharon, P., 140,000 year isotope climatic record from raised coral reefs in New Guinea, Nature, 304, 720-723, 1983.

Aharon, P., and J. Chappell, Carbon and oxygen isotope probes of reef environment histories, in Perspectives on Coral Reefs, edited by D.

J. Barnes, pp. 1-15, Brian Clouston Publ., Canberra, Australia, 1983.

Aharon, P., and H. H. Veeh, Isotope studies of insular phosphates explain atoll phosphatization, Nature, 309, 614-617, 1984.

Arrhenius, G., Late Cenozoic climatic changes as recorded by the equatorial current system, Tellus, 2, 83-88, 1950.

Bloom, A. L., W. S. Broecker, J. Chappell, R. K. Matthews, and K. J. Mesolella, Quaternary sea level fluctuations on a tectonic coast: New $^{230}Th/^{234}U$ data from the Huon Peninsula, New Guinea, Quat. Res., 4, 185-205, 1974.

Broecker, W. S., Glacial to interglacial changes in ocean chemistry, Prog. Oceanogr., 11, 151-197, 1982a.

Broecker, W. S., Ocean chemistry during glacial time, Geochim. Cosmochim. Acta, 46, 1689-1705, 1982b.

Broecker, W. S., and V. M. Oversby, Chemical Equilibria in the Earth, 318 pp., McGraw Hill, New York, 1971.

Chappell, J., Geology of coral terraces, Huon Peninsula, New Guinea: A study of Quaternary tectonic movements and sea-level changes, Geol. Soc. Am. Bull., 85, 553-570, 1974.

Chappell, J., and H. A. Polach, Some effects of partial recrystallization on $^{14}C$ dating late Pleistocene corals and molluscs, Quat. Res., 2, 244-252, 1972.

Craig, H., Isotopic standards for carbon and oxygen and correction factors for mass spectrometric analysis of carbon dioxide, Geochim. Cosmochim. Acta, 12, 133-149, 1957.

Crossland, C. J., Dissolved nutrients in coral reef waters, in Perspectives on Coral Reefs, edited by D. J. Barnes, pp. 56-68, Brian Clouston Publ., Canberra, Australia, 1983.

Crowley, T. J., and R. K. Matthews, Isotope-plankton comparisons in a late Quaternary core with a stable temperature history, Geology, 11, 275-278, 1983.

DeVooys, C. G. N., Primary production in aquatic environments, in The Global Carbon Cycle, SCOPE 13, edited by B. Bolin, E. T. Degens, S. Kempe, and P. Ketner, pp. 259-292, John Wiley, New York, 1979.

Dunbar, R. B., and G. M. Wellington, Stable isotopes in a branching coral monitor seasonal temperature variation, Nature, 293, 453-455, 1981.

Emiliani, C., Oxygen isotopes and paleotemperature determinations, Congr. Int. Assoc. Quat. Res. 4th, 831-843, 1956.

Emrich, K., D. H. Ehhalt, and J. C. Vogel, Carbon isotope fractionation during the precipitation of calcium carbonate, Earth Planet. Sci. Lett., 8, 363-371, 1970.

Epstein, S., and T. Mayeda, Variation of $^{18}O$ content of waters from natural sources, Geochim. Cosmochim. Acta, 4, 213-224, 1953.

Erez, J., Vital effect on stable isotope

composition seen in foraminifera and coral skeletons, Nature, 273, 199–202, 1978.

Gleason, J. D., I. Friedman, and B. B. Hanshaw, Extraction of dissolved carbonate species from natural water for carbon isotope analysis, Prof. Pap. U.S. Geol. Surv. 650-D, 248–250, 1969.

Kinsey, D. W., Productivity and calcification estimates using slack-water periods and field enclosures, in Coral Reefs: Research Methods, edited by D. R. Stoddart and R. E. Johannes, pp. 439–468, UNESCO, Paris, 1978.

Kinsey, D. W., Standards of performance in coral reef primary production and carbon turnover, in Perspectives on Coral Reefs, edited by D. J. Barnes, pp. 209–220, Brian Clouston Publ., Canberra, Australia, 1983.

Kinsey, D. W., and A. Domm, Effects of fertilization on a coral reef environment-Primary production studies, Proc. Int. Coral Reef Symp. 2nd, 1, 49–66, 1974.

Kroopnick, P., Correlations between $^{13}$C and $\Sigma CO_2$, Earth Planet. Sci. Lett., 22, 397–403, 1974.

Land, L. S., J. C. Lang, and B. N. Smith, Preliminary observations on the carbon isotopic composition of some reef coral tissues and symbiotic zooxanthellae, Limnol. Oceanogr., 20, 283–287, 1975.

Larkum, A. W. D., The primary productivity of plant communities on coral reefs, in Perspectives on Coral Reefs, edited by D. J. Barnes, pp. 221–230, Brian Clouston Publ., Canberra, Australia, 1983.

McCrea, J. M., On the isotopic chemistry of carbonates and a paleotemperature scale, J. Chem. Phys., 18, 849–857, 1950.

Morris, R. J., M. J. McCartney, and P. P. E. Weaver, Sapropelic deposits in a sediment from the Guinea Basin, South Atlantic, Nature, 309, 611–614, 1984.

Muller, P. J., and E. Suess, Productivity, sedimentation rate, and sedimentary organic matter in the oceans, I, Organic carbon preservation, Deep Sea Res., 26A, 1347–1362, 1979.

Newell, R. E., A. R. Navato, and J. Hsiung, Long-term global sea surface temperature fluctuations and their possible influence on atmospheric $CO_2$ concentrations, Pure Appl. Geophys., 116, 351–371, 1978.

Pedersen, T. F., Increased productivity in the eastern equatorial Pacific during the last glacial maximum (19,000 to 14,000 yr. B. P.), Geology, 11, 16–19, 1983.

Sarmiento, J. L., and J. R. Toggweiler, A new model for the role of the oceans in determining atmospheric $pCO_2$, Nature, 308, 621–624, 1984.

Shackleton, N. J., Carbon-13 in Uvigerina: Tropical rainforest history and the equatorial Pacific carbonate dissolution cycles, in The Fate of Fossil Fuel $CO_2$ in the Oceans, edited by N. R. A. Andersen and A. Malahoff, pp. 401–427, Plenum, New York, 1977.

Shackleton, N. J., M. A. Hall, J. Line, and Cang Shuxi, Carbon isotope data in core V 19–30 confirm reduced carbon dioxide concentration in the ice age atmosphere, Nature, 306, 319–322, 1983.

Smith, S. V., Carbon dioxide dynamics: A record of organic carbon production, respiration, and calcification in the Eniwetok reef flat community, Limnol. Oceanogr., 18, 106–120, 1973.

Smith, S. V., Coral reef carbon dioxide flux, Proc. Int. Coral Reef Symp. 2nd, 1, 77–85, 1974.

Smith, S. V., and D. W. Kinsey, Calcification and organic carbon metabolism as indicated by carbon dioxide, in Coral Reefs: Research Methods, edited by D. R. Stoddart and R. E. Johannes, pp. 469–484, UNESCO, Paris, 1978.

Smith, S. V., and P. Kroopnick, Carbon 13 isotopic fractionation as a measure of aquatic metabolism, Nature, 294, 252–253, 1981.

Veeh, H. H., and J. Chappell, Astronomical theory of climatic change: Support from New Guinea, Science, 167, 862–865, 1970.

Weber, J. N., $^{13}C/^{12}C$ ratios as natural isotopic tracers elucidating calcification processes in reef-building and non-reef-building corals, Proc. Int. Coral Reef Symp. 2nd, 2, 289–298, 1974.

Weber, J. N., and P. M. J. Woodhead, Diurnal variations in the isotopic composition of dissolved inorganic carbon in seawater from coral reef environments, Geochim. Cosmochim. Acta, 35, 891–902, 1971.

Wefer, G., and J. S. Killingley, Growth histories of Strombid snails from Bermuda recorded in their O-18 and C-13 profiles, Mar. Biol., 60, 129–135, 1980.

Wefer, G., and W. H. Berger, Stable isotope composition of benthic calcareous algae from Bermuda, J. Sediment. Petrol., 51, 459–465, 1981.

Woodwell, G. M., The carbon dioxide question, Sci. Am., 238, 34–43, 1978.

# DISTRIBUTION OF MAJOR VEGETATIONAL TYPES DURING THE TERTIARY

Jack A. Wolfe

U.S. Geological Survey, Denver, Colorado 80 225

Abstract. During the latest Paleocene-early
Eocene (about 50-55 million years ago (Ma)),
broad-leaved evergreen vegetation extended to
latitude 70°-75°, and multistratal evergreen
vegetation to 55°-60°. During cool intervals of
the Eocene, broad-leaved evergreen forests were
restricted to about 50°, with poleward areas
occupied by densely stocked coniferous forest.
Semideciduous tropical to paratropical forest
occupied southeastern Asia and southeastern North
America during the middle and late Eocene, re-
placing broad-leaved evergreen forests. Following the terminal Eocene event, dense coniferous forest occupied areas poleward of 50°-60°,
and microthermal broad-leaved deciduous forest
(unknown in the Eocene) occupied areas of the
northern hemisphere south to 35°. Broad-leaved
evergreen forests in both hemispheres were
equatorward of 35°, and multistratal forests
equatorward of 20°. In the southern hemisphere,
areas between 35° and 50° were occupied by a
mixed forest of conifers and broad-leaved evergreens. During the Neogene, broad-leaved evergreen vegetation slightly expanded (particularly
during the mid-Miocene warm interval), as did
coniferous forest at the expense of broad-leaved
deciduous forest. Antarctica was deforested by
the Miocene. Woodlands replaced forests in
southwestern North America and the Mediterranean
region during the Miocene. The rising Himalayas
produced steppe vegetation in central Asia during
the Miocene; in western North America, steppes
developed during the Pliocene. Grasslands in
central North America can be no older than late
Miocene and may be of younger origin. Taiga is
first present in the Arctic region at about 5 Ma
north of 65°-70° and has continued to expand.
Tundra is first recorded at about 2-3 Ma. Evidence of Tertiary desert is absent.

## Introduction

The major factors in the carbon cycle at any
point in time are the geographic distribution of
various vegetational types and the biomass of
each type. During parts of the Paleocene and
Eocene, the great poleward expansion of multistratal, humid, broad-leaved evergreen vegetation
must have produced a greatly different carbon
cycle than during the Holocene. Similarly, the
carbon cycle during the Oligocene and early Miocene would have been markedly different from that
of both the Eocene and Holocene: broad-leaved
deciduous forests covered large areas at middle
latitudes of the northern hemisphere, extensive
grasslands and deserts were absent, and multistratal evergreen forests were of more restricted
distribution than during the Eocene or Holocene.

Even if the past distributions of major vegetational types were accurately known, the variations of biomass within a given vegetational
type could be considerable. Where possible, I
will suggest either a modern analog for particular vegetational types during particular periods
or structural features of the vegetation from
paleobotanical data. The direct paleobotanical
data include in situ preservational situations,
where forests have been preserved by events such
as volcanic ash falls. In instances in which
direct paleobotanical data are lacking, modern
analogs are suggested, based on diversity; a
given forest type typically reaches maximal diversity as optimal growth conditions for the type
are approached.

In order to map the vegetation of the Tertiary, analyses should ideally be based on extensive sequences of assemblages from many
latitudes, altitudes, and orographic positions
and from continental and maritime climatic regimes. Such ideal conditions do not exist.
Further, many floral sequences yet remain to be
studied, and some assemblages, although in
critical geographic areas, are of uncertain age.

Of necessity, the following models of vegetation distribution are based on two major assumptions. First, when data are unavailable from
a given continental area, the vegetation may be
inferred from that of an analogous area on another continent. For example, nothing is known
of Miocene vegetation in eastern North America
north of latitude 40°. The present, markedly
different vegetation and climate north of this
latitude in eastern and western North America
indicate that it would be problematic to apply
the abundant western North American Miocene data
to the other side of the continent; on the other

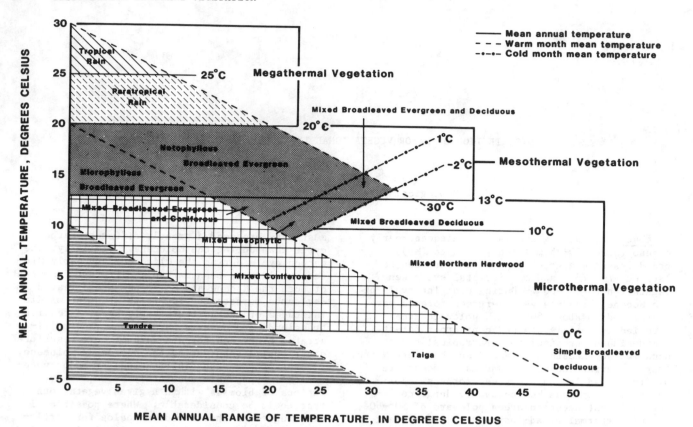

Fig. 1.  Temperature parameters of humid to mesic forests of eastern Asia
(adapted from Wolfe [1979]).  Note that the southern boundaries of Taiga and
Simple Broad-leaved Deciduous forest are shown as approximating a mean annual
temperature of 0°C rather than 3°C (as in the work by Wolfe [1979]).

hand, the climatic zones of northeastern Asia are
today similar to those of eastern North America,
and thus the data on Miocene vegetation in northeastern Asia can be applied to the Miocene of
northeastern North America.

A second major assumption is that the tem-
perature/vegetation relations based on extant
east Asian humid to mesic forests [Wolfe, 1979]
are applicable to the Tertiary (Figure 1).  For
example, if the boundary between Tropical and
Paratropical Rain forests in early Eocene maritime areas was at 50°, and if the latitudinal
temperature gradient is inferred to have been
about 0.25°C/deg latitude [Wolfe, 1978], then
Paratropical Rain forest should have extended to
latitude 70° and Microphyllous or Notophyllous
Broad-leaved Evergreen forests should have extended to the poles, if only temperature is considered.  In continental interiors, lower mean
annual temperatures would have resulted in a
somewhat lower latitudinal placement of various
vegetational belts relative to coastal areas.

The North American Tertiary plant megafossil
record is by far the most completely known for
any continent and will be the primary data base
for the models developed.  Paleogene rocks and

contained floral sequences are known in the
Mississippi Embayment, High Plains and adjacent
Rocky Mountain region of both the United States
and Canada, and along the Pacific coast from
California to Alaska.  Neogene floral sequences
are primarily in western North America from
California and Nevada north to Alaska and the
Canadian Arctic islands.  Scattered Neogene data
are available from the High Plains, Atlantic
coastal plain, southern coastal plain, and
southern Mexico.

Estimation of the areal extent of uplands in
the Paleogene is difficult.  Uplands would presumably have a cooler and less biologically productive vegetation than adjacent lowland areas.
In western North America, such upland areas are
well documented [Axelrod, 1966; Wolfe and Wehr,
in press] from at least the early Eocene on.
Generally, however, the other continents had no
known major upland areas during the Eocene, except Antarctica, which apparently had some alpine
glaciation during the Eocene [Matthews and Poore,
1980; Webb, 1983].

The following models are, wherever possible,
based primarily on physiognomic analyses of leaf
assemblages.  Foliar physiognomy is one of the

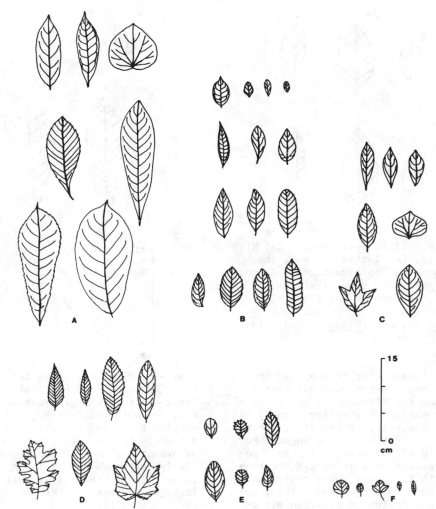

Fig. 2.  Foliar physiognomy characteristic of some major vegetational types.
(a) Tropical Rain forest.  (b) Tropical semideciduous forest.  (c) Notophyllous Broad-leaved Evergreen forest.  (d) Microthermal broad-leaved deciduous
forest.  (e) Mesothermal xerophyllous scrub.  (f) Arctic vegetation.  Leaf
size classes can generally be estimated from length:  less than 8 cm, microphyllous or smaller; 8-12 cm, notophyllous; more than 12 cm, mesophyllous or larger
(including megaphyllous and macrophyllous).

prime attributes of the classification of extant
vegetation [e.g., Webb, 1959].  Tropical and
Paratropical Rain forests (Figure 2), for example, are marked by many foliar characteristics.  These forests have a high percentage
(>60%) of species that have leaves that are
entire margined, the leaf size is large (notophyllous to typically mesophyllous), most leaves
have elongated apices (drip tips), lianes are
diverse (most liane leaves have a cordate base,
are broad, and have palmate venation), the density of foliar venation is high, and most leaves
are coriaceous (indicative of an evergreen habit).  In contrast, in Mixed Northern Hardwood
forest, most species have nonentire margins
(indeed, many have strongly lobed leaves), leaf

size is small (notophyllous to microphyllous),
drip tips are absent, lianes are not diverse, the
density of foliar venation is low, and the leaves
are typically thin (indicative of a deciduous
habit).  Analysis of foliar physiognomy, first
proposed for fossil assemblages by Bailey and
Sinnott [1915], has been used in a general way by
many paleobotanists [e.g., Chaney and Sanborn,
1933; MacGinitie, 1941; Axelrod and Raven, 1978].
Although recently criticized by some workers
[e.g., Dolph, 1979; Dolph and Dilcher, 1979], the
criticisms appear to be based largely on misin-
terpretations or inadequate sampling of modern
vegetation [Wolfe, 1981].

The advantage of physiognomic methodology is
that climatic interpretations are based directly

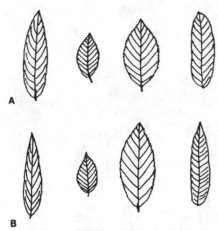

Fig. 3. Streamside foliar physiognomy. (a) High middle latitude. (b) Low latitude. Note that low-latitude (megathermal to mesothermal) leaves have physiognomy (generally nonentire margins) similar to that of high middle latitude (microthermal) streamside leaves. From left to right: willow (Salix), elm (Ulmus, s.l.), alder (Alnus), wingnut (Pterocarya).

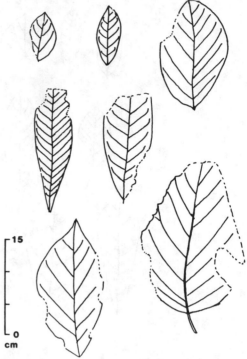

Fig. 4. Latest Paleocene leaves from the Tongue River Member of the Fort Union Formation, Powder River Basin, Wyoming. Most of the leaves are coriaceous (probably evergreen). Most of the species that have nonentire margins are either streamside types (e.g., Betulaceae, Juglandaceae) or probable lianes belonging to Vitidaceae (a family, although common in megathermal rain forests today, that invariably has toothed foliage). Compare to Figure 2a.

on fossils and are independent of supposed systematic affinities and the presumed climatic tolerances of extant relatives. Thus, species of extinct genera (or extinct higher-level taxa) can be included in the analysis, an important factor in dealing with early Tertiary assemblages. Further, the systematic affinities of many taxa described by early workers are, at best, questionable. Of course, in instances where reliable determinations of taxa have been made, these can provide a general validation of physiognomic analyses or provide data in the absence of physiognomic data.

The factor of plate tectonics has been analyzed in regard to the paleolatitudinal placement of the various assemblages. The base maps used are adapted from Smith et al. [1981]. Unless otherwise stated, latitudinal positions cited in the current report are the paleo-latitudinal positions.

A major problem in modeling the distribution of Tertiary vegetation is that of synchroneity.

The occurrence of major temperature fluctuations, particularly in the Paleocene and Eocene, is well documented, both from land plants and from marine planktics [e.g., Wolfe and Poore, 1982, and references cited therein]. Whether these fluctuations were cyclic cannot be determined from available radiometric data. Nevertheless, data from a given cool interval should be compared to data from the same interval; if a map is based on data from both warm and cool intervals, the re-

TABLE 1. Physiognomic Analyses of Two Leaf Assemblages From the Tongue River
Member of the Fort Union Formation, Powder River Basin, Wyoming

| | Number of Species | Percentage of Species | | | | |
|---|---|---|---|---|---|---|
| | | Meso-phyll | Noto-phyll | Micro-phyll | Entire Margined | Coria-ceous |
| Quarter Circle Hills | 28 | 29 | 36 | 36 | 86 | 93 |
| Welch Mine | 27 | 53 | 30 | 7 | 48 | 59 |
| Total | 55 | 45 | 33 | 22 | 67 | 75 |

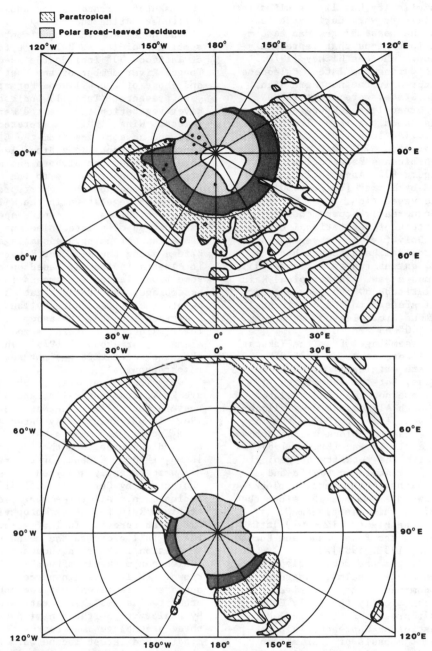

Fig. 5.  Suggested vegetational map of the latest Paleocene-early Eocene (about 50-55 Ma).  Solid circles represent some critical assemblages of latest Paleocene or early Eocene age; open circles represent some critical assemblages of earlier Paleocene age.  Continental positions are adapted from Smith et al. [1981].  Patterns correspond to those in Figure 1.

sulting map could have some very peculiar contradictions.

### Paleocene and Eocene Forests

The vegetation of the Paleocene (54-65 million years ago (Ma)) is the least known of any epoch,

despite the voluminous literature on Paleocene floras.  Almost all described North American Paleocene floras represent floodplain vegetation, which is a low-diversity type that is generally not sensitive to temperatures (Figure 3).  Thus, a large number of species are common to the Paleocene floras of southern Alaska and those of

Wyoming, but this similarity has little climatic significance. Preliminary work during the last 2 years has indicated that assemblages can be obtained from floodplain regions that represent interfluve vegetation. Whereas Hickey [1977, 1980] suggested that during the latest Paleocene mean annual temperatures in the northern High Plains of the United States were about 15°C (i.e., Notophyllous Broad-leaved Evergreen forest would have been present on the interfluves), analyses of recent collections indicate a mean annual temperature of about 22°–23°C and the presence of the high-biomass Paratropical Rain forest (Table 1; Figure 4). Apparently, most assemblages analyzed by Hickey [1977, 1980] represent streamside vegetation.

The latest Paleocene-early Eocene cool interval has a simple pattern of vegetational distribution (Figure 5). Only five major vegetational types are known. Tropical Rain forest had its greatest latitudinal extent (up to latitude 50°), except for mountainous areas. Paratropical Rain forest extended to latitude 60°–65°, except for a 70° limit in coastal areas. In the continental interiors, Notophyllous Broad-leaved Evergreen forest extended from 60°–65° to 70°.

The early Eocene assemblages from the Yakutat terrane [Wolfe, 1977] were at about 65°–70°N (these assemblages were originally dated by mollusks as middle Eocene, but the associated "Domengine" mollusks are now regarded as late early Eocene). Although Alaska has many allochthonous terranes [Howell et al., 1983], the interpretation that the Yakutat block was at a low latitude during the early Eocene [Bruns, 1983] is at strong variance with the totality of geologic [Plafker, 1984] and paleontologic [Wolfe and McCoy, 1984] data. Although the Yakutat block may have moved northward as much as 5° since the Eocene [Plafker, 1983], this dis-placement is more than equaled by the higher (12°–20°) latitudes for Alaska during the Paleo-cene and Eocene [Hillhouse and Gromme, 1982, 1983].

The Yakutat assemblages are dominantly broad-leaved evergreen and contain palms, mangroves, and other taxa now dominantly to exclusively megathermal. Physiognomically the early Eocene Yakutat assemblages represent Paratropical Rain forest. The somewhat older Kupreanof Island assemblage could, on the basis of leaf margin analysis, represent Notophyllous Broad-leaved Evergreen forest, or could, on the basis of leaf size analysis [Wolfe, 1980], represent Paratropical Rain forest. The Kupreanof Island assemblage is small (21 species), and experience indicates that leaf-margin percentages tend to increase when additions are made to small floras [Wolfe, 1971] (cf. the Susanville flora, which had a percentage of 68 based on 22 species and a percentage of 82 based on 40 species). The Kupreanof Island assemblage occurs inboard of the major transform fault system in southeastern Alaska and certainly indicates that the occurrence of broad-leaved evergreen assemblages on

the Yakutat terrane is not anomalous for high latitudes during the Paleogene.

The North American continental interior assemblages of latest Paleocene to early Eocene age at latitude 50° include the assemblages of the Tongue River Member of the Fort Union Formation and those of the Willwood Formation. The analyses presented (Table 1) indicate that the Tongue River vegetation represented Paratropical Rain forest. Wing [1981] interpreted the early Eocene Willwood assemblages from the Bighorn Basin to represent Notophyllous Broad-leaved Evergreen forest; either the Bighorn Basin was at a moderate altitude relative to the Powder River Basin, or the possible overrepresentation of streamside vegetation in the Willwood assemblages has given them a "too cool" aspect. A cursory examination of a late Paleocene (Tiffanian in the mammalian chronology) assemblage from Joffre Bridge along the Red Deer River (latitude 65°N) in Alberta indicates, based on the occurrence of some broad-leaved evergreens in this floodplain assemblage, that the vegetation was at least mesothermal (i.e., Notophyllous Broad-leaved Evergreen forest); the assemblage from Smokey Tower, Alberta, contains a moderate coniferous element [Christophel, 1976], which indicates that the vegetation was not megathermal (tropical to paratropical).

Only four megafossil assemblages of Eocene age are known from Australia, and all are from southeastern Australia. The Nerriga assemblage [Hill, 1982] is palynologically dated as middle or early late Eocene, the Maslin Bay and Vegetable Creek as middle Eocene, and the Anglesea as late Eocene [Kemp, 1978]. Except for the Vegetable Creek, all were recently analyzed for some physiognomic characters by Christophel [1981]. These analyses indicate that all three represent either Paratropical Rain forest or Notophyllous Broad-leaved Evergreen forest. In leaf size, all three assemblages fall close to the boundary between the two vegetational types, although the taphonomic bias (which can markedly affect leaf size) is unknown. In leaf margin percentages, the Nerriga appears to represent warm notophyllous, and the other two assemblages paratropical vegetation. By analogy to northern hemisphere trends, the three assemblages may represent the early, late middle, and latest Eocene warm intervals. In the northern hemisphere, the absolute amount of warmth decreases sequentially from the earliest to the latest of these Eocene warm intervals. The equatorward movement of Australia was about 5° during the Eocene, and this could explain the relative warmths indicated by the Australian Eocene assemblages in contrast to the northern hemisphere pattern. In any case, it is clear that large-leaved, broad-leaved evergreen vegetation extended to at least 55°–60°S in Australia during the Eocene.

The biomass of these Eocene multistratal rain forests was probably high. One criterion is diversity; generally the drier a rain forest cli-

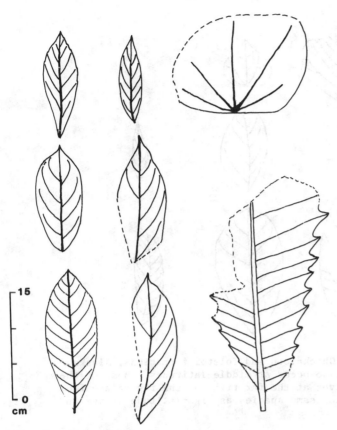

Fig. 6. Late middle Eocene leaves from Susanville, California. All the leaves are coriaceous
(probably evergreen). Compare to Figure 2a.

mate, the lower are both diversity and biomass
[Richards, 1952], and the Eocene rain forest assemblages are typically very diverse. In few
instances, however, have assemblages been preserved in situ; one such instance is the late
middle Eocene Susanville flora in northern California (Figure 6). This assemblage was obtained
from an interval of about 5-15 cm and appears to
represent (1) a forest floor litter entombed by a
crevasse-splay sandstone and (2) a leaf fall
caused by flooding and death of the forest vegetation. The leaf litter is largely composed of a
dentate, large-leaved (0.7-1 m long) plant (along
with a few fern fronds) that probably represents
a ground cover or low tufted plant. The leaf
fall horizon above contains at least 45 species
(a highly diverse assemblage for the limited
outcrop area of less than 20 m X 1 m). These
species include broad-leaved evergreens of megaphyllous size, some of which have finely serrate
margins (probably lower story trees), broad-
leaved evergreens of notophyllous to megaphyllous
size (probably upper story to crown or emergent
trees), and at least one probable liane. These
leaf size and leaf margin spectra closely parallel those in extant Tropical Rain forest [Rich-

ards, 1952]. The very large leaf size of the
inferred lower story plants indicates a high,
closed canopy that allowed little light to reach
the ground, as well as high levels of precipitation.

Areas poleward of 70°N were occupied by a
peculiar vegetation that has no modern analog.
The dominants in this vegetation were broad-
leaved deciduous and deciduous conifers (Glyptostrobus and Metasequoia); some broad-leaved
evergreens (including palms) were present between
latitudes 70 and 75° but are generally absent
poleward. Unlike extant microthermal broad-
leaved deciduous vegetation, few species in this
Paleogene vegetation had lobed leaves, and the
leaf size was exceedingly large (Figure 7).
Diversity was also low in this polar deciduous
forest, which probably gave the vegetation a
monotonous aspect. Although the inferred latitudinal temperature gradient during the early
Eocene indicates that Notophyllous Broad-leaved
Evergreen forest could have occurred poleward of
70°, the low winter light levels probably pro-
hibited the occurrence of broad-leaved evergreens. L. J. Hickey (oral communication, 1982)
has noted this type of large-leaved deciduous
vegetation in the Paleocene and early Eocene of
Ellesmere Island; other assemblages include those
from the early Eocene of Spitsbergen [Heer, 1868;
Schweitzer, 1974] and the Alaskan Cantwell [Wolfe
and Wahrhaftig, 1970] and Chickaloon [Wolfe,
1966, 1980] assemblages, both of Paleocene age.

Although the deciduousness of the polar broad-
leaved deciduous vegetation was probably an
adaptation to absence of winter light, the large
leaf size possibly represents adaptation to (1)
the low angle of incidence of solar radiation
during the summer (the leaves could be large,
because of the absence of an overheating problem
and a large size would allow the interception of
more radiation) and (2) the long growing season
days combined with high temperatures (allowing
long photosynthetic periods). Early and middle
Miocene assemblages at latitude 65° in Alaska
have a similarly large leaf size (see below). If
the crown trees of the polar deciduous forest had
such large foliage, it is unlikely that the forest had more than one tree stratum; undergrowth
was probably sparse.

The polar broad-leaved deciduous forest probably covered large land areas that are presently
occupied by the Arctic Ocean. The only evidence
of marine influence in the early Tertiary of the
Arctic Basin is occasional incursions of marine
tongues in the generally nonmarine rocks of areas
such as Banks Island [Doerenkamp et al., 1976].
Presumably, the Arctic Ocean was generally restricted to the deep part of the basin during the
early and middle Tertiary.

During warm intervals of the late middle and
latest Eocene (Figure 8), broad-leaved vegetation
again extended far poleward, although less so
than during the early Eocene. Generally, the
boundaries between respective vegetational zones

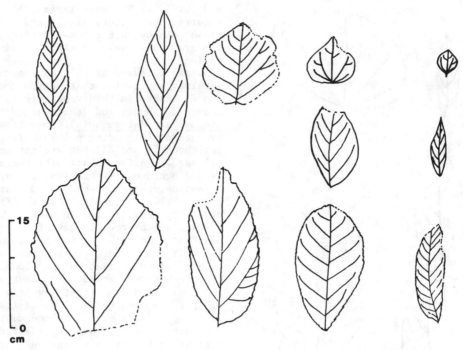

Fig. 7.  Latest Paleocene leaves from the Chickaloon and Tolstoi formations, Alaska, and comparison to leaves of species that also occur at middle latitudes.  Most of the leaves are thin (probably deciduous).  Leaves at the far right of the figure are from middle-latitude localities and represent the same species as high-latitude leaves in the column to the immediate left.

(except probably for the polar broad-leaved deciduous forest) were displaced 5°-10° equatorward from the early Eocene boundaries.  Assemblages in the Mississippi Embayment region (Figure 9) indicate a dry climate.  This tropical semideciduous vegetation has a leaf size comparable to that of semideciduous forest along the west coast of Mexico and in the Caribbean region today; generally, drip tips are absent, and the probable liane element is not diverse.  Legumes, which are particularly predominant in dry tropical vegetation today, are the most diverse family in the Mississippi Embayment middle and late Eocene assemblages.  The extent of this vegetation is not certain, but correlative middle Eocene assemblages in the continental interior include the subhumid upper Green River assemblages [MacGinitie, 1969].

MacGinitie [1969] interpreted the upper Green River assemblages as indicative of a distinctive Eocene savanna (i.e., shrubs and widely spaced trees but no grass on the interfluves).  The typical leaf size of the upper Green River assemblages, however, is like that of megathermal semideciduous forest and is considerably larger than leaf size in modern savanna or scrub vegetation.  Further, spinose leaves, which are characteristic of many species in extant megathermal and mesothermal scrub vegetation, are rare in the Green River assemblages.  Although I

agree with MacGinitie that the upper Green River assemblages are indicative of a subhumid climate, the assemblages are physiognomically most similar to megathermal (or very warm mesothermal) semideciduous forest.

Although the stratigraphic positions within the Eocene series are not given, Guo [1980] reported Chinese plant assemblages that represent subhumid, semideciduous vegetation south of latitude 45°N (Guo's "Paratropical Arid Floral Province"); to the north was humid vegetation containing both broad-leaved evergreen and deciduous plants (Guo's "Subtropical Humid Floral Province"), which Guo equated to the present vegetation of the Yangtze Valley (Notophyllous Broad-leaved Evergreen forest).  Guo [1980] probably included assemblages of both warm and cool intervals.  Iljinskaja [1963] also reported subhumid late Eocene assemblages from eastern Kazakhstan.  In southern European USSR [Dorofeyev, 1965] west to England [Chandler, 1964], the Eocene vegetation all represented rain forest; during warm intervals, the vegetation was Tropical Rain forest, and during cool intervals, the vegetation was Paratropical Rain forest.

Truswell and Harris [1982] have noted the presence of moderate amounts (up to 2%) of grass pollen in probable middle Eocene rocks in central Australia.  These authors suggested that, although the vegetation was generally closed for-

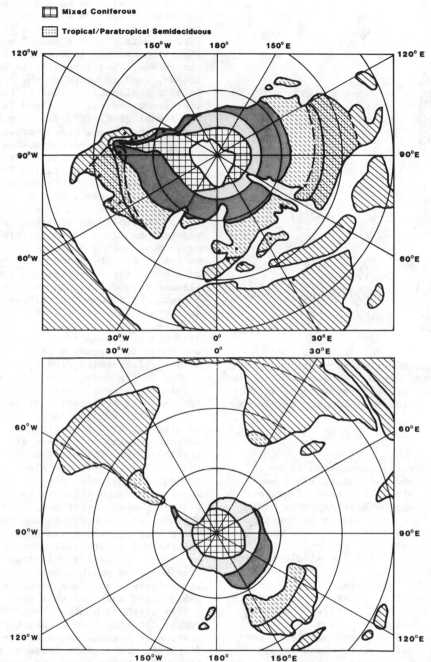

Fig. 8. Suggested vegetational map of the late middle Eocene. Solid circles represent some critical assemblages. Continental positions are adapted from Smith et al. [1981]. Patterns correspond to those in Figure 1.

est, some local edaphically dry areas may have supported open vegetation. The meager Australian data could conceivably represent an analog of the northern hemisphere subhumid vegetation.

During cool intervals of the early middle and early late Eocene, broad-leaved evergreen vegetation was generally equatorward of 50° (Figure 10). The area poleward of 50° was occupied by a diverse Mixed Coniferous forest, analogous to that of upland Taiwan. The described Japanese Eocene assemblages appear to all represent the cool interval of the early late Eocene (e.g., the mammal-dated Ube flora [Huzioka and Takahashi, 1970]). The "Oligocene" assemblages from the Kushiro coal field [Tanai, 1970] are stratigraphically below planktic faunas (the Poronai

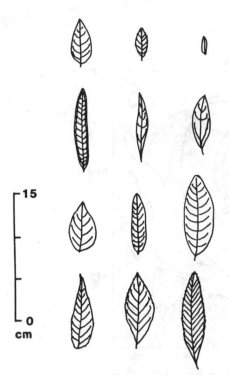

Fig. 9. Middle Eocene leaves from the Puryear assemblage, Claiborne Group, Tennessee. Included are both coriaceous (probably evergreen) and thin (probably deciduous) leaves. Compare to Figure 2b.

faunas) that are generally considered to be of latest Eocene age [Asano and Hatai, 1967] and thus probably the Kushiro plant assemblages are of early late Eocene age. These Japanese assemblages represent Microphyllous to Notophyllous Broad-leaved Evergreen forest.

Actual high-latitude megafossil records of Mixed Coniferous forest during the middle to late Eocene are generally lacking. Two possible exceptions are the Disko Island [Heer, 1883] and Rex Creek assemblages (the "Oligocene lower Healy Creek" of Wahrhaftig et al. [1969]). The Disko Island assemblage contains some pinaceous megafossils, and the Rex Creek assemblage has associated palynofloras that contain abundant and diverse Pinaceae (E. B. Leopold, unpublished data, 1970). The presence of high latitude coniferous forests can also be inferred from the fact that Mixed Coniferous forest occupied middle latitude upland areas (700–900 m and above [Wolfe and Wehr, in press]).

The biomass of the Eocene coniferous forests was probably high, particularly at high middle latitudes. The Yellowstone Park petrified forests fall at about the early/middle Eocene boundary and have been interpreted by Axelrod [1966] as vegetation of altitude intermediate between lowland assemblages (e.g., Kisinger Lakes [Mac-

Ginitie, 1974]) and upland assemblages (e.g., Republic [Wolfe and Wehr, in press]). Many of the coniferous trunks of the Yellowstone petrified forests are more than 3 m in diameter [Knowlton, 1899], indicative of optimal growing conditions. The diversity in both conifers and broad-leaved adjuncts of the Republic assemblage is also analogous to that in the high, dense coniferous forests of upland Taiwan.

Palynological data from the Eocene of Antarctica indicate a mixed forest of southern beech (Nothofagus) and podocarpaceous conifers [Kemp, 1972]. Nothofagus pollen, however, occurs as a common element in Australian Eocene beds that lack megafossils of this taxon [Christophel, 1981]. The apparent abundance of Nothofagus in palynofloras may be more a function of high pollen production than of actual abundance in the vegetation. Thus the Antarctic Eocene vegetation could have been dominantly coniferous. Alternatively, the abundance of Nothofagus may be related to the fact that the genus was a dominant element in a south polar broad-leaved deciduous forest. No Eocene leaf floras are known from Antarctica that would assist in resolving this problem. Probable middle or late Eocene leaf assemblages dominated by Nothofagus and podocarps are known from areas such as the South Shetland Islands [Zastawniak, 1981]; in the analyses presented in the current report, however, the age within the Eocene is critical, and I have thus not included such poorly dated assemblages.

The inferred distribtion of megathermal vegetation in high northern latitudes has an important bearing on present disjunctions of megathermal taxa. The early Tertiary boreotropical flora [Wolfe, 1975] was presumed to have largely spread via Beringia, although some dispersal across the North Atlantic land bridge was considered probable [Tiffney, 1980; Wolfe, 1980]. The map for the early Eocene indicates that the North Atlantic route was broad and continuous in megathermal vegetation. Beringia might have provided a route for a limited number of megatherms in the southernmost part of the polar broad-leaved deciduous forest zone or via islands immediately south of this zone.

The southern hemisphere offered no direct routes for migration of megathermal plants. Migration of megatherms must have resulted from dispersal via oceanic islands, such as those in the Indian Ocean [Kemp and Harris, 1975]. Mesothermal (as well as microthermal) taxa, however, probably could have migrated directly between Australia, Antarctica, and South America.

## Oligocene and Neogene Forests

At the end of the Eocene, a major cooling concomitant with a shift to high mean annual ranges of temperature is inferred (the terminal Eocene event). Large areas of the northern hemisphere were occupied by microthermal broadleaved deciduous forest (Figure 11). Paleo-

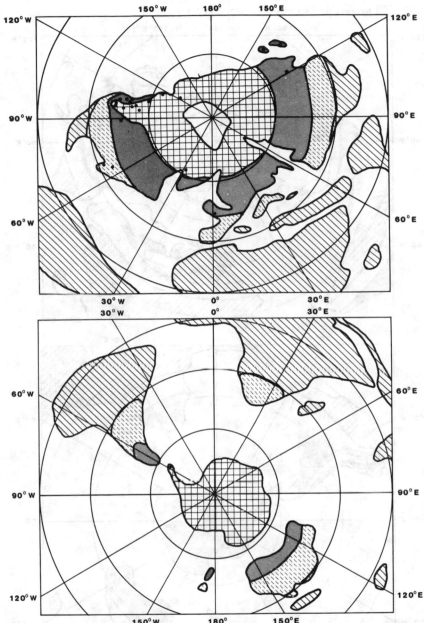

Fig. 10.  Suggested vegetational map of the Eocene cool intervals, early middle (about 46-50 Ma) and early late (about 37-41 Ma) Eocene.  Solid circles represent some critical assemblages of these two intervals.  Continental positions are adapted from Smith et al. [1981].  Patterns correspond to those in Figure 1.

botanical data for various Oligocene to early Miocene assemblages indicate that broad-leaved evergreen vegetation was restricted to equatorward of 35° (in comparison to 70° in the early Eocene and 40° today) and megathermal, multistratal vegetation to equatorward of 15° (in comparison to 60°-65° in the early Eocene and 20-25° today).  Mixed Coniferous forest also attained a wider distribution than previously, partly due to the uplift of various mountains.

The inference regarding restriction of megathermal and mesothermal vegetation to the most equatorward position in the Tertiary is based on scattered data, in part palynological. In equatorial Borneo, Muller [1966] recorded spruce (Picea), hemlock (Tsuga), and alder (Alnus) in the Oligocene; these taxa decreased in abundance during the Oligocene and disappeared during the early Miocene.  In Puerto Rico (latitude 15°), Graham and Jarzen [1969] recorded

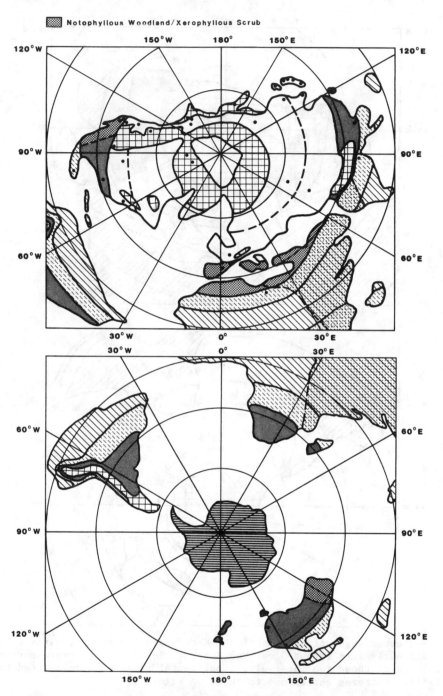

Fig. 11.  Suggested vegetational map of the early Miocene (about 18-22 Ma).  Solid circles represent some critical assemblages of Oligocene and early Miocene age.  Continental positions are adapted from Smith et al. [1981].  Patterns correspond to those in Figure 1.  The dashed line within the broad-leaved deciduous forest region delineates Mixed Northern Hardwood forest to the north from more southern broad-leaved deciduous forests.

beech (Fagus) and sweetgum (Liquidambar) in late Oligocene or early Miocene pollen assemblages. In all three instances, high mountains, for which no geologic data exist, have been suggested as the source for these cool elements.

The megafossil data are also meager regarding this cool interval.  In northern Vietnam (latitude 20°), Mchedlishvili [1960] recorded taxa characteristic of Notophyllous Broad-leaved Evergreen forest in a region today well within

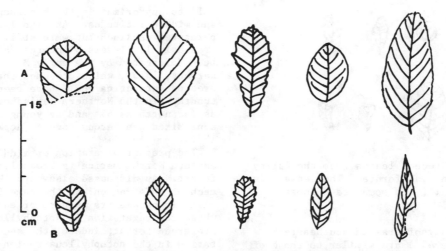

Fig. 12.  Comparison of high-latitude to middle-latitude early to middle Miocene
leaves.  (a) High-latitude leaves from the Tyonek and Healy Creek formations,
Alaska.  (b) Middle-latitude leaves from the Collawash assemblage, "Little Butte
Volcanic Series," Oregon.  From left to right:  hickory (Carya), alder (Alnus), oak
(Quercus), yellowwood (Cladrastis), and willow (Salix).

the Paratropical Rain forest; these assemblages
are unfortunately not well dated.  On Taiwan
(latitude 30°), Chaney and Chuang [1968] recorded
the presence of taxa (including Acer with lobed
leaves) characteristic of notophyllous vegetation
in a region now occupied by paratropical vegetation up to 500 to 700 m; this assemblage is
dated by planktics as early Miocene.  The most
significant set of leaf data come from Japan.
Broad-leaved evergreen vegetation (largely notophyllous) now extends in Japan to latitude 38°N
[Wolfe, 1979].  Southern Honshu and Kyushu were
between latitudes 35° and 40°N during the early
Miocene, but the vegetation was broad-leaved
deciduous [Tanai, 1961]; moreover, the type of
deciduous vegetation was beech dominated, like
that of the Mixed Northern Hardwood forest today.  I consider that the megafossil and microfossil data, when taken together, point to the
conclusion that the various types of broad-leaved
evergreen vegetation were restricted to lower
latitudes than at present.

The development of microthermal broad-leaved
deciduous forests following the terminal Eocene
event was a novelty.  During the Eocene, the
highly equable climates would not have allowed
the development of such vegetation, although
broad-leaved deciduous trees and shrubs were
important accessories in mesothermal evergreen
and microthermal coniferous vegetation.  The
Oligocene microthermal broad-leaved deciduous
forests were generally of low diversity in comparison to the descendant forests of the Miocene.  This initial low diversity of the broad-
leaved deciduous forests is thought to reflect
the severity and rapidity of the terminal Eocene
event [Wolfe, 1972]; a similar low diversity has
been observed in early Oligocene coniferous
palynofloras in southeastern Australia [Kemp,
1978].

Mixed Coniferous forests apparently occupied
large areas at high middle to high latitudes
following the terminal Eocene event; the areas
occupied were probably somewhat less than those
occupied by Mixed Coniferous forest during cool
intervals of the Eocene.  No direct information
is available regarding the biomasses of either
the broad-leaved deciduous or coniferous forests
during the Oligocene.  As noted above, the low
diversity is probably not indicative of subop-
timal growing conditions in the Oligocene but was
a function of the severity of the terminal Eocene
event.

In the southern hemisphere, Oligocene vegetation on Antarctica was inferred from palynofloras to have been dominated by southern beech
(Nothofagus) and podocarpaceous conifers [Kemp
and Barrett, 1975], although recently Truswell
[1983] questioned whether the supposed Oligocene
palynomorphs were redeposited from Eocene
rocks.  Hill and Macphail [1983] discussed the
probable overrepresentation of southern beech
pollen in an Oligocene site in Tasmania, where
the megafossil and microfossil assemblages can be
compared.  Despite the abundance of southern
beech pollen, Hill and Macphail concluded that
the highest stratum of the vegetation was formed
by a diverse coniferous element above a canopy of
southern beech, analagous to the present forests
on South Island, New Zealand.  The Antarctic
vegetation analyzed by Kemp and Barrett [1975]
could represent similar vegetation, although
thermophilic elements are absent that are present
to the north.  Kemp [1978, p. 194] suggested that
the Antarctic Oligocene vegetation could have
been analogous to "...the open, stunted deciduous
forests presently growing in Patagonia..."  The
northward extension of the mixed southern beech-
coniferous forests during the Oligocene was at
least to latitude 50° [Kemp, 1978].

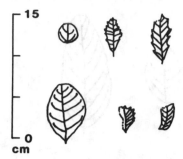

Fig. 13.  Early Miocene leaves from the Valley Springs Formation, California.  The leaves are evergreen and small, and some are spinose.

Early Miocene palynofloras in southeastern Australia (latitude 45°) are similar to those of the late Oligocene; i.e., the vegetation probably had a broad-leaved canopy with a diverse emergent stratum of conifers above.  Limited megafossil data indicate that the broad-leaved trees were evergreen (Lauraceae and Myrtaceae; J. A. Owen, 1975, cited in Kemp [1978]).

As in the Eocene polar broad-leaved deciduous forest, the Mixed Northern Hardwood forest that occupied high northern latitudes had a large leaf size.  Comparison of Miocene leaves of the same or closely related species between high and middle latitudes indicates that invariably the high-latitude leaves are larger (Figure 12).  In all other physiognomic features, however, the high-latitude Mixed Northern Hardwood forest of the early to middle Miocene is similar to the present Mixed Northern Hardwood forest of eastern Asia [Wolfe and Tanai, 1980].

Data on the biomasses of Miocene forests can generally be inferred from diversity.  The diversity in the microthermal broad-leaved deciduous forests was typically high.  For example, the early Miocene Collawash flora of Oregon (representative of Mixed Mesophytic forest) contains over 140 megafossil species, an exceedingly diverse flora, even allowing for taphonomic factors such as transport.  The middle Miocene Seldovia Point flora from Alaska [Wolfe and Tanai, 1980] is of comparable diversity to the densely stocked Mixed Northern Hardwood forest of Manchuria.  The coniferous forests of the Miocene are also typically of high diversity.

The thermal maximum of the Neogene occurred during the late early to early middle Miocene (13-18 Ma); this maximum is frequently referred to as the mid-Miocene warming.  Notophyllous Broad-leaved Evergreen forest expanded to at least 45°N along the Pacific coast of North America [Wolfe, 1981], in western Europe [Mai, 1981], and in Japan [Tanai and Huzioka, 1967].  Pollen assemblages from the mid-Miocene of New Jersey contain thermophilic taxa (e.g., Engel-hardia, Cyrilla, Sapotaceae, and Alangium; J. A. Wolfe, unpublished data, 1973).

At 13 Ma (late middle Miocene) temperatures fell to approximately their present levels at low and middle latitudes.  At high latitudes, temperatures declined but were still not at their presently low levels.  Antarctica appears to have been deforested by the early Miocene [Kemp and Barrett, 1975], which presumes that any ice-free area on Antarctica would have been occupied by tundra.  In the Northern Hemisphere, assemblages as far north as 65° and as young as 5 Ma represent Mixed Coniferous forest [Hopkins et al., 1971].

The present vegetation of middle latitude eastern North America is broad-leaved deciduous forest.  As discussed elsewhere [Wolfe, 1979], much of this region has the same fundamental climatic parameters as part of eastern Asia, where the vegetation is Notophyllous Broad-leaved Evergreen forest; indeed, the secondary vegetation in the notophyllous region taxonomically resembles the vegetation in eastern North America (Pinus, Liquidambar, lobed deciduous Quercus).  Despite the fact that the cold month means are the same in the subject parts of eastern Asia and eastern North America, the latter region is subject to occasional but highly intense outbreaks of Arctic air masses, and it is probably such outbreaks that result in the anomalous lack of broad-leaved evergreens in eastern North America (except in areas under strong marine influence, such as the islands off the Carolina coast).  As noted above, mid-Miocene pollen assemblages in New Jersey (and also in Maryland) contain taxa that are now exclusively broad-leaved evergreen.  Assemblages younger than 13 Ma generally lack such taxa, although Rachele [1976] recorded a few probable broad-leaved evergreens from early late Miocene beds in New Jersey.  A leaf assemblage from Maryland (probably 5-8 Ma), although represented by a small collection, contains only deciduous taxa, including lobed deciduous Quercus (J. A. Wolfe, unpublished data, 1973).  Apparently the anomalous broad-leaved deciduous vegetation of eastern North America developed at 13 Ma, which implies the beginning of intense Arctic fronts during the winter.

The inferred development of intense Arctic cold fronts at 13 Ma does not necessarily indicate the formation of a permanently frozen Arctic Ocean.  Indeed, the freezing of the Arctic appears to have started at about 2-3 Ma.  The sequence of assemblages from the Kolyma lowland [Sher et al., 1979] indicate a high (for this latitude) winter temperature until about 2-3 Ma, when the sediments contain definite evidence of ice wedges (and thus a drastically low winter temperature).  Such a drop in winter temperature would occur if the Arctic Ocean became frozen over during the winter.

During the Neogene, microthermal broad-leaved deciduous forest was gradually reduced in areal extent.  In both continental and maritime Siberia, broad-leaved deciduous forest generally persisted until sometime in the late Miocene, when most of the region was occupied by Mixed

Coniferous forest and ultimately by Taiga or Simple Broad-leaved Deciduous (birch) forest [Baranova et al., 1968]. In Alaska, Mixed Northern Hardwood forest was replaced by Mixed Coniferous forest at about 13 Ma [Wolfe and Tanai, 1980]. At middle latitudes, the low-altitude Mixed Mesophytic forest was replaced by Mixed Coniferous forest during the Pliocene in both northwestern United States (Wolfe, 1981) and in central and western Europe [Szafer, 1954]; the current broad-leaved deciduous forest of central and western Europe appears to be an anomaly resulting from Quaternary glaciation [Wolfe, 1979].

## Neogene Low-Biomass Vegetation

Of particular significance to the Tertiary history of the carbon cycle is the origin and spread of low-biomass vegetational types: savanna (including woodland and scrub), steppe, Taiga, tundra, and true desert. All these vegetational types appear to be of Miocene or younger origin.

Evidence of savanna during the Oligocene is absent, although Axelrod and Raven [1978] inferred the presence of "savanna-woodland and thorn scrub" in the Saharan-Libyan desert region between 25 and 30 Ma. However, recent investigations of the early Oligocene Fayum plant assemblage in northern Egypt indicate the presence of a humid gallery forest [Bown et al., 1982]. A small Oligocene leaf and wood assemblage from southwestern Libya, moreover, contains some taxa now endemic to the moist tropical region of West Africa [Louvet and Mouton, 1975]; the wood is semidiffuse porous, indicative of only mild seasonality. In the early Miocene (but not Oligocene) of Venezuela, grass pollen is moderately common and increased in abundance during the Miocene, indicating the appearance and spread of savanna [Germeraad et al., 1968]. Leaf and fructification assemblages in eastern Africa also indicate the growth of savanna during the Miocene [Axelrod and Raven, 1978], as do wood assemblages in western India [Prakash, 1972]. Savanna probably occupied a continuous belt from eastern Africa east to western India by the middle to late Miocene. Kemp [1978] inferred the presence of open vegetation (gallery forest with grassy interfluves) because of the abundance of grass pollen in central Australia during the middle Miocene. Early Miocene leaf assemblages in central and southern California, such as the Tehachapi [Axelrod, 1939] and Valley Springs [J. A. Wolfe, unpublished data, 1984], are xerophyllous (small, coriaceous leaves, many of which are spinose (Figure 13) and probably indicate a savanna-woodland or thorn scrub. An early to middle Miocene assemblage from the upland area of central New Mexico (now in the upper part of the Chihuahuan Desert) represents a sparse juniper woodland [H. Meyer, unpublished data, 1982]. All these occurrences are from equatorial regions up to about latitude 35°. The data indicate that during the early to middle Miocene, savanna

developed in low-latitude areas that are presently dry.

The history of middle-latitude grasslands (or steppes) is much more uncertain and variable. By the late Miocene, steppes developed in north-western China (H. Li, oral communication, 1984; Song et al. [1981]), probably from the orographic effect of the rising Himalayas. In eastern Kazakhstan and western Siberia, however, Miocene vegetation was still broad-leaved deciduous forest [Iljinskaja, 1963; Dorofeyev, 1963]. In areas such as Nebraska, late middle Miocene vegetation was apparently still also broad-leaved forest, although there is some evidence of at least some interfluve grassland [MacGinitie, 1962]. The latest Miocene fructification and leaf assemblages in Kansas [Chaney and Elias, 1936] indicate that widespread grasslands may have occurred in this region. The fructifications include abundant grass, and the leaf assemblages are the low-diversity type that typically occurs along streams in unforested grassland regions. Similar leaf assemblages occur in the latest Miocene of the southern Great Basin (H. E. Schorn, unpublished data, 1984) and in the early Pliocene of the Columbia Plateau [Chaney, 1938; Wolfe, 1969]. Palynofloras from the latest Miocene of western Wyoming [Barnosky, 1984] that are dominated by xeric shrubs (e.g., Ephedra, Sarcobatus, Artemesia) alternate with more mesic assemblages.

Early to middle Miocene fructification assemblages from 75° to 80°N on Banks Island in the Canadian Arctic [Hills et al., 1974] re-present Mixed Coniferous forest. Included are pine (Pinus), spruce (Picea), and hemlock (Tsuga), as well as broad-leaved trees such as butternut (Juglans). These forests were on the borderlands of the Arctic Ocean and indicate that Taiga could not have been present in the Arctic during the early half of the Miocene.

Taiga first definitely appears in what is probably a late Miocene assemblage from the coastal area of central Siberia [Sher et al., 1979]. However, temperatures appear to have remained sufficiently high in areas south of 65°-70°N in Alaska during the early Pliocene to support Mixed Coniferous forest. By the end of the Pliocene, Taiga appears to have attained about its present areal extent.

Tundra vegetation in the northern hemisphere is of Pliocene origin, at least insofar as any recognizable areal extent is concerned. Small patches of alpine tundra could have appeared associated with glaciation in southern Alaska during the late Miocene [Denton and Armstrong, 1969], and analogous situations could have developed in other Arctic or subarctic areas. The first fossil evidence for tundra, however, is in the late Pliocene of the Kolyma lowland (latitude 70°N), where a mosaic of Taiga and tundra has been recorded [Sher et al., 1979]. In the Bering Strait region, grass and sedge pollen became increasingly common in Pliocene palynofloras

[Wolfe, 1972]. As late as the early Quaternary, however, a Taiga of larch (Larix) persisted on the Arctic Coast of Alaska (J. A. Wolfe, unpublished data, 1982). Similarly, an early Quaternary assemblage from Kap Kobenhavn, northeast Greenland, also represents Taiga [Funder et al., in press]. As noted above, tundra may have occupied large areas of the unglaciated parts of Antarctica during the Neogene.

The origin of present deserts has been a subject of considerable speculation, largely because many systematists consider the modifications (and taxonomic isolation) of most desert plants to be extreme and indicative of a long adaptive history to an arid environment. On the other hand, the paleobotanical evidence for either Tertiary desert environments or an ancient history of now-desert taxa is absent, although Truswell and Harris [1982] suggested that some sclerophyllous vegetation could have occupied local edaphically dry environments during the Eocene. For example, the pollen of Cactaceae is easily recognizable, and cacti are a major element in the deserts of both South and North America (as well as a major epiphytic group in megathermal rain forests); yet, no pollen of Cactaceae is known in the Tertiary [Muller, 1981]. Such data argue against a slow adaptation to areally limited xeric conditions during the Tertiary. Where evidence is available from areas that are now desert, these areas appear to have had a savanna-woodland vegetation during the late Neogene. For example, a 5-6 Ma leaf assemblage from the Verde Valley of Arizona (now in the upper part of the Sonoran Desert) is composed of notophyllous broad-leaved evergreens (J. A. Wolfe, unpublished data, 1984), unlike the microphylls that occur along streams in the present Sonoran Desert. Available data indicate that present deserts are of post-Tertiary origin.

The origin and spread of low-biomass vegetational types during the Neogene is the result of two major temperature trends during this period. In polar, areas mean annual temperature (and summer temperature) has consistently declined during the Neogene. Taiga, which exists in areas of low (<0°C) mean annual temperature, and tundra, which exists in areas of low summer heat (warm month mean <10°C), would both gradually spread. At low latitudes, on the other hand, mean annual temperature has increased. These divergent trends resulted in an increased latitudinal temperature gradient, which, in turn, intensified the subtropical high-pressure systems and the summer drought along the western sides of continents. Further, even if the amount of precipitation at low latitudes remained constant, an increase in temperature would result in a greater need for water by plants. Thus at lower latitudes, savanna gradually spread during the Neogene. The origin and spread of mid-latitude steppes, on the other hand, may reflect the elevation of particular mountain ranges, as well as secular climatic change.

## Future Work

I have painted only a partial picture of the distribution of Tertiary vegetation (and certainly with too broad a brush). Much refinement can be done, particularly in regard to estimates of biomass. Many areas remain to be sampled for Tertiary floras, and many known assemblages await detailed physiognomic analyses. Hopefully, in the near future our data base will substantially increase and allow a series of more detailed and accurate maps of the distribution of Tertiary vegetation.

Acknowledgments. For many discussions and critiques of concepts presented above, I wish to thank James A. Doyle, Leo J. Hickey, Estella B. Leopold, Harry D. MacGinitie, Howard E. Schorn, Bruce H. Tiffney, Elizabeth M. Truswell, Wesley Wehr, and Scott L. Wing.

## References

Asano, K., and K. Hatai, Micro- and macropaleontological Tertiary correlations within Japanese islands and with planktonic foraminiferal sequences of foreign countries, in Tertiary Correlations and Climatic Changes in the Pacific, edited by K. Hatai, pp. 77-87, Sasaki, Sendai, Japan, 1967.

Axelrod, D. I., A Miocene flora from the western border of the Mohave Desert, Carnegie Inst. Washington Publ., 516, 129 pp., 1939.

Axelrod, D. I., The Eocene Copper Basin flora of northeastern Nevada, Univ. Calif. Berkeley Publ. Geol. Sci., 59, 125 pp., 1966.

Axelrod, D. I., and P. H. Raven, Late Cretaceous and Tertiary vegetation history of Africa, in Biogeography and Ecology of Southern Africa, edited by M. J. A. Werger, pp. 77-130, W. Junk, The Hague, 1978.

Bailey, I. W., and E. W. Sinnott, A botanical index of Cretaceous and Tertiary climates, Science, 41, 831-834, 1915.

Baranova, I. P., S. F. Biske, V. F. Goncharov, I. A. Kulkova, and A. S. Titkov, Cenozoic of the Northeast of the USSR (in Russian), Tr. Akad. Nauk SSSR Sib. Fil. Inst. Geol. and Geofiz., 38, 125 pp., 1968.

Barnosky, C. W., Late Miocene vegetational and climatic variations inferred from a pollen record in northwest Wyoming, Science, 223, 49-51, 1984.

Bown, T. M., M. J. Kraus, S. L. Wing, J. G. Fleagle, B. H. Tiffney, E. L. Simons, and C. F. Vondra, The Fayum primate forest revisited, J. Hum. Evol., 11, 603-632, 1982.

Bruns, T. R., Model for the origin of the Yakutat block, an accreting terrane in the northern Gulf of Alaska, Geology, 11, 718-721, 1983.

Chandler, M. E. J., The Lower Tertiary Floras of Southern England IV, 151 pp., British Museum (Natural History), London, 1964.

Chaney, R. W., The Deschutes flora of eastern

Oregon, Carnegie Inst. Washington Publ., 476, 187-216, 1938.

Chaney, R. W., and G. C. Chuang, An oak-laurel forest in the Miocene of Taiwan, I, Geol. Soc. China, no. II, 3-18, 1968.

Chaney, R. W., and M. K. Elias, Late Tertiary floras from the High Plains, Carnegie Inst. Washington Publ., 476, 1-72, 1936.

Chaney, R. W., and E. I. Sanborn, The Goshen flora of west-central Oregon, Carnegie Inst. Washington Publ., 439, 103 pp., 1933.

Christophel, D. C., Fossil floras of the Smokey Tower locality, Alberta, Palaeontographica, Abt. B, 157, 1-43, 1976.

Christophel, D. C., Tertiary megafossil floras of Australia as indicators of floristic associations and paleoclimate, in Ecological Biogeography of Australia, edited by A. Keast, pp. 379-390, W. Junk, The Hague, 1981.

Denton, G., and R. L. Armstrong, Miocene-Pliocene glaciations in southern Alaska, Am. J. Sci., 267, 1121-1142, 1969.

Doerenkamp, A., S. Jardine, and P. Moreau, Cretaceous and Tertiary palynomorph assemblages from Banks Island and adjacent areas (N.W.T), Bull. Can. Pet. Geol., 24, 372-417, 1976.

Dolph, G. E., Variation in leaf margin with respect to climate in Costa Rica, Bull. Torrey Bot. Club, 106, 104-109, 1979.

Dolph, G. E., and D. L. Dilcher, Foliar physiognomy as an aid in determining paleoclimate, Palaeontographica, Abt. B, 170, 151-172, 1979.

Dorofeyev, P. I., Tertiary Floras of Western Siberia (in Russian), 346 pp., Akademii Nauk SSSR, Leningrad, 1963.

Dorofeyev, P. I., On some problems of the history of flora (in Russian), Bot. Zh., 50, 1501-1522, 1965.

Funder, S., O. Bennike, G. S. Mogensen, B. Noe-Nygaard, S. A. Pedersen, and K. S. Petersen, The Kap København Formation, a late Cainozoic sedimentary sequence in north Greenland, Rep. Geol. Survey Greenland, in press.

Germeraad, J. H., C. A. Hopping, and J. Muller, Palynology of Tertiary sediments from tropical areas, Rev. Palaeobot. Palynol., 6, 189-348, 1968.

Graham, Alan, and D. M. Jarzen, Studies in neotropical paleobotany, I, The Oligocene communities of Puerto Rico, Ann. Mo. Bot. Gard., 56, 308-357, 1969.

Guo, S., Late Cretaceous and Eocene floral provinces, report, 9 pp., Inst. of Geol. and Paleontol., Academia Sinica, Nanjing, 1980.

Heer, O., Die fossile Flora der Polarlander, Flora Fossilis Arctica, vol. 1, 192 pp., J. Wurster, Zurich, 1868.

Heer, O., Den zweiten Theil der fossilen Flora Gronlands, 48-110, Flora Fossilis Arctica, vol. 7, 275 pp., J. Wurster, Zurich, 1883.

Hickey, L. J., Stratigraphy and paleobotany of the Golden Valley Formation (early Tertiary) of western North Dakota, Geol. Soc. Am. Mem., 150, 183 pp., 1977.

Hickey, L. J., Paleocene stratigraphy and flora of the Clark's Fork Basin, Univ. Mich. Pap. Paleontol., 24, 33-49, Univ. Michigan, 1980.

Hill, R. S., The Eocene megafossil flora of Nerriga, New South Wales Australia, Palaeontographica, Abt. B, 181, 44-77, 1982.

Hill, R. S., and M. K. Macphail, Reconstruction of the Oligocene vegetation at Pioneer, northeast Tasmania, Alcheringa, 7, 181-299, 1983.

Hillhouse J. W., and C. S. Gromme, Limits to northward drift of the Paleocene Cantwell Formation, central Alaska, Geology, 10, 552-556, 1982.

Hillhouse, J. W., and C. S. Gromme, Wrangellia in southern Alaska 50 million years ago (abstract), Eos Trans. AGU, 64, 687, 1983.

Hills, L. V., J. E. Klovan, and A. R. Sweet, Juglans eocinerea n. sp., Beaufort Formation (Tertiary) southwestern Banks Island, Arctic Canada, Can. J. Bot., 52, 65-90, 1974.

Hopkins, D. M., J. V. Mathews, J. A. Wolfe, and M. L. Silberman, A Pliocene flora and insect fauna from the Bering Strait region, Palaeogeogr. Palaeoclimatol. Palaeoecol., 9, 211-231, 1971.

Howell, D. G., E. R. Schermer, D. L. Jones, Z. Ben-Avraham, and E. Scheibner, Preliminary tectonostratigraphic terrane map of the circum-Pacific region, U.S. Geol. Surv. Open File Rep., 83-716, 18 pp., 1983.

Huzioka, K. and E. Takahashi, The Eocene flora of the Ube coal-field, southwest Honshu, Japan, pp. 1-88, J. Min. Coll. Akita Univ., Ser. A, 4, 1-88, 1970.

Iljinskaja, I. A., The fossil flora of the Mountain Kiin-Kerish Zaisan basin (in Russian), Tr. Akad. Nauk SSSR Inst. Bot. Komarova, Ser. 8, Palaeobotanika, 4, 141-187, 1963.

Kemp, E. M., Reworked palynomorphs from the west Ice Shelf area, East Antarctica, and their possible geological and paleoclimatological significance, Mar. Geol., 13, 145-157, 1972.

Kemp, E. M., and P. J. Barrett, Antarctic vegetation and early Tertiary glaciation, Nature, 258, 507-508, 1975.

Kemp, E. M., Tertiary climatic evolution and vegetation history in the southeast Indian Ocean region, Palaeogeogr. Palaeoclimatol. Palaeoecol., 24, 169-208, 1978.

Kemp, E. M., and W. K. Harris, The vegetation of Tertiary islands on the Ninetyeast Ridge, Nature, 258, 303-307, 1975.

Knowlton, F. H., Fossil flora of the Yellowstone National Park, in Monograph 32, part 2, pp. 651-882, U.S. Geological Survey, Reston, Va., 1899.

Louvet, P., and J. Mouton, La flore Oligocene du Djebel Coquin (Libye), Actes 95e Congr. Natl. Soc. Savantes, Sect. sci., 3, 79-96, 1975.

MacGinitie, H. D., A middle Eocene flora from the central Sierra Nevada, Carnegie Inst. Washington Publ., 543, 178 pp., 1941.

MacGinitie, H. D., The Kilgore flora, Univ. Calif. Berkeley Publ. Geol. Sci., 35, 67-158, 1962.

MacGinitie, H. D., The Eocene Green River flora of northwestern Colorado and northeastern Utah, Univ. Calif. Berkeley Publ. Geol. Sci., 83, 140 pp., 1969.

MacGinitie, H. D., An early middle Eocene flora from the Yellowstone-Absaroka volcanic province, northwestern Wind River basin, Wyoming, Univ. Calif. Berkeley Publ. Geol. Sci., 108, 103 pp., 1974.

Mai, D. H., Entwicklung und klimatische Differonzierung der Laubwaldflora Mitteleuropas im Tertiar, Flora Jena, 171, 525-582, 1981.

Matthews, R. K., and R. Z. Poore, Tertiary record and glacio-eustatic sea-level fluctuations, Geology, 8, 501-504, 1980.

Mchedlishvili, P. A., New data on the Tertiary flora of North Viet Nam (in Russian), Dokl. Akad. Nauk SSSR, 135, 694-697, 1960.

Muller J., Montane pollen from the Tertiary of northwestern Borneo, Blumea, 14, 231-235, 1966.

Muller, J., Fossil pollen records of extant angiosperms, Bot. Rev., 47, 1-142, 1981.

Plafker, G., The Yakutat block: An actively accreting tectonostratigraphic terrane in southern Alaska (abstract), Geol. Soc. Am. Abstr. Programs, 15, 406, 1983.

Plafker, G., Comment on "A model for the origin of the Yakutat block, an accreting terrane in the northern Gulf of Alaska" by T. R. Bruns, Geology, 562-563, 1984.

Prakash, U., Paleoenvironmental analysis of Indian Tertiary floras, Geophytology, 2, 178-205, 1972.

Rachele, L. D., Palynology of the Legler lignite: A deposit in the Tertiary Cohansey Formation of New Jersey, U.S.A., Rev. Palaeobot. Palynol., 22, 225-252, 1976.

Richards, P. W., The Tropical Rain Forest, 450 pp., Cambridge University Press, New York, 1952.

Schweitzer, H. J., Die "tertiären" Koniferen Spitzbergens, Palaeontographica, Abt. B, 149, 1-89, 1974.

Sher, A. V., T. N. Kaplina, R. E. Giterman, A. V. Lozhkin, A. A. Arkhangelov, S. V. Kiselyov, Y. P. Koutnetsov, E. I. Virina, and V. S. Zazhigin, Scientific excursion on problem "Late Cenozoic of the Kolyma Lowland," Pac. Sci. Congr. (Khabarovsk) 14th, Tour 11 Guidebook, 115 pp., 1979.

Smith, A. G., A. M. Hurley, and J. C. Briden, Phanerozoic Paleocontinental World Maps, 102 pp., Cambridge University Press, New York, 1981.

Song, Z., H. Li, Y. Zheng, and G. Liu, Miocene floristic regions of China, Geol. Soc. Amer. Spec. Paper, 187, 249-254, 1981.

Szafer, W., Pliocene flora from the vicinity of Czorsztyn (West Carpathians) and its relationship to the Pleistocene, Prace Inst. Geol., 11, 238 pp., 1954.

Tanai, T., Neogene floral change in Japan, J. Fac. Sci. Hokkaido Univ., Ser. 4, 11, 119-298, 1961.

Tanai, T., The Oligocene floras from the Kushiro coal field, Hokkaido, Japan, J. Fac. Sci. Hokkaido Univ., Ser. 4, 14, 383-514, 1970.

Tanai, T., and K. Huzioka, Climatic implications of Tertiary floras in Japan, in Tertiary Correlations and Climatic Changes in the Pacific, edited by K. Hatai, pp. 89-94, Sasaki, Sendai, 1967.

Tiffney, B. H., The Tertiary flora of eastern North America and the North Atlantic land bridge (abstract), Int. Congr. Syst. Evol. Biol. Abstr. 2nd, 373, 1980.

Truswell, E. M., Recycled Cretaceous and Tertiary pollen and spores in Antarctic marine sediments: A catalogue, Palaeontographica, Abt. B, 186, 121-174, 1983.

Truswell, E. M., and W. K. Harris, The Cainozoic palaeobotanical record in arid Australia: Fossil evidence for the origins of an arid-adapted flora, in Evolution of the Flora and Fauna of Arid Australia, edited by W. R. Barker and P. J. M. Greenslade, pp. 57-76, Peacock Publications, Frewville, South Australia, 1982.

Wahrhaftig, C., J. A. Wolfe, E. B. Leopold, and M. A. Lanphere, The coal-bearing group in the Nenana coal field, Alaska, U.S. Geol. Surv. Bull., 1274-D, D1-D30, 1969.

Webb, L. J., A physiognomic classification of Australian rain forests, J. Ecol., 47, 551-570, 1959.

Webb, P. N., Climatic, palaeo-oceanographic and tectonic interpretation of Paleogene-Neogene biostratigraphy from MSSTS-1 drillhole, McMurdo Sound, Antarctica, in Antarctic Earth Science, edited by R. L. Oliver, P. R. James, and J. B. Jago, p. 560, Australian Academy of Science, Canberra, 1983.

Wing, S. L., A study of paleoecology and paleobotany in the Willwood Formation (early Eocene, Wyoming), Ph.D thesis, Yale Univ., New Haven, Conn.

Wolfe, J. A., Tertiary plants from the Cook Inlet region, Alaska, U.S. Geol. Surv. Prof. Pap., 398-B, B1-B32, 1966.

Wolfe, J. A., Neogene floristic and vegetational history of the Pacific Northwest, Madrono, 20, 83-110, 1969.

Wolfe, J. A., Tertiary climatic fluctuations and methods of analysis of Tertiary floras, Palaeogeogr. Palaeoclimatol. Palaeoecol., 9, 25-57, 1971.

Wolfe, J. A., An interpretation of Alaskan Tertiary floras, in Floristics and Paleofloristics of Asia and Eastern North America, edited by A. Graham, pp. 201-233, Elsevier, New York, 1972.

Wolfe, J. A., Some aspects of plant geography of the Northern Hemisphere during the Late Cretaceous and Tertiary, Ann. Mo. Bot. Gard., 62, 264-279, 1975.

Wolfe, J. A., Paleogene floras from the Gulf of Alaska region, U.S. Geol. Surv. Prof. Pap., 997, 108 pp., 1977.

Wolfe, J. A., A paleobotanical interpretation of

Tertiary climates in the Northern Hemisphere, Am. Sci., 66, 694-703, 1978

Wolfe, J. A., Temperature parameters of humid to mesic forests of eastern Asia and relation to forests of other regions of the Northern Hemisphere and Australasia, U.S. Geol. Surv. Prof. Pap., 1106, 37 pp., 1979

Wolfe, J. A., Tertiary climates and floristic relationships at high latitudes in the Northern Hemisphere, Palaeogeogr. Palaeoclimatol. Palaeoecol., 30, 313-323, 1980.

Wolfe, J. A., Paleoclimatic significance of the Oligocene and Neogene floras of northwestern United States, in Paleobotany, Paleoecology, and Evolution, edited by K. Niklas, pp. 79-101, Praeger, New York, 1981.

Wolfe, J. A., and S. McCoy, jr., Comment on "A model for the origin of the Yakutat block, an accreting terrane in the northern Gulf of Alaska" by T. R. Bruns, Geology, 12, 564-565, 1984.

Wolfe, J. A., and R. Z. Poore, Tertiary marine and nonmarine climatic trends in, Climate in Earth History, Studies in Geophysics, edited by J. C. Crowell and W. Berger, pp. 154-158, National Research Council, Washington, D.C., 1982.

Wolfe, J. A., and T. Tanai, The Miocene Seldovia Point flora from the Kenai Group, Alaska, U.S. Geol. Surv. Prof. Pap., 1105, 52 pp., 1980.

Wolfe, J. A., and C. Wahrhaftig, The Cantwell Formation of the cental Alaska Range, U.S. Geol. Surv. Bull., 1294-A, A41-A46, 1970.

Wolfe, J. A., and W. Wehr, Middle Eocene dicotyledonous plants from Republic, northeastern Washington, U.S. Geol. Surv. Prof. Pap., in press.

Zastawniak, E., Tertiary leaf flora from the Point Hennequin Group of King George Island (South Shetland Islands, Antarctica), preliminary report, Stud. Geol. Pol., 72, part 2,. 98-108, 1981.

# CENOZOIC FLUCTUATIONS IN BIOTIC PARTS OF THE GLOBAL CARBON CYCLE

Jerry S. Olson

Environmental Sciences Division
Oak Ridge National Laboratory
Oak Ridge, Tennessee 37831

Abstract. The mass of organic carbon in land plants presumably decreased over much of Tertiary time. Global average cooling, mountain building, rain shadows, and other drying displaced forests and led to the expansion of shrublands, grasslands and deserts. Quaternary cold repeatedly stimulated the expansion of tundra and cold deserts. Lowering of sea level partly compensated with new areas for coastal and wetland vegetation. Interglacial and postglacial ice retreats opened new lands for boreal, mostly conifer forests (taiga), and for renewed storage of peat in mires. Early Holocene broad-leaved forests expanded again (along with temperate and humid tropical/subtropical climates) and constitute most of the world's plant carbon mass. Slightly less than 800 $10^9$ metric tons C is a plausible estimate in all live land plants in mid-Holocene time. Variations from this estimate by a factor of about 2 seem likely within the late Cenozoic Era: higher in early Miocene and lower in glacial times; 460-660 x $10^9$ metric tons C from A.D. 1980 to ∿1780 A.D.

## Introduction

The paper by Wolfe [this volume] explains the basis for inferring past vegetational and climatic patterns from fossil leaves [see also Dolph and Dilcher, 1979]. Here I will relate the distribution of Tertiary and Quaternary vegetational units to probable influences on the levels and trends of organic carbon. Some of these trends may be interpreted as responses to the general lowering of temperature and $CO_2$ in the atmosphere since the preceding Cretaceous period. Quantifying carbon inventories is difficult, but is necessary for better understanding of the natural carbon cycle and its effects on climate.

The shifting of Gondwanaland across the south pole in Paleozoic time had culminated in Pennsylvanian-Permian glaciations [Lloyd, 1984]. Boucot and Gray [1979, 1982, 1983] invoke evidence about climate, biogeography, and rock facies to complement magnetic indications, in their inference that the continental plates might have merged, perhaps in the southern hemisphere, as early as the Devonian period, when early vascular plants, pteridophytes, were rapidly evolving on land. Around 200 million years ago (Ma) [Gordon, 1975], the Triassic rocks with evaporites and other indications of aridity are logically related to the presumed continental climate [Green, 1961]. By 100 Ma, near mid-Cretaceous time, early flowering plants (angiosperms) had spread rapidly [Scott et al., 1960] over a world without glaciers, formerly dominated by the gymnosperms, or "naked-seeded" plants. How the Cenozoic changes discussed by Wolfe [this volume] and the rest of this paper can be related to inventories or trends of organic carbon and atmospheric carbon dioxide [Olson, 1970, 1981a, b, 1982] is a problem that needs to be evaluated by modern biogeography, paleoecology, and geochemistry.

## Estimating Plant Mass and Carbon in Tertiary Ecosystems

### Qualitative Changes: Paleogene

The necessarily simplified maps (Figure 1, [after Wolfe this volume, Figures 5 and 10]) of late Paleocene to Eocene vegetation reflect an equable climatic pattern, first only slightly cooler than that of the Cretaceous [e.g., MacGinitie, 1969, 1974]. The broad-leaved deciduous forest (around both poles) and associated animals [Estes and Hutchison, 1980] suggests moderate winter temperatures and low annual amplitude of temperature. Moving toward the equator, one encounters broad-leaved evergreen, paratropical, and tropical vegetation [Wolfe, this volume]. The broad-leaved evergreen forest, mixed with conifers in places, had the notophyll or microphyll leaf features which Wolfe [this volume] analyzed (see also caption of my Figure 1). Paratropical forest occurs today in slightly seasonal, humid climates of southern China and adjacent

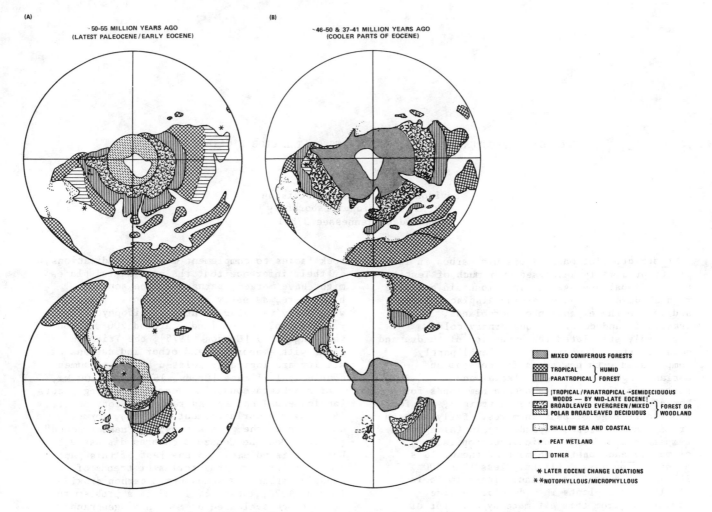

(A)
~50-55 MILLION YEARS AGO
(LATEST PALEOCENE / EARLY EOCENE)

(B)
~46-50 & 37-41 MILLION YEARS AGO
(COOLER PARTS OF EOCENE)

MIXED CONIFEROUS FORESTS

TROPICAL } HUMID
PARATROPICAL } FOREST

|TROPICAL / PARATROPICAL →SEMIDECIDUOUS
WOODS — BY MID-LATE EOCENE!
BROADLEAVED EVERGREEN / MIXED } FOREST OR
POLAR BROADLEAVED DECIDUOUS } WOODLAND

SHALLOW SEA AND COASTAL

• PEAT WETLAND

OTHER

✱ LATER EOCENE CHANGE LOCATIONS
✱✱ NOTOPHYLLOUS / MICROPHYLLOUS

Fig. 1.   Connection of early Cenozoic continental geography to major ecosystem groups
(after Wolfe [this volume], with shifted continental positions after Smith et al.
[1981]): (a) warm Paleocene/Eocene (~50-55 Ma); (b) cooler Eocene times (46-50 and
37-41 Ma).   Some coastal plain and continental plate boundary areas probably
fluctuated above and below sea level during and between these intervals (dots).   These
and other wetland and coastal landscape complexes probably had highly variable plant
biomass and carbon.   They contributed small plant pools but more than average inputs
to land and shallow water sediments of the geologic column because of their proximity
to depositional environments.   (According to Wolfe [1979], microphyllous broad-leaved
evergreen forest tends to have a high proportion of narrow, hard-leaved, i.e.,
sclerophyll, foliage, mainly today in zones of only a few degrees annual temperature
range, having a warmest month of 20°C or cooler, commonly also with light reduced by
fog in transitions to dwarfed montane and subalpine forest.   In zones with contrast of
monthly temperature [Wolfe, this volume, Figure 1] notophyllous forest is mostly
broad-leaved evergreen, typically 40 to 60% with entire margins, without drip tips, of
the "oak-laurel tree group:" Fagaceae, Lauraceae, plus Theaceae and Magnoliaceae, up
to altitudes and latitudes approximating annual mean of 13°C, and downward to zones
near a mean of 20°C in south and southeast Asia.   Paratropical forest, having these
families plus a wider assortment of others with temperate as well as tropical
affinities, is typically two-layered; Wolfe [this volume, Figure 1] suggest mean annual
temperature spanning 21° to 25°C belts in that region.   Tropical rain forests occur
above the latter mean, or even slightly below it, in hot, wet zones, where dry seasons
are absent, not long, or severe, or where locally favorable substrate moisture reserves
counteract drying; ideally three closed forest layers occur, typically with additional
emergents, with buttressed stems and more than 75% entire leaves with drip tips.)

**CARBON IN LIVE PLANTS ~20 MILLION YEARS AGO
(EARLY MIOCENE)**

⊠⊠ TROPICAL RAIN FOREST
‖‖‖ PARATROPICAL RAIN FOREST
NOTOPHYLLOUS BROADLEAVED
EVERGREEN FOREST
NOTOPHYLLOUS WOODLAND/DRY SCRUB
BROADLEAVED DECIDUOUS
FOREST (MINOR EVERGREEN)
MIXED CONIFER FOREST
MIXED CONIFER WOODLAND
TROPICAL-SEMI DECIDUOUS
PARATROPICAL-SEMI DECIDUOUS
NONFOREST
COASTAL, SEA

Fig. 2.  Estimated areas of vegetation in early Miocene time (~20 million years ago), by analogy with categories for modern major world ecosystem complexes [Olson et al., 1983], and Wolfe's paleobotany [1979; this volume].

highlands of modern southeast Asia or in Holdridge's [1964] "submontane rain forest" life zone.  The widespread "tropical humid forest" (called "rain forest" by Wolfe and many others) denotes fossil finds without indication of protective buds, and with leaf traits uncommon in regions with either cold or dry seasons. Semideciduous forest or open woodland is inferred to have developed only in middle Eocene time, presaging the monsoonal climate which is presently familiar on the south or southeast

Asian hemisphere (top right part of Figure 1a, adapted from Wolfe [this volume, Figure 8]).

Only the first and last of these ecosystems have sufficient seasonality, of light or temperature and of drought, respectively, to suggest a reduced level of productivity, i.e., input rate of organic carbon storage.  Over a large part of the continents, paratropical and tropical humid forests almost certainly built up steady state accumulations (i.e., when income equals loss rate) that carried high carbon in live plants.  Locally, in the wetlands, the turnover rate of dead residues was low enough to create a net storage of peat, which became lignite and soft coal, e.g., deposits in North Park, Colorado [Hail and Leopold, 1960], Wyoming and North Dakota (black circles of Figure 1a).

At least the two intervals of cooler Eocene climate (∿46–50 and 37–41 Ma), as inferred by Wolfe [this volume, Figure 10] and indicated in Figure 1b, led to the development and spread of mixed coniferous forest.  Antarctic collections may have overrepresentation of southern beech (Nothofagus) pollen; the extent of a mixed broad-leaved/conifer forest is unclear.  The conifers and broad-leaved genera are normally evergreen, but either may have deciduous forms at high latitudes:  Metasequoia and Glyptostrobus north of 70°N in early Eocene along with polar broad-leaved deciduous forest; southern deciduous or mixed Nothofagus east of the Andes today.

In the late Eocene/early Oligocene, tropical and paratropical forests became more widespread again.  However, they were displaced toward the equator by broad-leaved evergreen forests which Wolfe calls notophyllous or microphyllous (see further Neogene discussion below for more comment on definitions in Figure 1 caption).

Oligocene collections suggest a further restriction of hot- and warm-temperature-adapted species to more equatorial regions [e.g., Louvet and Mouton, 1975].  High-altitude sources for pollen of spruce (Picea), hemlock (Tauga), and alder (Alnus), in early Oligocene of Borneo, and of northern beech (Fagus) and sweet gum (Liquidambar) in late Oligocene/early Miocene of Puerto Rico have been suggested to account for their fossils at low latitudes (see discussion and references in the work by Wolfe [this volume]).  In Venezuela, Podocarpus (a gymnosperm now typical of mountains above 2000 m) occurred as pollen sporadically in the Eocene and more abundantly in the Oligocene and Miocene, but Scott et al. [1960] argue for long-distance river transport for this common microphyll genus and for Alnus in Orinoco sediments.  However, we still have no clear picture of the island and/or mountain paleogeography in places near many continental plate or shelf boundaries.

Wolfe is sceptical of Axelrod and Raven's [1978] inference of savanna woodland and low scrub in today's North African desert region as long ago as 25–30 Ma.  There was at least a gallery forest in the Oligocene in this area [Brown et al., 1982].  Above the basal (Paleocene) part of the Coalmont formation in Colorado, which has little or no grass pollen, Hail and Leopold [1960] found much grass pollen, along with that of Platocarya (Juglandaceae) and Tilia trees, which are also abundant in early and middle Eocene rocks of the Green River formation in Wyoming.

## Neogene Transformations

In Miocene time, however, an increasingly dry belt designated semideciduous woods at 20 Ma extended from East Africa to what is now northwestern India (Figure 2), with more grassy savanna aspect thereafter.  My map differs from Wolfe's [this volume] Figure 11 in locating an area of grassland north of the rising (coniferous) Himalayan upland, but the extent of this community is uncertain and surely varied during the Miocene.  Between late middle Miocene (interfluve grasslands in Nebraska according to MacGinitie [1962]) and late Miocene of Kansas [Chaney and Elias, 1936], i.e., later than my Figure 2 mapping, fossil evidence is accepted as indicating the main spread of North American grassland.  Figure 2 also indicates grassland somewhat schematically in today's Llanos region of Venezuela and in southern Africa, but the limited evidence from pollen is not sufficient to define their dominance, extent in space or time, and grassland relations with savannas or drought-deciduous forests.  Evolution of grasses could be active in many additional places, subordinated to woody plants, or in openings too small or poorly defined to outline on global maps.

Barnosky's [1984] analysis of the Teewinot Formation in Jackson Hole, Wyoming, indicates virtually complete elimination of earlier paratropical and moist temperate elements by late Miocene, (or Pliocene in the more inclusive scope of Leopold and MacGinitie [1960, Figure 3]), 9.5 Ma, in the increasingly continental climates (seasonally dry and cold) around the rising Rocky Mountains.  Of special interest is her resolution of dry interludes, of a few thousand to tens (or hundreds) of thousands of years, interrupting even briefer wet intervals.  Genera now associated with desert scrubs were represented in pollen records that were still dominated by pine, spruce, and fir (perhaps dispersed from the mountains, as from today's Sierra Nevada contribution to nearby desert communities).

By ∿5 Ma in Pliocene time , mountain building and rain shadows had further displaced forests and expanded shrubs, grassland, and desert communities [Leopold and MacGinitie, 1960].  Biomass and carbon in plants presumably underwent a major transition between the bounds which are discussed in the following sections.  Arctic climates still had comparatively mild winters [e.g., Hopkins et al., 1971].

## Preliminary Carbon Estimates for the Early Miocene Vegetation

The early Cenozoic vegetation types (Figure 1) may be sufficiently different from modern vegetation to raise uncertainties about analogy, besides other errors inherent in estimates about climate or stand characteristics for later times. For the trial case of early Miocene vegetation, based on Figure 2, an attempt is made in Table 1 to illustrate a way of approaching this problem: by comparison with recent vegetation considered in the rest of this paper. Tentative planimetered and dot-counted areas for the hemispheres and whole earth are estimated in the first three data columns of Table 1. The alternative area estimates in parentheses indicate credible higher alternative estimates of area for grasslands ($\sim9 \times 10^6$ km$^2$ compared to the fairly conservative $\sim3 \times 10^6$ km$^2$ taken for the area added to Figure 2). Such additional area allocated for grassland is compensated by the lower values for wooded areas in parentheses: subtracted from broad-leaved deciduous forest in the northern hemisphere and broad-leaved notophyllous-microphyllous "forest" (possibly already including some more open woodlands) in the southern hemisphere.

Translating the distributions of area (A) into total carbon in vegetation (C) requires appropriately averaged estimates of carbon density per unit area (D). My best estimates for D (without parentheses in Table 1) are drawn by analogy with categories of Olson et al. [1983]. Higher alternative values in parentheses for tropical and paratropical "rain forest" are compared, using the 20 and 16 kg m$^{-2}$ means for tropical rain forest and seasonal forest of Whittaker and Likens [1973], but these latter categories are not really equivalent and are now considered upper bounds. These numbers (four thirds of my preferred means) are probably inapplicable to wide areas; many kinds of interruption, disturbance, and large areas of regenerating forest and deficient site quality (sandy or thin soils) would be expected to lower average carbon densities, even prior to the evolution of Homo sapiens. Brown and Lugo [1984] are lowering their previous estimates of biomass and carbon in modern forests, but their values are relevant mainly to the issues of enhanced recent human depletion of the forests, as by cycles of shifting cultivation, enhanced burning, and tree harvest by humans, mostly in recent centuries in the tropics [cf. Myers, 1984] and over several millennia elsewhere.

Mediterranean-type woods with interruptions by scrub and grassland were allocated 7 kg m$^{-2}$ of carbon compared with 4 kg m$^{-2}$ from Olson et al. [1983]. An intermediate number (5) in parentheses reflects the possibility of severe fires in the climate of summer drought even before humans, i.e., without such extra grazing and other human disturbance as developed throughout Holocene times. A lower alternative number, like 14 instead of 16 g C m$^{-2}$, for "typical" conifer forest also reflects the possibility of lower average because of extensive fires, uneven stocking density, and varied age class and site quality mosaics, even prior to the later Neogene drying. On the high side, assumed open, mostly conifer woodland was separated from the rest in the Rocky Mountain region and subarctic zone as well as the eastern Andes, with a mean density of 6 kg C m$^{-2}$. On the high side, Pacific coastal conifer density of 20 kg C m$^{-2}$ may be high as a regional average for these reasons.

Tundra-like vegetation for Antarctica is assigned a nominal 1 kg C m$^{-2}$ above the value for the more severe Arctic tundra climates and polar deserts of today. Uncertainties in both A and D, for coastal and wetland systems and islands missed in the mapping and for grasslands, may not have a large effect on C. Nor may local (e.g., rain shadow) deserts. The main influence of all such omissions would be to preempt even more woody vegetation (and high carbon) from areas to which it is presently assigned, thereby lowering global carbon totals in Table 1.

Consider first the carbon numbers without parentheses in Table 1. The underlined subtotals give 1284 x 10$^9$ metric tons for broad-leaved, mostly closed-canopy forests alone. Adding other semideciduous and coniferous woods, my medium estimate for early Miocene woods total is 1518 x 10$^9$ tons, probably including some wooded swamps (10$^9$ tons = 10$^{15}$ g = petagram). Adding 29 x 10$^9$ tons for other vegetation, 1547 x 10$^9$ tons is my current, very tentative global estimate. (Even doubling the last increment to 59 x 10$^9$ tons would increase the total only to 1608 x 10$^9$ tons. Yet increasing the grassland and desert total would probably require a much greater compensating decrease in pools in various woods, as indicated by parenthetic values in the left column with parentheses. Other alternative total estimates and uncertainties can best be evaluated after a review of Holocene vegetation, for which we can make closer analogy with modern conditions, less interpolation and extrapolation, and more subdivisions and local detail than will ever be feasible by reconstruction for the distant geologic past.)

Quaternary Fluctuations of Ecosystems

Pleistocene cold repeatedly stimulated the expansion of tundra and cold desert communities, which were displaced outward along the fringes of successive glaciations. Broecker [1983] has adapted a generalized Pleistocene world "forest" map from files of G. Kukla and other Climatic Impact Assessment Program members [CLIMAP Project Members, 1976] for conditions of 18,000 years before the present. They show "forests" severely restricted, a small fraction of the area covered by present forests; most of the

TABLE 1. Preliminary Reconstruction of Carbon in Live Early Miocene Vegetation (~20 Million Years Ago)

| Life Forms of Major Vegetation[a] | Area[b] (A), $10^6$ km$^2$ | | | Carbon Density[c] (D), kg m$^{-2}$ | Carbon (C), $10^9$ metric tons |
|---|---|---|---|---|---|
| | Northern | Southern | Global | | |
| Tropical rain forest | 9 | 11 | 20 | 15 (to 20) | 300 (to 400) |
| Paratropical rain forest | 9 | 10 | 19 | 12 (to 16) | 228 (to 304) |
| Subtotal | | | 39 | | 528 (to 704) |
| Notophyllous/microphyllous broad-leaved "forest"[d] | (6 to) 9 | 8 | (15 to) 17 | 12 | 204 (180 to) |
| "Mediterranean" type (mostly evergreen temperate) | 7 | 5 | 12 | (5 to) 7 | 84 (30 to) |
| Subtotal | | | 29 | | 288 (240 to) |
| Broad-leaved deciduous[d] | (35 to) 39 | | (35 to) 39 | 12 | 468 (420 to) |
| Total of all the above | | | (101 to) 107 | | 1284 (1188 to) (to 1460) |
| Mixed conifer (closed and open) | | | | | |
| Pacific coastal forest | 1 | | 1 | 20 | 20 |
| Other closed forest | 7 | 1 | 8 | (14 to) 16 | 128 (112 to) |
| Open woodland (mostly conifer) | 5 | 1 | 6 | 6 | 36 |
| Subtotal | | | 15 | | 184 (168 to) |
| Semideciduous Woods | | | | | |
| Tropical | 4 | 1 | 5 | 7 (to 12) | 35 (to 60) |
| Paratropical semideciduous | 4 | 1 | 5 | 3 (to 12) | 15 (to 60) |
| Subtotal | | | 10 | | 50 (to 120) |
| Total of all above (woods) | | | 132 | | 1518 (1406 to) (to 1764) |
| Other vegetation | | | | | |
| Expanding grasslands[d] | 2 (to 6) | 1 (to 3) | 3 (to 9) | 1 | 3 (9 to) |
| Tundra-like | | 15 | 15 | 1 | 15 |
| Wetland, coastal | | | ~0 | 1 (to 10) | 11 (1 to 21) |
| Deserts | | | | ? | ~0 |
| Total C in live plants | 96 | 55 | 150 | | 1547 |
| Plausible estimate range | | | | | (1441 to 1803)[e] |

a Based on characterizations in text, Figure 2 legend, and Wolfe [1979, this volume].

b From areas based on planimetering Figure 2, and text.

c Carbon inferences were analogous with generalizations of Table 1, with Olson [1970a], and Olson et al. [1983] plus adjustments for conditions prior to land clearing history of Richards et al. [1983].

d May also include grassy openings in forest or more or less open woodlands, as mapped in Figure 2. Areas in parentheses allocate additional the southern hemisphere broad-leaved evergreen forest to grassland; also may include some of the northern hemisphere broad-leaved deciduous forest besides the minimum area estimate conjectured north of the Himalayan highland conifer forest and Llanos.

e Approximately -7% to 17% uncertainty from alternative estimates (in parentheses) as discussed in text and footnote d. Other errors of +20% or more might accumulate in geochemical totals due either to area or to carbon per unit area as discussed in references of footnote c.

world's open woodlands, grassy savannas, and northern taiga areas that are transitional to tundra complexes are not discriminated from "other exposed land."

As large volumes of water were locked up in each glacier, sea level was lowered. Coastal vegetation complexes and wetlands on flat coastal plains shifted accordingly. It is not clear to what degree these moist habitats were occupied by forests, or where these areas were dominated by scrubland, bog, or marsh communities of low biomass and live carbon.

Interglacial and postglacial ice retreats opened new lands for boreal (taiga) conifer forest and for storage of peat in mires. Early Holocene broad-leaved forests expanded, along with temperate and humid tropical/subtropical climates. Solomon and Tharp [this volume] illustrate how simulation of changing numbers and sizes of trees under changed climatic conditions may be analogous to some of the historic changes that took place in eastern North America since the last maximum glaciation.

Hot, dry (xerothermic or hypsithermal) climates again stimulated expansion of grassland and deserts, thereby diminishing forest mass. Whether the expansion or maturation of forest successions or locally moistened climates in other (e.g., subtropical) areas compensated for decreased wood in mid-latitudes is an interesting problem [Grove, 1984]. Expanding human agriculture, fuel, buildings, and other uses hastened the oxidation of wood and of many hundred petagrams for gigatons of stored live carbon, as explained below.

Expanding forest fires and grass burning for hunting and grazing also shortened the transient storage of carbon in live plants, standing or fallen dead trees, and litter, until recent decades of forest protection and fire control. Increased turnover rate lowered the level that could be maintained in balance with any given production level. Climatic changes also alter input to, and turnover from, large pools of peat and other humus. How much humus transfers to subfossil and fossil carbon are major issues requiring research combining ecology, pedology, and several specialties of earth science.

### Estimating Holocene Changes in Area, Plant Mass and Carbon

Most of the world's current land plant mass is in forests [Olson, 1974; Ajtay et al., 1979; Olson et al., 1978, 1983] and probably has been in forests or other woods throughout Cenozoic time (e.g., Table 1). Yet the area suitable for woods, i.e., closed forests, woodlands, tall scrub, and savanna, has fluctuated widely. We now work backward to infer the probable carbon content for past Holocene millennia by analogy with modern communities.

### Methods and Sources of Data

A recent analysis [Olson et al., 1983] evaluates the wide assortment of information sources and methods for estimating the areas and carbon amounts per unit area in modern vegetation. Our map base [Olson and Watts, 1982; Olson et al., 1983] locates prevalent ecosystem complexes within cells of 0.5° latitude x 0.5° longitude. Matthews [1983] has a similar map of typical vegetation Unesco types with 1°-cell resolution [Olson, 1970b; Unesco, 1973]. Sums of the areas in map cells of the same type of complex estimate total area A that is dominated by a particular type. Amounts of carbon in live vegetation per unit of ground area (density D) are inferred first from the type means for well-developed or preserved stands. There the vegetation shows a relatively natural or undisturbed aspect, or has had many decades for recovery. However, much vegetation has been disturbed by humans, blowdown, fire, or floods and supports timber volumes, biomass, and plant carbon far below this asymptotic condition; it may never recover under current regimes of frequent disturbance or management harvests. Quantifying the world's inventory of carbon for modern, widely disrupted landscape has been based partly on forest and rangeland surveys and general experience with regional patterns, and a wide variety of literature that is cited in the text and appendix of Olson et al. [1983].

### Modern Estimates

Active research aims to improve estimates of areal coverage of modern vegetation and rapid historic clearing for agriculture [Olson, 1981a; Richards et al., 1983; Houghton et al., 1983], forestry [Olson et al., 1978, 1983; Woodwell et al., 1983], and grazing. Reduction in the average amount of dry mass and carbon per unit area of vegetation, for disturbed regimes, has to be averaged over cycles of disturbance and recovery [Olson, 1981b].

It is not yet possible to use a single, statistically controlled procedure to untangle the random errors and biases for various regions and types. The distinction between local stand variations in D by types, and plausible variations among cells within the global area mapped in the same type has been retained in the table of Olson and Watts [pp. 436-437 in Watts, 1982]. Table 2 of Olson et al. [1983] gives inferred medium estimates of A, D, and the product C = A x D which was provisionally assigned for the main types matching the terminology of the colored map published with that report. Table 5 of Olson et al. [1983] gives a rearrangement that makes it easier to translate data from our categories into various other published terminology; it also gives low and high estimates (judgmentally, considering

TABLE 2.   Climatic Zonal Distribution of Major

| Typical Map Zone | Bioclimatic Complex and Terrain Temperature, Relief | Prevalent Climatic Moisture | Prevalent Vegetation Complex |
|---|---|---|---|

### Nonwoods, Besides Wetlands

| | Polar to subpolar | | |
|---|---|---|---|
| 1-3 | polar | arid to humid | tundra, barren |
| 3-9a | boreal alpine | humid to arid | highland meadow |
| 4 | subpolar | mostly humid | wooded tundra |
| | | | |
| | Grassland | | |
| 22 | other alpine | humid to arid | mountain meadow |
| 13 | temperate | subhumid | tallgrass, marsh |
| 13-14 | temperate, montane | semiarid | grass, shrub, puna |
| 22a | tropical subalpine | humid | paramo/scrubby |
| 26 | tropical/subtropic | semiarid | savanna grassland |
| | | | |
| | Desert (excluding local moist sites) | | |
| 19a,20 | temperate, tropical | sandscapes | bare to grass/shrub |
| 21 | cold highlands | arid, semiarid | "mountain desert" |
| 16-20 | other temperate | arid, semiarid | shrub steppe/bare |
| 27 | tropical/subtropic | very arid, arid | desert/semiarid |

Subtotal, excluding woods and major open wetlands

### Woods, Nontropical/Subtropical

| | Boreal, "semiboreal" and other conifer | | |
|---|---|---|---|
| 5-7 | boreal, continental | semiarid-humid | taiga (northern conifer) |
| 7a-9a | semiboreal or montane | mostly humid | mostly conifer |
| 8-11a | temperate/montane | more humid | conifer (mostly) |
| | | | |
| | Temperate broad-leaved and mixed forest | | |
| 10 | cool | humid | mostly deciduous |
| 11 | warm | humid | notophyll and deciduous |
| 11a-b | warm/montane | mostly subhumid | woodland sclerophyll |

Subtotal temperate and boreal

### Woods, Tropical/Subtropical/Paratropical

| | Closed, tall forest | | |
|---|---|---|---|
| 23 | hot (wet site)* | humid | rainforest, evergreen |
| | hot | mostly humid | mostly evergreen |
| | | | |
| | Tropical/subtropical seasonal forest | | |
| 24 | hot or high (dry) | humid and dry | monsoon/semideciduous |
| (32) | warm to hilly | humid | paratropical forest |
| | | | |
| | Hot-warm interrupted woodland and forest | | |
| 26 | hot | humid-dry | "forest" savanna |
| 25 | hot | dry-subhumid | scrub; tree savanna |

Subtotal of tropical forests and interrupted woods

Total major woods (besides wetlands)

Ecosystems Estimated for ∿6000 Years Before Present

| Area, $10^6$ km$^2$ | Plant Carbon, $10^{15}$ g ($10^9$ metric tons) | Density of Plant Carbon, kg m$^{-2}$ | Production, C $10^9$ metric tons yr$^{-1}$ | Turnover, yr$^{-1}$ |
|---|---|---|---|---|
| | | **Nonwoods, Besides Wetlands** | | |
| 10.79 | 14.6 | | 2.49 | |
| 7.86 | 6.0 | 0.8 | 0.9 | 0.11 |
| 1.58 | 2.5 | 1.6 | 0.71 | 0.29 |
| 1.35 | 6.1 | 4.5 | 0.88 | 0.14 |
| 22.99 | 32.1 | 1.4 | 10.27 | 0.33 |
| 1.81 | | | | |
| 1.04 | | | | |
| 9.43 | | | | |
| 0.94 | | | | |
| 9.77 | | | | |
| 29.35 | 8.58 | 0.3 | 3.37 | 0.4 |
| 5.77 | | | | |
| 3.20 | | | | |
| 10.45 | | | | |
| 9.93 | | | | |
| 63.1 | 55.3 | | 16 | |
| | | **Woods, Nontropical/Subtropical** | | |
| 20.78 | 223 | | 7.34 | |
| 10.1 | 91 | 9 | 3.33 | 0.037 |
| 6.91 | 64.1 | 9.2 | 1.93 | 0.030 |
| 3.77 | 68 | 18 | 2.08 | 0.030 |
| 13.01 | 148 (to 181) | | 5.66 | |
| 3.41 | 42 (to 58) | 14 (to 17) | 2.09 | 0.05 |
| 5.76 | 81 (to 98) | 14 (to 17) | 1.17 | 0.14 |
| 3.83 | 24.8 | 6.5 | 2.40 | 0.17 |
| 33.8 | 371 (to 404) | | 13 | |
| | | **Woods, Tropical/Subtropical/Paratropical** | | |
| 13.39 | 205 (to 250) | | 22 | |
| 4.56 | 73 | 16 | 11.2 | 0.15 |
| 8.83 | 132 (to 177) | 15 (to 20) | 10.8 | 0.082 |
| 3.60 | 50 | | 5.83 | |
| 1.18 | 16.5 | 14 | 1.83 | 0.11 |
| 2.42 | 33.8 | 14 | 4.0 | 0.12 |
| 14.05 | 90 (to 141) | | 11.2 | |
| 8.93 | 45 (to 80) | 5 (to 9) | 6.9 | 0.15 |
| 5.12 | 45 (to 61) | 9 (to 12) | 4.0 | 0.09 |
| 31 | 345 (to 441) | | 39 | |
| 64.8 | 717 (to 845) | | 52 | |

TABLE 2.

| Typical Map Zone | Bioclimatic Complex and Terrain Temperature, Relief | Prevalent Climatic Moisture | Prevalent Vegetation Complex |
|---|---|---|---|
| | **Wetlands, Besides Those Already Accounted for Above** | | |
| 30 | Floodplains of dry regions (arid, semiarid) | | |
| | temperate floodplains | wet/moist site | woods-herb mix |
| | hot floodplains | wet/moist site | herb-woods mosaic |
| | hot swamps | meadow soils | swamp savanna |
| | Cold wetlands | | |
| 2-10a | polar to boreal | wet peat | bog and fen (open) |
| 2-10a | polar to boreal | floodplains | scrub, meadows |
| 4-10a | boreal, semiboreal | mostly peat | bog woods and scrub |

Subtotal for these wetlands

Grand total, nonwoods, woods, wetlands of either (or mixed) types

many factors) for C and for D, which is the greatest single factor leading to this uncertainty. A typical range of uncertainty is $\pm20\%$; the span is wider for some types, but presumably compensating errors between the high and low side of the unknown mean leave its error in this general range: $\sim$460 to 660 x $10^9$ g C. Olson [1970a], Olson et al. [1978, 1983], and Ajtay et al. [1979] converged on a medium estimate near 560 x $10^9$ tons of C on the earth for recent decades. Overcoming sampling uncertainties and common biases (e.g., neglecting natural and artificial gaps in forests that are assumed to be continuous and fully occupying mapped areas) make it more likely that improved estimates for 1980 conditions will tend through the lower half of the uncertainty range instead of the higher.

Untenably high estimates of the extent of tall or dense forest vegetation have been widely used for geochemical estimations and calculations. This reflects the bias just mentioned, and a closely related tendency to attribute to modern times (1980 or 1950) a condition which might have been more plausible sometime before accelerating human population growth and land conversion in the 19th century [Richards et al., 1983]. Whittaker and Likens' [1973] estimates did try to provide for the removal of much natural vegetation in areas tabulated as modern arable land, but they made little or no provision for the above-mentioned reduction in plant carbon density D in vast areas of cut, burned, or naturally regenerating forests, nor in the wide areas of grazing land, other than planted pastures. Bazilevich et al. [1971], reprinted as Rodin et al. [1975], are

still cited as authorities advocating still higher modern plant biomass or carbon. However, they did not even attempt to reduce estimates of area for wild vegetation as a result of agricultural and other management.

Ca 3000 Years B.P.

Conferences with Rodin et al. suggested [Olson, 1974] that their maps and estimates were likely to approximate the biospheric conditions for some Holocene time, prior to the main expansion of civilized technology and agriculture. One modification of their map was called "Continental Ecosystem Patterns and Reconstructed Living Carbon Prior to the Iron Age" [Olson, 1970c], but that map did not separately treat the biomass reductions due to early cultivation. Otherwise that map may approximate geography of major zones within a few centuries of 3000 years before present. That attribution implies climates fairly close to those of the 20th century, and makes no pretense at treating land areas of forest harvest.

Ca 6000 Years B.P.: Ranges of Carbon Per Unit Area

Figure 3 is a further, fairly minor modification of the map just cited. It is projected tentatively for $\sim$6000 years B.P., when human populations and civilized agriculture were even more limited than for 3000 years B.P. The main changes shown are extension of transitional mosaics (10b, 11b): moving the edge of prairie eastward and steppe northward,

(Continued)

| Area, $10^6$ km$^2$ | Plant Carbon $10^{15}$ g ($10^9$ metric tons) | Density of Plant Carbon, kg m$^{-2}$ | Production, C $10^9$ metric tons yr$^{-1}$ | Turnover, yr$^{-1}$ |
|---|---|---|---|---|
| | | Wetlands, Besides Those Already Accounted for Above | | |
| 2.28 | 16.11 | | 4.25 | |
| 1.07 | 11.04 | 10.3 | 2.91 | 0.26 |
| 0.32 | 2.66 | 8.3 | 0.78 | 0.30 |
| 0.89 | 2.41 | 2.7 | 0.56 | 0.29 |
| 3.35 | 8.2 | | 0.68 | |
| 1.30 | 1.96 | 1.5 | 0.19 | 0.09 |
| 0.6 | 1.61 | 2.7 | 0.16 | 0.10 |
| 1.45 | 4.63 | 3.2 | 0.32 | 0.07 |
| 5.63 | 24.3 | | 4.93 | |
| 133.5 | 796 (to 925) | | 72.3 | |

in North America and Eurasia, respectively. Perhaps the Lebanese and Zagros Mountain areas already had reduced forest biomass at both these Holocene times, influenced by very early neighboring settlements. The map scale makes it difficult or unfeasible to show minor shifts related to other Holocene settlements or to climatic changes. Reductions in organic biomass had already commenced in some limited areas of Africa, India, China, and Europe under pressure from early shifting cultivation and irrigation. We do not know yet how closely such reductions may be offset by higher mid-Holocene carbon where subtropical deserts had more grass, shrub, or trees than their regions have now [Grove, 1984]. For example, at least the southern half of the Sahara-Libyan desert had more shrub and grass cover than now (area 27), and trees may have been far more continuous in savanna and dry woodland complexes (areas 26 and 25 on Figure 3).

The scale for quantities of organic carbon (lower left of Figure 3A) includes deliberately overlapping scales in two columns. The top row includes mosaics having no plant cover over wide areas, mixed with plants of low stature. Lichens on rocks alternate with sparse, low scrub in high Arctic tundra (range II, 0-400 g C m$^{-2}$) and in polar deserts and other very arid deserts (range I, 0-100 g C m$^{-2}$). The second row includes transitional semidesert and desert steppe and more substantial grassland and shrubland (range III, 100-1000 g C m$^{-2}$; range IV, 400-2000 g C m$^{-2}$). Some of the wetter low Arctic tundra and mountain (tussock) tundra also falls in these ranges. The third row of the carbon scale (range V, 1000-4000 g C m$^{-2}$; range VI, 2000-6000 g C m$^{-2}$) is high enough

that herbaceous growth must be lush, as on moist sites, or more commonly, tall shrubs and/or low or sparse trees add a woody component of biomass and carbon. These conditions typify various interrupted woods categories of the Olson and Watts [1982] map legend.

Distributed through the ocean areas of Figure 3 are legends for the vegetation zones (arabic numerals) which are associated with the plant mass ranges given with Roman numerals in the map boxes and paragraph comments above and below. In descending position (ascending numeric order) of these zones named in Figure 3, the first three lines of Table 2 of Olson [1975] for zones 1 to 3 correspond to the lines for polar desert, lowland tundra (tundra gley soils), and mountain tundra, respectively, in Table 3 of Rodin et al. [1975], summed in the top line in Table 2 of the present paper and Table 1 of Olson [1974]. The "open bog and mire" (second line of Table 1 of Olson [1974] of boreal as well as Arctic zones is moved to the last section in Table 2 of the present paper, along with some bog woods and floodplains. The subarctic zone 4 includes maritime meadows as well as wooded tundra. Zone 5 is northern taiga, but maritime taiga may also have an interrupted, savanna-like cover, as do some analogous subalpine zones that are difficult or impossible to map globally, even on the 1:30,000,000 scale of Olson and Watts' [1982] map.

Most of the savanna, scrub-woodland complex (zone 26) and parts of the dry woodland-savanna complex (zone 25) fall in this intermediate range (ranges V and VI) of carbon in biomass now. Burning, shifting cultivation, and drought [Grove, 1984] may have reduced the carbon from a

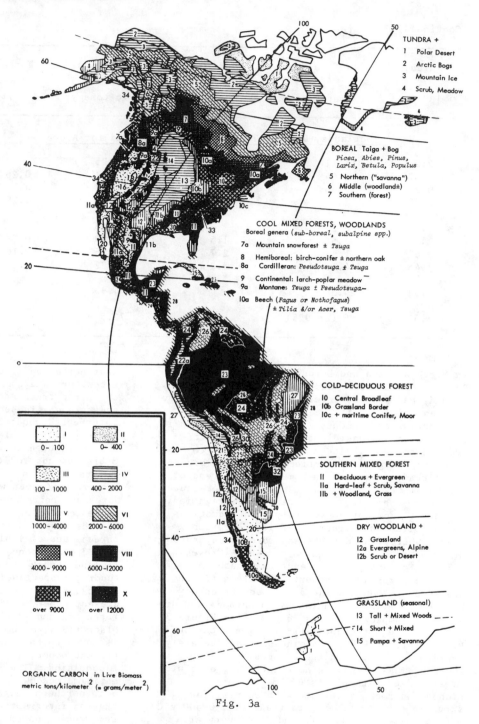

Fig. 3a

Fig. 3. Adjustment of Olson's [1970c] map "Continental Ecosystem Patterns and Reconstructed Living Carbon Prior to the Iron Age" (∿3000 years B.P.) for an interim estimate of conditions around 6000 years before present, sometimes called mid-Holocene time. Scale of overlapping ranges of organic carbon (lower left) are typical for extended landscape complexes prior to major clearing for farming and grazing, but allowing some biomass reduction by early shifting farming, grazing hunting, and burning on an extensive basis by lithic, mesolithic or late paleolithic cultures. Numbered zones are augmented from a simpler list by Bazilevich et al. [1969, 1971]. (Left) Western hemisphere. (Right) Eastern hemisphere.

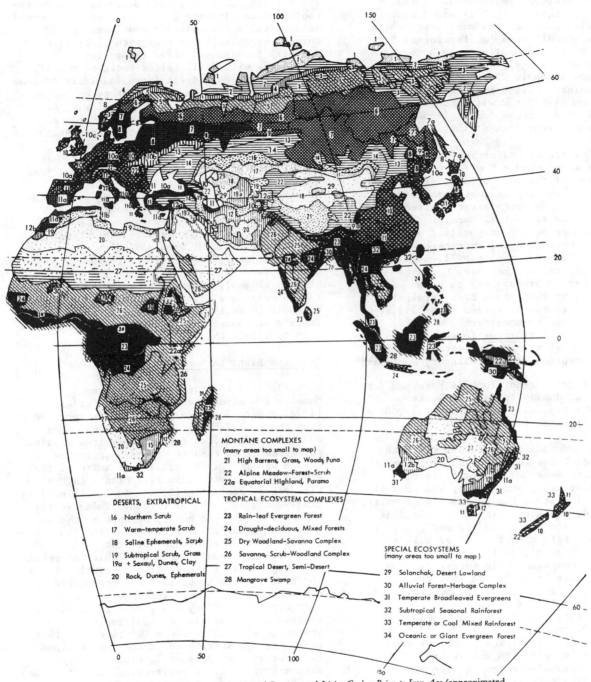

**MONTANE COMPLEXES**
(many areas too small to map)
21  High Barrens, Grass, Woods, Puna
22  Alpine Meadow-Forest-Scrub
22a  Equatorial Highland, Paramo

**DESERTS, EXTRATROPICAL**
16  Northern Scrub
17  Warm-temperate Scrub
18  Saline Ephemerals, Scrub
19  Subtropical Scrub, Grass
19a + Saxaul, Dunes, Clay
20  Rock, Dunes, Ephemerals

**TROPICAL ECOSYSTEM COMPLEXES**
23  Rain-leaf Evergreen Forest
24  Drought-deciduous, Mixed Forests
25  Dry Woodland-Savanna Complex
26  Savanna, Scrub-Woodland Complex
27  Tropical Desert, Semi-Desert
28  Mangrove Swamp

**SPECIAL ECOSYSTEMS**
(many areas too small to map )
29  Solonchak, Desert Lowland
30  Alluvial Forest-Herbage Complex
31  Temperate Broadleaved Evergreens
32  Subtropical Seasonal Rainforest
33  Temperate or Cool Mixed Rainforest
34  Oceanic or Giant Evergreen Forest

*Continental Ecosystem Patterns and Reconstructed Living Carbon Prior to Iron Age* (approximated by J. OLSON, after N. I. BAZILEVICH *et al.*, 1969, Basic Theoretical Problems of Biological Productivity. "Nauka", U.S.S.R. Acad. Sci., Leningrad.)

Fig. 3b

former higher range within the time interval between 3000 and 6000 years ago in the moister parts or gallery forest sites of these zones. A finer subdivision by climate or by archeologic records would be required to show just where and when.

The fourth row of the carbon legend (range VII, 4000–9000 g C m$^{-2}$; range VIII, 6000–12000 g C m$^{-2}$) covers most of the main and southern boreal climatic zone. This matches northern, mostly coniferous forest of the so-called taiga in the original Russian sense of that word, instead of the frequent American usage that appropriates "northern taiga" as noted above. In North America parts of the more continental the non-boreal Cordilleran conifer forest (zone 8a) may have been burned often enough to share these ranges; on average, however, the Pacific Coast Range and probably Cascades and montane zones of the Sierra Nevada include a higher range of carbon associated with the so-called oceanic or giant evergreen forest (zone 34 in the special ecosystems list at the lower right of Figure 3B).

In the fifth row of the carbon legend, range IX (generally above 9000 g C m$^{-2}$) probably applied to most of the cooler parts of the cold-deciduous, summer green, or nemoral forest (main zone 10) in the northeastern United States, and northwest-central Europe. The latter area and presently unspecified parts of the east Asian deciduous and evergreen broad-leaved ("laurel" or notophyll, mostly Pacific zone 11) forests may have had more reduction of carbon biomass due to human occupation even by 6000 years B.P. Limited temperate broad-leaved evergreen (zone 31) areas of southeastern and southwestern Australian Eucalyptus and the Nothofagus forests of New Zealand and Australia probably averaged in this range, with exceptions in the category mentioned next.

The lower right legend block is reserved for several kinds of unusually tall or dense forests. These typically develop over 12,000 g C m$^{-2}$, at least after recovery from disturbances (range X in Figure 3). They probably included cool mixed forest of North America's Great Lakes (northern hardwoods-white pine-hemlock, in zone 10a), southeastern deciduous and/or evergreen forest, and some of the cool rain forest of the Appalachian summits, southern Chile, and Australian Great Dividing Range (zones 11 and 31, including wet sclerophyll or "tall open forest" as well as closed temperate rain forest, and zone 33, "temperate or cool mixed rain forest"). Some of the "microphyllous broad-leaved evergreen forest" of Wolfe [1979, pp. 9-18] typically occurs in small belts of discontinuous high-altitude forest in the tropical montane complexes of Olson and Watts [1982a, b]; where montane grasslands and alpine, rock, or other open patches alternate with forest, they diminish the regional average of live biomass for many of these complexes.

What Wolfe [1979, this volume] calls "notophyllous broad-leaved evergreen forest," or "oak-laurel" life form, includes most Asiatic and Central American parts of my latitudinal zone 11, and some high- or medium-altitude zones in areas that I map as zone 32 (or parts of even zone 23) in Figure 3. The lowland or coastal parts of my zone 32 are the closest approximation shown in that figure to Wolfe's [1979, pp. 7-9] two-layered "paratropical"

forest; the latter may also occur in some of the cooler parts of zones 23 and 24. Zone 23 covers subtropical and tropical broad-leaved humid forest, especially the evergreen parts with drip-tip leaves, and three canopy layers (sometimes with still taller emergent trees). Tropical seasonal forest (zone 24) includes some deciduous elements in the overstory or understory, in African and particularly the Asiatic-North Australian monsoonal climates, where the contrast between wet and dry season is especially pronounced.

The map of Olson [1970c] attempted to separate extrahigh ranges of plant carbon (range XII above 20000 g m$^{-2}$, and range XI, above 16,000 g m$^{-2}$) which are probably valid for the best developed examples of these forest communities. While the zones so mapped have many stands reaching this potential standing crop, their regional mean may prove to be lower because of many less massive forests that occur on less favorable sites (leached sands or hilltop ridges). Many tropical and temperate rain forests are affected by hurricanes or typhoons [Chan 1982; Olson et al., 1978, 1983]. These disturbances and occasional fires also tend to dilute the average.

## Holocene Plant Carbon Totals

Table 2 of the present paper is my combination of area data from Rodin et al. [1975] Figure 3 and carbon estimates improved from those of Table 3 of Olson [1974]. The additional revisions from the previous Russian estimates cited tend to diminish estimates of carbon density D. My earlier estimate of the global pool of carbon in live plants in the terrestrial ecosystem [Olson, 1974, Table 3] was near 1070 x 10$^9$ tons of carbon. This clearly represents an upper bound, because of the points just made about the former overestimation. From my best current allowance for past overestimation and historical studies of recent clearing rates [Richards et al., 1983] , 716 x 10$^9$ tons C seems to be the most reasonable estimate for the forest and other woods categories of ∿6000 years before the present: see also upper middle parts of Table 3. Approximately 255 to 300 x 10$^9$ tons C of that total are now attributed to tropical lowland rain forest plus montane forest, and other seasonal tropical or paratropical forest (zones 23, 24, and 32), not counting perhaps 90 to 141 x 10$^9$ tons C originally in tropical scrub, open woodland, and associated savanna regions (zones 25 and 26). Tables 2 and 3 also give comparable breakdown for outside the tropics, for some additional wetlands not already included in forests, plus the additional ecosystems that are essentially nonwooded, totaling 796 x 10$^9$ tons of plant carbon.

From Table 3 (upper right, after Table 5 of Olson et al. [1983]) my medium estimate for "modern" tropical/subtropical wet and moist forest (160 x $10^9$ tons) plus tropical dry forest and woodland (69 x $10^9$ tons) adds up to only 229 x $10^9$ tons for the modern equivalent as carbon. The extra 14 x $10^9$ tons of C estimated for young second growth is not large by comparison. Biomass estimates based on forest volume inventories in Table 2 of Brown and Lugo [1984] are lower, and conversion from biomass to organic carbon (a conversion factor of 50% in their calculations; 45% to 47% for mine), make their estimates even lower than mine.

My medium "modern" estimate for C in forests, woodlands dense or tall shrublands and outside the hot zones was 233 x $10^9$ tons, not counting 8 x $10^9$ tons for wetlands.

Both the mid-Holocene and "modern" estimates of area and carbon are thus fairly similar for temperate plus boreal zones compared with the tropical, paratropical, and subtropical forests and interrupted woods (Table 3). For these latter, subtraction suggests net releases to the atmosphere near 103 x $10^9$ tons as elemental carbon, compared with 138 x $10^9$ tons (for the cooler zones) and 16 x $10^9$ tons for wetlands, over ∿6000 years. The sum of 257 x $10^9$ tons implies long-term average release at least near 0.043 x $10^9$ tons/yr. Clearly that smooths over much faster deforestation rates of $CO_2$ release in the current century (especially for the tropics and southern hemisphere) and for the 19th century and earlier for north temperate and boreal zones [Olson et al., 1978, 1983; Richards et al., 1983]. If the global plant total were already approaching 460 x $10^9$ tons of plant carbon instead of 560 x $10^9$ tons, the same mid-Holocene estimate as before implies almost 0.06 x $10^9$ tons released to the atmosphere per year. If our older, higher estimates of primordial vegetation prove applicable instead, then 0.1 x $10^9$ tons of C/yr is conceivable.

Such nonwooded landscapes as desert, tundra, grassland, and other shrubland account for 51 x $10^9$ tons C, compared with at least 55.3 x $10^9$ tons C for the mid-Holocene, perhaps not a significant difference. Including crops that are partly woody and their surroundings adds 21 x $10^9$ tons C. Thus gains of organic carbon from horticultural or wildland relics in this category could offset less than 10% of the losses (i.e., inputs mostly to atmospheric $CO_2$) from the changes of other kinds of landscapes.

## Further Discussion and Prospects

It may well become feasible to reconstruct plant carbon budgets for early Cenozoic time, by combining closer attention to climatic seasonality of modern vegetation with that inferred for past geographic areas outlined in Figure 1. Presumably, however, there will be additional uncertainties besides the obvious ones questioning analogy of plant form and gaps in the record of plant collections.

How much particular uncertainties could affect the world total was suggested by focusing on the few alternatives that follow from parenthetical values of estimates lower or higher than the preferred carbon totals for early Miocene vegetation in Table 1. If Whittaker and Likens' [1973] density estimates for tropical closed forests were substituted, the global high estimate would be near 1803 x $10^9$ tons C. Higher area and carbon values for coastal and wetland vegetation could contribute more forest cover and global carbon, but this category, by definition, is mainly intended to exclude those in forests that are already accounted for in the upper lines of the tables.

Reasons discussed earlier and in Olson et al. [1983] help explain common biases of overestimation of forest area, or carbon density. The bottom line of Table 1 suggests a range from 106 x $10^9$ tons below to 206 x $10^9$ tons above the intermediate estimate of 1547 x $10^9$ tons for all kinds of live vegetation, to cover the particular uncertainties or biases indicated by parentheses throughout the table. If another 20% uncertainty were superimposed, analogous with the more random errors of sampling, which we face in modern vegetation as well, an extra ±309 x $10^9$ tons of C might have to be added. However, it seems unlikely that all these uncertainties would pile up on one side or the other; some would compensate. Thus, there is little doubt that early Miocene vegetation had carbon contents well above the estimates near 800 x $10^9$ tons C for middle Holocene time. The inferred a twofold difference seems most likely, due mainly to the aridity and cold, which restrict both the areas of woody vegetation and density of plant carbon per unit area.

The complexities of unraveling vegetation for particular times of Pleistocene glacial landscapes force me to leave that complex problem outside the scope of this paper. Maps published by Broecker [1983] after collective estimates of various CLIMAP team members are provocative in showing areas of closed forest for 18,000 years ago much smaller than those of Olson et al. [1983] for modern forests and woodlands. The biggest uncertainties hinge upon my categories of interrupted woods, which must have occupied large parts of the area not shown as forest by Broecker. Presently, it seems plausible that several of the times of major glaciation could have had vegetation carbon pools as low as half that of middle Holocene time, lower even than our estimates of the "modern" landscape in which many forests have been removed or greatly reduced in numbers and masses of trees.

The present analysis suggests that plant carbon in 1980 is more likely to be in the range of 460 to 560 billion tons of carbon; within the past century or two, it could have ranged up to

TABLE 3. Summary Comparing Estimated Areas and Plant Carbon for Early Miocene, Mid-Holocene, and Modern World Landscapes: ~20,000,000, 6000, and 0 to 200 (or More) Years Before 1980

| Major Groups of Ecosystem Complexes and Associated Climate and/or Site | Reconstructed by Analogy — Early Miocene[a] Area | Early Miocene[a] Carbon | Mid-Holocene[b] Area | Mid-Holocene[b] Carbon | "Modern" 1780 to 1980[c] Area | "Modern" 1780 to 1980[c] Carbon | Change in ~6000 Years |
|---|---|---|---|---|---|---|---|
| **Forest, plus interrupted woods in which herbs and tall or dense shrubs may cover more area than trees** | | | | | | | |
| Tropical, paratropical, subtropical woods | | | | | | | |
| Rain forest and evergreen moist forest | 49 | 578 | 31 | 345 (to 440) | 29 | 243 | -102 (to -152) |
| Semideciduous/deciduous forest | 39 | 528 | 13.4 | 205 (to 250) | | | |
| Equivalent broad-leaved humid and montane woods (some well regrown) | 10 | 50 | 3.6 | 50 | 11.0 | 160 | |
| Dry woodland and woody savanna and thorn woods | | | 14.0 | 90 (to 141) | 15.4 | 69 | |
| Secondary woods and thickets from all above (crops, especially in shifting cultivation, may be included in this and preceding two categories) | | | ~0 | ? | 3.0 | 14 | |
| Mostly temperate and boreal woods | 92 | 940 | | | | | |
| Mostly coniferous forest and parkland | 15 | 184 | 33.8 | 371 (to 404) | 28.7 | 232 | -138 (to -188) |
| Broad-leaved evergreen woods (including mediterranean) | 29 | 288 | 20.8 | 223 | 15.0 | 143 | |
| Broad-leaved, mostly deciduous temperate woods | 39 | 468 | | | 9.5 | 72 | |
| Equivalent to last two (some mixed with conifer) | | | 13.0 | 148 (to 181) | | | |
| Secondary or other low woods and thickets, mostly nontropical | | | | | 4.2 | 18 | |
| **Floodplains and wetlands (swamps, bogs, fens), besides those already included in the above woods, plus coastal** | | | | | | | |
| Floodplains and swamps of semiarid, arid zones | ?[a] | 11? | 5.6[d] | 24 | 2.9[d] | 8+ | -16 |
| Floodplains and bogs of poplar to boreal zones | | | 2.3 | 16 | | | |
| Major wetlands and other coastal | | | 3.3 | 8 | 2.85 | 8 | |
| **Other "nonwoods" besides crops, etc., below** | | | 63.1[e] | 55.3 | 56[f] | 51 | -4.3 |
| Tundra, polar desert, "wooded tundra" | 18 | 18 | 10.8 | 14.6 | 13.0 | 13 | |
| Tropical subalpine (paramo, scrubby) | 13 | 13 | 0.9 | 1.6  16 | | | |
| Other grasslands equivalent to last two (with some shrub or parkland) | 3 | 3 | 22.1 | 30.5  39 | 24.8 | 32  38 | |
| Desert and semidesert (excluding moist sites) | ? | 0 | 29.3 | 8.6 | 18.2 | 6 | |
| **Cropped, residential, commercial, part** | | | | | 15.9 | 21 | +21 |
| Paddy and other irrigated land | 0 | 0 | ~0.001 | 0 | 3.6 | 9 | |
| Other crops, settlements, marginal lands | | | | | 12.3 | 12 | |
| World totals | 151 | 1547 | 133.5 | 823 (to 973) | 132.5 | 556[d] | -240 |
| | | | | | | | -0.04 to 0.6 per yr |

Areas in $10^6$ km² ($10^8$ ha). Plant carbon (estimating live above- and below-ground parts) in $10^9$ metric tons (gigatons) or $10^{15}$ g (petagrams). Rounding may account for minor differences compared with source tables cited below. Some nonsignificant figures are carried to minimize accumulation of rounding errors.

aFor assumed carbon per unit area and major uncertainties, see Table 1. Overestimate of approximate area due to maps counting submerged parts of continents as land may exceed underestimates related to missed islands and coastal fringes.

bSee Table 2 and text for derivation, and reasons why estimates are below those of Olson [1974, 1975], following Bazilevich et al. [1971] and Rodin et al. [1975]. Note that presently derived 6000 years B.P. total is close to Whittaker and Likens' [1973] estimate attributed to 1950, mainly because of more conservative estimates of biomass and plant carbon per unit area.

cWithin limits of rounding error and definitions, "modern" values are equivalent to medium levels of Olson et al. [1978, 1983] attributed to 1970 and 1980, respectively. However, because many information sources were not current, and there may still be high biases in expanding to large area estimates from limited source data, the last column may be more appropriate for earlier times. Judgments of uncertainty suggest a global range of 460 x $10^9$ to 660 x $10^9$ metric tons C in the last 200 years, with estimates likely to become refined toward the lower end of this range for 1980, and toward the upper end for 1780.

dAdditional floodplains and swamps that normally carry forest cover with only minor interruptions along channels and ponds are included in upper part of table instead of in these lines.

eArea estimates based on dot counting by Bazilevich et al. [1971] applied here to ∿6000 years before present and those based on computer map areas of Olson and Watts [1982] for "modern" times were essentially independent. "Nonwoods" definitions made the former source more inclusive, so the larger area for mid-Holocene desert and semidesert is probably an artifact of definition; deserts may have increased instead of declining.

660 billion tons, in view of the clearing rates of these times. Whittaker and Likens' [1973] estimate, near 830 x $10^9$ tons C, formerly was construed by them as appropriate for a modern date like 1950; now this level of plant carbon seems near or above this paper's estimate for a much earlier time as in the mid-Holocene, i.e., before most of the early farming and deforestation in cooler zones of the earth, as well as before the recently expanded tropical deforestation.

If the global totals for climates of the glacial maximum were near or below those of modern time, I infer a reduction to half of the levels for earlier Holocene times, within intervals of only a few thousand years. Clearly, this flux of several hundred gigatons of change, sustained over a few thousand years, is great enough to influence levels of atmospheric $CO_2$ substantially, but not as rapidly as the modern industrial releases of fossil carbon that were earlier laid down as sinks in the wetlands of the world.

In closing, please note the last two columns of Table 2 include, for convenient comparison, summary estimates of organic production and turnover. While several of my plant carbon pools were diminished substantially from those of Rodin, Bazilevich, and Rozov [1975], the net primary production (photosynthesis minus plant respiration) numbers are unchanged from or slightly below the estimates from their Table 3 and Figure 1 map (analogous with my Figure 3). These Holocene global production totals are at least 10 x $10^9$ tons C/yr above several others for the modern world [Deevey, 1960; Olson, 1970; Whittaker and Likens, 1973; several chapters of Lieth and Whittaker, 1975; Olson et al., 1978, 1983; Ajtay et al., 1979]. Yet we do not know whether the discrepancy is another artifact of method, or a very significant decrease in carbon production under stress from human technology. Agriculturalists would prefer to think the expensive investments of civilized culture have increased organic production. It remains to be seen whether extensive gains of this kind have been more than offset by absence of crop cover during much of the year, or by various kinds of biospheric deterioration over vast areas.

Turnover is calculated simply as the ratio of net primary production rate to plant carbon pool. This would be strictly appropriate only for those systems that were approximating a steady state on a regionally averaged basis. Such figures are useful for yardsticks of comparison with the modern ecosystems in which turnover rates have mostly been increased (residence times shortened: as by grazing, tree thinning, as well as harvest cuts, and increasing frequency and/or intensity of fires, as discussed by Olson [1981b]), even if the global net primary production rates had changed little for our "civilized" landscape. The essence of the grassland and desert ecosystems

that expanded so much in Pliocene and
Pleistocene epochs is the shortening of plant
residence time for carbon (increased turnover
fraction near 0.3 to 0.4). Some wetlands and
young or highly productive forests may approach
tis rapid carbon turnover, or its implied
recycling of nutrients. By contrast,
shrublands, open woodlands, and especially most
mature closed forests epitomize patterns of life
that, by definition, entail more storage of
carbon and nutrients within woody plants. That
pattern apparently dominated most of the
biosphere from carboniferous to Miocene times.
The abrupt fluctuations so typical of the
Quaternary period have not only altered the live
carbon storage in geologically brief times, but
have repeatedly switched the inputs to peat,
other soil profiles, humus of paleosols, and
carbonates in semiarid soils [Olson et al., in
press].

Acknowledgments. I am especially grateful to
Eric Sundquist for pairing my presentation with
Jack Wolfe's, and to Wolfe for substantial
information and clarification. Allen Solomon,
Dale Johnson, David Shriner, Ted Moore, and
anonymous reviewers also aided the formulation
and revision of the manuscript. The research
was sponsored by the Carbon Dioxide and Climate
Research Division, U.S. Department of Energy,
under contract DE-AC05-84OR21400 with Martin
Marietta Energy Systems, Inc. Environmental
Sciences Division Publication No. 2462.

## References

Ajtay, G. L., P. Ketner, and P. Duvigneaud,
Terrestrial primary production and phytomass,
in The Global Carbon Cycle, SCOPE 13, edited
by B. Bolin, E. T. Degens, S. Kempe, and
P. Ketner, pp. 129-181, John Wiley, New York,
1979.

Axelrod, D. I., and P. H. Raven, Late Cretaceous
and Tertiary vegetation history of Africa, in
Biogeography and Ecology of Southern Africa,
edited by M. J. A. Werger, pp. 77-130,
W. Junk, The Hague, 1978.

Barnosky, C. W., Late Miocene vegetational and
climatic variations inferred from a pollen
record in northwest Wyoming, Science, 223,
49-51, 1984.

Bazilevich, N. I., L. E. Rodin, and N. N. Rozov,
Geographical aspects of biological
productivity, in Soviet Geography: Review and
Translation, 12, pp. 293-317, Scripta, Silver
Spring, Maryland, 1971.

Bazilevich, N. I., T. K. Gordeeva,
O. V. Zalensky, L. E. Rodin, and J. K. Ross,
Basic Theoretical Problems of Biological
Productivity (in Russian), Nauka, Leningrad,
1969.

Boucot, A. J., and J. Gray, Epilogue: A Paleozoic
Pangaea?, in Historical Biogeography, Plate
Tectonics, and the Changing Environment,
edited by J. Bray and A. J. Boucot,
pp. 465-482, Oregon State University Press,
Corvallis, 1979.

Boucot, A. J., and J. Gray, Paleozoic data of
climatological significance and their use for
interpreting Silurian-Devonian climate, in
Climate in Earth History, pp. 189-198,
National Research Council, National Academy
Press, Washington, D.C., 1982.

Boucot, A. J., and J. Gray, A Paleozoic Pangaea,
Science, 222, 571-581, 1983.

Broecker, W., The ocean, Sci. Am., 245, 146-160,
1983.

Brown, S., and A. E. Lugo, Biomass of tropical
forests: A new estimate based on forest
volumes, Science, 223, 1290-1293, 1984.

Brown, T. M., M. J. Kraus, S. L. Wing,
J. G. Fleagle, B. H. Tiffney, E. L. Simons,
and C. F. Vondra, The Fayum primate forest
revisited, J. Human Evol., 11, 603-632, 1982.

Chan, Yip-Hoi, Storage and release of organic
carbon in Peninsular Malaysia, Int. J.
Environ. Stud., 18, 211-222, 1982.

Chaney, R. W., and M. K. Elias, Late Tertiary
Floras from the High Plains, Carnegie Inst.
Washington Publ., 476, 1-72, 1936.

CLIMAP Project Members, The surface of the
ice-age earth, Science, 191, 1131-1136, 1976.

Dasmann, R. F., Towards a system for classifying
natural regions of the world and their
representation by national parks and reserves,
Biol. Conserv., 4, 247-255, 1972.

Dolph, G. E., and D. L. Dilcher, Foliar
physiognomy as an aid in determining
paleoclimate, Palaeontographica, Abt. B, 170,
151-172, 1979.

Estes, R., and J. H. Hutchison, Eocene lower
invertebrates from Ellesmere Island, Canadian
Artic archipelago, Palaeogeogr. Palaeoclimatol.
Palaeoecol., 30, 325-347, 1980.

Gordon, W. A., Distribution by latitude of
Phanerozoic evaporite deposits, J. Geol., 83,
671-684, 1975.

Green, R., Palaeoclimatic significance of
evaporites, in Descriptive Palaeoclimatology,
edited by A. E. M. Nairn, pp. 61-88,
Wiley-Interscience, New York, 1961.

Grove, A. T., Changing climate, changing biomass,
and changing atmospheric $CO_2$, Prog.
Biometeorol., 3, 5-10, 1984.

Hail, W. J., and E. B. Leopold, Paleocene and
Ecocene age of the Coalmont Formation, North
Park, Colorado, U.S. Geol. Surv. Profess.
Papers, 400-B, 260-261, 1960.

Holdridge, L. R., Life Zone, Ecology, Tropical
Science Center, San Jose, Costa Rica, 1964.

Hopkins, D. M., J. V. Mathews, J. A. Wolfe, and
M. L. Silberman, A Pliocene flora and insect
fauna from the Bering Strait region,
Palaeogreogr. Palaeoclimatol. Palaeoecol., 9,
211-231, 1971.

Houghton, R. A., J. E. Hobbie, J. M. Melillo,
B. Moore, B. J. Peterson, G. R. Shaver, and
G. M. Woodwell, Changes in the carbon content

of terrestrial biota and soils between 1860 and 1980: A net release of $CO_2$ to the atmosphere, Ecological Monographs, 53, 235-262, 1983.

Leopold, E. B., and H. D. MacGinitie, Development and affinities of Tertiary floras in the Rocky Mountains, Floristics and Paleofloristics of Asia and Eastern North America, edited by A. Graham, pp. 147-200, Elsevier, Amsterdam, 1972.

Lieth, H., and R. H. Whittaker, Primary Productivity of the Biosphere, Springer-Verlag, New York, 1975.

Lloyd, C. R., Pre-Pleistocene paleoclimates: a summary of the geological and paleontological evidence, Adv. Geophys., 26, 35-140, 1984.

Louvet, P., and J. Mouton, La flore Oligocene du Djebel Coquin (Libye), Actes Congr. Natl. Soc. Savantes, 95th Sect. Sci., 3, 79-96, 1975.

MacGinitie, H. D., The Kilgore flora, Univ. Calif. Berkeley Publ. Geol. Sci., 35, 67-158, 1962.

MacGinitie, H. D., The Eocene Green River flora of northwestern Colorado and northeastern Utah, Univ. Calif. Berkeley Publ. Geol. Sci., 35, 67-158, 1969.

MacGinitie, H. D., An early middle Eocene flora from the Yellowstone-Absaroka volcanic province, northwestern Wind River Basin, Wyoming, Univ. Calif. Berkeley Publ. Geol. Sci., 108, 103, 1974.

Matthews, F., Global vegetation and land use: New high-resolution data bases for climate studies, J. Climate Appl. Meteorol., 22, 474-487, 1983.

Myers, N., The Primary Source: Tropical Forests and Our Future, W. W. Norton, New York, 1984.

Olson, J. S., Carbon cycles and temperate woodlands, in Analysis of Temperate Forest Ecosystem, Ecol. Stud., no. 1, edited by D. E. Reichle, pp. 226-241, Springer-Verlag, New York, 1970a.

Olson, J. S., Geographic index of world ecosystems, in Analysis of Temperate Forest Ecosystem, Ecol. Stud., no. 1, edited by D. E. Reichle, pp. 297-304, Springer-Verlag, New York, 1970b.

Olson, J. S., Continental ecosystem patterns and reconstructed living carbon prior to the Iron Age, in Analysis of Temperate Forest Ecosystem, Ecol. Stud., no. 1, edited by D. E. Reichle, back flyleaf, Springer-Verlag, New York, 1970c.

Olson, J. S., Terrestrial ecosystem, in Encyclopedia Britannica, 15th edition, pp. 144-149, Chicago, Illinois, 1974.

Olson, J. S., Productivity of forest ecosystems, Productivity of forest ecosystems, in Productivity of World Ecosystems, edited by D. E. Reichle, J. F. Franklin, and D. W. Goodwall, pp. 33-43, National Academy of Sciences, Washington, D.C., 1975.

Olson, J. S., The role of the biosphere in the carbon cycle, in Beyond the Energy Crisis,

edited by R. A. Fazzolare and C. B. Smith, vol. 4, pp. a125-a140, Pergamon, New York, 1981a.

Olson, J. S., Carbon balance in relation to fire regime, in Fire Regimes and Ecosystem Properties, edited by H. A. Mooney, T. M. Bonnicksen, N. L. Christensen, J. E. Lotan, and W. A. Reiners, USDA For. Serv. Tech. Rep., WO-26, 327-378, 1981b.

Olson, J. S., Earth's vegetation and atmospheric carbon dioxide, in Carbon Dioxide Review: 1982, edited by W. C. Clark, pp. 388-398, Oxford University Press, New York, 1982.

Olson, J. S., and J. A. Watts, Major world ecosystem complexes, ranked by carbon in live vegetation, in Carbon Dioxide Review: 1982, edited by W. C. Clark, color map, Oxford University Press, 1982.

Olson, J. S., H. A. Pfuderer, and Y.-H. Chan, Changes in the global carbon cycle and the biosphere, Rep. ORNL/EIS-109, Oak Ridge National Laboratory, Oak Ridge, Tennessee, 1978. (Available from the National Technical Information Service, Springfield, Virginia.)

Olson, J. S., J. A. Watts, and L. J. Allison, Carbon in live vegetation of major world ecosystems, Rep. ORNL-5862, Oak Ridge National Laboratory, Oak Ridge, Tennessee, 1983. (Available from the National Technical Information Service, Springfield, Virginia.)

Olson, J. S., R. M. Garrels, R. A. Berner, T. V. Armentano, M. I. Dyer, J. E. Lovelock, H. G. Ostlund, and D. H. Yaalon, How does (did) the natural global carbon cycle operate?, State-of-the-Art-Report on Global Carbon Cycle Research, U.S. Department of Energy, Washington, D.C., in press.

Richards, J. P., J. S. Olson, and R. M. Rotty, Development of a data base for carbon dioxide releases resulting from conversion of land to agricultural uses, Rep. ORAU/IEA-82-10(M), Oak Ridge National Laboratory, Oak Ridge, Tennessee, 1983. (Available from the National Technical Information Service, Springfield, Virginia.)

Rodin, L. E., N. I. Bazilevich, and N. N. Rozov, Productivity of the world's main ecosystem, in Productivity of World Ecosystems, edited by D. E. Reichle, J. F. Franklin, and D. W. Goodwall, pp. 13-26, National Academy of Sciences, Washington, D.C., 1975.

Scott, R. A., E. S. Barghoorn, and E. B. Leopold, How old are the angiosperms?, Amer. J. Sci., Bradley Volume, 258-A, 284-299, 1960.

Smith, A. G., A. M. Hurley, and J. C. Briden, Phanerozoic Paleocontinental World Maps, p. 102, Cambridge University Press, New York, 1981.

Solomon, A. M., and M. L. Tharp, Simulation experiments with late Quaternary carbon storage in mid-latitude forest communities, this volume.

Unesco, International Classification and Mapping of Vegetation, Paris, 1973.

Watts, J., The Carbon Dioxide Questions: A Data Sampler, in Carbon Dioxide Review: 1982, edited by W. C. Clark, pp. 429-269, Oxford University Press, New York, 1982.

Whittaker, R. H., and G. E. Likens, Carbon in the biota, Carbon and the Biosphere, edited by G. M. Woodwell, and E. V. Pecan, pp. 281-302, CONF-720510, U.S. Atomic Energy Commission, 1973. (Available from National Technical Information Service, Springfield, Virginia.)

Wolfe, J. A., Temperature parameters of humid to mesic forests of eastern Asia and relation to forests of other regions of the Northern Hemisphere and Australasia, U.S. Geol. Surv. Prof. Pap., 1106, 37 pp., 1979

Wolfe, J. A., Distribution of major vegetational types during the Tertiary, this volume.

Woodwell, G. M., J. E. Hobbie, R. A. Houghton, J. M. Melillo, B. Moore, B. J. Peterson, and G. R. Shaver, Global deforestation: Contribution to atmospheric carbon dioxide, Science, 222, 1081-1086, 1983.

# AN IMPROVED GEOCHEMICAL MODEL OF ATMOSPHERIC $CO_2$ FLUCTUATIONS OVER THE PAST 100 MILLION YEARS

Antonio C. Lasaga

Department of Geosciences, The Pennsylvania State University
University Park, Pennsylvania 16802

Robert A. Berner

Kline Geology Laboratory, Yale University
New Haven, Connecticut 06511

Robert M. Garrels

Department of Marine Science, University of South Florida
St. Petersburg, Florida 33701

Abstract. The carbonate-silicate geochemical cycle of Berner, Lasaga and Garrels [Berner et al., 1983] is extended to include reactions involving organic carbon and sulfur. To study the effect of the extended cycle on atmospheric $CO_2$, new reservoirs are added for organic carbon, pyrite, gypsum, and atmospheric oxygen. The present-day cycle is constructed taking into account oxidation of pyrite, diagenetic pyrite formation, sulfate reduction at mid-ocean ridges, organic carbon weathering and organic carbon burial, in addition to the fluxes included by Berner et al. [1983]. The organic carbon and reduced sulfur fluxes over the last 100 million years are obtained from a generalized isotope model and data on $\delta^{13}C$ and $\delta^{34}S$. The augmented model modifies the results of BLAG, Berner et al. [1983], while retaining the basic conclusions, specifically the link between $CO_2$, paleoclimate, and tectonism.

## Introduction

The distribution of elements within the ocean-atmosphere-lithosphere system is the result of a dynamic balance between many geochemical processes [Chamberlain, 1968; Budyko and Ronov, 1979]. To decipher the major processes in the balance, one must simplify the system by the introduction of broad reservoirs. The dynamics is then represented by the possible fluxes between these reservoirs. Recently, Berner, Lasaga, and Garrels (Berner et al. [1983]; henceforth this paper will be referred to as BLAG [1983]) described a model of the carbonate-silicate geochemical cycle that enabled a study of its effect on atmospheric carbon dioxide over the past 100 million years. This model elaborated on the classic "Urey" reactions:

$$CO_2 + CaSiO_3 \underset{\text{metamorphism}}{\overset{\text{weathering}}{\rightleftarrows}} CaCO_3 + SiO_2 \quad (1a)$$

$$CO_2 + MgSiO_3 \underset{\text{metamorphism}}{\overset{\text{weathering}}{\rightleftarrows}} MgCO_3 + SiO_2 \quad (1b)$$

Because the amount of $CO_2$ in the atmosphere is very small compared to the amount of carbon in carbonates or organic carbon, atmospheric $CO_2$ is expected to be quite dependent on any small imbalances between the fluxes represented by reactions (1). BLAG [1983] focused on 5 reservoirs: calcite, dolomite, oceans ($Mg^{2+}$, $Ca^{2+}$, $HCO_3^-$), Ca-Mg silicates, and the atmosphere. The fluxes between these reservoirs were based on the following geochemical processes: weathering, metamorphism-magmatism, calcite precipitation, and sea water-basalt reactions. A kinetic model was constructed, which took into account the effects of tectonic processes on the geochemical fluxes, and this model was used to deduce the evolution of the geochemical cycle and atmospheric $CO_2$ in particular over the last 100 million years.

BLAG [1983] pointed out some major conclusions regarding the relation between tectonic processes, $CO_2$, and paleoclimatic conditions. However, the model clearly can be extended in several directions. This paper presents some major additions, which will refine the previous model and, in

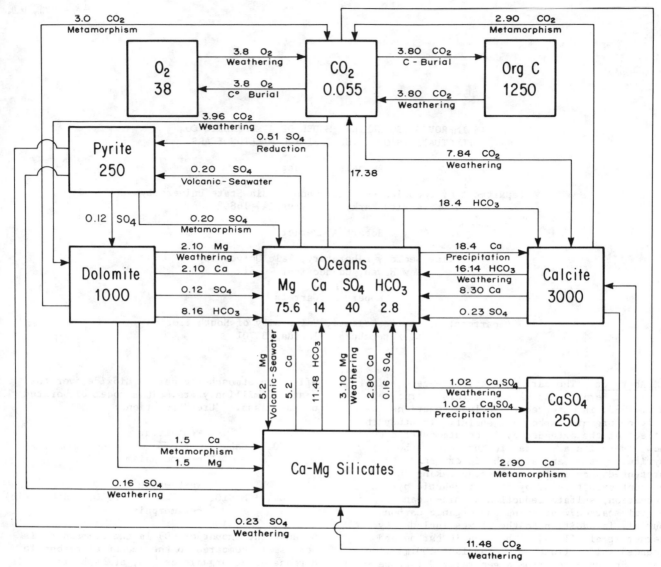

Fig. 1.   Enlarged present-day geochemical cycle.  Reservoir contents are in $10^{18}$ moles, and fluxes are given in units of $10^{18}$ moles per million years (see text).

fact, remove some of the problems that arose in the results of BLAG [1983].

The most important new reservoir considered is the organic carbon reservoir (see Figure 1). Besides the processes already mentioned, atmospheric $CO_2$ can also be influenced by imbalances in the overall reaction

$$CH_2O + O_2 \underset{\substack{\text{organic} \\ \text{matter burial}}}{\overset{\text{oxidation}}{\rightleftarrows}} CO_2 + H_2O \qquad (2)$$

where $CH_2O$ stands for the organic carbon in crustal rocks (which is approximated to have a zero-valence oxidation state).  This organic carbon includes the carbon in the biosphere, dead

organic matter, and the organic carbon in metamorphic and sedimentary rocks, the latter containing by far most of the carbon [Holland, 1978].  By adding the reduced carbon to the inorganic oxidized carbon, the treatment of carbon in the model now becomes much more complete.

In addition to the organic carbon reservoir, pyrite, gypsum, and oxygen reservoirs have been added to the model.  These additions are a consequence of including organic carbon in the model as discussed below.

Present-Day Overall Geochemical Cycle

The expanded geochemical cycle is shown in Figure 1.  All the reservoir contents and fluxes from BLAG [1983] have been incorporated into the

TABLE 1. Present-Day Weathering Fluxes in the Carbonate-Silicate Cycle Taken From Figure 1
and Changed to Include Weathering by $H_2SO_4$

| Flux | Species | Flux, $10^{18}$ mol m.y.$^{-1}$ | |
| --- | --- | --- | --- |
| | | Associated With $H_2SO_4$ Weathering | Associated With $CO_2$ Weathering |
| Dolomite $\longrightarrow$ ocean | Ca | 0.12 | 1.98 |
| Dolomite $\longrightarrow$ ocean | Mg | 0.12 | 1.98 |
| Dolomite $\longrightarrow$ ocean | $HCO_3$ | 0.24 | 7.92 |
| Calcite $\longrightarrow$ ocean | Ca | 0.46 | 7.84 |
| Calcite $\longrightarrow$ ocean | $HCO_3$ | 0.46 | 15.68 |
| Silicate $\longrightarrow$ ocean | Ca | 0.08 | 2.72 |
| Silicate $\longrightarrow$ ocean | $HCO_3$ | 0 | 11.48 |
| Atmosphere $\longrightarrow$ dolomite | $CO_2$ | 0 | 3.96 |
| Atmosphere $\longrightarrow$ calcite | $CO_2$ | 0 | 7.84 |
| Atmosphere $\longrightarrow$ silicate | $CO_2$ | 0 | 11.48 |

figure. Nonetheless, the addition of new reservoirs and fluxes must be accounted for. We will discuss these numbers in this section.

## Weathering Fluxes

The weathering fluxes for carbonates and silicates from our earlier model remain the same. However, the original weathering fluxes have now been divided into the portion due to $CO_2$ weathering and that due to sulfuric acid weathering, the latter arising from the oxidation of exposed pyrite on the continents. For example, calcium silicates can weather according to the reactions given by

$$CaSiO_3 + 2H_2CO_3 + H_2O \longrightarrow Ca^{2+} + 2HCO_3^- + H_4SiO_4 \qquad (3a)$$

or

$$CaSiO_3 + H_2SO_4 + H_2O \longrightarrow Ca^{2+} + SO_4^{2-} + H_4SiO_4 \qquad (3b)$$

A similar set of reactions can be written for $MgSiO_3$, $CaCO_3$, and $CaMg(CO_3)_2$. The partitioning of the weathering flux in Figure 1 into that due to $CO_2$ and that due to $H_2SO_4$ is given in Table 1. The assumption made is that the fraction of the total sulfuric acid production taken up by a reservoir is the same as the fraction of the total $CO_2$ weathering flux that is taken up by the same reservoir. This partition will be discussed later in establishing the proper kinetic rate constants. For a discussion of the weathering flux partitioning, consult BLAG [1983].

## Pyrite Reservoir

The total amount of pyrite involved in sedimentary cycling is estimated at $250 \times 10^{18}$ moles [Berner and Raiswell, 1983]. This pyrite resides

largely in sedimentary rocks, especially shales. If rocks containing pyrite are exposed at the surface, the pyrite will be oxidized, and sulfuric acid will be a weathering product. The flux of sulfuric acid due to the oxidation of pyrite balances the pyrite formation flux, which arises from sulfate reduction in marine sediments. The estimated flux for the deposition of pyrite in present marine sediments (preindustrial value) is around $0.51 \times 10^{18}$ moles S/m.y. [Berner and Raiswell, 1983] as shown in Figure 1.

Pyrite formation will induce fluxes to and from reservoirs other than pyrite. The overall reaction for the reduction of sulfate and the formation of pyrite in marine sediments can be written as [Garrels and Perry, 1974]

$$7.5\,CH_2O + 4SO_4^{2-} + 8Ca^{2+} + 8HCO_3^- + Fe_2O_3$$
$$\longrightarrow 2FeS_2 + 8CaCO_3 + 7.5CO_2 + 11.5H_2O \qquad (4)$$

Note that the formation of pyrite results in the precipitation of calcite (triggered by the $CH_2O$ oxidation), loss of $Ca^{2+}$, $HCO_3^-$ and $SO_4^{2-}$, and the liberation of $CO_2$ (see Garrels and Perry [1974] for further discussion). Thus, if $0.51 \times 10^{18}$ moles S/m.y. are buried as pyrite, $1.02 \times 10^{18}$ moles $CaCO_3$/m.y. will form and $1.02 \times 10^{18}$ moles/m.y. of $Ca^{2+}$ and $HCO_3^-$ will be lost from the oceans. This means that of the $18.40 \times 10^{18}$ moles/m.y. of calcite precipitated (as shown in Figure 1), only $17.38 \times 10^{18}$ moles come from ocean $CaCO_3$ precipitation and $1.02 \times 10^{18}$ moles are due to pyrite formation. The flux $17.38 \times 10^{18}$ moles $CaCO_3$/m.y. is used (rather than the previous $17.54 \times 10^{18}$) to calculate the kinetic constants ($k_{prep}$, $K_{eq}$) for calcite precipitation.

The weathering on the continents of $0.51 \times 10^{18}$ moles of pyrite sulfur per million years will also affect $O_2$ according to the overall reaction

$$2FeS_2 + 4H_2O + 7.5O_2 \longrightarrow Fe_2O_3 + 4H_2SO_4 \qquad (5)$$

Therefore, an $O_2$ output flux of $15/8 \times 0.51 \times$

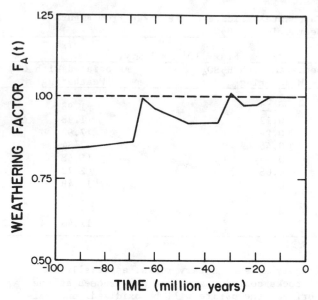

Fig. 2. Modified weathering area correction factor $f_A(t)$ as a function of time. Modified from data of Barron et al. [1980]. The dashed line represents present-day value.

$10^{18}$ or 0.96 x $10^{18}$ moles O$_2$/m.y. is induced by pyrite weathering as shown in Figure 1. Furthermore, the amount of organic matter oxidized by sulfate (not O$_2$) will then be counted as contributing to O$_2$ production (i.e., the reverse of reaction (2)). Thus, in Figure 1 there is an additional input flux of 15/8 x 0.51 or 0.96 x $10^{18}$ moles O$_2$/m.y. associated with pyrite formation.

The circulation of seawater through basalts in the mid-ocean ridge leads to the high-temperature reduction of sulfate and the formation of pyrite. This flux has been estimated at 0.2 x $10^{18}$ moles S/m.y., which represents 28% of total oceanic sulfate reduction, a reasonable estimate of this process based on presently available evidence (see Von Damm [1983] for a discussion). A balancing flux of 0.2 x $10^{18}$ moles S/m.y. is assumed for the loss of pyrite due to volcanic degassing.

## Organic Carbon Reservoir

The total amount of organic carbon in the crust is estimated at 1250 x $10^{18}$ moles C as shown in Figure 1 [Garrels et al., 1975]. Most of this carbon resides in shales [Holland, 1978]. The fluxes in and out of the reservoir reflect the overall reaction CH$_2$O + O$_2$ $\rightleftarrows$ CO$_2$. The present-day burial rate of organic carbon in sediments yields a preagricultural flux of 3.8 x $10^{18}$ moles C/m.y. based on the average organic content of modern sediments [Berner, 1982] and the average Cenozoic total sediment burial rate [Gregor, 1984]. This is balanced by an organic

carbon weathering (oxidation) flux of 3.8 x $10^{18}$ moles C/m.y. The organic carbon burial flux is a net flux; i.e., it takes into account the oxidation of part of the original carbon buried by diagenetic sulfate reduction (equation (4)).

The oxidation and burial of organic carbon will result in similar oxygen fluxes according to the overall reaction (2). Therefore, Figure 1 includes a flux of 3.8 x $10^{18}$ moles O$_2$/m.y. from the O$_2$ to the CO$_2$ reservoir (CH$_2$O weathering) and a flux of 3.8 x $10^{18}$ moles O$_2$/m.y. from the CO$_2$ to the O$_2$ reservoir (CH$_2$O burial). Obviously, imbalances in this subsystem would lead to changes in both the CO$_2$ and O$_2$ reservoirs.

## Gypsum-Anhydrite Reservoir

The total amount of gypsum and anhydrite in evaporites is taken as 250 x $10^{18}$ moles S [Berner and Raiswell, 1983]. This sulfate is weathered at a rate of 1.02 x $10^{18}$ moles S/m.y. based on the river flux [Berner and Raiswell, 1983]. There will be a balancing flux of 1.02 x $10^{18}$ moles S/m.y. resulting from the precipitation of gypsum in evaporite basins.

## Oxygen Reservoir

The total content of oxygen in the atmosphere is 38 x $10^{18}$ moles O$_2$ [Holland, 1978]. The fluxes in and out of this reservoir in Figure 1 have already been described in the earlier sections.

### Kinetic Treatment of the Cycle Over Geologic Time

In order to follow the evolution of the geochemical cycle with time, we must relate the fluxes in Figure 1 to particular rate laws. Some of these rate laws have already been discussed by BLAG [1983]. In this section we will concentrate on the new rate laws associated with the enlarged geochemical cycle. Because of the importance of stable isotopic data to the calculation of fluxes for organic carbon and pyrite sulfur, a separate subsection is devoted to these fluxes.

## Weathering Rate Laws

The rate laws for the weathering of silicates and carbonates by CO$_2$ remain the same as before, i.e., a linear rate law:

$$F_W = k_W A \tag{6}$$

where A stands for the reservoir content of the particular silicate or carbonate. The weathering of pyrite (equation (5)), organic matter (equation (2)), and gypsum (anhydrite) is also assumed to follow a linear rate law:

$$F_{W_{py}} = k_{W_{py}} A_{py} \tag{7}$$

$$F_{W_{gy}} = k_{W_{gy}} A_{gy} \qquad (8)$$

$$F_{W_{C^\circ}} = k_{W_{C^\circ}} C^\circ \qquad (9)$$

(For simplification, organic carbon is represented here as $C^\circ$, which is not to be confused with elemental carbon.) The rate constants $k_W$ in (6), (7), (8), and (9) will also depend on variations in continental land area and temperature-$CO_2$ variations as outlined by BLAG [1983] (see also discussion below), i.e.,

$$k_W = k_W^\circ f_A(t) f_B(CO_2) \qquad (10)$$

where $f_A(t)$ is the correction due to changes in total land area and $f_B(CO_2)$ is the $CO_2$-feedback function (see BLAG, 1983). The function $f_A(t)$ here is slightly modified from the $f_A(t)$ function used by BLAG [1983] to include some new interpolation points based on relative sea level fluctuation given by Vail [1979] (see Figure 2).

The weathering of pyrite (equation (5)) produces sulfuric acid, which in turn is used to weather silicates and carbonates (see Table 1). Each weathering flux associated with sulfuric acid will, therefore, depend on the total production given by $F_{W_{py}}$ in (8), i.e.,

$$F_W' = f F_{W_{py}} \qquad (11a)$$

where $F_W'$ stands for the weathering flux of carbonates or silicates caused by reaction with sulfuric acid and $f$ stands for the fraction of sulfuric acid used in weathering each reservoir. In deriving the numbers of Table 1 the assumption was made that the fraction $f$ of the sulfuric acid weathering a particular carbonate or silicate reservoir is the same as the fraction of the total $CO_2$ weathering assigned to that particular reservoir. If we write the weathering reactions as

$$2CaCO_3 + H_2SO_4 \longrightarrow SO_4^{2-} + 2HCO_3^- + 2Ca^{2+}$$

$$CaMg(CO_3)_2 + H_2SO_4 \longrightarrow SO_4^{2-} + 2HCO_3^- + Ca^{2+} + Mg^{2+}$$

$$H_2O + CaSiO_3 + H_2SO_4 \longrightarrow SO_4^{2-} + Ca^{2+} + H_4SiO_4$$

$$H_2O + MgSiO_3 + H_2SO_4 \longrightarrow SO_4^{2-} + Mg^{2+} + H_4SiO_4$$

we have

$$F_{W_C}' = 2F_{W_{py}} f_c \qquad (11b)$$

$$F_{W_D}' = F_{W_{py}} f_D \qquad (11c)$$

$$F_{W_{CaSi}}' = F_{W_{py}} \cdot f_{CaSi} \qquad (11d)$$

$$F_{W_{MgSi}}' = F_{W_{py}} \cdot f_{MgSi} \qquad (11e)$$

The factor of 2 in (11b) follows from the stoichiometry of the calcite reaction above. The present-day uptake of sulfuric acid by these reservoirs, based on each analogous $f$ value for $CO_2$ weathering, is (Table 1) $0.12 \times 10^{18}$ (dolomite), $0.23 \times 10^{18}$ (calcite), $0.08 \times 10^{18}$ (Ca-silicates), and $0.08 \times 10^{18}$ (Mg-silicates), accounting for the total flux, $0.51 \times 10^{18}$ moles of $H_2SO_4$ (Figure 1). For any time in the past,

$$f_C = \frac{0.23}{0.51} \times \frac{f_c'}{f_C'^\circ} \qquad (12a)$$

$$f_D = \frac{0.12}{0.51} \frac{f_D'}{f_D'^\circ} \qquad (12b)$$

$$f_{CaSi} = \frac{0.08}{0.51} \frac{f_{CaSi}'}{f_{CaSi}'^\circ} \qquad (12c)$$

$$f_{MgSi} = \frac{0.08}{0.51} \frac{f_{MgSi}'}{f_{MgSi}'^\circ} \qquad (12d)$$

where the $f'$ refers to the fraction of $CO_2$ weathering uptake due to that reservoir, for example,

$$f_C' = \frac{k_{W_C} C}{k_{W_C} C + 2k_{W_D} D + k_{W_{Mg}} S_{MgSi} + k_{W_{Ca}} S_{CaSi}} \qquad (13)$$

$f'^\circ$ refers to the present-day value of $f'$. Equations (6)-(13) and the fluxes in Figure 1 and Table 1 completely specify the various weathering fluxes needed in the kinetic description of the geochemical cycle.

## Fluxes Due to Volcanic-Seawater Reaction

The overall circulation of seawater in the hydrothermal systems at mid-ocean ridges produces a series of important chemical reactions [Holland, 1978; Edmond et al., 1979]. BLAG [1983] discussed the important depletion of seawater magnesium and the concomitant release of an equivalent amount of calcium. Also associated with the hydrothermal circulation of seawater is the high-temperature reduction of sulfate by basalt [Bischoff and Dickson, 1975; Mottl and Holland, 1978]. This basalt-seawater reaction would produce sulfides, mainly pyrite. This flux (see Figure 1) from seawater sulfate will be assumed to be proportional to the amount of seawater sulfate, $M_{SO_4}$, i.e.,

$$F_{py-hyd} = k_{py-hyd} M_{SO_4} \qquad (14)$$

for reasons similar to those discussed under Mg removal. The rate constant $k_{py-hyd}$ will be a function of the overall hydrologic circulation rate and thus will vary with variations in the

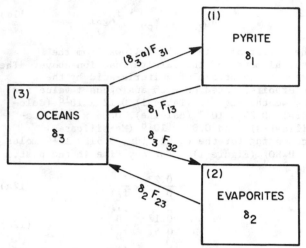

Fig. 3. Three-box model employed in computing fluxes from isotope data (see text).

global spreading rate just as discussed for Mg removal in BLAG [1983], i.e.,

$$k_{py-hyd} = k^°_{py-hyd} f_{SR}(t) \qquad (15)$$

where $f_{SR}(t)$ is the correction factor associated with variations in the global spreading rate, i.e., $SR(t)/SR(0)$.

## Fluxes Due to Magmatism-Metamorphism

BLAG [1983] discussed the role of decarbonation reactions, which result from metamorphism and magmatism. In this paper there is an additional flux from pyrite to ocean sulfate due to subsurface thermal decomposition of pyrite in rocks. This flux (Figure 1) is assumed to be linearly proportional to the amount of pyrite in the reservoir,

$$F_{met-py} = k_{M_{py}} A_{py} \qquad (16)$$

where $k_{M_{py}}$ depends on the spreading rate in the same way as the other metamorphic rate constants, for example,

$$k_{M_{py}} = k^°_{M_{py}} f_{SR}(t) \qquad (17)$$

The organic carbon reservoir is also affected by magmatism-metamorphism. Thermal breakdown of $CH_2O$ to $CH_4$ and $CO_2$ at depth is accompanied by loss of these gases to the atmosphere and subsequent oxidation of $CH_4$. The net effect is the same as that for organic carbon weathering as given in reaction (2) above. However, at present there is no simple way of distinguishing organic carbon breakdown at depth from surficial weathering, and thus the former process will be treated as part of weathering. However, it is understood that this is a defect of the model which awaits

further research on the topic. Fortunately, the effect of this simplification on atmospheric $CO_2$ is minor. Because of the abundance of carbonate rocks, most $CO_2$ involved in metamorphism-magmatism must be derived from the thermal decarbonation of calcite and dolomite.

## Isotope Data and Fluxes of Carbon and Sulfur

There are three fluxes unaccounted for in Figure 1: the burial rates of organic carbon in marine sediments, $F_{Cf}^°$; the formation of pyrite sulfur, $F_{py_f}$, in diagenesis; and the precipitation of gypsum-anhydrite in evaporite basins, $F_{gy_f}$. The use of isotope data ($\delta^{13}C$ and $\delta^{34}S$) to obtain information on these geochemical fluxes has been carried out by previous workers [e.g., Holland, 1973; Garrels and Lerman, 1981; Berner and Raiswell, 1983].

In this paper we will develop a modified approach to obtain the fluxes $F_{Cf}^°$, $F_{py_f}$, and $F_{gy_f}$ from data on $\delta^{13}C$ and $\delta^{34}S$ during the last 100 million years. Figure 3, which is similar to that of Garrels and Lerman [1981], illustrates the three-box model being used. Note that the figure can apply to either the sulfur or carbon cycles. Let us label the size of the reduced ($CH_2O$ or $FeS_2$) reservoir $S_1$, the size of the oxidized (carbonates, gypsum) reservoir $S_2$, and the size of the ocean-atmosphere reservoir $S_3$. The corresponding isotopic compositions ($\delta^{13}C$ or $\delta^{34}S$) will be labeled $\delta_1$, $\delta_2$, $\delta_3$. The amounts of the heavy isotope in each reservoir can be represented by $\delta_i S_i$. We will assume no isotopic fractionation in any of the fluxes except the flux from the ocean reservoir to the reduced

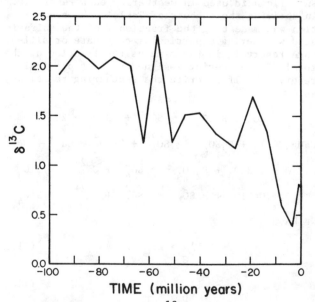

Fig. 4. Variations in $\delta^{13}C$ of carbonate rocks during the past 100 million years (data from Lindh [1983]).

reservoir, $F_{31}$. The latter flux will have a constant fractionation factor $\alpha$. Assuming linear rates for the $F_{13}$ and $F_{23}$ fluxes, the following kinetic equations are then obtained:

$$\frac{dS_1}{dt} = F_{31} - k_{13}S_1 \qquad (18)$$

$$\frac{dS_2}{dt} = F_{32} - k_{23}S_2 \qquad (19)$$

$$\frac{dS_3}{dt} = k_{13}S_1 + k_{23}S_2 - F_{31} - F_{32} \qquad (20)$$

$$\frac{d(\delta_1 S_1)}{dt} = (\delta_3 - \alpha) F_{31} - \delta_1 k_{13}S_1 \qquad (21)$$

$$\frac{d(\delta_2 S_2)}{dt} = \delta_3 F_{32} - \delta_2 k_{23}S_2 \qquad (22)$$

$$\frac{d(\delta_3 S_3)}{dt} = \delta_1 k_{13}S_1 + \delta_2 k_{23}S_2$$
$$- (\delta_3 - \alpha) F_{31} - \delta_3 F_{32} \qquad (23)$$

$$S_t = S_1 + S_2 + S_3 \qquad (24)$$

$$\delta_t = \delta_1 S_1 + \delta_2 S_2 + \delta_3 S_3 \qquad (25)$$

The last two equations represent the conservation of total element and of heavy isotopes. Note that equations (18)-(25) do not make the common assumption of steady state for the ocean-atmosphere reservoir. However, equations (18)-(25) are not all independent. For example,

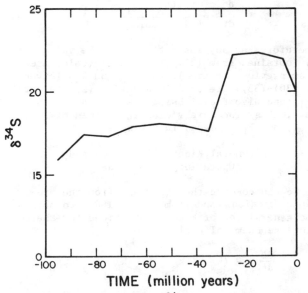

Fig. 5. Variations in $\delta^{34}S$ of sulfate rocks during the past 100 million years (data from Lindh [1983]).

TABLE 2. Isotopes: Initial Conditions

| Species | Value[a] |
|---|---|
| $S_{org\ C}$ | 1250 |
| $S_{carbonates}$ | 5000 |
| $\delta^{13}C_{org\ C}$ | $-25.4°/oo$ |
| $\delta^{13}C_{carb}$ | $0.6°/oo$ |
| $\delta^{13}C_{ocean}$ | $0.77°/oo$ |
| $S_{evap}$ | 250 |
| $S_{pyr}$ | 250 |
| $S_{sulfate}$ | 40 |
| $\delta^{34}S_{evap}$ | $+20°/oo$ |
| $\delta^{34}S_{pyr}$ | $-15°/oo$ |
| $\delta^{34}S_{ocean}$ | $+20°/oo$ |
| $\alpha_{org\ C}$ | $-26°/oo$ |
| $\alpha_{pyr}$ | $-35°/oo$ |

[a]Reservoir contents in $10^{18}$ moles.

equations (20) and (23) can be derived from the other equations. Therefore, if $\delta_3$ is assumed a known function of time, there are only six independent equations and seven unknowns: $S_1$, $S_2$, $S_3$, $\delta_1$, $\delta_2$, $F_{31}$, and $F_{32}$. To solve this problem, we need an additional input. One solution is to set $S_3$ equal to a constant, as Garrels and Lerman [1981] have done. However, we can generalize this solution to letting $S_3$ be an explicit function of time in order to investigate the effect of $S_3$ variations on the results. If $S_3$ and $\delta_3$ (and $k_{13}$ and $k_{23}$) are known from the present to 100 million years ago, equations (18)-(25) can be integrated backwards to obtain all the other parameters. The integration was done by an implicit finite-difference method which converged readily.

Figures 4 and 5 show the isotope data used. Table 2 gives the initial (present) conditions employed. Figure 6 compares the result of computing the pyrite flux assuming constant ocean sulfate to that assuming that ocean sulfate increases linearly from its present value (40 x $10^{18}$ moles) to 47 x $10^{18}$ at t = -25 m.y., then decreases linearly to 32 x $10^{18}$ at t = -75 m.y., and thereafter remains constant. Clearly, the constant ocean sulfate assumption is fairly good under these conditions. Of course, the gypsum fluxes must then be different in the two models and this is taken into account.

In general, it is found that during the last 100 million years the steady state ocean assumption for both heavy isotopes and total mass, while excellent for the carbon system, is acceptable for the sulfur system and especially

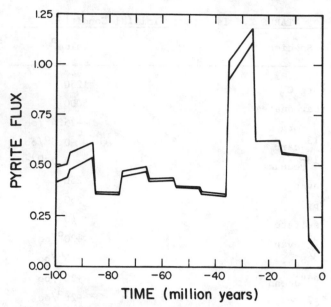

Fig. 6. Comparison of the pyrite flux (in $10^{18}$ moles per million years) computed from the isotope data and using the steady state assumption (lower curve) or a nonsteady state ocean sulfate model (upper curve) as described in the text.

so for the pyrite flux. If the steady state assumption is made, then

$$F_{31} + F_{32} = k_{13}S_1 + k_{23}S_2 \qquad (26)$$

Furthermore, if the oceanic isotopic changes are ignored (excellent assumptions for carbon and reasonable for sulfur), then a mean $\bar{\delta}$ defined by

$$\bar{\delta} = \frac{\delta_1 S_1 + \delta_2 S_2}{S_1 + S_2} \qquad (27)$$

will be a constant, and also

$$\delta_3 F_{32} + (\delta_3 - \alpha)F_{31} = \delta_1 k_{13}S_1 + \delta_2 k_{23}S_2 \quad (28)$$

We can solve these equations for $F_{31}$ to obtain

$$F_{31} = [\delta_3(k_{13}S_1 + k_{23}S_2) + \delta_1(k_{23} - k_{13})S_1$$
$$- k_{23}\bar{\delta}(S_1 + S_2)]\,\alpha^{-1} \qquad (29)$$

In the overall program, which computes the evolution of the geochemical cycle, the values of $S_1$ and $S_2$ are calculated at any one time, and $\delta_3$ is obtained from the isotope data. The rate constants in equation (29), however, are not fixed as in earlier work [e.g., Garrels and Lerman, 1981; Berner and Raiswell, 1983] but must change in accordance with the model developed in our earlier paper and in this paper. In particular, this means that for the carbon cycle the $k_{13}$ and $k_{23}$ in (29) are given by

$$k_{13} = k_{13}^{W^\circ} f_A(t)\, f_B(CO_2) \qquad (30)$$

$$k_{23} = k_{23}^{W^\circ} f_A(t)\, f_{CO_2}(t) + k_{23}^{M^\circ} f_{SR}(t) \quad (31)$$

where $k_{23}^{W^\circ}$ is the present total carbonate weathering constant (i.e., calcite plus dolomite), and so (see Figure 1)

$$k_{23}^{W^\circ} = \frac{12.5}{5000.0}$$

Likewise for metamorphism-magmatism,

$$k_{23}^{M^\circ} = \frac{5.9}{5000.0}$$

For the sulfur cycle, the rate constants are

$$k_{13} = k_{13}^{W^\circ} f_A(t)\, f_B(CO_2) \qquad (32)$$

$$k_{23} = k_{23}^{W^\circ} f_A(t)\, f_B(CO_2) \qquad (33)$$

where (see Figure 1)

$$k_{13}^{W^\circ} = \frac{0.51}{250} \qquad k_{23}^{W^\circ} = \frac{1.02}{250}$$

Finally, to obtain $\delta_1$ in equation (29) we need to go back to

$$\frac{d(\delta_1 S_1)}{dt} = F_{31}(\delta_3 - \alpha) - k_{13}S_1\delta_1$$

or

$$\delta_1\frac{dS_1}{dt} + S_1\frac{d\delta_1}{dt} = F_{31}(\delta_3 - \alpha) - k_{13}S_1\delta_1$$

which, using equation (18), yields

$$\frac{d\delta_1}{dt} = \frac{F_{31}(\delta_3 - \delta_1 - \alpha)}{S_1} \qquad (34)$$

Therefore, at any instant in time the values of $\delta_1$ (obtained from (34)), $S_1$ and $S_2$ (calculated from previous section), and $k_{13}$ and $k_{23}$ (given by (30)-(33)) are used in equation (29) to obtain $F_{31}$ (and also $F_{32}$). Using the $\delta^{13}C$ and $\delta^{34}S$ data, this procedure yields the values of $F_{C_f}$, $F_{py_f}$, and $F_{gy_f}$ needed in the model.

General Kinetic Equations and
Operation of the Model

We can combine the equations from the previous sections and from BLAG [1983] to obtain the general set of kinetic equations for the complete geochemical cycle:

$$\frac{dD}{dt} = -(k_{W_D} + k_{M_D})\,D - k_{W_{py}}A_{py} \cdot f_D \quad (35a)$$

$$\frac{dC}{dt} = -(k_{W_C} + k_{M_C})\,C + 2\,F_{py_f} - 2\,k_{W_{py}}A_{py} \cdot f_c$$
$$+ k_{prep}\,(M_{Ca}M_{HCO_3}^2 - K_{eq}A_{CO_2}) \qquad (35b)$$

$$\frac{dS_{CaSi}}{dt} = k_{M_C} C + k_{M_D} D - k_{W_{CaSi}} S_{CaSi} - k_{V-SW} M_{Mg}$$
$$- k_{W_{py}} A_{py} \cdot f_{CaSi} \qquad (35c)$$

$$\frac{dS_{MgSi}}{dt} = k_{M_D} D - k_{W_{MgSi}} S_{MgSi} + k_{V-SW} M_{Mg}$$
$$- k_{W_{py}} A_{py} \cdot f_{MgSi} \qquad (35d)$$

$$\frac{dA_{C^\circ}}{dt} = F_{C_f^\circ} - k_{W_{C^\circ}} A_{C^\circ} \qquad (35e)$$

$$\frac{dA_{py}}{dt} = F_{py_f} + k_{py-hyd} M_{SO_4} - k_{W_{py}} A_{py}$$
$$- k_{M_{py}} A_{py} \qquad (35f)$$

$$\frac{dA_{gy}}{dt} = F_{gy_f} - k_{W_{gy}} A_{gy} \qquad (35g)$$

$$\frac{dA_{O_2}}{dt} = F_{C_f^\circ} + \frac{15}{8} F_{py_f} - k_{W_{C^\circ}} A_{C^\circ}$$
$$- \frac{15}{8} k_{W_{py}} A_{py} \qquad (35h)$$

$$\frac{dM_{Mg}}{dt} = k_{W_D} D + k_{W_{MgSi}} S_{MgSi} - k_{V-SW} M_{Mg}$$
$$+ k_{W_{py}} A_{py} (f_D + f_{MgSi}) \qquad (35i)$$

$$\frac{dM_{Ca}}{dt} = k_{W_D} D + k_{W_C} C + k_{W_{CaSi}} S_{CaSi} + k_{V-SW} M_{Mg}$$
$$+ k_{W_{py}} A_{py} \cdot (f_D + 2f_C + f_{CaSi})$$
$$+ k_{W_{gy}} A_{gy} - 2F_{py_f} - F_{gy_f}$$
$$- k_{prep}(M_{Ca} M_{HCO_3}^2 - K_{eq} A_{CO_2}) \qquad (35j)$$

$$\frac{dM_{SO_4}}{dt} = k_{W_{gy}} A_{gy} + k_{W_{py}} A_{py} - F_{py_f} - F_{gy_f}$$
$$+ k_{M_{py}} A_{py} - k_{py-hyd} M_{SO_4} \qquad (35k)$$

$$\frac{dM_{HCO_3}}{dt} = 4 k_{W_D} D + 2 k_{W_C} C + 2 k_{W_{CaSi}} S_{CaSi}$$
$$+ 2k_{W_{MgSi}} S_{MgSi} - 2F_{py_f} + 2k_{W_{py}} A_{py}$$
$$\cdot (f_C + f_D)$$
$$- 2k_{prep}(M_{Ca} M_{HCO_3}^2 - K_{eq} A_{CO_2}) \qquad (35l)$$

$$\frac{dA_{CO_2}}{dt} = 2k_{M_D} D - 2k_{W_D} D - 2k_{W_{MgSi}} S_{MgSi}$$
$$- 2k_{W_{CaSi}} S_{CaSi} + k_{M_C} C - k_{W_C} C$$
$$- F_{C_f^\circ} + k_{W_{C^\circ}} A_{C^\circ}$$
$$+ k_{prep}(M_{Ca} M_{HCO_3}^2 - K_{eq} A_{CO_2}) \qquad (35m)$$

where M is the mass of $Ca^{2+}$, $Mg^{2+}$, $HCO_3^-$, or $SO_4^{2-}$ in oceans,

$$k_W = k_W^\circ f_A(t) f_B(CO_2) \qquad (36)$$

$$k_M = k_M^\circ f_{SR}(t) \qquad (37)$$

$$k_{V-SW} = k_{V-SW}^\circ f_{SR}(t) \qquad (38)$$

$$k_{py-hyd} = k_{py-hyd}^\circ f_{SR}(t) \qquad (39)$$

and $f_C$, $f_D$, $f_{CaSi}$, $f_{MgSi}$ are defined by equations (12) and (13).

The computation methods were very similar to those used by Berner et al. [1983], and the reader is referred to them for details. However, there was one major difference in the establishment of the initial conditions. In the present paper, the initial values of $M_{Ca}$, $M_{Mg}$, $M_{HCO_3}$, and $A_{CO_2}$ (i.e., $Ca^{2+}$, $Mg^{2+}$, and $HCO_3^-$ in the oceans and atmospheric $CO_2$) were obtained by assuming quasi-steady state, i.e.,

$$\frac{dM_{Ca}}{dt} = \frac{dM_{Mg}}{dt} = \frac{dM_{HCO_3}}{dt} = \frac{dA_{CO_2}}{dt} = 0 \qquad (40)$$

A recent paper by Kasting [1985] has already proposed such a scheme. We have generalized his scheme to include the new fluxes in the model. Equations (35i), (35j), (35m), and (40) can be recast as

$$0 = a + k_{V-SW} M_{Mg} - k_{prep}(M_{Ca} M_{HCO_3}^2 - K_{eq} A_{CO_2}) \qquad (41)$$

$$0 = b - k_{V-SW} M_{Mg} \qquad (42)$$

$$0 = d + k_{prep}(M_{Ca} M_{HCO_3}^2 - k_{eq} A_{CO_2}) \qquad (43)$$

where

$$a \equiv (k_{W_D}^\circ D + k_{W_C}^\circ C + k_{W_{CaSi}}^\circ S_{CaSi}) f(CO_2)$$
$$\equiv a^\circ f(CO_2) \qquad (44)$$

$$b \equiv (k_{W_D}^\circ D + k_{W_{MgSi}}^\circ S_{MgSi}) f(CO_2)$$
$$\equiv b^\circ f(CO_2) \qquad (45)$$

$$d = d_1 - d^\circ f(CO_2) - F_{C_f^\circ} \qquad (46)$$

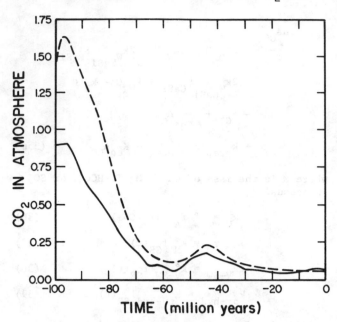

Fig. 7.  Comparison of the earlier BLAG [1983] $CO_2$ versus time curve (dashed) with the $CO_2$ curve obtained by the improved model.  Units for $CO_2$ content are $10^{18}$ moles.  (Both curves use the same Southam and Hay [1977] spreading rate curve corrected for loss by subduction.)

$$d_1 = 2k_{M_D} D + k_{M_C} C \tag{47}$$

$$d^\circ = 2k^\circ_{W_D} D + k^\circ_{W_C} C + 2k^\circ_{W_{MgSi}} S_{MgSi}$$
$$+ 2k^\circ_{W_{CaSi}} S_{CaSi} - k^\circ_{W_{C^\circ}} A_{C^\circ} \tag{48}$$

In (44), (45), and (48) the $k^\circ_W$ refer to the present-day $k_W$ corrected only for the land area term, $f_A(t)$, but not $f(CO_2)$, because we do not know $A_{CO_2}$ at this stage of the calculation.  Now from (29) we can write $F_{C^\circ_f}$ as

$$F_{C^\circ_f} = q + r\, f(CO_2) \tag{49}$$

where

$$q \equiv \frac{\delta_3 k^M_{63} S_{carb} + \delta_5 k^M_{63} A_{C^\circ} - k^M_{63} \bar{\delta}(S_{carb} + A_{C^\circ})}{\alpha}$$

$$r \equiv \{\delta_3 k^\circ_{53} A_{C^\circ} + \delta_3 S_{carb} k^\circ_{63} + \delta_5 A_{C^\circ}(k^\circ_{63} - k^\circ_{53})$$
$$- k^\circ_{63} \bar{\delta}(S_{carb} + A_{C^\circ})\} / \alpha$$

Adding (41), (42), and (43) yields

$$0 = a^\circ f(CO_2) + b^\circ f(CO_2) + d_1 - d^\circ f(CO_2)$$
$$- q - r\, f(CO_2)$$

or

$$f(CO_2) = \frac{d_1 - q}{d^\circ - a^\circ - b^\circ + r} \tag{50}$$

$d_1$, $d^\circ$, $a^\circ$, $b^\circ$, $r$, and $q$ are all assumed known.  Therefore, (50) can be used to solve for $A_{CO_2}$.  Once $A_{CO_2}$ is known, (42) can be used to obtain $M_{Mg}$.  Finally, $M_{Ca}$ and $M_{HCO_3}$ are obtained from

$$(M_{HCO_3} - M^\circ_{HCO_3}) = 2(M_{Mg} - M^\circ_{Mg}) + 2(M_{Ca} - M^\circ_{Ca}) \tag{51}$$

and (41).  Equation (51) is just a charge balance equation.  (Note that the equation $dM_{HCO_3}/dt = 0$ is redundant; i.e., it follows from $dM_{Ca}/dt = dM_{Mg}/dt = 0$.)

### Results and Discussion

The results of the model are shown in Figure 7. The solid curve gives the results for the evolution of atmospheric $CO_2$ using the modified Southam and Hay [1977] $f_{SR}(t)$ curve [see Berner et al., 1983].  For comparison the dashed curve gives the earlier results [BLAG, 1983], which neglected organic carbon fluxes.  It is clear that the overall qualitative behavior has not changed drastically.  However, an important difference is the lower increase in $CO_2$ for the Cretaceous period predicted by the extended model.  These

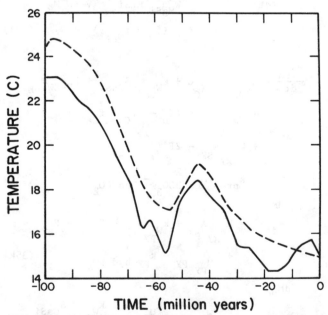

Fig. 8.  Comparison of the earlier BLAG [1983] temperature versus time curve (dashed) with the temperature curve obtained by the improved model. (Both curves use the same Southam and Hay [1977] $f_{SR}(t)$ curve.)

lower $CO_2$ values, 13 times present-day values
if t = -90 m.y. versus 26 times present-day
values if organic carbon is ignored, lead to
more reasonable temperatures (based on paleotemperature studies). The corresponding
temperature plots are shown in Figure 8. Mean
global temperatures close to 25°C, as predicted
by the earlier model, are on the high side
[Savin, 1977]. Therefore, the inclusion of the
organic carbon reservoir and the fluxes
associated with it has resulted in a favorable
modification of the $CO_2$-temperature results.

The temperature curve of the present model
exhibits slightly more fine structure than that
of the earlier model (Figure 8). This reflects
the added fluctuations due to the organic carbon
subsystem. There is paleobotanical evidence
[e.g., Savin, 1977] that there was an Eocene
temperature maximum. The earlier model yielded
such a maximum; however, in the present model
the Eocene temperature maximum is more clearly
delineated.

An important problem that arose in the earlier
calculations was the increase of $HCO_3^-$ in the
oceans relative to $Ca^{2+}$. Based on the sequence
of salts observed in evaporites during the
Phanerozoic, Holland [1972] concluded that the
amount of Ca in the oceans must be large enough
to precipitate $CaSO_4$ during evaporative concentration without all Ca being removed by earlier
$CaCO_3$ precipitation. Therefore, we require that

$$M_{Ca} > \frac{1}{2} M_{HCO_3} \qquad (52)$$

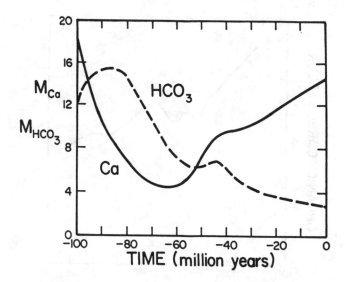

Fig. 9b. Same as Figure 9a but using the
earlier, BLAG [1983] model.

Equation (52) must hold at all times. BLAG
[1983] pointed out that their results mildly
violated (52). The results of the present model
remove this problem completely, as illustrated in
Figure 9a.

The reason for both the modification of the
$CO_2$-temperature curves and the removal of the
Ca-$HCO_3$ problem can be traced in part to the organic carbon burial flux $F_{C_f^\circ}$, shown in Figure
10a. Note the increase in $F_{C_f^\circ}$ as we go back to
the Cretaceous. This increase in $F_{C_f^\circ}$ (which
actually slightly outweighs the increase in
weathering rate) leads to a sink for $CO_2$ (hence
the modification) and concomitantly a sink for
$HCO_3$ (which now is kept at lower values in
Figure 9a). The calculated flux for the model
with constant kinetic coefficients is also given
for comparison in Figure 10b. Note also the
increase in $F_{C_f^\circ}$, albeit somewhat smaller than
that for varying coefficients. It should be
pointed out, however, that $F_{C_f^\circ}$ in Figure 10b increases even though the weathering rate constants
are fixed. Therefore, as in the more correct
model (Figure 10a), the increases in $F_{C_f^\circ}$ outstrip
the weathering rate increase.

Figure 11 gives the model results for Mg in
the ocean. Because of the leveling of the Ca
curve mentioned above, the Mg curve now shows
less variation than in the BLAG results (dashed
curve in Figure 11), although there is still a
slight decrease with time. Figure 12 shows that
the pH values are similar in the two models.

Another result of the higher $F_{C_f^\circ}$ values is
that the organic carbon reservoir was increasing
slightly during the last 100 million years. This
result is shown in Figure 13, which depicts the
amount, $A_{C^\circ}$, predicted from the model. As can be

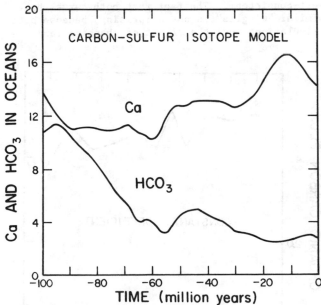

Fig. 9a. Computer results for oceanic composition of Ca and $HCO_3$ as a function of time using
the improved geochemical model and the Southam
and Hay [1977] $f_{SR}(t)$ curve.

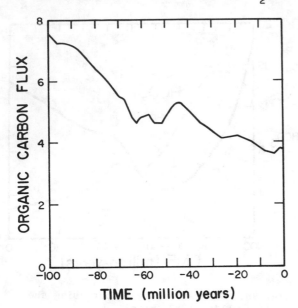

Fig. 10a.  Flux of organic carbon burial (in 10$^{18}$ moles per million years) derived from the $\delta^{13}$C data in Figure 4 and allowing for variation in the rate constants (see text).

seen, the reservoir grew from 1180 x 10$^{18}$ to the present value (1250 x 10$^{18}$).

A new problem in the extended model arises from the increase in $A_C$°. As the organic carbon reservoir grows, there is only a small change in the sulfide-sulfate reservoirs: during the last 100 million years the sulfide reservoir decreases by 6 x 10$^{18}$ moles and the sulfate reservoir increases by the same amount. Because the organic carbon increases are not offset by oxidation of sulfide to sulfate, there is an imbalance in the oxygen reservoir. This imbalance leads to unacceptable increases in oxygen during the last 100 million years. This can be seen in Figure 14, which shows an increase in O$_2$ of almost 300%. Part of the problem lies in our simplistic treatment of oxygen (which is a "passive" reservoir in our model), and part lies in possible adjustments of the isotope model.

It is important, at this point, to emphasize the sensitivity of the isotope model to the present-day conditions. The calculations discussed previously were all carried out utilizing the isotope data given in Table 2. In particular, the mean carbon isotopic composition for Tertiary sediments comes out as

$$\bar{\delta}=[M_{carb}\,\delta_{carbonates} + M_{org.C}\,\delta_{org.C}]/M_{total}$$

$$=[(5000)\,(0.6) + (1250)\,(-25.4)]/6250$$

$$= -4.6°/oo$$

There is some ambiguity regarding the correct average isotopic values to use for the present-day carbonate and organic carbon reservoirs. If a

higher value of +1.5°/oo is used for the carbonate and correspondingly a value of -24.5°/oo for the organic carbon, which gives a mean $\bar{\delta}$ = -3.7°/oo, the changes are quite significant, as can be seen from the dashed curves in Figures 13 and 14. With the new higher mean $\bar{\delta}$, the changes in both the oxygen and organic carbon reservoirs are minimized. Figure 15 compares the results for the temperature. The high mean $\bar{\delta}$ results give higher temperatures. However, the results with the higher mean $\bar{\delta}$ are not as self-consistent as those shown earlier (i.e., the end results do not closely match present-day values). If a value of +2°/oo is chosen for the carbonates, the oxygen curve now goes significantly lower with time, again in an unacceptable manner.

These variations caused by the changes in $\bar{\delta}$ and $\delta_{carb}$ should provide caution in using the isotope models, including others that have been presented elsewhere. Obviously, the isotope models must be anchored at several points to avoid the fluctuations discussed here. General models, such as the one developed here, can be very useful in anchoring the isotope models.

The model computes the carbonate flux according to the kinetic equation

$$Flux = k_{prep}\ (M_{Ca}\ M^2_{HCO_3} - K_{eq}\ A_{CO_2})\qquad (53)$$

derived by BLAG [1983]. However, once $F_{C_f}$° is calculated from the isotope data (equation (34)), equation (31) can be used to obtain independently an isotopically derived carbonate flux. It would be desirable to have both of these carbonate fluxes similar in magnitude. Figure 16 shows the two fluxes. Clearly, the model is reasonably self-consistent. The fact that both curves have similar "wiggles" is not surprising, because both calculations feel the effect of changing land

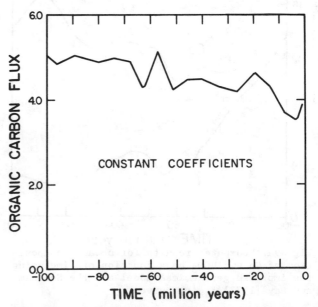

Fig. 10b.  Same as Figure 10a but with fixed rate constants.

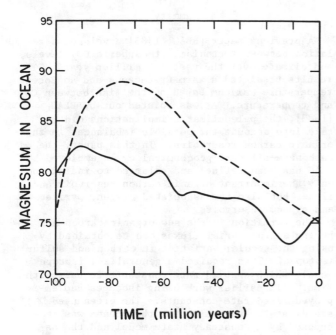

Fig. 11.  Computer results for Mg in the ocean as a function of time.  Mg content in $10^{18}$ moles. The dashed curve is the BLAG [1983] result using a modified Southam and Hay [1977] spreading curve. The solid curve is that calculated from the improved model.

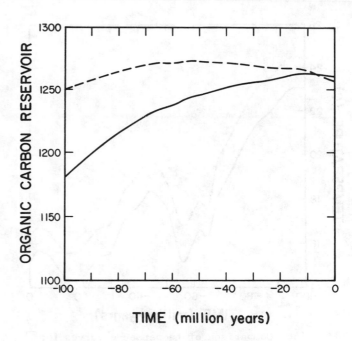

Fig. 13.  Computer results for the mass of organic carbon (in $10^{18}$ moles) during the past 100 million years.  The dashed curve is computed using the modified mean $\bar{\delta}$ (see text). The solid curve is that calculated from the improved model.

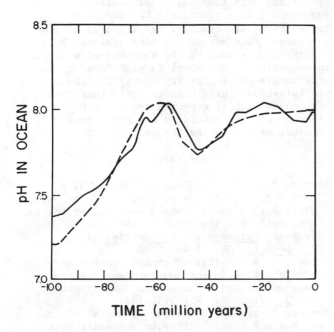

Fig. 12.  Computer results for pH of the ocean as a function of time.  The dashed curve is the BLAG [1983] result using a modified Southam and Hay [1977] spreading curve.  The solid curve is that calculated from the improved model.

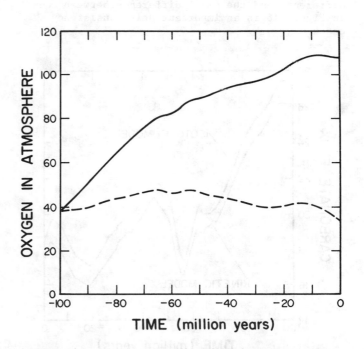

Fig. 14.  Computer results for the amount of oxygen in the atmosphere (in $10^{18}$ moles) during the past 100 million years.  The dashed curve is computed using the modified mean $\bar{\delta}$ (see text). The solid curve is that calculated from the improved model.

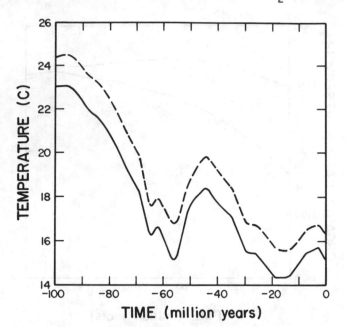

Fig. 15. Comparison of temperature curves for the last 100 million years computed with the improved model (solid curve) and with the modified mean $\bar{\delta}$ (dashed curve) (see text).

area and tectonism. However, the absolute magnitude of the fluxes could have been quite different, and the small difference between them in Figure 16 is an important self-consistency check.

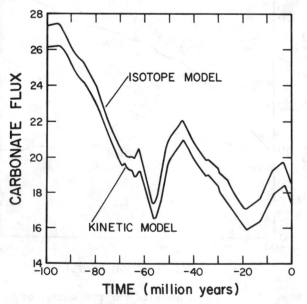

Fig. 16. Comparison of the carbonate flux (in units of 10$^{18}$ moles per million years) computed from the kinetic equation (53) and from the isotope model (see text).

Summary

A previous paper [BLAG, 1983] developed a relation between tectonism, atmospheric CO$_2$ levels, and climate over the past 100 million years. The results predicted a warm Cretaceous and an Eocene temperature maximum based on the link between CO$_2$ and temperature. As was pointed out by BLAG [1983], the paleoclimatic implications did not take into account the possible imbalances in the organic carbon reservoirs. In this paper, the carbonate-silicate geochemical cycle employed by BLAG has been refined and enlarged to include not only the important organic carbon reservoir and fluxes but also the associated oxygen, pyrite, and gypsum reservoirs.

The variation in the new organic carbon and pyrite sulfur burial fluxes can be obtained by using the secular variation in carbon and sulfur isotopes. This required a generalized isotope model, which enabled rate constants to vary with time. All earlier work using isotopes had employed fixed rate constants. The often used steady state assumption for the oceans was checked by a nonsteady state model and the assumption was found to be reasonable, at least for the last 100 million years.

The computational scheme used in calculating the evolution of the enlarged geochemical cycle was optimized by initializing the system in a self-consistent manner to ensure that CO$_2$ in the atmosphere and Mg, Ca, and HCO$_3$ in the oceans were at quasi-steady state.

The improved geochemical model corroborated the conclusions of BLAG, yet led to a significant optimization in the quality of the results. In particular, more reasonable CO$_2$ levels and temperatures are reached, and the evolution of the oceans does not at any time violate the constraints imposed by the sequence of minerals in evaporites. The model is also found to be self-consistent in its treatment of carbonate precipitation fluxes. In general, therefore, the improved model strengthens even more the link between CO$_2$ levels in the atmosphere, climate, and tectonism proposed by BLAG.

References

Barron, E. J., J. L. Sloan II, and C. G. A. Harrison, Potential significance of land-sea distribution and surface albedo variations as a climatic forcing factor: 180 m.y. to the present, Palaeogeogr. Palaeoclimatol. Palaeoecol., 30, 17-40, 1980.
Berner, R. A., Burial of organic carbon and pyrite sulfur in the modern ocean and its geochemical and environmental significance, Am. J. Sci., 282, 451-573, 1982.
Berner, R. A., and R. Raiswell, Burial of organic carbon and pyrite sulfur in sediments over Phanerozoic time: A new theory, Geochim. Cosmochim. Acta, 47, 855-862, 1983.
Berner, R. A., A. C. Lasaga, and R. M. Garrels, The carbonate-silicate geochemical cycle and

its effect on atmospheric carbon dioxide over the past 100 million years, Am. J. Sci., 283, 641-683, 1983.

Bischoff, J. L., and F. W. Dickson, Seawater-basalt interaction at 200°C and 500 bars: Implications for origin of sea-floor heavy metal deposits and regulation of seawater chemistry, Earth Planet. Sci. Lett., 25, 385-397, 1975.

Budyko, M. I., and A. B. Ronov, Chemical evolution of the atmosphere in the Phanerozoic, Geochem. Int., Engl. Transl., 1-9, 1979.

Chamberlain, T. C., The influence of great epochs of limestone formation upon the constitution of the atmosphere, J. Geol., 6, 609-621, 1968.

Edmond, J. M., C. Measures, R. E. McDuff, L. H. Chan, R. Collier, B. Grant, L. J. Gordon, and J. B. Corliss, Ridge crest hydrothermal activity and the balance of the major and minor elements in the ocean: The Galapagos data, Earth Planet. Sci. Lett., 46, 1-18, 1979.

Garrels, R. M., and A. Lerman, Phanerozoic cycles of sedimentary carbon and sulfur, Proc. Natl. Acad. Sci. USA, 78, 4652-4656, 1981.

Garrels, R. M., and E. A. Perry, Jr., Cycling of carbon, sulfur, and oxygen through geologic time, in The Sea, vol. 5, edited by G. E. Goldberg, pp. 303-336, Wiley-Interscience, New York, 1974.

Garrels, R. M., F. T. Mackenzie, and C. Hunt, Chemical Cycles and the Global Environment: Assessing Human Influences, Kaufmann, Los Angeles, Calif., 1975.

Gregor, C. B., The mass-age distribution of Phanerozoic sediments, in Geochronology and the Geological Record, edited by N. J. Snelling, J. Watson, J. Briden, and B. J. Bluck, Geological Society of London, in press, 1984.

Holland, H. D., The geologic history of sea water: An attempt to solve the problem, Geochim. Cosmochim. Acta, 36, 637-651, 1972.

Holland, H. D., Systematics of the isotopic composition of sulfur in the oceans during the Phanerozoic and its implications for atmospheric oxygen, Geochim. Cosmochim. Acta, 37, 2605-2616, 1973.

Holland, H. D., The Chemistry of the Atmospheres and Oceans, 351 pp., Wiley-Interscience, New York, 1978.

Kasting, J., Comments on "The carbonate-silicate geochemical cycle and its effect on atmospheric carbon dioxide over the past 100 million years" by R. A. Berner, A. C. Lasaga, and R. M. Garrels, Am. J. Sci., in press, 1985.

Lindh, T. B., Temporal variations in $^{13}C$, $^{34}S$ and global sedimentation during the Phanerozoic, M.S. thesis, Univ. of Miami, Coral Gables, Fla., 1983.

Mottl, M. J., and H. D. Holland, Chemical exchange during hydrothermal alteration of basalt by seawater, I, Experimental results for major components of seawater, Geochim. Cosmochim. Acta, 42, 1103-1115, 1978.

Savin, S., The history of the earth's surface temperature during the past 100 million years, Annu. Rev. Earth Planet. Sci., 5, 319-355, 1977.

Southam, J. R., and W. W. Hay, Time scales and dynamic models of deep-sea sedimentation, J. Geophys. Res., 82, 3825-3842, 1977.

Vail, P. R., and J. Hardenbol, Sea-level changes during the Tertiary, Oceanus, 22, 71-79, 1979.

Von Damm, K. L., Chemistry of submarine hydrothermal solutions at 21° North, East Pacific Rise and Guarymas Basin, Gulf of California, Ph.D. dissertation, 241 pp., Woods Hole Oceanogr. Inst./Mass. Inst. of Technol. Joint Program in Oceanography, Cambridge, 1983.

# OCEANIC CARBON ISOTOPE CONSTRAINTS ON OXYGEN AND CARBON DIOXIDE IN THE CENOZOIC ATMOSPHERE

N. J. Shackleton

Godwin Laboratory for Quaternary Research, University of Cambridge,
Cambridge CB2 3RS, England

Lamont-Doherty Geological Observatory, Palisades, N. Y. 10964

Abstract. The carbon isotope record of carbonate sediments deposited on the ocean floor during the past $10^8$ years may be intepreted in terms of changes in the global organic carbon reservoir, which has dimished by about $10^{19}$ moles over the past few tens of millions of years, drawing down the atmospheric oxygen content from 0.25 atmospheres to its present value of 0.2 atmospheres.

The history of $^{13}C$ gradients within the ocean, recorded in the $^{13}C$ content of foraminiferal tests, constrains the upper limit of possible excursions in the total dissolved $CO_2$ in the ocean, which was not appreciably higher in the early Cenozoic and Mesozoic than it is today, providing a useful constraint on the predictions of the Berner et al. (1983) model, and hence on the implication that that the cause of warm earth climate during the late Mesozoic and early Cenozoic was a high atmospheric $CO_2$ concentration.

## Introduction

It is becoming better appreciated that carbon isotope data from oceanic carbonates provide a powerful tool for evaluating the history of the global carbon system (Scholle and Arthur, 1980; Shackleton, 1977; Arthur, 1982; Shackleton, 1984). However, it is important to use $^{13}C$ data from the appropriate part of the ocean carbonate system if the data are to be fully exploited. For example, the history of the $^{13}C$ content of ocean-dissolved $CO_2$ is best obtained by analyzing benthic foraminiferal carbonate, because the average chemistry of the ocean is better represented at the seafloor than close to the surface. Curry and Lohmann (1982, 1983) and Shackleton et al. (1983b) have utilized vertical and horizontal gradients in the $^{13}C$ content of benthonic foraminifera to estimate past changes in deep-water circulation, while Shackleton et al. (1983a) used the gradient between planktonic

and benthonic species to infer changes in the $CO_2$ concentration in the atmosphere. In both these examples, it is the photosynthetic removal of isotopically light carbon from dissolved $CO_2$ in the photic zone and its regeneration by oxidation in deeper water that provide the observed $^{13}C$ gradients.

In contrast, in order to use the long-term (long compared with the $10^5$-year residence time of carbon in the ocean) $^{13}C$ record to evaluate the history of the global organic carbon reservoir, one must obtain data for the total carbonate (dominated by coccoliths) rather than for selected foraminiferal species. Such an approach was taken for the Mesozoic by Scholle and Arthur (1980); Shackleton (1984) has extended their approach through the Cenozoic, where we have a better paleo-oceanographic base on which to interpret the data.

## The Global Organic Carbon Reservoir

There is now a reasonable body of $^{13}C$ data available from Cenozoic sediments. Figure 1 shows two data sets for the Paleocene and lower Eocene. DSDP577 (Shackleton et al., 1984b) had a paleo-latitude of about 15°N in the west Pacific, while DSDP527 (Shackleton and Hall, 1984, also additional unpublished data, 1983) was from about 30°S in the Atlantic. Figure 1 shows that the two data sets are very similar, and justifies the use of data from a single area as representative of average deep-sea sediment.

It has been believed for several years that the sediment budget of the oceans has changed dramatically during the Cenozoic. This stems from the important study of Davies et al. (1977). However, it now appears that the very large fluctuations in global sedimentation rates postulated by Davies et al. (1977) and Davies and Worsley (1981) may have been an artifact of the geological time scale that was in use, since they largely vanish when the time scales of Berggren

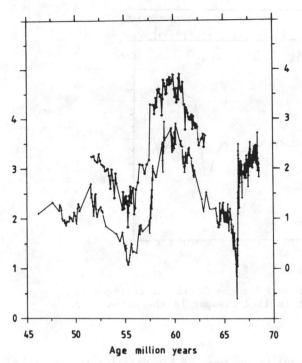

Fig. 1.    Carbon isotope data for bulk sediment in DSDP Site 527 (South Atlantic) and DSDP Site 577 (tropical Pacific) plotted on a common time scale. For ease of comparison the two records are slightly offset; the left scale applies to Site 527 (pluses) and the right scale to Site 577 (circles).

et al. (1984a, b) are used instead (Shackleton, 1984). Thus it may be a reasonable approximation to assume that the carbon input to the ocean has been constant. Using the approximation of a constant input (with an assumed constant $^{13}C$ content), Shackleton (1984) estimated long-term changes in the global organic carbon reservoir from the long-term $^{13}C$ record.

Figure 2 is based on a record of $^{13}C$ for marine sediments averaged in million-year increments. Although based on a limited geographical area, this is probably a good approximation to a global average record (Shackleton, 1984). The figure is scaled in terms of rate of change in the global organic carbon reservoir. An input to the ocean of $150 \times 10^{18}$ g carbon/m.y. is assumed. Steady state is defined as no change in the global organic carbon reservoir and is assumed to be represented by the accumulation of $120 \times 10^{18}$ g/m.y. of carbon as carbonate ($\delta^{13}C$ +2.25 per mil) and $30 \times 10^{18}$ g/m.y. of carbon as organic carbon ($\delta^{13}C$ -22.75 per mil). These figures are those used by Garrels and Perry (1974) and by Garrels and Lerman (1981).

The use of the limestone $^{13}C$ record to derive changes in the partitioning of carbon between the limestone and organic carbon reservoirs is generally accepted (Scholle and Arthur, 1980; Veizer et al., 1980; Garrels and Lerman, 1981). The outcome of applying the newer versions of the geological time scale to the data summarised by Davies et al. (1977) and by Whitman and Davies (1981) suggests that a model that assumes a constant input to the ocean basins, as we have done for the construction of Figure 2, is not unreasonable; much more work will be needed to produce a better model, because contrary to the conclusion of Davies et al. (1977) the changes appear to have been quite small.

It is widely agreed that through the Phanerozoic there must have been a mechanism which maintained the atmospheric oxygen concentration in an approximately constant range (Walker, 1977). Veizer et al., (1980) propose that oxygen removed by the oxidation of organic carbon is released by sulfate reduction and, further, that this is demonstrated by a covariance between the Phanerozoic records for $^{13}C$ and $^{34}S$. Garrels and Lerman (1981) have exploited this approach to model a predicted $^{13}C$ record from the observed $^{34}S$ record of oceanic sulfate. Figure 3 compares the output of their model with the ocean sediment $^{13}C$ data of Shackleton and Hall (1984) for the Cenozoic, extended into the Mesozoic with the data of Scholle and Arthur (1980). Figure 3 shows that the $^{34}S$ record does not imply the release since the Miocene of sufficient oxygen by sulphate reduction to compensate for that absorbed by the oxidation of organic carbon. This figure also suggests that for the late Cenozoic, the impact of the oxidation and reduction of carbon on the atmospheric oxygen balance has outweighed the contribution of sulfur to such an extent that sulfur may reasonably be ignored. Although the $^{34}S$ record of ocean-dissolved sulfate (Claypool et al., 1980) is not known in much detail for the Cenozoic, it has the strength that it is firmly anchored to measurements of actual dissolved sulfate at the recent end of the time scale. It could be argued that comparing our newer $^{13}C$ data with the predictions of the model produced by Garrels and Lerman (1981) only tests their model. However, it does appear that on the longer time scale, the $^{13}C$ record predicted by their model is consistent with the available data. Thus the divergence between the Garrels and Lerman (1981) model output and the observational data in the Cenozoic is probably due to special circumstances in the Cenozoic such as those that are proposed here.

Figure 4 (after Shackleton, 1984) shows the history of the global organic carbon reservoir during the Cenozoic that may be deduced from the ocean $^{13}C$ record. Figure 4 is obtained by compiling the change in the organic carbon reservoir in each million year increment (from Figure 2) back in time. Since this construction relies on the cumulation of departures from the steady state partitioning of carbon between the

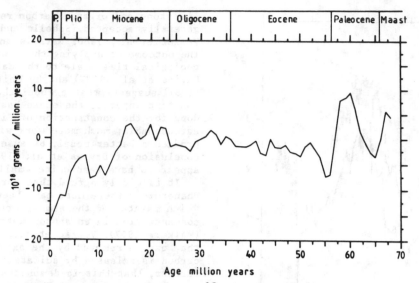

Fig. 2.   Bulk deep-sea sediment carbonate $^{13}$C record for the Cenozoic, averaged in million-year increments, and scaled in terms of implied changes in the global organic carbon reservoir (from Shackleton, 1984).

organic and limestone reservoirs, it is rather sensitive to the value chosen for the long-term steady state $^{13}$C content of the limestone pool. The value of +2.25 per mil used by Shackleton

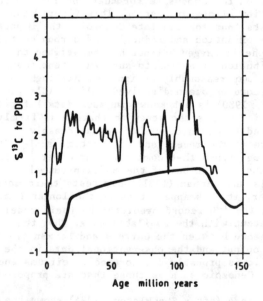

Fig. 3.   Carbon isotope record predicted by Garrels and Lerman (1981) on the basis of sulfur isotope data (heavy line), compared with data from Figure 2 (for the Cenozoic) and with data from Scholle and Arthur (1980) (for the Mesozoic). The horizontal dashed line shows the steady state value used in order to construct Figure 4.

(1984) is the mean of his data for the Cenozoic excluding the interval from the middle Miocene to present, and is shown by a dashed horizontal line in Figure 3.

Figure 4 also shows the changes in atmospheric oxygen resulting from changes in carbon burial. This calculation assumes, as is implied by Figure 3, that the sulfur system has made a small impact on the atmosphere during this time. Figure 4 shows that there has been a significant reduction in atmospheric oxygen, chiefly since the middle Miocene, and that an atmospheric oxygen pressure of 0.25 atmospheres may have characterized the later Mesozoic and early Cenozoic. The $^{13}$C data sets used by Veizer et al. (1980) and by Garrels and Lerman (1981) do not adequately record the isotopic history of global ocean carbonate during the Cenozoic, possibly for the reason that none of the data to which they had access had been gathered for that purpose.

Although the conclusion that there has been a significant draw-down in atmospheric oxygen during the later Cenozoic is at first sight surprising, there is evidence in its favor. If there is indeed a feedback mechanism whereby oxygen released by organic carbon formation is consumed by sulfide oxidation, and vice versa, the mechanism must operate mainly in marine sediments rather than on land, because the area of the sea-floor that is covered by reduced sediments is sensitive to small variations in atmospheric oxygen concentration. Thus it may be the oxygen concentration in the ocean, rather than the atmospheric oxygen pressure, that is maintained at a more-or-less constant level by mutual interaction among the various oxidation-reduction systems. This observation is not in

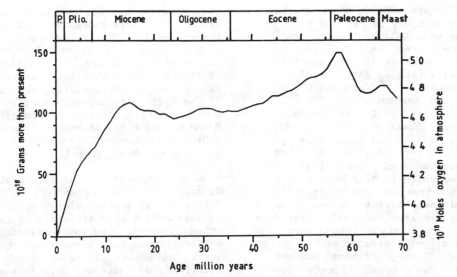

**Fig. 4.** Changes in the global organic carbon reservoir and of the atmospheric oxygen concentration (from Shackleton, 1984) estimated from the data in Figure 2 and ignoring the smaller variations in atmospheric oxygen that may have resulted from changes in the sulfate-sulfide system.

conflict with the model proposed by Berner and Raiswell (1983) whereby changes in the burial of organic carbon on the continents provide the initial impetus for shifts in the global balance of reduced carbon versus reduced sulfur. Viewed in this light, the late Cenozoic consumption of organic carbon may be seen as an expected consequence of the increase (by about 20%) in oceanic dissolved oxygen that must have come about with the drop in ocean deep-water temperatures from around $10^{\circ}$ to about $2^{\circ}$ that occurred during the Cenozoic. Although the precise timing of this temperature drop is somewhat contentious (Matthews and Poore, 1980), there is little doubt as to its reality.

### Implications of Within-Ocean $^{13}C$ Variations

The various species of planktonic foraminifera deposit their carbonate at different depths in the water column depending on the details of their life cycle (Emiliani, 1954), so that at times of high foraminiferal diversity, one has a good opportunity to monitor the total oceanic $^{13}C$ range that prevailed at that time by analyzing all available species (Williams et al., 1977; Boersma et al., 1979). Species that lived closest to the ocean surface, in the warmest water, have the lightest $^{18}O$ values and also the heaviest $^{13}C$ values, as is to be expected, since it is only close to the surface in the photic zone that isotopically light carbon can be extracted by photosynthesis.

Broecker (1974) has pointed out that the range of $^{13}C$ values in the ocean provides a measure of the P/C ratio in the ocean (where P represents the limiting nutrient but is not necessarily

phosphorus). This is because there is a more-or-less constant ratio of phosphorus to isotopically light organic carbon in the organic matter that is produced at the surface and subsequently returned to the deep water dissolved $CO_2$ pool by oxidation. However, Broecker (1974) also points out that there is a relationship between the dissolved $CO_2$ added to the deep water and its oxygen utilization. Since the dissolved-oxygen concentration can only lie in the range between its saturation value (where the deep water forms in contact with the atmosphere) and zero, the $^{13}C$ range also estimates the O/C ratio of newly formed deep water. The dissolved-oxygen content where the deep water forms is approximately predictable from the atmospheric oxygen concentration and the temperature. The $^{13}C$ range is a less good estimate of the O/C ratio of newly formed deep water than of the P/C ratio of upwelling deep water only in the sense that we expect the limiting nutrient to be totally depleted in warm, stratified surface water, whereas it is possible to imagine an ocean in which dissolved oxygen does not fall to near-zero values anywhere in the deep water. Thus the range in $^{13}C$ values within the ocean at any time in the geological past provides an estimate of the lower limit of the O/C ratio in newly formed deep water.

Berner et al. (1983) have developed a very interesting model for the operation of the global weathering system under the influence of changes in the input of $CO_2$ (e.g., from changes in sea floor spreading). These authors have explored the response of their model for various permutations of values for the parameters controlling the model. Their results may be summarized in terms of the level to which atmospheric $CO_2$ must rise before the rate at which $CO_2$ is removed by silicate weathering will balance a postulated rise in the $CO_2$ input from the ocean ridge system. Their favored version of this model predicts substantially (almost an order of magnitude) higher ocean total dissolved $CO_2$ in the early Cenozoic and Mesozoic.

The available [13]C data for the Paleocene (Boersma et al., 1979; Shackleton et al., 1984a) show that the [13]C in deep water (estimated by analyzing a variety of individually calibrated benthic species) was about +2.5 per mil at its most positive value, which it held betwen 58 and 60 million years ago. Specimens of Morozovella velascoensis attain most positive values close to +5.0 per mil during the same interval. These data constrain the O/C value in newly formed deep water to have been slightly greater at that time than today, since a [13]C range of about 2.0 per mil is observed in today's ocean. Since we have shown above that the atmospheric oxygen concentration may have been about 20% greater at that time than today, we conclude that the maximum possible value for ocean total dissolved $CO_2$ during the Paleocene was approximately its present level. If the change in atmospheric oxygen concentration that we have postulated is ignored, the [13]C data would constrain ocean-dissolved $CO_2$ in the Paleocene to about 20% below its present value, while a dissolved-$CO_2$ load of twice the present value would demand that the atmosphere contained more than twice the present oxygen content, far outside the range permitted by the data discussed above.

A continuous record of the oceanic [13]C gradient for the Cenozoic is shown by Arthur (1982). Although isotopic data for individual foraminiferal species from the Cretaceous are scarcer, Boersma and Shackleton (1981) have obtained sufficient [13]C data for benthic and planktonic foraminifera from the later part of the Cretaceous that one may be confident a [13]C range of at least 1.0 per mil was present. Thus ocean-dissolved $CO_2$ was certainly not more than twice its present level during the past $10^8$ years. It should be noted that since the time constant for changes in ocean dissolved $CO_2$ in the model of Berner et al. (1983) is over $10^7$ years, it is a sufficient test of their model prediction to select data from times of maximum foraminiferal diversity, or from samples that are in unusually pristine condition. It is fortunate that minor diagenesis is more likely to reduce the original [13]C difference between species in a sample than to increase it, so that it is unlikely that analytical data will exaggerate the constraint. It is quite possible that further analyses of a range of different species from selected time intervals will further restrict any possible upward excursions in ocean-dissolved $CO_2$.

This conclusion apparently provides one of the constraints that Berner et al. (1983) seek for

their models in their concluding paragraph. Although the more recent version of their model (Lasaga et al., this volume) does not require such high levels of dissolved $CO_2$ in the ocean as the version of Berner et al. (1983), they are still high enough that the constraint may be of value. This is particularly interesting in view of the difficulty in explaining the "abnormally" (by Pleistocene standards) warm climates of the Mesozoic and early Cenozoic without appealing to a high atmospheric $CO_2$ content (Barron and Washington, this volume).

Acknowledgments.  I am very grateful to Eric Sundquist for inviting me to this Chapman Conference at a particularly appropriate point in my research; I also sincerely appreciate the stimulus provided by the delightful and excellent scientists who were present and shared their ideas. Bob Berner was especially helpful in clarifying my concepts. Mike Arthur, Bill Curry, and Ted Moore generously reviewed a very hastily produced manuscript. NERC grant GR3/3606 has supported this work.

## References

Arthur, M. A.,  The carbon cycle--Controls on atmospheric $CO_2$ and climate in the geologic past, in Climate in Earth History, pp. 55-67 National Academy Press, Washington, D.C., 1982.

Barron, E. J. and W. M. Washington, Warm Cretaceous Climates: High Atmospheric $CO_2$ as a plausible mechanism, this volume.

Berggren, W. A., D. V. Kent and J. J. Flynn, Paleogene geochronology and chronostratigraphy, in Geochronology and the Geological Record, Geol. Soc. London Spec. Publ., edited by N. J. Snelling, in press 1984a.

Berggren, W. A., D. V. Kent and J. A. van Couvering, Neogene geochronology and chronostratigraphy, in Geochronology and the Geological Record. Geol. Soc. London. Spec. Publ., edited by N. J. Snelling, in press 1984.

Berner, R. A., A. C. Lasaga and R. M. Garrels, The carbonate-silicate geochemical cycle and its effect on atmospheric carbon dioxide over the past 100 million years, Am. J. Sci., 283, 641-683, 1983.

Berner, R. A. and R. Raiswell, Burial of organic carbon and pyrite sulfur in sediments over Phanerozoic time: A new theory, Geochim. Cosmochim. Acta, 47, 855-862, 1983.

Boersma, A. and N. J. Shackleton, Oxygen and carbon isotope variations and planktonic foraminifer depth habitats, Late Cretaceous to Paleocene, Central Pacific, Deep Sea Drilling Project Sites 463 and 465, Initial Rep. Deep Sea Drill. Proj., 62, 695-717, 1981.

Boersma, A., N. J. Shackleton, M. A. Hall, and Q. C. Given, Carbon and oxygen isotope records at DSDP Site 384 (North Atlantic) and some Paleocene paleotemperatures and carbon isotope variations in the Atlantic Ocean, Initial Rep. Deep Sea Drill. Proj., 43, 695-717, 1979.

Broecker, W. S., Chemical Oceanography, 214 pp., Harcourt, Brace, Jovanovich, New York, 1974.

Claypool, G., W. T. Holser, I. R. Kaplan, H. Sakai and I. Zak, The age curves of sulfur and oxygen isotopes in marine sulfate and their mutual interpretation, Chem. Geol., 28, 199-260, 1980.

Curry, W. B. and G. P. Lohmann, Carbon isotopic changes in benthic foraminifera from the western South Atlantic: Reconstruction of glacial abyssal circulation patterns, Quat. Res., 18, 218-235, 1982.

Curry, W. B. and G. P. Lohmann, Reduced advection into Atlantic Ocean deep eastern basins during last glaciation maximum. Nature, 306, 577-580, 1983.

Davies, T. A. and T. R. Worsley, Paleoenvironmental implications of oceanic carbonate sedimentation rates, in The Deep Sea Drilling Project: A Decade of Progress, SEPM Spec. Publ. 32, edited by J. E. Warme, R. G. Douglas, and E. L. Winterer., pp. 169-179, Society of Economic Paleontologists and Mineralogists, Tulsa, Okla., 1981. Tulsa: SEPM Special Publication No 32

Davies, T. A., W. W. Hay, J. R. Southam, and T. R. Worsley, Estimates of Cenozoic oceanic sedimentation rates, Science, 197, 53-55, 1977.

Emiliani, C., Depth habitats of some pelagic foraminifera as indicated by oxygen isotope ratios Amer. J. Sci., 252, 149-158, 1954.

Garrels, R. M., and A. Lerman, Phanerozoic cycles of sedimentary carbon and sulfur, Proc. Natl. Acad. Sci. USA, 78, 4652-4656, 1981.

Garrels, R. M., and E. A. Perry, Cycling of carbon, sulfur, and oxygen through geologic time, The Sea, 5, 303-336, 1974.

Lasaga, A. C., R. A. Berner, and R. M. Garrels, An improved model of atmospheric $CO_2$ fluctuations over the past 100 million years, this volume.

Matthews, R. K., and R. Z. Poore, Tertiary $\delta^{18}O$ record and glacio-eustatic sea-level fluctuations. Geology, 8, 501-504, 1980.

Scholle, P. A., and M. A. Arthur, Carbon isotope fluctuations in Cretaceous limestones: potential stratigraphic and petroleum exploration tool, Am. Assoc. Pet. Geol. Bull., 64, 67-87, 1980.

Shackleton, N. J., Carbon-13 in Uvigerina: Tropical rainforest history and the equatorial Pacific carbonate dissolution cycles, in The Fate of Fossil Fuel $CO_2$ in the Oceans, edited by N. R. Andersen and A. Malahoff, pp. 401-427, Plenum, New York, 1977.

Shackleton, N. J., The carbon isotope record of the Cenozoic. in Petroleum Source Rocks, edited by J. Brooks and A. J. Fleet, Blackwell, Oxford, in press, 1984.

Shackleton, N. J., and M. A. Hall, Carbon isotope data from leg 74 sediments, Initial Rep. Deep Sea Drill. Proj., 74, 613-619, 1984.

Shackleton, N. J., M. A. Hall, J. Line, and S. Cang, Carbon isotope data in core V19-30 confirm reduced carbon dioxide concentration in the ice age atmosphere, Nature, 306, 319-322, 1983.

Shackleton, N. J., J. Imbrie, and M. A. Hall, Oxygen and carbon isotope record of East Pacific core V19-30: implications for the formation of deep water in the late Pleistocene North Atlantic Earth Planet. Sci. Lett., 65, 233-244, 1984.

Shackleton, N. J., M. A. Hall, and A. Boersma, Oxygen and carbon isotope data from leg 74 foraminifers. Initial Rep. Deep Sea Drill. Proj., 74, 599-612, 1984.

Shackleton, N. J., M. A. Hall, and U. Bleil, Carbon isotope stratigraphy, Site 577, Initial Rep. Deep Sea Drill. Proj., 86, in press, 1984.

Veizer, J., W. T. Holser, and C. K. Wilgus, Correlation of $^{13}C/^{12}C$ and $^{34}S/^{32}S$ secular variations. Geochim. Cosmochim. Acta, 44, 579-587, 1980.

Walker, J. C. G., Evolution of the Atmosphere 318 pp. Macmillan, New York, 1977.

Whitman, J. M., and T. A. Davies, Cenozoic oceanic sedimentation rates: How good are the data?, Marine Geol., 30, 269-284, 1979.

Williams, D. F., M. A. Sommer, and M. L. Bender, Carbon isotope compositions of recent planktonic foraminifera of the Indian Ocean. Earth Planet. Sci. Lett., 36, 391-403, 1977.

# CHARCOAL FLUXES INTO SEDIMENTS OF THE NORTH PACIFIC OCEAN: THE CENOZOIC RECORD OF BURNING

James R. Herring

U.S. Geological Survey, Lakewood, Colorado 80226

Abstract. Charcoal particles occur in sediments of the North Pacific Ocean at least as old as the earliest Cenozoic. The particles have morphologies similar to those from modern trees and shrubs. The charcoal results from plant burning on land; particles subsequently were blown into the ocean. This study measured charcoal concentrations in Cenozoic sediments from 11 Deep Sea Drilling Project sites. Although the charcoal concentrations in Holocene sediments vary considerably, they are consistently greater at temperate and high latitudes than at low latitudes. At each site, Holocene sediment charcoal concentrations also are consistently greater than older sediments at each site. Fluxes of charcoal to the sediment (mass per unit area per unit time) were calculated from the charcoal concentrations and the sediment accumulation rates. Through the Paleogene the charcoal flux remained small; but in the Neogene (at most sites, in the Miocene) the fluxes began to increase, and since the late Neogene the fluxes have been approximately two orders of magnitude greater than in the Paleogene. The increase appears to be real: there has been no chemical or biological destruction of the charcoal, which would cause older fluxes to appear smaller. The size distribution of the charcoal particles, which would be reduced by such destruction, shows no consistent trend through the sediment column. Much of the increased charcoal flux into upper Cenozoic sediments appears to be due to increased plant burning on land; the rest to increased wind transport of the charcoal from land to the ocean. An increase in burning is consistent with paleobotanical and paleoclimatic evidence, which indicates that an increasingly cool climate throughout the Cenozoic increased the relative abundance of temperate, hence more burnable, plants. The amount of burning in turn changed because of changes in the abundance of burnable plants or changes in combustion conditions affected by relative humidity, temperature, or oxygen-nitrogen concentration in the atmosphere. Over the past 5 million years, the estimate for the permanent removal of carbon from the atmosphere as inert charcoal is approximately $10^{14}$ g, or about 0.01% of the present atmosphere carbon mass, per year.

## Introduction

The study of charcoal in sediments has several important aspects. The characterization of the charcoal component of the sediments is important for a complete understanding of the carbon cycle. Previous studies of the carbon cycle have little appreciated or quantitatively considered this component. For example, most analytical techniques for particulate carbon include charcoal, if present, in the carbon fraction, but these techniques ignore charcoal as a separate component. Or, determinations of stable carbon isotope ratios in the particulate carbon fraction of sediments includes a contribution from charcoal whose stable carbon isotope ratio is largely unknown; and this charcoal may compose up to 50% of the total noncarbonate carbon content of some Holocene pelagic sediments.

The study of charcoal is also important in understanding the role of fire as an ecological agent. Destructive effects of fire have been noted for thousands of years, but the impact of natural burning on plants, plant evolution, soils, and animals has been studied only comparatively recently [Cooper, 1961; Komarek, 1962]. No quantitative assessments have been made of the ecological role of fire throughout the Cenozoic, although the existence of ancient fires is recognized [Komarek, 1972; Harris, 1958; Cope and Chaloner, 1980].

No previous investigation has examined the temporal and spatial variations in the flux of charcoal to marine sediments throughout the Cenozoic. This study is based upon one made by Smith et al. [1973], who analyzed concentrations of charcoal in surficial, Quaternary marine sediments. They noted that the distribution of charcoal concentrations in Pacific deep-sea deposits corresponds to the distribution of modern nontropical forests on a latitudinal basis. The latitudinal distribution of charcoal concentrations was attributed to production of charcoal from forest fires on land and dispersion by the latitudinally zoned, major surface wind systems. However, no temporal or areal investigation of charcoal fluxes was made in their study, which limited interpretation. Without knowledge of the areal distribution of fluxes, the observed spatial trend in concentrations could have resulted from other processes, for example, dilution of the charcoal by sediment with differing rates of accumulation or from latitudinal differences in land to water area ratios or in charcoal production area.

The observation of Smith et al. [1973] concerning the correlation of charcoal concentration patterns with the latitudinal extent of nontropical forest appears reasonable. The climate of the middle north latitudes supports the abundant growth of temperate plants [Walter, 1973] and also promotes intense thunderstorm activity and associated lightning production [Komarek, 1966]. Lightning is the major natural ignition source for fire in the boreal environment today [Komarek, 1964, 1965, and 1968] and is also a reasonable possible ignition source in the past. Modern plant distribution and climate favor temperate, north latitudes for the greatest production of charcoal. In polar latitudes, vegetation is sparse, and in tropical latitudes, vegetation, while abundant, does not support extensive combustion due to high rainfall and high water content of the plants [Walter, 1973]. In the tropics, oxidation of plant debris occurs but is due primarily to bacterial decomposition of what otherwise could become fire-supporting litter [Kittredge, 1948]. In the southern hemisphere, temperate latitudes have between 10 and 20% of the land area of comparable latitudes in the northern hemisphere; as a result, the amount of temperate plant biomass is considerably less in this hemisphere than for the northern temperate latitudes [Walter, 1973]. In summary, the accumulation of burnable fuels is favored by temperate plants [Dodge, 1972], and the distribution of temperate plants is concentrated in middle latitudes of the northern hemisphere [Walter, 1973]. Consequently, it seems reasonable to expect that larger and more frequent fires occur in these latitudes than elsewhere.

Charcoal particles are transported from the continents to the oceans by the tropospheric winds. Initially, combustion injects charcoal particles into the atmosphere. Subsequent dispersion depends on the size and nature of the airborne charcoal particles, along with the velocity and direction of the winds, which in turn depend primarily on the altitude and latitude. Charcoal from modern fires is injected into the lower atmosphere and rarely exceeds heights of 10 km [Palmer and Northcutt, 1975; S. Countryman, personal communication, 1975]. Thus, few particles, if any, are injected into the upper troposphere or lower stratosphere at

heights necessary for nonlatitudinal or intra-hemispherical transport. At heights of 10 km or less, the charcoal will be distributed by the major tropospheric wind systems. These planetary winds are geostrophic and thus move along latitudes as a result of the fluid dynamic constraints imposed by a rotating earth. It is interesting to note that in the geological past these dynamic contraints still required that the winds remain latitudinal, even if they were stronger or weaker than those of today. Therefore, modern charcoal particles, and probably those of the past as well, are wind transported within the latitudinal zones from which they originate.

In order for latitudinal charcoal production patterns to be preserved in the sediment, charcoal particles must settle through the oceanic water column quickly as compared to their rates of lateral movement caused by oceanic circulation. But the calculated Stokes' settling velocities for charcoal particles of a few micrometers diameter would be only a few centimeters per day, far too slow to preserve latitudinal patterns. Clearly, if latitudinal production patterns are to be preserved in the sediment, the charcoal must settle much faster. The charcoal distribution patterns noted by Smith et al. [1973] and the close correspondence between these patterns and latitudinal plant distribution indeed suggest that something has accelerated the settling velocities of the charcoal particles. Schrader [1971] and Roth et al. [1975] identified fecal pellets of zooplankton as the rapid transfer vehicles of modern, sinking diatoms and nannoplankton. They observed that distribution patterns for these tiny, slow-sinking organisms in the surface sediments mimicked those of the surface waters, and they concluded that the same patterns occurred because of rapid transfer of the organisms through the water colunm in fecal pellets from filter feeding organisms. More recent studies, for example, those of Honjo and Roman [1978] and Deuser et al. [1983], continue to support the role of fecal pellets in transferring fine-grained particles from surface waters rapidly to the seafloor. These pellets likely also explain what appears to be the abnormally rapid settling of the charcoal.

Marine sediments seem to offer the only satisfactory repository for charcoal on time scales of several tens of millions of years. In nonmarine environments the persistence and fate of charcoal is as yet unresolved. One experiment has suggested metabolic degradation of charcoal particles in soils [Shneour, 1966], but no responsible organisms could be identified. Griffin and Goldberg [1975] measured charcoal concentrations in San Diego County soils and suggested that charcoal was not persistent in soils but that the lack of persistence could have been due to resuspension and removal. Persistence on land has been noted for charcoal particles found in Miocene rocks [Weide, 1968], where the rocks had been exposed to subaerial weathering and microbial conditions since at least the Pleistocene. In addition, the common existence of charcoaled remains found in conjunction with anthropological studies, for example, by Zohary and Spiegel-Roy [1975], implies that charcoal is stable in certain terrestrial environments over time periods of thousands to millions of years. In marine environments, in contrast to the apparent conflict over the persistence of charcoal on land, it appears that charcoal is stable for at least millions of years [Parkin et al., 1970; Jedwab, 1971]. Other nonmarine sediments, such as glacial or lake sediments, also may contain charcoal, but the record of these deposits covers at best only the late Quaternary. Thus, in order to study the Cenozoic record of combustion processes on land, one must go to sediments of the oceans, where charcoal is both persistent and has a record of sufficient length.

Because charcoal persists in sediments, the type of study like that of Smith et al. [1973] can be extended into the geological past. Furthermore, if charcoal fluxes can be determined, then spatial and temporal variations in the latitudinal zones of greatest burning can be explicitly delineated. The subsequent goal of this study will be to understand the significance of changes of fluxes of charcoal in terms of (1) changes in the number of fires over time (charcoal production), (2) changes in the transport capacity of the winds in carrying charcoal to the oceans (wind intensity and zonal variability), and (3) whether any apparent changes in the flux of charcoal to the sediments could have resulted from diagenetic effects or other artifacts.

## Nature of Charcoal

Charcoal as it occurs in nature is composed of elemental carbon, which may be associated with unburned or partially burned plant material if there has been incomplete combustion. Thus, charcoal may occur along with complex organic compounds which originate from the plant material, such as humic substances or organometallic compounds [Bailey et al., 1973], or from subsequent pyrolysis.

Herein, charcoal is defined as the carbon fraction of the particulate residue after extraction of mineral carbon compounds by acidification and removal of organic material by peroxidation in alkaline solution. Other terms for this substance have been used, for example, microcrystalline graphite [Smith et al., 1973], graphitic carbon [Shneour, 1966] and amorphous carbon. Names referring to crystallinity are rejected because the crystalline nature of the carbon is yet to be established. The charcoal contains no graphite structure discernible by X-ray diffraction, although some weak graphite

crystallinity can be detected using electron diffraction. The terms soot [Novakov et al., 1974], fusain [Muir, 1970], particulate carbon [Jedwab, 1971], and free carbon [Kukreja and Bove, 1976] are rejected because of their ambiguous meanings. While the proper chemical term for the residue is elemental carbon [Smith et al., 1973], the term charcoal has been selected for this study because of the genetic implication that the carbon originates from burned plants.

In sediments, charcoal occurs as individual dark brown to black granules. The morphology of the particles, as revealed by scanning electron microscopy, is intricate, with craters, holes, and tubes. The morphology ranges from structured particles, which are similar in form to charcoal from modern plants [Komarek et al., 1973] to amorphous, nearly spherical, particles.

## Charcoal Flux

The range of sedimentation rates of various constituents of marine sediment makes it impractical to compare concentrations of a single sedimentary component. Comparisons of concentrations, especially of minor or trace components, are too misleading because of variable rates of dilution by other, major components. Fluxes, the masses deposited within a unit area per unit time, are essential for meaningful comparisons. In this case the fluxes of charcoal have been calculated using the measured charcoal concentration in the sediment along with the reported sedimentation rate and sediment physical properties data in the Initial Reports of the Deep Sea Drilling Project. First, the dry bulk density of the sediment, corrected for interstitial water salinity, is calculated from the reported values of wet bulk density, interstitial water content, and interstitial fluid salinity. The product of the dry bulk density and the sedimentation rate produces the mass flux or mass accumulation rate. Once this mass flux is known, the flux of charcoal, or of any other sediment component, is obtained from the product of that component's concentration times the mass flux.

Uncertainties in the sedimentation rate produce the most significant error in the mass flux calculation. In the open ocean, sedimentation rates are generally a few meters per million years, but the rates can range over several orders of magnitude. The sedimentation rates used in this study are those reported in the Initial Reports of the Deep Sea Drilling Project and have been determined by relating the paleontological age, which is based on fossil assemblages, to an absolute time scale, generally that of Berggren and van Couvering [1974]. Other, more refined time scales, for example, that of Harland et al. [1982], have come into accepted use since this scale was used for calculation. However, any effect of a different

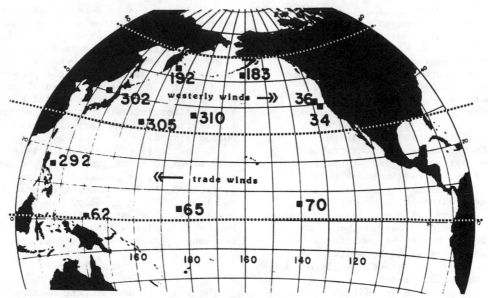

Fig. 1. Deep Sea Drilling Project site localities used in this study. Data on their position and water depth are given in Table 1. Approximate surface average wind zones are shown. Trade winds tend to be capricious in direction and intensity, while westerly winds are more uniform in direction and intensity.

time scale on the flux calculations is minor, never more than a factor of 2 difference. Some sedimentation rates have been recalculated in order to remove a correction for compaction or in order to take advantage of more accurate age information subsequent to publication.

Analytical Methods

The analytical method used for charcoal determinations is a modification of the technique developed by Smith et al. [1975] and consists of an extraction step followed by an analytical step. The extraction step removes diluting and interfering mineral matter using HF and hot HCl, followed by removal of organic material using hot

KOH and then 30% hydrogen peroxide in 6N KOH. Smith et al. [1975] showed that this treatment removes infrared-detectable organic matter from the residue but that the elemental carbon is not appreciably oxidized. After extraction the residue consists of charcoal in addition to any minerals that persist through the extraction and digestion. Typical resistant minerals, determined using X-ray diffraction, that occur in residua of Cenozoic North Pacific sediments are rutile, anatase, and, occasionally, tourmaline, zircon, pyrophyllite, barite, and pyrite.

The analytical step quantitatively determines the amount of elemental carbon in the residue. The analysis consists of intensely pulverizing the charcoal and resistant mineral mixture using vigorous shaking of ball bearings, at 20 oscillations per second, in a stainless steel vial for 24 hours. The impacts produce a partial oxidation of the carbon, which has a series of infrared absorbances at 1720, 1580, and 1240 $cm^{-1}$. These absorbances result from carbonyl bonds formed by the interaction of carbon with oxygen from the air. The 1580 $cm^{-1}$ absorbance has a linear relationship to the mass of the elemental carbon in the residue [Smith et al., 1975]. The charcoal concentration is measured by integrating the 1580 $cm^{-1}$ absorbance from 1480 to 1650 $cm^{-1}$ and comparing this area to a working curve obtained by using petroleum charcoal as a standard. The reproducibility on duplicate pellets of the same residue is within 5% or less.

The primary advantage of the infrared analytical technique, compared to many others that are used for carbon analysis, is that the

TABLE 1. Data for Deep Sea Drilling Project Sites

| Site | Latitude, $^{\circ}$N | Longitude | Water Depth, m |
|------|-----------|-----------|----------------|
| 34 | 39$^{\circ}$28′ | 127$^{\circ}$17′W | 384 |
| 36 | 40$^{\circ}$59′ | 130$^{\circ}$07′W | 116 |
| 183 | 52$^{\circ}$35′ | 161$^{\circ}$13′W | 4708 |
| 192 | 53$^{\circ}$01′ | 164$^{\circ}$43′E | 3000 |
| 302 | 40$^{\circ}$20′ | 136$^{\circ}$54′E | 2400 |
| 292 | 15$^{\circ}$49′ | 124$^{\circ}$39′E | 2900 |
| 62 | 1$^{\circ}$52′ | 141$^{\circ}$56′E | 2591 |
| 65 | 4$^{\circ}$21′ | 176$^{\circ}$59′E | 6130 |
| 70 | 6$^{\circ}$20′ | 140$^{\circ}$21′W | 5060 |
| 305 | 32$^{\circ}$00′ | 157$^{\circ}$51′E | 2903 |
| 310 | 36$^{\circ}$52′ | 176$^{\circ}$54′E | 3516 |

concentration of charcoal is determined and that small quantities of organic material, which could interfere with the detection or quantification of charcoal, also can be detected. This advantage does not exist for other common techniques used for the determination of carbonaceous substances in sediments.

If infrared spectra of the ground or unground residues contain C-H fundamental absorbance bands, the residue is considered to contain residual organic material, and the amount of charcoal is reported as an upper limit concentration. Upper limit concentrations also have been reported when insufficient charcoal occurred in the residue for infrared analysis. In this case, extracted residues were weighed after removal of the mineral matter using heavy liquid.

## Results

The present study comprises 186 samples, excluding duplicates, of marine sediment analyzed for its charcoal concentrations. The samples have been taken from seven marginal, two central, and two low-latitude central sites in the North Pacific Basin (Figure 1 and Table 1). The sites are representative of low latitudes (sites 292, 62, 65, and 70), mid-latitudes (sites 302, 305, 310, 34, and 36), and high latitudes (sites 192 and 183). These locations span a latitudinal range from near the equator to about $53^{\circ}N$. While the sample localities do not quite extend to the latitudes of the northernmost tropospheric surface wind system, the polar easterly (consequently, the most northern land flora are not sampled), the sites do span latitudes representative of 82% of the area of the northern hemisphere. The ages of the sediment samples span an interval of 65 million years (m.y.) and range from Late Cretaceous to Quaternary. From the known sedimentation rates of each of the sediment samples a deposition time has been integrated ranging from a minimum of several centuries to a maximum of several tens of millennia.

The primary criterion in site selection is the existence of a nearly complete sedimentary section with little interruption and with a fossil content that can produce a reliable geochronology. Unfortunately, most of the potentially usable sites presently lie along the equator in a latitudinal zone in which Quaternary sediments have a low concentration of wind-transported charcoal [Smith et al., 1973] or the usable sites are adjacent to continents. In the latter case the charcoal is diluted with terrigenous sediment, which may contain non-eolian charcoal. At the time of sampling in 1976, few acceptable, dated sites existed at the center of the North Pacific Basin. However, the sampled sites 305 and 310 lie toward the center of the basin and have relatively complete Cenozoic sediment sequences with sufficient fossils for age assignments.

A summary plot of the log of the charcoal flux versus the absolute age of the sediment is included in Figure 2, along with individual plots of the charcoal flux data for each site. Herring [1977] has presented a tabulation of the original data, and a summary of those data is given in Table 2. The log of the charcoal flux has been used because of the range in orders of magnitude of flux among the sites. The age scale is expanded to show the past 10 m.y. in greater detail. Note the rather consistent decrease in charcoal flux into older sediments at most sites. Fluxes reported as upper limit values, of course, permit no determination of trends in the charcoal flux over time.

Charcoal particles were examined for size distribution from one high-latitude site, two mid-latitude sites, and one low-latitude site [Herring, 1977]. Particle size analysis was performed on isolates from the residue after heavy liquid extraction to remove mineral matter but before the residue was dried and ground. None of the sites shows any uniform trend of particle sizes versus sediment age. Median particle diameters, around a few micrometers, at each site vary by a factor of 5 or less. Also, the variation for samples among the different sites is within a factor of 5 or less. Thus, there is relative uniformity of the size of the charcoal fraction both temporally and spatially.

## Morphology of Charcoal Particles

Figure 3 displays some of the representative morphology of the charcoal particles. The particles usually appear as porous fragments or, occasionally, spheres (Figures 3i and 3j). The structure of the particle in Figure 3b is similar to that of angiosperm structures, in particular to that of charred grass [Komarek et al., 1973]. Also, the particle in Figure 3h has a structure that is similar to that of gymnosperm tissue and is quite similar, in particular, to that of charred coniferous woods [Komarek et al., 1973]. Although fragments occasionally retain structure similar to that of plant charcoal, the detail is usually insufficient for specific identification of the plant. Charcoal particles retain more of the original plant morphology when found closer to land (Figures 3a, 3c, 3g, 3h, and 3j), compared to particles from open ocean sites (Figures 3e and 3f). Also, sites closer to land contain more larger charcoal particles than open ocean sites.

## Observations

The following conclusions can be drawn from this study:

1. Charcoal is stable in the marine environment for at least several tens of millions of years. In particular, this is shown by identifiable charcoal occurrence in the oldest sediments examined in this study, 65 m.y. in

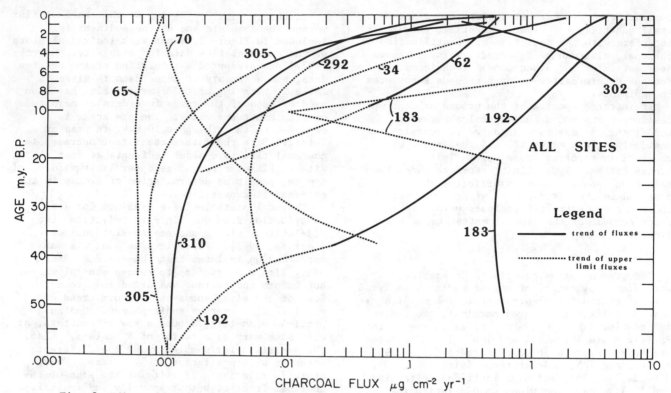

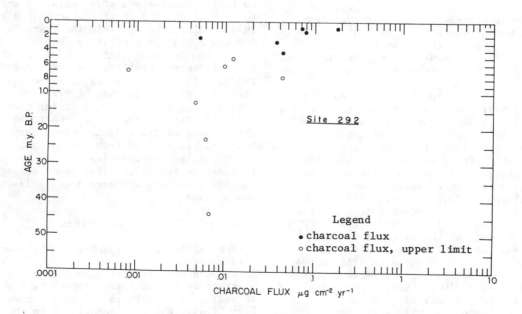

Fig. 2.  Charcoal fluxes as a function of sediment age at the sites analyzed in this study.  Note the log scale for the fluxes and dashed trend lines, which indicate upper limit values for the fluxes.  The first 10 million years is shown with an expanded age scale.  The first graph in the series is a composite of the remaining graphs for comparison of fluxes at all sites.  Trend lines of fluxes on this graph are approximate.  The original data are tabulated by Herring [1977].

Fig. 2.  (continued)

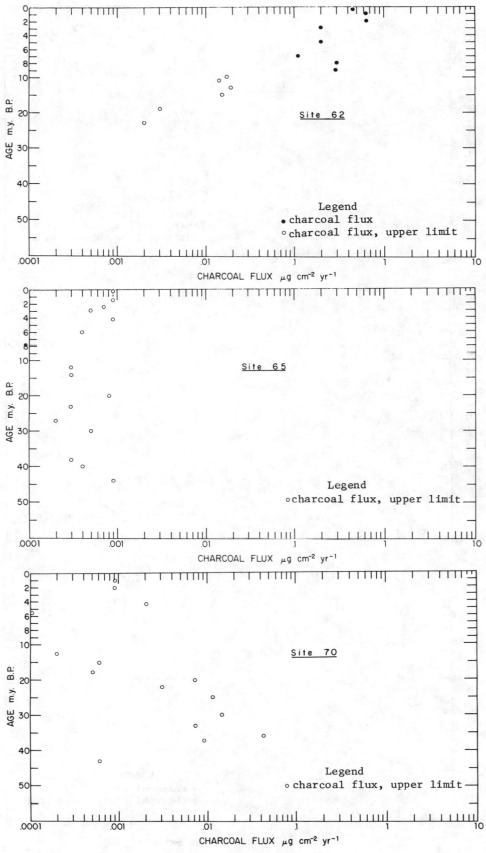

Fig. 2. (continued)

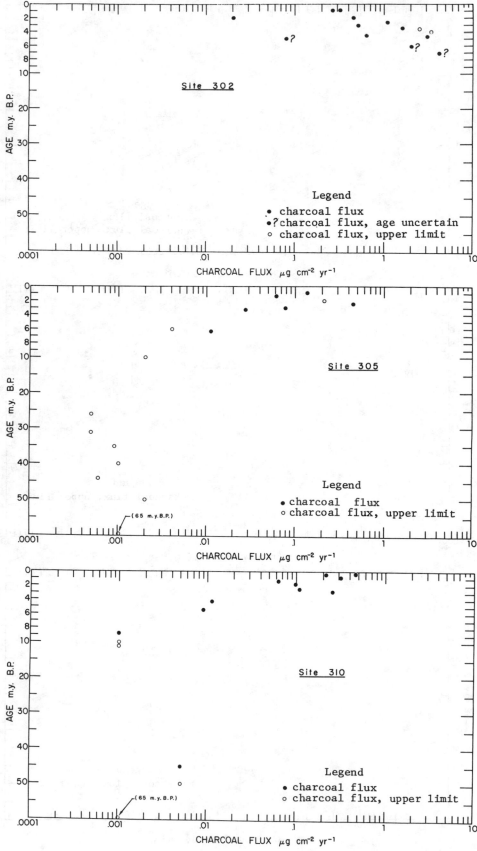

Fig. 2. (continued)

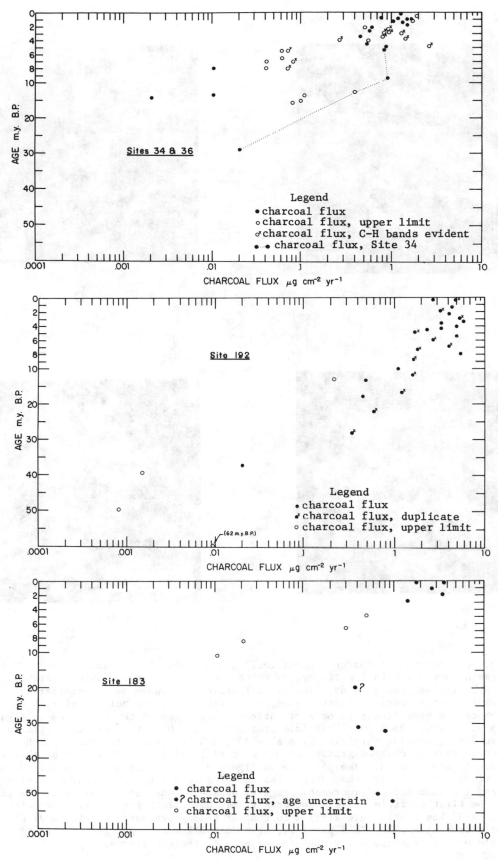

Fig. 2. (continued)

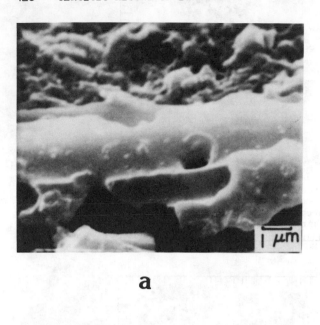

**a**

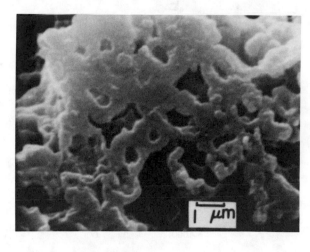

**b**

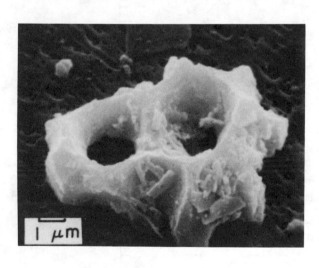

**c**

**d**

Fig. 3.  Micrographs of charcoal particles.  Holes in the background of the micrographs are pores in the filters on which the particles are mounted.  Figures 3a-3d show particles from site 62.  The axial features in Figure 3a are similar to tracheids or parenchmya in plant tissue, while the transverse hole is similar to a ray.  Note the very fine pits or perforations along edges of the particles in Figures 3c and 3d.  Figure 3e shows a particle from site 305.  Note the perforate structure in Figures 3f-3j, showing particles from site 302, and the axial structure, which is similar to that of charred grasses [Komarek et al., 1973].  Total length of the first particle is about 60 μm.  The particle in Figure 3g has a morphology similar to that of coniferous wood.  Note the large holes and pits in the tissue compared to other tissues, for example, in Figure 3h.  Figures 3i and 3j show spherical particles in which the first particle has a smoother surface compared to the other.  The particle in Figure 3j has a structure similar to that of some palynomorphs.  Figure 3k shows a particle from site 183.  This particle, of middle Oligocene age, is the oldest particle with recognizable perforations or pits on the exterior surface.

**e**

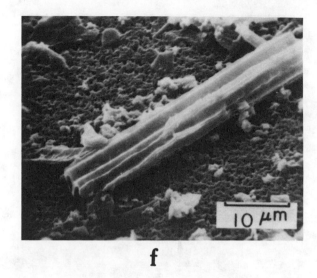

**f**

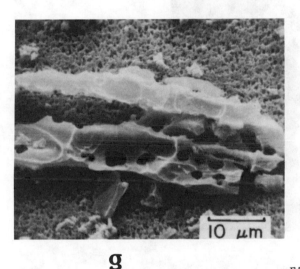

**g**

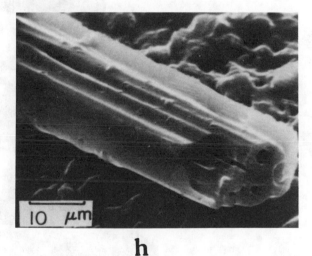

**h**

Fig. 3.    (continued)

age.  Charcoal particles with recognizable plant morphology occur in sediments as old as Oligocene.  Much older fossil charcoal has been tentatively identified by Cope and Chaloner [1980], and it is likely that charcoal has been produced since the occupation of land by vascular plants in the Paleozoic whenever atmospheric conditions produced ignition and permitted combustion.

2.  Charcoal can constitute a significant fraction of the noncarbonate particulate carbon component in sediments, especially those of late Neogene or Quaternary age.  For example, the charcoal concentrations in Pliocene age samples at site 192 compose up to 50 % of the organic carbon contents reported by Creager et al. [1973].

3.  Charcoal concentrations and distributions in Quaternary sediments of the North Pacific are similar to those obtained by Smith et al. [1973] for sediments of similar age in the North Pacific (Figure 4a).  In both cases, there are large concentrations in temperate to high-latitude sediments and smaller concentrations in lower-latitude sediments.  This study also indicates that this distributional pattern was maintained for at least 10 m.y. into the past (Figure 4b).

4.  When charcoal concentrations in Quaternary sediments are converted into charcoal fluxes, the flux distribution pattern is similar to the distribution pattern of charcoal concentrations, with higher fluxes in northern latitudes relative to the fluxes in low latitudes (Figure 5a).

i

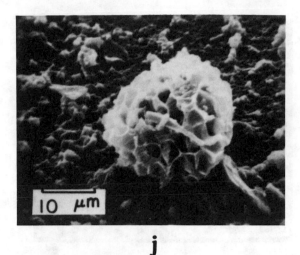

j

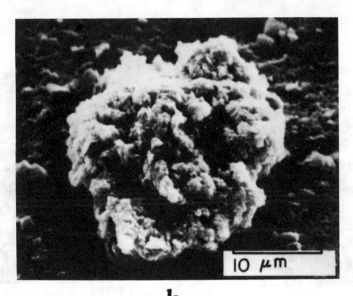

k

Fig. 3.  (continued)

Also, this pattern of charcoal fluxes was maintained for at least 10 m.y. into the past (Fig. 5b).  The spatial similarity between the pattern of Quaternary charcoal fluxes determined in this study and the pattern of Quaternary charcoal concentrations reported by Smith et al. [1973] suggests that their concentration pattern is not an artifact produced by varying degrees of sediment dilution.

5.  Sites near land masses have higher charcoal fluxes than open ocean sites at comparable latitudes, possibly as a result of charcoal contribution from river input and/or turbidite redeposition from shelf sediments.  For example, the Quaternary and Neogene charcoal fluxes to sites 34, 36, and 302 are greater than

those to sites 305 and 310 at approximately the same latitude.  Site 34 also has a greater charcoal flux than site 36, which is close to the same location but farther seaward.  The effect of proximal land also occurs in the low-latitude sites 62 and 292, which have Quaternary charcoal fluxes at least two orders of magnitude greater than the low-latitude, open ocean sites 65 and 70.  Clearly, it would be useful in any subsequent studies of eolian-borne charcoal in sediments to carefully avoid such continental margin sites where charcoal also may occur as a result of other transport processes.  Sampling should be done with consideration of lithologies that might indicate these processes, such as turbidite redeposition.

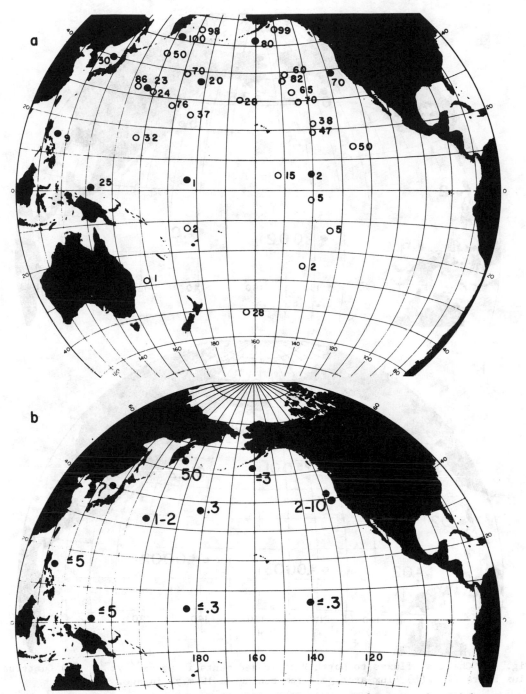

Fig. 4.   Charcoal concentrations in Pacific sediments.   Values are in weight percent
times 1000.   Solid circles are values taken from this study, and open circles are
values from that of Smith et al. [1973].   (a)  Quaternary averages of charcoal
concentrations.   (b)  Same as for Figure 4a except that concentrations are from
sediments around 10 m.y. in age.

6.   Mid- and high-latitude sites have charcoal
fluxes which increase in younger sediments by
about two orders of magnitude.   These increases
occur mostly in Neogene sediments.   Site 302 in

the Sea of Japan is an exception, with smaller
fluxes in youngest sediments.   However, this
likely resulted from the effect of proximal land
around most of the semi-enclosed marginal sea.

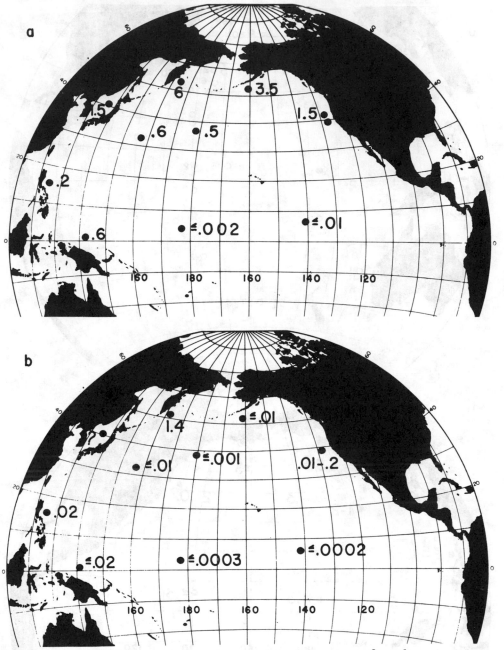

Fig. 5. Charcoal fluxes to North Pacific sediments in μg cm$^{-2}$ yr$^{-1}$. (a) Quaternary flux averages. (b) Flux averages from sediments 10 m.y. in age.

Large charcoal fluxes in older strata of site 183 also may have been due to an influx of terrigenous sediments and charcoal. Two of the low-latitude sites also exhibit an increase in charcoal flux into younger sediments. Over the past 5 m.y. and 10 m.y. at sites 292 and 62, respectively, there is a one order of magnitude increase in charcoal flux into the sediments. Prior to these ages, and in the other low-latitude sites, the trend in charcoal flux values cannot be determined due to inability to measure accurately the very low charcoal concentrations.

## Discussion

There are analytical and calculational errors in the flux measurement. The analytical precision on duplicates was given as 5% or less;

however, there also is an error associated with the assumptions and calculations of sedimentation rates that proportionally affects the charcoal flux calculation. A reasonable estimate of this error is a factor of 2; hence, neither the analytical nor the mass flux uncertainties are sufficient to explain the observed order of magnitude changes in charcoal flux.

This study shows that charcoal concentrations and flux distribution patterns in Quaternary sediments are maintained back at least through the middle Cenozoic. Also, for mid-latitude and high-latitude sites of the North Pacific there has been a significant increase in the flux of charcoal to younger sediments, and this increase has mainly occurred during the late Neogene and Quaternary. These results can be explained on the basis that there has been an increase in the ability of the surface winds to transport charcoal to the ocean basins and/or that there has been an increase in the amount of plant material burning on land during this time.

Other possible explanations for the increase of charcoal flux to younger sediments are not convincing. For example, the possibility of increased charcoal deterioration in older sediments due to inorganic or biochemical diagenetic reactions cannot be supported on the basis of the known chemical resistance of charcoal, well illustrated in this study by the rigor of the extraction step in the analytical method, which leaves charcoal intact. (Also see, for example, Mantell [1968] for a discussion of chemical inertness of elemental carbon.) Nor can charcoal degradation be supported on the basis of particle size analysis of the charcoal. For the latter consideration, any degradative reaction of the charcoal will first affect smaller particles, which have a higher surface area to mass ratio. This would produce a trend of increasing charcoal particle size with increasing sediment age. But the lack of a trend in the size analysis of the charcoal particles suggests little, if any, destruction of the charcoal.

Another alternative explanation for a substantial increase in charcoal flux through time involves movement of the sample sites into latitudes of increased charcoal production. Most of the sites lie on the Pacific plate, and this plate has been moving primarily northwest during the middle and late Cenozoic [Minster et al., 1974]. This could move former low-latitude, low-flux sites into temperate, high-flux latitudes and produce an apparent increase of flux into the younger sediments. In order to assess the potential significance of site movement the localities have been backtracked using Minster et al.'s [1974] rotational pole of 72°N and 83°W and an angular spreading rate of $0.81°$ $(m.y.)^{-1}$. When the sites on the Pacific plate are tracked back in time counterclockwise along small circles about this pole, the latitudinal movement is limited to only a few degrees during the time of greatest flux change. A similar reconstruction

of the site locations during the past 5 m.y. using the work of Van Andel et al. [1975] shows a similar location change of but a few degrees latitude. Thus, plate movement effects on the charcoal flux seem negligible. This observation is also in agreement with that of Creager et al. [1973], who indicated that latitudinal movement for sites 183 and 192 has been small. Latitudinal shifting of the continents, which are the source of the charcoal, has not been considered; however, Minster et al. [1974] suggested that North America and perhaps Eurasia have had only slight northerly components to their motion during the Cenozoic. Therefore, geographical relocation of the sample sites or the sources of charcoal is not considered to have significantly affected the temporal and spatial distribution of charcoal flux.

Could combustion conditions have been different in the past? In particular, could the oxygen content of the atmosphere have been sufficiently different to have produced differences in burning characteristics and to have altered the amounts of charcoal that could be produced from land plants? Combustion is extremely sensitive to oxygen content ($pO_2$) in the atmosphere, and alteration of the $pO_2$ by 10 to 20% of its present concentration has substantial effects on combustion. Only a modest increase in $pO_2$ greatly enhances combustion. The $pO_2/pN_2$ ratio also affects burning, but is of secondary importance to $pO_2$; in particular, lower ratios lead to less burning due to the suppressor effect of nitrogen. An increase in atmospheric $pO_2$ also can produce an interesting decrease in charcoal production, as higher $pO_2$ leads not only to greater incidence of ignition but also to greater efficiency of combustion. Thus, little charcoal, the incomplete oxidation product, remains, and most of the carbon is oxidized completely to carbon dioxide. Thus, past changes in $pO_2$ would drastically affect charcoal production. Also, any atmospheric oxygen changes would be synchronous and uniform worldwide due to the rapid mixing time of the gases in the troposphere. If smaller charcoal production and subsequent fluxes to older sediments result from different atmospheric oxygen contents in the past, the effect on charcoal flux at all sites should be similar and simultaneous. But the charcoal flux data do not support the theory of atmospheric oxygen changes, and the theory is rejected. For example, the charcoal flux into older sediments decreases by two orders of magnitude but at slightly different times during the past 10 m.y. for sites 305 and 310; the overall flux decrease for both sites is about four orders of magnitude. At site 192 during the past 10 m.y. the decrease is less than one order of magnitude. Watson et al. [1978] and Bartknecht [1981] provided some insight into the very substantial role of oxygen in combustion processes, although definitive experiments on charcoal production as a function of oxygen content have yet to be done.

Shackleton's [this volume] model proposes an increase of $pO_2$ in the early to middle Cenozoic. But at this time there is no concurrent increase in charcoal flux above the very low charcoal flux values as might be expected from higher $pO_2$ leading to increased burning. His model is somewhat controversial, as other interpretations do not require the increase of atmospheric $pO_2$ [e.g., Lasaga et al., this volume].

Thus, the principal reasons for the increased charcoal flux over time are due to possibly two factors: (1) an increase in the wind strength and subsequent transport of charcoal from land to ocean and/or (2) an increase in the amount of plant burning and subsequent charcoal production on land. Consider, first, the influence of changes in wind transport capacity. These changes have been used to explain concentration differences in various sedimentary components [Rex et al., 1969; Huang et al., 1975; Bowles, 1975; Leinen and Heath, 1981], but in the sedimentary record, no wind-transported species has been identified that indisputably can distinguish between changes in the intensity of wind transport of the species versus changes in the production or availability of that species. Also, these two factors may not be independent of each other if changes in climate, which determine wind strength, likewise influence the availability of a wind-transported component, for example, through enhanced weathering. In their study, Leinen and Heath [1981] noted a two order of magnitude increase in the flux of eolian-supplied quartz into sediments of the North Pacific at about 2.5 m.y. B.P. (before present). A similar increase in the flux of eolian quartz at site 310 was noted by Janecek and Rea [1983]. Both studies attributed the increase in quartz flux as due to increased wind strength beginning about 7 m.y. B.P. and the balance of the increase at about 2.5 m.y. B.P. as due to the onset of dry northern hemisphere glaciation. On the basis of their data and those of others, who also suggest an increased atmospheric circulation at this time, it is tempting to explain the increased charcoal fluxes as due to enhanced transport from the land via more vigorous wind circulation. However, in the quartz study, the climate changes that led to an enhancement of the winds also were a major factor in increasing the availability of quartz from glacial scour, so that the quartz flux increase was due to increased availability of quartz as well as increased wind transport. The relative contribution of two factors on the quartz flux cannot be determined, and consequently, not all of the two order of magnitude increase in quartz flux is due to increased wind transport.

A similar argument can be made for charcoal flux. Although it is not possible to independently eliminate a change in wind transport to explain charcoal flux increase through time, a change in wind transport does not appear to explain all of the observed magnitude and distribution of the charcoal flux change during the late Cenozoic. The reasonably uniform distribution of particle size of the charcoal during periods of maximum flux change suggests that the carrying capacity of the winds for charcoal was not appreciably different in the past or changing with a fixed pattern through time. Also, the offshore sites, particularly site 310, show a rather continual increase in charcoal flux over the past 10 m.y., rather than the sharp change observed at 2.5 m.y. B.P. for quartz. At least some major component of the flux increase, therefore, seems due to an increase in supply: burning on land. To some degree, then, it appears that the smaller charcoal flux into the older mid- and high-latitude sediments of the North Pacific is due to less burning on land prior to the Miocene and that the burning progressively increased during the late Neogene.

This conclusion of less burning in the past is consistent with data on global Cenozoic climate and resultant floral distributions. There is general accord that, as the Cenozoic climate became progressively cooler and drier through time, especially in middle and high latitudes [Savin et al., 1975; Dorf, 1964], the floral communities in North America changed as well, as described, for example, by Axelrod [1958], MacGinitie [1962], and Wolfe [1975, 1977, this volume]. The climate changes were global; hence, by analogy, the general considerations of floral succession in North America also apply to Eurasia, which would be the principal source of eolian-transported charcoal furnished to the Pacific Ocean via the westerly winds. The early Tertiary was very warm, and tropical flora extended from low to middle latitudes, while an arctic flora existed in the higher, cooler latitudes. No abundant analog of a modern temperate flora existed. These arctic and tropical, early to middle Tertiary plants would not have burned well as is the case with modern arctic and tropical floras, which are either too sparse [Walter, 1973] or too wet, respectively, to support extensive combustion. Thus, it is reasonable to expect that charcoal production would have been low in the early to middle Tertiary, when only arctic and tropical floral communities dominated. However, as the climate cooled during the middle to late Tertiary, a temperate flora with forests and grasslands proliferated and provided a more abundant, burnable substrate. There also may have been a symbiosis between increased fire frequency and temperate plant proliferation, as fire is known to influence modern floral community succession and establishment [Cooper, 1961] through the gradual establishment of fire-tolerant and fire-requiring species. The establishment of extensive communities of temperate, fire-tolerant plants, such as coniferous forest, chaparral, and grassland, ensured the continued supply of

abundant charcoal production and hence the
continued supply of greater charcoal fluxes to
the sediments toward the end of the Tertiary.

Increasing combustion of grass may have
contributed significantly to the increase in
charcoal flux to North Pacific sediments.
Neogene sediments at site 302 (Figures 3f and 3h)
contain particles that are similar in morphology
to modern grasses [Komarek et al., 1973].
Present grassland communities require frequent,
nearly annual burning to eliminate taller, light-
competing plants [Walter, 1973]. And extensive
mid-latitude prairie grasslands appeared sometime
during the Miocene in an environment cool and wet
enough for luxuriant growth yet also with annual
seasons, producing a dry part of the year during
which the grass could burn. While the extent of
these grasslands during the Neogene is not well
documented [MacGinitie, 1962], the proliferation
must have occurred about then and could be either
a consequence of or an accompaniment to the
increase in burning at the same time.

Finally, the charcoal fluxes in the past may
provide a measure of the importance of fire and
its interaction with the environment. The
possible association between plant succession and
fire frequency for grass has been mentioned.
Frequencies of burn can affect plant succession
and establishment of a stable floral community.
Grasslands, for example, require nearly annual
burns to eliminate taller, light-competing
plants. Likewise, a forest can tolerate burning
on the order of every decade or so without
alteration of its existing plant climax
community. From this study it appears that fire
has had a progressively more important role in
determining the evolution and establishment of
plant communities over the past 10 m.y.

Fire and/or fire products may have other, as
yet undetermined conseqences on past
environments. Blumer and Youngblood [1975]
suggested that plant burning is responsible for
the release of polycyclic aromatic hydrocarbons
to the atmosphere. Other possible chemical
interactions with the environment could involve
the adsorbant and catalytic properties of
charcoal. Other possible physical interactions
with the environment include the effect of
atmospheric charcoal particles in reducing the
albedo, acting as an atmospheric heat source
[Gray and Frank, 1974], or acting as cloud
condensation nuclei which could influence
rainfall patterns and abundance [Eagan et al.,
1974; Warner, 1968]. With more extensive
knowledge of the past temporal and spatial
variations in charcoal flux and knowledge or
assumption of the residence times of charcoal
particles in the atmosphere or hydrosphere, it
would be possible to estimate the atmospheric or
hydrospheric concentrations and residence times
of charcoal particles. It may then become
possible to estimate the amounts of fire-
associated products released from plant burning
as well as other effects on the environment over

time. Also, because charcoal is stable, and
hence chemically inert, on time scales of at
least several tens of millions of years, the
continual production of charcoal from plant
burning serves as a net sink or removal of carbon
from the plant cycle and thus of atmospheric
carbon dioxide [Seiler and Crutzen, 1980].

Some estimate can be made of how charcoal
participates in the global carbon cycle. First,
it is necessary to understand how open ocean
charcoal fluxes relate to global charcoal
production over time. However, it is somewhat
difficult to relate open ocean charcoal fluxes of
the North Pacific Basin to global charcoal
production, as only a minor amount of the
produced total amount of charcoal ultimately is
transported to the open ocean of the North
Pacific. Most of the charcoal must remain on
land near to where it was created. On land the
brittle charcoal is comminuted and ultimately is
incorporated into soil or eroded as
particulates. Large pieces of charcoal cannot be
moved very far by the winds, so that most
charcoal remains local and seldom is wind
transported far beyond the continent. Thus, most
of the charcoal produced by forest and brush
fires never travels beyond coastal sediments,
which probably are the primary reservoir for
charcoal. I have observed this coastal
deposition of charcoal from fires in the United
States, when coastal waters and beaches became
black with centimeter-long and larger pieces of
charcoal following a fire several tens of
kilometers distant. This observation is further
supported by an observed enhancement of a factor
of 1000 of modern charcoal flux into coastal
[Griffin and Goldberg, 1975] versus open ocean
[Smith et al., 1973] sediment. Additional
support is provided by comparing the Quaternary
depositional flux of charcoal to the assumed
production amount by assuming that the amount of
modern, natural burning has been maintained back
through the Quaternary. If an assumed northern
hemisphere charcoal production of approximately
$10^{14}$ g is totally dispersed into the oceanic
sediments of the northern hemisphere temperate
latitudes, then the observed and measured
Quaternary fluxes in this study, between 0.1 and
1 $\mu$g cm$^{-2}$ yr$^{-1}$, are low by a factor of close to
1000. Clearly, only a small fraction of the
charcoal produced on land reaches the sediments
of the open ocean; and most of the charcoal
produced throughout the history of land plants
has likely been buried in coastal sediments.

A highly simplified estimate can be made for
the global, permanent removal rate of atmospheric
carbon as charcoal, as well as for the change of
this removal rate over time. Estimates of modern
plant and other biomass burning range from 1 to
10 billion tons per year [e.g., Seiler and
Crutzen, 1980, and references therein] for the
release of carbon, roughly 1% of present
atmospheric carbon mass. The natural, or pre-
human, amount of carbon fixed as charcoal can be

TABLE 2.   Charcoal Fluxes

| Site | Sample (Core-Section) | Depth m | Age (m.y. B.P.) | Residue Recovery mg | Residue Recovery ppm | Sedimentation Rate m/m.y. | Mass Flux ($mg\ cm^{-2}\ yr^{-1}$) | Charcoal Concentration ppm | Charcoal Flux ($\mu g\ cm^{-2}\ yr^{-1}$) | Comments |
|---|---|---|---|---|---|---|---|---|---|---|
| 292 | 1-CC | 6 | 0.8 | 177.0 | 1238 | 8 | 0.71 | 251 | 0.178 | |
| | 1-CC | 6 | 0.8 | 17.7 | 538 | 8 | 0.71 | 99 | 0.070 | |
| | 2-3 | 10 | 1.3 | 83.1 | 993 | 8 | 0.74 | 106 | 0.078 | |
| | 3-1 | 18 | 2.3 | 19.5 | 668 | 8 | 0.64 | 8 | 0.0051 | |
| | 3-CC | 23 | 2.9 | 51.3 | 302 | 8 | 0.78 | 48 | 0.037 | |
| | 4-CC | 34 | 4.4 | 55.8 | 358 | 8 | 0.84 | 51 | 0.043 | |
| | 5-4 | 40 | 5.1 | 12.7 | 188 | 8 | 0.83 | <14 | <0.012 | |
| | 6-3 | 48 | 6.1 | 3.1 | 69 | 8 | 0.82 | <12 | <0.0098 | |
| | 6-CC | 54 | 6.9 | 34.1 | 150 | 8 | 0.78 | <1 | <0.0007 | |
| | 7-5 | 61 | 7.8 | 1.1 | 53 | 8 | 0.82 | <53 | <0.043 | |
| | 11-5 | 99 | 12.7 | 4.2 | 87 | 12.5 | 0.97 | <5 | <0.0047[a] | |
| | 15-6 | 138 | 23 | 1.1 | 26 | 12.5 | 1.42 | <4 | <0.0061 | |
| | 33-2 | 306 | 44 | 3.1 | 41 | | 1.94 | <4 | <0.0068 | |
| 62.1 | 1-2 | 8 | 0.3 | 48.7 | 832 | 25 | 1.8 | 251 | 0.45 | |
| | 2-3 | 19 | 1 | 52.4 | 847 | 25 | 2.0 | 317 | 0.63 | |
| | 4-1 | 34 | 2 | 66 | 904 | 25 | 2.0 | 320 | 0.64 | |
| | 7-2 | 65 | 3 | 18.9 | 216 | 25 | 2.4 | 84 | 0.20 | |
| | 11-6 | 107 | 5 | 23.3 | 247 | 25 | 2.5 | 80 | 0.20 | |
| | 15-2 | 140 | 7 | 12.5 | 113 | 25 | 2.7 | 41 | 0.11 | |
| | 20-5 | 193 | 8 | 49.0 | 413 | 25 | 2.8 | 110 | 0.30 | |
| | 24-6 | 231 | 9 | 46.1 | 394 | 25 | 2.9 | 102 | 0.29 | |
| | 28-2 | 265 | 10 | 2.3 | 22 | 25 | 2.8 | <6 | <0.017 | <0.06[a] |
| | 32-2 | 303 | 11 | 2.9 | 24 | 25 | 2.9 | <5 | <0.014 | <0.07[a] |
| | 37-2 | 346 | 13 | 5.0 | 44 | 25 | 2.9 | <7 | <0.019 | <0.12[a] |
| 62.0 | 4-3 | 399 | 15 | 8.8 | 61 | 25 | 3.4 | <4 | <0.015 | <0.21[a] |
| | 5-4 | 495 | 19 | 0.9 | 7 | 18 | 2.7 | <1 | <0.003 | <0.015[a] |
| | 6-2 | 522 | 23 | 0.5 | 6 | 11 | 1.9 | <1 | <0.002 | <0.009[a] |
| 65.0 | 1-1 | 1 | 0.1 | 1.5 | 66 | 8 | 0.15 | <9.1 | <0.0014 | |
| | 2-2 | 12 | 1.5 | 2.1 | 98 | 8 | 0.21 | <9.2 | <0.0019 | |
| | 2-CC | 19 | 2.5 | 3.8 | 78 | 8 | 0.21 | <7.5 | <0.0016 | |
| | 3-4 | 24 | 3.0 | 0.7 | 25 | 8 | 0.22 | <5.2 | <0.0011 | |
| | 4-5 | 35 | 4.3 | 1.7 | 48 | 8 | 0.20 | <9.1 | <0.0018 | |
| | 5-5 | 44 | 6.0 | 0.9 | 27 | 3 | 0.08 | <3.8 | <0.0003 | |
| | 7-4 | 61 | 12 | 0.6 | 19 | 3 | 0.08 | <2.8 | <0.0002 | |
| | 8-2 | 66 | 14 | 0.9 | 34 | 3 | 0.09 | <3.4 | <0.0003 | |
| | 10-1 | 84 | 20 | 1.3 | 50 | 3 | 0.09 | <7.6 | <0.0007 | |
| | 11-2 | 95 | 23 | 0.8 | 24 | 3 | 0.09 | <2.9 | <0.0003 | |
| | 12-4 | 106 | 27 | 2 g | 4.5 x $10^4$ | 3 | 0.14 | <2.1 | <0.0003 | (residue contaminated with $K_2SiF_6$) |
| | 13-4 | 115 | 30 | 0.9 | 23 | 3 | 0.09 | <4.8 | <0.0004 | |
| | 15-CC | 137 | 38 | 0.8 | 11 | 3 | 0.13 | <3.2 | <0.0004 | |
| | 16-5 | 144 | 40 | 0.5 | 15 | 3 | 0.10 | <3.9 | <0.0004 | |
| 65.1 | 4-1 | 155 | 44 | 1.1 | 36 | 3 | 0.10 | <8.5 | <0.0008 | |

| | | | | | | | | | | Notes |
|---|---|---|---|---|---|---|---|---|---|---|
| 70[b] | 1-2 | 2 | 1 | 11 | 510 | 1.6 | 0.05 | <20 | <0.0009 | |
| | 1-3 | 4 | 2 | 11 | 440 | 1.6 | 0.05 | <20 | <0.0009 | |
| | 1-5 | 7 | 4 | 6.3 | 210 | 1.6 | 0.05 | <50 | <0.002 | |
| | 1-6 | 9 | 6 | 162 | 7810 | 1.6 | 0.05 | <3 | <0.0001 | residue mostly barite |
| | 3-5 | 24 | 12 | 0.5 | 6 | 3.3 | 0.27 | <1 | <0.0002 | |
| | 4-4 | 33 | 15 | 3.6 | 90 | 3.3 | 0.15 | <4 | <0.0006 | |
| | 5-5 | 42 | 18 | 2.9 | 100 | 3.3 | 0.10 | <5 | <0.0005 | |
| | 8-5 | 69 | 20 | 1.0 | 9 | 14.3 | 1.46 | <5 | <0.007 | |
| 70A[b] | 11-6 | 103 | 22 | 0.4 | 5 | 14.3 | 1.38 | <2 | <0.003 | |
| | 4-3 | 144 | 25 | 4.2 | 40 | 14.3 | 1.22 | <9 | <0.011 | |
| | 11-3 | 205 | 30 | 1.0 | 10 | 14.3 | 1.71 | <8 | 0.014 | |
| | 16-2 | 252 | 33 | 0.5 | 4 | 14.3 | 1.63 | <4 | <0.007 | |
| | 22-2 | 290 | 36 | 5.1 | 50 | 14.3 | 1.39 | <30 | <0.042 | |
| | 25-2 | 308 | 37 | 0.9 | 9 | 14.3 | 1.52 | <6 | <0.009 | |
| | 28-1 | 327 | 43 | 0.4 | 10 | 1 | 0.06 | <10 | <0.0006 | |
| 302 | 2-4 | 24 | 0.8 | 32 | 2600 | 27 | 1.89 | 140-170 | 0.26-0.32 | duplicate pellet |
| | 3-6 | 46 | 1.7 | 25 | 2600 | 27 | 1.89 | 10 | 0.02 | |
| | 3-CC | 47 | 1.7 | 206 | 3487 | 27 | 1.89 | 244 | 0.46 | |
| | 5-5 | 83 | 2.5 | 4.1b | 400 | 60 | 5.0 | 101 | 0.51 | light fraction residue |
| | 7-5 | 120 | 3.0 | 3.1b | 360 | 84 | 5.0 | 101 | 0.51 | light fraction residue |
| | 8-CC | 142 | 3.3 | 51.5 | 681 | 84 | 4.6 | 354 | 1.6 | |
| | 9-2 | 157 | 3.5 | 3.6 | 541 | 84 | 4.6 | <541 | <2.5 | |
| | 11-2 | 193 | 3.9 | 5.0 | 736 | 84 | 4.6 | <736 | <3.4 | |
| | 14-5 | 254 | 4.4 | 1.4 | 161 | 194 | 11.6 | 55 | 0.63 | |
| | 14-CC | 256 | 4.4 | 97.6 | 1050 | 194 | 11.6 | 273 | 3.2 | |
| | 16-1 | 356 | 4.9(?) | 3.4 | 389 | 194 | 13.6 | 6 | 0.08 | |
| | 17-2 | 461 | 6.0(?) | 21.8 | 1154 | 100 | 7.5 | 281 | 2.1 | |
| | 18-1 | 530 | 7.0(?) | 28.0 | 1795 | 100 | 7.5 | 579 | 4.3 | |
| 305 | 1-CC | 8 | 0.9 | 68.1 | 781 | 6 | 0.58 | 230 | 0.133 | |
| | 2-1 | 12 | 1.2 | 36.9 | 607 | 6 | 0.56 | 109 | 0.061 | |
| | 2-CC | 15 | 2.0 | 29.6 | 379 | 6 | 0.57 | <379 | <0.22 | |
| | 3-2 | 22 | 2.5 | 35.3 | 306 | 41 | 4.2 | 107 | 0.449 | |
| | 4-2 | 31 | 3.0 | 28.5 | 253 | 41 | 4.3 | 18 | 0.077 | |
| | 5-2 | 38 | 3.2 | 28.3 | 208 | 4 | 0.38 | 70 | 0.027 | |
| | 5-CC | 44 | 6.0 | 8.2 | 151 | 4 | 0.40 | <10 | <0.004 | |
| | 6-2 | 47 | 6.3 | 18.2 | 157 | 4 | 0.41 | 27 | 0.011 | |
| | 6-5 | 52 | 10 | 36.4 | 261 | 4 | 0.46 | <5 | <0.002 | |
| | 6-CC | 54 | 26 | 2.7 | 21 | 2 | 0.25 | <2 | <0.0005 | |
| | 7-CC | 64 | 31 | 1.6 | 18 | 2 | 0.22 | <2 | <0.0005 | |
| | 8-CC | 73 | 35 | 1.9 | 25 | 2 | 0.24 | <4 | <0.0009 | |
| | 9-CC | 82 | 40 | 11.9 | 97 | 2 | 0.23 | <6 | <0.001 | |
| | 10-CC | 90 | 44 | 6.8 | 26 | 8 | 0.23 | <2 | <0.0006 | |
| | 11-CC | 101 | 50 | 3.1 | 19 | 8 | 0.96 | <3 | <0.002 | |
| | 18-CC | 167 | 65 | 18.8 | 84 | 4 | 0.47 | <3 | <0.001 | |
| 310 | 1-4 | 5 | 0.2 | 47.8 | 1038 | 30 | 2.4 | 190 | 0.456 | |
| | 2-1 | 7 | 0.4 | 51.7 | 564 | 22 | 1.9 | 111 | 0.211 | |
| | 2-CC | 12 | 0.7 | 112.3 | 736 | 22 | 1.9 | 174 | 0.331 | |
| | 3-CC | 24 | 1.5 | 21.4 | 244 | 13 | 1.1 | 56 | 0.062 | |
| | 4-2 | 26 | 1.6 | 31.8 | 310 | 13 | 1.2 | 82 | 0.098 | |
| | 4-CC | 34 | 2.5 | 51.0 | 448 | 10 | 1.0 | 108 | 0.108 | |

TABLE 2. (continued)

| Site | Sample (Core-Section) | Depth m | Age (m.y. B.P.) | Residue Recovery mg | Residue Recovery ppm | Sedimentation Rate m/m.y. | Mass Flux (mg cm$^{-2}$ yr$^{-1}$) | Charcoal Concentration ppm | Charcoal Flux (μg cm$^{-2}$ yr$^{-1}$) | Comments |
|---|---|---|---|---|---|---|---|---|---|---|
| 310 (cont) | 5-CC | 43 | 2.8 | 47.7 | 253 | 35 | 3.6 | 70 | 0.252 | |
| | 6-CC | 53 | 4.2 | 47.5 | 289 | 35 | 3.5 | 3 | 0.011 | |
| | 7-CC | 61 | 5.4 | 59.2 | 448 | 5 | 0.46 | 20 | 0.009 | |
| | 8-6 | 70 | 8.7 | 46.0 | 401 | 5 | 0.46 | 3 | 0.001 | |
| | 9-5 | 79 | 10 | 31.5 | 247 | 5 | 0.50 | <2 | <0.001 | |
| | 9-6 | 80 | 10.5 | 73.7 | 417 | 5 | 0.50 | <3 | <0.001 | |
| | 10-CC | 90 | 45 | 68.4 | 694 | 7 | 0.82 | 6 | 0.005 | |
| | 11-CC | 98 | 50 | 14.5 | 165 | 7 | 0.74 | <6 | <0.005 | |
| | 13-CC | 119 | 65 | 4.3 | 22 | 3 | 0.36 | <3 | <0.001 | |
| 36 | 1-1 | 1 | 0.04 | 38.1 | 2989 | 24 | 1.27 | 968 | 1.23 | |
| | 2-1 | 10 | 0.48 | 16.9 | 1047 | 24 | 1.82 | <1047 | <1.90 | |
| | 2-5 | 16 | 0.67 | 11.4 | 1279 | 24 | 1.61 | 476 | 0.77 | |
| | 2-6 | 17 | 0.71 | 38.8 | 2735 | 24 | 1.61 | 725 | 1.17 | |
| | 3-4 | 23 | 0.96 | 15.8 | 1349 | 24 | 1.66 | 923-1020 | 1.49-1.65 | duplicate pellet |
| | 3-4 | 23 | 0.96 | 6.9 | 1000 | 24 | 1.66 | <1000[c] | <1.66[c] | duplicate sample |
| | 3-6 | 26 | 1.1 | 19.3 | 1477 | 24 | 1.75 | 610 | 1.07 | |
| | 4-3 | 31 | 1.3 | 31.5 | 2726 | 24 | 1.51 | 867 | 1.31 | |
| | 5-3 | 40 | 1.7 | 20.0 | 1859 | 24 | 1.68 | 884 | 1.49 | |
| | 6-1 | 49 | 2.1 | 20 | 1869 | 15 | 0.91 | 672 | 0.61 | |
| | 6-1 | 49 | 2.1 | 2.8 | 602 | 15 | 0.91 | <602[c] | <0.52[c] | duplicate pellet |
| | 6-3 | 52 | 2.3 | 23.7 | 2557 | 15 | 0.92 | <1015[d] | <0.93[d] | |
| | 6-6 | 56 | 2.5 | 15.4 | 1357 | 15 | 1.02 | 567 | 0.58 | |
| | 8-1 | 66 | 3.2 | 30.0 | 2515 | 15 | 1.11 | <852-1169 | <0.94-1.30[d] | duplicate pellet |
| | 8-1 | 66 | 3.2 | 6.3 | 770 | 15 | 1.09 | <770[c] | <0.84[c] | |
| | 8-2 | 68 | 3.3 | 26.7 | 2494 | 15 | 1.01 | <783[c] | <0.79 | |
| | 8-4 | 71 | 3.5 | 12.5 | 1326 | 16 | 1.08 | 412 | 0.45 | |
| | 8-6 | 74 | 3.7 | 42.9 | 3388 | 16 | 1.15 | <1203[d] | <1.38[d] | |
| | 9-2 | 78 | 3.9 | 48.3 | 962 | 16 | 0.89 | <308[d] | <0.27[d] | |
| | 9-2 | 78 | 3.9 | 4.1 | 579 | 16 | 0.95 | <579[c] | <0.55[c] | duplicate sample |
| | 10-2 | 85 | 4.4 | 22.9 | 1890 | 16 | 0.99 | 544 | 0.54 | |
| | 10-5 | 90 | 4.7 | 41.8 | 3900 | 16 | 0.91 | <2863[d] | <2.61[d] | |
| | 11-2 | 94 | 4.9 | 33.4 | 2714 | 16 | 1.03 | 884 | 0.87 | |
| | 12-1 | 103 | 5.5 | 5.1 | 628 | 1.5 | 0.11 | <628[c] | <0.07[c] | |
| | 12-1 | 103 | 5.5 | 22.3 | 2019 | 1.5 | 0.10 | <627[d] | <0.06[d] | duplicate sample |
| | 12-2 | 105 | 6.5 | 5.2 | 605 | 1.5 | 0.10 | <605 | <0.06 | |
| | 12-3 | 106 | 7.0 | 23.0 | 1991 | 1.5 | 0.10 | <759[d] | <0.08[d] | |
| | 12-3 | 106 | 7.0 | 3.3 | 362 | 1.5 | 0.11 | <362[c] | <0.04[c] | duplicate sample |
| | 12-5 | 107 | 8.0 | 6.8 | 401 | 1.5 | 0.11 | <401 | <0.04[c] | sample at 10 cm |
| | 12-5 | 108 | 8.1 | 21.6 | 1716 | 1.5 | 0.11 | <611 | <0.07[d] | sample at 140 cm |
| | 12-5 | 108 | 8.1 | 5.5 | 330 | 1.5 | 0.11 | 98 | 0.01 | duplicate sample |
| | 12-6 | 110 | 13.6 | 6.2 | 471 | 3 | 0.24 | 24 | 0.01 | |
| | 12-6 | 110 | 13.6 | 8.5 | 446 | 3 | 0.24 | <446[c] | <0.11[c] | duplicate sample |
| | 13-2 | 113 | 14.6 | 7.9 | 948 | 3 | 0.17 | 9 | 0.002 | |

| Group | Sample | | | | | | | | | |
|---|---|---|---|---|---|---|---|---|---|---|
| 34 | 13-3 | 115 | 15 | 6.9 | 649 | 3 | 0.15 | <649 | <0.10c | |
| | 13-4 | 117 | 15.8 | 4.6 | 529 | 3 | 0.15 | <529 | <0.08 | |
| | 7-6 | 140 | 5.5 | 20.9 | 1953 | 19 | 1.06 | 781 | 0.83 | |
| | 9-3 | 215 | 9.4 | 14.9 | 1634 | 19 | 1.06 | 849 | 0.90 | |
| | 11-3 | 280 | 12.5 | 3.1 | 223 | 25 | 1.81 | <223 | <0.40 | |
| | 16-1 | 350 | 29 | 7.3 | 439 | 2.5 | 0.25 | 70 | 0.02 | |
| 192 | 2-4 | 5.5 | 0.11 | 18.1 | 1464 | 47 | 3.7 | 723 | 2.7 | |
| | 4-5 | 16 | 0.34 | 39.2 | 2609 | 47 | 4.3 | 1108 | 4.8e | 2.3% |
| | 6-2 | 60 | 1.3 | 28.0 | 2590 | 47 | 2.9 | 1495 | 4.3 | |
| | 7-4 | 80 | 1.7 | 23.8 | 1453 | 47 | 3.8 | 835 | 3.2e | 0.0% |
| | 8-6 | 100 | 2.1 | 32.0 | 2085 | 47 | 3.7 | 1145 | 4.2 | |
| | 9-4 | 127 | 2.7 | 48.1 | 3083 | 47 | 3.8 | 1431 | 5.4e | 0.3% |
| | 10-3 | 152 | 3.1 | 23.2 | 1908 | 82 | 5.2 | 1125 | 5.9 | |
| | 11-2 | 180 | 3.5 | 14.7 | 1296 | 82 | 5.2 | 640 | 3.3 | |
| | 12-3 | 210 | 3.9 | 20.5 | 1863 | 82 | 5.3 | 941 | 5.0 | |
| | 13-5 | 240 | 4.2 | 12.4 | 1038 | 82 | 5.3 | 624 | 3.3 | |
| | 14-3 | 256 | 4.4 | 9.5 | 929 | 82 | 4.8 | 475 | 2.3 | |
| | 15-3 | 275 | 4.7 | 9.2 | 714 | 82 | 5.3 | 315 | 1.7e | 3.8% |
| | 17-3 | 331 | 5.3 | 21.2 | 1863 | 82 | 5.0 | 1006 | 5.0 | |
| | 18-5 | 360 | 5.7 | 10.6 | 926 | 82 | 5.1 | 537 | 2.7e | 1.5% |
| | 20-2 | 434 | 6.6 | 17.4 | 1561 | 82 | 5.2 | 800 | 4.2e | 5.1% |
| | 21-2 | 480 | 7.2 | 9.1 | 1041 | 82 | 5.2 | 351 | 1.8e | 0.0% |
| | 22-1 | 526 | 7.7 | 11.7 | 1268 | 82 | 5.8 | 777 | 4.5 | |
| | 23-1 | 573 | 8.6 | 12.9 | 1065 | 39 | 2.8 | 570 | 1.6e | 2.1% |
| | 24-2 | 630 | 10 | 5.2 | 676 | 39 | 2.6 | 434 | 1.1 | |
| | 25-1 | 676 | 11.2 | 14.9 | 1170 | 39 | 2.8 | 556 | 1.6e | 2.5% |
| | 26-1 | 713 | 12.9 | 2.3 | 256 | 9 | 0.87 | <25b | <0.22b | |
| | 26-2 | 714 | 13.0 | 20.9 | 1122 | 9 | 0.97 | 504 | 0.49 | |
| | 27-1 | 748 | 16.7 | 29.9 | 1751 | 9 | 1.2 | 1100 | 1.2e | 2.9% |
| | 27-5 | 754 | 17.4 | 18.3 | 923 | 9 | 0.45 | 411 | 0.45 | |
| | 29-1 | 794 | 22 | 17.5 | 859 | 9 | 0.59 | 419 | 0.59e | 3.3% |
| | 30-1 | 852 | 28 | 14.5 | 787 | 9 | 0.34 | 245 | 0.34e | 2.0% |
| | 34-4 | 927 | 37 | 7.2 | 356 | 9 | 0.02 | 13 | 0.02 | |
| 192A | 1-3 | 946 | 39 | 0.028 | 1b | 8 | 1.5 | <1 | <0.0015b | |
| | 4-3 | 1023 | 49 | 0.012 | 0.5b | 8 | 1.5 | <0.5 | <0.0008b | |
| | 5-1 | 1043 | 62 | 0.210 | 11b | 4 | 0.78 | <11 | <0.0090b | |
| 183 | 1-1 | 1 | 0.02 | 45.0 | 1975 | 45 | 4.36 | 835 | 3.64 | |
| | 2-2 | 7 | 0.16 | 8.8 | 952 | 45 | 3.32 | 519 | 1.72 | |
| | 7-1 | 51 | 1.1 | 17.4 | 1265 | 45 | 3.87 | 665 | 2.58 | |
| | 10-2 | 83 | 1.8 | 23.6 | 1565 | 45 | 4.36 | 797 | 3.47 | |
| | 11-1 | 100 | 2.7 | 29.7 | 1742 | 16 | 1.69 | 836 | 1.41 | |
| | 12-5 | 125 | 4.7 | 5.4 | 426 | 16 | 1.12 | <426 | <0.48a | |
| | 15-4 | 152 | 6.5 | 2.9 | 293 | 16 | 0.99 | <293 | <0.29a | |
| | 18-3 | 178 | 8.5 | 2.3b | 280 | 12 | 0.68 | <30b | <0.02b | |
| | 20-2 | 197 | 11 | 2.3b | 241 | 4 | 0.26 | <24b | <0.01b | |
| | 23-1 | 221 | 20(?) | 50.8 | 2175 | 2.5 | 0.35 | 1057 | 0.37 | |
| | 25-CC | 255 | 31 | 17.5 | 978 | 14 | 1.89 | 214 | 0.40 | |

TABLE 2. (continued)

| Site | Sample (Core-Section) | Depth m | Age (m.y. B.P.) | Residue Recovery mg | Residue Recovery ppm | Sedimentation Rate m/m.y. | Mass Flux (mg cm$^{-2}$ yr$^{-1}$) | Charcoal Concentration ppm | Charcoal Flux (µg cm$^{-2}$ yr$^{-1}$) | Comments |
|---|---|---|---|---|---|---|---|---|---|---|
| 183 (cont) | 28-2 | 271 | 32 | 17.0 | 992 | 14 | 1.73 | 458 | 0.79 | |
| | 33-CC | 346 | 37 | 48.2 | 1953 | 9 | 1.29 | 439 | 0.56 | |
| | 38-1 | 475 | 50 | 19.6 | 1041 | 9 | 1.28 | 508 | 0.65 | |
| | 39-1 | 502 | 52 | 49.3 | 1614 | 9 | 1.43 | 656 | 0.94 | |

aFlux based on total residue.

bLight fraction of residue or fluxes based on light fraction. All site 70 and site 70A light residues except C25-S2 contain minerals as shown by their X-ray diffraction spectra; therefore the charcoal fluxes of these samples are even lower than the reported upper limits.

cGrinding vial leaked; upper limit based on total residue.

dC-H absorbances in infrared spectra.

eAverage of duplicate pellets; deviation from average given in percent of average.

estimated by assuming that half of present biomass burning is natural and that any plant burning results in 10% of the plant carbon mass converted to charcoal. Thus a reasonable estimate for the natural fixation and permanent removal of atmospheric carbon as inert charcoal, prior to human intervention, would be on the order of $10^{14}$ g/yr, or about 0.01% of the atmospheric carbon mass per year. Presumably, other sources of carbon resupply the atmosphere to balance this removal. Throughout the Cenozoic this rate of carbon removal and fixation should have paralleled the observed change in charcoal flux, assuming most of the flux change was due to supply changes rather than transport difference. Thus, carbon removal from the atmosphere as charcoal should decrease back into the geologic past.

## Conclusion

This study has shown that charcoal is a nontrivial component of the particulate, carbonate-free carbon fraction of some sediments, especially in those sediments of Quaternary or late Neogene age from boreal latitudes. I have shown that it is possible to obtain some insight into the record of past plant burning on land by measuring the temporal and spatial variation of charcoal flux to marine sediments. The conclusion, that the amount of plant burning has been less during the Cenozoic past, seems consistent with other evidence. Furthermore, the charcoal component may have had some as yet unassessed chemical or physical interaction with the atmosphere and/or hydrosphere.

There are ideal localities available for the extension and refinement of this work. The sediments of the North Atlantic Ocean should contain much of the charcoal derived from temperate latitudes of North America. In these sediments it generally is much easier to obtain accurate ages and sedimentation rates, compared to similar latitudes in the North Pacific Ocean. In addition, many sediment cores from the North Atlantic have had detailed isotopic stratigraphy worked out, so that a clear separation of Pleistocene glacial stages can be made. The comparison between glacial and nonglacial stages should show interesting and very dramatic changes in charcoal flux to the sediments, as marked changes in burning patterns on land during glacial, compared to interglacial, times would be expected. The variation in charcoal production between glacial and interglacial intervals may explain some of the scatter of the Quaternary charcoal fluxes at the offshore sites 305 and 310. Any comparisons of charcoal fluxes between the glacial and interglacial stages should include measurement not only of the charcoal but also of independent indicators of wind transport, charcoal particle size and morphology, and stable carbon isotopes, as an indicator of the combusted plant type, as well.

## References

Axelrod, D. I., Evolution of the Madro-Tertiary Geoflora, Bot Rev., 24, 433-509, 1958.

Bailey, J. M., R. Lee, P. C. Rankin, and T. W. Speir, Composition and properties of an age sequence of charcoals and associated humic acids, N. Z. J. Geol. Geophys., 16, 703-715, 1973.

Bartknecht, W., Explosions, translated from Russian by H. Burg and T. Almond, Springer-Verlag, New York, 1981.

Berggren, W. A., and J.A. van Couvering, The late Neogene, Palaeogeog. Palaeoclimatol. Palaeoecol., 16, 230 pp., 1974.

Blumer, M., and W. W. Youngblood, Polycyclic aromatic hydrocarbons in soils and Recent sediments, Science, 188, 53-55, 1975.

Bowles, F. A., Paleoclimatic significance of quartz/illite variations in cores from the eastern equatorial North Atlantic, Quat. Res., 5, 225-235, 1975.

Cooper, C. F, The ecology of fire, Sci. Am., 204, 150-160, 1961.

Cope, M. J., and W. G. Chaloner, Fossil charcoal as evidence of past atmospheric composition, Nature, 283, 647-649, 1980.

Creager, J. S., et al., Initial Reports of the Deep Sea Drilling Project, vol 19, 913 pp., U. S. Government Printing Office, Washington D. C., 1973.

Deuser, W. G., P. G. Brewer, T. D. Jickells, and R. F. Commeau, Biological control of removal of abiogenic particles from the surface ocean, Science, 219, 388-391, 1983.

Dodge, M., Forest fuel accumulation--A growing problem, Science, 177, 139-142, 1972.

Dorf, E., The use of fossil plants in paleoclimatic interpretation, in Problems in Paleoclimatology, edited by A. E. M. Nairn, pp. 13-31, Interscience, New York, 1964.

Eagan, R. C., P. V. Hobbs, and L. F. Radke, Measurements of cloud condensation nuclei and cloud droplet size distributions in the vicinity of forest fires, J. Appl. Meteorol., 13, 553-557, 1974.

Gray, W. M., and W. M. Frank, Feasibility of meso- and synoptic-scale weather modification from carbon particle seeding, Proceedings of WMO/IAMAP Scientific Conference on Weather Modification, Tashkent, 1-7 October, 1974. WMO Rep. 399, pp. 95-101, World Meterol. Organ., Geneva, 1974.

Griffin, J. J., and E. D. Goldberg, The fluxes of elemental carbon in coastal marine sediments, Limnol. Oceanogr., 20, 456-463, 1975.

Harland, W. B., A. V. Cox, P. G. Llewellyn, C. A. G. Pickton, A. G. Smith, and R. Walters, A Geologic Time Scale, 131 pp., Cambridge University Press, New York, 1982.

Harris, T. M., Forest fire in the Mesozoic, J. Ecol., 46, 447-453, 1958.

Herring, J. R., Charcoal fluxes to Cenozoic sediments of the North Pacific, Ph.D. thesis, 102 pp., Univ. of Calif., San Diego, May 1977.

Honjo, S., and M. R. Roman, Marine copepod fecal pellets: Production, preservation, and sedimentation, J. Mar. Res., 36, 45-57, 1978.

Huang, T. C., N. D. Watkins, and D. M. Shaw, Atmospherically transported volcanic glass in deep-sea sediments: Volcanism in sub-Antarctic latitudes of the South Pacific during the late Pliocene and Pleistocene time, Geol. Soc. Am. Bull., 86, 1305-1315, 1975.

Janecek, T. R., and D. K. Rea, Eolian deposition in the northeast Pacific Ocean: Cenozoic history of atmospheric circulation, Geol. Soc. Am. Bull., 94, 730-738, 1983.

Jedwab, J., Particules de matiere carbonee dans des nodules de manganese, C. R. Hebd Seances Acad. Sci., 272, 2968, 1971.

Kittredge, J., Forest Influences, 248 pp., McGraw-Hill, New York, 1948.

Komarek, E. V., Fire ecology, Proc. Annu. Tall Timbers Fire Ecol. Conf., 1st, 95-107, 1962.

Komarek, E. V., The natural history of lightning, Proc. Annu. Tall Timbers Fire Ecol. Conf., 3rd, 139-184, 1964.

Komarek. E. V., Fire ecology-Grasslands and man, Proc. Annu. Tall Timbers Fire Ecol. Conf., 4th, 169-220, 1965.

Komarek, E. V., The meterological basis for fire ecology, Proc. Annu. Tall Timbers Fire Ecol. Conf., 5th, 85-125, 1966.

Komarek, E. V., Lightning and lightning fires as ecological forces, Proc. Annu. Tall Timbers Fire Ecol. Conf., 8th, 69-199, 1968.

Komarek, E. V., Ancient fires, Proc. Annu. Tall Timbers Fire Ecol. Conf., 12th, 219-240, 1972.

Komarek, E. V., B. Komarek, and T. C. Carlysle, The ecology of smoke particulates and charcoal residues from forest and grassland fires: A preliminary atlas, Misc. Publ. 3, 49 pp.,Tall Timbers Res. Sta., Tallahassee, Fla., 1973.

Kukreja, V. P., and J. L. Bove, Determination of free carbon collected on high volume glass filter, Environ. Sci. Technol., 10, 187-189, 1976.

Lasaga, A. C., R. A. Berner, and R. M. Garrels, An improved geochemical model of atmospheric $CO_2$ fluctuations over the past 100 million years, this volume.

Leinen, M., and G. R. Heath, Sedimenatary indicators of atmospheric activity in the northern hemisphere during the Cenozoic, Palaeogeog. Palaeoclimatol. and Palaeoecol., 36, 1-21, 1981.

MacGinitie, H. D., The Kilgore Flora, Univ. Calif. Berkeley Publ. Geol. Sci., 35, 67-157, 1962.

Mantell, C. L., Carbon and Graphite Handbook, 538 pp., Wiley-Interscience, New York, 1968.

Minster, J. B., T. H. Jordan, P. Molnar, and E. Haines, Numerical modelling of instantaneous plate tectonics, Geophys. J. R. Astron. Soc., 36, 541-576, 1974.

Muir, M., A new approach to the study of fossil wood, in Scanning Electron Microscopy/1970,

IIT Research Institute, Chicago, Ill., 1970.

Novakov, T. S. G. Chang, and A. B. Harker, Sulfates as pollution particlulates; catalytic formation on carbon (soot) particles, Science, 186, 259-261, 1974.

Palmer, T. Y., and L. I. Northcutt, Convection columns above large experimental fires, Fire Technol., 11, 111-118, 1975.

Parkin, D. W., D. R. Phillips, R. A. L. Sullivan, and L. Johnson, Airborne dust collections over the North Atlantic, J. Geophys. Res., 75, 1782-1793, 1970.

Rex, R. W., J. K. Jackson, M. L. Jackson, and R. N. Clayton, Eolian origin of quartz in solids of Hawaiian Islands and in Pacific pelagic sediments, Science, 163, 277-279, 1969.

Roth, P. H., M. M. Mullin, and W. H. Berger, Coccolith sedimentation by fecal pellets: Laboratory experiments and field observations, Geol. Soc. Am. Bull., 86, 1079-1084, 1975.

Savin, S. M., R. G. Douglas, and F. G. Stehli, Tertiary marine paleotemperatures, Geol. Soc. Am. Bull., 86, 1499-1510, 1975.

Schrader, H. J., Fecal pellets: Role in sedimentation of pelagic diatoms, Science, 174, 55-57, 1971.

Seiler, W., and P. J. Crutzen, Estimates of gross and net fluxes of carbon between the biosphere and the atmosphere from biomass burning, Clim. Change, 2, 207-247, 1980.

Shackleton, N. J., Oceanic carbon isotope constraints on oxygen and carbon dioxide in the Cenozoic atmosphere, this volume.

Shneour, E., Oxidation of graphitic carbon in certain soils, Science, 151, 991-992, 1966.

Smith, D. M., J. J. Griffin, and E. D. Goldberg, Elemental carbon in marine sediments: A baseline for burning, Nature, 241, 268-270, 1973.

Smith, D. M., J. J. Griffin, and E. D. Goldberg, Spectrometric method for quantitative determination of elemental carbon, Anal. Chem., 47, 233-238, 1975.

Van Andel, T. H., G. R. Heath, and T. C. Moore, Jr., Cenozoic history and paleoceanography of the central equatorial Pacific, Geol. Soc. Am. Mem., 143, 134 pp., 1975.

Walter, H., Vegetation of the Earth, translated from German, 2nd ed., 237 pp., English University Press, London, 1973.

Warner, J., A reduction in rainfall associated with smoke from sugar cane fires--An inadvertent weather modification, J. Appl. Meteorol., 7, 247-251, 1968.

Watson, A., J. E. Lovelock, and L. Margulis, Methanogenesis, fires and the regulation of atmospheric oxygen, Biosystems, 10, 293-298, 1978.

Weide, D. L., The geography of fire in the Santa Monica Mountains, M.A. thesis, 180 pp., Calif. State Univ., Los Angeles, 1968.

Wolfe, J. A., Some aspects of plant geography of the northern hemisphere during the Late Cretaceous and Tertiary, Ann. Mo. Bot. Gard. 62, 264-279, 1975.

Wolfe, J. A., Paleogene flora from the Gulf of Alaska region, U.S. Geol. Surv., Prof. Pap., 997, 108 pp., 1977.

Wolfe, J. A., Distribution of major vegetational types during the Tertiary, this volume.

Zohary, D., and P. Spiegel-Roy, Beginnings of fruit growing in the old world, Science, 187, 319-327, 1975.

# CHANGES IN CALCIUM CARBONATE ACCUMULATION IN THE EQUATORIAL PACIFIC DURING THE LATE CENOZOIC: EVIDENCE FROM HPC SITE 572

Nicklas G. Pisias

College of Oceanography, Oregon State University, Corvallis, Oregon 97331

Warren L. Prell

Department of Geological Sciences, Brown University, Providence, Rhode Island 02901

**Abstract.** Hydraulic Piston Core Site 572, located at 1°N, 114°W (3903 m), recovered a continuous hydraulic piston cored section of late Miocene to late Pleistocene pelagic sediments. The sediment is composed of biogenic carbonate and silica, with nonbiogenic material being only a minor component. Detailed analysis of the calcium carbonate content shows that both the degree of variability in carbonate deposition and total carbonate accumulation changed markedly between the late Miocene and Pliocene at this equatorial Pacific site. During this interval, carbonate mass accumulation rates decrease from 2.6 to 0.8 g cm$^{-2}$ (1000 years)$^{-1}$. Comparison with other data shows that the marked change in accumulation of calcium carbonate at this site does not represent a change in the carbonate budget of the equatorial Pacific but rather a redistribution of carbonate in this region. The detailed carbonate records from the Pliocene section of Site 572 suggest that changing carbonate deposition controls the carbonate percent records. During the late Miocene, however, variations in biogenic silica accumulation play an important role in controlling the variations in calcium carbonate percentage at this site.

## Introduction

The deposition of calcium carbonate in the world's oceans is an important part of the global carbon cycle. Hence changes in the amount and rate of calcium carbonate deposition provide insight into the variability of carbon cycling. Variations in calcium carbonate content (referred to as carbonate) in marine sediments have received much attention since the work of Arrhenius [1952], who was the first to document high-resolution Pleistocene carbonate records from the equatorial Pacific and suggest their possible link to global climate. Hays et al.

[1969] extended the carbonate record into the Pliocene and showed that carbonate variations could be used as a powerful stratigraphic tool.

Although the nature of the variability in carbonate content in Pleistocene sediments of the Pacific is well defined, the processes controlling this variability are not well understood. Arrhenius hypothesized that carbonate content reflected changes in carbonate productivity which varied with changing global climate. Studies by Olausson [1965], Broecker [1971] and Berger [1973] suggest that changes in the rate of dissolution are also important in controlling carbonate deposition. However, analysis of cores from above the zone of marked carbonate dissolution, the lysocline, suggests that variation in carbonate productivity may be important [Adelseck and Anderson, 1978]. Boyle [1983] estimated that changes in local productivity of the eastern equatorial Pacific could account for over 50% of the changes in the local carbonate accumulation during the last 130,000 years. Thus changes in both the rate of carbonate supply and dissolution play important roles in controlling the variations in carbonate deposition.

Most studies have concentrated on the carbonate fraction, while few have quantitatively examined the noncarbonate fraction in equatorial Pacific sediments. In regions away from terrigenous sediment sources, the noncarbonate fraction in the equatorial Pacific is dominated by biogenic silica (opal) produced predominantly by diatoms and Radiolaria. Like calcium carbonate, the deposition of opal is controlled by changes in surface productivity and recycling by dissolution. Thus variations in the cycling of biogenic silica through changes in productivity and dissolution are also reflected in percent variation in the calcium carbonate content of deep-sea sediments.

In this study we present a detailed analysis of Deep Sea Drilling Site 572 from the equatorial

Pacific that allows us to define the degree of variability in both the amount and rate of carbonate deposition in pre-Pleistocene sections. Thus we can extend the high-resolution analysis of Arrhenius [1952], Hays et al. [1969], and others through the Pliocene and into the late Miocene with a sampling resolution of 3000 to 5000 years.

## High-Resolution Carbonate Data
## From HPC Site 572

Leg 85 of the Deep Sea Drilling Project drilled five long sections in the equatorial Pacific. Detailed sampling of Site 572 (01°26.09'N, 113°50.52'W; water depth 3903 m) provides data on the nature of carbonate variations during the latest Miocene and Pliocene with a sampling resolution of 3000 to 5000 years. At Site 572 the hydraulic piston corer recovered sediments to 154 m subbottom and provided a high-quality, undisturbed sediment section spanning the last 6 m.y. Site 572 has always been within the equatorial divergence zone. Hence the intervals studied at this site reflect depositional processes associated with equatorial divergence and related current systems.

To examine the nature of carbonate variability and its response to oceanographic and climatic variability in the late Miocene and Pliocene, selected intervals were sampled every 10 cm. The intervals were selected on the basis of (1) high sedimentation rates so that high-resolution records could be obtained; (2) shipboard evaluation of carbonate fossil preservation to ensure adequate numbers of foraminiferal tests for later stable isotopic analyses; (3) intervals which spanned major paleoclimatic or paleoceanographic events; and (4) at least 15 m of good continuous recovery so that it could be sampled to provide a long enough sequence for detailed statistical analysis in both time and frequency domains.

Three intervals from Site 572 were sampled in detail, 572C cores 4 to 7 (29 to 68 m subbottom), 572A cores 11 and 12 (92 to 110 m subbottom), and 572C cores 15 and 16 (121 to 134 m subbottom). The first samples the interval of 2 to 4.6 Ma and spans the time when major glaciation allegedly began in the northern hemisphere. A detailed carbonate stratigraphy and stable isotopic analysis of this section are presented by Prell [1984]. Calcium carbonate analyses were completed using a modified version of the Jones and Kaiteris [1983] vacuum-gasometric technique for samples from cores 11 and 12 and using the carbonate bomb technique as modified by Dunn [1980] for samples from cores 15 and 16. Analytic precisions for these techniques are 0.5% and 1%, respectively [Pisias and Prell, 1984].

Sediment accumulation rates were estimated for Site 572 from diatom and radiolarian biostratigraphic datums [Mayer et al., 1984]. Only datums which have been correlated to paleomagnetic time scale were used to estimate sedimentation rates.

The sedimentation rate in the interval between 29 and 68 m subbottom is about 15 m/m.y. so that the sampling resolution is on the order of 5000 years. The sedimentation rate below 68 m subbottom is estimated to be 50 m/m.y. [Mayer et al., 1984]. Thus the detailed intervals between 92 to 110 m and 121 to 134 m subbottom provide very high resolution records of carbonate variations in the late Miocene. These intervals sample the time periods of about 4.6 to 5 Ma and 5.2 to 5.3 Ma, respectively, with a sample resolution of 2000 years.

## Results

Analysis of the sediment accumulation rate data from Site 572 reveals that a marked change in the mean accumulation rate occurred at about 4 Ma, with carbonate mass accumulation rates decreasing by a factor of 3 and noncarbonate rates decreasing by a factor of 5 (Table 1). The detailed carbonate data reveal that a change in the variability of carbonate content is associated with the marked decrease in sedimentation rates. The depth series of percent carbonate variations are similar in their general appearance and statistical properties (Figures 1 and 2; Table 2). The mean values for four intervals range from 69 to 79% CaCO$_3$, with the standard deviations of the intervals ranging from 6.6 to 9.2% carbonate. The interval from 30 to 68 m was divided into two parts, with the number of data points in each approximately equal to the other depth intervals sampled. Differences between the extreme values of the mean and variance are significant at the 10% level only based on pair-wise Student's t and F tests (Table 2). The average wavelength of variations in all sections is on the order of 20 to 50 samples per cycle, with secondary fluctuations on the order of 10 samples per cycle.

Although the depth series are remarkably similar, the time series reveal striking differences in the frequency of carbonate variations between the intervals from 4.4 to 5.2 Ma and from 2 to 4 Ma (Figure 3). These differences are best quantified in the frequency domain. The variance spectra for the intervals 2 to 3 Ma and 4.7 to 5 Ma are shown in Figure 4. The results of analyses of all four intervals represented in Table 2 are summarized in Table 3. The sample resolution in the 2- to 3-Ma interval is on the order of 5000 to 6000 years, so that the highest frequency resolved is equivalent to a 10,000- to 12,000-year period. In this interval (2 to 3 Ma), the spectrum is dominated by three spectral components centered at periods of about 300,000, 67,000, and 35,000 years (Figure 4a). Analysis of both the 2- to 3-Ma and 3- to 4-Ma intervals shows that the amplitudes of the carbonate variation associated with these components (where amplitude is defined as 2 x [variance]$^{1/2}$ associated with the spectral component) are 8 to 12%, 7.5 to 9%, and 2 to 5%, respectively.

TABLE 1.    Sedimentation and Mass Accumulation Rate Estimates
for HPC Section of Site 572

| Subbottom Depth m | Age, Ma | Rate, m/my | Mean Density, g/cm³ | Mean CaCO₃, % | Mass Accumulation Rates, g cm⁻² (1000 years)⁻¹ CaCO₃ | Non-CaCO₃ |
|---|---|---|---|---|---|---|
| 9.5 | 0.55 | | | | | |
| | | 13 | 0.57 | 79 | 0.58 | 0.16 |
| 29.0 | 2.0 | | | | | |
| | | 21 | 0.61 | 73 | 0.93 | 0.35 |
| 49.0 | 2.95 | | | | | |
| | | 15.2 | 0.70 | 75 | 0.80 | 0.20 |
| 68.0 | 4.2 | | | | | |
| | | 50.0 | 0.72 | 73 | 2.60 | 1.00 |
| 208.0 | 6.95 | | | | | |

In the interval of 4.7 to 5.0 Ma the spectrum shows a significant amount of variance associated with very high frequencies (Figure 4b). From 12 to 20% of the total variance in the intervals 4.7 to 5 Ma and 5.2 to 5.6 Ma is found at frequencies equivalent to 16,000-year periods. The dominant spectral component, containing 50 to 60% of the variance, centers at periods of about 40,000 to 50,000 years. In these late Miocene records the resolution of long periods is hindered by the limited length of record (about 400,000 years); however, the fact that the spectra do not show a continual increase in variance with decreasing frequency of periods over 40,000 years suggests that long periods are not major contributors to the variance of the intervals analyzed.

To determine whether the differences between these intervals represent real changes in the variability of the carbonate system of the late Miocene and Pliocene, we must consider a number of problems. Sample aliasing (the effect of sampling a "high" frequency signal at too low a resolution) may be important, since the sampling resolution of the Site 572 data varies by a fac-

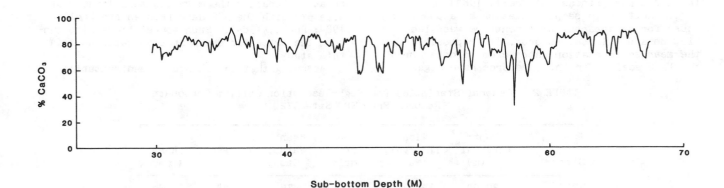

Fig. 1. Calcium carbonate versus depth in HPC Site 572. Detailed carbonate data from Hole 572C are shown as a solid line. The dashed line shows shipboard analyses from Hole 572A. Biostratigraphy is from Mayer et al. [1984].

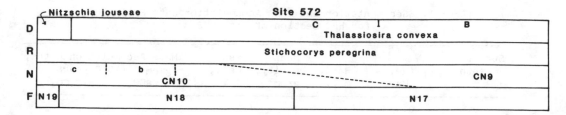

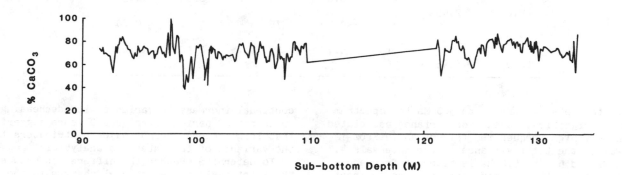

Fig. 2.   Calcium carbonate versus subbottom depth in HPC Site 572.   Detailed carbonate data from hole 572A (92 to 110 m) and 572C (121 to 134 m) are shown as solid line.   The dashed line shows shipboard analyses from Hole 572A.   Biostratigraphy is from Mayer et al. [1984].

tor of 3.  Theoretically, the sample spacing of 5000 years in the interval of 2 to 4 Ma should resolve periods on the order of 10,000 years. Since no evidence of important periods shorter than 35,000 years exists in the Pliocene records, either in terms of significant spectral peaks or in the amount of variance accounted for by the high frequencies, it is unlikely that aliasing is an important problem in the Pliocene section of HPC Site 572.  Additional detailed sampling in the intervals 30 to 33 m and 35 to 36.5 m subbottom also suggests that sample aliasing is not a major contributor to the observed differences in these records [Pisias and Prell, 1984]

In addition to sample aliasing as a possible cause for an apparent difference between the Miocene and Pliocene records, it is possible that the mean sedimentation rates used to determine the time scales for these records have signifi-

cant errors.  Magnetic stratigraphy is not available for Site 572, and sedimentation rate estimates were made from biostratigraphic datums which have been correlated to the geomagnetic time scale.  The rate estimates used here were made from a combination of radiolarian and diatom datums which have been correlated to the magnetic time scale by a number of investigators [see Barron et al., 1984].  Stratigraphic analysis of all sites drilled on Leg 85 suggests that correlations to within a few meters are possible throughout the Pliocene and Miocene sections of Sites 573 and 572.  Thus it is possible to correlate accurately the magnetic stratigraphy of Site 573 with the 572 data [Pisias and Prell, 1984; Prell, 1984].  Such correlations also support the estimated sedimentation rates used in this study.

Assuming that the change in sedimentation

TABLE 2.   General Statistics for High-Resolution Calcium Carbonate
Records From HPC Site 572

| Site | Depth Interval, (m) | Time Interval, (Ma) | Number of Samples | Mean Value, % CaCO$_3$, | Variance, $\%^2$ | Range, % CaCO$_3$ |
|------|------|------|------|------|------|------|
| 522C | 30-50 | 2-3 | 178 | 75 | 44 | 35 |
| 572C | 50-68 | 3-4 | 180 | 79 | 85 | 59 |
| 572A | 92-110 | 4.7-5 | 178 | 69 | 75 | 61 |
| 572C | 121-134 | 5.2-5.6 | 141 | 75 | 45 | 37 |

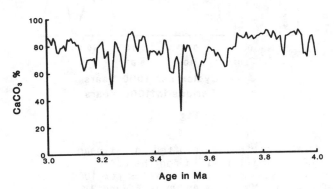

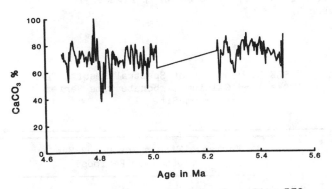

Fig. 3. Detailed carbonate data from Site 572 versus age. Age estimates are from Mayer et al. [1984].

rates estimated between the intervals of 68 to 208 m and 29 to 68 m subbottom is correct, the mass accumulation rate of calcium carbonate changes from 3.5 to 1 g cm$^{-2}$ (1000 years)$^{-1}$, respectively. Associated with this threefold decrease in mass accumulation rate (Table 1) is a marked decrease in high-frequency variations in carbonate content. Therefore if the carbonate variations are converted to mass accumulation rates, the late Miocene carbonate record at Site 572 requires not only high-frequency variations of carbonate content but also very high amplitude variations in the total mass of carbonate dissolved or preserved as compared to the Pliocene interval.

## Discussion

### Causes of High-Frequency Carbonate Fluctuations: A Mass Accumulation Rate Approach

To examine the processes related to the high-frequency variations in calcium carbonate, we have used a different approach for describing changes in sediment carbonate content from previous workers. A useful review of the hypotheses for variation in calcium carbonate concentrations is given by Volat et al. [1980]. The weight percent of calcium carbonate in pelagic sediments reflects the balance between the rate of input of calcium carbonate to the sediment surface, the rate of carbonate dissolution, and the rate of accumulation of non-$CaCO_3$ dilutants. The effects of the factors controlling carbonate deposition can be modeled by assuming the initial composition of the pelagic rain [Heath and Culberson, 1970; Berger, 1973; Dean et al., 1981; Gardner, 1982]. The simulation in Figure 5 assumes that the initial pelagic rain is 85% $CaCO_3$ and 15% noncarbonate. Variations in the concentration of calcium carbonate can result from changes in dilution by non-$CaCO_3$ or changes in the percentage of calcium carbonate dissolved. The percent of dilution, D, needed to produce the observed calcium carbonate, Cf, from the initial carbonate content, Ci [Heath and Culberson, 1970], is

$$D = \frac{100(Ci/Cf) - 1}{1 - Ci} \qquad (1)$$

The percent carbonate lost to dissolution, L, needed to produce the observed noncarbonate fraction, Nf, from the initial fraction, Ni, is

$$L = \frac{100[1 - (Ni/Nf)]}{1 - Ni} \qquad (2)$$

This formulation makes the assumption that dissolution and dilution are the primary controls of carbonate variations. As discussed below, this approach also makes important assumptions about the relationship between rate of dissolution and changes in total mass accumulation rate.

This model shows that a change in the observed carbonate content from the initial 85% to a value of 80% by dissolution only requires about 30% of the initial carbonate to be dissolved. If we compare two different sediment records, as in the case of the Pliocene and Miocene of Site 572, and calculate the percent dissolution required to describe the individual records, we have inherently assumed that dissolution is the only process responsible for changes in carbonate content (i.e., no change in supply or dilution by other sediment fractions) and that the system is steady state over each interval. This implies that, for a given change in $CaCO_3$ content, the proportion of $CaCO_3$ that has been dissolved is constant and is independent of the absolute rate of dissolu-

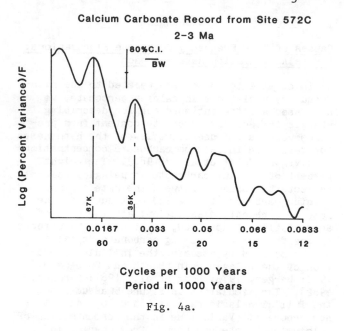

Fig. 4a.

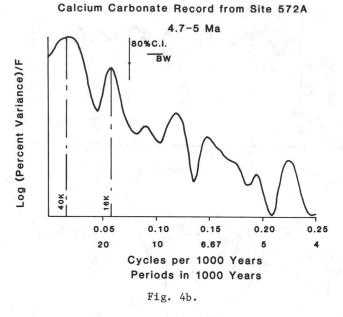

Fig. 4b.

Fig. 4. Variance spectra of Site 572 carbonate data. Spectra are plotted on a linear frequency scale and a log variance scale. The confidence interval is for the 80% level; BW gives band width of spectral calcuations. Frequencies are in cycles per 1000 years, and periods in 1000 years. (a) Spectra for 2 to 3 Ma (data shown in Figure 3). (b) Spectra for 4.7 to 5 Ma (data shown in Figure 3).

tion. For example, the change from 85 to 80% CaCO$_3$ would require a dissolution rate 3 times higher for a sediment column with a carbonate deposition rate of 3 g cm$^{-2}$ (1000 years)$^{-1}$ than for one with a rate of 1 g cm$^{-2}$ (1000 years)$^{-1}$. For the Site 572 data this assumption of dissolution-controlled variation suggests that the amplitude of variation in CaCO$_3$ percentages of the late Miocene, which is similar to the Pliocene data, represents a calcium carbonate dissolution rate 3 times greater than that needed to produce similar variations of the Pliocene. Thus not only was the wavelength of carbonate variations much shorter during the Miocene, but also the variation in the total mass of carbonate dissolved was significantly higher.

Since the mass of carbonate preserved or the rate of dissolution provides more information than CaCO$_3$ percentages, we suggest a different approach to describing changes in carbonate content. Here we describe changes in CaCO$_3$ concentration in terms of changing rates of dissolution (expressed in grams of CaCO$_3$ per square centimeter per 1000 years) rather than changing proportions of dissolution. This approach reflects our belief that changes from the mean concentration of calcium carbonate caused by specific changes in oceanographic processes are best described in terms of changing rates of dissolution rather than simply changes in the relative proportion of carbonate preserved at any one location.

TABLE 3. Summary of Spectral Results for Detailed Calcium Carbonate Time Series From Site 572

| Interval, Ma (m) | Variance Within Frequency Bands: (1000-Year Periods) | | | | |
|---|---|---|---|---|---|
| | ∞–142 | 142–42 | 42–30 | Total | Sample Interval Years |
| 2 to 3 (30–50) | 15 | 14 | 2 | 35 | 6000 |
| 3 to 4 (50–68) | 37 | 21 | 6 | 71 | 6000 |

| Interval, Ma (m) | Variance Within Frequency Bands: (1000-Year Periods) | | | | |
|---|---|---|---|---|---|
| | 142–23 | 23–12.5 | 12.5–7 | Total | Sample Interval Years |
| 4.7 to 5 (92–110) | 34 | 10 | 6 | 62 | 2000 |
| 5.2 to 5.6 (121–134) | 20 | 8 | 3 | 41 | 2000 |

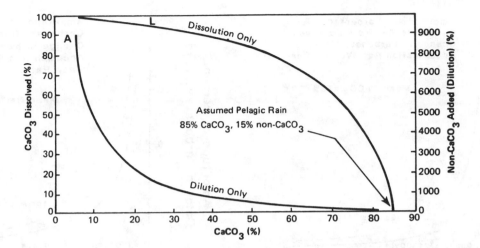

Fig. 5.  Curves for the loss of carbonate by dissolution (L) and addition of noncarbonate dilutants (A) to an initial carbonate rain of 85% [after Dean et al., 1981]

To examine an important consequence of this approach, consider the amplitude of carbonate variations associated with the 67,000-year spectral component of the 2- to 3-Ma carbonate record of Site 572 (Figure 4a). The amplitude of this frequency component is about 10% CaCO in an interval with a mean sediment accumulation rate of 1.0 g cm$^{-2}$ (1000 years)$^{-1}$ and a mean carbonate content of 75% (Table 1). Assuming that the supply of non-CaCO$_3$ sediment has remained constant, a change in the mass accumulation of CaCO$_3$ of 0.2 g cm$^{-2}$ (1000 years)$^{-1}$ is required to produce the observed 10% carbonate content signal. Unlike the approach described by equations (1) and (2), we have not made an assumption about the initial carbonate content of the pelagic rain. The value of 0.2 g cm$^{-2}$ (1000 years)$^{-1}$ is the amplitude of the change in accumulation rate associated with the 67,000-year frequency component and may reflect changes in dissolution as well as production. If the total mass accumulation rate was 3.5 g cm$^{-2}$ (1000 years)$^{-1}$, as in the 5.2- to 5.6-Ma interval, the 0.2 g cm$^{-2}$ (1000 years)$^{-1}$ change in carbonate deposition would produce a carbonate content change of only 4%.

The relationship between the amplitude of the change in the carbonate concentration and the mean mass accumulation rate can be calculated for a given mean carbonate percentage (75% is used in Figure 6a). Two values for the net change in carbonate mass accumulation rates (0.20 and 0.35 g cm$^{-2}$ [1000 years]$^{-1}$) are shown in Figure 6a. A mean of 75% was chosen because this is a typical value for the mean carbonate value of all sites drilled on Leg 85 in the equatorial Pacific [Mayer et al., 1984]. As shown in Figure 6a, for mass accumulation rates from 0.5 to 2.0 g cm$^{-2}$ (1000 years)$^{-1}$ (sedimentation rates on the order of 1 to 3 cm [1000 years]$^{-1}$), a 0.2 g cm$^{-2}$ (1000 years)$^{-1}$ rate change in carbonate deposition is

reflected as a change in carbonate concentration which varies by a factor of about 5.

Thus far we have assumed that changes in carbonate concentrations reflect changes in net carbonate mass accumulation rates (i.e., our supply-dissolution model). If we consider the alternate case that changes in calcium carbonate content reflect changes in noncarbonate (dilution), we can calculate the change in the mass accumulation necessary to produce carbonate fluctuations. Using the same starting values of a mean mass accumulation rate of 1.0 g cm$^{-2}$ (1000 years)$^{-1}$ and a mean carbonate content of 75%, a change in net noncarbonate accumulation of 0.07 g cm$^{-2}$ (1000 years)$^{-1}$ is required to produce a 10% CaCO$_3$ amplitude in the carbonate record. Although this value is less than half of the change needed if only variations in carbonate supply and dissolution are considered, it represents 30% of the mean noncarbonate accumulation. As shown in Figure 6b, the amplitude of carbonate variations caused by changes in noncarbonate deposition also varies as a function of total mean accumulation. An increase in the amplitude of carbonate content variations with decreasing sedimentation rates is produced by variations in noncarbonate accumulation rates.

Without detailed studies of the degree of carbonate dissolution in the sediment from Site 572 or detailed time control to constrain very short term changes in sedimentation rates, it is difficult to evaluate the ultimate cause of the carbonate content fluctuations in this equatorial Pacific site. One possible constraint, however, is supplied by the variance spectrum calculated from these data. If changes in the CaCO$_3$ records reflect changes in carbonate supply and/or dissolution, then we would expect significant changes in the short-term mass accumulation rates. In Figure 7 the percent change in total accumulation

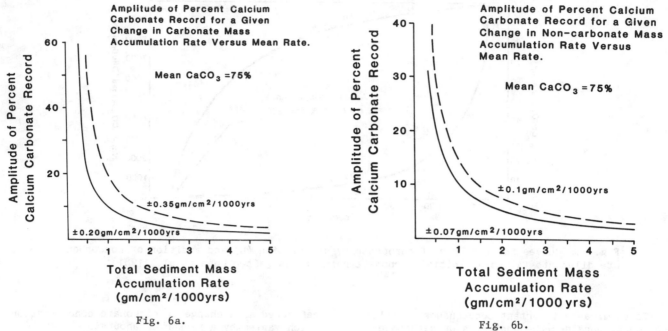

Fig. 6a.

Fig. 6b.

Fig. 6.   Amplitude of the carbonate concentration record versus total mean accumulation rates for a given response in the net accumulation rate of carbonate or noncarbonate. (a) Calculations made assuming a mean carbonate concentration of 75%, a constant noncarbonate flux and a change in net carbonate deposition of 0.20 (solid line) and 0.35 (dashed line) g cm$^{-2}$ (1000 years)$^{-1}$. (b) Calculations made assuming a mean carbonate concentration of 75%, a constant carbonate flux and a change in net noncarbonate accumulation of 0.07 (solid line) and 0.1 (dashed line) g cm$^{-2}$ (1000 years)$^{-1}$.

rate associated with the estimated 0.2 and 0.07 g cm$^{-2}$ (1000 years)$^{-1}$ rate changes in carbonate and noncarbonate accumulation is plotted as a function of mean sedimentation rates.  For a mean of 1 g cm$^{-2}$ (1000 years)$^{-1}$ changes in noncarbonate deposition would produce a 15% change in total accumulation as opposed to a 40% change if changes in carbonate deposition are controlling carbonate concentrations.

In Figure 3 the carbonate minima tend to occur over shorter depth intervals relative to maxima. This could reflect reduced sedimentation rates during intervals of decreased carbonate concentration.   This type of sedimentation rate distortion would produce harmonics in the variance spectra associated with important periodic components (R. Holman and N. Pisias, unpublished manuscript, 1984).   In the 2- to 3-Ma interval of the Site 572 data, the secondary spectral peak (35,000 years) is very close to the harmonic of the 67,000-year component.  If these are harmonic components and reflect changes in carbonate accumulation rates at the primary frequency of 67,000 years, then given the observed amplitudes of the primary frequency, the equations developed by R. Holman and N. Pisias (unpublished manuscript, 1984) predict a ratio of 2.63 between the amplitudes of the primary and harmonic components.  The observed ratio of 2.66 suggests that variation in carbonate accumulation rates in the

Pliocene section of Site 572 is the primary factor controlling the carbonate concentrations. Whether changes in carbonate accumulation at this site are controlled by variation in dissolution or productivity is yet to be determined.

The variance spectra of the late Miocene data show little evidence of harmonic components.  The two dominant spectral peaks are centered at 40,000 and 16,000 years, which are not harmonics. This suggests that the variations in carbonate concentration are not associated with marked distortion in sedimentation rates.  Although we cannot rule out variations in CaCO$_3$ accumulation rates, the absence of harmonic components and the calculations presented in Figure 7 suggest that variations in opal accumulation play an important part in controlling variations in carbonate content of the late Miocene section of Site 572.  In either case the data from Site 572 clearly show that the frequency of variation and the magnitude of fluctuations in carbonate accumulation were significantly different in the late Miocene as compared to the Pliocene.

Causes of Major Change in Carbonate Accumulation at Site 572:  Regional Versus Local Processes

To place the sediment accumulation rate change observed at Site 572 into a regional perspective, we have used the carbonate accumulation maps of

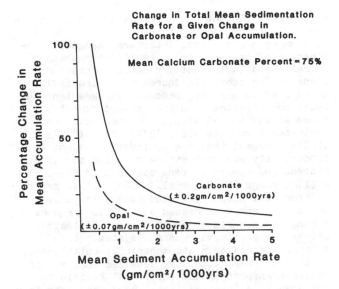

**Change in Total Mean Sedimentation Rate for a Given Change in Carbonate or Opal Accumulation.**

**Mean Calcium Carbonate Percent=75%**

Fig. 7. Percent change in mean accumulation rate versus mean accumulation rate for a 0.07 g cm⁻² (1000 years)⁻¹ change in the noncarbonate accumulation rate (dashed line) and a 0.2 g cm⁻² (1000 years)⁻¹ change in the carbonate accumulation rate (solid line).

van Andel et al. [1975] to examine changes in the total mass of carbonate deposited in the equatorial Pacific during the Cenozoic. The comparison of sedimentation rates estimated by van Andel et al. and the rate estimates for Site 572 must be viewed with some caution given the developments in geochronology since 1976. However, the pattern of sedimentation rate changes observed in the HPC drill sites of the equatorial Pacific [Mayer et al., 1984] is very similiar to the pattern estimated by van Andel et al. [1975].

In Table 4 we list the total carbonate accumulation in the equatorial Pacific as a function of time. These estimates were made from the isopleth maps of carbonate accumulation rates [van Andel et al., 1975, Figure 38]. Total carbonate mass accumulation rates were determined for the equatorial region between 5°N and 5°S and bounded east-west by 100°W and 155°W. For most of the late Cenozoic, very little carbonate accumulation is observed north or south of this latitudinal band. Prior to 14 Ma, however, significant carbonate accumulation is observed north and south of the latitude limits used here. Important areas of carbonate accumulation in the Pacific not included in these calculations are the western equatorial region, the southwest Pacific, the Panama Basin in the eastern equatorial Pacific, and the East Pacific Rise in the southeast Pacific. The North Pacific is not included in these calculations, but data from recent sedimentation studies suggest that carbonate accumulation in this region is much less important. Analysis of surface sediments north of 10°N shows

that over 80% of all samples contain less than 10% calcium carbonate [Berger et al., 1976].

Although the equatorial Pacific is a major area of carbonate accumulation, the total carbonate flux calculated for the equatorial region represents less than 10% of the total Pacific carbonate accumulation estimated by Worsley and Davies [1981]. Expanding the region integrated on the carbonate maps of van Andel et al. [1975] does not change this difference significantly. For example, the estimated carbonate flux for the total Pacific during the last 3 m.y. is 4 x 10¹⁴ g yr⁻¹, whereas we estimate that the flux in the equatorial regions is on the order of 0.3 x 10¹⁴ g yr⁻¹. Thus over 13 times more carbonate must be accumulating in other parts of the Pacific than has been observed in the equatorial region, assuming all these estimates are correct. To reconcile these data, we calculated the area of the deep Pacific basin where surface sediments contained greater than 50% calcium carbonate from the maps of Berger et al. [1976]. If the accumulation rates in the nonequatorial region averaged 2 g cm⁻² (1000 years)⁻¹, then the net carbonate flux in the Pacific would be on the order of 4 x 10¹⁴ g yr⁻¹. This average sediment accumulation rate is high for the southeast Pacific, the largest region of carbonate accumulation outside the equatorial band. Kendrick [1974] estimated a sediment accumulation rate of 1.5 g cm⁻² (1000 years)⁻¹ at 11°S on the crest of the East Pacific Rise. Away from the rise crest, however, rates fall off very rapidly to approximately 0.5 g cm⁻² (1000 years)⁻¹ [Leinen et al., 1984]. Without detailed rate data from the western and southwestern Pacific, it is difficult to evaluate whether the estimated Pacific carbonate flux of Worsley and Davies [1981] and the rate estimated for the equatorial Pacific can be reconciled. It is possible that the extrapolations made by Worsley and Davies produced significant errors and the net carboate flux in the Pacific has been overestimated.

The calculated fluxes within the equatorial band show that total carbonate accumulation has

TABLE 4. Total Carbonate Accumulation From 5°N to 5°S and 100°W to 155°W

| Interval, Ma | Total Carbonate Rate 10¹⁴ g (1000 years)⁻¹ |
|---|---|
| 0-1 | 337 |
| 3-5 | 295 |
| 6-7 | 281 |
| 10-11 | 126 |
| 14-15 | 916 |
| 18-19 | 437 |
| 21-22 | 755 |
| 25-26 | 318 |
| 28-30 | 1074 |

Data are from van Andel et al. [1975, Figure 38].

remained relatively constant since the late Miocene. Between the late Miocene (6 to 7 Ma) and the Pliocene (3 to 5 Ma), the total change is less than 5%. This is in marked contrast to the threefold decrease in carbonate accumulation observed at Site 572. Noting that the resolution provided by the van Andel et al. [1975] data is much lower than that from Site 572, we conclude from these data that the marked change in carbonate accumulation at Site 572 between late Miocene and Pliocene reflects changes in local oceanographic processes that affected the areal distribution of carbonate accumulation, rather than a change in the total ocean carbonate budget that would have resulted in a change in the total carbonate flux in the equatorial Pacific. The fact that the sedimentation rate change occurred in both carbonate and noncarbonate (mostly biogenic silica) suggests that surface oceanographic processes and biogenic production played an important role in controlling the sedimentation rate changes observed in Site 572.

With the available data we can only speculate on the causes of this rate change in the pattern of carbonate accumulation. The timing of the sedimentation rate change (4 Ma) is considerably older than the the major steps in global climate, the 3.2-Ma shift in benthic isotopes associated with a change in deep-water circulation [Prell, 1984] or the onset of major glaciation at 2.4 Ma [Shackleton et al., 1984]. The timing of the sedimentation rate change is very close to the estimated time of the closure of the Panamanian seaway. The closing of Panama has been estimated to have occurred between 3.2 and 3.8 Ma [Keigwin, 1982]. The effect of a shoaling land mass at the isthmus of Panama on equatorial circulation, however, is as uncertain as the nature of equatorial circulation prior to 3.8 Ma. Therefore the nature of sedimentation rate changes and available knowledge of equatorial circulation allow some speculation.

The strength and pattern of currents in the equatorial Pacific today reflect atmospheric circulation and its control of surface wind strength and the position of the Intertropical Convergence. In addition to the surface circulation pattern, an important feature of the equatorial Pacific circulation is the slope of the thermocline along the equator. In general, the thermocline is relatively shallow (~50 m) in the east and becomes deeper towards the west, where its depth is on the order of 150 m. The wind-induced divergence along the equator causes upwelling of water from the upper 100 m and combined with the sloping thermocline results in an east-west gradient in surface productivity. In the west, where upwelling involves nutrient-depleted water from above the thermocline, productivity is reduced compared to the east, where upwelling brings nutrient-rich water from below the thermocline into the surface region. The slope of the thermocline along the equator changes with the changing intensity of westward

transport along the equator that is controlled in turn by surface wind intensity.

Thus we might expect long-term changes in the strength of the equatorial Pacific circulation to be reflected in the pattern of sedimentation along the equator. An increase in circulation would lead to a strong productivity gradient and would be reflected in differing sedimentation rates at equatorial sites. Sedimentation rate estimates from Site 572 (113°W) and Site 573 (133°W) suggest that the east-west gradient in productivity was much stronger during the late Miocene and early Pliocene compared to the last million years [Mayer et al., 1984]. This may be related to a less restricted Panamanian seaway, but the inferred increased circulation requires strong atmospheric circulation that ultimately drives the equatorial current system. This inferred increase in atmospheric circulation must be tested. Unfortunately, detailed analysis of eolian transport in the equatorial Pacific does not span this interval [Rea, 1982], and data from the North Pacific lack the resolution neccessary to resolve this interval [Rea and Janecek, 1982]. Again we point out that, although it is not possible to state the processes associated with the changes in the carbonate accumulation pattern in the Pacific and their relation to changes in the circulation patterns, these changes did not result in a change in the total CaCO$_3$ accumulation in the equatorial Pacific. The sedimentation rate change at Site 572 reflects local processes and not necessarily a change in global carbonate deposition.

## Conclusions

Analysis of sedimentation rates and variations in the carbonate content of sediments at Site 572 lead to the following conclusions:

1. At about 4.2 Ma a significant change in both total and carbonate mass accumulation occurred at this equatorial site.

2. Associated with this change in sedimentation rates is a marked change in the dominant wavelengths of variation in the carbonate records. Between the late Miocene and Pliocene the wavelength of carbonate variations changes from frequencies on the order of 140,000-year periods to about 40,000-year periods.

3. Modeling carbonate variations in terms of mass accumulation rate changes combined with time series analysis shows that during the Pliocene the data are consistent with the hypothesis that changes in carbonate accumulation control the carbonate records. During the late Miocene, changing accumulation of noncarbonate (mostly biogenic silica) sediment plays an important role in controlling the carbonate records.

4. The change in accumulation rates at 4.2 Ma is not reflected in the total mass of calcium carbonate being deposited in the equatorial Pacific. Thus the accumulation rates reflect a change in the regional pattern of deposition of

calcium carbonate and not necessarily a change in the calcium carbonate budget of the oceans.

5.  The 4.2-Ma age of the transition in sediment accumulation suggests that the closing of the Panamanian seaway may have had an important role in controlling the observed changes in equatorial Pacific sedimentation, although the data do not allow definitive conclusions about the nature of oceanic circulation prior to the closing of the equatorial seaway.

Acknowledgments.  This research was supported by National Science Foundation grants ATM80-18897 and ATM82-45480 (SPECMAP) through the Climate Dynamic Section, Division of Atmospheric Sciences, and OCE83-15471 through the Submarine Geology and Geophysics Section of Ocean Sciences.

## References

Adelseck, C. G., Jr., and T. F. Anderson, The late Pleistocene record of productivity fluctuations in the eastern equatorial Pacific Ocean, Geology, 6, 388-391, 1978.

Arrhenius, G., Sediment cores from the east Pacific.  Properties of the sediment, Rep. Swed. Deep Sea Exped., 1947-1948, 1, 1-228, 1952.

Barron, J. A., C. A. Nigrini, A. Pujos, T. Saito, F. Theyer, E. Thomas, and N. Weinreich, Biostratigraphic summary and synthesis of the central equatorial Pacific, DSDP, Leg 85, Initial Rep. Deep Sea Drill. Proj., 85, in press, 1984.

Berger, W. H., Deep-sea carbonates: Pleistocene dissolution cycles, J. Foraminiferal Res., 3, 187-195, 1973.

Berger, W. H., C. G. Adelseck, and L. A. Mayer, Distribution of carbonate in surface sediments of the Pacific Ocean, J. Geophys. Res., 81, 2617-2640, 1976.

Boyle, E. A., Chemical accumulation variations under the Peru Current during the past 130,000 years, J. Geophys. Res., 88, 7667-7680, 1983.

Broecker, W. S., Calcite accumulation rates and glacial to interglacial changes in oceanic mixing, in Late Cenozoic Glacial Ages, edited by K. K. Turekian, pp. 239-266, Yale University Press, New Haven, Conn., 1971.

Dean, W. E., J. V. Gardner, and P. Cepek, Tertiary carbonate-dissolution cycles on the Sierra Leone Rise, eastern equatorial Atlantic Ocean, Mar. Geol., 39, 81-101, 1981.

Dunn, D. A., Revised technique for quantitative calcium carbonate analysis using the "Karbonat-Bombe" and correlations to other quantitative carbonate analysis methods, J. Sediment. Petrol., 50, 631-637, 1980.

Gardner, J. V., High-resolution carbonate and organic-carbon stratigraphies for the late Neogene and Quaternary from the eastern equatorial Pacific, Initial Rep. Deep Sea Drill. Proj., 68, 347-364, 1982.

Hays, J. D., T. Saito, N. D. Opdyke, and L. H. Burckle, Pliocene-Pleistocene sediments of the equatorial Pacific:  Their paleomagnetic, biostratigraphic, and climatic record, Geol. Soc. Am. Bull., 80, 1481-1514, 1969.

Heath, G. R., and C. Culberson, Calcite: Degree of saturation, rate of dissolution and compensation depth in the deep oceans, Geol. Soc. Am. Bull., 81, 3157-3160, 1970.

Jones, G. A., and P. Kaiteris, A vacuum-gasometric technique for rapid and precise analysis of calcium carbonate in sediments and soils, J. Sediment. Petrol., 53, 655-659, 1983.

Keigwin, L., Isotope paleoceangraphy of the Caribbean and east Pacific: Role of Panama uplift in late Neogene time, Science, 217, 350-353, 1982.

Kendrick, J. W., Trace element studies of metalliferous sediments in cores from the East Pacific Rise and Bauer Deep, 10°S, M.S. thesis, 117 pp., Oregon State Univ., Corvallis, 1974.

Leinen, M. S., et al. (Eds.), Initial Reports of the Deep Sea Drilling Project, vol. 92, U.S. Government Printing Office, Washington, D. C., in press, 1984.

Mayer, L., et al. (Eds.), Initial Reports of the Deep Sea Drilling Project, vol. 85, U.S. Government Printing Office, Washington, D. C., in press, 1984.

Olausson, E., Evidence of climatic changes in North Atlantic deep-sea cores, with remarks on isotopic paleotemperature analysis, Prog. Oceanogr., 3, 221-252, 1965.

Pisias, N. G., and W. L. Prell, High resolution carbonate records from HPC Site 572, Initial Rep. Deep Sea Drill. Proj., 85, in press, 1984.

Prell, W. L., Stable isotope and carbonate stratigraphy for the Pliocene of Hole 572C: Paleoceanographic data bearing on the question of Pliocene glaciation, Initial Rep. Deep Sea Drill. Proj., 85, in press, 1984.

Rea, D. K., Fluctuations in eolian sedimentation during the past five glacial-interglacial cycles: A preliminary examination of data from Deep Sea Drilling Project Hole 503B, eastern equatorial Pacific, Initial Rep. Deep Sea Drill. Proj., 68, 409-418, 1982.

Rea, D. K., and T. R. Janecek, Late Cenozoic changes in atmospheric circulation deduced from North Pacific eolian sediments, Mar. Geol., 49, 149-167, 1982.

Shackleton, N. J., et al., Oxygen isotope calibration on the onset of ice-rafting and history of glaciation in the North Atlantic, Nature, 307, 620-623, 1984.

van Andel, T., G. R. Heath, and T. C. Moore, Jr., Cenozoic history of the central equatorial Pacific Ocean, Mem. Geol. Soc. Am., 143, 1-134, 1975.

Volat, J. L., L. Pastouret, and C. Vergnaud-Grazzini, Dissolution and carbonate fluctuations in Pleistocene deep-sea cores: A Review, Mar. Geol., 34, 1-28, 1980.

Worsley, T., and T. Davies, Paleoenvironmental implications of oceanic carbonate sedimentation rates, in Deep Sea Drilling Project: A Decade of Progress, Spec. Publ. 32, edited by J. Warme, R. Douglas, and E. L. Winterer, pp. 169-180, Society of Economic Paleontologists and Mineralogists, Tulsa, Okla., 1981.

# CARBON DIOXIDE AND POLAR COOLING IN THE MIOCENE:
## THE MONTEREY HYPOTHESIS

Edith Vincent and Wolfgang H. Berger

Scripps Institution of Oceanography, University of California, San Diego
La Jolla, California 92093

Abstract. A pronounced shift in the $\delta^{13}C$ of foraminifera in the latest early Miocene has been proposed by various authors. Our data in the tropical Indian Ocean show an excursion of $\delta^{13}C$ signals toward heavier values, lasting for about 4 million years. The excursion is documented for benthic foraminifera as well as for deep-living and for shallow-dwelling planktonic species. The initial $\delta^{13}C$ shift occurs within Magnetic Chron 16, at about 17.5 Ma. It represents a change toward heavier $\delta^{13}C$ values by about 1o/oo in surface and bottom waters. The excursion terminates at approximately 13.5 Ma. The Chron 16 Carbon shift coincides with the cessation of an early Miocene warming trend, seen in the $\delta^{18}O$ signals. The mid-Miocene cooling step (presumably associated with Antarctic ice buildup, near 15 Ma) is centered on the carbon isotope excursion. We propose that the initial carbon shift was caused by rapid extraction of organic carbon from the ocean-atmosphere system. Subsequently, the excursion toward heavy values was maintained by continued extraction of organic carbon, into ocean-margin deposits. Beginning at the end of the early Miocene, fine-grained diatomaceous sediments rich in organic matter were deposited all around the margins of the northern Pacific. In California, these sediments are known as the Monterey Formation. This formation is the result of coastal upwelling, which arose because of the development of strong zonal winds and a strong permanent thermocline.

Zonal winds and thermocline evolution, in turn, depended on increasing temperature contrast between high and low latitudes. We hypothesize that a feedback loop was established, such that an initial increase in the planetary temperature gradient started thermocline development which led to organic carbon extraction at the ocean margins which resulted in a drop in atmospheric carbon dioxide concentration. Concomitant cooling (reverse greenhouse effect) strengthened thermocline development, leading to further cooling. The loop was broken when available nutrients were used up. The total amount of excess carbon buildup, according to the hypothesis, is between 40 and 80 atmospheric carbon masses for the duration of the Monterey carbon isotope excursion. This amount corresponds to that present in the ocean, that is, one ocean carbon mass.

## Introduction

Thanks to deep-sea drilling and the diligent work of a large number of dedicated earth scientists during the last 15 years, it is now possible to reconstruct the evolution of climate for the Tertiary with considerable accuracy. Among the many clues which are useful in such climate reconstruction, the oxygen isotopes in planktonic and benthic foraminifera have emerged as the master signal carriers. Savin et al. [1975] and Shackleton and Kennett [1975] were the first to provide a reasonably detailed picture of what the climate signals look like (Figure 1).

Perhaps the most striking feature of the record is the presence of a number of major cooling "steps," that is, substantial shifts toward heavier $\delta^{18}O$ values over relatively short time intervals. In high latitudes, the steps occur in both planktonic and benthic records, while in low latitudes they only show in benthic records. Thus, in tropical oceans, there is an ever-increasing difference between benthic and planktonic values which reflects an ever-increasing latitudinal temperature gradient on earth. One of the chief problems in Cenozoic paleoceanography is the nature of these steps. What sorts of changes in the ocean-atmosphere system do they signify? Which mechanisms are responsible for the rapidity of the changes? Here we propose that the carbon cycle must be taken into consideration if these questions are to be attacked.

## The Mid-Miocene Oxygen Shift

One of the most interesting "steps" is the mid-Miocene oxygen shift, which spans the time period from about 15.5 to 12.5 million years ago. This step, more that any other event, initiates

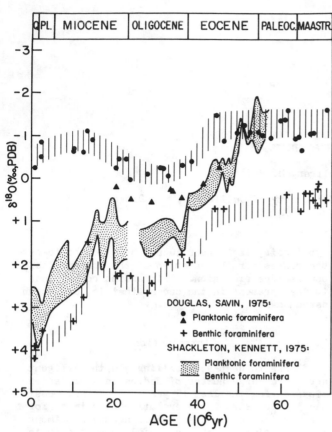

Fig. 1.   Oxygen isotopic composition of planktonic and benthic foraminifera in deep-sea sediments.  Data of Douglas and Savin [1975] and
Savin et al. [1975], central Pacific, and Shackleton and Kennett [1975], subantarctic Pacific.
After Berger et al. [1981].

the modern ocean-atmosphere system with cold
polar regions, a strong planetary temperature
gradient, and a strong thermocline.  From this
time on, also, the planktonic foraminiferal
assemblages look much like modern ones.  Savin et
al. [1975], Shackleton and Kennett [1975], Savin
et al. [1981], and Woodruff et al. [1981]
associated this oxygen shift with both high-
latitude cooling (and production of cold bottom
waters) and with ice buildup on Antarctica.
Matthews and Poore [1980], on the other hand,
drew attention to the lack of hard evidence for
ice mass formation on Antarctica, and suggested
that the mid-Miocene oxygen isotope shift
represents mainly a drop in temperature.  Large-
scale ice formation may in fact have started well
before the Miocene.  We have no new arguments on
this matter, which is being actively discussed in
the literature [Miller and Fairbanks, 1983;
Keigwin and Keller, 1984; Miller and Thomas,
1984].
    Considering the evidence and the arguments

offered by the authors cited, we shall adopt the
following working hypothesis.
    1.  The fact that the mid-Miocene step exists
suggests that the ocean-atmosphere system changed
permanently toward a colder state.  An obvious
mechanism providing for the irreversibility is
the buildup of a permanent ice body.
    2.  The low temperature of the Pacific bottom
water (within a few degrees of Pleistocene
values, see Figure 1) at that time suggests that
freezing temperatures existed not far poleward
from the location (or locations) of bottom water
formation, presumably within Antarctica.  Hence,
it is likely that some parts of Antarctica were
already covered by snow and ice in the early
Miocene.
    3.  A substantial buildup of ice should have
resulted in (1) a drop in sea level and (2)
increased amplitude of short-term climatic fluctuations after the step.  The evidence is weak,
but the sea level curve of Vail and Hardenbol
[1979] indicates a regression event at about 13
Ma at the termination of the mid-Miocene oxygen
step.  It is suggested, therefore, that cooling
and ice buildup were linked.  To explain the $\delta^{18}O$
step, we assume a combination of ice buildup and
temperature drop, with temperature delivering
well over one half of the effect, and tending to
precede the ice effect.
    The interpretation of the mid-Miocene $\delta^{18}O$
step in terms of temperature and ice volume
change is one thing, but finding a physical
explanation for the occurrence of the step is
another.
    Basically, there are two classes of arguments
in the literature regarding explanations for
step-like transitions in Cenozoic climate.  The
first class contains hypotheses which invoke
specific rearrangements in the configuration of
continents and oceans, usually the opening or
closing of a passage which initiates or terminates the transfer of heat across latitudes, by
ocean currents.  For example, Shackleton and Kennett [1975] proposed that the opening of Drake
Passage triggered the mid-Miocene event.  The
opening of this gate would have resulted in the
initiation of the Circumpolar Current, and hence
in the thermal isolation of the Antarctic continent.  Similarly, Blanc et al. [1980] and
Schnitker [1980] suggested that the subsidence of
the Iceland Ridge below a critical depth triggered the mid-Miocene event by causing a fundamental rearrangement of the North Atlantic circulation system linking the entire North Polar Sea
as a heat sink to the world ocean.  The second
class of arguments consists of hypotheses which
call on positive feedback mechanisms in order to
translate trends into steps, at certain thresholds.  For example, once the maximum summer temperature on Antarctica falls below a certain
value (owing to a long-term cooling trend), snow
and ice would have remained on the ground
throughout the year, providing strong albedo
feedback for cooling [Berger, 1981].

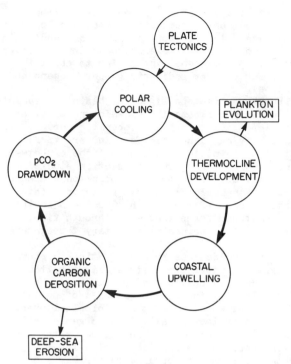

Fig. 2.  Feedback loop connecting oxygen isotope and carbon isotope records.

Historical arguments rely on co-occurrences. They make for narratives connecting seemingly unrelated events which occur in close association in the stratigraphic record. Purely physical explanations are essentially an exercise in deductive elementary climatology. Potentially, they can be modeled. In this paper we intend to exemplify a combined approach, using both historical and physical arguments to attack the problem of the origin of the mid-Miocene temperature step. We concentrate here on the possibility of positive climatic feedback from a change in the carbon dioxide content of the atmosphere. Various elements in such a feedback system are discussed by Berger [1982].

### Hypothesis

The Monterey Hypothesis runs as follows (see Figure 2):

1. The mid-Miocene oxygen isotope step occurs within a substantial $\delta^{13}C$ excursion toward heavier values, what we shall call the "Monterey Carbon Isotope Excursion." This excursion signifies excess extraction of organic matter, over and above the normal long-term conditions of dynamic equilibrium. The deposition of carbon-rich sediments in oxygen-poor environments can provide for the excess extraction.

2. The timing of the $\delta^{13}C$ excursion as seen in the deep-sea record and that of the deposition of the California Monterey Formation suggest that much of the called-for organic carbon may have been deposited around the edges of the northern Pacific. A strong increase in coastal upwelling is indicated, presumably driven by a threshold in global cooling. Coastal upwelling appears evident from high contents in organic carbon and phosphorus in Monterey rocks and from an abundance of laminated sediments indicating anaerobic deposition. Increased upwelling in the central equatorial Pacific has been proposed by Barron [1984] on the basis of changes in abundance of the diatom species Denticulopsis nicobarica coincidentally with the onset of the $\delta^{13}C$ excursion [Vincent and Killingley, 1984]. Evidence for increased equatorial upwelling is also evidence for increased coastal upwelling and mixing because equatorial upwelling is related to zonal wind stress, which also governs coastal upwelling.

3. The apparent increase in carbon deposition is such that it should have exerted a considerable "pull" on the ocean-atmosphere reservoir. This pull is here proposed to have lowered the $pCO_2$, and hence the global temperature, by the well-known "greenhouse" mechanisms [Hansen et al., 1981] working in reverse. An increase in the temperature gradient from the drop in $pCO_2$ would have further strengthened the thermocline-upwelling-carbon-deposition chain, providing for positive feedback.

4. The cycle was broken when the system became unable to sustain excess carbon deposition because of a decrease in available phosphorus. A drop in sea level, due to ice buildup, also would have generated negative feedback, through a decrease in area of deposition and an increase in the area of erosion of organic-rich soil and older marine sediments. Since increased delivery of phosphorus presumably would result from increased erosion, the negative feedback from this factor might not have been large.

Our hypothesis rests on circumstantial evidence: there can be little question of the association between $\delta^{18}O$ and $\delta^{13}C$ in the deep-sea record, and it does seem probable that this association signifies upwelling; recall that upwelling is increased during glacials, in the Pleistocene [CLIMAP, 1976]. However, there is no straight path from $\delta^{13}C$ signals to $pCO_2$ variations, and the phosphorus budget is not known for this time. A direct link between planktonic-benthic $\delta^{13}C$ gradients and atmospheric $pCO_2$ has been explored by Broecker [1982] and by Keir and Berger [1983] and has been accepted as fact by Shackleton et al. [1983]. Here we are not concerned with the productivity-$pCO_2$ link which is the subject of those and other articles (see review by Berger and Keir [1984]). Instead, on the time scale of $10^6$ years, trends involving large carbon reservoirs are at issue. For such trends, present-day steady state conditions are not directly relevant, and there is an increasingly complex network of assumptions which surrounds $pCO_2$ calculations [see Berner et al.,

1983]. We shall not attempt to link our hypothesis to this network. Thus, our suggestions regarding mid-Miocene pCO$_2$ changes remain explicitly qualitative.

In what follows, we discuss the stable isotope record from Deep-Sea Drilling Project (DSDP) Site 216 in the light of our hypothesis. The data obtained for benthic foraminifera, for deep-living and for shallow-dwelling planktonic species throughout the Miocene at this site as well as at three other DSDP sites (214, 237, 238) in the tropical Indian Ocean, have been published by Vincent et al. [1984]. They were collected within the framework of the Cenozoic Paleoceanography Project (CENOP). The various advantages of multispecies isotope signals, of which but few investigations have availed themselves, will become apparent in the course of the discussion.

### The Isotope Record of DSDP Site 216

We have selected Site 216 from the four sites for which multiple isotope stratigraphies are available [Vincent et al., 1983, 1984], because despite the condensed nature of the sedimentary sequence at this site it has the most complete and undisturbed record for the early and middle Miocene. It is also the only one which yielded siliceous biostratigraphic control throughout that interval. Similar trends are shown by the isotope stratigraphies at the four sites. Site 216 is located on the Ninetyeast Ridge (latitude 1°27'N, longitude 90°12'E) in a water depth of 2247 m, well above the present-day calcite compensation depth (CCD). In the early Miocene the site was presumably shallower by some 1000 m. The sediments consist of calcareous ooze typically containing more than 85% CaCO$_3$. Both calcareous and siliceous planktonic microfossils are abundant throughout the sequence. Foraminiferal faunas are well preserved in the upper and middle Miocene and moderately to poorly preserved in the lower Miocene.

Foraminiferal species selected for isotope measurements are (1) Dentoglobigerina altispira and Globigerinoides sacculifer, two shallow-dwelling planktonic forms, (2) Globoquadrina venezuelana, a deep-dwelling form, and (3) Oridorsalus umbonatus, a benthic foraminifera. Isotope data are plotted against depth below sea floor in Figure 3. Biostratigraphic zonations established on calcareous and siliceous planktonic microfossils by various authors were summarized in the synthesis by Vincent [1977]. They are shown on the right-hand side of Figure 3. There is good agreement between various fossil groups concerning the sequential order of paleontologic events (first appearance and disappearance of species). A number of biostratigraphic datums calibrated elsewhere to the paleomagnetic time scale (see references in work by Vincent et al. [1984]) are identified with triangular tick marks in Figure 3. Their numerical ages are those given by Barron et al. [1984] based on

Berggren et al.'s [1984] paleomagnetic time scale. Sediment accumulation rates are low with an average value of about 5 m/m.y. for sediments older than 7.5 m.y. (lower Miocene through lower part of upper Miocene) increasing to about 19 m/m.y. for younger sediments (upper Miocene and Pliocene).

The oxygen isotope signal of benthic foraminifera clearly reflects the mid-Miocene shift of about 1o/oo (Figure 3, O. umbonatus curve). The shallow-living planktonic foraminifera (SPF: D. altispira) maintain δ$^{18}$O compositions of between 0 and 0.5o/oo throughout the section studied. The signal of the deep-living planktonic species (DPF: G. venezuelana) follows an intermediate course. The difference in δ$^{18}$O between benthic and planktonic forms increases through time, reflecting the increasing planetary temperature gradient.

The carbon isotope record of Site 216 shows the expected separation between light bottom water and heavy shallow water values of about 2o/oo. This difference in carbon isotope values reflects the workings of the "biological pump," which removes isotopically light organic matter from sunlit waters to within and below the thermocline where it is oxidized, releasing light carbon dioxide. The organic matter has a δ$^{13}$C value near -20o/oo. A difference of -2o/oo, therefore, indicates that 10% of the total dissolved carbon in shallow water was removed through the settling of organic matter. The more efficient the pump, the greater the shallow-versus-deep difference (other factors being equal!). Changes in the difference, through geologic time, are interpreted as reflecting changes in ocean productivity, therefore. There is evidence in our data of increasing separation of δ$^{13}$C values and hence of increasing productivity upward in the section.

The δ$^{13}$C signal of the deeper-living planktonic foraminifera G. venezuelana may contain clues to the strength of the oxygen minimum. On the whole, the signal follows that of the shallow-dwelling species rather closely. A major separation between the two planktonic species is associated with the main carbon shift, which also marks the termination of the early Miocene warming trend, as seen in the oxygen isotopes. It appears that the strength of the oxygen minimum increased with the strength of the thermocline, at that time.

The most obvious feature of all carbon isotope records is the lower-to-middle Miocene excursion toward heavy values, which coincides in time with the deposition of the lower and central portion of the Monterey Formation of California. If this excursion were only known from a single foraminiferal signal (or a bulk signal, which is the same as the planktonic one, essentially), its significance would then not be clear. It could indicate either regional or global changes in the carbon chemistry of ocean waters. From the multispecies records there is no doubt that the changes are

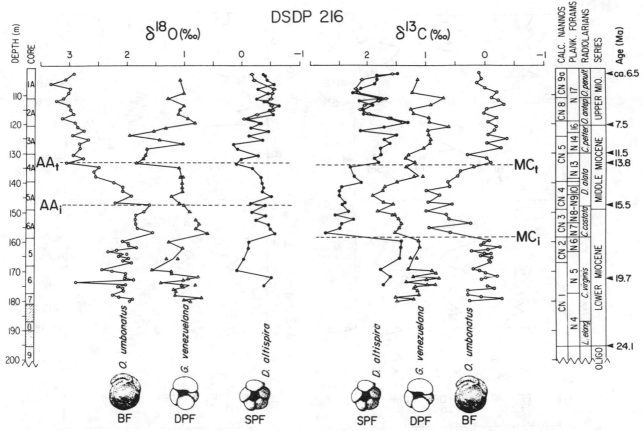

Fig. 3. Miocene oxygen and carbon isotope stratigraphies of the benthic foraminifera (BF) <u>Oridorsalis</u> <u>umbonatus</u> (open circles), the shallow-dwelling planktonic foraminifera (SPF) <u>Dentoglobigerina</u> <u>altispira</u> (solid circles), and <u>Globigerinoides</u> <u>sacculifer</u> (diamonds), and the deep-dwelling planktonic species (DPF) <u>Globoquadrina</u> <u>venezuelana</u> (triangles) at DSDP Site 216, tropical Indian Ocean, from 100 m to 200 m below seafloor. Data from Vincent et al. [1984]. Biostratigraphy after Vincent [1977]. Numerical ages at right are derived from the position in cores of biostratigraphic datums calibrated elsewhere to the paleomagnetic time scale (data of Vincent et al. [1984]). Horizons $MC_i$ and $MC_t$ mark the initiation and the termination respectively, of the "Monterey Carbon Isotope Excursion." The interval between $AA_i$ and $AA_t$ corresponds to the Antarctic cooling episode.

global and involve the entire water column.

The signals indicate a temporary increase in the $\delta^{13}C$ values of the dissolved carbon. To produce such an increase, $^{12}C$ must be preferentially withheld from the influx of carbon to the system, or preferentially extracted, or both. The process which is most readily visualized is a buildup of organic carbon, in marine sediments, below regions of high productivity. A link between changing ocean productivity, organic carbon deposition, and $\delta^{13}C$ change in the ocean was suggested some time ago by Tappan [1968]. More recently, Fischer and Arthur [1977], Arthur [1979], Broecker [1982], and Keir and Berger [1983] have elaborated on the subject, among others. Biosphere fluctuations, which may drive $\delta^{13}C$ fluctuations in the Pleistocene [Shackleton,

1977], are too small to produce a sustained excursion such as the one under discussion here, as we shall see.

## Time Scale and Correlation Analysis

Before we can proceed with correlating the stable isotope record of Site 216 to changes in sedimentation in the marginal North Pacific, or anywhere else, we must convert the depth scale in Figure 3 to a time scale. Four chief marker points for this conversion (numbered triangular tick marks in Figure 4) were selected from the paleomagnetically dated biostratigraphic datums. We have interpolated between these datum levels and replotted the signals on the appropriate scale in Figures 4a and 4b. Only those levels

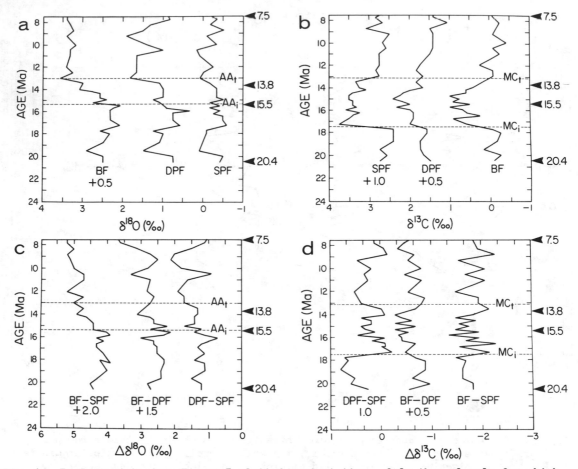

Fig. 4.   Isotopic data from Figure 3 plotted against time.  Only those levels for which data points are available in all six categories are shown (Table 1).  Data are shifted along the δ$^{18}$O and δ$^{13}$C axis so that curves do not overlap.  Ages are interpolated between four chief marker points (triangular tick marks at right).  (a) δ$^{18}$O signals. (b) δ$^{13}$C signals.  (c) Difference in δ$^{18}$O signals.  (d) Difference in δ$^{13}$C signals.

downcore are shown for which data points are available in all six categories (Table 1). Differences are plotted in Figures 4c and 4d.

The graphs in Figures 4c and 4d emphasize the importance of differences in δ$^{18}$O and δ$^{13}$C in the three original carriers.  In the δ$^{18}$O BF-SPF curve, the step increase within the AA event is clearly seen.  The later increase, near 10 Ma, signifies another change in the system toward a greater planetary temperature gradient.  This time it is by warming of tropical surface waters, rather than by cooling in high latitudes.  The curves involving DPF are less easily interpreted, because of the possible changes in water depth of DPF and the vicinity to the equator.

The δ$^{13}$C differences are not readily interpreted.  The BF-SPF curve could be taken as a measure of productivity and hence CO$_2$ partitioning between ocean and atmosphere [Broecker, 1982].  There is no clear shift or maximum asso-

ciated with the Monterey excursion.  This is somewhat surprising, given the various lines of evidence for increased productivity at this time, referred to earlier.

A correlation matrix has been prepared for the different curves shown in Figure 4 [Vincent et al., 1984].  The most remarkable correlations are as follows.  (1) Within the oxygen isotope signals, the low correlation between benthic and shallow planktonic signal (0.18) suggests that ice buildup and decay was of secondary importance in producing oxygen isotope variations.  Temperature fluctuations apparently dominated. (2) Within the carbon signals, the high positive correlations (0.73; 0.76; 0.85) indicate the dominance of whole water column phenomena.  That is, global changes in dissolved carbon composition are reflected, rather than changes in water mass fractionation.  (3) Correlations between oxygen and carbon isotope records are surprisingly high

TABLE 1.  Isotope Values at DSDP Site 216 for Levels Where Six
Numbers Are Available.

| DEPTH BELOW SEA FLOOR, (m) | SPF D. altispira | | DPF G. venezuelana | | BF O. umbonatus | | Age (Ma) |
|---|---|---|---|---|---|---|---|
| | $\delta^{18}O$ | $\delta^{13}C$ | $\delta^{18}O$ | $\delta^{13}C$ | $\delta^{18}O$ | $\delta^{13}C$ | |
| 102.69 | -0.36 | 1.47 | 1.06 | 1.13 | 2.93 | 0.08 | |
| 104.72 | -0.43 | 1.82 | 1.06 | 1.13 | 3.07 | 0.04 | |
| 109.32 | -0.45 | 2.26 | 0.99 | 1.23 | 3.02 | -0.22 | |
| 111.21 | -0.25 | 1.75 | 1.27 | 0.69 | 3.11 | -0.13 | |
| 113.56 | -0.64 | 1.89 | 0.99 | 0.91 | 3.01 | -0.34 | |
| 115.07 | -0.47 | 2.06 | 1.07 | 1.25 | 2.98 | -0.05 | |
| 116.57 | -0.54 | 2.06 | 1.10 | 1.20 | 2.93 | 0.00 | |
| 118.07 | -0.11 | 1.58 | 0.92 | 0.94 | 3.13 | -0.22 | |
| 119.52 | -0.33 | 1.80 | 0.80 | 0.93 | 2.90 | -0.07 | 7.7 |
| 121.11 | -0.15 | 1.70 | 1.31 | 0.60 | 2.86 | -0.21 | 8.3 |
| 122.33 | -0.46 | 2.12 | 1.52 | 0.96 | 2.75 | -0.18 | 8.77 |
| 123.83 | -0.12 | 1.49 | 1.93 | 0.95 | 2.95 | -0.07 | 9.34 |
| 125.33 | -0.36 | 1.70 | 1.41 | 0.91 | 2.67 | -0.39 | 9.91 |
| 126.72 | 0.15 | 1.60 | 1.00 | 0.93 | 2.83 | -0.05 | 10.44 |
| 128.36 | 0.08 | 1.76 | 1.63 | 1.03 | 2.82 | -0.29 | 11.07 |
| 130.63 | -0.28 | 1.75 | 1.66 | 1.29 | 2.94 | 0.27 | 11.93 |
| 131.83 | 0.02 | 1.86 | 1.70 | 1.34 | 2.76 | -0.05 | 12.39 |
| 133.33 | 0.10 | 1.79 | 1.82 | 1.18 | 3.05 | -0.12 | 12.96 |
| 134.83 | -0.18 | 2.30 | 1.07 | 1.33 | 2.49 | 0.16 | 13.54 |
| 137.83 | -0.33 | 2.23 | 0.99 | 1.19 | 2.55 | 0.44 | 14.01 |
| 139.84 | -0.35 | 2.10 | 1.02 | 1.48 | 2.23 | 0.60 | 14.31 |
| 141.31 | -0.37 | 2.45 | 1.02 | 1.68 | 2.08 | 0.57 | 14.52 |
| 142.71 | -0.35 | 2.49 | 1.01 | 1.72 | 2.08 | 0.97 | 14.73 |
| 144.33 | -0.51 | 2.49 | 1.21 | 1.89 | 1.93 | 0.74 | 14.96 |
| 146.68 | -0.15 | 2.39 | 1.06 | 1.70 | 2.21 | 0.94 | 15.31 |
| 147.33 | -0.4 | 2.44 | 0.99 | 1.54 | 1.61 | 0.53 | 15.4 |
| 149.66 | -0.17 | 2.33 | 0.89 | 1.80 | 1.71 | 0.97 | 15.81 |
| 150.83 | -0.41 | 2.56 | 0.31 | 1.51 | 1.85 | 0.63 | 16.03 |
| 152.33 | -0.24 | 2.24 | 0.79 | 1.45 | 1.86 | 0.58 | 16.3 |
| 153.82 | -0.46 | 2.45 | 0.72 | 1.40 | 1.85 | 0.21 | 16.58 |
| 155.31 | -0.49 | 2.48 | 0.78 | 1.44 | 1.82 | 0.92 | 16.86 |
| 156.83 | -0.57 | 2.72 | 0.57 | 1.51 | 1.55 | 0.56 | 17.14 |
| 159.74 | -0.12 | 1.41 | 1.25 | 1.11 | 2.08 | -0.9 | 17.68 |
| 161.87 | -0.11 | 1.42 | 1.00 | 1.09 | 1.83 | -0.27 | 18.08 |
| 165.47 | -0.03 | 1.44 | 1.09 | 1.25 | 1.97 | -0.13 | 18.74 |
| 169.24 | 0.09 | 1.76 | 1.54 | 1.29 | 2.43 | 0.18 | 19.44 |
| 171.67 | -0.51 | 1.59 | 0.73 | 1.20 | 2.03 | -0.23 | 19.89 |
| 174.39 | -0.38 | 1.71 | 0.78 | 0.96 | 2.02 | -0.06 | 20.4 |

(maximum of -0.66), signifying that climatic variations ($\delta^{18}O$) are closely linked to the carbon budget of the ocean and hence to the $CO_2$ exchange between ocean and atmosphere.

### The Monterey Formation

The name Monterey Formation or Monterey Shale has been applied to lithologically similar rocks of Miocene age exposed in the coastal regions of California. It constitutes one of the world's most remarkable sedimentary units. Long known for the considerable amount of its siliceous deposits (laminated diatomites and diagenetically derived porcelanites and cherts) it also contains distinctive kinds of organic-rich mudstones, phosphatic rocks, and authigenic carbonates. It is therefore an important repository for silica, phosphorus and carbon.

A type area for this formation was designated in the vicinity of Monterey, California (see Figure 5). Bramlette's [1946] treatise on the diatomaceous unit of the Monterey Formation still provides the most thorough description to date. Renewed interest in Monterey rocks as a major source and reservoir for petroleum has prompted considerable research interest over the past few decades. Recent symposia and summary articles on the Monterey Formation give details on various aspects of this rock assemblage [see Garrison et

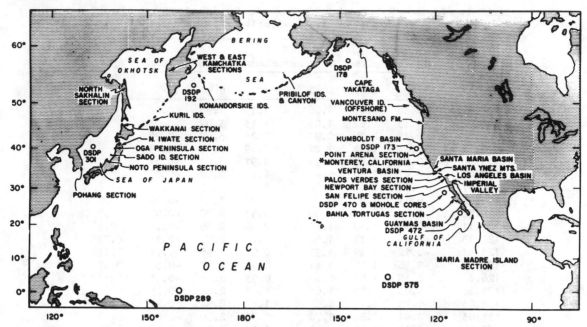

Fig. 5. Locations of important sites of the Deep Sea Drilling Project (DSDP) and stratigraphic sections around the North Pacific margin containing well-developed Neogene diatomite facies and genetically related porcellanous shales and cherts equivalent to the Monterey Formation (the type Monterey is indicated by an asterisk). After Ingle [1981]. DSDP sites in the equatorial Pacific are those with expanded lower and middle Miocene stable isotope records [Savin et al., 1981; Woodruff et al., 1981; Vincent and Killingley, 1984].

al., 1981; Isaacs, 1981, 1983; Isaacs et al., 1983].

Monterey-type deposits characterize not only California basins but many Neogene basinal sequences around the margin of the North Pacific, including the Gulf of California, Bering Sea, and Sea of Japan (Figure 5). In formulating a "Monterey Hypothesis" we use the name "Monterey" in a broad sense, referring to Monterey rocks and their equivalents rimming the Pacific. These deposits represent the combined product of tectonic, climatic, and oceanographic events.

A number of depositional models have been proposed [see Ingle, 1981, and references therein]. The synchronous formation of marginal basins around the Pacific rim may have been linked to an increased rate of spreading on the East Pacific Rise beginning in late Oligocene time (about 25-22 Ma). Deposition of Oligocene to lower Miocene volcanic continental and/or shallow marine clastics was followed by rapid subsidence. Silled basins developed, coincident with rising sea level, which created sediment-starved basins by late early Miocene time (about 18 Ma). Filling of these basins occurred during a deep marine phase from late early Miocene through late Miocene time (about 18-5.5 Ma), by hemipelagic sedimentation from highly productive coastal waters during an extended period of generally low terrigenous input. During Plio-Pleistocene time,

a rapidly deposited wedge of terrigenous clastics diluted and capped the earlier, predominantly biogenous sediments.

The deep marine phase is represented by a characteristic succession of facies within the Monterey Formation. Local variations aside, there typically is a distinct threefold vertical subdivision consisting of a lower calcareous facies, a middle transitional unit of phosphatic rocks, and an upper siliceous facies [Pisciotto and Garrison, 1981; Isaacs et al., 1983]. Correlative but thinner diatomaceous deposits are also present in deep-sea sedimentary sequences recovered at a number of DSDP sites in the marginal Pacific (see Figure 5).

### Correlation of Site 216 with the Monterey Formation

A correlation between the stratigraphies of the coastal deposits and those of the deep sea is by no means a trivial undertaking. Substantial differences in the ecologic environments hamper direct biostratigraphic correlation, while regional tectonics during and after deposition, as well as diagensis, complicate the comparison.

Early biostratigraphic studies on the Monterey Formation were based on Kleinpell's [1938] California stages defined on the basis of benthic foraminifera. The resulting provincial biostra-

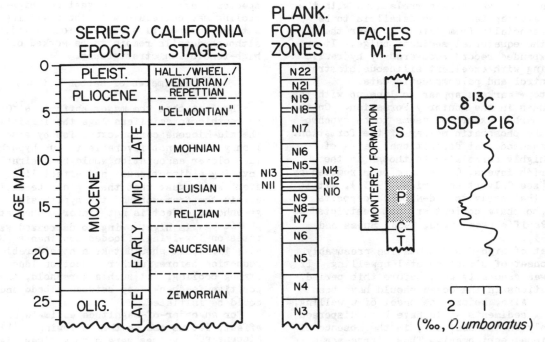

Fig. 6. Correlation of the provincial Neogene benthic foraminifera stages of California, standard Neogene planktonic foraminiferal zones, and the radiometric time scale. After Ingle [1981]. Generalized distributions in time of the Monterey Formation (M.F.) and its threefold facies adapted from Pisciotto and Garrison [1981]; C: calcareous facies, P: phosphatic facies; S: siliceous facies, T: terrigenous facies. Generalized $\delta^{13}C$ signal at DSDP Site 216. The $\delta^{13}C$ curve is constructed from average values between benthic and shallow-water planktonic foraminifera.

tigraphy provided only a very generalized correlation for placing the Monterey Formation into a worldwide time frame. Early studies of planktonic foraminifera were hindered by the paucity of biostratigraphic index species in the low-diversity, cool-temperate faunal assemblages of the Monterey Formation. Planktonic foraminiferal biostratigraphy in Monterey rocks and in the mid-latitudes of the Pacific was given new impetus by the use of quantitative analyses [Ingle 1973a, b]. Fluctuations in species abundance allowed Ingle to identify a sequence of cold and warm paleoclimatic events and fluctuations in the California Current system. Stratigraphic correlation by climatic fluctuation provided a tool to correlate mid-latitude and low-latitude foraminiferal zonations throughout the Pacific. Biostratigraphic correlation between the Monterey Formation and deep-sea sequences was further refined by the use of other planktonic microfossils, especially calcareous nannofossils and diatoms.

Integration of various biostratigraphic studies with radiometric dates obtained from volcanic units [Turner, 1970] (also see summaries by Obradovich and Naeser [1981] and Ingle [1981]) as well as paleomagnetic data recently obtained from dolomite in the Monterey Formation [Coe et al.,

1984; Hornafius and Lagle, 1984] has produced considerable refinement in the placement of California stages into chronologic schemes [Crouch and Bukry, 1979; Ingle, 1980, 1981; Blake, 1981; Keller and Barron, 1981; Obradovich and Naeser, 1981; Poore et al., 1981, 1983; Armentrout and Echols, 1983]. The position of the Monterey Formation as well as its generalized threefold facies succession into a time framework is given in Figure 6.

It is of interest, of course, to see whether the facies development in the Monterey Formation shows any parallelisms to the fluctuations in the stable isotope signals extracted from deep-sea sediments. An attempt to correlate the Monterey Formation with our deep-sea isotopic profiles is also shown in Figure 6. The isotope stratigraphies of Site 216 are correlated through biostratigraphy to those of other DSDP sequences in the equatorial Pacific (Site 289 [Woodruff et al., 1981]; Site 77 [Savin et al., 1981;] sites 573, 574, and 575 [Vincent and Killingley, 1984]). At all these sites, isotopic data clearly show the carbon excursion toward positive values and its stratigraphic relationship to the mid-Miocene oxygen isotope step.

At this point our correlations still remain tentative. We (and others) are working on refinements, through cross-correlations with the Pacific stratigraphies. Much detail is to be expected, especially from Site 575, near the center of the equatorial sediment bulge. It provides an expanded record recovered by hydraulic piston coring with excellent siliceous biostratigraphic control and paleomagnetic data.

Our isotope markers appear to line up with the facies changes in the Monterey Formation. On the whole, the carbon excursion seems to be synchronous with the phosphatic member of the formation. It may be assumed that depositional rates of P and C are highly correlated (although in the phosphate-rich layers, C may have been removed by winnowing [see Calvert and Price, 1983]). It is noteworthy that early to mid-Miocene deposits with high phosphate content are not restricted to the North Pacific [see Riggs, 1984; Suess and Thiede, 1983].

The onset of high-P sedimentation presumably marks the onset of strong coastal upwelling. If upwelling was generally weak before this period, P concentrations in the ocean should have been quite high. Also, before the onset of upwelling, P-containing sediments would have been dispersed rather widely on the seafloor, in the absence of strong regional enrichment. Thus, large-scale shallow erosional processes could have readily tapped an abundant P supply. The time of initiation of upwelling, therefore, would have been privileged with respect to P availability, especially if erosion did take place on the deep-sea floor.

There is evidence of hiatus formation just preceding carbon shift time, in the central equatorial Pacific based on isotope data and diatom biostratigraphy at DSDP Site 573 [Vincent and Killingley, 1984; Barron, 1984]. This short hiatus corresponds to NH1b of Barron and Keller [1982]. It is not present at all deep-sea sequences. However, in continuous sedimentary sequences it corresponds to the start of a pronounced CaCO$_3$ dissolution event (marked by a decrease in CaCO$_3$ content and foraminiferal preservation). The maximum dissolution is coeval with the carbon shift, but dissolution increases slightly before the change in $\delta^{13}$C. This is well documented in both continuous, expanded records at DSDP sites 574 and 575 in the equatorial Pacific [Vincent and Killingley, 1984].

We regard the newly active upwelling systems as pumps, pulling biogenic materials out of the deep sea to place them into the margins. This applies to phosphorus, as well as to carbon and silica. Carbonate also is involved initially, but in the course of increased upwelling the low pH produced by carbonic acid development in organic-rich surface sediments tends to discriminate against the preservation of carbonate.

The mid-Miocene oxygen shift, in our scale, lasts from about 15.2 to 13 Ma: it falls in the second half of the Monterey Carbon Excursion, which lasts from approximately 17.5 to 13.5 Ma. Fluctuations in the abundance of certain diatom species [Barron, 1984] suggest how high-latitude cooling may correlate with the upwelling history. A trend for positive correlation is indicated, although details remain to be worked out by high-resolution stratigraphy.

### Estimates of Carbon Buildup

The fact that the major shift in $\delta^{18}$O values in benthic foraminifera lags the initiation of the mid-Miocene carbon excursion by some 2 million years is not favorable to our hypothesis. A much closer association would be desirable for invoking a direct cause-and-effect link. However, we suggest that the lag can be explained within the framework of the hypothesis. The CO$_2$ greenhouse effect is not linear. When atmospheric CO$_2$ is high, owing to decreased weathering rates on low-lying, flooded continents [Berner et al., 1983], it should take a considerable CO$_2$ reduction before further reduction leads to strong cooling. At such a threshold, then, a positive feedback loop between albedo and pCO$_2$ could be initiated.

For an order-of-magnitude estimate of the lag effect, let us assume the following. (1) Early Miocene pCO$_2$ values were 3 to 4 times higher than present ones. (2) Partitioning between ocean and atmosphere followed thermodynamic rules (equilibrium with calcite). (3) Strong positive correlation between pCO$_2$ and global temperature sets in somewhere between 2 and 3 times present pCO$_2$. We must then remove enough carbon to lower pCO$_2$ permanently by about 1 unit to obtain the feedback condition. The corresponding amount, drawn mainly from the ocean, would be 30 to 60 (present) atmospheric carbon masses (1 ACM = 700 * 10$^9$t).

These crude assumptions are backed by the $\delta^{13}$C record. To show this, we produce a simple input-output scheme as illustrated in Figure 7. Organic carbon and carbonate carbon enter and leave the ocean-atmosphere system in a dynamic equilibrium. The $\delta^{13}$C is set at zero. If we are to produce an excursion toward heavy values, we must shut off a portion of the organic input, or increase the organic output at the expense of carbonate, or both.

Our hypothesis proposes that organic matter was preferentially extracted beginning 17.5 million years ago. If this is the reason for the $\delta^{13}$C excursion, we can calculate the necessary carbon buildup. It is given by

$$\Delta C(org)/\Delta t = \Delta \delta /\Delta t * R/\delta(org) \qquad (1)$$
$$+(\delta(t)-\delta(0))*F(\Delta t)/\delta(org)$$

where $\delta$ stands for $\delta^{13}$C of carbonate and $\delta(org)$ for $\delta^{13}$C of organic matter. R is the carbon reservoir (ocean + atmosphere), F is the carbon flux (units per $\Delta t$), and $\Delta t$ is the time step. Two terms yield the carbon buildup from the $\delta^{13}$C signal. The first term uses the difference

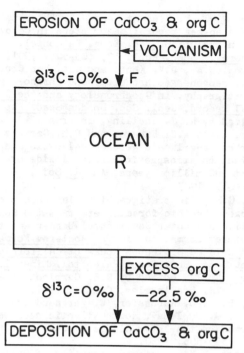

EROSION OF CaCO₃ & org C

VOLCANISM

$\delta^{13}C = 0‰$  F

OCEAN
R

EXCESS org C

$\delta^{13}C = 0‰$    −22.5 ‰

DEPOSITION OF CaCO₃ & org C

Fig. 7. Simple conceptual model of control of ocean's $\delta^{13}C$ values. F stands for flux, R for reservoir.

being flushed) and the (presumed) equilibrium value of $\delta^{13}C$. We have also run experiments with an adjustable equilibrium $\delta^{13}C$, under the assumption that the system changes the input ratio between carbonate and carbon, to minimize disequilibrium. Such adjustment quickly gets rid of the carbon buildup proposed and thus defeats the hypothesis.

We note that under the hypothesis, about one-half ocean carbon mass is extracted into organic carbon before the $pCO_2$ falls sufficiently for the greenhouse feedback to condition the planet for high-latitude ice formation, which presumably helped accelerate the oxygen isotope step. It remains to ask whether the amount of carbon sequestered in the Monterey Formation and associated deposits is within the general limits of our calculations. If we assume a content of 5% of organic carbon for these upwelling sediments, and a density of 2 g/cm³, we need a sediment thickness of 2 m over the surface of the earth. Monterey-type deposits cover about 600,000 km², which is 0.4% of the earth's surface. Thus, these deposits have to be roughly 500 m thick, which they are, more or less.

## Conclusions

We conclude that the Monterey Hypothesis, which relates the early Miocene carbon isotope

between successive $\delta^{13}C$ values in carbonate (= seawater) and compares it with reservoir size (R) and the $\delta^{13}C$ value of organic matter. For example, to shift the $\delta^{13}C$ of the ocean by 1o/oo (= $\Delta\delta/\Delta t$) about 3 ACM of organic carbon must be removed from the ocean's 60 ACM (= R), since the $\delta^{13}C(org)$ is near −20o/oo [Vincent et al., 1980]. The second term is more important in the long run, because it considers the excess removal necessary if disequilibrium is to be maintained. The ocean is being flushed by the input F (over the time interval $\Delta t$ considered), which is at a $\delta^{13}C$ set for equilibrium (= $\delta(0)$), and has to be reset to $\delta^{13}C(t)$ within the available time, using excess extraction of $\delta^{13}C(org)$. The value used for R is relatively unimportant; we used the present one. The value of F is rather important; we assumed a flux equivalent to a carbon residence time of 120,000 years, that is, F = R/120,000 [e.g., Garrels et al., 1973].

Results of stepwise calculations on various $\delta^{13}C$ curves and with various values of $\delta^{13}C(0)$ are shown in Figure 8. The amount of carbon extraction is typically one ocean carbon mass, and it varies by a factor of 1.5 depending upon the assumptions. It is relatively unimportant which of the three $\delta^{13}C$ signals is used for the calculations. Benthic values alone, or perhaps an average between surface and benthic values, may reflect best the actual composition of the ocean. The crucial parameters are the assumed flux (that is, the rate at which the ocean is

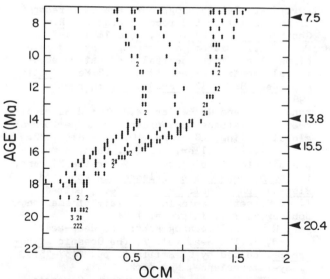

Fig. 8. Calculated carbon mass from model in Figure 7, using equation (1) and stepwise summation. The output is the excess organic carbon extracted from the system, in units of ocean carbon mass (OCM), beginning 20 million years ago. Six runs are shown. Inputs are the $\delta^{13}C$ curves of benthic foraminifera (with assumed equilibrium $\delta^{13}C$ of −0.15 and −0.05o/oo), shallow planktonic foraminifera (1.6o/oo and 1.8o/oo), and average signal (0.8o/oo and 1.0o/oo). The most likely estimate is 1 OCM (= 60 ACM = 60 * 700 * 10¹⁵gC).

shift to the onset of upwelling and to climatic developments, is a viable proposition.

Our conclusion is based on the correlation of the early to mid-Miocene carbon isotope excursion with the facies development in the Monterey Formation, and on the fact that calculations of carbon buildup produce reasonable values for the masses involved. The lag between the changes in the carbon system (beginning near 17.5 Ma) and the main climatic step involving cooling of high latitudes and ice buildup (15.5 Ma to 13 Ma) may be significant. It suggests that the carbon cycle is part of the climatic feedback system and had to be adjusted for cooling and ice buildup to take place. A drawdown of the atmospheric pCO$_2$, through increased upwelling from an initial cooling, is proposed as the crucial step preparing the way for late Neogene glaciations.

Acknowledgments. We thank Jim Ingle and John Barron for discussions. This work was supported by NSF grant OCE 83-10518.

## References

Armentrout, J.M., and R.J. Echols, Biostratigraphic-chronostratigraphic scale of the northeastern Pacific Neogene, in IGCP-114 International Workshop of Pacific Neogene Biostratigraphy Proc., pp. 7-27, Osaka, Japan, 1981.

Arthur, M.A., Paleoceanographic events - recognition, resolution and reconsideration, Rev. Geophys. Space Phys., 17, 1476-1494, 1979.

Barron, J.A., Diatom paleoceanography and paleoclimatology of the central and eastern equatorial Pacific between 18 and 6.2 Ma. Initial Rep. Deep Sea Drill. Proj., 85, in press, 1984.

Barron, J.A., and G. Keller, Widespread Miocene deep-sea hiatuses: coincidence with periods of global cooling, Geology, 10, 577-581, 1982.

Barron, J.A., G. Keller, and D.A. Dunn, A multiple microfossil biochronology for the Miocene, in The Miocene Ocean, Paleoceanography and Biogeography, Geol. Soc. Am. Mem., edited by J. P. Kennett, Geological Society of America, Boulder, Colo., in press, 1984.

Berger, W.H., Paleoceanography: the deep-sea record, in The Sea, Vol. 7, The Oceanic Lithosphere, edited by C. Emiliani, pp. 1437-1519, Wiley-Interscience, New York, 1981.

Berger, W.H., Deep-sea stratigraphy: Cenozoic climate steps and the search for chemo-climatic feedback, in Cyclic and Event Stratification, edited by G. Einsele and A. Seilacher, pp. 121-157, Springer-Verlag, New York, 1982.

Berger, W.H., and R.S. Keir, Glacial-Holocene changes in atmospheric CO$_2$ and the deep-sea record, in Climate Processes and Climate Sensitivity, Geophys. Monogr. Ser., Vol. 29 edited by J. Hansen and T. Takahashi, pp. 337-351, AGU, Washington, D.C., 1984.

Berger, W.H., E. Vincent, and H.R. Thierstein, The deep-sea record: major steps in Cenozoic ocean evolution, Soc. Econ. Paleontol. Mineral. Spec. Publ. 32, 489-504, 1981.

Berggren, W.A., D.V. Kent, and J.A. Van Couvering, Neogene geochronology and chronostratigraphy, in Geochronology and the Geological Record, Geol. Soc. London Spec. Pap., edited by N. J. Snelling, in press, 1984.

Berner, R.A., A.C. Lasaga, and R.M. Garrels, The carbonate-silicate geochemical cycle and its effect on atmospheric carbon dioxide over the past 100 million years, Am. J. Sci., 283, 641-683, 1983.

Blake, G.H., Biostratigraphic relationship of Neogene benthic foraminifera from the Southern California outer Continental Borderland to the Monterey Formation, in The Monterey Formation and Related Siliceous Rocks of California, Pac. Sect. Soc. Econ. Paleontol. Mineral. Spec. Publ., edited by R.E. Garrison et al., pp. 1-14, Los Angeles, Calif., 1981.

Blanc, P.L., D. Rabussier, C. Vergnaud-Grazzini, and J.C. Duplessy, North Atlantic deep water formed by the later middle Miocene, Nature, 289, 553-555, 1980.

Bramlette, M.N., Monterey Formation of California and origin of its siliceous rocks, U.S. Geol. Surv. Prof. Pap. 212, 1-57, 1946.

Broecker, W.S., Glacial to interglacial changes in ocean chemistry, Progr. in Oceanogr., 11, 151-197, 1982.

Calvert, S.E., and N.B. Price, Geochemistry of Namibian shelf sediments, in Coastal Upwelling, Part A edited by E. Suess and J. Thiede, pp. 337-375, Plenum, New York, 1983.

CLIMAP Project Members, The surface of the ice age earth, Science, 191, 1131-1137, 1976.

Coe, R.S., K.A. Mertz, Jr., and R.L. Pettit, Paleomagnetic results from dolomite rocks of the Monterey Formation, Arroyo Seco Area, California, in AAPG, SEPM, SEG, 59th Annual Meeting, Pacific Sections Program and Abstracts, pp. 53-54, American Association of Petroleum Geologists, Tulsa, Okla., 1984.

Crouch, J.K., and D. Bukry, Comparison of Miocene provincial foraminiferal stages to coccoliths in the California Continental Borderlands, Geology, 7, 211-215, 1979.

Douglas, R.G., and S.M. Savin, Oxygen and carbon isotope analyses of Tertiary and Cretaceous microfossils from Shatsky Rise and other sites in the North Pacific Ocean, Initial Rep. Deep Sea Drill. Proj., 32, 509-520, 1975.

Fischer, A.G., and M.A. Arthur, Secular variations in the pelagic realm, Soc. Econ. Paleontol. Mineral. Spec. Publ., 25, 19-50, 1977.

Garrels, R.M., F.T. MacKenzie, and C. Hunt, Chemical Cycles and the Global Environment, 206 pp., W. Kaufmann, Los Altos, Calif., 1973.

Garrison, R.E., R.G. Douglas, K.E. Pisciotto, C. M. Isaacs, and J. C. Ingle (Eds.), The Monterey Formation and Related Siliceous Rocks of California, Pac. Sect., Soc. Econ. Paleontol.

Mineral. Spec. Publ., 327 pp., Los Angeles, Calif., 1981.

Hansen, J., D. Johnson, A. Lacis, S. Lebedeff, P. Lee, D. Rind, and G. Russell, Climate impact of increasing atmospheric carbon dioxide, Science, 213, 957-966, 1981.

Hornafius, J.S., and C.W. Lagle, Magnetic stratigraphy and diatom biostratigraphy of the Monterey Formation at selected localities in the western transverse ranges, California, in AAPG, SEPM, SEG., 59th Annual Meeting Pacific Sections, Program and Abstracts, pp. 74-75, American Association of Petroleum Geologists, Tulsa, Okla., 1984.

Ingle, J.C., Jr., Neogene foraminifera from the northeastern Pacific Ocean, Leg 18, Deep Sea Drilling Project, Initial Rep. Deep Sea Drill. Proj., 18, 517-567, 1973a.

Ingle, J.C., Jr., Neogene foraminifera from the northeastern Pacific Ocean, Leg 18, DSDP, Initial Rep. Deep Sea Drill. Proj., 18, 949-959, 1973b.

Ingle, J.C., Jr., Cenozoic paleobathymetry and depositional history of selected sequences within the Southern California Continental Borderland, Cushman Found. Foraminiferal Res., Spec. Publ., 19, 163-195, 1980.

Ingle, J.C., Jr., Origin of Neogene diatomites around the north Pacific rim, in The Monterey Formation and Related Siliceous Rocks of California, Pac. Sect. Soc. Econ. Paleontol. Mineral. Spec. Publ., edited by R.E. Garrison et al., pp. 159-179, Los Angeles, Calif., 1981.

Isaacs, C.M. (Ed.), Guide to the Monterey Formation in the California Coastal Area, Ventura to San Luis Obispo, Pac. Sect. Am. Assoc. Pet. Geol. Spec. Publ. 52, pp. 1-91, Tulsa, Okla., 1981.

Isaacs, C.M., Compositional variation and sequence in the Miocene Monterey Formation, Santa Barbara coastal area, California, in Cenozoic Marine Sedimentation, Pacific Margin, U.S.A., Pac. Sect. Soc. Econ. Paleontol. Mineral. Spec. Pub., edited by D. Larue and R.J. Steel, pp. 117-132, Los Angeles, Calif., 1983.

Isaacs, C.M., K.A. Pisciotto, and R.E. Garrison, Facies and diagenesis of the Miocene Monterey Formation: a summary, in Siliceous Deposits in the Pacific Region, Dev. Sedimentol., vol. 36, edited by A. Iijima, J.R. Mein and R. Siever, pp. 247-282, Elsevier, New York, 1983.

Keigwin, L.D., and G. Keller, Middle Oligocene climatic change from equatorial Pacific DSDP Site 77, Geology, 12, 16-19, 1984.

Keir, R.S., and W.H. Berger, Atmospheric $CO_2$ content in the last 120,000 years: the phosphate extraction model, J. Geophys. Res., 88, 6027-6038, 1983.

Keller, G., and J.A. Barron, Integrated planktic foraminiferal and diatom biochronology for the Northeast Pacific and the Monterey Formation, in The Monterey Formation and Related Sili-

ceous Rocks of California, Pac. Sect., Soc. Econ. Paleontol. Mineral., Spec. Publ., edited by R.E. Garrison et al., pp. 43-54, Los Angeles, Calif., 1981.

Kleinpell, R.M., Miocene stratigraphy of California, 450 pp., American Association of Petroleum Geologists, Tulsa, Okla., 1938.

Matthews, R.K., and R.Z. Poore, Tertiary $\delta^{18}O$ record and glacio-eustatic sea-level fluctuations, Geology, 8, 501-504, 1980.

Miller, K.G., and R.G. Fairbanks, Evidence for Oligocene-middle Miocene abyssal circulation changes in the western North Atlantic, Nature, 306, 250-253, 1983.

Miller, K.G., and E. Thomas, Late Eocene to Oligocene benthic foraminiferal isotopic record, Site 574 equatorial Pacific, Initial Rep. Deep Sea Drill. Proj., 85, in press, 1984.

Obradovich, J.D., and C.W. Naeser, Geochronology bearing on the age of the Monterey Formation and siliceous rocks in California, in The Monterey Formation and Related Siliceous Rocks of California, Pac. Sect., Soc. Econ. Paleontol. Mineral., Spec. Publ., edited by R.E. Garrison et al., pp. 87-95, Los Angeles, Calif., 1981.

Pisciotto, K.A., and R.E. Garrison, Lithofacies and depositional environments of the Monterey Formation, California, in The Monterey Formation and Related Siliceous Rocks of California, Pac. Sect., Soc. Econ. Paleontol. Mineral., Spec. Publ., edited by R.E. Garrison et al., pp. 97-122, Los Angeles, Calif., 1981.

Poore, R.Z., K. McDougall, J.A. Barron, E.E. Brabb, and S.A. Kling, Microfossil biostratigraphy and biochronology of the type Relizian and Luisian stages of California, in The Monterey Formation and Related Siliceous Rocks of California, Pac. Sect., Soc. Econ. Paleontol. Mineral., Spec. Publ., edited by R.E. Garrison et al., pp. 15-41, Los Angeles, Calif., 1981.

Poore, R.Z., J.A. Barron, and W.O. Addicott, Biochronology of the northern Pacific Miocene, in IGCP-114 International Workshop of Pacific Neogene Biostratigraphy, pp. 91-97, Osaka, Japan, 1981.

Riggs, S.R., Paleoceanographic model of Neogene phosphorite deposition, U.S. Atlantic continental margin, Science, 223, 123-131, 1984.

Savin, S.M., R.G. Douglas, and F.G. Stehli, Tertiary marine paleotemperatures, Geol. Soc. Am. Bull., 86, 1499-1510, 1975.

Savin, S.M., R.G. Douglas, G. Keller, J.S. Killingley, L. Shaughnessy, M. A. Sommer, E. Vincent, and F. Woodruff, Miocene benthic foraminiferal isotope records: A synthesis, Mar. Micropaleontol., 6, 423-450, 1981.

Shackleton, N.J., Carbon-13 in Uvigerina: Tropical rainforest history and the equatorial Pacific carbonate dissolution cycles, in The Fate of Fossil Fuel $CO_2$ in the Oceans, edited by N.R. Andersen and A. Malahoff, pp. 401-427, New York, Plenum, 1977.

Shackleton, N.J., and J.P. Kennett, Paleotemperature history of the Cenozoic and the initia-

tion of Antarctic glaciation: Oxygen and carbon isotope analyses in DSDP sites 277, 279
and 281, Initial Rep. Deep Sea Drill. Proj.,
29, 743-756, 1975.

Shackleton, N.D., M.A. Hall, J. Line and Cang
Shuxi, Carbon isotope data in core V19-30 confirm reduced carbon dioxide concentration in
the ice age atmosphere, Nature, 306, 319-322,
1983.

Schnitker, D., North Atlantic oceanography as
possible cause of Antarctic glaciation and
eutrophication, Nature, 284, 615-616, 1980.

Suess, E., and J. Thiede, Coastal Upwelling, Its
Sedimentary Record, Part A, Responses of the
Sedimentary Regime to Present Coastal Upwel-
ling, 604 pp., Plenum, New York, 1983.

Tappan, H., Primary production, isotopes, extinctions and the atmosphere, Palaeogeogr. Palaeoclimatol. Palaeoecol, 4, 187-210, 1968.

Turner, D.L., Potassium-argon dating of Pacific
coast Miocene foraminiferal stages, Geol. Soc.
Am. Spec. Pap. 124, 91-129, 1970.

Vail, P.R., and J. Hardenbol, Sea-level change
during the Tertiary, Oceanus, 22, 71-79, 1979.

Vincent, E., Indian Ocean Neogene planktonic
foraminiferal biostratigraphy and its paleoceanographic implications, in Indian Ocean
Geology and Biostratigraphy, edited by J.R.
Heirtzler et al., pp. 469-584, AGU, Washington, D.C., 1977.

Vincent, E. and J.S. Killingley, Oxygen and carbon isotope record for the early and middle
Miocene in the central equatorial Pacific
(DSDP Leg 85) and paleoceanographic implications, Initial Rep. Deep Sea Drill. Proj. 85,
in press, 1984.

Vincent, E., J.S. Killingley, and W.H. Berger,
The Magnetic Epoch 6 carbon shift: A change in
the ocean's $^{13}C/^{12}C$ ratio 6.2 million years
ago, Mar. Micropaleontol., 5, 185-203, 1980.

Vincent, E., J.S. Killingley, and W.H. Berger,
Miocene oxygen and carbon isotope stratigraphy
of the tropical Indian Ocean, in The Miocene
Ocean, Paleoceanography and Biogeography,
Geol. Soc. Am. Mem., edited by J. Kennett, in
press, 1984.

Vincent, E., J.S. Killingley, and W.H. Berger,
The Chron-16 carbon event in the Miocene
Indian Ocean, Geol. Soc. Am., Abstr. Pro-
grams, 15, 712, 1983.

Woodruff, F., S.M. Savin, and R.G. Douglas,
Miocene stable isotope record: A detailed deep
Pacific Ocean study and its paleoclimatic
implications, Science, 212, 665-668, 1981.

# OLIGOCENE TO MIOCENE CARBON ISOTOPE CYCLES AND ABYSSAL CIRCULATION CHANGES

Kenneth G. Miller and Richard G. Fairbanks

Lamont-Doherty Geological Observatory of Columbia University
Palisades, New York 10964

Abstract. Three cycles of $\delta^{13}C$ occurred in Oligocene to Miocene benthic and planktonic foraminifera at western North Atlantic Sites 558 and 563. Intervals of high $\delta^{13}C$ occurred at about 35-33 Ma (early Oligocene), 25-22 Ma (across the Oligocene/Miocene boundary), and 18-14 Ma (across the early/middle Miocene boundary). Similar carbon isotopic fluctuations have been measured in benthic and planktonic foraminifera from the Atlantic, Pacific, and Indian oceans, suggesting that these cycles represent global changes in the $\delta^{13}C$ of mean ocean water. The average duration of the carbon cycles is 50 times greater than the residence time of carbon in the oceans. Therefore, the mechanism controlling these cycles must be tied to changes in the input ratio of organic carbon to carbonate from weathering rocks or to changes in the output ratio of organic carbon to carbonate in marine sediments. Following a strategy used to study modern and Pleistocene oceans, benthic foraminiferal $\delta^{13}C$ differences between the Atlantic and Pacific are used to infer Oligocene through Miocene abyssal circulation changes. The Atlantic was most enriched in $^{13}C$ relative to the Pacific from about 36-33 Ma (early Oligocene) and 26-10 Ma (late Oligocene to late Miocene). We interpret this as indicating supply of nutrient-depleted bottom water in the North Atlantic, perhaps analogous to modern North Atlantic Deep Water. High benthic foraminiferal $\delta^{18}O$ values at about 36-35 Ma, 31-28 Ma, 25-24 Ma, and younger than 15 Ma indicate the presence of ice sheets at these times. Covariance between benthic and planktonic foraminiferal $\delta^{18}O$ records of 0.3-0.5°/oo at 36 Ma, 31 Ma, and 25 Ma suggests that three periods of continental glaciation caused eustatic (global sea-level) lowerings of 30-50 m during the Oligocene epoch. The $\delta^{13}C$ cycles do not correlate with sea-level changes deduced from oxygen isotopic data, nor do they correlate with other proxy indicators for sea level.

## Introduction

Stable isotopic studies of the shells of microfossil foraminifera can provide information on changes in global carbon reservoirs, abyssal circulation, ice volume, and temperature. For example, synchronous planktonic and benthic foraminiferal $\delta^{13}C$ changes observed in different ocean basins and oceanographic settings provide evidence for changes in the $\delta^{13}C$ of mean ocean water. Deep-sea benthic foraminiferal $\delta^{13}C$ records can also provide a monitor of bottom-water $\delta^{13}C$ conditions, allowing reconstruction of Atlantic and Pacific carbon isotopic differences caused by abyssal circulation. Oligocene (36.6 to 23.7 Megannus (Ma) = $10^6$ years B.P.) to Miocene (23.7 to 5.3 Ma) North Atlantic isotopic records with good chronostratigraphic control were not available prior to drilling of Deep Sea Drilling Project (DSDP) Sites 558 and 563. In this study, we measured benthic and planktonic foraminiferal $\delta^{13}C$ and $\delta^{18}O$ from western North Atlantic Sites 558 and 563 in order to compare the Atlantic data with published Pacific isotopic records. Oxygen isotopic records are used as a proxy indicator of sea level, which we compared with the oceanic carbon cycles and changes in abyssal circulation deduced from $\delta^{13}C$ studies.

Previous studies of Tertiary carbon isotopes have been hampered by the fact that most existing data are based upon analyses of benthic assemblages or taxa which do not accurately record modern deep-sea $\delta^{13}C$ distribution. Studies have shown that the benthic foraminifera Cibicidoides accurately reflects deep-water $\delta^{13}C$ variations [Graham et al., 1981; Belanger et al., 1981]. Fortunately, this taxon is abundant in the Tertiary fossil record recovered by the DSDP. Deep- and bottom-water $\delta^{13}C$ variations may be caused by (1) changes in global carbon reservoirs [Broecker, 1970, 1982; Shackleton, 1977]; (2) abyssal circulation changes [Bender and Keigwin, 1979; Belanger et al., 1981; Curry and Lohmann, 1982, 1983; Boyle and Keigwin, 1982; Shackleton et al., 1983]; (3) local productivity changes, especially on continental margins [Sarnthein et al., 1982]; and (4) changes in the preformed chemistry of the source region [Mix and Fairbanks, 1984].

Miller and Fairbanks [1983] first delineated three cycles in Oligocene to middle Miocene $\delta^{13}C$

TABLE 1.   Oxygen and Carbon Isotopic Data, DSDP Sites 558 and 563

| Core-Section/ Depth, cm | Depth, msb | Age, Ma | $\delta^{18}O_{PDB}$ | $\delta^{13}C_{PDB}$ |
|---|---|---|---|---|
| | | Cibicidoides spp., Site 558 | | |
| 1-2/90-94 | 160.40 | 8.27 | 2.12 | 1.10 |
| 2-1/80-85 | 168.30 | 8.53 | 2.26 | 1.09 |
| 3-1/82-87 | 177.82 | 8.83 | 2.05 | 1.07 |
| 4-1/81-86 | 187.31 | 9.19 | 2.12 | 1.25 |
| 5-1/103-109 | 197.03 | 9.58 | 2.24 | 0.81 |
| 6-1/79-84 | 206.29 | 9.95 | 1.84 | 0.95 |
| 7cc | 215.80 | 10.93 | 2.14a | 1.09a |
| 8-1/82-86 | 225.32 | 11.02 | 1.66a | 0.69a |
| 9-1/16-22 | 234.16 | 11.74 | 1.63a | 0.96a |
| 11-1/103-109 | 254.03 | 13.36 | 1.56a | 1.06a |
| 12-1/80-85 | 263.30 | 14.11 | 1.44a | 1.35a |
| 13-1/103-109 | 273.03 | 14.91 | 1.11a | 1.21a |
| 14-1/67-71 | 282.17 | 15.65 | 1.21a | 1.43a |
| 15-1/131-135 | 292.31 | 16.48 | 1.56 | 2.08 |
| 16-1/79-83 | 301.29 | 17.53 | 1.24 | 1.01 |
| 16-4/68-73 | 305.68 | 18.52 | 1.21 | 0.85 |
| 17-1/128-132 | 311.28 | 20.60 | 1.52 | 0.61 |
| 17-4/56-61 | 315.06 | 21.54 | 1.03 | 0.89 |
| 18-1/92-96 | 320.42 | 22.88 | 1.33 | 1.50 |
| 18-4/16-21 | 324.16 | 23.66 | 1.71 | 1.49 |
| 18-4/81-86 | 324.81 | 23.77 | 1.57 | 0.58 |
| 18-5/23-28 | 325.73 | 23.94 | 2.09 | 1.34 |
| 19-1/2-6 | 329.02 | 24.64 | 1.10 | 1.03 |
| 19-1/141-145 | 330.41 | 24.78 | 2.15 | 1.04 |
| 19-2/34-39 | 330.84 | 24.86 | 1.18 | 0.60 |
| 19-3/69-74 | 332.69 | 25.19 | 1.08 | 0.66 |
| 19-4/118-122 | 334.68 | 25.55 | 1.40 | 0.93 |
| 19-5/110-115 | 336.10 | 25.80 | 1.57 | 0.59 |
| 20-1/118-122 | 339.68 | 26.44 | 1.29 | 0.86 |
| 20-2/66-71 | 340.66 | 26.62 | 0.74 | 0.44 |
| 20-3/14-19 | 341.64 | 26.79 | 1.31 | 0.55 |
| 20-4/122-126 | 344.22 | 27.26 | 1.66 | 0.89 |
| 21-1/46-49 | 348.46 | 28.02 | 1.40 | 0.80 |
| | | | 1.09 | 0.39 |
| 21-1/132-136 | 349.32 | 28.17 | 1.71a | 1.23a |
| 21-2/102-108 | 350.52 | 28.39 | 1.85 | 0.65 |
| 21-3/13-16 | 351.13 | 28.50 | 1.79 | 0.86 |
| 22-1/34-38 | 357.84 | 29.70 | 1.77 | 0.46 |
| 22-2/84-88 | 359.84 | 30.03 | 2.05a | 1.38a |
| 22-3/21-26 | 360.71 | 30.13 | 1.60 | 0.61 |
| 22-4/3-8 | 362.03 | 30.28 | 1.84 | 0.62 |
| 22-4/78-81 | 362.78 | 30.37 | 1.77 | 0.69 |
| 23-1/52-57 | 367.52 | 30.89 | 1.65 | 0.49 |
| 23-2/51-55 | 369.01 | 31.06 | 2.03 | 0.52 |
| 23-3/65-69 | 370.65 | 31.24 | 1.30b | 0.09b |
| 23-4/49-53 | 371.99 | 31.39 | 1.36b | 0.34b |
| 23-5/22-26 | 373.22 | 31.53 | 1.65b | 0.62b |
| 23-6/45-49 | 374.95 | 31.72 | 1.82b | 0.30b |
| 24-2/80-85 | 378.80 | 32.13 | 1.10b | 0.69b |
| 24-3/79-84 | 380.29 | 32.25 | 1.29b | 0.82b |
| 24-4/75-79 | 381.75 | 32.38 | 1.77b | 0.67b |
| 25-1/117-121 | 387.17 | 32.83 | 1.29b | 0.77b |
| 25-4/85-89 | 391.35 | 33.63 | 1.59b | 0.85b |
| 26-1/93-97 | 396.43 | 34.73 | 1.13b | 1.23b |

TABLE 1. (continued)

| Core-Section/ Depth, cm | Depth, msb | Age, Ma | $\delta^{18}O_{PDB}$ | $\delta^{13}C_{PDB}$ |
|---|---|---|---|---|
| | | | | |

Cibicidoides spp., Site 558 (continued)

| Core-Section/ Depth, cm | Depth, msb | Age, Ma | $\delta^{18}O_{PDB}$ | $\delta^{13}C_{PDB}$ |
|---|---|---|---|---|
| 26-4/22-26 | 400.22 | 35.51 | 1.51b | 1.13b |
| 27-1/81-85 | 405.81 | 36.50 | 1.04b | 1.48b |

Catapsydrax spp., Site 558

| Core-Section/ Depth, cm | Depth, msb | Age, Ma | $\delta^{18}O_{PDB}$ | $\delta^{13}C_{PDB}$ |
|---|---|---|---|---|
| 16-5/88-92 | 307.38 | 18.91 | 1.80 | 0.82 |
| 16-6/87-91 | 308.87 | 20.05 | 1.76 | 0.96 |
| 17-1/128-132 | 311.28 | 20.64 | 1.34 | 1.46 |
| 17-2/100-106 | 312.50 | 20.94 | 1.47 | 1.40 |
| 17-3/80-85 | 313.80 | 21.26 | 0.95 | 1.03 |
| 17-4/7-12 | 314.57 | 21.45 | 0.93 | 1.15 |
| 17-4/56-61 | 315.06 | 21.57 | 1.19 | 1.45 |
| 17-5/23-28 | 316.23 | 21.86 | 1.13 | 1.75 |
| 18-1/28-33 | 319.78 | 22.73 | 1.36 | 1.70 |
| 18-1/92-96 | 320.42 | 22.88 | 1.56 | 1.80 |
| 18-2/126-130 | 322.26 | 23.32 | 1.19 | 1.44 |
| 18-3/80-86 | 323.30 | 23.50 | 1.23 | 1.51 |
| 18-4/16-21 | 324.16 | 23.66 | 1.79 | 1.76 |
| 18-4/81-86 | 324.81 | 23.77 | 1.51 | 1.64 |
| 18-5/23-28 | 325.73 | 23.94 | 1.32 | 1.50 |
| 19-1/2-6 | 329.02 | 24.53 | 0.97 | 1.52 |
| 19-1/66-70 | 329.66 | 24.64 | 0.98 | 1.61 |
| 19-1/141-145 | 330.41 | 24.78 | 1.49 | 1.07 |
| 19-2/34-39 | 330.84 | 24.86 | 0.81 | 1.13 |
| 19-3/69-74 | 332.69 | 25.19 | 1.24 | 1.19 |
| 19-4/10-14 | 333.60 | 25.35 | 1.23 | 0.88 |
| 19-4/118-122 | 334.68 | 25.55 | 1.66 | 1.28 |
| 19-5/110-115 | 336.10 | 25.80 | 1.45 | 0.89 |
| 20-1/118-122 | 339.68 | 26.44 | 1.52 | 1.12 |
| 20-2/66-71 | 340.66 | 26.62 | 1.52 | 0.95 |
| 20-3/14-19 | 341.64 | 26.79 | 1.69 | 1.02 |
| 20-4/12-16 | 343.12 | 27.06 | 1.48 | 1.10 |
| 20-4/122-126 | 344.22 | 27.26 | 1.85 | 1.16 |
| 20-5/27-32 | 344.77 | 27.36 | 1.64 | 1.26 |
| 21-1/46-49 | 348.46 | 28.02 | 1.66 | 1.06 |
| 21-1/132-136 | 349.32 | 28.17 | 1.80 | 1.22 |
| 21-2/102-108 | 350.52 | 28.39 | 1.73 | 1.00 |
| 21-3/13-16 | 351.13 | 28.50 | 1.63 | 1.06 |
| 22-1/34-38 | 357.84 | 29.70 | 1.57 | 0.78 |
| 22-1/97-101 | 358.47 | 29.82 | 1.55 | 1.02 |
| 22-2/84-88 | 359.84 | 30.04 | 1.33 | 0.85 |
| 22-4/3-8 | 362.03 | 30.28 | 1.15 | 0.39 |
| 22-4/78-81 | 362.78 | 30.37 | 1.27 | 0.70 |
| 23-1/52-57 | 367.52 | 30.89 | 1.53 | 1.32 |
| 23-2/51-55 | 369.01 | 31.06 | 1.55 | 1.11 |
| 23-3/65-69 | 370.65 | 31.24 | 1.63 | 0.69 |
| 23-5/22-26 | 373.22 | 31.53 | 1.60 | 0.79 |
| 23-6/45-49 | 374.95 | 31.72 | 1.80 | 0.71 |
| 24-1/34-38 | 376.84 | 31.93 | 1.49 | 0.63 |
| 24-2/80-85 | 378.80 | 32.13 | 1.89 | 1.13 |
| 24-3/79-80 | 380.29 | 32.25 | 1.57 | 0.83 |
| 24-4/74-79 | 381.75 | 32.38 | 1.70 | 1.38 |
| 24cc | 382.40 | 32.73 | 1.43 | 1.12 |
| 25-1/117-121 | 387.17 | 32.83 | 1.62 | 1.19 |
| 26-4/22-26 | 400.22 | 35.56 | 1.68 | 1.52 |

TABLE 1.  (continued)

| Core-Section/ Depth, cm | Depth, msb | Age, Ma | $\delta^{18}O_{PDB}$ | $\delta^{13}C_{PDB}$ |
|---|---|---|---|---|
| | | | | |
| *Globorotalia opima nana*, Site 558 | | | | |
| | | | | |
| 18-1/92-96 | 320.42 | 22.88 | 1.02 | 1.73 |
| 18-4/16-21 | 324.16 | 23.66 | 1.29 | 1.81 |
| 19-1/141-145 | 330.41 | 24.78 | 1.22 | 0.87 |
| 19-3/69-74 | 332.29 | 25.12 | 0.72 | 1.14 |
| 19-4/118-122 | 334.68 | 25.55 | 0.38 | 0.93 |
| 20-2/66-71 | 340.66 | 26.62 | 0.47 | 0.69 |
| 20-4/122-126 | 344.22 | 27.26 | 0.71 | 0.84 |
| 21-1/132-136 | 349.32 | 28.17 | 0.76 | 0.69 |
| 21-2/102-108 | 350.52 | 28.39 | 0.85 | 0.79 |
| 22-1/34-38 | 357.84 | 29.70 | 1.14 | 0.87 |
| 22-4/3-8 | 362.03 | 30.28 | 1.36 | 0.84 |
| 23-1/52-57 | 367.52 | 30.89 | 1.40 | 0.95 |
| 24-4/75-79 | 381.75 | 32.38 | 1.25 | 1.22 |
| 25-1/117-121 | 387.17 | 32.83 | 1.05 | 1.20 |
| 25-4/85-89 | 391.35 | 33.63 | 1.06 | 0.98 |
| | | | | |
| *Globigerina angulisuturalis*, Site 558 | | | | |
| | | | | |
| 19-4/118-122 | 334.68 | 25.59 | 0.80 | 1.41 |
| 20-4/122-126 | 344.22 | 27.25 | 0.82 | 1.47 |
| | | | | |
| Concentrated Crystals From Benthic Foraminifera, Site 558 | | | | |
| | | | | |
| 26-4/22-26 | 400.22 | 35.35 | 1.89c | 1.32c |
| | | | | |
| *Cibicidoides* spp., Site 563 | | | | |
| | | | | |
| 1-1/119-125 | 157.69 | 8.06 | 2.27 | 1.29 |
| 1-3/18-24 | 159.68 | 8.29 | 2.28 | 1.03 |
| 2-1/33-38 | 166.33 | 8.73 | 2.30 | 1.05 |
| 2-4/34-39 | 170.84 | 9.08 | 2.41 | 1.04 |
| 3-1/35-40 | 175.85 | 9.47 | 2.40 | 1.23 |
| 3-4/36-41 | 180.36 | 9.82 | 2.23 | 1.25 |
| 4-1/37-42 | 185.37 | 10.22 | 2.03 | 0.84 |
| 4-4/30-36 | 189.80 | 10.58 | 1.61 | 0.94 |
| 5-2/130-136 | 197.30 | 11.25 | 2.24 | 1.44 |
| 5-5/130-136 | 201.80 | 11.66 | 2.25 | 1.36 |
| 6-2/130-136 | 206.80 | 12.10 | 2.24 | 0.87 |
| 6-5/121-127 | 211.21 | 12.50 | 1.91 | 1.07 |
| 7-2/131-136 | 216.31 | 12.95 | 2.19 | 1.15 |
| 7-5/110-116 | 220.60 | 13.33 | 2.12 | 1.10 |
| 8-2/114-120 | 225.64 | 13.77 | 2.27 | 1.12 |
| 8-5/114-120 | 230.14 | 14.17 | 2.14 | 1.19 |
| 8-1/21-26 | 232.71 | 14.40 | 2.17 | 1.70 |
| 9-2/114-120 | 235.14 | 14.62 | 1.96 | 1.72 |
| 9-3/104-109 | 236.54 | 14.74 | 0.97 | 1.01 |
| 9-4/109-114 | 238.09 | 14.88 | 1.26 | 1.00 |
| 9-5/114-120 | 239.64 | 15.02 | 1.28 | 1.32 |
| 9-6/113-118 | 241.13 | 15.15 | 1.32 | 1.21 |
| 10-2/90-96 | 244.40 | 15.48 | 1.53 | 1.95 |
| 10-3/105-113 | 246.05 | 15.67 | 1.15 | 1.72 |
| 10-4/95-101 | 247.45 | 15.82 | 1.05 | 1.82 |
| 10-5/93-98 | 248.93 | 15.99 | 1.64 | 2.07 |
| 10-5/110-116 | 249.10 | 16.01 | 1.22 | 1.43 |
| 10-6/90-95 | 250.40 | 16.15 | 1.44 | 1.77 |

TABLE 1.   (continued)

| Core-Section/ Depth, cm | Depth, msb | Age, Ma | $\delta^{18}O_{PDB}$ | $\delta^{13}C_{PDB}$ |
|---|---|---|---|---|
| | | *Cibicidoides* spp., Site 563 (continued) | | |
| 11-1/110-116 | 252.60 | 16.46 | 1.21 | 2.01 |
| 11-2/100-105 | 254.00 | 16.67 | 1.04 | 1.51 |
| 11-3/81-86 | 255.31 | 16.87 | 1.25 | 1.73 |
| 11-4/112-118 | 257.12 | 17.14 | 1.40 | 1.33 |
| 12-2/110-116 | 263.60 | 17.86 | 1.40 | 1.56 |
| 12-5/124-130 | 268.24 | 18.24 | 1.64 | 1.17 |
| 13-2/124-130 | 273.24 | 19.82 | 1.31 | 0.71 |
| 13-5/124-130 | 277.74 | 20.49 | 1.59 | 0.91 |
| 14-2/129-135 | 282.79 | 21.23 | 1.72 | 1.01 |
| 15-2/117-123 | 292.17 | 22.62 | 1.56 | 1.51 |
| 15-4/110-115 | 295.10 | 23.05 | 1.71 | 1.41 |
| 15-6/100-106 | 298.00 | 23.59 | 1.49 | 1.30 |
| 16-1/114-120 | 300.14 | 24.08 | 1.52 | 1.40 |
| 16-2/117-123 | 301.67 | 24.43 | 2.05 | 1.59 |
| 16-3/110-116 | 303.10 | 24.76 | 1.79 | 1.17 |
| | | | 1.84 | 1.30 |
| | | | 2.05 | 1.50 |
| 16-5/117-123 | 306.17 | 25.45 | 1.25 | 1.06 |
| 16-6/93-98 | 307.43 | 25.75 | 1.06 | 0.80 |
| 16-7/43-48 | 308.43 | 25.98 | 1.38 | 0.82 |
| 17-1/81-86 | 309.31 | 26.18 | 1.48 | 0.78 |
| 17-2/19-24 | 310.19 | 26.38 | 1.29 | 0.82 |
| 17-2/117-123 | 311.17 | 26.61 | 1.64a | 1.34a |
| 17-3/5-9 | 311.55 | 26.69 | 1.36 | 0.90 |
| 17-3/124-129 | 312.74 | 26.96 | 1.53 | 0.63 |
| 17-4/132-137 | 314.32 | 27.33 | 1.09 | 0.46 |
| 17-5/24-30 | 314.74 | 27.42 | 1.30 | 0.63 |
| 18-1/20-25 | 318.20 | 28.21 | 1.18 | 0.40 |
| 18-2/15-21 | 319.65 | 28.55 | 1.06 | 0.57 |
| 18-3/15-20 | 321.15 | 28.89 | 1.61 | 0.73 |
| 18-3/138-144 | 322.38 | 29.17 | 1.62 | 0.70 |
| 18-4/20-25 | 322.70 | 29.24 | 1.59 | 0.83 |
| 18-4/120-125 | 323.70 | 29.47 | 1.45 | 0.36 |
| 18-5/23-29 | 324.23 | 29.59 | 1.31 | 0.44 |
| 18-5/65-67 | 324.65 | 29.69 | 1.68 | 0.29 |
| 18cc | 324.70 | 29.70 | 1.51 | 0.55 |
| 19-1/12-18 | 327.62 | 30.36 | 1.58 | 0.46 |
| 19-2/7-12 | 329.07 | 30.68 | 1.37 | 0.48 |
| 19-3/12-18 | 330.62 | 31.03 | 1.35 | 0.55 |
| 20-1/24-28 | 337.24 | 32.51 | 1.59 | 0.79 |
| 20-2/41-47 | 338.91 | 32.88 | 1.40 | 0.93 |
| 20-4/20-26 | 341.70 | 33.10 | 1.59 | 0.70 |
| 20-7/4-10 | 346.04 | 33.51 | 1.55 | 0.93 |
| 21-1/110-116 | 347.60 | 33.55 | 1.75 | 1.29 |
| 21-2/112-116 | 349.12 | 33.67 | 1.15 | 0.63 |
| 21-3/90-95 | 350.40 | 33.76 | 1.47 | 1.06 |
| | | | 1.46 | 1.04 |
| 21-4/31-37 | 351.31 | 33.83 | 1.71 | 1.40 |
| | | | 1.58 | 1.18 |
| 21-5,20-25 | 352.70 | 33.91 | 1.33 | 1.79 |
| | | | 1.21 | 1.65 |
| 21-6/22-28 | 354.22 | 34.05 | 1.98 | 1.54 |
| 22-1/110-116 | 356.60 | 34.23 | 2.43 | 1.50 |
| 22-2/3-8 | 357.03 | 34.27 | 1.56 | 1.10 |
| | | | 1.63 | 1.16 |

TABLE 1.  (continued)

| Core-Section/ Depth, cm | Depth, msb | Age, Ma | $\delta^{18}O_{PDB}$ | $\delta^{13}C_{PDB}$ |
|---|---|---|---|---|
| | | Cibicidoides spp., Site 563 (continued) | | |
| 22-2/104-108 | 358.04 | 34.34 | 1.46 | 1.18 |
| 22-3/6-11 | 358.56 | 34.38 | 2.28 | 1.44 |
| 22-3/110-116 | 359.60 | 34.46 | 1.67 | 1.19 |
| | | | 1.74 | 1.06 |
| 22-3/145-149 | 359.95 | 34.49 | 1.74 | 0.95 |
| | | | 1.86 | 1.00 |
| 22CC | 360.10 | 34.50 | 1.63 | 1.03 |
| | | Catapsydrax spp., Site 563 | | |
| 13-2/124-130 | 273.24 | 19.82 | 0.82 | 1.04 |
| 13-5/124-130 | 277.74 | 20.49 | 0.78 | 1.07 |
| 14-2/129-135 | 282.79 | 21.24 | 0.99 | 1.02 |
| 15-2/117-123 | 292.17 | 22.63 | 0.97 | 1.62 |
| 15-5/117-123 | 296.67 | 23.31 | 1.05 | 1.63 |
| 16-2/117-123 | 301.67 | 24.45 | 1.22 | 1.69 |
| 16-5/117-123 | 306.17 | 25.47 | 1.05 | 1.01 |
| 17-2/117-123 | 311.17 | 26.62 | 1.26 | 1.17 |
| 17-5/24-30 | 314.74 | 27.43 | 1.48 | 1.12 |
| 18-2/23-29 | 319.73 | 28.57 | 1.51 | 0.98 |
| 18-5/23-29 | 324.23 | 29.60 | 1.55 | 0.76 |
| 19-1/12-18 | 327.62 | 30.36 | 1.76 | 0.94 |
| 19-3/12-18 | 330.62 | 31.03 | 1.14 | 1.06 |
| 20-2/41-47 | 338.91 | 32.88 | 1.23 | 1.26 |
| 20-4/20-26 | 341.70 | 33.10 | 1.30 | 1.26 |
| 20-7/4-10 | 346.04 | 33.51 | 1.08 | 1.27 |
| 21-1/140-146 | 347.60 | 33.55 | 0.91 | 1.21 |
| 21-4/31-37 | 351.31 | 33.83 | 1.08 | 1.42 |
| 22-1/110-116 | 356.60 | 34.23 | 0.84 | 1.30 |
| 22-3/110-116 | 359.60 | 34.46 | 0.91 | 1.30 |
| | | Globorotalia opima nana, Site 563 | | |
| 15-2/117-123 | 292.17 | 22.63 | 0.70 | 1.72 |
| 15-4/110-115 | 295.10 | 23.31 | 0.98 | 1.63 |
| 16-1/114-120 | 300.14 | 24.10 | 1.08 | 1.82 |
| 16-3/110-116 | 303.10 | 24.78 | 1.10 | 1.59 |
| 16-5/117-123 | 306.17 | 25.01 | 0.07 | 0.83 |
| 16-4/109-115 | 304.59 | 25.12 | 0.64 | 1.13 |
| 17-2/117-123 | 311.17 | 26.62 | 0.15 | 0.95 |
| 17-5/24-30 | 314.74 | 27.43 | 0.38 | 0.68 |
| 18-2/23-29 | 319.73 | 28.57 | 0.46 | 0.91 |
| 18-5/23-29 | 324.23 | 29.60 | 0.74 | 0.84 |
| 19-1/12-18 | 327.62 | 30.36 | 0.37 | 0.48 |
| 19-3/12-18 | 330.62 | 31.03 | 1.08 | 0.78 |
| 20-1/24-28 | 337.24 | 32.51 | 1.12 | 1.11 |
| 20-2/41-47 | 338.91 | 33.10 | 1.44 | 1.18 |
| 21-1/140-146 | 347.60 | 33.55 | 0.97 | 0.95 |
| 21-4/31-37 | 351.31 | 33.83 | 0.97 | 1.27 |

TABLE 1.  (continued)

| Core-Section/ Depth, cm | Depth, msb | Age, Ma | $\delta^{18}O_{PDB}$ | $\delta^{13}C_{PDB}$ |
|---|---|---|---|---|
| | | | | |

*Globorotalia opima opima*, Site 563

| Core-Section/ Depth, cm | Depth, msb | Age, Ma | $\delta^{18}O_{PDB}$ | $\delta^{13}C_{PDB}$ |
|---|---|---|---|---|
| 20-2/41-47 | 338.91 | 32.88 | 1.10 | 1.08 |
| 20-4/20-26 | 341.70 | 33.10 | 1.16 | 0.75 |
| 20-7/4-10 | 346.04 | 33.51 | 0.95 | 1.03 |
| 21-4/31-37 | 351.31 | 33.83 | 0.83 | 1.50 |

In samples with replicates, mean values were plotted in Figures 1-4. Age estimates were obtained by interpolating between biostratigraphic datum levels and magnetostratigraphic chron boundaries (Figures 1 and 2). Slight differences between $\delta^{18}O$ values reported here and those of Miller and Fairbanks [1983] are due to use of H. Craig's (personal communication, 1978, to Blattner and Hulston [1978]) correction to PDB standard.

a Possible recrystallization (see text), data not plotted.

b Large crystals noted; visible crystals removed from test by sonifying. Data plotted in Figures 2 and 3.

c Corresponding $\delta^{18}O$ values for this level: <u>Cibicidoides</u>, 1.5°/oo; <u>Catapsydrax</u>, 1.7°/oo.

records of <u>Cibicidoides</u> from the Atlantic and Pacific. (We use the term cycle here to refer to periodic or aperiodic sequences of changes in which the initial conditions are reestablished.) The $\delta^{13}C$ cycles also have been observed in planktonic foraminifera [Shackleton et al., 1984; Vincent et al., 1984; Vincent and Berger, this volume; this study], strongly suggesting that these $\delta^{13}C$ cycles resulted from global changes in the $\delta^{13}C$ of mean ocean water. These cycles occurred with periods of 7-12 m.y. and amplitudes of approximately 1.0°/oo, and are similar in amplitude to late Quaternary high-frequency ($10^4$-$10^5$ year) benthic and planktonic foraminiferal $\delta^{13}C$ cycles [Shackleton et al., 1983; Mix and Fairbanks, 1984; Fairbanks and Mix, 1984].

Differences in the amplitudes between Atlantic and Pacific benthic foraminiferal $\delta^{13}C$ cycles can be used to reconstruct abyssal circulation [e.g., Shackleton et al., 1983; Mix and Fairbanks, 1984]. Today, Atlantic deep and bottom waters are 1°/oo higher in $\delta^{13}C$ than are Pacific bottom waters, reflecting production of "young," $^{13}C$-enriched North Atlantic Deep Water (NADW)[Kroopnick et al., 1972; Kroopnick 1974, 1980, 1984]. Blanc et al. [1980], Blanc and Duplessy [1982], Miller and Fairbanks [1983] and Miller and Thomas [1984] attempted to isolate global from abyssal circulation $\delta^{13}C$ changes in the Tertiary by comparing carbon isotopic records from the Atlantic and the Pacific. This approach is similar to that of Curry and Lohmann [1982],

Shackleton et al. [1983], and Mix and Fairbanks [1984] for the late Quaternary, and is used in this study. We demonstrate that the Atlantic also was enriched in $^{13}C$ relative to the Pacific from 36-33 Ma (early Oligocene) and 26-10 Ma (late Oligocene to late Miocene), suggesting supply of $^{13}C$-enriched bottom water in the North Atlantic, perhaps in a mode analogous to modern NADW.

Tertiary paleoclimatic studies generally have assumed that the earth was substantially ice free prior to about 15 Ma (middle Miocene), and therefore $\delta^{18}O$ history prior to 15 Ma reflected only temperature changes [Shackleton and Kennett, 1975; Savin et al., 1975, 1981; Savin, 1977; Woodruff et al., 1981]. However, recent studies of benthic foraminifera indicate that continental ice sheets must have existed in the Oligocene at approximately 36-35 Ma and 31-28 Ma [Miller and Fairbanks, 1983; Keigwin and Keller, 1984; Miller and Thomas, 1984; Poore and Matthews, 1984; Shackleton et al., 1984]. We present further evidence that supports the existence of Oligocene ice sheets at 31-28 Ma, and new data which suggest glaciation at about 25-24 Ma (below the Oligocene/Miocene boundary). The $\delta^{13}C$ cycles are compared with the record of glaciation to assess whether the cycles were related to sea-level changes as they were in the Quaternary.

Methods and Samples

Isotopic measurements on benthic and planktonic foraminifera were obtained for the Oligocene to

Miocene sections at DSDP Sites 558 (37°46.24'N, 37°20.61'W; 3754 m present water depth) and 563 (33°38.53'N, 43°46.04'W; 3796 m present water depth) in the western North Atlantic (Figures 1-3; Table 1; see also Miller and Fairbanks [1983]). These two sites were selected for their relatively continuous records and for the excellent age control provided by magnetobiostratigraphy [Khan et al., 1984]. Our Atlantic data are compared with benthic foraminiferal isotopic data from the following/ Pacific sites: eastern equatorial Site 574 (03°59.24'N, 114°8.53'W; 4561 m present depth [Miller and Thomas, 1984]), eastern equatorial Site 77 (00°28.90'N, 133°13.70'W; 4291 m present depth [Savin et al., 1981; Keigwin and Keller, 1984]), and western equatorial Site 289 (00°29.92'S,158°30.69'E; 2206 m present depth [Savin et al., 1981]).

Both Site 558 and Site 563 are located beneath the subtropical gyre. The surface waters in subtropical gyres are generally low in nutrients and nearly uniform in carbon isotopic compostion [Kroopnick, 1984]. The carbon isotopic records of planktonic foraminifera from subtropical gyres are not subject to changes caused by variable upwelling rates of waters rich in nutrients and low in $\delta^{13}C$. Since the North Atlantic subtropical gyre was a stable feature throughout the Tertiary [Berggren and Hollister, 1974], Sites 558 and 563 are particularly suitable locations for obtaining Tertiary global carbon isotopic records.

All benthic foraminiferal isotopic analyses were performed on mixed species of Cibicidoides (including Planulina wuellerstorfi, which many authors place in Cibicidoides wuellerstorfi). Cibicidoides spp. (including P. wuellerstorfi) secrete calcite tests that are consistently offset from $\delta^{18}O$ equilibria by approximately 0.65°/oo [Shackleton and Opdyke, 1973; Duplessy et al., 1980; Woodruff et al., 1980; Belanger et al., 1981; Graham et al., 1981]. In addition, studies of Holocene core tops have demonstrated that Cibicidoides accurately reflects the distribution of $^{13}C$ of $\Sigma CO_2$ in the modern ocean [Belanger et al., 1981; Graham et al., 1981].

Planktonic foraminiferal isotopic analyses were performed on Globorotalia opima nana (Figures 1 and 2), the taxon which was generally most depleted in $^{18}O$ at Sites 558 and 563. Analyses of G. opima opima yielded values similar to those for G. opima nana, while analyses of Globigerina angulisuturalis yielded values similar to or more enriched than those for G. opima nana (Table 1). Poore and Matthews [1984] noted that compared with other species at Site 366 (equatorial Atlantic), G. opima nana exhibited intermediate $\delta^{18}O$ values; however at higher latitudes (Site 522, 26°S), similar to our locations, this taxon (and G. opima opima) appears to have been among the most depleted in $^{18}O$ [Poore and Matthews, 1984]. Therefore, we assume that it calcified in the surface mixed layer at Sites 558 and 563. Analyses of the planktonic foraminifera Catapsydrax spp. (Figure 3; Table 1) showed that

this was consistently the most enriched planktonic taxon at this location; Poore and Matthews [1984] similarly noted that this was the most enriched planktonic taxon at Site 366.

Samples were washed with sodium hexametaphosphate and/or hydrogen peroxide in tap water through a 63-$\mu$m sieve and air dried. Benthic foraminifera were ultrasonically cleaned in distilled water for 10-20 s, crushed, and roasted at 370°C in a vacuum to remove organic matter. Similar procedures were followed for planktonic foraminifera, although they were not sonified due to susceptibility to breakage. The $CaCO_3$ was reacted with $H_3PO_4$ at 50°C, and the gas was analyzed with a VG Micromass 903E mass spectrometer following standard procedures.

Ages are reported using the Berggren et al. [1984a, b] time scale by linearly interpolating between magnetostratigraphic chron boundaries where possible and using planktonic foraminiferal datum levels where magnetostratigraphic data are limited (Figures 1-4). With such magnetobiostratigraphic control [Khan et al., 1984], we believe that, in general, the Atlantic age models are reliable within 0.5 m.y. or better. An exception to this occurs in the lower Oligocene of Site 563, where the age models are uncertain. Pacific age models are based entirely upon biostratigraphy (Figure 4), and are therefore less accurate. Pacific isotope data are sparse, and the Pacific age model is tentative for the time interval 26-21 Ma.

Paleodepths for Sites 558 and 563 are estimated by "backtracking" [Sclater et al., 1971, 1977; Berger and Winterer, 1974] using the empirical western North Atlantic age versus subsidence curve [Tucholke and Vogt, 1979]. These sites subsided from approximately 2200-2400 m during the earliest Oligocene (36 Ma) to 3200 m during the latest Oligocene (24 Ma) and to 3600 m during the late middle Miocene (11 Ma).

The average benthic foraminiferal sampling intervals at Sites 558 and 563 are between 0.3 and 0.4 m.y. Planktonic foraminiferal sampling intervals are typically 0.7 m.y., although we anlayzed at 0.3 m.y. sampling interval for Catapsydrax spp. at Site 558 (Figure 3). If high-frequency changes similar to Quaternary isotopic fluctuations (on the order of $10^4$-$10^5$ years) occurred in the Oligocene-Miocene, our sampling may have introduced aliasing. However, the isotopic records from both sites, including planktonic and benthic data, show similar patterns (Figures 1-3), despite differences in sampling intervals. Therefore, the sampling interval is probably sufficient to resolve isotopic changes on the order of $10^6$-$10^7$ years.

## Results

### Oxygen Isotope Stratigraphy

Oxygen isotope data provide (1) constraints on ice-volume and temperature changes; (2)

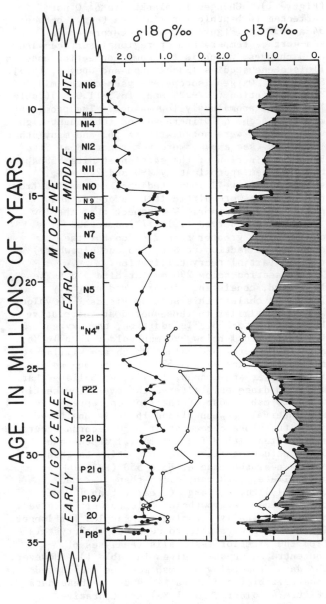

Fig. 1. Site 563 oxygen and carbon isotopic composition for the benthic foraminifera _Cibicidoides_ (solid circles) and the planktonic foraminifera _Globorotalia opima nana_ (open circles). The age model was obtained by interpolation between the following levels: base chron C4An, 159.5 msb, 8.2 Ma; base chron C5n, 188.0 msb, 10.4 Ma; top chron C5An, 204.5 msb, 11.9 Ma; base chron C5Bn, 242.5 msb, 15.3 Ma; top chron C5Cn, 251.0 msb, 16.2 Ma; top chron C5Dn, 260.0 msb, 17.6 Ma; base chron C5Dn, 267.0 msb, 18.1 Ma; unconformity 270.5 msb from 18.4 to 19.4 Ma; base chron C6n, 277.5 msb, 20.5 Ma; top chron C6Cn, 296.5 msb, 23.3 Ma; last occurrence (LO) of _Chiloguembelina_, 326.0 msb, 30.0 Ma; base chron C12n, 339.0 msb, 32.9 Ma; LO of _Pseudohastigerina_, 353.5 msb, 34.0 Ma.

supplementary stratigraphic control, especially in correlating sites without magnetostratigraphy (e.g., Sites 77, 289, 574); and (3) evidence for the suitability of our samples for carbon isotopic studies, allowing us to screen sections altered by diagensis.

Site 563. Rapid increases occurred in benthic foraminiferal $\delta^{18}O$ values (Figure 1) at 14.7 Ma (middle Miocene) and 25-24 Ma (immediately below the Oligocene/Miocene boundary). Peak values at these times were greater than 2.0°/oo. The 14.7 Ma event precisely correlates with a well-documented benthic foraminiferal increase noted in the Atlantic, Pacific, and southern oceans [e.g., Savin et al., 1975, 1981; Shackleton and Kennett, 1975; Boersma and Shackleton, 1977], providing control on stratigraphic correlations between basins. The 25-24 Ma increase has been detected elsewhere only at Site 558 [this study]. This apparently results from a previous lack of study of uppermost Oligocene sediments.

An interval of high benthic foraminiferal $\delta^{18}O$ values (greater than 2.0°/oo) occurred at other DSDP sites near the early/late Oligocene boundary (between about 31 and 28 Ma (Table 2)); such high values were not observed at Site 563 (Figure 1) (values only attained 1.7°/oo). However, at Site 563 some of the section representing the time 32-28 Ma probably is missing and/or mechanically disturbed as suggested by (1) poor recovery precluded the identification of magnetochrons C10 and C11 (32.6-29.7 Ma), and it is in these chrons that highest $\delta^{18}O$ values are expected (see below); (2) chrons C8n and C9n (29.7-26.9 Ma) are concatenated and a hiatus is indicated [Khan et al., 1984]; and (3) drilling disturbance between 335 meters subbottom (msb) and 324 msb (estimated age 31.5-29.5 Ma) may have mixed the samples, lowering the magnitude of peak values.

High benthic foraminiferal $\delta^{18}O$ values (greater than 2.0°/oo) occurred near the base of Site 563 (Figure 1). These values may correlate with peak $\delta^{18}O$ values noted at other DSDP sites at 36-35 Ma following the well-established increase that occurred just above the Eocene/Oligocene boundary [e.g., Savin et al., 1975; Shackleton and Kennett, 1975; Kennett and Shackleton, 1976; Keigwin, 1980; Miller and Curry, 1982; Belanger and Matthews, 1984; Keigwin and Keller, 1984; Miller and Thomas, 1984; Miller et al., 1984a; Oberhansli and Toumarkine, 1984; Poore and Matthews, 1984; Shackleton et al., 1984; Snyder et al., 1984]. Although the estimated age for these high $\delta^{18}O$ values at Site 563 is 34.5-34.0 Ma (Figure 1), we caution that the lower Oligocene age model (below 350 msb) at Site 563 is based upon extrapolation of sedimentation rates from biostratigraphic levels. We believe that these peak $\delta^{18}O$ values at Site 563 correlate with similar high values noted at 36-35 Ma elsewhere; this is within uncertainties in the age control at the base of Site 563.

The planktonic foraminiferal $\delta^{18}O$ record of _Globorotalia opima nana_ also exhibited high values

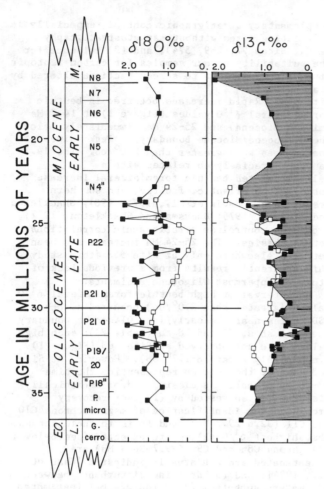

Fig. 2. Site 558 oxygen and carbon isotopic composition for the benthic foraminifera Cibicidoides (solid squares) and planktonic foraminifera Globorotalia opima nana (open squares). The age model was obtained by interpolation between the following levels: top chron C5n, 180.0 msb; 8.9 Ma; base chron C5n, 218 msb, 10.4 Ma; base chron C5Bn, 277.5 msb, 15.3 Ma; base chron C5Cn, 298.5 msb, 17.0 Ma; top chron C5Dn, 301.5 msb, 17.6 Ma; base chron C5Dn, 304.0 msb, 18.1 Ma; unconformity at 307.5 msb from 19.0 to 19.7 Ma; base chron C6n, 310.5 msb, 20.5 Ma; top chron C6Cn, 322.0 msb, 23.3 Ma; middle chron C10, 359.5 msb, 30.0 Ma; base chron C11n, 378.0 msb, 32.1 Ma; base chron C12n, 388.0 msb, 32.9 Ma; top chron C13n, 399.0 msb, 35.3 Ma; base Oligocene, 405.8 msb, 36.5 Ma.

at 25–24 Ma (Figure 1). Covariance of planktonic and benthic foraminiferal $\delta^{18}O$ at 25–24 Ma can be estimated as a minimum of $0.5°/oo$. Using a strategy applied to Quaternary studies [Shackleton and Opdyke, 1973], we interpret planktonic-benthic covariance as resulting from ice-volume changes. High $\delta^{18}O$ values also occurred in G. o. nana at about 34–31 Ma when benthic values were low

(Figure 1). Changes in planktonic $\delta^{18}O$ not reflected in benthic foraminifera (e.g., between 34 and 31 Ma (Figure 1)) are interpreted as sea-surface temperature or regional surface-water $\delta^{18}O$ changes, which is the same convention used in Pleistocene studies [Shackleton and Opdyke, 1973]. Site 558. Oxygen isotope analyses of the middle Miocene section (292–225 msb, 16.5–11.0 Ma)(Table 1) yielded anomalously low values. The prominent middle Miocene $\delta^{18}O$ increase and subsequent high values which were noted at Site 563 and many other locations (see above) were not recorded at Site 558. Examination of the section at Site 558 shows that between approximately 290 and 220 msb, sediments are well indurated chalks; samples from this section were difficult to process, and preservation is poor. We suspect that the low $\delta^{18}O$ values observed ($1.1-1.7°/oo$ versus consistently greater than $2.0°/oo$ for the correlative section at Site 563 and elsewhere) reflect partial recrystallization.

The section below 290 msb at Site 558 is less lithified, consisting of oozes and slightly indurated chalks; this section yields $\delta^{18}O$ values which are similar to those noted at correlative levels at Site 563. In addition, the oxygen isotopic record below 290 msb (older than 16 Ma) reveals patterns and values which are similar to records at Sites 77 and 574 [Keigwin and Keller, 1984; Miller and Thomas, 1984]. We believe that although diagenesis altered the overlying section (290–220 msb), most of the section representing the interval between 31 and 16 Ma is unaltered, since the close agreement of $\delta^{18}O$ records occurs among sites with different burial depths and diagenetic histories. The exception to this occurs near the base of Hole 558 (below approximately 370 msb; older than 31.2 Ma (Table 1)); here, large (greater than 150 μm) crystalline overgrowths of calcite were observed. By ultrasonically cleaning the samples for longer intervals we were able to separate the crystals from the exterior test walls (analyses of concentrated crystals given in Table 1). However, the data from below 370 msb must be considered less reliable, and comparisons using these data (Figure 4, older than 31 Ma) are tentative.

The benthic foraminiferal $\delta^{18}O$ record at Site 558 (Figure 2) shows that high values occurred at 25–24 Ma; high values also occurred immediately above and below the early/late Oligocene boundary (Figure 2; values exceeding $2.0°/oo$ in both intervals). Planktonic foraminiferal (Globorotalia opima nana) $\delta^{18}O$ values also were high at 25–24 Ma and near the early/late Oligocene boundary (Figure 2). The high benthic and planktonic foraminiferal $\delta^{18}O$ values noted between 25 and 24 Ma at Site 558 correlate with similar peak values at Site 563 (Figure 1). Although planktonic and benthic records may exhibit covariance of greater than $0.5°/oo$ at 25 Ma (Figure 1; Table 2), the increase in planktonics at about 25 Ma apparently began prior to the benthic foraminiferal $\delta^{18}O$ increase (Figure 2).

This apparent lead-lag relationship may be an artifact of sampling and/or aliasing of the record.

The record recovered at Site 558 representing the time interval 32-28 Ma is much more complete than that same interval at Site 563. Magnetochrons C9 through C12 (approximately 32-28 Ma) are well represented at Site 558 [Khan et al., 1984], suggesting that the record is complete. High benthic and planktonic foraminiferal $\delta^{18}O$ values occurred near the early/late Oligocene boundary (Figure 2); oxygen isotopic values increased prior to 31 Ma at Site 558 and reached maximum values from 31 to 30 Ma, with benthic values remaining high until 28 Ma (Figure 2). At Sites 574 and 77 in the Pacific, similar high $\delta^{18}O$ values in Cibicidoides were noted at 29-30 Ma [Keigwin and Keller, 1984; Miller and Thomas, 1984]. This slight apparent time discrepancy can be attributed to uncertainties in Pacific biostratigraphy. Given such uncertainties, we estimate the timing of this $\delta^{18}O$ event as follows. High values at Site 558 are associated with chrons C11n (top) to C9n and planktonic foraminiferal Zones P21a and P21b, constraining peak values to about 31-28 Ma. This is consistent with the observation of high values in Zones P21a/P21b at Sites 77 and 574. The timing of the change from lower to higher values at Site 558 apparently occurred in chron C11 at about 32-31 Ma (Figure 2), although the data older than 31 Ma may have been affected by diagnesis. However, this estimate is supported by data from Site 77 [Keigwin and Keller, 1984], where the change began within Zone P21a, constraining the increase to about 31±0.5 Ma.

At both Site 558 and Site 563, Catapsydrax spp. are consistently the most isotopically heavy planktonic foraminifera. The $\delta^{18}O$ record of this taxon (Figure 3) shows little correspondence to the record of benthic foraminifera or to Globorotalia opima nana (Figures 1-3).

## Global $\delta^{13}C$ Cycles

Three cycles of benthic foraminiferal $\delta^{13}C$ occurred at Site 563 in the Oligocene-Miocene (Figure 1). High values occurred at 36-33 Ma (within the early Oligocene), 26-22 Ma (across the Oligocene/Miocene boundary), and 18-14 Ma (across the early/middle Miocene boundary (Figure 1)). Nearly the same pattern occurred in planktonic foraminifera (in both Catapsydrax spp. and Globorotalia opima nana) at both subtropical gyre Sites 558 and 563 during the Oligocene to early Miocene (Figures 1-3). At Site 558, the same pattern can be seen in Oligocene to middle Miocene benthic foraminifera (Figure 2; Table 1).

The early to late Oligocene cycle also has been noted in benthic foraminifera from the Pacific [Keigwin and Keller, 1984; Miller and Thomas, 1984], South Atlantic [Poore and Matthews, 1984; Shackleton et al., 1984], Gulf of Mexico [Belanger and Matthews, 1984], and Bay of Biscay [Miller and

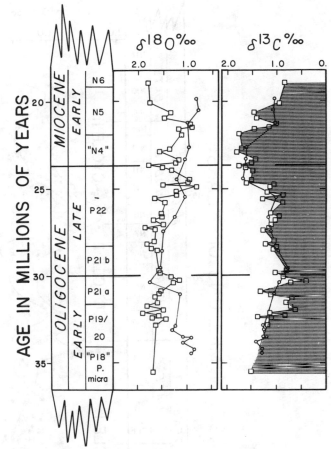

Fig. 3. Site 558 (open squares) and Site 563 (open circles) oxygen and carbon isotopic composition for the planktonic foraminifera Catapsydrax. Age models are as in Figures 1 and 2.

Curry, 1982]. Shackleton et al. [1984] also noted a similar change in Oligocene South Atlantic planktonic foraminifera. In the Miocene, the $\delta^{13}C$ cycles can be observed in both benthic and planktonic foraminifera from the Pacific [Savin et al., 1981; Vincent et al., 1984], Indian [Vincent and Berger, this volume], and South Atlantic [Shackleton et al., 1984] oceans.

An increase in benthic foraminiferal $\delta^{13}C$ also occurred across the Eocene/Oligocene boundary at Pacific and North Atlantic sites [Miller and Curry, 1982; Keigwin and Keller, 1984; Miller and Thomas, 1984; Miller et al., 1984a]. Keigwin [1980] and Oberhansli and Toumarkine [1984] noted a $\delta^{13}C$ increase in both benthic and planktonic foraminifera across the Eocene/Oligocene boundary in the Pacific and South Atlantic oceans, respectively.

It is apparent that benthic and planktonic foraminifera recorded three worldwide Oligocene-Miocene $\delta^{13}C$ cycles with amplitudes of approximately $0.5-1.0^{o}/oo$. Choosing $\delta^{13}C$ maxima as the boundaries (Figure 4), cycle 1 lasted about

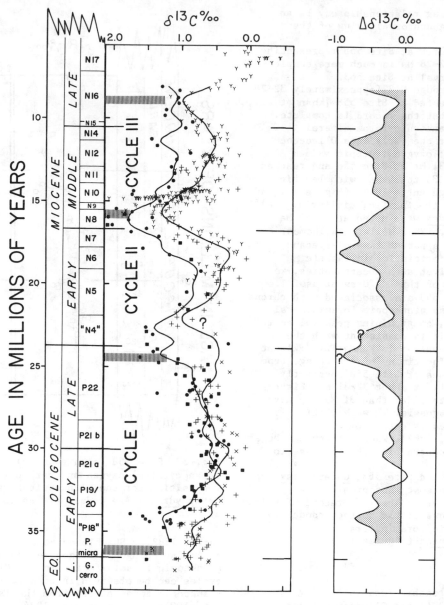

Fig. 4.  Comparison of benthic foraminiferal (Cibicidoides)  $\delta^{13}C$ from Atlantic Sites
558 (solid squares) and 563 (solid circles) with Pacific Sites 574 (crosses (data from
Miller and Thomas [1984])), 77 (pluses (data from Keigwin and Keller [1984] and Savin
et al. [1981])), and 289 (letter Y (data from Savin et al. [1981])).  Curves were
obtained by separately interpolating Atlantic and Pacific data to constant time
intervals (0.2 m.y.) and smoothing with an 11-point Gaussian convolution filter,
eliminating frequencies less than approximately 1/m.y.  Global cycle boundaries were
chosen as the most enriched interval constrained by the Pacific benthic record, except
the cycle boundary at about 24.5 Ma, which was chosen on the basis of the Atlantic
record (Pacific age model is unreliable at this time).  The difference curve was
obtained by subtracting the smoothed Atlantic   $\delta^{13}C$ record from the smoothed Pacific
record.  The solid line is drawn through the 0.0°/oo difference, and regions less than
zero are shaded.  Pacific isotope data are sparse, and Pacfic age models are
considered unreliable in the time interval 26–21 Ma.  Age models for Atlantic sites
are from Figures 1 and 2.  Those for Pacific sites are as follows: Site 574, after
Mayer, Theyer et al. [1984]; Site 77, after Miller and Fairbanks [1983] and Miller and
Thomas [1984]; Site 289, model obtained by interpolating between last occurrence of
Globorotalia margaritae (156.5 msb, 5.6 Ma)[Srinivasan and Kennett, 1981] and other
biostratigraphic datum levels cited by Miller and Fairbanks [1983].

12 m.y. (from about 36.5 to 24.5 Ma), cycle 2 lasted about 8.5 m.y. (from about 24.5 to 16 Ma), and cycle 3 lasted about 7 m.y. (from about 16 to 9 Ma).

Since the cycles delineated here occurred in benthic and planktonic foraminifera from many locations, they must have resulted from global changes in $\delta^{13}C$ of mean ocean water. Several mechanisms have been proposed to explain Pleistocene $\delta^{13}C$ variations with shorter periodicities ($10^3$–$10^4$ years), including changes in the size of the terrestrial organic carbon reservoir [Shackleton, 1977] and amount of organic carbon deposited on continental margins, presumably in response to sea-level changes [Broecker, 1982]. However, since the residence time of carbon in the oceans is approximately 1.8 x $10^5$ years [Froelich et al., 1982], these mechanisms cannot be invoked to explain the cycles delineated here. For example, a one-time transfer of organic carbon from the terrestrial biosphere to the ocean would result in a rapid $\delta^{13}C$ change in planktonic and benthic foraminifera, such as that measured in the Pleistocene. However, the average carbon isotopic composition of carbonate sediments would eventually be restored to the average carbon isotopic composition of incoming carbon derived from weathering rocks. The time it takes to replace the ocean carbon inventory is short (1.8 x $10^5$ years) compared with the average period of the carbon isotopic cycles (9 x $10^6$ years) delineated here. Thus, Oligocene-Miocene $\delta^{13}C$ cycles were not controlled by transfer of organic carbon between the terrestrial biosphere and ocean, and therefore must represent changes in $\delta^{13}C$ of riverine bicarbonate (input) or changes in the ratio of the amount of carbon buried as calcium carbonate to the amount buried as organic carbon (output).

Ronov [1968] estimated that 18% of sedimentary carbon is stored in the geological record as organic carbon and that the remaining 82% of carbon is stored in the form of carbonate. Sedimentary organic carbon is approximately $-25^o$/oo in $\delta^{13}C$ [Craig, 1953], and carbonate carbon is approximately $0.5^o$/oo [Keith and Weber, 1964], resulting in the following estimate for riverine (riv) $CO_2$:

$$input = output$$
$$\delta^{13}C_{CO_2}(riv) = f(\delta^{13}C_{CaCO_3}) + 1-f(\delta^{13}C_{CHO_2}) = 4./1^o/oo$$

where f is the fraction of carbon stored in the form of carbonate.

The $1^o$/oo changes noted in the Oligocene-Miocene $\delta^{13}C$ record (Figures 1-4) can be explained by changes in the ratio of $CaCO_3$ carbon to organic carbon in riverine input. Alternatively, the $1^o$/oo enrichment can be explained by a decrease in the fraction of $CaCO_3$ carbon deposited in marine sediments from 0.82 to 0.79. These estimates assume that organic carbon $\delta^{13}C$ remained constant. A decrease in $\delta^{13}C$ of organic carbon from $-25^o$/oo to approximately $-30^o$/oo would also explain a $1^o$/oo increase in $\delta^{13}C$ in $CaCO_3$. We know of no

measurements of organic carbon variability which support this possibility. The carbon isotope mass balance is determined here by the ratio of organic carbon to carbonate carbon and implies nothing about the total amount of carbon stored in the oceans.

## Discussion

We attempt to determine if there was an underlying oceanographic cause for the Tertiary global carbon cycles by comparing proxy indicators for sea-level and climatic changes with the $\delta^{13}C$ cycles. Such proxy indicators are provided by Tertiary records of ice volume-$\delta^{18}O$ history [this study], changes in depositional patterns on continental margins [Vail et al., 1977], sea-floor spreading rates [Pitman, 1978], and climatic changes recorded by terrestrial vegetation [Wolfe, 1978].

## Glacio-eustasy and the $\delta^{18}O$ Record

Based upon warm bottom-water temperatures ($5^o$–$9^oC$) inferred from analyses of mixed assemblages of benthic foraminifera, initial studies concluded that the earth was substantially ice free prior to the middle Miocene (about 15 Ma)[Shackleton and Kennett, 1975; Savin et al., 1975, 1981; Savin, 1977; Woodruff et al., 1981]. Matthews and Poore [1980] questioned the assumption of an ice-free world prior to 15 Ma, and Oligocene isotopic data presented here and elsewhere support their contention. The $\delta^{18}O$ data discussed here from Atlantic and Pacific DSDP sites show that high benthic foraminiferal values (greater than $2.0^o$/oo in Cibicidoides) occurred at 36-35 Ma, 31-28 Ma, 25-24 Ma, and younger than 15 Ma (Figure 1; Table 2). These $\delta^{18}O$ values indicate that in the Oligocene, bottom-water temperatures were as cold as or colder than at present (typically $2^oC$), assuming an ice-free world [Shackleton and Kennett, 1975; Graham et al., 1981]. Such cold bottom waters require a cold source region and are incompatible with an ice-free world. Therefore, significant continental ice volume must have existed in the Oligocene and from the middle Miocene to present [Miller and Fairbanks, 1983; Keigwin and Keller, 1984; Miller and Thomas, 1984; Poore and Matthews, 1984; Shackleton et al., 1984].

Covariance of planktonic and benthic foraminiferal $\delta^{18}O$ is used in the Quaternary as an ice-volume indicator [Shackleton and Opdyke, 1973]. Increases of at least $0.3$–$0.5^o$/oo apparently occurred in both planktonic and benthic foraminifera at about 36 Ma, 31 Ma, and 25 Ma (Table 2), suggesting that ice volume increased at these times. Using the Quaternary sea level-$\delta^{18}O$ calibration [Fairbanks and Matthews, 1978], sea level fell by 30-50 m at 36 Ma, 31 Ma, and 25 Ma. This implies the development of ice sheets at these times equal to one half of the size of modern ice sheets. Savin et al. [1975] suggested that the major middle Miocene (15 Ma) benthic foraminiferal $\delta^{18}O$ increase (Figure 1) may not

TABLE 2. Magnitude of Peak $\delta^{18}O$ Values for <u>Cibicidoides</u> (except as noted) and Planktonic-Benthic $\delta^{18}O$ Covariance

| DSDP Site | Location | Peak $\delta^{18}O$ <sup>o</sup>/oo, PDB | Covariance | Reference |
|---|---|---|---|---|
| | | Latest Oligocene (24-25 Ma event) | | |
| Site 558 | western North Atlantic | 2.1 | 0.5-0.9 | this study |
| Site 563 | western North Atlantic | 2.1 | 0.4-0.9 | this study |
| | | "Middle" Oligocene (28-31 Ma Event) | | |
| Site 77 | equatorial Pacific | 2.3 | ?0.4 | Keigwin and Keller [1984] |
| Site 574 | equatorial Pacific | 2.3 | a | Miller & Thomas [1984] |
| Site 558 | western North Atlantic | 2.1 | ?0.4[b] | this study |
| | | Earliest Oligocene (35-36 Ma Event) | | |
| Site 292 | Phillipine Sea | 2.3[c] | ?0.3 | Keigwin [1980] |
| Site 277 | southern ocean | 1.6[c] | 1.0 | Keigwin [1980] |
| Site 522 | South Atlantic | 2.5[d] | 0.5 | Oberhansli and Toumarkine [1984] |
| Site 119 | Bay of Biscay | 2.0 | a | Miller & Curry [1982] |

[a] no planktonic foraminiferal data available.
[b] based, in part, upon data with some recrystallization.
[c] = <u>Oridorsalis</u> spp.
[d] = <u>Stilostomella</u> spp.

have been accompanied by an increase in planktonic foraminiferal $\delta^{18}O$. However, Vincent and Berger [this volume] and Vincent et al. [1984] noted a synchronous, albeit smaller, increase in planktonic $\delta^{18}O$ at low latitude Pacific and Indian Ocean sites. Shackleton and Kennett [1975] noted covariance between benthic and planktonic foraminifera at high latitudes at about 15 Ma, and suggested that ice sheets developed at this time. We tentatively accept the premise that the middle Miocene (15 Ma) benthic foraminiferal increase was associated, in part, with an increase in ice volume. Thus, we infer that ice-volume increases caused sea-level falls at about 36, 31, 25, and 15 Ma.

Comparison of our inferred glacio-eustatic history with the carbon isotopic cycles shows no consistent response of carbon isotopes to sea-level changes:

1. The global $\delta^{13}C$ increases that occurred at the beginning of the Oligocene (about 36 Ma) and near the Oligocene/Miocene boundary (about 25 Ma) were associated with $\delta^{18}O$ increases interpreted as sea-level lowerings.

2. The inferred glacio-eustatic lowering of 31-28 Ma is not associated with a change in $\delta^{13}C$.

3. The global $\delta^{13}C$ increase beginning at 18 Ma leads a major $\delta^{18}O$ increase (inferred sea-level fall) by 2-3 million years (see also Vincent and Berger [this volume]). This inconsistent relationship precludes a simple causal relationship between sea-level changes and the $\delta^{13}C$ cycles.

Other Proxy Indicators of Sea level

Pitman [1978] used variations in the rate of sea-floor spreading to determine a tectono-eustatic record. Comparisons of the carbon isotopic cycles with the record of tectono-eustasy show that the carbon cycles were too rapid to be explained by this mechanism.

Vail et al. [1977] presented a record of changes in coastal onlap observed in seismic profiles from passive continental margins. They interpreted the onlap record as a record of sea level. Although Pitman [1978] and Watts [1982] questioned a direct causal relationship between onlap and sea-level records, we acknowledge that there may be a relationship between the glacio-eustatic record and the onlap record, albeit an indirect one [see Miller et al., 1984b]. In the Oligocene, major sea-level falls inferred by Vail et al. [1977] and Vail and Hardenbol [1979] occurred at 31 Ma and 25 Ma. Therefore, there is little correlation between the sea level record inferred from seismic stratigraphy and the carbon cycles.

We conclude that the low-frequency ($10^6$ years) Tertiary $\delta^{13}C$ cycles were not directly tied to global sea-level changes like the high-frequency ($10^4$-$10^5$ years) Pleistocene cycles.

## Terrestrial Climate

The timing of the global carbon isotope cycles appears to have been similar to changes in inferred terrestrial climate. Based upon changes in terrestrial vegetation, Wolfe [1978] suggested that periods of lowered mean annual temperature occurred in the early Oligocene, across the Oligocene/Miocene boundary, and in the middle Miocene, in good agreement with periods of high global $\delta^{13}C$ (Figure 4).

We suggest that the apparent correlation of the global $\delta^{13}C$ cycles with the terrestrial plant record may be invalid due to problems in correlation of the terrestrial record. For example, the period of greatest Paleogene at about 31-32 Ma, as constrained by radiometric dates [Wolfe and Hopkins, 1967]. Wolfe and Hopkins assigned this cool interval to the middle of the Oligocene, which agrees with the time scale of Berggren et al. [1984b]. Wolfe [1978] later recorrelated this same cooling with the Eocene/Oligocene boundary, suggesting that the terrestrial cooling is synchronous with the global increase in $\delta^{13}C$ noted here. However, we suggest that the terrestrial cooling occurred near the early/late Oligocene boundary as indicated by the radiometric dates of Wolfe and Hopkins [1967]. In this interpretation, the $\delta^{13}C$ cycles do not correlate well with the terrestrial plant record.

## Abyssal Circulation History: The Pacific-Atlantic $\delta^{13}C$ Difference

Comparison of the carbon isotopic record at Atlantic Sites 558 and 563 with Pacific Sites 77, 289, and 574 (Figure 4) shows that the Atlantic sites were generally higher than the Pacific sites, with greatest differences between oceans from about 36 to 33 Ma (early Oligocene) and from 26 to 10 Ma (latest Oligocene-Miocene). We interpret this as reflecting supply of $^{13}C$-enriched bottom water to the North Atlantic, perhaps in a manner analogous to NADW. The difference between smoothed Pacific minus smoothed

Atlantic $\delta^{13}C$ values (Figure 4) illustrates gross changes in production of bottom water. However, the difference curve is highly sensitive to the age models used and to stratigraphic uncertainties. The mean difference for the interval examined was about one half of the present-day difference (0.4 versus $1.0^o/oo$); peak differences that occurred at about 12-10 Ma, 18-16 Ma, and 25-22 Ma approached $1^o/oo$ (Figure 4). We caution that the variations in the difference curve may be entirely due to slight inaccuracies in Pacific biochronology. Still, the peaks at 12-10 Ma and 18-16 Ma correlate with periods of bottom-water erosion noted in the North Atlantic [Miller and Tucholke, 1983; Mountain and Tucholke, 1984]. We speculate that increased flow of North Atlantic deep water resulted in bottom-water erosion at these times; more Atlantic data and better stratigraphic control in the Pacific are needed to test this conclusion. The peak difference noted at 22-25 Ma (Figure 4) is insignificant due to sparse Pacific isotopic data and uncertainties in the Pacific age models for this interval. The reduced $\delta^{13}C$ difference noted in the Oligocene (33-26 Ma) can be interpreted as reflecting reduced production of North Atlantic deep water [Miller and Thomas, 1984] or increased preformed nutrient levels in bottom-water source regions [Mix and Fairbanks, 1984]. Unlike in the Pleistocene, there apparently are no correlations among the Atlantic-Pacific $\delta^{13}C$ difference record, global changes in $\delta^{13}C$, and eustatic changes during the Oligocene-Miocene.

## Conclusions

1. Three cycles of benthic and planktonic foraminiferal $\delta^{13}C$ occurred in the Oligocene to Miocene with periods of 7-12 m.y.; these cycles resulted from global changes in the $\delta^{13}C$ of mean ocean water. The $\delta^{13}C$ cycles were independent of changes in sea level indicated by changes in spreading rates (ocean volume), seismic stratigraphy, and oxygen isotopic stratigraphy. The most likely causes of these cycles were small changes in the input or burial ratio of carbonate carbon to organic carbon.

2. Pacific benthic foraminifera were depleted in $^{13}C$ relative to the Atlantic from 36-33 Ma (early Oligocene) and 26-12 Ma (late Oligocene to middle Miocene); we interpret this difference between basins as reflecting supply of $^{13}C$-enriched bottom waters to the deep North Atlantic at these times. Reduced $\delta^{13}C$ differences between 33 and 26 Ma (early to late Oligocene) may reflect reduced production of "NADW" or an increase in preformed nutrient content of "NADW" source regions.

3. High benthic foraminiferal oxygen isotopic values noted at about 15 Ma (middle Miocene), 25-24 Ma (near Oligocene/Miocene boundary), 31-28 Ma (near early/late Oligocene boundary), and 36-35 Ma (just above the Eocene/Oligocene boundary) suggest the existence of significant continental

ice at these times; covariance between planktonic and benthic foraminiferal $\delta^{18}O$ records is used to estimate that development of ice sheets lowered sea level at 36 Ma, 31 Ma, and 25 Ma by 30-50 m. Unlike those in the Pleistocene, Tertiary global $\delta^{13}C$ changes did not correlate with glacio-eustatic changes.

Acknowledgments. We thank W.B. Curry, L.D. Keigwin, and A.C. Mix for discussions, W.B. Curry and A.C. Mix for reviewing the manuscript, A.C. Mix for data-smoothing programs, A.D. Chave for plotting programs, I. Ivanciu and M. Katz for technical assistance, and G. Kolibas, A. Ivanciu, D. King, and B. Gruder for performing isotopic analyses. Samples were provided by the DSDP. We thank E. Sundquist and W. Broecker for inviting us to participate in the Chapman Conference and the American Geophysical Union for travel support. This work was supported by National Science Foundation grants OCE 83-10086 (K.G.M.), OCE 82-08784 and OCE 82-00717 (R.G.F.), and a grant from the ARCO Foundation. This is Lamont-Doherty Geological Observatory of Columbia University contribution 3742.

## References

Belanger, P.E., W.B. Curry, and R.K. Matthews, Core-top evaluation of benthic foraminiferal isotopic ratios for paleo-oceanographic interpretations, Palaeogeogr. Palaeoclimatol. Palaeoecol., 33, 205-220, 1981.

Belanger, P.E., and R.K. Matthews, The foraminiferal isotopic record across the Eocene/Oligocene boundary at Site 540, Initial Rep. Deep Sea Drill. Proj., 77, in press, 1984.

Bender, M.L., and L.D. Keigwin, Speculations about the upper Miocene change in abyssal Pacific dissolved bicarbonate $\delta^{13}C$, Earth Planet. Sci. Lett., 45, 383-393, 1979.

Berger, W.H., and E.L. Winterer, Plate stratigraphy and the fluctuating carbonate line, in Pelagic Sediments on Land and Under the Sea, Spec. Pub. Int. Assoc. Sediment. 1, edited by K.L. Hsu and H.C. Jenkyns, pp. 11-48, Blackwell, Oxford, England, 1974.

Berggren, W.A., and C.D. Hollister, Paleogeography, paleobiogeography and the history of circulation in the oceans, in Studies in Paleo-oceanography, Spec. Publ. 20, edited by W.W. Hay, pp. 126-186, Society Economic Paleontologists and Mineralogists, Tulsa, Okla., 1974.

Berggren, W.A., D.V. Kent, and J. Flynn, Paleogene geochronology and chronostratigraphy, in Geochronology and the Geologic Time Scale, edited by N.J., Snelling, Geological Society London, in press, 1984a.

Berggren, W.A., D.V. Kent, and J.A. Van Couvering, Neogene geochronology and chronostratigraphy, in Geochronology and the Geologic Time Scale, edited by N.J., Snelling, Geological Society London, in press, 1984b.

Blanc, P.-L., D. Rabussier, C. Vergnaud-Grazzini, and J.-C. Duplessy, North Atlantic deep water formed by the later middle Miocene, Nature, 283, 553-555, 1980.

Blanc, P.-L., and J.-C. Duplessy, The deep-water circulation during the Neogene and the impact of the Messinian salinity crisis, Deep Sea Res., 29A, 1391-1414, 1982.

Blattner, P., and J.R. Hulston, Proportional variations of geochemical $\delta^{18}O$ scales--An interlaboratory comparison. Geochim. Cosmochim. Acta, 42, 59-62, 1978.

Boersma, A., and N.J. Shackleton, Tertiary oxygen and carbon isotope stratigraphy, Site 357 (mid-latitude South Atlantic), Initial Rep. Deep Sea Drill. Proj., 39, 911-924, 1977.

Boyle, E.A., and L.D. Keigwin, Deep circulation of the North Atlantic over the last 200,000 years: Geochemical evidence, Science, 218, 784-787, 1982.

Broecker, W.S., A boundary condition on the evolution of atmospheric oxygen, J. Geophys. Res., 75, 3553-3557, 1970.

Broecker, W.S. Ocean chemistry during glacial time, Geochim. Cosmochim. Acta, 46, 1689-1705, 1982.

Craig, H., The geochemistry of the stable carbon isotopes, Geochim. Cosmochim. Acta, 18, 53, 1953.

Curry, W.B., and G.P. Lohmann, Carbon isotopic changes in benthic foraminifera from the western South Atlantic: Reconstruction of glacial abyssal circulation patterns, Quat. Res., 18, 218-235, 1982.

Curry, W.B., and G.P. Lohmann, Reduced advection into Atlantic Ocean deep eastern basins during last glacial maximum, Nature, 306, 577-580, 1983.

Duplessy, J.-C., J. Moyes, and C. Pujol, Deep water formation in the North Atlantic during the last ice age, Nature, 286, 479-482, 1980.

Fairbanks, R.G., and R.K. Matthews, The marine oxygen isotopic record in Pleistocene coral, Barbados, West Indies, Quat. Res., 10, 181-196, 1978.

Fairbanks, R.G., and A.C. Mix, Carbon isotope variability of the southern ocean during the last 30,000 years, Nature, in press, 1984.

Froelich, P.N., M.L. Bender, N.A. Luedthe, G.R. Heath, and T. DeVries, The marine phosphorous cycle, Am. J. Sci., 282, 474-511, 1982.

Graham, D.W., B.H. Corliss, M.L. Bender, and L.D. Keigwin, Carbon and oxygen isotopic disequilibria of Recent benthic foraminifera, Mar. Micropaleontol., 6, 483-497, 1981.

Keith, M.L., and J.N. Weber, Carbon and oxygen isotopic composition of mollusk shells from marine and fresh-water environments, Geochim. Cosmochim. Acta, 28, 1787, 1964.

Khan, M.J, D.V. Kent, and K.G. Miller, Magnetostratigraphy of Oligocene to Pleistocene sediments of DSDP Leg 82, Initial Rep. Deep Sea Drill. Proj., 82, in press, 1984.

Keigwin, L.D., Palaeoceanographic change in the Pacific at the Eocene-Oligocene boundary, Nature, 287, 722-725, 1980.

Keigwin, L.D., and G. Keller, Middle Oligocene climatic change from equatorial Pacific DSDP Site 77, Geology, 12, 16-19, 1984.

Kennett, J.P., and N.J. Shackleton, Oxygen isotope evidence for the development of the psychrosphere 38 Myr ago, Nature, 260, 513-515, 1976.

Kroopnick, P., Correlation between $^{13}C$ and $\Sigma CO_2$ in surface waters and atmospheric $CO_2$, Earth Planet. Sci. Lett., 22, 397-403, 1980.

Kroopnick, P., The distribution of $^{13}C$ in the Atlantic Ocean, Earth Planet. Sci. Lett., 49, 469-484, 1972.

Kroopnick, P., The distribution of C-13 of $TCO_2$ in the world oceans, Deep Sea Res., in press, 1984.

Kroopnick, P., R.F. Weiss, and H. Craig, Total $CO_2$, $^{13}C$, and dissolved oxygen-$^{18}O$ at GEOSECS II in the North Atlantic, Earth Planet. Sci. Lett., 16, 103-110, 1972.

Matthews, R.K., and R.Z. Poore, Tertiary $\delta^{18}O$ record and glacio-eustatic sea-level fluctuations, Geology, 8, 501-504, 1980.

Mayer, L., Theyer, F., et al., Initial Rep. Deep Sea Drill. Proj. 85, in press, 1984.

Miller, K.G., and W.B. Curry, Eocene to Oligocene benthic foraminiferal isotopic record in the Bay of Biscay, Nature, 296, 347-350, 1982.

Miller, K.G., and R.G. Fairbanks, Evidence for Oligocene-Middle Miocene abyssal circulation changes in the western North Atlantic, Nature, 306, 250-253, 1983.

Miller, K.G., and E. Thomas, Late Eocene to Oligocene benthic foraminiferal isotopic record, Site 574 equatorial Pacific, Initial Rep. Deep Sea Drill. Proj., 85, in press, 1984.

Miller, K.G., and B.E. Tucholke, Development of Cenozoic abyssal circulation south of the Greenland-Scotland Ridge, in Structure and Development of the Greenland-Scotland Ridge, NATO Conference Ser., edited by M.H.P. Bott, S. Saxov, M. Talwani, and J. Thiede, 549-589, Plenum, New York, 1983.

Miller, K.G., W.B. Curry, and D.R. Ostermann, Late Paleogene benthic foraminiferal paleoceanography of the Goban Spur Region, DSDP Leg 80, Initial Rep. Deep Sea Drill. Proj., 80, in press, 1984a.

Miller, K.G., G.S. Mountain, and B.E. Tucholke, Oligocene glacio-eustasy and erosion on the margins of the North Atlantic, Geology, in press, 1984b.

Mix, A.C., and R.G. Fairbanks, Late Pleistocene history of North Atlantic surface and deep ocean circulation, Earth Planet. Sci. Lett., in press, 1984.

Mountain, G.S., and B.E. Tucholke, Mesozoic and Cenozoic geology of U.S. Atlantic continental slope and rise, in Geologic Evolution of the United States Atlantic Margin, edited by C.W. Poag, Van Nostrand Reinhold, Stroudsberg, Penn., in press, 1984.

Oberhansli, H., and M. Toumarkine, The Paleogene oxygen and carbon isotope history of Site 522, 523, and 524 from the central South Atlantic, in Symposium on South Atlantic Paleoceanography,

edited by K.J. Hsu and H. Weissert, in press, 1984.

Pitman, W. C., Relationship between eustacy and stratigraphic sequences of passive margins, Geol. Soc. Am. Bull., 89, 1389-1403, 1978.

Poore, R.Z., and R.K. Matthews, Late Eocene-Oligocene oxygen and carbon isotope record from South Atlantic Ocean DSDP Site 522, Initial Rep. Deep Sea Drill. Proj., 73, 725-736, 1984.

Ronov, A.B., Probable changes in the composition of sea water during the course of geological time, Sedimentology, 10, 25, 1968.

Sarnthein, M., J. Thiede, U. Pflaumann, H. Erlenkeuser, D. Futterer, B. Koopmann, H. Lange, and E. Seibold, Atmospheric and oceanic circulation patterns off northwest Africa during the past 25 million years, in Geology of the Northwest African Continental Margin, edited by U. von Rad, K. Hinz, M. Sarnthein, and E. Seibold, 545-604, Springer-Verlag, New York, 1982.

Savin, S.M., The history of the earth's surface temperature during the last 100 million years, Annu. Rev. Earth Planet. Sci., 5, 319-355, 1977.

Savin, S.M., R.G. Douglas, G. Keller, J.S. Killingley, L. Shaughnessy, M.A. Sommer, E. Vincent, and F. Woodruff, Miocene benthic foraminiferal isotope records: A synthesis, Mar. Micropaleontol., 6, 423-450, 1981.

Savin, S.M., R.G. Douglas, and F.G. Stehli, Tertiary marine paleotemperatures, Geol. Soc. Am. Bull., 86, 1499-1510, 1975.

Sclater, J.G., R.N. Anderson, and N.L. Bell, Elevation of ridges and evolution of the central eastern Pacific, J. Geophys. Res., 76, 7888-7915, 1971.

Sclater, J.G., D. Abbott, and J. Thiede, Paleobathymetry of sediments of the Indian Ocean, in Indian Ocean Geology and Biostratigraphy edited by J.R. Heirtzler et al., 25-59, AGU, Washington, D.C., 1977.

Shackleton, N.J., Carbon-13 in Uvigerina: Tropical rain forest history and the equatorial Pacific carbonate dissolution cycles, in The Fate of Fossil Fuel $CO_2$ in the Oceans, edited by N.R. Anderson and Malahoff, 401-427, Plenum, New York, 1977.

Shackleton, N.J., and N.D. Opdyke, Oxygen isotopic and palaeomagnetic stratigraphy of equatorial Pacific core V28-238: Oxygen isotope temperatures and ice volume on a $10^5$ and $10^6$ year scale, Quat. Res., 3, 39-55, 1973.

Shackleton, N.J., and J.P. Kennett, Paleotemperature history of the Cenozoic and the initiation of Antarctic glaciation: oxygen and carbon isotopic analyses in DSDP Sites 277, 279, and 281, Initial Rep. Deep Sea Drill. Proj., 29, 743-755, 1975.

Shackleton, N.J., M.A. Hall, and A. Boersma, Oxygen and carbon isotope data from Leg 74 foraminifers, Initial Rep. Deep Sea Drill. Proj., 74, 599-612, 1984.

Shackleton, N.J., J. Imbrie, and M.A. Hall, Oxygen and carbon isotope record of East Pacific core

V19-30: Implications for the formation of deep water in the late Pleistocene, Earth Planet. Sci. Lett., 65, 233-244, 1983.

Snyder, S.W., C. Muller, and K.G. Miller, Biostratigraphy and paleoceanography across the Eocene/Oligocene boundary at Deep Sea Drilling Project Site 549, Geology, 12, 112-115, 1984.

Srinivasan, M.S., and J.P. Kennett, Neogene planktonic foraminiferal biostratigraphy and evolution: Equatorial to subantarctic, South Pacific, Mar. Micropaleontol., 6, 499-534, 1981.

Tucholke, B.E., and P.R. Vogt, Western North Atlantic: Sedimentary evolution and aspects of tectonic history, Initial Rep. Deep Sea Drill. Proj., 43, 791-825, 1979.

Vail, P.R., R.M. Mitchum, Jr., R.G. Todd, J.M. Widmier, S. Thompson, III, J.B. Sangree, J.N., Bubb, W.G. Hatelid, Seismic stratigraphy and global changes of sea level, in Seismic Stratigraphy--Applications to Hydrocarbon Exploration, Mem. 26, edited by C.E. Payton, pp. 49-205, American Association of Petroleum Geologists, Tulsa, Okla., 1977.

Vail, P.R., and J. Hardenbol, Sea-level changes during the Tertiary, Oceanus, 22, 71-79, 1979.

Vincent, E., and W.H. Berger, Carbon dioxide and polar cooling in the Miocene: The Monterey hypothesis, this volume.

Vincent, E., Killingley, J.S., and W.H. Berger, Miocene oxygen and carbon isotope stratigraphy of the tropical Indian Ocean, in Miocene paleoceanography and biogeography, edited by J.P. Kennett, Geological Society America, in press, 1984.

Watts, A.B., Tectonic subsidence, flexure, and global changes in sea level, Nature, 297, 469-474, 1982.

Wolfe, J.A., A paleobotanical interpretation of Tertiary climates in the northern hemisphere, Am. Sci., 66, 694-703, 1978.

Wolfe, J.A., and D.M. Hopkins, Climatic changes recorded by land floras in northwestern North America, in Tertiary Correlations and Climatic Changes in the Pacific, Proceedings of the 11th Pacific Science Congress, pp. 67-76, Science Council of Japan, Tokyo, Japan, 1967.

Woodruff, F., S.M. Savin, and R.G. Douglas, Biological fractionation of oxygen and carbon isotopes by Recent benthic foraminifera, Mar. Micropaleontol., 5, 3-13, 1980.

Woodruff, F., S.M. Savin, and R.G. Douglas, Miocene stable isotopic record: A detailed deep Pacific Ocean study and its paleoclimatic implications, Science, 212, 665-668, 1981.

# A "STRANGELOVE" OCEAN IN THE EARLIEST TERTIARY

Kenneth J. Hsü and Judith A. McKenzie

Geological Institute, Swiss Federal Institute of Technology, CH-8092 Zürich

Abstract. A decrease of up to $3^o/oo$ in the $\delta^{13}C$ values of planktic skeletons has been systematically observed across the Cretaceous/Tertiary boundary; the benthic skeletons show no corresponding changes. We interpret this decrease as a manifestation of the elimination of the surface-to-bottom carbon isotope gradient in ocean waters at a time when carbon fractionation by a photosynthesis respiration mechanism became ineffective. A concurrent release of excess $CO_2$ from a nearly barren ocean to the atmosphere could have caused global warming.

## Carbon Isotope Anomalies Across C/T Boundary

Analyses of carbon isotope ratios in pelagic sediments across the Cretaceous/Tertiary (C/T) boundary indicated a remarkable change in the chemistry of the ocean water in the very beginning of the Tertiary, as revealed by a negative $\delta^{13}C$ anomaly ranging from 1 to $3^o/oo$. First reported by Brennecke and Anderson [1977] the anomaly has been detected at many sites where the sedimentation across the C/T boundary was continuous (e.g. Figure 1). Perch-Nielsen et al. [1982] suggested the existence of possibly two more negative anomalies after the boundary, but the magnitude and timing of these are not the same everywhere. At DSDP Site 524 [Hsü et al., 1982; He et al., 1984], for example, we noted that the first negative anomaly of about $3^o/oo$ occurred at a horizon where the maximum iridium concentration was detected (Figure 2). This anomaly was registered in the bulk sample, which consisted mainly of nannoplankton. A second depletion occurs in both the bulk and the fine fraction at a horizon 1.2 m above the "iridium spike" after the first appearance of Globigerina pseudobulloides in sediments 50,000 years younger as estimated from sedimentation rates. Fossil tests of the benthic species Gavelinella beccariformis were analyzed above and below the boundary sediments. The $\delta^{13}C$ values of the benthic skeletons from the Cretaceous are less positive (by 1 or $2^o/oo$) than those

of the bulk or those of a fine fraction, which consists almost exclusively of nannoplankton (Figure 2). This is the normal oceanic pattern of fractionation of carbon isotopes between the surface and bottom waters, as will be discussed later. However, the $\delta^{13}C$ values of the benthic skeletons from the earliest Tertiary are about the same as, or even more positive than, those of the co-existing planktic fossils. This is an abnormal pattern, previously recorded by Boersma et al. [1979] in an earlier study. Investigations of carbon isotope values across the C/T boundary at DSDP Site 465A revealed the same pattern [Boersma and Shackleton, 1981]. Whereas the carbon isotope value of the benthic species shows no systematic variation across the boundary, the $\delta^{13}C$ of the planktic species indicates a negative shift of about -1.5 per mil across the boundary. As a consequence, the value for the earliest Tertiary planktonics was slightly more negative than that of the benthic from the sample, even though the latest Cretaceous gradient was normal; planktonic foraminifers of that age have $\delta^{13}C$ values more than 1 per mil more positive than that of the contemporaneous benthic species.

As it is generally recognized, the carbon isotope composition of the oceans is primarily controlled by the photosynthesis-respiration processes [Broecker and Peng, 1982]; the consequence of large decreases in productivity at the C/T boundary and in the earliest Tertiary oceans should have been recorded in the carbon isotope content of pelagic sediments from this period. It is, therefore, interesting to evaluate the carbon isotope composition of the sedimentary record as a reflection of fertility changes in the earliest Tertiary oceans.

## Carbon Isotope Fractionation

In the marine environments, fractionation of carbon isotopes between the surface and bottom waters is essentially accomplished via a photosynthesis-respiration mechanism [Broecker and Peng, 1982]. In the photic zone, plants pre-

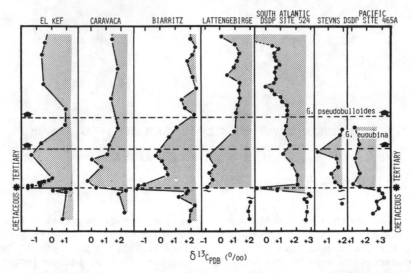

Fig. 1.  The carbon isotope stratigraphy of pelagic marine sediments from seven sections across the Cretaceous/Tertiary boundary. Note the carbon isotope shift at the boundary. Foraminiferal datum levels are first-appearance horizons (modified from Perch-Nielsen et al. [1982]).

ferentially incorporate carbon 12 from the dissolved inorganic carbon reservoir into the organic matter produced during photosynthesis. This organic matter subsequently sinks into the underlying water masses, where it is oxidized, enriching the carbon 12 in the dissolved inorganic carbon. In other words, carbon 12 is transferred from the surface to the bottom waters, establishing a carbon isotope gradient, which can be 2°/oo to 3°/oo [Kroopnick, 1974]. A "typical" profile [Kroopnick et al., 1977] for the carbon isotope composition of dissolved inorganic carbon is shown in Figure 3. In this profile from a mid-latitude Pacific Ocean site, the $\delta^{13}C$ value ranges from about 2.0°/oo in the near-surface waters to a low of about 0.2°/oo, which corresponds to the oxygen-minimum zone.

If for some reason the plankton production in the ocean should be completely suppressed, a mixing of ocean water would eliminate the carbon isotope gradient in a very short time, of the order of decades. We propose to call such a sterile ocean, as possibly envisioned for the earliest Tertiary, a "Strangelove" ocean, after the fictitious character who had visions of bringing about mass mortality with a nuclear war. A sudden destruction of oceanic life, or the "Strangelove" effect, has been suggested as a mechanism to produce the negative carbon shift of about 3°/oo in the planktic skeletons of the earliest Tertiary, while retaining the same isotopic composition for the benthic skeletons [Broecker and Peng, 1982; Hsü and McKenzie, 1983]. Hence mass mortality at the C/T boundary resulted in a collapse of the surface-to-bottom, carbon isotope gradient.

The effect of mass mortality upon the carbon isotope gradient in the oceans can be experimentally studied in freshwater lakes. McKenzie [1985] has shown that carbon isotope fractionation in aquatic environments behaves in a similar manner in both marine and lacustrine basins. Therefore, modern lakes can be used to model isotopic events which occurred in oceans of the geologic past. For example, a carbon isotope study of Lake Greifen, a highly productive lake in northeastern Switzerland, reveals that in summer, when photosynthesis is at a maximum, the $\delta^{13}C$ value of the dissolved carbonate in the surface waters is about 4°/oo more positive than in the underlying waters (Figure 4). A surface-to-bottom, carbon-isotope gradient is established as a consequence of photosynthesis-respiration activity, and a carbon isotope profile similar to those measured in the oceans (Figure 3) results. This transfer of carbon 12 from the surface to bottom waters essentially ceases in the winter months when productivity approaches a minimum. The vertical gradient collapses, and the $\delta^{13}C$ value of the dissolved bicarbonate is the same from top to bottom (Figure 4) [McKenzie, 1982].

The summer to winter cooling of the surface waters in Lake Greifen by as much as 22°C is an annual catastrophic event. Each summer the lake recovers from the previous mass mortality, and a new carbon isotope gradient is quickly reestablished. An analogous situation for the first carbon isotope anomaly at the C/T boundary is obvious. Mass mortality of marine plankton resulting from a catastrophic event causes the collapse of the surface-to-bottom carbon isotope gradient in

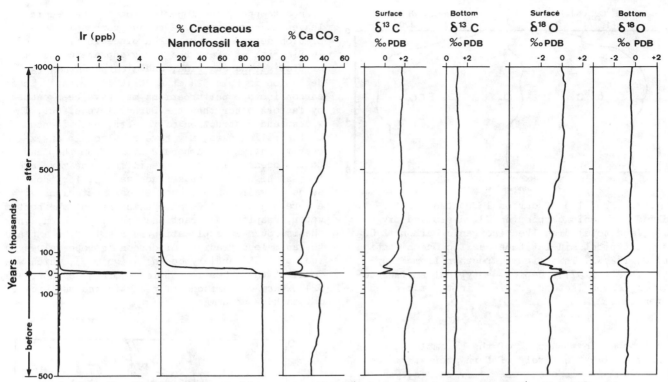

Fig. 2.  Geochemical anomalies across the C/T boundary at DSDP Site 524. Note the
two sharp perturbations represented by the iridium anomaly and by the calcium carbonate content; those perturbations were restricted to a boundary deposited in less
than 1000 years. The disappearance of the Cretaceous nannofossil taxa took place
within a time interval of about 40000 years. The variation of $\delta^{13}C$ and of $\delta^{18}O$ in
surface water is based upon analysis of the nannoplankton, and that in bottom water
is based upon analysis of <u>Gavelinella beccariformis</u>. Note the "instantaneous response" of the carbon isotope anomaly in surface waters and the delayed response
of the oxygen isotope anomaly. Note also the absence of a $\delta^{13}$ perturbation in
bottom waters (based on data of Hsü et al. [1982]). The numerical ages are magneto-
stratigraphical estimates.

the oceans. The rapid, negative $\delta^{13}C$ excursion
recorded in the calcareous plankton living in the
earliest Tertiary oceans reflects this breakdown.
Within a short period afterwards, hundreds to
thousands of years, the surface water productivity
gradually recovered, as seen in the El Kef section
(Figure 1), and the vertical gradient was restored.

A "Strangelove" Ocean

Considerable evidence has been presented that
a meteor impact was responsible for mass extinction at the end of the Cretaceous [Alvarez et al.,
1980; Hsü, 1980]. Computer modeling has revealed
the possible destruction of terrestrial environments by the fall of a solid bolide with a mass
of $10^{17}$ or $10^{18}$ g [Silver and Schultz, 1982]. One
scenario suggested that the high temperature
($10^{5}$ °K) of the mushroom cloud rising from the
impact site should produce considerable $NO_x$

through the combination of atmospheric nitrogen
and oxygen. This should in turn (1) destroy the
ozone layer of the stratosphere, (2) cause large-
scale defoliation of terrestrial plants, and (3)
be oxidized and fall eventually as acid rain to
change the pH of the oceans. At the same time the
ejecta envelope around the earth may have reduced
the transmission of sunlight sufficiently to
suppress the photosynthesis of phytoplanktons for
at least several months.

The evidence of forest destruction has been
clearly shown by the pollen record across the C/T
boundary of San Juan Basin, New Mexico [Orth et
al., 1982]. The percentage of tree pollen ranged
from 70% to 85% in the late Cretaceous and in the
early Tertiary assemblages of the region. The
only exception was the assemblage found in sediments with a large iridium anomaly, which probably signified the time of a large meteor impact. In this boundary sediment the tree pollen

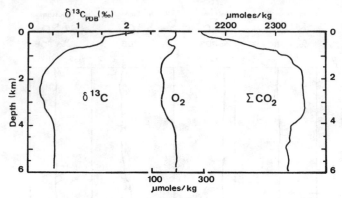

Fig. 3. Profiles of carbon 13 content of dissolved inorganic carbon ($\delta^{13}C$), dissolved oxygen
($O_2$), and total dissolved inorganic carbon ($\Sigma\ CO_2$)
from a typical mid-latitude Pacific Ocean station
(17°S, 172°W). This figure, redrawn from Kroopnick et al. [1977], depicts a typical surface-to-
bottom carbon isotope gradient in the marine environment.

constitutes less than 1% of the assemblage, which
consists almost entirely of fern spores.

The changes in ocean chemistry caused by temporary suppression of photosynthesis and catastrophic environmental pollution could have resulted in an ocean almost devoid of plankton, or
a "Strangelove" ocean. Lewis et al. [1982] calculated that an impact of a $10^{18}g$ asteroid could
produce $2 \times 10^{37}$ NO molecules and that a cometary
event of equal-iridium magnitude ($10^{19}g$, 40 km/s
entry velocity) could produce a NO yield of $2 \times 10^{40}$
molecules. Adding all the oxidized molecules
($2 \times 10^{40}$) to a 75-m-deep surface layer of the
ocean could reduce the pH of the seawater in this
layer by half a unit. We should recall that the
pH gradient of the oceans, like the carbon isotope gradient, is also a product of plankton
photosynthesis; when dissolved $CO_2$ in surface
waters is no longer being extracted by the plankton, the pH fractionation of the ocean waters
stops. Adding all this additional acid ($HNO_3$) to
a Strangelove ocean could drastically suppress
plankton fertility until the acid-rich surface
layer was destroyed by ocean mixing.

We interpret the carbon isotope shift across
the C/T boundary as largely a manifestation of
the Strangelove effect. The Strangelove ocean
probably did not exist for more than thousands
of years because the carbon isotope signal at
those localities returned quickly to normal when
plankton production of the oceans resumed, as the
record of nannoplankton blooms (Thoracosphaera
spp., Braarudosphaera spp.) shows [Perch-Nielsen
et al., 1982].

Although a tentative correlation of carbon isotope changes during the earliest Tertiary has

been attempted, the imperfection of the record
leaves much uncertain. A second carbon 12 depletion seemed to have taken place in El Kef and
in Biarritz at about the first appearance datum
of Globigerina eugubina, and a third carbon 12
depletion in Hole 524 at a still later date. Those
carbon isotope perturbations may have been caused
by factors other than the Strangelove effect. The
widespread destruction of forests, as evidenced
by the pollen data, may have resulted in an excessive influx of carbon 12 atoms from the land
to the oceans. In fact, the influx rate may have
been so high that the surface waters, for thousands of years, had dissolved carbonate ions with
a more negative $\delta^{13}C$ value than that of the bottom
water, despite the fractionation by plankton synthesis. Such an explanation is required to explain
the anomalous gradient of carbon isotope values in
the earliest Tertiary oceans (Figure 5), which was
found at the three localities (DSDP sites 356, 465,
524) where comparisons of planktic and benthic
values have been made.

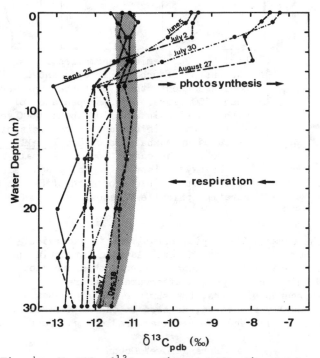

Fig. 4. Monthly $\delta^{13}C$ profiles of the dissolved
inorganic carbon in Lake Greifen. The shaded area
represents the range of $\delta^{13}C$ values found between
December and May, the period of minimum productivity. The carbon 13 increase in surface waters
due to photosynthesis and the carbon 13 depletion
in deep waters resulting from the respiration of
the sinking organic matter establish a surface-
to-bottom carbon isotope gradient during the summer months from May to September. Modified from
McKenzie [1982].

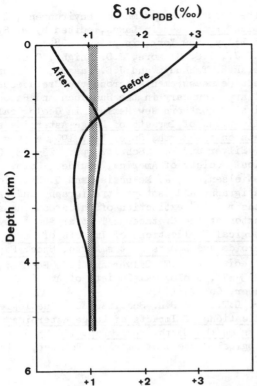

$\delta\,^{13}C_{PDB}(‰)$

Fig. 5.  Carbon isotope gradient in the earliest Tertiary. The anomalous carbon isotope gradient of dissolved carbonate in the earliest Tertiary ocean was first noted by Boersma et al. [1979] in their study of DSDP Site 356 samples. It was subsequently found by Hsü et al. [1982] at Site 524 and by Boersma and Shackleton [1981] at Site 465.

A corollary of the postulate of a Strangelove ocean is the release of $CO_2$ from the oceans to the atmosphere [Broecker and Peng, 1982]. Equilibration with the Strangelove ocean could increase the partial pressure of $CO_2$ in the atmosphere by a factor of 3. A greenhouse effect resulting from the increased atmospheric $CO_2$ may have caused a global warming. A second corollary of the postulate is the accumulation of nutrients in ocean waters as a result of mass mortality and decreased productivity. The nannoplankton of the earliest Tertiary oceans are mainly the products of unusual, monospecific blooms. We suggest that these blooms were possible because of the plentitude of nutrients in the Strangelove ocean. The interplay of $CO_2$ release from, and of repeated blooms in, a Strangelove ocean may have caused rapid fluctuations in the $CO_2$ content of the earliest Tertiary atmosphere. The consequence might be a very unstable climate, as suggested by the oxygen isotope record for the first 50,000 years after the terminal Cretaceous event (Figure 2); there were

rapid fluctuations with an overall warming trend [Hsü et al., 1982].

Conclusions

Environmental disasters could cause catastrophic decrease of plankton productivity, resulting in a Strangelove ocean. $CO_2$ not utilized during photosynthesis could be transferred from the oceans to the atmosphere, leading to a marked increase in the atmospheric $CO_2$ content and consequently major climatic changes.

Acknowledgements. Contribution 241 of the Laboratory of Experimental Geology, Zürich, and a contribution to I.G.C.P. 199.

References

Alvarez, L.W., W. Alvarez, F. Asaro, and H.V. Michel, Extraterrestrial cause for the Cretaceous-Tertiary extinction, Science 208, 1095-1108, 1980.
Boersma, A., and N. Shackleton, Oxygen- and carbon-isotope variations and planktonic depth habitats, late Cretaceous to Paleocene, central Pacific, Deep Sea Drilling Project sites 463 and 465, Intitial Rep. Deep Sea Drill. Proj., 62, 513-526, 1981.
Boersma, A., N. Shackleton, M. Hall, and Q. Given, Carbon and oxygen isotope records at DSDP Site 384 (North Atlantic) and some Paleocene paleotemperatures and carbon isotope variations in the Atlantic Ocean, Initial Rep. Deep Sea Drill. Proj., 43, 695-717, 1979.
Brennecke, J.C., and T.F. Anderson, Carbon isotope variation in pelagic carbonates (abstract), Eos Trans. AGU, 58, 415, 1977.
Broecker, W.S., and T.-H. Peng, Tracers in the Sea, 690 pp., Eldigo, New York, 1982.
He, Q., J.A. McKenzie, and H. Oberhänsli, Stable-isotope and percentage-of-carbonate data for upper Cretaceous/lower Tertiary sediments from Deep Sea Drilling Project Site 524, Cape Basin, South Atlantic, Initial Rep. Deep Sea Drill. Proj., 73, 749-754, 1984.
Hsü, K.J., Terrestrial catastrophe caused by cometary impact at the end of Cretaceous, Nature, 285, 201-203, 1980.
Hsü, K.J., and J.A. McKenzie, The Strangelove effect-carbon-isotope perturbation as indication of mass mortality (abstract), Terra Cognita, 3, 222, 1983.
Hsü, K.J., Q. He, J.A. McKenzie, H. Weissert, K. Perch-Nielsen, H. Oberhänsli, K. Kelts, J. LaBrecque, L. Tauxe, U. Krähenbühl, S.F. Percival, Jr., R. Wright, A.M. Karpoff, N. Petersen, P. Tucker, R.Z. Poore, A.M. Gombos, K. Pisciotto, M.F. Carman, Jr., and E. Schrei-

ber, Mass mortality and its environmental and evolutionary consequences, Science, 216, 249-256, 1982.

Kroopnick, P., Correlations between [13]C and $\Sigma$ $CO_2$ in surface waters and atmospheric $CO_2$, Earth Planet. Sci. Lett., 22, 397-403, 1974.

Kroopnick, P., S.V. Margolis, and C.S. Wong, $\delta^{13}C$ variations in marine carbonate sediments as indicators of the $CO_2$ balance between the atmosphere and oceans, in The Fate of Fossil Fuel $CO_2$ in the Oceans, edited by N.R. Anderson and A. Malakoff, pp. 295-321, Plenum, New York, 1977.

Lewis, J.S., G.H. Watkins, H. Hartman, and R.G. Prinn, Chemical consequences of major impact events on earth, in Geological Implications of Impacts of Large Asteroids and Comets on the Earth, Spec. Pap. 190, edited by L.T. Silver and P.H. Schultz, pp. 215-221, Geological Society of America, Boulder, Colo., 1982.

McKenzie, J.A., Carbon-13 cycle in Lake Greifen: A model for restrictive ocean basins, in Nature and Origin of Cretaceous Carbon-Rich Facies, edited by S.O. Schlanger and M.B. Cita, pp. 197-207, Academic Press, New York, 1982.

McKenzie, J.A., Carbon isotopes and productivity in the lacustrine and marine environment, in Chemical Processes in Lakes, edited by W. Stumm, John Wiley, New York, in press, 1985.

Orth, C.J., J.S. Gilmore, J.D. Knight, C.L. Pillmore, R.H. Tschudy, and J.E. Fasset, Iridium abundance measurements across the Cretaceous/ Tertiary boundary in the San Juan and Raton basins of northern New Mexico, in Geological Implications of Impacts of Large Asteroids and Comets on the Earth, Spec. Pap. 190, edited by L.T. Silver and P.H. Schultz, pp. 423-433, Geological Society of America, Boulder, Colo., 1982.

Perch-Nielsen, K., J. McKenzie, and Q. He, Biostratigraphy and isotope stratigraphy and the "catastrophic" extinction of calcareous nannoplankton at the Cretaceous/Tertiary boundary, in Geological Implications of Impacts of Large Asteroids and Comets on the Earth, Spec. Pap. 190, edited by L.T. Silver and P.H. Schultz, p. 353-371, Geological Society of America, Boulder, Colo., 1982.

Silver, L.T., and P.H. Schultz (Ed.), Geological Implications of Impacts of Large Asteroids and Comets on the Earth, Spec. Pap. 190, 528 pp., Geological Society of America, Boulder, Colo., 1982.

# MANTLE DEGASSING INDUCED DEAD OCEAN IN THE CRETACEOUS-TERTIARY TRANSITION

Dewey M. McLean

Department of Geology, Virginia Polytechnic Institute and
State University, Blacksburg, Virginia 24061

**Abstract.** Prior to the terminal Cretaceous marine extinctions about 65 m.y. ago (polarity chron R29), ecological stability prevailed suggesting steady state between rate of mantle $CO_2$ degassing and uptake by surficial sinks. The extinctions were characterized by ecological instability that persisted into the Early Tertiary; instability was coeval with the Deccan Traps flood basalt volcanism in India that flooded earth's surface with $2.6 \times 10^6$ $km^3$ of lavas; radiometric ages concentrate around 66-60 m.y., with the main volcanic activity around 65 m.y. during a reversed-to-normal polarity sequence; most Deccan Traps basalts were erupted during the Deccan Reversed Magnetic Polarity Epoch, with some activity into the overlying Nipani normal. Using the geomagnetic time scale of Harland et al. the Deccan Reversed interval is R29, and the Nipani N29. Ages of lower and upper boundaries of R29 are 65.39 and 64.86 m.y. ($5.3 \times 10^5$ years); based on this duration for the bulk of Deccan Traps volcanism, the rate of basalt production was 4.91 $km^3$/yr versus modern mid-ocean ridge basalt production of 1.2 $km^3$/yr. Deccan Traps total mantle $CO_2$ release was about $5 \times 10^{17}$ moles $CO_2$, released at a rate of $9.6 \times 10^{11}$ moles/yr, versus the modern release rate of about $4.1 \times 10^{12}$ moles/yr. Nearly 25% of total annual $CO_2$ release from all other sources, Deccan Traps $CO_2$ addition would have upset steady state, with release exceeding uptake, leading to $CO_2$ buildup in the atmosphere and marine mixed layer, triggering ecological instability. Reduced mixed layer photosynthesis and $CaCO_3$ production indicated in the record evidence of failure of the Williams-Riley pump (productivity plus gravity pump of $CO_2$ from shallow into deep oceans), producing dead ocean conditons which would have triggered additional $CO_2$ release and global ecological instability. The dinosaurs seem to have become extinct during magnetic polarity chron R28.

## Introduction

About 66-63 m.y. ago there were extinctions on our planet so notable they are used to subdivide two of the three eras of the Phanerozoic Eon; these are the Mesozoic and Cenozoic eras and, on a finer scale, the Cretaceous and Tertiary periods. In the oceans nearly 90% of the genera of calcareous planktonic foraminifera and coccolithophorids became extinct; their range truncations define the marine Cretaceous-Tertiary (K-T) contact; its age is estimated at 65 m.y. [Harland et al., 1982] and 66.0 m.y. (J. Hazel, Amoco, personal communication, 1984, based on Shaw's graphic correlation). On land the dinosaurs became extinct; long assumed to have become extinct suddenly at the end of the Cretaceous, their last remains are instrumental in picking the terrestrial K-T contact. Its age in eastern Montana is 63.7-63.9 m.y. [Baadsgaard and Lerbekmo, 1983], and 63.0-63.5 m.y. in the Red Deer Valley, Alberta [Lerbekmo, et al., 1979]. Long the subject of intense controversy, suggested causes of the extinctions span the gamut from extraterrestrial-based abrupt global catastrophe to natural earthly gradual bioevolutionary turnover. The extinctions were coeval with Deccan Traps flood basalt volcanism in India. Prior to the trans-K-T extinctions, ecological stability prevailed [Thierstein, 1981], suggesting steady state between mantle $CO_2$ production and uptake by surficial sinks. Deccan Traps mantle degassing addition to all other mantle $CO_2$ release elevated atmosphere/marine mixed layer $pCO_2$ above normal background levels, perturbing the Williams-Riley pump (productivity-gravity pump of $CO_2$ from atmosphere/mixed layer into deep oceans), contributing to $CO_2$ buildup, and severely reducing mixed layer photosynthesis and $CaCO_3$ production (the dead ocean concept [after Baes, 1982]). Development of a dead ocean would, in itself, have released vast amounts of $CO_2$ and, in concert with degassing, produced trans-K-T global ecological instability.

## Deccan Traps Volcanism

For geologic age of the Deccan Traps, Kaneoka [1980] notes that radiometric ages based on little altered samples "concentrate around 66-60 m.y.," and that "volcanic activity(ies) more than 70 m.y. ago may correspond to precursory one(s) for the main volcanic activity around 65 m.y. ago in the Deccan Traps." Klootwijk [1979] cites an age range of 65-60 m.y., noting that the lower age limit of the Deccan Traps is probably not much older than 65 m.y. Wensink et al. [1979] notes that most dates fall in the lowermost Tertiary or "around the Tertiary/Cretaceous boundary."

Covering about $5 \times 10^5$ $km^2$ of western and central India (Figure 1) with volume exceeding $10^6$ $km^3$ [Choubey, 1973], outliers on the Indian east coast

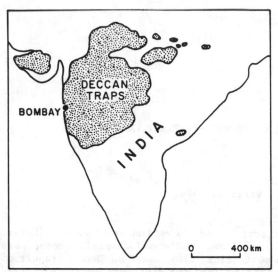

Fig. 1. India, showing present-day distribution of Deccan Traps basalts. Much of the original lava coverage has been destroyed by erosion as indicated by isolated outliers in eastern India.

suggest original coverage was greater than $10^6$ $km^2$ [Krishnan, 1960]; Pascoe [1964] suggests original coverage of Deccan and related volcanics of about 2.6 x $10^6$ $km^2$. The thickness is from 2000 m in western India to 100-200 m in central India [Bose, 1972].

Klootwijk [1979] notes that most of the Deccan Trap lavas were extruded during a relatively short period which spanned an older reversed (R1) and a younger normal (N1) polarity epoch, and that the reversed-to-normal polarity sequence did not last for longer than about 4 m.y. Wensink et al. [1979] note that most lavas were extruded during an older Deccan Reversed Magnetic Polarity Epoch, with volcanism continuing into the overlying Nipani normal epoch. Noting that the main Deccan Traps volcanism was about 65 m.y. ago, and following the geomagnetic time scale of Harland et al. [1982], the bulk of Deccan Traps volcanism occurred in reversed polarity chron R29. Estimates of the lower and upper boundaries of R29 are 65.39 and 64.86 m.y., respectively [Harland et al., 1982]. Thus, the bulk of the Deccan Traps lavas could have been extruded over the very short time interval of about 5.3 x $10^5$ years.

The Deccan Traps basalts represent vast outpourings of liquid lavas through fissure systems that produced thick, extensive tholeiitic basalt flows [Marathe et al., 1981]. Most were erupted along two major lineaments: the E-W trending Narmada-Son-Tapti (Figure 2) and a N-S lineament from Cutch to Bombay [Sukheswala, 1981]. The Narmada-Son-Tapti rift (Figure 2) that separates the Aravali and Deccan shields is perhaps early Paleozoic to Permo-Carboniferous in origin [Auden, 1949; Krishnan, 1961; West, 1962]. Volcanism in India is related to periodic activation of this rift [Krishnaswamy, 1981]. Part of a triple junction formed by its convergence with the N-S trending Konkan and off-shore faulted belt, elements of the Narmada and Konkan cross the Moho at a depth of 30-40 km [Krishnaswamy, 1981].

Other lavas erupted along these lineaments are olivine tholeiites, alkali olivine basalts, rhyolite/trachytes, nepheline syenites, and nephelinites, along with carbonatites, lamprophyres, tinguaites and picrites; they are minor in quantity relative to the tholeiites.

### Deccan Traps—Mantle Plume Connection

Deccan volcanism is linked to distention of the Indian crust via rifting and drift. Chandrasekharem and Parthasarathy [1978] relate the volcanism to convection currents that heated the Indian plate from below, creating fractures that facilitated extrusion of the lavas; others link the Deccan Traps to hot spot activity [Morgan, 1972a, 1972b; Sukheswala, 1981].

Morgan [1981] suggests that continental flood basalts are related to hot spot tracks, and that the track exists under the area to be flooded for tens of millions of years before the eruption. Morgan [1981] relates the Deccan Traps to the Reunion hot spot, which today exists as a volcanic island east of Madagascar. The hot spot is first indicated beneath India about 80 m.y. ago, with the bulk of Deccan Traps erupting about the time of the marine K-T contact. The track progresses along the Laccadive, Maldive, and Chagos Islands; the Carlsberg Rise migrates over the hot spot; and the track makes the southern part of the Mascarene Plateau, and the islands of Mauritius, and Reunion. Sukheswala [1981] also relates the Deccan volcanism to hot spot activity.

Plume origin lavas are enriched in halogens, light rare-earth elements, $^{87}Sr/^{86}Sr$, and other incompatible elements of varying volatility including K, Ba, Rb, Cs, and $CO_2$ [Moore, 1979]; thus, plumes may represent sites of upwelling of partial melt from a deep mantle source

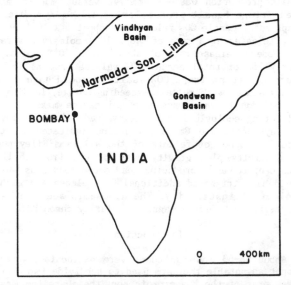

Fig. 2. The Narmada-Son lineament is an ancient rift system whose periodic activation has produced most of the basalts on the Indian subcontinent, including the Deccan Traps. Elements of the rift cross the Moho at a depth of 30-40 km.

relatively undepleted in volatiles, and other incompatible large ion elements. In contrast, normal ridge segments are from a shallower mantle zone previously depleted in these elements. The Deccan Traps do not compare chemically with any other plateau volcanics [Chandrasekharem and Parthasarathy, 1978]; seemingly derived from a less depleted mantle source they are a unique province resembling basalts associated with crustal fragmentation and oceanic ridges. Their evidence includes $K_2O/K_2O + Na_2O$ versus $TiO_2/P_2O_5$ plots that fall within the rift volcanic field, low $K_2O$ concentrations (0.16-0.56%), and $K_2O/K_2O + Na_2O$ ratios (0.04-0.16), and low $^{87}Sr/^{86}Sr$ ratios (0.7058).

Pearce et al. [1975] also cite oceanic affinities of the Deccan Traps. Their work on $TiO_2-K_2O-P_2O_5$ (for discriminating between oceanic and continental basalts) on 222 ocean floor and ocean ridge basalts separates 93% into the oceanic field and, of continental basalts, more than 80% into the continental field. Two exceptions to the latter are the Tertiary basalts of Greenland, and the Deccan Traps, both with oceanic affinities.

### Trans-K-T Carbon Cycle Perturbation

In volcanic emanations, carbon gases ($CO_2$, $CO$, $COS$, and $CH_4$) rank second in abundance to water vapor; of the carbon species, $CO_2$ usually dominates. Sulphur gases ($SO_2$, $H_2S$, $S_2$, and $SO_3$) are next. Other acid gases are $HCl$ and $HF$, etc. This work focusses upon $CO_2$; other gases will be addressed later.

Volcanic $CO_2$ release can be calculated in several ways if rates and duration of $CO_2$ liberated or total gas volume and the $CO_2$ fraction are known [Leavitt, 1982]. These data are not available for the Deccan Traps; $CO_2$ release is calculated from estimates of volume of original Deccan Traps lavas, total $CO_2$ dissolved, and fractionation of $CO_2$ released during eruption via

$$\text{moles } CO_2 = \sum_1 \frac{(s)\ (dg)\ (v)\ (d)}{mwCO_2}$$

where

| | |
|---|---|
| $mCO_2$ | moles of $CO_2$ released; |
| s | original weight fraction of $CO_2$ in the magma |
| dg | degassing fraction during eruption; |
| v | volume of erupted material; |
| d | solid density of the erupted material; |
| $mwCO_2$ | molecular weight of $CO_2$ (=44); |
| i | each eruption. |

In calculating Deccan Traps mantle $CO_2$ release, I use s = 0.005, dg = 0.6, v = 2.6 X $10^6$ $km^3$, d = 2.9 X $10^{15}$ g $km^{-3}$, and $mwCO_2$ = 44.0. For s, 0.002 would be a "best estimate," and 0.005 a "liberal" one [Leavitt, 1982]; I use the liberal value. For dg, 0.6 is a "best estimate" [Leavitt, 1982]. For v, using Pascoe's [1964] estimate of 2.6 x $10^6$ $km^2$ original coverage of the Deccan Traps with average original thickness of 1 km provides an estimate of original volume of 2.6 x $10^6$

$km^3$. Estimated $mCO_2$ release calculated on the entire Deccan Traps lava pile is 5 x $10^{17}$ ($mCO_2$ of the modern atmosphere is 5.6 x $10^{16}$).

Before the time of the Deccan Traps volcanism and the terminal Cretaceous marine extinctions, ecological stability prevailed [Thierstein, 1981], and marine mixed layer productivity was high. Thus, equilibrium likely existed between rates of mantle $CO_2$ degassing and uptake by surficial reservoirs. Excess $CO_2$ production over uptake would have triggered instability. The Deccan Traps mantle degassing was in addition to $CO_2$ release by all other sources. Assuming a pre-Deccan Traps steady state, Deccan Traps $CO_2$ contribution, if of sufficient magnitude, would have triggered disequilibrium, and perturbation of the carbon cycle.

Modern ocean ridges have a total length of about 6 x $10^4$ km. Based on a half spreading rate of 1 cm/yr and a basalt thickness of about 1 km, modern ridge basalt eruption rate is about 1.2 $km^3$/yr. I will now attempt to gain some estimate of the rate of Deccan Traps basalt production relative to modern ocean ridge production and assume that trans-K-T ridge production was at least equivalent. According to Klootwijk [1979] the Deccan Traps basalts were extruded during a reversed-to-normal polarity sequence. Noting that Kaneoka [1980] cites the "main volcanic activity around 65 m.y. ago in the Deccan Traps," and following the geomagnetic time scale of Harland et al. [1982], the reversed polarity chron is R29, and the normal N29. Klootwijk [1979] states that the sequence did not last longer than about 4 m.y. However, according to the time scale of Harland et al. [1982], R29 began and ended at 65.39 and 64.86 m.y., respectively, and N29 at 64.86 and 64.03. Thus, R29 and N29 together spanned a total of about 1.36 m.y.

However, the duration of extrusion of the bulk to the Deccan Traps basalts may have been much less. Wensink et al. [1979] note that the bulk of the Deccan Traps was extruded during the reversed magnetic polarity epoch that they designate the Deccan Reversed Magnetic Polarity Epoch (=R29). Using the time scale of Harland et al. [1982], the bulk of the Deccan Traps were extruded over the relatively short time of about 5.3 x $10^5$ years.

As noted above, an estimated volume of Deccan Traps basalts of 2.6 x $10^6$ $km^3$ is reasonable. Averaged over 1.36 m.y., the rate of Deccan Traps basalt production would have been 1.9 $km^3$/yr, a rate 1.6 times that of the modern MORB production rate of 1.2 $km^3$/yr. However, averaged over 5.3 x $10^5$ years, production would have been 4.91 $km^3$/yr, a rate 4 times modern mid-ocean ridge basalt production.

Next, I will calculate the rate of Deccan Traps annual $CO_2$ production relative to that of all other sources from Table 1 [after Leavitt, 1982]. The average of all estimates in Table 1 is 4.1 x $10^{12}$ moles $CO_2$ released per year. $CO_2$ release by the Deccan Traps considered over the 1.36 m.y. time span of magnetic polarity chrons R29 and N29 is 3.9 x $10^{11}$ moles $CO_2$/yr, about 10% of annual production from all other sources. However, $CO_2$ release calculated over the 5.3 x $10^5$ year duration of polarity chron R29, the equivalent of the Deccan Reversed Magnetic Polarity Epoch of Wensink et al. [1979], is 9.6 x $10^{11}$ moles $CO_2$/yr, nearly 25% of

TABLE 1.  $CO_2$ Input Into Atmosphere From Earth's Interior

| Source of $CO_2$ | $CO_2$ Released moles $yr^{-1}$ | Reference |
|---|---|---|
| Igneous and metamorphic | $6.7 \times 10^{12}$ | Borchert [1951] |
| Igneous only | $1.3 \times 10^{12}$ | |
| Total $CO_2$ since earth's origin | $5.0 \times 10^{11}$ | Rubey [1951] |
| $CO_2$ released from earth's interior | $2.0 \times 10^{12}$ | Plass [1956] |
| Total $CO_2$ on earth's surface | $1.0 \times 10^{12}$ | Li [1972] |
| Volcanic $CO_2$ | $1.7 \times 10^{10}$ | Libby and Libby [1972] |
| Outgassing of oceanic crust | $2.3 \times 10^{11}$ | Anderson [1975] |
| Hawaiian estimate extrapolated to worldwide | $2.0 \times 10^{13}$ | Buddemeier and Puccetti [1974] |
| Volcanoes, fumeroles, and hot springs | $1.7-8.3 \times 10^{12}$ | Baes et al. [1976] |

annual $CO_2$ production from all other sources.

Thus, if the Deccan Traps basalts were erupted mostly during the short duration of $5.3 \times 10^5$ years, the 25% increase in $CO_2$ production above steady state should have perturbed the previous equilibrium between mantle $CO_2$ production and uptake by surface sinks indicated by climatic stability prior to the marine extinctions [Thierstein, 1981]. Because the deep oceans were warm ($14°-15°C$), $CO_2$ uptake by that sink would have been reduced. Also, so far as is known, polar regions lacked ice caps, and the marine cold water circulation drive would have been less vigorous than today; Berger [1979] notes that salinity drive was likely predominant. Also, that reducing conditions prevailed on the seafloor at the time of the K-T contact seems to indicate sluggish circulation. Thus, the effects of the Deccan Traps $CO_2$ production addition to all other sources would have been cumulative, and would have been expressed as $CO_2$ buildup in the atmosphere and marine mixed layer. For point of argument, Deccan Traps $CO_2$ production of $9.6 \times 10^{11}$ moles $CO_2$/yr would, by itself, have accumulated to $5.8 \times 10^{16}$ moles $CO_2$ in only $6 \times 10^4$ years, an amount equivalent to that of the modern atmosphere.

Clearly, the Deccan Traps $CO_2$ addition to all other sources would have triggered disequilibrium between mantle release and uptake by sinks. That geological conditions were conducive to $CO_2$ accumulation in the atmosphere and the marine mixed layer would have further perturbed earth's carbon cycle.

Biological Perturbation—Dead Sea Linkage to Deccan Traps Degassing

Prior to the terminal Cretaceous marine extinctions, ecological stability seems to have prevailed, and productivity was high in the marine mixed layer [Thierstein, 1981]. During the K-T transition, instability and reduced productivity prevailed. This instability was coeval with Deccan Traps mantle degassing.

Evidence of perturbation at and about the marine K-T contact is the marine extinctions themselves together with coeval drops in $^{13}C$ and $^{18}O$ isotopic values, and mixed layer biogenic $CaCO_3$ production; these are coeval with the Ir/clay deposition and early stages of Deccan Traps volcanism. The duration of the iridium enrichment of about $10^5$ years (C. Officer, personal communication, 1984) may reflect duration of the most vigorous degassing.

Immediately following the marine extinctions, which seem to have spanned at least $5 \times 10^4$ years [Hsu et al., 1982a], Strangelove conditions [Hsu, 1984] existed in the mixed layer, in which thoracospherids, braarudosphaerids and small planktonic foraminifera dominated Early Tertiary plankton; duration of Strangelove conditions ranges between Hsu's repopulation of the oceans $10^5$ years after Ir/clay deposition, and the Haq et al. [1977] dominance of thoracospherids, etc., for about $1-2 \times 10^6$ years. Arthur and Dean [1982] cite low surface water productivity for 1-2 m.y. following the crisis. Thierstein [1981] indicates ecological

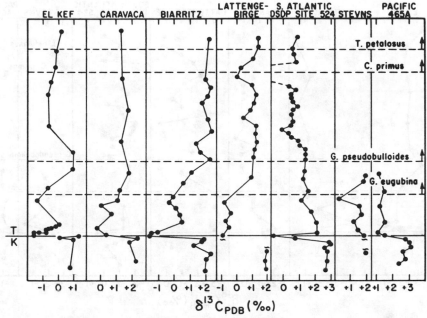

Fig. 3. Carbon 13 record at relatively complete trans-K-T sections showing a depletion of $^{13}C$ beginning before deposition of the K-T boundary Ir/clay layer [after Perch-Nielsen et al., 1982].

instability for hundreds of thousands of years following the marine extinction event. Hsu et al. [1982b] suggested return to normal surface water enriched in $^{13}C$ after $5 \times 10^5$ years with complete recovery of mixed layer fertility.

During the K-T transition, the temperature of the deep oceans was $14°-15°C$ (versus $2°-3°C$ today) [Thompson and Barron, 1981], reducing the capacity of earth's major sink for $CO_2$ uptake; mantle $CO_2$ flooding onto earth's surface would thus have accumulated in the atmosphere and mixed layer. Sluggish ocean circulation would have caused $CO_2$ and other acid gases to accumulate in the atmosphere/mixed layer faster than they would have circulated into the deep oceans by the slow process of diffusion; $CO_2$ would have accumulated in the mixed layer, affecting pH. Marine circulation was sluggish; deep waters derived their density from high salinity, and anaerobism developed at times [Berger, 1979]. That Maastrichtian oceans were less stratified than modern ones is indicated by low-latitude top-bottom temperature gradients of one-third modern values [Douglas and Savin, 1973]. Also, reducing conditions seem to have prevailed on the seafloor during the time of K-T contact deposition.

Reduction of terrestrial floras (and thus global photosynthesis) should cause marine $CaCO_3$ dissolution via $CO_2$ uptake by the oceans [Shackleton, 1977]. By analogy, the terminal Cretaceous coccolithophorid extinctions should have triggered $CO_2$ buildup. Algae remove $CO_2$ from the atmosphere/mixed layer, converting it to particulates that settle into and store $CO_2$ in the deep oceans maintaining low $pCO_2$ in the atmosphere and mixed layer; this is the Williams-Riley pump [Dyson, 1982]; its failure today would raise atmospheric $CO_2$ levels severalfold [National Research Council, 1977].

The concept of a dead ocean [after Baes, 1982] is relevant to the trans-K-T scenario. By definition this is where photosynthesis and $CaCO_3$ precipitation are turned off. According to Baes, a modern dead ocean at steady state would produce a substantially higher (nearly doubled) partial pressure of $CO_2$ in the atmosphere. The terminal Cretaceous failure of coccolithophorids would certainly have produced some degree of dead ocean and consequent $CO_2$ release. And, because of the continued Deccan Traps $CO_2$ contribution, steady state conditions did not prevail, leading to even greater $CO_2$ buildup. To make a point concerning the potential of the Williams-Riley pump for $CO_2$ release, modern total failure of the pump with no $CO_2$ uptake by other sinks, based on a modern mixed layer net primary productivity of $2.1 \times 10^{15}$ moles carbon/year, would release $5.3 \times 10^{16}$ moles of carbon in 25 years, doubling atmospheric $CO_2$. Even partial failure during a time of reduced marine circulation and warm deep oceans would trigger $CO_2$ buildup in the atmosphere and mixed layer.

At and about the K-T contact as defined by the Ir/clay, marine mixed layer photosynthesis and $CaCO_3$ production were severely reduced, indicating dead sea conditions. Certainly, the coccolithophorid collapse is indicative of failure of the Williams-Riley pump. Hsu et al. [1982b] note mass mortality in the oceans followed by nearly total suppression of photosynthetic activities by planktonic organisms. Perch-Nielsen et al. [1982] suggest that a $^{13}C$ decrease of about 3 per mil in Tertiary sediments at the K-T contact (Figure 3) is the result of greatly reduced photosynthesis and the increase of isotopically lighter bicarbonate ions in

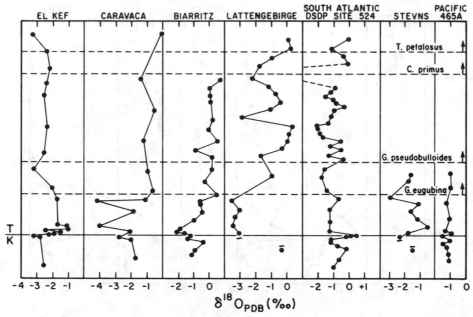

Fig. 4. Oxygen 18 record at relatively complete trans-K-T sections showing depletion of $^{18}O$ beginning before deposition of the K-T boundary Ir/clay layer [after Perch-Nielsen et al., 1982]. The $^{18}O$ record, which seems latitude dependent, indicates Early Tertiary climatic warming at latitudes of the final dinosaurian record in the North American Rocky Mountain region.

the surface water; $^{12}C$ is preferentially used during photosynthesis. Thierstein [1980] noted that the collapse of the Late Cretaceous nannoplankton community resulted in a sudden decrease of carbonate supply to the ocean floor. Trans-K-T cut off of $CaCO_3$ is seen in sections with nearly complete trans-K-T stratigraphic sections. At El Kef, $CaCO_3$ content decreases sharply at the K-T boundary from 37 to 0% in the boundary clay [Perch-Nielsen et al., 1982]. At DSDP Site 524, $CaCO_3$ decreases to 2% in the boundary clay [Hsu et al., 1982b]. According to J. Hazel (Amoco, personal communication, 1984), based on Shaw's graphic correlation, El Kef and DSDP Site 524 are among the roughly half-dozen nearly complete trans-K-T stratigraphic sections yet discovered.

At and about the K-T contact as defined by the Ir/clay were shifts in $^{13}C$ and $^{18}O$ isotopic values (Figures 3 and 4); they were coeval with drops in photosynthesis and mixed layer $CaCO_3$ production, and with iridium enrichment. Isotopic shifts, iridium enrichment, and drop in $CaCO_3$ occur stratigraphically lower than the K-T boundary Ir/clay and persist into Early Tertiary. These are not evidence of a sudden catastrophic event but, instead, gradual event spanning tens of thousands to hundreds of thousands of years. As noted previously, C. Officer (personal communication, 1984) suggests a duration for the iridium enrichment of about $10^5$ years. Sections with hiatuses at the K-T contact have created the illusion of catastrophe [McLean, 1981a]. Pre-Ir/clay $^{13}C$ depletion is recorded at: El Kef, Caravaca, Biarritz, DSDP Site 524, and DSDP Site 465A (Figure 3). The depletion may reflect mantle degassing. Because (1) surface water $^{13}C$ is determined

by exchange with the atmosphere, and by the amount of carbon removed versus that replenished [Kroopnick et al., 1977], (2) the $^{13}C$ record is preserved in biogenic $CaCO_3$ produced in the mixed layer coevally with Deccan Traps mantle degassing, and (3) mantle $CO_2$ is depleted in $^{13}C$, I propose that at least some of the $^{13}C$ depletion is reflective of mantle degassing. I also propose that the K-T boundary iridium enrichment is reflective of mantle release (discussed herein).

The Trans-K-T Marine Extinctions

The terminal Cretaceous marine extinctions were restricted in both habitat (mixed layer of the oceans) and organisms involved (calcareous microplankton). Calcareous microplankton includes planktonic foraminifera and coccolithophorids, each of which lost about 87% of their genera [Emiliani et al., 1981]. Claims that other marine groups became extinct simultaneously must be proved from demonstrably complete trans-K-T stratigraphic sections; this has not been done. For organisms at or near the peak of their evolutionary diversification, near-catastrophic extinctions occurred only among calcareous microplankton of warm water habitats; noncalcareous plankton, and most other groups that died out at or near the end of the Cretaceous, show graded extinction histories [Kauffman, 1979, 1982]. These include the ammonoids, belemnoids, inoceramid and rudistid bivalves, hermatypic corals, and larger foraminifera, all of which were in evolutionary decline and were represented by a few generalized genera and species at the end of the Cretaceous. The most severe terminal Cretaceous marine extinctions were in Tethyan or marginal Tethyan regions.

For the calcareous microplankton, most planktonic foraminifera live in near-surface waters between 10 and 50 m [Be, 1977], as do most coccolithophorids, and both seem sensitive to modest pH change (discussed below). Thus, a $CO_2$-induced mixed layer pH change may explain the restricted nature of the extinctions. Restriction is also seen in that planktonic foraminifera which were affected more than benthonics have never made up more than 1% of the total foraminiferal faunas [Lipps, 1979].

For planktonic foraminifera there is correlation between low pH and low foraminiferal abundance [Boltovskoy and Wright, 1976]. Myers, [1943] notes that small specimens survive better in acid waters than do large ones. For the Salton Sea, pH is the most important factor affecting distribution [Arnal, 1961].

Boltovskoy and Wright [1976] note that pH plays a different role in the calcareous test than in the protoplasm. Below pH 7.8, calcareous tests begin to dissolve [Boltovskoy and Wright, 1976]. In a Mission Bay, California, marsh with daily pH range from 6.8 to 8.5, tests dissolve below 7.6 unless the organisms expend energy to maintain them [Phleger and Bradshaw, 1966]. Some species can be decalcified at pH 8.0, with recovery of the foraminifera [McEnery and Lee, 1970]. For the protoplasm, specimens of Elphidium macellum and Quinqueloculina become less active as the pH drops below that of normal sea water, and all activity ceases at a pH of 7.0.

Coccolithophorids are also sensitive to pH change. Coccolithus huxleyi loses its coccoliths at pH 7.3-7.0 (C. S. Sikes, personal communication, 1984). Culture studies of the modern Coccolithus huxleyi and Cricosphaera carterae [Sikes et al., 1980; Sikes and Wilbur, 1980, 1982] suggest that coccolithophorids are unlike other algae in lacking a mechanism for concentrating $CO_2$ before fixation. Carbon enters the cells as $HCO_3^-$ and is supplied to the calcification site as the source for coccolith production; its incorporation into $CaCO_3$ dehydrates $HCO_3^-$, producing $CO_2$ that is used in photosynthesis.

That coccolithophorids are unlike other algae may make them more vulnerable to mixed layer pH change than groups that suffered less during trans-K-T carbon cycle perturbtion. Dinoflagellate studies in my graduate program show little composition change across the K-T contact due to unusual extinction rates. For diatoms, A. Gombos (Exxon, personal communication, 1984) showed over 80% of the Cretaceous genera crossing the K-T contact into Tertiary.

Trans-K-T extinctions of some invertebrate groups can also be explained via mixed layer pH change. For cephalopods, both the ammonoids and belemnoids became extinct at (or near) the end of the Cretaceous. Eggs of the former were miniscule [after Emiliani et al., 1981], and an epipelagic larval stage must have followed. For the latter, their closest living relative, Loligo, inhabits shallow waters and fastens its eggs on shallow (10-30 m) bottoms; both groups would have been affected by trans-K-T mixed layer pH change. Conversely, modern Nautilus, as representative of the surviving nautiloids, fastens it eggs to bedrock at depths

below 100 m. Terminal Cretaceous brachiopod extinctions in Denmark [Alvarez et al., 1984] are also explained via mixed layer pH change; they lived on a $CaCO_3$ substrate produced by mixed layer coccolithophorids; trans-K-T cutoff of coccolithophorid $CaCO_3$ to the seafloor would have also affected benthonic brachiopods.

The terminal Cretaceous marine record does not support the concept of catastrophe, even for the organisms most affected: planktonic foraminifera and coccolithophorids. Hsu et al. [1982a] note that at DSDP Site 524 Cretaceous species did not become extinct until in Early Tertiary about $5 \times 10^4$ years after deposition of the K-T boundary Ir/clay; evidence that Cretaceous taxa in Tertiary strata were survivors and are not reworked is that they have the same isotopic values as Tertiary taxa, showing that they lived in the same Early Tertiary waters [Perch-Nielsen et al., 1982]. As an estimate of the duration of the marine extinctions, Kent's [1977] estimate of $10^4$ years at the Gubbio, Italy, section cannot be correct in light of $7.8 \times 10^5$ years of Upper Cretaceous strata missing at Gubbio (J. Hazel, Amoco, personal communication, 1984).

Trans-K-T Global Warming and Terrestrial Extinctions

A number of studies have indicated trans-K-T global warming that would have been coeval with Deccan Traps mantle degassing. Margolis et al. [1977], interpreting the isotope record from several DSDP sites, state that the K-T boundary occurs during a time of significant global warming of bottom and surface waters. The Boersma et al. [1979] study of over 200 samples of K-T sediments from Atlantic Ocean DSDP sites 384, 86, 95, 152, 144, 20C, 21, 356, 357, and 329 indicates a significant temperature rise across the K-T boundary both at the surface and in deep waters of the Atlantic Ocean; their evaluation of Douglas and Savin's [1973] data indicates warming across the K-T boundary for the Pacific as well.

Oxygen-18 data from nearly complete stratigraphic sections indicate $CO_2$-induced global warming beginning in the K-T transition and peaking in the Early Tertiary. At DSDP Site 524, Hsu et al. [1982b], noting that ocean temperatures increased substantially after the end of the Crecaceous, record a $5^\circ C$ temperature increase of bottom waters (from $16^\circ C$ to $21^\circ C$). Maximum temperature, reached in the Early Tertiary about $3 \times 10^4$ years after deposition of the K-T boundary Ir/clay, was about $10^\circ C$ above the background. Hsu et al. [1982a] indicate $2^\circ C$ to $10^\circ C$ warming of the surface water. Hsu et al. [1982a] attribute the warming to $CO_2$-induced greenhouse conditions.

Other evidence of trans-K-T warming is Smit's [1982] $^{18}O$ data from the Gredero and Biarritz sections that support "catastrophic temperature rise" after the K-T boundary event. Perch-Nielsen et al. [1982] show variable results (Figure 4). They note that geographic position plus changes in paleocirculation produced the oxygen isotope stratigraphy at the various locations, noting also that changes in global temperatures affect the higher-latitude waters more than the lower.

Early Tertiary Deccan Traps mantle degassing coeval with global warming is significant to the terrestrial dinosaurian extinctions; time-rock data indicate that the dinosaurs lived into the Early Tertiary. The global K-T isochron is defined on the basis of extinctions in marine strata; age estimates of the marine K-T contact are 65 m.y. [Harland et al., 1982] and 66.0 m.y. (J. Hazel, Amoco, personal communication, 1984). In the Red Deer Valley, Alberta, the last dinosaurs occur in a geomagnetic reversed interval 6 m below the pollen-based K-T contact at the top of the Nevis coal, and 5 m below a bentonite with transparent, unaltered sanidine which yielded radiometric dates of 63.0, 63.3, 63.5, and 63.5 m.y. [Lerbekmo et al., 1979]. Based on sedimentation rates, an estimate of the duration of time between the last dinosaurs and the K-T contact is $9 \times 10^4$ years [Lerbekmo et al., 1979]. Thus, on the basis of radiometric dates, the dinosaurs disappeared between 63.0 and 64.0 m.y. years ago in the Red Deer Valley. In eastern Montana, Baadsgaard and Lerbekmo [1983] reported an age of the "Z" coal K-T contact of 63.7-63.9 m.y. According to the geomagnetic time scale of Harland et al. [1982], the K-T contact in both areas would seem to be in polarity chron R28, whereas the marine K-T contact is in R29. Thus, if the radiometric dates are correct, the last dinosaur remains are younger than the marine extinctions that define the global K-T isochron and, by definition, the dinosaurs became extinct during the Early Tertiary, seemingly in the late stages of Strangelove ocean conditions. For attempts to use the dinosaurs for age dating purposes, I have suggested previously that the dinosaurs disappeared at different times in different places in the Rocky Mountain region [McLean, 1981b, 1982] and are thus useless for chronostratigraphic purposes.

Several years ago [McLean, 1978], I proposed $CO_2$-induced "greenhouse" conditions as a factor in the dinosaurian extinctions, evoking coupling between climatic warming and repoductive dysfunction. Confirmation of global warming at the time of the dinosaurian extinctions would seem to strengthen my case. Erben et al. [1983] have reported pathological dinosaurian egg shell thinning in Early Tertiary strata from France and Spain; Maastrichtian shells were normal. Fassett [1982] reports dinosaur bones above the pollen-based K-T contact in the San Juan Basin of New Mexico.

## K-T Boundary Iridium

Iridium enrichment at and about the marine K-T contact, is the basis for terminal Cretaceous asteroid impact global catastrophe theory [Alvarez et al., 1980]. Impact dust injected into the stratosphere supposedly plunged earth into darkness for several months, cutting off photosynthesis and simultaneously killing both marine and terrestrial plants, and also the terrestrial dinosaurs. Fallout of the dust is presumed to have produced a global Ir/clay layer. Recently, a thin clay layer (or layers) containing iridium has been found in the Rocky Mountain region; some researchers assume it to be isochronous with the marine Ir/clay [Smit and Van Der Kaars, 1984]. In fact, neither marine nor terrestrial records support extraterrestrial based catastrophe. The marine Ir/clay is more plausibly explained via mantle degassing. The terrestrial Ir/clay

and its diaplectic (shocked) quartz [Bohor et al., 1984] are younger than the marine Ir/clay K-T contact.

Zoller et al. [1983] recently reported iridium in the gas phase at Kilauea at concentrations up to $6 \times 10^5$ times that in the basalt, demonstrating that iridium can be released from the mantle. This is important for the origin of the K-T Ir/clay. During Deccan Traps volcanism, India was east of Africa and slightly south of the equator. Fissure flow volcanism produces thermal plumes that can inject particulates into the stratosphere. India's location during Deccan Traps volcanism was such that prevailing wind directions in both hemispheres would have carried iridium-bearing particulates to where the K-T boundary Ir/clay is best developed: western Europe and the eastern south Atlantic [Officer and Drake, 1983]. That the Ir/clay is rich in smectite shows its volcanic origin.

During earth's differentiation, iridium is believed to have been swept into the core. Cousins [1973] states that greater abundance of platinum group elements in basic and ultrabasic rocks, believed to form at great depths, suggests that siderophiles are more concentrated in the core than in the crust. Thus, if the Deccan Traps represents mantle plume activity originating near earth's mantle/core boundary [G. Wasserburg, California Institute of Technology, personal communication, 1982, based on Nd/Sm), (or shallower source enriched in platinum group elements), it provides a mechanism for conveying siderophiles to earth's surface, to then be distributed along with volcanic dust by wind and ocean currents. Stumpel [1974] notes that Ir can be mobilized, as does Cousins [1973]. Alvarez et al. [1980] report iridium in a warm spring on Mt. Hood, Oregon. Coevality of the marine K-T boundary Ir/clay and Deccan Traps volcanism is too striking to ignore. Accumulation of volcanic dust carrying iridium would, over a long time interval, produce an Ir/clay. Barker and Anders' [1968] work on iridium and osmium in deep-sea sediments shows an increase with decreased sedimentation rate, and that locally high iridium concentrations could result from terrestrial sources as trace-element-rich particles from land or submarine volcanism. In the K-T transition while planktonic $CaCO_3$ supply was cut off to the sea floor, volcanic clays that ordinarily would be admixed with $CaCO_3$ would have settled out as a discrete layer. Iridium would also have accumulated via accretion of $10^5$ kg of interplanetary dust onto earth daily [Grieve and Sharpton, 1981]. As noted by C. Officer (personal communication, 1984) the marine iridium represents deposition over an interval of at least $10^5$ years.

Inferences that the Rocky Mountain terrestrial iridium spikes are the same age as the marine, or that they represent accumulation of dust following an asteroid, are not supported by the record. The terrestrial iridium spike(s) is (are) younger than the marine K-T boundary Ir/clay. The Raton Basin iridium spikes occur in magnetic polarity chron N29 (determined by J. Hazel, Amoco, personal communication, 1984, based on Shaw's graphic correlation). For the eastern Montana site, dating of a bentonite in the "Z" coal K-T boundary marker [Baadsgaard and Lerbekmo, 1983] indicates an age of 63.7-63.9 m.y. According to the geomagnetic time scale of Harland et al. [1982], this is most likely in R28. Also, Luck and Turekian [1983] have shown based on study of $^{187}Os/^{186}Os$ that the marine and terrestrial Ir/clay layers did not have common origin. Distribution of iridium in the Rocky Mountain terrestrial Ir/clay does not support its origin via bolide impact fallout. If the Ir/clay resulted from dust settling out after impact, the iridium would be admixed uniformly throughout the clay. This is not the case. In the Raton Basin sites [Pillmore et al., 1984], the iridium occurs at the top of the 1-4 cm thick clay; at the York Canyon site, the iridium does not even occur in the clay, but in the overlying coal. For the Hell Creek, Montana, Ir/clay spike, Smit and Van Der Kaars [1984] reported that the iridium anomaly "occurs in the top of a thin clay layer at the base of the Z coal."

Last, by asteroid theory, dust-induced darkness killed off most terrestrial plants; plant life should have been extremely sparse during a long recovery period. However, coal(s) rests immediately upon the Rocky Mountain Ir/clay at all localities, showing vigorous plant growth following deposition of the Ir/clay. Asteroid impact is not supported by the record. Chemical concentration of iridium associated with swamp and marsh development seems operative in the Rocky Mountain iridium anomalies.

## Conclusions

Mantle degassing perturbation of earth's surficial carbon reservoirs was a factor in the terminal Cretaceous marine extinctions that define the global K-T isochron, and in the Early Tertiary terrestrial dinosaurian extinctions. Deccan Traps volcanism in India, the greatest episode of continental flood basalt volcanism in the Phanerozoic, began erupting coevally with the marine extinctions (65 m.y.). Mantle $CO_2$ release at a time while trans-K-T deep oceans were warm (14°-15°C) and not a sink for $CO_2$ uptake would have triggered $CO_2$ accumulation in the atmosphere/mixed layer, causing failure of the Williams-Riley productivity-gravity pump of $CO_2$ from the atmosphere/mixed layer into the deep oceans, creating a trans-K-T dead ocean (severely reduced mixed layer photosynthesis and $CaCO_3$ production); dead ocean conditions are supported by the sedimentary record. Main Deccan Traps activity was in the Early Tertiary, coeval with Early Tertiary global warming indicated by depletion of $^{18}O$ and with the dinosaurian extinctions 63 to 64 m.y. ago in the Red Deer Valley, Alberta, Canada; the dinosaurs may have disappeared in late stages of the Early Tertiary Strangelove oceans [Hsu, 1984] in which plankton was dominated by thoracosphaerids, braarudosphaerids, and small foraminifera.

## References

Alvarez, L. W., W. Alvarez, F. Asaro, and H. V. Michel, Extraterrestrial cause for the Cretaceous-Tertiary extinction, Science, 208, 1095-1108, 1980.

Alvarez, W., E. G. Kauffman, F. Surlyk, L. W. Alvarez, F. Asaro, and H. V. Michel, Impact theory of mass extinctions and the invertebrate fossil record, Science, 223, 1135-1141, 1984.

Anderson, A. T., Some basaltic and andesitic gases, Rev. Geophys. Space Phys., 13, 37-55, 1975.

Arnal, R. E., Limnology, sedimentation, and microorganisms of the Salton Sea, California, Geol. Soc. Am., 72, 427-478, 1961.

Arthur, M. A., and W. E. Dean, Changes in deep ocean circulation and carbon cycling at the Cretaceous-Tertiary boundary (abstract), in American Association for the Advancement of Science, 148th National Meeting, p. 48, AAAS, Washington, D.C., 1982.

Auden, J. B., A geological discussion of the Satpura Hypothesis and Garo-Rajmahal Gap, Proc. Natl. Inst. Sci. India, 15, 313-340, 1949.

Baadsgaard, H., and J.F. Lerbekmo, Rb-Sr and U-Pb dating of bentonites, Can. J. Earth Sci., 20, 1282-1290, 1983.

Baes, C. F., H. E. Goeller, J. S. Olsen, and R. M. Rotty, The global carbon dioxide problem, Oak Ridge Natl. Lab Rep., ORNL-5194, 1976.

Baes, C. F., Jr., Effects of ocean chemistry and biology on atmospheric carbon dioxide, in The Carbon Review: 1982, edited by W. C. Clark, pp. 189-204, Clarendon Press, Oxford, 1982.

Barker, J. L., and Anders, E., Accretion rate of cosmic from iridium and osmium contents of deep-sea sediments, Geochim. Cosmochim. Acta, 32, 627-645, 1968.

Be, A. W. H., An ecological, zoogeographic and taxonomic review of Recent planktonic foraminifera, in Oceanic Micropaleontology, edited by A. T. S. Ramsay, pp. 1-100, Academic, New York, 1977.

Berger, W. H., Preservation of foraminifera, in Foraminiferal Ecology and Paleoecology, SEPM Short Course 6, edited by J. H. Lipps, pp. 105-155, Society of Economic Paleontologists and Mineralogists, Tulsa, Okla., 1979.

Boersma, A., N. Shackleton, M. Hall, and Q. Given, Carbon and oxygen isotope records at DSDP site 384 (North Atlantic) and some Paleocene paleotemperatures and carbon isotope variations in the Atlantic Ocean, Initial Rep. Deep Sea Drill. Proj., 43, 695-717, 1979.

Bohor, B. F., E. F. Foord, P. J. Modreski, and D. M. Triplehorn, Mineralogical evidence for an impact event at the Cretaceous-Tertirary bundary, Science, 224, 867-869, 1984.

Boltovskoy, E., and R. Wright, Recent Foraminifera, 515 pp., Junk, The Hague, 1976.

Borchert, H., Zur Geochemie des Kohlenstoffs, Geochim. Cosmochim. Acta, 2, 62-75, 1951.

Bose, M. K., Deccan basalts, Lithos, 5, 131-145, 1972.

Buddemeier, R. W., and A. Puccetti, C-14 dilution estimates of vocanic $CO_2$ emission rates (abstract), Eos Trans. AGU, 55, 488, 1974.

Chandrasekharam, D. and A. Parthasarathy, Geochemical and tectonic studies on the coastal and inland Deccan volcanics and a model for the evolution of Deccan Trap volcanism, Neues Jahrb. Mineral. Abh., 132, 214-229, 1978.

Choubey, V. D., Long-distance correlation of Deccan flows, central India, Geol. Soc. Am. Bull., 84, 2785-2780, 1973.

Cousins, C. A., Notes on the geochemistry of the platinum group elements, Trans. Geol. Soc. South Africa, 76, 77-81, 1973.

Douglas, R. G., and S. M. Savin, Oxygen and carbon isotope analyses of Cretaceous and Tertiary foraminifera from the central North Pacific, Initial Rep. Deep Sea Drill. Proj., 17, 591-605, 1973.

Dyson, F. J., Balance sheets of the carbon and oxygen cycles, in The Long Term Impacts of Increasing Atmospheric Carbon Dioxide Levels, edited by G. J. MacDonald, pp. 43-56, Ballinger, Cambridge, Mass., 1982.

Emiliani, C., E. B. Kraus, and E. M. Shoemaker, Sudden death at the end of the Mesozoic, Earth Planet. Sci. Lett., 55, 317-334, 1981.

Erben, H. K., A. R. Ashraf, K. Krumsiek, and J. Thein, Some dinosaurs survived the Cretaceous "final event" (abstract), Terra Cognita, 3, KA 6, 1983.

Fassett, J. E., Dinosaurs in the San Juan Basin, New Mexico, may have survived the event that resulted in creation of an iridium-enriched zone near the Cretaceous/Tertiary boundary, in Geological Implications of Impacts of Large Asteroids and Comets on the Earth, Geol. Soc. Am. Spec. Pap. 190, edited by L. T. Silver and P. H. Schultz, pp. 435-447, Geological Society of America, Boulder, Colo., 1982.

Grieve. R. A. F., and V. L. Sharpton, The Cretaceous-Tertiary extinction event, Lithos, 9, 65-73, 1981.

Harland, W. B., A. V. Cox, P. G. Llewellyn, C. A. G. Pickton, A. G. Smith, and R. Walters, A Geologic Time Scale, 131 pp., Cambridge University Press, New York, 1982.

Haq, B. U., K. Perch-Nielsen, and G. P. Lohman, Contribution to the Paleocene nannofossil biogeography of the central and southwest Atlantic Ocean (Ceara Rise and Sao Paulo Plateau, DSDP Leg 39), Initial Rep. Deep Sea Drill. Proj., 39, 841-848, 1977.

Hsu, K. J., Q. X. He, J. A. McKenzie, H. Weissert, K. Perch-Nielsen, H. Oberhansli, K. Kelts, J. LaBrecque, L. Tauxe, U. Krahenbuhl, S. F. Percival, R. Wright, A. M. Karpoff, N. Petersen, P. Tucker, R. Z. Poore, A. M. Gombos, K. Pisciotto, M. F. Carman, and E. Schreiber, Mass mortality and its environmental and evolutionary consequences, Science, 216, 249-256, 1982a.

Hsu, K. J., J. A. McKenzie, and Q. X. He, Terminal Cretaceous environmental and evolutionary changes, in Geological Implications of Impacts of Large Asteroids and Comets on the Earth, Geol. Soc. Am. Spec. Pap. 190, edited by L. T. Silver and P. H. Schultz, pp. 317-328, Geological Society of America, Boulder, Colo., 1982b.

Hsu, K. J., Strangelove ocean at the beginning of Tertiary, paper presented at Chapman Conference on Natural Variations in Carbon Dioxide and the Carbon Cycle, American Geophysical Union, Tarpon Springs, Jan. 9 to 13, 1984.

Kaneoka, I., Ar$^{40}$/Ar$^{39}$ dating on volcanic rocks of the Deccan Traps, India, Earth Planet. Sci. Lett., 46, 233-243, 1980.

Kauffman, E. G., The ecology and biogeography of the Cretaceous-Tertiary event, in Cretaceous-Tertiary Boundary Event Symposium II, edited by T. V. Birkelund and R. G. Bromley, pp. 29-37, University of Copenhagen, 1979.

Kauffman, E. G., Environmental deterioration and graded

extinction at the end of the Cretaceous (abstract), in American Association for the Advancement of Science, 14 8th National Meeting, p. 48, AAAS, Washington, D.C., 1982.

Kent, D. V., An estimate of the duration of the faunal change at the Cretaceous-Tertiary boundary, Geology, 5, 769-771, 1977.

Klootwijk, C. T., A review of paleomagnetic data from the Indo-Pakistani fragments of Gondwanaland, in Geodynamics of Pakistan, edited by A. Farah and K. A. DeJong, pp. 41-80, Geological Survey of Pakistan, Quetta, 1979.

Krishnan, M. S., Geology of India and Burma, Higginbothams (Private), Ltd., Madras, 1960. Krishnan, M. S., A structural and tectonic history of India, Geol. Surv. India Mem. 81, 1-137, 1961.

Krishnaswamy, V. S., The Deccan volcanic episode, related tectonism and geothermal manifestations, in Deccan Volcanism and Related Basalt Provinces in Other Parts of the World, Geol. Soc. India Mem. 3, edited by K. V. Subbarao and R. N. Sukheswala, pp. 1-7, Rashtrothana Press, Bangalore, 1981.

Kroopnick, P. M., S. V. Margolis, and C. S. Wong, Carbon-13 variations in marine carbonate sediments as indicators of the $CO_2$ balance between the atmosphere and oceans, in The Fate of Fossil Fuel $CO_2$ in the Oceans, edited by N. R. Andersen and A. Malahoff, pp. 295-321, Plenum, New York, 1977.

Leavitt, S. W., Annual volcanic carbon dioxide emission: an estimate from eruption chronologies, Environ. Geol., 4, 15-21, 1982.

Lerbekmo, J. F., M. E. Evans, and H. Baadsgaard, Magnetostratigraphy, biostratigraphy and geochronology of Cretaceous-Tertiary boundary sediments, Red Deer Valley, Nature, 279, 26-30, 1979.

Li, Y. H., Geochemical mass balance among lithosphere, hydrosphere, and atmosphere, Am. J. Sci., 272, 119-137, 1972.

Libby, L. M., and W. F. Libby, Vulcanism and radiocarbon dates, Proc. Int. Conf. Radiocarbon Dating 8th, A72-A75, 1972.

Lipps, J. H., The ecology and paleoecology of planktic foraminifera, in Foraminiferal Ecology and Paleoecology, SEPM Short Course 6, edited by J. H. Lipps, pp. 62-104, Society of Economic Paleontologists and Mineralogists, Tulsa, Okla., 1979.

Luck, J. M. and K. K. Turekian, Osmium-187/osmium-186 in manganese nodules and the Cretaceous-Tertiary boundary, Science, 222, 613-615, 1983.

Margolis, S. V., P. M. Kroopnick, and D. E. Goodney, Cenozoic and late Mesozoic paleooceanographic and paleoglacial history recorded in circum-Antarctic deep sea sediments, Mar. Geol., 25, 131-147, 1977.

McEnery, M. E., and J. J. Lee, J. Protozool., 17, 184-195, 1970.

McLean, D. M., A terminal Mesozoic "greenhouse": lessons from the past, Science, 201, 401-406, 1978.

McLean, D. M., A test of terminal Mesozoic "catastrophe", Earth Planet. Sci. Lett., 53, 103-108, 1981a.

McLean, D. M., Diachroneity of the K-T contact and of the dinosaurian extinctions, paper presented at Conference on Large Body Impacts and Terrestrial Evolution: Geological, Climatological, and Biological Implications, p. 36, Lunar and Planetary Institute, Houston, Tex., 1981b.

McLean, D. M., Diachroneity of the Cretaceous/Tertiary contact and of the dinosaurian extinctions, Biol. Extinction, 1, 23, 1982.

Moore, J. G., Vesicularity and $CO_2$ in mid-ocean ridge basalt, Nature, 282, 250-253, 1979.

Morgan, W. J., Deep mantle convection plumes and plate motion, Am. Assoc. Pet. Geol. Bull., 56, 203-213, 1972a.

Morgan, W. J., Plate motions and deep mantle convection, Geol. Soc. Am. Mem., 132, 7-22, 1972b.

Morgan, W. J., Hotspot tracks and the opening of the Atlantic and Indian oceans, in The Oceanic Lithosphere, edited by C. Emiliani, pp. 443-487, Wiley-Interscience, New York, 1981.

Myers, E. H., Life activities of foraminifera in relation to marine ecology, Proc. Am. Philos. Soc., 86, 439-458, 1943.

National Research Council Geophysics Study Committee, in Energy and Climate, p. 17, National Academy of Sciences, Washington, D. C., 1977.

Officer, C. B., and C. L. Drake, The Cretaceous-Tertiary Transition, Science, 219, 1383-1390, 1983.

Pascoe, E. H., A Manual of the Geology of India and Burma, vol. III. Publication of the Government of India, Calcutta, 1964.

Pearce, T. H., B. E. Gorman, and T. C. Birkett, The $TiO_2-K_2O-P_2O_5$ diagram: a method of discriminating between oceanic and non-oceanic basalts, Earth Planet. Sci. Lett., 24, 419-426, 1975.

Perch-Nielsen, K., J. McKenzie, and Q. X. He, Biostratigraphy and isotope stratigraphy and the "catastrophic" extinction of calcareous nannoplankton at the Cretaceous/Tertiary boundary, in Geological Implications of Impacts of Large Asteroids and Comets on the Earth, Geol. Soc. Am. Spec. Pap. 190, edited by L. T. Silver and P. H. Schultz, pp. 353-371, Geological Society of America, Boulder, Colo., 1982.

Phleger, F. B., and J. S. Bradshaw, Science, 154, 1551-1553, 1966.

Pillmore, C. L., R. H. Tschudy, C. J. Orth, J. S. Gilmore, and J. D. Knight, Geologic framework of non-marine Cretaceous-Tertiary boundary sites, Raton Basin, New Mexico, Science, 223, 1180-1183, 1984.

Plass, G. N., The carbon dioxide theory of climate change, Tellus, 8, 140-154, 1956.

Rubey, W. M., Geologic history of the sea, Geol. Soc. Am. Bull., 63, 1111-1148, 1951.

Shackleton, N. J., Carbon-13 in Uvigerina: Tropical rainforest history and the equatorial Pacific carbonate dissolution cycles, in The Fate of Fossil Fuel $CO_2$ in the Oceans, edited by N. R. Andersen and A. Malahoff, pp. 401-428, Plenum, New York, 1977.

Sikes, C. S., R. D. Roer, and K. M. Wilbur, Photosynthesis and coccolith formation: inorganic carbon sources and net inorganic reaction of deposition, Limnol. Oceanogr., 25, 248-261 1980.

Sikes, C. S., and K. M. Wilbur, M., Calcification by coccolithophorids: effects of pH and Sr, J. Phycol., 16, 433-436 1980.

Sikes, C. S., and K. M. Wilbur, Function of coccolith formation, Limnol. Oceanogr., 27, 18-26, 1982.

Smit, J., Extinction and evolution of planktonic foraminifera after a major impact at the Cretaceous/Tertiary boundary, in Geological Implications of Impacts of Large Asteroids and Comets on the Earth, Geol. Soc. Am. Spec. Pap. 190, edited by L. T. Silver and P. H. Schultz, pp. 329-352, Geological Society of America, Boulder, Colo., 1982.

Smit, J., and S. Van Der Kaars, Terminal Cretaceous extinctions in the Hell Creek area: Compatible with catastrophic extinctions, Science, 223, 1177-1179, 1984.

Stumpel, E. F., The genesis of platinum deposits: further thoughts, Sci. and Engineering, 6, 120-141, 1974.

Sukheswala, R. N., Deccan basalt volcanism, in Deccan Volcanism and Related Basalt Provinces in Other Parts of the World, Geol. Soc. India Mem. 3, edited by K. V. Subbarao and R. N. Sukheswala, pp. 8-18, Rashtrothana Press, Bangalore, 1981.

Thierstein, H. R., Cretaceous oceanic catastrophism, Paleobiology, 6, 244-247, 1980.

Thierstein, H. R., Late Cretaceous nannoplankton and the change at the Cretaceous-Tertiary boundary, Soc. Econ. Paleontol. Mineral. Spec. Publ., 32, 55-394, 1981.

Thompson, S. L., and Barron, E. J., Comparison of Cretaceous and present earth albedos: implications for the causes of paleoclimates, Jour. Geol., 89, 143-167, 1981.

Wensink, H., N. A. I. M. Boelrijk, E. H. Hebeda, H. N. A. Priem, E. A. T. Verdurmen, and R. H. Verschure, Paleomagnetism and radiometric age determinations of the Deccan Traps, India, in Fourth International Gondwana Symposium: Papers, vol. II, edited by B. Laskar and C. S. Raja Rao, pp. 832-849, Hindustan Publishing Corporation (India), Delhi, 1979.

West, W. D., The line of the Narmada-Son Valley, Current Sci., 31, 143-144, 1962.

Zoller, W. H., J. R. Parrington, and J. M. Kotra, Iridium enrichment in airborne particles from Kilauea volcano: January 1983, Science, 222, 1118-1120, 1983.

# VARIATIONS IN THE GLOBAL CARBON CYCLE DURING THE CRETACEOUS RELATED TO CLIMATE, VOLCANISM, AND CHANGES IN ATMOSPHERIC $CO_2$

M. A. Arthur

Graduate School of Oceanography, University of Rhode Island, Narragansett, Rhode Island 02882

W. E. Dean

U.S. Geological Survey, Denver Federal Center, Denver, Colorado 80225

S. O. Schlanger

Department of Geology, Northwestern University, Evanston, Illinois 60201

Abstract. The stratigraphic record from both deep-sea and shallow-water depositional environments indicates that during late Aptian through Cenomanian time (1) global climates were considerably warmer than at present; (2) latitudinal gradients of atmospheric and oceanic temperatures were considerably less than at present; (3) rates of accumulation of organic matter of both marine and terrestrial origin were as high as or higher than during any other interval in the Mesozoic or Cenozoic; (4) the rate and volume of accumulation of $CaCO_3$ in the deep sea were reduced in response to a marked shoaling of the carbonate compensation depth; (5) seafloor spreading rates were somewhat more rapid than at any other time in the Cretaceous or Cenozoic; (6) off-ridge volcanism was intense and widespread, particularly in the ancestral Pacific Ocean basin; and (7) sea level was relatively high, forming widespread areas of shallow shelf seas. A marked increase in the rate of $CO_2$ outgassing due to volcanic activity between about 110 and 70 m.y. ago may have resulted in a buildup of atmospheric $CO_2$. A significant fraction of this atmospheric $CO_2$ may have been reduced by an increase in the production and burial of terrestrial organic carbon. Some excess $CO_2$ may have been consumed by marine algal photosynthesis, but marine productivity apparently was low during the Aptian-Albian relative to terrestrial productivity. Terrestrial productivity also may have been stimulated by increased rainfall that resulted from a warm global climate and increased marine transgression as well as by the higher $CO_2$.

## Introduction

There has been considerable interest recently in the effects of increased anthropogenic $CO_2$ on global climate. Comparatively little effort, however, has been made to document natural changes in the concentration of atmospheric $CO_2$ over geologic time [but see Budyko and Ronov, 1979; Mackenzie and Pigott, 1981; Fischer, 1982; Sandberg, 1983]. Recently, Berner et al. [1983] constructed a numerical model to calculate levels of atmospheric $CO_2$ in response to the carbonate-silicate weathering cycle. They concluded that the concentration of atmospheric $CO_2$ is highly sensitive to changes in the seafloor spreading rate and land area. Spreading rate is important in that it controls the rate of outgassing of mantle-derived $CO_2$; land area is important in that it controls the rate of continental weathering and hence the carbonate-silicate cycle. Berner et al. [1983] assumed that the carbon cycle (photosynthesis-respiration) has been in balance over the past 100 m.y. [Garrels and Lerman, 1981], and that the rate of burial of organic carbon (Corg) has been small compared with rates of carbon fixing by weathering and outgassing of $CO_2$ by metamorphism and volcanism. They obtained their estimate of the rate of outgassing of $CO_2$ by metamorphism, volcanism, and geothermal activity as a remainder in balancing their weathering and precipitation model.

The purpose of this paper is to present evidence that during the middle to late Cretaceous (110 to 70 m.y. ago), high rates of production and burial of Corg, and intense and widespread volcanism in both the ancestral Pacific Ocean Basin and in circum-Pacific terranes may have had profound effects on the global $CO_2$ cycle that were not accounted for originally by the model of Berner et al. [1983]. Lasaga et al. [this volume] have subsequently incorporated the carbon and sulfur cycles in a revised model which independently establishes the importance of Corg burial.

## Importance of Middle Cretaceous Volcanism

### The Modern Baseline

Javoy et al. [1982] computed a range of values of the present-day $CO_2$ flux from the mantle based on carbon isotopic data. In Table 1 we have computed values of the present-day yearly flux of $CO_2$ using the extreme parameters of Javoy et al. [1982] but using a half spreading rate of 3 cm $yr^{-1}$ based on the more detailed data of Parsons [1981], rather than the value of 4 cm $yr^{-1}$ used by Javoy et al. Javoy et al. [1982] also estimated $CO_2$ degassing from both a 2-km-thick upper basalt layer of oceanic crust and a 5-km thick layer of basalt plus gabbro, and these two possibilities are included in the calculations in Table 1. They used initial carbon concentrations of mantle magma ranging from 0.2 to 1.0% based on Rayleigh distillation calculations [Pineau and Javoy, 1983]. The extremes of the Javoy et al. [1982] data give extreme carbon fluxes by mid-ocean ridge volcanism of 0.43 x $10^{14}$ g C $yr^{-1}$ and 5.4 x $10^{14}$ g C $yr^{-1}$ (Table 1). From the range of their calculations, they chose an average value of 4 x $10^{14}$ g C $yr^{-1}$ for the carbon flux from present-day ridge volcanism, and modified this value to 2.4 x $10^{14}$ g C $yr^{-1}$ in their response [Javoy et al., 1983] to comments by Walker [1983]. Des Marais and Moore [1984] consider that isotopically light carbon in ridge crest basaltic glass is contamination, and therefore the flux rates of Javoy et al. [1982, 1983] are too high. On the basis of $^3$He and carbon ratios in submarine hydrothermal fluids, Des Marais and Moore [1984] calculated flux rates of $CO_2$ from mantle associated with spreading centers of about $10^{12}$ moles C $yr^{-1}$, or about 0.1 x $10^{14}$ g C $yr^{-1}$. Thus, there is more than an order of magnitude difference between estimates of the flux of $CO_2$ from the mantle, ranging from 0.1 x $10^{14}$ g C $yr^{-1}$ to 2.4 x $10^{14}$ g C $yr^{-1}$ as the preferred mean flux rate of Javoy et al. [1983].

The mantle carbon fluxes calculated by Javoy et al. [1982, 1983] and Des Marais and Moore [1984] include only outgassing of $CO_2$ by ridge crest volcanism. Some unknown amount of $CO_2$ also will be released to the atmosphere by island arc and circum-oceanic continental volcanism at the subducting leading edges of plates, and we have assumed that this amount is equal to the amount released at the ridges. Presumably, the "metamorphic-magmatic" $CO_2$ flux calculated by Berner et al. [1983] as a remainder from their weathering and precipitation model includes arc as well as ridge volcanism. The metamorphic-magmatic remainder of their model is 5.9 x $10^{18}$ moles $CO_2$ m.y.$^{-1}$, or 0.71 x $10^{14}$ g C $yr^{-1}$. For comparison, Holland [1978] calculated a flux of 0.8 x $10^{14}$ g C $yr^{-1}$ for arc plus ridge volcanism. Therefore, assuming that the $CO_2$ flux from island arc and continental borderland volcanism is equal to that from ridge volcanism, we have ranges of fluxes from these two sources today of 0.2 x $10^{14}$ g C $yr^{-1}$ to 4.8 x $10^{14}$ g C $yr^{-1}$, with intermediate values of 0.71 x $10^{14}$ g C $yr^{-1}$ [Berner et al., 1983] and 0.8 x $10^{14}$ g C $yr^{-1}$ [Holland, 1978].

Berner et al. [1983], using their present-day metamorphic-magmatic flux rate of 5.9 x $10^{18}$ moles C m.y.$^{-1}$ (0.71 x $10^{14}$ g C $yr^{-1}$), and the spreading rate model of Southam and Hay [1977] corrected for $CO_2$ loss by subduction of carbonate sediments, calculated that the flux of $CO_2$ to the atmosphere from metamorphic-magmatic sources was between 1.2 and 1.65 times higher 100 m.y. ago than at present. Berner et al. [1983] tried different land area and spreading rate models and obtained different values of $CO_2$ mass through time, but for each model they tried, the concentration of $CO_2$ in the atmosphere during the middle Cretaceous was always at least several times higher than it is today.

### Estimates of Cretaceous Volcanism

Whatever values are used for present-day volcanic outgassing of $CO_2$, and no matter what assumptions are made about spreading rates and land areas, we wish to emphasize that volcanic activity was considerably greater during the Cretaceous than it is today. In addition to increased ridge and plate margin volcanism in the Cretaceous, there also was widespread mid-plate volcanic activity that may have had a significant effect on the addition of $CO_2$ to the atmosphere. During the middle and Late

TABLE 1.  Estimates of $CO_2$ Outgassing From Volcanic-Metamorphic Sources

| Ridge Length,[a] $10^9$ cm | Crust Thickness,[b] $10^5$ cm | Spreading Rate,[c] cm $yr^{-1}$ | Basalt Created per year, $10^{14}$ $cm^3$ | Basalt Mass, at $\rho=3$ g $cm^{-3}$, $10^{14}$ g | $CO_2$ Outgassed | |
|---|---|---|---|---|---|---|
| | | | | | % C[d] | $10^{14}$ g C yr |
| 6 | 2 | 6 | 72 | 216 | 0.2 | 0.43 |
| 6 | 2 | 6 | 72 | 216 | 1.0 | 2.16 |
| 6 | 5 | 6 | 180 | 540 | 0.2 | 1.08 |
| 6 | 5 | 6 | 180 | 540 | 1.0 | 5.4 |

| Other Literature Values | Source | $CO_2$ Outgassed, $10^{14}$ g C $yr^{-1}$ |
|---|---|---|
| Ridge and arc | Holland [1978] | 0.8 $\pm$ 0.4 |
| "Metamorphism" of calcite and dolomite | Berner et al. [1983, Table 2] | 0.7 |
| Ridge only | Javoy et al. [1982] | 4.0 |
| | Javoy et al. [1983] | 2.4 |
| | Des Marais and Moore [1984] | 0.1 |

[a]Javoy et al. [1982].
[b]Thickness of basalt layer of oceanic crust:  2 km.  Thickness of basalt layer and gabbro layers of oceanic crust:  5 km [Javoy et al., 1982].
[c]Half spreading rate of 3 cm $yr^{-1}$ [Parsons, 1981].
[d]Minimum and maximum values of initial carbon concentration in mantle magma: 0.2% and 1.0% [Javoy et al; 1982].

Cretaceous, the area of the present western Pacific Ocean experienced a complex history of seafloor spreading [Larson and  Chase, 1972; Larson, 1976; Winterer, 1976; Hilde et al., 1977; Watts et al., 1980; Rea and Dixon, 1983] and mid-plate volcanism [Jackson and Schlanger, 1976; Schlanger et al., 1981; Vallier et al., 1983; Rea and Vallier, 1983].  The intense volcanic activity during this time resulted in numerous oceanic islands, seamounts and plateaus that cover an area of more than $3 \times 10^7$ $km^2$ in the western Pacific away from spreading ridges [Rea and Vallier, 1983].

Most of these volcanic features formed between about 70 and 110 m.y. ago [Schlanger et al., 1981].  On the basis of dredged and drilled basalts [Schlanger et al., 1981] and the age distribution of volcanogenic sediments in Deep Sea Drilling Project (DSDP) holes in the western Pacific (Figure 1) [Rea and Vallier, 1983], there appear to have been two main pulses of off-ridge volcanism in the Pacific basin, one about 100 m.y. ago and another about 80 m.y. ago.  These

two periods of volcanism correlate with periods of volcanic intensity associated with the emplacement of granitoid plutons in the present continental borderlands of the Pacific basin (Figure 2).

Menard [1964] estimated that there were more than 10,000 volcanoes higher than 1 km above the seafloor in the Pacific.  Most of these volcanoes occur in the western Pacific in an area of the Pacific plate west of the 90-m.y. isochron (Figures 3 and 4).  Menard concluded that the rate of volcanic effusion in the western Pacific Ocean was about the same as that of continental plateau basalts, but the area affected by this volcanism was hundreds of times greater.  If the extent of volcanism that we observe for that time in the Cretaceous on the  western part (>90 m.y.) of the Pacific plate, the "Darwin Rise" of Menard [1964], was as extensive on the Kula, Farallon, and Phoenix plates (Figure 5), which are now mostly subducted, as proposed by Schlanger et al. [1981], (see also Rea and Dixon [1983]), then the amount of mid-plate volcanic effusion that

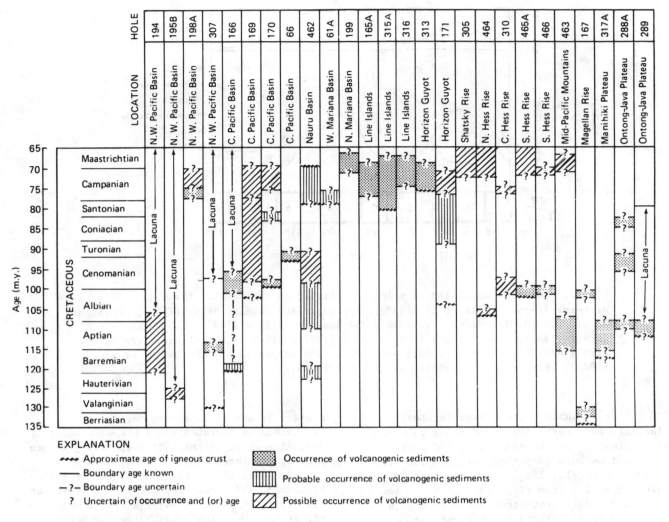

Fig. 1. Occurrence of volcanogenic sediments in DSDP holes in the western Pacific Ocean. From Thiede et al. [1981]; modified from Rea and Vallier [1983].

occurred within a few tens of millions of years must have been enormous. Major mid-plate volcanism also occurred in the Caribbean about 80 m.y. ago [Burke et al., 1978], and in the South Atlantic about 90 to 100 m.y. ago with the formation of the Rio Grande Rise-Walvis Ridge system [Kumar, 1979]. This extensive mid-plate volcanism must have had a great effect on atmospheric $CO_2$.

Unfortunately, it is difficult to determine the volume of this Cretaceous volcanic pile, but we have attempted a rough estimate of the volume of Cretaceous mid-plate volcanism in the western Pacific. The area of the Pacific plate that was involved in this volcanic activity is about 3 x $10^7$ km$^2$ [Rea and Vallier, 1983]. The thickened oceanic crust under volcanic plateaus on the

Pacific plate typically is 12-18 km thick [Vallier et al., 1983]. If the average thickness of basalt extruded on top of older crust during the episodes of Cretaceous mid-plate volcanism on the Pacific plate is assumed to be about one third of this 12-18 km, or about 5 km, then the volume of the Cretaceous volcanic pile in the western Pacific is 15 x $10^7$ km$^3$, or 15 x $10^{22}$ cm$^3$. If this volume formed within 40 m.y., then the average rate of formation was 38 x $10^{14}$ cm$^3$ yr$^{-1}$. At a density of 3 g cm$^{-3}$, and an initial $CO_2$ content of 0.2% (lowest estimate of Javoy et al [1982]), the rate of $CO_2$ evolved by extrusion of this basalt would have been 0.23 x $10^{14}$ g C yr$^{-1}$. Using the minimum rate of $CO_2$ evolution from ridge volcanism today based on the data of Javoy et al. [1982] and Parsons [1981] (0.43 x

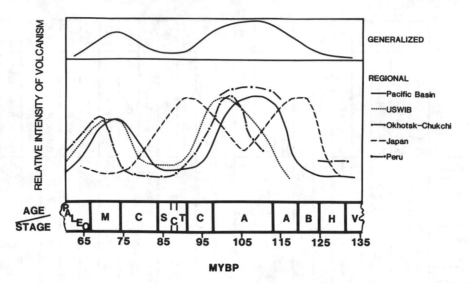

Fig. 2.  Relative intensity of volcanic activity in the Pacific Basin data from
Winterer [1976], Watts et al. [1980], Schlanger and Premoli Silva [1981], Schlanger
et al. [1981], Thiede et al. [1981], and Rea and Vallier [1983]; U.S. Western
Interior Basin (USWIB) data from Kauffman [1977]; Okhotsk-Chukchi data from Lebeder
[1983] and Shilo et al. [1983]; Japan data from Nozawa [1983] and Shilo  et al.
[1983]; Peru data from Cobbing and Pitcher [1983].

$10^{14}$ g C $yr^{-1}$; Table 1), and assuming that
island-arc and continental borderland volcanism
contributes $CO_2$ at the same rate, and that both
contributions were 1.65 times greater 100 m.y.
ago [Berner et al., 1983], then the rate of
evolution of $CO_2$ from ridge and arc volcanism
during the middle Cretaceous would have been 1.42
x $10^{14}$ g C $yr^{-1}$. Using a present rate of
evolution of $CO_2$ from ridge and arc-associated
volcanism of 0.71 x $10^{14}$ g C $yr^{-1}$ calculated as a
remainder in the weathering and precipitation
model of Berner et al. [1983], and applying the
same 1.65 times increase for the middle
Cretaceous, the rate of $CO_2$ evolution from ridge
and arc volcanism would have been 1.17 x $10^{14}$ g
C $yr^{-1}$. The calculated rate then of $CO_2$
evolution from Cretaceous volcanism on the
Pacific plate alone (Figure 5) is 0.23 x $10^{14}$ g
C $yr^{-1}$, which is about 20% of the estimated $CO_2$
evolved from ridge and arc volcanism. As shown
in Figure 5, the Pacific basin at about 110 m.y.
ago was occupied by large segments of at least
four plates; at the present time the Farallon
and Kula plates have been almost entirely
subducted. We argue here that both the Farallon
and Kula plates were the sites of mid-plate
volcanism as widespread and intense as that on
the Pacific plate where the volcanic edifices
have been preserved. Major volcanic plateaus,

similar in age to Hess Rise, are now
allochthonous terranes along the west coast of
Central America [Schmidt-Effing, 1979], and the
present Caribbean plate is thought to be a
stranded part of the Farallon plate [Malfait and
Dinkleman, 1972; Mattson, 1969]. Henderson et
al. [1984] showed that major aseismic volcanic
ridges have been subducted below the North
American borderland. Schlanger et al. [1981] and
Schlanger and Premoli Silva [1981] suggested that
volcanic edifices must have been strewn across
the Farallon plate to serve as "stepping stones"
for the migration of the shallow-water Late
Cretaceous faunas from the Caribbean to the
Pacific plate. We believe that the 0.23 x $10^{14}$ g
C $yr^{-1}$ contributed by degassing of $CO_2$ from the
Pacific plate is a minimum figure. Actually,
considering the sizes of the Farallon and Kula
plates, the carbon contributed from the ancestral
Pacific may have been 3 times as large, or 0.69 x
$10^{14}$ g C $yr^{-1}$ which would amount to about 60% of
the estimated $CO_2$ evolved from ridge and arc
volcanism.

The intensity of ridge crest volcanic activity
during the middle Cretaceous also is indicated by
changes in sea level [e.g. Russell, 1968; Menard,
1964; Hays and Pitman, 1973; Hallam, 1977;
Pitman, 1978; Matsumoto, 1980; Mörner, 1980] and
in the ratio of $^{87}Sr/^{86}Sr$ in marine carbonate

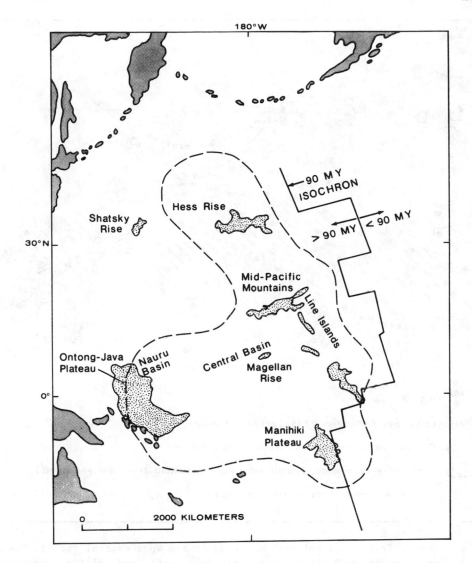

Fig. 3. Area of the western Pacific (within dashed line) characterized by excessive mid-plate volcanism during the middle Cretaceous (120 to 90 m.y. ago) according to Watts et al. [1980]. The position of the 90-m.y. crustal isochron also is indicated.

sediments [Burke et al., 1982] (Figure 6). The quantitative relationships between volcanic activity, as expressed in spreading rate, and changes in sea level are discussed by Hays and Pitman [1973] and Pitman [1978]. The overall sea level curve of Pitman [1978] for the past 90 m.y. shown in Figure 6 is essentially a curve of the relative spreading rate. The age curve of $^{87}Sr/^{86}Sr$ (Figure 6) also is strongly influenced by spreading rate [Brass, 1976; Graham et al., 1982]. The $^{87}Sr/^{86}Sr$ ratio of active-margin basalts is 0.704, and that of continental sialic rocks is 0.72. Marine carbonate sediments

monitor the $^{87}Sr/^{86}Sr$ ratio of seawater, and those with a higher $^{87}Sr/^{86}Sr$ ratio indicate that there was a relatively greater proportion of Sr supplied to the oceans by weathering of continental rocks. Sediments with a lower ratio indicate that there was a greater proportion of oceanic Sr supplied by submarine alteration of oceanic rocks. The $^{87}Sr/^{86}Sr$ curve in Figure 6 shows that beginning in the Late Jurassic and continuing through the Cretaceous, there was a large relative increase in the proportion of oceanic Sr supplied from weathering of oceanic basalt. Two minor minima at about 100 and 80

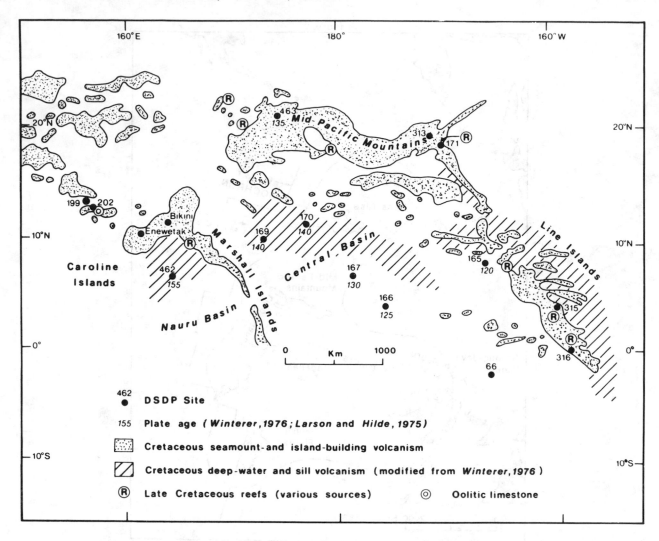

Fig. 4.  Map of Cretaceous volcanic features in the west central Pacific Ocean
showing locations of DSDP sites, Late Cretaceous reefs, and oolitic limestone.
Modified from Winterer [1976], incorporating data from Larson and Hilde [1975].

m.y. are interpreted to be a result of the two
main pulses of Cretaceous mid-plate volcanism
(Figure 2).

Importance of Organic Carbon Burial

The Modern Baseline

Establishing a reasonable Corg burial rate for
the Holocene is important because the value is
used as a baseline in calculating Corg burial
rates in the past together with data on secular
variations in the carbon isotopic composition of
marine carbonates [e.g., Garrels and Lerman,
1981; Berner and Raiswell, 1983].  The model
presented by Berner et al. [1983] incorporated

only the effects of weathering and metamorphism
within the carbonate-silicate geochemical cycle
in accommodating increases and decreases in rates
of $CO_2$ outgassing.  They assumed that over the
last 100 m.y. "...changes in the rate of
sedimentary organic carbon burial [Garrels and
Lerman, 1981] have been distinctly less than
changes in the rate of $CO_2$ uptake by weathering
or $CO_2$ outgassing by metamorphic-volcanic
activity" [Berner et al., 1983, p. 644].  The
evidence upon which this conclusion is based is a
model of secular changes in average $^{13}C$ of
marine carbonate carbon by Garrels and Lerman
[1981], who used data originally compiled by
Veizer et al. [1980].  In their model, Garrels
and Lerman [1981] assumed that the annual Corg

## CRETACEOUS PACIFIC OCEAN
### (110 m.y. B.P.)

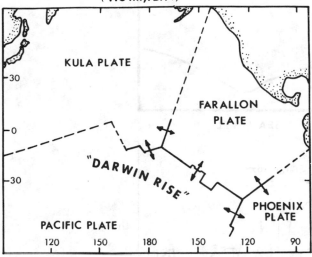

Fig. 5.  Plate configuration of the Cretaceous Pacific Ocean 110 my B.P. [Larson and Pitman, 1975] showing the paleoposition of the nascent "Darwin Rise" of Menard [1964].

burial rate in the world ocean at present is about $0.46 \times 10^{14}$ g C yr$^{-1}$.  They consider this rate to be near the global average through time.  Berner and Raiswell [1983] also adopted the value of $0.46 \times 10^{14}$ g C yr$^{-1}$ in their correction to and analysis of the Garrels and Lerman [1981] model.  This Corg burial rate is significantly lower than that estimated by Gershanovich et al. [1974], Walker [1977], Holland [1978], and Berner [1982] (Table 2). These authors' estimates are all in close agreement at about $1.2 \times 10^{14}$ g C yr$^{-1}$.  Other workers have provided even higher estimates of rates of Corg burial; for example, Wollast and Billen [1981] suggested that $2 \times 10^{14}$ g C yr$^{-1}$ of terrestrial particulate organic carbon (POC) and dissolved organic carbon (DOC) are buried in the coastal zone alone.  Meybeck [1981, p. 219] estimated that rivers supply about $3.95 \times 10^{14}$ g C yr$^{-1}$ as POC and DOC (POC about $1.80 \times 10^{14}$ g C yr$^{-1}$).  Berner [1982] calculated a burial rate of Corg in deltaic-shelf sediments of $1.3 \times 10^{14}$ g C yr$^{-1}$, which after correction of a 20% loss during diagenesis equals about $1.04 \times 10^{14}$ g C yr$^{-1}$.  Mackenzie [1981] estimated a burial rate of Corg on shelves alone of over $2.8 \times 10^{14}$ g C yr$^{-1}$.

Another approach to estimating global Corg burial rates is through rates of supply of preindustrial "reactive" phosphate to oceans by rivers.  Some recent estimates of phosphorus available for burial in the oceans, which at steady state is equal to the amount supplied by rivers, are shown in Table 2.  Using a value of $1.08 \times 10^{12}$ g P yr$^{-1}$ for preindustrial (i.e. nonanthropogenic) phosphorus input to the oceans [Froelich et al., 1982], assuming that all of that phosphorus is buried as organic phosphorus (Porg), and assuming a Redfield C:P atomic ratio of 106:1 in living plankton, then that amount of phosphorus would allow sedimentation and burial, at a minimum, of $0.47 \times 10^{14}$ g C yr$^{-1}$.  Froelich et al. [1982] by independent means calculated a value of $0.43 \times 10^{14}$ for net marine Corg burial.  The phosphorus calculation probably gives a reasonable rate for marine Corg burial because on the one hand phosphorus is recycled during early diagenesis of Corg, which reduces the P:C ratio in buried organic matter and allows further production and sedimentation of Corg [e.g., Suess and Müller, 1980].  On the other hand, some "reactive" phosphorus is buried in other important sedimentary sinks, particularly with $CaCO_3$ and iron oxyhydroxides, and is not available for Corg production [Froelich et al., 1982].  Using Holland's [1978] estimate of the rate of supply of reactive phosphorus to the oceans by rivers (Table 2), the same calculation using the Redfield ratio gives a marine Corg burial rate of $0.98 \times 10^{14}$ g C yr$^{-1}$.  This phosphorus flux rate, however, may include an anthropogenic contribution of as much as 50% [e.g. Meybeck, 1982], and therefore overestimates the steady state preindustrial marine Corg burial rate.

Garrels and Lerman [1981] and Berner and Raiswell [1983] used rates of Corg burial for their models that are at the low end of the range of estimates.  Garrels and Mackenzie [1971] assumed a burial rate of $0.30 \times 10^{14}$ g C yr$^{-1}$ as Corg in their mass balance calculations.  Garrels and Perry [1974] estimated that $0.305 \times 10^{14}$ g C yr$^{-1}$ of Corg burial occurs today; their estimate was derived from the product of the average Corg content of marine sediments (0.5% according to them) and the rate of supply of sediment from rivers of $61 \times 10^{14}$ g yr$^{-1}$. However, Garrels and Perry's [1974] estimated sediment supply is 3 times smaller than that cited by Holland [1978], Berner [1982], and Meybeck [1982], and their average sedimentary Corg content also was low.  Garrels et al. [1975] also used $0.3 \times 10^{14}$ g C yr$^{-1}$ as the rate of burial of Corg.  Budyko and Ronov [1979] used a value of $0.29 \times 10^{14}$ g C yr$^{-1}$ for modern Corg burial, based on an estimated Pliocene rate.  In their original model, which was a precurser to

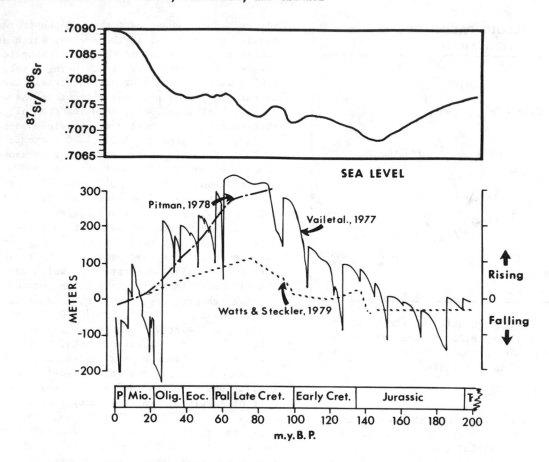

Fig. 6.  Curves showing estimates of relative sea level height in meters above present sea level for the Mesozoic and Cenozoic.  From Vail et al. [1977], Pitman [1978], and Watts and Steckler [1979].  Curve at top of diagram is of $^{87}Sr/^{86}Sr$ in marine sediments from Burke et al. [1982].

that of Berner et al. [1983] and encompassed the entire Phanerozoic, Budyko and Ronov [1979] did not estimate any Corg burial rate higher than $0.29 \times 10^{14}$ g C $yr^{-1}$ for any part of the Phanerozoic; the unweighted Phanerozoic average Corg burial rate cited by them was $0.013 \times 10^{14}$ g C $yr^{-1}$.  Their average Corg accumulation rate for the Phanerozoic is at least an order of magnitude too low because the values used in its calculation were not corrected for loss of sediment mass by erosion, and because the sediment masses and Corg concentrations they used ignored the fact that much sediment is deposited on continental slopes and rises.

R. A. Berner (personal communication, 1983) stated that Berner and Raiswell [1983] (see also Lasaga et al. [1984]) used a low rate of global Corg burial ($0.46 \times 10^{14}$ g C $yr^{-1}$) for the "modern" rate because they felt that it better represented a preindustrial value.  However,

Berner [1982] reasoned that his estimated global Corg burial rate of $1.26 \times 10^{14}$ g C $yr^{-1}$ after diagenesis did not reflect the influence of man, and that Corg burial in marine sediments today must account for much less than 10% of "missing" anthropogenic $CO_2$.  In general, those who allow for anthropogenic influences on modern Corg burial rates provide estimates of total rates of at least twice those reported by Walker [1977], Holland [1978], and Berner [1982], and 6 to 10 times that used by Garrels and Lerman [1981].  For these reasons we consider the values of Walker [1977], Holland [1978], and Berner [1982] (i.e., about $1.2 \times 10^{14}$ g C $yr^{-1}$; Table 2) as good estimates of modern Corg burial rates with no influence of man.

We have attempted to balance the carbon cycle for the present, both in terms of carbon mass and stable-isotope composition (Table 3 and Figure 7).  We incorporated the constraints imposed on

**TABLE 2.** Estimated Fluxes and Reservoirs for Consideration in Carbon Cycle

### Selected Dissolved River and Hydrothermal Fluxes

| Source | $Ca^{2+}$ Input, $10^{14}$ g yr$^{-1}$ Rivers | $Ca^{2+}$ Input, $10^{14}$ g yr$^{-1}$ Hydrothermal | $HCO_3^-$, $10^{14}$ g C yr$^{-1}$ Rivers | Reactive $PO_4$-P, $10^{14}$ g P yr$^{-1}$ Rivers | Total (DOC + POC) Terrestrial Corg, $10^{14}$ g C yr$^{-1}$ Rivers |
|---|---|---|---|---|---|
| Holland [1978] | 6.9 | 2.0 | 5.1 | 0.023 | 1.4 |
| Berner et al. [1983] | 5.8 | 2.1 | 4.21[a] | – | – |
| Meybeck [1981, 1982] | 5.01[b] | – | 3.84 | 0.02 | 3.95 |
| Walker [1977] | – | – | 4.8 | – | – |
| Froelich et al. [1982] | – | – | 2.15 | 0.011[b] | – |
| Kempe [1979] | 5.64 | – | 4.45 | – | 1.89 |
| Garrels and Mackenzie [1971] | – | – | 3.90 | – | 3.50 |
| Schlesinger and Melak [1981] | – | – | – | – | 3.7–4.1 |

### Major Carbon Reservoirs[c]

| Reservoir | Mass, $10^{15}$ g C | Concentration | $\delta^{13}C$, °/oo |
|---|---|---|---|
| Atmospheric $CO_2$ | 615±20 | 290±10 ppm | −6[b] |
| Ocean mixed layer (top 75 m) | 670 (DIC) | 2.05 mol m$^{-3}$ | +2 |
| Total ocean | 37,400 (DIC) | [90% as $HCO_3$] | 0 |
|  | 1,000 (DOC) |  | −24 |
| Deep-sea |  | 2.31 mol m$^{-3}$ |  |
| Terrestrial biota | 800+100 |  | −25 |
| Humus, soil | 1,500+200 |  | −25 |

### Carbon Sedimentation Fluxes

| Source | $CaCO_3$ Burial ($HCO_3$-Consumption), $10^{14}$ g C yr$^{-1}$ | Corg Burial, $10^{14}$ g C yr$^{-1}$ | Old Corg Oxidized, $10^{14}$ g C yr$^{-1}$ | Total Riverine Sediment Supply, $10^{14}$ g yr$^{-1}$ |
|---|---|---|---|---|
| Berner et al. [1983] | 2.10 (2.20)[d] | not given | – | – |
| Holland [1978] | 2.20 | 1.20 | 0.9 | 200 |
| Walker [1977] | 2.40 | 1.20 (0.4; p. 131) | 0.4 | 180 |
| Berner [1982] | – | 1.26 | – | 180 |
| Garrels et al. [1975] | – | 0.29 | – | – |
| Garrels and Mackenzie [1971]; Garrels and Perry [1974] | 1.50±0.3 | 0.30 | 0.3 | 61 |
| Budyko and Ronov [1979] (Ronov [1976]) | – | 0.29 | – | – |
| Kempe [1979, p. 357] | 1.6 | 0.4[b]–7.2 | – | – |
| Mopper and Degens [1979, p. 294] | – | 1.1 | – | – |
| Mackenzie [1981]; Wollast and Billen [1981] | – | 2.3[e] | – | – |

[a]Does not include oxidized Corg or contributions of excess acid weathering due to pyrite oxidation.
[b]Preindustrial.
[c]Source: mainly Bolin et al. [1981, p. 10].
[d]If all river $Ca^{2+}$ used.
[e]Largely terrestrial Corg.

**TABLE 3.   A Balanced Carbon Cycle (Halocene)**

| | Inputs | | | Outputs | |
|---|---|---|---|---|---|
| Source | Amount, $10^{14}$ g C $yr^{-1}$ | $\delta^{13}C$, o/oo PDB | Source | Amount, $10^{14}$ g C $yr^{-1}$ | $\delta^{13}C$, o/oo PDB |
| Oxidized "old" Corg | 0.9[a] | −25.0 | Net burial of Corg | 1.15 | −25.0 |
| Carbonate rock weathering | 1.4[b] | +0.4[c] | CaCO₃ deposition | 2.1[b] | +0.4 |
| Atmospheric CO₂ consumed in rock weathering | 2.1[b] | −6.0[d] | Return of CO₂ to atmosphere by CaCO₃ deposition | 2.1[b] | −6.0 |
| "Mantle" CO₂ degassing | 0.95 | −6.0[e] | | | |
| Total | 5.35 | −7.6 | | 5.35 | −7.6 |

[a]Holland [1978].
[b]Berner et al. [1983].
[c]Veizer and Hoefs [1976].
[d]Bolin et al. [1981] (preindustrial).
[e]Javoy et al. [1982]; Des Marais and Moore

carbon cycling by the carbonate-silicate cycle as outlined by Berner et al. [1983]. However, because they ignored the subcycle of oxidation of "old" Corg and burial of "new" Corg, it was necessary for us to add these to their model. They have subsequently revised their model and independently have incorporated Corg burial into their revised model [Lasaga et al., this volume]. Holland [1978] provided an estimate of the rate of oxidation of "old" sedimentary Corg of 0.9 x 10$^{14}$ g C yr$^{-1}$ based on estimates of the relative masses of rock types exposed to erosion [e.g. Blatt and Jones, 1975], estimates of erosion rates [e.g., Garrels and Mackenzie, 1971], and his estimate of the average Corg content of sedimentary rocks. This estimated oxidation rate would allow for burial of at least an equal amount of Corg in order to maintain mass balance and to help maintain carbon isotopic mass balance in the system. Burial of amounts of Corg above those allowed by Corg oxidation demands a further source of CO$_2$ to the atmosphere, as does consideration of the balance between CO$_2$ consumed by weathering and CO$_2$ liberated by carbonate sedimentation in the model of Berner et al. [1983]. This additional source of CO$_2$ is from

volcanic degassing as suggested by Holland [1978]. The model of Berner et al. [1983] required an addition of about 0.7 x 10$^{14}$ g C yr$^{-1}$ from metamorphic-magmatic CO$_2$, calculated as a remainder. Our version of a balanced cycle (Table 3 and Figure 7) requires an addition of about 0.9 x 10$^{14}$ g C yr$^{-1}$, which is near the amount originally suggested by Holland [1978] and very close to the calculated value of 0.86 x 10$^{14}$ g C yr$^{-1}$ based on the volcanic degassing model of Javoy et al. [1982]. The complete mass and carbon isotope balances (Table 3) require annual deposition of Corg of 1.15 x 10$^{14}$ g C yr$^{-1}$, which is close to that proposed by Walker [1977], Holland [1978], and Berner [1982] (Table 2). The rates of global burial of Corg and CaCO$_3$ at present (excluding the influence of man) suggest a ratio of Corg/CaCO$_3$ carbon deposition of 0.55, which is much higher than the approximate 0.20 depositional ratio of Broecker [1970]. Broecker's value was derived from consideration of the amount and carbon isotopic composition of the bicarbonate supplied to the ocean by rivers, and a partitioning of that carbon into the Corg and CaCO$_3$ reservoirs assuming that the average isotopic fractionation of carbon during

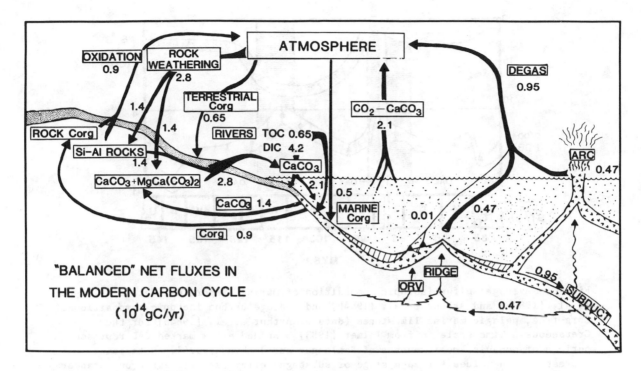

Fig. 7.  Balanced net fluxes of carbon in the modern carbon cycle.  All fluxes are in $10^{14}$ g C yr$^{-1}$.  ORV, off-ridge volcanism; TOC, organic carbon; DIC, dissolved inorganic carbon.

photosynthesis is $-25^o$/oo.  The partitioning of carbon into the two reservoirs is constrained so as to not result in a net change in the $\delta^{13}$C of oceanic total dissolved carbon.  Our model and that of Holland [1978] suggest a steady state depositional Corg/CaCO$_3$ ratio of about 0.6 and an erosional ratio of about 0.6 as well.  This contrasts with a Corg/CaCO$_3$ erosional ratio of 0.46 suggested by Junge et al. [1975]; they also calculated a long-term Corg/CaCO$_3$ burial ratio of 0.27, which is close to that postulated by Broecker [1970].  We emphasize, however, that our Corg burial rate includes burial of both marine and terrestrial Corg, whether deposited in marine sediments or in nonmarine environments.  On the basis of the limitations imposed by fluvial supply of phosphate on marine Corg burial, we estimate that approximately 60% of the global carbon burial at present is in the form of terrestrial Corg.

## Estimates of Rates of Corg Burial During the Cretaceous

The value for the rate of Corg burial in the modern oceans can be used together with carbon isotope data from carbonates of Cretaceous age to estimate Cretaceous rates of Corg burial.  The

methods of calculation are those described by Garrels and Lerman [1981] as modified by Berner and Raiswell [1983], except that we use a present steady state rate of Corg deposition of 1.15 x $10^{14}$ g C yr$^{-1}$ (Table 3; Figure 7).  We used an average $\delta^{13}$C of modern seawater of 0 $^o$/oo and of 0.4 $^o$/oo for modern carbonate output.

The carbon isotope data for Cretaceous carbonates (Figure 8) are those of M. A. Arthur et al. (unpublished data, 1984).  Averages of $\delta^{13}$C by stage or substage were calculated on the basis of a data set of over 3000 analyses of the stable isotopic composition of pelagic carbonates.  Those samples having Corg contents greater than 0.7% were eliminated because of the possibility that carbon isotope values were shifted towards more negative values during Corg degradation and cementation [e.g., Scholle and Arthur, 1980].  The stage or substage values are plotted on the Harland et al. [1983] time scale (Figure 8), with error bars representing one standard deviation about the mean.  The time-weighted Cretaceous average $\delta^{13}$C in CaCO$_3$ is 2.05 $^o$/oo.

The $\delta^{13}$C data were used to calculate the trends in Corg deposition during the Cretaceous assuming constant mass of the oceanic total dissolved carbon reservoir.  The calculated Corg burial rates are shown in Figure 9.  The average

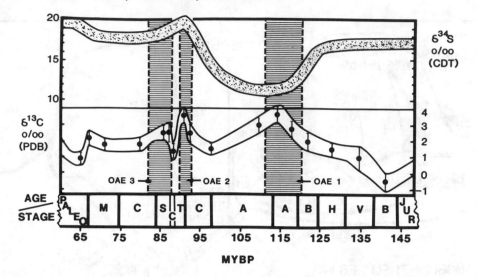

Fig. 8.   Average sulfur isotopic composition of marine evaporites (data of Claypool et al. [1980b] and Arthur et al. [1984b]) and average carbon isotopic composition of Corg-free, pelagic marine limestones (data of Arthur et al. [1984b]) for the Cretaceous.   Time scale is from Palmer [1983].   Shaded areas marked OAE represent periods of oceanic anoxic events of Schlanger and Jenkyns [1976].   Data points represent mean values for each stage or substage; error bars represent one standard deviation greater and less than the mean.

Cretaceous carbonate $\delta^{13}C$ of 2.05 °/oo corresponds to an average Corg burial rate of about $1.8 \times 10^{14}$ g C yr$^{-1}$, a factor of 1.57 times the estimated modern burial rate.   However, the pattern of $\delta^{13}C$ changes suggests that there were major temporal variations in Corg burial rates through the Cretaceous.   Three episodes of apparently higher $\delta^{13}C$ and Corg accumulation correspond to "oceanic anoxic events" (OAEs) recognized on the basis of the distribution of Corg-rich strata (Figure 9) [Schlanger and Jenkyns, 1976; Arthur and Schlanger, 1979; Jenkyns, 1980] and increases in carbonate carbon isotope ratios [Scholle and Arthur, 1980].   Periods between OAEs are characterized by relatively low Corg burial rates (Figure 9).

Cretaceous Corg burial differs from that at present in that apparently a larger proportion of Corg was buried in the deep sea during the Cretaceous.   Black shales of Early to middle Cretaceous age are widespread in the North and South Atlantic Ocean basins, in the Tethys, and in the central Pacific [e.g., Schlanger and Jenkyns, 1976; Thiede and van Andel, 1977; Fischer and Arthur, 1977; Ryan and Cita, 1977; Thierstein, 1979; Arthur and Natland, 1979; Summerhayes, 1981; Weissert, 1981; de Graciansky et al., 1982; Thiede et al., 1982; Dean et al., 1984; Arthur et al., 1984a].   The estimated deep-

sea Corg accumulation rates for stages or substages during the Cretaceous shown in Figure 9 are based on the average Corg contents and mass accumulation rates of deep-sea strata recovered during DSDP legs 1-48 [Arthur, 1982] calculated over the area of the present seafloor using the time scale of Harland et al. [1983].   Corg accumulation rates for pre-Barremian intervals are not reported because rocks of that age have been recovered at few DSDP sites.   During the Aptian-Albian and Cenomanian-Turonian OAEs the deep-sea Corg accumulation rates (for the paleoseafloor below about 2000 m because at DSDP sites, drilling rarely penetrates strata deposited at depths shallower than this) are as great as $0.85 \times 10^{14}$ g C yr$^{-1}$ or about 33% of the estimated global total Corg accumulation rates for those times.   The burial rate of Corg in the deep sea today is about $0.06 \times 10^{14}$ g C yr$^{-1}$ or about 5% of the global total [Berner, 1982].

Therefore, it appears that substantial amounts of Corg were buried during the Cretaceous, and that the deep-sea basins were important repositories for Corg during certain intervals of the Cretaceous.   The question of the relative importance of increased marine primary productivity versus enhanced marine Corg preservation as explanations for the unusually high rates of Corg burial during parts of the

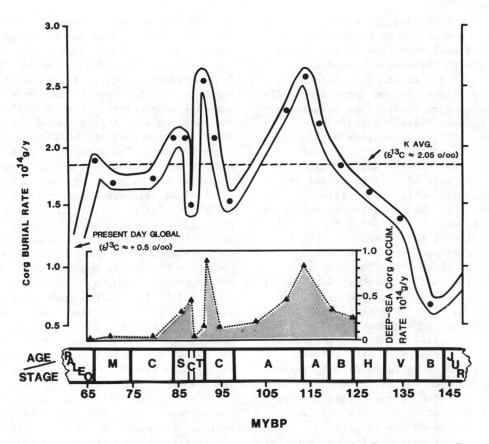

Fig. 9. Burial rates of Corg by stage or substage during the Cretaceous. Top curve is burial rate in $10^{14}$ g C $yr^{-1}$, a present $\delta^{13}C$ of marine carbonate of +0.5 °/oo, the stage or substage average $\delta^{13}C$ values for pelagic limestones from Figure 8, and computed using the model of Garrels and Lerman [1981] as corrected by Berner and Raiswell [1983]. Inset shows average deep-sea Corg burial rates for stages or substages computed from Corg data for DSDP holes from Legs 1 through 48 [Arthur, 1982]. Time scale is from Palmer [1983].

Cretaceous has been explored in a number of papers [e.g., Summerhayes, 1981; Waples, 1983; Jenkyns, 1980; Thiede et al., 1982; Habib, 1982; Arthur et al.,1984a; Arthur and Dean, 1985]. Preservation of terrestrial Corg in the deep sea, particularly during the Aptian-Albian Corg burial episode, is a further important factor. On the basis of optical studies of kerogen in Aptian-Albian sediments from DSDP sites in the North Atlantic [e.g., Habib, 1979, 1982; Summerhayes, 1981; Summerhayes and Masran, 1983] and geochemical characteristics of kerogen [Tissot et al., 1979, 1980; Herbin and Deroo, 1982; Summerhayes, 1981] it appears that a substantial proportion of the kerogen in most samples of Corg-rich sediments consists of terrigenous Corg. Perhaps more that 50% of the Corg buried in the deep North Atlantic during the Aptian-

Albian was land derived (ca. 0.4-0.5 x $10^{14}$ g C $yr^{-1}$). The Cenomanian-Turonian OAE appears to have resulted in burial of mainly marine Corg. This requires higher-than-normal marine productivity and enhanced Corg preservation mainly within expanded and intensified midwater oxygen-minimum zones [e.g., Schlanger et al., 1984; Arthur et al., 1984b].

Data from sulfur isotopes also support the idea that a large proportion of "unreactive" Corg was buried during the Aptian-Albian. Although the carbon isotope curve suggests high rates of Corg burial during the Aptian-Albian, the sulfur isotope curve (Figure 8) suggests that this was a time of relatively low rates of sulfide-sulfur burial according to the reasoning outlined by Garrels and Lerman [1981] and Berner and Raiswell [1983]. The sulfur isotope curve shown in Figure

8 incorporates data from isotopic studies of marine evaporites summarized by Claypool et al. [1980], with additional data from sulfur isotopic studies of sulfate in Cretaceous inoceramid bivalves by M. A. Arthur et al. (unpublished data, 1984). The sulfur isotope curve supposedly reflects secular changes in the sulfur isotopic composition of oceanic dissolved sulfate [Claypool et al., 1980]. Excursions toward lighter $\delta^{34}S$ values reflect net transfer of reduced sulfur to the oxidized sulfur reservoir by decreases in the rate of sulfide burial or increases in evaporite sulfate deposition. Heavier $\delta^{34}S$ values reflect relative increases in rate of sulfide formation. Berner [1982, 1984] and Berner and Raiswell [1983] suggest that a typical reduced S/reduced C ratio in modern marine sediments is about 0.4. Therefore, increased rates of burial of reactive Corg should result in increased rates of sulfide deposition in marine sediments, with a resulting increase in $\delta^{34}S$ values as monitored by marine evaporites. Increased rates of sulfide deposition (higher S/C ratios) may occur in an anoxic water column [Leventhal, 1983], which would tend to increase oceanic $\delta^{34}S$ values even further. This clearly did not happen during the Aptian-Albian because $\delta^{34}S$ values decrease, which suggests that the Corg supplied to the ocean was largely unreactive, thereby not contributing to bacterial sulfate reduction and pyrite formation. An alternative, of course, is that the rate of sulfide oxidation due to weathering increased. However, during the Cenomanian-Turonian, $\delta^{34}S$ values increased rapidly, and this suggests that there was a increase in the rate of sulfate reduction and pyrite formation that accompanied increased rates of burial of reactive marine Corg. Such considerations are important because Corg burial is nutrient limited. The extreme rates of Cretaceous Corg burial would require particularly high nutrient supplies, more effective recycling of nutrients in the marine realm, or both. If the increased rates of burial of Corg were due to burial of terrestrial Corg, then an increased nutrient supply to the oceans would not be necessary in order to accommodate some of the postulated $pCO_2$ increase. Bralower and Thierstein [1984] suggested that mid-Cretaceous marine productivity was significantly lower than that of today.

## Calcium Carbonate Burial Rates, Mineralogy, Oceanic $TCO_2$ and the CCD

We have thus far argued that volcanic and hydrothermal $CO_2$ output may have been greater in the Cretaceous than considered previously. We also have indicated, however, that Cretaceous Corg burial rates were significantly greater in the middle Cretaceous than today. Therefore, a large proportion of the increased $CO_2$ outgassing may have been buffered by increased Corg burial. Nonetheless, it is possible that $pCO_2$ was substantially higher during the Cretaceous as argued by Berner et al. [1983] and Lasaga et al. [this volume]. The magnitude of the Cretaceous $pCO_2$ is at question, but there is no easy way to constrain the value using available geologic or geochemical evidence. Bender [1984] calculated the effects of changing different seawater chemical parameters (CCD; degree of calcite saturation at the CCD; Ca-ion concentration; deep-ocean $TCO_2$ concentration; deep-ocean phosphate concentration; deep-ocean temperature; rain ratio of $CaCO_3/CH_2O$) on $pCO_2$ using a two-box model and fixing various combinations of parameters at their present values, within limits imposed by available geochemical analyses of Cenozoic pelagic sediments. He argued that with the proper data sets we may be able to invert the sedimentary record in order to detect past changes in $pCO_2$. Is there any evidence from the Cretaceous record that would indicate that there was increased $pCO_2$, and would also help to estimate the magnitude of the increase?

Estimated values for modern fluxes of calcium and bicarbonate to the oceans from various sources, excluding internal recycling, and the partitioning of calcium carbonate burial fluxes in the ocean are given in Table 4. Burial of all of the annual supply of calcium to the oceans, disregarding recycled carbonates from within the basins (erosion, reexposure and dissolution of previously deposited pelagic carbonates, [e.g., Moore and Heath, 1977]), results in deposition of a maximum of $18.4 \times 10^{14}$ g $CaCO_3$ yr$^{-1}$, or $2.2 \times 10^{14}$ g C yr$^{-1}$ [Berner et al., 1983]. This value is close to the estimates of Holland [1978] and Broecker [1974], but is higher than that accepted by Hay and Southam [1977] by a factor of 1.5 because the latter authors were not aware of potential hydrothermal sources of calcium at the time.

Milliman [1974] and Hay and Southam [1977] estimated the rate of burial of calcite on the shelves. Milliman [1974] postulated a long-term (post-Miocene) rate of $3.62 \times 10^{14}$ g $CaCO_3$ yr$^{-1}$ buried on the continental shelves, which today have an area of about $0.26 \times 10^{18}$ cm$^2$. He allowed for burial of about $4.6 \times 10^{18}$ g $CaCO_3$ yr$^{-1}$ on the continental slope. These estimates leave about $10.2 \times 10^{14}$ g $CaCO_3$ yr$^{-1}$ for burial in the deep sea, mainly as pelagic carbonate.

TABLE 4.  Burial Rate Estimates for CaCO3:
"Modern" and Cretaceous

| Age and Location | CaCO3 Burial Rate, $10^{14}$ g CaCO3 yr$^{-1}$ |
|---|---|
| Total Modern | 18.4[a] |
| Long-term Cenozoic deep-sea pelagic rate | 13.0[b] |
| Post-Miocene | |
| Continental shelves | 3.6[c] |
| Continental slope | 4.6[c] |
| Deep sea | 10.2[d] |
| Holocene | |
| Shelf-slope total | 13.1[e] |
| Deep-sea | 11.8[c] (5.3[e]) (8.0[f]) |
| Mid-Cretaceous | |
| Shelf | 7.2[g] |
| Slope and deep-sea | 0.6-5.0[h] |

[a]Maximum of Berner et al., [1983].
[b]Davies and Worsley [1981].
[c]Milliman [1974].
[d]Calculated as remainder of the estimates of Milliman [1974].
[e]Milliman [1974] as corrected by Hay and Southam [1977].
[f]Minimum calculated using data of Broecker and Peng [1982].
[g]Minimum calculated by doubling Milliman's [1974] average post-Miocene rate for doubling of shelf area during mid-Cretaceous.
[h]Range of reasonable extimates (see text).

However, as Hay and Southam [1977] point out, Milliman's [1974] estimates of Holocene carbonate sedimentation rates are significantly higher and result in burial rates of about 13.1 x $10^{14}$ g CaCO3 yr$^{-1}$ on the shelves and slope. Using this latter estimate for shelf and slope burial leaves only 5.3 x $10^{14}$ g CaCO3 yr$^{-1}$ for deposition in the deep sea if the system is to remain in steady state. Independently, however, Milliman [1974] estimated that 11.8 x $10^{14}$ g CaCO3 yr$^{-1}$ was buried in the deep sea during the Holocene. If the latter estimates are valid, then the oceanic carbonate system presently is not in steady state and (or) there have been major changes in mass of oceanic total dissolved carbon [e.g. Broecker, 1982] from glacial to interglacial. Using the value of 1.2 g CaCO3 cm$^{-2}$ $10^{-3}$ yr$^{-1}$ for carbonate accumulation above the lysocline today [Broecker

and Peng, 1982], and a value of 20% of the area of the deep-sea floor above the lysocline, we obtain a value of at least 8.0 x $10^{14}$ g CaCO3 yr$^{-1}$ burial for pelagic carbonates. In consideration of the above cited estimates, we suggest a maximum of 8.5 x $10^{14}$ g CaCO3 yr$^{-1}$ burial at present on the shelves, and a minimum value of 9.9 x $10^{14}$ g CaCO3 yr$^{-1}$ on the slopes and deep-sea floor.

If atmospheric $pCO_2$ was significantly higher during the Cretaceous, as Berner et al. [1983] suggested, then we might expect to see an imprint on the marine carbonate system in the form of increased rates of CaCO3 dissolution and consequent decreased rates of carbonate accumulation on the deep-sea floor, possible shoaling of the carbonate compensation depth (CCD), and perhaps evidence of increased dissolution of more soluble carbonate minerals (aragonite and high-Mg calcite) in shoal-water environments. However, a consequence of the revised model of Lasaga et al. [this volume] is that the increased $pCO_2$ obtained in their simulation for the Cretaceous results in a significant increase in weathering and, therefore, an increase in the flux rates of calcium and bicarbonate to the oceans. As a result, the predicted burial rates of CaCO3 are nearly 1.5 times those of today, even though the greater $pCO_2$ probably resulted in a decrease in oceanic pH. We suspect that the weathering and burial rates predicted by the model of Lasaga et al. [this volume] are too high on the basis of estimates of CaCO3 burial that follow.

Figure 10 shows that the CCD rose during the Early to middle Cretaceous in all ocean basins [Thierstein, 1979] and remained shallow from the Aptian through most of the Campanian. This time interval with a shallow CCD coincides closely with timing of major volcanism and organic carbon burial as outlined above. It is surprising that the CCD would have shoaled as it did, considering the large increases in supply of calcium and bicarbonate predicted by Lasaga et al. [this volume]. Bender [1984] pointed out that the depth of the CCD is sensitive to $pCO_2$ such that an increase of about 13% in $pCO_2$ results in about a 1.1 km rise, all other intensive variables remaining constant. The implications of the Cretaceous CCD rise for $pCO_2$ are uncertain, but the CCD rise does correspond with decreased CaCO3 accumulation rates in the deep sea.

Our average calculated deep-sea CaCO3 accumulation rate for the Albian (100 m.y. B.P.) from DSDP data for legs 1-48 is 5.0 x $10^{-4}$ g CaCO3 cm$^{-2}$ yr$^{-1}$. If this value is extrapolated to the entire area of the seafloor below shelf

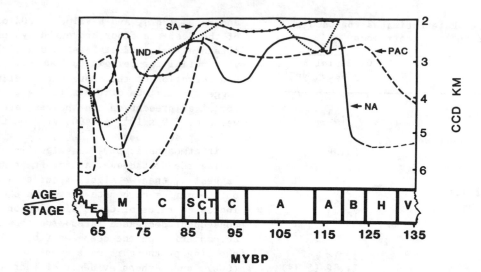

Fig. 10.  Variation through time of the carbonate compensation depth (CCD; taken as the depth at which pelagic sediments contain <20% $CaCO_3$) for the four major ocean basins (NA and SA, North and South Atlantic; IND, Indian; PAC, Pacific). Modified from Thierstein [1979].  Time scale from Palmer [1983].

depth during the Cretaceous, assuming no major changes from today [see Davies and Worsley, 1981], about $16.6 \times 10^{14}$ g $CaCO_3$ yr$^{-1}$ would have been deposited on the deep-sea floor.  However, because the CCD was much shallower, we should calculate the rate only for the area of the seafloor above about 3 km, which is approximately $0.5 \times 10^{18}$ cm$^{-2}$.  The result is burial of only about $2.5 \times 10^{14}$ g $CaCO_3$ yr$^{-1}$.  Similarly, an average Coniacian-Santonian rate of $2.6 \times 10^{-4}$ g $CaCO_3$ cm$^{-2}$ yr$^{-1}$ for the deep sea yields a global deep-sea carbonate accumulation rate of only $1.3 \times 10^{14}$ g $CaCO_3$ yr$^{-1}$.  If we use the average accumulation rate of $1.2 \times 10^{-4}$ g $CaCO_3$ cm$^{-2}$ yr$^{-1}$ from DSDP sites 417 and 418 [Arthur and Dean, 1985], which were at the Mid-Atlantic ridge crest during the Albian, for the global deep-sea carbonate accumulation rate above the CCD during the middle Cretaceous we obtain an even lower estimate of $0.6 \times 10^{14}$ g $CaCO_3$ yr$^{-1}$.  Considering the uncertainties in our calculations, the estimated global $CaCO_3$ burial rate during the Albian probably was no more than $5 \times 10^{14}$ g $CaCO_3$ yr$^{-1}$, or roughly one half of that estimated for today.

Hay and Southam [1977], Fischer and Arthur [1977], Sclater et al. [1979], and Davies and Worsley [1981], among others, noted the apparent correlation between shallower CCD levels and higher sea level.  As pointed out by Hay and Southam [1977] and Southam and Hay [1977], there are at least two  explanations for this

correlation:  1) decreasing land area in some way decreased the amounts of dissolved calcium and bicarbonate supplied to the oceans by rivers, thereby decreasing the steady state burial rate of $CaCO_3$ in the oceans;  and (2) increasing shelf area perhaps promoted more $CaCO_3$ burial in shallow-water environments, thus depriving the deep sea of carbonate and perhaps causing a rise in the CCD.  The last explanation seems unlikely for the Cretaceous in light of the results by Berner et al. [1983] and Lasaga et al. [this volume] which indicate that rates of supply of calcium and bicarbonate to the oceans by rivers increased.  Sclater et al. [1979] constructed a simple model of the oceanic carbonate system and examined the effects that changing hypsography in the Cretaceous (higher spreading rates, greater ridge volume, expanded area of shelf seas) would have on the CCD.  They suggest that in the absence of changes in the input of dissolved calcium and bicarbonate, a relative CCD rise of 1.4 km would result from a transfer of the main areas of carbonate deposition from deep-sea basins to shallow shelf seas.  Davies and Worsley [1981] also found a general antithetic relationship between global deep-sea carbonate accumulation rates and sea level in the Cenozoic.  They suggested that either more carbonate was trapped on the shelves during highstands of sea level, or fluvial input of calcium and bicarbonate to the ocean was decreased, or both.  However, much of the variation in deep-sea CaCO3 burial rates found by these authors (particularly in the Paleogene) may have been a function of nonlinearities in the older time scale used, which has recently been revised [e.g., Harland et al., 1983; Palmer, 1983; Berggren et al., 1984]. The average long-term Cenozoic deep-sea carbonate accumulation rate calculated by Davies and Worsley [1981], and corrected for internal recycling using the Moore and Heath [1977] hiatus curve, is $13.0 \times 10^{14}$ g $CaCO_3$ $yr^{-1}$. This rate is even larger than the estimates of the modern burial rate of $CaCO_3$ cited above. Therefore it is possible that the observed shoaling of the CCD in the Cretaceous of perhaps 2 km is related mainly to a shift in carbonate deposition from the deep sea to the shelves during transgressions.

Simply doubling the estimated $CaCO_3$ accumulation rate for shelf seas for the approximate doubling of the area of shallow shelf seas during the Cretaceous gives a shelf $CaCO_3$ burial rate of about $7.2 \times 10^{14}$ g $CaCO_3$ $yr^{-1}$. This rate plus the $5 \times 10^{14}$ g $CaCO_3$ $yr^{-1}$ estimate of Albian deep-sea carbonate burial suggests that there was a major global decrease in carbonate accumulation during the Albian, with the major component of the decrease attributable to the deep sea. As an upper limit for the shelf burial rate, we can use the average $CaCO_3$ accumulation rate for Upper Cretaceous chalks of Great Britain of about $4.8 \times 10^{-4}$ g $CaCO_3$ $cm^{-2}$ $yr^{-1}$ extrapolated over about 75% of the Albian shelf-sea area estimated by Barron et al. [1981a] to obtain $18 \times 10^{14}$ g $CaCO_3$ $yr^{-1}$. This rate would consume all of today's river and hydrothermal influx of calcium, and most likely overestimates the shelf burial rate by a factor of 2. Despite the large uncertainties in these calculations, we infer that the Cretaceous deep-sea $CaCO_3$ accumulation rates were much lower than those expected.

Better $CaCO_3$ burial estimates are necessary because they will help constrain models such as that presented by Lasaga et al. [this volume]. Although a $pCO_2$ increase in the Cretaceous is likely, the magnitude predicted by their model and/or the impact on weathering and carbonate burial rates is likely overestimated on the basis of evidence presented here.

Mackenzie and Pigott [1981] suggested that the primary mineralogy of ancient ooids, which can with care be interpreted from petrographic textures, might indicate something about atmospheric $pCO_2$ (and oceanic pH) and/or the magnesium content of seawater through time. A preponderance of primary calcitic ooids would suggest that aragonite was undersaturated, and therefore seawater had a higher $pCO_2$ (lower pH)

and/or a higher magnesium concentration. Sandberg [1983] further constrained the secular trends in ooid mineralogy. The data suggest that calcitic ooid formation predominated during the Jurassic and Cretaceous, giving way to primary aragonite precipitation during the Cenozoic. Sandberg favored elevated $pCO_2$ during the Cretaceous as the cause of this pattern, in agreement with Fischer [1982], who also proposed elevated $pCO_2$ during the Cretaceous on the basis of increased volcanism and warmer climate. Again, although suggestive, these data do not prove a higher Cretaceous $pCO_2$, nor do they constrain $pCO_2$ levels. It is possible that the inhibition of aragonite precipitation is due to a higher Mg/Ca ratio in seawater, a factor that is predicted by the Lasaga et al. [this volume] model.

One other constraint on $pCO_2$ can be suggested. Knowledge of the gradient in $\delta^{13}C$ between surface- and deep-water masses can provide information on oceanic $[TCO_2]$ or the ratio $[TCO_2]/[PO_4{}^{3-}]$ in deep water. The detailed arguments are developed by Broecker [1982], Shackleton et al. [1983] and Bender [1984]. The method essentially depends on the fact that deepwater $\delta^{13}C$ decreases and $PO_4{}^{3-}$ increases at a predictable rate because of the addition of $PO_4{}^{3-}$ and isotopically light $CO_2$ during organic matter degradation in the water column. Therefore, the average surface- to deep-water $\delta^{13}C$ gradient at any time is determined by the average $[PO_4{}^{3-}]$ in deep water (which determines the rate of production of Corg in surface waters) and the average $[TCO_2]$ in surface water sinking to become deep water. Any increase in $[TCO_2]$ at constant $[PO_4{}^{3-}]$ as the result of an increase in $pCO_2$ will diminish the effect of organic matter degradation on the $\delta^{13}C$ of $TCO_2$ of deep water, thereby decreasing the $\delta^{13}C$ gradient. The average modern $\Delta\delta^{13}C$ between surface and bottom waters is about 2 $^o/oo$. Although there are a variety of obstacles to determining past $\delta^{13}C$ gradients (e.g., lack of knowledge of "vital effects" on $\delta^{13}C$ of now extinct calcareous organisms and paucity of suitable diagenetically unaltered material), available data [Boersma et al., 1979; Boersma and Shackleton, 1981; Zachos et al., 1984] suggest the $\delta^{13}C$ gradient during the Late Cretaceous (85 to 67 m.y. B.P.) was about 1.0 $^o/oo$. A halving of the $\delta^{13}C$ gradient represents a doubling of oceanic $TCO_2$, which corresponds to a doubling of $pCO_2$ if $[PO_4{}^{3-}]$ and other factors remain approximately constant. The data are therefore suggestive of a higher Late Cretaceous $pCO_2$, but not of the magnitude suggested by Lasaga et al. [this volume].

### Evidence From Cretaceous Climate

Oxygen isotope paleotemperature data indicate that mid-Cretaceous ocean surface-water temperatures were significantly higher than those at the end of the Cretaceous and particularly those at present [e.g., Savin, 1977, 1982]. Figure 11 summarizes some of the available oxygen isotope data converted to paleotemperatures for northwest Europe at about $40^{\circ}N$ latitude during the Cretaceous. The trends show that there was warming during the Aptian-Albian followed by cooling during the Late Cretaceous, cooling with secondary warming in the Coniacian-early Campanian, and further cooling during the Campanian-Maestrichtian. These trends for higher-latitude regions are essentially those summarized by Savin [1977, 1982] based on planktonic foraminifera and nannofossil data from low latitudes, and by Stevens and Clayton [1971] based on data from belemnites from a variety of localities. The Campanian-Maestrichtian cooling was also recognized by Boersma and Shackleton [1981] in planktonic foraminifera $\delta^{18}O$ data from several ocean basins. The peak warm periods coincide closely with volcanic episodicity shown in Figure 2.

Barron [1983] and Crowley [1983] reviewed other relevant faunal and floral data which suggest that mid-Cretaceous paleoclimates were warmer and more equable than today's climate. Global latitudinal thermal gradients were substantially lower during the Cretaceous, and the poles probably were essentially ice free. Barron [1983] suggested a possible poleward latitudinal shift of major carbonate buildups relative to today of about $5^{\circ}$, which may indicate a poleward shift of the $21^{\circ}C$ isotherm. The extrapolated polar temperature for 100 m.y. B.P. is about $5^{\circ}C$ based on isotope paleotemperature gradients from low through middle latitudes, or as high as $15^{\circ}C$ if based on deep-water temperatures derived from benthic foraminifera, assuming a polar source for deep-water masses in the Cretaceous. However, other considerations suggest that deeper water masses could have been derived from saline low-latitude sources during the Cretaceous [e.g., Brass et al., 1982; Arthur and Natland, 1979; Saltzman and Barron, 1982]. The global mean temperature was probably about $6^{\circ}C$ higher than that of today [Barron, 1983], but could have been as much as $14^{\circ}C$ higher.

Barron and colleagues [e.g., Thompson and Barron, 1981; Barron et al, 1981b; Barron and Washington, 1982] have attempted to simulate mid-Cretaceous climates using realistic paleo-geography and the National Center for Atmospheric

Research (NCAR) global climate model. Although the model produced warmer Cretaceous climate, a series of reasonable assumptions did not produce the desired distribution of temperature with latitude. Some other factor, such as a vastly increasing latent heat transport from low to high latitudes, more effective surface- and deep-water circulation, or higher atmospheric $pCO_2$, must be called upon to explain the temperature discrepancies. Barron [1983] suggested that, in the extreme, an estimated six-fold to eight-fold increase in $pCO_2$ could bring about better agreement between the simulations and observations.

Therefore, there is evidence for the possible role of increased atmospheric $pCO_2$ in promoting warmer global temperatures, as postulated by Budyko and Ronov [1979] and Berner et al. [1983]. The timing of the warmest Cretaceous climatic episodes corresponds closely to periods of major Cretaceous volcanism. We suggest, therefore, that there is some evidence for elevated $pCO_2$, but that the mass was probably less than about 8 times that at present because of the drawdown by much higher rates of Corg burial during the mid-Cretaceous.

### Discussion:  Additional Effects

The two main effects of variations in atmospheric $CO_2$ concentration would be on $CaCO_3$ equilibria and global temperature, which were discussed above. There are additional effects, however, that could have important impact on the global carbon cycle. Sea level during the Cretaceous was as high as or higher than at any other period in the Phanerozoic. The importance of a higher sea level in reducing the land area [Barron et al., 1981a] available for weathering and erosion, and hence on the carbonate-silicate cycle, was discussed by Berner et al. [1983], who found, however, that weathering rates apparently increased in the Cretaceous over those at present because of the higher $pCO_2$ predicted by their model. Land area decreased no more than 15% during maximum Cretaceous sea level. On the other hand, higher sea level increased the area covered by shallow shelf areas which, with the distribution of the continents in the Cretaceous, probably increased the area of evaporation. Greater evaporation, coupled with warmer mean global temperature and consequently higher evaporation rates, would have increased the amount of water vapor in the atmosphere [e.g., Manabe and Wetherald, 1975; Barron and Washington, 1982]. Increased water vapor in the atmosphere most likely would have increased the

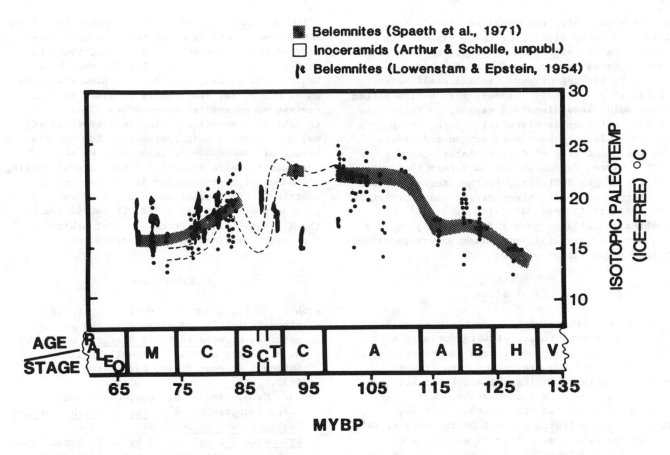

■ Belemnites (Spaeth et al., 1971)
□ Inoceramids (Arthur & Scholle, unpubl.)
🔥 Belemnites (Lowenstam & Epstein, 1954)

Fig. 11. Ocean surface water paleotemperatures for the Cretaceous computed from oxygen isotopic compositions of belemnite and inoceramid samples from localities in northwestern Europe (~40° north latitude). Time scale from Palmer [1983].

amount of precipitation over land areas. Increased precipitation would have several important effects. First, it would have allowed for greater terrestrial plant growth. Additional terrestrial plant growth also would have resulted from greater areas of coastal wetlands related to a global rise in sea level. The rapid evolution and radiation of angiosperms also occurred during the middle Cretaceous [e.g., Hickey and Doyle, 1977]. Greater plant growth would have resulted in greater rate of fixing of terrestrial Corg and ultimately in a greater rate of burial of terrestrial Corg on land, in coastal wetlands, and in the deep sea. The combination of more available moisture and increased plant growth probably also resulted in an increase in rate of chemical weathering [e.g., Holland, 1978] even though the actual area available for weathering was less as indicated by the models of Berner et al. [1983] and Lasaga et al. [this volume]. Van Houten [1982] outlined evidence suggesting that the intensity of weathering (lateritization),

indicating warmer and more humid climates, had greater latitudinal expanse during the Cretaceous than during the late Cenozoic to Holocene. We have already mentioned the unusual abundance of terrestrial Corg in Aptian-Albian deep-sea black shales, and Cretaceous coals also are widespread. In the absence of sufficient nutrients in the oceans to sustain productivity and allow increased net burial of marine Corg, the increased burial of terrestrial Corg may have been a buffer to limiting $CO_2$ accumulation in the atmosphere [e.g., Arthur, 1982].

An increase in the areas of warm, shallow shelf seas with increased evaporation also may have resulted in greater production of warm, saline bottom water as suggested by Brass et al. [1982]. This saline bottom water may have produced salinity stratification in the deep sea and formed the dominant deep-water mass in the world ocean [e.g., Arthur and Natland, 1979; Wilde and Berry, 1982; Brass et al., 1982]. This bottom-water mass may have been more stagnant, or, alternatively, thermohaline circulation, driven today by sinking of cold, oxygen-rich surface waters at high latitudes, may have been more vigorous during the Cretaceous, driven mainly by sinking of warm, saline shelf water [Brass et al., 1982]. Warmer, more saline bottom water holds less dissolved oxygen, so regardless of the bottom-water circulation rate, the bottom waters would have been more oxygen deficient. The increased flux of Corg probably placed a larger oxygen demand on bottom waters that were already oxygen deficient, further decreasing the concentration of dissolved oxygen and increasing the concentration of dissolved CO2. The combination of increased rate of supply of Corg and more oxygen-deficient bottom waters permitted better preservation of Corg.

## Conclusions

We propose that increased outgassing of mantle $CO_2$ and increased burial of Corg during the Cretaceous had profound effects on the cycling of $CO_2$. Increased $CO_2$ production from greater volcanic activity during the middle to Late Cretaceous tended to increase the $CO_2$ concentration of the atmosphere. This $CO_2$ increase was buffered in part by the greater rate of Corg burial. Our preferred estimates of volcanic $CO_2$ outgassing and Corg burial for the preindustrial modern world are $0.9 \times 10^{14}$ g C $yr^{-1}$ and $1.15 \times 10^{14}$ g C $yr^{-1}$, respectively. $CO_2$ outgassing during the Cretaceous by ridge-crest plus island-arc and continental-borderland volcanism, plus the $0.23 \times 10^{14}$ g C $yr^{-1}$ that we assume is a minimum for additional mid-plate volcanism gives a total rate of volcanic $CO_2$ outgassing of about $1.7 \times 10^{14}$ g C $yr^{-1}$, or an increase of about $0.8 \times 10^{14}$ g C $yr^{-1}$ between the present and the middle Cretaceous. Our calculation of the average middle Cretaceous Corg burial rate is $1.8 \times 10^{14}$ g C $yr^{-1}$, or an increase of about $0.65 \times 10^{14}$ g C $yr^{-1}$ between present and middle Cretaceous rates. These results do suggest that there may have been a small net addition of $CO_2$ to the atmosphere, between $CO_2$ added by volcanic outgassing and that consumed by additional Corg burial, of less than $0.15 \times 10^{14}$ g C $yr^{-1}$ during the middle Cretaceous. Such an estimate also ignores the buffering of $pCO_2$ increases by increased weathering rates. The available geologic evidence, although not compelling, indicates that $CO_2$ degassing and increased $pCO_2$ affected the chemical, biological, and physical conditions of the Cretaceous world.

Acknowledgments. We thank the convenors of the AGU Chapman Conference for inviting us to present our paper at the meeting and to contribute to this volume. M.L. Bender, R.A. Berner, D.L. Gautier, A. Lerman, P.M. Meyers, L.M. Pratt and an anonymous reviewer kindly provided us with reviews of an earlier manuscript. Their thoughtful comments are sincerely appreciated; we, of course, remain responsible for any bias or errors of commission or omission. We are indebted to Maria Burdett for her careful editing and manuscript preparation and to her and Tom Kostick for illustrations. This work was, in part, supported by NSF grant OCE 840062 and by the U.S.G.S. Climate Program. Chet Atkins and Doc Watson provided musical inspiration.

## References

Arthur, M. A., The carbon cycle - Controls on atmospheric $CO_2$ and climate in the geologic past, in Climate in Earth History, edited by W. H. Berger and J. C. Crowell, pp. 55-67, National Academy Press, Washington, D.C., 1982.

Arthur, M. A., and W. E. Dean, Cretaceous paleoceanography, in Decade of North American Geology, Western North Atlantic Basin Synthesis Volume, edited by B. E. Tucholke and P. R. Vogt, Geological Society of America, Boulder, Colo., in press, 1985.

Arthur, M. A., and J. H. Natland, Carbonaceous sediments in North and South Atlantic: The role of salinity in stable stratification of early Cretaceous basins, in Deep Drilling Results in the Atlantic Ocean: Continental Margins and Paleoenvironment, Maurice Ewing Ser., vol. 3, edited by M. Talwani, W. W. Hay, and W. B. F. Ryan, pp. 375-401, AGU, Washington, D.C., 1979.

Arthur, M. A., and S. O. Schlanger, Cretaceous "oceanic anoxic events" as causal factors in development of reef-reservoired oilfields, Am. Assoc. Pet. Geol. Bull., 63, 870-885, 1979.

Arthur, M. A., W. E. Dean, and D. A. V. Stow, Models for the deposition of Mesozoic-Cenozoic fine-grained organic-carbon-rich sediment in the deep sea, in Fine-Grained Sediments: Processes and Products, edited by D. A. V. Stow and D. Piper, Geological Society of London, in press, 1984a.

Arthur, M. A., S. O. Schlanger, and H. C. Jenkyns, The Cenomanian-Turonian oceanic anoxic event, II, Paleoceanographic controls on organic matter production and preservation, in Marine Petroleum Source Rocks, edited by J.

Brooks and A. Fleet, Geological Society of London, in press, 1984b.

Barron, E. J., A warm, equable Cretaceous: The nature of the problem, Earth Sci. Rev., 19, 305-338, 1983.

Barron, E. J., and W. M. Washington, Cretaceous climate: A comparison of atmospheric simulations with the geologic record, Palaeogeogr. Palaeoclimatol. Palaeoecol., 40, 103-133, 1982.

Barron, E. J., C. G. A. Harrison, J. L. Sloan II, and W. W. Hay, Paleogeography, 180 million years ago to the present, Eclogae Geol. Helv., 74, 443-469, 1981a.

Barron, E. J., S. L. Thompson, and S. H. Schneider, An ice-free Cretaceous? Results from climate model simulations, Science, 212, 501-508, 1981b.

Bender, M. L., On the relationship between ocean chemistry and atmospheric $CO_2$ during the Cenozoic, in Climate Processes and Climate Sensitivity, Geophys. Monogr., vol. 29, edited by J. E. Hansen and T. Takahashi, pp. 352-359, AGU, Washington, D.C., 1984.

Berggren, W. A., D. V. Kent, and J. J. Flynn, Paleogene geochronology and chrono-stratigraphy, in Geochronology and the Geologic Record, edited by N. J. Snelling, Geological Society of London, in press, 1984.

Berner, R. A., Burial of organic carbon and pyrite sulfur in the modern ocean: Its geochemical and environmental significance, Am. J. Sci., 282, 451-473, 1982.

Berner, R. A., Sedimentary pyrite formation: An update, Geochim. Cosmochim. Acta, 48, 605-615, 1984.

Berner, R. A., and R. Raiswell, Burial of organic carbon and pyrite sulfur in sediments over Phanerozoic time: A new theory, Geochim. Cosmochim. Acta, 47, 855-862, 1983.

Berner, R. A., A. C. Lasaga, and R. M. Garrels, The carbonate-silicate geochemical cycle and its effect on atmospheric carbon dioxide over the past 100 million years, Am. J. Sci., 283, 641-683, 1983.

Blatt, H., and R. L. Jones, Proportions of exposed igneous, metamorphic, and sedimentary rocks, Geol. Soc. Amer. Bull., 86, 1085-1088, 1975.

Boersma, A., and N. J. Shackleton, Oxygen- and carbon-isotope variations and planktonic-foraminifer depth habitats, Late Cretaceous to Paleocene, central Pacific, Deep Sea Drilling Project Sites 463 and 465, Initial Rep. Deep Sea Drill. Proj., 62, 513-526, 1981.

Boersma, A., N. J. Shackleton, M. A. Hall, and Q. C. Given, DSDP Site 384 (North Atlantic) and some Paleocene paleotemperatures and carbon isotope variations in the Atlantic Ocean, Initial Rep. Deep Sea Drill. Proj., 43, 695-717, 1979.

Bolin, B., C. D. Keeling, R. B. Bacastow, A. Björkström, and U. Siegenthaler, Carbon cycle modelling, in Carbon Cycle Modelling, SCOPE 16, edited by B. Bolin, pp. 1-28, John Wiley, New York, 1981.

Bralower, T. J., and H. R. Thierstein, Low-productivity and slow deep-water circulation in mid-Cretaceous oceans, Geology, 12, 614-618, 1984.

Brass, G. W., The variation of the marine $^{87}Sr/^{86}Sr$ ratio during Phaneozoic time: Interpretation using a flux model; Geochim. Cosmochim. Acta, 40, 721-730, 1976.

Brass, G. W., J. R. Southam, and W. H. Peterson, Warm saline bottom water in the ancient ocean, Nature, 296, 620-623, 1982.

Broecker, W. S., A boundary condition on the evolution of atmospheric oxygen, J. Geophys. Res., 75, 3553-3557, 1970.

Broecker, W. S., Chemical Oceanography, 214 pp., Harcourt Brace Jovanovich, New York, 1974.

Broecker, W. S., Ocean chemistry during glacial times, Geochim. Cosmochim. Acta, 46, 1689-1705, 1982.

Broecker, W. S. and T.-H. Peng, Tracers in the Sea, 892 pp, Eldigio, New York, 1982.

Budyko, M. I., and A. B. Ronov, Chemical evolution of the atmosphere in the Phanerozoic, Geochem. Int., 16, 1-9, 1979.

Burke, K., P. J. Fox, and A. M. C. Sengör, Buoyant ocean floor and the evolution of the Caribbean, J. Geophys. Res., 83, 3949-3954, 1978.

Burke, W. H., R. E. Denison, E. A. Hetherington, R. B. Koepnick, H. F. Nelson, and J. B. Otto, Variation of seawater $^{87}Sr/^{86}Sr$ throughout Phanerozoic time, Geology, 10, 516-519, 1982.

Claypool, G. E., W. T. Holser, I. R. Kaplan, H. Sakai, and I. Zak, The age curves of sulfur and oxygen isotopes in marine sulfate and their mutual interpretation, Chem. Geol., 28, 199-260, 1980.

Cobbing, E. J., and W. S. Pitcher, Andreas plutonism in Peru and its relationship to volcanism and metallogenesis at a segmented plate edge, in Circum-Pacific Plutonic Terranes, edited by J. A. Roddick, pp. 277-292, Geol. Soc. Am. Mem. 159, Geological Society of America, Boulder, Colo., 1983.

Crowley, T. J., The geologic record of climatic change, Rev. Geophys. Space Phys. 21, 828-877, 1983.

Davies, T. A., and T. R. Worsley,

Paleoenvironmental implications of oceanic carbonate sedimentation rates, Soc. Econ. Paleontol. Mineral. Spec. Publ., 32, 169-179, 1981.

Dean, W. E., M. A. Arthur, and D. A. V. Stow, Origin and geochemistry of Cretaceous deep-sea black shales and multicolored claystones, with emphasis on Deep Sea Drilling Project Site 530, southern Angola Basin, Initial Rep. Deep Sea Drill. Proj., 75, 819-844, 1984.

de Graciansky, P. C., E. Brosse, G. Deroo, J-P. Herbin, L. Montadert, C. Müller, J. Sigal, and A. Schaaf, Les formations d'age Crétacé de l'Atlantique Nord et leur Matière organique: Paléogéographie et milieus de dépôt, Rev. Inst. Fr. du Pet., 37, 275-337, 1982.

Des Marais, D. J., and J. G. Moore, Carbon and its isotopes in mid-oceanic basaltic glasses, Geochim. Cosmochim. Acta, in press, 1984.

Fischer, A. G., Long-term climatic oscillations recorded in stratigraphy, in Climate in Earth History, edited by W. H. Berger and W. C. Crowell, pp. 97-104, National Academy Press, Washington, D.C., 1982.

Fischer, A. G., and M. A. Arthur, Secular variations in the pelagic realm, in Deep Water Carbonate Environments, edited by H. E. Cook and P. Enos, Spec. Publ. 25, pp. 19-50, Society of Economic Paleontologists and Mineralogists, Tulsa, Okla., 1977.

Froelich, P. N., M. L. Bender, N. A. Luedtke, G. R. Heath, and P. DeVries, The marine phosphorus cycle, Am. J. Sci., 282, 484-511, 1982.

Garrels, R. M., and A. Lerman, Phanerozoic cycles of sedimentary carbon and sulfur, Proc. Natl. Acad. Sci. USA, 78, 4652-4656, 1981.

Garrels, R. M., and F. T. Mackenzie, Evolution of Sedimentary Rocks, 397 pp., W. W. Norton, New York, 1971.

Garrels, R. M., and E. A. Perry, Cycling of carbon, sulfur and oxygen throughout geologic time, in The Sea, vol. 5, edited by E. D. Goldberg, pp. 303-336, Wiley-Interscience, New York, 1974.

Garrels, R. M., A. Lerman, and F. T. Mackenzie, Controls of atmospheric $O_2$ and $CO_2$: Past, present, and future, Am. Sci., 64, 306-315, 1975.

Gershanovich, D. E., T. I. Gorshkova, and A. I. Konyukov, Organic matter of modern submarine sediments of the continental margins, in Organic Matter of Modern and Ancient Sediments and Methods of Its Investigation (in Russian), pp. 63-80, Nauka, Moscow, 1974.

Graham, D. W., M. L. Bender, D. F. Williams, and L. D. Keigwin, Jr., Strontium-calcium ratios ni Cenozoic planktonic foraminifera, Geochim. Cosmochim. Acta, 46, 1281-1292, 1982.

Habib, D., Sedimentary origin of North Atlantic Cretacaeous palynofacies, in Deep Drilling Results in the Atlantic Ocean: Continental Margins and Paleoenvironments, Maurice Ewing Ser., vol. 3, edited by M. Talwani, W. W. Hay, and W. B. F. Ryan, pp. 420-437, AGU, Washington, D.C., 1979.

Habib, D., Sedimentary supply origin of Cretaceous black shales, in Nature and Origin of Cretaceous Carbon-Rich Facies, edited by S. O. Schlanger and M. Cita, pp. 113-128, Academic, New York, 1982.

Hallam, A., Secular changes in marine inundation of USSR and North America through the Phanerozoic, Nature, 268, 769-772, 1977.

Harland, W. B., A. V. Cox, P. G. Llewellyn, C. A. G. Pickton, A. G. Smith, and R. Walters A Geologic Time Scale, 128 pp., Cambridge University Press, New York, 1983.

Hay, W. W., and J. R. Southam, Modulation of marine sedimentation by the continental shelves, in The Fate of Fossil Fuel $CO_2$ in the Oceans, edited by N. R. Anderson, and A. Malahoff, pp. 569-604, Plenum, New York, 1977.

Hays, J. D., and W. C. Pitman III, Lithospheric plate motion, sea level changes, and climatic and ecological consequences, Nature, 246, 18-22, 1973.

Henderson, L. J., R. G. Gordon, and D. C. Engebretson, Mesozoic aseismic ridges on the Farallon plate and southward migration of shallow subduction during the Laramide orogeny, Tectonics, 3, 121-132, 1984.

Herbin, J.-P., and G. Deroo, Sedimentologie de la matière organique dans les formations du Mesozoique de l'Atlantique Nord, Bull. Soc. Geol. Fr., 24, 497-510, 1982.

Hickey, L. J., and J. A. Doyle, Early Cretaceous fossil evidence for angiosperm evolution, Bot. Rev., 43, 3-104, 1977.

Hilde, T. W. C., S. Uyeda, and L. Kroenke, Evolution of the western Pacific and its margin, Tectonophysics, 38, 145-165, 1977.

Holland, H. D., The Chemistry of the Atmosphere and Oceans, 351 pp., John Wiley, New York, 1978.

Jackson, E. D., and S. O. Schlanger, Regional syntheses, Line Islands chain, Tuamotu Islands chain, and Manihiki Plateau, central Pacific Ocean, Initial Rep. Deep Sea Drill. Proj., 33, pp. 915-927, 1976.

Javoy, M., F. Pineau, and C.-J. Allegre, Carbon geodynamic cycle, Nature, 300, 171-173, 1982.

Javoy, M., F. Pineau, and C.-J. Allegre, Carbon geodynamic cycle--Reply, Nature, 303, 731,1983.

Jenkyns, H. C., Cretaceous anoxic events:  From continents to oceans, J. Geol. Soc. London, 137, 171-188, 1980.

Junge, C. E., M. Schidlowski, R. Eichmann, and H. Pietrek, Model calculations for the terrestrial carbon cycle:  Carbon isotope geochemistry and evolution of photosynthetic oxygen, J. Geophys. Res., 80, 4542-4552, 1975.

Kauffman, E. G., Geological and biological overview:  Western interior Cretaceous basin, Mt. Geol., 14, 75-100, 1977.

Kempe, S., Carbon in the freshwater cycle, in The Global Carbon Cycle, edited by E.T. Degens, S. Kempe, and P. Ketner, pp. 317-342, John Wiley, New York, 1979.

Kumar, N., Origin of "paired" aseismic rises: Ceara and Sierra Leone Rises in the equatorial Atlantic, and the Rio Grande Rise and Walvis Ridge in the South Atlantic, Mar. Geol., 30, 175-191, 1979.

Larson, R. L., Late Jurassic and Early Cretaceous evolution of the western central Pacific Ocean, J. Geomagn. Geoelectr., 28, 219-236, 1976.

Larson, R. L., and C. G. Chase, Late Mesozoic evolution of the western Pacific, Geol. Soc. Am. Bull., 38, 3627-3644, 1972.

Larson, R. L., and T. W. Hilde, A revised time scale of magnetic reversals for the Early Cretaceous and Late Jurassic, J. Geophys. Res., 80, 2586-2594, 1975.

Larson, R. L., and W. C. Pitman III, World-wide correlations of Mesozoic magnetic anomalies and its implications, Geol. Soc. Am. Bull., 83, 3645-3661, 1975.

Lebeder, Y. L., Volcanism and the Cretaceous climate, Int. Geol. Rev., 25, 407-414, 1983.

Leventhal, J. S., An interpretation of carbon and sulfur relationships in Black Sea sediments as indicators of environments of deposition, Geochim. Cosmochim. Acta, 47, 133-138, 1983.

Lowenstam, H. A., and S. Epstein, Paleotemperatures of the post-Aptian Cretaceous as determined by the oxygen isotope method, J. Geol., 62, 207-248, 1954.

Mackenzie, F. T., Global carbon cycle:  Some minor sinks for $CO_2$, in Carbon Dioxide Effects Research and Assessment Program Conf., Flux of Organic Carbon by Rivers to the Ocean, pp. 359-384, U.S. Department of Energy, Washington, D.C., 1981.

Mackenzie, F. T., and J. D. Pigott, Tectonic controls of Phanerozoic sedimentary rock cycling, J. Geol Soc. London, 138, 183-196, 1981.

Malfait, B. T., and M. G. Dinkelman, Circum-Caribbean tectonic and igneous activity and the evolution of the Caribbean plate, Geol. Soc. Am. Bull., 83, 251-272, 1972.

Manabe, S., and R. T. Wetherald, The effects of doubling the $CO_2$ concentration on the climate of a general circulation model, J. Atmos. Sci., 32, 3-15, 1975.

Matsumoto, T., Inter-regional correlation of transgressions and regressions in the Cretaceous period, Cretaceous Res., 1, 359-373, 1980.

Mattson, P. H., The Caribbean:  A detached relic of the Darwin Rise (abstract), EOS Trans. AGU, 50, 317, 1969.

Menard, H. W., Marine Geology of the Pacific, 271 pp., McGraw-Hill, New York, 1964.

Meybeck, M., River transport of organic carbon to the ocean, in Carbon Dioxide Effects Research and Assessment Program Conf., Flux of Organic Carbon by Rivers to the Ocean, pp. 219-269, U.S. Department of Energy, Washington, D.C., 1981.

Meybeck, M., Carbon, nitrogen, and phosphorus transport by world rivers, Amer. J. Sci., 282, 401-450, 1982.

Milliman, J. D., Marine Carbonates, 375 pp., Springer-Verlag, New York, 1974.

Moore, T. C., and G. R. Heath, Survival of deep-sea sedimentary sections, Earth Planet. Sci. Lett., 37, 71-80, 1977.

Mopper, K., and E. T. Degens, Organic carbon in the ocean:  Nature and cycling, in The Global Carbon Cycle, SCOPE 13, edited by B. Bolin et al., pp. 293-316, John Wiley, New York, 1979.

Mörner, N.-A., Relative sea-level, tectono-eustasy, geoidal-eustasy and geodynamics during the Cretaceous, Cretaceous Res., 1, 329-340, 1980.

Nozawa, T., Felsic plutonism in Japan, in Circum-Pacific Plutonic Terranes, Geol. Soc. Am. Mem. 159, edited by J. A. Roddick, pp. 105-122, Geological Society of America, Boulder, Colo., 1983.

Palmer, A. R., The DNAG time scale, Geology, 11, 182, 1983.

Parsons, B., The rates of plate creation and consumption, Geophys. J. R. Astron. Soc., 67, 437-448, 1981.

Pineau, F., and M. Javoy, Carbon isotopes and concentrations in mid-oceanic ridge basalts, Earth Planet. Sci. Lett., 62, 239-257, 1983.

Pitman, W. C., III, Relationship between eustacy and stratigraphic sequences of passive margins, Geol Soc. Am. Bull., 89, 1389-1403, 1978.

Rea, D. K. and J. M. Dixon, Late Cretaceous and

Paleogene tectonic evolution of the North Pacific Ocean, Earth Planet. Sci. Lett., 65, p. 145-166, 1983.

Rea, D. K., and T. Vallier, Two Cretaceous volcanic episodes in the western Pacific Ocean, Geol. Soc. Am. Bull., 94, 1430-1437, 1983.

Ronov, A. B., Global carbon geochemistry, volcanism, carbonate accumulation, and life, Geochem. Int., 8, Engl. Transl., 1252-1277, 1976.

Russell, K. L., Oceanic ridges and eustatic changes in sea level, Nature, 218, 861-862, 1968.

Ryan, W. B. F., and M. B. Cita, Ignorance concerning episodes of ocean-wide stagnation, Mar. Geol., 23, 197-215, 1977.

Saltzman, E. S., and E. J. Barron, Deep circulation in the Late Cretaceous: Oxygen isotope paleotemperatures from remains in D.S.D.P. cores, Palaeogeogr., Palaeoclimatol., Palaeoecol., 40, 167-181, 1982.

Sandberg, P. A., An oscillating trend in Phanerozoic non-skeletal carbonate mineralogy, Nature, 305, 19-22, 1983.

Savin, S. M., The history of the earth's surface temperature during the last 100 million years, in Annu. Rev. Earth Planet. Sci., 5, 319-355, 1977.

Savin, S. M., Stable isotopes in climatic reconstructions, in Climate in Earth History, edited by W. H. Berger and J. C. Crowell, pp. 164-171, National Academy Press, Washington, D.C., 1982.

Schlanger, S. O., and H. C. Jenkyns, Cretaceous oceanic anoxic events--Causes and consequences, Geol. Mijnbouw, 55, 179-184, 1976.

Schlanger, S. O., and I. Premoli Silva, Tectonic, volcanic, and paleogeographic implications of redeposited reef faunas of Late Cretaceous and Tertiary age from the Nauru Basin and the Line Islands, Initial Rep. Deep Sea Drill. Proj., 61, 817-828, 1981.

Schlanger, S. O., H. C. Jenkyns, and I. Premoli Silva, Volcanism and vertical tectonics in the Pacific Basin related to global Cretaceous transgressions, Earth Planet. Sci. Lett., 52, 435-449, 1981.

Schlanger, S. O., M. A. Arthur, H. C. Jenkyns, and P. A. Scholle, The Cenomanian-Turonian oceanic anoxic event, I, Stratigraphy and distribution of organic carbon-rich beds and the marine $\delta^{13}C$ excursion, in Marine Petroleum Source Rocks, edited by J. Brooks and A. Fleet, Geological Society of London, in press, 1984.

Schlesinger, W. H., and J. M. Melack, Transport of organic carbon in the world's rivers, Tellus, 33, 172-187, 1981.

Schmidt-Effing, R., Alter und Genese des Nicoya-Komplexes, einer ozeanischen Paleokruste (Ober Jura bis Eozärs) en sudlichen Zentralamerika, Geol. Rundsch., 68, 457-494, 1979.

Scholle, P. A., and M. A. Arthur, Carbon isotopic fluctuations in pelagic limestones; potential stratigraphic and petroleum exploration tool, Am. Assoc. Pet. Geol. Bull., 64, 67-87, 1980.

Sclater, J. G., E. Boyle, and J. M. Edmond, A quantitative analysis of some factors affecting carbonate sedimentation in the oceans, in Deep Drilling Results in the Atlantic Ocean: Continental Margins and Paleoenvironment, Maurice Ewing Ser., vol. 3, edited by M. Talwani, W. W. Hay, and W. B. F. Ryan, pp. 235-248, AGU, Washington, D.C., 1979.

Shackleton, N. J., M. A. Hall, J. Line, and S. Cang, Carbon isotope data in core V-10-30 confirm reduced carbon dioxide concentration in ice age atmosphere, Nature, 306, 319-322, 1983.

Shilo, N. A., A. P. Milov, and A. P. Sobolev, Mesozoic granitoids of northeast Asia, in Circum-Pacific Plutonic Terranes, Geol. Soc. Am. Mem., 159, edited by J. A. Roddick, pp. 149-158, Geological Society of America, Boulder, Colo., 1983.

Southam, J. R., and W. W. Hay, Time scales and dynamic models of deep-sea sedimentation, J. Geophys. Res., 82, 3825-3842, 1977.

Stevens, G. R., and R. N. Clayton, Oxygen isotope studies on Jurassic and Cretaceous belemnites from New Zealand and their biogeographic significance, N. Z. J. Geol. Geophys., 14, 829-897, 1971.

Suess, E. and P. J. Müller, Productivity, sedimentation rate, and sedimentary organic matter in the oceans, II, Elemental fractionation, in Biogeochimie de la Matière Organique a L'Interface Eau-Sediment Marin, Coll. Int. C.N.R.S., 293, pp. 17-26, Centre Nationale de la Recherche Scientifique, Paris, 1980.

Summerhayes, C. P., Organic facies of mid-Cretaceous black shales in the deep North Atlantic, Am. Assoc. Pet. Geol. Bull., 65, 2364-2380, 1981.

Summerhayes, C. P., and T. C. Masran, Organic facies of Cretaceous and Jurassic sediments from Deep Sea Drilling Project Site 534 in the Blake-Bahama Basin, western North Atlantic, Initial Rep. Deep Sea Drill. Proj., 76, 469-480, 1983.

Thiede, J., and T. H. van Andel, The paleo-environment of anaerobic sediments in the late Mesozoic South Atlantic Ocean, Earth Planet. Sci. Lett., 33, 301-309, 1977.

Thiede, J., W. E. Dean, and G. E. Claypool, Oxygen-deficient depositional paleo-environments in the mid-Cretaceous tropical and subtropical Pacific Ocean, in Nature and Origin of Cretaceous Carbon-Rich Facies, edited by S. O. Schlanger and M. B. Cita, pp. 79-100, Academic, New York, 1982.

Thiede, J., W. E. Dean, D. K. Rea, T. L. Vallier, and C. G. Adelseck, The geologic history of the mid-Pacific Mountains in the central North Pacific Ocean--A synthesis of deep-sea drilling studies, Initial Rep. Deep Sea Drill. Proj., 62, 1073-1120, 1981.

Thierstein, H. R., Paleoceanographic implications of organic carbon and carbonate distribution in Mesozoic deepsea sediments, in Deep Drilling Results in the Atlantic Ocean: Continental Margins and Paleoenvironment, Maurice Ewing Ser., vol. 3, edited by M. Talwani, W. W. Hay, and W. B. F. Ryan, pp. 249-274, AGU, Washington, D.C., 1979.

Thompson, S. L., and E. J. Barron, Comparison of Cretaceous and present earth albedos: Implications for the causes of paleoclimates, J. Geol., 89, 143-167, 1981.

Tissot, B., G. Deroo, and J. P. Herbin, Organic matter in Cretaceous sediments of the North Atlantic: Contribution to sedimentology and paleogeography, in Deep Drilling Results in the Atlantic Ocean: Continental Margins and Paleoenvironment, Maurice Ewing Ser., vol. 3, edited by M. Talwani, W. W. Hay, and W. B. F. Ryan, pp. 362-374, AGU, Washington, D.C., 1979.

Tissot, B., G. Demaison, P. Masson, J. R. Delteil, and A. Combaz, Paleoenvironment and petroleum potential of Middle Cretaceous black shales in Atlantic basins, Am. Assoc. Pet. Geol. Bull., 64, 2051-2063, 1980.

Vail, P. R., R. M. Mitchum, Jr., and S. Thompson III, Global cycles of relative changes of sea level, 4, in Seismic Stratigraphy--Applications to Hydrocarbon Exploration, Mem. 26, pp. 83-98, American Association of Petroleum Geologists, Tulsa, Okla., 1977.

Vallier, T. L., W. E. Dean, D. K. Rea, and J. Thiede, Geologic evolution of Hess Rise, central North Pacific Ocean, Geol. Soc. Am. Bull., 94, 1289-1307, 1983.

Van Houten, F. B., Ancient soils and ancient climates, in Climate in Earth History, pp. 112-117, National Academy Press, Washington, D.C., 1982.

Veizer, J., and J. Hoefs, The nature of $^{18}O/^{16}O$ and $^{13}C/^{12}C$ secular trends in sedimentary carbonate rocks, Geochim. Cosmochim. Acta, 40, 1387-1395, 1976.

Walker, J. C. G., Evolution of the Atmosphere, 318 pp., Macmillan, New York, 1977.

Walker, J. C. G., Carbon geodynamic cycle--Comment, Nature, 303, 730-731, 1983.

Waples, D. W., Reappraisal of anoxia and organic richness, with emphasis on Cretaceous of North Atlantic, Am. Assoc. Pet. Geol. Bull., 67, 963-978, 1983.

Watts, A. B. and M. S. Steckler, Subsidence and eustacy at the continental margin of Eastern North America, in Deep Drilling Results in the Atlantic Ocean: Continental Margins and Paleoenvironment, Maurice Ewing Ser., vol. 3, edited by M. Talwani, W. W. Hay and W. B. F. Ryan, pp. 218-234, AGU, Washington D. C., 1979.

Watts, A. B., J. H. Bodine, and N. M. Ribe, Observations of flexure and the geological evolution of the Pacific Ocean Basin, Nature, 283, 532-537, 1980.

Weissert, H., The environment of deposition of black shales in the Early Cretaceous: An ongoing controversy, in The Deep Sea Drilling Project: A Decade of Progress, Spec. publ. 32, edited by J. F. Warme, R. G. Douglas, and E. L. Winterer, pp. 547-560, Society of Economic Paleontologists and Mineralogists, Tulsa, Okla., 1981.

Wilde, P., and W. B. N. Berry, Progressive ventilation of the oceans - Potential for return to anoxic conditions in the post-Paleozoic, in Nature and Origin of Cretaceous Carbon-Rich Facies, edited by S. O. Schlanger and M. B. Cita, pp. 209-224, Academic, New York, 1982.

Winterer, E. L., Bathymetry and regional tectonic setting of the Line Islands chain, Initial Rep. Deep Sea Drill. Proj., 33, 731-748, 1976.

Wollast, R., and G. Billen, The fate of terrestrial organic carbon in the coastal area, in Carbon Dioxide Effects Research and Assessment Program, Conf., Flux of Organic Carbon by Rivers to the Oceans, pp. 331-359, U.S. Department of Energy, Washington, D.C., 1981.

Zachos, J. C., M. A. Arthur, R. C. Thunell, and D. F. Williams, Stable isotope and geochemical studies across the Cretaceous/Tertiary boundary at DSDP Site 577, Shatsky Rise, Leg 86, Initial Rep. Deep Sea Drill. Proj., 86, in press, 1985.

# NATURAL VARIATIONS IN THE CARBON CYCLE DURING THE EARLY CRETACEOUS

H. J. Weissert and J. A. McKenzie

Geological Institute, Swiss Federal Institute of Technology, CH-8092 Zürich

J. E. T. Channell

Department of Geology, University of Florida, Gainesville, Florida 32611

Abstract. Lower Cretaceous pelagic sediments
deposited in the Central Tethys Ocean and along
its southern continental margin and outcropping
in the Swiss and Italian Alps were chosen for a
study of Tethyan paleoceanography. The analyzed
sediments document bottom water evolution in
different bathymetric and paleogeographic environments. Bottom waters of the Berriasian
Tethys were well oxygenated and allowed the deposition of white nannofossil ooze; first indications of a rise in the calcium carbonate compensation depth due to poorly oxygenated bottom
waters are seen in Valanginian shales of the Central Tethys. A progressive shallowing of the low-
oxygen or anoxic water mass is seen in cycles of
Barremian pelagic limestones and organic carbon-
rich shales from basinal settings along the continental margin. A period of extensive anoxia
occurred during the middle Aptian, a time when
organic carbon-rich sediments were deposited on
a global scale. The "black-shale"-limestone cycles
analyzed in a Barremian basinal section appear to
have a duration of $10^5$ years. They reflect periodic changes in the deep-water circulation pattern.
Anoxic events caused by changes in regional hydrography did not disturb the global carbon budget as
read from the C isotope record, but low-amplitude
and low-frequency fluctuations in the C isotope
record are superimposed on Barremian and Aptian
lithologic cycles. These fluctuations of $\delta^{13}C$
values of up to 1°/oo occurring in cycles of about
$10^6$ years may be explained by ocean-wide variations in surface water productivity and burial
of organic matter in restricted basinal environments. A large $\delta^{13}C$ spike of +4°/oo, dated as middle to Late Aptian, reflects a pronounced increase
in Lower Cretaceous oceanic productivity coupled
with massive burial of organic matter of marine
and terrestrial origin in basinal environments.
The Aptian oceanic event was not restricted to
the Tethys but also occurred in other major
Cretaceous oceans. We conclude that the positive
C isotope spikes preserved in Lower Cretaceous
pelagic sediments are not simply indicators of
global anoxia but signal periods of high productivity which cause the formation of organic
carbon-rich sediments in environments with
limited deep-water renewal.

## Introduction

The paleogeographic map of the Early Cretaceous shows a fragmented Pangea continent with a
small Atlantic Ocean separating the Americas from
Africa and Eurasia, a narrow Tethys Ocean between
Africa and Eurasia, and the young Indian Ocean
splitting India from Africa [Smith et al., 1981].
A gigantic Pacific-Tethys Ocean covers the opposite side of the globe. An equable climate with
temperature gradients between equatorial and polar regions half as large as today's and high
atmospheric $pCO_2$ governed atmospheric conditions
during this period [Barron, 1983; Berner et al.,
1983] . Within this paleotectonic and paleoclimatic framework, ocean history is marked by a
series of peculiar events, widely known as
"oceanic anoxic events" [Schlanger and Jenkyns,
1976; Ryan and Cita, 1977; Arthur and Natland,
1979; Jenkyns, 1980; Weissert, 1981], where in
various bathymetric and paleogeographic settings
organic carbon-rich sediments were preserved.
Within the Early Cretaceous stratigraphic column,
Barremian-Albian pelagic or hemipelagic sediments
show alternations between black, bituminous and
oxydized beds [e.g., Arthur and Fischer, 1977;
Dean and Gardner, 1982]. Numerous scenarios have
been proposed to explain the mode of formation of
these carbon-rich sediments; a periodic expansion
of the oxygen-minimum zone [Schlanger and Jenkyns,
1976; Fischer and Arthur, 1977; Thiede and van
Andel, 1977], repeated density stratification due
to injection of high- or low-salinity water
[Ryan and Cita, 1977; Thierstein and Berger, 1978;
Arthur and Natland, 1979] and periodic massive

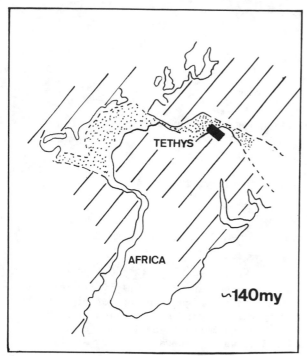

Fig. 1.  Tentative reconstruction of the Tethys-Atlantic Ocean near the Jurassic-Cretaceous boundary. The location of our area of study is indicated by the bar.

burial of terrestrial organic matter [Habib, 1979; Graciansky et al., 1979] are among the most popular mechanisms. Other authors emphasize the unique paleogeographic conditions within the Early Cretaceous oceans, where anoxic conditions could develop in a series of small basins without depleting the global deep-water reservoir in oxygen [Waples, 1983]. Whether these oceanic anoxic events are simply linked to global anoxia or are coupled with fluctuations in surface water productivity remains controversial [de Boer, 1982; Roth, 1983; Thierstein and Bralower, 1983; Calvert, 1984].

In this contribution we begin by tracing the evolution of a "normal" oxygenated Late Jurassic ocean to a periodically anoxic ocean during the Early Cretaceous. For our study we selected one small segment of the Cretaceous Tethyan sediments which now is exposed in the Alps and Appennines (Figure 1). We consider this segment, which represents a cross section from the central oceanic Tethys to the proximal continental margin of the Adriatic carbonate platform, to be a typical sequence representative of the Tethys and Atlantic oceans [Bernoulli, 1972; Bernoulli and Jenkyns, 1974]. This cross section allowed us to compare the sedimentological evolution of various bathymetric and paleogeographic settings throughout the Early Cretaceous. We compared sediments from

the deepest part of the Tethys, underlain by ophiolitic basement now preserved in alpine thrust nappes (Platta nappe [Trümpy, 1975]),with those from environments along the continental margin. In the latter area, we analyzed a section from a deep, distal basin (Lombardian Basin), one from an adjacent submarine high (Trento Plateau) and one from a more proximal basin (Belluno Basin) (Figure 2). Sedimentological and carbon isotope data helped us to trace bottom water evolution during the Early Cretaceous and to evaluate the regional extent of periodic anoxic events, their effect on the global carbon cycle and the consequences of changes in surface water productivity on the oceanic carbon system. A comparison of sections with differing diagenetic history allowed us to evaluate possible effects of diagenesis on the carbon isotope record. With the help of a detailed magnetostratigraphy in the Barremian and Aptian sections we further could correlate the carbon isotope stratigraphy and use carbon isotopes as indicators for perturbations in the oceanic carbon system.

From a "Normal" to a Periodically Anoxic Ocean

## Progressive Lithostratigraphic Changes

During the Late Jurassic, red radiolarian chert and limestone were deposited on the deeply subsided southern continental margin of the alpine Tethys. These radiolarian sediments overlay a drowned continental margin sequence ultimately floored by continental basement [Winterer and Bosellini, 1981]. An analogous radiolarite sequence was deposited on ophiolitic basement of the central Tethys. These radiolarites probably reflect conditions of heightened fertility in Tethyan surface water and an elevated calcium carbonate compensation depth (CCD) [Jenkyns and Winterer, 1982]. Bottom water was well oxygenated at depositional depths of 2.5 to 4.0 km [Bosellini and Winterer, 1975]. During the Late Tithonian the radiolarian ooze was replaced by a white nannofossil ooze. Hsü [1975] and Weissert [1979] proposed that this facies change is an indicator for a decrease of productivity in Tethyan surface water. This proposed change in surface water productivity was associated with a progressive alteration of Tethyan bottom water characteristics, which we will document in more detail.

At the Jurassic-Cretaceous boundary (Tithonian-Berriasian), Calpionellid-bearing nannofossil ooze with rare radiolaria was deposited in the central Tethys as well as in basins and on submarine highs of the continental margin (Figure 3). These deposits form the base of the widely known Maiolica Formation or Calpionellid Limestone Formation, which closely resembles the pelagic limestone sequence recovered from the central Atlantic Ocean [Bernoulli, 1972]. These

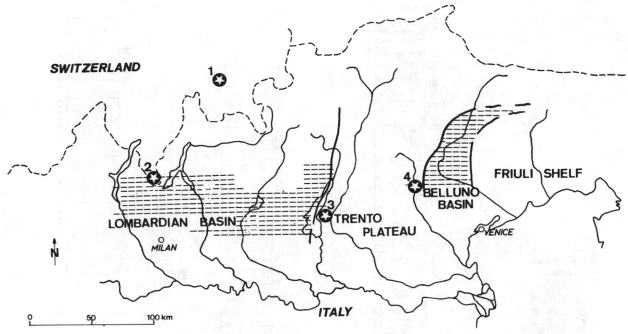

Fig. 2.  Localities of the stratigraphic sections studied in the Swiss and Italian Alps: (1) Platta, (2) Breggia, (3) Spiazzi/Valle Aviana, (4) Cismon. Cretaceous paleogeographic elements of the southern continental margin are labeled along the Southern Alps.

sediments were everywhere homogenized by burrowing fauna.

Only a few million years later, during the Valanginian, the limestone sequence of the deep central Tethys was replaced by grey and black siliceous shales alternating with dark pelagic limestones and chert beds of up to 10 cm thickness (Palombini Formation). Less distinct changes occur in Valanginian deposits of Lombardian Basin sequences. The thick-bedded (up to 1 m) white pelagic limestone of the Tithonian-Berriasian is replaced by thin-bedded (10 cm) dark and strongly bioturbated nannofossil limestone with millimeter intercalations of dark marl separating individual limestone beds. In the sections from the Trento Plateau and the Belluno Basin we did not recognize any lithologic changes throughout the Berriasian-Valanginian.

In Hauterivian-Aptian sediments of the central Tethys (Platta nappe), dated only indirectly because the overlying hemipelagic marls are of Albian age [Dietrich, 1970], dark siliceous shales dominate with only an occasional intercalation of pelagic limestones (Figure 3). The basinal environments along the continental margin were apparently more susceptible to changes in Hauterivian-Aptian oceanography in the Lombardian Basin, the Hauterivian and Barremian are characterized by a sequence of alternating nannofossil limestones and dark siliceous marls. The limestones

are bioturbated, and individual beds, up to 15 cm thick, are separated by thin seams of clay. In intervals of a few tens of centimeters up to 1 m, the limestone deposits are interrupted by layers of black, siliceous marl. These beds, up to 20 cm thick, lack bioturbation and show millimeter lamination structures [Weissert et al., 1979]. In the transition from limestones to marls the $CaCO_3$ content drops from values near 90% to values near 40-75%. The main carbonate constituents in both lithologies are nannofossils and diagenetic calcite. Diagenetic quartz, clay minerals, pyrite and organic matter of marine and terrestrial origin [Arthur and Premoli-Silva, 1982] are identified as additional constituents in the "black shales". On the nearby Trento Plateau, no organic carbon-rich sediments were deposited during this time interval, and in the sediments of the Belluno Basin, the observed dark shale intercalations rarely exceed centimeters in thickness. Hardgrounds on the top of the Maiolica Formation are thought to be a result of strong erosional or non-depositional events within the Lombardian Basin during Late Barremian and Early Aptian. The paleomagentically dated Barremian-Aptian interval at the Cismon section in the Belluno Basin is more complete [Channell et al., 1979]. The change from nannofossil limestone to grey, green and red foraminiferal marls is dated as middle Aptian. These deposits compare well

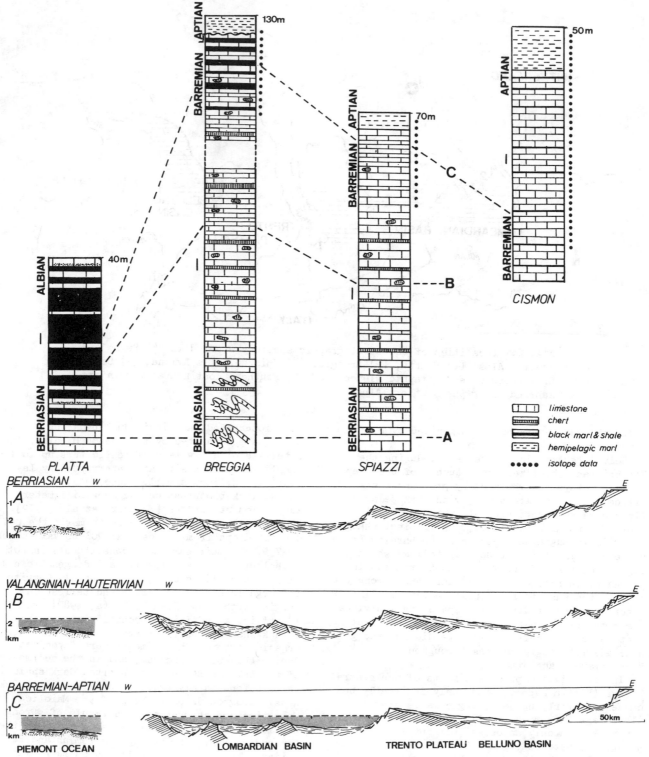

Fig. 3.  Stratigraphic sections of the Early Cretaceous central Tethys (Platta) and the southern continental margin (Breggia-Lombardian Basin, Spiazzi-Trento Plateau, Cismon-Belluno Basin). Three time slices (A,B,C) through the central Tethys and its southern margin show evolution of bottom water conditions during the Early Cretaceous. Anoxic bottom water is indicated by the intensity of the shading.

with the Aptian-Albian foraminiferal marls of the Lombardian Basin, where again bituminous shales are intercalated [Arthur and Premoli-Silva, 1982]. Foraminiferal marls, grey in color and dated as Aptian/Albian [Cita, 1964; Arthur and Premoli-Silva, 1982], overlie the nannofossil limestones on the Trento Plateau.

Not all of the described dark shales and marls are rich in organic carbon. Less than 1% organic carbon was measured in some samples from the Valanginian and Hauterivian siliceous shales of the central Tethys [Green, 1981]. In the Barremian black shales of the Lombardian Basin, organic carbon contents range between 1 and 3.4% [Bitterli, 1965]. High organic carbon contents of up to 15% are reported from Aptian horizons [Bitterli, 1965: Arthur and Premoli-Silva, 1982]. In all of these beds a mixed terrestrial and marine source can be assigned to the organic matter.

## An Oceanographic Interpretation

At the end of the Jurassic, surface water productivity in the Tethys decreased as radiolaria, which are limited by the supply of silica, were gradually replaced by nannofossils. The reduced input of organic matter into deeper water masses is regarded as a major cause lowering the CCD [Hsü, 1976; Weissert, 1979; Jenkyns and Winterer, 1982]. In addition, oxygenation of bottom water became less effective than during times of radiolarian deposition, due, perhaps, to changes in bottom water circulation. A common hypothesis is that deep-water circulation in Early Cretaceous oceans was "sluggish" due to salinity-controlled stratification of the water masses [e.g., Thierstein and Berger, 1978]. Southam et al. [1982] present an alternative model where oxygen-depleted intermediate water masses are linked with intensified thermohaline circulation and overturn of water masses. Hsü [1976] and Weissert [1979, 1981] suggested that the mechanism of deep water foramtion in the present-day Mediterranean could serve as a possible model for the circulation pattern in the Early Cretaceous Tetyhs Ocean. In the eastern Mediterranean, saline surface water is formed during the summer season by evaporation. During winter months, cold continental winds cause a cooling of the surface water in shallow marginal seas, such as the Adriatic Sea. The cold water mixes with the dense, saline water and produces the deep water of the eastern Mediterranean [Lacombe and Tchernia, 1972]. At the beginning of the Cretaceous, dense water masses formed by this mechanism could have reached the deepest parts of the Tethys, continuously renewing their oxygen content.

A diversification of the hydrographic conditions of the Tethyan deep water occurred during the Valanginian as indicated by the onset of black shale deposition. We regard these dark, siliceous shales of the central Tethys as an indication of a shallowing CCD, due, first, to an increase in dissolved $CO_2$ caused by slow regeneration of bottom water and, second, to progressive deepening of the central Tethys [Winterer and Bosellini, 1981]. These low-oxygen or even anoxic conditions were limited to the deepest part of the Tethys and could be explained by decreased generation of deep water due to changes in the summer or winter climate according to the analogy with the modern Mediterranean. If we regard the few data on organic carbon content (less than 1%) from the Valanginian dark, siliceous shales as typical values, we can assume that surface water productivity remained low, particularly when compared with the high organic carbon contents in the Aptian shales (up to 15%).

A few million years later, during the Hauterivian and Barremian, low-oxygen/anoxic bottom water conditions were not limited to the central

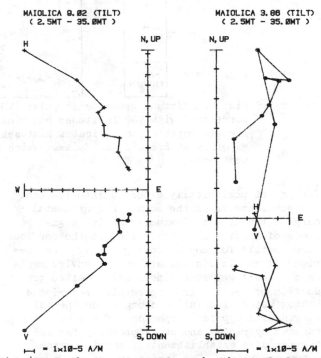

Fig. 4.  Typical orthogonal projections of alternating field demagnetization data of a normally (left) and reversely magnetized sample of Maiolica limestone. Crosses indicate projection on the horizontal plane, and circles projection on the vertical plane. The range of peak fields is indicated in milliteslas. The directions are corrected for the tilt of the beds. Note that reversely magnetized samples often carry a normal overprint which is removed by peak fields of greater than 10 milliteslas.

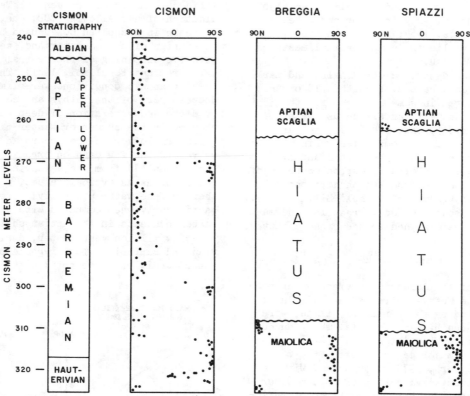

Fig. 5. Virtual geomagnetic polar (VGP) latitudes from Breggia and from Spiazzi, correlated with VGP latitudes and biostratigraphy from Cismon. This magnetostratigraphic correlation indicates hiatuses between the Aptian Scaglia and underlying Maiolica at Breggia and Spiazzi which are not identical in duration for the two sections.

Tethys but periodically extended into the basinal environments of the southern continental margin. Bituminous, laminated shales signal periods of anoxia, when conditions on the seafloor were hostile for any burrowing fauna. These periods of anoxia did not affect the environments on the shallower submarine highs. Limited production of deep water coupled with restricted bottom water circulation along a continental margin with typical morphology of basins separated by ridges and banks probably favored the periodic establishment of anoxic bottom water conditions. The laminated shales formed during periods of anoxia show considerable variations in their organic carbon content [Bitterli, 1965]. If we assume that sediment accumulation rates during periods of anoxia did not vary greatly, we can construe that these fluctuations, superimposed on circulation changes, reflect changes in the rate of organic carbon burial due to changes in biological surface water productivity. Calvert [1984] concludes that increasing preservation of organic

matter in deep-sea sediments is facilitated by increasing accumulation rates and/or surface water productivity. Anoxic conditions alone do not result in the accumulation of organic matter.

The highest organic carbon contents were measured in siliceous shales of Late Aptian age. A low carbonate content in these "black shales" may be caused either by a low carbonate supply or by increased carbonate dissolution near the Aptian CCD [Arthur and Premoli-Silva, 1982]. Either increased dissolution or decreased primary carbonate supply would result in low accumulation rates if the clay sedimentation rate remained relatively constant. Under these circumstances the extremely high percentage of marine and terrestrial organic matter preserved in the sediments could only be the result of an increased biological productivity along the southern continental margin of the Aptian Tethys. High accumulation rates of marine organic matter along the eastern North Atlantic were also related to increased biological productivity in the Aptian [Dean and Gardner, 1982].

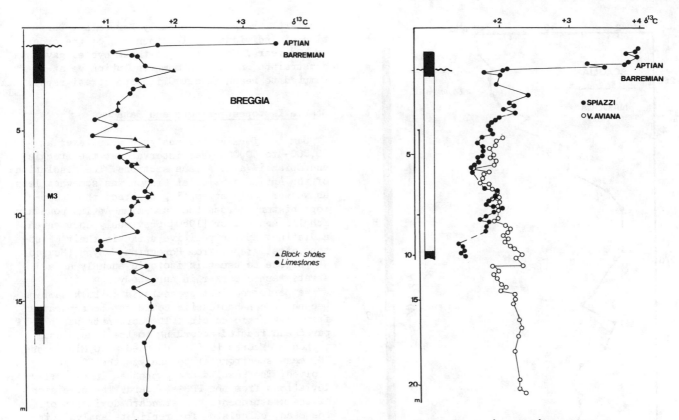

Fig. 6.   Carbon isotope and magnetostratigraphy of the Barremian-Aptian at
Breggia (Lombardian Basin) and Spiazzi/Valle Aviana (Trento Plateau).

Carbon Isotope Stratigraphy in Barremian-Aptian
Sediments of the Southern Continental Margin

Biostratigraphy and Magnetostratigraphy

Three sections at Breggia, Spiazzi/Valle
Aviana and Cismon (Figure 2) were chosen for
their different diagenetic histories and paleo-
environments. Magnetostratigraphies were derived
from the uppermost Maiolica at Spiazzi and
Breggia and compared with that from the Cismon
locality, where the most continuous Early Aptian
section is preserved and where magnetostratigraphy has been established [Channell et al.,
1979]. The magnetic properties of the Maiolica
limestones are discussed in the Cismon study. At
Breggia and Spiazzi the Maiolica limestones have
magnetic properties similar to those at Cismon,
but at Valle Aviana the magnetization of the
rocks was too weak for paleomagnetic investigations. Orthogonal projections of alternating
field (AF) demagnetization data (Figure 4) indicate a low-coercivity magnetization component
which is probably carried by detrital magnetite.
This simple magnetization can be used to generate
magnetostratigraphies which can be used to correlate between Maiolica sections. Based on such
magnetostratigraphic correlation (Figure 5) there

appears to be a significant hiatus between Aptian
Scaglia and Barremian Maiolica both at Spiazzi
and at Breggia. The unconformity is marked by a
distinct lithologic break at these two sections
but is not apparent at Cismon. The duration of
this hiatus is probably not identical at Breggia
and Spiazzi, as the normal magnetozone immediately below the unconformity (Figure 5) has a
thickness of 10 cm and 2.5 m at Spiazzi and
Breggia, respectively. The reversed magnetozone
below this normal zone, which is observed both
at Spiazzi and at Breggia, has a thickness of
10.5 m and 13 m at the two localities, respectively, and the correlative magnetozone at Cismon
has a comparable thickness.

The magnetostratigraphic correlation shown in
Figure 5 is consistent with available paleontological data. The Cismon biostratigraphy is based
on planktonic foraminifera for the Aptian/Albian
and on calcareous nannofossils for the Barremian
[Channell et al., 1979]. The biostratigraphy of
the Aptian Scaglia above the unconformity at
Breggia and Spiazzi is based on planktonic
foraminifera [Cita, 1964; Premoli-Silva and Luterbacher, 1965; Arthur and Premoli-Silva, 1982].
The biostratigraphy below the unconformity at
these two localities is not well constrained.

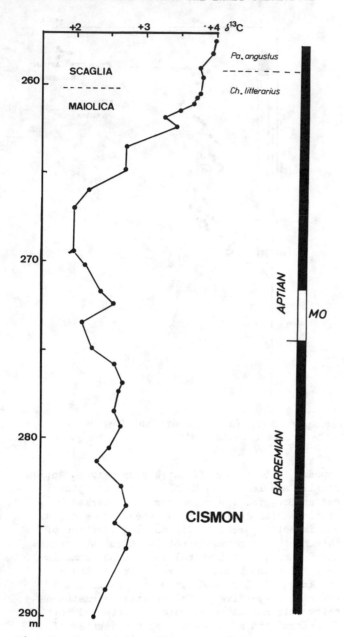

Fig. 7. Barremian-Aptian carbon isotope stratigraphy and magnetostratigraphy of the Cismon section (Belluno Basin).

The only control available is at Breggia, where
a Barremian ammonite fauna has been discovered
in a black marl horizon located 4.7 m below the
unconformity [Rieber, 1977].

The magnetostratigraphic information allowed
us to estimate the length of the limestone-black
marl cycles in the Barremian Maiolica. We first
assigned the chron M3 to the reversed magnetozone observed at the Breggia locality. We further assumed that the limestone sedimentation
rate and the clay sedimentation rate remained

constant throughout the Early Barremian. Duration of individual cycles then was of the order
of $10^5$ years. The length of these cycles gave us
constraints on the minimum resolution we attempted to reach in our C isotope stratigraphy.

Stable Isotopes: Methods and Data

Samples for isotope analysis were taken at
20,000- to 50,000-year intervals at the Breggia
and Spiazzi/Valle Aviana sections. The resolution
of the Aptian section at Cismon was somewhat less,
as we were only attempting to trace the C isotope history across the Barremian-Aptian boundary.
Scholle and Arthur [1980] previously documented
a distinct Aptian positive spike in their C isotope stratigraphy from Peregrina Canyon (Mexico).
We propose to use this isotopic anomaly as a
stratigraphic marker in our study.

We performed isotope analysis on bulk samples
because the nannofossils could not be separated
from diagenetic calcite. The carbonate was prepared and reacted according to the traditional
method of McCrea [1950]. We used a VG Micromass-
903 mass spectrometer to analyze the $CO_2$ gas
evolved. The results are expressed in per mil
deviations from the PDB-standard. The precision
of the measurements, as standard deviation of
the mean, calculated for replicate analysis is
$\pm 0.10^o/oo$ for $\delta^{13}C$ and $\pm 0.20^o/oo$ for $\delta^{18}O$. The
results are tabulated in Table 1 and graphically
presented in Figures 6 and 7.

A Diagenetic or an Oceanographic Signal?

During progressive lithification by burial
diagenesis the nannofossil ooze experiences significant changes in texture and geochemistry.
Diagenetic calcite is added to nannofossil calcite through precipitation from a $CaCO_3$-over-
saturated pore fluid [Schlanger and Douglas,
1974]. New calcite cement may alter significantly
the original isotope signature of bulk carbonate.
Most suscpetible to diagenetic changes is the
oxygen isotope composition of pelagic sediments.
A highly temperature-dependent fractionation
factor between $H_2O$ and calcite is responsible
for a progressively more negative oxygen isotope
signature during increasing lithification with
depth [Matter et al., 1975; McKenzie et al., 1978;
Garrison, 1982]. In lithified sediments, therefore, the oxygen isotope composition is not a
reliable tracer for ocean history, and we will
focus our discussion on the carbon isotope data.

The basinal Lombardian section and the Trento
Plateau section both have a similar pattern in
the C isotope curve, but the values at the Breggia section are shifted by $0.5^o/oo$ to more negative values (Figure 8). We assume that this
relative difference expresses different dia-

genetic pathways for the two studied sections. The Trento Plateau section, built up of nanno-fossil limestone, underwent classical limestone diagenesis [Matter et al., 1975]; i.e., dia-genetic calcite with marine-derived $CO_2$ as its source was precipitated under continuously in-creasing temperatures up to about 50°C. The bulk C isotope composition should differ from the original nannofossil ooze value only as a result of the temperature-dependent fractionation in the system $HCO_3$-$CaCO_3$ [Emrich et al., 1970]. Since this factor is very small, we assume that the measured values on the Trento Plateau closely reflect the $\delta^{13}C$ value of the dissolved carbon in the Cretaceous ocean. In the Breggia section, the numerous bituminous shales could have in-fluenced the C isotope composition of bituminous shales and pelagic limestones. During bacterial oxidation and sulfate reduction, $CO_2$ derived from organic matter could have been added to the dissolved $CO_2$ in the pore fluid. This organic $CO_2$ with $\delta^{13}C$ values near -20°/oo is clearly dif-ferent in its isotopic composition from the marine derived $CO_2$ [Irwin et al., 1977; Irwin, 1980]. Pore water enriched in isotopically light $CO_2$ would not be restricted to organic carbon-rich layers but could be expelled during burial through the overlying nannofossil ooze. Calcite cement precipitated from the isotopically lighter dissolved $CO_2$ could have added a more negative $^{13}C$ signal to the bulk carbonate C isotope strati-graphy of the Barremian Breggia section.

If, indeed, the basinal carbon isotope record experienced a significant diagenetic overprint, we must evaluate whether the observed fluctua-tions are of oceanic origin or represent a dia-genetic phenomenon. We approached this question by comparing and correlating the basinal and plateau C isotope records for the Barremian.

## Fluctuations in the Barremian C Isotope Record

Constant $\delta^{13}C$ values characterize the lower part of both measured sequences (Figure 6). With-in magnetozone M4 the constancy in the isotope pattern is replaced by a low-amplitude and low-frequency fluctuations of up to 1.2°/oo in the Breggia record and up to 0.8°/oo in the Spiazzi/ Valle Aviana. While in the lower part of the re-versed magnetozone correlated to M3 a matching of individual $^{13}C$ spikes is still speculative, we recognize a distinct period of low $\delta^{13}C$ values succeeded by a positive spike at the top of M3. A return to decreased $\delta^{13}C$ values in the following normally magnetized interval is measured in both the basinal and plateau sections.

Do the recorded C isotope fluctuations cor-relate with changes in sedimentology? The high-frequency pattern of black shale-limestone cycles of the order of $10^5$ years is not mirrored in the C isotope record (Figure 9), but we can correlate

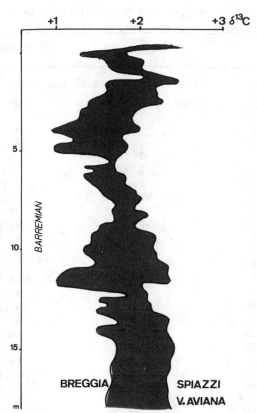

Fig. 8. Carbon isotope curves of the sections Breggia and Spiazzi/Valle Aviana, showing the diagenetically induced negative shift of the Breggia curve.

the most distinct positive spikes with layers of organic carbon-rich sediments. The C isotope re-cord with its low-frequency fluctuations of the order of $10^6$ years can be superimposed upon se-cond-order variations, i.e., upon high-frequency sedimentary cycles. We therefore must modify an earlier scenario where we assigned positive $\delta^{13}C$ spikes to all the organic carbon-rich marls [Weissert et al., 1979]. Not all black shales cor-respond to increases in carbon 13 composition of the pelagic sediments. The periodic anoxic events extending into the Lombardian Basin during the Barremian frequently had no impact on the oceanic carbon budget, because they were not accompanied by increased biogenic productivity in the sur-face water. The major factor controlling these anoxic events was periodic oxygen depletion in the bottom water of the narrow Tethys, which had a rugged morphology and numerous silled basins, a situation comparable to the present-day Cali-fornia borderlands [Douglas, 1981].

On the other hand, the controlling factor on the carbon isotope distribution and, hence, the carbon cycle in the Barremian oceans was surface

TABLE 1.  Oxygen and Carbon Isotope Data, Maiolica Formation, Southern Alps

| Sample | Oxygen 18 | Carbon 13 |
|--------|-----------|-----------|

Breggia, Lombardian Basin

| Sample | Oxygen 18 | Carbon 13 |
|--------|-----------|-----------|
| BR-0 | | |
| 0.00 | -1.34 | +1.76 |
| 0.40 | -2.39 | +1.10 |
| 0.70 | -1.70 | +1.38 |
| 0.75* | -2.37 | +1.45 |
| 0.86 | -2.91 | +1.06 |
| 1.20 | -2.19 | +1.53 |
| 1.49 | -1.98 | +2.00 |
| 1.75 | -2.16 | +1.64 |
| 2.00 | -1.72 | +1.48 |
| 2.18 | -1.85 | +1.52 |
| 2.45 | -1.56 | +1.57 |
| 2.65 | -1.63 | +1.40 |
| 2.90 | -1.80 | +1.36 |
| 3.35* | -0.71 | +0.85 |
| 3.80 | -1.49 | +1.19 |
| 4.50 | -2.01 | +0.85 |
| 4.55 | -2.26 | +0.69 |
| 4.65 | -1.55 | +1.34 |
| 4.78 | -1.61 | +1.17 |
| 5.50 | -1.80 | +0.67 |
| 5.55* | -1.87 | +1.42 |
| 5.80 | -2.50 | +0.00 |
| 6.14 | -0.03 | +1.59 |
| 6.22 | -1.42 | +1.08 |
| 6.28 | -2.28 | +1.47 |
| 6.70 | -2.06 | +1.20 |
| 7.08 | -2.27 | +1.30 |
| 7.13 | -2.28 | +1.47 |
| 7.20 | -2.62 | +1.34 |
| 7.55* | -3.91 | +0.50 |
| 8.10 | -1.69 | +1.67 |
| 8.85 | -1.71 | +1.57 |
| 8.90* | -1.92 | +1.60 |
| 9.05 | -1.65 | +1.38 |
| 9.22* | -2.17 | +1.45 |
| 9.55 | -1.65 | +1.36 |
| 9.82 | -2.29 | +0.95 |
| 10.00 | -1.85 | +1.38 |
| 11.10 | -1.53 | +1.45 |
| 11.55 | -1.67 | +0.85 |
| 11.90 | -1.64 | +0.84 |
| 12.10* | -1.77 | +0.86 |
| 12.40 | -- -- | +1.24 |
| 12.45* | -2.05 | +1.85 |
| 12.80 | -1.52 | +1.23 |
| 13.50 | -1.55 | +1.42 |
| 14.00 | -1.67 | +1.56 |
| 14.30 | -1.98 | +1.38 |
| 15.00 | -1.44 | +1.63 |
| 15.50 | -1.39 | +1.64 |
| 16.60 | -1.02 | +1.61 |
| 16.70 | -1.54 | +1.64 |
| 19.00 | -1.37 | +1.61 |

TABLE 1 (continued)

| Sample | Oxygen 18 | Carbon 13 |
|--------|-----------|-----------|

Spiazzi, Trento Plateau

| Sample | Oxygen 18 | Carbon 13 |
|--------|-----------|-----------|
| SP-0.80 | -1.38 | +3.93 |
| -0.60 | -1.59 | +3.72 |
| -0.50 | -0.87 | +3.86 |
| -0.45 | -1.18 | +3.83 |
| -0.40 | -1.84 | +3.27 |
| -0.30 | -1.08 | +3.62 |
| -0.20 | -1.38 | +3.29 |
| -0.05 | -1.16 | +2.05 |
| 0.00 | -1.21 | +2.11 |
| +0.10 | -1.04 | +2.07 |
| +0.20 | -1.14 | +2.10 |
| +0.30 | -1.70 | +1.76 |
| +0.40 | -1.05 | +1.97 |
| +0.50 | -1.26 | +2.00 |
| +0.60 | -1.02 | +2.04 |
| +0.70 | -0.92 | +2.04 |
| +0.80 | -0.87 | +2.10 |
| +0.90 | -1.04 | +2.05 |
| +1.00 | -1.06 | +1.96 |
| +1.20 | -1.16 | +2.01 |
| +1.40 | -0.88 | +1.88 |
| +1.60 | -0.71 | +2.46 |
| +1.80 | -1.07 | +2.13 |
| +2.00 | -0.88 | +2.01 |
| +2.20 | -1.00 | +2.07 |
| +2.40 | -0.71 | +2.25 |
| +2.60 | -1.01 | +2.01 |
| +2.80 | -0.96 | +1.94 |
| +3.00 | -0.96 | +1.88 |
| +3.20 | -1.15 | +1.81 |
| +3.40 | -1.09 | +1.86 |
| +3.60 | -1.05 | +1.89 |
| +3.80 | -0.95 | +1.94 |
| +4.00 | -1.05 | +1.79 |
| +4.20 | -1.04 | +1.79 |
| +4.40 | -1.21 | +1.70 |
| +4.60 | -0.92 | +1.75 |
| +4.80 | -0.94 | +1.80 |
| +5.00 | -1.57 | +1.83 |
| +5.20 | -0.98 | +1.81 |
| +5.40 | -1.20 | +1.74 |
| +5.60 | -1.08 | +1.74 |
| +5.80 | -1.16 | +1.69 |
| +6.00 | -1.32 | +1.65 |
| +7.00 | -1.05 | +1.89 |
| +7.20 | -0.94 | +1.97 |
| +7.40 | -1.20 | +1.93 |
| +7.60 | -1.21 | +1.91 |
| +7.80 | -1.39 | +1.83 |
| +8.00 | -0.76 | +1.90 |
| +8.20 | -0.79 | +1.99 |
| +8.40 | -0.99 | +1.90 |
| +8.60 | -0.93 | +1.79 |
| +8.80 | -1.09 | +1.70 |
| +9.00 | -0.58 | +1.85 |
| +9.20 | -0.85 | +1.82 |

TABLE 1 (continued)

| Sample | Oxygen 18 | Carbon 13 |
|--------|-----------|-----------|

### Valle Aviana, Trento Plateau

| Sample | Oxygen 18 | Carbon 13 |
|--------|-----------|-----------|
| AV 0.00 | -1.25 | +2.16 |
| 0.20 | -1.54 | +2.04 |
| 0.40 | -1.39 | +2.05 |
| 0.60 | -1.49 | +1.94 |
| 0.80 | -1.61 | +1.83 |
| 1.00 | -1.40 | +2.02 |
| 1.20 | -1.33 | +1.98 |
| 1.40 | -1.43 | +2.07 |
| 1.60 | -1.39 | +2.06 |
| 1.80 | -1.65 | +1.82 |
| 2.00 | -1.56 | +1.98 |
| 2.20 | -1.46 | +1.91 |
| 2.40 | -1.79 | +1.90 |
| 2.60 | -1.61 | +1.85 |
| 2.80 | -1.42 | +1.97 |
| 3.00 | -1.54 | +2.00 |
| 3.20 | -1.48 | +2.10 |
| 3.40 | -1.40 | +2.12 |
| 3.60 | -1.55 | +2.03 |
| 3.80 | -1.50 | +2.13 |
| 4.00 | -1.37 | +2.11 |
| 4.20 | -1.19 | +2.13 |
| 4.40 | -1.18 | +2.13 |
| 4.60 | -1.76 | +2.15 |
| 4.80 | -1.15 | +2.15 |
| 5.00 | -1.26 | +2.06 |
| 5.20 | -1.98 | +2.22 |
| 5.40 | -1.15 | +2.29 |
| 5.60 | -1.13 | +2.28 |
| 5.80 | -1.22 | +2.22 |
| 6.00 | -1.37 | +2.25 |
| 6.20 | -3.21 | +2.26 |
| 6.40 | -1.25 | +2.31 |
| 6.60 | -1.24 | +2.34 |
| 6.80 | -1.60 | +2.45 |
| 7.00 | -1.44 | +2.51 |
| 7.20 | -1.36 | +2.41 |
| 7.40 | -1.27 | +2.46 |
| 7.60 | -1.29 | +2.51 |
| 7.80 | -1.25 | +2.08 |
| 8.00 | -1.07 | +2.20 |
| 8.20 | -1.41 | +2.13 |
| 8.40 | -1.27 | +2.14 |
| 8.60 | -1.15 | +2.15 |
| 8.80 | -1.18 | +2.15 |
| 9.00 | -1.25 | +2.23 |
| 9.20 | -1.27 | +2.21 |
| 9.40 | -1.25 | +2.35 |
| 9.60 | -1.25 | +2.38 |
| 9.80 | -1.34 | +2.34 |
| 10.00 | -1.31 | +2.35 |
| 11.00 | -1.14 | +2.40 |
| 11.60 | -1.24 | +2.42 |
| 12.00 | -1.31 | +2.42 |

TABLE 1 (continued)

### Valle Aviana, Trento Plateau (continued)

| Sample | Oxygen 18 | Carbon 13 |
|--------|-----------|-----------|
| 13.00 | -1.09 | +2.36 |
| 13.50 | -1.30 | +2.56 |
| 14.40 | -1.31 | +2.63 |
| 14.90 | -1.36 | +2.59 |
| 15.20 | -1.37 | +2.65 |
| 16.50 | -1.12 | +2.77 |
| 17.00 | -1.10 | +3.02 |
| 18.00 | -1.08 | +3.03 |
| 18.70 | -1.21 | +2.97 |
| 19.70 | -1.98 | -2.48 |

### Cismon, Belluno Basin

| Sample | Oxygen 18 | Carbon 13 |
|--------|-----------|-----------|
| C 257.50 | -0.39 | +3.92 |
| 258.20 | -0.49 | +3.91 |
| 259.00 | -0.51 | +3.71 |
| 260.50 | -0.76 | +3.72 |
| 260.65 | -2.17 | +3.66 |
| 261.10 | -1.45 | +3.61 |
| 261.50 | -2.47 | +3.42 |
| 261.90 | -2.89 | +3.22 |
| 263.50 | -2.28 | +2.67 |
| 264.80 | -2.07 | +2.64 |
| 266.00 | -2.45 | +2.12 |
| 267.00 | -2.77 | +1.90 |
| 269.45 | -1.39 | +1.91 |
| 270.20 | -2.77 | +2.08 |
| 271.20 | -0.73 | +1.90 |
| 272.40 | -0.72 | +2.44 |
| 273.40 | -0.85 | +2.05 |
| 274.90 | -2.58 | +2.17 |
| 275.85 | -1.74 | +2.50 |
| 276.85 | -0.40 | +2.62 |
| 277.35 | -0.57 | +2.56 |
| 278.50 | -1.77 | +2.50 |
| 279.30 | -0.54 | +2.59 |
| 280.55 | -2.26 | +2.13 |
| 281.30 | -0.81 | +2.24 |
| 282.70 | -0.83 | +2.58 |
| 283.75 | -0.51 | +2.67 |
| 284.75 | -1.87 | +2.50 |
| 285.40 | -0.50 | +2.70 |
| 286.20 | -0.81 | +2.67 |
| 288.50 | -0.44 | +2.37 |
| 290.00 | -0.97 | +2.21 |

The sample numbers at Breggia, Spiazzi, and Valle Aviana equal the meter distance from the top of the Maiolica formation. The sample numbers of the Cismon section are equal to the arbitrary depth values of the plotted section of Channell et al. [1979].

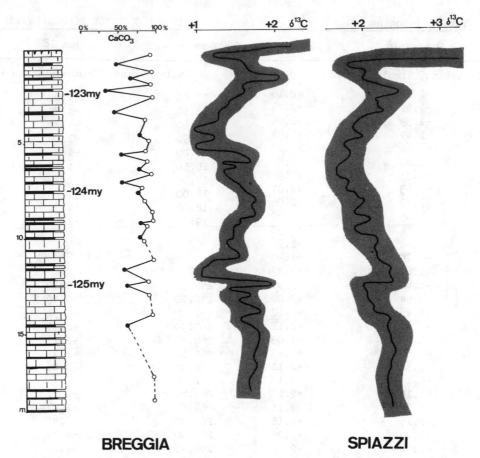

**BREGGIA**                    **SPIAZZI**

Fig. 9.   Tentative correlation of the carbon isotope stratigraphy with the
sedimentary cycles recorded in the uppermost part of the Maiolica Formation at
Breggia. Calcium carbonate contents of the Breggia sediments were analyzed with
a Ströhlein Coulomat. The absolute ages are according to GSA-1983 geological
time scale.

water productivity. During blooms of marine plankton, the light $^{12}C$ isotopes are preferentially
removed from the surface water and transferred
to the deep water [e.g. Kroopnick, 1974]. At
times when high productivity coincided with
anoxia in bottom water, burial of isotopically
light organic matter in the sediments was most
effective. This burial essentially removes carbon 12 from the oceanic reservoir, allowing carbon 13 concentrations to increase. Therefore,
most positive $^{13}C$ spikes do coincide with
organic carbon-rich black shales. Fluctuations
in organic matter content of the bituminous
shales could also be explained by varying rates
of surface water productivity.

## The C Isotope Event in the Aptian

Superimposed on the sedimentary cycles and the
low-amplitude C isotope fluctuations is the 4°/oo
excursion in the Aptian C isotope record. The

evolution of the Barremian and Early Aptian isotope record is best documented in the extended
pelagic limestone sequence of Cismon (Figure 7).
The middle Aptian part of the section missing at
Breggia and Spiazzi is preserved at Cismon as
foraminiferal marl. The resolution in the C isotope record established for the time interval
of interest is not as detailed as in the previous
sections studied, but we can recognize the low-
frequency and low-amplitude fluctuations which
continue up into the Early Aptian above the magnetic interval MO. Within only a few hundred
thousand years the $\delta^{13}C$ value shifts from +2°/oo
to +4°/oo, as mentioned above. We correlate this
event with the spikes analyzed in the top of the
Spiazzi section and the Breggia section. The
sharper change in these latter curves is interpreted as artificial due to discontinuous sedimentation. We correlate this Aptian spike further
with the signals Scholle and Arthur [1980] measured in the sequence of the Peregrina Canyon

(Mexico). While they tentatively dated their spike as Late Aptian-Albian, we can assign an age of Late Early Aptian to the beginning of the event, as determined from our paleomagnetic investigation.

The C isotope event of the Aptian is useful as a global stratigraphic marker, as it has been recognized in the Pacific [Coplen and Schlanger, 1973], in the Caribbean area [Scholle and Arthur, 1980], and in the alpine Tethys. It also reflects one of the major disturbance of the global carbon-cycle in the last 200 million years. Enhanced removal of the $^{12}C$ isotope from the aquatic system by excessive burial of organic matter over millions of years is regarded as a major cause for shifting the C isotope curve to more positive values [Scholle and Arthur, 1980]. We are again confronted with the question, is excessive burial of organic matter linked with increased productivity and/or with widespread oceanic anoxia? Modelling lacustrine systems, McKenzie [1982, 1985] has demonstrated that increasing productivity together with the removal of $^{12}C$ from the aquatic resevoir through the burial of organic matter produces positive carbon isotope shifts in the pelagic carbonates. She also showed that increasing surface water productivity together with a recycling of the organic carbon produces negative carbon isotope shifts in the pelagic carbonates. Apparently, increasing productivity and rapid burial of the organic matter can combine to promote positive carbon-isotope shifts. If the sedimented organic matter is not removed from the zones of bacterial oxidation or sulfate reduction, recycling of the $^{13}C$-depleted $CO_2$ can occur, effectively minimizing or eliminating the possibilitiy of maintaining a positive carbon isotope shift over geologically long periods of time. By analogy, therefore, we interpret the long Aptian carbon isotope shift to be a result of increased production and rapid burial of organic matter which was facilitated by anoxia.

According to our lithological data the deepest part of the Tethys remained poorly oxygenated throughout the Aptian. Periodically, the Aptian anoxic deep water extended into the continental margin basins, causing the formation of cyclic organic carbon-rich deposits [Arthur and Premoli-Silva, 1982]. The lithological cycles seem to record changes in regional hydrography, while the C isotope stratigraphy traces major geochemical changes in the oceans. In addition to our proposed changes in bottom water production due to thermohaline circulation, we must also postulate rate changes in surface water productivity in order to explain the positive C isotope anomaly in the Aptian. If periods of increased organic productivity coincided with times of limited deep-water renewal, we would expect a further expansion of the anoxic deep-water layer, increased carbonate dissolution due to high contents of dissolved $CO_2$, and exceptional conditions for the preservation of organic matter as postulated by Dean and Gardner [1982]. During times of intensified deep-water production, only the deepest basins of the Tethys and Atlantic oceans would remain as sinks for organic matter [e.g., Waples, 1983], while along the continental margins, foraminiferal marls and siliceous shales would be deposited.

The organic carbon-rich sediments, with Corg contents of up to 15%, of the Aptian Tethys-Atlantic oceans were preferentially formed in deep, restricted basins. In the Pacific Ocean an expanded oxygen-minimum layer caused the formation of organic carbon-rich shales on submarine plateaus, according to a model by Schlanger and Jenkyns [1976]. They, too, relate the expansion of the oxygen-minimum layer to increased productivity in Aptian oceans.

The Aptian, we conclude, was a time of both widespread anoxia and increased organic productivity. The Aptian also is recognized as a time of high global sea level, rapid formation of new oceanic crust [Le Pichon and Huchon, 1984], and periodically fluctuating climate belts [Hochuli, 1981]. It remains a question how these changes interact with the postulated oceanic changes and if we will find tectonic or climatic arguments for the sudden end of the positive C isotope excursion somewhere in the Albian.

## Summary

Organic carbon-rich sediments periodically were deposited in various bathymetric and paleogeographic settings of all the major Early Cretaceous oceans. Along a section through the central Tethys and its southern continental margin in the Swiss and Italian Alps we traced the evolution of the Early Cretaceous Tethys from an oxygenated ocean to an ocean marked by periodic oxygen depletion of the bottom water. Restricted renewal of the deep water first occurred in the deepest realms of the Tethys, depleting bottom water of oxygen. Dark siliceous shales were deposited near the Early Cretaceous CCD. During the Barremian and Aptian the poorly oxygenated deep-water mass periodically extended onto the southern continental margin, where, in deep basinal environments, organic carbon-rich marls were formed. Submarine highs and shallower basins, more proximal to the continental margin, remained virtually unaffected by the anoxic water masses. Apparently, the area covered by the anoxic bottom water mass was growing throughout the Early Cretaceous.

In the Barremian the organic carbon-rich marls in the deep continental margin basins alternate with pelagic limestones in cycles of about $10^5$ years duration, reflecting periodic changes in bottom water circulation along the African continental margin. Superimposed on these lithologic cycles are fluctuations in the C isotope record in intervals of up to $10^6$ years. These positive $^{13}$C events generally coincide with the deposition of organic carbon-rich sediments. Since these fluctuations are recorded both in the deep basins and on submarine highs, we regard them to be a result of ocean-wide fluctuations in surface water chemistry. Further, we propose that a large, positive $\delta^{13}$C excursion of up to +4°/oo in the Aptian reflects extremely widespread anoxic conditions combined with a high global surface water productivity and the burial of large amounts of organic matter.

Acknowledgements. We thank D. Bernoulli (University of Basel) and two anonymous reviewers for their stimulating criticisms of an earlier draft of this paper. This study was supported by the Swiss National Science Foundation (grant 2.855-0.80) and by NSF (grant EAR 8212535). This is contribution no.249 of the Laboratory for Experimental Geology, ETH, Zürich.

## References

Arthur, M.A., and A.G. Fischer, Upper Cretaceous-Paleocene magnetic stratigraphy at Gubbio, Italy, I, Lithostratigraphy and sedimentology, Geol. Soc. Am. Bull., 88, 367-389, 1977.

Arthur, M.A., and J.H. Natland, Carbonaceous sediments in the North and South Atlantic: The role of salinity in stable stratification of Early Cretaceous basins, in Deep Drilling Results in the Atlantic Ocean: Continental Margins and Paleoenvironment, Maurice Ewing Ser., vol.3, edited by M. Talwani, W. Hay, and W.B.F. Ryan, pp.375-401, AGU, Washington, D.C., 1979

Arthur, M.A., and I. Premoli-Silva, Development of widespread organic carbon-rich strata in the Mediterranean Tethys, in Nature and Origin of Cretaceous Carbon-Rich Facies, edited by S. O. Schlanger and M.B. Cita, pp.7-54, Academic Press, New York, 1982.

Barron, E., A warm, equable Cretaceous: The nature of the problem, Earth Sci. Rev., 19, 305-338, 1983.

Berner, R.A., A.C. Lasaga, and R.G. Garrels, The carbonate-silicate geochemical cycle and its effect on atmospheric carbon dioxide over the past 100 million years, Am. J. Sci., 283, 641-683, 1983.

Bernoulli, D., North Atlantic and Mediterranean Mesozoic facies: A comparison, Initial Rep. Deep Sea Drill. Proj., 11, 801-822, 1972.

Bernoulli, D., and H.C. Jenkyns, Alpine, Mediterranean and central Atlantic Mesozoic facies in relation to the early evolution of the Tethys, Soc. Econ. Paleontol. Mineral. Spec. Publ., 19, 129-160, 1974.

Bitterli, P., Bituminous intercalations in the Cretaceous of the Breggia River, S. Switzerland, Bull. Ver. Schweiz. Pet. Geol. Ing., 31, 179-185, 1965.

Bosellini, A., and E. Winterer, Pelagic limestone and radiolarite: A genetic model, Geology, 3, 279-282, 1975.

Calvert, S.A., Oceanographic controls on the accumulation of organic matter in marine sediments, in Marine Petroleum Source Rocks, edited by J. Brooks and A. Fleet, pp. xx-xx, Blackwell Scientific Publications, London, in press, 1984.

Cita, M.B., Ricerche micropaleontologiche e stratigrafiche sui sedimenti pelagici del Giurassico superiore e del Cretacico inferiore nella catena del M. Baldo, Mem. Riv. Ital. Paleontol. Stratigr., 10, 184 pp., 1964.

Channell, J.E.T., W. Lowrie, and F. Medizza, Middle and Early Cretaceous magnetic stratigraphy from the Cismon section, northern Italy, Earth Planet. Sci. Lett., 42, 153-166, 1979.

Coplen, T.B., and S.O. Schlanger, Oxygen and carbon isotope studies of carbonate sediments from Site 167, Magellan Rise, Leg 17, Initial Rep. Deep Sea Drill, Proj., 17, 505-509, 1973.

Dean, W.E., and J.V. Gardner, Origin and geochemistry of redox cycles of Jurassic to Eocene age, Cap Verde Basin (DSDP Site 367), continental margin of Northwest Africa, in Nature and Origin of Cretaceous Carbon-Rich Facies, edited by S.O. Schlanger and M.B. Cita, pp. 55-78, Academic Press, New York, 1982.

de Boer, P.L., Some remarks about stable isotope composition of cyclic pelagic sediments from the Cretaceous in the Appenines (Italy), in Nature and Origin of Cretaceous Carbon-Rich Facies, edited by S.O. Schlanger and M.B. Cita, pp. 129-143, Academic Press, New York, 1982.

Irwin, H., Early diagenetic carbonate precipitation and pore fluid migration in the Kimmeridge clay of Dorset, England, Sedimentology, 27, 577-591, 1980.

Irwin, H., C.D. Curtis, and M. Coleman, Isotopic evidence for source of diagenetic carbonate formed during burial of organic-rich sediments, Nature, 269, 209-213, 1977.

Jenkyns, H.C., Cretaceous anoxic events: From continents to oceans, J. Geol. Soc. London, 137, 137-188, 1980.

Jenkyns, H.C., and E.L. Winterer, Paleoceanography of Mesozoic ribbon radiolarites, Earth Planet. Sci. Lett., 60, 351-375, 1982.

Kroopnick, P., The dissolved $O_2$-$CO_2$-$^{13}$C system

in the eastern equatorial Pacific, Deep Sea
Res., 21, 211-227, 1974.

Lacombe, H., and P. Tchernia, Caractères hydro-
logiques et circulation des eaux de Mediter-
ranée, in The Mediterranean Sea, A Natural
Sedimentation Laboratory, edited by D.J.
Stanley, pp. 25-36, Dowden, Hutchinson and
Ross, Stroudsburg, Pa., 1972.

LePichon, X., and P. Huchon, Geoid, Pangea and
convection, Earth Planet. Sci., Lett., 64,
123-135, 1984.

Matter, A., R.G. Douglas, and K. Perch-Nielsen,
Fossil preservation, geochemistry and diagenesis of pelagic carbonates from Shatsky Rise,
northwest Pacific, Intitial Rep. Deep Sea
Drill. Proj., 32, 891-921, 1975.

McCrea, J.M., The isotopic chemistry of carbonates and a paleotemperature scale, J. Chem.
Phys., 18, 849-857, 1950.

McKenzie, J.A., Carbon-13 cycle in Lake Greifen,
a model for restricted ocean basin, in Nature
and Origin of Cretaceous Carbon-Rich Facies,
edited by S.O. Schlanger and M.B. Cita, pp.
197-207, Academic Press, New York, 1982.

McKenzie, J.A., Carbon isotopes and productivity
in the lacustrine and marine environments, in
Chemical Processes in Lakes, edited by W.
Stumm, John Wiley, New York, in press, 1985.

McKenzie, J.A., D. Bernoulli, and R.E. Garrison,
Lithification of pelagic-hemipelagic sediments
at DSDP Site 372: Oxygen isotope alteration
with diagenesis, Initial Rep. Deep Sea Drill.
Proj., 42A, 473-478, 1978.

Premoli-Silva, I., and H. Luterbacher, Kreide
und Pliozän der Umgebung von Balerna (Süd-
Tessin), Bull. Ver. Schweiz. Pet. Geol. Ing.,
31, 160-178, 1965.

Rieber, H., Eine Ammonitenfauna aus der oberen
Maiolica der Breggia-Schlucht (Tessin/Schweiz),
Eclogae, Geol. Helv., 70, 777-787, 1977.

Roth, P.H., Cretaceous paleoceanography: Calcareous nannofossil evidence (abstract), in
First International Conference on Paleoceanography, ETH, Zürich, p.50, 1983.

Ryan, W.B.F., and M.B. Cita, Ignorance concerning episodes of oceanwide stagnation, Mar.
Geol. 23, 197-215, 1977.

Schlanger, S.O., and R.G. Douglas, The pelagic
ooze-chalk-limestone transition and its implication for marine stratigraphy, Int. Assoc.

Sedimentol. Spec. Publ., 1, 117-148, 1974.

Schlanger, S.O., and H.C. Jenkyns, Cretaceous
oceanic anoxic events: Causes and consequences,
Geol. Mijnbouw, 55, 179-184, 1976.

Scholle, P.A., and M.A. Arthur, Carbon isotopic
fluctuations in pelagic limestones: Potential
stratigraphic and petroleum exploration tool,
Am. Assoc. Pet. Geol. Bull., 64, 67-87, 1980.

Smith, A.G., R.M. Hurley, and J.C. Briden,
Phanerozoic continental world maps, Cambridge
Earth Sci. Ser., 102 pp., Cambridge University
Press, New York, 1981.

Southam, J.R., W.H. Peterson, and G.W. Brass,
Dynamics of anoxia, Palaeogeogr. Palaeocli-
matol.Palaeocol., 40, 183-198, 1982.

Thiede, J., and T.H. van Andel, The paleoenvironment of anaerobic sediments in the late Mesozoic South Atlantic Ocean, Earth Planet. Sci.
Lett., 33, 301-309, 1977.

Thierstein, H., and T.J. Bralower, Estimates of
surface water fertility in mid-Cretaceous
Oceans (abstract), in First International
Conference on Paleoceanography, Zürich, p.62,
1983.

Thierstein, H., and W.H. Berger, Injection
events in ocean history, Nature, 276, 461-466,
1978.

Trümpy, R., Penninic-Austroalpine boundary in
the Swiss Alps: A presumed former continental
margin and its problems, Am. J. Sci., 275-A,
209-238, 1975.

Waples, D.W., Reappraisal of anoxia and organic
richness with emphasis on Cretaceous North
Atlantic, Am. Assoc. Pet., Geol. Bull., 67,
963-978, 1983.

Weissert, H., Die Paläozeanographie der südwest-
lichen Tethys in der Unterkreide, Mitt. Geol.
Inst. Eidg. Tech. Hochsch. Univ. Zürich, 226,
174 pp., 1979.

Weissert, H., The environment of deposition of
black shales in the Early Cretaceous: An ongoing controversy, Soc. Econ., Paleontol.
Mineral. Spec. Publ., 32, 547-560, 1981.

Weissert, H., J.A. McKenzie, and P. Hochuli,
Cyclic anoxic events in the Early Cretaceous
Tethys Ocean, Geology, 7, 147-151, 1979.

Winterer, E.L., and A. Bosellini, Subsidence
and sedimentation on Jurassic passive continental margin, Southern Alps, Italy, Am.
Assoc. Pet. Geol. Bull., 65, 394-421, 1981.

## WARM CRETACEOUS CLIMATES: HIGH ATMOSPHERIC CO$_2$ AS A PLAUSIBLE MECHANISM

Eric J. Barron and Warren M. Washington

National Center for Atmospheric Research, Boulder, Colorado 80303

Abstract. Sensitivity experiments with a general circulation model of the atmosphere coupled to a simple ocean model are the basis for an investigation of whether changing geography is a sufficient mechanism to explain warm Cretaceous (≈100 Ma) climates or whether other mechanisms, such as a higher atmospheric CO$_2$ concentration, are required. Although Cretaceous geography results in a substantial warming in comparison with the present day, the warming is insufficient to explain the geologic data. Several lines of evidence suggest that an estimated two to tenfold increase in CO$_2$ with respect to present values is a plausible explanation of this problem. Higher values of CO$_2$ result in additional climate problems. These model experiments have implications for geochemical models with climate-dependent weathering rates.

## Introduction

Glacial climates, such as the present-day climate, are representative of the cold end of the recorded range of paleoclimates [e.g., Frakes, 1979]. In contrast, the climate of the Cretaceous period, with globally averaged surface temperatures 6°-14°C higher than at present [Barron, 1983], is representative of the opposite end of the spectrum. The explanation of this variability in the earth's climatic state is problematic. Since the formulation of continental drift and plate tectonics, changing geography has become the most frequently cited explanation of the contrast between the Cretaceous warmth and the present glacial climate [e.g., Cox, 1968; Irving and Robertson, 1968; Kemp, 1972; Donn and Shaw, 1977]. Other forcing factors have received much less attention. However, recently, the importance of variable atmospheric carbon dioxide as proposed by Chamberlin [1899], Budyko and Ronov [1979], and Fischer [1982] has received credible quantitative support from the geochemical model of Berner et al. [1983]. It is now plausible that both geography and atmospheric CO$_2$ were important climatic forcing factors during Phanerozoic earth history.

Climate model sensitivity studies offer an important perspective in examining the potential of various forcing factors as explanations of past warm climates. The following questions can be considered in this manner: (1) Is changing geography a sufficient mechanism to explain Cretaceous climates, or is higher atmospheric CO$_2$ required? (2) If higher atmospheric CO$_2$ is required, what level of CO$_2$ in the Cretaceous atmosphere is estimated from climate sensitivity studies? (3) Since continental weathering rates are climate dependent, what are the implications of climate sensitivity studies for geochemical models? These questions are addressed here by performing sensitivity studies with a general circulation model (GCM) of the atmosphere coupled to a simple ocean model and by considering the results of previous model simulations.

The climate model studies suggest that Cretaceous geography resulted in a substantial warming compared to present geography but that this warming is insufficient to explain fully the paleoclimatic data. An estimated two to tenfold increase in atmospheric CO$_2$ would be required to achieve a more "realistic" Cretaceous simulation. Finally, the model results have implications for geochemical models of CO$_2$ evolution with temperature- and precipitation-dependent weathering rates. The globally averaged temperature and precipitation show fairly consistent relationships with respect to changes in CO$_2$ level regardless of either present-day or Cretaceous geography. However, there are latitudinal, geographic, and intermodel variations which indicate that global changes in climate variables due to higher CO$_2$ will be of limited value for inclusion in geochemical models.

## Climate Model Description

The model used in this study is a version of the National Center for Atmospheric Research Community Climate Model (CCM) which evolved from the spectral model of Bourke et al. [1977] and McAvaney et al. [1978]. The model uses a sigma vertical coordinate system with nine levels and the spectral transform method for computing nonlinear horizontal transport terms. The model has an associated grid of 40 latitudes and 48 longitudes over the entire globe.

The major modification to the model is a new radiation-cloudiness parameterization described by Ramanathan et al. [1983]. The absorptance formulation for $CO_2$ is in excellent agreement with observations [Kiehl and Ramanathan, 1983]. The version of the model used here has a predictive cloud parameterization based on model-derived relative humidity and static stability. Model-derived precipitation and evaporation rates are used to simulate the change in soil moisture and snow cover, following Washington and Williamson [1977]. The formation and decay of sea ice is dependent on ocean surface temperatures. Land surfaces have an albedo of 0.13 unless a desert albedo of 0.25 is specified. If the surface becomes snow covered, the albedo for short solar wavelengths is 0.80 and for long solar wavelengths is 0.55.

The simulations described here employ an energy balance ocean formulation which does not include heat storage, transport, or diffusion. Experiments with this type of ocean formulation are usually restricted to simulations with annual average insolation. The major advantage of the model is that it reaches equilibrium relatively quickly; however, the deficiencies suggest that it should be used primarily to examine first-order effects. Comparisons of the model climatology with present-day boundary conditions and present observations can be found in work by Washington and Meehl [1983]. The model satisfactorily simulates many climatic characteristics, including the observed annual mean temperature structure and the zonal wind distribution. Ramanathan et al. [1983] have also made detailed comparisons for present-day predicted and observed clouds.

## Warm Cretaceous Climate:
## Is Geography the Solution?

Estimates of mean annual surface temperatures with respect to latitude for the mid-Cretaceous have been derived by Barron [1983] based on a variety of data. Cretaceous temperatures can be determined only within the two broad limits reproduced in Figure 1. Mean annual polar surface temperatures in the northern hemisphere are estimated to be on the order of 15°-25°C warmer, while equatorial surface temperatures were similar or slightly warmer (5°C) than at present. In addition to the constraints on mean annual surface temperatures, the distribution of tropical paleofloras [e.g., Vakhrameev, 1975] suggests that continental interiors at latitudes of at least 45°N did not experience subfreezing winters. Although the limits are fairly broad, these factors constrain the degree of warming which must be explained by changes in geography or other factors.

Based solely on correlations between geographic variables and paleoclimate data, plate tectonic processes appear to be a logical explanation of the cooling between the Cretaceous and

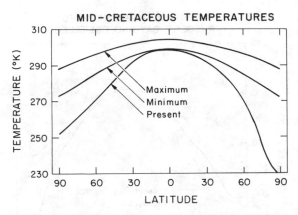

Fig. 1. Present-day zonally averaged surface temperatures (degrees Kelvin) with respect to latitude in comparison with temperature limits for the Cretaceous period derived from geologic data.

the present. Two aspects of geography are frequently cited. First, the degree of high-latitude continentality is of major importance [Cox, 1968; Frakes and Kemp, 1972; Donn and Shaw, 1977]. High-latitude land serves as a surface for the accumulation of high-albedo snow and also limits poleward heat transport by the ocean. Second, one of the best correlations between paleotemperatures and geography is with global land area or global sea level. Barron et al. [1980] have suggested that this mechanism follows from simple surface albedo arguments.

The model results of Barron and Washington [1984] confirm that geography is an important mechanism of climatic change on long time scales. In the CCM experiment, Cretaceous geography resulted in a 4.8° warming in comparison with a present-day control. As can be seen in Figure 2, the warming was relatively symmetric with respect to latitude, and the polar regions were more sensitive than equatorial regions. These results are in substantial agreement with the paleoclimatic data; however, surface temperatures are too cool in comparison with the geologic record everywhere but in the tropics.

Barron and Washington [1984] performed additional sensitivity experiments to determine which geographic variables (continental positions, topography, ice cover, and sea level) were responsible for the model warming. Interestingly, continental positions explained almost the entire northern hemisphere model warming, while topography and absence of ice cover on Antarctica (both ice related) explained almost the entire southern hemisphere model warming. There was no model sensitivity to sea level variations, because changes in cloud cover and evaporation compensated for the changes in surface albedo in the subtropics, where the greatest continental flooding occurred. There are two aspects of the simulations which are problematic. First, the total warming due to geography is insufficient,

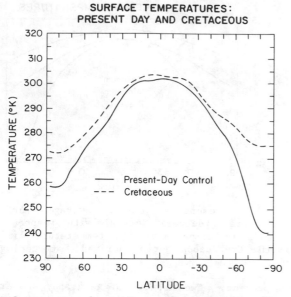

**SURFACE TEMPERATURES:
PRESENT DAY AND CRETACEOUS**

Fig. 2. A comparison of zonally averaged surface temperatures (degrees Kelvin) with respect to latitude for a present-day control simulation and a Cretaceous geography sensitivity experiment.

and second, although the geologic record shows a good correlation between sea level and climate, the model indicated no sensitivity of climate to sea level.

There are three plausible explanations of these two discrepancies. First, the paleoclimatic data may be overinterpreted. Second, the model sensitivity may be inappropriate because a feedback process has not been included or a parameterization may be in error. Finally, another forcing factor in addition to geography may be required. The Cretaceous temperature limits are rather broad, and since most evidence seems to support a warmer estimate rather than the cooler estimate, an error in interpretation of the paleoclimatic data seems the least probable of the three explanations. Before considering alternative mechanisms, several drawbacks to the GCM model must be evaluated.

There are three major limitations to the mean annual energy balance ocean version of the CCM. The first limitation is the lack of ocean heat transport and storage, the second is the lack of a seasonal cycle, and the third is the accuracy of the interactive cloud parameterization. The importance of these limitations has not been examined explicitly; however, additional sensitivity experiments suggest that these limitations are insufficient to solve the discrepancy between the Cretaceous model simulation and the Cretaceous paleoclimatic data.

Neglecting ocean heat transport in a climate model has the potential to be a serious omission. GCM models with energy balance oceans typically account for an "implied" ocean heat flux by tuning some albedo parameter (e.g., surface or cloud albedos) in a present simulation (see Spelman and Manabe [1984] for some discussion). This modification may not result in the correct model sensitivity to a specific perturbation. However, two factors suggest that the addition of ocean heat transport will not solve the discrepancy between the Cretaceous model simulation and the paleoclimatic data.

First, Spelman and Manabe [1984] show in a similar model that the sensitivity is strongly dependent on the area-averaged surface air temperature (due to the important role of ice albedo feedback). If the surface air temperatures in the present control simulation are reasonable, as is the case for the CCM [Washington and Meehl, 1983], the sensitivity of the model may not be altered by the addition of an explicit oceanic heat transport formulation.

Second, Barron and Washington [1982] performed Cretaceous January simulations with ocean heat transport implied by adjusting sea surface temperatures. Even with an implied enormous ocean heat transport, continental interior temperatures were not maintained above freezing. Additional CCM experiments with extreme limits of oceanic heat transport including implied instantaneous oceanic mixing also indicate that continental interior temperatures are actually colder for cases of very large ocean heat transport. The increased ocean heat transport is compensated by a decreased atmospheric heat transport, and the advection of warm air into continental interiors decreases as the intensity of atmospheric motions decreases. These two factors, that addition of an explicit ocean heat transport parameterization will not necessarily alter the model sensitivity and that addition of extreme implied ocean heat transport may even exacerbate the problem of maintaining continental interiors above freezing, suggest that the ocean heat transport deficiency of the model is not a complete solution to the problem.

The lack of a seasonal cycle is also a limitation. There is evidence from CCM simulations for differences in sensitivity between mean annual and seasonal versions of the model. Washington and Meehl [1984] also suggest that differences in area-weighted surface air temperature may be a major component of the explanation of these differences in sensitivity. More important for the arguments here, the limits to atmospheric advection of warm air into continental interiors and the small thermal intertia of the continents will prevent large land masses which experience low winter light conditions from maintaining above freezing conditions. Certainly, seasonal models have bearing on whether permanent snow or ice may accumulate, but the addition of a seasonal cycle also seems unlikely to result in substantially warmer mean conditions.

The cloudiness parameterization in the model is more difficult to evaluate. Present-day predicted and observed clouds compare favorably [Ramanathan et al., 1983], but beyond using the present-day seasonal cycle as a surrogate climate change the cloud-climate sensitivity cannot be evaluated.

In addition to the specific limitations of the CCM described above there is a more general question about the validity of climate sensitivity estimates derived from models [Dickinson, 1984]. Estimates from different models vary widely, and there is substantial uncertainty concerning the explanation of the variations. Based on experiments with other climatic forcing factors the sensitivity of the CCM is less than that of many other GCM's due to a smaller ice-albedo feedback. One explanation (noted above) is that the CCM is better tuned; in other words, the area-averaged surface air temperature in the CCM is warmer and is a closer representation of modern observations than are many other models. Consequently, the ice-albedo feedback will be smaller. However, until appropriate experiments are completed, we cannot be sure that this is a full explanation.

The conclusion from the series of sensitivity experiments and inferences described above is that it is more likely that the discrepancy between the Cretaceous model simulation and the paleoclimatic data is due to an additional forcing factor rather than overinterpretation of the geologic record or some limitation of the climate model. Both of these aspects limit our ability to determine the magnitude of the additional forcing but apparently not the necessity of considering an additional mechanism.

### Atmospheric $CO_2$ as a Logical Alternative

Although variations in atmospheric carbon dioxide through geologic time have been proposed for a variety of reasons [Chamberlin, 1899; Budyko and Ronov, 1979; Fischer, 1982], the quantitative calculations of Berner et al. [1983] based on the carbonate-silicate geochemical cycle lend substantial credibility to this proposal. Essentially, the Berner et al. [1983] model suggests that the most important influence on atmospheric $CO_2$ was tectonics, specifically the rate of seafloor spreading. Rapid seafloor spreading, characteristic of the mid-Cretaceous, results in a greater $CO_2$ flux to the atmosphere through volcanic activity and also results in a higher global sea level by altering the ocean basin volume [Pitman, 1978]. Weathering of silicate minerals, which is largely dependent on continental area, apparently cannot compensate for the increased $CO_2$ flux to the atmosphere.

Although Berner et al. [1983] point out a number of assumptions and limitations to the geochemical model, their results would explain the correlation between climatic warmth and sea level variations and the lack of sensitivity to sea level variations in the climate simulations. Because a portion of the heat loss from the earth's surface would be reradiated back to the surface, higher atmospheric $CO_2$ may also partially solve the problem of cold continental interior temperatures in the January climate simulations. For these reasons, a higher atmospheric $CO_2$ content, combined with geography, is a logical mechanism to explain the warmth of the Cretaceous period.

The next question, and an additional constraint in estimating the natural variation of carbon dioxide during earth history, is the level of atmospheric $CO_2$ required to explain Cretaceous climates based on climate model sensitivity studies.

### Cretaceous $CO_2$ Climate Sensitivity

Most estimates of the increase in globally averaged surface temperature for a quadrupling of atmospheric $CO_2$ range from approximately $3°-4°K$ [e.g., Augustsson and Ramanathan, 1977; Manabe and Wetherald, 1980; Manabe and Stouffer, 1980; Washington and Meehl, 1983]. However, some models [e.g., Hansen et al., 1984] have a substantially greater sensitivity implied by $CO_2$ doubling experiments. The differences in sensitivity between different models has been difficult to evaluate partly due to the very complexity of the models. At least one major factor, as mentioned earlier, is the results of Spelman and Manabe [1984], which indicate relatively large differences in sensitivity due to differences in area-weighted surface temperatures. Washington and Meehl explain much of the difference between an annually averaged solar forcing experiment [Washington and Meehl, 1983] and annual solar cycle experiment [Washington and Meehl, 1984] as due to this factor. In the control simulations the annual average experiment has a globally averaged surface temperature which is more than $1°$ higher than the seasonal experiment. The seasonal and annual experiments had $3.5°$ and $1.3°$ sensitivities, respectively, for a twofold increase in $CO_2$. Including the seasonal cycle must also explain part of this difference.

The Cretaceous geography climate simulation has a globally averaged surface temperature which is $1.2°-8.2°K$ lower than the limits determined from the paleoclimatic data [Barron, 1983]. The range in $CO_2$ required in addition to geography to satisfy the Cretaceous temperature limits can be estimated to range from a doubling to a tenfold increase.

However, there are two factors which might suggest that the climate sensitivity derived from previous model experiments may be inappropriate for the Cretaceous earth. First, the Cretaceous geography resulted in a substantially warmer model simulation than the present geography. The arguments described previously suggest that the climate sensitivity on a warmer planet with a reduced importance of ice albedo feedback due to less ice and snow will have a reduced sensitivity to $CO_2$ variations. Second, if the $CO_2$ sensitivity

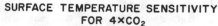
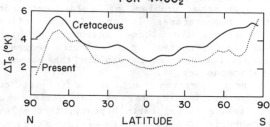

SURFACE TEMPERATURE SENSITIVITY
FOR 4×$CO_2$

Fig. 3.  A comparison of zonally averaged surface
temperature (degrees Kelvin) sensitivity for a
quadrupling of $CO_2$ for present and Cretaceous
geography.

is land-sea dependent, Cretaceous sensitivity may
be different from the present sensitivity.
Consequently, two additional climate model sensitivity studies were performed with the CCM by
quadrupling the atmospheric $CO_2$ in both the
present-day control and Cretaceous geography
experiments.  A quadrupling is of the right order
to place the Cretaceous globally averaged surface
temperature well within the two limits derived
from the paleoclimatic data.

The model sensitivity for the Cretaceous quadrupling of atmospheric $CO_2$ is actually greater
than in the present-day control (3.6° compared to
2.7°K).  Two factors contribute to this difference.  First, in the ice albedo parameterization
the albedos are not allowed to change until all
snow or ice disappears at a particular grid
point.  Ice albedo feedback is still of some
importance in the Cretaceous simulation because
polar temperatures are close to the temperature/
snow limits.  The ice albedo feedback in the
present simulation may also be somewhat
restricted because permanent ice is specified on
Antarctica and Greenland.  Second, water vapor
climate feedback is of greater importance in the
Cretaceous experiment than in the present-day
experiment.  This occurs because of the strong
nonlinear relationship between surface temperature and saturation vapor pressure.  Both of
these factors are reflected in greater surface
temperature sensitivity in the Cretaceous simulation than in the present-day simulation at almost
every latitude (Figure 3).

The conclusions from Spelman and Manabe [1984]
and Washington and Meehl [1984] suggest that the
area-weighted surface temperature in the control
has a major influence on model sensitivity due to
the importance of ice albedo feedback.  A colder
planet with more ice and snow at lower latitudes
has a greater sensitivity than a warmer planet.
The Cretaceous experiments suggest a refinement
of this interpretation due to the nature of the
ice albedo parameterization in the CCM.  The
Cretaceous experiments are evidence that there
may be a sensitivity threshold in the CCM.  With
warmer temperatures than at present but polar

temperatures near temperature/snow limits, the
combination of water vapor climate feedbacks and
ice albedo feedbacks may result in greater sensitivity than for some cooler temperatures or
warmer temperatures in which ice albedo feedback
does not play an important role.

A comparison of the Cretaceous paleotemperature limits with the model-derived zonally averaged surface temperatures from the geography
experiment and the geography plus $CO_2$ quadrupling
experiment is given in Figure 4.  The zonally
averaged surface temperatures for the $CO_2$ experiment are close to or within the limits of the
paleoclimatic data.  However, the model-derived
temperatures are at the upper limit for the
tropics and the lower limit for polar regions.
This result is not necessarily desirable and
introduces an additional climate problem.  If we
were to increase the $CO_2$ concentration above a
quadrupling in order to increase polar temperatures, it is likely that the tropical temperature
limit derived from oxygen isotope paleotemperatures (see Barron [1983] for a review) would be
exceeded.  For $CO_2$ increases on the order of 8 to
10 times the present value it is conceivable that
the temperature tolerances of some tropical
organisms may be exceeded; for instance, Vaughn
and Wells [1943] estimated the maximum temperature tolerances of corals to be approximately
308°-310°K.

In the Cretaceous $CO_2$ experiment there is a
substantial increase in poleward latent heat flux
which does not quite compensate for the decreased
sensible heat flux due to the reduced equator-to-
pole temperature gradients.  Mechanisms which
would maintain the total poleward heat flux would
certainly reduce the problem of overheating the
tropics with greater $CO_2$ concentration.  Whereas
ocean heat transport could not solve the temperature discrepancies for the Cretaceous geography
experiment, it is plausible that it could reduce

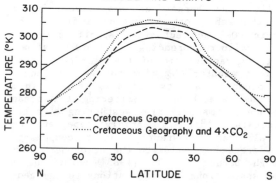

COMPARISON OF CRETACEOUS TEMPERATURES:
MODEL AND LIMITS

Fig. 4.  Cretaceous zonally averaged surface
temperature limits (degrees Kelvin) in comparison
with Cretaceous model-derived surface temperatures for the geography and geography plus $CO_2$
quadrupling experiments.

this tropical temperature problem. Additionally, the climate simulations lack equatorial upwelling, which would reduce tropical temperatures, perhaps by a few degrees. However, this would not have an important effect at latitudes away from the equator, and the problem (Figure 4) exists up to 30° in latitude; nor would it be important in all tropical environments. Even if $CO_2$ levels greater than 4 times the present value could be rationalized based on the limitations described above, $CO_2$ amounts greater than 10 times the present value are still not reasonable for the last 100 million years given the temperature tolerances of organisms, tropical isotopic paleotemperatures, and the model sensitivity to $CO_2$ variations.

## Implications for Geochemical Models

Changes in weathering fluxes due to increased $CO_2$ are climate dependent in the geochemical model of Berner et al. [1983]. Both temperature and runoff (parameterized as a function of temperature) are important variables in determining the rate at which $CO_2$ is removed from the atmosphere by weathering silicate rocks. Increased $CO_2$ serves to increase both surface temperatures and the intensity of the hydrologic cycle. The question is whether different paleogeographies result in substantially different sensitivities in these climatic variables. The CCM experiments are the only simulations to date which have bearing on this problem.

In comparison with the present-day control the CCM experiment with Cretaceous geography has a somewhat greater sensitivity in surface temperature. This was interpreted as the result of temperature water vapor feedback and the fact that ice albedo feedback was still important despite the warmer temperatures because of the albedo parameterization. Without additional

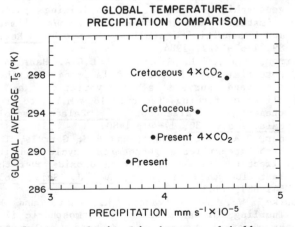

Fig. 5. The relationship between globally averaged surface temperature (degrees Kelvin) and globally averaged precipitation (millimeter per second).

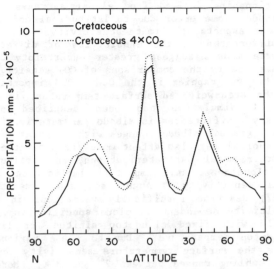

Fig. 6. The latitudinal sensitivity of precipitation rate (millimeter per second) for a quadrupling of $CO_2$ with Cretaceous geography.

experiments it is not possible to evaluate whether this difference in sensitivity has any relationship to the different geographies or whether it is a function only of the characteristics of the surface temperature distributions as has been interpreted here. The only point which is clear is that geochemical models which consider $CO_2$ as the only means of global temperature change are ignoring the important role of geography.

For the CCM sensitivity experiments described here there is a consistent relationship between globally averaged surface temperature and global precipitation, with greater precipitation for higher temperatures (Figure 5). In both the present-day and Cretaceous experiments the increased precipitation is latitudinally dependent, with the increased precipitation occurring within the tropics and poleward of 40° in latitude (Figure 6). Although global averages show consistent simple relationships, for different paleogeographies, continental precipitation can vary greatly. Present-day $CO_2$ sensitivity experiments with the CCM indicate a net drying of the continents for a quadrupling of $CO_2$ and runoff declines with increased $CO_2$ and increased global precipitation. In contrast, global runoff rates in the Cretaceous simulations are proportional to the increased global precipitation. Washington and Meehl [1984] found more soil moisture in higher-precipitation areas with associated runoff but less in regions like the downward branches of the Hadley circulation. It is noteworthy that Cretaceous geography has substantially less land area in the subtropical regions associated with the downgoing branch of the Hadley circulation than does that of the present day.  The results suggest that the application of relationships derived from global averages will introduce some error due to latitudinal and geographic aspects of climate sensitivity.

Differences in the sensitivity of different climate models suggest greater uncertainty than is implied by the comparisons of $CO_2$ sensitivity for two geographies in the CCM.  The importance of the area-weighted surface temperature in the control simulation has been described previously.  Differences in albedo parameterizations (i.e., gradual albedo changes with temperature or thresholds) may also affect model sensitivities.

Other model parameterizations, in particular for cloudiness, have important implications for model sensitivity.  Augustsson and Ramanathan [1977] describe specifically variations in $CO_2$ sensitivity dependent on cloud specifications in models with fixed cloud-top altitude or fixed cloud-top temperature.  Due to these variables alone the surface temperature sensitivity for a $CO_2$ doubling ranged between $2°$ and $3°K$.  Models with more advanced interactive cloud cover based on relative humidity and static stability still evidence differences in sensitivity.  For instance, Hansen et al. [1984] find a large sensitivity in the tropics for warmer temperatures due to decreased total tropical cloudiness but an increase in cirrus-type clouds.  The increase in cirrus-type clouds leads to more emitted downward flux from the upper troposphere, which in turn causes greater tropical ocean temperature sensitivity.  Other model studies, including the work of Washington and Meehl [1984], fail to demonstrate the same effect.

Differences in parameterizations and the complexity of models hamper intercomparisons.  The result is that the relatively large differences in published estimates of $CO_2$-climate sensitivity are not always fully understood.  With the exception of the warming due to geography, discrepancies between models may be a greater source of error for geochemical models than differences due to $CO_2$ sensitivity for different geographies.

## Summary

Although changing geography is a substantial climatic forcing factor, it is insufficient to explain the warmth of the Cretaceous period based on the sensitivity of a general circulation model of the atmosphere.  This point is dependent on the actual global climate sensitivity, which could plausibly be either greater than or less than the model sensitivity derived here.  The total sensitivity of the model to changing geography is an important factor, but additionally, the model lacked sensitivity to global sea level variations despite the strong correlation of sea level with paleoclimatic data.  A variety of additional sensitivity experiments and recent geochemical models suggest that increased atmospheric $CO_2$ during the Cretaceous is a logical

explanation of this discrepancy between the model simulation and the paleoclimatic data.

$CO_2$-climate sensitivity studies suggest that a two to tenfold increase in $CO_2$ relative to the present would be sufficient to explain the above discrepancies.  However, even levels of $CO_2$ greater than a quadrupling of the present values introduce an additional climate problem, overheating of tropical regions.

Finally, globally averaged temperature and precipitation show fairly consistent relationships with respect to changes in $CO_2$ within the limited range of $CO_2$ sensitivity experiments described here and may have application to geochemical models.  However, latitudinal, geographic, and intermodel variations introduce substantial uncertainties.  In addition, the climatic variation due to changes in geography is not negligible in comparison to potential $CO_2$ effects and is therefore a limitation of present geochemical models.

Acknowledgments.  The National Center for Atmospheric Research is sponsored by the National Science Foundation.  The authors gratefully acknowledge comments by Starley Thompson and Robert Dickinson and the programming aid of Lynda Verplank.

## References

Augustsson, T., and V. Ramanathan, A radiative-convective model study of the $CO_2$ climate problem, J. Atmos. Sci., 34, 448-451, 1977.

Barron, E.J., A warm, equable Cretaceous:  The nature of the problem, Earth Sci. Rev., 19, 305-338, 1983.

Barron, E.J., and W.M. Washington, Cretaceous climate:  A comparison of atmospheric simulations with the geologic record, Palaeogeogr. Palaeoclimatol. Palaeoecol., 40, 103-133, 1982.

Barron, E.J., and W.M. Washington, The role of geographic variables in explaining paleoclimates:  Results from Cretaceous climate model sensitivity studies, J. Geophys. Res., 89, 1267-1279, 1984.

Barron, E.J., J.L. Sloan, and C.G.A. Harrison, Potential significance of land-sea distribution and surface albedo variations as a climatic forcing factor; 180 M.Y. to the present, Palaeogeogr. Palaeoclimatol. Palaeoecol., 30, 17-40, 1980.

Berner, R.A., A.C. Lasaga, and R.M. Garrels, The carbonate-silicate geochemical cycle and its effect on atmospheric carbon dioxide over the last 100 million years, Am. J. Sci., 283, 641-683, 1983.

Bourke, W., B. McAvaney, K. Puri, and R. Thurling, Global modeling of atmospheric flow by spectral methods, in Methods in Computational Physics, General Circulation Models of the Atmosphere, vol. 17, edited by J. Chang, pp. 267-324, Academic, New York, 1977.

Budyko, M., and A. Ronov, Chemical evolution of the atmosphere in the Phanerozoic, Geochem. Int., 16, 1-9, 1979.

Chamberlin, T.C., An attempt to frame a working hypothesis of the cause of glacial periods on an atmospheric basis, J. Geol., 7, 545-585, 1899.

Cox, A., Polar wandering, continental drift, and the onset of Quaternary glaciation, in Causes of Climatic Change, Meteorol. Monogr. vol. 8, edited by J.M. Mitchell, pp. 112-125, American Meteorological Society, Boston, Mass., 1968.

Dickinson, R.E., Climate sensitivity, in Smagorinsky Festschrift, Academic, New York, 1984.

Donn, W., and D. Shaw, Model of climate evolution based on continental drift and polar wandering, Geol. Soc. Am. Bull., 88, 390-396, 1977.

Fischer, A.G., Long-term climatic oscillations recorded in stratigraphy, in Climate in Earth History, edited by W.H. Berger and J.C. Crowell, pp. 97-104, National Academy Press, Washington, D.C., 1982.

Frakes, L.A., Climates Throughout Geologic Time, 310 pp., Elsevier, New York, 1979.

Frakes, L.A., and E. Kemp, Influence of continental positions on early Tertiary climates, Nature, 240, 97-100, 1972.

Hansen, J., A. Lacis, D. Rind, G. Russell, P. Stone, I. Fung, R. Ruedy and J. Lerner, Climate sensitivity: Analysis of feedback mechanisms, in Climate Processes and Climate Sensitivity, Geophys. Monogr., vol. 29, edited by J.E. Hansen and T. Takahashi, pp. 130-163, AGU, Washington, D.C., 1984.

Irving, E., and W. Robertson, The distibution of continental crust and its relation to ice ages, in The History of the Earth's Crust, edited by R. Phinney, pp. 168-177, Princeton University Press, Princeton, N.J., 1968.

Kiehl, J.T., and V. Ramanathan, $CO_2$ radiative parameterization used in climate modes: Comparison with narrow band models and with laboratory data, J. Geophys. Res., 88, 5191-5202, 1983.

Manabe, S., and R.J. Stouffer, Sensitivity of a global climate model to an increase of $CO_2$ concentration in the atmosphere, J. Geophys. Res., 85, 5529-5554, 1980.

Manabe, S., and R. Wetherald, On the distribution of climate change resulting from an increase in $CO_2$ content of the atmosphere, J. Atmos. Sci., 37, 99-118, 1980.

McAvaney, B.J., W. Bourke, and K. Puri, A global spectral model for simulation of the general circulation, J. Atmos. Sci., 35, 1557-1582, 1978.

Pitman, W., Relationship between eostasy and stratigraphic sequences of passive margins, Geol. Soc. Am. Bull., 89, 1389-1403, 1978.

Ramanathan, V., E.J. Pitcher, R.C. Malone, and M.L. Blackmon, The response of a spectral general circulation model to refinements in radiative processes, J. Atmos. Sci., 40, 605-630, 1983.

Spelman, M.J., and S. Manabe, Influence of oceanic heat transport upon the sensitivity of a model climate, J. Geophys. Res., 89, 571-586, 1984.

Vakhrameev, V., Main features of phytogeography of the globe in Jurassic and Early Cretaceous time, Paleontol. J., 2, 123-133, 1975 (in Russian).

Vaughn, T., and J. Wells, Revision of the suborders, families and genera of Scleractinia, Geol. Soc. Am., Spec. Pap., 44, 363 pp., 1943.

Washington, W.M., and G.A. Meehl, General circulation model experiments on the climatic effects due to a doubling and quadrupling of carbon dioxide concentration, J. Geophys. Res., 88, 6600-6610, 1983.

Washington, W.M., and G.A. Meehl, Seasonal cycle experiment on the climate sensitivity due to a doubling of $CO_2$ with an atmospheric general circulation model coupled to a simple mixed-layer ocean model, J. Geophys. Res., 89, in press, 1984.

Washington, W.M., and D.L. Williamson, A description of the NCAR global circulation models, in Methods in Computational Physics, General Circulaton Models of the Atmosphere, vol. 17, edited by J. Chang, pp. 111-172, Academic, New York, 1977.

# MID-CRETACEOUS CONTINENTAL SURFACE TEMPERATURES: ARE HIGH CO$_2$ CONCENTRATIONS NEEDED TO SIMULATE ABOVE-FREEZING WINTER CONDITIONS?

Stephen H. Schneider, Starley L. Thompson, and Eric J. Barron

National Center for Atmospheric Research, Boulder, Colorado 80307

Abstract. We have found that mid-Cretaceous paleogeography (for 100 million years ago) and very efficient oceanic mixing (e.g., holding all oceanic surface temperatures fixed at 293°K) together are still insufficient to produce above-freezing (or even near-freezing) surface temperatures in our model in mid- to high-latitude continental interiors in midwinter. For mean annual solar forcing and a no-heat capacity, no-ocean transport "swamp" ocean, Barron and Washington (1984) had obtained high northern latitude land and oceanic surface temperatures in the 270°-280°K range, warmer in the high-latitude midcontinents in the northern hemisphere than the much warmer all-293°K oceanic case presented here. In our all-293°K case, the equator-to-pole ocean surface temperature gradient is very weak, as is the vigor of atmospheric circulation, which, in turn, provides insufficient ocean to land atmospheric heat transport to mitigate midwinter continental radiative cooling. Increasing the atmospheric air temperature gradient from equator to pole can increase atmospheric circulation and moderate somewhat the midcontinental cold temperatures. Nevertheless, all the cases we have considered suggest that midwinter high-latitude freezing is still present, particularly in regions of assumed higher topography.

## Introduction

Understanding the causes of the warm equable climates of the mid-Cretaceous (about 100 million years ago) has been a problem of considerable interest to paleoclimatologists. Although the proposed explanations have included a diverse variety of so-called "external" climatic forcing factors, the effect of a different paleogeography has received the most attention [e.g., Frakes and Kemp, 1972; Donn and Shaw, 1977; Barron et al., 1981; Barron and Washington, 1984]. More recently, the importance of increased levels of CO$_2$ has

also been supported from several research approaches [e.g., Budyko and Ronov, 1979; Berner et al., 1983; Barron and Washington, this volume].

Considerable paleoclimatic data [e.g., Frakes, 1979; Barron, 1983] suggest warm mid-latitude and polar conditions during much of the Cretaceous. For Antarctica these data are restricted to coastal regions, but for both North America-Greenland and Asia, the distribution of tropical-subtropical floras and temperature-sensitive vertebrates suggests that even mid- to high-latitude continental interior localities were above freezing in winter. Temperate floras did exist in the Siberian region above a paleolatitude of 60°N [e.g., Vakhrameev, 1964, 1975]. The evidence for warm winters in mid- to high-latitude continental interiors is an important constraint in explaining the climate of the mid-Cretaceous.

Physical models of the climatic systems are required to separate out quantitatively the relative importance of physical processes which could explain such warm Cretaceous conditions. Barron et al. [1981], using a simple zonal energy balance climate model (for example, see Thompson and Schneider [1979]), showed that paleogeographic changes alone, even with strong positive cloud- and albedo-temperature feedback processes assumed, would probably not be sufficient to prevent midwinter freezing in middle to high latitudes unless radically altered oceanic heat transport, different external forcings (e.g., increased CO$_2$) or both were also operative. Unfortunately, the Barron et al. [1981] zonal energy balance model could not explicitly resolve temperatures over continents and oceans. Nor could it explicitly account for atmospheric heat transport between air over land and air over oceans, the direct process by which oceanic heat transport communicates its influence on land surface temperatures (for a simple treatment, see, for example, the energy balance models of North et al. [1983] and Harvey and Schneider [1985]). Moreover, oceanic heat transport has a

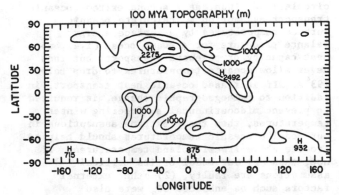

Fig. 1.   100 Ma topography (meters) from Barron and Washington [1982] as used in all our simulations. Note that the higher-elevation areas in paleo-North America, for example, will tend to be centers for subfreezing midcontinental surface temperatures, such as in Figures 2 and 3. However, by lowering topography to sea level (an extreme assumption), Barron and Washington [1984] found that average surface temperatures would still be increased by only about 6°K.

substantial indirect influence on midcontinental surface temperatures. By controlling the geographic distribution of sea surface temperature, ocean heat transport affects the global hydrologic cycle and, particularly, the global atmospheric circulation patterns. (This will be shown below.) It is obvious that knowledge of oceanic heat transport patterns is important in understanding even midcontinental surface temperatures.

Three-dimensional atmospheric circulation model studies which explicitly resolve land and sea also explicitly calculate atmospheric motions [Barron and Washington, 1982, 1984]. Although exhibiting greater sensitivity to paleogeography, such three-dimensional models do confirm the basic inferences of the Barron et al. [1981] energy balance climate model experiments. Barron and Washington [1982], for example, describe experiments with an assumed minimum sea surface temperature of 10°C (an implied large poleward ocean heat transport) and found that mid-latitude continental interior regions, such as Asia, still became cold in winter. This result brings into question whether greater oceanic heat transport alone is a solution to the problem of keeping continental interiors warm and whether other external forcings are required. Here we build on these model studies using a version of the NCAR Community Climate Model (CCM), a spectral general circulation model of the atmosphere [see Washington and Meehl, 1983; Barron and Washington, 1984]. We examine the possible effects of enhanced oceanic heat transports on

midcontinental surface temperatures in the mid-Cretaceous and conclude that enhanced atmospheric forcing (such as increase in $CO_2$ concentrations) may also be necessary to explain the relative warmth of continental interiors during this period.

The Cretaceous paleocontinental positions and topography used [from Barron and Washington, 1984] are shown in Figure 1. Their surface temperature results for a "swamp" ocean [see Washington and Meehl, 1983] with the paleogeography of Figure 1 and for mean annual solar forcing were given in Figures 3 and 4 of Barron and Washington [1984]. The Cretaceous case was 4.8°C warmer in surface temperature than the control; and the poles warmed up by several-fold more. Surface temperatures over land and oceans in the higher northern latitudes were typically between 270° and 280°K, and well below freezing in the Antarctic. However, in this swamp ocean case, no heat is stored or transported by oceans poleward, and it is natural to inquire whether increased poleward transport of heat due to oceanic currents might not alter this result.

Model Experiments

Since we do not use a dynamical oceanic circulation model coupled to our atmospheric circulation model, we instead assume how oceanic dynamics could alter sea surface temperatures. We first make the extreme assumption that mid-Cretaceous oceanic motions were perfectly efficient in transporting heat within the oceans; that is, despite a large latitudinal gradient in solar forcing, we assume a globally uniform oceanic surface temperature fixed at 293°K (20°C), a reasonable warm-world average surface temperature.

When the oceanic surface temperatures are prescribed, rather than explicitly calculated, the total climatic system is not constrained to conserve energy. Thus no claim is made that the ocean surface temperature distribution is realistic. However, the aim of this study is not to produce believable ocean surface temperature distributions, but simply to assume a wide range of possible distributions in order to assess the relative impacts on midcontinental temperatures. To produce believable ocean surface temperatures would require coupling to a believable oceanic model. At best, we will have bracketed the actual ocean surface temperature distributions of 100 Ma; at the least, we will have gained insight into the role of oceanic surface temperature distributions on mid-Cretaceous, midcontinental, winter surface temperatures. Ironically, perhaps, the latter issue is very similar in physical character to an earth perturbed by a transient, thick, high-altitude aerosol, which obscures sunlight and allows substantial land surface cooling relative to

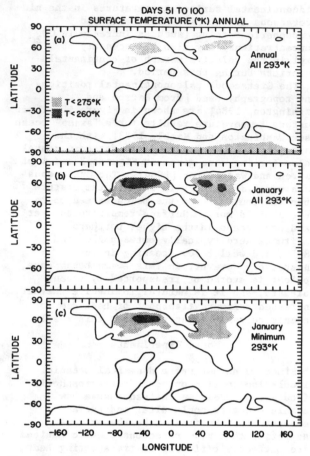

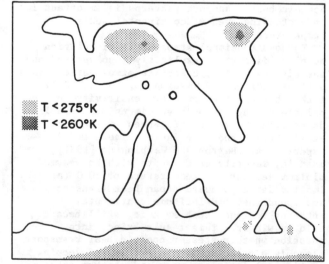

Fig. 2.  Surface temperature for annually averaged solar forcing (Figure 2a) computed by averaging the model-generated weather statistics over a 50-day period.  For all cases, $CO_2$ levels and the latitudinal distribution of solar insolation are based on today's values, and for Figures 2a and 2b all oceanic surface temperatures are held fixed at $293^\circ K$ ($20^\circ C$), equivalent to a superefficient heat-transporting ocean.  Note that subfreezing surface temperatures are computed in areas of mid- to high-latitude zones, even for the mean annual solar conditions.  Surface temperatures for January solar forcing are also shown in Figures 2b and 2c.  However, in Figure 2c, ocean temperatures are first calculated by a "swamp" ocean model [see Washington and Meehl, 1983], but are then never allowed to drop below $293^\circ K$.

warmer surrounding oceanic areas (for example, see Alvarez et al. [1981], Turco et al. [1983], or Covey et al. [1984]).
    The reduction in horizontal surface temperature contrasts resulting from the all-$293^\circ K$ ocean surface temperature assumption will drastically reduce the vigor of the atmospheric

circulation.  Thus, as a second extreme oceanic transport assumption we calculate oceanic surface temperatures by a surface energy balance procedure for a swamp ocean (i.e., no heat capacity or poleward transport) but then never allow oceanic temperatures to drop below $293^\circ K$.  If increased oceanic heat transport, in addition to paleogeographic change, is required to prevent midcontinental subfreezing winter temperatures, then these extreme assumptions in our model's oceanic temperatures should help to produce above-freezing land temperatures.  If not, then (1) our model or its critical assumptions are faulty, (2) other external factors such as enhanced $CO_2$ were also operative, or (3) subfreezing temperatures in midwinter, midcontinent areas did occur at 100 Ma.

Results

    Figure 2a shows the average surface temperature result assuming  mean annual latitudinal distribution of solar radiative forcing and all oceanic temperatures held fixed at $293^\circ K$.  In continental interiors in middle to high latitudes and (especially) in higher topographic regions, surface temperatures are still significantly below freezing.  Despite the very high implicit oceanic heat transport implied by the all-$293^\circ K$ oceans, the northern hemisphere midcontinental temperatures have

Fig. 3.  Same as Figure 2a, but for one particular instant of time.  Note that the instantaneous weather patterns produce a different, and sometimes more locally severe (e.g., some surface temperatures are less than $260^\circ K$), detailed map of surface temperatures than for the 50-day average on Figure 2a.

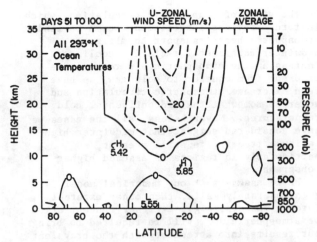

Fig. 4. Eastward component of wind velocity (meters per second) averaged around latitude circles and over a 50-day period shown as a function of latitude and height for the all-293°K ocean temperature case in Figure 2a. The all-293°K assumption has so weakened the equator-to-pole lower atmospheric temperature gradient that mean zonal westerly mid-latitude "jet streams" have been severely reduced relative to present (where jet core speeds are 20-40 m/s) or to the case in Figure 5, where tropical oceanic temperatures are allowed to go over 293°K.

actually decreased relative to the swamp ocean case (no horizontal heat transport) in Figure 3 of Barron and Washington [1984]. The reason our warmer ocean case has colder land surface temperatures is associated with atmospheric circulation changes. The simulated vigor of atmospheric motions in the all-293°K case is insufficient to export the extremely warm mid- to high-latitude oceanic surface temperatures to midcontinental areas. Moreover, a daily temperature map (Figure 3) shows that the effect of averaging 50 days, as in Figure 2, is only to reduce the severity of local cold spots by averaging out such locally extreme weather events.

Both Figure 2a and Figure 3 are results from simulations using annually averaged solar energy. It is, of course, likely that January solar forcing will only produce lower temperatures in northern continental interiors, as is indeed evident in Figure 2b. Even eliminating all topography, an implausible assumption, would probably do very little to avoid deep subfreezing average conditions in midcontinent in midwinter (and daily weather events would, by analogy to Figure 3, produce even more severe local freezes). Although actual topography for 100 Ma is speculative, Barron and Washington [1984, Figure 16] showed that, for annual average solar insolation

forcing, surface temperatures in mid-latitude, higher-altitude areas were only about 6°C warmer when all topography was reduced to sea level heights. Quite simply, surface radiative energy balance in midwinter keeps high-latitude continental interior surface temperatures well below freezing.

Although our all-293°K ocean surface temperature assumption is designed as an extreme exaggeration of efficient oceanic heat transport, it could be argued that the extreme reduction in equator-to-pole atmospheric temperature gradient generated by such efficient oceanic mixing would so weaken atmospheric circulation as to render it incapable of transporting realistic amounts of heat from the assumed warm oceans to the midwinter continental interiors. Indeed, Figure 4 does show that atmospheric winds are very weak in the zonal average in our all-293°K case corresponding to Figure 2a. Therefore, we performed an additional set of simulations in which oceanic surface temperatures are explicitly calculated but never allowed to drop below 293°K, even in polar night waters. With tropical oceanic temperatures now greater than 300°K, a substantial atmospheric temperature gradient is set up on a very warm planet. Figure 5 shows that the atmospheric circulation is much more vigorous in this case, as might be expected. But, as Figure 2c still shows, despite the much enhanced atmospheric circulation strength and the much warmer average planetary surface temperatures, subfreezing continental interiors in midwinter

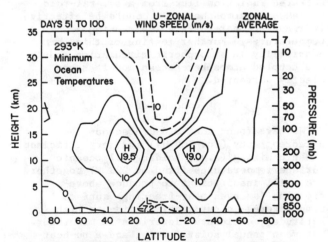

Fig. 5. Same as Figure 4, except that oceanic temperatures are allowed to exceed 293°K in a "swamp" ocean. Tropical surface temperatures thus exceed 300°K. The enhanced equator-to-pole lower atmospheric temperature gradient helps to create a much more vigorous atmospheric circulation, which partially mitigates the radiative cooling of midwinter, midcontinental surfaces.

are still widespread, even though there has been a considerable warming everywhere. Clearly, even these exaggerated, high-transport warm oceans are unable to route enough heat via atmospheric motions to offset the midwinter, continental interior radiative cooling which creates subfreezing land surface temperatures. However, in the no-ocean transport case of Barron and Washington [1984], midcontinental temperatures at high northern latitudes were typically warmer than our all-293°K very warm ocean case, primarily, we have argued, because of the effects of the oceanic surface temperatures on atmospheric motions. Thus, the implicit effects of oceanic heat transport on large-scale atmospheric motions will be critical in determining surface temperatures in continental interiors.

This conclusion should be valid even though it comes from our non-energy-conserving prescribed oceanic surface temperature experiments. For midwinter solar forcings, it appears from all our cases that additional forcing mechanisms are required to produce near- or above-freezing temperatures on land, of which a much enhanced atmospheric "greenhouse effect" from greatly increased $CO_2$ concentration is a leading candidate (for example, see Budyko and Ronov [1979], Barron et al. [1981, p. 506], Berner et al. [1983], or Barron and Washington [this volume]). In particular, since cold winter surface temperatures are often associated with increased atmospheric vertical stability near the surface (e.g., atmospheric temperature inversions), a large increase in downward infrared radiation flux from a several-fold increase in atmospheric $CO_2$ could be effective in reducing near-surface subfreezing temperatures. Further testing of this possibility in atmospheric circulation models and coupled atmosphere/ocean models seems clearly warranted.

## Conclusions

We have found that mid-Cretaceous paleogeography (for 100 Ma) and very efficient oceanic mixing (e.g., holding all oceanic surface temperatures fixed at 293°K) together are still insufficient to produce above-freezing (or even near-freezing) surface temperatures in our model in mid- to high latitude, continental interiors in midwinter. For mean annual solar forcing and a no-heat capacity, no-ocean transport swamp ocean, Barron and Washington [1984] had obtained high northern latitude land and oceanic surface temperatures in the 270°-280°K range, warmer in the high-latitude midcontinents in the northern hemisphere than the much warmer all-293°K oceanic case (Figure 2a) presented here. In our all-293°K case, the equator-to-pole ocean surface temperature gradient is very weak, as

is the vigor of atmospheric circulation, which, in turn, provides insufficient ocean-to-land atmospheric heat transport to mitigate midwinter continental radiative cooling. Increasing the atmospheric air temperature gradient from equator to pole can, we have seen, increase atmospheric circulation and moderate somewhat the midcontinental cold temperatures. Nevertheless, all the cases we have considered suggest that midwinter high-latitude freezing is still present, particularly in regions of assumed higher topography.

If we assume that our numerical model is even moderately valid, then other external forcing factors, e.g., a higher concentration of atmospheric $CO_2$, will be required to bring our results into agreement with the prevalent interpretation of the geologic record, which suggests that subfreezing surface temperatures at 100 Ma were rare. Although, as we explained earlier, our quantitative results should not be taken literally because of many questionable approximations and assumptions in both the data and models, we believe that they reinforce the likelihood that an enhanced atmospheric $CO_2$ concentration was present at mid-Cretaceous times.

Acknowledgments . The National Center for Atmospheric Research is operated by the University Corporation for Atmospheric Research and is sponsored by the National Science Foundation.

## References

Alvarez, L. W., W. Alvarez, F. Asaro, and H. V. Michel, Asteroid extinction hypothesis, Science, 211, 654-656, 1981.

Barron, E. J., A warm, equable Cretaceous: The nature of the problem, Earth Sci. Rev., 19, 305-338, 1983.

Barron, E. J., and W. M. Washington, Cretaceous climate: A comparison of atmospheric simulations with the geologic record, Palaeogeogr. Palaeoclimatol. Palaeoecol., 40, 103-133, 1982.

Barron, E. J., and W. M. Washington, The role of geographic variables in explaining paleoclimates: Results from Cretaceous climate model sensitivity studies, J. Geophys. Res., 89, 1267-1279, 1984.

Barron, E. J., and W. M. Washington, Warm Cretaceous climates: High atmospheric $CO_2$ as a plausible mechanism, this volume.

Barron, E. J., S. L. Thompson, and S. H. Schneider, An ice free Cretaceous? Results from climate model simulations, Science, 212, 501-508, 1981.

Berner, R. A., A. C. Lasaga, and R. M. Garrels, The carbonate-silicate geochemical cycle and its effect on atmospheric carbon dioxide over the last 100 million years, Am. J. Sci., 283, 641-683, 1983.

Budyko M. I., and A. B. Ronov, Atmospheric evolution in the Phanerozoic (in Russian), Geokhimiya, 5, 643-653, 1979.

Budyko, M. I., and A. B. Ronov, Chemical evolution of the atmosphere in the Phanerozoic, Geochem. Int., Engl. Transl., 16, 1-9, 1980.

Covey, C., S. H. Schneider, and S. L. Thompson, Global atmospheric effects of massive smoke injections from a nuclear war: Results from general circulation model simulations, Nature, 308, 21-25, 1984.

Donn, W., and D. Shaw, Model of climate evolution based on continental drift and polar wandering, Geol. Soc. Am. Bull., 88, 390-396, 1977.

Frakes, L. A., Climates Throughout Geologic Time, 310 pp., Elsevier, New York, 1979.

Frakes, L. A., and E. Kemp, Influence of continental positions in early Tertiary climates, Nature, 240, 97-100, 1972.

Harvey, L. D. D., and S. H. Schneider, Transient climate response to external forcing on $10^0$-$10^3$ year time scales, I, Experiments with globally averaged, coupled, atmosphere and ocean energy balance models, J. Geophys. Res., in press, 1985.

North, G. R., J. G. Mengel, and D. A. Short, Simple energy balance model resolving the seasons and the continents: Application to the astronomical theory of the ice ages, J. Geophys. Res., 88, 6576-6586, 1983.

Thompson, S. L., and S. H. Schneider, A seasonal zonal energy balance climate model with an interactive lower layer, J. Geophys. Res., 84, 2401-2414, 1979.

Turco, R. P., O. B. Toon, T. Ackerman, J. B. Pollack, and C. Sagan, Global atmospheric consequences of nuclear war, Science, 222, 1283-1292, 1983.

Vakhrameev, V. A., Jurassic and Cretaceous floras of Eurasia and the paleofloristic provinces of this period, (in Russian), Trans. Geol. Inst., Academy of Sciences of the USSR, 102, 1-263, 1964.

Vakhrameev, V. A., Main features of phytogeography of the globe in Jurassic and early Cretaceous time, Paleontol. J., (English translation), 2, 123-133, 1975.

Washington, W. M., and G. A. Meehl, General circulation model experiments on the climatic effects due to a doubling and quadrupling of carbon dioxide concentration, J. Geophys. Res., 88, 6600-6610, 1983.

# PROTEROZOIC TO RECENT TECTONIC TUNING OF BIOGEOCHEMICAL CYCLES

T. R. Worsley

Department of Geological Sciences, Ohio University, Athens, Ohio 45701

J. B. Moody

ONWI-Battelle, Columbus, Ohio 43201

R. D. Nance

Department of Geological Sciences, Ohio University, Athens, Ohio 45701

Abstract. An approximately 0.5-Ga plate tectonic cycle (Worsley et al., 1984) of continental dispersion and accretion has been correlated with tectonic and platform sediment trends, climate, evolution of life forms, and stable isotope distributions for the last 3.0 Ga. Orogenic peaks center on 2.6, 2.1, 1.8-1.6, 1.1, 0.6, and 0.25 Ga, with the beginnings of the peaks marking the onset of supercontinent-producing collisions and the ends of the peaks corresponding to the thermal elevation that ultimately results in continental breakup. Continental rifting is also related to episodes of mafic-dike swarm production and subsequent rapid seafloor spreading. The major effect of the plate tectonic megacycle on platform sedimentation is to foster marine platform sedimentation during fragmentation-induced submergence and to inhibit sedimentation during assembly-produced emergence. Episodes of platform biogeochemical precipitation (magnetite, hematite, chert, bitumens, dolomite, calcite, phosphorite) and organic and inorganic lag deposits (detrital pyrite and uranitite, redbeds, coal) can be integrated with this plate tectonic cycle of continental fragmentation and assembly. Carbon dioxide and water vapor have controlled the earth's climate throughout its geologic history. The first strong suggestion for a drop in $CO_2$ levels occurs at the onset of Proterozoic glaciation at 2.1 Ga because orogenic quiescence between 2.5 and 2.2 Ga would suggest drowned fragmented continents that would be difficult to glaciate. Continental assembly and emergence favor but do not mandate glacial intervals. The appearance of minute traces of oxygen in the atmosphere is related to the appearance of oxyphobic procaryotic photosynthesizers present in the late Archean ocean. $O_2$ levels elevated dramatically as $O_2$-tolerant eucaryotes first evolved and then achieved aerobic metabolism. The continued evolution of life is related not only to oxygen levels in the atmosphere/ocean system but also to marine transgressions caused by rifting and formation of passive-margined oceans. The carbon, sulfur, and strontium isotopic record can be correlated to platform sedimentation and the Phanerozoic freeboard record. The above correlations suggest that a ~0.5-Ga plate tectonic episodicity is a driving variable responsible for orogenic, eustatic, stable isotope, platform-sediment, biogenetic, and carbon reservoir megacycles that can be recognized at least as far back as 2.5 Ga.

## Introduction

It has long been apparent to earth scientists studying the evolution of the continents, ocean, and atmosphere that each of the diverse endogenic and exogenic processes that power the planet's evolution has always interacted to yield a preserved rock record that confirms the long-term steady state of the system as it approaches maximum entropy. The use of this principle has facilitated understanding of the genetic relationships coupling the histories of such seemingly unrelated as plutonism, volcanism, orogeny, metamorphism, sedimentation, atmosphere-ocean chemistry, and life into a system collectively termed the rock cycle. However, the seemingly smooth, simple, and predictable radiogenic power sources that drive exogenic and endogenic processes have failed to generate an equally smooth earth history. Instead we see an earth history characterized by intervals for which the system did run smoothly that alternate with sharply contrasting episodes of extremely rapid change. If, as we believe, the episodicity is not an artifact of preservational bias, either the radiogenic heat production must oscillate (most unlikely) or the earth itself must act in such a way as to alternately retard and accelerate loss of its internal radiogenic heat to produce the episodes.

Umbgrove [1947], in a startlingly prescient synthesis of global geologic phenomena entitled The Pulse of the Earth, ascribed the alternating periods to

deep-seated forces [that are] the paramount
source of all subcrustal energy, which manifests
itself with a [∿250 Ma] periodicity observed
in a whole series of phenomena in the earth's
crust and on is surface, viz. the alternating
decrease and increase of compression of the
earth's crust and the closely related epochs
of folding, the process of mountain-building
and submersion of borderlands, the periodic
formation of basins and dome-shaped elevations, the magmatic cycles and rhythmic
cadence of world-wide transgressions and
regressions, the pulsation of climate, and -
lastly -the pulse of life.

Umbgrove assigned a ∿250-Ma period to the two Phanerozoic cycles he observed using an age of 470 Ma for the base of the Cambrian. Because Pulse of the Earth was written during an interval before the plate tectonic paradigm captured geologic imagination, and when the theory of continental drift was in strong eclipse due to the absence of a continent-moving mechanism, Umbgrove was not able to ascribe a cause to the global tectonic cycle he so clearly saw, and the book never attained the prominence it deserves. Fischer [1981] revived the Umbgrove model in plate tectonic guise (suggesting a ∿300-Ma period for each of two complete Phanerozoic supercycles based on a 570-Ma duration for the Phanerozoic), thereby providing the mechanism by which pulsation could occur. However, like Umbgrove, he offered no explanation as to why plate tectonics should be cyclic. Worsley et al. [1984], partly on the basis of Precambrian extrapolation, suggested that the Phanerozoic exhibits only slightly more than one cycle and that each cycle actually has an apparent 0.5-Ga periodicity. They also offered a speculative explanation of why plate tectonic processes are intrinsically oscillatory and proposed a simple model that accounts for long-term quasi-periodic episodicity in plutonic, volcanic, orogenic, and metamorphic history back to 2.5 Ga. These plate tectonic episodes are propagated into the sedimentary and biotic record by the atmosphere-ocean system.

As plate-tectonic-driven crustal recycling provides the ultimate nutrient-recycling mechanism for the evolution of life on earth, an understanding of long-term plate tectonic episodicity is fundamental to tracing the earth's biogeochemical evolution. Needless to say, such an understanding is critical if one wishes to accurately reconstruct the long, complex history of the carbon cycle. Carbon is unique among the elements because it is able to form the long polymers upon which life is based,

and its compounds behave as acids or bases, solids or volatiles, or oxygen sources or sinks as needed to buffer the effects of organic evolution. Therefore, carbon has played a critical role in determining the state of the atmosphere-ocean system and that system's effects on the history of life and the life-mediated sedimentation.

In the following article, we offer a new interpretation of the known histories of tectonic, platform sediment, climate, and the evolution of life by tuning the geologic time scale (within allowable dating error) to accommodate the close coupling that must exist between the timing of major changes within these four processes. We restrict ourselves to the platform record because other present-day environments (i.e., slope, rise, or abyssal) have no pre-Mesozoic analogs with which to compare them. We begin with a short summary of our model for tectonic episodicity and its calibration to the Phanerozoic record [Worsley et al., 1984] and offer additional supporting evidence from stable isotopes. We then apply the model to the Proterozoic and "tune" the record to account for the major events in the history of the carbon and other biogeochemical cycles in terms of first-order plate tectonic processes that mediate their evolution.

In constructing our quasi-periodic plate tectonic model [Worsley et al., 1984] and in this article reconciling the model to the biogeochemical history of the earth, we were greatly influenced by, and made extensive use of, the excellent syntheses and summaries of Umbgrove [1947], Garrels and MacKenzie [1971], Veizer [1976], Windley [1977], Holland [1978], Condie [1982a, b], Anderson [1982], and Berner et al. [1983]. We refer the reader to those sources and their extensive reference lists for information that is only briefly presented in this paper.

Finally, before we begin our discussion, we offer the caveat that the seductively simple scenerio we present makes heavy-handed use of Ockham's razor. Perhaps we even dulled the razor's edge by cutting too many corners in our intentional bias toward simplicity. We think that for the purpose at hand namely, to discover order and coherence in earth history, the benefits to be gained from the Ockham's razor approach outweigh the disadvantages.

## The Model

To account for the growing body of evidence of long-term episodicity in tectonic processes [Williams, 1980], dramatically expressed in the recurrence of supercontinents, we have proposed a simple model in which quasi-periodic episodicity is attributed to the influence on terrestrial heat flow of the repeated assembly of successive supercontinents [Anderson, 1982]. In this section we briefly review this model but refer the reader to Worsley et al. [1984] for details of the methods, assumptions, and sources implicit in its calibration.

Average mantle heat flow through oceanic crust exceeds that through continental platform by a fac-

tor of 4 [Sclater et al., 1980], while the world
ocean basin occupies some 60% of the earth's surface
[i.e., Condie, 1982b]. Therefore, the total mantle
heat dissipated through ocean basin is about 6 times
that dissipated through the continents. The hemis-
pherical asymmetry in the distribution of ocean that
has persisted throughout the entire Phanerozoic [i.e.,
Bambach et al., 1980] becomes especially pronounced
during periods of continental assembly. Consequently,
the ocean hemisphere (Panthalassa) tends to remain
an area of general mantle upwelling, the net effect
of which will be the tendency to maintain an antipodal
supercontinent (Pangea) in an approximately stationary position. However, heat buildup beneath an
assembled, stationary Pangea will cause an upward
displacement of its geothermal gradient. This dis-
placement will initially result in uplift of the
supercontinent through thermal expansion and volume-
increasing phase changes, and ultimately, in hot-
spot-induced rifting followed by dispersal of the
continental fragments down the flanks of the thermal
dome [Anderson, 1982, 1984].

During dispersal of a supercontinent, growth of
passive-margined ("Atlantic-type") ocean between
the Pangea fragments is accommodated by subduction
of active-margined ("Pacific-type") Panthalassa.
However, we speculate that this process can only
continue until the margins of the aging Atlantic-
type oceans become sufficiently old and dense to
spontaneously self-subduct [Worsley et al., 1984].
As self-subduction must occur within 200 Ma of rifting [Hynes, 1982], growth of passive-margined oceans
at today's spreading rates can only acheive a
√33% reduction in the size of Panthalassa.
Panthalassa will, therefore, remain the principal
mantle heat dissipator and, with the onset of subduction in the passive-margined oceans, they will
again close, thereby reassembling a supercontinent.

## Model Calibration to the
## Phanerozoic Geologic Record

### Freeboard and Platform Flooding Megacycles

To calibrate our plate tectonic megacycle, we
first gauged its effect on continental freeboard
with time. In this context, freeboard, or the position of the world shelf break with respect to sea
level, is a two-component system determined by the
total volume of the world ocean basin as governed
by seafloor elevation and by the relative elevation
of continental platform as governed by its area and
degree of subcrustal heating. In deriving the two
components, we assumed constant or uniformly accreting continental crust and ocean water volume in an
ice-free world and used as our initial calibration
point today's ice-free water depth at the world shelf
break of +200 m.
Calibration of the seafloor component was based
on the age-depth relationship for oceanic crust
[Parsons and Sclater, 1977] gauged to the shelf
break and on the seafloor model of Berger and
Winterer [1974], in which the age of the world ocean

floor is related to the percentage of Atlantic-type
ocean, assuming today's spreading rates. Calibration
of the continental platform component utilized the
present shelf break elevation of stationary Africa
to yield a minimum estimate of 400 m for the Pangea
thermal dome, a figure that closely approximates
the prediction of Hay and Southam [1977]. Our model
shelf break curve for the Phanerozoic (Figure 1a)
represents the algebraic addition of the two components as constrained by an estimated Pangea duration
of 120 Ma based on that of the Permo-Carboniferous
example, and the 520- and 80-Ma highstand ages of
the first-order eustatic sea level curve of Vail
et al. [1977]. The duration of this one established
megacycle is 440 Ma.
To test the validity of this calculated curve,
we compared its predictions to the established records
of long-term eustatic sea level fluctuations, the
percentage platform flooding, and the number of continents through time (Figure 1a). These variables
are strongly modulated by plate tectonics, and each
strongly covaries with the model's predictions
throughout the Phanerozoic.
Predicted shelf water depth (Figure 1a) closely
matches continental freeboard as reflected in both
the "Vail curves" and the platform flooding curves
of Strakov and the Termiers as compiled by Hallam
[1981]. We do not claim our shelf water depths as
"absolute", but we do claim that relative amplitudes and phases are in correct proportion.
Continent numbers were derived from Ziegler et al.
[1979], Bambach et al. [1980], and Barron et al.
[1981], with the sole modification being the lumping
of closely adjacent continents in the Mesozoic-
Cenozoic (Neozoic). This deliberate bias serves
to decrease the resolution of the Neozoic to that
of the Paleozoic and hence make the two counts comparable. The striking inverse correlation between continent number and freeboard reflects the direct relationship between continental area and average elevation established for today's continents by Hay
and Southam [1977] and Harrison et al. [1981]. Dispersed continents with relatively small areas, therefore, have low elevations and hence correspond to
highstands of sea level, whereas assembled continents
(Pangeas) are elevated and therefore correspond to
lowstands. We believe (as does Anderson, [1982,
1984]) that the underlying cause for this observed
relationship is the increasing tendency toward stasis
with respect to the mantle, and hence thermal elevation, with increasing continental size.

### Carbon, Sulfur, and Strontium Stable
### Isotope Megacycles

A strong correlation exists between the stable
isotope records of carbon [Veizer et al., 1980;
MacKenzie and Pigott, 1981], sulfur [Claypool et
al., 1980; MacKenzie and Pigott, 1981], and strontium [Burke et al., 1982; Veizer et al., 1983] in
platform sediments and the Phanerozoic record of
freeboard that is also explainable in terms of our
Pangea assembly-fragmentation model (Figure 1b).

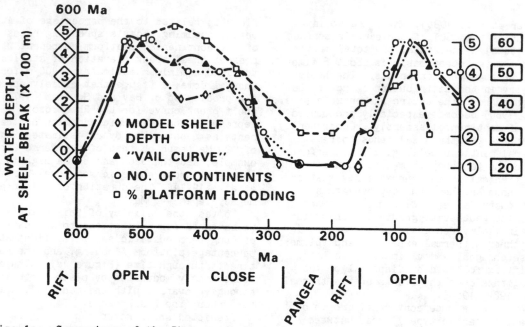

Fig. 1a.  Comparison of the Phanerzoic tectonic-eustatic model to known paleogeography and measured platform drowning and emergence [after Worsley et al., 1984]

This correlation strongly suggests that plate tectonic episodicity [MacKenzie and Pigott, 1981] is the driving variable responsible for Phanerozoic stable isotope megacycles, all of which strongly influence the carbon cycle.

Life processes preferentially sequester isotopically light carbon into reduced compounds (organic matter), leaving the heavier isotope to accumulate in oxidized form as carbonate [i.e., MacKenzie and Pigott, 1981]. Evaporite formation, on the other hand, actively sequesters heavy sulfur [MacKenzie and Pigott, 1981]. Times of globally high rates of reduced carbon accumulation in nonmarine deposits, shales, and other paralic sediments (all enriched in organic matter) should correspond to isotopically heavy records of oxidized carbon in the reciprocal marine carbonate reservoir, whereas times of low reduced carbon accumulation should correspond to

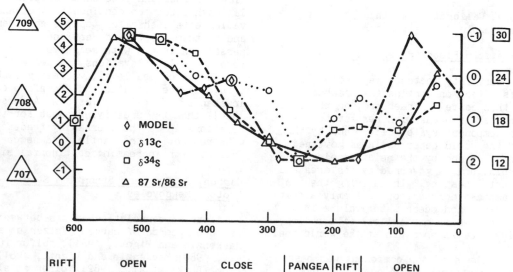

Fig. 1b.  Comparison of the Worsley et al., [1984] model to measured stable isotope trends:  $\delta^{13}C$ [Veizer et al., 1980]; $\delta^{34}S$ [Claypool et al., 1980]; $^{87}Sr/^{86}Sr$ [Burke et al., 1982; and Veizer et al., 1983].

the opposite.  In a similar fashion, evaporites should be enriched in heavy sulfur compared to open marine carbonates.  If the evaporite record alone is considered [i.e., Claypool et al, 1980], times of sparse evaporite formation should yield evaporites rich in isotopically heavy sulfur in the few evaporites forming, whereas periods of extensive evaporite formation should immediately draw down the total supply of heavy sulfur, thereby subsequently producing isotopically light evaporite deposits as long as production rate remains high.  The two most recent Pangeas ($\sim$700 and 250 Ma) correspond to high production rates of nonmarine deposits, shale, and evaporite, all of which should combine to yield an isotopically heavy marine carbonate carbon record and an isotopically light sulfur evaporite record. Fragmentation leading to maximum dispersion would yield the opposite.  Figure 1b shows this to be the case.

Strontium isotopes are not significantly fractionated by life processes or selective evaporation as are carbon and sulfur, so that we must look to another mechanism to explain their observed trend (Figure 1b).  Continental crust, because of its higher Rb/Sr ratio, provides a $^{87}$Sr-enriched flux into the oceans relative to mantle-derived Sr (see Veizer et al. [1983] for a review).  Consequently, times of continental assembly and emergence (or, on a shorter time scale, individual orogenic events and glacial episodes) should correspond to high $^{87}$Sr/$^{86}$Sr ratios, whereas fragmenting Pangeas corresponding to rising sea levels and faster seafloor spreading rates (i.e. increased mantle-derived $^{86}$Sr) should show low $^{87}$Sr/$^{86}$Sr ratios.  The general agreement of Sr isotope ratios in carbonates therefore once again lends support to our plate tectonic megacycle (Figure 1b).

## Protero-Phanerozoic Tectonic and Biogeochemical Trends

For the Phanerozoic, a diverse suite of tectonic, eustatic, and geochemical trends correlate to an apparent plate tectonic megacycle of $\sim$440 Ma for which we proposed [Worsley et al., 1984] a simple, endogenic heat-driven model of plate tectonic episodicity.  This plate-tectonic episodicity in turn appears to be the driving variable that controls episodic, secular, and recurring shifts in platform sedimentation, climate, and the punctuated evolution of life.  Episodic trends common to both the Proterozoic and the Phanerozoic and some exclusive to the Proterozoic that exhibit a periodicity compatible with that of the Worsley et al., [1984] megacycle suggest that the model is also applicable to the Precambrian .  If we ascribe orogenic peaks and coinciding glacial episodes to periods of supercontinent assembly and emergence, and lagging mafic dike swarms coupled with rapid biotic diversification to periods of continental breakup and subsidence (as is the case for the Permo-Carboniferous and latest Precambrian), a $\sim$500 Ma periodicity is revealed that persists back to the Archean [Worsley et al., 1984].

In the following section, we will show that the model is also applicable to the more poorly and less quantitatively known Proterozoic platform record by using published qualitative information about tectonic, platform sediment, climatic, biotic, and stable isotope trends.  To accomplish this, we briefly review the diverse geologic evidence pertaining to the timing of the onsets, cessations, and major changes in rates of the trends shown in Figure 2 and re-interpret the punctuated nature of the trends in terms of our supercontinent assembly-fragmentation model.  In constructing Figure 2, we assumed that dating error increases linearly with age from an error of zero today to + 75 or - 75 Ma at 3 Ga.  We then adjust (or "tune") published estimates of the timing of changes that we believe are genetically related to supercontinent assembly-fragmentation episodes to coincide with assembly or fragmentation and with each other within the allowable error.

## Tectonic Trends

We believe that there is a genetic relationship between orogenic peaks [Condie, 1982b], supercontinents [Veizer et al., 1983], and mafic emanation episodes [Halls, 1982] that persists at least as far back as the advent of the granitic continental crust at about 2.6 Ga [Windley, 1977], and that constitutes the controlling variable of all other trends. Orogenic peaks occur at about 0.5-Ga intervals and center on 2.6, 2.1, 1.8-1.6, 1.1, 0.6, and 0.25 Ga (Figure 2).  We ascribe the beginnings of the peaks to the onset of supercontinent-producing continental collisions, and the demises of the peaks to the stasis-induced thermal elevation that ultimately results in supercontinent rifting and breakup. Rifting, in turn, corresponds to episodes of mafic-dike swarm production [Halls, 1982] and subsequent rapid seafloor spreading as the thermally unstable supercontinent fragments and flounders.  This pattern persists despite major changes in tectonic style due to crustal evolution and the exponential decline in the earth's heat production [Worsley et al., 1984], perhaps  because the changes have influenced seafloor spreading processes to a much greater extent than continental tectonics.  However, higher geothermal gradients would have caused earlier continents to behave in a more ductile fashion, thereby inhibiting their rifting sufficiently to form wide oceans. We speculate that the world consisted of one or two supercontinents that stretched and recoiled on a rapidly cycling, more komatiite-rich world ocean crust [Arndt, 1983; Nisbet and Fowler, 1983].  Areas of ensialic stretching are recorded as Archean greenstone belts [Condie, 1981, 1982b] where mantle-derived ultramafic and mafic magma upwelled to the surface, and subsequently, as the ensialic orogenic belts of the Proterozoic [Windley, 1977; Condie, 1982b].  The first major change in this scenerio occurred with the advent of ophiolites (peridotite/ gabbro/basalt/sediment sequences postulated to be obducted slivers of oceanic crust) that appears to

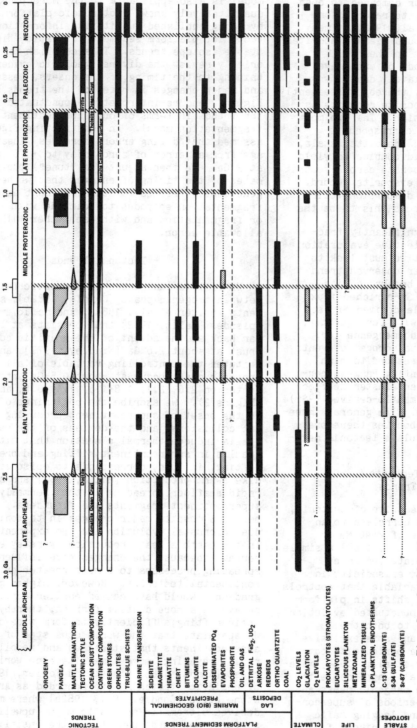

Fig. 2.  Qualitative summary of late Archean to Recent tectonic, platform sediments, climate, life, and stable isotope trends.  Lag deposits are subaerial or littoral deposits; mantle emanations are material added to the crust or water/atmospheric system derived from the mantle.  In constructing this chart, we took liberties with published information to adjust the timing of events (within allowable dating errors) in order to emphasize the close coupling that we propose exists between the trends and the quasi-periodic 0.5-Ga plate tectonic cycle.  Please see the text for references.  Trends are as follows:  abundant, intense or heavy shown by solid bars (documented) and stippled bars (speculative); common or moderate shown by solid lines (documented) and dot-grouped lines (speculative).  Proposed supercontinent fragmentation is indicated by vertical hatched bars.

coincide with the demise of komatiite-enriched basaltic oceanic crust somewhere around 1 Ga. A dense, komatiite-rich ocean crust (if it ever did exist) would have been universally subductable and, therefore, not preserved in the geologic record [Arndt, 1983]. Continued lowering of worldwide thermal gradient caused a concomitant decrease in the proportion of undepleted mantle melted to produce ocean crust from greater than 60% (komatiite) to the 15-20% (tholeiite) that we see today. At about 0.6 Ga, the first blueschists (low-temperature/ high- pressure metamorphic assemblages found at convergent margins) appear, heralding the onset of thermal gradients similar to those found at today's convergent margins [Ernst, 1972].

## Platform Sediment Trends

The primary coupling between tectonic trends and associated geochemical, climatic, and biospheric trends related to the carbon cycle, in addition to the obvious sealevel effects, occurs because supercontinent fragmentation episodes increase the input rate of mantle material to ensialic basins or new mid-ocean ridges, whereas supercontinent assembly decreases them [Veizer et al., 1983]. An increase in the rate of mantle-derived material intensifies juvenile element input to the lithosphere/hydrosphere (we consider the atmosphere to be a negligibly small reservoir on these time scales). At the same time, rapid subduction and metamorphism of seafloor-precipitated carbonate minerals releases carbon dioxide that is then deposited on the concurrently subsiding platforms as "new" carbonate and/or organic matter [Berner et al., 1983]. During supercontinent accretion, erosion from collision-thickened, thermally domed continental crust allows transfer of the continent-sequestered carbon back to the deep sea.

The overriding effect of the plate tectonic megacycle is to foster marine platform sediment preservation during fragmentation-induced submergence and inhibit it during assembly-produced emergence. As shown in Figure 2, there is remarkably good agreement between the predicted [Worsley et al., 1984] and observed [Condie, 1981] record of marine sediments for the last 2.5 Ga. However, superimposed upon this episodic sediment accumulation trend are a series of secular trends in the various sedimentary components that are clearly related to biogenesis and its effects on the carbon-oxygen budget of the hydrosphere.

Increasing $O_2$ levels in the atmosphere/hydrosphere system have caused continued stripping of redox-sensitive elements from seawater, thereby greatly influencing the nature of platform sedimentation through time [Veizer, 1976]. Geochemical evidence [i.e., Holland, 1978] demands a neutral hydrosphere until about 2 Ga, when photosynthetic $O_2$ overwhelmed the iron buffering system of the oceans [Cloud, 1974; Veizer, 1976]. The terrestrial sedimentary record is more ambiguous in that weathering profiles indicate $Fe^{2+}$ leaching from Archean soils but not thereafter, suggesting minor free oxygen

appearing in the atmosphere by about 2.5 Ga [Veizer, 1976]. The presence of minor free oxygen in the atmosphere prior to its appearance in the hydrosphere requires that the continents ceased to be a significant source of dissolved iron about 2.5 Ga. A submarine hydrothermal source of reduced iron is therefore required to maintain continuous precipitation of photosynthetically produced oxygen in the hydrosphere, offering indirect evidence of a continuously active seafloor spreading system that emanates dissolved reduced iron. By 1.8 Ga, however, photosynthetic $O_2$ production overwhelmed the ridge system seawater buffer, and iron was thereafter insoluble in seawater, precipitating as oxyhydroxides near submarine hydrothermal vents as it does today.

## Marine Biogeochemical Precipitates

Siderite/magnetite/hematite. The above-described secular trend in platform sedimentation in which ferrous iron is progressively replaced by ferric iron conforms to the increase in photosynthetic oxygen levels within the atmosphere/hydrosphere that began at about 3 Ga. The extremely low oxygen stress that probably characterized the middle Archean ocean resulted in abundant dissolved $Fe^{2+}$ that must have behaved much like Ca and Mg do today in forming carbonates. Siderite is indeed a common accessory mineral in middle Archean sediments [Veizer, 1976]. As photosynthetic $O_2$ production increased around 3 Ga [Cloud, 1983], siderite became less abundant, and the excess $O_2$ was used to produce marine precipitates enriched in magnetite between 3 and 2.5 Ga [Veizer, 1976]. Further, $O_2$ stress rendered hematite the dominant precipitate between 2.5 and 2 Ga, after which the soluble iron buffer must have become depleted, allowing $O_2$ to begin accumulating in the hydrosphere. As Figure 2 shows, the preserved record of this secular iron trend is strongly punctuated by the postulated supercontinent accretion and dispersion cycle.

Silica. The solubility of silica in seawater is not greatly pH or Eh sensitive [Holland, 1978]. It is therefore probable that dissolved silica has never exceeded 25 ppm in seawater [Holland, 1978], so that the trend of decreasing platform precipitation of silica (chert) at about 1.5 Ga [Veizer, 1976] observed in Figure 2. is not explainable in terms of decreased absolute silica flux through the oceans as was the case for iron. We believe that the trend represents a change in the partitioning between chert deposition on the platforms and the deep sea. From the early Archean through the early Proterozoic, chert was a common component of platform precipitates, perhaps mediated by benthic procaryotes. At about 1.5 Ga, an already flourishing eucaryotic [Cloud, 1968] or plastid-symbiotic [Margulis, 1981] procaryotic plankton became evident in the geologic record. We suggest (as does Cloud [1983]) that components of this plankton developed the ability to secrete silica at this time. A siliceous plankton community would transfer the major geochemical sink for this material from the

shelf to the deep sea. The record of this postulated event has not yet been found preserved on
the platforms and cannot be documented from deep-
sea sediments because the currently oldest known
preserved evidence of deep seafloor (ophiolites)
occurs at around 1 Ga. It is interesting to note,
however, that these oldest ophiolites are capped
by cherts [Windley, 1977], that we believe will
prove to be the remains of these siliceous
plankton. Intensive study of 0.6-to 1.5-Ga
old-cherts using appropriate preparation
techniques may clarify the problem of the origins
of the relatively advanced siliceous biota already
apparent in the Cambrian.

Bitumens. The geologic record of bitumens
(disseminated organic matter) exhibits a weak
trend of decrease with time that corresponds
mainly to transfer of organic carbon to ever more
specific geochemical reservoirs as life evolved
[Veizer, 1976]. The only noticeable break in the
abundance occurs with the demise of the iron
buffer corresponding to the advent of $O_2$ at 2 Ga.
Possible further reductions corresponding to the
development of bioturbation cannot be documented
with available data.

Dolomite and calcite. The total quantity of
carbonate (calcite plus dolomite plus siderite)
deposited through geologic time [Garrels and
MacKenzie, 1971; Condie, 1981] conforms quite well
to the supercontinent assembly-fragmentation model
in that long term (i.e., 0-3 Ga) deposition rate
is fairly constant but accelerates during
dispersals and decreases during assemblies.
However, the search for reasons behind an observed
long-term secular trend beginning at about 1 Ga in
which Ca steadily replaces Mg in platform
carbonates has inspired continued and ongoing
debate that has contributed significant volume to
the geochemical literature and is known as the
"dolomite problem" [see Holland, 1978]. Simply
stated, the problem concerns finding suitable geochemical sinks for Mg that account for ever
decreasing amounts of dolomite in the geologic
record since about 1 Ga assuming near-constant
riverine input through time. Currently
fashionable explanations include (1) the "time
jeopardy" model, in which we simply have not
waited long enough for currently preserved calcite
to be converted into dolomite by groundwater; (2)
the "$CO_2$ inhibition" model, in which steadily
dropping $CO_2$ levels in the hydrosphere
progressively inhibit dolomite precipitation; and
(3) the "ridge hydrothermal" model, in which an increasing rate of mid-ocean ridge hydrothermal circulation through time has allowed for increasing Mg-
Ca exchange with ocean crust. Models 1 and 2 encounter serious mass balance objections, and model
3 encounters an equally serious thermodynamic objection in that heat flow and therefore, supposedly,
seafloor spreading rates and hydrothermal circulation rates are continually decreasing.

While we do not attempt to resolve the dolomite
problem here, we suggest that the key to understanding this dilemma involves the transition from ductile to brittle tectonics that occurred between 1

and 2 Ga [Windley, 1977; Condie, 1982]. A komatiite-
rich ocean crust with a higher thermal gradient and
a much higher coefficient of expansion upon hydra-
tion ($\sim$30) than tholeiite ($\sim$10) might offer
a poorer Mg sink because the depth to which hydrothermal circulation could penetrate and the length
of time the system would remain open should both
be much smaller. Total Ca-Mg exchange for such a
system would be much lower despite faster seafloor
spreading rates. We suggest here that the Mg drawdown effect that is so prominent today began about
at 1.5 Ga and became strongly manifested during the
0.6-Ga fragmentation cycle.

Phosphorus. The Precambrian record of sedimentary phosphorus is obscure even though P is
thought to be the limiting nutrient of life on earth
[Broecker, 1983]. Long-term geochemical flux of
the element through the hydrosphere has probably
not changed appreciably with time, so that the absolute biomass which depends on it has not changed
either. What has changed is partitioning to geochemical sinks. As Figure 2 shows, disseminated
$PO_4$ has been a consistent but minor component of
platform precipitates since at least 2 Ga [Cook and
McElhinny, 1979], when the hydrosphere first became
oxygenated. Unlike carbon, phosphorus is essentially insoluble in seawater, ($\sim$0.1 ppm) and
does not form volatile or soluble compounds. Once
precipitated, it is relatively difficult to reflux
into seawater, and its geologic history tracks the
ever expanding effort of organisms to keep it in
the hydrosphere. Prior to the advent of the metazoans with their ability to burrow, life was restricted to the water column and the sediment-water
interface. Hence, we suggest here that any relatively insoluble element not dissolved back into
seawater within the upper 1 cm of the sediment
column was lost to the hydrosphere [see Emerson,
this volume and Moody et al., 1981]. Such a system
would, therefore, bury such elements in proportion
to their precipitation in the overlying water column.
It would then seem reasonable that all pre-metazoan
P accumulated in disseminated form because there
was no way to reflux it for subsequent concentration
into specialized sinks. The arrival of the metazoans
changed this simple picture because their burrowing
ability gave ever deeper sediment layers access to
seawater, meaning that element reflux to seawater
has become continually more effecient. The phosphorus record clearly records this process. The
advent of the metazoans at 1 Ga [Kauffman, 1983]
heralds the first small, highly concentrated deposits of phosphorite rock in that refluxing of P to
seawater allowed for selected concentration at fast-
sedimentation upwelling sites at the expense of more
sterile and slowly sedimenting sites because
bioturbation-refluxing efficiency is inversly
proportional to sedimentation rate. The invention
of skeletons at 0.6 Ga allowed even more efficient
burrowing and corresponds to the first occurrence
of common phosphorite deposits (see Cook and Shergold
[1984] for a review).

Evaporites. Evidence of evaporitic halides and
sulfates preserved as hopper crystals are known from
the earliest geologic record [Dimroth, 1981], but

owing to their short depositional half-life, the actual minerals are not found preserved until the late Proterozoic (Figure 2). Since then, however, periods of major evaporite preservation coincide with the early stages of supercontinent breakup.

Oil and gas. Oil and gas are dominantly marine deposits whose geologic record is closely tied to the evolution of marine plankton and to the increasingly oxygenated hydrosphere. In essence, they represent the ever more restricted sink for reduced carbon that must balance the continually increasing $O_2$ content of the hydrosphere. Oil and gas make their first common appearance in the record at the 0.6-Ga breakup phase and become common during the 0.2-Ga fragmentation event. Like evaporites, oil and gas have a short geologic half-life, and we cannot be certain whether their observed geologic history owes more to evolution or preservation.

## Subaerial and Littoral Sediments

Detrital Uraninite and pyrite. Uraninite and pyrite (Figure 2) are stable only under reducing or neutral atmospheric conditions [Veizer, 1976], so that their presence as detritus in aeolian, fluvial, or deltaic sediments is a clear indication of a neutral or at most very slightly oxidizing (assuming rapid mechanical erosion and burial) atmosphere. The disappearance of detrital uraninite and pyrite from the platform record at about 1.8 Ga [Veizer, 1976] indicates platform drowning coupled with their increasing instability with the appearance of a moderately oxygenated atmosphere, both of which suggest a supercontinent fragmentation.

Arkose. Arkose is a fluvial or deltaic sediment deposited mechanically under at least slightly oxidizing conditions and which has endured only minor chemical alteration during transport and deposition [Garrels and MacKenzie, 1971]. The prerequisites for arkose formation are similar to those for detrital pyrite and uraninite except that arkose will tolerate higher $O_2$ levels. The appearance of abundant arkose in the platform record at about 2.5 Ga (Figure 2) is consistent with rapid mechanical denudation of the first highly elevated granitic supercontinent, while its decline at 1.5 Ga conforms to relatively high oxygen levels and hence strong chemical weathering.

Redbeds. Redbeds (Figure 2) represent subaerial deposition in the presence of moderate amounts of free $O_2$ [Garrels and MacKenzie, 1971]. They seem to make their first appearance in the record during rifting of an emergent Pangea at 2 Ga [Veizer, 1976] when atmospheric $O_2$ levels were first high enough to form them, and have been a conspicuous although minor component of the sedimentary record ever since. Redbeds are most abundant during the early stages of Pangea "rifting without drifting" where subaerial grabens favor their preservation (i.e., Triassic redbeds).

Orthoquartzites. Quartz represents the stable end-member of the felsic magmatic reaction series and likewise represents the stable end member to prolonged and repeated erosion cycles on the earth's surface [Garrels and MacKenzie, 1971]. Rare Archean orthoquartzites [Condie, 1981] probably result from shale-derived gneiss and metamorphosed chert, as the Archean contained few if any quartz-containing igneous rocks. With the advent of granitic continents at 2.5 Ga, orthoquartzite becomes common in the geologic record and appears to become increasingly abundant at each marine transgression following a supercontinent breakup (Figure 2).

Coal. Coal (Figure 2) is a terrestrial biochemical precipitate of volatiles and carbon produced by land plants and therefore makes its first appearance when land plants evolved [Veizer, 1976]. Coal becomes common in the record as the seas regressed during the assembly of the most recent supercontinent. As with the marine hydrocarbons, coal represents a large but much restricted reservoir of reduced carbon that resulted from the continued photosynthetic extraction of $O_2$ from the crust.

## Climate, Life, and the Carbon Cycle

Insofar as sedimentary rocks are in part the product of life processes which in turn greatly influence climate, we have already discussed to some extent the history of climate and life on earth for the last 3 Ga. The purpose of this section is to review the earth's climatic history in terms of what can be inferred about the history of $CO_2$ and $O_2$ levels in the atmosphere as they relate to supercontinent assembly-fragmentation episodes. In doing this, we assume that feedback systems operated in such a way as to keep the world ocean in aqueous form so that past climates have never been very different from today's. Because astronomical calculations imply a solar constant about 25% lower than today's shortly after the origin of the solar system, a stronger greenhouse effect was required to keep the world ocean aqueous. Many gases are good candidates for maintaining the greenhouse, but in the absence of compelling evidence for a reducing Archean atmosphere, we prefer the simplest choice that $CO_2$ and water vapor controlled the earth's climate throughout its geologic history.

## Carbon Dioxide Levels

In the absence of photosynthesis, the main control of $CO_2$ levels in the atmosphere would be the rate it is produced by mantle outgassing versus the rate of consumption through carbonate precipitation. Archean levels were probably high [i.e., Holland, 1978], but lack of a quantitative means of estimating them hinders quantitative modeling. The first suggestion of a drop in $CO_2$ levels occurs with the onset of early Proterozoic glaciation between 2.5 and 2.1 Ga [Christie-Blick, 1982]. We would place the date close to 2.1 Ga because orogenic quiescence between 2.5 and 2.2 Ga [Condie, 1982b] would suggest drowned, fragmented continents that would be difficult to glaciate. This first evidence of a large ice volume indicates that atmospheric $CO_2$ finally dropped to levels low enough to weaken the greenhouse effect sufficiently to produce climates equivalent to today's. A strong, delicately balanced life-mediated feedback system between decreasing $CO_2$ levels and increasing solar constant has kept climatic excursions within the liquid range of water ever since. In this scenerio, subsequent glacials are viewed as times of an emergent polar (at least with respect to the plane of the ecliptic) continent. Continental assembly and emergence (Figure 2), therefore, favor but do not mandate glacial intervals as evidenced by an emergent present-day Antarctica situated at the south pole.

## Oxygen Levels and Life

Oxygen levels appear to broadly reciprocate $CO_2$ levels throughout earth history, although the two are only indirectly coupled via metabolic processes. Photosynthesizing but oxyphobic procaryotes probably existed in the middle Archean, and the oxygen they produced was neutralized in the production of banded iron formation [Cloud, 1968], thereby keeping the atmosphere essentially neutral and the deep sea weakly reducing. Oxyphobic procaryotic photosynthesizers flourished unchallenged until the development of plastid-containing organisms with oxygen-mediating enzymes (perhaps by symbiosis [see Margulis, 1981]) precipitated the "iron crisis" at about 2 Ga. At this time, $O_2$ production permanently overwhelmed the iron buffer, dealing a blow to the procaryotes from which they never recovered [Cloud, 1968]. The appearance of significant oxygen in the atmosphere-hydrosphere quickly led to the energy-efficient aerobic respiration characteristic of eucaryotes. Direct fossil evidence suggests the presence of eucaryotes at the 1.5-Ga fragmentation. However, the conspicuous presence of redbeds at 1.8 Ga suggests that some organism of which we have no current record was already tolerant of, and perhaps metabolized, oxygen by that time. We believe it to be a eucaryote or symbiotic eucaryotic predecessor that evolved during the period of ecological stress just prior to the 2-Ga Pangea breakup.

$O_2$ levels continued to build as eucaryotes became ever more efficient and oxygen tolerant and, finally, oxygen obligate. We speculate here that the drawdown of the platform chert reservoir at about 1.5 Ga noted by Veizer [1976] corresonds to the development of siliceous eucaryotc plankton (an event not yet found in the fossil record). The appearance of metazoan trace fossils at the 1-Ga Pangea fragmentation [Kauffman, 1983] and their sparse preservation as body fossils by 0.8 Ga indicate a hydrosphere rich enough in oxygen (>1% of present atmospheric level) to support collagen-forming metabolism [Towe, 1970]. The imprint of the metazoans on the geologic record is profound because their burrowing activities allowed for efficient scavenging and refluxing to seawater of newly deposited elements contained within the upper 10 cm of the marine sediment column for the first time in geologic history, thereby ending forever the equatable biogenic-production-to-burial scenerio characterizing the earlier record. The consequences of efficient refluxing result in more slowly sedimenting areas being "picked clean" of their organic matter by scavenging metazoans. The refluxed elements are then ultimately transformed in high concentration (through numerous repetitions of the refluxing process) to the narrowly focused upwelling areas. Ironically, the high biologic production and rapid sedimentation of the upwelling areas maintains the reducing conditions at the sediment-water interface that locally restrict the refluxing activity of the very metazoans responsible for the nutrient transfer that allowed for the prolific productivity in the first place. The significant reduction of disseminated bitumens and phosphate [Veizer, 1976] and the final decline in the stromatolites [i.e., Windley, 1977] coupled with the advent of common phosphorites at 1 Ga correspond to the advent of the metazoans. The appearance of abundant phosphorite in the record of upwelling areas at about 0.6 Ga [Cook and Shergold, 1984], coinciding with the appearance of hydrocarbons, marks the increase in burrowing and refluxing efficiency permitted by the development of shells. Invasion of the low-lying (fragmented continents) terrestrial environment by land plants at about 0.45 Ga served to increase the contrast between reduced and oxidized carbon and sulfur reservoirs by sequestering continually larger portions of reduced carbon and sulfur onto the continents (coal). This sequestering left the marine productivity belts to capture the remainder, thereby producing the virtually completely oxidized shell of terrestrial and marine sediments interspersed with small but highly concentrated pockets of reduced carbon and sulfur that we see today.

The last major tectonic event in the history of the carbon cycle is the fragmentation of Permo-Carboniferous Pangea at about 0.2 Ga. This event corresponds to the proliferation of hydrocarbons, endotherms, and calcareous plankton [Cloud, 1974]. The rise of the endotherms represents the response to the diurnal, seasonal, and latitudinal thermal contrasts developed during the most recent drawdown of atmospheric $CO_2$ corresponding to the assembly of the latest Pangea [Fischer, 1981]. The development of calcareous plankton during the Pangea fragmentation fostered very high sedimentation rates at passive margins, which resulted in the enhanced burial of organic carbon at these sites that corresponds to the abundant hydrocarbons there now.

## Stable Isotopes

Stable isotope trends for the latest Proterozoic to Recent have already been discussed earlier, and little additional information is available for earlier times, with the exception of a [87]Sr record supporting the fragmentation of a supercontinent at 1 Ga [Veizer et al., 1983]. Therefore, the Proterozoic enrichment-depletion episodes for these isotopes plotted in Figure 2 are predictive speculations of how the Proterozoic record might look based on similar coupling of these isotopes to assembly-fragmentation episodes for the Phanerozoic.

## Conclusions

The earth is a heat engine powered by an internal radiogenic heat source that continuously differentiates and density-stratifies its elements. The endogenic process is expressed at the surface by the process of plate tectonics that leaves mountains as direct evidence of the heat dissapation process. Counteracting the endogenic mountain-building process is the sun-powered exogenic process that uses phase changes between solid, liquid, and gaseous water to destroy all elevated structures. The long-term interaction between the two competing processes produces a record of steady state equilibrium that is punctuated by excursions in the endogenic cycle that alternately retard and then accelerate heat loss around steady state values. These excursions are clearly decipherable in the rock record as ∿0.5-Ga first-order oscillations in platform sedimentation in which assembled supercontinents correspond to low sea levels and vice versa. A detailed record for the last 0.6 Ga has allowed for the construction and testing of a heat-driven plate tectonic episodicity model that faithfully reproduces all major Phanerozoic tectonic, sedimentological, and biogenic trends. Application of the model to the Proterozoic record offers intriguing clues to how the system operated then. If the inferences we have made are correct, application of the model could greatly simplify interpretation of how plate tectonic processes have orchestrated the life-mediated history of the earth's atmosphere-hydrosphere and crust.

Acknowledgments. The writing of this article is based on the efforts of a large fraction of the geologic community over a long period of time. Special thanks are offered to two compassionate anonymous reviewers, who greatly improved the manuscript by making us aware of our responsibility to separate fact from speculation. We fervently hope that the product stimulates enough new thinking on the problems of crustal cycling to justify the liberties we took in writing it. Battelle Memorial Institute is thanked for their generous support of manuscript preparation.

## References

Anderson, D.L., Hotspots, polar wander Mesozoic convection and the geoid, Nature, 297, 391-393, 1982.

Anderson, D.L., The earth as a planet: Paradigms and paradoxes, Science, 223, 347-355, 1984.

Arndt, N.T., Role of a thin, komatiite-rich oceanic crust in the Archean plate-tectonic process, Geology, 11, 372-375, 1983.

Bambach, R.K., C.R. Scotese, and A.M. Zeigler, Before Pangea: The geographics of the Paleozoic world, Am. Sci., 68, 26-38, 1980.

Barron, E.J., C.G.A. Harrison, J.L. Sloan, and W.W. Hay, Paleogeography, 180 million years ago to the present, Eclogae Geol. Helv., 74, 443, 1981.

Berger, W.H., and E.L. Winterer, Plate stratigraphy and the fluctuating carbonate line, in Pelagic Sediments: On Land and Under the Sea, Spec. Publ. Int. Assoc. Sediment. 1, edited by K.J. Hsu and H.C. Jenkyns, pp. 11-98, Blackwell, Oxford, England, 1974.

Berner, R. A., A. C. Lasaga, and R. M. Garrels, The carbonate-silicate geochemical cycle and its effect on atmospheric carbon dioxide over the past 100 million years, Am. J. Sci., 283, 641-683, 1983.

Broecker, W. S., Tracers in the Sea, 690 pp., Eldigio, 1983.

Burke, W.H., R.E. Denison, E.A. Hetherington, R.B. Koepnick, H.F. Nelson and, J.B. Otto, Variation of seawater $^{87}Sr/^{86}Sr$ throughout Phanerozoic time, Geology, 10, 516-519, 1982.

Christie-Blick, N., Pre-Pleistocene glaciation on earth: Implications for climatic history of Mars, Icarus, 50, 423-443, 1982.

Claypool, G.E., W.T. Holser, I.R. Kaplan, H. Sakai, and I. Zak, The age curves of sulfur and oxygen isotopes in marine sulfate and their mutual interpretation, Chem. Geol., 28, 199-259, 1980.

Cloud, P.E., Pre-metazoan evolution and the origins of the metazoa, in Evolution and Environment, edited by E.T. Drake, pp. 1-72, Yale University Press, New Haven, Conn., 1968.

Cloud, P.E., Evolution and ecosystems, Am. Sci., 62, 54-67, 1974.

Cloud, P. E., The biosphere, Sci. Am., 249, 176-189, 1983.

Condie, K.C., Archean Greenstone Belts, 434 pp., Elsevier, New York, 1981.

Condie, K.C., Early and middle Proterozoic supracrustal successions and their tectonic settings, Am. J. Sci., 282, 341-357, 1982a.

Condie, K.C., Plate Tectonics and Crustal Evolution, 310 pp., Pergamon, New York, 1982b.

Cook, P.J., and M.W. McElhinny, A reevaluation of the spatial and temporal distribution of sedimentary phosphate deposits in the light of plate tectonics, Econ. Geol., 74, 315-330, 1979.

Cook, P.J., and J.H. Shergold, Phosphorus, phosphorites and skeletal evolution at the Precambrian-Cambrian boundary, Nature, 308, 231-236, 1984.

Dimroth, E., The carbon-oxygen system at present and in the Precambrian, in Evolution of the Earth, Geodyn. Ser., Vol. 5, edited by R.J. O'Connell and W.S. Fyfe, pp. 69-74, AGU, Washington, D.C., 1981.

Emerson, S.R., Organic carbon preservation in Marine Sediments, this volume.

Ernst, W.G., Occurrence and mineralogic evolution of blueschist belts with time, Am. J. Sci., 272, 657-668, 1972.

Fischer, A. G., Climatic oscillations in the biosphere, in Biotic Crises in Ecological and Evolutionary Time, edited by M. H. Nitecki, pp. 102-131, Academic, New York, 1981.

Garrels, R.M., and F.T. MacKenzie, Evolution of

Sedimentary Rocks:  A Geochemical Approach, 397 pp., Norton, New York, 1971.

Hallam, A., Facies Interpretation and the Stratigraphic Record, 291 pp., W.H. Freeman, San Francisco, Calif., 1981.

Halls, H.C., The importance and potential of mafic dyke swarms in studies of geodynamic processes, Geosci. Can., 9, 145-154, 1982.

Harrison, C.G.A., G.W. Brass, E. Saltzman, J. Sloan, J. Southam, and J.M. Whitman, Sea level variations, global sedimentation rates and the hypsographic curve, Earth Planet. Sci. Lett., 54, 1-16, 1981.

Hay, W.W., and J.R. Southam, Modulation of marine sedimentation by the continental shelves, in The Role of Fossil Fuel $CO_2$ in the Oceans, edited by N.R. Anderson and A. Malahoff, pp. 569-605, Plenum, New York, 1977.

Holland, H.D., The Chemistry of the Atmosphere and Oceans, 351 pp., Wiley-Interscience, New York, 1978.

Hynes, A., Stability of the oceanic tectosphere-A model for early Proterozoic intercratonic orogeny, Earth Planet. Sci. Lett., 61, 333-345, 1982.

Kauffman, E.G., Brooksella canyonensis:  A billion year old complex metazoan trace fossil from the Grand Canyon, Geol. Soc. Am. Abstr. Programs, 15, 608, 1983.

MacKenzie, F.T., and J.D. Pigott, Tectonic controls of Phanerozoic sedimentary rock cycling, J. Geol. Soc. London, 138, 183-196, 1981.

Margulis, L., Symbiosis in Cell Evolution, 419 pp., W.H. Freeman, San Francisco, Calif., 1981.

Moody, J.B., T.R. Worsley, and P.R. Manoogian, Long-term phosphorus flux to deep-sea sediments, J. Sediment. Petro., 51, 307-312, 1981.

Nisbet, E.G., and C.M.R. Fowler, Model for Archean plate tectonics, Geology, 11, 376-379, 1983.

Parsons, B., and J.G. Sclater, An analysis of the variation of ocean floor bathymetry and heat flow with age, J. Geophys. Res., 82, 803-827, 1977.

Sclater, J.G., C. Jaupart, and D. Galson, The heat flow through oceanic and continental crust and the heat loss of the earth, Rev. Geophys. Space Phys., 18, 269-311, 1980.

Towe, K.M., Oxygen-collagen priority and the early metazoan fossil record, Proc. N.Y. Acad. Sci., 65, 781-788, 1970.

Umbgrove, J.H.F., The Pulse of the Earth, 358 pp., Martinus Nijhoff, The Hague, 1947.

Vail, P.R., R.M. Mitchum, and S. Thompson, Global cycles of relative changes of sea level, in Seismic Stratigraphy-Applications to Hydrocarbon Exploration, Mem. 26, edited by C.E. Payton, pp. 83-97, American Association of Petroleum Geologists, Tulsa, Okla., 1977.

Veizer, J., Evolution of ores of sedimentary affiliation through geologic history; relations to the general tendencies in evolution of the crust, hydrosphere, atmosphere and biosphere, in Handbook of Strata-Bound and Stratiform Ore Deposits-III, edited by K.H. Wolf, pp. 1-41, Elsevier, New York, 1976.

Veizer, J., W.T. Holser, and C.K. Wilgus, Correlation of $^{13}C/^{12}C$ and $^{34}S/^{32}S$ secular variations, Geochim. Cosmochim. Acta, 44, 579-587, 1980.

Veizer, J., W. Compston, N. Clauer, and M. Schidlowski, $^{87}Sr/^{86}Sr$ in late Proterozoic carbonates:  Evidence for a "mantle" event at .900 Ma ago, Geochim. Cosmochim. Acta, 47, 295-302, 1983.

Williams, H., Structural telescoping across the Appalachian orogeny and the minimum width of the Iapetus Ocean, in The Continental Crust and Its Mineral Deposits, Spec. Pap. 20, edited by C.A. Burk and C.L. Drake, pp. 421-440, Geological Association of Canada, 1980.

Windley, B.F., The Evolving Continents, 385 pp., John Wiley, New York, 1977.

Worsley, T.R., D. Nance, and J.B. Moody, Global tectonics and eustasy for the past 2 billion years, Mar. Geol., 58, 373-400, 1984.

Ziegler, A.M., C.R. Scotese, W.S. McKenow, M.E. Johnson, and R.K. Bamback, Paleozoic paleogeography, Annu. Rev. Earth Planet. Sci., 7, 473-502, 1979.

# POTENTIAL ERRORS IN ESTIMATES OF CARBONATE ROCK ACCUMULATING THROUGH GEOLOGIC TIME

William W. Hay

Museum, University of Colorado, Boulder, Colorado 80309

Abstract. One of the arguments for changes in atmospheric $CO_2$ on geologic time scales rests on consideration of the rates of accumulation of carbonate rock through time. Estimates of the volumes of different rock types, including carbonate, deposited over geologic time have generally assumed that sedimentary rocks can be divided into three major assemblages, one characteristic of platform areas, another characteristic of geosynclines, and a third characteristic of pelagic sediment. The latter assemblage is frequently omitted from the estimate computations. While it may be reasonable to assume that shelf sediments beneath the sea are probably similar to the platform sediments exposed on land, it may be unreasonable to expect that the same proportions of sediment types can be extrapolated to the deposits of the slope and rise. Cratonic sediments form about 14% of the sedimentary mass, passive margin shelf sediments 8%, and passive margin slope and rise sediments 24%. Because the mass of slope and rise sediments is larger than that of cratonic and shelf sediments, there may be significant errors in estimates of total mass of carbonate rock formed at a given time. Passive margin slope and rise areas may be carbonate sinks early in their development and may exclude carbonate later in their development. The apparent ratio of clastics to carbonates accumulating at any time might also vary significantly according to the overall composition of the slope and rise deposits. The apparent decline in rate of accumulation of carbonate since the Mesozoic is in part a function of the shift in site of deposition of carbonate from solely shallow water environments to both shallow and deep water environments as a result of the development of the oceanic calcareous plankton.

## Introduction

Budyko and Ronov [1979] have proposed that the chemical evolution of the atmosphere in the Phanerozoic can be interpreted from the record of formation of sediment. They suggested that the $O_2$ content of the atmosphere is proportional to the amount of organic carbon buried in sediments and that the mass of $CO_2$ in the atmosphere at any given time is related to the overall consumption of $CO_2$ in the formation of sediments; for the latter relation, they assumed that the simplest approximation is direct proportionality. Further, they noted a correlation between the amounts of volcanogenic and carbonate rocks formed at any given time. Their conclusion was that volcanic activity raised the $CO_2$ concentrations in the atmosphere and that the increased mass of $CO_2$ was consumed in weathering and an increased mass of carbonate rock being formed.

The argument rests upon the assumption that output of various sediment components for intervals of geologic time is well enough known to be extrapolated to estimates of total sediment output. The purpose of this paper is to provide a critical evaluation of this assumption with respect to carbonate.

## Estimating the Masses of Different Sediment Types Through Time

The most detailed estimates of masses of different sediment types through time have been prepared by Ronov [1949, 1959, 1964, 1968, 1972, 1976, 1980], Ronov and Khain [1954, 1955, 1956, 1961, 1962, 1970], Ronov and Yaroshevskiy [1967, 1976, 1977], Ronov and Migdisov [1970], Ronov et al.[1969, 1972a, b, 1973, 1974, 1976, 1977, 1978], and Khain et al. [1975, 1977, 1979], Vinogradov and Ronov [1956]. The results of these compilations have been presented by Ronov [1980] based on Phanerozoic sedimentary sequences of the continents and "in less detail on those of the oceans" [Ronov, 1980 (Engl. trans. 1982, p. 1365)]. The data are from the Russian platform, the Scythian Plate and Caucasus, the Carpathians, the Urals, the North American and Siberian platforms, the West Siberian Plain, the Tien Shan, the Cordillera, the Appalachians and "several of the platformal and geosynclinal regions" [Ronov, 1980 (Engl. trans. 1982, p. 1368)]. The distribution of the most important types of continental sediment and volcanogenic associations expressed as volumes was presented in Ronov's [1980] Table 13; the distribution through time of the mass of Phanerozoic sedimentary and volcanogenic rocks of the continents and the masses of carbon dioxide and carbon buried in the deposits was presented in Ronov's [1980b] Table 14. Although Ronov is now a "mobilist" and supporter of the theory of plate tectonics, all of these mass balance studies were conducted in a "fixist" geological context, assuming permanence of continents and ocean basins.

Southam and Hay [1981] have discussed global sedimentary cycling in a plate tectonic context and have suggested that the deposits formed on passive margins since the breakup of Pangea constitute about one third of the total sedimentary mass. The passive margin shelf deposits are becoming known through exploration for petroleum, and as a first approximation resemble shallow water marine deposits exposed on the continent in terms of lithologic associations. However, the passive margin slope and rise deposits are virtually unknown, and the assumption that they have lithologic associations similar to those of shallow water deposits is probably invalid. The pelagic sediments of the ocean basins compose about one sixth of the total mass of sediments, and thanks to 15 years of Deep Sea Drilling Project operations their composition is relatively well known. However, the ocean floor is in motion, and the pelagic sediments are continuously removed through subduction, having a half-life of only 50-60 m.y. [Worsley et al. 1984]. Because the compositions of the continental slope and rise and the pelagic sediment reservoirs are different from those of the continental sediments discussed by Budyko and Ronov [1979], a different estimate of the mass of carbonates accumulating at any given time may result.

## Proportions of Carbonate Rock in the Major Sediment Reservoirs

Table 1 is an extract from the data of Ronov [1980] presenting the volume of carbonate rock, total volume of sediments, and the percentage of carbonate rock in the total sediment for platforms, geosynclines, and the continents as a whole. These data show an interesting pattern over intervals of era length.

For any era, the proportions of carbonate rock to total sediment are higher in platform rocks than in geosynclinal rocks, although the volume of carbonate in geosynclinal rocks is greater on an absolute basis. More significant, there is a decline in the percentage of carbonate rock as a part of the total sediment from Paleozoic to Mesozoic to Cenozoic. Cenozoic sediments on the continents contain only about half as much carbonate rock as Paleozoic sedimentary rocks.

Ronov's [1980] estimate of the total volume of sedimentary rock on the continents is $640 \times 10^6$ $km^3$. This is in close agreement with the estimates of Southam and Hay [1981] for the total of sedimentary rock on the cratons ($165 \times 10^6$ $km^3$) plus Precambrian-Paleozoic geosynclines ($211 \times 10^6$ $km^3$) plus Mesozoic-Cenozoic geosynclines ($203 \times 10^6$ $km^3$) plus the sediment of the passive margin shelves exposed on land ($1/2 \times 95 \times 10^6$ $km^3$): $626.5 \times 10^6$ $km^3$. Except for the geosynclinal regions, the estimates were derived using different data and assumptions.

Table 2 presents Southam and Hay's [1981] estimate of the volume of sediment in each of the major sedimentary reservoirs, an estimate of the percentage of carbonate rock in each of the reservoirs, and the volume of carbonate rock in each of the reservoirs. The data of Ronov [1980] were used to calculate the percentage of carbonate rock on the craton (31.7%; assumed to be dominantly Paleozoic by Southam and Hay), in Precambrian-Paleozoic geosynclines (24.9%), and in Mesozoic-Cenozoic geosynclines (18.9%, as given in Table 1). Active margins are largely devoid of carbonate rock except for coral reef development in the tropics, so that the percentage of carbonate rock in them is given as 11%, corresponding to the Cenozoic average for geosynclines.

The passive margin shelves probably have the same proportions as the Mesozoic-Cenozoic platform deposits (19.0%). Partitioning of carbonate between the shelves and the deep sea probably depends on sea level changes [Milliman, 1974; Hay and Southam, 1977].

The passive margin slope and rise deposits are largely unknown. The only data set which may be applicable is the Ocean Margin Drilling Program atlases of the east coast of North America [Bryan and Heirtzler, 1984; Ewing and Rabinowitz, 1984; Shor and Uchupi, 1984; Uchupi and Shor, 1984]. These have interpreted lithofacies and sediment thickness data for a large segment of this classic passive margin, but the interpretation is largely inferred from seismic data and has not been verified by observation. Inspection of the maps suggests that the proportion of carbonate rock in the slope and rise probably resembles that of the platforms for the Mesozoic (19.5%) but is significantly less (10%?) for the Cenozoic, resembling the proportion in geosynclines.

Judging from the Ocean Margin Drilling Program Atlases for the east coast passive margin of North America, about two thirds of the sediment is Mesozoic and one third Cenozoic, so that the proportion of carbonate might be estimated to be 16%. However, other passive margins may have had different histories, i.e., separation occurred later, the regional climatology was different, etc., so that extrapolation of an estimate based on the east coast of North America is inappropriate. The average age of all presently existing passive margins is 130 m.y. [Southam and Hay, 1981, Figure 12], so that an overall proportion of sediment closer to half Mesozoic, half Cenozoic is more likely for passive margins on a global scale. Hence, I used 15% as an estimate for the proportion of carbonate rock in all passive margin slope and rise deposits.

The proportion of carbonate ooze in pelagic sediments has been thoroughly investigated by the Deep Sea Drilling Project; a recent average of several thousand analyses indicated that the carbonate ooze component is 74% of the total pelagic sediment by volume and about 52% by weight [Southam and Hay, 1981].

Changes in volumes of carbonate rocks (expressed as volumes of $CO_2$ in carbonate rocks) were plotted by Ronov [1976, 1980] and shown to closely parallel the changes in the area of shelf seas with time. This is to be expected because the data are based largely on rocks exposed on the continents, and carbonates are deposited preferentially in shallow epeiric seas. The question remains whether the changes in volumes of carbonate rocks will still show the same pattern after information from the ocean basins and their margins is added.

## Proportions of Carbonate in Other Sedimentary Rock Types

Not all carbonate is present as carbonate rock; some carbonate occurs in other sedimentary rocks as fossils, cements, and minor interbeds. Ronov [1980] has calculated the mass of carbon dixoide present as carbonate in other sedimentary rock types (clays, sands, and evaporites). These data were presented only as total masses for all continental rocks. Although there is presently no basis for estimating the amount of carbonate in noncarbonate rocks in the slope and rise deposits, it is likely that any error introduced by neglecting this factor would be small. The content of carbon in pelagic sediments other than carbonate oozes is included in the estimated proportion of 52% carbonate. Table 3 presents Ronov's data for masses of $CO_2$ in carbonate rocks and as carbonate in other sedimentary rocks for intervals of geologic time, the percentage of total carbonate present as carbonate rock, and the percentage of total carbonate present as carbonate in other rock types. The percentage of total carbonate on the continents which occurs as carbonate rock varies with time, as seen in column 4 of Table 3, but is generally higher than 75%. The Late Permian and Pliocene have the lowest proportions of carbonate rock to total carbonate, 48.5% and 38.5%, respectively; during these times, significantly more carbonate was deposited as carbonate in other rock types than was deposited as carbonate rock. Through much of geologic time, carbonate is well segregated into carbonate rock, and during the Paleozoic it is common that more than 85% of the total carbonate on the continents is segregated into carbonate rocks.

## Reevaluation of the Output of $CO_2$ Into Sediment Through Time

Table 3 also shows the apparent $CO_2$ accumulation rates for rocks on the continents [after Ronov, 1980], expressed in units of $10^{21}$ g/m.y.; these apparent accumulation rates are obtained by dividing the $CO_2$ mass in existing rocks representing each time interval by the length of the interval. Column 6 represents the data from which Budyko and Ronov [1979] estimated the $CO_2$ content of the atmosphere.

Ronov's platform deposits correspond to the cratonic reservoir and the presently emergent portion of the passive margin shelf reservoirs of Southam and Hay. Ronov's geosynclines correspond to the Precambrian-Paleozoic and Mesozoic-Cenozoic geosynclines of Southam and Hay. Not included by Ronov were the passive margin slope and rise deposits and the pelagic sediments. Clearly, Ronov's estimates of changes of carbonate rock volumes will not change for the Paleozoic. However, for the Mesozoic-Cenozoic the total volume of carbonate accounted for by Ronov is $51.7 \times 10^6$ km$^3$ (very close to the $56.4 \times 10^6$ km$^3$ estimated to be in Mesozoic-Cenozoic geosynclines and passive margin shelves in Table 2), whereas the total volume of Mesozoic-Cenozoic carbonate estimated in Table 2 is 3 times as large, $169.4 \times 10^6$ km$^3$. The passive margin slope and rise and pelagic sediment reservoirs contain about $113 \times 10^6$ km$^3$ of carbonate; the detailed distribution of this in time is not presently known. Hence it is likely that the Mesozoic-Cenozoic variations in volume of carbonate rock plotted by Ronov will be significantly altered as the age distribution of carbonates in the ocean basins and margins becomes known.

The $CO_2$ in carbonate in passive margin slope and rise deposits and in pelagic sediments should be added to Ronov's estimate based on rocks on the continents. Assuming from Table 2 a volume of carbonate rock in the slope and rise of $42 \times 10^6$ km$^3$, and using Ronov's value of 2.52 for the density of carbonate rock, yields a mass of $105.8 \times 10^{21}$ g. Calculating the mass of $CO_2$ as 44% of $CaCO_3$ yields $46.6 \times 10^{21}$ g. Assuming equal masses of carbonate in the Mesozoic (Jurassic-Cretaceous) and Cenozoic portions of the slope and rise, it is immediately evident that for the Cenozoic, carbonate accumulation in this province is nearly equivalent to that on the continental blocks. For the Mesozoic, slope and rise sediment accumulation could be half that on the continental blocks. It is apparent that changing the assumptions could significantly alter estimates of the total $CO_2$ in sediments.

Following analogous calculations, the pelagic sediment reservoir contains a mass of approximately $57 \times 10^{21}$ g of $CO_2$. This presently

TABLE 1. Volumes of Carbonate Rock and Total Sediments

| | | | Platforms | | |
|---|---|---|---|---|---|
| | Age, Ma | Length, m.y. | Volume of Carbonate Rock $10^6$ $km^3$ | % | Total Volume of Sediment $10^6$ $km^3$ |
| | 1 | 2 | 3 | | 4 |
| Pliocene | 2–9 | 7 | 0.12 | 4.7 | 2.56 |
| Miocene | 9–25 | 16 | 0.70 | 13.8 | 5.07 |
| Oligocene | 25–37 | 12 | 0.10 | 3.2 | 3.08 |
| Eocene | 37–58 | 21 | 2.05 | 24.8 | 8.28 |
| Paleocene | 58–66 | 8 | 0.85 | 25.6 | 3.32 |
| Tertiary | 2–66 | 64 | 3.82 | 17.12 | 22.31 |
| Upper Cretaceous | 66–100 | 34 | 5.36 | 28.1 | 19.09 |
| Lower Cretaceous | 100–132 | 32 | 2.77 | 13.4 | 20.69 |
| Upper Jurassic | 132–153 | 21 | 3.01 | 32.3 | 9.33 |
| Middle Jurassic | 153–168 | 15 | 1.80 | 23.1 | 7.80 |
| Lower Jurassic | 168–185 | 17 | 1.31 | 19.3 | 6.78 |
| Upper Triassic | 185–210 | 25 | 0.47 | 4.4 | 10.70 |
| Middle Triassic | 210–220 | 10 | 0.58 | 31.5 | 1.85 |
| Lower Triassic | 220–235 | 15 | 0.50 | 10.4 | 4.81 |
| Mesozoic | 66–235 | 169 | 15.80 | 19.49 | 81.05 |
| Mesozoic and Tertiary | 2–235 | 232 | 19.61 | 18.98 | 103.36 |
| Upper Permian | 235–255 | 20 | 0.33 | 6.9 | 4.79 |
| Lower Permian | 255–280 | 25 | 2.24 | 26.2 | 8.55 |
| Pennsylvanian | 280–320 | 40 | 1.62 | 18.5 | 8.76 |
| Mississippian | 320–345 | 25 | 2.40 | 50.3 | 4.77 |
| Upper Devonian | 345–360 | 15 | 2.06 | 34.3 | 6.00 |
| Middle Devonian | 360–376 | 16 | 1.11 | 21.2 | 5.23 |
| Lower Devonian | 376–400 | 24 | 0.19 | 3.0 | 6.44 |
| Silurian | 400–435 | 35 | 1.41 | 24.7 | 5.70 |
| Ordovician | 435–490 | 55 | 3.45 | 37.6 | 9.17 |
| Upper Cambrian | 490–515 | 25 | 3.21 | 53.2 | 6.03 |
| Middle Cambrian | 515–545 | 30 | 3.30 | 61.2 | 5.39 |
| Lower Cambrian | 545–570 | 25 | 3.54 | 46.0 | 7.81 |
| Paleozoic | 235–570 | 335 | 24.91 | 31.68 | 78.64 |
| Phanerozoic | 2–570 | 568 | 44.53 | 24.5 | 182.03 |

existing mass rests on ocean crust of varying age. Converting the areas of seafloor of given age tabulated by Sclater et al. [1980] to the geologic time intervals used by Budyko and Ronov [1979] and given here in Table 1, the area of the existing seafloor of Pliocene age is 26.8 x $10^6$ $km^2$, Miocene 46.0 x $10^6$ $km^2$, Oligocene 32.8 x $10^6$ $km^2$, Eocene 56.3 x $10^6$ $km^2$, Paleocene 18.5 x $10^6$ $km^2$, Late Cretaceous 71.0 x $10^6$ $km^2$, Early Cretaceous 39.5 x $10^6$ $km^2$, Late Jurassic 14.3 x $10^6$ $km^2$, and Middle Jurassic 6.3 x $10^6$ $km^2$. The carbonate accumulates only above the carbonate compensation depth, which varies with time and ocean basin but averages about 4 km, a depth reached by the seafloor about 25 m.y. after formation. Global pelagic carbonate accumulation rates have varied with time (Davies and Worsley, 1981), but when averaged over the time intervals used by Budyko and Ronov are approximately equal. Subsequent erosion and removal as

hiatuses are formed reduces the proportion of surviving sediment significantly (Moore and Heath, 1977), so that for the Pliocene, about 80% of the sediments originally deposited have survived, for the Miocene 60%, Oligocene 40%, Eocene 30%, Paleocene 20%, and older sediments perhaps 10%. Further, it has been assumed that the influx of $CO_2$ as carbonate into pelagic sediment began with the rise of calcareous plankton 100 m.y. ago [Poldervaart 1955; Hay and Southam, 1976; Sibley and Vogel, 1976]. From these considerations the existing total of 57 x $10^{21}$ g of $CO_2$ can be estimated to be divided among time intervals as shown in column 9 of Table 3. Clearly, for epochs of the Cenozoic, the pelagic sediment reservoir dominates the global system, having significantly larger accumulation rates than the continental blocks and passive margin slope and rise combined.

on the Continents [After Ronov, 1980]

| Geosynclines | | | Continents | | |
| Volume of Carbonate Rock $10^6$ km$^3$ | % | Total Volume of Sediment $10^6$ km$^3$ | Volume of Carbonate Rock $10^6$ km$^3$ | % | Total Volume of Sediment $10^6$ km$^3$ |
| 5 | | 6 | 7 | | 8 |
| --- | --- | --- | --- | --- | --- |
| 0.06 | 0.9 | 6.96 | 0.18 | 1.9 | 9.52 |
| 1.17 | 8.7 | 13.44 | 1.87 | 10.1 | 18.51 |
| 0.52 | 8.0 | 6.50 | 0.62 | 6.5 | 9.58 |
| 2.54 | 20.7 | 12.27 | 4.59 | 22.3 | 20.55 |
| 0.42 | 13.5 | 3.11 | 1.27 | 19.8 | 6.43 |
| 4.71 | 11.14 | 42.28 | 5.53 | 13.21 | 64.59 |
| 5.37 | 18.6 | 28.88 | 10.73 | 22.4 | 47.94 |
| 5.74 | 18.8 | 30.47 | 8.51 | 16.6 | 51.16 |
| 3.42 | 21.5 | 15.89 | 6.43 | 25.5 | 25.22 |
| 2.64 | 18.2 | 14.53 | 4.44 | 19.9 | 22.33 |
| 3.01 | 21.1 | 14.25 | 4.32 | 20.3 | 21.03 |
| 4.22 | 18.3 | 23.06 | 4.69 | 19.7 | 23.76 |
| 1.86 | 16.7 | 11.16 | 2.44 | 18.8 | 13.01 |
| 1.10 | 16.9 | 6.50 | 1.60 | 14.1 | 11.31 |
| 27.36 | | 144.74 | 43.16 | 10.1 | 215.76 |
| 32.07 | 18.4 | 187.02 | 51.69 | 20.0 | 280.35 |
| 1.49 | 11.2 | 13.29 | 1.82 | 33.1 | 18.08 |
| 8.88 | 35.5 | 25.03 | 11.12 | 27.7 | 33.58 |
| 7.29 | 31.2 | 23.38 | 8.91 | 41.6 | 32.14 |
| 11.58 | 49.2 | 28.81 | 13.98 | 24.1 | 35.58 |
| 5.61 | 21.7 | 25.80 | 7.67 | 24.1 | 31.80 |
| 6.62 | 24.7 | 26.83 | 7.73 | 13.6 | 32.06 |
| 3.50 | 16.9 | 20.70 | 3.69 | 17.5 | 27.14 |
| 3.66 | 15.7 | 23.30 | 5.07 | 20.6 | 29.00 |
| 5.65 | 16.1 | 35.02 | 9.10 | 34.2 | 44.19 |
| 3.41 | 25.5 | 13.35 | 6.62 | 35.3 | 19.38 |
| 4.45 | 26.9 | 16.56 | 7.75 | 34.2 | 21.95 |
| 5.31 | 29.1 | 18.23 | 8.90 | 26.47 | 26.04 |
| 67.36 | 24.92 | 270.30 | 92.36 | 22.5 | 348.94 |
| 99.52 | 21.76 | 457.32 | 144.05 | | 639.35 |

### Potential Transfer of C Between Ccarb and Corg Reservoirs

Ronov [1980] also presented an inventory of organic carbon (Corg) in the sediments on the continental blocks. Budyko and Ronov [1979] noted that the change in mass of atmospheric oxygen with time must equal the rate of input of oxygen minus the rate of consumption of oxygen by oxidation of minerals and assumed that the rate of input of oxygen can be calculated from the amounts of organic carbon in sediments. Their argument for determining atmospheric $O_2$ through time will not be discussed here, but Table 4 presents a summary of Ronov's data on the amount of carbon present as carbonate (Ccarb) and as organic carbon on the continents. Also given are estimates of the masses of organic and carbonate carbon in passive margin and pelagic deposits along with a new global total. In making these estimates I have used data from Tables 2 and 4, have assumed that the organic carbon content of passive margin deposits is the same as that in rocks on the continents and that the organic carbon content of pelagic sediments is zero. Adding in the passive margin slope and rise deposits significantly increases the estimated total of organic carbon for the Mesozoic and Cenozoic. Ronov's data taken alone suggest a secular trend in favor of increasing the proportion of carbon preserved as organic carbon through time. However, when the passive margin and pelagic reservoirs are added in, the proportions remain much more nearly constant, with about 15% of the total carbon being buried as organic carbon.

It seems likely that almost all carbon that is recycled is at some stage $CO_2$, and that partitioning between the organic carbon and carbonate carbon reservoirs is established

TABLE 2.   Distribution of Sedimentary and Carbonate Rocks among Reservoirs

| Sediment Reservoir | Volume of Sediment, $10^6$ km$^3$ 1 | % Carbonate Rock 2 | Volume of Carbonate Rock $10^6$ km$^3$ 3 |
|---|---|---|---|
| Cratonic | 165 | 31.7% | 52.3 |
| Paleozoic geosynclines | 211 | 24.9% | 52.5 |
| Mesozoic-Cenozoic geosynclines | 203 | 18.9% | 38.4 |
| Active margins | 19 | 11.1% | 1.9 |
| Passive margin shelves | 95 | 19.0% | 18.0 |
| Passive margin slope and rise | 280 | 15.0% | 42.0 |
| Pelagic | 142 | 50.0% | 71.0 |

TABLE 3.   Distribution and Apparent Accumulation

| | $CO_2$ in Carbonate Rocks (Continents) $10^{21}$ g 1 | $CO_2$ as Carbonate in Other Rocks (Continents) $10^{21}$ g 2 | Total $CO_2$ (Continents) $10^{21}$ g 3 | % $CO_2$ in Carbonate Rocks (Continents) 4 | % $CO_2$ as Carbonate in Other Rocks (Continents) 5 |
|---|---|---|---|---|---|
| Pliocene | 0.20 | 0.32 | 0.52 | 38.5 | 61.5 |
| Miocene | 2.07 | 0.75 | 2.82 | 73.4 | 26.6 |
| Oligocene | 0.68 | 0.22 | 0.90 | 75.6 | 24.4 |
| Eocene | 5.09 | 1.46 | 6.55 | 77.7 | 22.3 |
| Paleocene | 1.41 | 0.21 | 1.62 | 87.0 | 13.0 |
| Tertiary | 9.45 | 2.96 | 12.41 | 76.1 | 23.9 |
| Late Cretaceous | 11.89 | 3.02 | 14.91 | 79.7 | 20.3 |
| Early Cretaceous | 9.42 | 1.70 | 11.12 | 84.7 | 15.3 |
| Late Jurassic | 7.11 | 1.35 | 8.46 | 84.0 | 16.0 |
| Middle Jurassic | 4.91 | 0.88 | 5.79 | 84.8 | 15.2 |
| Early Jurassic | 4.79 | 0.54 | 5.33 | 89.9 | 10.1 |
| Late Triassic | 5.22 | 1.74 | 6.96 | 75.0 | 25.0 |
| Middle Triassic | 2.70 | 0.68 | 3.38 | 79.9 | 20.1 |
| Early Triassic | 1.76 | 1.02 | 2.78 | 63.3 | 36.7 |
| Mesozoic | 47.80 | 10.93 | 58.73 | 81.4 | 18.6 |
| Mesozoic & Tertiary | 57.25 | 13.89 | 71.14 | 80.5 | 19.5 |
| Late Permian | 2.19 | 2.33 | 4.52 | 48.5 | 51.5 |
| Early Permian | 13.18 | 1.75 | 15.13 | 88.4 | 11.6 |
| Pennsylvanian | 10.71 | 1.18 | 11.89 | 90.1 | 9.9 |
| Mississippian | 16.78 | 0.72 | 17.50 | 95.9 | 4.1 |
| Late Devonian | 9.22 | 1.04 | 10.26 | 89.9 | 10.1 |
| Middle Devonian | 9.30 | 1.25 | 10.55 | 88.2 | 11.8 |
| Early Devonian | 4.43 | 1.30 | 5.73 | 77.3 | 22.7 |
| Silurian | 6.10 | 1.71 | 7.81 | 78.1 | 21.9 |
| Ordovician | 10.94 | 3.67 | 14.61 | 74.9 | 25.1 |
| Late Cambrian | 7.96 | 0.13 | 8.09 | 98.4 | 1.6 |
| Middle Cambrian | 9.30 | 0.33 | 9.63 | 96.6 | 3.4 |
| Early Cambrian | 10.68 | 0.62 | 11.30 | 94.5 | 5.5 |
| Paleozoic | 110.79 | 16.03 | 127.02 | 85.1 | 14.9 |
| Phanerozoic | 168.04 | 29.92 | 198.16 | 82.6 | 17.4 |

directly or indirectly by life processes. Taking organic carbon in the offshore passive margin shelf, slope, and rise deposits into account suggests total organic carbon fluxes for the Mesozoic and Cenozoic about 50% higher than those presented by Ronov [1980].

### The Effect of Sedimentary Cycling and Losses From the System

Ronov and his colleagues inventoried existing sediment volumes and masses. They made no attempt to estimate the actual masses that would have existed at times in the past, which must include sediments subsequently eroded, strongly metamorphosed, or subducted.

Figure 1 shows the accumulations of $CO_2$ as carbonate on the continents as presented by Ronov

[1980] and the increments of $CO_2$ present as carbonate in the offshore passive margin deposits and in pelagic sediments. Ronov's estimated volumes of carbonate do not vary greatly through geologic time, but adding the increments for the marine sediment reservoirs results in the significantly larger volumes of young sediment that would be expected as a result of sediment cycling, as discussed by Garrels and Mackenzie [1971] and Veizer and Jansen [1979]. Because such large volumes of carbonate rock are present in the pelagic sediment reservoir, the pattern for total sediment volumes would be slightly different. It is evident that the rate of sediment recycling is in part a function of sediment type.

Loss of sediment from the system through subduction or metamorphism has not been taken

Rates of $CO_2$ in Sedimentary Reservoirs

| Apparent $CO_2$ Accumulation Rate (Continents) $10^{21}$ g/m.y. 6 | $CO_2$ in Passive Margin Slope and Rise $10^{21}$ g 7 | $CO_2$ Accumulation Rate Passive Margin Slope and Rise $10^{21}$ g/m.y. 8 | $CO_2$ in Pelagic Sediments $10^{21}$ g 9 | $CO_2$ Accumulation Rate Pelagic Sediments $10^{21}$ g/m.y. 10 | Apparent Global $CO_2$ Accumulation Rate $10^{21}$ g/m.y. 11 |
|---|---|---|---|---|---|
| 0.074 | | 0.364 | 10.9 | 1.563 | 2.001 |
| 0.108 | | 0.364 | 16.9 | 1.058 | 1.530 |
| 0.075 | 23.3 | 0.364 | 8.1 | 0.675 | 1.114 |
| 0.312 | | 0.364 | 10.3 | 0.491 | 1.167 |
| 0.203 | | 0.364 | 2.2 | 0.271 | 0.838 |
| 0.194 | | 0.364 | 48.4 | 0.756 | 1.314 |
| 0.439 | | 0.196 | 8.6 | 0.253 | 0.888 |
| 0.348 | | 0.196 | | | 0.544 |
| 0.403 | 23.3 | 0.196 | | | 0.599 |
| 0.386 | | 0.196 | | | 0.582 |
| 0.314 | | | | | 0.314 |
| 0.264 | | | | | 0.264 |
| 0.338 | | | | | 0.338 |
| 0.185 | | | | | 0.185 |
| 0.311 | | | | | 0.311 |
| 0.281 | | | | | 0.281 |
| 0.226 | | | | | 0.226 |
| 0.605 | | | | | 0.605 |
| 0.297 | | | | | 0.297 |
| 0.700 | | | | | 0.700 |
| 0.684 | | | | | 0.684 |
| 0.659 | | | | | 0.659 |
| 0.239 | | | | | 0.239 |
| 0.223 | | | | | 0.223 |
| 0.266 | | | | | 0.266 |
| 0.324 | | | | | 0.324 |
| 0.321 | | | | | 0.321 |
| 0.452 | | | | | 0.452 |
| 0.379 | | | | | 0.379 |
| 0.349 | | | | | 0.531 |

TABLE 4.   Masses of Carbon

| | On Continents[a] | | In Passive Margins | | In Pelagic Sediment | | Total | |
|---|---|---|---|---|---|---|---|---|
| | Corg | Ccarb | Corg | Ccarb | Corg | Ccarb | Corg | Ccarb |
| | 1 | 2 | 3 | 4 | 5 | 6 | 7 | 8 |
| Cenozoic | 1.13 | 3.39 | 2.78 | 6.36 | 0 | 13.21 | 3.91 | 22.96 |
| Mesozoic | 3.05 | 16.03 | 2.29 | 6.36 | 0 | 2.35 | 5.34 | 24.74 |
| Devonian–Permian | 1.98 | 20.63 | | | | | | |
| Cambrian–Silurian | 1.11 | 14.04 | | | | | | |
| Phanerozoic | 7.27 | 54.09 | | | | | | |

Carbon masses are given as $10^{21}$ g.
[a]After Ronov [1980].

into account.  Using a half-life of 100 m.y. for pelagic sediment, and assuming a constant rate of accumulation of red clay/siliceous ooze in the ocean basins, Southam and Hay [1981] estimated that about 50% of all sediment formed from the weathering of igneous rocks may have been lost through subduction over the past 4.5 billion years.  With a shorter half-life, the amount lost through subduction may be significantly larger.  However, for carbonate the loss through subduction is probably a minor factor because significant carbonate output into the deep sea did not start until 100 m.y. ago.  Clearly, significant volumes of sedimentary rock have been lost through metamorphism, but it is not presently possible to evaluate the magnitude of these losses.

## Discussion

Why did Budyko and Ronov [1979] omit marine sediments from their calculations?  Based on the estimates of volume and composition of major sedimentary reservoirs by Ronov and Yaroshevskiy [1967, 1976, 1977], it would not have been apparent to Budyko and Ronov that adding the passive margin slope and rise deposits and pelagic sediments would alter the pattern of carbonate deposition they observed.  Ronov and Yaroshevskiy's estimates indicated a small, relatively insignificant volume for slope and rise deposits; further, they considered the ocean basins to be permanent features, so that the passive margin deposits would have been expected to represent all of Phanerozoic time, not just the Mesozoic and Cenozoic.  For the pelagic sediments, they calculated a composition assuming the masses to be directly proportional to the area of each sediment type, as had El Wakeel and Riley [1961].  This results in an average composition of pelagic sediment not very different from that of platform or geosynclinal rocks.  It is only if the composition is calculated taking volume into account that

pelagic sediment is found to be very different from all other assemblages [Poldervaart, 1955; Sibley and Wilband, 1977; Southam and Hay, 1977, 1981].  Thus, because the volume of passive margin deposits had been underestimated, because its age distribution had not been understood, and because the composition of pelagic sediment had been incorrectly assumed to resemble that of platform and geosynclinal sediments, it was possible for Budyko and Ronov to disregard these complications, believing that they would have little effect on their conclusions.

Did Budyko and Ronov find evidence for higher $CO_2$ levels in the atmosphere in the geologic past?  Their argument that the $CO_2$ content of the atmosphere would be proportional to the $CO_2$ content of rocks formed at a given time may be correct, but their data represent only rocks presently existing on the continents.  Because of recycling and the shift in site of carbonate deposition to the deep sea, their data do not provide the required value of the $CO_2$ content of rocks formed at a given time.  However, in addition to the volumes and masses of carbonate, Budyko and Ronov determined the volumes and masses of volcanogenic rocks, primarily associated with active margins.  Ronov [1980] found correlations between carbonate rock volume, sea level, and volume of volcanics.  The relation between volume of carbonate rocks on the continents and sea level is readily understood as discussed above.  The volume of volcanics is undoubtedly related to rate and volume of material subducted per unit time, and hence to the ultimate control of sea level: the rate of production of young oceanic crust.  Budyko and Ronov assumed that incremental $CO_2$ inputs to the atmosphere were a result of increased volcanic activity, so that the more volcanism that occurred in a given time period, the higher the atmospheric $CO_2$ level.  The recent geochemical model of Berner et al. [1983] also relates higher $CO_2$ levels in the Cretaceous to higher rates of subduction and volcanism.  Thus, Budyko and

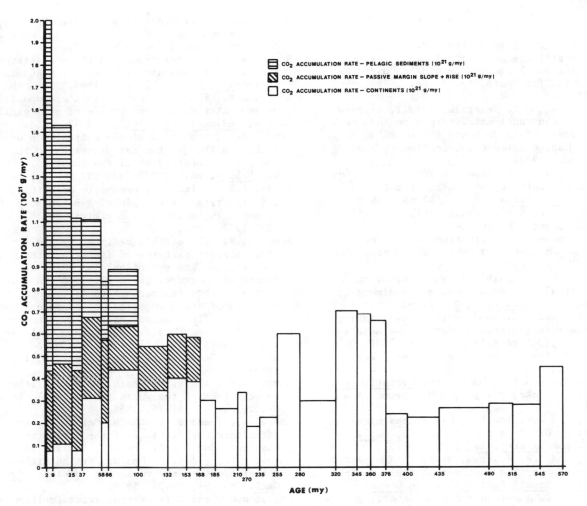

Fig. 1.  Accumulations of $CO_2$ as carbonate on the continents [from Ronov 1980], in passive margin slope and rise deposits, and in pelagic sediments.

Ronov's argument for the importance of volcanics as a clue to the rate of $CO_2$ introduction into the atmosphere is valid, and their data, although again affected by recycling and hence not entirely accurate as a quantitative estimate, are a good reflection of the geologic record.  In this sense, Budyko and Ronov did find qualitative evidence for higher $CO_2$ levels in the past.

A complicating factor in the relationship between subduction, volcanism, and atmospheric $CO_2$ is the time-dependent changing nature of the pelagic sediment being subducted.  If pelagic sediments were largely devoid of carbonate prior to 100 m.y. ago, $CO_2$ inputs from subduction-related volcanism should have been much reduced. This matter needs further investigation to

clarify the atmospheric $CO_2$/volcanism/carbonate relations prior to the mid—Cretaceous.

The work of Ronov and his colleagues in analyzing sediment quantitatively has provided an invaluable data base which shows many significant trends which need to be explained and understood.  It is important that further studies extend the data base so that the rocks beneath the sea can be included to provide a complete inventory of existing sediment.

Acknowledgements.  This work was supported by grant OCE 8218194 from the Ocean Sciences Section of the National Science Foundation.  The figure was drafted by Michael Rosol and the manuscript typed by Rebecca Phillips.

Scott W. Starratt
Dept. of Paleontology
U. C. Berkeley
Berkeley, Ca. 94720

## References

Berner, R.A., A.C. Lasaga, and R.M. Garrels, The carbonate-silicate geochemical cycle and its effect on atmospheric carbon dioxide over the past 100 million years, Am. J. Sci., 283, 641-683, 1983.

Bryan, G.M., and J.R. Heirtzler (Eds.), Eastern North American Continental Margin and Adjacent Ocean Floor, 28° to 36°N and 70° to 82°W, 41 + a-h pp., Marine Science International, Woods Hole, Mass., 1984.

Budyko, M.I., and A.B. Ronov, Chemical evolution of the atmosphere in the Phanerozoic (in Russian), Geokhimiya, no. 5, 643-653, 1979. (Geochem. Int., Engl. Transl., 15, 1-9, 1979.)

Davies, T.A., and T.R. Worsley, Paleoenvironmental implications of oceanic carbonate sedimentation rates, SEPM Spec. Pub. 32 169-179, 1981.

El Wakeel, S.K., and J.R. Riley, Chemical and mineralogical studies of deep sea sediments, Geochim. Cosmochim. Acta, 25, 110-146, 1961.

Ewing, J.I., and P.D. Rabinowitz (Eds.), Eastern North American Continental Margin and Adjacent Ocean Floor, 34° to 41°N and 68° to 78°W, 32 + a-h pp., Marine Science International, Woods Hole, Mass., 1984.

Garrels, R.M., and F.T. Mackenzie, Evolution of Sedimentary Rocks, 397 pp., W.W. Norton, New York, 1971.

Hay, W.W., and J.R. Southam, Calcareous plankton and loss of CaO from the continents, Geol. Soc. Am. Abstr. Programs, 7, 1105, 1975.

Hay, W.W., and J.R. Southam, Modulation of marine sedimentation by the continental shelves, in The Fate of Fossil Fuel $CO_2$ in the Oceans, edited by N.R. Anderson and A. Malahoff, pp. 569-604, Plenum, New York, 1977.

Khain, V.Ye., A.B. Ronov, and A.N. Balukhovskiy, Cretaceous lithologic associations of the world (in Russian), Sov. Geol., no. 11, 10-39, 1975.

Khain, V.Ye., A.B. Ronov, and K.B. Seslavinskiy, Silurian lithologic associations of the world (in Russian), Sov. Geol., no 5, 21-42, 1977.

Khain V.Ye., A.B. Ronov, and A.N. Balukhovskiy, Neogene lithologic associations of the world (in Russian), Sov. Geol., no. 10, 15-23, 1979.

Milliman, J.D., Marine carbonates, in Recent Sedimentary Carbonates, vol. 1, edited by J.D. Milliman, G. Mueller, and U. Foerster, xv + 375 pp., Springer-Verlag, New York, 1974.

Moore, T.C., and G.R. Heath, Survival of deep-sea sedimentary sections, Earth Planet Sci. Lett. 37, 71-80.

Poldervaart, A., Chemistry of the earth's crust, Geol. Soc. Amer. Spec. Pap., 62, 110-144, 1955.

Ronov, A.B., A history of the sedimentation and epeirogenic movements of the European part of the USSR (based on the volumetric method) (in Russian) Tr. Geofiz. Inst. Akad Nauk SSR, 3(130), 390, 1949.

Ronov, A.B., Post-Precambrian geochemical history of the atmosphere and hydrosphere (in Russian), Geokhimiya, no. 5, 379-409, 1959.

Ronov, A.B., General trends in the composition of the crust, ocean, and atmosphere (in Russian), Geokhimiya, no. 8, 715-743, 1964.

Ronov, A.B., Probable changes in the composition of sea water during the course of geological time, Sedimentology, 10(1), 25-43, 1968.

Ronov, A.B., Evolution of the composition of the rocks and the geochemical processes in the earth's sedimentary shell (in Russian), Geokhimiya, no. 2, 137-147, 1972.

Ronov, A.B., Volcanism, carbonate deposition, and life (patterns of the global geochemistry of carbon) (in Russian), Geokhimiya, no. 8, 1252-1277, 1976.

Ronov, A.B., The earth's sedimentary shell (quantitative patterns of its structures, compositions and evolution) The 20th V.I. Vernadskiy Lecture, March 12, 1978 (in Russian) in The earth's sedimentary shell (quantitative patterns of its structures, compositions and evolution) (in Russian), edited by Yaroshevskiy pp. 1-80 Nauka, Moscow, 1980 (Int. Geol. Rev., Engl. transl., 24, 1313-1363, 1365-1338, 1982; also; Am. Geol. Inst. Reprint ser., 5, 1-73, 1983).

Ronov, A.B., and V.Ye. Khain, Devonian lithologic associations of the world (in Russian), Sov. Geol., no. 41, 47-76, 1954.

Ronov, A.B., and V.Ye. Khain, Carboniferous lithologic associations of the world (in Russian), Sov. Geol., no. 48, 92-117, 1955.

Ronov, A.B., and V.Ye. Khain, Permian lithologic associations of the world (in Russian), Sov. Geol., no. 54, 20-36, 1956.

Ronov, A.B., and V.Ye. Khain, Triassic lithologic associations of the world (in Russian), Sov. Geol., no. 1, 27-48, 1961.

Ronov, A.B., and V.Ye. Khain, Jurassic lithologic associations of the world (in Russian), Sov. Geol., no. 1, 9-34, 1962.

Ronov, A.B., and V.Ye. Khain, Fundamental problems of the relationships between sedimentation and tectonics, (in Russian), in State and Objectives of Soviet Lithology (in Russian), vol. 1, pp. 67-709, Nauka, Moscow, 1970.

Ronov, A.B., and A.A. Migdisov, Evolution of the chemical composition of the rocks of the shields and the sedimentary cover of the Russian and North American platforms (in Russian), Geokhimiya, no. 4, 403-438, 1970.

Ronov, A.B., and A.A. Yaroshevskiy, Chemical structure of the earth's crust (in Russian), Geokhimiya, no. 11, 1285-1309, 1967.

Ronov, A.B., and A.A. Yaroshevskiy, A new model for the chemical structure of the earth's crust (in Russian), Geokhimiya, no. 12, 25-38, 1976. (Geochem. Int., Engl. Transl., 13, 89-121, 1977).

Ronov, A.B., A.A. Migdisov, and N.V. Barskaya,

Evolutionary patterns of the sedimentary rocks and the paleogeographic conditions of sedimentation on the Russian Platform (An attempt at a quantitative study) (in Russian), Litol. Polezn. Iskop., no. 6, 3-36, 1969.

Ronov, A.B., A.A. Migdisov, and A.A. Yaroshevskiy, Geochemical history of the earth's outer shells, in Reviews of Current Geochemistry and Analytical Chemistry (in Russian), pp. 88-98, Nauka, Moscow, 1972a.

Ronov, A.B., A.A. Migdisov, and V.Ye. Khain, Reliability of quantitative methods of investigation in lithology and geochemistry (in Russian), Litol. Polezn. Iskop., no. 1, 3-26, 1972b.

Ronov, A.B., A.A. Migdisov, and V.Ye. Khain, Capabilities and limitations of the volumetric method (as exemplified by the Russian platforms and the deep trenches surrounding it) (in Russian), Litol.Polezn. Iskop., no. 4, 3-14, 1973.

Ronov, A.B., K.B. Seslavinskiy, and V.Ye. Khain, Cambrian lithologic associations of the world (in Russian), Sov. Geol., no. 12, 10-33, 1974. (Int. Geol. Rev., Engl. Transl. 4, 373-394, 1974).

Ronov, A.B., V.Ye. Khain, and K.B. Seslavinskiy, Ordovician lithologic associations of the world (in Russian), Sov. Geol., no. 1, 7-27, 1976a. ed in (Int. Geol. Rev., Engl. Transl., 12, 1395-1412, 1976.)

Ronov, A.B., A.A. Migdisov, and S.B. Lobach-Zhuchenko, Problems of the evolution of the chemical composition of the sedimentary rocks, and regional metamorphism (in Russian), Geokhimiya, no. 2, 163-186, 1977.

Ronov, A.B., V.Ye. Khain, and A.N. Balukhovskiy, Paleogene lithologic associations of the continents (in Russian), Sov. Geol., no. 3, 10-42, 1978. (Int. Geol. Rev., Engl. Transl., 4, 415-446, 1979.)

Sclater, J.G., C. Jaupart, and D. Galson, The heat flow through oceanic and continental crust and the heat loss of the earth, Rev. Geophys. Space Phys., 18, 269-311, 1980.

Shor, A.N., and E. Uchupi, (Eds.), Eastern North American Continental Margin and Adjacent Ocean Floor, 39° to 46°N and 54° to 64°W, 24 + a-h pp., Marine Science International, Woods Hole, Mass., 1984.

Sibley, D.F., and T.A. Vogel, Chemical mass balance of the earth's crust: The calcium dilemma and the role of pelagic sediments, Science, 192, 551-553, 1976.

Sibley, D.F., and J.T. Wilband, Chemical balance of the earth's crust, Geochim. Cosmochim. Acta, 41, 545-554, 1977.

Southam, J.R., and W.W. Hay, Time scales and dynamic models of deep-sea sedimentation, J. Geophys. Res., 82, 3825-3842, 1977.

Southam, J.R., and W.W. Hay, Global sedimentary mass balance and sea level changes, in The Oceanic Lithosphere, The Sea, vol. 7, edited by C. Emiliani, pp. 1617-1684, John Wiley, New York, 1981.

Uchupi, E., and A.N. Shor, (Eds.), Eastern North American Continental Margin and Adjacent Ocean Floor, 39° to 46°N and 64° to 74°W, 30 + a-h pp., Marine Science International, Woods Hole, Mass., 1984.

Veizer, J., and S.L. Jansen, Basement and sedimentary recycling and continental evolution, J. Geol., 87, 341-370, 1979.

Vinogradov, A.P., and A.B. Ronov, Composition of the sedimentary rocks of the Russian Platform in relation to the history of its tectonic movements (in Russian), Geokhimiya, no. 6, 3-24, 1956.

Worsley, T.R., D. Nance, and J.B. Moody, Global tectonics and eustasy for the past 2 billion years, Mar. Geol., 58: 373-400.

# NONSKELETAL ARAGONITE AND pCO$_2$ IN THE PHANEROZOIC AND PROTEROZOIC

Philip A. Sandberg

Department of Geology, University of Illinois, Urbana, Illinois 61801

Abstract. Calcitized, originally aragonitic
nonskeletal carbonates may be recognized
by a combination of textural and compositional
properties including: disruptive replacement
textures, relic aragonite inclusions, and
elevated strontium content. Temporal distribution
of once-aragonitic constituents appears
to be discontinuous during the Phanerozoic.
That suggests an alternation of "aragonite-
inhibiting" conditions, in which nonskeletal
carbonates formed only as calcite (of uncertain,
but perhaps lower Mg$^{2+}$ content), and "aragonite-
facilitating" conditions, in which both
aragonite and calcite (mainly high-magnesium
calcite) formed. Fischer (1981, 1982)
and Mackenzie and Pigott (1981) suggested
that pCO$_2$ changes are coupled to changes
in plate tectonic mode. Control of nonskeletal carbonate mineralogy by the influence
of pCO$_2$ or some covariant factor most probably oceanic Mg/Ca) is suggested by the
close agreement between sea level curves
and the presence-absence curves for nonskeletal aragonite. Times of aragonite inhibition
by elevated pCO$_2$ would probably also have
been times of diminished non-skeletal carbonate precipitation and of some submarine
occurrence of processes commonly regarded
as meteoric, such as carbonate mud stabilization and molds or void-fill calcite casts
of aragonite skeletons. A few examples
of apparent nonskeletal aragonite occur
in Precambrian rocks at least as old as
-1.9 Ga. The inferred levels of pCO$_2$ associated with apparent inhibition of nonskeletal
aragonite precipitation in the Phanerozoic
make it difficult to reconcile possible
occurrences of Precambrian aragonite cements
and ooids with standard interpretations
of very high pCO$_2$ throughout the Precambrian.

## Introduction

Literature on carbonates has generally
suggested that originally aragonitic nonskeletal carbonates were present throughout
the Phanerozoic. A wide range of nonskeletal
carbonates with diverse states of textural
preservation or disruption have been taken
to have been originally aragonite, largely
because of an assumed equivalence between
ancient ooids and submarine cements and
their modern counterparts, as in the Bahamas
or Persian Gulf. This includes constituents
such as radial calcite ooids and fibrous,
isopachous calcite cements which retain
fine texture and regular optic orientations, purportedly as the result of pseudomorphic or paramorphic calcitization. More
recent detailed study of aragonite diagenesis
indicates that calcitization produces distinctive, disrupted textures, commonly with
aragonite relics and/or elevated Sr$^{2+}$ content
[Sandberg, 1975, 1984]. The common thread
I observe among diverse aragonites of widely
different geologic ages is that pseudomorphic
or paramorphic calcitization does not appear
to be a potential diagenetic fate for aragonite, at least not in natural settings. This
does not refer to preservation of the morphology of a grain despite replacement (e.g.,
a calcite "pseudomorph" of an aragonite
gastropod shell, equivalent to saying "a
calcitized shell"). I mean that pseudomorph-
ing of fine crystals does not occur.

It has recently been suggested that
original mineralogy of nonskeletal marine
carbonates (ooids, cements) has changed
throughout the Phanerozoic [Folk, 1974;
Sandberg, 1975; Milliken and Pigott, 1977;
Mackenzie and Pigott, 1981], probably in
a nonlinear, oscillatory manner [Sandberg
and Popp, 1981; Sandberg, 1983].

Evidence for presence of original aragonite
mineralogy varies in its reliability (see
discussion by Sandberg [1983, 1984]).
Unequivocal states include preservation
as aragonite, or calcitized with aragonite
relics. Disrupted texture without aragonite
relics and with or without elevated Sr$^{2+}$
content is less secure. Moldic preservation
is equivocal. External morphologies (botryoids, square-end crystals) may also be

useful supplemental criteria. Using, as much as possible, the most reliable compositional and textural criteria, nonskeletal aragonite appears (Figure 1) to be restricted to times of emergent continents, low rates of sea floor spreading and subduction, and thus low pCO$_2$ and perhaps also lower oceanic Mg/Ca.

Three reports of possible aragonite in the Precambrian suggest that the trend may be extendable into even earlier earth history. Because of the scarcity of data, Precambrian oscillations cannot be defined. Nevertheless, the potential presence of aragonite (particularly at ~1.9 Ga) poses some problems for standard interpretations of Precambrian atmosphere composition.

## Discussion

### Carbonate Data for Phanerozoic Cycles

A variety of Phanerozoic nonskeletal carbonates (both ooids and cements) were accepted, on the basis of relatively secure criteria [Sandberg, 1983], as originally aragonitic. The temporal distribution of those inferred aragonites suggested an oscillatory trend through the Phanerozoic, with nonskeletal aragonite only intermittently present (Figure 1). A number of references not included in that study warrant discussion here. Rich [1982] reported some Late Mississippian ooids which were composed of coarse, irregular calcite traversed by relics of concentric laminations. Strontium content and possible presence of aragonite relics were not investigated. Although no fine-textured, radial-concentric ooids occurred in the unit, Rich reported that such ooids occurred in beds both above and below.

Controls over such short-term variation in ooid mineralogy constitute one of the more intriguing questions of ooid history. The fact that both types of ooids can occur in the same bed [e.g., Sandberg, 1983, Figure 4] suggests that the differences in environment are probably not extreme. Deposits of mixed mineralogy could result from later mixing of temporally successive layers of aragonitic and of calcitic ooids, or through transport from different environments of formation into a common site of accumulation. In the former case, one might expect that the mixing would diminish vertically away from layers deposited near the time of mineralogic transition. In the latter case, one might expect contemporaneous, geographically separated deposits of each of the two mineralogies. In addition to mixed deposits of aragonitic ooids and calcitic ooids, there are deposits of bimineralic ooids (inner cortex originally aragonitic, outer cortex originally calcite,

or vice versa) [Sandberg, 1983], as in the Pennsylvanian of Kansas.

In addition to the "secure" aragonite cements of the mid-Phanerozoic cataloged earlier [Sandberg, 1983], Dauphin and Cuif [1980] found unaltered aragonite cement on Norian (Late Triassic), still-aragonitic cephalopods and bivalves. Peryt [1981] found what appeared to be calcitized botryoidal aragonite cements in the upper Permian of Poland.

Wright [1981] argued that facicular optic calcite (FOC) layers in some Lower Carboniferous pisoids had been originally aragonitic. Both FOC and its common companion, radiaxial fibrous calcite (RFC), retain a regular, fan-like extinction sweep, reflecting crystals or subcrystals which are elongate and generally perpendicular to the substrate. Neither FOC nor RFC is reasonably derivable from calcitization [Sandberg, 1984]. The optic regularity and uniformity of crystal size, shape and orientation which they typically possess does not characterize calcitized known aragonites. Elongate clear regions in the FOC, regarded by Wright as relics, generally have morphologic and extinction orientations appropriate for their position in the bundle of elongate FOC crystals. Aragonite relics in calcitization mosaics typically contrast strongly in both size and optic orientation with the surrounding calcite. The "square-end terminations" in Wright's material are not valid as indicators of aragonite mineralogy for the reasons stated by Folk [1977].

Ooids of the Burgsvik oolite (Silurian, Gotland, Sweden) were included by Richter [1983] among those ancient ooids with textural properties indicative of original aragonite mineralogy. In thin section, they are composed of one or a few calcite crystals in which there are uncommon but clear concentric growth increments, generally with a nucleus evident. The Sr$^{2+}$ content has not yet been analyzed, but preliminary study by means of scanning electron microscopy and energy dispersive x-ray has not shown any aragonite relics. There are, however, abundant clays intermingled throughout the calcite of the ooids, sometimes delineating concentric structure and sometimes irregularly transecting parts of the ooid. Clays are most common at or near the ooid peripheries. Perhaps significantly, the clays are not present in the intergranular cement. All this does not preclude original aragonite mineralogy, but points out some distinctive dissimilarities between these ooids and typical alteration textures in calcitized ooids. Clays both within and around the ooids serve to outline regions of calcite with apparent growth faces. Those regions include overgrow-

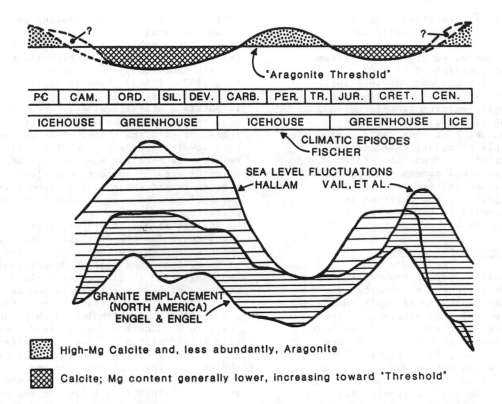

Fig. 1.  Inferred first-order cycles in nonskeletal carbonate mineralogy, showing correlation to cycles of climate, sea level (from Vail et al. [1977] and Hallam [1977] and granite emplacement (from Engel and Engel [1964]), as given by Fischer [1981, 1982]. See Mackenzie and Pigott [1981] for comments on relationship of $pCO_2$ and $Mg^{2+}$ content in calcite.  The presence of aragonite correlates well with times of low sea level. From Sandberg [1983], reproduced with permission of the publishers of Nature.

ths with relatively large crystal termina-tions (10-50 μm wide) which can give the ooid a rather dentate margin.  In some cases, there is apparent growth of such crystals into the nucleus region, suggesting a metastable nucleus which dissolved, and whose volume is being infilled.

Unaltered aragonitic skeletons are relatively common in middle and later Mesozoic rocks, certainly relative to the rare examples (in impermeable lithologies) in older rocks. It is therefore reasonable to infer that, if aragonitic cements (or ooids) had formed during that part of the Mesozoic, they should have at least the same chance of preservation and eventual discovery.  That interpretation is strengthened by empirical evidence that nonskeletal aragonite is less susceptible to dissolution or calciti-zation than are skeletal aragonites (see discussion by Sandberg [1984]).  There are rather common instances of calcitized aragonitic skeletons with overlying, unalter-ed aragonite cement [e.g., Land, 1967; Sandberg and Popp, 1981].  The difference

may relate to relative abundances of lattice defects, which appear to be quite high in skeletal aragonite (R. J. Reeder, per-sonal communication, 1984).

Unequivocal submarine cements of middle to late Mesozoic age have been described, as in the Jurassic ooid hardgrounds of the Paris Basin [Purser, 1969, 1971]. However, they are all calcite and nearly always fine radial (but see Wilkinson, et al., [1984]).

Kahle [1965] published a curve of temporal variation in $Sr^{2+}$ content in ooids, based on 62 oolitic samples nonuniformly distributed from Cambrian to Recent.  Although Kahle did not have data on Permian or Triassic ooids, his mean value for Pennsylvanian (four samples) was 950 ppm (range: 610-1430 ppm).  It is probably not fortuitous that the overall form of his curve is quite similar to the temporal curve for nonskeletal aragonite (Figure 1).  Given the roughly 9,000-10,000 ppm $Sr^{2+}$ in modern Bahamian ooids [Milliman, 1974], the value of 280 ppm $Sr^{2+}$ reported by Kahle from the predominantly oolitic Miami Limestone (Pleistocene, south Florida) is unusually low. That is true, even if one assumes a completely open, water-dominated diagenetic system and a partition coefficient of 0.05. That very low Sr$^{2+}$ value had considerable influence on interpretations of strontium distribution in aragonite-calcite transformation. It is important to note, however, that the sample from which it came was a very quartzose, partly peloidal, essentially nonoolitic part of the Miami Limestone (C. F. Kahle, personal communication, 1980), composed largely of calcite cement. The value is thus probably not relevant to aragonite diagenesis in the way that has been implied in the literature.

## Global Tectonic-Climatic Cycles

In a long-delayed paper, Brouwer [1983] pointed out the influence of conditions beyond the local depositional environment, i.e., global controls, in determining the nonrandom temporal distribution of sediments of special character in the geologic record. He suggested [Brouwer, 1983, Figure 2] a link to mantle convection and plate tectonic mode, and identified chains of interrelated events, correlating plate tectonic influence on sea level with changes in pCO$_2$, erosion and sedimentation rates, extinction and evolution of shallow marine organisms, and extent of carbonate sedimentation. Other authors, mentioned above, have also correlated plate tectonic mode with temporal changes in pCO$_2$, glaciation, and rates of evolution and extinction and with nonskeletal carbonate mineralogy.

As a first-order approximation, the occurrence of "aragonite plus calcite" versus "calcite only" tracks fairly well the first-order sea level curves (Figure 1). These first-order tectonic-eustatic cycles in the Phanerzoic [e.g., Vail et al., 1977; Hallam, 1977] and potentially also in the Precambrian [Worsley et al., 1984] are modified by second order oscillations, such as the late Jurassic and mid-Cretaceous highs and the late Ordovician (Ashgillian) low. Sheridan [1983] interpreted the secondary oscillations as products of pulsation tectonics, an influence on spreading rates and hence eustatic levels by cyclic mantle plume eruptions. Such secondary oscillations in factors ultimately influencing pCO$_2$ could conceivably cause a temporary reversal of the first-order trend in nonskeletal carbonate mineralogy. The most easily recognizable type of "reversed subepisode" would be the presence of aragonite in the midst of an aragonite-inhibiting episode. The reverse condition, absence of aragonite in any given unit deposited

during an aragonite-facilitating episode, is not conclusive evidence of a reversed trend. Even subrecent or modern nonskeletal carbonates may locally be high-magnesium calcite (HMC) only (see discussion by Sandberg [1975, 1984]). Thus for delineation of temporal trends in mineralogy, the problem is partly a statistical one. For any given age, the significance of aragonite absence increases with the number of cement and/or ooid samples studied. An obvious point here is the need for more careful determination of ages of ancient nonskeletal aragonites to see if their distribution reflects any influence by second-order tectonic-eustatic cycles.

Leggett et al. [1981] presented a sea level curve for the Ordovician and Silurian which showed an emergent state during most of Ashgillian time (latest Ordovician). The Ashgillian is also characterized by evidence of near-polar glaciation in the North African region [Fairbridge, 1964] and lacks the black shales which predominate in much of the underlying Ordovician (Llandeillian to Lower Ashgillian) and in the overlying Silurian. The black shales were taken as evidence for the presence, at those times, of the "greenhouse" state of Fischer [1981] (submergent mode of Mackenzie and Pigott [1981]). Those "greenhouse" states, with the intervening "icehouse" state of the Ashgillian, exemplify the second-order tectonic-eustatic oscillations for which corresponding second-order oscillations might exist in the trend of nonskeletal carbonate mineralogy. Carbonates of Ashgillian age are a good place to look for such evidence, in the form of originally aragonitic ooids or cements. In the few samples of Ashgillian age which I have seen, those constituents have textures indicative of original calcite mineralogy.

## Possible Mechanisms for Mineralogic Changes

Nonskeletal aragonite precipitation may have been enhanced since the Jurassic by changes in oceanic Mg/Ca ratio due to decrease in extent of dolomitization [Holland, 1984] and to the shift of major carbonate deposition to low-magnesium calcite (LMC) plankton in the deep sea (see discussion by Sandberg [1975]). However, those factors cannot account for the apparent earlier, intermittent Phanerozoic occurrence of nonskeletal aragonite. Control by cyclic changes in pCO$_2$ or related effects is in accord with the timing of mineralogic changes extending back at least to the early Paleozoic (Figure 1), with presence of aragonite temporally linked to low pCO$_2$. The most likely related effect is change in the oceanic Mg/Ca ratio (only partly from the effect of LMC plankton and the post-Paleozoic decrease in dolomitization), acting synergistically with $pCO_2$ changes to promote or inhibit aragonite. Increase in magnesium removal by oceanic basalt alteration is likely during times of high spreading rates. The sea level curve may reflect rate of seawater cycling through ridges, with high sea level corresponding to high spreading rates and larger ridge volumes and thus higher probability of high-temperature seawater-basalt reactions. However, Holland [1984] has also suggested that the temporal curve of $\delta^{87}Sr$ variation [Burke, et al., 1982] may reflect changes in seawater cycling through mid-ocean ridge (MOR) systems. As pointed out by Holland [1984], basalt-seawater interaction is not simply a matter of ridge volume or spreading rate. The amount of high-temperature, fast-spreading ridge is of prime importance. Holland noted that the good correlation between time variations of $\delta^{87}Sr$ and $\delta^{34}S$ supports the interpretation that variation of both may be the result of changes in rates of seawater cycling through MOR systems. Sea-level curves compiled for the Phanerozoic by different authors vary, but none shows close covariance with the $\delta^{87}Sr$ curve of Burke, et al. [1982]. There is thus a quandary regarding the timing of the plate tectonically induced changes in factors likely to influence nonskeletal carbonate mineralogy. At least to a first approximation, the temporal distribution of apparent oscillations in carbonate mineralogy more closely correlates with the sea level fluctuations.

A proposal that the apparent mineralogical oscillations in the Phanerozoic are related to changes in $pCO_2$ is not without its problems. A possible easing of the dilemma is afforded by a concurrent effect of changes in oceanic Mg/Ca ratio on carbonate mineralogy. However, it is not clear that a lowered Mg/Ca ratio alone would inhibit aragonite in the manner implied by the geologic record.

If times of aragonite inhibition are correlated to high $pCO_2$, they should also be times when inorganic carbonate pecipitation in general would be less favored. One might therefore expect that ooids and submarine cements might be less common in the submergent episodes. The ooid distribution curve of Sandberg [1983, Figure 2] is not necessarily a good measure of true Phanerozoic ooid abundances, but it does suggest a diminishment in ooids for the mid-Paleozoic submergent episode. The lack of such a correlation for the Mesozoic submergent episode may reflect a lesser departure from present-day conditions in the Jurassic-Cretaceous than was true for the Ordovician. This in accord with the differences in aragonite preservation between those two times (noted elsewhere in this paper). The VFC preservation of molluscan shells which is so characteristic of the Ordovician could even be partly the product of submarine dissolution and sparry cement infilling. Thus the sea floor may at times have been the location for diagenetic phenomena we tend to think of as exclusively meteoric, such as equant spar formation [Folk, 1974; Purser, 1971] and carbonate mud stabilization. Intraformational conglomerates in micritic limestones are fairly common in the early Paleozoic and in the Mississippian. A statistical analysis of temporal distribution of such indicators of submarine stabilization would be quite useful for evaluation of the speculation presented above.

L. M. Walter (personal communication, 1984) has found experimentally that, although elevated $SO_4^{2-}$ concentration inhibits both polymorphs, the effect is many times stronger on calcite than on aragonite. There is no direct link between $pCO_2$ and $SO_4^{2-}$ concentration, but there may be some connection through mass balance relationships of carbonates and evaporites and $Ca^{2+}$ concentration. Walter also found that $PO_4^{3-}$ more strongly inhibited growth of aragonite than calcite. The relationship of these experimental results to the geologic record is yet unclear. There are cements in the early Cambrian [James and Klappa, 1983] which appear to have been aragonitic. However, abundant shallow water phosphorites of that age suggest a rather high $PO_4^{3-}$ concentration [Cook and Shergold, 1984].

Given and Wilkinson [1984] suggest that $CO_3^{2-}$ levels may influence the mineralogy of cements even in pore systems in which aragonite and calcite may be intimately associated. They further suggest that fibrous and equant morphology are related to high and low $CO_3^{2-}$ concentrations, respectively.

For the apparent Phanerozoic oscillations in $CaCO_3$ mineralogy, the $pCO_2$ itself may be a significant direct factor. Elevated $pCO_2$ would not only lower the $CO_3^{2-}$ concentration but could perhaps also produce a marine system below aragonite saturation, but still supersaturated with respect to calcite. The likely $Mg^{2+}$ content of nonskeletal calcite during such times may not always have been as low as would have been thought recently. The work of Walter [1984] suggests that, if aragonite were absent, nonbiogenic marine calcite need only have been less than 12 mol% $MgCO_3$ rather than <8 mol%, as suggested by the work of Plummer and Mackenzie [1974]. Under those conditions, such "relatively LMC" could be the normal marine precipitate. Equant morphology, perhaps associated with lower Mg$^{2+}$ content, could occur, as suggested by Folk [1974] and found in bored Jurassic hardgrounds by Purser [1971] and also recently by Wilkinson, et al., [1982, 1984] in the Ordovician and Jurassic. Modern submarine cements may vary locally from aragonite to HMC, possibly in response to differences in CO$_3^{2-}$ [Given and Wilkinson, 1984]. Fluctuations among the somewhat lower CO$_3^{2-}$ concentrations during times of high pCO$_2$ could produce coeval submarine cements composed only of calcite which vary locally from fibrous [e.g., Purser, 1969; Sandberg, 1984] to equant [Wilkinson, et al., 1982, 1984]. The work of L. M. Walter (personal communication, 1984) suggests that aragonite-HMC cement successions could relate to bacterial sulfate reduction and organic phosphate buildup in restricted pore waters. During times when higher pCO$_2$ and lower CO$_3^{2-}$ levels might result in only calcitic submarine cements, the potential effect of SO$_4^{2-}$ and PO$_4^{3-}$ is uncertain. Could they also influence morphology in calcite?

Ooid mineralogy might be a better indicator of ancient ocean chemistry than is submarine cement mineralogy. Ooids are formed more nearly in the open ocean, but in the relatively restricted intergranular cavities in which cements form, the water composition may be quite different. The picture is complicated, however, by the possible influence of microorganisms on tangential ooid formation in modern marine environments.

If elevated pCO$_2$ influenced nonskeletal carbonate mineralogy in the geologic past, what about possible influence on diagenetic behavior of carbonates? One might postulate that diagenetic environments would have been more corrosive to carbonates during times of high atmospheric pCO$_2$. However, soil gas CO$_2$ content can be elevated by soil microbe activity to values many times present atmospheric level (PAL) and probably above values likely to be achieved by long-term, plate tectonically influenced mechanisms. Thus, the extent of organic-rich soil development, as a function of terrestrial vegetation [see Holland, 1984; James and Choquette, 1984], may influence meteoric (especially vadose) diagenetic environments more than long term oscillations in atmosphere-hydrosphere pCO$_2$.

Mode of preservation of aragonitic shells shows a change through the Paleozoic that does not fit well with a simple model of increasing influence of soil CO$_2$ on diagenetic fluids as terrestrial plant communities evolved. Prior to the Devonian-Mississippian inception of extensive terrestrial floras, soil zone development should have been minimal. It is therefore intriguing that

Ordovician material I have seen is nearly universally characterized by aragonite shell dissolution, suggestive of strongly corrosive diagenetic fluids. In rocks of Ordovician age, molluscan shells are common or even dominant constituents of many limestones. Only a single sample I have studied showed calcitization, with relics of the growth laminations preserved in the replacement calcite. In all others, molluscan shells are preserved as void-fill calcite (VFC) casts. VFC preservation of some shells still occurs in many younger Paleozoic rocks. However, in Devonian and especially Carboniferous rocks, calcitized aragonite shells become more common and may even be locally the dominant mode of preservation.

Preservation states of once-aragonitic constituents from the Mesozoic submergent episode are even more varied. Jurassic and Cretaceous corals and mollusks and other aragonite skeletons are commonly preserved as VFC. They appear to be less frequently calcitized, but may even sometimes be preserved as unaltered aragonite. Because of that divergence from preservation states of the earlier submergent episode, it is difficult to argue that the long-term oscillations in pCO$_2$ are the only control on aragonite diagenetic mode, at least if alteration is mainly meteoric. If alteration were largely either submarine or a burial diagenetic phenomenon (as implied by some carbon isotope data [Hudson, 1975]), the role of soil gas CO$_2$ would be less important.

## Possible Precambrian Aragonites

Although the inference is presently supported only by diagenetic textures and cement morphology, original aragonite mineralogy is possible for three Precambrian nonskeletal carbonates. These are the ooids from the very late Proterozoic of Norway (about -0.7 Ga, M. E. Tucker, personal communication, 1982), somewhat older ooids (-1.2 Ga [Tucker, 1984]) from Montana, and botryoidal cements from the Wopmay Orogen of NW Canada (-1.9 Ga [Grotzinger and Read, 1983]). The coexisting disrupted and well preserved ooids from -1.2 Ga [Tucker, 1984] are very like those in some Pennsylvanian (upper Carboniferous) oolites in which aragonite ooids have been identified on the basis of aragonite relics [P.A. Sandberg and B. N. Popp, unpublished manuscript, 1984]. Tucker's [1984] material also had rather elevated strontium content. The botryoids of Grotzinger and Read [1983] are dolomitized. However, their morphologies and alteration textures are very like those in Phanerozoic botryoids in which original aragonite mineralogy can be confidently inferred on the basis of high $Sr^{2+}$ content or aragonite relics. Such "secure" criteria may not be present in any of the Precambrian examples just discussed. Strontium content is a function of the openness of the diagenetic system, and even some unequivocal Mesozoic and Cenozoic examples of calcitized aragonites contain no aragonite relics [Sandberg and Hudson, 1983]. Survival of aragonite relics after dolomitization has not yet been observed (see, however, Gebelein et al. [1980, fig. 11]), but I have seen aragonite relics in gypsum-replaced Cenozoic Mediterranean corals.

Those three Precambrian data points are obviously insufficient for demonstrating any oscillatory trend. At best, they imply that conditions favorable for aragonite deposition could have occurred at least several times, as far back as -1.9 Ga. That possibility alone has rather significant implications for the nature of Precambrian environments. Distribution of nonskeletal aragonite in the Phanerozoic suggests that its precipitation is favored by one or more factors related to lower $pCO_2$. That relationship is at odds with the common inference of high $pCO_2$ throughout the Precambrian (as high as 30 times PAL at about -1.9 Ga) [Hart, 1978; Owen et al., 1979].

Except for the ooids at -0.7 Ga, the inferred Precambrian occurrences of aragonite are not close in age to widespread glaciogenic rocks, even given the considerable uncertainty in the ages. However, even in the Phanerozoic, occurrence of nonskeletal aragonite is not always closely linked to times of glaciation. The temporal range of mid-Phanerozoic aragonite (Late Mississippian - ?Early Jurassic) is considerably beyond the time of major glaciation (Pennsylvanian-Early Permian). That range includes the Triassic, which (with the late Cambrian and pre-Ashgillian Ordovician) has perhaps the least evidence of glaciation among Phanerozoic periods [Hambrey and Harland, 1981, pp. 947, 952].

## Precambrian Climate and Aragonite Occurrences

Studies of the "faint early sun" suggest solar luminosity 25% below the present value at -4.5 Ga [Owen et al., 1979]. Evidence (waterlain sediments and stromatolites) of liquid water indicates that, by about -3.5 Ga, any lower surface temperature which would result from lower luminosity must have been compensated by a greenhouse gas. Considerably lower solar luminosity may have been a major contribution to the cooling which produced the early Proterozoic glaciation, despite possibly elevated $CO_2$. However, by late Proterozoic time (-0.75 to -0.5 Ga), luminosity had essential-

ly reached the present value. Furthermore, a steady increase in luminosity cannot explain the oscillations between warmer, nonglacial intervals and the intervening glaciations (-2.6 Ga ?, Witwatersrand; -2.3 to -2.1 Ga, Gowgonda; -0.9 to -0.6 Ga). Fluctuation in the other major factor, greenhouse gas, is implied. For the interval we are concerned with, that is $CO_2$.

Owen et al. [1979] noted that inclusion of the effect of weak absorption bands of $CO_2$ increased the greenhousing potential of that gas sufficiently so that it could alone counter the cooling effect of inferred lesser early solar luminosity. Their model predicted (for surface temperatures in the range of 284°-290°K) $pCO_2$ values of 56 x PAL at -2.5 Ga and 9 x PAL at -1.5 Ga (PAL not reached until about -0.5 Ga). Best fit computer runs of Hart [1978], in contrast, predicted PAL $pCO_2$ back to -0.75 Ga, although his values for -1.5 and -2.5 Ga were comparable to those of Owen et al. [1979].

Frakes [1979] mentioned several possible biologic influences on $CO_2$ abundance in the Precambrian. Such influence would be effected through photosynthetic or related reactions but maintained over long times through the burial of organic carbon or of carbonates precipitated by photosynthetically induced local lowering of $pCO_2$. Carbonate-silicate cycles have a major influence on the $CO_2$ levels and temperatures in the atmosphere-hydrosphere because the "carbon fluxes between reservoirs are very large compared to the amount of $CO_2$ in the atmosphere" [Berner et al., 1983, pp. 679-680].

Funnell [1981, p. 179] pointed out that elevation of $pCO_2$ at various times in the Phanerozoic indicated that $CO_2$ was being "added to the system faster than the biosphere can sequester it." Conversely, inception of widespread photosynthesis in the Precambrian could have played a significant role in decreasing $pCO_2$, particularly at the time of the early Proterozoic glaciations [Frakes, 1979]. Frakes also commented on the conflict between presence of any Proterozoic nonskeletal carbonates at all and the generally accepted inference of contemporaneous high $pCO_2$. His implication that ooid formation is incompatible with conditions in which glaciogenic sediments would form is certainly true on a local or even regional scale. However, extrapolation of that observation to suggest that ooids and glaciogenic sediments cannot be contemporaneous on a global scale is refuted by the presence of numerous oolitic deposits in the Pleistocene and Permo-Carboniferous [Sandberg, 1983]. The very widespread low latitude [Piper, 1973] or even equatorial [Harland, 1981] glaciation which appears to have occurred during latest Precambrian time certainly suggests significant decrease in $CO_2$ and its greenhousing effect. The apparent presence of aragonite at about -0.7 Ga [M. E. Tucker, unpublished manuscript, 1984] and in the early Cambrian [James and Klappa, 1983] supports that interpretation.

Manabe and Wetherald [1980] modeled an atmosphere with doubled and quadrupled $CO_2$. They found that $CO_2$ increase of that magnitude resulted in significant increase in poleward latent heat transport and strong reduction not only in meridional temperature gradient, but also in variability of the model climate.

During the mid-Paleozoic and middle to late Mesozoic, there was an apparent inhibition of nonskeletal aragonite. $CO_2$ levels at those times were well below those which most authors envision for the Precambrian. If the apparent Proterozoic aragonite occurrences and the inferred $CO_2$-related control on aragonite precipitation are both valid, they are difficult to reconcile with conventional views on the nature of the Precambrian atmosphere.

Using an approach based on constraints of the known sedimentary record (like the study of Holland, [1972]), Walker [1983] calculated possible limits on the composition of the Archean ocean, including the proposed high levels of $pCO_2$ implied by greenhousing requirements. He found [Walker, 1983, p. 518] no inconsistencies between the postulated higher $pCO_2$ and the Archean record, "provided that high partial pressure was accompanied by generally lower pH for seawater, higher concentrations of calcium and bicarbonate, and lower concentration of carbonate and sulphate ions." As noted by the work of L. M. Walter (personal communication, 1984) and of Given and Wilkinson [1984], lower sulfate and carbonate concentrations would make aragonite precipitation less likely. If $pCO_2$ levels had been consistently high during the Precambrian, then nonskeletal aragonite probably should not have occurred.

## Summary Statement

In assessing the geologic history of aragonite, we need data on two major topics: (1) diagenetic fate of aragonite and recognizability of its former presence in ancient rocks, and (2) the factors which control the likelihood of nonskeletal aragonite deposition. The first is critical for delineation of the temporal trends in nonskeletal carbonates on which an environmental history can be based. Knowledge of the second helps us to make interpretations of ancient environmental conditions, based on the carbonate mineralogy. Clearly, there is a need for more experimental studies on factors influencing the relative precipitation rates of the various carbonate phases.

In the Phanerozoic, there appears to be a good correlation between times of occurrence of nonskeletal aragonite and times of inferred lower $pCO_2$ (and possible higher oceanic Mg/Ca), based on the sea level curves. Nonskeletal aragonite may have been deposited at several times in the Precambrian. Those two observations are in conflict with standard views on the nature of the Precambrian atmosphere, suggesting a need for careful reexamination of assumptions of high $pCO_2$ throughout the Precambrian.

Acknowledgements. This paper has profited from discussions with numerous participants at the AGU Chapman Conference on $CO_2$, January 1984. It represents the culmination of part of the work supported by grant DES 7515208 from the National Science Foundation. I am grateful to T. F. Anderson and Lynn M. Walter for valuable reviews of the manuscript. The support of the facilities and staff of the Department of Geological Sciences, Cornell University, during my time there as a Visiting Scientist, Spring 1984, is gratefully acknowledged.

References

Berner, R. A., A. C. Lasaga, and R. M. Garrels, The carbonate-silicate geochemical cycle and its effect on atmospheric carbon dioxide over the past 100 Million Years, Am. J. Sci., 283, 641-683, 1983.

Brouwer, A., Global controls in the Phanerozoic sedimentary record, Newsl. Stratigr., 12, 166-174, 1983.

Burke, W. H., R. E. Denison, E. A. Hetherington, R. B. Koepnick, H. F. Nelson, and J. B. Otto, Variation of seawater $^{87}Sr/^{86}Sr$ throughout Phanerozoic time, Geology, 10, 516-519, 1982.

Cook, P.J., and J. H. Shergold, Phosphorus, phosphorites and skeletal evolution at the Precambrian-Cambrian boundary, Nature, 308, 231-236, 1984.

Dauphin, Y., and J. P. Cuif, Implications systémtiques de l'analyse microstructurale des rostres de trois genres d'aulacocéridés triasiques (Cephalopoda-Coleoidea), Palaeontographica, Abt. A, 169, 28-50, 1980.

Engel, A. E. J., and C. G. Engel, Continental accretion and the evolution of North America, in Advancing Frontiers in Geology and Geophysics, edited by A. P. Subramaniam and S. Balakrishna, Indian Geophysical Union, pp. 17-37, 1964.

Fairbridge, R. W., Glacial grooves and

periglacial features in the Saharan Ordovician, in Glacial Geomorphology, edited by D. R. Coates, pp. 317-327, State University of New York, Binghamton, 1964.

Fischer, A.G., Climatic oscillations in the biosphere, in Biotic Crises in Ecological and Evolutionary Time, edited by M. Nitecki, pp. 103-131, Academic Press, New York, 1981.

Fischer, A. G., Long-term oscillations recorded in stratigraphy, in Climate in Earth History, edited by W. Berger, pp. 97-104, National Academy Press, Washington, D.C., 1982.

Folk, R. L., The natural history of crystalline calcium carbonate: effect of magnesium content and salinity, J. Sediment. Petrol., 44, 40-53, 1974.

Folk, R. L., Reply: Calcite and aragonite fabrics Carlsbad Caverns, J. Sediment. Petrol., 47, 1400-1401, 1977.

Frakes, L. A., Climate throughout geologic time, Elsevier, New York, 1979.

Funnell, B. M., Mechanisms of autocorrelation, J. Geol. Soc. London, 138, 177-181, 1981.

Gebelein, C. D., R. P. Steinen, P. Garrett, E. J. Hoffman, J. M. Queen, and L. N. Plummer, Subsurface dolomitization beneath the tidal flats of central west Andros Island, Bahamas, in Concepts and Models of Dolomitization, Spec. Publ. 28, edited by D. H. Zenger, J. B. Dunham, and R. L. Ethington, pp. 31-49, Society of Economic Paleontologists and Mineralogists, Tulsa, Okla., 1980.

Given, R. K., and B. H. Wilkinson, Morphology, composition, and mineralogy of abiotic sedimentary carbonates: Control by kinetic reaction parameters and physical environment, J. Sediment. Petrol., in press, 1984.

Grotzinger, J. P., and J. F. Read, Evidence for primary aragonite precipitation, Lower Proterozoic (1.9 Ga) Rocknest Dolomite, Wopmay Orogen, Northwest Canada, Geology, 11, 710-713, 1983.

Hallam, A., Secular changes in marine inudation of USSR and North America throughout the Phanerozoic, Nature, 269, 769-772, 1977.

Hambrey, M. J., and W. B. Harland, Summary of earth's pre-Pleistocene glacial record, in Earth's Pre-Pleistocene Glacial Record, edited by M. J. Hambrey and W. B. Harland, pp. 943-954, Cambridge University Press, New York, 1981.

Harland, W. B., Chronology of earth's glacial and tectonic record, J. Geol. Soc. London, 138, 197-203, 1981.

Hart, M. H., The evolution of the atmosphere of the earth, Icarus, 33, 23-29, 1978.

Holland, H. D., The geologic history of seawater -- An attempt to solve the problem, Geochim. Cosmochim. Acta, 36, 637-651, 1972.

Holland, H. D., The chemical evolution of the atmosphere and oceans, 582 pp., Princeton University Press, Princeton, N.J., 1984.

Hudson, J. D., Carbon isotopes and limestone cement, Geology, 3, 19-22, 1975.

James, N. P., and P. W. Choquette, Limestones-- The meteoric diagenetic environment, Geosci. Can., in press, 1984.

James, N. P., and C. F. Klappa, Petrogenesis of early Cambrian reef limestones, Labrador, Canada, J. Sediment. Petrol., 53, 1051-1096, 1983.

Kahle, C. F., Strontium in oolitic limestones, J. Sediment. Petrol., 35, 846-856, 1965.

Land, L. S., Diagenesis of skeletal carbonates, J. Sediment. Petrol., 47, 914-930, 1967.

Leggett, J. K., W. S. McKerrow, L. R. Cocks, and R. B. Rickards, Periodicity in the early Paleozoic marine realm, J. Geol. Soc. London, 138, 167-176, 1981.

Mackenzie, F. T., and J. D. Pigott, Tectonic controls of Phanerozoic sedimentary rock cycling, J. Geol. Soc. London, 138, 183-196, 1981.

Manabe, S., and R. T. Wetherald, On the distribution of climate change resulting from an increase in $CO_2$ content of the atmosphere, J. Atmo. Sci., 37, 99-118, 1980.

Milliken, K. L., and J. D. Pigott, Variation of oceanic Mg/Ca ratio through time -- Implication of the calcite sea, Geol. Soc. Am. Abstr. Programs, 9, 64-65, 1977.

Milliman, J. D., Marine Carbonates, 375 pp., Springer-Verlag, New York, 1974.

Owen, T., R. D. Cess, and V. Ramanathan, Enhanced $CO_2$ greenhouse to compensate for reduced solar luminosity on early earth, Nature, 227, 640-642, 1979.

Peryt, T., Former aragonitic submarine hemispheroids associated with vadose deposits, Zechstein Limestone (Upper Permian), Fore-Sudetic Area, Western Poland, Neues Jahrb. Geol. Palaeontol., Monatsh., 1981, 559-570, 1981.

Piper, J. D. A., Latitudinal extent of late Precambrian glaciations, Nature, 244, 342-344, 1973.

Plummer, L. N., and Mackenzie, F. T., Predicting mineral solubility from rate data. Application to the dissolution of magnesium calcites, Am. J. Sci., 274, 61-83, 1974.

Purser, B. H., Synsedimentary marine lithification of Middle Jurassic limestones in the Paris Basin, Sedimentology, 12, 205-230, 1969.

Purser, B. H., Middle Jurassic synsedimentary marine cements from the Paris Basin, in Carbonate Cements, edited by O. P.

Bricker, pp. 178-184, Johns Hopkins University Press, Baltimore, Md., 1971.

Rich, M., Ooid cortices composed of neomorphic pseudospar: Possible evidence for ancient originally aragonitic ooids, J. Sediment. Petrol., 52, 843-847, 1982.

Richter, D. K., Calcareous ooids: A synopsis, in Coated Grains, edited by T. Peryt, pp. 71-99, Springer-Verlag, New York, 1983.

Sandberg, P. A., New interpretations of Great Salt Lake ooids and of ancient non-skeletal carbonate mineralogy, Sedimentology, 22, 497-537, 1975.

Sandberg, P. A., An oscillating trend in Phanerozoic non-skeletal carbonate mineralogy, Nature, 305, 19-22, 1983.

Sandberg, P. A., Aragonite cements and their occurrences in ancient limestones, in Carbonate Cements, Spec. Pub. 36, edited by N. Schneidermann and P. M. Harris, Society of Economic Paleontologists and Mineralogists, Tulsa, Okla., in press, 1984.

Sandberg, P. A., and J. D. Hudson, Aragonite relic preservation in Jurassic calcite-replaced bivalves, Sedimentology, 30, 879-892, 1983.

Sandberg, P. A., and B. N. Popp, Pennsylvanian aragonites from southeastern Kansas-- Environmental and diagenetic implications, Am. Assoc. Pet. Geol. Bull., 65, 985, 1981.

Sheridan, R. E., Phenomena of pulsation tectonics related to the breakup of the eastern North American continental margin, Tectonophysics, 94, 169-185, 1983.

Tucker, M. E., Calcitic, aragonitic and mixed calcitic-aragonitic ooids from the mid-Proterozoic Belt Supergroup, Montana, Sedimentology, 31, 627-644, 1984.

Vail, P. R., R. M. Mitchum, Jr., and S. Thompson, Seismic stratigraphy and global changes in sea level, Part 4, in Seismic Stratigraphy, Mem. 26, edited by C. E. Peyton, pp. 83-97, American Association of Petroleum Geologists, Tulsa, Okla., 1977.

Walker, J.C.G., Possible limits on the compositions of the Archean ocean, Nature, 302, 518-520, 1983.

Walter, L. M., Revised data on the relative solubility of carbonate minerals: Implications for diagenesis, in Carbonate Cements, Spec. Pub. 36, edited by N. Schneidermann and P. M. Harris, Society of Economic Paleontologists and Mineralogists, Tulsa, Okla., in press, 1984.

Wilkinson, B.H., S.U. Janecke, and C.E. Brett, Low-magnesian calcite marine cement in Middle Ordovician hardgrounds from Kirkfield, Ontario, J. Sediment. Petrol., 52, 47-57, 1982.

Wilkinson, B.H., A. L. Smith, and K.C. Lohmann, Sparry calcite marine cements in the Upper Jurassic limestones of southeastern Wyoming, J. Sediment. Petrol., in press, 1984.

Worsley, T. R., D. Nance, and J. B. Moody, Global tectonics and eustasy for the past 2 billion years, Marine Geol., 58, 373-400, 1984.

Wright, V.P., Algal aragonite-encrusted pisoids from a Lower Carboniferous schizohaline lagoon, J. Sediment. Petrol., 51, 479-489, 1981.

# CARBONATES AND ANCIENT OCEANS: ISOTOPIC AND CHEMICAL RECORD ON TIME SCALES OF $10^7$-$10^9$ YEARS

Ján Veizer

Derry Laboratory, Department of Geology, University of Ottawa
Ottawa, Ontario, Canada, K1N 6N5

## Introduction

The nature, and $CO_2$ content, of the terrestrial atmosphere in the course of its geologic evolution has been intimately linked with that of the hydrosphere, biosphere, and lithosphere (e.g., Garrels and Mackenzie, 1971; Walker, 1977; Holland, 1978, 1984). However, the interacting fluxes are complex and operate on different time scales. To compound the difficulty, the lack of unequivocal samples of pre-Pleistocene atmospheres enables only indirect inferences concerning their attributes. With respect to possible variations in the past atmospheric $CO_2$ content, the most promising constraints are emerging from understanding of the carbon cycle. It is the aim of this extended abstract to point out factors controlling the carbon cycle on $10^7$-$10^9$ years time scales, as emerging from chemical and isotopic studies of ancient marine carbonates.

## Isotopic Record on $10^9$-Year Time Scales

The exogenic terrestrial reservoir of carbon is composed of oxidized (carbonates, dissolved marine C species, $CO_2$) and reduced (living and buried organic matter) subreservoirs, with average $\delta^{13}C$ values of ~ +1.0 ± 1.5 and −25 ± $5^0/oo$ PDB (Peedee Belemnite), respectively. The isotopic dichotomy is a consequence of the ~4:1 partitioning of carbon between these two subreservoirs (e.g., Broecker, 1970). The outstanding feature of the carbon isotope record is that it remains basically unmodified as far back as 3.5 (and possibly 3.8) billion years ago (Ga), that is, as far back as the available rock record (e.g., Schidlowski et al., 1975, 1983; Veizer and Hoefs, 1976). This observation argues for great antiquity and essential conservatism of the biosphere (e.g., Schopf, 1983). Furthermore, it indicates that the fundamental partitioning of carbon into inorganic (oxidized) and "living" (reduced) subreservoirs is set and maintained by the internal working of the earth. The exact scenario is as yet not available, but the rate of nutrient supply due to tectonism is one possible alternative.

The $\delta^{13}C$ record notwithstanding, application of the present-day steady state to $10^9$-year time scales is not supported by other isotopic pairs. The redox balance in the carbon cycle is very effective, and the cycle may be viewed as essentially cannibalistic on time scales of <$10^7$ years (e.g., Bolin, 1981; Bolin and Cook, 1983). On longer time scales, however, it is imperative that the carbon cycle be coupled to a cycle of another redox sensitive element(s), in order to counteract the compounding effect of "leaks" on the oxygen balance. It will be shown below that sulfur has been this redox couple for at least the Phanerozoic era. However, consideration of the S isotopic record on a time scale of $10^9$ years (e.g., Schidlowski et al., 1983; Thode and Goodwin, 1983) shows that, prior to ~2.7 Ga, the isotopic fractionation of S between its oxidized and reduced subreservoirs was relatively minor, with $\delta^{34}S$ of sulfides and sulfates clustering around +5 ± $5^0/oo$ CDM (Canon Diablo Meteorite), that is, close to the expected mantle value of ~$0^0/oo$ (CDM). This contrasts with the Phanerozoic record, where the $\delta^{34}S_{sulfate}$ is +20 ± $10^0/oo$ and the coeval $\delta^{34}S_{sulfide}$ is ~40 ± $20^0/oo$ lighter (Claypool et al., 1980). The incomplete Precambrian record suggests that the transition from the relatively unfractionated to the strongly fractionated pattern evolved at ~2.3 ± 0.4 Ga. This transition has been equated with bacterial dissimilatory sulfate reduction (the cause of the pronounced isotopic fractionation) becoming geologically important during the same time interval (cf. Schidlowski et al., 1983). An alternative, or more likely complementary, explanation is that this transition reflects a decrease in the rate of sulfur cycling through oceanic crust due to a decline in generation and dissipation of the earth's internal heat (Veizer et al., 1982; Cameron, 1982; Veizer, 1983a). Whatever the cause, the C-O-S coupling during the period before 2.3 ± 0.4 Ga differed from that of the Phanerozoic steady state. Furthermore, the lack of a clear isotopic fractionation in $\delta^{34}S$ between sulfate and sulfide indicates that sulfur was not the dominant redox couple to carbon during early terrestrial history. It is possible to argue

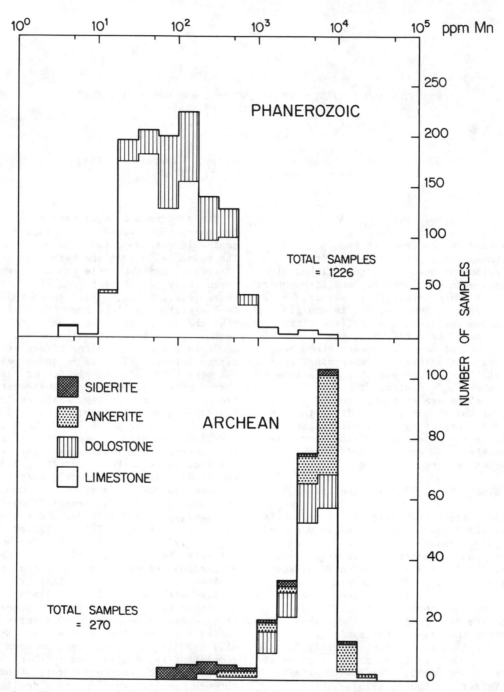

Fig. 1.  Histograms of Mn distributions in Archean and Phanerozoic carbonates. Note that these data are all for the carbonate (acid soluble) fraction of the rocks and not whole rock analyses.  All results were obtained in the same laboratory (University of Ottawa) by the same preparation and analytical techniques.  Sources of data for the Archean are Veizer et al. (1982, and unpublished work, 1980), and for the Phanerozoic they are Al-Aasm (1984), Brand and Veizer (1980), Majid (1983), and Veizer et al.  (1978).

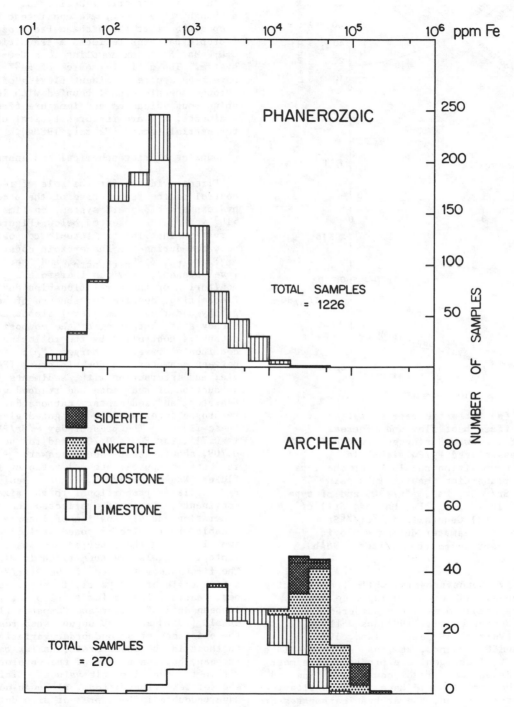

Fig. 2. Histograms of Fe distributions in Archean and Phanerozoic carbonates. Explanations and sources of data are as in Figure 1.

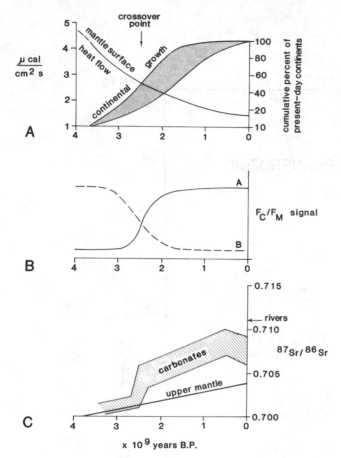

Fig. 3. (a) Schematic representation of mantle surface heat flow and of areal continental growth during geologic history. (b) The associated $F_C/F_M$ signal in seawater composition should be of the type A if the transition results in a gain (i.e., $^{87}Sr/^{86}Sr$, Na?, $^{18}O/^{16}O$) and of type B if in a loss (i.e., Fe, Mn, Ba, Sr?) of an entity. (c) Generalized $^{87}Sr/^{86}Sr$ variations in seawater during geologic history. Reproduced from Veizer (1984b).

corroborative results for clastic sediments (Ronov and Migdisov, 1971) (see Veizer (1978) for a review), show unequivocally an enrichment in Fe and Mn in early terrestrial sediments. The data, although not a proof, are consistent with a proposition that the Archean-Proterozoic change in isotopic patterns reflects a transition from C-O-FeMn-(S) redox coupling to a C-O-S-(FeMn) system. The ultimate source of $Fe^{2+}$ and $Mn^{2+}$ could have been a more prominent fluxing of seawater through oceanic crust, coupled with leaching of ubiquitous volcanics and immature first-cycle sediments; all are features typical of the early terrestrial record (Veizer, 1983a).

Coupling of Biogeochemical and Inorganic Cycles

Direct evidence for the role of tectonism as a control of the redox state of the atmosphere-hydrosphere-biosphere system, on time scales of $>10^7$ years, is presented below (Figure 4). The observation that the Sr isotopic composition of seawater during the Phanerozoic covaries with $\delta^{34}S_{sulfate}$, $\delta^{13}C_{carbonate}$, and "sea level stands", provides therefore a tool for monitoring of these relationships during the Precambrian, despite the absence of seismic stratigraphy and thus of sea level stands data. The steady state buffering of the composition of the oceans is controlled by the following fluxes: continental river discharges ($F_C$); interaction between seawater and oceanic crust ($F_M$); efflux via, and interaction with, sediments ($F_S$); and subduction of sediments and trapped water ($F_D$). Seawater, and sedimentary carbonates, therefore record an integrated and weighted global signal of these fluxes. The present-day $^{87}Sr/^{86}Sr$ for $F_C$ is 0.711, for $F_M$ is 0.703, and for seawater is 0.709, showing that modern seawater $^{87}Sr/^{86}Sr$ reflects an approximate 4:1 ratio of $F_C/F_M$ fluxes (Wadleigh, 1982). In the geological past, $F_C$ was likely proportional to the size of the continents, whereas $F_M$ reflected the intensity of generation and dissipation of internal heat through oceanic crust. The proposed variations in these two fluxes and the expected and observed Sr isotopic signals are summarized in Figure 3. The first-order feature of the $^{87}Sr/^{86}Sr$ curve, on a time scale of $10^9$ years, is the preponderance of near-mantle values prior to ~2.5 Ga. The existing Archean data of Veizer and Compston (1976), Veizer et al. (1982) and ~130 unpublished results confirm the existence of second-order variations, similar to those in the Phanerozoic, as far back as the Archean, but the bottom of the envelope is well within the coeval mantle values. Similar shorter-term fluctuations, although not resolved, have undoubtedly been present also during the Proterozoic. Nevertheless, the bottom of the envelope becomes significantly more radiogenic than the coeval mantle since ~2.5 Ga. This is again consistent with the proposed transition from a tectonically controlled $F_M$-dominated atmosphere-hydrosphere-biosphere system into an

that the early photosynthesis, while fractionating carbon isotopes, was not producing free oxygen. However, this scenario is not considered likely on biological (Schopf et al., 1983) as well as geochemical (Veizer, 1983a; Holland, 1984, chapters 7 and 8) grounds, at least since the late Archean times. If oxygen were produced, the most likely candidates for a redox couple are iron and manganese; the latter would have been quantitatively subordinate. Indeed, the Archean carbonates are markedly enriched in these elements compared to their Phanerozoic counterparts (Figures 1 and 2). Veizer's (1978) and ~150 unpublished results showed that this was the case also for carbonates of the early Proterozoic age. Contrary to assertions by Holland (1984, chapters 5, 6, and 8), these and

esearch Council of Canada, University
d the American Geophysical Union;
f figures by W.F. Schmiedel and E.
 by Julie Hayes and Heather Egan; and
T. Sundquist.  Publication 10-84 of
rleton Centre for Geoscience Studies.

## References

, Stabilization of aragonite to low-Mg
race elements and stable isotopes in
.D. thesis, pp. 306, University of
awa, Ont., 1984.
 W.E. Dean, and S.O. Schlanger,
in the global carbon cycle during the
 related to climate, volcanism, and
 atmospheric $CO_2$, this volume.
d.), Carbon Cycle Modeling, 390 pp.,
 New York, 1981.
d R.B. Cook (Eds.), The Major
cal Cycles and Their Interactions, 532
iley, New York, 1983.
d J. Veizer, Chemical diagenesis of a
ent carbonate system, 1, Trace
. Sediment. Petrol., 50, 1219-1236,

., A boundary condition on the
f atmospheric oxygen, J. Geophys.
553-3557, 1970.
., Ocean chemistry during glacial
him.  Cosmochim. Acta, 46, 1689-1705,

, Sulfate and sulfate reduction in the
mbrian oceans, Nature, 296, 145-148,

., W.T. Holser, I.R. Kaplan, H. Sakai,
 The age curves of sulfur and oxygen
 marine sulfate and their mutual
ions, Chem.  Geol., 28, 199-260,

, and A. Lerman, Phanerozoic cycles of
 carbon and sulfur, Proc. Natl. Acad.
., 78, 72-78, 1981.
, and F.T. Mackenzie, The Evolution of
 Rocks, 397 pp., Norton, New York,

, and E.A. Perry, Cycling of carbon,
oxygen through geologic time, in The
, edited by E.D. Goldberg, pp.
ley-Interscience, New York, 1974.
nd W.C. Pitman, Lithospheric plate
 level changes and climatic and
consequences, Nature, 246, 18-22,

, The Chemistry of the Atmosphere and
 pp., John Wiley, New York, 1978.
, The Chemical Evolution of the
and Ocean, 582 pp., Princeton
Press, Princeton, N.J., 1984.
 Gradual and abrupt shifts in ocean
uring Phanerozoic time, in Patterns of
arth Evolution, edited by H.D.
 A.J. Trendal, pp. 123-143,
rlag, New York, 1984.
 R.A. Berner, and R.M. Garrels, An

of the
n (Veizer, 1983a, b),
om consideration of

vity, it must be
 Phanerozoic-like
land, 1984).  These
question the
sting data base (an
es of the controversy)
Phanerozoic versus
thermore, they are
stry of the multi-
 where the provenance
ling factor.  The
of such sediments
s 5-8) can also be
on that they reflect a
first-cycle sediments
ing geologic age (cf.
 1979).  Such
ly pronounced at the
 cratonization of the
-Proterozoic
, however, that the
utually exclusive, but
ents are therefore

$10^8$-Year Time Scales

n this time scale is
ine basis for the

posed that fast
her "sea level
a level" curve for the
7) is believed to
 Others (Watts et
ue that this sea level
bsidence at the
n either case, the sea
 of tectonism.  Figure
rge transgressions
 were usually
y derived, radiogenic
e, and heavy
tion to this
, and I do not under-
s anomaly.  These
$10^8$-year time scale,
factor of at least the
 a $10^7$-year time
c parameters can vary
itions (upper and
e tectonic controls.
$\delta^{13}C$ are also the
versa, not only in
t also in terms of the
these general trends.
, are distributed
rend.  This indicates
 be the dominant
ions on time scales of

$<10^7$ years. The internal dynamics of the atmosphere-hydrosphere-biosphere system accounts therefore for the covariance of $\delta^{34}S$ and $\delta^{13}C$, although the direct geological meaning of this negative correlation (Garrels and Perry, 1974; Garrels and Lerman, 1981; Veizer et al., 1980; Holser, 1984) is somewhat enigmatic.

## Summary

Consideration of isotopic and elemental composition of sedimentary carbonates during geologic history suggests that, on time scales of $>10^7$ years, tectonic causes have governed the operation of the terrestrial exogenic cycles of C-O-S-FeMn-Sr. On a $10^9$-year time scale, the atmosphere-hydrosphere-biosphere system appears to have been dominated by interactions with mantle-derived components (volcanics, sediments, emanations) during the early stages and continental crustal components during the later stages of planetary evolution, the transition occurring during the late Archean. This led to a transition from C-O-FeMn-(S) redox coupling to a redox balance dominated by C-O-S-(FeMn) interactions.

The tectonic control on $10^7$- to $10^8$-year time scales must have been exercised through coupling of seafloor spreading with mountain building and erosion, both having comparable time constants (Veizer, 1984a). Thus faster spreading apparently results in orogeny and enhanced erosion. As a consequence, continentally derived radiogenic Sr dominates seawater chemistry at high "sea level stands." Simultaneously, such conditions are characterized also by light $\delta^{13}C$, implying less sequestering of carbon into organic matter (less production and/or burial). These observations clearly pose problems for models suggesting large creation and/or burial of biomass at the times of large epicontinental seas (high sea level stands) (Mackenzie and Pigott, 1981; Arthur, et al., this volume; Lasaga et al., this volume). The basis for all these models is the data set observed for the last 100 Ma, with a disproportionate weight given to the Cretaceous. Yet, the Cretaceous appears to represent an exception rather than a norm, if the whole Phanerozoic era is considered (cf. Figure 4). These data may also pose problems for models proposing biomass control through supplies of nutrients, such as phosphorus (Broecker, 1982). Times of high erosional rates (radiogenic $^{87}Sr/^{86}Sr$) should also be the times of maximal nutrient supply and not vice versa.

The control of the carbon cycle on time scales of $<10^7$ years likely occurs through internal dynamics of the atmosphere-hydrosphere-biosphere system (e.g., oceanic circulation patterns, glaciations, temperature and salinity gradients) (Shackleton, this volume), and the causes of fluctuation in atmospheric $CO_2$ should be therefore sought in the internal operation of this system as well.

Acknowledgments. The author acknowledges financial support by the Natural Sciences and Engineering [...] of Ottawa, a[...] preparation [...] Hearn; typin[...] editing by E[...] the Ottawa-C[...]

Al-Aasm, I.S[...]
    calcite:[...]
    rudists, P[...]
    Ottawa, Ot[...]
Arthur, M.A.[...]
    Variations[...]
    Cretaceous[...]
    changes in[...]
Bolin, B. ([...]
    John Wiley[...]
Bolin, B., a[...]
    Biogeochem[...]
    pp., John [...]
Brand, U., a[...]
    multicompo[...]
    elements, [...]
    1980.
Broecker, W.[...]
    evolution[...]
    Res., 75,[...]
Broecker, W.[...]
    times, Geo[...]
    1982.
Cameron, E.M[...]
    early Prec[...]
    1982.
Claypool, G.[...]
    and I. Zak[...]
    isotopes i[...]
    interpreta[...]
    1980.
Garrels, R.M[...]
    sedimentar[...]
    Sci. U.S.[...]
Garrels, R.M[...]
    Sedimentar[...]
    1971.
Garrels, R.M[...]
    sulfur and[...]
    Sea, vol. [...]
    303-336, W[...]
Hays, J.D.,[...]
    motion, se[...]
    ecological[...]
    1973.
Holland, H.D[...]
    Oceans, 35[...]
Holland, H.D[...]
    Atmosphere[...]
    University[...]
Holser, W.T.[...]
    chemistry[...]
    Change in[...]
    Holland an[...]
    Springer-V[...]
Lasaga, A.C.[...]

improved geochemical model of atmospheric $CO_2$ fluctuations over the past 100 million years, this volume.

Mackenzie, F.T., and J.D. Pigott, Tectonic control of Phanerozoic sedimentary rock cycling, J. Geol. Soc., 138, 183-196, 1981.

Majid, A.H., Lithofacies, chemical diagenesis, and dolomitization of the Tertiary carbonates of the Kirkuk oil field, 269 pp., Ph.D. thesis, University of Ottawa, Ottawa, Ont., 1983.

Pitman, W.C., Passive continental margins: A review, Rev. Geophys. Space Phys., 21, 1520-1527, 1983.

Ronov, A.B., and A.A. Migdisov, Geochemical history of the crystalline basement and the sedimentary cover of the Russian and North American platforms, Sedimentology, 16, 137-185, 1971.

Schidlowski, M., R. Eichmann, and C.E. Junge, Precambrian sedimentary carbonates: Carbon and oxygen isotope geochemistry and implications for the terrestrial oxygen budget, Precambrian Res., 2, 1-69, 1975.

Schidlowski, M., J.M. Hayes, and I.R. Kaplan, Isotopic inferences of ancient biochemistries: Carbon, sulfur, hydrogen, and nitrogen, in Earth's Earliest Biosphere: Its Origin and Evolution, edited by J.W. Schopf, pp. 149-186, Princeton University Press, Princeton, N.J., 1983.

Schopf, J.W. (Ed.), Earth's Earliest Biosphere:Its Origin and Evolution, 543 pp., Princeton University Press, Princeton, N.J., 1983.

Schopf, J.W., J.M. Hayes, and M.R. Walter, Evolution of earth's earliest ecosystems: Recent progress and unresolved problems, in Earth's Earliest Biosphere: Its Origin and Evolution, edited by J.W. Schopf, pp. 361-384, Princeton University Press, Princeton, N.J., 1983.

Shackleton, N.J., Oceanic carbon isotope constraints on oxygen and carbon dioxide in the Cenozoic atmosphere, this volume.

Thode, H.G., and A.M. Goodwin, Further sulfur and carbon isotopic studies of late Archean iron-formations of the Canadian Shield and the role of sulfate-reducing bacteria, Precambrian Res., 20, 337-356, 1983.

Vail, P.R., and R.M. Mitchum, Global cycles of relative changes of sea level from seismic stratigraphy, Am. Assoc. Petl. Geol. Mem., 29, 469-472, 1979.

Vail, P.R., R.M. Mitchum, and S. Thompson, Seismic stratigraphy and global changes of sea level, 4, Global cycles of relative changes of sea level, Am. Assoc. Petl. Geol. Mem., 26, 83-97, 1977.

Veizer, J., Secular variations in the composition of sedimentary carbonate rocks, II, Fe, Mn, Ca, Mg, Si and minor constituents, Precambrian Res., 6, 381-413, 1978.

Veizer, J., The Archean-Proterozoic earth, in Earth's Earliest Biosphere: Its Origin and Evolution, edited by J.W. Schopf, pp. 240-259, Princeton University Press, Princeton, N.J., 1983a.

Veizer, J., Trace elements and isotopes in sedimentary carbonates, Rev. Mineral., 11, 265-300, 1983b.

Veizer, J., Recycling on the evolving Earth: Geochemical record in sediments, Int. Geol. Congr. Proc. 27th, 11, pp. 325-345, VNU Science Press, Utrecht, 1984a.

Veizer, J., The evolving Earth: Water tales, Precambrian Res., 25. 5-12, 1984b.

Veizer, J., and W. Compston, $^{87}Sr/^{86}Sr$ in Precambrian carbonates as an index of crustal evolution, Geochim. Cosmochim. Acta, 40, 905-915, 1976.

Veizer, J., and J. Hoefs, The nature of $^{18}O/^{16}O$ and $^{13}C/^{12}C$ secular trends in sedimentary carbonate rocks, Geochim. Cosmochim. Acta, 40, 1387-1395, 1976.

Veizer, J., and S.L. Jansen, Basement and sedimentary recycling and continental evolution, J. Geol., 87, 341-370, 1979.

Veizer, J., J. Lemieux, B. Jones, M.R. Gibling, and J. Savelle, Paleosalinity and dolomitization of a lower Paleozoic carbonate sequence: Somerset and Prince of Wales Islands, Arctic Canada, Can. J. Earth Sci., 15, 1448-1461, 1978.

Veizer, J., W.T. Holser, and C.K. Wilgus, Correlation of $^{13}C/^{12}C$ and $^{34}S/^{32}S$ secular variations, Geochim. Cosmochim. Acta, 44, 579-587, 1980.

Veizer, J., W. Compston, J. Hoefs, and H. Nielsen, Mantle buffering of the early oceans, Naturwissenschaften, 69, 173-180, 1982.

Veizer, J., W. Compston, N. Clauer, and M. Schidlowski, $^{87}Sr/^{86}Sr$ in late Proterozoic carbonates: Evidence for a mantle event at ~900 m.y. ago, Geochim. Cosmochim. Acta, 47, 295-302, 1983.

Wadleigh, M.A., Marine geochemical cycle of strontium, 187 pp., M.Sc. thesis, University of Ottawa, Ottawa, Ont., 1982.

Walker, J.C.G., Evolution of the Atmosphere, 318 pp., Macmillan, New York, 1977.

Watts, A.B., G.D. Karner, and M.S. Steckler, Lithospheric flexure and the evolution of sedimentary basins, Philos. Trans. R. Soc. London, Ser. A, 305, 249-281, 1982.

# CARBON EXCHANGE BETWEEN THE MANTLE AND THE CRUST, AND ITS EFFECT UPON THE ATMOSPHERE: TODAY COMPARED TO ARCHEAN TIME

David J. Des Marais

NASA Ames Research Center, Moffett Field, California 94035

Abstract. Over geologic time scales of tens of millions to billions of years, atmospheric $CO_2$ levels are influenced by the exchange of carbon between the earth's surface and the mantle. Estimates of the mid-ocean ridge mantle carbon flux can be made by estimating concentration ratios of carbon to helium in hydrothermal fluids and tholeiitic glasses and by multiplying these by the oceanic primordial $^3$He flux. The estimated ranges of carbon fluxes for the earth today and 3 b.y. ago are $1 \times 10^{12}$ to $8 \times 10^{12}$ moles $yr^{-1}$ and $3 \times 10^{12}$ to $48 \times 10^{12}$ moles $yr^{-1}$, respectively. These fluxes are comparable in magnitude to the present-day flux estimated for carbonate metamorphism, $6 \times 10^{12}$ moles $yr^{-1}$. A net carbon flux from the mantle of $10 \times 10^{12}$ moles $yr^{-1}$ would require less than 700 m.y. to generate the present-day crustal carbon inventory. Given present-day geothermal gradients, carbonate sediments can be at least partly subducted into the mantle. Perhaps between 3 and 50 percent of the sedimentary carbon presently transported by the lithospheric plates to subduction zones will eventually be injected into the upper mantle. A hotter upper mantle 3 b.y. ago would have made this injection process less efficient. During the late Archean, the earth's crustal carbon inventory very likely equaled or even exceeded the present crustal inventory. This circumstance, together with the liklihood that the land area at that time was considerably less than it is today, suggests that 3 b.y. ago the atmosphere contained at least two orders of magnitude more $CO_2$ than it does today.

## Introduction

Over geologic time scales of tens of millions to billions of years, the crustal carbon (C) inventory has been influenced by the processes which govern the exchange of C between the earth's surface and the deep crust and upper mantle. Rubey [1951] first proposed that plutonic gas emanations furnished the "excess" volatiles found within the earth's crust, oceans, and atmosphere. Mid-ocean ridge volcanism is the single most important source of mantle volatiles, and crustal production rates along the ridges have apparently varied over the past 100 m.y. [Hays and Pitman, 1973]. Even the total heat flow from the earth's interior, which is the driving force behind tectonics, has varied in the past and was substantially greater during the early Precambrian [Lambert, 1976]. Such variations have probably affected the abundance of atmospheric $CO_2$; therefore a closer inspection of thermally mediated C fluxes merits attention.

The purpose of this paper is to examine the exchange of C between the mantle and the earth's surface. The "metamorphic" flux of $CO_2$, that is, the $CO_2$ produced by the thermal decomposition of carbonates and organic matter in heated sediments [e.g., Barnes et al., 1978], is not considered here in detail. How great is the mantle C flux from mid-oceanic ridges? Is C subducted and reinjected into the upper mantle? If so, how has the net C flux between the earth's interior and its surface varied over geologic time scales during which the earth's heat flow has changed markedly?

Questions such as these were considered by Javoy et al. [1982], who calculated a mid-oceanic ridge mantle C flux and discussed its implications. Using a Rayleigh distillation model to help explain the isotopically light total C that they released by heating the mid-ocean ridge tholeiitic glasses, they estimated that the tholeiitic melt initially contains between 0.2 and 1 percent C. Such high concentrations require that the present-day mid-ocean ridge mantle C flux be very large. A mean estimate is $20 \times 10^{12}$ moles $yr^{-1}$ [Javoy et al., 1983]. Such a large flux would deliver a mass of C equal to the entire crustal C inventory in only 335 m.y. Therefore Javoy et al. [1982] concluded that C is being subducted back into the mantle. Large C fluxes between the mantle and crust imply that the mantle has played an active role in controlling the size of the crustal C reservoir and thus the atmospheric $CO_2$ level. The broader significance of elemental recycling through the mantle is discussed by Allegre [1982].

Des Marais and Moore [1984] measured C in mid-oceanic ridge basalts by combusting the samples sequentially at increasing temperatures. They identified a [13]C-depleted component which was apparently added to the samples after their eruption. If this isotopically light component is indeed nonmantle in origin, then the basalts' C isotopic composition is quite uniform and similar to the value inferred for mantle C [see Hunt, 1979; Pineau et al., 1976]. Therefore the initial C contents of tholeiitic magmas cannot be estimated using an isotopic Rayleigh distillation model.

It is still worthwhile to calculate the mid-ocean ridge mantle flux by some other, independent method. If this flux is large enough to indicate that subduction of C must occur, then the proposal of Javoy et al. [1982] that C is being cycled between the mantle and crust remains viable.

The purpose of this paper is to calculate a present-day mid-ocean ridge mantle C flux by correlating it with the flux of juvenile $^3$He. The behavior of carbonate sediments during subduction is also considered. The sizes of these fluxes, as well as their effects on atmospheric $CO_2$ levels, will be estimated for the Precambrian earth.

## Present-Day Mid-Ocean Ridge Carbon Flux

The purpose of this section is to outline how a present-day mid-ocean ridge flux of mantle C might be calculated, to estimate the range of values for the flux using presently available literature data for C and $^3$He, and to indicate where more data are needed to improve the calculation.

The isotope $^3$He has been established as a reliable tracer of mantle degassing for a variety of reasons [see Clarke et al., 1969; Lupton and Craig, 1975; Weiss et al., 1977; Jenkins et al., 1978, and references therein]. The earth is indeed still outgassing $^3$He which it acquired during its initial accretion. Radiogenic sources of $^3$He in the crust and mantle are minimal. The atmosphere does not contain much $^3$He, in part because helium is being continually lost into space. Little $^3$He is reinjected into the mantle during subduction, because marine sediments contain essentially no $^3$He and the $^6$Li contribution of $^3$He is very small. The mantle contribution of $^3$He to the deep ocean has been successfully resolved from the other major $^3$He source: the radioactive decay of tritium produced in the atmosphere.

The estimate of a mantle C flux into the oceans requires that both the mantle $^3$He flux and the mantle C to $^3$He elemental ratio be determined. The flux of mantle $^3$He into the oceans is estimated, with perhaps a 20 percent uncertainty, to be 4 atoms cm$^{-2}$ s$^{-1}$, averaged over the entire surface of the earth [Craig et al., 1975]. Our knowledge of the average global mantle C to $^3$He elemental ratio for the mid-ocean ridges is much less certain. Des Marais and Moore [1984] assumed a C to $^3$He molar ratio of 7.8x10$^8$ from measurements of hydrothermal fluids from the East Pacific Rise at 21$^o$N [Craig et al., 1980] and the Galapagos Rift. However, mantle C and $^3$He are injected into the ocean not only by hydrothermal activity but also during eruptive events. Furthermore, the above estimate of the C to $^3$He ratio is obtained only from rift systems in the eastern Pacific. The C to $^3$He ratio might vary between localities or between hydrothermal and eruptive events. These issues are perhaps more effectively addressed by considering, in addition, the C to $^3$He ratios in the tholeiitic glasses. Such glasses contain a residuum of the volatiles exhaled during volcanic eruptions. Furthermore, these glasses are available from ridge systems worldwide, whereas corresponding analyses of hydrothermal fluids are frequently difficult, if not impossible, to obtain. On the other hand, if elemental fractionation occurs as these volatiles enter and traverse the ridge system, the elemental C to $^3$He ratios in glasses might differ from the ratio of C to $^3$He which is expelled directly into the ocean during an eruption. It is essentially impossible to analyze directly those volatiles which are eruptively injected into the ocean. However, a comparison of C to $^3$He ratios for basaltic glasses and hydrothermal fluids from the same site can offer at least a general impression of the magnitude of any elemental fractionation which might occur.

At the Galapagos and East Pacific Rise 21$^o$N sites, C and $^3$He have been analyzed both in hydrothermal fluids and in glasses. Representative data are summarized in Table 1. Hydrothermal fluids probably inherit their mantle volatiles from the magma chamber and/or the hydrothermal alteration of hot rock above the chamber. The volatiles in basaltic glasses represent that fraction which was not released from the erupting magma. Therefore, if fractionation occurs, the volatiles in the glasses will contain a relatively larger amount of that element which is more strongly retained in the magma, whereas the hydrothermal fluids should contain a relatively larger amount of the more volatile component. The C to $^3$He ratio is lower in the hydrothermal fluids than it is in the basalts (Table 1); therefore $^3$He is apparently more readily degassed. However, the elemental fractionation is not very large.

The fractionation which accompanies the degassing of an erupting magma can also be assessed by comparing the percentages of C and $^3$He in the glass vesicles to those in the groundmass of a given sample. The more readily degassed element should have a larger percentage of its total abundance occupying the vesicles. Table 2 gives vesicle data for two sets of samples, each set coming from the same field locality. The fraction of C present in the vesicles as $CO_2$ actually exceeds the fraction of $^3$He in the vesicles. This result suggests that C is released

TABLE 1.   C and He Data from Eastern Pacific Spreading Ridges

| Sample | Galapagos | East Pacific Rise |
|---|---|---|
| | Hydrothermal Fluids | |
| $[^3He]$, $\mu$m g$^{-1}$ | $2.0\times10^{-10}$ [a] | $6.14\times10^{-9}$ [b] |
| $[C]$, $\mu$m g$^{-1}$ | $0.25$ [c] | $3.4$ [d] |
| $[C]/[^3He]$ | $1.24\times10^9$ $(1.0\times10^9)$ [d] | $0.56\times10^9$ |
| | Tholeiitic Glasses | |
| $[^3He]$, $\mu$m g$^{-1}$ | $6.22\times10^{-9}$ [e] | $7.51\times10^{-9}$ [b] |
| $[C]$, $\mu$m g$^{-1}$ | $8.75$ [f] | $10.0$ [g] |
| $[C]/[^3He]$ | $1.4\times10^9$ | $1.3\times10^9$ |

[a]Jenkins et al. [1978], for 12° water, excluding Garden of Eden.
[b]Craig et al. [1980], 21°N.
[c]Corliss et al. [1979], for 12°C water.
[d]H. Craig (personal communication, 1984).
[e]Kurz and Jenkins [1981], sample 714-6-4.
[f]Des Marais and Moore [1984], sample 714-1-A.
[g]This work, sample ALV 981-R23 from 21°N, analyzed by the method of Des Marais and Moore [1984].

a little more readily than $^3He$, which is in contrast to the observation made when the glasses are compared with the hydrothermal fluids. However, the comparison of the fluids and glasses is based on at least a few more data and perhaps reflects a greater spectrum of processes associated with the degassing magma.  Therefore, it is assumed in the following calculation that the "true" C to $^3He$ ratio for a given ridge locality lies somewhere between 0.5 and 1.0 times the ratio measured in the glasses from that locality.  Clearly, more simultaneous measurements

TABLE 2.   Percentage of the C and $^3He$ in Tholeiitic Glasses
Which  Resides in the Vesicles

| Sample | Vesicle Volume, Percent | Percent C in Vesicles[a] | Percent He in Vesicles[b] |
|---|---|---|---|
| | Mid-Atlantic Ridge (FAMOUS) | | |
| 519 4-1 | 1.0 | 58 | |
| 519 2-1 | 2.0 | | 47 |
| | Galapagos Rift | | |
| 714 1-A | 0.1 | 17 | |
| 714 G-4 | 0.1 | | 15 |

[a]Des Marais and Moore [1984].
[b]Kurz and Jenkins [1981].

of C and $^3$He in glasses are required for a critical assessment of these effects.

At this moment, no individual glass samples have been reliably analyzed simultaneously for both their C and $^3$He contents. The ratios listed in Table 3 were calculated using data from different samples which were from approximately the same ridge locality and eruption depth. One uncertainty associated with using data from different samples is that samples vary with respect to their volume percent of vesicles. Furthermore, as Table 2 indicates, vesicles often contain a considerable percentage of a glass' $^3$He and C content. For the Cayman and FAMOUS localities, the samples analyzed for $^3$He had a different volume percent of vesicles than those analyzed for C. Fortunately, the $^3$He and C abundances in both the groundmass and the vesicles in these samples has been estimated [Kurz and Jenkins, 1981; Des Marais and Moore, 1984], so an approximate correction can be made. For example, if the sample analyzed for C has 1.0 percent vesicles and the sample analyzed for $^3$He has 2.0 percent, then the C sample's carbon content can be recalculated to reflect a 2.0 percent vesicle content. Using the $^3$He and C abundance values corresponding to a 2.0 percent vesicle content, a molar ratio can be calculated. Conversely, a sample's $^3$He content can be recalculated to reflect a 1.0 percent vesicle content, and a second molar ratio can be calculated. When the molar ratios are estimated by adjusting either the C abundances or the $^3$He abundances, the two estimates are found to differ by 20 percent or less. This agreement is good, given the uncertainties associated with estimating vesicle volumes. Averages of such pairs of estimates for both the Cayman and FAMOUS samples are given in Table 3. For Galapagos, the $^3$He and C data were both from samples having 0.1 percent vesicles, so no correction was needed. No vesicle data are available for the East Pacific Rise samples.

The C to $^3$He ratios in the glasses range from $1.4\times10^9$ to $7\times10^9$. The ratios for the mid Atlantic and Cayman ridges exceed those for the East Pacific sites. Does this difference reflect variations in elemental fractionation during eruptions at these sites? If not, then the different ratios might reflect spatial heterogeneity within the oceanic mantle. Kurz et al. [1982] observed heterogeneities in the $^3$He to $^4$He ratio along the central North Atlantic Ridge. They suggest that such variations might be caused by subducted crust which, along with its $^4$He-producing nuclides, is recycled back into the mantle. It will be important to investigate whether samples having higher C to $^3$He ratios have a higher component of subducted and recycled C.

TABLE 3.  C and $^3$He Data for Glasses from Different Spreading Ridges

| Sample | Eruption Depth, m | Vesicle Volume, Percent | $[^3\text{He}]$, $\mu\text{m g}^{-1}$ | $[\text{C}]$, $\mu\text{m g}^{-1}$ | $[\text{C}]/[^3\text{He}]$ |
|---|---|---|---|---|---|
| Mid-Cayman Rise | | | | | |
| KNR 54 47-24[a] | 5400 | 0.3 | $1.31\times10^{-9}$ | | |
| | | | | | $7\times10^9$ [c] |
| KN 54 52-32[b] | 5393 | 1.4 | | 12.5 | |
| East Pacific Rise, 21°N | | | | | |
| From Table 1 | 2600 | | | | $1.3\times10^9$ |
| Mid Atlantic Ridge, FAMOUS | | | | | |
| ALV 519 2-1-b[a] | 2700 | 2.0 | $2.75\times10^{-9}$ | | |
| | | | | | $7\times10^9$ |
| ALV 519 4-1[b] | 2690 | 1.0 | | 12.4 | |
| Galapagos Ridge | | | | | |
| From Table 1 | 2500 | 0.1 | | | $1.4\times10^9$ [c] |

[a] Kurz and Jenkins [1981].
[b] Des Marais and Moore [1984].
[c] For the computation of the ratios, the differences in vesicle contents between samples were considered (see text).

This might be accomplished by correlating higher C to $^3$He ratios with other potential indicators of subducted crust, such as low $^3$He/$^4$He, high $^{87}$Sr/$^{86}$Sr, and high concentrations of other elements which tend to be enriched in the crust, relative to the mantle.

The figure for the $^3$He flux and the range of values for the C to $^3$He ratio can be combined to obtain a rough estimate of the C flux. The global oceanic magmatic $^3$He flux is estimated to be 4 atoms cm$^{-2}$ s$^{-1}$ [Craig et al., 1975], or 1.07x10$^3$ moles yr$^{-1}$. The range of C to $^3$He ratios to be used must take into account the range of values actually measured in the glasses (Table 3), as well as the elemental fractionation effects discussed above. The low end of the range is therefore assumed to equal 0.5 times 1.3x10$^9$; the high end of the range is assumed to be 1.0 times 7x10$^9$. This range of ratios, 0.65x10$^9$ to 7x10$^9$, together with the $^3$He flux, gives the following range of values for the C flux: 0.7x10$^{12}$ moles yr$^{-1}$ to 7.5x10$^{12}$ moles yr$^{-1}$.

An estimate of the mantle C flux to the earth's surface should perhaps include that C which remains in mid-ocean ridge basalts after their eruption. Conceivably, some portion of this C is released during subsequent chemical alteration of these basalts (e.g., during subduction). This C contribution can be estimated by calculating the rate at which C is added to new basaltic crust formed at the ridges. Assuming a crustal production rate of 3 km$^2$ yr$^{-1}$, a crustal thickness of 2 to 5 km [Javoy et al., 1982], a basalt density of 3 g cm$^{-2}$, and a basalt C content equal to 200 µg C/g [Des Marais and Moore, 1984], this additional component of the C flux is estimated to be between 0.3x10$^{12}$ moles yr$^{-1}$ and 0.75x10$^{12}$ moles yr$^{-1}$. If this component is included, the total flux estimate ranges from 1x10$^{12}$ to about 8x10$^{12}$ moles C yr$^{-1}$.

It is useful to calculate how much time this range of fluxes would require to produce an amount of C equal to that which currently resides in the atmosphere and oceans and in the igneous, metamorphic, and sedimentary rocks of the earth's crust. This inventory has been estimated recently to be 8x10$^{22}$ g [Javoy et al., 1982], or 6.7x10$^{21}$ moles. The ridge flux would require between 0.8 and 7 b.y. to provide this much C. Another approach is to calculate how much C the ridges would produce at this rate during the earth's lifetime (4.5x10$^9$ years). This amounts to between 4.5x10$^{21}$ and 3.6x10$^{22}$ moles C, which ranges from about 0.7 to about 5 times the present size of the crustal C reservoir. If the mid-ocean ridges have delivered C to the crust at a rate which exceeds that needed to equal the crustal inventory over the earth's history, then C must have been reinjected into the mantle via subduction. Given the range of values estimated for the present-day flux, we cannot conclude with certainty that subduction of C is required. However, crustal production rates have apparently varied over geologic time; therefore, the rates of C

outgassing and, possibly, C subduction have also varied. It is instructive to ask whether the balance between the outgassing and subduction of C may have been different in the past. If major changes did occur, then they very likely affected the atmospheric $CO_2$ content.

### The Precambrian Mid-Ocean Ridge Mantle Carbon Flux

Lupton et al. [1980] noted that the ratio of heat flow to mantle $^3$He flux is relatively constant both in the Galapagos hydrothermal vents and East Pacific Rise vents at 21$^O$N. This ratio is expected to vary between localities where the ratio of convective to conductive heat loss is different (see, for example, Jenkins et al. [1978]). Despite such variations, a greater heat flow along ridges would cause greater $^3$He fluxes. Given that the heat flow controls the rate of mantle outgassing and that mantle carbon is also being delivered to the ridges, a higher heat flow also gives larger C fluxes.

That global heat flows were greater during the Precambrian has been inferred both by model calculations [e.g., Lambert, 1976; Turcotte, 1980] and by the high degree of mantle melting as inferred from komatiites [Green, 1975; Arndt, 1977]. Turcotte [1980] estimated that the global heat flow 3 b.y. ago was about 2.2 times its present value. Heat loss can occur either by conduction through the solid crust or by convection associated with volcanism and plate tectonics. Archean supracrustal rocks apparently experienced geothermal gradients similar to present-day continental gradients [Bickle, 1978; Boak and Dymek, 1982], implying that continental heat flow may not have been much higher then than it is today. The higher Archean heat flows must therefore have been accommodated principally by the ocean basins. As much as 40 to 45 percent of the present-day global heat loss may be due to plate motions [Bickle, 1978]. Plate motions can be quantified by estimating the global production rate of new crust along mid-ocean ridges. Presently, that rate is estimated to be 3 km$^2$ yr$^{-1}$ [Chase, 1972; Garfunkel, 1975; Minster et al., 1974; Parsons, 1981].

It appears likely that a greater percentage of the Precambrian heat flow was dissipated by plate motions and production of new crust [Bickle, 1978] and, possibly, by increased mantle convection. Modelers have attempted to predict crustal production rates in the Precambrian. These rates were controlled both by the radionuclide ($^{40}$K, U, and Th) content of the mantle and core and by the efficiency with which heat and mass are exchanged between the core, lower mantle, and upper mantle, and are generally considered to have exceeded present-day rates. Among the models proposed, those which invoke relatively high radionuclide concentrations require high oceanic crustal production rates to dissipate the additional heat and to avoid generating continental thermal gradients steeper than those indicated by the Archean supracrustal assemblages. The efficiency with which the mantle is mixed is also important. Models which invoke relatively efficient mantle-wide convection must limit the lower mantle's radionuclide concentrations, and thus the lower mantle's heat production, in order to avoid injecting more $^{40}$Ar into the atmosphere than is presently observed. For example, the model of Sleep [1979] assumes relatively efficent mantle convection and therefore must limit the lower mantle's $^{40}$K concentrations and therefore the Precambrian crustal production rates. Sleep's model concludes that crustal production rates varied between 3 km$^2$ yr$^{-1}$ (the present-day rate) and 6 km$^2$ yr$^{-1}$ during the past 3 b.y. Models which assume relatively high rates of heat production in the Archean require that the production rate 3 b.y. ago was much greater than at present (for example, 18 km$^2$ yr$^{-1}$ [Bickle, 1978]). Some models (e.g., that of Bickle [1978]) do not consider the constraints imposed by the atmospheric $^{40}$Ar inventory. However, O'Nions and Oxburgh [1983] suggest that the exchange of helium and, by implication, $^{40}$Ar between the upper and lower mantle is more restricted than is the heat flow. This restriction allows greater heat production to occur without producing greater-than-observed atmospheric $^{40}$Ar levels. If heat flow between the lower and upper mantle is indeed more facile than is rare gas transport, then a crustal production rate for 3 b.y. ago which lies near the middle of the range between 3 and 18 km$^2$ yr$^{-1}$ is reasonable.

Because the solid-state diffusivities of most volatiles are lower than the thermal diffusivity within the earth, volatile transport is dominated by large-scale movements of volatile-rich phases, rather than by molecular diffusion [O'Connell and Houseman, 1983]. Such transport within the earth is therefore enhanced by greater mantle convection and crustal production rates. Furthermore, higher temperatures in the upper mantle favor greater abundances of the more mobile volatile phases. Even in the model of Sleep [1979], where the crustal production rate 3 b.y. ago could have been as low as the present-day rate, a hotter mantle would have caused eutectic melting to commence at 95 km depth in the Archean oceanic crust, compared to 30 km depth in today's ocean crust. Eutectic melting promotes upward migration of mobile volatile phases, and an Archean melting depth which was approximately 3 times the present-day melting depth certainly would have increased the rate of volatile outgassing, perhaps by a factor of 3. Thus higher heat flow in the Precambrian earth was very likely accompanied by more rapid volatile transport and outgassing. The mid-ocean ridge carbon flux 3 b.y. ago is assumed to have been between 3 and 6 times the present value.

Given the range of values estimated earlier for the present-day carbon flux (1x10$^{12}$ to 8x10$^{12}$ moles yr$^{-1}$), the rates estimated for 3 b.y. ago range from 3x10$^{12}$ to 48x10$^{12}$ moles yr$^{-1}$. Could an Archean mantle sustain such high rates? If the carbon found in the present-day crust constitutes most of the C in the crust/mantle system, and if most of the C issuing along the ridges has been recycled repeatedly via subduction, then such high Archean outgassing rates could not have occurred. On the other hand, if the upper mantle is a large reservoir of C, then such high rates are likely. Given that the solubility of C in magmas increases with pressure and, therefore, depth in the earth [Mysen et al., 1975], the mantle is apparently capable of storing vast quantities of C. Even if the earth had accreted with minimal volatiles in its mantle [Arrhenius, 1981], the upper mantle has very likely been stirred several times during earth history [O'Connell, 1982; Houseman, 1983], and therefore it should have acquired C relatively early.

It is possible to estimate the C content of the upper mantle, defined here to be the outermost 700 km of the mantle (see evidence summarized by Houseman [1983]), if one assumes that the C concentration of the magma erupting at the ridges represents the C concentration of the entire upper mantle. This concentration is estimated by dividing the present-day ridge C flux by the volume of new crust which is being degassed annually (between 6 and 15 km$^3$ yr$^{-1}$ [Javoy et al., 1982, 1983]). Assuming that the upper mantle has a volume of 3.17x10$^{11}$ km$^3$, its C inventory is between 2.1x10$^{22}$ moles and 4.2x10$^{23}$ moles. This inventory is between 3.2 and 69 times the crustal C inventory; therefore, markedly higher Archean outgassing rates appear to be possible.

These outgassing rates for 3 b.y. ago require between 0.14 and 2.2 b.y. to produce enough C to equal the present-day crustal inventory. If such rates persisted during all of Archean time, then between 0.9 and 14 times the present crustal C inventory was outgassed during the Archean. Such outgassing rates require that C must have been subducted. Conceivably, C was actively exchanged between the Archean mantle and upper crust.

## Subduction of Carbon

Our understanding of the fate of sedimentary C during subduction is far from complete. Even the fate of oceanic sediment during subduction is a controversial topic. At issue is the extent to which sediments are subducted as opposed to being scraped off the descending slab and accreting laterally at continental margins or island arcs. Huang et al. [1980] have summarized the evidence regarding carbonate subduction in today's earth. They discuss the chemical transformations of siliceous carbonates which accompany the increasing temperatures and pressures at depth. Figure 1, which is adapted from Figures 8 and 9 of Huang et al. [1980], shows some of the reactions involving siliceous limestones, together with estimates of temperatures along the upper surface of a subducted slab. The dissociation curve for dry siliceous limestone lies well above the

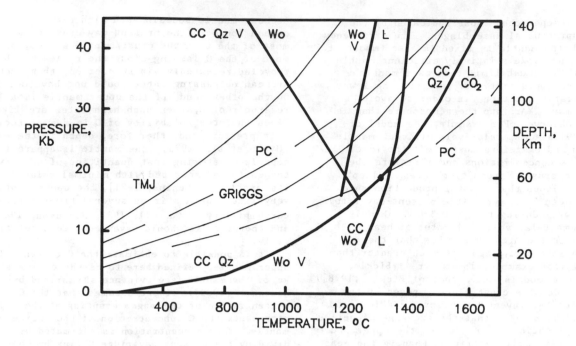

Fig. 1.   Selected reactions involving calcite-quartz assemblages situated on the upper surface of a subducted slab, modified from Figures 8 and 9 of Huang et al. [1980]. Heavy lines delineate transformations between the following:   CC - calcite, $CO_2$, L - liquid, Qz - Quartz, V - Vapor, and Wo - wollastonite, $Ca(SiO_3)$.   The two heavy curved lines correspond to anhydrous conditions; the two heavy straight lines correspond to reactions occurring in the presence of $CO_2$-$H_2O$ mixtures.   The straight line shown for CC+Qz+V = Wo assumes $P_{CO2}$ = 15 kbar.   The straight line for Wo+V = L assumes $P_{H2O}$ = 15 kbar.   The two solid light lines are estimates of the present-day temperature along the surface of subducted oceanic lithosphere (GRIGGS - Griggs [1972]; TMJ - Toksoz et al.[1971]).   The pair of dashed light lines labeled PC correspond to this author's estimates of temperature along the surface of subducted oceanic lithosphere 3 b.y. ago, assuming that the temperature at 100 km depth was 300°C hotter then than it is today [Sleep, 1979].   This figure illustrates that subducted carbonate sediments were more likely to form mobile, volatile phases 3 b.y. ago than they are today.

estimated present-day temperatures for subducted sediments.  For wet siliceous carbonates (refer to the straight lines in Figure 1), wollastonite is first formed.  Subsequently, for one of the two estimates of modern day slab temperatures, wollastonite melts, and a liquid of carbonatitic composition is formed.  However, the marbles produced during subduction may permit only limited access to migrating $H_2O$.  Therefore, as Huang et al. [1980] conclude, "...much of the subducted calcite would probably escape dissociation and melting and be carried down to considerable depths, possibly for long-term storage of carbon in the mantle, either as aragonite, or as diamond if the carbonate is reduced at whole or in part. Aragonite masses should react marginally with mantle peridotite to produce dolomite or magnesite, depending on depth."

If the mass of C in the crust either does not change or changes relatively slowly with time, then the rate at which C is subducted must approximately equal the rate at which mantle C is

exhaled.   Let us assume that, because submarine volcanism accounts for more than 90 percent of the mantle [3]He exhaled [O'Nions and Oxburgh, 1983], such volcanism therefore delivers more than 90 percent of the mantle C.   Therefore, the C subduction rate on today's earth should approximately equal the value estimated earlier for the mid-ocean ridge mantle flux, between $1 \times 10^{12}$ and $8 \times 10^{12}$ moles C $yr^{-1}$.   It is interesting to compare this range of values with an estimate of the rate at which C in pelagic sediments arrives at the subduction zones.   The mass of injected C which reaches the mantle, divided by the mass of sedimentary C which arrives at the subduction zones but is not yet injected, should give a rough estimate of the efficiency with which sedimentary C reaches the mantle.   That C which does not reach the mantle presumably resides within sediments which are not subducted, or it returns toward the earth's surface by volcanism. Javoy et al. [1982] calculated how much C in a 500- to 1000-m thick layer of pelagic sediments arrives at subduction zones, assuming both a reasonable sediment composition and a value for the crustal subduction rate which equals the accepted literature value for the crustal production rate. If one assumes a crustal subduction rate of 3 $km^2$ $yr^{-1}$ and a sediment composition identical to that used by Javoy et al. [1982], the amount of C arriving at subduction zones is between $15 \times 10^{12}$ and $29 \times 10^{12}$ moles $yr^{-1}$. Therefore, somewhere between 3 and 50 percent of the C which arrives at subduction zones might actually reach the mantle and be stored there. What was the subducted C flux 3 b.y. ago? There is disagreement as to whether subduction was even operative at that time [e.g., Baer, 1977], because the hotter crustal temperatures would have prevented the formation of eclogite, whose high density is thought by some to be necessary for subduction. If subduction did not occur in the Archean earth, then the C released by thermal processes could have accumulated in the crust and attained a very high concentration in the atmosphere.

Assuming for the moment that subduction did occur, how readily could C have been subducted 3 b.y. ago, relative to present-day circumstances? Figure 1 illustrates the temperatures and depths experienced by subducting sediments in an Archean crust and mantle (see dashed light lines). The lines were drawn assuming that the earth's surface temperature 3 b.y. ago was the same as it is today and that the temperature at 100 km depth was $300^{\circ}C$ greater than it is today (see evidence summarized by Sleep [1979]). Note that, relative to the present-day situation, 3-b.y.-old subducted sediments followed a trajectory which, for a given pressure, subjected them to much greater temperatures. These greater temperatures enhanced the liklihood that subducted minerals reacted to form mobile C phases which migrated upward into volcanos and therefore escaped injection into the mantle.

The qualitative arguments presented above cannot be translated into quantitative estimates of Archean C subduction rates. However, geophysicists do agree that subduction of basaltic crust into a hotter mantle was more difficult [e.g., Sleep, 1979]. Subduction into a hotter mantle favored more efficient mobilization of sedimentary C and therefore a greater probability that this C returned to the earth's surface via volcanism.

Precambrian Atmospheric Carbon Dioxide Levels

Over geologic time scales of a million years or more, the exchange of C between the atmosphere and ocean and the crust and mantle has exerted an important influence upon atmospheric $CO_2$ levels. This interaction was outlined eloquently by Rubey [1951], who first recognized that thermal as well as weathering and depositional processes were important.

Variations in crustal exchange processes and atmospheric $CO_2$ during the past 100 m.y. have been modeled recently [Berner et al., 1983; Lasaga et al., this volume]. Weathering and organic matter burial are major sinks for atmospheric $CO_2$, whereas carbonate precipitation and $CO_2$ production by geothermal processes are major sources. Berner et al. [1983] suggested that the rate of seafloor spreading exerted a strong influence upon atmospheric $CO_2$ levels during the past 100 m.y. Their model uses the seafloor spreading rate as a measure of mid-ocean ridge hydrothermal activity (calcium and magnesium reactions) and rates of sediment metamorphism. The rate of $CO_2$ release by sediment metamorphism exerts a strong effect on atmospheric $CO_2$ levels. The calculations described in the present paper suggest that, on today's earth, the ridge C flux is comparable to the metamorphic C flux. Thus the flux of mantle C from ridges might also significantly affect the atmosphere. If this is so, then the Cretaceous period might have witnessed atmospheric $CO_2$ levels which were actually somewhat higher than the levels predicted by Lasaga et al. [this volume].

The hotter Archean mantle must have influenced significantly the inventory of C in the crust and atmosphere. Consider the mantle-crust-ocean-atmosphere system as consisting of a mantle C reservoir and a crust-ocean-atmosphere C reservoir. The amount of C in the mantle is controlled in part by the rate at which C is delivered to the crust and the rate at which C is returned by subduction. During the Archean, hotter mantle temperatures conferred greater mobility to C, due to more vigorous convection and to the greater abundances of mobile liquid and gaseous C phases. Higher rates of crustal production were accompanied by higher rates of C outgassing. This outgassing is estimated to have been 3 to 6 times larger 3 b.y. ago, and was probably even greater during the early Archean. At the same time, the hotter mantle retained subducted C with greater difficulty. These considerations are consistent with an Archean earth in which the crustal C inventory actually exceeded the present-day crustal C inventory, perhaps by a substantial amount.

The rates of weathering, carbonate precipitation, organic matter preservation, etc. during the Archean exerted important controls on atmospheric $CO_2$ levels. These processes must be considered before even an approximate estimate of Archean atmospheric $CO_2$ levels can be made. A detailed consideration of such processes is beyond the scope of this paper. However, for purposes of illustration, conditions during the Cretaceous earth 100 m.y. ago as inferred by the model of Berner et al. [1983] can be compared with the earth 3 b.y. ago. In an improved version of the Berner et al. model, Lasaga et al. [this volume] calculate that a greater seafloor spreading rate (about 1.6 times present, which they assumed caused the metamorphic $CO_2$ flux to be 1.6 times present), and lower continental area (about 0.84

times present) were responsible for producing atmospheric $CO_2$ levels of about 13 times present. Less land area favors higher $CO_2$ concentrations, because rock weathering is the major sink for atmospheric $CO_2$. Rates of mantle $CO_2$ outgassing 3 b.y. ago are estimated to have been 3 to 6 times the present rate, whereas subduction of C was less efficient. Several current models for the evolution of the continents assert that land area was less than half the present area (see Veizer and Jansen [1979] for a discussion). These considerations imply that Archean atmospheric $CO_2$ levels greatly exceeded those during the Cretaceous, and very likely were at least two orders of magnitude above present levels.

Several other investigators have suggested that the Precambrian witnessed elevated atmospheric $CO_2$ concentrations. For example, Holland [1976] noted that the widespread deposition of dolomite during the Precambrian was favored by higher rates of $CO_2$ input into the ocean-atmosphere system. He proposed that this $CO_2$ derived both from greater outgassing of juvenile C and from a greater rate of carbonate sediment metamorphism due to a higher geothermal gradient. Walker [1978] used these concepts to help explain the evolution of the atmospheres of the earth, Mars, and Venus. Sagan and Mullen [1972] recognized that, because the early sun was cooler, the earth's atmosphere must have exerted a stronger greenhouse effect to prevent the oceans from freezing. Carbon dioxide appears to have been the most suitable greenhouse gas [Hart, 1978; Owen et al., 1979]. Kasting et al. [1984] calculated that an atmospheric $CO_2$ concentration approximately 600 times greater than present is required to sustain early Archean ocean temperatures similar to those of today. Walker et al. [1981] proposed a mechanism whereby declining atmospheric $CO_2$ concentrations stabilized the earth's surface temperature over geologic time.

This paper has attempted to outline how mantle C influenced the early earth's crustal C cycle. The role played by mantle C will be better defined both by improved estimates of crust-mantle exchange and by ongoing studies of early crustal and biological evolution.

## References

Allegre, C. J., Chemical geodynamics, Tectonophysics, 81, 109-132, 1982.

Arndt, N. T., Ultrabasic magmas and high-degree melting of the mantle, Contrib. Mineral. Petrol., 64, 205-221, 1977.

Arrhenius, G., Interaction of ocean-atmosphere with planetary interior, Adv. Space Res., 1, 37-48, 1981.

Baer, A. J., Speculations on the evolution of the lithosphere, Precambrian Res., 5, 249-260, 1977.

Barnes, I., W. P. Irwin, and D. E. White, Global distribution of carbon dioxide discharges and major zones of seismicity, U.S. Geol. Surv. Water Resour. Invest. 78-39, 12 pp., 1978.

Berner, R. A., A. C. Lasaga, and R. M. Garrels, The carbonate-silicate geochemical cycle and its effect on atmospheric carbon dioxide over the past 100 million years, Am. J. Sci., 283, 641-683, 1983.

Bickle, M. J., Heat loss from the earth: A constraint on Archaean tectonics from the relation between geothermal gradients and the rate of plate production, Earth Planet. Sci. Lett., 40, 301-315, 1978.

Boak, J. L., and R. F. Dymek, Metamorphism of the ca. 3800 Ma supracrustal rocks at Isua, West Greenland: Implications for early Archaean crustal evolution, Earth Planet. Sci. Lett., 59, 155-176, 1982.

Chase, C. G., The N plate problem of plate tectonics, Geophys. J. R. Astron. Soc., 29, 117-122, 1972.

Clarke, W. B., W. A. Beg, and H. Craig, Excess $^3$He in the sea: Evidence for terrestrial primordial helium, Earth Planet. Sci. Lett., 6, 213-220, 1969.

Corliss, J. B., J. Dymond, L. I. Gordon, J. M. Edmond, R. P. von Herzen, R. D. Ballard, K. Green, D. Williams, A. Bainbridge, K. Crane, and T. H. van Andel, Submarine thermal springs on the Galapagos Rift, Science, 203, 1073-1082, 1979.

Craig, H., W. B. Clarke, and M. A. Beg, Excess $^3$He in deep water on the East Pacific Rise, Earth Planet. Sci. Lett., 26, 125-132, 1975.

Craig, H., J. A. Welhan, K. Kim, R. Poreda, and J. E. Lupton, Geochemical studies of the 21°N EPR hydrothermal fluids (abstract), Eos Trans. AGU, 61, 992, 1980.

Des Marais, D. J., and J. G. Moore, Carbon and its isotopes in mid-oceanic basaltic glasses, Earth Planet. Sci. Lett., 69, 43-57, 1984.

Garfunkel, Z., Growth, shrinking, and long-term evolution of plates and their implications for the flow pattern in the mantle, J. Geophys. Res., 80, 4425-4432, 1975.

Green, D. H., Genesis of Archean peridotitic magmas and constraints on Archean geothermal gradients and tectonics, Geology, 3, 15-18, 1975.

Griggs, D. T., The sinking lithosphere and the focal mechanism of deep earthquakes, in The Nature of the Solid Earth, edited by E. C. Robertson, pp. 361-384, McGraw-Hill, New York, 1972.

Hart, M. H., The evolution of the atmosphere of the earth, Icarus, 33, 23-39, 1978.

Hays, J. D., and W. C. Pitman, Lithospheric plate motion, sea level changes, and ecological consequences, Nature, 246, 18-22, 1973.

Holland, H. D., The evolution of seawater, in The Early History of the Earth, edited by B. F. Windley, pp. 559-567, John Wiley, New York, 1976.

Houseman, G., Large aspect ratio convection cells in the upper mantle, Geophys. J. R. astr. Soc., 75, 309-334, (1983).

Huang, W. -L., P. J. Wyllie, and C. E. Nehru, Subsolidus and liquidus phase relationships in the system $CaO-SiO_2-CO_2$ to 30 kbar with geological applications, Am. Mineral., 65, 285-301, 1980.

Hunt, J. M., Petroleum Geochemistry and Geology, 617 pp., W. H. Freeman, San Francisco, Calif., 1979.

Javoy, M., F. Pineau, and C. J. Allegre, Carbon geodynamic cycle, Nature, 300, 171-173, 1982.

Javoy, M., F. Pineau, and C. J. Allegre, Carbon geodynamic cycle--Reply, Nature, 303, 731, 1983.

Jenkins, W. J., J. M. Edmond, and J. B. Corliss, Excess $^3$He and $^4$He in Galapagos submarine hydrothermal waters, Nature, 272, 156-158, 1978.

Kasting, J. F., J. B. Pollack, and D. Crisp, Effects of high $CO_2$ levels on surface temperature and atmospheric oxidation state of the early earth, J. Atmos. Chem., in press, 1984.

Kurz, M. D., and W. J. Jenkins, The distribution of helium in oceanic basalt glasses, Earth Planet. Sci. Lett., 53, 41-54, 1981.

Kurz, M. D., W. J. Jenkins, J. G. Schilling, and S. R. Hart, Helium isotopic variations in the mantle beneath the central North Atlantic Ocean, Earth Planet. Sci. Lett., 58, 1-14, 1982.

Lambert, R. S. J., Archean thermal regimes, crustal and upper mantle temperatures, and a progressive evolutionary model for the earth, in The Early History of the Earth, edited by B. F. Windley, pp. 363-373, John Wiley, New York, 1976.

Lasaga, A. C., R. A. Berner, and R. M. Garrels, An improved geochemical model of atmospheric $CO_2$ fluctuations over the past 100 million years, this volume.

Lupton, J. E., and H. Craig, Excess $^3$He in oceanic basalts: Evidence for terrestrial primordial helium, Earth Planet. Sci. Lett., 26, 133-139, 1975.

Lupton, J. E., G. P. Klinkhammer, W. R. Normark, R. Haymon, K. C. McDonald, R. F. Weiss, and H. Craig, Helium-3 and manganese at the 21°N East Pacific Rise hydrothermal site, Earth Planet. Sci. Lett, 50, 115-127, 1980.

Minster, J. B., T. H. Jordon, P. Molnar, and E. Haines, Numerical modelling of instantaneous plate tectonics, Geophys. J. R. Astron. Soc., 36, 541-576, 1974.

Mysen B. O., R. J. Arculus, and D. H. Eggler, Solubility of carbon dioxide in melts of andesite, tholeiite, and olivine nephelinite composition to 30kbar pressure, Contrib. Mineral. Petrol., 53, 227-239, 1975.

O'Connell, R. J., Heat and volatile transport in planetary interiors, paper presented at Conference on Planetary Volatiles, Lunar and Planet. Inst., Alexandria, Minn., Oct. 9-12, 1982.

O'Connell, R. J., and G. A. Houseman, Heat and volatile transport in planetary interiors, in Conference on Planetary Volatiles, NASA Tech. Rep. 83-01, compiled by R. O. Pepin and R. O'Connell, p. 131, Lunar and Planetary Institute, Houston, Tex., 1983.

O'Nions, R. K., and E. R. Oxburgh, Heat and helium in the earth, Nature, 306, 429-431, 1983.

Owen, T., R. D. Cess, and V. Ramanathan, Enhanced $CO_2$ greenhouse to compensate for reduced solar luminosity on early earth, Nature, 277, 640-642, 1979.

Parsons, B., The rates of plate creation and consumption, Geophys. J. R. Astron. Soc., 67, 437-448, 1981.

Pineau, F., M. Javoy, and Y. Bottinga, $^{13}C/^{12}C$ ratios of rocks and inclusions in popping rocks of the Mid-Atlantic Ridge and their bearing on the problem of isotopic composition of deep-seated carbon, Earth Planet. Sci. Lett., 29, 413-421, 1976.

Rubey, W. W., Geologic history of sea water--An attempt to state the problem, Geol. Soc. Am. Bull., 62, 1111-1148, 1951.

Sagan, C., and G. Mullen, Earth and Mars: evolution of atmospheres and surface temperatures, Science, 177, 52-56, 1972.

Sleep, N. H., Thermal history and degassing of the earth: Some simple calculations, J. Geol., 87, 671-686, 1979.

Toksoz, M. N., J. W. Minear, and B. R. Julian, Temperature field and geophysical effects of a downgoing slab, J. Geophys. Res., 76, 1113-1138, 1971.

Turcotte, D. L., On the thermal evolution of the earth, Earth Planet. Sci. Lett., 48, 53-58, 1980.

Veizer, J., and S. L. Jansen, Basement and sedimentary recycling and continental evolution, J. Geol., 87, 341-370, 1979.

Walker, J. C. G., Atmospheric evolution on the inner planets, in Comparative Planetology, edited by C. Ponnamperuma, pp. 141-163, Academic, New York, 1978.

Walker, J. C. G., P. B. Hays, and J. F. Kasting, A negative feedback mechanism for the long-term stabilization of earth's surface temperature, J. Geophys. Res., 86, 9776-9782, 1981.

Weiss, R. F., P. Lonsdale, J. E. Lupton, A. E. Bainbridge, and H. Craig, Hydrothermal plumes in the Galapagos Rift, Nature, 267, 600-603, 1977.

# PHOTOCHEMICAL CONSEQUENCES OF ENHANCED CO$_2$ LEVELS IN EARTH'S EARLY ATMOSPHERE

James F. Kasting

NASA Ames Research Center, Moffett Field, California 94035

Abstract. Greatly enhanced atmospheric CO$_2$ concentrations are the most likely mechanism for offsetting the effects of reduced solar luminosity early in the earth's history. CO$_2$ levels of 80 to 600 times the present value could have maintained a mean surface temperature of 0°C to 15°C, given a 25% decrease in solar output. Such high CO$_2$ levels are at least qualitatively consistent with our present understanding of the carbonate-silicate geochemical cycle. The presence of large amounts of CO$_2$ has important implications for the composition of the earth's prebiotic atmosphere. The hydrogen budget of a high-CO$_2$ primitive atmosphere would have been strongly influenced by rainout of H$_2$O$_2$ and H$_2$CO. The reaction of H$_2$O$_2$ with dissolved ferrous iron in the early oceans could have been a major sink for atmospheric oxygen. The requirement that this loss of oxygen be balanced by a corresponding loss of hydrogen (by escape to space and rainout of H$_2$CO) implies that the atmospheric H$_2$ mixing ratio was greater than $2 \times 10^{-5}$ and the ground level O$_2$ mixing ratio was below $10^{-12}$, even if other surface sources of H$_2$ were small. These results are only weakly dependent on changes in solar UV flux, rainout rates, and vertical mixing rates in the primitive atmosphere.

## Introduction

An outstanding challenge in understanding the earth's history is to try to simulate the environmental conditions under which life originated. Two factors which would almost certainly have influenced this process were the climate and atmospheric oxidation state of the primitive earth. Minimum requirements for chemical and biological evolution include the presence of liquid water over at least part of the earth's surface and the absence of free O$_2$, which would have acted as a poison to unprotected organisms [Walker, 1977]. The purpose of this paper is to show that both the climate and atmospheric oxidation state of the primitive earth were probably strongly influenced by greatly enhanced carbon dioxide levels in the early atmosphere.

Kasting et al. [1984b] (henceforth, KPC) have recently summarized the theoretical arguments for higher carbon dioxide levels during the earth's early history. The most convincing argument, from an atmospheric scientist's viewpoint, is that CO$_2$ is the only greenhouse gas which seems capable of resolving the "faint early sun" paradox. The problem, which was first recognized by Sagan and Mullen [1972], is that of maintaining the earth's surface temperature above freezing despite a 25% or greater reduction in solar luminosity early in solar system history [Newman and Rood, 1977; Walker, 1982]. Water vapor, the most effective greenhouse gas available, cannot solve the faint sun problem by itself because of the strong temperature dependence of its saturation vapor pressure. Other infrared-active gases, such as NH$_3$ or CH$_4$, have relatively short photochemical lifetimes and no obvious large surface or atmospheric source [Kasting, 1982; Kasting et al., 1983]. Carbon dioxide, on the other hand, has a long photochemical lifetime and a large source associated with the process of silicate reconstitution [Holland, 1978; Walker et al., 1981]. Several investigators, including Owen et al. [1979], Kuhn and Kasting [1983], and KPC, have shown that a CO$_2$ increase by a factor of roughly 100–1000 would be sufficient to restore the mean surface temperature to approximately its present value.

An alternative solution to the faint sun problem, proposed by Rossow et al. [1982], is to assume that cloud albedo feedback played a major role in stabilizing the earth's climate. Rossow et al. have formulated a one-dimensional radiative-convective model of the atmosphere in which cloudiness is assumed to decrease with decreasing surface temperature. For reduced solar fluxes, the primary effect is to decrease the planetary albedo and thereby increase the fraction of the sun's radiation which is absorbed by the earth. The resultant warming of the surface would ensure that temperatures remained above freezing at low latitudes for as much as a 20% decrease in solar luminosity.

This hypothesis, however, is unattractive for a variety of reasons. First, the 20% solar flux reduction assumed by Rossow et al. is below the range of values (-22% to -29%) thought to apply during the earth's early history [Newman and Rood, 1977; Gough, 1981]. Second, their calculation has neglected the effect of ice albedo feedback, which should tend to amplify the temperature drop associated with a decrease in solar flux. As discussed by KPC, it seems probable that the albedo increase due to polar ice formation would have tended to offset any albedo decrease related to clouds. Consequently, the procedure followed in this paper, as in several previous calculations [Owen et al., 1979; Kuhn and Kasting, 1983; KPC], is to assume that the earth's albedo has remained constant throughout the planet's history.

The idea that $CO_2$ concentrations were greatly enhanced in the early atmosphere is supported by additional theoretical arguments. Walker et al. [1981] argued that the rate of consumption of $CO_2$ by silicate weathering should decrease sharply at lower surface temperatures, largely as a result of a decrease in the global average precipitation rate. This effect, by itself, should produce a big enough enhancement in $CO_2$ to prevent the earth from freezing over. They also showed that the $CO_2$ partial pressure should depend strongly on the "volcanic" or "metamorphic" source of $CO_2$, which results from conversion of carbonate minerals to silicate minerals deep within the earth's crust. A younger and more tectonically active earth should have had a larger volcanic source and, consequently, a higher atmospheric $CO_2$ level. This same concept is parameterized differently in the model of Berner et al. [1983], in which the volcanic $CO_2$ flux is assumed to be proportional to the rate of seafloor spreading. In addition, Berner et al. argued that the $CO_2$ partial pressure should depend inversely on the amount of continental surface area exposed to weathering. Since the fraction of the earth's surface covered by continents may have originally been much smaller than it is today [Hargraves, 1977; Windley, 1977; Veizer et al., 1982], this factor should also have favored higher $CO_2$ levels in the distant past. Finally, Lovelock and Whitfield [1982] and Lovelock and Watson [1982] have pointed out that land plants speed up the rate of silicate weathering by enhancing the $CO_2$ partial pressure in soils above its atmospheric value. The absence of land plants prior to about 420 million years ago is yet another factor favoring high $CO_2$ concentrations in the early atmosphere.

Unfortunately, the physical processes which influence the carbonate-silicate geochemical cycle cannot be quantified well enough to allow us to estimate early $CO_2$ levels with any degree of confidence. The best that we can do is to approach the problem in an inverse sense, that is, to ask how much $CO_2$ would have been needed to maintain a temperate climate on the early earth. Since we do not presently know whether the early climate was warmer or colder than today's, one can only hope to derive a range of plausible $CO_2$ concentrations. One of the purposes of this paper is to determine that range for a realistic set of assumptions about the early atmosphere.

The other purpose of this paper is to investigate the effects of high $CO_2$ concentrations on atmospheric photochemistry. KPC have already shown that one of the major effects is to increase the photochemical production rates of hydrogen peroxide ($H_2O_2$) and formaldehyde ($H_2CO$). These gases are formed from the by-products of $CO_2$ and $H_2O$ photolysis and, hence, become more abundant as $CO_2$ levels increase. KPC pointed out that rainout of these gases should have affected the atmospheric oxidation state, since the loss of one $H_2O_2$ molecule involves a net loss of one free oxygen atom, while the loss of one $H_2CO$ molecule implies a net loss of four free hydrogen atoms. Prior to the advent of oxygenic photosynthesis, the hydrogen budget of the early atmosphere should have been regulated by the relation [KPC, equation (4)].

$$\Phi_{esc}(H) + 4 \times \text{Rain}\,(H_2CO) = \Phi_{out}(H) + 2 \times \text{Rain}\,(H_2O_2) \qquad (1)$$

where Rain ($H_2CO$) and Rain ($H_2O_2$) are the net rainout rates of the two gases, $\Phi_{esc}(H)$ is the rate of hydrogen escape to space, and $\Phi_{out}(H)$ represents the surface source of hydrogen. The escape rate is given by the limiting flux relationship

$$\Phi_{esc}(H) \simeq 2.5 \times 10^{13}\, f_t(H) \quad H \text{ atoms cm}^{-2}\,\text{s}^{-1} \qquad (2)$$

where $f_t(H)$ is the sum of the stratospheric volume mixing ratios of all the hydrogen-bearing species weighted by the number of H atoms they contain. The proportionality constant in (2) is taken from Walker [1977] and is appropriate for an atmosphere in which most of the hydrogen at the homopause is in the form of $H_2$. KPC, by comparison, used a proportionality constant of $4 \times 10^{13}$, which is appropriate for pure atomic hydrogen. Equation (2) probably gives a more realistic estimate for the escape rate from the primitive atmosphere, except perhaps during the first 100 million years of earth history when the solar EUV flux was very high, in which case the conversion of $H_2$ into H might have been effected at relatively low altitudes.

The surface hydrogen source $\Phi_{out}(H)$ in (1) includes volcanic outgassing of $H_2$ and CO and photostimulated iron oxidation in the oceans. The volcanic outgassing rate was probably greater than or equal to the present release rate of $4 \times 10^8$ ($H_2$+CO) molecules cm$^{-2}$ s$^{-1}$ estimated by Kasting et al. [1979]. The rate of photostimulated iron oxidation, in which hydrogen is generated by absorption of near-UV photons by Fe(OH)[+] [Braterman et al., 1983], is difficult to estimate but could have been up to two orders of magnitude greater than the volcanic hydrogen flux, based on the availability of ferrous iron [KPC]. Since only a small part of the iron should have been oxidized, however, and some of that by reaction with dissolved H$_2$O$_2$, it may be unfair to treat this as an independent hydrogen source. More realistically, the sum of the rate of photo-stimulated iron oxidation plus rainout of H$_2$O$_2$ should have remained equal to some fraction of the rate of supply of ferrous iron. In this paper we shall assume that all of the ferrous iron oxidation resulted from reaction with H$_2$O$_2$ and none from Braterman's mechanism. We shall also assume that the volcanic source of hydrogen was negligible. (This assumption can be justi-fied for the case of high atmospheric P$_{CO_2}$ and present-day outgassing rates.) The effect of these assumptions is to produce a lower limit on the H$_2$ mixing ratio and, hence, an upper limit on O$_2$ concentrations in the early atmosphere.

This paper extends the work of KPC by con-sidering the consequences of changes in rainout rates, solar UV fluxes, and vertical mixing rates. The emphasis is on constructing self-consistent models of the early atmosphere. To achieve this goal, we use a radiative-convective model to pro-vide inputs to the photochemical model, so that CO$_2$ levels, temperature, and H$_2$O profiles are varied in a consistent manner. Finally, we will briefly mention how data from paleosols may provide an indirect measure of CO$_2$ and H$_2$O$_2$ con-centrations as far back as 2.9 billion years ago.

### Numerical Models

The radiative-convective and photochemical models used in this study are described by Kasting et al. [1984a] and KPC. As such, only the essential features of the models and a few important changes will be described here.

The radiative-convective model employs the method of Rodgers and Walshaw [1966] to parame-terize line absorption by H$_2$O, using an approxi-mate Voigt line shape taken from Fels [1979]. H$_2$O continuum absorption was treated by the method of Roberts et al. [1976]. The parameteri-zation of CO$_2$ absorption is described by KPC. The amount of greenhouse warming caused by CO$_2$ agrees well with that calculated in an inde-pendent study by Kuhn and Kasting [1983] but is substantially smaller than that predicted by Owen et al. [1979]. The present model utilizes the moist adiabatic lapse rate in the tropo-sphere and assumes a tropospheric relative humid-ity profile similar to that of Manabe and Wetherald [1967]. The treatment of H$_2$O in the stratosphere follows the procedure of Visconti [1982], whereby the relative humidity is allowed to increase above the cloud-top level (the top of the convective region) provided that the H$_2$O mixing ratio does not increase with altitude. This maximizes the stratospheric H$_2$O content, yielding an upper limit on the hydrogen escape rate and, hence, an upper limit on the rate of

abiotic production of oxygen. By contrast, KPC fixed the stratospheric relative humidity at its cloud-top value, which could have caused them to underestimate stratospheric H$_2$O mixing ratios by a factor of 5 or 6.

The photochemical model solves the vertical transport and continuity equations for 12 differ-ent species: O, O$_2$, H, OH, HO$_2$, H$_2$O$_2$, H$_2$, H$_2$O, CO, CO$_2$, HCO, and H$_2$CO. For numerical reasons, O$_3$ and O($^1$D) were assumed to be photochemical equilibrium at each height step. Little accu-racy is lost by this approximation since both species would be extremely short-lived in a low-O$_2$ atmosphere. The chemical reactions and (updated) rate constants linking these species are shown in Table 1. Of particular importance for this study is a change in the rate constant for H$_2$O$_2$ production (R15). Photolysis rates were calculated by the method of Yung [1976], which includes the effects of Rayleigh scattering. Rainout of H$_2$O$_2$ was parameterized by assuming a constant rainout rate of $2.4 \times 10^{-6}$ s$^{-1}$ below 6 km, decreasing to $4.6 \times 10^{-7}$ s$^{-1}$ at 10 km and to zero above that height [Fishman and Crutzen, 1977]. The assumed boundary conditions are the same as those of KPC. The numerical solver, however, was replaced with a time-dependent rou-tine based on the reverse Euler method [Dahlquist and Bjorck, 1974].

Two additional changes from the model of KPC were made to improve self-consistency. First, the surface pressure (in bars) in both the radiative-convective and photochemical models was assumed to be given by

$$P_s = 0.77 \left[ 1 + \frac{32 f_{O_2} + 44\, f_{CO_2}}{28 \left( 1 - f_{O_2} - f_{CO_2} \right)} \right] \qquad (3)$$

where $f_{O_2}$ and $f_{CO_2}$ are the ground level volume mixture ratios of O$_2$ and CO$_2$. Equation (3) accounts for the pressure variation due to large increases in atmospheric CO$_2$, given a fixed back-ground pressure of 0.77 bar of N$_2$. (By contrast, KPC assumed a constant surface pressure of 1 bar.)

The second change was in the method of param-eterizing condensation of water vapor in the photochemical model. Instead of specifying H$_2$O all the way up to the cold trap (the height where saturation H$_2$O mixing ratio is at a minimum), the H$_2$O content was held fixed only up to the top of the convective region. Above that height, H$_2$O was treated as a gas which obeys the flux and continuity equations. If the calculated mixing ratio exceeded saturation, however, the excess water vapor was assumed to condense out in a period of a few hours. The resultant H$_2$O dis-tributions were thus consistent with both eddy mixing and condensation processes.

TABLE 1.  Reactions and Rate Constants

| | Reaction | Rate Constant, $cm^3$ $s^{-1}$ | Reference |
|---|---|---|---|
| (R 1) | $H_2O + O(^1D) \rightarrow OH + OH$ | $2.2 \times 10^{-10}$ | JPL [1983] |
| (R 2) | $H_2 + O(^1D) \rightarrow OH + H$ | $1 \times 10^{-10}$ | JPL [1983] |
| (R 3) | $H_2 + O \rightarrow OH + H$ | $3 \times 10^{-14} T \exp(-4480/T)$ | Hampson and Garvin [1977] |
| (R 4) | $H_2 + O \rightarrow OH + H$ | $6.1 \times 10^{-12} \exp(-2030/T)$ | JPL [1983] |
| (R 5) | $H + O_3 \rightarrow OH + O_2$ | $1.4 \times 10^{-10} \exp(-470/T)$ | JPL [1983] |
| (R 6) | $H + O_2 + M \rightarrow HO_2 + M$ | 3-body rate[a] | JPL [1983] |
| (R 7) | $H + HO_2 \rightarrow H_2 + O_2$ | $0.09 \times 7.4 \times 10^{-11}$ | JPL [1983] |
| (R 8) | $H + HO_2 \rightarrow H_2O + O$ | $0.04 \times 7.4 \times 10^{-11}$ | JPL [1983] |
| (R 9) | $H + HO_2 \rightarrow OH + OH$ | $0.87 \times 7.4 \times 10^{-11}$ | JPL [1983] |
| (R10) | $OH + O \rightarrow H + O_2$ | $2.2 \times 10^{-11} \exp(117/T)$ | JPL [1983] |
| (R11) | $OH + HO_2 \rightarrow H_2O + O_2$ | $1 \times 10^{-11} (7 + 4 \text{ Patm})$ | JPL [1983] |
| (R12) | $OH + O_3 \rightarrow HO_2 + O_2$ | $1.6 \times 10^{-12} \exp(-940/T)$ | JPL [1983] |
| (R13) | $HO_2 + O \rightarrow OH + O_2$ | $3 \times 10^{-11} \exp(200/T)$ | JPL [1983] |
| (R14) | $HO_2 + O_3 \rightarrow OH + 2 O_2$ | $1.4 \times 10^{-14} \exp(-580/T)$ | JPL [1983] |
| (R15) | $HO_2 + HO_2 \rightarrow H_2O_2 + O_2$ | $2.3 \times 10^{-13} \exp(590/T)$ $+ 1.7 \times 10^{-33} \exp(1000/T)[M]$ | JPL [1983] |
| (R16) | $H_2O_2 + OH \rightarrow HO_2 + H_2O$ | $3.1 \times 10^{-12} \exp(-187/T)$ | JPL [1983] |
| (R17) | $H_2O_2 + O \rightarrow HO_2 + OH$ | $1.4 \times 10^{-12} \exp(-2000/T)$ | JPL [1983] |
| (R18) | $OH + OH + M \rightarrow H_2O_2 + M$ | 3-body rate[a] | JPL [1983] |
| (R19) | $O + O_3 \rightarrow 2 O_2$ | $8 \times 10^{-12} \exp(-2060/T)$ | JPL [1983] |
| (R20) | $OH + OH \rightarrow H_2O + O$ | $4.2 \times 10^{-12} \exp(-242/T)$ | JPL [1983] |
| (R21) | $O(^1D) + N_2 \rightarrow O + N_2$ | $1.8 \times 10^{-11} \exp(107/T)$ | JPL [1983] |
| (R22) | $O(^1D) + O_2 \rightarrow O + O_2$ | $3.2 \times 10^{-11} \exp(67/T)$ | JPL [1983] |
| (R23) | $O + O + M \rightarrow O_2 + M$ | $2.76 \times 10^{-34} \exp(710/T)[M]$ | Campbell and Thrush [1967] |
| (R24) | $O + O_2 + M \rightarrow O_3 + M$ | 3-body rate[a] | JPL [1983] |
| (R25) | $H + H + M \rightarrow H_2 + M$ | $2.6 \times 10^{-33} \exp(375/T)[M]$ | Liu and Donahue [1974] |
| (R26) | $H + OH + M \rightarrow H_2O + M$ | $6.1 \times 10^{-26} [M]/T^2$ | McEwan and Phillips [1975] |
| (R27) | $CO + OH \rightarrow CO_2 + H$ | $1.5 \times 10^{-13} (1 + 0.6 \text{ Patm})$ | JPL [1983] |
| (R28) | $CO + O + M \rightarrow CO_2 + M$ | $6.5 \times 10^{-33} \exp(-2180/T)[M]$ | Hampson and Garvin [1977] |
| (R29) | $H + CO + M \rightarrow HCO + M$ | $2 \times 10^{-33} \exp(-850/T)[M]$ | Baulch et al. [1976] |
| (R30) | $H + HCO \rightarrow H_2 + CO$ | $1.2 \times 10^{-10}$ | Hochanadel et al. [1980] |
| (R31) | $HCO + HCO \rightarrow H_2CO + CO$ | $2.3 \times 10^{-11}$ | Hochanadel et al. [1980] |
| (R32) | $OH + HCO \rightarrow H_2O + CO$ | $5 \times 10^{-11}$ | Baulch et al. [1976] |
| (R33) | $O + HCO \rightarrow H + CO_2$ | $1 \times 10^{-10}$ | Hampson and Garvin [1977] |
| (R34) | $O + HCO \rightarrow OH + CO$ | $1 \times 10^{-10}$ | Hampson and Garvin [1977] |
| (R35) | $O_2 + HCO \rightarrow HO_2 + CO$ | $3.5 \times 10^{-12} \exp(140/T)$ | JPL [1983] |
| (R36) | $H_2CO + H \rightarrow H_2 + HCO$ | $2.8 \times 10^{-11} \exp(-1540/T)$ | JPL [1979] |
| (R37) | $H_2CO + OH \rightarrow H_2O + HCO$ | $1 \times 10^{-11}$ | JPL [1983] |
| (R38) | $H_2CO + O \rightarrow HCO + OH$ | $3 \times 10^{-11} \exp(-1550/T)$ | JPL [1983] |
| (R39) | $O_2 + h\nu \rightarrow O + O(^1D)$ | $1.14 \times 10^{-7}$ $s^{-1}$ [b] | Thompson et al. [1963] |
| (R40) | $O_2 + h\nu \rightarrow O + O$ | $7.42 \times 10^{-8}$ $s^{-1}$ [b] | Thompson et al. [1963] |
| (R41) | $H_2O + h\nu \rightarrow H + OH$ | $2.86 \times 10^{-6}$ $s^{-1}$ [b] | Thompson et al. [1963] |
| (R42) | $O_3 + h\nu \rightarrow O_2 + O(^1D)$ | $7.65 \times 10^{-3}$ $s^{-1}$ [b] | DeMore and Raper [1966] |
| (R43) | $O_3 + h\nu \rightarrow O_2 + O$ | $2.37 \times 10^{-4}$ $s^{-1}$ [b] | DeMore and Raper [1966] |
| (R44) | $H_2O_2 + h\nu \rightarrow 2OH$ | $9.06 \times 10^{-5}$ $s^{-1}$ [b] | JPL [1983] |
| (R45) | $HO_2 + h\nu \rightarrow OH + O$ | $5.78 \times 10^{-4}$ $s^{-1}$ [b] | JPL [1983] |
| (R46) | $CO_2 + h\nu \rightarrow CO + O(^1D)$ | $6.62 \times 10^{-9}$ $s^{-1}$ [b] | Thompson et al. [1963] |
| (R47) | $CO_2 + h\nu \rightarrow CO + O$ | $1.74 \times 10^{-9}$ $s^{-1}$ [b] | Shemansky [1972] |
| (R48) | $H_2CO + h\nu \rightarrow H_2 + CO$ | $5.55 \times 10^{-5}$ $s^{-1}$ [b] | JPL [1983] |
| (R49) | $H_2CO + h\nu \rightarrow HCO + H$ | $6.31 \times 10^{-5}$ $s^{-1}$ [b] | JPL [1983] |
| (R50) | $HCO + h\nu \rightarrow H + CO$ | $1 \times 10^{-2}$ $s^{-1}$ | Pinto et al. [1980] |

[a] See expression in JPL [1983].
[b] Calculated from solar fluxes [Mount et al., 1980; Ackerman, 1970] and cross sections; diurnally averaged values at 63.5 km for atmosphere containing 600 PAL of $CO_2$.

## Results

Before examining high-CO$_2$ atmospheres, we first considered a case in which the surface temperature was assumed to be high (288 K) even though the atmospheric CO$_2$ content was low (0.8 times present atmospheric level (PAL)). While not entirely self-consistent, this situation corresponds to that assumed by Kasting et al. [1979] and Kasting and Walker [1981] and is at least of academic interest. Temperature and water vapor profiles for this calculation were taken from the model of KPC. An upper limit on the O$_2$ level was then derived by adjusting the H$_2$ content in the photochemical model so as to balance equation (1), under the assumption that $\Phi_{out}$(H) was equal to zero (see Introduction). The ground level H$_2$ mixing ratio which satisfied this constraint was $5 \times 10^{-7}$, as shown in Table 2. The calculated ground level O$_2$ mixing ratio was $1.6 \times 10^{-14}$, which may be compared with the value of $10^{-8}$ estimated by Kasting and Walker [1981] for similar circumstances. The difference between these estimates is caused by including rainout of H$_2$O$_2$ in the atmospheric hydrogen budget in the present model.

This experiment implies that the ground level O$_2$ mixing ratio should have been very low, even if the primitive atmosphere contained only present amounts of CO$_2$. Our conclusion rests solely upon the availability of ferrous iron in the oceans and its ability to react with dissolved H$_2$O$_2$. We shall see that this process should have been even more effective if the atmosphere had contained enhanced levels of CO$_2$.

We turn now to the construction of self-consistent models of the primitive atmosphere. The solar luminosity is assumed to be 25% lower than it is today, which is about midway through the range of estimated values [Newman and Rood, 1977; Gough, 1981]. We further assume that the mean global surface temperature T$_s$ was between 0°C and the present value of 15°C. The higher value corresponds to a sea ice line at 72° latitude and a global surface albedo of 0.10, according to Wang and Stone [1980]. Temperatures warmer than this cannot be ruled out but seem improbable given the large decrease in solar constant. The lower temperature should produce an ice line near 51° latitude and a global surface albedo of 0.16. As discussed earlier, we assume that the increase in surface albedo would be balanced by a decrease in cloud albedo. It is unlikely that such compensation could occur at temperatures below freezing, so 0°C is probably a realistic lower limit on T$_s$.

The CO$_2$ mixing ratios required by our model to reproduce these surface temperatures are $3.3 \times 10^{-2}$ (for 0°C) and 0.2 (for 15°C). From equation (3), we see that this corresponds to surface pressure increments of 0.04 bar and 0.3 bar of CO$_2$, respectively. The CO$_2$ partial pressure at present

($f_{CO_2}$ = $3.3 \times 10^{-4}$) is $5 \times 10^{-4}$ bar, so the required CO$_2$ increases correspond to 80 PAL and 600 PAL, respectively. The CO$_2$ enhancements calculated in this fashion may be compared with the estimates of 100-1000 PAL quoted by Owen et al. [1979], Kuhn and Kasting [1983], and KPC, based on maintaining the present surface temperature given a 24-30% decrease in solar flux.

Vertical temperature profiles calculated with the radiative-convective model are shown in Figure 1. The solid horizontal lines mark the cloud-top level, and dashed lines show the position of the cold trap. The saturation H$_2$O mixing ratio at the cold trap is $2.4 \times 10^{-7}$ for the case of 80 PAL of CO$_2$, or $6.3 \times 10^{-7}$ for 600 PAL of CO$_2$. Both values are significantly lower than the H$_2$O mixing ratio in the earth's present stratosphere, which is about 3-5 ppm [Walker, 1977]. The low stratospheric H$_2$O mixing ratios are the result of extremely cold temperatures caused by the absence of UV absorption by O$_3$. As mentioned earlier, this implies a low rate of oxygen production from photodissociation of H$_2$O followed by hydrogen escape.

The next step was to run the photochemical model using these temperatures and water vapor profiles as inputs. The results of doing so are presented in Figures 2 and 3, which show mixing ratio profiles of selected long-lived species. Other characteristics of the 80 PAL and 600 PAL CO$_2$ cases, including rainout rates and hydrogen escape rates, are listed in Table 2. The surface source of hydrogen $\Phi_{out}$(H) has in each case been set equal to zero, so these model atmospheres represent the most oxidizing environments possible. Figures 2 and 3 show that the low-CO$_2$ and high-CO$_2$ cases are qualitatively similar, especially with regard to ground level O$_2$ concentrations. Quantitatively, the high-CO$_2$ case has higher rainout rates of H$_2$O$_2$ and H$_2$CO and, consequently, requires a higher equilibrium H$_2$ mixing ratio in order to balance equation (1). The CO concentration is also higher as a consequence of the increased CO$_2$ photolysis rate. The CO concentration is of interest because of its possible involvement in synthesis of methane and other hydrocarbon compounds within the atmosphere [Bar-Nun and Chang, 1983].

Let us pause here to consider how these results would have changed had we assumed a finite surface source of hydrogen. The present-day volcanic outgassing rate of $4 \times 10^8$ (H$_2$ + CO) molecules cm$^{-2}$ s$^{-1}$ estimated by Kasting et al. [1979] corresponds to a net hydrogen source of $8 \times 10^8$ H atoms cm$^{-2}$ s$^{-1}$. (Note that CO and H$_2$O can react indirectly to give CO$_2$ plus H$_2$.) By comparison, the H$_2$O$_2$ rainout rates for the 80 PAL and 600 PAL CO$_2$ cases are $1.2 \times 10^9$ cm$^{-2}$ s$^{-1}$ and $3.6 \times 10^9$ cm$^{-2}$ s$^{-1}$, respectively. These numbers must also be multiplied by a factor of 2 to get the effective hydrogen source in H atoms cm$^{-2}$ s$^{-1}$, since the excess oxygen atom in H$_2$O$_2$ must ulti-

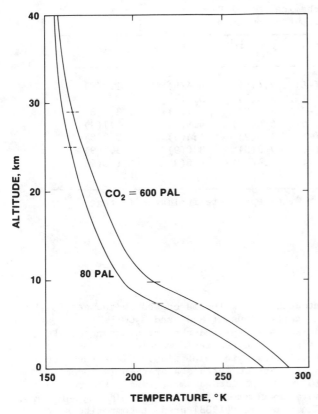

Fig. 1.  Calculated temperature profiles for 80 and 600 PAL of $CO_2$.  The solid horizontal lines mark the cloudtop level, and the dashed lines indicate the position of the cold trap.

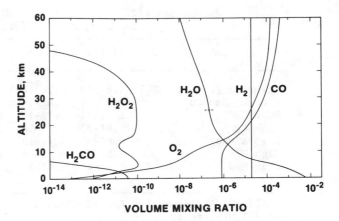

Fig. 2.  Volume mixing ratios of long-lived species for 80 PAL of $CO_2$.

mately derive from $H_2O$.  The present-day volcanic hydrogen source is therefore equivalent to roughly 30% of the assumed hydrogen source in the low-$CO_2$ case, or 10% in the high-$CO_2$ case. Inclusion of this outgassing term in equation (1) would thus result in an increase of about 10–30% in the equilibrium $H_2$ mixing ratio in these two model atmospheres.  The photochemical consequences of such a modest change in $H_2$ concentration would be minimal.  The ground level $O_2$ concentration, for example, would presumably decrease by roughly this same amount.

Suppose, on the other hand, that the volcanic outgassing rate of hydrogen was much greater than it is today or, equivalently, that there was a large additional supply of hydrogen from photo-stimulated iron oxidation in the oceans.  Then the atmospheric $H_2$ mixing ratio should have risen until equilibrium was achieved.  Other workers [Yung and McElroy, 1979; Pinto et al., 1980] have speculated that the resultant $H_2$ mixing ratio could have been as high as $10^{-3}$.  The effect of such an $H_2$ increase on our "standard" 600 PAL

$CO_2$ atmosphere is shown by the "high $H_2$" case in Table 2.  As might be expected, the concentrations of $O_2$ and $H_2O_2$ decrease, while those of $CO$ and $H_2CO$ become larger.  The surface hydrogen source required to sustain this $H_2$ concentration, assuming that the escape rate remains diffusion-limited, is about $6 \times 10^{10}$ H atoms $cm^{-1}$ $s^{-2}$.

The primary consequence of a large surface hydrogen source would thus have been to decrease the concentrations of oxidants and increase the concentrations of reduced species in the early atmosphere.  With this in mind we now return to the case where $\Phi_{out}(H)$ is assumed to be negligible, so that the calculated $O_2$ concentrations represent an upper bound.  An interesting question with respect to chemical or biological evolution is whether changes in any of our assumed

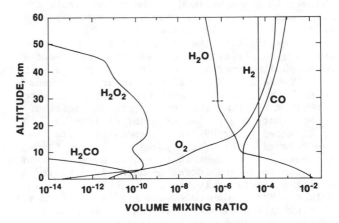

Fig. 3.  Volume mixing ratios of long-lived species for 600 PAL of $CO_2$ (standard case).

TABLE 2.   Sensitivity of Photochemical Model Results
to Changes in Different Input Parameters

| Case | $f(H_2)$[a] | $f(O_2)$[a] | $f(CO)$[a] | $\Phi_{esc}(H)$[b] | $Rain(H_2O_2)$[b] | $Rain(H_2CO)$[b] |
|---|---|---|---|---|---|---|
| Standard (600 PAL $CO_2$) | 4.8(−5) | 8.2(−14) | 9.6(−6) | 2.4(9) | 3.6(9) | 1.2(9) |
| Present-day $CO_2$ | 5.0(−7) | 1.6(−14) | 1.5(−10) | 7.0(7) | 3.5(7) | 0 |
| 80 PAL $CO_2$ | 2.0(−5) | 1.9(−13) | 8.0(−7) | 1.0(9) | 1.2(9) | 3.5(8) |
| High $H_2$ | 1.0(−3) | 1.8(−18) | 5.2(−5) | 5.0(10) | 4.6(7) | 2.1(9) |
| Low rainout | 3.0(−5) | 1.6(−13) | 6.0(−6) | 1.5(9) | 1.4(9) | 3.3(8) |
| High UV | 1.5(−5) | 5.6(−14) | 2.5(−6) | 7.8(8) | 8.0(9) | 3.8(9) |
| High Keddy | 2.8(−4) | 2.6(−13) | 1.3(−4) | 1.4(10) | 9.0(9) | 1.3(9) |

All models assume 600 PAL of $CO_2$ and $\Phi_{out}(H) = 0$, except where indicated in text. Read 1.0(9) as $1.0 \times 10^9$.

[a] Ground level volume mixing ratios.
[b] Column-integrated rates (molecules $cm^{-2}$ $s^{-1}$).

model parameters could have led to significant increases in the ground level abundance of $O_2$.

An obvious source of uncertainty in our calculations results from the steep gradients in $H_2O_2$ and $H_2CO$ concentrations in the lower troposphere. This is illustrated more clearly in Figure 4, which shows number density profiles for the high-$CO_2$ model atmosphere. Nearly all of the $H_2CO$ which rains out comes from the lowest 3 km of the atmosphere, whereas most of the dissolved $H_2O_2$ originates between 3 and 7 km. Our model assumes a constant rainout rate below 6 km, which may not be sufficiently accurate to describe this situation. Clearly, a more sophisticated parameterization of precipitation patterns could alter the balance between rainout of $H_2O_2$ and $H_2CO$ and thereby change the equilibrium $H_2$ mixing ratio.

Simple linear changes in the rainout rate, on the other hand, are unlikely to result in major changes in atmospheric $H_2$. The rainout rate in our standard model corresponds to an atmospheric lifetime for soluble species of about 5 days. Tripling this lifetime to 15 days for the high-$CO_2$ case resulted in a 50% reduction in $H_2$ and a corresponding increase in $O_2$ by less than a factor of 2 (Table 2). Thus, simple linear variations in rainout lifetimes should not greatly influence estimates of primitive atmospheric composition.

Somewhat larger changes might be induced by enhanced solar UV fluxes associated with the T-Tauri phase of the young sun. Such flux increases are thought to be caused by more rapid rotation rates and increased magnetic activity during a star's early history [Zahnle and Walker, 1982]. Canuto et al. [1982] and KPC have investigated possible photochemical effects of a wavelength-independent flux increase. Canuto et al. [1983] looked at the effects of a more realistic increase which is more pronounced at short wavelengths. Of these investigators, only KPC took into account the possible change in

atmospheric $H_2$ levels caused by changes in the concentrations of $H_2O_2$ and $H_2CO$.

We have repeated the calculations of KPC using a more realistic wavelength dependence for the increase in solar flux. Since the time scale for the earth's accretion is about $10^8$ years [Canuto et al., 1983], the relevant solar spectrum is that for the 100-m.y.-old sun. Canuto et al. [1983] present composite spectra of three T-Tauri stars estimated to be between 1 and 10 m.y. old which show UV enhancements by a factor of 3 at 3000 Å, increasing to ~$10^4$ at 1000 Å. The maximum flux enhancement is estimated to decrease to a factor of ~$10^2$ at an age of 50 m.y. By comparison, Zahnle and Walker [1982] estimate that the flux between 1000 Å and 2000 Å was enhanced by a factor of 5–10 for the 100-m.y.-old sun. As a rough composite of these estimates, we have assumed that the UV

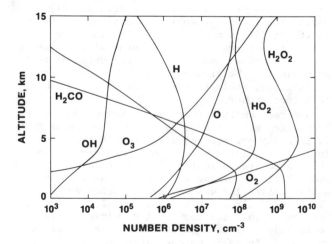

Fig. 4. Tropospheric number densities for 600 PAL of $CO_2$.

flux just after the earth was formed was given by

$$F(\lambda) = F_o(\lambda) \times 10^{(3000-\lambda)/1000}$$

$$\lambda < 3000 \text{ Å} \quad (4)$$

where $F_o(\lambda)$ is the present solar maximum flux. This expression yields a factor of 10 enhancement at 2000 Å (where $CO_2$ and $H_2O$ photolysis commence) and a factor of 60 enhancement at Ly α. The photolysis rate of $H_2O_2$ is increased by a factor of 2-3, while photolysis of $H_2CO$ is only marginally affected.

The resulting compositional changes for the high-$CO_2$ atmosphere are shown in Table 2 and Figure 5. The equilibrium $H_2$ mixing ratio is reduced by a factor of 3 compared to the standard case, mostly due to a large increase in the rainout rate of $H_2CO$. The rainout rate of $H_2O_2$ also increases, but not by as much as that of $H_2CO$. This is because $H_2O_2$ is photolyzed more rapidly, whereas the $H_2CO$ photolysis rate does not change. The $O_2$ content of the upper atmosphere (near 60 km) is enhanced by a factor of 20, but the ground level $O_2$ concentration actually declines by about 30%. The column depth of ozone remains far too small (by a factor of ~500) to provide effective shielding from solar UV.

The response of $O_2$ and $H_2$ in this model is qualitatively similar to that found by KPC but occurs for somewhat different reasons. The wavelength-independent factor of 10 UV increase assumed by KPC caused a decrease in the rainout rates of both $H_2O_2$ and $H_2CO$. The decrease in $H_2O_2$ was more pronounced, so the net result was a slight (40%) decrease in atmospheric $H_2$. Both models agree that increases in solar UV should lead to small decreases in the ground level $O_2$ mixing ratio. These results contradict the findings of Canuto et al. [1982, 1983], who predicted that ground level $O_2$ mixing ratios would rise. Possible reasons for this discrepancy are

discussed by KPC. The conclusion to be drawn from this experiment is that increases in the solar UV flux should not have substantially altered ground level $O_2$ concentrations in the primitive atmosphere, because of the requirement that the atmospheric hydrogen budget remain balanced.

A final sensitivity study was to investigate the possible effect of increased vertical mixing rates in the lower stratosphere. The absence of the temperature inversion caused by ozone heating should have resulted in a decrease in static stability in this region and, hence, an increase in the rate of eddy mixing. Because of the pronounced gradient in $O_2$ mixing ratio at these heights, this could have caused a large increase in the downward flux of $O_2$ into the troposphere. An obvious question is whether such a change could have raised the $O_2$ concentration at the surface.

To investigate this possibility, we increased the eddy diffusion coefficient in the standard high-$CO_2$ model to $2 \times 10^5$ cm$^{-2}$ s$^{-1}$ (the assumed tropospheric value) throughout the lower stratosphere. This corresponds to a factor of 40 increase in vertical mixing rates near 14 km. The resulting changes in atmospheric composition are shown in Table 2 and Figure 6. The ground level $O_2$ concentration does indeed increase, but by only a factor of 3. (Only an enhancement by several orders of magnitude would be significant.) The reason that the increase is this small, however, is the strong feedback provided by rainout of $H_2O_2$. The increased downward flux of $O_2$ into the troposphere results in a dramatic rise in the photochemical production rate of $H_2O_2$. Since the hydrogen budget must remain balanced, however, the model responds by raising the atmospheric $H_2$ mixing ratio by a factor of 6. The increased $H_2$ then moderates the rise of both $H_2O_2$ and $O_2$. The conclusion is that changes in vertical mixing rates, like changes in rainout rates or solar UV fluxes, should not have significantly altered the atmospheric oxidation state.

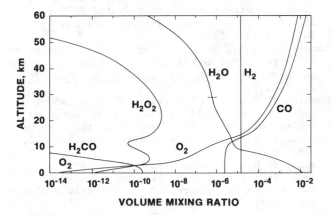

Fig. 5. Same as Figure 3 but for solar UV enhancement given by equation (4).

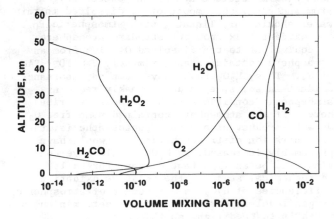

Fig. 6. Same as Figure 3 but for increased vertical mixing rates in the stratosphere.

## Discussion

A frustrating aspect of doing the type of calculations described here is the lack of observational evidence to confirm or deny one's predictions. This is likely to remain a problem for prebiotic atmosphere simulations, since the rock record is nonexistent prior to 3.8 b.y. ago, by which time life had probably already evolved. The latter part of the Archean, however, may prove to be more susceptible to testing of model hypotheses. Inferences about atmospheric oxygen can be drawn from observations of uraninite deposits, banded iron formations, and redbeds [Walker, 1977]. The data, while not always consistent, indicate relatively low atmospheric $O_2$ levels prior to 2 b.y. ago. Now, new types of evidence are being uncovered from studies of paleosols, or ancient soils. By examining the degree of silicate weathering in such samples, along with the $Fe^{2+}/Fe^{3+}$ ratio, one may estimate the $O_2/CO_2$ ratio in the air to which the soil was originally exposed. The oldest paleosols which have been studied date back to 3.1 b.y. ago. They are consistent with $CO_2$ levels of the order of 20-30 PAL and $O_2$ levels of a few percent times present levels [Holland, 1984]. Such $CO_2$ concentrations, while lower than those considered here, are approximately what is needed to compensate for the lesser (~15%) decrease in solar luminosity at that time [KPC].

A possible complication, which is suggested by the model results presented here, is that the ferric iron observed in the soil profiles could have been oxidized by $H_2O_2$ rather than $O_2$. (The data are not specific as to the nature of the oxidant (H. D. Holland, personal communication, 1984).) The present globally-averaged rainout rate of $O_2$ is about $6 \times 10^{11}$ $O_2$ molecules $cm^{-2}$ $s^{-1}$, assuming a dissolved $O_2$ concentration of 7.2 ml/l (at 15°C) and a precipitation rate of 100 cm $yr^{-1}$ [Sellers, 1965]. By comparison, the estimated $H_2O_2$ rainout rate in our standard high-$CO_2$ atmosphere is $3.6 \times 10^9$ molecules $cm^{-2}$ $s^{-1}$. Since each $H_2O_2$ molecule contains one excess oxygen atom and since the amount of $O_2$ dissolved in rainwater should vary linearly with atmospheric $P_{O_2}$, the oxidant flux from our standard atmosphere is equivalent to the dissolved $O_2$ flux from an atmosphere containing approximately $3 \times 10^{-3}$ PAL of $O_2$. Thus, $H_2O_2$ could have been an important oxidant in soils, even under weakly reducing atmospheric conditions such as those described here. If the atmosphere contained some free $O_2$ as a by-product of oxygenic photosynthesis, the concentration of tropospheric $H_2O_2$ would have presumably increased, so $H_2O_2$ should have continued to be an important oxidant for soils. A photochemical study of the relative oxidation efficiencies of $H_2O_2$ as a function of atmospheric oxygen level is currently being undertaken and may help to shed light on this matter.

## Conclusion

The faint young sun problem is most readily solved by assuming large increases in atmospheric $CO_2$. $CO_2$ levels of 80 to 600 PAL could have kept the mean global surface temperature between 0°C and 15°C, given a 25% decrease in solar flux. Large $CO_2$ enhancements can be justified theoretically on the basis of several different geochemical arguments.

High $CO_2$ levels would have influenced the atmospheric oxidation state by increasing photochemical production of $H_2O_2$ and $H_2CO$. Rainout of these gases was an important factor in the atmospheric hydrogen budget. For a $CO_2$ level of 600 PAL, rainout of $H_2O_2$ should have kept the $H_2$ mixing ratio above $2 \times 10^{-5}$ even if no additional surface sources of hydrogen were present. Predicted ground level $O_2$ mixing ratios are between $10^{-12}$ and $10^{-14}$. Changes in solar UV fluxes or in rates of precipitation or vertical mixing would have changed atmospheric composition slightly, but should not have substantially altered the weakly reducing character of the prebiotic atmosphere.

## References

Ackerman, M., Ultraviolet solar radiation related to mesospheric processes, Inst. Aeronaut. Spatiale Belge Brussels, A-77, 149-159, 1970.

Bar-Nun, A., and S. Chang, Photochemical reactions of carbon monoxide and water in earth's primitive atmosphere, J. Geophys. Res., 88, 6662-6672, 1983.

Baulch, D. L., D. D. Drysdale, J. Duxbury, and S. J. Grant, Evaluated Kinetic Data for High Temperature Reactions, vol. 3, Butterworth's, London, 1976.

Berner, R. A., A. C. Lasaga, and R. M. Garrels, The carbonate-silicate geochemical cycle and its effect on atmospheric carbon dioxide over the past 100 million years, Am. J. Sci., 283, 641-683, 1983.

Braterman, P. S., A. G. Cairns-Smith, and R. W. Sloper, Photo-oxidation of hydrated $Fe^{+2}$ — Significance for banded iron formations, Nature, 303, 163-164, 1983.

Campbell, I. M., and B. A. Thrush, The association of oxygen atoms and their combination with nitrogen atoms, Proc. R. Soc. London, Ser. A., 296, 222-232, 1967.

Canuto, V. M., J. S. Levine, T. R. Augustsson, and C. L. Imhoff, UV radiation from the young sun and oxygen and ozone levels in the prebiological paleoatmosphere, Nature, 296, 816-820, 1982.

Canuto, V. M., J. S. Levine, T. R. Augustsson, C. L. Imhoff, and M. S. Giampapa, The young sun and the atmosphere and photochemistry of the early earth, Nature, 305, 281-286, 1983.

Dahlquist, G., and A. Bjorck, Numerical Methods,

translated from Swedish by N. Anderson, Prentice-Hall, Englewood Cliffs, New Jersey, 573 pp., 1974.

DeMore, W. B., and O. F. Raper, Primary processes in ozone photolysis, J. Chem. Phys., 44, 1780-1783, 1966.

Fels, S. B., Simple strategies for inclusion of Voigt effects in infrared cooling rate calculations, Appl. Opt., 18, 2634-2637, 1979.

Fishman, J., and P. Crutzen, A numerical study of tropospheric photochemistry using a one-dimensional model, J. Geophys. Res., 82, 5897-5906, 1977.

Gough, D. O., Solar interior structure and luminosity variations, Sol. Phys., 74, 21-34, 1981.

Hampson, R. F., and D. Garvin, Reaction rate and photochemical data for atmospheric chemistry-1977, NBS Spec. Pub., 513, 1977.

Hargraves, R. B., Precambrian geologic history, Science, 193, 363-374, 1976.

Hochanadel, C. J., T. J. Sworski, and P. J. Ogren, Ultraviolet spectrum and reaction kinetics of the formyl radical, J. Phy. Chem., 84, 231-235, 1980.

Holland, H. D., The Chemistry of the Atmosphere and Oceans, J. Wiley, New York, 1978.

Holland, H. D., The Chemical Evolution of the Atmosphere and Oceans, Princeton University Press, Princeton, New Jersey, 1984.

JPL (Jet Propulsion Laboratory), Chemical kinetic and photochemical data for use in stratospheric modelling, JPL Publ., 79-27, 1979.

JPL (Jet Propulsion Laboratory), Chemical kinetic and photochemical data for use in stratospheric modelling, JPL Publ., 83-62, 1983.

Kasting, J. F., Stability of ammonia in the primitive terrestrial atmosphere, J. Geophys. Res., 87, 3091-3098, 1982.

Kasting, J. F., and J. C. G. Walker, Limits on oxygen concentration in the prebiological atmosphere and the rate of abiotic fixation of nitrogen, J. Geophys. Res., 86, 1147-1158, 1981.

Kasting, J. F., S. C. Liu, and T. M. Donahue, Oxygen levels in the prebiological atmosphere, J. Geophys. Res., 84, 3097-3107, 1979.

Kasting, J. F., K. J. Zahnle, and J. C. G. Walker, Photochemistry of methane in the earth's early atmosphere, Precambrian Res., 20, 121-148, 1983.

Kasting, J. F., J. B. Pollack, and T. P. Ackerman, Response of earth's atmosphere to increases in solar flux and implications for loss of water from Venus, Icarus, 57, 335-355, 1984a.

Kasting, J. F., J. B. Pollack, and D. Crisp, Effects of high $CO_2$ levels on surface temperature and atmospheric oxidation state of the early earth, J. Atmos. Chem., in press, 1984b.

Kuhn, W. R., and J. F. Kasting, The effects of increased $CO_2$ concentrations on surface temperature of the early earth, Nature, 301, 53-55, 1983.

Liu, S. C., and T. M. Donahue, Realistic model

of hydrogen constituents in the lower atmosphere and escape flux from the upper atmosphere, J. Atmos. Sci., 31, 2238-2242, 1974.

Lovelock, J. E., and A. J. Watson, The regulation of carbon dioxide and climate; Gaia or or geochemistry, Planet. Space Sci., 30, 795-802, 1982.

Lovelock, J. E., and M. Whitfield, Life span of the biosphere, Nature, 296, 561-563, 1982.

Manabe, S., and R. T. Wetherald, Thermal equilibrium of the atmosphere with a given distribution of relative humidity, J. Atmos. Sci., 24, 241-259, 1967.

McEwan, M. J., and L. F. Phillips, Chemistry of the Atmosphere, J. Wiley, New York, 1975.

Mount, G. H., G. J. Rottman, and J. G. Timothy, The solar spectral irradiance 1200-2550 Å at solar maximum, J. Geophys. Res., 85, 4271-4274, 1980.

Newman, M. J., and R. T. Rood, Implications of solar evolution for the earth's early atmosphere, Science, 198, 1035-1037, 1977.

Owen, T., R. D. Cess, and V. Ramanathan, Early earth: An enhanced carbon dioxide greenhouse to compensate for reduced solar luminosity, Nature, 277, 640-642, 1979.

Pinto, J. P., C. R. Gladstone, and Y. L. Yung, Photochemical production of formaldehyde in the earth's primitive atmosphere, Science, 210, 183-185, 1980.

Roberts, R. E., J. E. A. Selby, and L. M. Biberman, Infrared continuum absorption by atmospheric water vapor in the 8-12 μm window, Appl. Opt., 15, 2085-2090, 1976.

Rodgers, C. D., and C. D. Walshaw, The computation of infrared cooling rate in planetary atmospheres, Q. J. R. Meteorol. Soc., 92, 67-92, 1966.

Rossow, W. B., A. Henderson-Sellers, and S. K. Weinreich, Cloud feedback: A stabilizing effect for the early earth?, Science, 217, 1245-1247, 1982.

Sagan, C., and G. Mullen, Earth and Mars: Evolution of atmospheres and surface temperatures, Science, 177, 52-56, 1972.

Sellers, W. D., Physical Climatology, University of Chicago Press, Chicago, Ill., 1965.

Shemansky, D. E., $CO_2$ extinction coefficient 1700-3000 Å, J. Chem. Phys., 56, 1582-1587, 1972.

Thompson, B. A., P. Harteck, and R. R. Reeves, Jr., Ultraviolet absorption coefficients of $CO_2$, CO, $O_2$, $H_2O$, $N_2O$, $NH_3$, NO, $SO_2$, and $CH_4$ between 1850 and 4000 Å, J. Geophys. Res., 68, 6431-6436, 1963.

Veizer, J., W. Compston, J. Hoefs, and H. Nielson, Mantle buffering of the early oceans, Naturwissenschaften, 69, 173-180, 1982.

Visconti, G., Radiative-photochemical models of the primitive terrestrial atmosphere, Planet. Sci., 30, 785-793, 1982.

Walker, J. C. G., Evolution of the Atmospheres, Macmillan, New York, 1977.

Walker, J. C. G., Climatic factors on the Archean

earth, Pure Appl. Geophys., 40, 1-11, 1982.

Walker, J. C. G., P. B. Hays, and J. F. Kasting, A negative feedback mechanism for the long-term stabilization of earth's surface temperature, J. Geophys. Res., 86, 9776-9782, 1981.

Wang, W. C., and P. H. Stone, Effect of ice-albedo feedback on global sensitivity in a one-dimensional radiative-convective climate model, J. Atmos. Sci., 37, 545-552, 1980.

Windley, B. F., Timing of continental growth and emergence, Nature, 270, 426-428, 1977.

Yung, Y. L., A numerical method for calculating the mean intensity in an inhomogeneous Rayleigh scattering atmosphere, J. Quant. Spectrosc. Radiat. Transfer, 16, 755-761, 1976.

Yung, Y. L., and M. B. McElroy, Fixation of nitrogen in the prebiotic atmosphere, Science, 203, 1002-1004, 1979.

Zahnle, K. J., and J. C. G. Walker, The evolution of solar ultraviolet luminosity, Rev. Geophys. Space Phys., 20, 280-292, 1982.

# ARE INTERPRETATIONS OF ANCIENT MARINE TEMPERATURES CONSTRAINED BY THE PRESENCE OF ANCIENT MARINE ORGANISMS?

James W. Valentine

Department of Geological Sciences, University of California
Santa Barbara, California 93106

Abstract. The thermal tolerances of most organisms are probably not selected per se but arise from molecular structures selected for their kinetic properties -- for capacity rather than resistance adaptations. Organisms contain molecules which display activities at temperatures far beyond any encountered in life. A few prokaryotes can tolerate boiling water; prokaryotic phototrophs live to about 73°C, eukaryotes to about 62°C, eukaryotic phototrophs to about 56°C, and metazoans to about 50°C; all groups are represented at temperatures near freezing. Living marine metazoans are adapted to today's temperatures. If late Precambrian marine temperatures were at, say, 50°C or so, we probably could not tell it from fossil evidence.

## Introduction

When dealing with conditions which may have prevailed on earth in the remote geological past, one is faced with evidence that life has been continuously present since before about 3.5 billion years (b.y.) ago [Awramik, 1981]. This evidence places significant constraints upon the past ranges of environmental variables, which must have remained continuously within the limits which life will tolerate. Marine temperature is an important factor in regulating the chemistry and composition of the oceans and the atmosphere, and it is also one of the chief limiting factors in the distribution of organisms. It is therefore of particular interest to examine the question as to just what constraints the history of appearance of organisms of different grades and types does place upon the history of ocean temperatures.

## Empirical Temperature Limits in Living Organisms

The mechanisms by which temperature proves limiting to living organisms are known in a general way at gross physiological and molecular levels [e.g., Prosser and Brown, 1961; Hochachka and Somero, 1984]. However, our understanding of the factors which limit the ability of organisms to become adapted to extreme temperatures is rudimentary; knowledge in this area remains largely empirical. As I am concerned here with the marine environments, the temperatures of interest fall between the freezing and boiling temperatures of ocean water.

At the higher temperature extremes, there is a clear correlation between the grade of complexity of organisms and the temperature extreme to which some living members of each grade are adapted. The organisms which have achieved the most extreme high-temperature adaptations are chemautotrophic and chemoheterotrophic prokaryotes, some of which can live at temperatures exceeding 100°C [Brock, 1978; Castenholz, 1979]. Volcanic springs are commonly acidic, and some of the extreme thermophiles are acidophilic as well [Degryse, 1976; Brock, 1978]. The highest temperatures known to be tolerated by photoautotrophic prokaryotes are around 73°C [Castenholz, 1973]. This is probably the uppermost temperature at which photosynthesis of any sort occurs today. The extremely thermophilic prokaryote groups are not particularly closely related, and the different groups each have low-temperature relatives, thus making it appear that extreme thermal resistance has been achieved a number of times independently. Some high-temperature organisms are known from only small areas, and high-temperature environments are certainly spottily distributed, so that there is a strong suspicion that the extreme thermophiles which are now living have evolved relatively recently, rather than representing ancient lineages continuously endemic to high-temperature environments.

The most temperature-resistant eukaryotes are fungi, which can evidently tolerate 62°C or so [Tansey and Brock, 1972]. The highest temperature at which eukaryotic photosynthesis is known to be carried forward is 56°C [Tansey and

Brock, 1978]. Finally, animals are not known to tolerate temperatures above about 50°C [Brock, 1978]. All of these record-setting thermophiles live in terrestrial environments, and most are associated with hot springs.

In the shallow sea, temperatures above about 40°C are not continuously available. Intertidal organisms, which are sometimes exposed to direct sunlight on warm days, have relatively high thermal tolerances. Even among temperate-zone organisms, experiments demonstrate that many common intertidal species tolerate temperatures well above 40°C when submerged [Gunter, 1957]. In the deep sea, temperatures of hundreds of degrees Celsius are found in the vent emissions in active rift zones, best studied in the eastern Pacific [Corliss et al., 1979]. Although communities of high biomass are associated with these submarine hot springs, the Metazoa do not appear to be tolerant of extreme high temperatures, living in the cool waters surrounding the vent springs. Reports of high-temperature prokaryotes from vent samples [Baross and Deming, 1983] have not been confirmed [Trent et al., 1984].

### Physiological and Molecular Bases of Thermal Limits in Living Organisms

In the sea, poikilothermous species appear to range to very near the limits of their thermal tolerance. In other words, temperature is probably a major factor in restricting their distributions; at least, they are not limited by some parameter which does not closely correlate with temperature. Nevertheless, it is a question whether their temperature tolerance is selected per se, or whether their ranges are simply extended to the limits of their tolerance [Valentine, 1984]. The latter case is strongly suggested by studies of the thermostabilities of enzymes and of other biologically important molecules.

In order for life as we know it to exist, reactions must occur which involve the breakage and formation of covalent bonds at rates which are from 8 to 12 orders of magnitude higher than occur in noncatalyzed reactions in earth environments. These high rates are achieved by enzyme catalysis; the enzymes have the effect of lowering the activation energy of the reactions to the point where the ambient thermal energy of the organisms is sufficient to permit the reactions to go forward. In order to operate properly the enzymes themselves must undergo rapid reversible changes in conformation which permit interactions with their substrates. Therefore, by contrast with the strong covalent bonding which characterizes the metabolized substances, the enzymes function via weak bonds (hydrogen bonds and van der Waals and hydrophobic interactions), some of which are broken at kinetic energies which are well within a single order of magnitude of the biological temperature of the organisms. There is thus the threat that a rise

in ambient temperature will destabilize the enzymes.

The evolutionary stabilization of enzymes at higher temperatures seems to be easily accomplished in a number of ways; it often involves a small increase in the number of weak bonds affecting Tertiary and Quaternary enzyme structures, and minimal changes in the primary structure. A variety of even more subtle changes have been identified also. Hydrophobic bonds actually increase in strength with temperature, at least to the boiling point. In general, stiffening of the appropriate enzyme structures for stability at higher temperatures reduces the flexibility and utility of the enzymes at lower temperatures, though there are exceptions. A particularly clear account of these principles is by Hochachka and Somero [1984].

Measured at some intermediate condition, an enzyme from a cold-adapted species will commonly display more activity than the corresponding enzyme from a warm-adapted species. In such cases we are evidently dealing with allozymes, and although there are as yet few data, allozymes which appear from experimental evidence to be thermally adaptive have been shown to occur in fish, both in warmer- versus cooler-adapted populations of the same species [Place and Powers, 1979] and in warmer- versus cooler-adapted species which are closely related [Graves and Somero, 1982].

A different story is presented by many other enzymes from marine poikilotherms, which when studied experimentally have proven to have thermostabilities which not only exceed the physiological tolerances of the whole organism, but which also sometimes retain activities at temperatures not ordinarily encountered in the ocean. Read [1963] has studied the thermostability of aspartic-glutamic transaminase in a number of bivalve molluscan species. Mytilus edulis, the lethal temperature for which is about 27°C, has an inactivation temperature for this enzyme of 50°C; corresponding temperatures for Placopecten magellanicus are 28°C and 45°C, and for Aequipecten irradians, 33°C and 45°. Even though the physiological activities of some enzymes may be seriously impaired at temperatures below the experimental inactivation temperatures, it is clear that in many cases the thermostability of enzymes has not been directly involved in limiting the thermal tolerance of organisms.

Physiologists distinguish two sorts of thermal adaptation [Christophersen and Precht, 1953]. (1) Resistance adaptations are those which permit an organism to cope with temperature extremes, whether during larval stages, as an adult, at reproduction, or during some other critical function. (2) Capacity adaptions are those which produce useful activities, biochemical or physiological, within the temperature extremes. Read's studies on bivalves led him to postulate that the structure of many enzymes, from which their thermostabilities derive, is selected as a capacity adaptation and that the temperature limits of enzymatic activity are not selected per se. For example, Read [1963] found that the enzymes of some bivalve species which are tolerant of low oxygen concentrations have broad thermostabilities; selection for anoxic or microaerophilic conditions evidently produces enzymes with structures that happen to be broadly thermostable. Another capacity adaptation is for enzymes which have high activation energies at temperatures near the lower end of a species' tolerance, so that peak enzyme activity may occur even when temperatures have only barely risen above some low point. Some enzymes with this capacity confer potential activity at temperatures so high that they will never be encountered (see review by Read [1967]). Even when enzymes have thermostabilities which lie well within the range of ocean temperatures, they are probably selected for their kinetic properties and thus do not represent resistance adaptations for temperature tolerance.

If enzymes which are stable at higher temperatures are so easily discovered by selection for capacity adaptations, it is a question as to why high-temperature enzymes are not regularly evolved or, put more generally, why species are physiologically limited, and therefore confined in their distributions, by temperature barriers. There are two suggestions which seem quite plausible. One, due to Read [1967], is that species are simply limited by their less stable enzymes. It may be noted that the higher-temperature forms of some enzymes have correspondingly less activity at lower temperatures, and it might not behoove some species to replace a tolerance for lower temperatures with one for higher temperatures. Another suggestion, due to Bullock [1955], is that since the activity coefficients of different enzymes are different, temperature changes will evoke differential shifts in enzyme activities and thus imbalances in enzyme-mediated functions can result. The viable physiological temperature range is thus visualized as an envelope within which harmonious enzymatic activities are maintained.

Substances other than enzymes have been suggested as creating temperature limits. Lipids have been studied from this standpoint; there are series of lipid structures which vary in lability at some standard temperature; the less labile occur in high-temperature forms, and the more labile in low-temperature forms, so that each form has lipids of appropriate "runniness" under its normal conditions [see Brock, 1978]. Plasma membranes have also been implicated as a possible source of temperature limitation. Composed of lipids, proteins, and ions, they can be exceedingly complex, with over 150 chemically different lipid molecules harbored in a single membrane [Esser, 1979]. Membranes play the leading role in controlling the flow of substances between the interior and exterior of the cell, qualitatively and quantitatively, and adaptations which permit such control at different temperature ranges have been identified [Brock, 1978].

## Are There Evolutionary Limits to Thermal Adaptation?

With this general background we may now ask whether there are indications of constraints upon temperature adaptations which would preclude the evolution of thermal resistance to some temperatures. The most striking piece of data which bears upon this problem is the viability of prokaryotes at extreme high temperatures. Prokaryotes have large complements of enzymes, lipids, nucleic acids, and other molecular species which they contain in common with the most advanced metazoans. Indeed, 93% of the enzyme activities identified in mammals are also known in prokaryotes [Britten and Davidson, 1971]. This fact does not make it appear likely that there are simple compositional restrictions to thermal stability in eukaryotes when compared to prokaryotes.

Nevertheless, in attempting to account for the observed greater thermal restriction of eukaryotes, an obvious strategy is to find some feature of eukaryotes which is not shared by prokaryotes and then to blame the restriction on that feature. For example, Tansey and Brock [1972] point out that the eukaryotic endoplasmic membranes are structurally distinct from any of the prokaryotic membranes, for, among other differences, they must permit the passage of macromolecules and are quite porous at that level. Therefore they suggested that these endoplasmic membranes may be responsible for the restriction of eukaryotes to temperatures below about 62°C. This is a perfectly legitimate suggestion, but we must not lose sight of the fact that there is no evidence that the membranes in question are actually responsible for the temperature limits observed, and even if they were proven to be so limiting, there is no evidence that natural selection could not produce eukaryotic endoplasmic membranes which are thermostable at high temperatures.

Still another feature restricted largely to higher organisms is collagen, one of the most abundant structural proteins in metazoans. Some of the various collagens have melting temperatures consistently close to the upper thermal limits of tolerance of the species in which they occur [Rigby, 1968]. The most labile collagen form, tropocollagen, is a building block for the assembly of other collagen structures which in fact can withstand temperatures significantly in excess of species' tolerance limits. Variation in the melting point of tropocollagen is achieved at least in part via variation in the amino acid hydroxyproline, which is more abundant and perhaps differently placed in the tropocollagen structure at higher temperatures. It has been suggested that the reason for the lability of tropocollagen is to facilitate the formation of mature collagen structures [Hochachka and Somero, 1984]; in other words, selection keeps this molecule as flexible as is possible in any given species. At any rate, there is at this time no way to specify an upper limit to the potential thermostability even of tropocollagen with confidence. So long as our knowledge of thermal limitations remains on an empirical basis without an underpinning of theoretical principles, we cannot assert that there have been any particular limits to marine thermal adaptations.

Today, marine poikilotherms appear to be well adapted to the entire range of temperatures that are actually present in the shallow sea. The fact that diversities are lower in some situations than in others does not appear to be temperature related at least, not temperature per se. For example, although low-temperature waters in high latitudes support only low species diversities, species packing within equally cold-water communities in the deep sea compares favorably with, and may well surpass, that in comparable tropical communities from shallow water [Hessler and Sanders, 1967; Hessler and Jumars, 1974]. In warm-water localities with restricted faunas there are salinity or other variations which are more likely than temperature to be associated with the biotic depauperation. It is true that no organisms seem adapted to the high vent water temperatures, but that is not surprising considering the magnitude and abruptness of the change from deep-sea to vent waters. Furthermore, and perhaps more important the vent emissions are in fact being very well exploited by a community which includes abundant metazoan forms of diverse phyla, the members of which are not at all required to adapt to the higher temperatures in order to realize the benefits of the thermal springs. Exploitation of terrestrial hot springs also occurs downstream from the spring vents in cooler waters [Brock, 1978]. It is a fair question whether or not there is anything for eukaryotes to gain by invading the hottest spots in localized spring environments.

There is, however, much evidence that when broad opportunities for thermal adaptation are gradually opened, or when new thermal adaptations are required by a changing environment, evolution can easily respond to meet the challenges. All our examples are from within the usual temperature range of the present oceans, naturally enough. There is general agreement that the thermal history of the deep sea during the last 50 million years or so has been one of falling temperatures, perhaps at the outside lowering bottom water from over 12°C to near 2°C at present [Shackleton, 1979]. High-latitude surface temperatures may have fallen similarly. At a possible average change of 1°C per 5 m.y., there has evidently been no difficulty in thermal adaptation among the deep-sea biota. As previously mentioned, the now frigid deep-sea environment supports one of the highest levels of species packing known in comparable habitats, and the shelves poleward of the 12°C isotherm, though of relatively low diversity, nevertheless support many thousands of species of cryophilic poikilotherms. Some interpretations of tropical surface water paleotemperatures suggest a rise during the last 20 m.y. or so [see Shackleton, 1979]; however, the extent of the rise is in dispute [Matthews and Poore, 1980]. It is nevertheless symptomatic of the state of paleotemperature interpretation that the composition of the invertebrate fauna itself has not provided any definitive evidence on this point. If the tropical temperatures have or have not risen at an average rate of 1°C per few million years for the last 20 m.y., we have not been able to tell the difference.

## Conclusions

That metazoan species could be adapted to much higher temperatures than those which are ordinarily present today in the shallow sea seems possible. The restriction of any given species to a particular thermal envelope is sufficiently explained by general physiological knowledge and by such hypotheses as those of Bullock [1955] and Read [1967]. If we imagine the ocean warming by 1°C each few million years into the future, it is perfectly plausible that the marine biota would adapt without obvious temperature-related compositional changes for quite a long while. Would the biota remain well adapted at, say, 50°C, 60°C, or higher? We have no way to be sure.

On the basis of the oxygen isotopic compositions of ancient marine cherts, Knauth and Epstein [1976] estimated paleotemperatures for Precambrian oceans far in excess of values found in today's oceans; for example, the estimates for 3-b.y. oceans was 70°C, and for 1.3-b.y. oceans, 52°C. These estimates might be adjusted today owing to improved knowledge of corrections for ancient isotopic compositions, but even as they stand, they cannot be ruled out because of the life forms known from those times. Even by empirical standards, prokaryotic microbiotas are possible at about 70°C, and it can be argued that the eukaryotes identified by about 1.3 b.y., which are interpreted as types of marine algae, might have lived at over 50°C. For that matter, if the metazoan fauna at the Cambrian-Precambrian boundary diversified at marine temperatures of 50°C or more, we might not be able to tell it from fossil evidence so long as subsequent cooling proceeded slowly.

In conclusion, we do not now seem able to rule out the possibility that ordinary-appearing marine metazoans could withstand water temperatures much higher than those in the sea today (excepting hot vent emissions). Perhaps we will learn of principles which in fact do constrain potential metazoan thermal environments, but at present the evidence is empirical only and is, naturally enough, based on the present fauna, which is adapted to present ocean temperatures. For the moment at least, interpretations of very ancient marine temperatures would seem to be relatively unfettered by constraints from fossil evidence, so long as the postulated conditions are commensurate with the presence of an ocean.

Acknowledgments. S. M. Awramik and J. J. Childress, University of California, Santa Barbara, provided criticism, expertise, and references to significantly improve this paper; the research on which it is based was supported in part by NSF grant EAR81-21212 and NASA-Ames grant NAG 2-73.

## References

Awramik, S. M., The pre-Phanerozoic biosphere -- Three billion years of crises and opportunities, in Biotic, Ecological and Evolutionary Time, edited by M. H. Nitecki, pp. 83-102, Academic, New York, 1981.

Baross, J. A., and J. W. Deming, Growth of "black smoker" bacteria at temperatures of at least 250°C, Nature, 303, 423-426, 1983.

Britten, R. J., and E. H. Davidson, Repetitive and non-repetitive DNA sequences and a speculation on the origins of evolutionary novelty, Q. Rev. Biol., 46, 111-138, 1971.

Brock, T. D., Thermophilic Microorganisms and Life at High Temperatures, 465 pp., Springer-Verlag, New York, 1978.

Bullock, T. H., Compensation for temperature in the metabolism and activity of poikilotherms, Biol. Rev. Cambridge Philos. Soc., 30, 311-342, 1955.

Castenholz, R. W., Ecology of blue-green algae in hot springs, in Biology of Blue-Green Algae, edited by N. G. Carr and B. A. Whitton, pp. 379-414, Blackwell, Oxford, 1973.

Castenholz, R. W., Evolution and ecology of thermophilic microorganisms, in Strategies of Microbial Life in Extreme Environments, edited by M. Shilo, pp. 373-392, Verlag Chemie, New York, 1979.

Christophersen, J., and H. Precht, Die Bedeutung des Wassergehaltes der Zelle fur Temperaturanpassungen, Biol. Zentralbl., 72, 104-119, 1953.

Corliss, J. B., J. Dymond, L. I. Gordon, J. M. Edmond, R. P. von Herzen, R. D. Ballard, K. Green, D. Williams, A. Bainbridge, K. Crane, and T. H. van Andel, Submarine thermal springs on the Galapagos Rift, Science, 203, 1073-1083, 1979.

Degryse, E., Bacterial diversity at high temperature, in Enzymes and Proteins From Thermophilic Microorganisms, edited by H. Zuber, pp. 401-410, Birkhauser, Basel 1976.

Esser, A. F., Physical chemistry of Thermostable membranes, in Stretegies of Microbial Life in Extreme Environments, edited by M. Shile, pp. 433-454, Verlag Chimie, New York, 1979.

Graves, J. E., and G. N. Somero, Electrophoretic and functional enzymatic evolution in four species of eastern Pacific barracudas from different thermal environments, Evolution, 36, 97-106, 1982.

Gunter, G., Temperature, in Treatise on Marine Ecology and Paleoecology, vol. 1, Ecology, Geol. Soc. Am. Mem. 67, edited by J. W. Hedgpeth, pp. 159-184, Geological Society of America, Boulder, Colo., 1957.

Hessler, R. R., and P. A. Jumars, Abyssal community analysis from replicate box cores in the central North Pacific, Deep Sea Res., 21, 185-209, 1974.

Hessler, R. R., and H. L. Sanders, Faunal diversity in the deep-sea, Deep Sea Res., 14, 65-79, 1967.

Hochachka, P. W., and G. N. Somero, Biochemical Adaptation, 537 pp., Princeton University Press, Princeton, N. J., 1984.

Knauth, L. P., and S. Epstein, Hydrogen and oxygen isotope ratios in nodular and bedded cherts, Geochim. Cosmochim. Acta, 40, 1095-1108, 1976.

Matthews, R. K., and R. Z. Poore, Tertiary $\delta^{18}O$ record and glacioeustatic sea-level fluctuations, Geology, 8, 501-504, 1980.

Place, A. R., and D. A. Powers, Genetic variation and relative catalytic efficiencies: Lactate dehydrogenase B allozymes of Dundulus heteroclitus, Proc. Natl. Acad. Sci. USA, 76, 2354-2358, 1979.

Prosser, C. L., and F. A. Brown, Jr., Comparative Animal Physiology, 688 pp., Saunders, Philadelphia, Pa., 1961.

Read, K. R. H., Thermal inactivation of preparations of aspartic/glutamic transaminase from species of bivalved molluscs from the sublittoral and intertidal zones, Comp. Biochem. Physiol., 9, 161-180, 1963.

Read, K. R. H., Thermostability of proteins in poikilotherms, in Molecular Mechanisms of Temperature Adaptation, edited by C. L. Prosser, pp. 93-106, American Association Advances of Science, Washington, D. C., 1967.

Rigby, B. J., Temperature relationships of poikilotherms and the melting temperature of molecular collagen, Biol. Bull. Woods Hole, Mass., 135, 223-229, 1968.

Shackleton, N. J., Evolution of the earth's climate during the Tertiary era, in Evolution of Planetary Atmospheres and Climatology of the Earth, edited by D. Gautier, pp. 49-58, Centre National d'Etudes Spatiale, Toulouse, 1979.

Tansey, M. R., and T. D. Brock, The upper temperature limit for eukaryotic organisms, Proc. Natl. Acad. Sci. USA, 69, 2426-2428, 1972.

Tansey, M. R., and T. D. Brock, Microbial life at high temperatures: Ecological aspects, in Microbial Life in Extreme Environments, edited by D. J. Kushner, pp. 159-216, Academic, New York, 1978.

Trent, J. D., R. A. Chastain, and A. A. Yayanos, Possible artefactual basis for apparent bacterial growth at 250°C, Nature, 307, 737-740, 1984.

Valentine, J. W., Climate and evolution in the shallow sea, in Fossils and Climate, edited by P. J. Brenchley, John Wiley, New York, in press, 1984.